American Prisoners of War
held at Dartmoor
during the War of 1812

Society of the War of 1812 in the State of Ohio

Prisoners of War Book Series

Harrison Scott Baker II

American Prisoners of War Held at Barbados, Newfoundland and New Providence During the War of 1812

American Prisoners of War Held at Bermuda, Cape of Good Hope and Jamaica During the War of 1812

American Prisoners of War Held at Halifax During the War of 1812
Volume 1

American Prisoners of War Held at Halifax During the War of 1812
Volume 2

Eric Eugene Johnson

American Prisoners of War Held at Quebec During the War of 1812:
8 June 1813 – 11 December 1814

American Prisoners of War held in Montreal and Quebec During the War of 1812

American Prisoners of War Paroled at Dartmouth, Halifax, Jamaica and Odiham during the War of 1812

# American Prisoners of War
## Held at
## Dartmoor during the War of 1812

Transcribed by
Eric Eugene Johnson

Society of the War of 1812
in the
State of Ohio

HERITAGE BOOKS
2016

# HERITAGE BOOKS
*AN IMPRINT OF HERITAGE BOOKS, INC.*

### Books, CDs, and more—Worldwide

For our listing of thousands of titles see our website
at
www.HeritageBooks.com

Published 2016 by
HERITAGE BOOKS, INC.
Publishing Division
5810 Ruatan Street
Berwyn Heights, Md. 20740

Copyright © 2016 Society of the War of 1812 in the State of Ohio

Heritage Books by the Society of the War of 1812 in the State of Ohio:

Transcribed by Harrison Scott Baker

*American Prisoners of War Held at Bermuda, Cape of Good Hope and Jamaica During the War of 1812*
*American Prisoners of War Held at Barbados, Newfoundland and New Providence During the War of 1812*
*American Prisoners of War Held at Halifax, During the War of 1812, Volume I and II*

Transcribed by Eric Eugene Johnson

*American Prisoners of War Held at Dartmoor during the War of 1812*
*American Prisoners of War Held in Montreal and Quebec during the War of 1812*
*American Prisoners of War Held at Quebec During the War of 1812, 8 June 1813–11 December 1814*
*American Prisoners of War Paroled at Dartmouth, Halifax, Jamaica and Odiham during the War of 1812*
*Black Regulars in the War of 1812*
*Ohio and the War of 1812: A Collection of Lists, Musters and Essays*
*Ohio's Regulars in the War of 1812*

All rights reserved. No part of this book may be reproduced or transmitted in any form or by any means, electronic or mechanical, including photocopying, recording or by any information storage and retrieval system without written permission from the author, except for the inclusion of brief quotations in a review.

International Standard Book Numbers
Paperbound: 978-0-7884-5720-3
Clothbound: 978-0-7884-6447-8

## - Table of Contents -

| | |
|---|---:|
| Introduction | 1 |
| History of Dartmoor Depot | 5 |
| The Honored Dead | 7 |
| Alphabetical listing of names | 9 |
| Numeric listing by prisoner number | 362 |
| Prisoner listing by ship or regiment | 410 |
| Definitions | 488 |
| Bibliography | 493 |

American Prisoners of War held at Dartmoor during the War of 1812

# Introduction

This is a transcription of American prisoner of war records from the U.S. Navy, privateers and merchant vessels (plus some civilians) who were captured and then interned by the British Empire at the Dartmoor Depot in England during the War of 1812. There are also some U.S. Army, volunteers and militia soldiers included in these transcriptions.

This volume was compiled from copies of the *General Entry Book of American Prisoners of War* ledger of the British Admiralty made by the Public Records Office in London, Great Britain (ADM 103 series). These ledgers contain the information on 6,553 American prisoners of war who were interned between 2 April 1813 and 26 March 1815 at the Dartmoor Depot.

The *General Entry Book* (GEB) records are composed of lines for the recording of names and personal information of those incarcerated. The record of each prisoner is found on two facing pages. The clerk making the entries wrote the page number on the upper right side page of each page. The names and information of nine men can be found on each double page.

The records from Dartmoor Depot are contained in five ledgers which have been microfilmed. The names of these men were checked against the names appearing on the Alphabetical Index of American Prisoners of War at Dartmoor (ADM 103/511) for corrective spelling. The dates of death from the ledgers were checked against the Certificates of Death records which are found in ADM 103/465 and 640. The majority of the Certificates of Death were located and the causes of these deaths were added to each man's record.

| Microfilm | Dates | Number of prisoners |
|---|---|---|
| ADM 103/87 | April 1813 – May 1814 | 1,089 |
| ADM 103/88 | May 1814 – August 1814 | 1,034 |
| AMD 103/89 | August 1814 – September 1814 | 1,295 |
| ADM 103/90 | September 1814 – December 1814 | 2,162 |
| ADM 103/91 | December 1814 – March 1815 | 699 |

| ADM 103/511 | Alphabetical Index of POWs | |

| ADM 103/465 | Certificates of Death | |
| ADM 103/640 | Certificates of Death | |

Below are the column headers from the GEB ledgers used at the Dartmoor Depot. Titles in brackets, "[" and "]", indicates that the transcriber has changed the column headers from the original to a more meaningful header. Some of the columns have been eliminated in this book while other columns have been combined.

    Column 1 –Number [Nbr]
        Each prisoner of war arriving at a prisoner facility was assigned a number.

    Column 2 – By what Ship or how taken [How taken]
        This column lists the Royal Naval ship or privateer which captured the prisoner. Also included in this column are the men who gave themselves up as a prisoner, who were impressed by the Royal Navy, taken ashore or captured by the Royal Army.

American Prisoners of War held at Dartmoor during the War of 1812

Column 3 – Time when [When Taken]
   The date the prisoner was taken into custody.

Column 4 – Where Taken
   This column lists the location of capture which could indicate latitudes and longitudes if a sea, a port, or a geographic region or location.

Column 5 – Name of Prize [Prize]
   The name of the ship or vessel of the prisoner of war, or if from land forces, the name of the regiment or service.

Column 6 - Whether Man of War, Privateer, or Merchant Vessel [Ship Type]
   The type of ship or vessel, that is: man of war (warship), revenue cutter, privateer, letter of marque, merchant vessel, or prize of a privateer. Land forces indicate regular army, volunteers or militia.

Column 7 – Prisoner's Names [Title not used]
   The prisoner names given in the GEB are first name then last name, e.g. John Smith, but in this book the last name is given first then the first name.

Column 8 – Quality [Title not used]
   This column gives the rank of the prisoner. In addition to naval, privateer and merchant vessel ranks, there are also civilians, merchants, supercargoes and passengers found in these records.

Column 9 – Time when received into custody [Date Received]
   The date when a prisoner arrived at a prisoner facility.

Column 10 – From what ship, or whence received [From what ship]
   The location of the prisoner before being received at the current prisoner of war facility. This could be the capturing ship, the ship that transported the prisoner to England, a hospital, another prison facility or parole location.

Column 11 – Place of Nativity [Born]
   Lists the birth place of a prisoner.

Column 12 – Age
   The age of the prisoner in years, most likely, from his last birthday.

Columns 13 through 18 are not used in this book [Race]
   These columns indicates the height, hair color, color of eyes, type of complexion and any body marks (including tattoos) or wounds of a prisoner. Included in these physical descriptions are the races of non-Caucasians, which are Black, Negro, Mulatto, Creole or Chinese. The author has created a new column entitled Race to indicate non-Caucasians.

Columns 19 through 32 are not used in this book
   The personal items of a prisoner were inventoried upon arrival at a prisoner of war facility and the required missing items were replaced. These items include hammocks, beds, straw mattresses, cushions, blankets, hats, jackets, waistcoats, trousers, shirts, shoes, stockings and handkerchiefs.

Columns 33 through 35 have been combined into a single entry [Title not used]

# American Prisoners of War held at Dartmoor during the War of 1812

This new column indicates when a prisoner died at a prisoner facility, when he escaped from a prisoner facility, when he was discharged to another prison facility, or when he was released and sent back to the United States.

> Column 33 – Exchanged, Discharged, Died or Escaped [Field not used]
> "E" or "R" indicates that the prisoner escaped from the parole location while "D" indicates that he was discharged. "DD" indicates that he died while assigned to a parole location.
>
> Column 34 – Time When [Field not used]
> Contains the date of the event from column 33.
>
> Column 35 - Whither, and by what order if discharged. [Field not used]
> This column shows the place or ship that the prisoner was sent when he was discharged from the prison facility. Orders were given by the His Majesty's Transport Board.

British prisoner of war camps were called 'depots' and the British maintained three depots in England at Stapleton (near Bristol), at Dartmoor (near Plymouth), and at Norman Crossing (near Peterborough). The Americans were only interned at Stapleton and Dartmoor while all three depots housed French POWs and their allies.

The Mill Prison in Plymouth was initially used as a temporary depot. After the Dartmoor Depot was opened, the prisoners at the Mill Prison were sent to Dartmoor Depot. The prison was then used as a holding area for prisoners arriving in England, before they were assigned to a prison facility, and later as a holding area for prisoners being exchanged or released back to the United States.

Depots were actually forts which contained more than two prisons within their walls. Each prison was a barracks, housing approximately 1,000 POWs and each having its own courtyard for the prisoners. Each prison was separate from the other prisons by walls and each barracks normally contained prisoners from the same country. British army home regiments (militia) served as guards at each of the depots on a two to three month rotational basis.

Dartmoor Depot was organized into seven prisons. The French were assigned six barracks and the Americans one barracks at the beginning of the War of 1812. At its height, Dartmoor Depot contained over 10,000 French and American POWs within its walls. After the fall of Napoleon in 1814, the British sent the French POWs home and they then relocated nearly all of Americans from Stapleton and the prison ships to Dartmoor Depot. By March 1815, there were approximately 6,000 Americans at Dartmoor Depot.

The Stapleton Depot, outside of Bristol, had over 5,000 French prisoners but only 421 Americans. This depot was organized into three prisons with a third of the Americans assigned to each barracks.

Three Royal Navy bases had prison ships for POWs. There were located at Chatham, Portsmouth and Plymouth. Prison ships were obsolete warships and transports, ranging in size from ships-of-the-line to frigates and cargo ships. These ships were stripped down to the main deck and all naval stores and armaments were removed from the hull. All openings in the hull were boarded over. At the end of the War of 1812, most of these prison ships were scrapped. Britain used separate prison ships to house both civilian convicts and military POWs.

Finally, there were facilities for paroling officers, warrant officers, senior mates, and some civilians. Normally, the men who came from ships which had fourteen or more guns were eligible to be paroled.

These men were assigned to a village or city where they could live away from the depots. These villages and towns were called 'parole stations.'

The officers were given a daily allowance for room and board, and they were restricted to within a mile of their parole station. They were mustered twice a week. They were free to live among the population and to obtain work if they so desired. Any violations of a parole agreement would cause a man to be sent back to a prison depot. At Dartmoor Depot there was an eighth prison, called the Petty Officer's Prison, where officers who violated their paroles were kept. Officers and enlisted men were not housed together.

French officers were paroled at Ashburton, Okehampton, Moretonhampstead, Tauistock, Bodmin, Launceston, Callington, Roscron and Regilliack, while the American officers were paroled at Ashburton, Dartmouth, Oliham and Reading.

Forty-one Americans, who had been captured by the British while serving as crewmen on French warships and merchant vessels, were transferred on 4 June 1814 from the French prisons at Dartmoor Depot to the American prison at Dartmoor Depot. The names of these men were dropped from the French GEB ledgers and their names were entered in the American GEB (with new prisoner numbers). These new prisoner numbers were 1173 through 1213. Some of these men had been prisoners in England since 1805.

867 of these Americans are listed as Negroes, Blacks, Colored, Men of Color or Mulattos on the GEB ledgers, which amounts to 13% of the total American POW population. Sixteen men are listed as Creoles and there was one Chinese American.

Fifty-nine American men entered British service, that is, they enlisted in the British army, navy or merchant marines to avoid internment in Dartmoor Depot. Fifty-three Americans escaped from Dartmoor Deport, while 272 died and were buried in the American Cemetery near Dartmoor Deport.

No American POWs were exchanged back to the United States directly from Dartmoor Deport during the War of 1812. Some POWs were sent to parole locations or to prison ships, and from these facilities they were exchanged and returned home. The majority of men at Dartmoor Depot were released between April and July 1815 and returned home.

The penmanship in these ledgers was fair to very good. The spelling of non-familiar names was done phonetically. Definitions are included in back of this book.

Any errors or omissions are regretted and are the fault of the transcriber.

Eric Eugene Johnson

President (2008-2011)
Society of the War of 1812 in the State of Ohio

Historian General (2011-2014)
General Society of the War of 1812

## History of the Dartmoor Depot

Great Britain was involved in the Napoleonic Wars between 1793 and 1815. The Napoleonic Wars were a series of seven wars between France, and its allies, and Great Britain, Prussia, Austria-Hungary and Russia, plus their allies. Each segment of the Napoleonic Wars was called a coalition.

During the first four coalitions, Great Britain was fighting a naval war with the French Empire while supplying money and arms to its allies on the continent. During the Fourth Coalition, France invaded its ally Spain and threatened Portugal. England entered the land war in 1807 to aid Spain and Portugal. This war was called the Peninsular War and it would be fought continuously from the end of the Fourth Coalition to the end of the Sixth Coalition in 1814.

The Sixth Coalition brought an end to the French Empire and Napoleon abdicated his throne and went into exile. In 1815, he escaped from Elba and went back to France to form a new empire. He was defeated at Waterloo and was sent into permanent exile. This segment of the Napoleonic Wars was called the Seventh Coalition or the Hundred Days War.

At the beginning of these conflicts, England housed French naval prisoners of war in prison hulks at three of its naval bases: Plymouth, Portsmouth and Chatham. As the number of prisoners of war increased, England constructed two land prisons (called depots) at Stapleton and Norman Crossing.

A need for a third deport was realized in 1806 and the foundation stone for Dartmoor Depot was laid on 20 March 1806. It was opened in December 1808. The facility was built by both English laborers and French POWs (who were paid for their work).

Dartmoor Depot contained seven 1,000-man barracks, called prisons. Each building was enclosed with a palisade of iron bars, forty feet around each building, creating a courtyard. The seven buildings were surrounded by two walls, each fourteen feet high and twenty-two feet apart.

The floors of the buildings resembled the decks of a prisoner ship. The bottom two floors were for sleeping while the top floor was a living space during the day. The French prisoners included Eurasians, Malays, Chinese, Italians, Danes, Norwegians and Swedes.

The daily ration for each man was one-pound of bread, a half-pound of beef, a half-pint of peas, and a third of an ounce of salt. Fish was available on Fridays for those who wanted fish. There was also a local farmer's market established within the walls of the depot. The men were given a daily allowance and they could purchase fruits, vegetables and even clothing from the local populace. Outside of the walls were the prison administration facilities, prison hospital, and barracks for 500 guards.

Escapes were common throughout the history of this depot. Due to excellent soil conditions, tunnels could be dug under the walls while during the winter months, many scaled the walls. Snow drifts against the walls helped escapees while the bitter winters kept many guards from doing their duties. The winter of 1813-1814 was the worst recorded winter in fifty years.

The men who were caught escaping or who committed crimes within the prisons were housed in the Dungeon. This was a 30' by 20' building made of stone walls and a stone floor. The building contained no light within its walls.

During April 1813, the first 1,700 Americans arrived at Dartmoor Depot. 700 men came from the prison ships *Meteor* and *Le Brave* while the rest came from prison facility overseas, particularly from Halifax, Nova Scotia.

Once the Americans came, the prison guards were increased to 1,200 militia men because the English felt that most of the Americans were actually Englishmen and that they would cause more mischief than the French prisoners. In part, this was true, but most of the enlisted Frenchmen and their allies did not know English, which would cause a problem when escaping and trying to get back to their homelands. The Americans, on the other hand, could easily blend into the local population.

The French officers who were paroled caused more problems than the American officers. Most of the French officers knew English and if they were lucky, could make it back to France. The French officers were needed in order to free English officers in prisoner of war exchanges. The American officers on parole behaved themselves, most were exchanged and sent back to the United States.

In May 1813, 250 more Americans arrived at Dartmoor Depot, but 200 of them were sent to Stapleton Depot while the other fifty were sent on to the Chatham prison ships. July saw 120 prisoners sent to Chatham for exchange back to the United States. In August, fifty-nine Americans enlisted in the British service and were released from Dartmoor Depot.

The British began to have problems with the American Blacks. Most white Americans did not want to associate with or live with the Blacks in the barracks. Many of the whites accused the Blacks of stealing and committing other crimes. The British cleared Number 4 Prison (barracks) and assigned this building to the Blacks.

During April and May of 1814, the French and their allies were sent home after the end of the Sixth Coalition. The depot became an American prisoner of war facility. The first draft of men from Stapleton Depot and the prison ships at Chatham arrived in June. Eventually, the prison ships at Plymouth, Portsmouth and Chatham were emptied and the Americans were sent to Dartmoor Depot.

With the end of the War of 1812 on 17 February 1815, the United States and England began preparations to release these men and turned the American POWs back to the United States. The last of the Americans left on 11 July 1815. In the same month, 4,000 French POWs arrived at Dartmoor Depot. These men were captured by the British during the Seventh Coalition. They would be returned to France in 1816.

There were 1,478 deaths records at the depot including 272 Americans. The depot was closed in 1816 and it would remain dormant until 1850 when it was reopened as a civilian prison.

On 6 April 1815, the American prisoners of war rioted at Dartmoor Deport. The frustration of knowing that the war was over and that you were still a prisoner of war was too great for many men to handle. The men stormed the prison gate and the guards fired upon the men. Seven Americans were killed, two died from wounds, and another fifty-four men were wounded. The British government compensated the men, or their families, who were wounded or killed during the riot.

American Prisoners of War held at Dartmoor during the War of 1812

## - *In memory of those who did not return* -

## The Honored Dead

| | | |
|---|---|---|
| Adams, James | Carter, Daniel | Gladding, Joseph |
| Adams, John | Cateret, James | Graves, Thomas |
| Adams, Robert | Chandler, Simon | Grey, John |
| Adams, William (1) | Chult, David | Gunn, Charles William |
| Adams, William (2) | Clark, William | Gwynn, Josiah |
| Addigo, Henry | Clarke, Simon | Hall, Thomas |
| Allan, Asa | Cobbs, James | Harman, Isaac |
| Allen, Archibald | Coffee, Ram'l | Harris, Simeon |
| Allen, John Baptiste | Cole, John | Harris, William |
| Almeno, Jose | Coleman, William | Harrison, Silas |
| Amos, Peter | Collins, John | Hart, James |
| Anderson, Jacob | Congdon, James | Hawley, Frederick |
| Andrews, Joshua | Conklin, Vertius | Haycock, Joseph |
| Appleton, Daniel | Cook, Benjamin | Haywood, John |
| Archer, Daniel | Cooper, Thomas | Henderson, Alexander |
| Aubry, John Martinalias | Cornish, Charles | Henry, James |
| Babb, Benjamin | Curren, Nathaniel | Hentey, Jacob |
| Badson, Jacob | Cussar, Jacob O. | Heny, Daniel |
| Bailey, Moses | Davenport, John | Holbrook, Ebenezer |
| Baker, Charles | Davis, John | Holding, Henry |
| Baldwin, John | DeBates, Amos | Holford, Elisha |
| Barnett, James | Denham, Silas | Holiday, Francis |
| Barren, Thomas | Denning, Joseph | Holstein, Richard |
| Barry, Peter | Devinas, John | Hosson, John |
| Bateman, John | Diamond, William | Hydra, Dempsey |
| Bean, William | Dillon, William | Jack, John |
| Beck, William | Diluo, Benjamin | Jackson, Thomas (1) |
| Bella, Darius | Dyer, Jonathan | Jackson, Thomas (2) |
| Beverly, Henry | Edgar, William | Jarvis, Thomas |
| Birch, Peter | Erwin, William | Jenkins, Nathaniel |
| Bisley, Horace | Evans, Edward | Jennings, John |
| Blasdon, Philip | Fernald, William | Johannes, John |
| Bradly, Samuel | Fisher, Charles | Johnson, John |
| Bray, Isacher | Fletcher, William B. | Johnson, Joseph Toker |
| Brien, Lewis | Flowers, John | Johnson, William (1) |
| Brisons, John | Fogerty, Archibald | Johnson, William (2) |
| Brown, Charles | Fowler, Joshua | Jones, George |
| Brown, George | Francis, John | Jones, Isaac |
| Brown, William | Freely, Henry | Jones, James |
| Burbage, Henry | Gardner, Francis | Jones, Stephen |
| Burts, Joseph | Gardner, Jerry | Jones, Thomas |
| Butler, John | Gardner, Timothy | Jose, Emanuel |
| Campbell, Henry | Gatewood, James | Joseph, Pedro |
| Campbell, James | Gayler, James | Kelley, John |
| Campreche, St. Jago | Gennison, Michael | King, Uriel |
| Carson, John | Gibson, William | Knabbs, James |

American Prisoners of War held at Dartmoor during the War of 1812

| | | |
|---|---|---|
| Lackey, Joseph | Perigo, Joel | Smithers, Richard |
| Lamb, Anthony | Perkins, John | Snell, Shaderick |
| Larkin, Amos | Perkins, Joseph | Squibb, Silas |
| Larkin, Lewis | Peters, Aaron | Stacey, Stephen |
| Lawson, James | Peterson, Jacob | Stanwood, Timothy |
| Lee, Richard | Peterson, John | Steel, John |
| Lee, Richard Robert | Peterson, Lawrence | Stone, John |
| Leurand, Ambrose | Pettingall, Joseph | Stout, John |
| Lilley, Samuel | Pierce, Samuel | Stove, Lewis |
| Lippart, Thomas D. | Pinkham, Jacob | Studdy, Richard |
| Long, Joseph | Polland, John | Sutton, Martin |
| Louis, John | Porter, Gideon | Taylor, David |
| Lovely, Placid | Potter, John | Thomas, John |
| Leveridge, William | Queenwell, Peter | Thompson, Abraham |
| Mann, James | Ranson, Joseph | Thompson, Henry |
| March, Jesse | Rawlinson, Thomas | Thompson, Martin |
| Marshall, Benjamin | Raysden, John | Thompson, Thomas |
| Marshall, John | Read, David | Thompson, William |
| Marshall, Solomon | Read, William | Timmerman, Mathew |
| Martin, Daniel | Resmabin, Benjamin | Toley, Elisha |
| Mead, William | Ricks, Thomas | Tomkins, Abraham |
| Mendoza, Caesar | Roberts, Francis | Tophouse, Samuel |
| Menillo, John | Roberts, John | Tremerin, Joseph |
| Miller, Edward | Robertson, James | Tucker, James |
| Miller, Richard | Robinson, Samuel | Tulford, Joseph |
| Mills, William | Robinson, William | Turner, David |
| Mingo, Albert | Rogers, Luke | Turner, John |
| Mista, Sullivan | Roth, James | Tuttle, French |
| Mitchell, Ezekiel | Rowland, William | Tyre, William |
| Mitchell, Reuben | Salisbury, Joseph | Vaughan, Nicholas |
| Modge, Daniel | Saul, Francis | Washington, John |
| Monte, Charles | Saunders, William | Wescott, Edward |
| Montgomery, John | Sawyer, Jacob | West, George (1) |
| Moore, George | Schew, Richard | West, George (2) |
| Moore, Henry | Seapatch, John | Whettan, John |
| Morrell, Jacob | Shaw, William | Widger, Joseph |
| Murray, James | Shelton, Smith | Williams, Charles |
| Nash, Daniel | Sherridan, Henry | Williams, Edward |
| Parish, Samuel | Simmons, David | Williams, John |
| Norton, Edward | Simmons, Ebenezer | Williams, Thomas |
| Osborne, John L. | Simmons, Thomas | Williams, Joseph |
| Parker, Thomas | Simondson, Isaac | Williams, Samuel |
| Parker, William | Smith, Andrew | Williams, William |
| Paul, Jonathan | Smith, Nicholas | Young, William |
| Peck, Thomas | Smith, Richard | |

*- Those who die in service to the United States should not be forgotten –*

American Prisoners of War Held at Dartmoor during the War of 1812

## Alphabetical listing of names

Medical terms used in death descriptions

| Medical Terms | Translation |
|---|---|
| febris | fever |
| variola | pox |
| phthisis pulmonalis | tuberculosis |
| perissneuoniria | pneumonia |
| mania | depression |
| erysipelatous inflammation | skin infection |
| erysipelas | skin infection |

Abbert, Solomon - Seaman - Nbr: 2161 - Prize: Rattlesnake - Ship Type: P - How taken: HMS Hyperion - When taken: 26 Jun 1814 - Where taken: off Cape Finisterre - Date Received: 16 Aug 1814 - From what ship: HMS Dublin, Halifax - Born: Shapley - Age: 25 - Released on 2 May 1815.

Abbey, Obadiah - Seaman - Nbr: 933 - Prize: Fanny - Ship Type: MV - How taken: HMS Eurotas - When taken: 23 Dec 1813 - Where taken: at sea - Date Received: 31 Jan 1814 - From what ship: Plymouth - Born: Hanfield - Age: 31 - Released on 27 Apr 1815.

Abbott, Benjamin - Seaman - Nbr: 2488 - Prize: US Sloop Frolic - Ship Type: MW - How taken: HMS Orpheus - When taken: 20 Apr 1814 - Where taken: off Cuba - Date Received: 16 Aug 1814 - From what ship: HMT Queen, Halifax - Born: Massachusetts - Age: 45 - Released on 3 May 1815.

Abbott, Daniel - Boy - Nbr: 1862 - Prize: Vivid - Ship Type: P - How taken: HMS Nymphe - When taken: 20 Apr 1813 - Where taken: off Cape Cod - Date Received: 29 Jul 1814 - From what ship: HMS Ville de Paris, Chatham Depot - Born: Boston - Age: 17 - Released on 2 May 1815.

Abbott, Enoch - Seaman - Nbr: 4031 - Prize: US Brig Rattlesnake - Ship Type: MW - How taken: HMS Leander - When taken: 13 Jul 1814 - Where taken: off Shelburne - Date Received: 6 Oct 1814 - From what ship: HMT Chesapeake, Halifax - Born: Arundel - Age: 25 - Released on 9 Jun 1815.

Abbott, Ephraim - Seaman - Nbr: 1064 - Prize: Fair American - Ship Type: MV - How taken: HMS Andromache - When taken: 19 Jan 1814 - Where taken: Bay of Biscay - Date Received: 10 May 1814 - From what ship: Plymouth - Born: Boston - Age: 25 - Released on 10 Apr 1815.

Abbott, Timothy - Seaman - Nbr: 543 - Prize: Orders in Council - Ship Type: MV - How taken: HMS Surveillante - When taken: 1 Jun 1813 - Where taken: off Cape Ortegal (Spain) - Date Received: 8 Sep 1813 - From what ship: Plymouth - Born: Hampshire - Age: 25 - Released on 26 Apr 1815.

Abbott, William - Seaman - Nbr: 5197 - Prize: Enterprize - Ship Type: P - How taken: HMS Tenedos - When taken: 21 May 1813 - Where taken: off Cape Cod - Date Received: 31 Oct 1814 - From what ship: HMT Mermaid, Chatham - Born: Ipswich - Age: 22 - Released on 11 Jul 1815.

Abman, John - Seaman - Nbr: 1645 - Prize: Paul Jones - Ship Type: P - How taken: HMS Leonidas - When taken: 23 May 1813 - Where taken: Channel - Date Received: 23 Jun 1814 - From what ship: Stapleton - Born: Baltimore - Age: 26 - Race: Colored - Released on 1 May 1815.

Achro, Joseph - Seaman - Nbr: 4816 - Prize: President - Ship Type: P - How taken: HMS Pique - When taken: 7 May 1814 - Where taken: off Porto Rico - Date Received: 9 Oct 1814 - From what ship: HMT Freya, Halifax - Born: Cartagena - Age: 25 - Race: Mulatto - Released on 21 Jun 1815.

Adams, Abijah - 2nd Mate - Nbr: 5407 - Prize: Rattlesnake - Ship Type: LM - How taken: HMS Rhin - When taken: 12 Mar 1814 - Where taken: off Bermuda - Date Received: 31 Oct 1814 - From what ship: HMT Leyden, Chatham - Born: Boston - Age: 27 - Released on 1 Jul 1815.

Adams, Henry - Mate - Nbr: 5221 - Prize: Fly - Ship Type: P - How taken: HMS Dover - When taken: 27 Jun 1813 - Where taken: off Newfoundland - Date Received: 31 Oct 1814 - From what ship: HMT Mermaid, Chatham - Born: New York - Age: 24 - Released on 11 Jul 1815.

Adams, Isaac - Carpenter - Nbr: 4865 - Prize: Olive Branch - Ship Type: MV - How taken: HMS Acasta - When taken: 16 Oct 1814 - Where taken: Chesapeake Bay - Date Received: 9 Oct 1814 - From what ship: HMT Freya, Halifax - Born: Maryland - Age: 27 - Released on 21 Jun 1815.

Adams, J. - Seaman - Nbr: 5633 - How taken: Gave himself up from HMS Racehorse - When taken: Jul 1813 - Date rec'd: 24 Dec 1814 - From what ship: HMT Tay - Born: Boston - Age: 38 - Released on 28 Apr 1815.

Adams, James - Seaman - Nbr: 2218 - Prize: Hussar - Ship Type: P - How taken: HMS Saturn - When taken: 25

American Prisoners of War Held at Dartmoor during the War of 1812

## Alphabetical listing of names

May 1814 - Where taken: off Sandy Hook - Date Received: 16 Aug 1814 - From what ship: HMS Dublin, Halifax - Born: New York - Age: 25 - Released on 2 May 1815.

Adams, James - Passenger - Nbr: 1729 - Prize: Helen and Emeline - Ship Type: LM - How taken: HMS Telegraph - When taken: 15 Aug 1813 - Where taken: off St. Anders - Date Received: 17 Jul 1814 - From what ship: Exeter, having broken parole at Ashburton (England) - Born: New York - Age: 35 - Released on 10 Apr 1815.

Adams, James - Seaman - Nbr: 4851 - Prize: Greyhound - Ship Type: MV - How taken: Amity (Privateer) - When taken: 5 May 1814 - Where taken: off Bermuda - Date Received: 9 Oct 1814 - From what ship: HMT Freya, Halifax - Born: North Carolina - Age: 20 - Died on 6 Nov 1814.

Adams, Jesse - Seaman - Nbr: 4888 - How taken: Gave himself up from HMS Centaur - When taken: 16 Oct 1814 - Date Received: 24 Oct 1814 - From what ship: HMT Salvador del Mundo - Born: New Hampshire - Age: 23 - Released on 21 Jun 1815.

Adams, John - Sailing master - Nbr: 5089 - Prize: Ida - Ship Type: LM - How taken: HMS Newcastle - When taken: 9 Aug 1814 - Where taken: Long 34 - Date Received: 28 Oct 1814 - From what ship: HMT Alkbar, Halifax - Born: Boston - Age: 27 - Died on 3 Dec 1814 from febris.

Adams, John - Seaman - Nbr: 444 - Prize: Magdalen - Ship Type: MV - How taken: HMS Superb - When taken: 15 Apr 1813 - Where taken: off Belle Isle, France - Date Received: 1 Jul 1813 - From what ship: Plymouth - Born: Pennsylvania - Age: 16 - Released on 20 Apr 1815.

Adams, John - Seaman - Nbr: 873 - Prize: Agnes - Ship Type: LM - How taken: Jane (Cutter) - When taken: 29 Nov 1813 - Where taken: Bay of Biscay - Date Received: 31 Jan 1814 - From what ship: Plymouth - Born: Boston - Age: 17 - Released on 27 Apr 1815.

Adams, John - Seaman - Nbr: 1219 - Prize: Rambler - Ship Type: MV - How taken: Morley (Transport) - When taken: 13 Jan 1813 - Where taken: off Isle of France - Date Received: 14 Jun 1814 - From what ship: Mill Prison (Plymouth, England) - Born: Boston - Age: 25 - Released on 26 Apr 1815.

Adams, Peter W. - Cook - Nbr: 1163 - How taken: Gave himself up from HMS Hebrius - Date Received: 10 May 1814 - From what ship: Plymouth - Born: Norfolk - Age: 24 - Race: Black - Released on 28 Apr 1815.Adams, Robert - Seaman - Nbr: 6290 - How taken: Gave himself up from HMS Voluntaire - Date Received: 17 Feb 1815 - From what ship: HMT Ganges, Plymouth - Born: Massachusetts - Age: 24 - Released on 13 Jun 1815.

Adams, Robert - Seaman - Nbr: 5656 - Prize: Herald - Ship Type: P - How taken: HMS Endymion - When taken: 15 Aug 1814 - Where taken: off Nantucket - Date Received: 24 Dec 1814 - From what ship: HMT Impregnable - Born: Not listed - Died on 5 Feb 1815 from pneumonia.

Adams, Samuel - Seaman - Nbr: 6203 - Prize: Prince de Neufchatel - Ship Type: P - How taken: Leander (Newcastle Acasta) - When taken: 20 Dec 1814 - Where taken: Lat 38 Long 56 - Date Received: 30 Jan 1815 - From what ship: HMT Pheasant - Born: Salem - Age: 20 - Released on 5 Jul 1815.

Adams, Samuel - Seaman - Nbr: 2268 - Prize: Experiment - Ship Type: MV - How taken: HMS Bulwark - When taken: 15 Apr 1814 - Where taken: off Newport - Date Received: 16 Aug 1814 - From what ship: HMS Dublin, Halifax - Born: Barnstable - Age: 19 - Released on 3 May 1815.

Adams, Thomas - Seaman - Nbr: 2935 - How taken: Gave himself up from HMS Tremendous - Date Received: 24 Aug 1814 - From what ship: HMT Alpheus, Chatham - Born: Charleston - Age: 48 - Race: Black - Released on 19 May 1815.

Adams, Thomas - Seaman - Nbr: 3137 - How taken: Gave himself up from HMS Ganymede - When taken: 6 Apr 1813 - Date Received: 11 Sep 1814 - From what ship: HMT Freya, Chatham - Born: New York - Age: 48 - Released on 28 May 1815.

Adams, William - Seaman - Nbr: 4848 - Prize: Hawk - Ship Type: P - How taken: HMS Pique - When taken: 26 Apr 1814 - Where taken: off Porto Rico - Date Received: 9 Oct 1814 - From what ship: HMT Freya, Halifax - Born: North Carolina - Age: 22 - Died on 24 Apr 1815 from variola.

Adams, William - Maters at Arms - Nbr: 5926 - Prize: Harlequin - Ship Type: P - How taken: HMS Bulwark - When taken: 23 Nov 1814 - Where taken: off Halifax - Date Received: 27 Dec 1814 - From what ship: HMT Penelope - Born: Maine - Age: 26 - Released on 3 Jul 1815.

Adams, William - Seaman - Nbr: 4755 - How taken: Gave himself up from HMS Africa - When taken: 4 Oct 1813 - Date Received: 9 Oct 1814 - From what ship: HMT Freya, Chatham - Born: Colchester - Age: 23 - Race: Black - Died on 15 Mar 1815 from variola.

Addigo, Henry - Soldier - Nbr: 739 - Prize: US Brig Argus - Ship Type: MW - How taken: HMS Pelican - When taken: 14 Aug 1813 - Where taken: Irish Channel - Date Received: 3 Nov 1813 - From what ship: Plymouth -

American Prisoners of War Held at Dartmoor during the War of 1812

## Alphabetical listing of names

Born: New York - Age: 41 - Died on 23 Dec 1813 from pneumonia.

Addy, Francis - Seaman - Nbr: 5794 - Prize: James - Ship Type: MV - How taken: HMS Harpy - When taken: 18 Dec 1813 - Where taken: off Isle of France - Date Received: 26 Dec 1814 - From what ship: HMT Argo - Born: Philadelphia - Age: 30 - Released on 27 Apr 1815.

Adivoe, Henry - Seaman - Nbr: 2065 - How taken: Apprehended at London - When taken: 11 Jun 1813 - Date Received: 3 Aug 1814 - From what ship: HMS Bittern, Chatham Depot - Born: New York - Age: 34 - Released on 2 May 1815.

Adrianne, Jose - Seaman - Nbr: 4808 - Prize: President - Ship Type: P - How taken: HMS Pique - When taken: 7 May 1814 - Where taken: off Porto Rico - Date Received: 9 Oct 1814 - From what ship: HMT Freya, Halifax - Born: St. Martins - Age: 25 - Race: Mulatto - Released on 21 Jun 1815.

Ainsley, William - Seaman - Nbr: 2243 - Prize: Carbineer - Ship Type: Prize - How taken: HMS Ringdove - When taken: 24 Apr 1814 - Where taken: off Bermuda - Date Received: 16 Aug 1814 - From what ship: HMS Dublin, Halifax - Born: Philadelphia - Age: 21 - Released on 3 May 1815.

Akens, William - Seaman - Nbr: 238 - Prize: Charlotte - Ship Type: MV - How taken: HMS Warspite - When taken: 3 Mar 1813 - Where taken: Bay of Biscay - Date Received: 2 Apr 1813 - From what ship: Plymouth - Born: New Orleans - Age: 23 - Released on 20 Apr 1815.

Akerman, William - Seaman - Nbr: 1223 - How taken: Sent into custody from MV Havannah - Date Received: 14 Jun 1814 - From what ship: Mill Prison (Plymouth, England) - Born: New York - Age: 28 - Released on 27 Apr 1815.

Albert, Hezekiah - Seaman - Nbr: 4632 - Prize: Orbit - Ship Type: MV - How taken: HMS Achates - When taken: 29 Jan 1813 - Where taken: Lat 44N Long 13W - Date Received: 9 Oct 1814 - From what ship: HMT Leyden, Chatham - Born: Rhode Island - Age: 20 - Released on 15 Jun 1815.

Albert, John - Seaman - Nbr: 1944 - How taken: Gave himself up from HMS Royal William - When taken: 3 Feb 1813 - Date Received: 3 Aug 1814 - From what ship: HMS Alceste, Chatham Depot - Born: New Jersey - Age: 28 - Released on 2 May 1815.

Albro, George - Seaman - Nbr: 2756 - How taken: Gave himself up from HMS Blake - When taken: 10 Dec 1812 - Date Received: 24 Aug 1814 - From what ship: HMT Liverpool, Chatham - Born: Newport - Age: 23 - Released on 26 Apr 1815.

Alder, Clough - Seaman - Nbr: 5511 - Prize: Ann Dorothy - Ship Type: Prize - How taken: HMS Maidstone - When taken: 30 Oct 1814 - Where taken: off Cape Sable Island (Canada) - Date Received: 17 Dec 1814 - From what ship: HMT Loire, Halifax - Born: Hamburg, Germany - Age: 39.

Aldridge, Richard - Seaman - Nbr: 3432 - Prize: Mary - Ship Type: Prize - How taken: HMS Bellerophon - When taken: 16 Dec 1813 - Where taken: off Land's End - Date Received: 13 Sep 1814 - From what ship: HMT Niobe, Chatham - Born: Pennsylvania - Age: 24 - Released on 28 May 1815.

Alexander, George - Seaman - Nbr: 6547 - How taken: Gave himself up from HMS Theodoria - When taken: 22 Feb 1815 - Date Received: 7 Mar 1815 - From what ship: HMT Ganges, Plymouth - Born: Maryland - Age: 32 - Race: Mulatto - Released on 11 Jul 1815.

Alexander, George - Seaman - Nbr: 2868 - How taken: Gave himself up from HMS Scorpion - When taken: 27 May 1813 - Date Received: 24 Aug 1814 - From what ship: HMT Alpheus, Chatham - Born: Philadelphia - Age: 26 - Released on 19 May 1815.

Alexander, J. B. - Seaman - Nbr: 1206 - Prize: Peggy - Ship Type: MV - How taken: HMS Indefatigable - When taken: 17 Sep 1810 - Where taken: off Bordeaux - Date Received: 4 Jun 1814 - From what ship: Dartmouth - Born: New Orleans - Age: 23 - Sent to Plymouth on 8 Jul 1814.

Alexander, James - Seaman - Nbr: 5080 - Prize: Ida - Ship Type: LM - How taken: HMS Newcastle - When taken: 9 Aug 1814 - Where taken: Long 34 - Date Received: 28 Oct 1814 - From what ship: HMT Alkbar, Halifax - Born: Baltimore - Age: 23 - Released on 29 Jun 1815.

Alexander, James - Seaman - Nbr: 6337 - Prize: Prince de Neufchatel - Ship Type: P - How taken: Leander (Newcastle Acasta) - When taken: 20 Dec 1814 - Where taken: Lat 38 Long 56 - Date Received: 19 Feb 1815 - From what ship: HMT Ganges, Plymouth - Born: Portland - Age: 27 - Released on 6 Jul 1815.

Alexander, Joseph - Seaman - Nbr: 2343 - Prize: Snap Dragon - Ship Type: P - How taken: HMS Martin - When taken: 10 Jun 1814 - Where taken: off Halifax - Date Received: 16 Aug 1814 - From what ship: HMS Dublin, Halifax - Born: Brazil - Age: 17 - Race: Creole - Released on 3 May 1815.

Alexander, Pedro - Seaman - Nbr: 5129 - Prize: Calabria - Ship Type: MV - How taken: HMS Castilian - When taken: 29 Sep 1814 - Where taken: off Ireland - Date Received: 31 Oct 1814 - From what ship: HMT Castillian - Born: Havre de Grace - Age: 36 - Released on 29 Jun 1815.

American Prisoners of War Held at Dartmoor during the War of 1812

## Alphabetical listing of names

Alexander, R. - Seaman - Nbr: 4535 - Prize: Wolf Cove - Ship Type: MV - How taken: HMS Briton - When taken: 1 Dec 1812 - Where taken: off Brest - Date Received: 8 Oct 1814 - From what ship: HMT Leyden, Chatham - Born: Massachusetts - Age: 22 - Released on 27 Apr 1815.

Alford, William - Seaman - Nbr: 5373 - Prize: James - Ship Type: MV - How taken: HMS Harpy - When taken: 16 Dec 1812 - Where taken: off Isle of France - Date Received: 31 Oct 1814 - From what ship: HMT Leyden, Chatham - Born: New Hampshire - Age: 24 - Released on 17 Jul 1815.

Allan, Asa - Seaman - Nbr: 4956 - Prize: Herald - Ship Type: P - How taken: HMS Endymion - When taken: 15 Aug 1814 - Where taken: off Nantucket - Date Received: 28 Oct 1814 - From what ship: HMT Alkbar, Halifax - Born: New Bedford - Age: 37 - Died on 14 Nov 1814.

Allen, Ambrose - Seaman - Nbr: 4009 - Prize: US Brig Rattlesnake - Ship Type: MW - How taken: HMS Leander - When taken: 13 Jul 1814 - Where taken: off Shelburne - Date Received: 6 Oct 1814 - From what ship: HMT Chesapeake, Halifax - Born: Marblehead - Age: 54 - Released on 9 Jun 1815.

Allen, Archibald - Seaman - Nbr: 5706 - Prize: Amazon - Ship Type: Prize - How taken: HMS Bulwark - When taken: 22 Sep 1814 - Where taken: Georges Bank - Date Received: 24 Dec 1814 - From what ship: HMT Penelope - Born: Morristown - Age: 20 - Died on 2 Mar 1815 from phthisis pulmonalis.

Allen, Arthur - Seaman - Nbr: 5035 - Prize: Betsey - Ship Type: Prize - How taken: HMS Pylades - When taken: 7 Sep 1814 - Where taken: Canten - Date Received: 28 Oct 1814 - From what ship: HMT Alkbar, Halifax - Born: Portland - Age: 16 - Released on 29 Jun 1815.

Allen, Charles - Seaman - Nbr: 3647 - Prize: Ulysses - Ship Type: MV - How taken: HMS Majestic - When taken: 29 Jun 1813 - Where taken: off Western Islands (England) - Date Received: 30 Sep 1814 - From what ship: HMT President - Born: Gloucester - Age: 27 - Released on 4 Jun 1815.

Allen, David - Seaman - Nbr: 4414 - Prize: Fire Fly - Ship Type: LM - How taken: HMS Revolutionnaire - When taken: 19 Oct 1813 - Where taken: at sea - Date Received: 8 Oct 1814 - From what ship: HMT Leyden, Chatham - Born: Massachusetts - Age: 20 - Released on 14 Jun 1815.

Allen, Edward - 2nd Mate - Nbr: 2540 - Prize: Sea Nymphe - Ship Type: MV - How taken: HMS Thraster - When taken: 4 Mar 1814 - Where taken: North Sea - Date Received: 16 Aug 1814 - From what ship: HMT Salvador del Mundo - Born: Nantucket - Age: 21 - Released on 3 May 1815.

Allen, Edward D. - Prize master - Nbr: 4522 - Prize: Yankee - Ship Type: P - How taken: HMS Shannon - When taken: 20 Aug 1813 - Where taken: at sea - Date Received: 8 Oct 1814 - From what ship: HMT Leyden, Chatham - Born: Connecticut - Age: 23 - Released on 15 Jun 1815.

Allen, Eveshia - Prize Master - Nbr: 806 - Prize: Collin - Ship Type: MV - How taken: HMS Helicon and HMS Whiting - When taken: 26 Oct 1813 - Where taken: Channel - Date Received: 3 Nov 1813 - From what ship: Plymouth - Born: Massachusetts - Age: 27 - Released on 26 Apr 1815.

Allen, George - Seaman - Nbr: 5205 - Prize: Porcupine - Ship Type: LM - How taken: HMS Acasta - When taken: 18 Jul 1813 - Where taken: off Halifax - Date Received: 31 Oct 1814 - From what ship: HMT Mermaid, Chatham - Born: New York - Age: 19 - Race: Black - Released on 29 Jun 1815.

Allen, Henry - Clerk - Nbr: 6153 - Prize: Lion - Ship Type: P - How taken: HMS Granicus - When taken: 2 Dec 1814 - Where taken: off Lisbon - Date Received: 18 Jan 1815 - From what ship: HMT Impregnable - Born: Salem - Age: 24 - Released on 27 Apr 1815.

Allen, Henry - Mate - Nbr: 1684 - Prize: Polly - Ship Type: P - How taken: HMS Barbados - When taken: 2 Apr 1814 - Where taken: off San Domingo - Date Received: 2 Jul 1814 - From what ship: Plymouth - Born: Salem - Age: 24 - Escaped on 23 Oct 1814.

Allen, Henry - Seaman - Nbr: 217 - Prize: Mars - Ship Type: MV - How taken: HMS Warspite - When taken: 26 Feb 1813 - Where taken: off Basque Roads (France) - Date Received: 2 Apr 1813 - From what ship: Plymouth - Born: Vermont - Age: 21 - Sent to Plymouth on 15 Jun 1813.

Allen, Isaac - Seaman - Nbr: 2034 - How taken: Gave himself up from HMS Impeteux - When taken: 2 Dec 1812 - Date Received: 3 Aug 1814 - From what ship: HMS Lyffey, Chatham Depot - Born: Philadelphia - Age: 20 - Released on 26 Apr 1815.

Allen, James - Seaman - Nbr: 662 - Prize: Joel Barlow - Ship Type: LM - How taken: HMS Briton - When taken: 3 Jul 1813 - Where taken: off Bordeaux - Date Received: 8 Sep 1813 - From what ship: Plymouth - Born: New London - Age: 44 - Released on 26 Apr 1815.

Allen, John - Seaman - Nbr: 4557 - How taken: Gave himself up from HMS Astrea - When taken: 13 Apr 1813 - Date Received: 8 Oct 1814 - From what ship: HMT Leyden, Chatham - Born: Delaware - Age: 36 - Released on 15 Jun 1815.

Allen, John - Seaman - Nbr: 1947 - How taken: Gave himself up from HMS Cornwall - When taken: 21 May 1813 -

American Prisoners of War Held at Dartmoor during the War of 1812

## Alphabetical listing of names

Date Received: 3 Aug 1814 - From what ship: HMS Alceste, Chatham Depot - Born: Barnstable - Age: 24 - Released on 2 May 1815.

Allen, John - Seaman - Nbr: 1353 - Prize: Paul Jones - Ship Type: P - How taken: HMS Leonidas - When taken: 23 May 1813 - Where taken: Channel - Date Received: 19 Jun 1814 - From what ship: Stapleton - Born: Massachusetts - Age: 37 - Released on 28 Apr 1815.

Allen, John - Seaman - Nbr: 165 - How taken: Impressed at Liverpool - When taken: 10 Jan 1813 - Date Received: 2 Apr 1813 - From what ship: Plymouth - Born: Richmond - Age: 23 - Sent to Dartmouth on 30 Jul 1813.

Allen, John - Seaman - Nbr: 98 - Prize: St. Martin's Planter - Ship Type: P - How taken: HMS Dublin - When taken: 9 Feb 1813 - Where taken: Lat 43 N, Long 33 50 W - Date Received: 2 Apr 1813 - From what ship: Plymouth - Born: Gloucester - Age: 21 - Released on 20 Apr 1815.

Allen, John - Boy - Nbr: 2789 - Prize: Prompt - Ship Type: MV - How taken: Chance (Privateer) - When taken: 28 Mar 1813 - Where taken: Bay of Biscay - Date Received: 24 Aug 1814 - From what ship: HMT Liverpool, Chatham - Born: Boston - Age: 14 - Released on 19 May 1815.

Allen, John - Seaman - Nbr: 5776 - Prize: William Penn - Ship Type: MV - How taken: HMS Acorn - When taken: 27 Oct 1812 - Where taken: Lat 14 - Date Received: 26 Dec 1814 - From what ship: HMT Argo - Born: Nantucket - Age: 19 - Released on 27 Apr 1815.

Allen, John Baptiste - Seaman - Nbr: 4967 - Prize: Herald - Ship Type: P - How taken: HMS Endymion - When taken: 15 Aug 1814 - Where taken: off Nantucket - Date Received: 28 Oct 1814 - From what ship: HMT Alkbar, Halifax - Born: Africa - Age: 40 - Race: Negro - Died on 21 Nov 1814.

Allen, John D. - Seaman - Nbr: 2805 - How taken: Apprehended at London - When taken: 17 Jul 1813 - Date Received: 24 Aug 1814 - From what ship: HMT Liverpool, Chatham - Born: New York - Age: 29 - Released on 19 May 1815.

Allen, Joseph - Seaman - Nbr: 609 - Prize: US Brig Argus - Ship Type: MW - How taken: HMS Pelican - When taken: 14 Aug 1813 - Where taken: Irish Channel - Date Received: 8 Sep 1813 - From what ship: Plymouth - Born: Boston - Age: 37 - Sent to Dartmouth on 2 Nov 1814.

Allen, Joseph - Seaman - Nbr: 2672 - Prize: Orders in Council - Ship Type: MV - How taken: HMS Surveillante - When taken: 1 Jun 1813 - Where taken: off Cape Ortegal (Spain) - Date Received: 21 Aug 1814 - From what ship: HMT Freya, Chatham - Born: New York - Age: 24 - Race: Mulatto - Released on 19 May 1815.

Allen, Lark - Seaman - Nbr: 6058 - Prize: William - Ship Type: Prize - How taken: HMS Armide - When taken: 11 Oct 1814 - Where taken: off Newport - Date Received: 28 Dec 1814 - From what ship: HMT Penelope - Born: Not legible - Age: 19 - Race: Black - Released on 3 Jul 1815.

Allen, Philip - Seaman - Nbr: 652 - Prize: Marmion - Ship Type: MV - How taken: HMS President - When taken: 14 Aug 1813 - Where taken: off Nantes - Date Received: 8 Sep 1813 - From what ship: Plymouth - Born: Maryland - Age: 28 - Race: Negro - Released on 26 Apr 1815.

Allen, Thomas - Seaman - Nbr: 5985 - Prize: McDonough - Ship Type: P - How taken: HMS Bacchante - When taken: 1 Nov 1814 - Where taken: Lat 42 Long 67 - Date Received: 27 Dec 1814 - From what ship: HMT Penelope - Born: Maine - Age: 32 - Released on 27 Apr 1815.

Allen, Thomas - Seaman - Nbr: 409 - Prize: Viper - Ship Type: MV - How taken: HMS Superb - When taken: 15 Apr 1813 - Where taken: Bay of Biscay - Date Received: 1 Jul 1813 - From what ship: Plymouth - Born: Alexandria - Age: 23 - Released on 20 Apr 1815.

Allen, William - Seaman - Nbr: 5306 - Prize: Elbridge Gerry - Ship Type: P - How taken: HMS Crescent - When taken: 25 Aug 1813 - Where taken: off St. Johns - Date Received: 31 Oct 1814 - From what ship: HMT Leyden, Chatham - Born: Boston - Age: 53 - Released on 29 Jun 1815.

Allen, William - Seaman - Nbr: 5971 - Prize: Halifax Packet - Ship Type: Prize - How taken: HMS Bulwark - When taken: 22 Sep 1814 - Where taken: Georges Bank - Date Received: 27 Dec 1814 - From what ship: HMT Penelope - Born: Newport - Age: 21 - Released on 3 Jul 1815.

Allen, William - Seaman - Nbr: 171 - Prize: Brazilian - Ship Type: MV - How taken: Olive Branch - When taken: 18 Dec 1812 - Where taken: off Porto Rico - Date Received: 2 Apr 1813 - From what ship: Plymouth - Born: Boston - Age: 35 - Sent to Dartmouth on 30 Jul 1813.

Allen, William - Seaman - Nbr: 2083 - Prize: True Blooded Yankee - Ship Type: P - How taken: HMS Hope - When taken: 24 Jun 1813 - Where taken: off Brest - Date Received: 3 Aug 1814 - From what ship: HMS Bittern, Chatham Depot - Born: Georgetown - Age: 36 - Released on 2 May 1815.

Allerd, Erick - Seaman - Nbr: 2177 - Prize: Rattlesnake - Ship Type: P - How taken: HMS Hyperion - When taken: 26 Jun 1814 - Where taken: off Cape Finisterre - Date Received: 16 Aug 1814 - From what ship: HMS Dublin, Halifax - Born: New Hampshire - Age: 37 - Released on 2 May 1815.

American Prisoners of War Held at Dartmoor during the War of 1812

## Alphabetical listing of names

Alley, Jacob - Seaman - Nbr: 4005 - Prize: US Brig Rattlesnake - Ship Type: MW - How taken: HMS Leander - When taken: 13 Jul 1814 - Where taken: off Shelburne - Date Received: 6 Oct 1814 - From what ship: HMT Chesapeake, Halifax - Born: Boston - Age: 25 - Released on 9 Jun 1815.

Alley, Samuel - Seaman - Nbr: 454 - Prize: Eliza - Ship Type: MV - How taken: HMS Surveillante - When taken: 27 Mar 1813 - Where taken: Bay of Biscay - Date Received: 8 Sep 1813 - From what ship: Plymouth - Born: New Haven - Age: 35 - Released on 26 Apr 1815.

Allison, J. B. - Mate - Nbr: 3861 - Prize: Thorn - Ship Type: MV - How taken: HMS Bulwark - When taken: 9 Jul 1814 - Where taken: off Nantucket - Date Received: 5 Oct 1814 - From what ship: HMT Orpheus, Halifax - Born: Alexandria - Age: 44 - Released on 10 May 1815.

Allison, Thomas - Seaman - Nbr: 3852 - Prize: Dominique - Ship Type: LM - How taken: HMS Dotterel - When taken: 21 May 1814 - Where taken: off Charleston - Date Received: 5 Oct 1814 - From what ship: HMT Orpheus, Halifax - Born: Charleston - Age: 24 - Released on 9 Jun 1815.

Allison, William R. - Seaman - Nbr: 4592 - Prize: Prize to the Diomede - Ship Type: P - How taken: HMS Sapphire - When taken: 24 Feb 1814 - Where taken: off America - Date Received: 9 Oct 1814 - From what ship: HMT Leyden, Chatham - Born: Alexandria - Age: 28 - Released on 15 Jun 1815.

Allister, Isaac - Seaman - Nbr: 570 - Prize: US Brig Argus - Ship Type: MW - How taken: HMS Pelican - When taken: 14 Apr 1813 - Where taken: Irish Channel - Date Received: 8 Sep 1813 - From what ship: Plymouth - Born: Hobeanos, America - Age: 26 - Released on 6 May 1814.

Almeno, Jose - Seaman - Nbr: 4825 - Prize: President - Ship Type: P - How taken: HMS Pique - When taken: 7 May 1814 - Where taken: off Porto Rico - Date Received: 9 Oct 1814 - From what ship: HMT Freya, Halifax - Born: Cartagena - Age: 16 - Died on 3 Nov 1814.

Alston, Andrew - Seaman - Nbr: 6084 - Prize: David Porter - Ship Type: LM - How taken: HMS Pylades - When taken: 13 Sep 1814 - Where taken: Georges Bank - Date Received: 28 Dec 1814 - From what ship: HMT Penelope - Born: Not legible - Age: 17 - Released on 3 Jul 1815.

Alston, Richard - Seaman - Nbr: 5156 - Prize: Volante - Ship Type: P - How taken: HMS Curlew - When taken: 25 Nov 1813 - Where taken: off Halifax - Date Received: 31 Oct 1814 - From what ship: HMT Mermaid, Chatham - Born: Marblehead - Age: 21 - Released on 29 Jun 1815.

Alton, Peter - Marine - Nbr: 1276 - How taken: Gave himself up - Date Received: 14 Jun 1814 - From what ship: Mill Prison (Plymouth, England) - Born: Balston - Age: 21 - Released on 28 Apr 1815.

Amerson, Charles - Seaman - Nbr: 1518 - Prize: Essex - Ship Type: MV - How taken: HMS Pyramus - When taken: 20 Apr 1813 - Where taken: Bay of Biscay - Date Received: 23 Jun 1814 - From what ship: Stapleton - Born: Massachusetts - Age: 18 - Released on 1 May 1815.

Amos, Cheney - Seaman - Nbr: 544 - Prize: Orders in Council - Ship Type: MV - How taken: HMS Surveillante - When taken: 1 Jun 1813 - Where taken: off Cape Ortegal (Spain) - Date Received: 8 Sep 1813 - From what ship: Plymouth - Born: Litchfield - Age: 21 - Released on 26 Apr 1815.

Amos, Elijah - Seaman - Nbr: 6051 - Prize: Enterprize - Ship Type: MV - How taken: HMS Niemen - When taken: 30 Aug 1814 - Where taken: off Bahamas - Date Received: 28 Dec 1814 - From what ship: HMT Penelope - Born: Washington - Age: 27 - Race: Mulatto - Released on 3 Jul 1815.

Amos, Isaac - Seaman - Nbr: 1946 - How taken: Gave himself up from HMS Sultan - When taken: 3 Feb 1813 - Date Received: 3 Aug 1814 - From what ship: HMS Alceste, Chatham Depot - Born: Boston - Age: 20 - Released on 2 May 1815.

Amos, Peter - Passenger - Nbr: 3821 - Prize: Nimble - Ship Type: Prize - How taken: HMS Arab - When taken: 5 Apr 1814 - Where taken: Lat 37 Long 65 - Date Received: 5 Oct 1814 - From what ship: HMT Orpheus, Halifax - Born: Martha's Vineyard - Age: 22 - Race: Mulatto - Died on 18 Feb 1815 from variola.

Anders, John - Seaman - Nbr: 3204 - How taken: Gave himself up from HMS America - When taken: 16 Jul 1813 - Date Received: 11 Sep 1814 - From what ship: HMT Freya, Chatham - Born: Alexandria - Age: 24 - Released on 28 May 1815.

Anders, John - Seaman - Nbr: 3200 - How taken: Gave himself up from HMS America - When taken: 16 Jul 1813 - Date Received: 11 Sep 1814 - From what ship: HMT Freya, Chatham - Born: Charleston - Age: 27 - Race: Black - Released on 28 May 1815.

Anderson, Alexander - Seaman - Nbr: 5067 - Prize: Ida - Ship Type: LM - How taken: HMS Newcastle - When taken: 9 Aug 1814 - Where taken: Long 34 - Date Received: 28 Oct 1814 - From what ship: HMT Alkbar, Halifax - Born: Maryland - Age: 29 - Released on 29 Jun 1815.

Anderson, Daniel - Seaman - Nbr: 1440 - Prize: Leo - Ship Type: LM - How taken: HMS Magicienne - When taken: 4 Jun 1813 - Where taken: off France - Date Received: 19 Jun 1814 - From what ship: Stapleton - Born:

American Prisoners of War Held at Dartmoor during the War of 1812

## Alphabetical listing of names

Massachusetts - Age: 22 - Released on 28 Apr 1815.
Anderson, David - Cook - Nbr: 4675 - Prize: Portsmouth Packet - Ship Type: Prize - How taken: HMS Fantome - When taken: 5 Oct 1813 - Where taken: off Portland - Date Received: 9 Oct 1814 - From what ship: HMT Leyden, Chatham - Born: New York - Age: 31 - Race: Mulatto - Released on 15 Jun 1815.
Anderson, David - Seaman - Nbr: 1557 - Prize: Caroline - Ship Type: MV - How taken: HMS Medusa - When taken: 12 Apr 1813 - Where taken: Bay of Biscay - Date Received: 23 Jun 1814 - From what ship: Stapleton - Born: Maryland - Age: 31 - Race: Negro - Released on 1 May 1815.
Anderson, Ebenezer - Seaman - Nbr: 4044 - Prize: US Brig Rattlesnake - Ship Type: MW - How taken: HMS Leander - When taken: 13 Jul 1814 - Where taken: off Shelburne - Date Received: 6 Oct 1814 - From what ship: HMT Chesapeake, Halifax - Born: Norfolk - Age: 29 - Released on 13 Jun 1815.
Anderson, Edward - Seaman - Nbr: 992 - Prize: Apparencen - Ship Type: MV - How taken: HMS Castilian - When taken: 27 Jan 1814 - Where taken: off Ushant (France) - Date Received: 31 Jan 1814 - From what ship: Plymouth - Born: Baltimore - Age: 26 - Released on 27 Apr 1815.
Anderson, Edward - Seaman - Nbr: 2465 - Prize: US Sloop Frolic - Ship Type: MW - How taken: HMS Orpheus - When taken: 20 Apr 1814 - Where taken: off Cuba - Date Received: 16 Aug 1814 - From what ship: HMT Queen, Halifax - Born: Newbury - Age: 28 - Released on 3 May 1815.
Anderson, Francis - Seaman - Nbr: 3952 - Prize: Rolla - Ship Type: P - How taken: HMS Loire - When taken: 10 Dec 1813 - Where taken: off Bull Island (South Carolina) - Date Received: 5 Oct 1814 - From what ship: HMT President, Halifax - Born: Connecticut - Age: 17 - Released on 9 Jun 1815.
Anderson, George - Seaman - Nbr: 1042 - Prize: US Brig Argus - Ship Type: MW - How taken: HMS Pelican - When taken: 16 Aug 1813 - Where taken: Irish Channel - Date Received: 10 May 1814 - From what ship: Plymouth - Born: Liddy - Age: 24 - Sent to Dartmouth on 19 Oct 1814.
Anderson, George - Seaman - Nbr: 5289 - Prize: Volunteer - Ship Type: MV - How taken: Victoria (Privateer) - When taken: 26 Dec 1813 - Where taken: at sea - Date Received: 31 Oct 1814 - From what ship: HMT Leyden, Chatham - Born: Philadelphia - Age: 36 - Race: Black - Released on 29 Jun 1815.
Anderson, Henry - Seaman - Nbr: 5273 - Prize: Ajax - Ship Type: MV - How taken: HMS Resolution - When taken: 13 Apr 1813 - Where taken: off Western Islands (England) - Date Received: 31 Oct 1814 - From what ship: HMT Leyden, Chatham - Born: Addington - Age: 20 - Released on 29 Jun 1815.
Anderson, Jacob - Seaman - Nbr: 2225 - Prize: Hussar - Ship Type: P - How taken: HMS Saturn - When taken: 25 May 1814 - Where taken: off Sandy Hook - Date Received: 16 Aug 1814 - From what ship: HMS Dublin, Halifax - Born: Portland - Age: 24 - Died on 26 Jan 1815 from phthisis pulmonalis.
Anderson, James - Seaman - Nbr: 747 - How taken: Sent into custody at Plymouth - Date Received: 3 Nov 1813 - From what ship: Plymouth - Born: Boston - Age: 27 - Released on 26 Apr 1815.
Anderson, James - Seaman - Nbr: 2773 - How taken: Gave himself up from HMS Leonidas - Date rec'd: 24 Aug 1814 - From what ship: HMT Liverpool, Chatham - Born: Long Island - Age: 39 - Released on 19 May 1815.
Anderson, John - Seaman - Nbr: 6471 - Prize: Decatur - Ship Type: P - How taken: HMS Rhin - When taken: 5 Jun 1814 - Where taken: off San Domingo - Date Received: 3 Mar 1815 - From what ship: HMT Ganges, Plymouth - Born: Charleston - Age: 20 - Race: Creole - Released on 14 Jun 1815.
Anderson, John - Seaman - Nbr: 6388 - How taken: Gave himself up from HMS Iris - When taken: Feb 1813 - Date Received: 24 Feb 1815 - From what ship: HMT Ganges, Plymouth - Born: Charlestown - Age: 32 - Race: Mulatto - Released on 11 Jul 1815.
Anderson, John - Seaman - Nbr: 5121 - Prize: Scourge - Ship Type: P - How taken: HMS Ringdove - When taken: 12 Aug 1814 - Where taken: off Portland - Date Received: 28 Oct 1814 - From what ship: HMT Alkbar, Halifax - Born: Cape Francis - Age: 20 - Race: Mulatto - Released on 29 Jun 1815.
Anderson, John - Seaman - Nbr: 5030 - Prize: Landrail - Ship Type: Prize - How taken: HMS Wasp - When taken: 27 Jul 1814 - Where taken: Georges Bank - Date Received: 28 Oct 1814 - From what ship: HMT Alkbar, Halifax - Born: Norway - Age: 31 - Released on 21 Jun 1815.
Anderson, John - Seaman - Nbr: 4885 - Prize: Saratoga - Ship Type: P - How taken: HMS Barracouta - When taken: 9 Oct 1814 - Where taken: off Western Islands (England) - Date Received: 24 Oct 1814 - From what ship: HMT Salvador del Mundo - Born: Rhode Island - Age: 33 - Released on 21 Jun 1815.
Anderson, John - Seaman - Nbr: 4553 - Prize: Pilot - Ship Type: LM - How taken: Victoria (Privateer) - When taken: 28 Jan 1814 - Where taken: off Bordeaux - Date Received: 8 Oct 1814 - From what ship: HMT Leyden, Chatham - Born: Maryland - Age: 24 - Race: Black - Released on 15 Jun 1815.
Anderson, Joseph - Seaman - Nbr: 1341 - Prize: Paul Jones - Ship Type: P - How taken: HMS Leonidas - When taken: 23 May 1813 - Where taken: Channel - Date Received: 19 Jun 1814 - From what ship: Stapleton -

American Prisoners of War Held at Dartmoor during the War of 1812

## Alphabetical listing of names

Born: Maryland - Age: 26 - Race: Negro - Released on 28 Apr 1815.

Anderson, Joseph - Seaman - Nbr: 2607 - How taken: Gave himself up from HMS Diomede - When taken: 12 Oct 1812 - Date Received: 21 Aug 1814 - From what ship: HMT Freya, Chatham - Born: Baltimore - Age: 33 - Released on 26 Apr 1815.

Anderson, Robert - Seaman - Nbr: 110 - How taken: Gave himself up from HMS Foxhound - Date Received: 2 Apr 1813 - From what ship: Plymouth - Born: Rhode Island - Age: 26 - Sent to Chatham on 27 May 1813.

Anderson, Thomas - Seaman - Nbr: 3472 - How taken: Gave himself up from HMS Prince - When taken: 12 Sep 1814 - Date Received: 19 Sep 1814 - From what ship: HMT Salvador del Mundo - Born: Savannah - Age: 24 - Released on 4 Jun 1815.

Anderson, William - Seaman - Nbr: 729 - Prize: Ned - Ship Type: LM - How taken: HMS Royalist - When taken: 6 Sep 1813 - Where taken: Bay of Biscay - Date Received: 27 Sep 1813 - From what ship: Plymouth - Born: Virginia - Age: 19 - Released on 26 Apr 1815.

Anderson, William - Seaman - Nbr: 2528 - Prize: Diomede - Ship Type: P - How taken: HMS Rifleman - When taken: 28 Jul 1814 - Where taken: off Halifax - Date Received: 16 Aug 1814 - From what ship: HMT Queen, Halifax - Born: Massachusetts - Age: 20 - Released on 3 May 1815.

Andrews, Asa - Seaman - Nbr: 2396 - Prize: US Sloop Frolic - Ship Type: MW - How taken: HMS Orpheus - When taken: 20 Apr 1814 - Where taken: off Cuba - Date Received: 16 Aug 1814 - From what ship: HMT Queen, Halifax - Born: Beverly - Age: 23 - Sent to Dartmouth on 19 Oct 1814.

Andress, Daniel - Seaman - Nbr: 1613 - Prize: Fox - Ship Type: LM - How taken: HMS Pheasant - When taken: 23 Apr 1813 - Where taken: Bay of Biscay - Date Received: 23 Jun 1814 - From what ship: Stapleton - Born: Philadelphia - Age: 26 - Released on 1 May 1815.

Andress, George - Seaman - Nbr: 812 - Prize: Sybille - Ship Type: MV - How taken: HMS Zenobia - When taken: 27 Jun 1813 - Where taken: off Cape St. Mary's - Date Received: 3 Nov 1813 - From what ship: Plymouth - Born: Philadelphia - Age: 30 - Released on 26 Apr 1815.

Andrews, Benjamin - Seaman - Nbr: 896 - Prize: Squirrel - Ship Type: MV - How taken: HMS Belle Poule - When taken: 14 Dec 1813 - Where taken: at sea - Date Received: 31 Jan 1814 - From what ship: Plymouth - Born: Massachusetts - Age: 23 - Released on 27 Apr 1815.

Andrews, Charles - Seaman - Nbr: 678 - Prize: VA Planter - Ship Type: MV - How taken: HMS Pyramus - When taken: 18 Mar 1813 - Where taken: off Nantes - Date Received: 14 Sep 1813 - From what ship: Plymouth - Born: Newport - Age: 36 - Released on 20 Apr 1815.

Andrews, Charles - Seaman - Nbr: 381 - Prize: VA Planter - Ship Type: MV - How taken: HMS Pyramus - When taken: 18 Mar 1813 - Where taken: off Nantes - Date Received: 1 Jul 1813 - From what ship: Plymouth - Born: Newport - Age: 31 - Sent to Mill Prison (Plymouth, England) on 3 Aug 1813.

Andrews, James - Boy - Nbr: 5492 - Prize: General Putnam - Ship Type: P - How taken: HMS Leander - When taken: 8 Nov 1814 - Where taken: Long 65 Lat 42 - Date Received: 17 Dec 1814 - From what ship: HMT Loire, Halifax - Born: Beverly - Age: 17 - Released on 1 Jul 1815.

Andrews, James - Seaman - Nbr: 3305 - Prize: Porcupine - Ship Type: LM - How taken: British squadron - When taken: 3 Jun 1813 - Where taken: off Cape Sable Island (Canada) - Date Received: 13 Sep 1814 - From what ship: HMT Niobe, Chatham - Born: Boston - Age: 21 - Released on 28 May 1815.

Andrews, Jeremiah - Seaman - Nbr: 5607 - Prize: Mary - Ship Type: MV - How taken: Tamer - When taken: Sep 1814 - Date Received: 24 Dec 1814 - From what ship: HMT Tay - Born: Long Island - Age: 29 - Race: Mulatto - Released on 5 Jul 1815.

Andrews, Joshua - Seaman - Nbr: 5108 - Prize: David Porter - Ship Type: LM - How taken: HMS Pylades - When taken: 12 Sep 1814 - Where taken: Georges Bank - Date Received: 28 Oct 1814 - From what ship: HMT Alkbar, Halifax - Born: Ipswich - Age: 21 - Died on 21 Nov 1814.

Andrews, Nathaniel - Boy - Nbr: 2974 - Prize: Frolic - Ship Type: P - How taken: HMS Heron - When taken: 25 Jan 1814 - Where taken: off St. Thomas - Date Received: 29 Aug 1814 - From what ship: HMT Bittern - Born: Beverly - Age: 16 - Released on 19 May 1815.

Andrews, Peter - Seaman - Nbr: 4088 - Prize: Nelly - Ship Type: Prize - How taken: HMS Bulwark - When taken: 10 Jul 1814 - Where taken: Georges Bank - Date Received: 6 Oct 1814 - From what ship: HMT Chesapeake, Halifax - Born: Pennsylvania - Age: 24 - Released on 13 Jun 1815.

Andrews, Samuel - Seaman - Nbr: 6375 - Prize: US Brig Syren - Ship Type: MW - How taken: HMS Medway - When taken: 12 Jul 1814 - Where taken: off Cape of Good Hope - Date Received: 24 Feb 1815 - From what ship: HMT Ganges, Plymouth - Born: Norway - Age: 21 - Released on 11 Jul 1815.

Andrews, Thomas - Seaman - Nbr: 129 - Prize: Good Intent - Ship Type: P - How taken: HMS Rota - When taken:

American Prisoners of War Held at Dartmoor during the War of 1812

## Alphabetical listing of names

26 Jan 1813 - Where taken: Lat 43'30" N, Long 20 W - Date Received: 12 Apr 1813 - From what ship: Plymouth - Born: Marblehead - Age: 21 - Sent to Dartmouth on 30 Jul 1813.

Angel, Silvester - Seaman - Nbr: 2702 - How taken: Gave himself up from HMS Illustrious - When taken: 1 Dec 1813 - Date Received: 21 Aug 1814 - From what ship: HMT Freya, Chatham - Born: New London - Age: 26 - Released on 19 May 1815.

Anon, John - Seaman - Nbr: 5417 - How taken: Impressed at London - When taken: 13 Oct 1814 - Date Received: 31 Oct 1814 - From what ship: HMT Leyden, Chatham - Born: Philadelphia - Age: 22 - Race: Black - Released on 1 Jul 1815.

Anthony, James - Seaman - Nbr: 5279 - Prize: Blockade - Ship Type: P - How taken: HMS Recruit - When taken: 17 Aug 1813 - Where taken: off America - Date Received: 31 Oct 1814 - From what ship: HMT Leyden, Chatham - Born: Orleans - Age: 28 - Released on 29 Jun 1815.

Anthony, John - Seaman - Nbr: 921 - Prize: US Frigate Chesapeake - Ship Type: MW - How taken: HMS Shannon - When taken: 1 Jun 1813 - Where taken: off Boston - Date Received: 31 Jan 1814 - From what ship: Plymouth - Born: Beyo, South America - Age: 20 - Sent to Dartmouth on 19 Oct 1814.

Anthony, John - Seaman - Nbr: 5167 - Prize: Volante - Ship Type: P - How taken: HMS Curlew - When taken: 25 Nov 1813 - Where taken: off Halifax - Date Received: 31 Oct 1814 - From what ship: HMT Mermaid, Chatham - Born: New Orleans - Age: 49 - Released on 29 Jun 1815.

Antoine, John - Seaman - Nbr: 4002 - Prize: US Brig Rattlesnake - Ship Type: MW - How taken: HMS Leander - When taken: 13 Jul 1814 - Where taken: off Shelburne - Date Received: 6 Oct 1814 - From what ship: HMT Chesapeake, Halifax - Born: Bedford - Age: 28 - Released on 9 Jun 1815.

Antoni, Francis - Seaman - Nbr: 5425 - Prize: Volunteer - Ship Type: MV - How taken: Victoria (Privateer) - When taken: 26 Dec 1813 - Where taken: at sea - Date Received: 31 Oct 1814 - From what ship: HMT Leyden, Chatham - Born: Portugal - Age: 23 - Released on 1 Jul 1815.

Antonia, Peter - Seaman - Nbr: 5539 - Prize: Ann - Ship Type: Prize - How taken: HMS Hamadryad - When taken: 10 Sep 1814 - Where taken: Lat 41 Long 23 - Date Received: 17 Dec 1814 - From what ship: HMT Loire, Halifax - Born: Vigo (Spain) - Age: 25 - Released on 1 Jul 1815.

Antony, Stephen - Seaman - Nbr: 1515 - Prize: Courier - Ship Type: LM - How taken: HMS Rover - When taken: 14 Mar 1813 - Where taken: Bay of Biscay - Date Received: 23 Jun 1814 - From what ship: Stapleton - Born: Maryland - Age: 28 - Race: Colored - Released on 1 May 1815.

Appene - Chinese servant - Nbr: 590 - Prize: US Brig Argus - Ship Type: MW - How taken: HMS Pelican - When taken: 14 Aug 1813 - Where taken: Irish Channel - Date Received: 8 Sep 1813 - From what ship: Plymouth - Born: Canton, China - Age: 21 - Race: Chinese - Sent to Dartmouth on 2 Nov 1814.

Appleton, Daniel - Seaman - Nbr: 2425 - Prize: US Sloop Frolic - Ship Type: MW - How taken: HMS Orpheus - When taken: 20 Apr 1814 - Where taken: off Cuba - Date Received: 16 Aug 1814 - From what ship: HMT Queen, Halifax - Born: Ipswich - Age: 28 - Died on 4 Jan 1815 from variola.

Appleton, John - Gunner - Nbr: 4391 - Prize: Portsmouth Packet - Ship Type: Prize - How taken: HMS Fantome - When taken: 5 Oct 1813 - Where taken: off Portland - Date Received: 8 Oct 1814 - From what ship: HMT Leyden, Chatham - Born: Virginia - Age: 34 - Released on 14 Jun 1815.

Archer, Benjamin - Seaman - Nbr: 6127 - Prize: Betsey - Ship Type: Prize - How taken: HMS Bellerophon - When taken: 2 Nov 1814 - Where taken: Long 61 - Date Received: 6 Jan 1814 - From what ship: HMT Impregnable - Born: Salem - Age: 18 - Released on 5 Jul 1815.

Archer, Daniel - Prize master - Nbr: 5698 - Prize: McDonough - Ship Type: P - How taken: HMS Bacchante - When taken: 7 Nov 1814 - Where taken: Georges Bank - Date Received: 24 Dec 1814 - From what ship: HMT Penelope - Born: Salem - Age: 22 - Died on 14 Jan 1815 from pneumonia.

Archer, James - Seaman - Nbr: 5147 - Prize: Thorn - Ship Type: P - How taken: HMS Shannon - When taken: 7 Nov 1813 - Where taken: off Newfoundland - Date Received: 31 Oct 1814 - From what ship: HMT Mermaid, Chatham - Born: Berwick - Age: 22 - Released on 29 Jun 1815.

Archer, Joseph - Seaman - Nbr: 886 - Prize: General Kempt - Ship Type: P - How taken: HMS Foxhound - When taken: 18 Dec 1813 - Where taken: Lat 48'60" Long 5'7" - Date Received: 31 Jan 1814 - From what ship: Plymouth - Born: Salem - Age: 28 - Released on 27 Apr 1815.

Archer, Samuel - Steward - Nbr: 3767 - Prize: Fame - Ship Type: P - How taken: HMS Thistle - When taken: 10 Apr 1814 - Where taken: after being cast ashore on a seal island - Date Received: 30 Sep 1814 - From what ship: HMT President, Halifax - Born: Salem - Age: 24 - Released on 9 Jun 1815.

Argent, William - Seaman - Nbr: 2209 - Prize: Hussar - Ship Type: P - How taken: HMS Saturn - When taken: 25 May 1814 - Where taken: off Sandy Hook - Date Received: 16 Aug 1814 - From what ship: HMS Dublin,

American Prisoners of War Held at Dartmoor during the War of 1812

## Alphabetical listing of names

Halifax - Born: Boston - Age: 29 - Released on 2 May 1815.

Ariskins, Samuel - Seaman - Nbr: 2507 - Prize: General Stark - Ship Type: P - How taken: HMS Sophie - When taken: 24 Apr 1814 - Where taken: off Bermuda - Date Received: 16 Aug 1814 - From what ship: HMT Queen, Halifax - Born: Pennsylvania - Age: 25 - Released on 3 May 1815.

Armstead, Edward - Seaman - Nbr: 1104 - Prize: Diamond - Ship Type: P - How taken: HMS Vengeur - When taken: 6 Mar 1814 - Where taken: Lat 47.4 Long 5.60 - Date Received: 10 May 1814 - From what ship: Plymouth - Born: Norfolk - Age: 21 - Released on 20 Mar 1815.

Armstrong William - Seaman - Nbr: 76 - Prize: Rolla - Ship Type: MV - How taken: HMS Surveillante - When taken: 11 Feb 1813 - Where taken: Bay of Biscay - Date Received: 2 Apr 1813 - From what ship: Plymouth - Born: New York - Age: 27 - Sent to Plymouth on 15 Jun 1813.

Armstrong, Andrew - Prize Master - Nbr: 2242 - Prize: Rolla - Ship Type: Prize - How taken: HMS Loire - When taken: 9 Dec 1813 - Where taken: off Newport - Date Received: 16 Aug 1814 - From what ship: HMS Dublin, Halifax - Born: Delaware - Age: 31 - Released on 2 May 1815.

Armstrong, Charles - 2nd Mate - Nbr: 6430 - Prize: Chance - Ship Type: P - How taken: HMS Statira - When taken: 1 Apr 1814 - Where taken: Lat 38 Long 24 - Date Received: 3 Mar 1815 - From what ship: HMT Ganges, Plymouth - Born: Hampton - Age: 34 - Released on 4 Jun 1815.

Armstrong, David - Seaman - Nbr: 1498 - Prize: Margaret - Ship Type: Prize - How taken: HMS Foxhound - When taken: 27 May 1814 - Where taken: off Isles of Scilly - Date Received: 20 Jun 1814 - From what ship: Mill Prison (Plymouth, England) - Born: Baltimore - Age: 19 - Released on 1 May 1815.

Armstrong, Elisha - Seaman - Nbr: 2143 - How taken: Gave himself up from HMS Victory - When taken: 5 Dec 1812 - Date Received: 8 Aug 1814 - From what ship: HMT Raven, Chatham - Born: Harford - Age: 22 - Released on 26 Apr 1815.

Armstrong, George - Seaman - Nbr: 2205 - Prize: Hussar - Ship Type: P - How taken: HMS Saturn - When taken: 25 May 1814 - Where taken: off Sandy Hook - Date Received: 16 Aug 1814 - From what ship: HMS Dublin, Halifax - Born: New Castle - Age: 23 - Released on 2 May 1815.

Armstrong, John - Seaman - Nbr: 1452 - How taken: Impressed at Liverpool - When taken: 19 Mar 1813 - Date Received: 19 Jun 1814 - From what ship: Stapleton - Born: Virginia - Age: 26 - Released on 28 Apr 1815.

Armstrong, Joseph - Seaman - Nbr: 1277 - How taken: Sent into custody from HMS Egmont - Date rec'd: 14 Jun 1814 - From what ship: Mill Prison (Plymouth, England) - Born: Not listed - Age: 45 - Race: Black - Released on 28 Apr 1815.

Armstrong, Thomas - Seaman - Nbr: 1852 - How taken: Gave himself up from HMS Swiftsure - When taken: 26 Dec 1812 - Date Received: 21 Jul 1814 - From what ship: HMT Redbeard & Pincher, Chatham Depot - Born: Lancaster - Age: 26 - Released on 26 Apr 1815.

Armstrong, William - Seaman - Nbr: 1473 - Prize: Governor Gerry - Ship Type: MV - How taken: HMS Lyra - When taken: 29 May 1813 - Where taken: Bay of Biscay - Date Received: 19 Jun 1814 - From what ship: Stapleton - Born: Philadelphia - Age: 36 - Race: Negro - Released on 28 Apr 1815.

Arnandez, Justine - Seaman - Nbr: 4882 - Prize: Saratoga - Ship Type: P - How taken: HMS Barracouta - When taken: 9 Oct 1814 - Where taken: off Western Islands (England) - Date Received: 24 Oct 1814 - From what ship: HMT Salvador del Mundo - Born: Leghorn, Italy - Age: 26 - Released on 21 Jun 1815.

Arnold, Alfred - Seaman - Nbr: 425 - Prize: Viper - Ship Type: MV - How taken: HMS Superb - When taken: 15 Apr 1813 - Where taken: Bay of Biscay - Date Received: 1 Jul 1813 - From what ship: Plymouth - Born: Rhode Island - Age: 23 - Sent to Plymouth on 7 Dec 1813.

Arnold, Benjamin - Marine - Nbr: 4145 - Prize: Prize to the Diomede - Ship Type: Prize - How taken: HMS Sapphire - When taken: 27 Feb 1814 - Where taken: at sea - Date Received: 6 Oct 1814 - From what ship: HMT Niobe, Chatham - Born: Massachusetts - Age: 23 - Released on 13 Jun 1815.

Arnold, James - Seaman - Nbr: 5426 - Prize: Rattlesnake - Ship Type: LM - How taken: HMS Rhin - When taken: 12 May 1814 - Where taken: off Bermuda - Date Received: 31 Oct 1814 - From what ship: HMT Leyden, Chatham - Born: Providence - Age: 30 - Released on 1 Jul 1815.

Arnold, James - Master - Nbr: 1777 - How taken: Apprehended at Plymouth - When taken: 13 Jul 1813 - Date Received: 20 Jul 1814 - From what ship: HMS Milford, Plymouth - Born: Weymouth - Age: 27 - Released on 1 May 1815.

Arnold, John - Seaman - Nbr: 6016 - Prize: McDonough - Ship Type: P - How taken: HMS Bacchante - When taken: 1 Nov 1814 - Where taken: Lat 42 Long 67 - Date Received: 28 Dec 1814 - From what ship: HMT Penelope - Born: Lebanon - Age: 33 - Released on 3 Jul 1815.

Arnold, Obadiah - Seaman - Nbr: 4579 - Prize: Bunker Hill - Ship Type: P - How taken: HMS Pomone - When

American Prisoners of War Held at Dartmoor during the War of 1812

## Alphabetical listing of names

taken: 2 Mar 1814 - Where taken: Channel - Date Received: 9 Oct 1814 - From what ship: HMT Leyden, Chatham - Born: Not legible - Released on 15 Jun 1815.

Arnold, William - Seaman - Nbr: 1906 - Prize: Tom Thumb - Ship Type: MV - How taken: Lyon (Privateer) - When taken: 15 Feb 1813 - Where taken: Bay of Biscay - Date Received: 3 Aug 1814 - From what ship: HMS Alceste, Chatham Depot - Born: Baltimore - Age: 18 - Released on 2 May 1815.

Arthur, John - Seaman - Nbr: 2371 - Prize: Snap Dragon - Ship Type: P - How taken: HMS Martin - When taken: 10 Jun 1814 - Where taken: off Halifax - Date Received: 16 Aug 1814 - From what ship: HMT Queen, Halifax - Born: Nantucket - Age: 23 - Released on 3 Apr 1815.

Artis, William - Seaman - Nbr: 5323 - Prize: Juliana Smith - Ship Type: P - How taken: HMS Nymphe - When taken: 11 May 1813 - Where taken: off Cape Sable Island (Canada) - Date Received: 31 Oct 1814 - From what ship: HMT Leyden, Chatham - Born: Dunley - Age: 56 - Race: Black - Released on 29 Jun 1815.

Ash, Oliver - Seaman - Nbr: 6516 - How taken: Gave himself up from HMS Orontes - When taken: Feb 1815 - Date Received: 3 Mar 1815 - From what ship: HMT Ganges, Plymouth - Born: Norfolk - Age: 23 - Race: Mulatto - Released on 11 Jul 1815.

Ash, Sampson - Boy - Nbr: 645 - Prize: US Brig Argus - Ship Type: MW - How taken: HMS Pelican - When taken: 14 Aug 1813 - Where taken: Irish Channel - Date Received: 8 Sep 1813 - From what ship: Plymouth - Born: Baltimore - Age: 19 - Sent to Dartmouth on 30 Jul 1813.

Ashby, William - 2nd Lieutenant - Nbr: 2198 - Prize: Hussar - Ship Type: P - How taken: HMS Saturn - When taken: 25 May 1814 - Where taken: off Sandy Hook - Date Received: 16 Aug 1814 - From what ship: HMS Dublin, Halifax - Born: Connecticut - Age: 32 - Released on 2 May 1815.

Ashfield, Henry - Seaman - Nbr: 2793 - Prize: Weazel - Ship Type: MV - How taken: HMS Foxhound - When taken: 25 Mar 1813 - Where taken: Bay of Biscay - Date Received: 24 Aug 1814 - From what ship: HMT Liverpool, Chatham - Born: New York - Age: 19 - Released on 19 May 1815.

Ashmore, Edward - Seaman - Nbr: 716 - Prize: Ned - Ship Type: LM - How taken: HMS Royalist - When taken: 6 Sep 1813 - Where taken: Bay of Biscay - Date Received: 27 Sep 1813 - From what ship: Plymouth - Born: Laberton - Age: 26 - Released on 26 Apr 1815.

Ashton, Joseph - Seaman - Nbr: 4001 - Prize: US Brig Rattlesnake - Ship Type: MW - How taken: HMS Leander - When taken: 13 Jul 1814 - Where taken: off Shelburne - Date Received: 6 Oct 1814 - From what ship: HMT Chesapeake, Halifax - Born: Norfolk - Age: 29 - Released on 9 Jun 1815.

Ashton, William - Mate - Nbr: 3604 - Prize: Blanche - Ship Type: MV - How taken: HMS Barbados - When taken: 24 Jan 1814 - Where taken: off St. Bartholomew - Date Received: 30 Sep 1814 - From what ship: HMT Sybella - Born: Marblehead - Age: 26 - Released on 4 Jun 1815.

Astropp, Hans Christopher - Seaman - Nbr: 859 - Prize: Amiable - Ship Type: LM - How taken: HMS Magnificent - When taken: 30 Oct 1813 - Where taken: off Bordeaux - Date Received: 4 Dec 1813 - From what ship: Plymouth - Born: Bergen, Norway - Age: 29 - Released on 7 Feb 1814 (Danish citizen).

Athroun, Samuel - 2nd Mate - Nbr: 108 - Prize: Rolla - Ship Type: MV - How taken: HMS Surveillante - When taken: 11 Feb 1813 - Where taken: Bay of Biscay - Date Received: 2 Apr 1813 - From what ship: Plymouth - Born: Philadelphia - Age: 30 - Released on 20 Apr 1815.

Atkins, Henry - Seaman - Nbr: 5799 - Prize: US Frigate Superior - Ship Type: MW - How taken: British gunboats - When taken: 26 Aug 1814 - Where taken: Lake Ontario - Date Received: 26 Dec 1814 - From what ship: HMT Argo - Born: Virginia - Age: 22 - Released on 16 Jun 1815.

Atkins, Joseph - Seaman - Nbr: 1379 - Prize: Courier - Ship Type: P - How taken: HMS Rover - When taken: 14 May 1813 - Where taken: Bay of Biscay - Date Received: 19 Jun 1814 - From what ship: Stapleton - Born: Massachusetts - Age: 21 - Released on 28 Apr 1815.

Atkins, Uriah - Seaman - Nbr: 23 - How taken: Apprehended at Gibraltar - When taken: 8 Aug 1813 - Date rec'd: 2 Apr 1813 - From what ship: Plymouth - Born: Buxton - Age: 26 - Sent to Plymouth on 15 Jun 1813.

Atkins, William - Seaman - Nbr: 2437 - Prize: US Sloop Frolic - Ship Type: MW - How taken: HMS Orpheus - When taken: 20 Apr 1814 - Where taken: off Cuba - Date Received: 16 Aug 1814 - From what ship: HMT Queen, Halifax - Born: Salem - Age: 26 - Released on 3 May 1815.

Atkinson, Charles - Seaman - Nbr: 5368 - Prize: Rambler - Ship Type: MV - How taken: Morley (Transport) - When taken: 10 Feb 1813 - Where taken: off Isle of France - Date Received: 31 Oct 1814 - From what ship: HMT Leyden, Chatham - Born: Salem - Age: 33 - Released on 1 Jul 1815.

Atkinson, Henry - Seaman - Nbr: 2154 - Prize: Rattlesnake - Ship Type: P - How taken: HMS Hyperion - When taken: 26 Jun 1814 - Where taken: off Cape Finisterre - Date Received: 16 Aug 1814 - From what ship: HMS Dublin, Halifax - Born: Baltimore - Age: 32 - Released on 2 May 1815.

American Prisoners of War Held at Dartmoor during the War of 1812

## Alphabetical listing of names

Atkinson, William - Seaman - Nbr: 361 - Prize: Two Brothers - Ship Type: MV - How taken: Beetle (LM) - When taken: 18 Mar 1813 - Where taken: off Western Islands (England) - Date Received: 1 Jul 1813 - From what ship: Plymouth - Born: Maryland - Age: 36 - Released on 20 Apr 1815.

Atwood, Edward - Seaman - Nbr: 1925 - How taken: Gave himself up from HMS Royal William - When taken: 3 Feb 1813 - Date Received: 3 Aug 1814 - From what ship: HMS Alceste, Chatham Depot - Born: Putney - Age: 26 - Released on 2 May 1815.

Atwood, Elisha - Seaman - Nbr: 5746 - Prize: US Schooner Ohio - Ship Type: MW - How taken: British gunboats - When taken: 12 Aug 1814 - Where taken: Fort Erie - Date Received: 26 Dec 1814 - From what ship: HMT Argo - Born: Massachusetts - Age: 25 - Released on 3 Jul 1815.

Atwood, James - Seaman - Nbr: 5345 - Prize: John & Mary - Ship Type: Prize - How taken: HMS Borer - When taken: 20 Oct 1813 - Where taken: off Long Island - Date Received: 31 Oct 1814 - From what ship: HMT Leyden, Chatham - Born: Brighton - Age: 20 - Released on 1 Jul 1815.

Atwood, Jesse - Seaman - Nbr: 4344 - Prize: Union - Ship Type: LM - How taken: HMS Curlew - When taken: 1 Apr 1814 - Where taken: off Halifax - Date Received: 7 Oct 1814 - From what ship: HMT Salvador del Mundo, Halifax - Born: Wellfleet - Age: 24 - Released on 14 Jun 1815.

Atwood, John - Seaman - Nbr: 2332 - Prize: Snap Dragon - Ship Type: P - How taken: HMS Martin - When taken: 10 Jun 1814 - Where taken: off Halifax - Date Received: 16 Aug 1814 - From what ship: HMS Dublin, Halifax - Born: New York - Age: 26 - Released on 3 May 1815.

Atwood, John - Seaman - Nbr: 1271 - How taken: Sent into custody from MV Young William - Date Received: 14 Jun 1814 - From what ship: Mill Prison (Plymouth, England) - Born: Middletown - Age: 28 - Released on 28 Apr 1815.

Atwood, Nathaniel - Seaman - Nbr: 4455 - Prize: Juliana Smith - Ship Type: P - How taken: HMS Nymphe - When taken: 11 May 1813 - Where taken: off Cape Sable Island (Canada) - Date Received: 8 Oct 1814 - From what ship: HMT Leyden, Chatham - Born: Massachusetts - Age: 22 - Released on 15 Jun 1815.

Atwood, P. - Seaman - Nbr: 2239 - Prize: Dispatch - Ship Type: Prize - How taken: HMS Narcissus - When taken: 4 Oct 1813 - Where taken: off Nantucket - Date Received: 16 Aug 1814 - From what ship: HMS Dublin, Halifax - Born: Massachusetts - Age: 23 - Released on 2 May 1815.

Atwood, Thomas - Seaman - Nbr: 2865 - Prize: Prompt - Ship Type: MV - How taken: Change (Privateer) - When taken: 28 Mar 1813 - Where taken: Bay of Biscay - Date Received: 24 Aug 1814 - From what ship: HMT Alpheus, Chatham - Born: Kingston - Age: 29 - Released on 19 May 1815.

Aubry, John Martinalias - Seaman - Nbr: 4826 - Prize: President - Ship Type: P - How taken: HMS Pique - When taken: 7 May 1814 - Where taken: off Porto Rico - Date Received: 9 Oct 1814 - From what ship: HMT Freya, Halifax - Born: Cartagena - Age: 19 - Race: Mulatto - Died on 17 Feb 1815 from variola.

Augustin, Anthony - Seaman - Nbr: 80 - Prize: Rolla - Ship Type: MV - How taken: HMS Surveillante - When taken: 11 Feb 1813 - Where taken: Bay of Biscay - Date Received: 2 Apr 1813 - From what ship: Plymouth - Born: New Orleans - Age: 47 - Sent to Mill Prison (Plymouth, England) on 10 Jul 1813.

Augustus, Amos - Seaman - Nbr: 125 - Prize: Governor McKean - Ship Type: LM - How taken: HMS Rover - When taken: 26 Jan 1813 - Where taken: off Bordeaux - Date Received: 2 Apr 1813 - From what ship: Plymouth - Born: Wilmington - Age: 17 - Race: Black - Sent to Dartmouth on 30 Jul 1813.

Augustus, Benjamin - Seaman - Nbr: 2836 - How taken: Gave himself up from HMS Royal George - When taken: 29 Oct 1812 - Date Received: 24 Aug 1814 - From what ship: HMT Liverpool, Chatham - Born: Portland - Age: 22 - Released on 27 Apr 1815.

Austen, William - Seaman - Nbr: 3441 - How taken: Gave himself up from HMS Clorinda - When taken: 18 Dec 1813 - Date Received: 13 Sep 1814 - From what ship: HMT Niobe, Chatham - Born: Philadelphia - Age: 32 - Released on 28 May 1815.

Austin, Jonathan - Seaman - Nbr: 1805 - How taken: Gave himself up from HMS Clarence - Date Received: 20 Jul 1814 - From what ship: HMS Milford, Plymouth - Born: Nerystoris - Age: 29 - Released on 1 May 1815.

Austin, Joseph - Seaman - Nbr: 3022 - Prize: US Sloop Frolic - Ship Type: MW - How taken: HMS Orpheus - When taken: 20 Apr 1814 - Where taken: off Cuba - Date Received: 2 Sep 1814 - From what ship: Naval Hospital, Plymouth - Born: Rhode Island - Age: 26 - Released on 3 May 1815.

Averell, Loring - Seaman - Nbr: 243 - Prize: William Bayard - Ship Type: MV - How taken: HMS Warspite - When taken: 3 Mar 1813 - Where taken: Bay of Biscay - Date Received: 2 Apr 1813 - From what ship: Plymouth - Born: Connecticut - Age: 19 - Released on 20 Apr 1815.

Averill, Joseph - Seaman - Nbr: 5940 - Prize: Harlequin - Ship Type: P - How taken: HMS Bulwark - When taken: 23 Nov 1814 - Where taken: off Halifax - Date Received: 27 Dec 1814 - From what ship: HMT Penelope -

American Prisoners of War Held at Dartmoor during the War of 1812

## Alphabetical listing of names

Born: Alfred - Age: 20 - Released on 3 Jul 1815.

Averill, Samuel - Seaman - Nbr: 5428 - Prize: Rattlesnake - Ship Type: LM - How taken: HMS Rhin - When taken: 12 May 1814 - Where taken: off Bermuda - Date Received: 31 Oct 1814 - From what ship: HMT Leyden, Chatham - Born: West-Cappel (France) - Age: 21 - Released on 1 Jul 1815.

Avery, Charles - Seaman - Nbr: 2099 - How taken: Gave himself up from HMS Strombolo - When taken: 25 Apr 1812 - Date Received: 3 Aug 1814 - From what ship: HMS Bittern, Chatham Depot - Born: New York - Age: 34 - Released on 2 May 1815.

Avery, John - Seaman - Nbr: 5177 - Prize: Vivid - Ship Type: P - How taken: HMS Nymphe - When taken: 20 Apr 1813 - Where taken: off Cape Cod - Date Received: 31 Oct 1814 - From what ship: HMT Mermaid, Chatham - Born: Massachusetts - Age: 18 - Released on 29 Jun 1815.

Avies, Samuel - Seaman - Nbr: 6229 - Prize: Prince de Neufchatel - Ship Type: P - How taken: Leander (Newcastle Acasta) - When taken: 20 Dec 1814 - Where taken: Lat 38 Long 56 - Date Received: 30 Jan 1815 - From what ship: HMT Pheasant - Born: Boston - Age: 56 - Released on 5 Jul 1815.

Avis, James - Seaman - Nbr: 1540 - Prize: Zebra - Ship Type: LM - How taken: HMS Pyramus - When taken: 20 Apr 1813 - Where taken: Bay of Biscay - Date Received: 23 Jun 1814 - From what ship: Stapleton - Born: New York - Age: 20 - Released on 1 May 1815.

Avlyn, Lawrence - Seaman - Nbr: 5364 - How taken: Gave himself up at Chatham - When taken: 23 Jan 1814 - Date Received: 31 Oct 1814 - From what ship: HMT Leyden, Chatham - Born: Boston - Age: 18 - Released on 1 Jul 1815.

Avril, Ebenezer - Seaman - Nbr: 5696 - Prize: McDonough - Ship Type: P - How taken: HMS Bacchante - When taken: 7 Nov 1814 - Where taken: Lat 42 Long 67 - Date Received: 24 Dec 1814 - From what ship: HMT Penelope - Born: Arundel - Age: 25 - Released on 3 Jul 1815.

Ayres, Henry - Seaman - Nbr: 3144 - Prize: Dolphin - Ship Type: P - How taken: HMS Curlew - When taken: 1 Feb 1812 - Where taken: off Barbados - Date Received: 11 Sep 1814 - From what ship: HMT Freya, Chatham - Born: Maryland - Age: 30 - Released on 28 May 1815.

Ayres, John - Seaman - Nbr: 5556 - Prize: Levant - Ship Type: MV - How taken: HMS Forester - When taken: 4 Jan 1814 - Where taken: Bahamas Banks - Date Received: 17 Dec 1814 - From what ship: HMT Loire, Halifax - Born: Boston - Age: 28 - Released on 1 Jul 1815.

Ayres, Parker - Seaman - Nbr: 5100 - Prize: David Porter - Ship Type: LM - How taken: HMS Pylades - When taken: 12 Sep 1814 - Where taken: Georges Bank - Date Received: 28 Oct 1814 - From what ship: HMT Alkbar, Halifax - Born: Dumbarton - Age: 22 - Released on 29 Jun 1815.

Ayres, Robert - Seaman - Nbr: 634 - Prize: US Brig Argus - Ship Type: MW - How taken: HMS Pelican - When taken: 14 Aug 1813 - Where taken: Irish Channel - Date Received: 8 Sep 1813 - From what ship: Plymouth - Born: New Jersey - Age: 28 - Sent to Dartmouth on 2 Nov 1814.

Ayres, William - Seaman - Nbr: 5243 - How taken: Gave himself up from HMS Progress - When taken: 18 Aug 1813 - Date Received: 31 Oct 1814 - From what ship: HMT Mermaid, Chatham - Born: Providence - Age: 33 - Released on 29 Jun 1815.

Azumus, Jerome - Seaman - Nbr: 4819 - Prize: President - Ship Type: P - How taken: HMS Pique - When taken: 7 May 1814 - Where taken: off Porto Rico - Date Received: 9 Oct 1814 - From what ship: HMT Freya, Halifax - Born: Cartagena - Age: 20 - Race: Black - Released on 21 Jun 1815.

Babb, Benjamin - Seaman - Nbr: 1922 - How taken: Gave himself up from HMS Victory - When taken: 3 Feb 1813 - Date Received: 3 Aug 1814 - From what ship: HMS Alceste, Chatham Depot - Born: Barrington - Age: 34 - Died on 29 Jan 1815 from phthisis pulmonalis.

Babb, Nathaniel - Seaman - Nbr: 5006 - Prize: Charlotte - Ship Type: Prize - How taken: HMS Wasp - When taken: 31 Aug 1814 - Where taken: off Cape Sable Island (Canada) - Date Received: 28 Oct 1814 - From what ship: HMT Alkbar, Halifax - Born: West-Cappel (France) - Age: 19 - Released on 21 Jun 1815.

Babcock, Charles - Seaman - Nbr: 2053 - How taken: Impressed at Gravesend - When taken: 27 Apr 1813 - Date Received: 3 Aug 1814 - From what ship: HMS Lyffey, Chatham Depot - Born: Ingram - Age: 43 - Released on 2 May 1815.

Babcock, Daniel - Seaman - Nbr: 3045 - How taken: Gave himself up from HMS Galatea - When taken: 24 Aug 1814 - Date rec'd: 2 Sep 1814 - What ship: HMT Sultan - Born: Rhode Island - Age: 25 - Released on 28 Apr 1815.

Babson, John - Seaman - Nbr: 5070 - Prize: Ida - Ship Type: LM - How taken: HMS Newcastle - When taken: 9 Aug 1814 - Where taken: Long 34 - Date Received: 28 Oct 1814 - From what ship: HMT Alkbar, Halifax - Born: Cape Ann - Age: 14 - Released on 3 May 1815.

American Prisoners of War Held at Dartmoor during the War of 1812

## Alphabetical listing of names

Bacchus, John - Seaman - Nbr: 4388 - Prize: Norfolk - Ship Type: P - How taken: HMS Fantome - When taken: 29 Jun 1813 - Where taken: off Delaware - Date Received: 8 Oct 1814 - From what ship: HMT Leyden, Chatham - Born: New London - Age: 23 - Race: Black - Released on 14 Jun 1815.

Backley, Walter - Prize master - Nbr: 6481 - Prize: Mary - Ship Type: P - How taken: Lapphin - When taken: 15 Jun 1814 - Where taken: off Saint Marys - Date Received: 3 Mar 1815 - From what ship: HMT Ganges, Plymouth - Born: Fairfield - Age: 28 - Released on 11 Jul 1815.

Backman, Charles - Seaman - Nbr: 2936 - How taken: Gave himself up from HMS Prince of Wales - Date Received: 24 Aug 1814 - From what ship: HMT Alpheus, Chatham - Born: Philadelphia - Age: 20 - Released on 19 May 1815.

Backman, Isaac - Seaman - Nbr: 5109 - Prize: David Porter - Ship Type: LM - How taken: HMS Pylades - When taken: 12 Sep 1814 - Where taken: Georges Bank - Date Received: 28 Oct 1814 - From what ship: HMT Alkbar, Halifax - Born: Gothenburg - Age: 27 - Released on 29 Jun 1815.

Bacock, Mathew - Seaman - Nbr: 5952 - Prize: Harlequin - Ship Type: P - How taken: HMS Bulwark - When taken: 23 Nov 1814 - Where taken: off Halifax - Date Received: 27 Dec 1814 - From what ship: HMT Penelope - Born: Virginia - Age: 30 - Race: Black - Released on 3 Jul 1815.

Bacon, Elisha - Seaman - Nbr: 5082 - Prize: Ida - Ship Type: LM - How taken: HMS Newcastle - When taken: 9 Aug 1814 - Where taken: Long 34 - Date Received: 28 Oct 1814 - From what ship: HMT Alkbar, Halifax - Born: Barnstable - Age: 22 - Released on 29 Jun 1815.

Bacon, James - Seaman - Nbr: 6193 - Prize: Prince de Neufchatel - Ship Type: P - How taken: Leander (Newcastle Acasta) - When taken: 20 Dec 1814 - Where taken: Lat 38 Long 56 - Date Received: 30 Jan 1815 - From what ship: HMT Pheasant - Born: Baltimore - Age: 28 - Released on 5 Jul 1815.

Baday, Edward - Seaman - Nbr: 737 - Prize: Hannah & Eliza - Ship Type: MV - How taken: HMS Lyra - When taken: 29 May 1813 - Where taken: off Bayonne - Date Received: 3 Nov 1813 - From what ship: Plymouth - Born: New York - Age: 24 - Released on 26 Apr 1815.

Badcock, Bradley - Seaman - Nbr: 3466 - How taken: Gave himself up from HMS Prince - When taken: 1 Sep 1814 - Date Received: 19 Sep 1814 - From what ship: HMT Salvador del Mundo - Born: Massachusetts - Age: 35 - Released on 4 Jun 1815.

Baddington, Asa - Seaman - Nbr: 2609 - How taken: Gave himself up from HMS Stag - When taken: 11 Sep 1812 - Date Received: 21 Aug 1814 - From what ship: HMT Freya, Chatham - Born: New London - Age: 47 - Released on 26 Apr 1815.

Badger, John - Seaman - Nbr: 2970 - How taken: Gave himself up from HMS Gorgon - Date Received: 24 Aug 1814 - From what ship: HMT Hannibal - Born: New York - Age: 47 - Race: Mulatto - Released on 19 May 1815.

Badger, Peter - Seaman - Nbr: 2322 - Prize: Hussar - Ship Type: P - How taken: HMS Saturn - When taken: 25 May 1814 - Where taken: off Sandy Hook - Date Received: 16 Aug 1814 - From what ship: HMS Dublin, Halifax - Born: Boston - Age: 28 - Released on 3 May 1815.

Badson, Jacob - Seaman - Nbr: 1131 - Prize: Young Dixon - Ship Type: MV - How taken: HMS Fly & HMS Avon - When taken: 3 Apr 1814 - Where taken: at sea - Date Received: 10 May 1814 - From what ship: Plymouth - Born: Boston - Age: 31 - Died on 22 Mar 1815 from variola.

Bagley, William - Seaman - Nbr: 6143 - How taken: Gave himself up from HMS Ajax - Date Received: 17 Jan 1815 - From what ship: HMT Impregnable - Born: Portland - Age: 36 - Released on 5 Jul 1815.

Bagley, William - Seaman - Nbr: 369 - How taken: Impressed at sea - When taken: 4 Apr 1813 - Date Received: 1 Jul 1813 - From what ship: Plymouth - Born: Portland - Age: 34 - Sent to Dartmouth on 30 Jul 1813.

Bailey, Aaron - Seaman - Nbr: 2521 - Prize: Snap Dragon - Ship Type: P - How taken: HMS Martin - When taken: 10 Jun 1814 - Where taken: off Halifax - Date Received: 16 Aug 1814 - From what ship: HMT Queen, Halifax - Born: Boston - Age: 19 - Released on 3 May 1815.

Bailey, Daniel - Seaman - Nbr: 3399 - Prize: Thomas - Ship Type: P - How taken: HMS Nymphe - When taken: 24 Jun 1813 - Where taken: off Halifax - Date Received: 13 Sep 1814 - From what ship: HMT Niobe, Chatham - Born: Portsmouth - Age: 22 - Released on 28 May 1815.

Bailey, Isaac - Seaman - Nbr: 5164 - Prize: Volante - Ship Type: P - How taken: HMS Curlew - When taken: 25 Nov 1813 - Where taken: off Halifax - Date Received: 31 Oct 1814 - From what ship: HMT Mermaid, Chatham - Born: Baltimore - Age: 29 - Race: Mulatto - Released on 29 Jun 1815.

Bailey, John - Seaman - Nbr: 2796 - Prize: Weazel - Ship Type: MV - How taken: HMS Foxhound - When taken: 25 Mar 1813 - Where taken: Bay of Biscay - Date Received: 24 Aug 1814 - From what ship: HMT Liverpool, Chatham - Born: Gloucester - Age: 29 - Released on 19 May 1815.

American Prisoners of War Held at Dartmoor during the War of 1812

## Alphabetical listing of names

Bailey, Joseph - Seaman - Nbr: 5964 - Prize: Amazon - Ship Type: Prize - How taken: HMS Bulwark - When taken: 22 Sep 1814 - Where taken: Georges Bank - Date Received: 27 Dec 1814 - From what ship: HMT Penelope - Born: Gloucester - Age: 24 - Released on 3 Jul 1815.

Bailey, Moses - Seaman - Nbr: 5819 - Prize: US Schooner Scorpion - Ship Type: MW - How taken: British gunboats - When taken: 6 Sep 1814 - Where taken: Lake Erie - Date Received: 26 Dec 1814 - From what ship: HMT Argo - Born: Pennsylvania - Age: 21 - Race: Black - Died on 17 Feb 1815 from variola.

Bailey, Peter - Seaman - Nbr: 3415 - Prize: Thomas - Ship Type: P - How taken: HMS Nymphe - When taken: 24 Jun 1813 - Where taken: off Halifax - Date Received: 13 Sep 1814 - From what ship: HMT Niobe, Chatham - Born: Bristol - Age: 19 - Released on 28 May 1815.

Bailey, S. - Passenger - Nbr: 6497 - Prize: Dolores - Ship Type: MV - How taken: HMS Abercrombie - When taken: 23 Oct 1814 - Where taken: St. Jago (Jamaica) - Date Received: 3 Mar 1815 - From what ship: HMT Ganges, Plymouth - Born: Baltimore - Age: 18 - Released on 11 Jul 1815.

Bailey, Samuel - Seaman - Nbr: 1788 - Prize: Ferox - Ship Type: MV - How taken: HMS Medusa & HMS Lyra - When taken: 28 Mar 1813 - Where taken: off Cape Ortegal (Spain) - Date Received: 20 Jul 1814 - From what ship: HMS Milford, Plymouth - Born: Portland - Age: 32 - Released on 1 May 1815.

Bailey, Thomas - Seaman - Nbr: 2960 - Prize: Blanche - Ship Type: MV - How taken: HMS Barbados - When taken: 24 Jan 1814 - Where taken: off St. Bartholomew - Date Received: 24 Aug 1814 - From what ship: HMT Hannibal - Born: Massachusetts - Age: 49 - Released on 19 May 1815.

Bailey, Warren - Seaman - Nbr: 3919 - Prize: Rolla - Ship Type: P - How taken: HMS Loire - When taken: 10 Dec 1813 - Where taken: off Bull Island (South Carolina) - Date Received: 5 Oct 1814 - From what ship: HMT President, Halifax - Born: Rhode Island - Age: 18 - Released on 9 Jun 1815.

Bailey, William - Seaman - Nbr: 2893 - How taken: Gave himself up from HMS Union - When taken: 27 May 1813 - Date Received: 24 Aug 1814 - From what ship: HMT Alpheus, Chatham - Born: Philadelphia - Age: 29 - Released on 19 May 1815.

Baird, James - Seaman - Nbr: 3923 - Prize: Rolla - Ship Type: P - How taken: HMS Loire - When taken: 10 Dec 1813 - Where taken: off Bull Island (South Carolina) - Date Received: 5 Oct 1814 - From what ship: HMT President, Halifax - Born: New York - Age: 21 - Released on 9 Jun 1815.

Baird, Martin - Seaman - Nbr: 3935 - Prize: Rolla - Ship Type: P - How taken: HMS Loire - When taken: 10 Dec 1813 - Where taken: off Bull Island (South Carolina) - Date Received: 5 Oct 1814 - From what ship: HMT President, Halifax - Born: Staten Island - Age: 37 - Released on 9 Jun 1815.

Baisley, Abraham - Marine - Nbr: 4600 - Prize: York Town - Ship Type: P - How taken: HMS Maidstone - When taken: 18 Jul 1813 - Where taken: Grand Banks - Date Received: 9 Oct 1814 - From what ship: HMT Leyden, Chatham - Born: New York - Age: 22 - Released on 15 Jun 1815.

Bakeman, Robert - Seaman - Nbr: 6059 - Prize: William - Ship Type: Prize - How taken: HMS Armide - When taken: 11 Oct 1814 - Where taken: off Newport - Date Received: 28 Dec 1814 - From what ship: HMT Penelope - Born: Albany - Age: 23 - Race: Black - Released on 3 Jul 1815.

Baker, Basil - Seaman - Nbr: 2532 - Prize: Hussar - Ship Type: P - How taken: HMS Saturn - When taken: 25 May 1814 - Where taken: off Sandy Hook - Date Received: 16 Aug 1814 - From what ship: HMT Queen, Halifax - Born: Philadelphia - Age: 22 - Released on 3 May 1815.

Baker, Benjamin - Seaman - Nbr: 2160 - Prize: Rattlesnake - Ship Type: P - How taken: HMS Hyperion - When taken: 26 Jun 1814 - Where taken: off Cape Finisterre - Date Received: 16 Aug 1814 - From what ship: HMS Dublin, Halifax - Born: New York - Age: 24 - Released on 2 May 1815.

Baker, Charles - Seaman - Nbr: 2942 - Prize: Atalanta - Ship Type: MV - How taken: HMS Barbados - When taken: 19 Jan 1814 - Where taken: off St. Bartholomew - Date Received: 24 Aug 1814 - From what ship: HMT Hannibal - Born: Virginia - Age: 19 - Died on 30 Jan 1815 from phthisis pulmonalis.

Baker, Daniel - Seaman - Nbr: 2892 - How taken: Gave himself up from HMS Union - When taken: 27 May 1813 - Date Received: 24 Aug 1814 - From what ship: HMT Alpheus, Chatham - Born: Massachusetts - Age: 21 - Released on 19 May 1815.

Baker, Edward - Seaman - Nbr: 5913 - Prize: Harlequin - Ship Type: P - How taken: HMS Bulwark - When taken: 23 Nov 1814 - Where taken: off Halifax - Date Received: 27 Dec 1814 - From what ship: HMT Penelope - Born: New York - Age: 19 - Released on 15 Jul 1815.

Baker, Edward E. - Seaman - Nbr: 5084 - Prize: Ida - Ship Type: LM - How taken: HMS Newcastle - When taken: 9 Aug 1814 - Where taken: Long 34 - Date Received: 28 Oct 1814 - From what ship: HMT Alkbar, Halifax - Born: Duxbury - Age: 17 - Released on 29 Jun 1815.

Baker, Henry - Seaman - Nbr: 5023 - Prize: Landrail - Ship Type: Prize - How taken: HMS Wasp - When taken: 27

American Prisoners of War Held at Dartmoor during the War of 1812

## Alphabetical listing of names

Jul 1814 - Where taken: Georges Bank - Date Received: 28 Oct 1814 - From what ship: HMT Alkbar, Halifax - Born: New York - Age: 27 - Escaped on 1 Jun 1815.

Baker, Henry - Seaman - Nbr: 4051 - Prize: US Brig Rattlesnake - Ship Type: MW - How taken: HMS Leander - When taken: 13 Jul 1814 - Where taken: off Shelburne - Date Received: 6 Oct 1814 - From what ship: HMT Chesapeake, Halifax - Born: Virginia - Age: 39 - Released on 13 Jun 1815.

Baker, James - Seaman - Nbr: 4329 - Prize: Martha - Ship Type: Prize - How taken: HMS Belviders - When taken: 25 Feb 1814 - Where taken: off New London - Date Received: 7 Oct 1814 - From what ship: HMT Salvador del Mundo, Halifax - Born: Boston - Age: 19 - Released on 14 Jun 1815.

Baker, Jesse - Seaman - Nbr: 6166 - Prize: Lion - Ship Type: P - How taken: HMS Granicus - When taken: 2 Dec 1814 - Where taken: off Lisbon - Date Received: 18 Jan 1815 - From what ship: HMT Impregnable - Born: Baltimore - Age: 30 - Released on 5 Jul 1815.

Baker, John - Seaman - Nbr: 214 - Prize: Mars - Ship Type: MV - How taken: HMS Warspite - When taken: 26 Feb 1813 - Where taken: off Basque Roads (France) - Date Received: 2 Apr 1813 - From what ship: Plymouth - Born: Baltimore - Age: 25 - Released on 20 Apr 1815.

Baker, John - Seaman - Nbr: 1199 - Prize: Ocean - Ship Type: P - How taken: HMS Achates - When taken: 14 Jun 1810 - Where taken: Channel - Date Received: 4 Jun 1814 - From what ship: Dartmouth - Born: Baltimore - Age: 32 - Released on 28 Apr 1815.

Baker, John - Seaman - Nbr: 421 - Prize: Viper - Ship Type: MV - How taken: HMS Superb - When taken: 15 Apr 1813 - Where taken: Bay of Biscay - Date Received: 1 Jul 1813 - From what ship: Plymouth - Born: Pennsylvania - Age: 19 - Released on 20 Apr 1815.

Baker, Richard - Purser steward - Nbr: 3883 - Prize: US Brig Rattlesnake - Ship Type: MW - How taken: HMS Leander - When taken: 11 Jul 1814 - Where taken: off Shelburne - Date Received: 5 Oct 1814 - From what ship: HMT Orpheus, Halifax - Born: Providence - Age: 32 - Released on 9 Jun 1815.

Baker, Robert - Seaman - Nbr: 3183 - How taken: Gave himself up from HMS Hibernia - When taken: 27 Jul 1813 - Date Received: 11 Sep 1814 - From what ship: HMT Freya, Chatham - Born: Virginia - Age: 29 - Released on 28 May 1815.

Baker, Stephen - Seaman - Nbr: 1325 - How taken: Impressed - When taken: 7 Jul 1813 - Date Received: 19 Jun 1814 - From what ship: Stapleton - Born: Rhode Island - Age: 23 - Released on 28 Apr 1815.

Baker, Thomas - Captain - Nbr: 6389 - Prize: Nellenville - Ship Type: MV - How taken: HMS Onyx - When taken: 25 Dec 1814 - Where taken: off San Domingo - Date Received: 3 Mar 1815 - From what ship: HMT Ganges, Plymouth - Born: Rhode Island - Age: 26 - Sent to Ashburton (England) on 10 Mar 1815.

Balch, George W. - Master - Nbr: 5566 - Prize: Harlequin - Ship Type: P - How taken: HMS Bulwark - When taken: 23 Nov 1814 - Where taken: off Halifax - Date Received: 17 Dec 1814 - From what ship: HMT Loire, Halifax - Born: Massachusetts - Age: 37 - Released on 10 Apr 1815.

Balch, Samuel - Officer Marines - Nbr: 5465 - Prize: General Putnam - Ship Type: P - How taken: HMS Leander - When taken: 8 Nov 1814 - Where taken: Long 65 Lat 42 - Date Received: 17 Dec 1814 - From what ship: HMT Loire, Halifax - Born: Wakefield - Age: 26 - Released on 1 Jul 1815.

Balch, William - Seaman - Nbr: 6206 - Prize: Prince de Neufchatel - Ship Type: P - How taken: Leander (Newcastle Acasta) - When taken: 20 Dec 1814 - Where taken: Lat 38 Long 56 - Date Received: 30 Jan 1815 - From what ship: HMT Pheasant - Born: Boston - Age: 25 - Released on 5 Jul 1815.

Baldwin, James - Seaman - Nbr: 5119 - Prize: Adams - Ship Type: MV - How taken: Taken by the 62nd British Regiment at Hampton - Date Received: 28 Oct 1814 - From what ship: HMT Alkbar, Halifax - Born: Charleston - Age: 25 - Released on 29 Jun 1815.

Baldwin, John - Seaman - Nbr: 1608 - Prize: Fox - Ship Type: LM - How taken: HMS Pheasant - When taken: 23 Apr 1813 - Where taken: Bay of Biscay - Date Received: 23 Jun 1814 - From what ship: Stapleton - Born: Boston - Age: 24 - Died on 5 Dec 1814 from pneumonia.

Baldwin, John - Seaman - Nbr: 1065 - Prize: Fair American - Ship Type: MV - How taken: HMS Andromache - When taken: 19 Jan 1814 - Where taken: Bay of Biscay - Date Received: 10 May 1814 - From what ship: Plymouth - Born: Boston - Age: 23 - Released on 27 Apr 1815.

Baldwin, Theophilus - Seaman - Nbr: 6525 - How taken: Gave himself up from MV Triton - When taken: Nov 1814 - Date Received: 3 Mar 1815 - From what ship: HMT Ganges, Plymouth - Born: Connecticut - Age: 25 - Released on 11 Jul 1815.

Bale, Charles - Seaman - Nbr: 3360 - How taken: Gave himself up from HMS Union - When taken: 17 Oct 1813 - Date Received: 13 Sep 1814 - From what ship: HMT Niobe, Chatham - Born: Hampshire - Age: 25 - Released on 28 May 1815.

American Prisoners of War Held at Dartmoor during the War of 1812

## Alphabetical listing of names

Baley, John - Seaman - Nbr: 2028 - How taken: Gave himself up from HMS Castilian - When taken: 26 Apr 1813 - Date Received: 3 Aug 1814 - From what ship: HMS Lyffey, Chatham Depot - Born: Bristol - Age: 26 - Race: Black - Released on 2 May 1815.

Ball, James - Seaman - Nbr: 685 - Prize: Joel Barlow - Ship Type: LM - How taken: HMS Briton - When taken: 3 Jul 1813 - Where taken: off Bordeaux - Date Received: 27 Sep 1813 - From what ship: Plymouth - Born: Pennsylvania - Age: 19 - Released on 26 Apr 1815.

Ball, Peter - Seaman - Nbr: 5026 - Prize: Landrail - Ship Type: Prize - How taken: HMS Wasp - When taken: 27 Jul 1814 - Where taken: Georges Bank - Date Received: 28 Oct 1814 - From what ship: HMT Alkbar, Halifax - Born: Not legible - Age: 35 - Released on 21 Jun 1815.

Ball, Stephen - Seaman - Nbr: 6231 - Prize: Prince de Neufchatel - Ship Type: P - How taken: Leander (Newcastle Acasta) - When taken: 20 Dec 1814 - Where taken: Lat 38 Long 56 - Date Received: 30 Jan 1815 - From what ship: HMT Pheasant - Born: North Carolina - Age: 19 - Released on 5 Jul 1815.

Ball, William - Seaman - Nbr: 4943 - Prize: Herald - Ship Type: P - How taken: HMS Endymion - When taken: 15 Aug 1814 - Where taken: off Nantucket - Date Received: 28 Oct 1814 - From what ship: HMT Alkbar, Halifax - Born: Portsmouth - Age: 22 - Released on 21 Jun 1815.

Ballace, Daniel - Seaman - Nbr: 6129 - Prize: Betsey - Ship Type: Prize - How taken: HMS Bellerophon - When taken: 2 Nov 1814 - Where taken: Long 61 - Date Received: 6 Jan 1814 - From what ship: HMT Impregnable - Born: Bristol - Age: 22 - Race: Black - Released on 5 Jul 1815.

Ballard, John - Seaman - Nbr: 3987 - Prize: US Brig Rattlesnake - Ship Type: MW - How taken: HMS Leander - When taken: 13 Jul 1814 - Where taken: off Shelburne - Date Received: 6 Oct 1814 - From what ship: HMT Chesapeake, Halifax - Born: Wiscasset - Age: 17 - Released on 9 Jun 1815.

Ballard, John - Seaman - Nbr: 2115 - How taken: Gave himself up from HMS Zenobia - When taken: 25 Aug 1812 - Date Received: 8 Aug 1814 - From what ship: HMT Raven, Chatham - Born: Georgetown - Age: 29 - Released on 26 Apr 1815.

Ballent, John - Seaman - Nbr: 5766 - Prize: Young William - Ship Type: Prize - How taken: HMS Plover - When taken: 10 Sep 1814 - Where taken: off Newfoundland - Date Received: 26 Dec 1814 - From what ship: HMT Argo - Born: Porto Rico - Age: 27 - Race: Mulatto - Released on 3 Jul 1815.

Bancroft, James - Mate - Nbr: 5088 - Prize: Ida - Ship Type: LM - How taken: HMS Newcastle - When taken: 9 Aug 1814 - Where taken: Long 34 - Date Received: 28 Oct 1814 - From what ship: HMT Alkbar, Halifax - Born: Boston - Age: 29 - Sent to Ashburton (England) on 2 Dec 1814.

Bancroft, Samuel - Seaman - Nbr: 470 - Prize: Essex - Ship Type: MV - How taken: HMS Pyramus - When taken: 2 Apr 1813 - Where taken: Bay of Biscay - Date Received: 8 Sep 1813 - From what ship: Plymouth - Born: Marblehead - Age: 29 - Released on 20 Apr 1815.

Bang, George - 5th Lieutenant - Nbr: 6307 - Prize: Prince de Neufchatel - Ship Type: P - How taken: Leander (Newcastle Acasta) - When taken: 20 Dec 1814 - Where taken: Lat 38 Long 56 - Date Received: 19 Feb 1815 - From what ship: HMT Ganges, Plymouth - Born: Boston - Age: 21 - Released on 14 Apr 1815.

Banker, Robert - Seaman - Nbr: 1552 - Prize: Zebra - Ship Type: LM - How taken: HMS Pyramus - When taken: 20 Apr 1813 - Where taken: Bay of Biscay - Date Received: 23 Jun 1814 - From what ship: Stapleton - Born: New York - Age: 19 - Released on 1 May 1815.

Banks, James - Seaman - Nbr: 4075 - Prize: New Zealander - Ship Type: Prize - How taken: HMS Belviders - When taken: 21 Apr 1814 - Where taken: off Delaware - Date Received: 6 Oct 1814 - From what ship: HMT Chesapeake, Halifax - Born: Massachusetts - Age: 50 - Released on 13 Jun 1815.

Banks, Perry - Seaman - Nbr: 1500 - Prize: Margaret - Ship Type: Prize - How taken: HMS Foxhound - When taken: 27 May 1814 - Where taken: off Isles of Scilly - Date Received: 20 Jun 1814 - From what ship: Mill Prison (Plymouth, England) - Born: Baltimore - Age: 23 - Released on 1 May 1815.

Bankstone, William - Seaman - Nbr: 3631 - Prize: Monarch - Ship Type: MV - How taken: HMS Dotterel - When taken: 14 Dec 1813 - Where taken: off Charleston - Date Received: 30 Sep 1814 - From what ship: HMT Sybella - Born: Waterford - Age: 18 - Released on 4 Jun 1815.

Bann, John - Seaman - Nbr: 2171 - Prize: Rattlesnake - Ship Type: P - How taken: HMS Hyperion - When taken: 26 Jun 1814 - Where taken: off Cape Finisterre - Date Received: 16 Aug 1814 - From what ship: HMS Dublin, Halifax - Born: Philadelphia - Age: 30 - Released on 2 May 1815.

Bannell, James - Seaman - Nbr: 1305 - How taken: Sent into custody from HMS Minden - Date Received: 14 Jun 1814 - From what ship: Mill Prison (Plymouth, England) - Born: New Jersey - Age: 32 - Released on 28 Apr 1815.

Banning, Peter - Seaman - Nbr: 6289 - How taken: Gave himself up from HMS Beaver - Date Received: 17 Feb

American Prisoners of War Held at Dartmoor during the War of 1812

## Alphabetical listing of names

1815 - From what ship: HMT Ganges, Plymouth - Born: Talbot County - Age: 32 - Race: Black - Released on 5 Jul 1815.

Bannister, George - Seaman - Nbr: 89 - Prize: Cashier - Ship Type: MV - How taken: HMS Reindeer - When taken: 3 Feb 1813 - Where taken: Bay of Biscay - Date Received: 2 Apr 1813 - From what ship: Plymouth - Born: New England - Age: 23 - Sent to Dartmouth on 30 Jul 1813.

Bannister, Joshua - Seaman - Nbr: 763 - How taken: Impressed at Liverpool - Date Received: 3 Nov 1813 - From what ship: Plymouth - Born: Philadelphia - Age: 26 - Released on 26 Apr 1815.

Banta, John - Seaman - Nbr: 2712 - How taken: Gave himself up from HMS Sterling Castle - When taken: 12 Jun 1813 - Date Received: 21 Aug 1814 - From what ship: HMT Freya, Chatham - Born: New Jersey - Age: 31 - Released on 26 Apr 1815.

Baptist, John - Cook - Nbr: 115 - Prize: Governor McKean - Ship Type: LM - How taken: HMS Rover - When taken: 26 Jan 1813 - Where taken: off Bordeaux - Date Received: 2 Apr 1813 - From what ship: Plymouth - Born: New Orleans - Age: 21 - Race: Negro - Sent to Dartmouth on 30 Jul 1813.

Baptiste, John - Cook - Nbr: 6365 - Prize: St. Johanna - Ship Type: Prize - How taken: HMS Sabine - When taken: 25 Oct 1814 - Where taken: off Newfoundland Banks - Date Received: 24 Feb 1815 - From what ship: HMT Ganges, Plymouth - Born: New Orleans - Age: 19 - Race: Black - Released on 6 Jul 1815.

Baptiste, John - Seaman - Nbr: 6394 - Prize: Nellenville - Ship Type: MV - How taken: HMS Onyx - When taken: 25 Dec 1814 - Where taken: off San Domingo - Date Received: 3 Mar 1815 - From what ship: HMT Ganges, Plymouth - Born: New Orleans - Age: 27 - Released on 11 Jul 1815.

Baptiste, John - Seaman - Nbr: 4634 - Prize: Harriett - Ship Type: MV - How taken: HMS Thistle - When taken: 24 Feb 1813 - Where taken: off St. Bartholomew - Date Received: 9 Oct 1814 - From what ship: HMT Leyden, Chatham - Born: Bordeaux - Age: 22 - Released on 15 Jun 1815.

Baptiste, John - Seaman - Nbr: 1870 - Prize: Elbridge Gerry - Ship Type: P - How taken: HMS Crescent - When taken: 16 Sep 1813 - Where taken: off St. George's - Date Received: 29 Jul 1814 - From what ship: HMS Ville de Paris, Chatham Depot - Born: Portland - Age: 22 - Race: Black - Released on 2 May 1815.

Barbadoes, Robert - Seaman - Nbr: 994 - Prize: Apparencen - Ship Type: MV - How taken: HMS Castilian - When taken: 27 Jan 1814 - Where taken: off Ushant (France) - Date Received: 31 Jan 1814 - From what ship: Plymouth - Born: Boston - Age: 29 - Released on 27 Apr 1815.

Barber, John H. - Seaman - Nbr: 3967 - Prize: Rolla - Ship Type: P - How taken: HMS Loire - When taken: 10 Dec 1813 - Where taken: off Bull Island (South Carolina) - Date Received: 5 Oct 1814 - From what ship: HMT President, Halifax - Born: Berkley - Age: 17 - Released on 9 Jun 1815.

Barber, William - Seaman - Nbr: 6183 - Prize: Lion - Ship Type: P - How taken: HMS Granicus - When taken: 2 Dec 1814 - Where taken: off Lisbon - Date Received: 21 Jan 1815 - From what ship: HMT Impregnable - Born: Philadelphia - Age: 28 - Released on 5 Jul 1815.

Barber, William - Seaman - Nbr: 1397 - Prize: Zebra - Ship Type: LM - How taken: HMS Pyramus - When taken: 20 Apr 1813 - Where taken: Bay of Biscay - Date Received: 19 Jun 1814 - From what ship: Stapleton - Born: Newport - Age: 40 - Race: Negro - Released on 28 Apr 1815.

Barden, Abel - Mate - Nbr: 6060 - Prize: High Flyer - Ship Type: MV - How taken: HMS Loire - When taken: 12 Sep 1814 - Where taken: off Delaware - Date Received: 28 Dec 1814 - From what ship: HMT Penelope - Born: Rhode Island - Age: 26 - Released on 3 Jul 1815.

Barker, A. L. - 2nd Lieutenant - Nbr: 3607 - Prize: Fiere Facia - Ship Type: P - How taken: HMS Ramillies - When taken: 26 Feb 1814 - Where taken: off NY - Date Received: 30 Sep 1814 - From what ship: HMT Sybella - Born: New York - Age: 23 - Released on 4 Jun 1815.

Barker, Charles - Seaman - Nbr: 6298 - Prize: John - Ship Type: Prize - How taken: Leander (Newcastle Acasta) - When taken: 30 Jan 1815 - Date Received: 19 Feb 1815 - From what ship: HMT Ganges, Plymouth - Born: New Haven - Age: 18 - Released on 5 Jul 1815.

Barker, Charles G. - Seaman - Nbr: 920 - How taken: Impressed at Cove of Cork - When taken: 20 Dec 1813 - Date rec'd: 31 Jan 1814 - From what ship: Plymouth - Born: Boston - Age: 23 - Released on 27 Apr 1815.

Barker, George - Seaman - Nbr: 5188 - Prize: Catherine - Ship Type: P - How taken: HMS La Hogue - When taken: 2 May 1813 - Where taken: off Cape Sable Island (Canada) - Date Received: 31 Oct 1814 - From what ship: HMT Mermaid, Chatham - Born: Africa - Age: 19 - Race: Negro - Released on 11 Jul 1815.

Barker, Israel - Seaman - Nbr: 732 - Prize: Ned - Ship Type: LM - How taken: HMS Royalist - When taken: 6 Sep 1813 - Where taken: Bay of Biscay - Date Received: 27 Sep 1813 - From what ship: Plymouth - Born: New York - Age: 24 - Released on 26 Apr 1815.

Barker, John - Seaman - Nbr: 2302 - Prize: Hussar - Ship Type: P - How taken: HMS Saturn - When taken: 25 May

American Prisoners of War Held at Dartmoor during the War of 1812

## Alphabetical listing of names

1814 - Where taken: off Sandy Hook - Date Received: 16 Aug 1814 - From what ship: HMS Dublin, Halifax - Born: Connecticut - Age: 23 - Released on 3 May 1815.

Barker, Joseph - Seaman - Nbr: 3684 - Prize: Alfred - Ship Type: P - How taken: HMS Epervier - When taken: 23 Feb 1812 - Where taken: off Newfoundland - Date Received: 30 Sep 1814 - From what ship: HMT President - Born: Marblehead - Age: 21 - Released on 4 Jun 1815.

Barker, Thomas - Seaman - Nbr: 6211 - Prize: Prince de Neufchatel - Ship Type: P - How taken: Leander (Newcastle Acasta) - When taken: 20 Dec 1814 - Where taken: Lat 38 Long 56 - Date Received: 30 Jan 1815 - From what ship: HMT Pheasant - Born: Boston - Age: 21 - Released on 5 Jul 1815.

Barker, Thomas - Surgeon - Nbr: 3488 - Prize: Snap Dragon - Ship Type: P - How taken: HMS Martin - When taken: 30 Jun 1814 - Where taken: off Halifax - Date Received: 19 Sep 1814 - From what ship: HMT Salvador del Mundo - Born: Virginia - Age: 22 - Sent to Dartmouth on 25 Oct 1814.

Barkman, Henry - 2nd Mate - Nbr: 434 - Prize: Magdalen - Ship Type: MV - How taken: HMS Superb - When taken: 15 Apr 1813 - Where taken: off Belle Isle, France - Date Received: 1 Jul 1813 - From what ship: Plymouth - Born: Pennsylvania - Age: 36 - Released on 20 Apr 1815.

Barlow, John - Seaman - Nbr: 574 - Prize: US Brig Argus - Ship Type: MW - How taken: HMS Pelican - When taken: 14 Apr 1813 - Where taken: Irish Channel - Date Received: 8 Sep 1813 - From what ship: Plymouth - Born: Amsterdam, Holland - Age: 22 - Sent to Plymouth on 8 Dec 1814.

Barlow, Robert - Seaman - Nbr: 245 - Prize: William Bayard - Ship Type: MV - How taken: HMS Warspite - When taken: 3 Mar 1813 - Where taken: Bay of Biscay - Date Received: 2 Apr 1813 - From what ship: Plymouth - Born: Pennsylvania - Age: 33 - Released on 20 Apr 1815.

Barnard, John - Seaman - Nbr: 4743 - How taken: Gave himself up from HMS Phoenix - When taken: 17 Jul 1814 - Date Received: 9 Oct 1814 - From what ship: HMT Freya, Chatham - Born: New Orleans - Age: 36 - Released on 15 Jun 1815.

Barnes, Edward L. - Quartermaster - Nbr: 4278 - Prize: Elbridge Gerry - Ship Type: MV - How taken: HMS Crescent - When taken: 16 Feb 1813 - Where taken: at sea - Date Received: 7 Oct 1814 - From what ship: HMT Niobe, Chatham - Born: Connecticut - Age: 41 - Released on 14 Jun 1815.

Barnes, Henry - Seaman - Nbr: 2260 - Prize: Prize to the Scourge - Ship Type: P - How taken: HMS Martin - When taken: 4 May 1814 - Where taken: off Newfoundland - Date Received: 16 Aug 1814 - From what ship: HMS Dublin, Halifax - Born: Norway - Age: 17 - Released on 3 May 1815.

Barnes, James - Seaman - Nbr: 3932 - Prize: Rolla - Ship Type: P - How taken: HMS Loire - When taken: 10 Dec 1813 - Where taken: off Bull Island (South Carolina) - Date Received: 5 Oct 1814 - From what ship: HMT President, Halifax - Born: New Haven - Age: 26 - Released on 9 Jun 1815.

Barnes, Joseph - Seaman - Nbr: 1174 - Prize: Cupidon - Ship Type: P - How taken: HMS Amazon - When taken: 23 Mar 1811 - Where taken: Lat 47N Long 7W - Date Received: 4 Jun 1814 - From what ship: Dartmouth - Born: Virginia - Age: 35 - Released on 28 Apr 1815.

Barnes, Nathaniel - Seaman - Nbr: 1369 - Prize: Paul Jones - Ship Type: P - How taken: HMS Leonidas - When taken: 23 May 1813 - Where taken: Channel - Date Received: 19 Jun 1814 - From what ship: Stapleton - Born: New York - Age: 35 - Released on 28 Apr 1815.

Barnes, Nathaniel - Seaman - Nbr: 4957 - Prize: Herald - Ship Type: P - How taken: HMS Endymion - When taken: 15 Aug 1814 - Where taken: off Nantucket - Date Received: 28 Oct 1814 - From what ship: HMT Alkbar, Halifax - Born: New York - Age: 47 - Released on 21 Jun 1815.

Barnes, Thomas - Seaman - Nbr: 5999 - Prize: McDonough - Ship Type: P - How taken: HMS Bacchante - When taken: 1 Nov 1814 - Where taken: Lat 42 Long 67 - Date Received: 28 Dec 1814 - From what ship: HMT Penelope - Born: Arundel - Age: 19 - Released on 3 Jul 1815.

Barnes, William - Seaman - Nbr: 39 - How taken: Apprehended at Gibraltar - When taken: 8 Jul 1813 - Date rec'd: 2 Apr 1813 - What ship: Plymouth - Born: Rehoboth - Age: 26 - Sent to Dartmouth on 30 Jul 1813.

Barnett, James - Mate - Nbr: 3862 - Prize: Buzi - Ship Type: MV - How taken: HMS Dragon - When taken: 20 Jul 1814 - Where taken: off VA - Date Received: 5 Oct 1814 - From what ship: HMT Orpheus, Halifax - Born: Pennsylvania - Age: 56 - Died on 8 Dec 1814 from pneumonia.

Barnett, Thomas - Seaman - Nbr: 3638 - Prize: Bordeaux Packet - Ship Type: LM - How taken: HMS Niemen - When taken: 28 Jan 1814 - Where taken: off Delaware - Date Received: 30 Sep 1814 - From what ship: HMT Sybella - Born: Lancaster - Age: 29 - Released on 19 May 1815.

Barnett, Tobias - Seaman - Nbr: 4800 - Prize: President - Ship Type: P - How taken: HMS Pique - When taken: 7 May 1814 - Where taken: off Porto Rico - Date Received: 9 Oct 1814 - From what ship: HMT Freya, Halifax - Born: Delaware - Age: 41 - Race: Black - Released on 21 Jun 1815.

American Prisoners of War Held at Dartmoor during the War of 1812

## Alphabetical listing of names

Baron, William - Seaman - Nbr: 144 - Prize: Criterion - Ship Type: MV - How taken: HMS Belle Poule - When taken: 14 Feb 1813 - Where taken: Bay of Biscay - Date Received: 2 Apr 1813 - From what ship: Plymouth - Born: New York - Age: 25 - Released on 20 Apr 1815.

Barraso, John - Seaman - Nbr: 1410 - Prize: Miranda - Ship Type: Prize - How taken: HMS Unicorn - When taken: 21 May 1813 - Where taken: off Ushant (France) - Date Received: 19 Jun 1814 - From what ship: Stapleton - Born: Rochelle - Age: 22 - Sent to Mill Prison (Plymouth, England) on 21 Jun 1814.

Barren, Thomas - Servant - Nbr: 587 - Prize: US Brig Argus - Ship Type: MW - How taken: HMS Pelican - When taken: 14 Aug 1813 - Where taken: Irish Channel - Date Received: 8 Sep 1813 - From what ship: Plymouth - Born: Norfolk - Age: 21 - Race: Negro - Died on 2 Nov 1813.

Barrentt, John - Seaman - Nbr: 3737 - Prize: Lizard - Ship Type: P - How taken: HMS Prometheus - When taken: 5 May 1814 - Where taken: off Halifax - Date Received: 30 Sep 1814 - From what ship: HMT President, Halifax - Born: Portsmouth - Age: 18 - Released on 4 Jun 1815.

Barrester, Peter - Seaman - Nbr: 1561 - Prize: Caroline - Ship Type: MV - How taken: HMS Medusa - When taken: 12 Apr 1813 - Where taken: Bay of Biscay - Date Received: 23 Jun 1814 - From what ship: Stapleton - Born: Philadelphia - Age: 33 - Released on 1 May 1815.

Barrett, Anthony - Seaman - Nbr: 213 - Prize: Mars - Ship Type: MV - How taken: HMS Warspite - When taken: 26 Feb 1813 - Where taken: off Basque Roads (France) - Date Received: 2 Apr 1813 - From what ship: Plymouth - Born: New Orleans - Age: 58 - Sent to Plymouth on 8 Jul 1814.

Barrett, George - Seaman - Nbr: 2310 - Prize: Hussar - Ship Type: P - How taken: HMS Saturn - When taken: 25 May 1814 - Where taken: off Sandy Hook - Date Received: 16 Aug 1814 - From what ship: HMS Dublin, Halifax - Born: Philadelphia - Age: 19 - Released on 3 May 1815.

Barrett, George - Seaman - Nbr: 3059 - How taken: Gave himself up from HMS Galatea - When taken: 28 Apr 1813 - Date Received: 2 Sep 1814 - From what ship: HMT Hydra, Chatham - Born: Lancaster - Age: 39 - Released on 28 Apr 1815.

Barrett, James - Seaman - Nbr: 1940 - How taken: Gave himself up from HMS Albacore - When taken: 3 Feb 1813 - Date Received: 3 Aug 1814 - From what ship: HMS Alceste, Chatham Depot - Born: Delaware - Age: 27 - Released on 2 May 1815.

Barrett, Thomas - Seaman - Nbr: 6207 - Prize: Prince de Neufchatel - Ship Type: P - How taken: Leander (Newcastle Acasta) - When taken: 20 Dec 1814 - Where taken: Lat 38 Long 56 - Date Received: 30 Jan 1815 - From what ship: HMT Pheasant - Born: Virginia - Age: 23 - Released on 5 Jul 1815.

Barrett, Thomas - Mate - Nbr: 5548 - Prize: Hornet - Ship Type: MV - How taken: HMS Surprize - When taken: 19 Aug 1814 - Where taken: Lat 35 Long 24 - Date Received: 17 Dec 1814 - From what ship: HMT Loire, Halifax - Born: Norfolk - Age: 38 - Released on 1 Jul 1815.

Barron, John - Seaman - Nbr: 4796 - Prize: Martin - Ship Type: MV - How taken: HMS Swaggerer - When taken: 13 Mar 1814 - Where taken: off St. Thomas - Date Received: 9 Oct 1814 - From what ship: HMT Freya, Halifax - Born: New York - Age: 28 - Released on 21 Jun 1815.

Barrows, John - Seaman - Nbr: 6037 - Prize: Regent - Ship Type: LM - How taken: HMS Forth - When taken: 19 Sep 1814 - Where taken: off Egg Harbor (New Jersey) - Date Received: 28 Dec 1814 - From what ship: HMT Penelope - Born: Philadelphia - Age: 23 - Released on 3 Jul 1815.

Barry, James - Seaman - Nbr: 603 - Prize: US Brig Argus - Ship Type: MW - How taken: HMS Pelican - When taken: 14 Aug 1813 - Where taken: Irish Channel - Date Received: 8 Sep 1813 - From what ship: Plymouth - Born: Annapolis - Age: 29 - Sent to Dartmouth on 2 Nov 1814.

Barry, Peter - Seaman - Nbr: 3222 - How taken: Gave himself up from HMS Jalousie - When taken: 3 Aug 1813 - Date Received: 11 Sep 1814 - From what ship: HMT Freya, Chatham - Born: Salem - Age: 39 - Race: Black - Died on 27 Nov 1814, found dead in prison.

Bartell, William - Seaman - Nbr: 3426 - Prize: Industry - Ship Type: P - How taken: HMS Heron - When taken: 3 Nov 1813 - Where taken: off Halifax - Date Received: 13 Sep 1814 - From what ship: HMT Niobe, Chatham - Born: Marblehead - Age: 38 - Released on 28 May 1815.

Bartholomew, B. - Mate - Nbr: 6086 - Prize: Foria - Ship Type: MV - How taken: HMS Loire - When taken: 26 Sep 1814 - Where taken: off Bahamas - Date Received: 28 Dec 1814 - From what ship: HMT Penelope - Born: Philadelphia - Age: 21 - Released on 3 Jul 1815.

Bartis, John - Seaman - Nbr: 2863 - Prize: Prompt - Ship Type: MV - How taken: Change (Privateer) - When taken: 28 Mar 1813 - Where taken: Bay of Biscay - Date Received: 24 Aug 1814 - From what ship: HMT Alpheus, Chatham - Born: New Orleans - Age: 26 - Race: Black - Released on 19 May 1815.

Bartlett, Caleb - Seaman - Nbr: 1433 - Prize: Leo - Ship Type: LM - How taken: HMS Magicienne - When taken: 4

American Prisoners of War Held at Dartmoor during the War of 1812

## Alphabetical listing of names

Jun 1813 - Where taken: off France - Date Received: 19 Jun 1814 - From what ship: Stapleton - Born: Massachusetts - Age: 34 - Released on 28 Apr 1815.

Bartlett, Enoch - Marine - Nbr: 5946 - Prize: Harlequin - Ship Type: P - How taken: HMS Bulwark - When taken: 23 Nov 1814 - Where taken: off Halifax - Date Received: 27 Dec 1814 - From what ship: HMT Penelope - Born: Nottingham - Age: 23 - Released on 3 Jul 1815.

Bartlett, Henry - Seaman - Nbr: 6093 - Prize: Black Swan - Ship Type: MV - How taken: HMS Maidstone - When taken: 24 Oct 1814 - Where taken: Georges Bank - Date Received: 28 Dec 1814 - From what ship: HMT Penelope - Born: Plymouth - Age: 22 - Released on 3 Jul 1815.

Bartlett, John - Seaman - Nbr: 2674 - Prize: Joseph - Ship Type: MV - How taken: HMS Iris - When taken: 8 Jun 1813 - Where taken: off Spain - Date Received: 21 Aug 1814 - From what ship: HMT Freya, Chatham - Born: Marblehead - Age: 24 - Released on 19 May 1815.

Bartlett, John - Seaman - Nbr: 4598 - Prize: Elbridge Gerry - Ship Type: P - How taken: HMS Crescent - When taken: 13 Nov 1813 - Where taken: off St. Johns - Date Received: 9 Oct 1814 - From what ship: HMT Leyden, Chatham - Born: Hampshire - Age: 23 - Released on 10 Apr 1815.

Bartlett, Robert - Seaman - Nbr: 3374 - Prize: Growler - Ship Type: P - How taken: HMS Electra - When taken: 7 Jul 1813 - Where taken: at sea - Date Received: 13 Sep 1814 - From what ship: HMT Niobe, Chatham - Born: Philadelphia - Age: 26 - Race: Black - Released on 28 May 1815.

Bartlett, Samuel - Boy - Nbr: 5630 - Prize: Mary - Ship Type: Prize - How taken: Not legible - When taken: 18 Jun 1814 - Date Received: 24 Dec 1814 - From what ship: HMT Tay - Born: Beverly - Age: 15 - Released on 2 May 1815.

Bartlett, Scorpio - Seaman - Nbr: 4246 - Prize: Requin - Ship Type: LM - How taken: HMS Venus - When taken: 6 Mar 1814 - Where taken: off Bordeaux - Date Received: 7 Oct 1814 - From what ship: HMT Niobe, Chatham - Born: Boston - Age: 64 - Race: Black - Released on 13 Jun 1815.

Bartlett, Thomas - Seaman - Nbr: 57 - Prize: Spitfire - Ship Type: MV - How taken: HMS Achates - When taken: 14 Feb 1813 - Where taken: off Ushant (France) - Date Received: 2 Apr 1813 - From what ship: Plymouth - Born: Marblehead - Age: 21 - Released on 20 Apr 1815.

Barton, Isaac - Seaman - Nbr: 843 - Prize: Chesapeake - Ship Type: LM - How taken: HMS Hotspur & HMS Pyramus - When taken: 26 Oct 1813 - Where taken: off Nantes - Date Received: 29 Nov 1813 - From what ship: Plymouth - Born: Rhode Island - Age: 54 - Released on 26 Apr 1815.

Barton, James - Seaman - Nbr: 1872 - Prize: Elbridge Gerry - Ship Type: P - How taken: HMS Crescent - When taken: 16 Sep 1813 - Where taken: off St. George's - Date Received: 29 Jul 1814 - From what ship: HMS Ville de Paris, Chatham Depot - Born: Massachusetts - Age: 17 - Released on 2 May 1815.

Barton, Mathew - Seaman - Nbr: 6293 - How taken: Gave himself up from HMS Sabine - Date Received: 17 Feb 1815 - From what ship: HMT Ganges, Plymouth - Born: Kent County - Age: 39 - Race: Mulatto - Released on 5 Jul 1815.

Barton, Robert - Seaman - Nbr: 5878 - Prize: Harlequin - Ship Type: P - How taken: HMS Bulwark - When taken: 23 Nov 1814 - Where taken: off Halifax - Date Received: 27 Dec 1814 - From what ship: HMT Penelope - Born: Philadelphia - Age: 24 - Released on 3 Jul 1815.

Barton, Samuel - Seaman - Nbr: 4127 - Prize: Bordeaux Packet - Ship Type: LM - How taken: HMS Niemen - When taken: 28 Jun 1814 - Where taken: off Delaware - Date Received: 6 Oct 1814 - From what ship: HMT Chesapeake, Halifax - Born: New York - Age: 33 - Released on 29 Jun 1815.

Bartrom, Lewis - Seaman - Nbr: 2246 - Prize: Carbineer - Ship Type: Prize - How taken: HMS Ringdove - When taken: 24 Apr 1814 - Where taken: off Bermuda - Date Received: 16 Aug 1814 - From what ship: HMS Dublin, Halifax - Born: Salem - Age: 26 - Released on 3 May 1815.

Basset, Gorham - Seaman - Nbr: 1284 - Prize: Indian Lass - Ship Type: Prize - How taken: Not listed - When taken: 29 Apr 1814 - Date Received: 14 Jun 1814 - From what ship: Mill Prison (Plymouth, England) - Born: Barnstable - Age: 22 - Released on 28 Apr 1815.

Bassett, John - Seaman - Nbr: 4478 - Prize: Enterprize - Ship Type: P - How taken: HMS Tenedos - When taken: 21 May 1813 - Where taken: off Cape Cod - Date Received: 8 Oct 1814 - From what ship: HMT Leyden, Chatham - Born: Massachusetts - Age: 25 - Released on 15 Jun 1815.

Bassett, William - Seaman - Nbr: 4233 - Prize: Teazer - Ship Type: P - How taken: HMS Boyce - When taken: 24 Mar 1814 - Where taken: at sea - Date Received: 7 Oct 1814 - From what ship: HMT Niobe, Chatham - Born: Philadelphia - Age: 20 - Race: Black - Released on 13 Jun 1815.

Bassington, John - Seaman - Nbr: 1190 - Prize: Jenny - Ship Type: MV - How taken: HMS Druid - When taken: 14 Jun 1812 - Date Received: 4 Jun 1814 - From what ship: Dartmouth - Born: Salem - Age: 45 - Released on

American Prisoners of War Held at Dartmoor during the War of 1812

## Alphabetical listing of names

26 Apr 1815.

Basson, John - Seaman - Nbr: 1212 - Prize: Imperatrice Reine - Ship Type: P - How taken: HMS Hotspur - When taken: 15 Feb 1813 - Where taken: off St. Antonio - Date Received: 4 Jun 1814 - From what ship: Dartmouth - Born: Philadelphia - Age: 31 - Released on 28 Apr 1815.

Bassonet, Charles - Seaman - Nbr: 830 - Prize: Chesapeake - Ship Type: LM - How taken: HMS Hotspur & HMS Pyramus - When taken: 26 Oct 1813 - Where taken: off Nantes - Date Received: 29 Nov 1813 - From what ship: Plymouth - Born: Pennsylvania - Age: 25 - Released on 26 Apr 1815.

Bastard, Walter - Seaman - Nbr: 3884 - Prize: US Brig Rattlesnake - Ship Type: MW - How taken: HMS Leander - When taken: 11 Jul 1814 - Where taken: off Shelburne - Date Received: 5 Oct 1814 - From what ship: HMT Orpheus, Halifax - Born: Marblehead - Age: 20 - Released on 9 Jun 1815.

Bateman, John - Seaman - Nbr: 3459 - Prize: Prize to Chasseur - How taken: HMS Tartarus - When taken: 1 Sep 1814 - Where taken: Atlantic - Date Received: 19 Sep 1814 - From what ship: HMT Salvador del Mundo - Born: Baltimore - Age: 18 - Died on 23 Nov 1814.

Bateman, Michael - Seaman - Nbr: 5175 - Prize: Vivid - Ship Type: P - How taken: HMS Nymphe - When taken: 20 Apr 1813 - Where taken: off Cape Cod - Date Received: 31 Oct 1814 - From what ship: HMT Mermaid, Chatham - Born: Salem - Age: 54 - Released on 29 Jun 1815.

Bates, Joseph - Seaman - Nbr: 3195 - How taken: Gave himself up from HMS Swiftsure - When taken: 26 Dec 1812 - Date Received: 11 Sep 1814 - From what ship: HMT Freya, Chatham - Born: Massachusetts - Age: 21 - Released on 27 Apr 1815.

Bates, Josiah - Seaman - Nbr: 5389 - Prize: Sister - Ship Type: MV - How taken: HMS Unicorn - When taken: 3 Jul 1814 - Where taken: off Christian Land - Date Received: 31 Oct 1814 - From what ship: HMT Leyden, Chatham - Born: Marblehead - Age: 22 - Released on 1 Jul 1815.

Batman, C. P. - Seaman - Nbr: 1450 - Prize: Tickler - Ship Type: LM - How taken: HMS Magicienne - When taken: 5 Jun 1813 - Where taken: Bay of Biscay - Date Received: 19 Jun 1814 - From what ship: Stapleton - Born: Massachusetts - Age: 24 - Released on 28 Apr 1815.

Batterson, John - Seaman - Nbr: 6077 - Prize: Daedalus - Ship Type: MV - How taken: HMS Niemen - When taken: 20 Sep 1814 - Where taken: off Delaware - Date Received: 28 Dec 1814 - From what ship: HMT Penelope - Born: New York - Age: 27 - Race: Black - Released on 3 Jul 1815.

Battes, John - Seaman - Nbr: 4477 - Prize: Enterprize - Ship Type: P - How taken: HMS Tenedos - When taken: 21 May 1813 - Where taken: off Cape Cod - Date Received: 8 Oct 1814 - From what ship: HMT Leyden, Chatham - Born: Massachusetts - Age: 43 - Released on 15 Jun 1815.

Battle, John - Seaman - Nbr: 6506 - How taken: Apprehended at Kingston - Date Received: 3 Mar 1815 - From what ship: HMT Ganges, Plymouth - Born: Farmington - Age: 25 - Released on 11 Jul 1815.

Batts, William - Boy - Nbr: 5891 - Prize: Harlequin - Ship Type: P - How taken: HMS Bulwark - When taken: 23 Nov 1814 - Where taken: off Halifax - Date Received: 27 Dec 1814 - From what ship: HMT Penelope - Born: Massachusetts - Age: 17 - Released on 11 Jul 1815.

Bauld, J. P. - Seaman - Nbr: 4835 - Prize: President - Ship Type: P - How taken: HMS Pique - When taken: 7 May 1814 - Where taken: off Porto Rico - Date Received: 9 Oct 1814 - From what ship: HMT Freya, Halifax - Born: Guadeloupe - Age: 21 - Released on 21 Jun 1815.

Baxter, David - Seaman - Nbr: 4594 - Prize: Caroline - Ship Type: MV - How taken: HMS Moselle - When taken: 12 Aug 1813 - Where taken: off Charleston - Date Received: 9 Oct 1814 - From what ship: HMT Leyden, Chatham - Born: Yarmouth - Age: 22 - Released on 15 Jun 1815.

Baxter, David - Seaman - Nbr: 3998 - Prize: US Brig Rattlesnake - Ship Type: MW - How taken: HMS Leander - When taken: 13 Jul 1814 - Where taken: off Shelburne - Date Received: 6 Oct 1814 - From what ship: HMT Chesapeake, Halifax - Born: Nantucket - Age: 23 - Released on 14 Jun 1815.

Baxter, Francis - Seaman - Nbr: 4742 - How taken: Gave himself up from HMS Phoenix - When taken: 17 Jul 1814 - Date Received: 9 Oct 1814 - From what ship: HMT Freya, Chatham - Born: Barnstable - Age: 24 - Released on 15 Jun 1815.

Baxter, John - Seaman - Nbr: 5502 - Prize: US Gunboat Nbr 2 - Ship Type: MW - How taken: British forces - When taken: 22 Aug 1814 - Where taken: Chesapeake Bay - Date Received: 17 Dec 1814 - From what ship: HMT Loire, Halifax - Born: Philadelphia - Age: 40 - Released on 29 Jun 1815.

Baxter, Marion - Seaman - Nbr: 6050 - Prize: Swift - Ship Type: MV - How taken: HMS Niemen - When taken: 8 Aug 1814 - Where taken: off Bahamas - Date Received: 28 Dec 1814 - From what ship: HMT Penelope - Born: Yarmouth - Age: 20 - Released on 11 Jul 1815.

Bean, Amos - Seaman - Nbr: 5438 - How taken: Gave himself up from HMS Mars - When taken: 31 Jan 1813 -

American Prisoners of War Held at Dartmoor during the War of 1812

## Alphabetical listing of names

Date Received: 11 Nov 1814 - From what ship: HMT Impregnable - Born: Brentwood - Age: 24 - Released on 1 Jul 1815.

Bean, John - Seaman - Nbr: 1859 - Prize: Volante - Ship Type: LM - How taken: HMS Valiant - When taken: 25 Mar 1813 - Where taken: Georges Bank - Date Received: 29 Jul 1814 - From what ship: HMS Ville de Paris, Chatham Depot - Born: Providence - Age: 29 - Released on 2 May 1815.

Bean, William - Seaman - Nbr: 2096 - How taken: Gave himself up from HMS Malta - When taken: 1 Jan 1813 - Date Received: 3 Aug 1814 - From what ship: HMS Bittern, Chatham Depot - Born: Virginia - Age: 34 - Died on 28 Nov 1814 from pneumonia.

Beans, James - Seaman - Nbr: 2312 - Prize: Hussar - Ship Type: P - How taken: HMS Saturn - When taken: 25 May 1814 - Where taken: off Sandy Hook - Date Received: 16 Aug 1814 - From what ship: HMS Dublin, Halifax - Born: Virginia - Age: 36 - Released on 3 May 1815.

Beard, Francis - Seaman - Nbr: 1612 - Prize: Fox - Ship Type: LM - How taken: HMS Pheasant - When taken: 23 Apr 1813 - Where taken: Bay of Biscay - Date Received: 23 Jun 1814 - From what ship: Stapleton - Born: Haiti - Age: 32 - Sent to Mill Prison (Plymouth, England) on 26 Jun 1814.

Beattie, James - Seaman - Nbr: 1903 - Prize: Quebec - Ship Type: Prize - How taken: HMS Derwent - When taken: 29 Jan 1813 - Where taken: off Lisbon - Date Received: 3 Aug 1814 - From what ship: HMS Alceste, Chatham Depot - Born: Frybourg - Age: 36 - Released on 2 May 1815.

Beatty, John - Seaman - Nbr: 4518 - Prize: Lash - Ship Type: P - How taken: HMS Tenedos - When taken: 26 May 1813 - Where taken: off Cape Cod - Date Received: 8 Oct 1814 - From what ship: HMT Leyden, Chatham - Born: New Jersey - Age: 24 - Released on 13 Jun 1815.

Bebe, Joseph - Seaman - Nbr: 825 - Prize: Chesapeake - Ship Type: LM - How taken: HMS Hotspur & HMS Pyramus - When taken: 26 Oct 1813 - Where taken: off Nantes - Date Received: 29 Nov 1813 - From what ship: Plymouth - Born: Massachusetts - Age: 22 - Released on 26 Apr 1815.

Becamp, Jacques - Seaman - Nbr: 4996 - Prize: Invincible - Ship Type: LM - How taken: HMS Armide - When taken: 15 Aug 1814 - Where taken: off Nantucket - Date Received: 28 Oct 1814 - From what ship: HMT Alkbar, Halifax - Born: Toulouse - Age: 31 - Released on 21 Jun 1815.

Beck, Francis - Seaman - Nbr: 6470 - Prize: Decatur - Ship Type: P - How taken: HMS Rhin - When taken: 5 Jun 1814 - Where taken: off San Domingo - Date Received: 3 Mar 1815 - From what ship: HMT Ganges, Plymouth - Born: Charleston - Age: 20 - Race: Creole - Released on 14 Jun 1815.

Beck, Henry - Steward - Nbr: 5664 - Prize: Harlequin - Ship Type: P - How taken: HMS Bulwark - When taken: 23 Nov 1814 - Where taken: off Halifax - Date Received: 24 Dec 1814 - From what ship: HMT Penelope - Born: Portsmouth - Age: 38 - Released on 28 May 1815.

Beck, Steward - Seaman - Nbr: 1638 - Prize: Henry Clements - Ship Type: MV - How taken: HMS Orestes - When taken: 15 Apr 1813 - Where taken: Bay of Biscay - Date Received: 23 Jun 1814 - From what ship: Stapleton - Born: Maryland - Age: 19 - Released on 1 May 1815.

Beck, William - Seaman - Nbr: 3069 - How taken: Impressed at Hull - When taken: 26 Feb 1814 - Date Received: 2 Sep 1814 - From what ship: HMT Hydra, Chatham - Born: New Hampshire - Age: 25 - Released on 28 May 1815.

Beck, William - Seaman - Nbr: 1934 - How taken: Gave himself up from HMS Royal William - When taken: 3 Feb 1813 - Date Received: 3 Aug 1814 - From what ship: HMS Alceste, Chatham Depot - Born: Portsmouth - Age: 50 - Died on 18 Jan 1814 from variola.

Beckett, William - Seaman - Nbr: 6142 - How taken: Gave himself up from HMS Ajax - Date Received: 17 Jan 1815 - From what ship: HMT Impregnable - Born: Virginia - Age: 28 - Race: Black - Released on 5 Jul 1815.

Beckford, John - Seaman - Nbr: 5406 - How taken: Impressed at London - When taken: 6 Oct 1814 - Date Received: 31 Oct 1814 - From what ship: HMT Leyden, Chatham - Born: Salem - Age: 23 - Released on 1 Jul 1815.

Beckwith, James - Seaman - Nbr: 4701 - How taken: Gave himself up from HMS Leopard - When taken: 25 Dec 1813 - Date Received: 9 Oct 1814 - From what ship: HMT Leyden, Chatham - Born: Maryland - Age: 25 - Released on 15 Jun 1815.

Beecher, William - Seaman - Nbr: 2788 - Prize: Prompt - Ship Type: MV - How taken: Chance (Privateer) - When taken: 28 Mar 1813 - Where taken: Bay of Biscay - Date Received: 24 Aug 1814 - From what ship: HMT Liverpool, Chatham - Born: New Haven - Age: 16 - Released on 19 May 1815.

Beeston, Robert - Seaman - Nbr: 5856 - Prize: Lion - Ship Type: P - How taken: HMS Granicus - When taken: 2 Dec 1814 - Where taken: off Lisbon - Date Received: 26 Dec 1814 - From what ship: HMT Impregnable - Born: Alexandria - Age: 26 - Race: Black - Released on 3 Jul 1815.

American Prisoners of War Held at Dartmoor during the War of 1812

## Alphabetical listing of names

Beets, Thomas - Seaman - Nbr: 5050 - Prize: Ida - Ship Type: LM - How taken: HMS Newcastle - When taken: 9 Aug 1814 - Where taken: Long 34 - Date Received: 28 Oct 1814 - From what ship: HMT Alkbar, Halifax - Born: Boston - Age: 24 - Released on 29 Jun 1815.

Behon, Simon - Seaman - Nbr: 4454 - Prize: Juliana Smith - Ship Type: P - How taken: HMS Nymphe - When taken: 11 May 1813 - Where taken: off Cape Sable Island (Canada) - Date Received: 8 Oct 1814 - From what ship: HMT Leyden, Chatham - Born: Massachusetts - Age: 18 - Released on 15 Jun 1815.

Belavoine, L. - Seaman - Nbr: 990 - Prize: Harvest - Ship Type: P - How taken: HMS Orestes - When taken: 20 Jan 1814 - Where taken: Bay of Biscay - Date Received: 31 Jan 1814 - From what ship: Plymouth - Born: Nantes - Age: 17 - Sent to Mill Prison (Plymouth, England) on 21 Jun 1814.

Belcour, James - Seaman - Nbr: 5093 - Prize: David Porter - Ship Type: LM - How taken: HMS Pylades - When taken: 12 Sep 1814 - Where taken: Georges Bank - Date Received: 28 Oct 1814 - From what ship: HMT Alkbar, Halifax - Born: France - Age: 28 - Released on 29 Jun 1815.

Belfast, Richard - Seaman - Nbr: 433 - Prize: Viper - Ship Type: MV - How taken: HMS Superb - When taken: 15 Apr 1813 - Where taken: Bay of Biscay - Date Received: 1 Jul 1813 - From what ship: Plymouth - Born: Connecticut - Age: 24 - Race: Black - Released on 20 Apr 1815.

Bell, George - Seaman - Nbr: 2386 - Prize: US Sloop Frolic - Ship Type: MW - How taken: HMS Orpheus - When taken: 20 Apr 1814 - Where taken: off Cuba - Date Received: 16 Aug 1814 - From what ship: HMT Queen, Halifax - Born: Boston - Age: 23 - Sent to Dartmouth on 19 Oct 1814.

Bell, Jacob - Seaman - Nbr: 5648 - Prize: Lion - Ship Type: P - How taken: HMS Granicus - When taken: 2 Dec 1814 - Where taken: off Lisbon - Date Received: 24 Dec 1814 - From what ship: HMT Tay - Born: Virginia - Age: 33 - Released on 3 Jul 1815.

Bell, James - Seaman - Nbr: 534 - Prize: Union - Ship Type: LM - How taken: HMS Goldfinch - When taken: 17 Jul 1813 - Where taken: Bay of Biscay - Date Received: 8 Sep 1813 - From what ship: Plymouth - Born: Virginia - Age: 27 - Released on 26 Apr 1815.

Bell, Joshua - Seaman - Nbr: 5460 - How taken: Gave himself up from HMS Constant - Date rec'd: 10 Dec 1814 - From what ship: HMT Impregnable - Born: Philadelphia - Age: 39 - Race: Black - Released on 1 Jul 1815.

Bell, Richard - Seaman - Nbr: 5315 - Prize: Thomas - Ship Type: P - How taken: HMS Nymphe - When taken: 24 Jun 1813 - Where taken: off Halifax - Date Received: 31 Oct 1814 - From what ship: HMT Leyden, Chatham - Born: Portland - Age: 19 - Race: Black - Released on 29 Jun 1815.

Bella, Darius - Seaman - Nbr: 2418 - Prize: US Sloop Frolic - Ship Type: MW - How taken: HMS Orpheus - When taken: 20 Apr 1814 - Where taken: off Cuba - Date Received: 16 Aug 1814 - From what ship: HMT Queen, Halifax - Born: New Providence - Age: 29 - Died on 25 Jan 1815 from apoplexy.

Bellinger, William - Seaman - Nbr: 255 - Prize: Pert - Ship Type: MV - How taken: HMS Warspite - When taken: 1 Mar 1813 - Where taken: off Basque Roads (France) - Date Received: 28 Jun 1813 - From what ship: Plymouth - Born: New Jersey - Age: 21 - Released on 20 Apr 1815.

Benedict, William - Marine - Nbr: 595 - Prize: US Brig Argus - Ship Type: MW - How taken: HMS Pelican - When taken: 14 Aug 1813 - Where taken: Irish Channel - Date Received: 8 Sep 1813 - From what ship: Plymouth - Born: New York - Age: 21 - Sent to Dartmouth on 2 Nov 1814.

Benjamin, Joseph - Seaman - Nbr: 2606 - How taken: Gave himself up from HMS Royal William - When taken: 1 Feb 1812 - Date Received: 21 Aug 1814 - From what ship: HMT Freya, Chatham - Born: Philadelphia - Age: 34 - Race: Black - Released on 26 Apr 1815.

Benjamin, P. - Seaman - Nbr: 2851 - Prize: Pallas - Ship Type: MV - How taken: Rebuff - When taken: 23 Jan 1813 - Where taken: off Cadiz - Date Received: 24 Aug 1814 - From what ship: HMT Alpheus, Chatham - Born: Stratford - Age: 17 - Released on 19 May 1815.

Bennett, Abisa - Seaman - Nbr: 3949 - Prize: Rolla - Ship Type: P - How taken: HMS Loire - When taken: 10 Dec 1813 - Where taken: off Bull Island (South Carolina) - Date Received: 5 Oct 1814 - From what ship: HMT President, Halifax - Born: Tiverton - Age: 22 - Race: Negro - Released on 9 Jun 1815.

Bennett, Charles - 2nd Mate - Nbr: 649 - Prize: Marmion - Ship Type: MV - How taken: HMS President - When taken: 14 Aug 1813 - Where taken: off Nantes - Date Received: 8 Sep 1813 - From what ship: Plymouth - Born: New York - Age: 27 - Released on 26 Apr 1815.

Bennett, D. C. - Seaman - Nbr: 5809 - Prize: US Schooner Scorpion - Ship Type: MW - How taken: British gunboats - When taken: 6 Sep 1814 - Where taken: Lake Erie - Date Received: 26 Dec 1814 - From what ship: HMT Argo - Born: New York - Age: 23 - Released on 3 Jul 1815.

Bennett, David - Seaman - Nbr: 5595 - Prize: Albion - Ship Type: Prize - How taken: HMS Jaseur - When taken: 21 Sep 1814 - Where taken: off Halifax - Date Received: 17 Dec 1814 - From what ship: HMT Loire, Halifax -

American Prisoners of War Held at Dartmoor during the War of 1812
## Alphabetical listing of names

Born: Alexandria - Age: 29 - Released on 1 Jul 1815.

Bennett, James - Seaman - Nbr: 295 - Prize: Ducornau - Ship Type: MV - How taken: HMS Pheasant - When taken: 15 Mar 1813 - Where taken: Bay of Biscay - Date Received: 28 Jun 1813 - From what ship: Plymouth - Born: Frankfort - Age: 26 - Released on 20 Apr 1815.

Bennett, James - Seaman - Nbr: 4104 - Prize: Mary - Ship Type: MV - How taken: HMS Junon - When taken: 6 Jun 1814 - Where taken: off Cape Ann - Date Received: 6 Oct 1814 - From what ship: HMT Chesapeake, Halifax - Born: Massachusetts - Age: 26 - Released on 13 Jun 1815.

Bennett, John - Seaman - Nbr: 4301 - How taken: Gave himself up from HMS Rigle - When taken: 9 May 1814 - Date Received: 7 Oct 1814 - From what ship: HMT Niobe, Chatham - Born: Shrewsbury - Age: 34 - Released on 14 Jun 1815.

Bennett, Peleg - Seaman - Nbr: 3615 - Prize: Monarch - Ship Type: MV - How taken: HMS Dotterel - When taken: 14 Dec 1813 - Where taken: off Charleston - Date Received: 30 Sep 1814 - From what ship: HMT Sybella - Born: Charleston - Age: 44 - Released on 4 Jun 1815.

Bennett, Robert - Seaman - Nbr: 5379 - Prize: Rose - Ship Type: MV - How taken: HMS Racehorse - When taken: 3 Feb 1813 - Where taken: off Isle of France - Date Received: 31 Oct 1814 - From what ship: HMT Leyden, Chatham - Born: Nantucket - Age: 29 - Released on 1 Jul 1815.

Bennett, Stephen - Seaman - Nbr: 2216 - Prize: Hussar - Ship Type: P - How taken: HMS Saturn - When taken: 25 May 1814 - Where taken: off Sandy Hook - Date Received: 16 Aug 1814 - From what ship: HMS Dublin, Halifax - Born: New Jersey - Age: 25 - Released on 2 May 1815.

Benny, David - Boy - Nbr: 1269 - Prize: Adeline - Ship Type: MV - How taken: HMS Magicienne - When taken: 16 Mar 1814 - Where taken: off Cape Finisterre - Date Received: 14 Jun 1814 - From what ship: Mill Prison (Plymouth, England) - Born: New York - Age: 17 - Released on 28 Apr 1815.

Benson, George - Seaman - Nbr: 1985 - How taken: Gave himself up from HMS Colossus - When taken: Apr 1813 - Date Received: 3 Aug 1814 - From what ship: HMS Lyffey, Chatham Depot - Born: Nottingham - Age: 39 - Released on 2 May 1815.

Benson, James - Seaman - Nbr: 610 - Prize: US Brig Argus - Ship Type: MW - How taken: HMS Pelican - When taken: 14 Aug 1813 - Where taken: Irish Channel - Date Received: 8 Sep 1813 - From what ship: Plymouth - Born: Delaware - Age: 30 - Sent to Dartmouth on 2 Nov 1814.

Benson, James - Seaman - Nbr: 2045 - How taken: Gave himself up from HMS North Star - When taken: 16 May 1812 - Date Received: 3 Aug 1814 - From what ship: HMS Lyffey, Chatham Depot - Born: Sacafras - Age: 23 - Released on 26 Apr 1815.

Benson, John - Seaman - Nbr: 6512 - How taken: Gave himself up from HMS Orontes - When taken: Feb 1815 - Date Received: 3 Mar 1815 - From what ship: HMT Ganges, Plymouth - Born: New York - Age: 22 - Race: Black - Released on 11 Jul 1815.

Benson, Joseph - Seaman - Nbr: 2408 - Prize: US Sloop Frolic - Ship Type: MW - How taken: HMS Orpheus - When taken: 20 Apr 1814 - Where taken: off Cuba - Date Received: 16 Aug 1814 - From what ship: HMT Queen, Halifax - Born: Marblehead - Age: 20 - Sent to Dartmouth on 19 Oct 1814.

Benson, Leven - Seaman - Nbr: 2042 - How taken: Gave himself up from HMS Sterling Castle - When taken: 5 Jun 1812 - Date Received: 3 Aug 1814 - From what ship: HMS Lyffey, Chatham Depot - Born: Maryland - Age: 36 - Released on 26 Apr 1815.

Benson, Mingo - Seaman - Nbr: 3119 - How taken: Gave himself up from HMS Derwent - When taken: Sep 1814 - Date Received: 11 Sep 1814 - From what ship: HMT Salvador del Mundo - Born: New York - Age: 33 - Race: Negro - Released on 28 May 1815.

Benson, Samuel - Seaman - Nbr: 5483 - Prize: General Putnam - Ship Type: P - How taken: HMS Leander - When taken: 8 Nov 1814 - Where taken: Long 65 Lat 42 - Date Received: 17 Dec 1814 - From what ship: HMT Loire, Halifax - Born: Beverly - Age: 24 - Released on 1 Jul 1815.

Bensted, John - Seaman - Nbr: 1055 - How taken: Impressed at Cork - Date Received: 10 May 1814 - From what ship: Plymouth - Born: Newbury - Age: 33 - Released on 27 Apr 1815.

Bent, Joseph - Seaman - Nbr: 5097 - Prize: David Porter - Ship Type: LM - How taken: HMS Pylades - When taken: 12 Sep 1814 - Where taken: Georges Bank - Date Received: 28 Oct 1814 - From what ship: HMT Alkbar, Halifax - Born: Portugal - Age: 24 - Released on 29 Jun 1815.

Benton, Samuel - Seaman - Nbr: 6420 - Prize: Farmer's Daughter - Ship Type: MV - How taken: HMS Leviathan - When taken: 29 May 1814 - Date Received: 3 Mar 1815 - From what ship: HMT Ganges, Plymouth - Born: Kent County - Age: 30 - Released on 11 Jul 1815.

Benyman, James - Seaman - Nbr: 4105 - Prize: Stark - Ship Type: P - How taken: HMS Sophie - When taken: 20

American Prisoners of War Held at Dartmoor during the War of 1812

## Alphabetical listing of names

Apr 1814 - Where taken: off Bermuda - Date Received: 6 Oct 1814 - From what ship: HMT Chesapeake, Halifax - Born: New London - Age: 19 - Released on 13 Jun 1815.

Berdick, Simon - Seaman - Nbr: 668 - Prize: Joel Barlow - Ship Type: LM - How taken: HMS Briton - When taken: 3 Jul 1813 - Where taken: off Bordeaux - Date Received: 8 Sep 1813 - From what ship: Plymouth - Born: Rhode Island - Age: 27 - Released on 26 Apr 1815.

Bernard, John - Seaman - Nbr: 1358 - Prize: Paul Jones - Ship Type: P - How taken: HMS Leonidas - When taken: 23 May 1813 - Where taken: Channel - Date Received: 19 Jun 1814 - From what ship: Stapleton - Born: New Orleans - Age: 26 - Sent to Mill Prison (Plymouth, England) on 26 Jun 1814.

Bernard, William - Mate - Nbr: 3912 - Prize: Nonsuch - Ship Type: MV - How taken: HMS Dotterel - When taken: 14 Dec 1813 - Where taken: off Block Island - Date Received: 5 Oct 1814 - From what ship: HMT President, Halifax - Born: Philadelphia - Age: 26 - Released on 4 Jun 1815.

Berry, Anton - Seaman - Nbr: 3585 - How taken: Gave himself up from MV Maria - When taken: May 1814 - Date Received: 30 Sep 1814 - From what ship: HMT Sybella - Born: Cape Ann - Age: 21 - Released on 4 Jun 1815.

Berry, Brook - Seaman - Nbr: 30 - How taken: Apprehended at Gibraltar - When taken: 8 Aug 1813 - Date Received: 2 Apr 1813 - From what ship: Plymouth - Born: Prince George's County - Age: 20 - Sent to Dartmouth on 30 Jul 1813.

Berry, Jesse - Seaman - Nbr: 6047 - Prize: Perry - Ship Type: MV - How taken: HMS Endymion - When taken: 3 Dec 1813 - Where taken: off NY - Date Received: 28 Dec 1814 - From what ship: HMT Penelope - Born: Plymouth - Age: 33 - Released on 11 Jul 1815.

Berry, Joseph - Seaman - Nbr: 6191 - Prize: Prince de Neufchatel - Ship Type: P - How taken: Leander (Newcastle Acasta) - When taken: 20 Dec 1814 - Where taken: Lat 38 Long 56 - Date Received: 30 Jan 1815 - From what ship: HMT Pheasant - Born: Massachusetts - Age: 48 - Released on 5 Jul 1815.

Berry, William - Seaman - Nbr: 5225 - Prize: York Town - Ship Type: P - How taken: HMS Maidstone - When taken: 18 Jul 1813 - Where taken: Grand Banks - Date Received: 31 Oct 1814 - From what ship: HMT Mermaid, Chatham - Born: Long Island - Age: 24 - Released on 29 Jun 1815.

Berryman, John - Seaman - Nbr: 1406 - Prize: Courier - Ship Type: P - How taken: HMS Rover - When taken: 14 May 1813 - Where taken: Bay of Biscay - Date Received: 19 Jun 1814 - From what ship: Stapleton - Born: Maryland - Age: 25 - Race: Colored - Released on 28 Apr 1815.

Berto, John - Seaman - Nbr: 1046 - Prize: Prince of Wales - Ship Type: P - How taken: Nelson (Transport) - When taken: 4 Feb 1814 - Where taken: at sea - Date Received: 10 May 1814 - From what ship: Plymouth - Born: New Orleans - Age: 40 - Sent to Plymouth on 8 Jul 1814.

Bertol, Samuel - Seaman - Nbr: 1258 - Prize: Adeline - Ship Type: MV - How taken: HMS Magicienne - When taken: 16 Mar 1814 - Where taken: off Cape Finisterre - Date Received: 14 Jun 1814 - From what ship: Mill Prison (Plymouth, England) - Born: Freeport - Age: 23 - Released on 28 Apr 1815.

Besson, Phillipe - Seaman - Nbr: 1026 - Prize: Joseph - Ship Type: MV - How taken: HMS Royalist - When taken: 18 Jan 1814 - Where taken: Bay of Biscay - Date Received: 31 Jan 1814 - From what ship: Plymouth - Born: Marblehead - Age: 18 - Released on 27 Apr 1815.

Bessop, John - Boy - Nbr: 3688 - Prize: Alfred - Ship Type: P - How taken: HMS Epervier - When taken: 23 Feb 1812 - Where taken: off Newfoundland - Date Received: 30 Sep 1814 - From what ship: HMT President - Born: Marblehead - Age: 16 - Released on 4 Jun 1815.

Best, John - Seaman - Nbr: 1938 - How taken: Gave himself up from HMS Albacore - When taken: 3 Feb 1813 - Date Received: 3 Aug 1814 - From what ship: HMS Alceste, Chatham Depot - Born: New Jersey - Age: 30 - Released on 2 May 1815.

Best, Robert - Seaman - Nbr: 1447 - Prize: Tickler - Ship Type: LM - How taken: HMS Magicienne - When taken: 5 Jun 1813 - Where taken: Bay of Biscay - Date Received: 19 Jun 1814 - From what ship: Stapleton - Born: Jersey - Age: 22 - Released on 28 Apr 1815.

Beter, John - Seaman - Nbr: 6485 - Prize: John - Ship Type: MV - How taken: Variable - When taken: 11 Aug 1814 - Where taken: off Cuba - Date Received: 3 Mar 1815 - From what ship: HMT Ganges, Plymouth - Born: Philadelphia - Age: 30 - Released on 11 Jul 1815.

Beverly, Henry - Seaman - Nbr: 2733 - How taken: Gave himself up at London - When taken: 5 Feb 1813 - Date Received: 24 Aug 1814 - From what ship: HMT Liverpool, Chatham - Born: New Market - Age: 21 - Died on 2 Dec 1814 from phthisis pulmonalis.

Beverley, Richard - Seaman - Nbr: 3402 - Prize: Thomas - Ship Type: P - How taken: HMS Nymphe - When taken: 24 Jun 1813 - Where taken: off Halifax - Date Received: 13 Sep 1814 - From what ship: HMT Niobe,

American Prisoners of War Held at Dartmoor during the War of 1812
## Alphabetical listing of names

Chatham - Born: Portsmouth - Age: 20 - Released on 28 May 1815.

Bevin, William - Seaman - Nbr: 620 - Prize: Betsy - Ship Type: MV - How taken: HMS Leonidas - When taken: 12 Aug 1813 - Where taken: Channel - Date Received: 8 Sep 1813 - From what ship: Plymouth - Born: Chatham - Age: 37 - Sent to Dartmouth on 2 Nov 1814.

Beymer, George - Seaman - Nbr: 815 - How taken: Impressed at Cork - When taken: 25 Oct 1813 - Date rec'd: 3 Nov 1813 - From what ship: Plymouth - Born: Philadelphia - Age: 27 - Released on 26 Apr 1815.

Bicker, Charles - Seaman - Nbr: 6536 - Prize: US Brig Syren - Ship Type: MW - How taken: HMS Medway - When taken: 12 Jul 1814 - Where taken: off Cape of Good Hope - Date Received: 7 Mar 1815 - From what ship: HMT Ganges, Plymouth - Born: Boston - Age: 23 - Released on 11 Jul 1815.

Bickwith, Benjamin - Seaman - Nbr: 1656 - Prize: Tom - Ship Type: LM - How taken: HMS Surveillante - When taken: 24 Apr 1813 - Where taken: Bay of Biscay - Date Received: 23 Jun 1814 - From what ship: Stapleton - Born: Pennsylvania - Age: 25 - Released on 1 May 1815.

Bidbee, Joseph - Prize master - Nbr: 6459 - Prize: Decatur - Ship Type: P - How taken: HMS Rhin - When taken: 5 Jun 1814 - Where taken: off San Domingo - Date Received: 3 Mar 1815 - From what ship: HMT Ganges, Plymouth - Born: Not legible - Age: 39 - Released on 4 Jun 1815.

Bids, Thomas - Seaman - Nbr: 5191 - Prize: Lark - Ship Type: MV - How taken: HMS Bream - When taken: 12 Apr 1813 - Where taken: off Cape Sable Island (Canada) - Date Received: 31 Oct 1814 - From what ship: HMT Mermaid, Chatham - Born: Salem - Age: 44 - Released on 29 Jun 1815.

Bidson, Thomas - Seaman - Nbr: 145 - Prize: Criterion - Ship Type: MV - How taken: HMS Belle Poule - When taken: 14 Feb 1813 - Where taken: Bay of Biscay - Date Received: 2 Apr 1813 - From what ship: Plymouth - Born: New York - Age: 22 - Released on 20 Apr 1815.

Bienfaux, Allen - Prize Master - Nbr: 783 - Prize: Betsy - Ship Type: MV - How taken: HMS Eurotas - When taken: 26 Oct 1813 - Where taken: off Ushant (France) - Date Received: 3 Nov 1813 - From what ship: Plymouth - Born: Brest - Age: 43 - Sent to Mill Prison (Plymouth, England) on 21 Jun 1814.

Bignell, Peter - Seaman - Nbr: 3871 - Prize: Tickler - Ship Type: MV - How taken: HMS Saturn - When taken: 13 Jul 1814 - Where taken: off America - Date Received: 5 Oct 1814 - From what ship: HMT Orpheus, Halifax - Born: Boston - Age: 50 - Released on 9 Jun 1815.

Bigsby, Samuel - Marine - Nbr: 4024 - Prize: US Brig Rattlesnake - Ship Type: MW - How taken: HMS Leander - When taken: 13 Jul 1814 - Where taken: off Shelburne - Date Received: 6 Oct 1814 - From what ship: HMT Chesapeake, Halifax - Born: New Hampshire - Age: 21 - Released on 9 Jun 1815.

Bill, Zachariah - Carpenter - Nbr: 3563 - Prize: Hawk - Ship Type: P - How taken: HMS Pique - When taken: 26 Apr 1814 - Where taken: off Bermuda - Date Received: 30 Sep 1814 - From what ship: HMT Sybella - Born: Washington - Age: 23 - Released on 4 Jun 1815.

Billings, Daniel - Seaman - Nbr: 5894 - Prize: Harlequin - Ship Type: P - How taken: HMS Bulwark - When taken: 23 Nov 1814 - Where taken: off Halifax - Date Received: 27 Dec 1814 - From what ship: HMT Penelope - Born: Massachusetts - Age: 23 - Released on 3 Jul 1815.

Billings, Richard - Seaman - Nbr: 3397 - Prize: Thomas - Ship Type: P - How taken: HMS Nymphe - When taken: 24 Jun 1813 - Where taken: off Halifax - Date Received: 13 Sep 1814 - From what ship: HMT Niobe, Chatham - Born: Massachusetts - Age: 21 - Released on 28 May 1815.

Billings, Robert - Seaman - Nbr: 5963 - Prize: Amazon - Ship Type: Prize - How taken: HMS Bulwark - When taken: 22 Sep 1814 - Where taken: Georges Bank - Date Received: 27 Dec 1814 - From what ship: HMT Penelope - Born: Massachusetts - Age: 18 - Released on 3 Jul 1815.

Billows, Charles - Seaman - Nbr: 6148 - How taken: Impressed at Cork - When taken: 1814 - Date Received: 17 Jan 1815 - From what ship: HMT Impregnable - Born: Camptown - Age: 28 - Race: Black - Released on 5 Jul 1815.

Bin, Peter - Soldier - Nbr: 5253 - Prize: 13th US Infantry - Ship Type: Troops - How taken: British Army - When taken: 13 Oct 1812 - Where taken: Canada - Date Received: 31 Oct 1814 - From what ship: HMT Leyden, Chatham - Born: Londonderry - Age: 24 - Released on 29 Jun 1815.

Bingham, Little - Seaman - Nbr: 3478 - How taken: Gave himself up from HMS Prince - When taken: 12 Sep 1814 - Date Received: 19 Sep 1814 - From what ship: HMT Salvador del Mundo - Born: Virginia - Age: 36 - Race: Mulatto - Released on 4 Jun 1815.

Birch, Andrew - Seaman - Nbr: 10 - Prize: Cashier - Ship Type: LM - How taken: HMS Reindeer - When taken: 3 Feb 1813 - Where taken: Bay of Biscay - Date Received: 2 Apr 1813 - From what ship: Plymouth - Born: Baltimore - Age: 19 - Sent to Dartmouth on 30 Jul 1813.

Birch, Peter - Seaman - Nbr: 3900 - Prize: Prosperity - Ship Type: MV - How taken: Not listed - When taken: 1 Jul

American Prisoners of War Held at Dartmoor during the War of 1812

## Alphabetical listing of names

1814 - Where taken: Chesapeake Bay - Date Received: 5 Oct 1814 - From what ship: HMT Orpheus, Halifax - Born: Philadelphia - Age: 57 - Race: Negro - Died on 13 Mar 1815 from pneumonia.

Bird, Comfort - Seaman - Nbr: 1078 - How taken: Impressed at Liverpool - When taken: 1 Feb 1814 - Date rec'd: 10 May 1814 - From what ship: Plymouth - Born: Dorchester - Age: 24 - Released on 27 Apr 1815.

Bird, James - Seaman - Nbr: 2823 - How taken: Gave himself up from HMS Philomel - When taken: 28 Dec 1812 - Date Received: 24 Aug 1814 - From what ship: HMT Liverpool, Chatham - Born: Kennebec - Age: 36 - Released on 26 Apr 1815.

Bird, John - Seaman - Nbr: 4927 - Prize: Herald - Ship Type: P - How taken: HMS Endymion - When taken: 15 Aug 1814 - Where taken: off Nantucket - Date Received: 28 Oct 1814 - From what ship: HMT Alkbar, Halifax - Born: New York - Age: 26 - Released on 21 Jun 1815.

Bird, Joseph - Seaman - Nbr: 6057 - Prize: William - Ship Type: Prize - How taken: HMS Armide - When taken: 11 Oct 1814 - Where taken: off Newport - Date Received: 28 Dec 1814 - From what ship: HMT Penelope - Born: Charlestown - Age: 22 - Released on 11 Jul 1815.

Bird, Thomas - Seaman - Nbr: 3768 - Prize: Fame - Ship Type: P - How taken: HMS Thistle - When taken: 10 Apr 1814 - Where taken: after being cast ashore on a seal island - Date Received: 30 Sep 1814 - From what ship: HMT President, Halifax - Born: Long Island - Age: 26 - Released on 9 Jun 1815.

Bisbee, Elijah - Seaman - Nbr: 4614 - Prize: Argus - Ship Type: MV - How taken: HMS San Domingo - When taken: 1 Mar 1814 - Where taken: off Savannah - Date Received: 9 Oct 1814 - From what ship: HMT Leyden, Chatham - Born: Plymouth - Age: 19 - Released on 15 Jun 1815.

Bisbee, J. D. - Seaman - Nbr: 4306 - Prize: Argus - Ship Type: MV - How taken: HMS San Domingo - When taken: 1 Mar 1814 - Where taken: off Savannah - Date Received: 7 Oct 1814 - From what ship: HMT Niobe, Chatham - Born: Middleborough - Age: 22 - Released on 4 Jun 1815.

Bisett, Robert - Seaman - Nbr: 6415 - Prize: Enterprize - Ship Type: P - How taken: HMS Argo - When taken: 16 Nov 1814 - Where taken: off San Domingo - Date Received: 3 Mar 1815 - From what ship: HMT Ganges, Plymouth - Born: Albany - Age: 25 - Released on 11 Jul 1815.

Bishop, Edward - Seaman - Nbr: 2861 - Prize: Dick - Ship Type: MV - How taken: HMS Dispatch - When taken: 17 Mar 1812 - Where taken: at sea - Date Received: 24 Aug 1814 - From what ship: HMT Alpheus, Chatham - Born: New York - Age: 27 - Released on 27 Apr 1815.

Bishop, William - Steward - Nbr: 55 - Prize: Spitfire - Ship Type: MV - How taken: HMS Achates - When taken: 14 Feb 1813 - Where taken: off Ushant (France) - Date Received: 2 Apr 1813 - From what ship: Plymouth - Born: Danvers - Age: 17 - Released on 10 Jul 1813.

Bisley, Horace - Seaman - Nbr: 202 - Prize: Star - Ship Type: MV - How taken: HMS Superb - When taken: 9 Feb 1813 - Where taken: Bay of Biscay - Date Received: 2 Apr 1813 - From what ship: Plymouth - Born: Rockhill - Age: 18 - Died on 11 Apr 1813 from pneumonia.

Biss, Daniel W. - Seaman - Nbr: 1383 - Prize: Courier - Ship Type: P - How taken: HMS Rover - When taken: 14 May 1813 - Where taken: Bay of Biscay - Date Received: 19 Jun 1814 - From what ship: Stapleton - Born: Massachusetts - Age: 25 - Released on 28 Apr 1815.

Bissen, John - Seaman - Nbr: 5855 - Prize: Lion - Ship Type: P - How taken: HMS Granicus - When taken: 2 Dec 1814 - Where taken: off Lisbon - Date Received: 26 Dec 1814 - From what ship: HMT Impregnable - Born: Massachusetts - Age: 27 - Released on 3 Jul 1815.

Bitters, John - Seaman - Nbr: 1296 - Prize: John & Frances - Ship Type: Prize - How taken: HMS Sterling Castle - When taken: 10 May 1814 - Where taken: off Cape Clear - Date Received: 14 Jun 1814 - From what ship: Mill Prison (Plymouth, England) - Born: Philadelphia - Age: 23 - Released on 28 Apr 1815.

Bisett, Robert - Seaman - Nbr: 6415 - Prize: Enterprize - Ship Type: P - How taken: HMS Argo - When taken: 16 Nov 1814 - Where taken: off San Domingo - Date Received: 3 Mar 1815 - From what ship: HMT Ganges, Plymouth - Born: Albany - Age: 25 - Released on 11 Jul 1815.

Black, Charles - Seaman - Nbr: 6176 - Prize: US Schooner Somers - Ship Type: MW - How taken: British gunboats - When taken: 6 Sep 1814 - Where taken: Lake Erie - Date Received: 21 Jan 1815 - From what ship: HMT Impregnable - Born: Philadelphia - Age: 18 - Race: Black - Released on 5 Jul 1815.

Black, Charles - Boatswain's Mate - Nbr: 3068 - Prize: Bunker Hill - Ship Type: P - How taken: HMS Pomona - When taken: 2 Jun 1814 - Where taken: Channel - Date Received: 2 Sep 1814 - From what ship: HMT Hydra, Chatham - Born: Philadelphia - Age: 33 - Released on 28 May 1815.

Black, George - Seaman - Nbr: 671 - Prize: Joel Barlow - Ship Type: LM - How taken: HMS Briton - When taken: 3 Jul 1813 - Where taken: off Bordeaux - Date Received: 8 Sep 1813 - From what ship: Plymouth - Born: Not legible - Age: 29 - Released on 26 Apr 1815.

American Prisoners of War Held at Dartmoor during the War of 1812

## Alphabetical listing of names

Black, James - Seaman - Nbr: 5296 - Prize: Fox - Ship Type: P - How taken: HMS Shannon - When taken: 7 Nov 1813 - Where taken: off Newfoundland - Date Received: 31 Oct 1814 - From what ship: HMT Leyden, Chatham - Born: Briston - Age: 27 - Released on 11 Jul 1815.

Black, Nicholas - Seaman - Nbr: 4361 - How taken: Gave himself up from HMS Hannibal - When taken: 2 Oct 1814 - Date Received: 7 Oct 1814 - From what ship: HMT Salvador del Mundo, Halifax - Born: Pennsylvania - Age: 31 - Released on 10 Apr 1815.

Black, Philip - Seaman - Nbr: 3090 - Prize: York Town - Ship Type: P - How taken: British squadron - When taken: 16 Jul 1813 - Where taken: off Halifax - Date Received: 3 Sep 1814 - From what ship: HMT Hydra, Chatham - Born: New York - Age: 22 - Released on 27 Apr 1815.

Black, Ruddick - Seaman - Nbr: 5347 - Prize: Sally - Ship Type: MV - How taken: HMS Maidstone - When taken: 17 Jul 1813 - Where taken: Grand Banks - Date Received: 31 Oct 1814 - From what ship: HMT Leyden, Chatham - Born: Barnstable - Age: 18 - Released on 1 Jul 1815.

Black, William F. - Mate - Nbr: 6364 - Prize: St. Johanna - Ship Type: Prize - How taken: HMS Sabine - When taken: 25 Oct 1814 - Where taken: off Newfoundland Banks - Date Received: 24 Feb 1815 - From what ship: HMT Ganges, Plymouth - Born: Gloucester - Age: 21 - Released on 6 Jul 1815.

Blackford, Henry - Seaman - Nbr: 6535 - Prize: US Brig Syren - Ship Type: MW - How taken: HMS Medway - When taken: 12 Jul 1814 - Where taken: off Cape of Good Hope - Date Received: 7 Mar 1815 - From what ship: HMT Ganges, Plymouth - Born: Massachusetts - Age: 26 - Released on 11 Jul 1815.

Blackler, Bernard - Seaman - Nbr: 275 - Prize: Cannoneer - Ship Type: MV - How taken: HMS Warspite - When taken: 14 Mar 1813 - Where taken: Bay of Biscay - Date Received: 28 Jun 1813 - From what ship: Plymouth - Born: Newburgh - Age: 26 - Sent to Plymouth on 8 Jul 1814.

Blackler, William - Prize Master - Nbr: 3663 - Prize: Alfred - Ship Type: P - How taken: HMS Epervier - When taken: 23 Feb 1812 - Where taken: off Newfoundland - Date Received: 30 Sep 1814 - From what ship: HMT President - Born: Marblehead - Age: 28 - Released on 10 Apr 1815.

Blacklidge, John - Marine - Nbr: 596 - Prize: US Brig Argus - Ship Type: MW - How taken: HMS Pelican - When taken: 14 Aug 1813 - Where taken: Irish Channel - Date Received: 8 Sep 1813 - From what ship: Plymouth - Born: New Jersey - Age: 25 - Sent to Dartmouth on 2 Nov 1814.

Blackston, Edward - Seaman - Nbr: 13 - Prize: Cashier - Ship Type: LM - How taken: HMS Reindeer - When taken: 3 Feb 1813 - Where taken: Bay of Biscay - Date Received: 2 Apr 1813 - From what ship: Plymouth - Born: New Sharon - Age: 23 - Sent to Dartmouth on 30 Jul 1813.

Blackston, William - Seaman - Nbr: 6397 - Prize: Nellenville - Ship Type: MV - How taken: HMS Onyx - When taken: 25 Dec 1814 - Where taken: off San Domingo - Date Received: 3 Mar 1815 - From what ship: HMT Ganges, Plymouth - Born: Baltimore - Age: 26 - Race: Mulatto - Released on 11 Jul 1815.

Bladen, John - Seaman - Nbr: 583 - Prize: US Brig Argus - Ship Type: MW - How taken: HMS Pelican - When taken: 14 Apr 1813 - Where taken: Irish Channel - Date Received: 8 Sep 1813 - From what ship: Plymouth - Born: Fairfax - Age: 23 - Sent to Dartmouth on 2 Nov 1814.

Blagcon, Stephen - Seaman - Nbr: 2598 - Prize: Tom Thumb - Ship Type: MV - How taken: Lyon (Privateer) - When taken: 15 Feb 1813 - Where taken: Bay of Biscay - Date Received: 21 Aug 1814 - From what ship: HMT Freya, Chatham - Born: Baltimore - Age: 23 - Released on 3 May 1815.

Blair, Benjamin - Seaman - Nbr: 5325 - Prize: Enterprize - Ship Type: P - How taken: HMS Tenedos - When taken: 21 May 1813 - Where taken: off Cape Cod - Date Received: 31 Oct 1814 - From what ship: HMT Leyden, Chatham - Born: Marblehead - Age: 28 - Released on 29 Jun 1815.

Blair, David - Seaman - Nbr: 5324 - Prize: Juliana Smith - Ship Type: P - How taken: HMS Nymphe - When taken: 11 May 1813 - Where taken: off Cape Sable Island (Canada) - Date Received: 31 Oct 1814 - From what ship: HMT Leyden, Chatham - Born: Marblehead - Age: 22 - Released on 29 Jun 1815.

Blair, John - Seaman - Nbr: 2485 - Prize: US Sloop Frolic - Ship Type: MW - How taken: HMS Orpheus - When taken: 20 Apr 1814 - Where taken: off Cuba - Date Received: 16 Aug 1814 - From what ship: HMT Queen, Halifax - Born: Worcester - Age: 26 - Released on 3 May 1815.

Blair, John R. - Master - Nbr: 1689 - Prize: Circe - Ship Type: MV - How taken: HMS Acteon - When taken: 22 Oct 1813 - Where taken: off Cape Hatteras - Date Received: 2 Jul 1814 - From what ship: Plymouth - Born: New York - Age: 27 - Sent to Ashburton (England) on 15 Aug 1814.

Blair, Robert - Seaman - Nbr: 4667 - Prize: Industry - Ship Type: P - How taken: HMS Heron - When taken: 3 Nov 1813 - Where taken: off Halifax - Date Received: 9 Oct 1814 - From what ship: HMT Leyden, Chatham - Born: Marblehead - Age: 22 - Released on 15 Jun 1815.

Blaird, David - Seaman - Nbr: 4715 - Prize: Hannah - Ship Type: MV - How taken: HMS Conquistador - When

American Prisoners of War Held at Dartmoor during the War of 1812

## Alphabetical listing of names

taken: 15 Jan 1814 - Where taken: at sea - Date Received: 9 Oct 1814 - From what ship: HMT Freya, Chatham - Born: Pelham - Age: 29 - Released on 15 Jun 1815.

Blake, Alexander - Seaman - Nbr: 399 - Prize: Young Holkar - Ship Type: MV - How taken: HMS Superb - When taken: 10 Apr 1813 - Where taken: off Belle Isle, France - Date Received: 1 Jul 1813 - From what ship: Plymouth - Born: Charlestown - Age: 19 - Released on 20 Apr 1815.

Blake, Charles - Seaman - Nbr: 3323 - Prize: York Town - Ship Type: P - How taken: HMS Nimrod - When taken: 17 Jul 1813 - Where taken: off St. Johns - Date Received: 13 Sep 1814 - From what ship: HMT Niobe, Chatham - Born: Boston - Age: 21 - Released on 28 May 1815.

Blake, Ebenezer - Sergeant Marines - Nbr: 6310 - Prize: Prince de Neufchatel - Ship Type: P - How taken: Leander (Newcastle Acasta) - When taken: 20 Dec 1814 - Where taken: Lat 38 Long 56 - Date Received: 19 Feb 1815 - From what ship: HMT Ganges, Plymouth - Born: Boston - Age: 29 - Released on 5 Jul 1815.

Blake, Philip - Seaman - Nbr: 332 - Prize: Charlotte - Ship Type: P - How taken: HMS Warspite - When taken: 3 Dec 1813 - Where taken: Bay of Biscay - Date Received: 28 Jun 1813 - From what ship: Plymouth - Born: Boston - Age: 21 - Released on 20 Apr 1815.

Blake, Thomas - Cook - Nbr: 866 - Prize: Amiable - Ship Type: LM - How taken: HMS Magnificent - When taken: 30 Oct 1813 - Where taken: off Bordeaux - Date Received: 4 Dec 1813 - From what ship: Plymouth - Born: Virginia - Age: 27 - Released on 27 Apr 1815.

Blake, William - Seaman - Nbr: 2884 - How taken: Gave himself up from HMS Undaunted - When taken: 28 May 1813 - Date Received: 24 Aug 1814 - From what ship: HMT Alpheus, Chatham - Born: Massachusetts - Age: 41 - Released on 19 May 1815.

Blakely, Michael - Seaman - Nbr: 5584 - Prize: Nonsuch - Ship Type: LM - How taken: HMS Dotterel - When taken: 14 Dec 1813 - Where taken: off Block Island - Date Received: 17 Dec 1814 - From what ship: HMT Loire, Halifax - Born: Chester - Age: 32 - Released on 1 Jul 1815.

Blanchard, Carvan - Seaman - Nbr: 26 - How taken: Apprehended at Gibraltar - When taken: 8 Aug 1813 - Date Received: 2 Apr 1813 - From what ship: Plymouth - Born: Milford - Age: 21 - Sent to Dartmouth on 30 Jul 1813.

Blanchard, George - Seaman - Nbr: 241 - Prize: William Bayard - Ship Type: MV - How taken: HMS Warspite - When taken: 3 Mar 1813 - Where taken: Bay of Biscay - Date Received: 2 Apr 1813 - From what ship: Plymouth - Born: Elizabethtown - Age: 21 - Sent to Plymouth on 15 Jun 1813.

Blanchard, John - Seaman - Nbr: 4508 - Prize: America - Ship Type: P - How taken: HMS Shannon - When taken: 23 May 1813 - Where taken: off Cape Cod - Date Received: 8 Oct 1814 - From what ship: HMT Leyden, Chatham - Born: Boston - Age: 18 - Released on 15 Jun 1815.

Blanchard, Samuel - Seaman - Nbr: 3632 - Prize: Monarch - Ship Type: MV - How taken: HMS Dotterel - When taken: 14 Dec 1813 - Where taken: off Charleston - Date Received: 30 Sep 1814 - From what ship: HMT Sybella - Born: Pennsylvania - Age: 47 - Race: Negro - Released on 4 Jun 1815.

Blanchet, Simon - Seaman - Nbr: 1587 - Prize: Price - Ship Type: MV - How taken: HMS Pyramus - When taken: 6 Apr 1813 - Where taken: Bay of Biscay - Date Received: 23 Jun 1814 - From what ship: Stapleton - Born: Charlestown - Age: 22 - Released on 1 May 1815.

Blandel, Jonathan - Seaman - Nbr: 756 - How taken: Impressed at Liverpool - Date Received: 3 Nov 1813 - From what ship: Plymouth - Born: Cannyborak - Age: 20 - Released on 26 Apr 1815.

Blaney, Stephen - Seaman - Nbr: 4504 - Prize: Enterprize - Ship Type: P - How taken: HMS Tenedos - When taken: 21 May 1813 - Where taken: off Cape Cod - Date Received: 8 Oct 1814 - From what ship: HMT Leyden, Chatham - Born: Massachusetts - Age: 27 - Released on 15 Jun 1815.

Blarney, Henry - Soldier - Nbr: 5260 - Prize: 13th US Infantry - Ship Type: Troops - How taken: British Army - When taken: 13 Oct 1812 - Where taken: Canada - Date Received: 31 Oct 1814 - From what ship: HMT Leyden, Chatham - Born: Mayo - Age: 29 - Released on 29 Jun 1815.

Blasdon, Philip - Soldier - Nbr: 5839 - Prize: 4th US Rifles - Ship Type: Troops - How taken: British Army - When taken: 17 Sep 1814 - Where taken: Fort Erie - Date Received: 26 Dec 1814 - From what ship: HMT Argo - Born: New Hampshire - Age: 35 - Died on 17 Jan 1815 from variola.

Blasker, Charles - Seaman - Nbr: 3627 - Prize: Monarch - Ship Type: MV - How taken: HMS Dotterel - When taken: 14 Dec 1813 - Where taken: off Charleston - Date Received: 30 Sep 1814 - From what ship: HMT Sybella - Born: Newark - Age: 18 - Race: Mulatto - Released on 4 Jun 1815.

Blasse, John M. - Seaman - Nbr: 1125 - Prize: Hope - Ship Type: P - How taken: HMS Sea Horse - When taken: 22 Mar 1814 - Where taken: at sea - Date Received: 10 May 1814 - From what ship: Plymouth - Born: Brest - Age: 21 - Sent to Mill Prison (Plymouth, England) on 21 Jun 1814.

American Prisoners of War Held at Dartmoor during the War of 1812

## Alphabetical listing of names

Blesdale, Jacob - Seaman - Nbr: 2376 - Prize: US Sloop Frolic - Ship Type: MW - How taken: HMS Orpheus - When taken: 20 Apr 1814 - Where taken: off Cuba - Date Received: 16 Aug 1814 - From what ship: HMT Queen, Halifax - Born: Massachusetts - Age: 26 - Sent to Dartmouth on 19 Oct 1814.

Bliss, Frederick - Seaman - Nbr: 4182 - How taken: Impressed at London - When taken: 7 Jul 1813 - Date Received: 7 Oct 1814 - From what ship: HMT Niobe, Chatham - Born: Norfolk - Age: 26 - Released on 13 Jun 1815.

Blisset, James - Seaman - Nbr: 612 - Prize: US Brig Argus - Ship Type: MW - How taken: HMS Pelican - When taken: 14 Aug 1813 - Where taken: Irish Channel - Date Received: 8 Sep 1813 - From what ship: Plymouth - Born: Gloucester - Age: 24 - Sent to Dartmouth on 2 Nov 1814.

Block, Edward - Seaman - Nbr: 3624 - Prize: Monarch - Ship Type: MV - How taken: HMS Dotterel - When taken: 14 Dec 1813 - Where taken: off Charleston - Date Received: 30 Sep 1814 - From what ship: HMT Sybella - Born: Pennsylvania - Age: 23 - Released on 4 Jun 1815.

Blodget, Caleb - Seaman - Nbr: 1520 - Prize: Essex - Ship Type: MV - How taken: HMS Pyramus - When taken: 20 Apr 1813 - Where taken: Bay of Biscay - Date Received: 23 Jun 1814 - From what ship: Stapleton - Born: New Hampshire - Age: 36 - Released on 1 May 1815.

Bloom, Joseph - Seaman - Nbr: 85 - Prize: Cashier - Ship Type: MV - How taken: HMS Reindeer - When taken: 3 Feb 1813 - Where taken: Bay of Biscay - Date Received: 2 Apr 1813 - From what ship: Plymouth - Born: Pennsylvania - Age: 25 - Sent to Dartmouth on 30 Jul 1813.

Blossom, Seth - Seaman - Nbr: 1882 - How taken: Gave himself up from HMS Belvidera - When taken: 15 Nov 1813 - Date Received: 29 Jul 1814 - From what ship: HMS Ville de Paris, Chatham Depot - Born: Baltimore - Age: 45 - Released on 2 May 1815.

Bloventon, John - Seaman - Nbr: 4214 - Prize: Devon - Ship Type: Prize - How taken: HMS Fly - When taken: 2 Jan 1814 - Where taken: at sea - Date Received: 7 Oct 1814 - From what ship: HMT Niobe, Chatham - Born: Albany - Age: 23 - Released on 20 Mar 1815.

Blue, Peter - Carpenter - Nbr: 1778 - Prize: Hope - Ship Type: MV - How taken: Chance (Privateer) - When taken: 15 Feb 1813 - Where taken: off Bordeaux - Date Received: 20 Jul 1814 - From what ship: HMS Milford, Plymouth - Born: Lisbon - Age: 46 - Released on 1 May 1815.

Blumbhouser, Samuel - Seaman - Nbr: 2827 - How taken: Gave himself up from HMS Monmouth - When taken: 15 Jul 1813 - Date Received: 24 Aug 1814 - From what ship: HMT Liverpool, Chatham - Born: Marblehead - Age: 40 - Released on 19 May 1815.

Blyth, Jonathan - 2nd Lieutenant - Nbr: 3766 - Prize: Fame - Ship Type: P - How taken: HMS Thistle - When taken: 10 Apr 1814 - Where taken: after being cast ashore on a seal island - Date Received: 30 Sep 1814 - From what ship: HMT President, Halifax - Born: Massachusetts - Age: 20 - Released on 9 Jun 1815.

Boardman, John - Prize master - Nbr: 5471 - Prize: General Putnam - Ship Type: P - How taken: HMS Leander - When taken: 8 Nov 1814 - Where taken: Long 65 Lat 42 - Date Received: 17 Dec 1814 - From what ship: HMT Loire, Halifax - Born: Ipswich - Age: 33 - Released on 1 Jul 1815.

Bodfish, William - Seaman - Nbr: 958 - Prize: Siro - Ship Type: LM - How taken: HMS Pelican - When taken: 13 Jan 1814 - Where taken: at sea - Date Received: 31 Jan 1814 - From what ship: Plymouth - Born: Boston - Age: 26 - Released on 27 Apr 1815.

Bodkin, William - Seaman - Nbr: 5217 - Prize: Thomas - Ship Type: P - How taken: HMS Nymphe - When taken: 24 Jun 1813 - Where taken: off Halifax - Date Received: 31 Oct 1814 - From what ship: HMT Mermaid, Chatham - Born: Portsmouth - Age: 55 - Race: Black - Released on 29 Jun 1815.

Bogart, K. W. - Seaman - Nbr: 579 - Prize: US Brig Argus - Ship Type: MW - How taken: HMS Pelican - When taken: 14 Apr 1813 - Where taken: Irish Channel - Date Received: 8 Sep 1813 - From what ship: Plymouth - Born: New York - Age: 21 - Sent to Dartmouth on 2 Nov 1814.

Boggart, John - Seaman - Nbr: 4731 - How taken: Gave himself up from HMS Salvador - When taken: 10 Dec 1813 - Date Received: 9 Oct 1814 - From what ship: HMT Freya, Chatham - Born: King's County - Age: 37 - Released on 15 Jun 1815.

Boggs, James - 2nd Mate - Nbr: 114 - Prize: Governor McKean - Ship Type: LM - How taken: HMS Rover - When taken: 26 Jan 1813 - Where taken: off Bordeaux - Date Received: 2 Apr 1813 - From what ship: Plymouth - Born: Londonderry Island - Age: 52 - Sent to Chatham on 5 Jul 1813.

Boggs, James - 2nd Mate - Nbr: 2684 - Prize: Governor McKean - Ship Type: MV - How taken: HMS Rover - When taken: 26 Jan 1813 - Where taken: off Bordeaux - Date Received: 21 Aug 1814 - From what ship: HMT Freya, Chatham - Born: Philadelphia - Age: 63 - Released on 20 Apr 1815.

Boillet, John - Seaman - Nbr: 3127 - How taken: Gave himself up from HMS Raven - When taken: 15 Aug 1813 - Date Received: 11 Sep 1814 - From what ship: HMT Freya, Chatham - Born: Charleston - Age: 29 -

American Prisoners of War Held at Dartmoor during the War of 1812

## Alphabetical listing of names

Released on 28 May 1815.
Boisseau, Joseph - Seaman - Nbr: 2575 - Prize: Rattlesnake - Ship Type: P - How taken: HMS Hyperion - When taken: 25 Jun 1814 - Where taken: off Cape Finisterre - Date Received: 21 Aug 1814 - From what ship: HMS Hyperion - Born: St. Malo - Age: 33 - Released on 3 May 1815.
Boivie, James - Seaman - Nbr: 372 - Prize: Independence - Ship Type: MV - How taken: HMS Superb - When taken: 16 Mar 1813 - Where taken: Bay of Biscay - Date Received: 1 Jul 1813 - From what ship: Plymouth - Born: Maryland - Age: 27 - Released on 20 Apr 1815.
Bolding, Garret - Seaman - Nbr: 3548 - Prize: Hawk - Ship Type: P - How taken: HMS Pique - When taken: 26 Apr 1814 - Where taken: off Bermuda - Date Received: 30 Sep 1814 - From what ship: HMT Sybella - Born: Argyle - Age: 23 - Released on 4 Jun 1815.
Bolton, John - Seaman - Nbr: 5005 - Prize: Invincible - Ship Type: LM - How taken: HMS Armide - When taken: 15 Aug 1814 - Where taken: off Nantucket - Date Received: 28 Oct 1814 - From what ship: HMT Alkbar, Halifax - Born: Newcastle - Age: 24 - Released on 21 Jun 1815.
Bond, Peter - Seaman - Nbr: 1723 - How taken: Impressed at Cork - When taken: 3 May 1814 - Date Received: 8 Jul 1814 - From what ship: Labrador - Born: Portsmouth - Age: 30 - Released on 1 May 1815.
Bond, Samuel - Seaman - Nbr: 2899 - How taken: Gave himself up from HMS Ocean - When taken: 28 May 1813 - Date Received: 24 Aug 1814 - From what ship: HMT Alpheus, Chatham - Born: Baltimore - Age: 40 - Race: Black - Released on 19 May 1815.
Bond, William - Seaman - Nbr: 2441 - Prize: US Sloop Frolic - Ship Type: MW - How taken: HMS Orpheus - When taken: 20 Apr 1814 - Where taken: off Cuba - Date Received: 16 Aug 1814 - From what ship: HMT Queen, Halifax - Born: New Bedford - Age: 19 - Released on 3 May 1815.
Bone, Charles - Seaman - Nbr: 3364 - How taken: Gave himself up from HMS Vixon - When taken: 12 Sep 1813 - Date Received: 13 Sep 1814 - From what ship: HMT Niobe, Chatham - Born: Rhode Island - Age: 22 - Race: Black - Released on 28 May 1815.
Bonfonce, Anthony - 2nd Lieutenant - Nbr: 6458 - Prize: Decatur - Ship Type: P - How taken: HMS Rhin - When taken: 5 Jun 1814 - Where taken: off San Domingo - Date Received: 3 Mar 1815 - From what ship: HMT Ganges, Plymouth - Born: Charleston - Age: 30 - Released on 13 Jun 1815.
Bonner, John - Seaman - Nbr: 6494 - Prize: James - Ship Type: MV - How taken: HMS Niemen - When taken: 6 Sep 1814 - Where taken: off Bahamas - Date Received: 3 Mar 1815 - From what ship: HMT Ganges, Plymouth - Born: Washington - Age: 20 - Released on 6 May 1815.
Bonny, John - Seaman - Nbr: 3902 - Prize: Henry Guilder - Ship Type: MV - How taken: HMS Niemen - When taken: 14 Jul 1814 - Where taken: at sea - Date Received: 5 Oct 1814 - From what ship: HMT Orpheus, Halifax - Born: Rhode Island - Age: 40 - Race: Negro - Released on 9 Jun 1815.
Boose, Abraham - Seaman - Nbr: 1700 - Prize: Fanny - Ship Type: Prize - How taken: HMS Sceptre - When taken: 12 May 1814 - Date Received: 2 Jul 1814 - From what ship: Plymouth - Born: Salem - Age: 32 - Released on 1 May 1815.
Booth, Charles - Seaman - Nbr: 882 - Prize: Charlotte - Ship Type: MV - How taken: HMS Dwarf - When taken: 4 Nov 1813 - Where taken: off Bordeaux - Date Received: 31 Jan 1814 - From what ship: Plymouth - Born: New York - Age: 30 - Released on 27 Apr 1815.
Booth, Thomas - Seaman - Nbr: 1927 - How taken: Gave himself up from HMS Royal William - When taken: 3 Feb 1813 - Date Received: 3 Aug 1814 - From what ship: HMS Alceste, Chatham Depot - Born: Maryland - Age: 30 - Released on 2 May 1815.
Bordage, Raymond - Seaman - Nbr: 4753 - How taken: Gave himself up from HMS Owen Glendower - When taken: 28 Jun 1813 - Date Received: 9 Oct 1814 - From what ship: HMT Freya, Chatham - Born: New Orleans - Age: 32 - Released on 15 Jun 1815.
Bordley, George - Seaman - Nbr: 2878 - How taken: Gave himself up from HMS Barfleur - When taken: 27 May 1813 - Date Received: 24 Aug 1814 - From what ship: HMT Alpheus, Chatham - Born: Rhode Island - Age: 30 - Race: Black - Released on 19 May 1815.
Borgin, Gabriel - Seaman - Nbr: 192 - Prize: Star - Ship Type: MV - How taken: HMS Superb - When taken: 9 Feb 1813 - Where taken: Bay of Biscay - Date Received: 2 Apr 1813 - From what ship: Plymouth - Born: Somerset County - Age: 17 - Sent to Mill Prison (Plymouth, England) on 10 Jul 1813.
Boriesa, Pierre - Seaman - Nbr: 1402 - Prize: Good Friends - Ship Type: MV - How taken: HMS Andromache - When taken: 2 Apr 1813 - Where taken: Bay of Biscay - Date Received: 19 Jun 1814 - From what ship: Stapleton - Born: Maryland - Age: 23 - Race: Negro - Released on 28 Apr 1815.
Borsdell, Justice - Seaman - Nbr: 1148 - Prize: Bunker Hill - Ship Type: P - How taken: HMS Pomone - When

American Prisoners of War Held at Dartmoor during the War of 1812

## Alphabetical listing of names

taken: 8 Mar 1814 - Where taken: at sea - Date Received: 10 May 1814 - From what ship: Plymouth - Born: New York - Age: 17 - Released on 28 Apr 1815.

Boss, Charles - Seaman - Nbr: 2417 - Prize: US Sloop Frolic - Ship Type: MW - How taken: HMS Orpheus - When taken: 20 Apr 1814 - Where taken: off Cuba - Date Received: 16 Aug 1814 - From what ship: HMT Queen, Halifax - Born: South Kingston - Age: 29 - Released on 3 May 1815.

Bosset, David - Seaman - Nbr: 1470 - Prize: Governor Gerry - Ship Type: MV - How taken: HMS Lyra - When taken: 29 May 1813 - Where taken: Bay of Biscay - Date Received: 19 Jun 1814 - From what ship: Stapleton - Born: Baltimore - Age: 43 - Race: Negro - Released on 28 Apr 1815.

Bostell, James - Seaman - Nbr: 2479 - Prize: US Sloop Frolic - Ship Type: MW - How taken: HMS Orpheus - When taken: 20 Apr 1814 - Where taken: off Cuba - Date Received: 16 Aug 1814 - From what ship: HMT Queen, Halifax - Born: Baltimore - Age: 31 - Released on 3 May 1815.

Boston, John - Seaman - Nbr: 4601 - Prize: Sally - Ship Type: MV - How taken: HMS Derwent - When taken: 21 Jan 1814 - Where taken: Grand Banks - Date Received: 9 Oct 1814 - From what ship: HMT Leyden, Chatham - Born: Salem - Age: 36 - Released on 15 Jun 1815.

Boswell, John - Seaman - Nbr: 5304 - Prize: Yankee - Ship Type: P - How taken: HMS Shannon - When taken: 20 Aug 1813 - Where taken: at sea - Date Received: 31 Oct 1814 - From what ship: HMT Leyden, Chatham - Born: New Orleans - Age: 18 - Released on 11 Jul 1815.

Botellio, Anthony - Seaman - Nbr: 969 - Prize: Siro - Ship Type: LM - How taken: HMS Pelican - When taken: 13 Jan 1814 - Where taken: at sea - Date Received: 31 Jan 1814 - From what ship: Plymouth - Born: St. Michael - Age: 19 - Released on 27 Apr 1815.

Bourdinon, Elijah - Seaman - Nbr: 5817 - Prize: US Schooner Scorpion - Ship Type: MW - How taken: British gunboats - When taken: 6 Sep 1814 - Where taken: Lake Erie - Date Received: 26 Dec 1814 - From what ship: HMT Argo - Born: Bordeaux - Age: 19 - Released on 3 Jul 1815.

Bourdon, Almond - Passenger - Nbr: 6271 - Prize: Lion - Ship Type: P - How taken: HMS Granicus - When taken: 2 Dec 1814 - Where taken: off Lisbon - Date Received: 17 Feb 1815 - From what ship: HMT Ganges, Plymouth - Born: L-Orient - Age: 25 - Escaped on 26 Mar 1815.

Bourdon, John - Seaman - Nbr: 1627 - Prize: Tom - Ship Type: LM - How taken: HMS Surveillante - When taken: 24 Apr 1813 - Where taken: Bay of Biscay - Date Received: 23 Jun 1814 - From what ship: Stapleton - Born: Maryland - Age: 39 - Released on 1 May 1815.

Bourns, George - Seaman - Nbr: 2029 - How taken: Gave himself up from HMS Impeteux - When taken: 26 Mar 1813 - Date Received: 3 Aug 1814 - From what ship: HMS Lyffey, Chatham Depot - Born: Philadelphia - Age: 30 - Released on 2 May 1815.

Bourns, John - Seaman - Nbr: 2009 - Prize: Dictator - Ship Type: MV - How taken: HMS Derwent - When taken: 7 May 1813 - Where taken: off Nantes - Date Received: 3 Aug 1814 - From what ship: HMS Lyffey, Chatham Depot - Born: Philadelphia - Age: 28 - Released on 2 May 1815.

Bourton, George - Seaman - Nbr: 1324 - How taken: Impressed at Bristol - When taken: 30 Jun 1813 - Date Received: 19 Jun 1814 - From what ship: Stapleton - Born: Massachusetts - Age: 23 - Released on 28 Apr 1815.

Boussell, Samuel - Seaman - Nbr: 3242 - Prize: Hepsey - Ship Type: MV - How taken: HMS Tenedos - When taken: 22 Jun 1813 - Where taken: off Lisbon - Date Received: 11 Sep 1814 - From what ship: HMT Freya, Chatham - Born: Connecticut - Age: 22 - Released on 28 May 1815.

Bouton, John - Seaman - Nbr: 4020 - Prize: US Brig Rattlesnake - Ship Type: MW - How taken: HMS Leander - When taken: 13 Jul 1814 - Where taken: off Shelburne - Date Received: 6 Oct 1814 - From what ship: HMT Chesapeake, Halifax - Born: Virginia - Age: 26 - Released on 9 Jun 1815.

Bovey, Benjamin - Seaman - Nbr: 1901 - Prize: Quebec - Ship Type: Prize - How taken: HMS Derwent - When taken: 29 Jan 1813 - Where taken: off Lisbon - Date Received: 3 Aug 1814 - From what ship: HMS Alceste, Chatham Depot - Born: New Orleans - Age: 18 - Released on 2 May 1815.

Bowden, Benjamin - Boy - Nbr: 3757 - Prize: Alfred - Ship Type: P - How taken: HMS Epervier - When taken: 23 Feb 1814 - Where taken: off Newfoundland - Date Received: 30 Sep 1814 - From what ship: HMT President, Halifax - Born: Marblehead - Age: 14 - Released on 28 May 1815.

Bowden, Frederick - Gunner's Mate - Nbr: 3671 - Prize: Alfred - Ship Type: P - How taken: HMS Epervier - When taken: 23 Feb 1812 - Where taken: off Newfoundland - Date Received: 30 Sep 1814 - From what ship: HMT President - Born: Marblehead - Age: 30 - Released on 4 Jun 1815.

Bowden, John - Boy - Nbr: 1868 - Prize: USRC Surveyor - Ship Type: MW - How taken: HMS Narcissus - When taken: 12 Jun 1813 - Where taken: Chesapeake Bay - Date Received: 29 Jul 1814 - From what ship: HMS

American Prisoners of War Held at Dartmoor during the War of 1812

## Alphabetical listing of names

Ville de Paris, Chatham Depot - Born: Virginia - Age: 17 - Sent to Dartmouth on 2 Nov 1814.

Bowden, Thomas - Seaman - Nbr: 4046 - Prize: US Brig Rattlesnake - Ship Type: MW - How taken: HMS Leander - When taken: 13 Jul 1814 - Where taken: off Shelburne - Date Received: 6 Oct 1814 - From what ship: HMT Chesapeake, Halifax - Born: Marblehead - Age: 29 - Released on 13 Jun 1815.

Bowden, William - Seaman - Nbr: 5473 - Prize: General Putnam - Ship Type: P - How taken: HMS Leander - When taken: 8 Nov 1814 - Where taken: Long 65 Lat 42 - Date Received: 17 Dec 1814 - From what ship: HMT Loire, Halifax - Born: Salem - Age: 18 - Released on 1 Jul 1815.

Bowden, William - Seaman - Nbr: 3375 - Prize: Growler - Ship Type: P - How taken: HMS Electra - When taken: 7 Jul 1813 - Where taken: at sea - Date Received: 13 Sep 1814 - From what ship: HMT Niobe, Chatham - Born: Marblehead - Age: 17 - Released on 28 May 1815.

Bowden, William - Mate - Nbr: 3381 - Prize: Growler - Ship Type: P - How taken: HMS Electra - When taken: 7 Jul 1813 - Where taken: at sea - Date Received: 13 Sep 1814 - From what ship: HMT Niobe, Chatham - Born: Marblehead - Age: 21 - Released on 28 May 1815.

Bowen, John - Seaman - Nbr: 6473 - Prize: Decatur - Ship Type: P - How taken: HMS Rhin - When taken: 5 Jun 1814 - Where taken: off San Domingo - Date Received: 3 Mar 1815 - From what ship: HMT Ganges, Plymouth - Born: Warren - Age: 34 - Released on 11 Jul 1815.

Bowen, Oliver - Seaman - Nbr: 5424 - Prize: Portsmouth Packet - Ship Type: P - How taken: HMS Maidstone - When taken: 18 Jul 1813 - Where taken: Grand Banks - Date Received: 31 Oct 1814 - From what ship: HMT Leyden, Chatham - Born: New York - Age: 23 - Released on 1 Jul 1815.

Bowen, Silvester - Seaman - Nbr: 4714 - Prize: Yankee - Ship Type: P - How taken: HMS Shannon - When taken: 20 Aug 1813 - Where taken: at sea - Date Received: 9 Oct 1814 - From what ship: HMT Freya, Chatham - Born: Rhode Island - Age: 26 - Released on 15 Jun 1815.

Bowers, Jonathan - Carpenter - Nbr: 1728 - Prize: Charlotte - Ship Type: LM - How taken: HMS Dwarf - When taken: 6 Nov 1813 - Where taken: off Bordeaux - Date Received: 17 Jul 1814 - From what ship: Exeter, having broken parole at Ashburton (England) - Born: Massachusetts - Age: 29 - Released on 2 Apr 1815.

Bowes, David - Servant - Nbr: 586 - Prize: US Brig Argus - Ship Type: MW - How taken: HMS Pelican - When taken: 14 Aug 1813 - Where taken: Irish Channel - Date Received: 8 Sep 1813 - From what ship: Plymouth - Born: Boston - Age: 15 - Race: Black - Sent to Dartmouth on 2 Nov 1814.

Bowland, William - Seaman - Nbr: 4916 - Prize: US Brig Rattlesnake - Ship Type: MW - How taken: HMS Leander - When taken: 13 Jul 1814 - Where taken: off Shelburne - Date Received: 28 Oct 1814 - From what ship: HMT Alkbar, Halifax - Born: Boston - Age: 45 - Released on 21 Jun 1815.

Bowles, John - 1st Lieutenant - Nbr: 5565 - Prize: Harlequin - Ship Type: P - How taken: HMS Bulwark - When taken: 23 Nov 1814 - Where taken: off Halifax - Date Received: 17 Dec 1814 - From what ship: HMT Loire, Halifax - Born: Portsmouth - Age: 37 - Released on 10 Apr 1815.

Bowmer, Isaac - Seaman - Nbr: 1144 - Prize: Bunker Hill - Ship Type: P - How taken: HMS Pomone - When taken: 8 Mar 1814 - Where taken: at sea - Date Received: 10 May 1814 - From what ship: Plymouth - Born: Philadelphia - Age: 26 - Released on 20 Mar 1815.

Boyce, Abraham - Seaman - Nbr: 4168 - How taken: Gave himself up from HMS Laspidon - When taken: 22 Oct 1812 - Date Received: 7 Oct 1814 - From what ship: HMT Niobe, Chatham - Born: New York - Age: 26 - Released on 29 Jun 1815.

Boyd, Andrew - Seaman - Nbr: 5333 - Prize: Elbridge Gerry - Ship Type: P - How taken: HMS Crescent - When taken: 16 Sep 1813 - Where taken: at sea - Date Received: 31 Oct 1814 - From what ship: HMT Leyden, Chatham - Born: Greenwich - Age: 24 - Released on 11 Jul 1815.

Boyd, Andrew - Seaman - Nbr: 3239 - Prize: Hepsey - Ship Type: MV - How taken: HMS Tenedos - When taken: 22 Jun 1813 - Where taken: off Lisbon - Date Received: 11 Sep 1814 - From what ship: HMT Freya, Chatham - Born: Baltimore - Age: 22 - Released on 28 May 1815.

Boyd, Ephraim - Seaman - Nbr: 5594 - Prize: Albion - Ship Type: Prize - How taken: HMS Jaseur - When taken: 21 Sep 1814 - Where taken: off Halifax - Date Received: 17 Dec 1814 - From what ship: HMT Loire, Halifax - Born: Middleburgh - Age: 25 - Released on 1 Jul 1815.

Boyd, John - Seaman - Nbr: 3346 - How taken: Gave himself up from HMS Bombay - When taken: 29 Oct 1812 - Date Received: 13 Sep 1814 - From what ship: HMT Niobe, Chatham - Born: North Carolina - Age: 40 - Released on 27 Apr 1815.

Boyd, Stephen - Seaman - Nbr: 5276 - Prize: Hindortar - Ship Type: MV - How taken: HMS Tenedos - When taken: 25 Jun 1813 - Where taken: off Lisbon - Date Received: 31 Oct 1814 - From what ship: HMT Leyden, Chatham - Born: Hansbro - Age: 30 - Released on 29 Jun 1815.

American Prisoners of War Held at Dartmoor during the War of 1812

## Alphabetical listing of names

Boyd, William - Seaman - Nbr: 1214 - How taken: Sent into custody from a Prussian ship - Date Received: 14 Jun 1814 - From what ship: Mill Prison (Plymouth, England) - Born: Salem - Age: 36 - Released on 28 Apr 1815.

Boyd, William - Seaman - Nbr: 4346 - Prize: Union - Ship Type: LM - How taken: HMS Curlew - When taken: 1 Apr 1814 - Where taken: off Halifax - Date Received: 7 Oct 1814 - From what ship: HMT Salvador del Mundo, Halifax - Born: Boston - Age: 21 - Released on 14 Jun 1815.

Boyer, Joseph - Seaman - Nbr: 1392 - Prize: Courier - Ship Type: P - How taken: HMS Rover - When taken: 14 May 1813 - Where taken: Bay of Biscay - Date Received: 19 Jun 1814 - From what ship: Stapleton - Born: Maryland - Age: 28 - Sent to Mill Prison (Plymouth, England) on 26 Jun 1814.

Boyle, John - Seaman - Nbr: 4079 - Prize: New Zealander - Ship Type: Prize - How taken: HMS Belviders - When taken: 21 Apr 1814 - Where taken: off Delaware - Date Received: 6 Oct 1814 - From what ship: HMT Chesapeake, Halifax - Born: Baltimore - Age: 34 - Released on 13 Jun 1815.

Boyle, Joseph - Seaman - Nbr: 5166 - Prize: Volante - Ship Type: P - How taken: HMS Curlew - When taken: 25 Nov 1813 - Where taken: off Halifax - Date Received: 31 Oct 1814 - From what ship: HMT Mermaid, Chatham - Born: Ellenton - Age: 40 - Released on 11 Jul 1815.

Boyleston, Zebediah - 4th Lieutenant - Nbr: 6306 - Prize: Prince de Neufchatel - Ship Type: P - How taken: Leander (Newcastle Acasta) - When taken: 20 Dec 1814 - Where taken: Lat 38 Long 56 - Date Received: 19 Feb 1815 - From what ship: HMT Ganges, Plymouth - Born: Springfield - Age: 31 - Released on 4 Jun 1815.

Boys, Hugh - Seaman - Nbr: 2298 - Prize: Hussar - Ship Type: P - How taken: HMS Saturn - When taken: 25 May 1814 - Where taken: off Sandy Hook - Date Received: 16 Aug 1814 - From what ship: HMS Dublin, Halifax - Born: New Boston - Age: 50 - Released on 3 May 1815.

Boziman, Ralph - Seaman - Nbr: 2350 - Prize: Snap Dragon - Ship Type: P - How taken: HMS Martin - When taken: 10 Jun 1814 - Where taken: off Halifax - Date Received: 16 Aug 1814 - From what ship: HMS Dublin, Halifax - Born: SC - Age: 49 - Released on 3 May 1815.

Bracket, James - Prize master - Nbr: 6404 - Prize: Nancy - Ship Type: P - How taken: Papillion - When taken: 27 Dec 1814 - Where taken: Lat 29 Long 20 - Date Received: 3 Mar 1815 - From what ship: HMT Ganges, Plymouth - Born: New Hampshire - Age: 25 - Released on 11 Jul 1815.

Bracket, John - Seaman - Nbr: 5155 - Prize: Revenge - Ship Type: P - How taken: HMS Shannon - When taken: 5 Nov 1813 - Where taken: off Halifax - Date Received: 31 Oct 1814 - From what ship: HMT Mermaid, Chatham - Born: Portland - Age: 19 - Released on 29 Jun 1815.

Bradbury, John H. - Seaman - Nbr: 1691 - How taken: Gave himself up from U.S. Ardent - When taken: 14 Dec 1813 - Date rec'd: 2 Jul 1814 - From what ship: Plymouth - Born: New York - Age: 22 - Released on 1 May 1815.

Bradford, Charles - Seaman - Nbr: 1483 - Prize: Napoleon - Ship Type: LM - How taken: HMS Belle Poule - When taken: 3 Apr 1813 - Where taken: off Cape Ortegal (Spain) - Date Received: 19 Jun 1814 - From what ship: Stapleton - Born: Massachusetts - Age: 19 - Released on 1 May 1815.

Bradford, Elisha - Marine - Nbr: 4034 - Prize: US Brig Rattlesnake - Ship Type: MW - How taken: HMS Leander - When taken: 13 Jul 1814 - Where taken: off Shelburne - Date Received: 6 Oct 1814 - From what ship: HMT Chesapeake, Halifax - Born: Freeport - Age: 20 - Released on 9 Jun 1815.

Bradford, George - 2nd Lieutenant - Nbr: 4696 - Prize: Growler - Ship Type: P - How taken: HMS Electra - When taken: 7 Jul 1813 - Where taken: at sea - Date Received: 9 Oct 1814 - From what ship: HMT Leyden, Chatham - Born: Massachusetts - Age: 30 - Race: Mulatto.

Bradford, James - Cook - Nbr: 699 - Prize: Hero - Ship Type: MV - How taken: HMS Tenedos - When taken: Sep 1812 - Where taken: Western ocean - Date Received: 27 Sep 1813 - From what ship: Plymouth - Born: Maryland - Age: 22 - Race: Black - Released on 26 Apr 1815.

Bradie, John - Seaman - Nbr: 4746 - How taken: Gave himself up from HMS Hussar - When taken: 26 Jul 1814 - Date Received: 9 Oct 1814 - From what ship: HMT Freya, Chatham - Born: Boston - Age: 30 - Race: Black - Released on 15 Jun 1815.

Bradie, Thomas - Cook - Nbr: 3606 - Prize: Flash - Ship Type: MV - How taken: HMS Acasta and HMS Loire - When taken: 8 Feb 1814 - Where taken: off NY - Date Received: 30 Sep 1814 - From what ship: HMT Sybella - Born: Pennsylvania - Age: 28 - Race: Mulatto - Released on 4 Jun 1815.

Bradley, William - Seaman - Nbr: 5701 - Prize: David Porter - Ship Type: LM - How taken: HMS Pylades - When taken: 12 Sep 1814 - Where taken: Georges Bank - Date Received: 24 Dec 1814 - From what ship: HMT Penelope - Born: Massachusetts - Age: 25 - Released on 3 Jul 1815.

Bradly, Samuel - Seaman - Nbr: 4356 - Prize: Fiere Facia - Ship Type: P - How taken: HMS Ramillies - When

American Prisoners of War Held at Dartmoor during the War of 1812

## Alphabetical listing of names

taken: 26 Feb 1814 - Where taken: off NY - Date Received: 7 Oct 1814 - From what ship: HMT Salvador del Mundo, Halifax - Born: Baltimore - Age: 29 - Race: Black - Died on 29 Mar 1814 from variola.

Bradt, Francis - Marine - Nbr: 592 - Prize: US Brig Argus - Ship Type: MW - How taken: HMS Pelican - When taken: 14 Aug 1813 - Where taken: Irish Channel - Date Received: 8 Sep 1813 - From what ship: Plymouth - Born: New York - Age: 35 - Sent to Dartmouth on 2 Nov 1814.

Brady, Hugh - Seaman - Nbr: 423 - Prize: Viper - Ship Type: MV - How taken: HMS Superb - When taken: 15 Apr 1813 - Where taken: Bay of Biscay - Date Received: 1 Jul 1813 - From what ship: Plymouth - Born: Philadelphia - Age: 31 - Released on 20 Apr 1815.

Brady, Jason - Seaman - Nbr: 1022 - How taken: Sent from Mill Prison (Plymouth, England) - Date Received: 31 Jan 1814 - From what ship: Plymouth - Born: New York - Age: 23 - Released on 27 Apr 1815.

Brady, John - Seaman - Nbr: 5444 - Prize: Ida - Ship Type: LM - How taken: HMS Newcastle - When taken: 9 Aug 1814 - Where taken: Long 34 - Date Received: 28 Oct 1814 - From what ship: HMT Alkbar, Halifax - Born: New York - Age: 27 - Released on 29 Jun 1815.

Bragden, James - Seaman - Nbr: 5180 - Prize: Montgomery - Ship Type: P - How taken: HMS Nymphe - When taken: 1 May 1813 - Where taken: off Cape Cod - Date Received: 31 Oct 1814 - From what ship: HMT Mermaid, Chatham - Born: New York - Age: 27 - Released on 29 Jun 1815.

Brain, William - Seaman - Nbr: 4875 - Prize: US Brig Rattlesnake - Ship Type: MW - How taken: HMS Leander - When taken: 13 Jul 1814 - Where taken: off Shelburne - Date Received: 24 Oct 1814 - From what ship: Royal Hospital, Plymouth - Born: Long Island - Age: 35 - Released on 21 Jun 1815.

Bramant, Laurence - Seaman - Nbr: 6333 - Prize: Prince de Neufchatel - Ship Type: P - How taken: Leander (Newcastle Acasta) - When taken: 20 Dec 1814 - Where taken: Lat 38 Long 56 - Date Received: 19 Feb 1815 - From what ship: HMT Ganges, Plymouth - Born: New Orleans - Age: 17 - Race: Black - Released on 5 Jul 1815.

Branblen, Andrew - Seaman - Nbr: 3809 - Prize: Stark - Ship Type: P - How taken: HMS Sophie - When taken: 20 Apr 1814 - Where taken: off Bermuda - Date Received: 5 Oct 1814 - From what ship: HMT Orpheus, Halifax - Born: New York - Age: 21 - Released on 9 Jun 1815.

Branch, Anthony - Seaman - Nbr: 3175 - How taken: Gave himself up from HMS Centaur - When taken: 10 Sep 1813 - Date Received: 11 Sep 1814 - From what ship: HMT Freya, Chatham - Born: Lancaster - Age: 35 - Released on 28 May 1815.

Branger, John - Seaman - Nbr: 696 - Prize: Montgomery - Ship Type: P - How taken: HMS Nymphe - When taken: 5 May 1813 - Where taken: Boston Bay - Date Received: 27 Sep 1813 - From what ship: Plymouth - Born: Gothenburg - Age: 26 - Released on 26 Apr 1815.

Brannen, Alexander - Seaman - Nbr: 2162 - Prize: Rattlesnake - Ship Type: P - How taken: HMS Hyperion - When taken: 26 Jun 1814 - Where taken: off Cape Finisterre - Date Received: 16 Aug 1814 - From what ship: HMS Dublin, Halifax - Born: Baltimore - Age: 22 - Released on 2 May 1815.

Brant, Thomas - Seaman - Nbr: 1545 - Prize: Zebra - Ship Type: LM - How taken: HMS Pyramus - When taken: 20 Apr 1813 - Where taken: Bay of Biscay - Date Received: 23 Jun 1814 - From what ship: Stapleton - Born: New York - Age: 20 - Released on 1 May 1815.

Brantre, James - Seaman - Nbr: 2611 - How taken: Gave himself up from HMS Romulus - When taken: 14 Aug 1812 - Date Received: 21 Aug 1814 - From what ship: HMT Freya, Chatham - Born: Philadelphia - Age: 27 - Released on 26 Apr 1815.

Bray, Andrew - Seaman - Nbr: 5068 - Prize: Ida - Ship Type: LM - How taken: HMS Newcastle - When taken: 9 Aug 1814 - Where taken: Long 34 - Date Received: 28 Oct 1814 - From what ship: HMT Alkbar, Halifax - Born: Cape Ann - Age: 20 - Released on 29 Jun 1815.

Bray, Issachar - Seaman - Nbr: 5053 - Prize: Ida - Ship Type: LM - How taken: HMS Newcastle - When taken: 9 Aug 1814 - Where taken: Long 34 - Date Received: 28 Oct 1814 - From what ship: HMT Alkbar, Halifax - Born: Cape Ann - Age: 23 - Died on 17 Nov 1814.

Bray, Zachariah - Seaman - Nbr: 3776 - Prize: Fame - Ship Type: P - How taken: HMS Thistle - When taken: 10 Apr 1814 - Where taken: after being cast ashore on a seal island - Date Received: 30 Sep 1814 - From what ship: HMT President, Halifax - Born: Salem - Age: 24 - Race: Negro - Released on 9 Jun 1815.

Brayden, Theodore - Seaman - Nbr: 4056 - Prize: US Brig Rattlesnake - Ship Type: MW - How taken: HMS Leander - When taken: 13 Jul 1814 - Where taken: off Shelburne - Date Received: 6 Oct 1814 - From what ship: HMT Chesapeake, Halifax - Born: Mansfield - Age: 20 - Released on 13 Jun 1815.

Brazel, James - Seaman - Nbr: 2489 - Prize: US Sloop Frolic - Ship Type: MW - How taken: HMS Orpheus - When taken: 20 Apr 1814 - Where taken: off Cuba - Date Received: 16 Aug 1814 - From what ship: HMT Queen,

American Prisoners of War Held at Dartmoor during the War of 1812

## Alphabetical listing of names

Halifax - Born: Beverly - Age: 27 - Released on 3 May 1815.
Brazier, Thomas - Seaman - Nbr: 3003 - Prize: St. Lawrence - How taken: HMS Aquilon - When taken: 9 Aug 1814 - Where taken: off Western Islands (England) - Date Received: 29 Aug 1814 - From what ship: HMT Bittern - Born: New York - Age: 21 - Released on 28 May 1815.
Brenton, York - Seaman - Nbr: 1926 - How taken: Gave himself up from HMS Royal William - When taken: 3 Feb 1813 - Date Received: 3 Aug 1814 - From what ship: HMS Alceste, Chatham Depot - Born: New Jersey - Age: 33 - Released on 2 May 1815.
Bressy, Charles - Seaman - Nbr: 4719 - How taken: Gave himself up from HMS Hussar - When taken: 5 Jul 1813 - Date Received: 9 Oct 1814 - From what ship: HMT Freya, Chatham - Born: Dorchester - Age: 28 - Released on 15 Jun 1815.
Bretade, E. F. - Seaman - Nbr: 4832 - Prize: President - Ship Type: P - How taken: HMS Pique - When taken: 7 May 1814 - Where taken: off Porto Rico - Date Received: 9 Oct 1814 - From what ship: HMT Freya, Halifax - Born: Nantes - Age: 40 - Released on 3 May 1815.
Brewster, Jacob - Seaman - Nbr: 131 - Prize: Mars - Ship Type: MV - How taken: HMS Warspite - When taken: 26 Feb 1813 - Where taken: off Basque Roads (France) - Date Received: 12 Apr 1813 - From what ship: Plymouth - Born: Duxbury - Age: 22 - Released on 20 Apr 1815.
Briant, Moses - Seaman - Nbr: 2017 - How taken: Gave himself up from HMS Royal William - When taken: 5 May 1813 - Date Received: 3 Aug 1814 - From what ship: HMS Lyffey, Chatham Depot - Born: Middleburg - Age: 23 - Released on 2 May 1815.
Brickman, John - Carpenter - Nbr: 5411 - Prize: Pilot - Ship Type: LM - How taken: Victoria (Privateer) - When taken: 23 Jan 1814 - Where taken: off Bordeaux - Date Received: 31 Oct 1814 - From what ship: HMT Leyden, Chatham - Born: Virginia - Age: 25 - Race: Mulatto - Released on 1 Jul 1815.
Bridge, Francis - Seaman - Nbr: 63 - Prize: Spitfire - Ship Type: MV - How taken: HMS Achates - When taken: 14 Feb 1813 - Where taken: off Ushant (France) - Date Received: 2 Apr 1813 - From what ship: Plymouth - Born: Marblehead - Age: 32 - Released on 20 Apr 1815.
Bridge, Jeremiah - Seaman - Nbr: 4259 - Prize: Argus - Ship Type: MV - How taken: HMS San Domingo - When taken: 1 Mar 1814 - Where taken: off Savannah - Date Received: 7 Oct 1814 - From what ship: HMT Niobe, Chatham - Born: Not legible - Age: 27 - Released on 13 Jun 1815.
Bridges, John - Seaman - Nbr: 363 - Prize: Two Brothers - Ship Type: MV - How taken: Beetle (LM) - When taken: 18 Mar 1813 - Where taken: off Western Islands (England) - Date Received: 1 Jul 1813 - From what ship: Plymouth - Born: Castine - Age: 23 - Sent to Liverpool on 25 Aug 1813.
Bridges, John - Seaman - Nbr: 4663 - Prize: Wasp - Ship Type: P - How taken: HMS Bream - When taken: 10 Jun 1813 - Where taken: off Halifax - Date Received: 9 Oct 1814 - From what ship: HMT Leyden, Chatham - Born: Beverly - Age: 27 - Released on 15 Jun 1815.
Bridson, Thomas - Seaman - Nbr: 4358 - Prize: Bordeaux Packet - Ship Type: LM - How taken: HMS Niemen - When taken: 28 Jan 1814 - Where taken: off Delaware - Date Received: 7 Oct 1814 - From what ship: HMT Salvador del Mundo, Halifax - Born: Newfoundland - Age: 22 - Released on 14 Jun 1815.
Brien, Lewis - Seaman - Nbr: 3549 - Prize: Hawk - Ship Type: P - How taken: HMS Pique - When taken: 26 Apr 1814 - Where taken: off Bermuda - Date Received: 30 Sep 1814 - From what ship: HMT Sybella - Born: North Carolina - Age: 24 - Died on 5 Nov 1814.
Briennis, John - Seaman - Nbr: 2451 - Prize: US Sloop Frolic - Ship Type: MW - How taken: HMS Orpheus - When taken: 20 Apr 1814 - Where taken: off Cuba - Date Received: 16 Aug 1814 - From what ship: HMT Queen, Halifax - Born: Bremen - Age: 33 - Released on 3 May 1815.
Briggs, B. - Seaman - Nbr: 3960 - Prize: Rolla - Ship Type: P - How taken: HMS Loire - When taken: 10 Dec 1813 - Where taken: off Bull Island (South Carolina) - Date Received: 5 Oct 1814 - From what ship: HMT President, Halifax - Born: Rhode Island - Age: 19 - Released on 9 Jun 1815.
Briggs, Benjamin - Seaman - Nbr: 3508 - Prize: Chasseur - Ship Type: P - How taken: HMS Whiting - When taken: Aug 1814 - Where taken: off the Western Isles (England) - Date Received: 24 Sep 1814 - From what ship: HMT Salvador del Mundo - Born: Salisbury - Age: 19 - Released on 4 Jun 1815.
Briggs, Boileau - Seaman - Nbr: 2077 - Prize: True Blooded Yankee - Ship Type: P - How taken: HMS Hope - When taken: 24 Jun 1813 - Where taken: off Brest - Date Received: 3 Aug 1814 - From what ship: HMS Bittern, Chatham Depot - Born: Virginia - Age: 31 - Released on 2 May 1815.
Briggs, Frank - Seaman - Nbr: 5344 - Prize: John & Mary - Ship Type: Prize - How taken: HMS Borer - When taken: 20 Oct 1813 - Where taken: off Long Island - Date Received: 31 Oct 1814 - From what ship: HMT Leyden, Chatham - Born: Freetown - Age: 21 - Released on 1 Jul 1815.

American Prisoners of War Held at Dartmoor during the War of 1812

## Alphabetical listing of names

Briggs, John - Seaman - Nbr: 2375 - Prize: US Sloop Frolic - Ship Type: MW - How taken: HMS Orpheus - When taken: 20 Apr 1814 - Where taken: off Cuba - Date Received: 16 Aug 1814 - From what ship: HMT Queen, Halifax - Born: New Bedford - Age: 38 - Sent to Dartmouth on 19 Oct 1814.

Briggs, John - Seaman - Nbr: 5551 - Prize: Minerva - Ship Type: MV - How taken: Lunenburg (Privateer) - When taken: 5 Sep 1814 - Where taken: Irish Shoals - Date Received: 17 Dec 1814 - From what ship: HMT Loire, Halifax - Born: West-Cappel (France) - Age: 23 - Released on 1 Jul 1815.

Briggs, Samuel - First Lieutenant - Nbr: 2527 - Prize: Diomede - Ship Type: P - How taken: HMS Rifleman - When taken: 28 Jul 1814 - Where taken: off Halifax - Date Received: 16 Aug 1814 - From what ship: HMT Queen, Halifax - Born: Salem - Age: 28 - Released on 3 May 1815.

Briggs, Thomas - Seaman - Nbr: 4457 - Prize: Juliana Smith - Ship Type: P - How taken: HMS Nymphe - When taken: 11 May 1813 - Where taken: off Cape Sable Island (Canada) - Date Received: 8 Oct 1814 - From what ship: HMT Leyden, Chatham - Born: Massachusetts - Age: 26 - Released on 15 Jun 1815.

Briggs, William - Seaman - Nbr: 1090 - How taken: Impressed at Cork - When taken: 9 Mar 1814 - Date Received: 10 May 1814 - From what ship: Plymouth - Born: New York - Age: 23 - Released on 27 Apr 1815.

Briggs, William - Boy - Nbr: 5182 - Prize: Montgomery - Ship Type: P - How taken: HMS Nymphe - When taken: 1 May 1813 - Where taken: off Cape Cod - Date Received: 31 Oct 1814 - From what ship: HMT Mermaid, Chatham - Born: Salem - Age: 14 - Released on 29 Jun 1815.

Bright, Samuel - Seaman - Nbr: 77 - Prize: Rolla - Ship Type: MV - How taken: HMS Surveillante - When taken: 11 Feb 1813 - Where taken: Bay of Biscay - Date Received: 2 Apr 1813 - From what ship: Plymouth - Born: Forsham - Age: 26 - Released on 20 Apr 1815.

Brightman, Joseph - Seaman - Nbr: 1117 - Prize: Bunker Hill - Ship Type: P - How taken: HMS Pomone & HMS Cadmus - When taken: 4 Mar 1814 - Where taken: Bay of Biscay - Date Received: 10 May 1814 - From what ship: Plymouth - Born: Dartmouth - Age: 24 - Released on 20 Mar 1815.

Brightman, Joseph - Seaman - Nbr: 2746 - How taken: Gave himself up from HMS Andromache - When taken: 5 Nov 1812 - Date Received: 24 Aug 1814 - From what ship: HMT Liverpool, Chatham - Born: Boston - Age: 36 - Released on 26 Apr 1815.

Brights, George - Seaman - Nbr: 1617 - Prize: Fox - Ship Type: LM - How taken: HMS Pheasant - When taken: 23 Apr 1813 - Where taken: Bay of Biscay - Date Received: 23 Jun 1814 - From what ship: Stapleton - Born: New Jersey - Age: 23 - Released on 1 May 1815.

Brill, John - Boy - Nbr: 2798 - Prize: Weazel - Ship Type: MV - How taken: HMS Foxhound - When taken: 25 Mar 1813 - Where taken: Bay of Biscay - Date Received: 24 Aug 1814 - From what ship: HMT Liverpool, Chatham - Born: New York - Age: 13 - Released on 19 May 1815.

Brisons, John - Seaman - Nbr: 4231 - Prize: Bunker Hill - Ship Type: P - How taken: HMS Pomone - When taken: 2 Mar 1814 - Where taken: Channel - Date Received: 7 Oct 1814 - From what ship: HMT Niobe, Chatham - Born: Baltimore - Age: 32 - Race: Black - Died on 24 Jan 1815 from phthisis pulmonalis.

Broadwater, Samuel - Seaman - Nbr: 1568 - Prize: Messenger - Ship Type: MV - How taken: HMS Iris - When taken: 10 Mar 1813 - Where taken: Bay of Biscay - Date Received: 23 Jun 1814 - From what ship: Stapleton - Born: Maryland - Age: 22 - Released on 1 May 1815.

Broden, Norman - Seaman - Nbr: 1905 - Prize: Rachael - Ship Type: MV - How taken: HMS Herring - When taken: 9 Feb 1813 - Where taken: at sea - Date Received: 3 Aug 1814 - From what ship: HMS Alceste, Chatham Depot - Born: Marblehead - Age: 30 - Released on 2 May 1815.

Bron, James - Seaman - Nbr: 1037 - Prize: US Brig Argus - Ship Type: MW - How taken: HMS Pelican - When taken: 16 Aug 1813 - Where taken: Irish Channel - Date Received: 10 May 1814 - From what ship: Plymouth - Born: Hampton - Age: 27 - Sent to Dartmouth on 19 Oct 1814.

Brook, David - Boy - Nbr: 3556 - Prize: Hawk - Ship Type: P - How taken: HMS Pique - When taken: 26 Apr 1814 - Where taken: off Bermuda - Date Received: 30 Sep 1814 - From what ship: HMT Sybella - Born: Washington - Age: 12 - Released on 28 Apr 1815.

Brooks, John - Cook - Nbr: 5624 - Prize: Rambler - Ship Type: MV - How taken: HMS Lion - When taken: 1 Feb 1813 - Where taken: Cape - Date Received: 24 Dec 1814 - From what ship: HMT Tay - Born: Salem - Age: 23 - Race: Black - Released on 3 Jul 1815.

Brooks, John - Seaman - Nbr: 1834 - Prize: Lightning - Ship Type: MV - How taken: HMS Medusa - When taken: 2 Apr 1813 - Where taken: Bay of Biscay - Date Received: 21 Jul 1814 - From what ship: HMT Redbeard & Pincher, Chatham Depot - Born: North Carolina - Age: 34 - Released on 1 May 1815.

Brooks, John - Seaman - Nbr: 5183 - Prize: Montgomery - Ship Type: P - How taken: HMS Nymphe - When taken: 1 May 1813 - Where taken: off Cape Cod - Date Received: 31 Oct 1814 - From what ship: HMT Mermaid,

American Prisoners of War Held at Dartmoor during the War of 1812

## Alphabetical listing of names

Chatham - Born: Africa - Age: 21 - Race: Negro - Released on 29 Jun 1815.

Brooks, Philip - Seaman - Nbr: 6159 - Prize: Lion - Ship Type: P - How taken: HMS Granicus - When taken: 2 Dec 1814 - Where taken: off Lisbon - Date Received: 18 Jan 1815 - From what ship: HMT Impregnable - Born: Maryland - Age: 28 - Race: Black - Released on 5 Jul 1815.

Brooks, Russell - Seaman - Nbr: 1094 - Prize: Bunker Hill - Ship Type: P - How taken: HMS Pomone - When taken: 8 Mar 1814 - Where taken: at sea - Date Received: 10 May 1814 - From what ship: Plymouth - Born: East Haddam - Age: 33 - Released on 27 Apr 1815.

Brooks, Thomas - Seaman - Nbr: 1087 - How taken: Impressed at Cork - When taken: 14 Feb 1814 - Date Received: 10 May 1814 - From what ship: Plymouth - Born: Virginia - Age: 23 - Released on 27 Apr 1815.

Broughton, Glover - Seaman - Nbr: 764 - How taken: Sent into custody from HMS Pallas - Date Received: 3 Nov 1813 - From what ship: Plymouth - Born: Marblehead - Age: 17 - Released on 26 Apr 1815.

Broughton, John - Prize Master - Nbr: 3668 - Prize: Alfred - Ship Type: P - How taken: HMS Epervier - When taken: 23 Feb 1812 - Where taken: off Newfoundland - Date Received: 30 Sep 1814 - From what ship: HMT President - Born: Marblehead - Age: 22 - Released on 4 Jun 1815.

Brower, John - Carpenter - Nbr: 1894 - Prize: Argus - Ship Type: MV - How taken: HMS San Domingo - When taken: 1 Mar 1814 - Where taken: off Savannah - Date Received: 29 Jul 1814 - From what ship: HMS Ville de Paris, Chatham Depot - Born: New York - Age: 42 - Released on 2 May 1815.

Brown, Aaron - Seaman - Nbr: 498 - Prize: Grand Napoleon - Ship Type: MV - How taken: HMS Goldfinch - When taken: 17 Apr 1813 - Where taken: Bay of Biscay - Date Received: 8 Sep 1813 - From what ship: Plymouth - Born: Rhode Island - Age: 38 - Released on 20 Apr 1815.

Brown, Abraham - Seaman - Nbr: 4729 - How taken: Gave himself up from Garmane - When taken: 3 Sep 1814 - Date Received: 9 Oct 1814 - From what ship: HMT Freya, Chatham - Born: Massachusetts - Age: 53 - Released on 15 Jun 1815.

Brown, Alexander - Marine - Nbr: 600 - Prize: US Brig Argus - Ship Type: MW - How taken: HMS Pelican - When taken: 14 Aug 1813 - Where taken: Irish Channel - Date Received: 8 Sep 1813 - From what ship: Plymouth - Born: Norway - Age: 37 - Sent to Dartmouth on 2 Nov 1814.

Brown, Amos - Seaman - Nbr: 4352 - Prize: Plutus - Ship Type: P - How taken: HMS Curlew - When taken: 1 Apr 1814 - Where taken: off Halifax - Date Received: 7 Oct 1814 - From what ship: HMT Salvador del Mundo, Halifax - Born: Salem - Age: 21 - Released on 14 Jun 1815.

Brown, Anthony - Seaman - Nbr: 3102 - Prize: Mary - Ship Type: Prize - How taken: Taken by pilot boat - When taken: 27 Aug 1814 - Where taken: Bristol Channel - Date Received: 9 Sep 1814 - From what ship: HMS Abercrombie - Born: Portugal - Age: 18 - Released on 28 May 1815.

Brown, Asher - Seaman - Nbr: 627 - Prize: US Brig Argus - Ship Type: MW - How taken: HMS Pelican - When taken: 14 Aug 1813 - Where taken: Irish Channel - Date Received: 8 Sep 1813 - From what ship: Plymouth - Born: Middletown - Age: 23 - Sent to Dartmouth on 2 Nov 1814.

Brown, Benjamin - Seaman - Nbr: 1461 - Prize: Revenge - Ship Type: LM - How taken: HMS Belle Poule - When taken: 10 Mar 1813 - Where taken: off Cornwall - Date Received: 19 Jun 1814 - From what ship: Stapleton - Born: Westborough - Age: 21 - Released on 28 Apr 1815.

Brown, Benjamin - Seaman - Nbr: 303 - Prize: Ducornau - Ship Type: MV - How taken: HMS Pheasant - When taken: 15 Mar 1813 - Where taken: Bay of Biscay - Date Received: 28 Jun 1813 - From what ship: Plymouth - Born: New Haven - Age: 29 - Released on 20 Apr 1815.

Brown, Benjamin - Seaman - Nbr: 1855 - Prize: Jason - Ship Type: MV - How taken: HMS Venerable - When taken: 1 Jun 1814 - Where taken: off Tenerife - Date Received: 21 Jul 1814 - From what ship: HMT Redbeard, Chatham Depot - Born: Nashville - Age: 17 - Released on 2 May 1815.

Brown, Benjamin - Clerk - Nbr: 3598 - Prize: Frolic - Ship Type: P - How taken: HMS Heron - When taken: 25 Jan 1814 - Where taken: off St. Thomas - Date Received: 30 Sep 1814 - From what ship: HMT Sybella - Born: Salem - Age: 21 - Released on 1 May 1815.

Brown, Benjamin - Seaman - Nbr: 5210 - Prize: Porcupine - Ship Type: LM - How taken: HMS Acasta - When taken: 18 Jul 1813 - Where taken: off Halifax - Date Received: 31 Oct 1814 - From what ship: HMT Mermaid, Chatham - Born: Thomastown - Age: 28 - Released on 11 Jul 1815.

Brown, Charles - Seaman - Nbr: 1350 - Prize: Paul Jones - Ship Type: P - How taken: HMS Leonidas - When taken: 23 May 1813 - Where taken: Channel - Date Received: 19 Jun 1814 - From what ship: Stapleton - Born: Virginia - Age: 23 - Race: Negro - Died on 17 Feb 1815 from variola.

Brown, Charles - Seaman - Nbr: 2560 - How taken: Gave himself up from HMS Bacchus - When taken: 12 Aug 1814 - Date Received: 16 Aug 1814 - From what ship: HMT Salvador del Mundo - Born: Salem - Age: 27 -

American Prisoners of War Held at Dartmoor during the War of 1812

## Alphabetical listing of names

Race: Negro - Released on 3 May 1815.

Brown, Charles - Seaman - Nbr: 3494 - How taken: Gave himself up from HMS Opossium - When taken: 3 Aug 1814 - Date Received: 19 Sep 1814 - From what ship: HMT Salvador del Mundo - Born: Philadelphia - Age: 26 - Race: Negro - Released on 4 Jun 1815.

Brown, David - Seaman - Nbr: 1172 - Prize: Zephyr - Ship Type: MV - How taken: HMS Pyramus - When taken: 30 Nov 1813 - Where taken: off L'Orient (France) - Date Received: 10 May 1814 - From what ship: Plymouth - Born: Maryland - Age: 35 - Race: Negro - Released on 28 Apr 1815.

Brown, David - Seaman - Nbr: 3139 - How taken: Impressed at London - When taken: 14 Sep 1813 - Date rec'd: 11 Sep 1814 - From what ship: HMT Freya, Chatham - Born: New York - Age: 21 - Released on 28 May 1815.

Brown, Ebenezer - Seaman - Nbr: 673 - Prize: Joel Barlow - Ship Type: LM - How taken: HMS Briton - When taken: 3 Jul 1813 - Where taken: off Bordeaux - Date Received: 8 Sep 1813 - From what ship: Plymouth - Born: Rhode Island - Age: 19 - Released on 26 Apr 1815.

Brown, Edward - Seaman - Nbr: 2423 - Prize: US Sloop Frolic - Ship Type: MW - How taken: HMS Orpheus - When taken: 20 Apr 1814 - Where taken: off Cuba - Date Received: 16 Aug 1814 - From what ship: HMT Queen, Halifax - Born: Ipswich - Age: 29 - Released on 3 May 1815.

Brown, Elisha - Seaman - Nbr: 2767 - How taken: Gave himself up from HMS Franchise - When taken: 20 Dec 1812 - Date Received: 24 Aug 1814 - From what ship: HMT Liverpool, Chatham - Born: Georgetown - Age: 25 - Released on 19 May 1815.

Brown, Francis - Seaman - Nbr: 556 - Prize: Godfrey and Mary - Ship Type: MV - How taken: Robert Todd - When taken: 23 Jun 1813 - Where taken: Lat 21, 30 Long 53 - Date Received: 8 Sep 1813 - From what ship: Plymouth - Born: New London - Age: 22 - Released on 26 Apr 1815.

Brown, Francis - Seaman - Nbr: 3246 - How taken: Impressed at Cork - When taken: 26 Sep 1813 - Date Received: 11 Sep 1814 - From what ship: HMT Freya, Chatham - Born: Pennsylvania - Age: 28 - Race: Mulatto - Released on 8 May 1815.

Brown, Frederick - Seaman - Nbr: 4585 - Prize: Bunker Hill - Ship Type: P - How taken: HMS Pomone - When taken: 2 Mar 1814 - Where taken: Channel - Date Received: 9 Oct 1814 - From what ship: HMT Leyden, Chatham - Born: Boston - Age: 23 - Race: Black - Released on 15 Jun 1815.

Brown, George - Seaman - Nbr: 5612 - How taken: Gave himself up from the Cape - When taken: 8 May 1812 - Date rec'd: 24 Dec 1814 - What ship: HMT Tay - Born: Baltimore - Age: 34 - Released on 27 Mar 1815.

Brown, George - Seaman - Nbr: 5788 - How taken: Gave himself up from East Indiaman Ocean - When taken: May 1814 - Date Received: 26 Dec 1814 - From what ship: HMT Argo - Born: Pennsylvania - Age: 31 - Race: Mulatto - Died on 11 Feb 1815 from pneumonia.

Brown, George - Seaman - Nbr: 1992 - How taken: Gave himself up from HMS President - Date Received: 3 Aug 1814 - From what ship: HMS Lyffey, Chatham Depot - Born: Baltimore - Age: 31 - Released on 3 Jul 1815.

Brown, George - Prize Master - Nbr: 162 - Prize: Margaret - Ship Type: P - How taken: HMS Nimrod - When taken: 9 Mar 1813 - Where taken: off Morlaix - Date Received: 2 Apr 1813 - From what ship: Plymouth - Born: Wellfleet - Age: 37 - Sent to Plymouth on 6 Jun 1813.

Brown, George - Seaman - Nbr: 3458 - Prize: Prize to Chasseur - How taken: HMS Tartarus - When taken: 1 Sep 1814 - Where taken: Atlantic - Date Received: 19 Sep 1814 - From what ship: HMT Salvador del Mundo - Born: Pennsylvania - Age: 27 - Released on 4 Jun 1815.

Brown, George - Seaman - Nbr: 3357 - How taken: Gave himself up from HMS Ocean - When taken: 29 Oct 1812 - Date Received: 13 Sep 1814 - From what ship: HMT Niobe, Chatham - Born: Not legible - Age: 35 - Released on 27 Apr 1815.

Brown, George - Seaman - Nbr: 3948 - Prize: Rolla - Ship Type: P - How taken: HMS Loire - When taken: 10 Dec 1813 - Where taken: off Bull Island (South Carolina) - Date Received: 5 Oct 1814 - From what ship: HMT President, Halifax - Born: Norfolk - Age: 25 - Released on 9 Jun 1815.

Brown, Henry - Seaman - Nbr: 6524 - How taken: Gave himself up from HMS Ashea - Date rec'd: 3 Mar 1815 - What ship: HMT Ganges, Plymouth - Born: Huntingdon - Age: 22 - Race: Black - Released on 11 Jul 1815.

Brown, Henry - Seaman - Nbr: 5821 - Prize: US Schooner Scorpion - Ship Type: MW - How taken: British gunboats - When taken: 6 Sep 1814 - Where taken: Lake Erie - Date Received: 26 Dec 1814 - From what ship: HMT Argo - Born: New York - Age: 22 - Race: Black - Released on 3 Jul 1815.

Brown, Henry - Boy - Nbr: 149 - Prize: Criterion - Ship Type: MV - How taken: HMS Belle Poule - When taken: 14 Feb 1813 - Where taken: Bay of Biscay - Date Received: 2 Apr 1813 - From what ship: Plymouth - Born: New York - Age: 17 - Sent to Mill Prison (Plymouth, England) on 10 Jul 1813.

Brown, Isaac - Seaman - Nbr: 3355 - How taken: Gave himself up from HMS Shearwater - When taken: 24 May

American Prisoners of War Held at Dartmoor during the War of 1812

## Alphabetical listing of names

1813 - Date Received: 13 Sep 1814 - From what ship: HMT Niobe, Chatham - Born: Not legible - Age: 30 - Race: Black - Released on 28 May 1815.

Brown, Jacob - Seaman - Nbr: 2826 - How taken: Gave himself up from HMS Onyx - When taken: 23 May 1813 - Date Received: 24 Aug 1814 - From what ship: HMT Liverpool, Chatham - Born: Germantown - Age: 29 - Released on 19 May 1815.

Brown, James - Seaman - Nbr: 5646 - Prize: Lion - Ship Type: P - How taken: HMS Granicus - When taken: 2 Dec 1814 - Where taken: off Lisbon - Date Received: 24 Dec 1814 - From what ship: HMT Tay - Born: Boston - Age: 27 - Race: Mulatto - Released on 3 Jul 1815.

Brown, James - Boatswain - Nbr: 321 - Prize: Pallas - Ship Type: MV - How taken: Rebuff - When taken: 23 Dec 1812 - Where taken: off Cadiz - Date Received: 28 Jun 1813 - From what ship: Plymouth - Born: Wethersfield - Age: 31 - Sent to Dartmouth on 30 Jul 1813.

Brown, James - Seaman - Nbr: 2644 - How taken: Gave himself up from HMS Caledonia - When taken: 5 Dec 1812 - Date Received: 21 Aug 1814 - From what ship: HMT Freya, Chatham - Born: Maryland - Age: 30 - Released on 26 Apr 1815.

Brown, James - Seaman - Nbr: 2705 - How taken: Gave himself up from HMS Serapis - When taken: 12 Dec 1813 - Date Received: 21 Aug 1814 - From what ship: HMT Freya, Chatham - Born: Virginia - Age: 26 - Released on 26 Apr 1815.

Brown, James - Seaman - Nbr: 3105 - How taken: Gave himself up from HMS Hannibal - When taken: 23 Aug 1814 - Date Received: 11 Sep 1814 - From what ship: HMT Salvador del Mundo - Born: Watertown - Age: 27 - Released on 28 May 1815.

Brown, James - Seaman - Nbr: 3628 - Prize: Monarch - Ship Type: MV - How taken: HMS Dotterel - When taken: 14 Dec 1813 - Where taken: off Charleston - Date Received: 30 Sep 1814 - From what ship: HMT Sybella - Born: Pennsylvania - Age: 53 - Race: Mulatto - Released on 4 Jun 1815.

Brown, Jean - Seaman - Nbr: 1586 - Prize: Price - Ship Type: MV - How taken: HMS Pyramus - When taken: 6 Apr 1813 - Where taken: Bay of Biscay - Date Received: 23 Jun 1814 - From what ship: Stapleton - Born: New Jersey - Age: 24 - Released on 1 May 1815.

Brown, Jesse - Seaman - Nbr: 335 - Prize: Star - Ship Type: MV - How taken: HMS Superb - When taken: 9 Feb 1813 - Where taken: Bay of Biscay - Date Received: 28 Jun 1813 - From what ship: Plymouth - Born: Connecticut - Age: 26 - Released on 2 Apr 1815.

Brown, Jesse - Seaman - Nbr: 1227 - How taken: Sent into custody from HMS Trident - When taken: 28 Nov 1812 - Date Received: 14 Jun 1814 - From what ship: Mill Prison (Plymouth, England) - Born: New York - Age: 27 - Race: Mulatto - Released on 26 Apr 1815.

Brown, John - Seaman - Nbr: 6213 - Prize: Prince de Neufchatel - Ship Type: P - How taken: Leander (Newcastle Acasta) - When taken: 20 Dec 1814 - Where taken: Lat 38 Long 56 - Date Received: 30 Jan 1815 - From what ship: HMT Pheasant - Born: New Hampshire - Age: 21 - Released on 5 Jul 1815.

Brown, John - Seaman - Nbr: 4378 - Prize: Growler - Ship Type: P - How taken: HMS Electra - When taken: 7 Jul 1813 - Where taken: at sea - Date Received: 8 Oct 1814 - From what ship: HMT Leyden, Chatham - Born: North Carolina - Age: 20 - Race: Black - Released on 28 May 1815.

Brown, John - Seaman - Nbr: 5311 - How taken: Sent into custody from HMS Belvidere - When taken: 12 Nov 1813 - Date Received: 31 Oct 1814 - From what ship: HMT Leyden, Chatham - Born: New York - Age: 20 - Released on 29 Jun 1815.

Brown, John - Seaman - Nbr: 452 - Prize: St. Martin's Planter - How taken: HMS Dublin - When taken: 9 Feb 1813 - Where taken: Lat 63 N, Long 33'50"W - Date Received: 8 Sep 1813 - From what ship: Plymouth - Born: New York - Age: 19 - Released on 26 Apr 1815.

Brown, John - Seaman - Nbr: 5669 - Prize: Harlequin - Ship Type: P - How taken: HMS Bulwark - When taken: 23 Nov 1814 - Where taken: off Halifax - Date Received: 24 Dec 1814 - From what ship: HMT Penelope - Born: Massachusetts - Age: 40 - Released on 3 Jul 1815.

Brown, John - Gunner - Nbr: 4424 - Prize: Elbridge Gerry - Ship Type: LM - How taken: HMS Crescent - When taken: 16 Sep 1813 - Where taken: at sea - Date Received: 8 Oct 1814 - From what ship: HMT Leyden, Chatham - Born: Charleston - Age: 20 - Released on 14 Jun 1815.

Brown, John - Prize master - Nbr: 3804 - Prize: Stark - Ship Type: P - How taken: HMS Sophie - When taken: 20 Apr 1814 - Where taken: off Bermuda - Date Received: 5 Oct 1814 - From what ship: HMT Orpheus, Halifax - Born: Boston - Age: 24 - Released on 9 Jun 1815.

Brown, John - Seaman - Nbr: 2901 - How taken: Gave himself up from HMS Ocean - When taken: 28 May 1813 - Date Received: 24 Aug 1814 - From what ship: HMT Alpheus, Chatham - Born: Massachusetts - Age: 20 -

American Prisoners of War Held at Dartmoor during the War of 1812

## Alphabetical listing of names

Race: Black - Released on 19 May 1815.

Brown, John W. - Seaman - Nbr: 1621 - Prize: Tom - Ship Type: LM - How taken: HMS Surveillante - When taken: 24 Apr 1813 - Where taken: Bay of Biscay - Date Received: 23 Jun 1814 - From what ship: Stapleton - Born: Albany - Age: 28 - Released on 1 May 1815.

Brown, Joseph - Master - Nbr: 4437 - Prize: Growler - Ship Type: P - How taken: HMS Electra - When taken: 7 Jul 1814 - Where taken: at sea - Date Received: 8 Oct 1814 - From what ship: HMT Leyden, Chatham - Born: Marblehead - Age: 29 - Released on 14 Jun 1815.

Brown, Joseph - Seaman - Nbr: 1853 - How taken: Gave himself up from HMS Swiftsure - When taken: 26 Dec 1812 - Date Received: 21 Jul 1814 - From what ship: HMT Redbeard & Pincher, Chatham Depot - Born: Old Providence - Age: 32 - Race: Negro - Released on 26 Apr 1815.

Brown, Ludwig - Seaman - Nbr: 305 - Prize: Ducornau - Ship Type: MV - How taken: HMS Pheasant - When taken: 15 Mar 1813 - Where taken: Bay of Biscay - Date Received: 28 Jun 1813 - From what ship: Plymouth - Born: Virginia - Age: 38 - Released on 20 Apr 1815.

Brown, Mark - Seaman - Nbr: 2633 - How taken: Gave himself up from HMS Leviathan - When taken: 22 Oct 1813 - Date Received: 21 Aug 1814 - From what ship: HMT Freya, Chatham - Born: Maryland - Age: 28 - Released on 3 May 1815.

Brown, Michael - Seaman - Nbr: 4552 - How taken: Gave himself up from HMS Cumberland - When taken: 16 Feb 1814 - Date Received: 8 Oct 1814 - From what ship: HMT Leyden, Chatham - Born: Virginia - Age: 55 - Released on 15 Jun 1815.

Brown, Reuben - Seaman - Nbr: 2829 - How taken: Gave himself up from HMS Berwick - When taken: 30 Nov 1812 - Date Received: 24 Aug 1814 - From what ship: HMT Liverpool, Chatham - Born: Boston - Age: 27 - Released on 26 Apr 1815.

Brown, Richard - Cook - Nbr: 821 - How taken: Impressed at Liverpool - When taken: 14 Sep 1813 - Date Received: 29 Nov 1813 - From what ship: Plymouth - Born: Baltimore - Age: 38 - Race: Negro - Released on 26 Apr 1815.

Brown, Robert - Seaman - Nbr: 3313 - Prize: Globe - Ship Type: P - How taken: HMS Victorious - When taken: 8 Jun 1813 - Where taken: Chesapeake Bay - Date Received: 13 Sep 1814 - From what ship: HMT Niobe, Chatham - Born: Baltimore - Age: 30 - Released on 25 Mar 1815.

Brown, Samuel - Seaman - Nbr: 307 - Prize: Ducornau - Ship Type: MV - How taken: HMS Pheasant - When taken: 15 Mar 1813 - Where taken: Bay of Biscay - Date Received: 28 Jun 1813 - From what ship: Plymouth - Born: New Orleans - Age: 38 - Sent to Plymouth on 8 Jul 1814.

Brown, Samuel - Seaman - Nbr: 1665 - Prize: Essex - Ship Type: MV - How taken: HMS Pyramus - When taken: 2 Apr 1813 - Where taken: Bay of Biscay - Date Received: 26 Jun 1814 - From what ship: Exeter - Born: New Haven - Age: 19 - Released on 1 May 1815.

Brown, Samuel - Seaman - Nbr: 3401 - Prize: Thomas - Ship Type: P - How taken: HMS Nymphe - When taken: 24 Jun 1813 - Where taken: off Halifax - Date Received: 13 Sep 1814 - From what ship: HMT Niobe, Chatham - Born: Massachusetts - Age: 25 - Released on 28 May 1815.

Brown, Samuel - Seaman - Nbr: 2819 - How taken: Gave himself up from HMS Malta - When taken: 18 Mar 1813 - Date Received: 24 Aug 1814 - From what ship: HMT Liverpool, Chatham - Born: Georgetown - Age: 30 - Released on 19 May 1815.

Brown, Sandy - Seaman - Nbr: 965 - Prize: Siro - Ship Type: LM - How taken: HMS Pelican - When taken: 13 Jan 1814 - Where taken: at sea - Date Received: 31 Jan 1814 - From what ship: Plymouth - Born: New York - Age: 25 - Released on 27 Apr 1815.

Brown, Sawyer - Seaman - Nbr: 3256 - How taken: Gave himself up from HMS Invincible - When taken: 1 Feb 1813 - Date Received: 11 Sep 1814 - From what ship: HMT Freya, Chatham - Born: New York - Age: 37 - Race: Black - Released on 28 May 1815.

Brown, Seth - Seaman - Nbr: 2086 - Prize: True Blooded Yankee - Ship Type: P - How taken: HMS Hope - When taken: 24 Jun 1813 - Where taken: off Brest - Date Received: 3 Aug 1814 - From what ship: HMS Bittern, Chatham Depot - Born: Bristol - Age: 23 - Released on 2 May 1815.

Brown, Stephen - 4th Lieutenant - Nbr: 938 - Prize: Siro - Ship Type: LM - How taken: HMS Pelican - When taken: 13 Jan 1814 - Where taken: at sea - Date Received: 31 Jan 1814 - From what ship: Plymouth - Born: Laban - Age: 22 - Released on 27 Apr 1815.

Brown, Thomas - Seaman - Nbr: 2195 - Prize: Olio - Ship Type: Prize - How taken: HMS Cyane - When taken: 22 May 1814 - Where taken: off NY - Date Received: 16 Aug 1814 - From what ship: HMS Dublin, Halifax - Born: Pennsylvania - Age: 24 - Released on 2 May 1815.

American Prisoners of War Held at Dartmoor during the War of 1812

## Alphabetical listing of names

Brown, Thomas - Seaman - Nbr: 1225 - How taken: Sent into custody from HMS Trident - When taken: 28 Oct 1812 - Date Received: 14 Jun 1814 - From what ship: Mill Prison (Plymouth, England) - Born: Philadelphia - Age: 34 - Released on 26 Apr 1815.

Brown, Thomas - Seaman - Nbr: 5238 - Prize: Polly - Ship Type: MV - How taken: HMS Maidstone - When taken: 18 Jul 1813 - Where taken: at sea - Date Received: 31 Oct 1814 - From what ship: HMT Mermaid, Chatham - Born: Rhode Island - Age: 24 - Released on 29 Jun 1815.

Brown, Thomas - Seaman - Nbr: 3876 - Prize: Fame - Ship Type: MV - How taken: HMS Niemen - When taken: 21 May 1814 - Where taken: off Cape Hatteras - Date Received: 5 Oct 1814 - From what ship: HMT Orpheus, Halifax - Born: Newburyport - Age: 38 - Released on 9 Jun 1815.

Brown, Thomas - Seaman - Nbr: 3220 - How taken: Gave himself up from HMS Orpheus - When taken: 10 Dec 1812 - Date Received: 11 Sep 1814 - From what ship: HMT Freya, Chatham - Born: New Jersey - Age: 31 - Released on 27 Apr 1815.

Brown, Thomas - Seaman - Nbr: 3417 - Prize: Fox - Ship Type: P - How taken: Manley - When taken: 1 Aug 1813 - Where taken: off Halifax - Date Received: 13 Sep 1814 - From what ship: HMT Niobe, Chatham - Born: New York - Age: 23 - Released on 28 May 1815.

Brown, Thomas - Boy - Nbr: 3061 - Prize: Polly - Ship Type: MV - How taken: HMS Maidstone - When taken: 20 Jul 1813 - Where taken: Grand Banks - Date Received: 2 Sep 1814 - From what ship: HMT Hydra, Chatham - Born: Salem - Age: 16 - Released on 28 May 1815.

Brown, Thomas - Seaman - Nbr: 2693 - Prize: Fox - Ship Type: MV - How taken: HMS Manly - When taken: 1 Aug 1813 - Where taken: off Halifax - Date Received: 21 Aug 1814 - From what ship: HMT Freya, Chatham - Born: New York - Age: 23 - Released on 19 May 1815.

Brown, Thomas - Seaman - Nbr: 2883 - How taken: Gave himself up from HMS Repulse - When taken: 27 May 1813 - Date Received: 24 Aug 1814 - From what ship: HMT Alpheus, Chatham - Born: Rhode Island - Age: 40 - Released on 19 May 1815.

Brown, William - Cook - Nbr: 1727 - Prize: Tom - Ship Type: LM - How taken: HMS Surveillante - When taken: 24 Apr 1813 - Where taken: Bay of Biscay - Date Received: 12 Jul 1814 - From what ship: Bristol - Born: New York - Age: 34 - Race: Negro - Released on 1 May 1815.

Brown, William - Seaman - Nbr: 4628 - How taken: Gave himself up from HMS Ulysses - When taken: 13 Jan 1813 - Date Received: 9 Oct 1814 - From what ship: HMT Leyden, Chatham - Born: New York - Age: 20 - Died on 20 Jul 1815 from phthisis pulmonalis.

Brown, William - Seaman - Nbr: 6479 - Prize: Mary - Ship Type: P - How taken: Lapphin - When taken: 15 Jun 1814 - Where taken: off Saint Marys - Date Received: 3 Mar 1815 - From what ship: HMT Ganges, Plymouth - Born: Virginia - Age: 19 - Released on 11 Jul 1815.

Brown, William - Seaman - Nbr: 694 - How taken: Impressed - When taken: 27 Aug 1813 - Date Received: 27 Sep 1813 - From what ship: Plymouth - Born: Philadelphia - Age: 27 - Released on 26 Apr 1815.

Brown, William - Seaman - Nbr: 228 - Prize: Charlotte - Ship Type: MV - How taken: HMS Warspite - When taken: 3 Mar 1813 - Where taken: Bay of Biscay - Date Received: 2 Apr 1813 - From what ship: Plymouth - Born: New York - Age: 30 - Released on 20 Apr 1815.

Brown, William - Carpenter - Nbr: 1967 - Prize: Orbit - Ship Type: MV - How taken: HMS Achates - When taken: 29 Jan 1813 - Where taken: Lat 44N Long 13W - Date Received: 3 Aug 1814 - From what ship: HMS Lyffey, Chatham Depot - Born: New London - Age: 43 - Released on 2 May 1815.

Brown, William - Seaman - Nbr: 692 - How taken: Impressed at Liverpool - When taken: 8 Sep 1813 - Date rec'd: 27 Sep 1813 - From what ship: Plymouth - Born: New Castle - Age: 25 - Released on 26 Apr 1815.

Brown, William - Seaman - Nbr: 1709 - How taken: Sent into custody from HMS Leyden - Date Received: 8 Jul 1814 - From what ship: Labrador - Born: Boston - Age: 28 - Released on 1 May 1815.

Brown, William - Seaman - Nbr: 539 - Prize: Matilda - Ship Type: MV - How taken: HMS Revolutionnaire - When taken: 27 Jul 1813 - Where taken: off L'Orient (France) - Date Received: 8 Sep 1813 - From what ship: Plymouth - Born: Virginia - Age: 20 - Sent to Plymouth on 7 Dec 1813.

Brown, William - Seaman - Nbr: 3692 - Prize: Alfred - Ship Type: P - How taken: HMS Epervier - When taken: 23 Feb 1814 - Where taken: off Newfoundland - Date Received: 30 Sep 1814 - From what ship: HMT President, Halifax - Born: Marblehead - Age: 21 - Released on 4 Jun 1815.

Brown, William - Seaman - Nbr: 3030 - Prize: US Sloop Frolic - Ship Type: MW - How taken: HMS Orpheus - When taken: 20 Apr 1814 - Where taken: off Cuba - Date Received: 2 Sep 1814 - From what ship: Naval Hospital, Plymouth - Born: New Haven - Age: 24 - Released on 3 May 1815.

Brown, William - Seaman - Nbr: 3332 - Prize: Polly - Ship Type: P - How taken: HMS Ringdove - When taken: 27

American Prisoners of War Held at Dartmoor during the War of 1812

## Alphabetical listing of names

Jul 1813 - Where taken: at sea - Date Received: 13 Sep 1814 - From what ship: HMT Niobe, Chatham - Born: Salem - Age: 21 - Released on 28 May 1815.

Brown, William - Seaman - Nbr: 2438 - Prize: US Sloop Frolic - Ship Type: MW - How taken: HMS Orpheus - When taken: 20 Apr 1814 - Where taken: off Cuba - Date Received: 16 Aug 1814 - From what ship: HMT Queen, Halifax - Born: Philadelphia - Age: 23 - Released on 3 May 1815.

Brown, William - Seaman - Nbr: 2480 - Prize: US Sloop Frolic - Ship Type: MW - How taken: HMS Orpheus - When taken: 20 Apr 1814 - Where taken: off Cuba - Date Received: 16 Aug 1814 - From what ship: HMT Queen, Halifax - Born: Baltimore - Age: 25 - Released on 3 May 1815.

Brown, William - Seaman - Nbr: 3951 - Prize: Rolla - Ship Type: P - How taken: HMS Loire - When taken: 10 Dec 1813 - Where taken: off Bull Island (South Carolina) - Date Received: 5 Oct 1814 - From what ship: HMT President, Halifax - Born: Rhode Island - Age: 19 - Released on 9 Jun 1815.

Brown, Charles - Seaman - Nbr: 3330 - Prize: York Town - Ship Type: P - How taken: HMS Nimrod - When taken: 17 Jul 1813 - Where taken: off St. Johns - Date Received: 13 Sep 1814 - From what ship: HMT Niobe, Chatham - Born: New York - Age: 28 - Released on 28 May 1815.

Brownell, Jonathan - Seaman - Nbr: 5645 - Prize: Lion - Ship Type: P - How taken: HMS Granicus - When taken: 2 Dec 1814 - Where taken: off Lisbon - Date Received: 24 Dec 1814 - From what ship: HMT Tay - Born: Newport - Age: 29 - Released on 3 Jul 1815.

Brownell, Paul - Lieutenant - Nbr: 3941 - Prize: Rolla - Ship Type: P - How taken: HMS Loire - When taken: 10 Dec 1813 - Where taken: off Bull Island (South Carolina) - Date Received: 5 Oct 1814 - From what ship: HMT President, Halifax - Born: Rhode Island - Age: 46 - Released on 9 Jun 1815.

Brownell, Richard - Seaman - Nbr: 299 - Prize: Ducornau - Ship Type: MV - How taken: HMS Pheasant - When taken: 15 Mar 1813 - Where taken: Bay of Biscay - Date Received: 28 Jun 1813 - From what ship: Plymouth - Born: Massachusetts - Age: 29 - Released on 20 Apr 1815.

Bruce, Joseph - Marine - Nbr: 4039 - Prize: US Brig Rattlesnake - Ship Type: MW - How taken: HMS Leander - When taken: 13 Jul 1814 - Where taken: off Shelburne - Date Received: 6 Oct 1814 - From what ship: HMT Chesapeake, Halifax - Born: Marblehead - Age: 47 - Released on 13 Jun 1815.

Bruce, Peter - Seaman - Nbr: 6040 - Prize: Regent - Ship Type: LM - How taken: HMS Forth - When taken: 19 Sep 1814 - Where taken: off Egg Harbor (New Jersey) - Date Received: 28 Dec 1814 - From what ship: HMT Penelope - Born: New York - Age: 18 - Released on 3 Jul 1815.

Brumbelcum, Merrit - Seaman - Nbr: 3679 - Prize: Alfred - Ship Type: P - How taken: HMS Epervier - When taken: 23 Feb 1812 - Where taken: off Newfoundland - Date Received: 30 Sep 1814 - From what ship: HMT President - Born: Marblehead - Age: 48 - Released on 4 Jun 1815.

Brush, Able - Seaman - Nbr: 5226 - Prize: York Town - Ship Type: P - How taken: HMS Maidstone - When taken: 18 Jul 1813 - Where taken: Grand Banks - Date Received: 31 Oct 1814 - From what ship: HMT Mermaid, Chatham - Born: Boston - Age: 28 - Released on 29 Jun 1815.

Brush, Caesar - Seaman - Nbr: 3875 - Prize: Tickler - Ship Type: MV - How taken: HMS Saturn - When taken: 13 Jul 1814 - Where taken: off America - Date Received: 5 Oct 1814 - From what ship: HMT Orpheus, Halifax - Born: Middleton - Age: 27 - Race: Negro - Released on 9 Jun 1815.

Brush, Samuel - Seaman - Nbr: 2950 - Prize: Rattlesnake - Ship Type: LM - How taken: HMS Rhin - When taken: 10 Mar 1814 - Where taken: West Indies - Date Received: 24 Aug 1814 - From what ship: HMT Hannibal - Born: New York - Age: 27 - Released on 10 Apr 1814.

Brush, Thomas - Seaman - Nbr: 2694 - Prize: Industry - Ship Type: MV - How taken: HMS Heron - When taken: 2 Nov 1813 - Where taken: off Halifax - Date Received: 21 Aug 1814 - From what ship: HMT Freya, Chatham - Born: Marblehead - Age: 54 - Released on 19 May 1815.

Brutus, Mario - Seaman - Nbr: 1013 - Prize: Rachel & Ann - Ship Type: P - How taken: HMS Cydnus - When taken: 6 Jan 1814 - Where taken: at sea - Date Received: 31 Jan 1814 - From what ship: Plymouth - Born: New Orleans - Age: 29 - Sent to Plymouth on 8 Jul 1814.

Bryan, John - Seaman - Nbr: 5741 - Prize: US Schooner Somers - Ship Type: MW - How taken: British gunboats - When taken: 12 Aug 1814 - Where taken: Fort Erie - Date Received: 26 Dec 1814 - From what ship: HMT Argo - Born: New York - Age: 17 - Released on 3 Jul 1815.

Bryant, Benjamin - Seaman - Nbr: 6288 - How taken: Gave himself up from HMS Quebec - Date Received: 17 Feb 1815 - From what ship: HMT Ganges, Plymouth - Born: Massachusetts - Age: 18 - Released on 5 Jul 1815.

Bryant, James - Seaman - Nbr: 1845 - Prize: Polly - Ship Type: MV - How taken: HMS Surveillante - When taken: 23 Mar 1813 - Where taken: Bay of Biscay - Date Received: 21 Jul 1814 - From what ship: HMT Redbeard & Pincher, Chatham Depot - Born: Beverly - Age: 18 - Released on 2 May 1815.

American Prisoners of War Held at Dartmoor during the War of 1812

## Alphabetical listing of names

Bryant, Stephen - Seaman - Nbr: 1322 - How taken: Impressed at Bristol - When taken: 28 Jun 1813 - Date Received: 19 Jun 1814 - From what ship: Stapleton - Born: Not listed - Age: 22 - Released on 28 Apr 1815.

Buchan, John - Seaman - Nbr: 706 - Prize: Ned - Ship Type: LM - How taken: HMS Royalist - When taken: 6 Sep 1813 - Where taken: Bay of Biscay - Date Received: 27 Sep 1813 - From what ship: Plymouth - Born: Boston - Age: 24 - Released on 26 Apr 1815.

Buchannan, Robert - Seaman - Nbr: 2994 - Prize: Frolic - Ship Type: P - How taken: HMS Heron - When taken: 25 Jan 1814 - Where taken: off St. Thomas - Date Received: 29 Aug 1814 - From what ship: HMT Bittern - Born: Philadelphia - Age: 30 - Race: Black - Released on 11 Jul 1815.

Buckley, Joseph - Seaman - Nbr: 3996 - Prize: US Brig Rattlesnake - Ship Type: MW - How taken: HMS Leander - When taken: 13 Jul 1814 - Where taken: off Shelburne - Date Received: 6 Oct 1814 - From what ship: HMT Chesapeake, Halifax - Born: Maryland - Age: 23 - Released on 9 Jun 1815.

Buckley, William - Lieutenant - Nbr: 3558 - Prize: Hawk - Ship Type: P - How taken: HMS Pique - When taken: 26 Apr 1814 - Where taken: off Bermuda - Date Received: 30 Sep 1814 - From what ship: HMT Sybella - Born: Maryland - Age: 54 - Released on 4 Jun 1815.

Buckling, Warren - Seaman - Nbr: 4106 - Prize: Stark - Ship Type: P - How taken: HMS Sophie - When taken: 20 Apr 1814 - Where taken: off Bermuda - Date Received: 6 Oct 1814 - From what ship: HMT Chesapeake, Halifax - Born: Rhode Island - Age: 35 - Released on 13 Jun 1815.

Budd, John - Seaman - Nbr: 4086 - Prize: New Zealander - Ship Type: Prize - How taken: HMS Belviders - When taken: 21 Apr 1814 - Where taken: off Delaware - Date Received: 6 Oct 1814 - From what ship: HMT Chesapeake, Halifax - Born: New York - Age: 31 - Released on 13 Jun 1815.

Budwell, Alpheus - Seaman - Nbr: 2352 - Prize: General Starks - Ship Type: P - How taken: HMS Sophie - When taken: 24 Apr 1814 - Where taken: off Bermuda - Date Received: 16 Aug 1814 - From what ship: HMS Dublin, Halifax - Born: Massachusetts - Age: 22 - Released on 3 May 1815.

Bull, Henry - Supercargo - Nbr: 4925 - Prize: Herald - Ship Type: P - How taken: HMS Endymion - When taken: 15 Aug 1814 - Where taken: off Nantucket - Date Received: 28 Oct 1814 - From what ship: HMT Alkbar, Halifax - Born: Connecticut - Age: 20 - Released on 17 Apr 1815.

Bumbelcomb, Sewell - Seaman - Nbr: 3693 - Prize: Alfred - Ship Type: P - How taken: HMS Epervier - When taken: 23 Feb 1814 - Where taken: off Newfoundland - Date Received: 30 Sep 1814 - From what ship: HMT President, Halifax - Born: Marblehead - Age: 22 - Released on 4 Jun 1815.

Bump, William - Seaman - Nbr: 5445 - How taken: Impressed at Liverpool - When taken: Sep 1814 - Date rec'd: 10 Dec 1814 - From what ship: HMT Impregnable - Born: Lime - Age: 28 - Released on 1 Jul 1815.

Bunharm, Jeremiah - Seaman - Nbr: 5962 - Prize: Amazon - Ship Type: Prize - How taken: HMS Bulwark - When taken: 22 Sep 1814 - Where taken: Georges Bank - Date Received: 27 Dec 1814 - From what ship: HMT Penelope - Born: Massachusetts - Age: 34 - Released on 3 Jul 1815.

Bunker, Isaiah - Prize Master - Nbr: 776 - Prize: Betsy - Ship Type: MV - How taken: HMS Eurotas - When taken: 26 Oct 1813 - Where taken: off Ushant (France) - Date Received: 3 Nov 1813 - From what ship: Plymouth - Born: Newcastle - Age: 23 - Escaped on 12 Mar 1815.

Bunker, James - Seaman - Nbr: 257 - Prize: William Bayard - Ship Type: MV - How taken: HMS Warspite - When taken: 3 Mar 1813 - Where taken: Bay of Biscay - Date Received: 28 Jun 1813 - From what ship: Plymouth - Born: Massachusetts - Age: 27 - Released on 20 Apr 1815.

Bunker, Robert - Seaman - Nbr: 3029 - Prize: US Sloop Frolic - Ship Type: MW - How taken: HMS Orpheus - When taken: 20 Apr 1814 - Where taken: off Cuba - Date Received: 2 Sep 1814 - From what ship: Naval Hospital, Plymouth - Born: Charleston - Age: 36 - Released on 28 May 1815.

Bunker, Thomas - Seaman - Nbr: 3156 - Prize: Fame (Whaler) - Ship Type: MV - How taken: HMS Cressy - When taken: 20 Jul 1813 - Where taken: at sea - Date Received: 11 Sep 1814 - From what ship: HMT Freya, Chatham - Born: Nantucket - Age: 19 - Released on 28 May 1815.

Bunkerson, John - Seaman - Nbr: 6173 - Prize: Lion - Ship Type: P - How taken: HMS Granicus - When taken: 2 Dec 1814 - Where taken: off Lisbon - Date Received: 18 Jan 1815 - From what ship: HMT Impregnable - Born: Savannah - Age: 20 - Released on 5 Jul 1815.

Bunkerson, John - Seaman - Nbr: 995 - Prize: Apparencen - Ship Type: MV - How taken: HMS Castilian - When taken: 27 Jan 1814 - Where taken: off Ushant (France) - Date Received: 31 Jan 1814 - From what ship: Plymouth - Born: New Orleans - Age: 18 - Sent to Mill Prison (Plymouth, England) on 26 Jun 1814.

Bunnel, Benjamin - Seaman - Nbr: 1463 - Prize: Revenge - Ship Type: LM - How taken: HMS Belle Poule - When taken: 10 Mar 1813 - Where taken: off Cornwall - Date Received: 19 Jun 1814 - From what ship: Stapleton - Born: New Haven - Age: 34 - Released on 28 Apr 1815.

American Prisoners of War Held at Dartmoor during the War of 1812

## Alphabetical listing of names

Bunyan, John - Seaman - Nbr: 3718 - Prize: Alfred - Ship Type: P - How taken: HMS Epervier - When taken: 23 Feb 1814 - Where taken: off Newfoundland - Date Received: 30 Sep 1814 - From what ship: HMT President, Halifax - Born: Marblehead - Age: 23 - Released on 4 Jun 1815.

Burbage, Henry - Seaman - Nbr: 3582 - Prize: Greyhound - Ship Type: MV - How taken: Amith (Privateer) - When taken: 5 Jun 1814 - Where taken: off Bermuda - Date Received: 30 Sep 1814 - From what ship: HMT Sybella - Born: Washington - Age: 26 - Died on 25 Dec 1814 from perissneuoniria.

Burbalk, Moses - 1st Lieutenant - Nbr: 5573 - Prize: McDonough - Ship Type: P - How taken: HMS Bacchante - When taken: 1 Nov 1814 - Where taken: Lat 42 Long 67 - Date Received: 17 Dec 1814 - From what ship: HMT Loire, Halifax - Born: Arundel - Age: 28 - Released on 10 Apr 1815.

Burchstad, John - Seaman - Nbr: 6212 - Prize: Prince de Neufchatel - Ship Type: P - How taken: Leander (Newcastle Acasta) - When taken: 20 Dec 1814 - Where taken: Lat 38 Long 56 - Date Received: 30 Jan 1815 - From what ship: HMT Pheasant - Born: Beverly - Age: 28 - Released on 5 Jul 1815.

Burdeen, Joseph - Seaman - Nbr: 5897 - Prize: Harlequin - Ship Type: P - How taken: HMS Bulwark - When taken: 23 Nov 1814 - Where taken: off Halifax - Date Received: 27 Dec 1814 - From what ship: HMT Penelope - Born: New Castle - Age: 20 - Released on 3 Jul 1815.

Burdge, Samuel - Cabin boy - Nbr: 311 - Prize: Ducornau - Ship Type: MV - How taken: HMS Pheasant - When taken: 15 Mar 1813 - Where taken: Bay of Biscay - Date Received: 28 Jun 1813 - From what ship: Plymouth - Born: New York - Age: 20 - Released on 9 Apr 1815.

Burdock, Enos - Seaman - Nbr: 3164 - How taken: Impressed at Gravesend - When taken: 11 Aug 1813 - Date rec'd: 11 Sep 1814 - From what ship: HMT Freya, Chatham - Born: Rhode Island - Age: 22 - Released on 11 Jul 1815.

Burford, William - Seaman - Nbr: 1279 - How taken: Sent into custody from HMS Nerus - Date Received: 14 Jun 1814 - From what ship: Mill Prison (Plymouth, England) - Born: Philadelphia - Age: 26 - Released on 28 Apr 1815.

Burgen, William - Seaman - Nbr: 5495 - Prize: General Putnam - Ship Type: P - How taken: HMS Leander - When taken: 8 Nov 1814 - Where taken: Long 65 Lat 42 - Date Received: 17 Dec 1814 - From what ship: HMT Loire, Halifax - Born: Marblehead - Age: 46 - Released on 1 Jul 1815.

Burger, Francis - Seaman - Nbr: 5146 - Prize: Thorn - Ship Type: P - How taken: HMS Shannon - When taken: 7 Nov 1813 - Where taken: off Newfoundland - Date Received: 31 Oct 1814 - From what ship: HMT Mermaid, Chatham - Born: Marblehead - Age: 19 - Released on 29 Jun 1815.

Burgess, John - Seaman - Nbr: 2430 - Prize: US Sloop Frolic - Ship Type: MW - How taken: HMS Orpheus - When taken: 20 Apr 1814 - Where taken: off Cuba - Date Received: 16 Aug 1814 - From what ship: HMT Queen, Halifax - Born: Massachusetts - Age: 20 - Released on 3 May 1815.

Burke, David - Seaman - Nbr: 2579 - How taken: Gave himself up at La Hague - When taken: 8 Feb 1813 - Date Received: 21 Aug 1814 - From what ship: HMT Salvador del Mundo - Born: Providence - Age: 24 - Race: Black - Released on 3 May 1815.

Burke, James - Seaman - Nbr: 4067 - Prize: Leicester - Ship Type: Prize - How taken: HMS Prometheus - When taken: Jun 1814 - Where taken: off Halifax - Date Received: 6 Oct 1814 - From what ship: HMT Chesapeake, Halifax - Born: Beverly - Age: 32 - Released on 13 Jun 1815.

Burke, John - Seaman - Nbr: 4733 - How taken: Gave himself up from HMS Bulwark - When taken: 28 Dec 1813 - Date Received: 9 Oct 1814 - From what ship: HMT Freya, Chatham - Born: Washington - Age: 23 - Released on 15 Jun 1815.

Burn, Reuben - Seaman - Nbr: 467 - Prize: Meteor - Ship Type: MV - How taken: HMS Briton - When taken: 12 Mar 1813 - Where taken: Bay of Biscay - Date Received: 8 Sep 1813 - From what ship: Plymouth - Born: New York - Age: 16 - Released on 20 Apr 1815.

Burn, Rufus - Prize master - Nbr: 4087 - Prize: Nelly - Ship Type: Prize - How taken: HMS Bulwark - When taken: 10 Jul 1814 - Where taken: Georges Bank - Date Received: 6 Oct 1814 - From what ship: HMT Chesapeake, Halifax - Born: Rhode Island - Age: 50 - Released on 13 Jun 1815.

Burnes, Robert - Seaman - Nbr: 40 - How taken: Apprehended at Gibraltar - When taken: 8 Jul 1813 - Date rec'd: 2 Apr 1813 - From what ship: Plymouth - Born: Rehoboth - Age: 23 - Sent to Plymouth on 15 Jun 1813.

Burnet, Charles - Seaman - Nbr: 1496 - Prize: Price - Ship Type: MV - How taken: HMS Iris - When taken: 13 Apr 1813 - Where taken: Bay of Biscay - Date Received: 19 Jun 1814 - From what ship: Mill Prison (Plymouth, England) - Born: New York - Age: 24 - Released on 1 May 1815.

Burnham, Abraham - Surgeon - Nbr: 6309 - Prize: Prince de Neufchatel - Ship Type: P - How taken: Leander (Newcastle Acasta) - When taken: 20 Dec 1814 - Where taken: Lat 38 Long 56 - Date rec'd: 19 Feb 1815 -

American Prisoners of War Held at Dartmoor during the War of 1812

## Alphabetical listing of names

From what ship: HMT Ganges, Plymouth - Born: Ipswich - Age: 30 - Sent to Ashburton (England) on 24 Feb 1815.

Burnham, Benjamin - Seaman - Nbr: 4126 - Prize: Bordeaux Packet - Ship Type: LM - How taken: HMS Niemen - When taken: 28 Jun 1814 - Where taken: off Delaware - Date Received: 6 Oct 1814 - From what ship: HMT Chesapeake, Halifax - Born: North Carolina - Age: 33 - Released on 13 Jun 1815.

Burnham, David - Seaman - Nbr: 6164 - Prize: Lion - Ship Type: P - How taken: HMS Granicus - When taken: 2 Dec 1814 - Where taken: off Lisbon - Date Received: 18 Jan 1815 - From what ship: HMT Impregnable - Born: Ipswich - Age: 50 - Released on 5 Jul 1815.

Burnham, David - Seaman - Nbr: 1941 - How taken: Gave himself up from HMS Albacore - When taken: 3 Feb 1813 - Date Received: 3 Aug 1814 - From what ship: HMS Alceste, Chatham Depot - Born: Ipswich - Age: 50 - Released on 2 May 1815.

Burnham, Enoch - Seaman - Nbr: 1491 - Prize: Essex - Ship Type: MV - How taken: HMS Pyramus - When taken: 2 Apr 1813 - Where taken: Bay of Biscay - Date Received: 19 Jun 1814 - From what ship: Stapleton - Born: Boston - Age: 22 - Released on 1 May 1815.

Burnham, James - Seaman - Nbr: 3971 - Prize: Rolla - Ship Type: P - How taken: HMS Loire - When taken: 10 Dec 1813 - Where taken: off Bull Island (South Carolina) - Date Received: 5 Oct 1814 - From what ship: HMT President, Halifax - Born: Vermont - Age: 22 - Race: Negro - Released on 9 Jun 1815.

Burnham, John - Seaman - Nbr: 2594 - How taken: Gave himself up from HMS Bold - When taken: 15 Feb 1813 - Date Received: 21 Aug 1814 - From what ship: HMT Freya, Chatham - Born: Boothbay - Age: 22 - Released on 3 May 1815.

Burns, Alexander - Boatswain - Nbr: 2279 - Prize: Diomede - Ship Type: P - How taken: HMS Rifleman - When taken: 28 May 1814 - Where taken: off Sable Island - Date Received: 16 Aug 1814 - From what ship: HMS Dublin, Halifax - Born: New York - Age: 23 - Released on 3 May 1815.

Burns, Benjamin - Boy - Nbr: 5986 - Prize: McDonough - Ship Type: P - How taken: HMS Bacchante - When taken: 1 Nov 1814 - Where taken: Lat 42 Long 67 - Date Received: 27 Dec 1814 - From what ship: HMT Penelope - Born: Portland - Age: 16 - Released on 19 May 1815.

Burns, Charles - Seaman - Nbr: 1563 - Prize: Caroline - Ship Type: MV - How taken: HMS Medusa - When taken: 12 Apr 1813 - Where taken: Bay of Biscay - Date Received: 23 Jun 1814 - From what ship: Stapleton - Born: Salem - Age: 22 - Race: Negro - Released on 1 May 1815.

Burr, Isaac - Seaman - Nbr: 397 - Prize: Young Holkar - Ship Type: MV - How taken: HMS Superb - When taken: 10 Apr 1813 - Where taken: off Belle Isle, France - Date Received: 1 Jul 1813 - From what ship: Plymouth - Born: Connecticut - Age: 19 - Released on 20 Apr 1815.

Burrell, Jesse - Seaman - Nbr: 4669 - Prize: Industry - Ship Type: P - How taken: HMS Heron - When taken: 3 Nov 1813 - Where taken: off Halifax - Date Received: 9 Oct 1814 - From what ship: HMT Leyden, Chatham - Born: Massachusetts - Age: 22 - Released on 15 Jun 1815.

Burrell, Michael - Boy - Nbr: 2391 - Prize: US Sloop Frolic - Ship Type: MW - How taken: HMS Orpheus - When taken: 20 Apr 1814 - Where taken: off Cuba - Date Received: 16 Aug 1814 - From what ship: HMT Queen, Halifax - Born: Salem - Age: 18 - Sent to Dartmouth on 19 Oct 1814.

Burrell, Ryal - Seaman - Nbr: 2058 - How taken: Gave himself up from HMS Latond - When taken: 21 Mar 1812 - Date Received: 3 Aug 1814 - From what ship: HMS Bittern, Chatham Depot - Born: Boston - Age: 24 - Released on 26 Apr 1815.

Burridge, Robert - Gunner - Nbr: 3423 - Prize: Industry - Ship Type: P - How taken: HMS Heron - When taken: 3 Nov 1813 - Where taken: off Halifax - Date Received: 13 Sep 1814 - From what ship: HMT Niobe, Chatham - Born: Marblehead - Age: 46 - Released on 28 May 1815.

Burrow, Charles - Seaman - Nbr: 3087 - Prize: Eliza (Whaler) - Ship Type: MV - How taken: HMS Charles - When taken: 20 Jul 1813 - Where taken: off Peter Head - Date Received: 3 Sep 1814 - From what ship: HMT Hydra, Chatham - Born: New York - Age: 25 - Released on 28 May 1815.

Burt, Thomas - Seaman - Nbr: 3918 - Prize: Rolla - Ship Type: P - How taken: HMS Loire - When taken: 10 Dec 1813 - Where taken: off Bull Island (South Carolina) - Date Received: 5 Oct 1814 - From what ship: HMT President, Halifax - Born: Newport - Age: 21 - Released on 9 Jun 1815.

Burton, John - Seaman - Nbr: 319 - Prize: Pallas - Ship Type: MV - How taken: Rebuff - When taken: 23 Dec 1812 - Where taken: off Cadiz - Date Received: 28 Jun 1813 - From what ship: Plymouth - Born: Trieste - Age: 29 - Sent to Dartmouth on 30 Jul 1813.

Burton, William - Seaman - Nbr: 6286 - How taken: Gave himself up from HMS Namur - Date Received: 17 Feb 1815 - From what ship: HMT Ganges, Plymouth - Born: Roxborough - Age: 55 - Released on 28 Apr 1815.

American Prisoners of War Held at Dartmoor during the War of 1812

## Alphabetical listing of names

Burton, William - Seaman - Nbr: 2757 - How taken: Gave himself up from HMS Blake - When taken: 10 Dec 1812 - Date Received: 24 Aug 1814 - From what ship: HMT Liverpool, Chatham - Born: Kent County - Age: 52 - Released on 26 Apr 1815.

Burts, Joseph - Seaman - Nbr: 1060 - Prize: Fair American - Ship Type: MV - How taken: HMS Andromache - When taken: 19 Jan 1814 - Where taken: Bay of Biscay - Date Received: 10 May 1814 - From what ship: Plymouth - Born: New York - Age: 22 - Died on 2 Dec 1814, hanged himself in prison Nbr 5.

Bush, George - Seaman - Nbr: 3626 - Prize: Monarch - Ship Type: MV - How taken: HMS Dotterel - When taken: 14 Dec 1813 - Where taken: off Charleston - Date Received: 30 Sep 1814 - From what ship: HMT Sybella - Born: Virginia - Age: 31 - Released on 4 Jun 1815.

Bushfield, James - Seaman - Nbr: 4563 - Prize: Volunteer - Ship Type: MV - How taken: Victoria (Privateer) - When taken: 26 Dec 1813 - Where taken: at sea - Date Received: 8 Oct 1814 - From what ship: HMT Leyden, Chatham - Born: New York - Age: 27 - Released on 15 Jun 1815.

Bushnell, Joseph - Seaman - Nbr: 4965 - Prize: Herald - Ship Type: P - How taken: HMS Endymion - When taken: 15 Aug 1814 - Where taken: off Nantucket - Date Received: 28 Oct 1814 - From what ship: HMT Alkbar, Halifax - Born: Norwich - Age: 28 - Released on 21 Jun 1815.

Buskell, William - Seaman - Nbr: 5384 - Prize: Rattlesnake - Ship Type: LM - How taken: HMS Rhin - When taken: 17 Mar 1814 - Where taken: off Bermuda - Date Received: 31 Oct 1814 - From what ship: HMT Leyden, Chatham - Born: Baltimore - Age: 40 - Released on 1 Jul 1815.

Bustin, John - Seaman - Nbr: 1832 - Prize: Lightning - Ship Type: MV - How taken: HMS Medusa - When taken: 2 Apr 1813 - Where taken: Bay of Biscay - Date Received: 21 Jul 1814 - From what ship: HMT Redbeard & Pincher, Chatham Depot - Born: New York - Age: 20 - Released on 1 May 1815.

Butcher, Jacob - Seaman - Nbr: 6416 - Prize: Enterprize - Ship Type: P - How taken: HMS Argo - When taken: 16 Nov 1814 - Where taken: off San Domingo - Date Received: 3 Mar 1815 - From what ship: HMT Ganges, Plymouth - Born: Massachusetts - Age: 23 - Race: Mulatto - Released on 11 Jul 1815.

Butler, Benjamin - Seaman - Nbr: 2530 - Prize: Hussar - Ship Type: P - How taken: HMS Saturn - When taken: 25 May 1814 - Where taken: off Sandy Hook - Date Received: 16 Aug 1814 - From what ship: HMT Queen, Halifax - Born: Baltimore - Age: 25 - Race: Black - Released on 3 May 1815.

Butler, Benjamin - Seaman - Nbr: 2484 - Prize: US Sloop Frolic - Ship Type: MW - How taken: HMS Orpheus - When taken: 20 Apr 1814 - Where taken: off Cuba - Date Received: 16 Aug 1814 - From what ship: HMT Queen, Halifax - Born: Salem - Age: 33 - Released on 3 May 1815.

Butler, Charles - Seaman - Nbr: 3019 - How taken: Gave himself up from HMS Zealous - When taken: 27 Aug 1814 - Date Received: 2 Sep 1814 - From what ship: HMT Centaur - Born: New York - Age: 27 - Released on 28 May 1815.

Butler, David - Prize master - Nbr: 6432 - Prize: Chance - Ship Type: P - How taken: HMS Statira - When taken: 1 Apr 1814 - Where taken: Lat 38 Long 24 - Date Received: 3 Mar 1815 - From what ship: HMT Ganges, Plymouth - Born: Boston - Age: 41 - Released on 11 Jul 1815.

Butler, George - Seaman - Nbr: 1863 - Prize: Governor Plumer - Ship Type: P - How taken: HMS Bold - When taken: 17 May 1813 - Where taken: off Newfoundland - Date Received: 29 Jul 1814 - From what ship: HMS Ville de Paris, Chatham Depot - Born: Berwick - Age: 17 - Released on 2 May 1815.

Butler, George - Seaman - Nbr: 1414 - Prize: Orders in Council - Ship Type: LM - How taken: HMS Surveillante - When taken: 1 Jun 1813 - Where taken: off Cape Ortegal (Spain) - Date Received: 19 Jun 1814 - From what ship: Stapleton - Born: Marblehead - Age: 28 - Released on 28 Apr 1815.

Butler, George - Seaman - Nbr: 2745 - How taken: Gave himself up from HMS Ceres - When taken: 20 Dec 1812 - Date Received: 24 Aug 1814 - From what ship: HMT Liverpool, Chatham - Born: Charleston - Age: 24 - Released on 26 Apr 1815.

Butler, Henry - Seaman - Nbr: 6528 - Prize: Vulture - Ship Type: MV - How taken: HMS Aurora - When taken: Mar 1814 - Where taken: off the Banks - Date Received: 7 Mar 1815 - From what ship: HMT Ganges, Plymouth - Born: Massachusetts - Age: 35 - Race: Black - Released on 11 Jul 1815.

Butler, Henry - Seaman - Nbr: 3248 - Prize: Wiley Reynard - Ship Type: P - How taken: HMS Shannon - When taken: 15 Aug 1812 - Where taken: off Halifax - Date Received: 11 Sep 1814 - From what ship: HMT Freya, Chatham - Born: New York - Age: 28 - Race: Mulatto - Released on 27 Apr 1815.

Butler, James - Seaman - Nbr: 5933 - Prize: Harlequin - Ship Type: P - How taken: HMS Bulwark - When taken: 23 Nov 1814 - Where taken: off Halifax - Date Received: 27 Dec 1814 - From what ship: HMT Penelope - Born: Virginia - Age: 36 - Released on 3 Jul 1815.

Butler, John - Seaman - Nbr: 2821 - How taken: Gave himself up from HMS Philomel - When taken: 28 Dec 1812 -

American Prisoners of War Held at Dartmoor during the War of 1812

## Alphabetical listing of names

Date Received: 24 Aug 1814 - From what ship: HMT Liverpool, Chatham - Born: Trallen - Age: 23 - Released on 26 Apr 1815.

Butler, John - Seaman - Nbr: 5626 - How taken: Gave himself up from HMS Semiramis - When taken: 20 Sep 1813 - Date Received: 24 Dec 1814 - From what ship: HMT Tay - Born: Pennsylvania - Age: 52 - Race: Black - Died on 23 Feb 1815 from variola.

Butler, Michael - Seaman - Nbr: 5910 - Prize: Harlequin - Ship Type: P - How taken: HMS Bulwark - When taken: 23 Nov 1814 - Where taken: off Halifax - Date Received: 27 Dec 1814 - From what ship: HMT Penelope - Born: Arundel - Age: 32 - Released on 3 Jul 1815.

Butler, Robert - Seaman - Nbr: 3634 - Prize: Monarch - Ship Type: MV - How taken: HMS Dotterel - When taken: 14 Dec 1813 - Where taken: off Charleston - Date Received: 30 Sep 1814 - From what ship: HMT Sybella - Born: Pennsylvania - Age: 36 - Race: Negro - Released on 4 Jun 1815.

Butler, Thomas - Seaman - Nbr: 4274 - How taken: Gave himself up from HMS Bulwark - When taken: 28 Dec 1813 - Date Received: 7 Oct 1814 - From what ship: HMT Niobe, Chatham - Born: Maryland - Age: 37 - Released on 13 Jun 1815.

Butler, William - Seaman - Nbr: 1995 - Prize: Kergettos - Ship Type: MV - How taken: HMS Castilian - When taken: 12 Mar 1813 - Where taken: off Cape Ortegal (Spain) - Date Received: 3 Aug 1814 - From what ship: HMS Lyffey, Chatham Depot - Born: Baltimore - Age: 40 - Released on 2 May 1815.

Butler, William - Seaman - Nbr: 4588 - Prize: Bunker Hill - Ship Type: P - How taken: HMS Pomone - When taken: 2 Mar 1814 - Where taken: Channel - Date Received: 9 Oct 1814 - From what ship: HMT Leyden, Chatham - Born: Newburyport - Age: 45 - Released on 15 Jun 1815.

Butnell, William - Seaman - Nbr: 644 - Prize: US Brig Argus - Ship Type: MW - How taken: HMS Pelican - When taken: 14 Aug 1813 - Where taken: Irish Channel - Date Received: 8 Sep 1813 - From what ship: Plymouth - Born: Bell Haven - Age: 17 - Sent to Dartmouth on 30 Jul 1813.

Buttman, Nathaniel - Seaman - Nbr: 4064 - Prize: US Brig Rattlesnake - Ship Type: MW - How taken: HMS Leander - When taken: 13 Jul 1814 - Where taken: off Shelburne - Date Received: 6 Oct 1814 - From what ship: HMT Chesapeake, Halifax - Born: Salem - Age: 22 - Released on 13 Jun 1815.

Buttman, Thomas - Seaman - Nbr: 4111 - Prize: Holstein - Ship Type: MV - How taken: HMS Belviders - When taken: 28 Mar 1813 - Where taken: off Newport - Date Received: 6 Oct 1814 - From what ship: HMT Chesapeake, Halifax - Born: Newbury - Age: 17 - Released on 13 Jun 1815.

Butts, Joseph - Seaman - Nbr: 3353 - How taken: Gave himself up from HMS Centaur - When taken: 10 Sep 1813 - Date Received: 13 Sep 1814 - From what ship: HMT Niobe, Chatham - Born: Boston - Age: 23 - Released on 28 May 1815.

Byard, Joseph - Seaman - Nbr: 1328 - Prize: Paul Jones - Ship Type: P - How taken: HMS Leonidas - When taken: 23 May 1813 - Where taken: Channel - Date Received: 19 Jun 1814 - From what ship: Stapleton - Born: Charouste, France - Age: 28 - Sent to Mill Prison (Plymouth, England) on 21 Jun 1814.

Byers, Sharp - Boy - Nbr: 2215 - Prize: Fiere Facia - Ship Type: P - How taken: HMS Ramillies - When taken: 27 Feb 1814 - Where taken: off NY - Date Received: 16 Aug 1814 - From what ship: HMS Dublin, Halifax - Born: New Jersey - Age: 17 - Race: Black - Released on 2 May 1815.

Byron, Ebenezer - Seaman - Nbr: 5330 - Prize: Portsmouth Packet - Ship Type: P - How taken: HMS Maidstone - When taken: 17 Jul 1813 - Where taken: Grand Banks - Date Received: 31 Oct 1814 - From what ship: HMT Leyden, Chatham - Born: Not legible - Age: 25 - Released on 29 Jun 1815.

Byron, John - Seaman - Nbr: 5965 - Prize: Amazon - Ship Type: Prize - How taken: HMS Bulwark - When taken: 22 Sep 1814 - Where taken: Georges Bank - Date Received: 27 Dec 1814 - From what ship: HMT Penelope - Born: Kent County - Age: 34 - Released on 3 Jul 1815.

Caban, Samuel - Seaman - Nbr: 2187 - Prize: Diomede - Ship Type: P - How taken: HMS Rifleman - When taken: 28 Jun 1813 - Where taken: off Halifax - Date Received: 16 Aug 1814 - From what ship: HMS Dublin, Halifax - Born: Salem - Age: 32 - Released on 2 May 1815.

Cadwell, Abraham - Seaman - Nbr: 3341 - How taken: Gave himself up from HMS Scorpion - When taken: 2 Dec 1812 - Date Received: 13 Sep 1814 - From what ship: HMT Niobe, Chatham - Born: Pennsylvania - Age: 36 - Released on 27 Apr 1815.

Cadwell, James - Seaman - Nbr: 1795 - Prize: Ferox - Ship Type: MV - How taken: HMS Medusa & HMS Lyra - When taken: 28 Mar 1813 - Where taken: off Cape Ortegal (Spain) - Date Received: 20 Jul 1814 - From what ship: HMS Milford, Plymouth - Born: Hadford - Age: 25 - Released on 1 May 1815.

Cadwell, James - Seaman - Nbr: 2782 - Prize: Tiger - Ship Type: MV - How taken: HMS Scylla - When taken: 22 May 1813 - Where taken: Bay of Biscay - Date Received: 24 Aug 1814 - From what ship: HMT Liverpool,

American Prisoners of War Held at Dartmoor during the War of 1812

## Alphabetical listing of names

Chatham - Born: New York - Age: 19 - Released on 19 May 1815.

Cadwell, John - Seaman - Nbr: 1307 - How taken: Sent into custody from HMS Minden - When taken: 17 May 1813 - Date Received: 14 Jun 1814 - From what ship: Mill Prison (Plymouth, England) - Born: Newport - Age: 20 - Released on 28 Apr 1815.

Cain, Enos - Seaman - Nbr: 2601 - How taken: Gave himself up from HMS Tigre - Date rec'd: 21 Aug 1814 - What ship: HMT Freya, Chatham - Born: Philadelphia - Age: 35 - Race: Mulatto - Released on 3 May 1815.

Cain, John - Seaman - Nbr: 2120 - How taken: Gave himself up from HMS Frederickstein - When taken: 28 Oct 1812 - Date Received: 8 Aug 1814 - From what ship: HMT Raven, Chatham - Born: Connecticut - Age: 26 - Released on 26 Apr 1815.

Cain, Joseph - Seaman - Nbr: 3557 - Prize: Hawk - Ship Type: P - How taken: HMS Pique - When taken: 26 Apr 1814 - Where taken: off Bermuda - Date Received: 30 Sep 1814 - From what ship: HMT Sybella - Born: Washington - Age: 16 - Released on 4 Jun 1815.

Cairns, Thomas - Seaman - Nbr: 5780 - Prize: William Penn - Ship Type: MV - How taken: HMS Acorn - When taken: 27 Oct 1812 - Where taken: Lat 14 - Date Received: 26 Dec 1814 - From what ship: HMT Argo - Born: Baltimore - Age: 20 - Released on 27 Apr 1815.

Cairns, Thomas - Seaman - Nbr: 3017 - How taken: Gave himself up from HMS Zealous - When taken: 27 Aug 1814 - Date Received: 2 Sep 1814 - From what ship: HMT Centaur - Born: Boston - Age: 25 - Released on 28 May 1815.

Calaman, John - Seaman - Nbr: 4314 - How taken: Gave himself up from HMS Armide - When taken: 8 Jun 1813 - Date Received: 7 Oct 1814 - From what ship: HMT Niobe, Chatham - Born: New York - Age: 33 - Released on 15 Jul 1815.

Calder, J. H. - Seaman - Nbr: 2777 - Prize: Tiger - Ship Type: MV - How taken: HMS Scylla - When taken: 22 May 1813 - Where taken: Bay of Biscay - Date Received: 24 Aug 1814 - From what ship: HMT Liverpool, Chatham - Born: New York - Age: 22 - Released on 19 May 1815.

Caldwell, Charles - Seaman - Nbr: 4304 - Prize: Raccoon - Ship Type: LM - How taken: HMS Venus - When taken: 6 Mar 1814 - Where taken: off Bordeaux - Date Received: 7 Oct 1814 - From what ship: HMT Niobe, Chatham - Born: Boston - Age: 20 - Released on 4 Jun 1815.

Caldwell, James - Seaman - Nbr: 215 - Prize: Mars - Ship Type: MV - How taken: HMS Warspite - When taken: 26 Feb 1813 - Where taken: off Basque Roads (France) - Date Received: 2 Apr 1813 - From what ship: Plymouth - Born: West Nottingham - Age: 28 - Released on 20 Apr 1815.

Calebs, Lewis - Seaman - Nbr: 4162 - How taken: Gave himself up from HMS Minos - When taken: 28 Dec 1813 - Date Received: 6 Oct 1814 - From what ship: HMT Niobe, Chatham - Born: New Orleans - Age: 25 - Race: Black - Released on 13 Jun 1815.

Calentin, Samuel - Seaman - Nbr: 2133 - How taken: Gave himself up from HMS Namur - When taken: 2 Dec 1812 - Date Received: 8 Aug 1814 - From what ship: HMT Raven, Chatham - Born: New York - Age: 44 - Released on 26 Apr 1815.

Calfax, William - Seaman - Nbr: 316 - Prize: Pallas - Ship Type: MV - How taken: Rebuff - When taken: 23 Dec 1812 - Where taken: off Cadiz - Date Received: 28 Jun 1813 - From what ship: Plymouth - Born: New Jersey - Age: 21 - Sent to Dartmouth on 30 Jul 1813.

Calhoun, Richard - Seaman - Nbr: 1731 - Prize: Fox - Ship Type: MV - How taken: HMS Scylla - When taken: 23 Apr 1813 - Where taken: off Bordeaux - Date Received: 17 Jul 1814 - From what ship: having escaped from Stapleton - Born: New York - Age: 20 - Released on 1 May 1815.

Call, George - Seaman - Nbr: 3818 - Prize: Nimble - Ship Type: Prize - How taken: HMS Arab - When taken: 5 Apr 1814 - Where taken: Lat 37 Long 65 - Date Received: 5 Oct 1814 - From what ship: HMT Orpheus, Halifax - Born: Massachusetts - Age: 22 - Released on 9 Jun 1815.

Callaghan, James - Seaman - Nbr: 6257 - How taken: Impressed at Liverpool - When taken: 21 Dec 1813 - Date Received: 4 Feb 1815 - From what ship: HMT Ganges - Born: Pennsylvania - Age: 23 - Race: Mulatto - Released on 5 Jul 1815.

Callam, John - Seaman - Nbr: 4712 - Prize: Montgomery - Ship Type: P - How taken: HMS Nymphe - When taken: 1 May 1813 - Where taken: off Cape Cod - Date Received: 9 Oct 1814 - From what ship: HMT Freya, Chatham - Born: Salem - Age: 26 - Released on 15 Jun 1815.

Callow, William - Seaman - Nbr: 3576 - Prize: William - Ship Type: MV - How taken: HMS Hasty - When taken: 14 May 1814 - Where taken: off Charleston - Date Received: 30 Sep 1814 - From what ship: HMT Sybella - Born: Southampton - Age: 20 - Released on 4 Jun 1815.

Calpry, Joseph - Seaman - Nbr: 4947 - Prize: Herald - Ship Type: P - How taken: HMS Endymion - When taken: 15

American Prisoners of War Held at Dartmoor during the War of 1812

## Alphabetical listing of names

Aug 1814 - Where taken: off Nantucket - Date Received: 28 Oct 1814 - From what ship: HMT Alkbar, Halifax - Born: Salem - Age: 22 - Released on 21 Jun 1815.

Cameron, Daniel - Seaman - Nbr: 4759 - Prize: York Town - Ship Type: P - How taken: British squadron - When taken: 10 Jul 1813 - Where taken: off Halifax - Date Received: 9 Oct 1814 - From what ship: HMT Freya, Chatham - Born: Virginia - Age: 35 - Released on 15 Jun 1815.

Cameron, William - Seaman - Nbr: 3792 - Prize: James - Ship Type: Prize - How taken: Rebecca (LM) - When taken: 13 Apr 1814 - Where taken: off Delaware - Date Received: 5 Oct 1814 - From what ship: HMT President, Halifax - Born: New Jersey - Age: 35 - Released on 9 Jun 1815.

Cammon, Robert - Seaman - Nbr: 391 - Prize: John & Frances - Ship Type: MV - How taken: HMS Belle Poule - When taken: 13 Mar 1813 - Where taken: Bay of Biscay - Date Received: 1 Jul 1813 - From what ship: Plymouth - Born: New York - Age: 27 - Released on 15 Apr 1815.

Camp, Tobias - Seaman - Nbr: 4359 - Prize: Fiere Facia - Ship Type: P - How taken: HMS Ramillies - When taken: 26 Feb 1814 - Where taken: off NY - Date Received: 7 Oct 1814 - From what ship: HMT Salvador del Mundo, Halifax - Born: New York - Age: 49 - Race: Negro - Released on 14 Jun 1815.

Campbell, Alexander - Seaman - Nbr: 820 - Prize: Alexander - Ship Type: MV - How taken: HMS Andromache - When taken: 23 Oct 1813 - Where taken: off Brest - Date Received: 29 Nov 1813 - From what ship: Plymouth - Born: Loubry - Age: 18 - Released on 26 Apr 1815.

Campbell, Henry - Seaman - Nbr: 6504 - How taken: Gave himself up in the Bay of Honduras - Date Received: 3 Mar 1815 - From what ship: HMT Ganges, Plymouth - Born: Delaware - Age: 28 - Died on 22 March 1815 from pneumonia.

Campbell, James - Quarter Gunner - Nbr: 3978 - Prize: US Brig Rattlesnake - Ship Type: MW - How taken: HMS Leander - When taken: 13 Jul 1814 - Where taken: off Shelburne - Date Received: 6 Oct 1814 - From what ship: HMT Chesapeake, Halifax - Born: Marblehead - Age: 46 - Released on 9 Jun 1815.

Campbell, James - Boatswain - Nbr: 5043 - Prize: Ida - Ship Type: LM - How taken: HMS Newcastle - When taken: 9 Aug 1814 - Where taken: Long 34 - Date Received: 28 Oct 1814 - From what ship: HMT Alkbar, Halifax - Born: New York - Age: 32 - Released on 29 Jun 1815.

Campbell, James - Seaman - Nbr: 2647 - How taken: Gave himself up from HMS Voluntaire - When taken: 25 Oct 1812 - Date Received: 21 Aug 1814 - From what ship: HMT Freya, Chatham - Born: New York - Age: 36 - Died on 7 Apr 1815 from gunshot in prison.

Campbell, James - Seaman - Nbr: 2428 - Prize: US Sloop Frolic - Ship Type: MW - How taken: HMS Orpheus - When taken: 20 Apr 1814 - Where taken: off Cuba - Date Received: 16 Aug 1814 - From what ship: HMT Queen, Halifax - Born: Boston - Age: 36 - Released on 3 May 1815.

Campbell, Jesse - Soldier - Nbr: 5837 - Prize: 4th US Rifles - Ship Type: Troops - How taken: British Army - When taken: 17 Sep 1814 - Where taken: Fort Erie - Date Received: 26 Dec 1814 - From what ship: HMT Argo - Born: Virginia - Age: 31 - Released on 29 Jun 1815.

Campbell, John - Seaman - Nbr: 4547 - Prize: Volunteer - Ship Type: MV - How taken: Victoria (Privateer) - When taken: 26 Dec 1813 - Where taken: at sea - Date Received: 8 Oct 1814 - From what ship: HMT Leyden, Chatham - Born: Maryland - Age: 26 - Released on 15 Jun 1815.

Campbell, John - Seaman - Nbr: 861 - Prize: Amiable - Ship Type: LM - How taken: HMS Magnificent - When taken: 30 Oct 1813 - Where taken: off Bordeaux - Date Received: 4 Dec 1813 - From what ship: Plymouth - Born: Delaware - Age: 18 - Released on 26 Apr 1815.

Campbell, John - Seaman - Nbr: 432 - Prize: Viper - Ship Type: MV - How taken: HMS Superb - When taken: 15 Apr 1813 - Where taken: Bay of Biscay - Date Received: 1 Jul 1813 - From what ship: Plymouth - Born: Norfolk - Age: 46 - Released on 20 Apr 1815.

Campbell, John - Seaman - Nbr: 3041 - How taken: Gave himself up from HMS Galatea - When taken: 24 Aug 1814 - Date Received: 2 Sep 1814 - From what ship: HMT Sultan - Born: New York - Age: 38 - Released on 28 May 1815.

Campbell, Moses - Lieutenant Marines - Nbr: 4926 - Prize: Herald - Ship Type: P - How taken: HMS Endymion - When taken: 15 Aug 1814 - Where taken: off Nantucket - Date Received: 28 Oct 1814 - From what ship: HMT Alkbar, Halifax - Born: New Jersey - Age: 28 - Released on 21 Jun 1815.

Campbell, Nathaniel - Seaman - Nbr: 3347 - How taken: Gave himself up from HMS Bombay - When taken: 29 Oct 1812 - Date Received: 13 Sep 1814 - From what ship: HMT Niobe, Chatham - Born: Albany - Age: 30 - Released on 27 Apr 1815.

Campbell, Thomas - Seaman - Nbr: 2536 - Prize: Hussar - Ship Type: P - How taken: HMS Saturn - When taken: 25 May 1814 - Where taken: off Sandy Hook - Date Received: 16 Aug 1814 - From what ship: HMT Queen,

American Prisoners of War Held at Dartmoor during the War of 1812

## Alphabetical listing of names

Halifax - Born: Philadelphia - Age: 32 - Race: Negro - Released on 11 Jul 1815.

Campbell, William - Cook - Nbr: 4144 - Prize: Minerva - Ship Type: MV - How taken: HMS Conquistador - When taken: 19 Jan 1814 - Where taken: Bay of Biscay - Date Received: 6 Oct 1814 - From what ship: HMT Niobe, Chatham - Born: L'Orient (France) - Age: 17 - Released on 13 Jun 1815.

Campreche, St. Jago (Jamaica) - Seaman - Nbr: 4820 - Prize: President - Ship Type: P - How taken: HMS Pique - When taken: 7 May 1814 - Where taken: off Porto Rico - Date Received: 9 Oct 1814 - From what ship: HMT Freya, Halifax - Born: Cartagena - Age: 18 - Race: Black - Died on 16 Jan 1815 from variola.

Canabon, J. - Seaman - Nbr: 1611 - Prize: Fox - Ship Type: LM - How taken: HMS Pheasant - When taken: 23 Apr 1813 - Where taken: Bay of Biscay - Date Received: 23 Jun 1814 - From what ship: Stapleton - Born: Corunna - Age: 31 - Sent to Mill Prison (Plymouth, England) on 30 Jun 1814.

Canada, William - Seaman - Nbr: 2121 - How taken: Gave himself up from HMS Rolus - When taken: 10 Nov 1812 - Date Received: 8 Aug 1814 - From what ship: HMT Raven, Chatham - Born: Abberville - Age: 34 - Released on 26 Apr 1815.

Canfield, Arthur - Seaman - Nbr: 4389 - Prize: Norfolk - Ship Type: P - How taken: HMS Fantome - When taken: 29 Jun 1813 - Where taken: off Delaware - Date Received: 8 Oct 1814 - From what ship: HMT Leyden, Chatham - Born: Middleton - Age: 23 - Released on 14 Jun 1815.

Cannon, Thomas - Seaman - Nbr: 4273 - How taken: Gave himself up from HMS Bulwark - When taken: 28 Dec 1813 - Date Received: 7 Oct 1814 - From what ship: HMT Niobe, Chatham - Born: New York - Age: 30 - Released on 13 Jun 1815.

Cannon, William - Seaman - Nbr: 6182 - Prize: Lion - Ship Type: P - How taken: HMS Granicus - When taken: 2 Dec 1814 - Where taken: off Lisbon - Date Received: 21 Jan 1815 - From what ship: HMT Impregnable - Born: Trenton - Age: 34 - Released on 5 Jul 1815.

Cannon, William - Seaman - Nbr: 1988 - How taken: Impressed at Yarmouth - When taken: 6 Mar 1813 - Date Received: 3 Aug 1814 - From what ship: HMS Lyffey, Chatham Depot - Born: Pennsylvania - Age: 35 - Released on 2 May 1815.

Cantelo, James - Seaman - Nbr: 3832 - Prize: Hussar - Ship Type: P - How taken: HMS Saturn - When taken: 24 May 1814 - Where taken: off Sandy Hook - Date Received: 5 Oct 1814 - From what ship: HMT Orpheus, Halifax - Born: New York - Age: 29 - Released on 9 Jun 1815.

Capewill, Bartholomew - Seaman - Nbr: 117 - Prize: Governor McKean - Ship Type: LM - How taken: HMS Rover - When taken: 26 Jan 1813 - Where taken: off Bordeaux - Date Received: 2 Apr 1813 - From what ship: Plymouth - Born: New Orleans - Age: 25 - Sent to Dartmouth on 30 Jul 1813.

Cappel, John - Seaman - Nbr: 2794 - Prize: Weazel - Ship Type: MV - How taken: HMS Foxhound - When taken: 25 Mar 1813 - Where taken: Bay of Biscay - Date Received: 24 Aug 1814 - From what ship: HMT Liverpool, Chatham - Born: Newport - Age: 18 - Released on 19 May 1815.

Capps, Denis - Seaman - Nbr: 4021 - Prize: US Brig Rattlesnake - Ship Type: MW - How taken: HMS Leander - When taken: 13 Jul 1814 - Where taken: off Shelburne - Date Received: 6 Oct 1814 - From what ship: HMT Chesapeake, Halifax - Born: North Carolina - Age: 23 - Released on 9 Jun 1815.

Carbenett, John - Seaman - Nbr: 581 - Prize: US Brig Argus - Ship Type: MW - How taken: HMS Pelican - When taken: 14 Apr 1813 - Where taken: Irish Channel - Date Received: 8 Sep 1813 - From what ship: Plymouth - Born: New York - Age: 16 - Sent to Plymouth on 8 Jul 1814.

Carberry, Thomas - 2nd Mate - Nbr: 6070 - Prize: Daedalus - Ship Type: MV - How taken: HMS Niemen - When taken: 20 Sep 1814 - Where taken: off Delaware - Date Received: 28 Dec 1814 - From what ship: HMT Penelope - Born: New York - Age: 23 - Released on 20 Apr 1815.

Card, Israel - Seaman - Nbr: 5423 - How taken: Gave himself up from East Indiaman Webster - When taken: 29 Aug 1813 - Date Received: 31 Oct 1814 - From what ship: HMT Leyden, Chatham - Born: Rhode Island - Age: 38 - Released on 1 Jul 1815.

Card, James - Mate - Nbr: 3580 - Prize: North Star - Ship Type: MV - How taken: HMS Crane - When taken: 10 Jun 1814 - Where taken: off St. Bartholomew - Date Received: 30 Sep 1814 - From what ship: HMT Sybella - Born: Newport - Age: 41 - Released on 4 Jun 1815.

Card, John - Seaman - Nbr: 3307 - Prize: Thomas - Ship Type: P - How taken: HMS Nymphe - When taken: 28 Jun 1813 - Where taken: off Halifax - Date Received: 13 Sep 1814 - From what ship: HMT Niobe, Chatham - Born: New Castle - Age: 18 - Released on 28 May 1815.

Carey, John - Seaman - Nbr: 1441 - Prize: Leo - Ship Type: LM - How taken: HMS Magicienne - When taken: 4 Jun 1813 - Where taken: off France - Date Received: 19 Jun 1814 - From what ship: Stapleton - Born: Brunswick - Age: 24 - Released on 28 Apr 1815.

American Prisoners of War Held at Dartmoor during the War of 1812

## Alphabetical listing of names

Carey, Liberty - Seaman - Nbr: 6108 - Prize: Garland - Ship Type: Prize - How taken: HMS Hamadryad - When taken: 22 Nov 1814 - Where taken: Georges Bank - Date Received: 6 Jan 1814 - From what ship: HMT Impregnable - Born: New Hampshire - Age: 35 - Race: Mulatto - Released on 5 Jul 1815.

Carland, Lewis - Seaman - Nbr: 787 - Prize: Avon - Ship Type: MV - How taken: HMS Eurotas - When taken: 27 Oct 1813 - Where taken: off Ushant (France) - Date Received: 3 Nov 1813 - From what ship: Plymouth - Born: LA - Age: 28 - Sent to Plymouth on 8 Jul 1814.

Carle, James - Seaman - Nbr: 6425 - Prize: Farmer's Daughter - Ship Type: MV - How taken: HMS Leviathan - When taken: 29 May 1814 - Date Received: 3 Mar 1815 - From what ship: HMT Ganges, Plymouth - Born: Boston - Age: 32 - Released on 11 Jul 1815.

Carleau, Daniel - Seaman - Nbr: 2608 - How taken: Gave himself up from HMS Partridge - When taken: 12 Sep 1812 - Date Received: 21 Aug 1814 - From what ship: HMT Freya, Chatham - Born: Philadelphia - Age: 29 - Released on 26 Apr 1815.

Carles, John - Seaman - Nbr: 3249 - Prize: Wiley Reynard - Ship Type: P - How taken: HMS Shannon - When taken: 15 Aug 1812 - Where taken: off Halifax - Date Received: 11 Sep 1814 - From what ship: HMT Freya, Chatham - Born: New Hampshire - Age: 50 - Released on 27 Apr 1815.

Carleton, Christopher - Seaman - Nbr: 6343 - Prize: Prince de Neufchatel - Ship Type: P - How taken: Leander (Newcastle Acasta) - When taken: 20 Dec 1814 - Where taken: Lat 38 Long 56 - Date Received: 19 Feb 1815 - From what ship: HMT Ganges, Plymouth - Born: Norway - Age: 25 - Released on 11 Jul 1815.

Carlton, William N. - Seaman - Nbr: 2811 - How taken: Impressed at Bristol - When taken: 12 Jul 1813 - Date Received: 24 Aug 1814 - From what ship: HMT Liverpool, Chatham - Born: Philadelphia - Age: 23 - Released on 19 May 1815.

Carman, George - Seaman - Nbr: 4293 - Prize: Commodore Perry - Ship Type: Schooner - How taken: Sent into custody from a cutter - When taken: 25 Feb 1814 - Where taken: off Bordeaux - Date Received: 7 Oct 1814 - From what ship: HMT Niobe, Chatham - Born: New Jersey - Age: 25 - Released on 14 Jun 1815.

Carnass, William - Seaman - Nbr: 5029 - Prize: Landrail - Ship Type: Prize - How taken: HMS Wasp - When taken: 27 Jul 1814 - Where taken: Georges Bank - Date Received: 28 Oct 1814 - From what ship: HMT Alkbar, Halifax - Born: Baltimore - Age: 24 - Escaped on 27 Jun 1815.

Carnell, George - Carpenter - Nbr: 5751 - Prize: US Schooner Ohio - Ship Type: MW - How taken: British gunboats - When taken: 12 Aug 1814 - Where taken: Fort Erie - Date Received: 26 Dec 1814 - From what ship: HMT Argo - Born: Rhode Island - Age: 22 - Released on 12 Jun 1815.

Carnes, John - Seaman - Nbr: 5287 - Prize: Wolf Cove - Ship Type: MV - How taken: HMS Briton - When taken: 1 Dec 1813 - Where taken: off Brest - Date Received: 31 Oct 1814 - From what ship: HMT Leyden, Chatham - Born: Copenhagen - Age: 31 - Released on 29 Jun 1815.

Carnes, Joseph - Seaman - Nbr: 385 - Prize: John & Frances - Ship Type: MV - How taken: HMS Belle Poule - When taken: 13 Mar 1813 - Where taken: Bay of Biscay - Date Received: 1 Jul 1813 - From what ship: Plymouth - Born: Boston - Age: 18 - Released on 20 Apr 1815.

Carnes, William - Seaman - Nbr: 6354 - Prize: Prince de Neufchatel - Ship Type: P - How taken: Leander (Newcastle Acasta) - When taken: 20 Dec 1814 - Where taken: Lat 38 Long 56 - Date Received: 19 Feb 1815 - From what ship: HMT Ganges, Plymouth - Born: Salem - Age: 30 - Released on 28 May 1815.

Carney, Samuel - Seaman - Nbr: 5726 - How taken: Gave himself up from HMS Warrior - Date Received: 26 Dec 1814 - From what ship: HMT Argo - Born: Philadelphia - Age: 32 - Released on 5 Jul 1815.

Carney, William - Seaman - Nbr: 2581 - How taken: Gave himself up from HMS Royal William - When taken: 5 Feb 1813 - Date Received: 21 Aug 1814 - From what ship: HMT Freya, Chatham - Born: Crossroads - Age: 48 - Released on 11 Jul 1815.

Caroline, Tobias - Seaman - Nbr: 2735 - How taken: Gave himself up from HMS Orion - When taken: 5 Sep 1812 - Date Received: 24 Aug 1814 - From what ship: HMT Liverpool, Chatham - Born: Albany - Age: 30 - Race: Black - Released on 26 Apr 1815.

Carpenter, C. R. - Marine - Nbr: 5945 - Prize: Harlequin - Ship Type: P - How taken: HMS Bulwark - When taken: 23 Nov 1814 - Where taken: off Halifax - Date Received: 27 Dec 1814 - From what ship: HMT Penelope - Born: Salem - Age: 33 - Released on 3 Jul 1815.

Carpenter, Henry - Seaman - Nbr: 509 - How taken: Impressed at Dublin - When taken: 10 Jun 1813 - Date Received: 8 Sep 1813 - From what ship: Plymouth - Born: New York - Age: 31 - Released on 26 Apr 1815.

Carpenter, Nathaniel - Passenger - Nbr: 2069 - How taken: Impressed at Hull - When taken: 18 May 1813 - Date Received: 3 Aug 1814 - From what ship: HMS Bittern, Chatham Depot - Born: Newport - Age: 27 - Released on 2 May 1815.

American Prisoners of War Held at Dartmoor during the War of 1812

## Alphabetical listing of names

Carr, James - Seaman - Nbr: 1004 - Prize: Zephyr - Ship Type: MV - How taken: HMS Surveillante - When taken: 6 Jan 1814 - Where taken: Bay of Biscay - Date Received: 31 Jan 1814 - From what ship: Plymouth - Born: Cumberland - Age: 19 - Released on 27 Apr 1815.

Carr, Jonathan - Seaman - Nbr: 2118 - How taken: Gave himself up from HMS Alceste - When taken: 14 Sep 1812 - Date Received: 8 Aug 1814 - From what ship: HMT Raven, Chatham - Born: Wallingford - Age: 24 - Released on 26 Apr 1815.

Carr, Thomas - Seaman - Nbr: 3870 - Prize: US Gunboat Nbr 51 - Ship Type: MW - How taken: HMS Nimrod - When taken: 22 May 1814 - Where taken: East passage - Date Received: 5 Oct 1814 - From what ship: HMT Orpheus, Halifax - Born: Charleston - Age: 19 - Released on 10 May 1815.

Carroll, John - Seaman - Nbr: 3043 - How taken: Gave himself up from HMS Galatea - When taken: 24 Aug 1814 - Date Received: 2 Sep 1814 - From what ship: HMT Sultan - Born: New York - Age: 38 - Released on 28 Apr 1815.

Carroll, Michael - Seaman - Nbr: 2410 - Prize: US Sloop Frolic - Ship Type: MW - How taken: HMS Orpheus - When taken: 20 Apr 1814 - Where taken: off Cuba - Date Received: 16 Aug 1814 - From what ship: HMT Queen, Halifax - Born: Marblehead - Age: 23 - Sent to Dartmouth on 19 Oct 1814.

Carroll, Samuel - Boy - Nbr: 3724 - Prize: Alfred - Ship Type: P - How taken: HMS Epervier - When taken: 23 Feb 1814 - Where taken: off Newfoundland - Date Received: 30 Sep 1814 - From what ship: HMT President, Halifax - Born: Marblehead - Age: 15 - Released on 4 Jun 1815.

Carroll, William - Seaman - Nbr: 5603 - Prize: Alexander - Ship Type: Prize - How taken: HMS Wasp - When taken: 14 Sep 1814 - Where taken: off Cape Sable Island (Canada) - Date Received: 17 Dec 1814 - From what ship: HMT Loire, Halifax - Born: Baltimore - Age: 28 - Released on 5 Jul 1815.

Carrot, Charles - Seaman - Nbr: 1079 - How taken: Impressed at Liverpool - When taken: 1 Feb 1814 - Date Received: 10 May 1814 - From what ship: Plymouth - Born: New York - Age: 25 - Race: Negro - Released on 27 Apr 1815.

Carson, John - Seaman - Nbr: 4326 - Prize: Fiere Facia - Ship Type: P - How taken: HMS Ramillies - When taken: 26 Feb 1814 - Where taken: off Long Island - Date Received: 7 Oct 1814 - From what ship: HMT Salvador del Mundo, Halifax - Born: New Orleans - Age: 26 - Race: Negro - Died on 16 Oct 1814 from pneumonia.

Carswell, John - Seaman - Nbr: 3686 - Prize: Alfred - Ship Type: P - How taken: HMS Epervier - When taken: 23 Feb 1812 - Where taken: off Newfoundland - Date Received: 30 Sep 1814 - From what ship: HMT President - Born: Marblehead - Age: 25 - Released on 4 Jun 1815.

Carswell, William - Seaman - Nbr: 4506 - Prize: America - Ship Type: P - How taken: HMS Shannon - When taken: 23 May 1813 - Where taken: off Cape Cod - Date Received: 8 Oct 1814 - From what ship: HMT Leyden, Chatham - Born: Massachusetts - Age: 21 - Released on 15 Jun 1815.

Carter, Daniel - Seaman - Nbr: 1395 - Prize: Zebra - Ship Type: LM - How taken: HMS Pyramus - When taken: 20 Apr 1813 - Where taken: Bay of Biscay - Date Received: 19 Jun 1814 - From what ship: Stapleton - Born: Virginia - Age: 26 - Died on 6 Oct 1814 from pneumonia.

Carter, Edward - Seaman - Nbr: 1512 - Prize: Zebra - Ship Type: LM - How taken: HMS Pyramus - When taken: 20 Apr 1813 - Where taken: Bay of Biscay - Date Received: 23 Jun 1814 - From what ship: Stapleton - Born: Norfolk - Age: 23 - Released on 1 May 1815.

Carter, Edward - Seaman - Nbr: 1394 - Prize: Zebra - Ship Type: LM - How taken: HMS Pyramus - When taken: 20 Apr 1813 - Where taken: Bay of Biscay - Date Received: 19 Jun 1814 - From what ship: Stapleton - Born: Boston - Age: 27 - Released on 28 Apr 1815.

Carter, Enoch - Seaman - Nbr: 4213 - Prize: Devon - Ship Type: Prize - How taken: HMS Fly - When taken: 2 Jan 1814 - Where taken: at sea - Date Received: 7 Oct 1814 - From what ship: HMT Niobe, Chatham - Born: Massachusetts - Age: 37 - Released on 13 Jun 1815.

Carter, George - Seaman - Nbr: 4880 - Prize: Saratoga - Ship Type: P - How taken: HMS Barracouta - When taken: 9 Oct 1814 - Where taken: off Western Islands (England) - Date Received: 24 Oct 1814 - From what ship: HMT Salvador del Mundo - Born: New York - Age: 20 - Released on 21 Jun 1815.

Carter, Henry - Seaman - Nbr: 4203 - Prize: Harvest - Ship Type: MV - How taken: HMS Orestes - When taken: 21 Jan 1814 - Where taken: at sea - Date Received: 7 Oct 1814 - From what ship: HMT Niobe, Chatham - Born: New York - Age: 26 - Race: Mulatto - Released on 13 Jun 1815.

Carter, James Wilson - Seaman - Nbr: 757 - How taken: Impressed at Liverpool - Date Received: 3 Nov 1813 - From what ship: Plymouth - Born: Long Island - Age: 26 - Released on 26 Apr 1815.

Carter, Jesse - Seaman - Nbr: 1685 - How taken: Impressed at Greenock - When taken: 18 Apr 1814 - Date Received: 2 Jul 1814 - From what ship: Plymouth - Born: Baltimore - Age: 23 - Race: Black - Released on 1

American Prisoners of War Held at Dartmoor during the War of 1812

## Alphabetical listing of names

May 1815.

Carter, John - Cook - Nbr: 2799 - Prize: Weazel - Ship Type: MV - How taken: HMS Foxhound - When taken: 25 Mar 1813 - Where taken: Bay of Biscay - Date Received: 24 Aug 1814 - From what ship: HMT Liverpool, Chatham - Born: New York - Age: 25 - Race: Black - Released on 19 May 1815.

Carter, Thomas - Seaman - Nbr: 1850 - Prize: True Blooded Yankee - Ship Type: P - How taken: HMS Fame - When taken: 24 Jun 1813 - Where taken: off Brest - Date Received: 21 Jul 1814 - From what ship: HMT Redbeard & Pincher, Chatham Depot - Born: New York - Age: 25 - Released on 2 May 1815.

Carton, Thomas - Carpenter - Nbr: 54 - Prize: Spitfire - Ship Type: MV - How taken: HMS Achates - When taken: 14 Feb 1813 - Where taken: off Ushant (France) - Date Received: 2 Apr 1813 - From what ship: Plymouth - Born: Pembroke - Age: 26 - Released on 29 Apr 1815.

Cartwright, George - Seaman - Nbr: 877 - How taken: Impressed at Liverpool - When taken: 1 Nov 1813 - Date Received: 31 Jan 1814 - From what ship: Plymouth - Born: Rhode Island - Age: 34 - Released on 27 Apr 1815.

Carty, Joseph - Seaman - Nbr: 4941 - Prize: Herald - Ship Type: P - How taken: HMS Endymion - When taken: 15 Aug 1814 - Where taken: off Nantucket - Date Received: 28 Oct 1814 - From what ship: HMT Alkbar, Halifax - Born: New London - Age: 17 - Released on 3 May 1815.

Casey, Henry - Seaman - Nbr: 453 - Prize: Eliza - Ship Type: MV - How taken: HMS Surveillante - When taken: 27 Mar 1813 - Where taken: Bay of Biscay - Date Received: 8 Sep 1813 - From what ship: Plymouth - Born: Washington - Age: 24 - Released on 26 Apr 1815.

Cash, John - Seaman - Nbr: 4373 - Prize: Growler - Ship Type: P - How taken: HMS Electra - When taken: 7 Jul 1813 - Where taken: at sea - Date Received: 8 Oct 1814 - From what ship: HMT Leyden, Chatham - Born: Massachusetts - Age: 38 - Released on 14 Jun 1815.

Cashman, Ansel - Passenger - Nbr: 5092 - Prize: David Porter - Ship Type: LM - How taken: HMS Pylades - When taken: 12 Sep 1814 - Where taken: Georges Bank - Date Received: 28 Oct 1814 - From what ship: HMT Alkbar, Halifax - Born: Boston - Age: 23 - Released on 14 Apr 1815.

Caslow, John - Marine - Nbr: 5653 - Prize: US Brig Rattlesnake - Ship Type: MW - How taken: HMS Leander - When taken: 13 Jul 1814 - Where taken: off Shelburne - Date Received: 24 Dec 1814 - From what ship: HMT Impregnable - Born: New York - Age: 29 - Released on 3 Jul 1815.

Casper, William - Boy - Nbr: 6385 - Prize: US Brig Syren - Ship Type: MW - How taken: HMS Medway - When taken: 12 Jul 1814 - Where taken: off Cape of Good Hope - Date Received: 24 Feb 1815 - From what ship: HMT Ganges, Plymouth - Born: Charlestown - Age: 19 - Released on 11 Jul 1815.

Cassam, Michael - Seaman - Nbr: 1080 - How taken: Impressed at Liverpool - When taken: 1 Feb 1814 - Date Received: 10 May 1814 - From what ship: Plymouth - Born: New York - Age: 28 - Race: Negro - Released on 4 Jun 1815.

Castagne, John - Seaman - Nbr: 3930 - Prize: Rolla - Ship Type: P - How taken: HMS Loire - When taken: 10 Dec 1813 - Where taken: off Bull Island (South Carolina) - Date Received: 5 Oct 1814 - From what ship: HMT President, Halifax - Born: Philadelphia - Age: 21 - Released on 9 Jun 1815.

Castana, John Baptist - Seaman - Nbr: 999 - Prize: Hall - Ship Type: MV - How taken: Not listed - When taken: 10 Jan 1814 - Where taken: off Ushant (France) - Date Received: 31 Jan 1814 - From what ship: Plymouth - Born: Bordeaux - Age: 58 - Sent to Mill Prison (Plymouth, England) on 21 Jun 1814.

Cate, Joseph - Seaman - Nbr: 5924 - Prize: Harlequin - Ship Type: P - How taken: HMS Bulwark - When taken: 23 Nov 1814 - Where taken: off Halifax - Date Received: 27 Dec 1814 - From what ship: HMT Penelope - Born: Boston - Age: 19 - Released on 3 Jul 1815.

Cateret, James - Seaman - Nbr: 3901 - Prize: Mary - Ship Type: Packet - How taken: HMS Loire - When taken: 9 Jul 1814 - Where taken: Chesapeake Bay - Date Received: 5 Oct 1814 - From what ship: HMT Orpheus, Halifax - Born: Talbot - Age: 22 - Died on 11 Nov 1814.

Catley, William - Seaman - Nbr: 5394 - Prize: Hawke - Ship Type: P - How taken: HMS Pique - When taken: 27 Apr 1814 - Where taken: West Indies - Date Received: 31 Oct 1814 - From what ship: HMT Leyden, Chatham - Born: Baltimore - Age: 24 - Escaped on 27 Jun 1815.

Cato, John - Seaman - Nbr: 1342 - Prize: Paul Jones - Ship Type: P - How taken: HMS Leonidas - When taken: 23 May 1813 - Where taken: Channel - Date Received: 19 Jun 1814 - From what ship: Stapleton - Born: New London - Age: 35 - Race: Negro - Released on 28 Apr 1815.

Catona, Manuel - Seaman - Nbr: 2287 - Prize: Diomede - Ship Type: P - How taken: HMS Rifleman - When taken: 28 May 1814 - Where taken: off Sable Island - Date Received: 16 Aug 1814 - From what ship: HMS Dublin, Halifax - Born: Leghorn, Italy - Age: 17 - Released on 3 May 1815.

American Prisoners of War Held at Dartmoor during the War of 1812

## Alphabetical listing of names

Catwood, Zenos - Seaman - Nbr: 1154 - How taken: Impressed at Cork - When taken: 20 Mar 1814 - Date Received: 10 May 1814 - From what ship: Plymouth - Born: Wellfleet - Age: 19 - Released on 28 Apr 1815.

Caulfield, E. - Seaman - Nbr: 5614 - How taken: Gave himself up from HMS Semiramis - When taken: 28 Sep 1813 - Date rec'd: 24 Dec 1814 - What ship: HMT Tay - Born: New York - Age: 35 - Released on 27 Mar 1815.

Cayler, Charles - Seaman - Nbr: 4102 - Prize: Mary - Ship Type: MV - How taken: HMS Junon - When taken: 6 Jun 1814 - Where taken: off Cape Ann - Date Received: 6 Oct 1814 - From what ship: HMT Chesapeake, Halifax - Born: Massachusetts - Age: 19 - Released on 13 Jun 1815.

Cayler, Henry - Seaman - Nbr: 4103 - Prize: Mary - Ship Type: MV - How taken: HMS Junon - When taken: 6 Jun 1814 - Where taken: off Cape Ann - Date Received: 6 Oct 1814 - From what ship: HMT Chesapeake, Halifax - Born: Massachusetts - Age: 38 - Released on 13 Jun 1815.

Cayler, Jacob - Master - Nbr: 4101 - Prize: Mary - Ship Type: MV - How taken: HMS Junon - When taken: 6 Jun 1814 - Where taken: off Cape Ann - Date Received: 6 Oct 1814 - From what ship: HMT Chesapeake, Halifax - Born: Massachusetts - Age: 31 - Released on 13 Jun 1815.

Ceaser, Joseph - Seaman - Nbr: 313 - Prize: Paulina - Ship Type: MV - How taken: HMS Lavinia - When taken: 9 Aug 1812 - Where taken: off Gibraltar - Date Received: 28 Jun 1813 - From what ship: Plymouth - Born: Massachusetts - Age: 37 - Sent to Dartmouth on 30 Jul 1813.

Cecil, Francis - Seaman - Nbr: 3869 - Prize: Dominique - How taken: HMS Nimrod - When taken: 5 Jul 1814 - Where taken: off Block Island - Date Received: 5 Oct 1814 - From what ship: HMT Orpheus, Halifax - Born: New York - Age: 23 - Released on 10 May 1815.

Cedus, Francis - Seaman - Nbr: 5834 - Prize: US Schooner Tigress - Ship Type: MW - How taken: British gunboats - When taken: 2 Sep 1814 - Where taken: Lake Erie - Date Received: 26 Dec 1814 - From what ship: HMT Argo - Born: Cadiz - Age: 24 - Released on 3 Jul 1815.

Celutan, Alexander - Seaman - Nbr: 4992 - Prize: Invincible - Ship Type: LM - How taken: HMS Armide - When taken: 15 Aug 1814 - Where taken: off Nantucket - Date Received: 28 Oct 1814 - From what ship: HMT Alkbar, Halifax - Born: Bordeaux - Age: 34 - Released on 21 Jun 1815.

Chace, George - Seaman - Nbr: 6106 - Prize: William Penn - Ship Type: MV - How taken: HMS Acorn - When taken: 27 Oct 1812 - Where taken: Lat 14 - Date Received: 6 Jan 1814 - From what ship: HMT Impregnable - Born: Nantucket - Age: 16 - Released on 27 Apr 1815.

Chain, John - Seaman - Nbr: 900 - Prize: Squirrel - Ship Type: MV - How taken: HMS Belle Poule - When taken: 14 Dec 1813 - Where taken: at sea - Date Received: 31 Jan 1814 - From what ship: Plymouth - Born: Maryland - Age: 23 - Released on 27 Apr 1815.

Chalk, John - Seaman - Nbr: 4170 - Prize: Quebec - Ship Type: Prize - How taken: HMS Derwent - When taken: 29 Jan 1813 - Where taken: off Lisbon - Date Received: 7 Oct 1814 - From what ship: HMT Niobe, Chatham - Born: Emden - Age: 38 - Released on 13 Jun 1815.

Chambers, Elias - Boy - Nbr: 1027 - Prize: Joseph - Ship Type: MV - How taken: HMS Royalist - When taken: 18 Jan 1814 - Where taken: Bay of Biscay - Date Received: 10 May 1814 - From what ship: Plymouth - Born: Marblehead - Age: 13 - Released on 1 May 1815.

Chambers, Henry - Seaman - Nbr: 140 - Prize: Criterion - Ship Type: MV - How taken: HMS Belle Poule - When taken: 14 Feb 1813 - Where taken: Bay of Biscay - Date Received: 2 Apr 1813 - From what ship: Plymouth - Born: Oldenburg - Age: 24 - Sent to Plymouth on 15 Jun 1813.

Chambers, Joseph - Seaman - Nbr: 2291 - Prize: Hussar - Ship Type: P - How taken: HMS Saturn - When taken: 25 May 1814 - Where taken: off Sandy Hook - Date Received: 16 Aug 1814 - From what ship: HMS Dublin, Halifax - Born: Petersburg - Age: 26 - Race: Negro - Released on 3 May 1815.

Chambers, William - Seaman - Nbr: 2436 - Prize: US Sloop Frolic - Ship Type: MW - How taken: HMS Orpheus - When taken: 20 Apr 1814 - Where taken: off Cuba - Date Received: 16 Aug 1814 - From what ship: HMT Queen, Halifax - Born: Marblehead - Age: 22 - Released on 3 May 1815.

Champlin, Thomas - Seaman - Nbr: 5219 - Prize: Rolla - Ship Type: P - How taken: HMS Victorious - When taken: 2 Jun 1813 - Where taken: off Halifax - Date Received: 31 Oct 1814 - From what ship: HMT Mermaid, Chatham - Born: Baltimore - Age: 48 - Released on 29 Jun 1815.

Chandler, Enoch - Seaman - Nbr: 4379 - Prize: Growler - Ship Type: P - How taken: HMS Electra - When taken: 7 Jul 1813 - Where taken: at sea - Date Received: 8 Oct 1814 - From what ship: HMT Leyden, Chatham - Born: Massachusetts - Age: 26 - Released on 14 Jun 1815.

Chandler, Samuel - Seaman - Nbr: 5542 - Prize: Lion - Ship Type: MV - How taken: HMS Brilliant - When taken: 4 Jan 1814 - Where taken: Bahamas Banks - Date Received: 17 Dec 1814 - From what ship: HMT Loire, Halifax - Born: Hillsborough - Age: 26 - Released on 1 Jul 1815.

American Prisoners of War Held at Dartmoor during the War of 1812

## Alphabetical listing of names

Chandler, Simon - Seaman - Nbr: 1523 - Prize: Essex - Ship Type: MV - How taken: HMS Pyramus - When taken: 20 Apr 1813 - Where taken: Bay of Biscay - Date Received: 23 Jun 1814 - From what ship: Stapleton - Born: Massachusetts - Age: 19 - Died on 22 Oct 1814 from phthisis pulmonalis.

Chandler, William - Seaman - Nbr: 5541 - Prize: Lion - Ship Type: MV - How taken: HMS Brilliant - When taken: 4 Jan 1814 - Where taken: Bahamas Banks - Date Received: 17 Dec 1814 - From what ship: HMT Loire, Halifax - Born: Hillsborough - Age: 22 - Released on 1 Jul 1815.

Chane, Daniel - Seaman - Nbr: 1281 - How taken: Sent into custody from HMS Nerus - Date Received: 14 Jun 1814 - From what ship: Mill Prison (Plymouth, England) - Born: Boston - Age: 25 - Released on 28 Apr 1815.

Channing, John - Seaman - Nbr: 1114 - Prize: Bunker Hill - Ship Type: P - How taken: HMS Pomone & HMS Cadmus - When taken: 4 Mar 1814 - Where taken: Bay of Biscay - Date Received: 10 May 1814 - From what ship: Plymouth - Born: Petersburg - Age: 33 - Released on 27 Apr 1815.

Chapel, John - Seaman - Nbr: 2223 - Prize: Hussar - Ship Type: P - How taken: HMS Saturn - When taken: 25 May 1814 - Where taken: off Sandy Hook - Date Received: 16 Aug 1814 - From what ship: HMS Dublin, Halifax - Born: Connecticut - Age: 32 - Released on 2 May 1815.

Chapley, John - Cook - Nbr: 2917 - How taken: Impressed at Chatham - When taken: 15 Aug 1813 - Date Received: 24 Aug 1814 - From what ship: HMT Alpheus, Chatham - Born: Boston - Age: 45 - Released on 19 May 1815.

Chapman, Abraham - Seaman - Nbr: 1681 - How taken: Gave himself up from HMS Orestes - Date Received: 2 Jul 1814 - From what ship: Plymouth - Born: Cape Cod - Age: 24 - Released on 1 May 1815.

Chapman, Daniel - Seaman - Nbr: 3717 - Prize: Alfred - Ship Type: P - How taken: HMS Epervier - When taken: 23 Feb 1814 - Where taken: off Newfoundland - Date Received: 30 Sep 1814 - From what ship: HMT President, Halifax - Born: Marblehead - Age: 54 - Released on 4 Jun 1815.

Chapman, James - Seaman - Nbr: 3589 - Prize: Salley & Betsey - Ship Type: MV - How taken: HMS Pique - When taken: 6 Feb 1814 - Where taken: off Porto Rico - Date Received: 30 Sep 1814 - From what ship: HMT Sybella - Born: Philadelphia - Age: 30 - Released on 4 Jun 1815.

Chappel, Edward - Seaman - Nbr: 2054 - How taken: Impressed at Gravesend - When taken: 27 Apr 1813 - Date Received: 3 Aug 1814 - From what ship: HMS Lyffey, Chatham Depot - Born: Newport - Age: 20 - Released on 2 May 1815.

Chapple, John - Seaman - Nbr: 2466 - Prize: US Sloop Frolic - Ship Type: MW - How taken: HMS Orpheus - When taken: 20 Apr 1814 - Where taken: off Cuba - Date Received: 16 Aug 1814 - From what ship: HMT Queen, Halifax - Born: Boston - Age: 21 - Released on 3 May 1815.

Chapple, Samuel - Seaman - Nbr: 4052 - Prize: US Brig Rattlesnake - Ship Type: MW - How taken: HMS Leander - When taken: 13 Jul 1814 - Where taken: off Shelburne - Date Received: 6 Oct 1814 - From what ship: HMT Chesapeake, Halifax - Born: Marblehead - Age: 20 - Released on 10 Apr 1815.

Chargne, Robert - Quarter Gunner - Nbr: 5748 - Prize: US Schooner Ohio - Ship Type: MW - How taken: British gunboats - When taken: 12 Aug 1814 - Where taken: Fort Erie - Date Received: 26 Dec 1814 - From what ship: HMT Argo - Born: Charleston - Age: 28 - Released on 12 Jun 1815.

Chariton, James - Seaman - Nbr: 4938 - Prize: Herald - Ship Type: P - How taken: HMS Endymion - When taken: 15 Aug 1814 - Where taken: off Nantucket - Date Received: 28 Oct 1814 - From what ship: HMT Alkbar, Halifax - Born: New York - Age: 23 - Released on 21 Jun 1815.

Charles, John - Seaman - Nbr: 6161 - Prize: Lion - Ship Type: P - How taken: HMS Granicus - When taken: 2 Dec 1814 - Where taken: off Lisbon - Date Received: 18 Jan 1815 - From what ship: HMT Impregnable - Born: Washington - Age: 24 - Race: Black - Released on 5 Jul 1815.

Charles, John - Seaman - Nbr: 987 - How taken: Impressed at Dublin - When taken: 1 Oct 1813 - Date rec'd: 31 Jan 1814 - What ship: Plymouth - Born: Boston - Age: 22 - Race: Negro - Sent to Plymouth on 8 Jul 1814.

Charles, Philip - Seaman - Nbr: 4872 - How taken: Impressed at Halifax - Date Received: 24 Oct 1814 - From what ship: Royal Hospital, Plymouth - Born: Not listed - Released on 11 Jul 1815.

Charles, Samuel - Seaman - Nbr: 5989 - Prize: McDonough - Ship Type: P - How taken: HMS Bacchante - When taken: 1 Nov 1814 - Where taken: Lat 42 Long 67 - Date Received: 27 Dec 1814 - From what ship: HMT Penelope - Born: Philadelphia - Age: 33 - Race: Mulatto - Released on 11 Jul 1815.

Charles, Samuel - Passenger - Nbr: 658 - Prize: Marmion - Ship Type: MV - How taken: HMS President - When taken: 14 Aug 1813 - Where taken: off Nantes - Date Received: 8 Sep 1813 - From what ship: Plymouth - Born: North Carolina - Age: 21 - Sent to Dartmouth on 4 Jul 1814.

Charlotte, William - Seaman - Nbr: 3039 - Prize: Commerce - How taken: HMS Superb - When taken: 23 Jun 1814 - Where taken: off Sandy Hook - Date Received: 2 Sep 1814 - From what ship: HMT Sultan - Born:

American Prisoners of War Held at Dartmoor during the War of 1812

## Alphabetical listing of names

Philadelphia - Age: 20 - Race: Creole - Released on 28 May 1815.

Charter, James - Passenger - Nbr: 503 - Prize: Governor Gerry - Ship Type: MV - How taken: HMS Royalist - When taken: 31 May 1813 - Where taken: Bay of Biscay - Date Received: 8 Sep 1813 - From what ship: Plymouth - Born: Maryland - Age: 30 - Sent to Dartmouth on 16 Jun 1814.

Chase, Allen - Prize master - Nbr: 4089 - Prize: Tyger - Ship Type: Prize - How taken: HMS Bulwark - When taken: 20 Jul 1814 - Where taken: Georges Bank - Date Received: 6 Oct 1814 - From what ship: HMT Chesapeake, Halifax - Born: Massachusetts - Age: 30 - Released on 1 May 1815.

Chase, Constant - Mate - Nbr: 4252 - Prize: Argus - Ship Type: MV - How taken: HMS San Domingo - When taken: 1 Mar 1814 - Where taken: off Savannah - Date Received: 7 Oct 1814 - From what ship: HMT Niobe, Chatham - Born: England - Age: 28 - Released on 13 Jun 1815.

Chase, Jacob - Seaman - Nbr: 2825 - How taken: Gave himself up from HMS Philomel - When taken: 28 Dec 1812 - Date Received: 24 Aug 1814 - From what ship: HMT Liverpool, Chatham - Born: Portland - Age: 29 - Released on 26 Apr 1815.

Chase, John - Seaman - Nbr: 4548 - Prize: Volunteer - Ship Type: MV - How taken: Victoria (Privateer) - When taken: 26 Dec 1813 - Where taken: at sea - Date Received: 8 Oct 1814 - From what ship: HMT Leyden, Chatham - Born: Massachusetts - Age: 23 - Released on 15 Jun 1815.

Chase, Joseph - Seaman - Nbr: 1248 - How taken: Sent into custody from HMS Edinburgh - When taken: 28 Oct 1812 - Date rec'd: 14 Jun 1814 - What ship: Mill Prison (Plymouth, England) - Born: Rhode Island - Age: 26 - Released on 26 Apr 1815.

Chase, Nathaniel - Seaman - Nbr: 2864 - Prize: Prompt - Ship Type: MV - How taken: Change (Privateer) - When taken: 28 Mar 1813 - Where taken: Bay of Biscay - Date Received: 24 Aug 1814 - From what ship: HMT Alpheus, Chatham - Born: Cape Cod - Age: 24 - Released on 19 May 1815.

Chase, Nathaniel - Seaman - Nbr: 3151 - Prize: Kitty - Ship Type: Prize - How taken: Dart of Guernsey (Privateer) - When taken: 20 Jun 1813 - Where taken: off the Western Isles (England) - Date Received: 11 Sep 1814 - From what ship: HMT Freya, Chatham - Born: Marblehead - Age: 23 - Sent to Dartmouth on 23 Sep 1814.

Chase, Nathaniel - Prize master - Nbr: 4390 - Prize: Portsmouth Packet - Ship Type: Prize - How taken: HMS Fantome - When taken: 5 Oct 1813 - Where taken: off Portland - Date Received: 8 Oct 1814 - From what ship: HMT Leyden, Chatham - Born: Cape Cod - Age: 30 - Released on 2 May 1815.

Chase, Oliver - Seaman - Nbr: 4527 - How taken: Gave himself up from HMS Racehorse - When taken: 25 Oct 1813 - Date Received: 8 Oct 1814 - From what ship: HMT Leyden, Chatham - Born: Massachusetts - Age: 29 - Released on 15 Jun 1815.

Chase, Samuel - Seaman - Nbr: 6162 - Prize: Lion - Ship Type: P - How taken: HMS Granicus - When taken: 2 Dec 1814 - Where taken: off Lisbon - Date Received: 18 Jan 1815 - From what ship: HMT Impregnable - Born: Maryland - Age: 45 - Race: Black - Released on 5 Jul 1815.

Chase, William - Seaman - Nbr: 5018 - Prize: Alexander - Ship Type: Prize - How taken: HMS Wasp - When taken: 6 Sep 1814 - Where taken: off Cape Sable Island (Canada) - Date Received: 28 Oct 1814 - From what ship: HMT Alkbar, Halifax - Born: Massachusetts - Age: 39 - Released on 20 Mar 1815.

Chasson, Thomas - Seaman - Nbr: 5849 - Prize: James - Ship Type: MV - How taken: HMS Harpy - When taken: 20 Dec 1812 - Where taken: off Isle of France - Date Received: 26 Dec 1814 - From what ship: HMT Argo - Born: Bengal, India - Age: 40 - Race: Mulatto - Released on 11 Jul 1815.

Check, Stephen - Seaman - Nbr: 166 - How taken: Impressed at Liverpool - When taken: 10 Jan 1813 - Date Received: 2 Apr 1813 - From what ship: Plymouth - Born: Kent County - Age: 31 - Sent to Dartmouth on 30 Jul 1813.

Cheney, Daniel - Seaman - Nbr: 932 - Prize: Fanny - Ship Type: MV - How taken: HMS Eurotas - When taken: 23 Dec 1813 - Where taken: at sea - Date Received: 31 Jan 1814 - From what ship: Plymouth - Born: New York - Age: 20 - Released on 27 Apr 1815.

Cherry, David - Gunner's mate - Nbr: 5914 - Prize: Harlequin - Ship Type: P - How taken: HMS Bulwark - When taken: 23 Nov 1814 - Where taken: off Halifax - Date Received: 27 Dec 1814 - From what ship: HMT Penelope - Born: Providence - Age: 29 - Released on 15 Jul 1815.

Cheslie, Amos - Seaman - Nbr: 4668 - Prize: Industry - Ship Type: P - How taken: HMS Heron - When taken: 3 Nov 1813 - Where taken: off Halifax - Date Received: 9 Oct 1814 - From what ship: HMT Leyden, Chatham - Born: Portsmouth - Age: 20 - Released on 11 Jul 1815.

Chessebrough, Benjamin F. - Seaman - Nbr: 1130 - Prize: Young Dixon - Ship Type: MV - How taken: HMS Fly & HMS Avon - When taken: 3 Apr 1814 - Where taken: at sea - Date Received: 10 May 1814 - From what ship: Plymouth - Born: New York - Age: 32 - Released on 28 Apr 1815.

American Prisoners of War Held at Dartmoor during the War of 1812

## Alphabetical listing of names

Chews, Thomas - Seaman - Nbr: 5804 - Prize: US Frigate Superior (Gig) - Ship Type: MW - How taken: British gunboats - When taken: 26 Aug 1814 - Where taken: Lake Ontario - Date Received: 26 Dec 1814 - From what ship: HMT Argo - Born: New York - Age: 24 - Released on 3 Jul 1815.

Chick, Moses - Seaman - Nbr: 6373 - Prize: US Brig Syren - Ship Type: MW - How taken: HMS Medway - When taken: 12 Jul 1814 - Where taken: off Cape of Good Hope - Date Received: 24 Feb 1815 - From what ship: HMT Ganges, Plymouth - Born: Wells - Age: 38 - Released on 11 Jul 1815.

Childs, Adam - Carpenter - Nbr: 4921 - Prize: Bordeaux Packet - Ship Type: LM - How taken: HMS Niemen - When taken: 28 Jan 1814 - Where taken: off Delaware - Date Received: 28 Oct 1814 - From what ship: HMT Alkbar, Halifax - Born: New York - Age: 26 - Released on 21 Jun 1815.

Childs, William - Seaman - Nbr: 5329 - Prize: Portsmouth Packet - Ship Type: P - How taken: HMS Maidstone - When taken: 17 Jul 1813 - Where taken: Grand Banks - Date Received: 31 Oct 1814 - From what ship: HMT Leyden, Chatham - Born: Warren - Age: 29 - Released on 29 Jun 1815.

Chine, Samuel - Seaman - Nbr: 2580 - Prize: Print - Ship Type: MV - How taken: HMS Colossus - When taken: 21 Jan 1813 - Where taken: off Long Island - Date Received: 21 Aug 1814 - From what ship: HMT Freya, Chatham - Born: Marblehead - Age: 25 - Released on 3 May 1815.

Chiney, Amos - Seaman - Nbr: 474 - Prize: Zebra - Ship Type: LM - How taken: HMS Pyramus - When taken: 20 Apr 1813 - Where taken: Bay of Biscay - Date Received: 8 Sep 1813 - From what ship: Plymouth - Born: New York - Age: 24 - Released on 20 Apr 1815.

Chip, Charles - Seaman - Nbr: 2853 - Prize: Dick - Ship Type: MV - How taken: HMS Dispatch - When taken: 17 Mar 1812 - Where taken: at sea - Date Received: 24 Aug 1814 - From what ship: HMT Alpheus, Chatham - Born: Norfolk - Age: 27 - Released on 27 Apr 1815.

Chipman, Christian - Seaman - Nbr: 669 - Prize: Joel Barlow - Ship Type: LM - How taken: HMS Briton - When taken: 3 Jul 1813 - Where taken: off Bordeaux - Date Received: 8 Sep 1813 - From what ship: Plymouth - Born: New London - Age: 29 - Released on 26 Apr 1815.

Chiseldine, John - Seaman - Nbr: 1481 - Prize: Hebe - Ship Type: MV - How taken: HMS Stag - When taken: 18 Apr 1813 - Where taken: Bay of Biscay - Date Received: 19 Jun 1814 - From what ship: Stapleton - Born: Newbury - Age: 20 - Released on 1 May 1815.

Chivers, Cantab - Seaman - Nbr: 3057 - How taken: Gave himself up from HMS Andromache - Date Received: 2 Sep 1814 - From what ship: HMT Hydra, Chatham - Born: New Haven - Age: 31 - Race: Black - Released on 11 Jul 1815.

Chivers, Joseph - Seaman - Nbr: 4536 - Prize: Wolf Cove - Ship Type: MV - How taken: HMS Briton - When taken: 1 Dec 1812 - Where taken: off Brest - Date Received: 8 Oct 1814 - From what ship: HMT Leyden, Chatham - Born: Massachusetts - Age: 25 - Released on 27 Apr 1815.

Choete, Thomas - Seaman - Nbr: 4613 - Prize: Argus - Ship Type: MV - How taken: HMS San Domingo - When taken: 1 Mar 1814 - Where taken: off Savannah - Date Received: 9 Oct 1814 - From what ship: HMT Leyden, Chatham - Born: Newburyport - Age: 18 - Released on 15 Jun 1815.

Christian, John - Seaman - Nbr: 5285 - Prize: Enterprize - Ship Type: P - How taken: HMS Tenedos - When taken: 21 May 1813 - Where taken: off Cape Cod - Date Received: 31 Oct 1814 - From what ship: HMT Leyden, Chatham - Born: Massachusetts - Age: 30 - Released on 29 Jun 1815.

Christian, John - Seaman - Nbr: 1145 - Prize: Bunker Hill - Ship Type: P - How taken: HMS Pomone - When taken: 8 Mar 1814 - Where taken: at sea - Date Received: 10 May 1814 - From what ship: Plymouth - Born: New Orleans - Age: 32 - Sent to Plymouth on 8 Jul 1814.

Christian, Nicholas - Seaman - Nbr: 6029 - Prize: Regent - Ship Type: LM - How taken: HMS Forth - When taken: 19 Sep 1814 - Where taken: off Egg Harbor (New Jersey) - Date Received: 28 Dec 1814 - From what ship: HMT Penelope - Born: Denmark - Age: 24 - Released on 3 Jul 1815.

Christian, Tyrel - Seaman - Nbr: 4750 - How taken: Gave himself up from HMS Minden - When taken: 4 Oct 1813 - Date Received: 9 Oct 1814 - From what ship: HMT Freya, Chatham - Born: Charleston - Age: 41 - Released on 15 Jun 1815.

Christie, James - Seaman - Nbr: 1444 - Prize: Tickler - Ship Type: LM - How taken: HMS Magicienne - When taken: 5 Jun 1813 - Where taken: Bay of Biscay - Date Received: 19 Jun 1814 - From what ship: Stapleton - Born: New York - Age: 23 - Released on 28 Apr 1815.

Christie, John - Seaman - Nbr: 2079 - Prize: True Blooded Yankee - Ship Type: P - How taken: HMS Hope - When taken: 24 Jun 1813 - Where taken: off Brest - Date Received: 3 Aug 1814 - From what ship: HMS Bittern, Chatham Depot - Born: Philadelphia - Age: 28 - Released on 2 May 1815.

Christie, Robert - Seaman - Nbr: 2444 - Prize: US Sloop Frolic - Ship Type: MW - How taken: HMS Orpheus -

American Prisoners of War Held at Dartmoor during the War of 1812

## Alphabetical listing of names

When taken: 20 Apr 1814 - Where taken: off Cuba - Date Received: 16 Aug 1814 - From what ship: HMT Queen, Halifax - Born: New Bedford - Age: 23 - Released on 3 May 1815.

Christy, Alexander - Seaman - Nbr: 232 - Prize: Charlotte - Ship Type: MV - How taken: HMS Warspite - When taken: 3 Mar 1813 - Where taken: Bay of Biscay - Date Received: 2 Apr 1813 - From what ship: Plymouth - Born: Charlestown - Age: 25 - Released on 14 Apr 1815.

Chugler, John - Marine - Nbr: 598 - Prize: US Brig Argus - Ship Type: MW - How taken: HMS Pelican - When taken: 14 Aug 1813 - Where taken: Irish Channel - Date Received: 8 Sep 1813 - From what ship: Plymouth - Born: New York - Age: 23 - Sent to Dartmouth on 2 Nov 1814.

Chult, David - Seaman - Nbr: 2740 - How taken: Gave himself up from HMS Salvador - When taken: 4 Jan 1814 - Date Received: 24 Aug 1814 - From what ship: HMT Liverpool, Chatham - Born: Salem - Age: 21 - Died on 3 Mar 1815 from variola.

Church, Benjamin - Seaman - Nbr: 1921 - How taken: Gave himself up from HMS Victory - When taken: 3 Feb 1813 - Date Received: 3 Aug 1814 - From what ship: HMS Alceste, Chatham Depot - Born: Newport - Age: 39 - Released on 2 May 1815.

Church, Jeremiah - Seaman - Nbr: 2067 - How taken: Apprehended at Gravesend - When taken: 16 May 1813 - Date Received: 3 Aug 1814 - From what ship: HMS Bittern, Chatham Depot - Born: Rhode Island - Age: 26 - Released on 2 May 1815.

Church, Richard - Seaman - Nbr: 2585 - How taken: Gave himself up from HMS Ulysses - When taken: 13 Jan 1813 - Date Received: 21 Aug 1814 - From what ship: HMT Freya, Chatham - Born: Baltimore - Age: 32 - Released on 3 May 1815.

Church, William - Prize Master - Nbr: 1289 - Prize: Traveler - Ship Type: Prize - How taken: HM Schooner Canso - When taken: 11 May 1814 - Where taken: off Cape Clear - Date Received: 14 Jun 1814 - From what ship: Mill Prison (Plymouth, England) - Born: Rhode Island - Age: 34 - Released on 28 Apr 1815.

Churchill, Henry - Seaman - Nbr: 2778 - Prize: Tiger - Ship Type: MV - How taken: HMS Scylla - When taken: 22 May 1813 - Where taken: Bay of Biscay - Date Received: 24 Aug 1814 - From what ship: HMT Liverpool, Chatham - Born: Massachusetts - Age: 37 - Released on 19 May 1815.

Churchill, Joseph - Seaman - Nbr: 6353 - Prize: Prince de Neufchatel - Ship Type: P - How taken: Leander (Newcastle Acasta) - When taken: 20 Dec 1814 - Where taken: Lat 38 Long 56 - Date Received: 19 Feb 1815 - From what ship: HMT Ganges, Plymouth - Born: Plymouth - Age: 22 - Released on 11 Jul 1815.

Churchill, Manuel - Seaman - Nbr: 5087 - Prize: Ida - Ship Type: LM - How taken: HMS Newcastle - When taken: 9 Aug 1814 - Where taken: Long 34 - Date Received: 28 Oct 1814 - From what ship: HMT Alkbar, Halifax - Born: Plymouth - Age: 29 - Released on 29 Jun 1815.

Churchill, Men'd. - Seaman - Nbr: 727 - Prize: Ned - Ship Type: LM - How taken: HMS Royalist - When taken: 6 Sep 1813 - Where taken: Bay of Biscay - Date Received: 27 Sep 1813 - From what ship: Plymouth - Born: Plymouth - Age: 24 - Released on 26 Apr 1815.

Claby, Martin - Seaman - Nbr: 4682 - Prize: Taken from an English whaler - Ship Type: MV - How taken: HMS Illustrious - When taken: 22 Oct 1813 - Where taken: at sea - Date Received: 9 Oct 1814 - From what ship: HMT Leyden, Chatham - Born: Boston - Age: 24 - Escaped on 1 Jun 1815.

Clamberg, Peter - Seaman - Nbr: 5934 - Prize: Harlequin - Ship Type: P - How taken: HMS Bulwark - When taken: 23 Nov 1814 - Where taken: off Halifax - Date Received: 27 Dec 1814 - From what ship: HMT Penelope - Born: Holland - Age: 41 - Released on 3 Jul 1815.

Clapp, George - Seaman - Nbr: 6220 - Prize: Prince de Neufchatel - Ship Type: P - How taken: Leander (Newcastle Acasta) - When taken: 20 Dec 1814 - Where taken: Lat 38 Long 56 - Date Received: 30 Jan 1815 - From what ship: HMT Pheasant - Born: Boston - Age: 19 - Released on 5 Jul 1815.

Clapp, George - Seaman - Nbr: 1062 - Prize: Fair American - Ship Type: MV - How taken: HMS Andromache - When taken: 19 Jan 1814 - Where taken: Bay of Biscay - Date Received: 10 May 1814 - From what ship: Plymouth - Born: Boston - Age: 19 - Released on 27 Apr 1815.

Clark, Abraham - Lieutenant - Nbr: 1813 - Prize: True Blooded Yankee - Ship Type: P - How taken: HMS Fame - When taken: 24 Jun 1813 - Where taken: off Brest - Date Received: 20 Jul 1814 - From what ship: HMS Milford, Plymouth - Born: New York - Age: 29 - Released on 20 Apr 1815.

Clark, Amos - Seaman - Nbr: 2617 - How taken: Gave himself up from HMS Tigre - When taken: 15 Aug 1812 - Date Received: 21 Aug 1814 - From what ship: HMT Freya, Chatham - Born: New York - Age: 23 - Released on 26 Apr 1815.

Clark, Elisha - Seaman - Nbr: 1963 - How taken: Gave himself up from HMS Andromache - When taken: 24 Dec 1812 - Date Received: 3 Aug 1814 - From what ship: HMS Lyffey, Chatham Depot - Born: New Bedford -

American Prisoners of War Held at Dartmoor during the War of 1812

## Alphabetical listing of names

    Age: 35 - Released on 26 Apr 1815.

Clark, Francis Marie - Passenger - Nbr: 6272 - Prize: Lion - Ship Type: P - How taken: HMS Granicus - When taken: 2 Dec 1814 - Where taken: off Lisbon - Date Received: 17 Feb 1815 - From what ship: HMT Ganges, Plymouth - Born: L'Orient (France) - Age: 23 - Escaped on 1 Jun 1815.

Clark, Jacob - Seaman - Nbr: 923 - How taken: Gave himself up from HMS Basilisk - Date Received: 31 Jan 1814 - From what ship: Plymouth - Born: Philadelphia - Age: 38 - Race: Negro - Released on 27 Apr 1815.

Clark, John - Seaman - Nbr: 6447 - Prize: Jemmell - Ship Type: MV - How taken: HMS Rhin - When taken: 28 May 1814 - Where taken: off Bermuda - Date Received: 3 Mar 1815 - From what ship: HMT Ganges, Plymouth - Born: Harwich - Age: 26 - Released on 11 Jul 1815.

Clark, John - Prize Master - Nbr: 2251 - Prize: Frederick - Ship Type: Prize - How taken: HMS Curlew - When taken: 4 May 1814 - Where taken: off Halifax - Date Received: 16 Aug 1814 - From what ship: HMS Dublin, Halifax - Born: Massachusetts - Age: 36 - Released on 3 May 1815.

Clark, John - Seaman - Nbr: 2394 - Prize: US Sloop Frolic - Ship Type: MW - How taken: HMS Orpheus - When taken: 20 Apr 1814 - Where taken: off Cuba - Date Received: 16 Aug 1814 - From what ship: HMT Queen, Halifax - Born: Maryland - Age: 27 - Sent to Dartmouth on 19 Oct 1814.

Clark, John - Seaman - Nbr: 614 - Prize: US Brig Argus - Ship Type: MW - How taken: HMS Pelican - When taken: 14 Aug 1813 - Where taken: Irish Channel - Date Received: 8 Sep 1813 - From what ship: Plymouth - Born: Gloucester - Age: 24 - Sent to Dartmouth on 2 Nov 1814.

Clark, John - Seaman - Nbr: 3963 - Prize: Rolla - Ship Type: P - How taken: HMS Loire - When taken: 10 Dec 1813 - Where taken: off Bull Island (South Carolina) - Date Received: 5 Oct 1814 - From what ship: HMT President, Halifax - Born: Philadelphia - Age: 27 - Released on 9 Jun 1815.

Clark, John - Soldier - Nbr: 5265 - Prize: 13th US Infantry - Ship Type: Troops - How taken: British Army - When taken: 13 Oct 1812 - Where taken: Canada - Date Received: 31 Oct 1814 - From what ship: HMT Leyden, Chatham - Born: Galway - Age: 46 - Released on 29 Jun 1815.

Clark, John D. - Seaman - Nbr: 2729 - How taken: Gave himself up from HMS Royal William - When taken: 18 Nov 1812 - Date Received: 24 Aug 1814 - From what ship: HMT Liverpool, Chatham - Born: New York - Age: 30 - Released on 26 Apr 1815.

Clark, Joseph - Seaman - Nbr: 1874 - Prize: Elbridge Gerry - Ship Type: P - How taken: HMS Crescent - When taken: 16 Sep 1813 - Where taken: off St. George's - Date Received: 29 Jul 1814 - From what ship: HMS Ville de Paris, Chatham Depot - Born: Cape Elizabeth - Age: 19 - Released on 2 May 1815.

Clark, Peleg - Seaman - Nbr: 3610 - Prize: Hornsby - Ship Type: P - How taken: HMS Sceptre - When taken: 4 Feb 1814 - Where taken: off Block Island - Date Received: 30 Sep 1814 - From what ship: HMT Sybella - Born: Newport - Age: 39 - Released on 4 Jun 1815.

Clark, Philip - Carpenter - Nbr: 1730 - How taken: Gave himself up from Bristol - When taken: 7 Jul 1814 - Date rec'd: 17 Jul 1814 - What ship: Bristol - Born: Monmouth County - Age: 44 - Released on 1 May 1815.

Clark, Samuel - Seaman - Nbr: 3440 - How taken: Gave himself up from HMS Clorinda - When taken: 18 Dec 1813 - Date Received: 13 Sep 1814 - From what ship: HMT Niobe, Chatham - Born: Pennsylvania - Age: 32 - Released on 28 May 1815.

Clark, Samuel - Seaman - Nbr: 4631 - Prize: Dart - Ship Type: MV - How taken: HMS Petrel - When taken: 5 Mar 1813 - Where taken: at sea - Date Received: 9 Oct 1814 - From what ship: HMT Leyden, Chatham - Born: Alfred - Age: 28 - Released on 15 Jun 1815.

Clark, Thomas - Cook - Nbr: 4972 - Prize: Invincible - Ship Type: LM - How taken: HMS Armide - When taken: 15 Aug 1814 - Where taken: off Nantucket - Date Received: 28 Oct 1814 - From what ship: HMT Alkbar, Halifax - Born: Guadeloupe - Age: 26 - Race: Negro - Released on 21 Jun 1815.

Clark, Titus - Mate - Nbr: 3083 - Prize: Eliza (Whaler) - Ship Type: MV - How taken: HMS Charles - When taken: 20 Jul 1813 - Where taken: off Peter Head - Date Received: 3 Sep 1814 - From what ship: HMT Hydra, Chatham - Born: Connecticut - Age: 22 - Sent to Ashburton (England) on 16 Oct 1814.

Clark, William - Seaman - Nbr: 2208 - Prize: Hussar - Ship Type: P - How taken: HMS Saturn - When taken: 25 May 1814 - Where taken: off Sandy Hook - Date Received: 16 Aug 1814 - From what ship: HMS Dublin, Halifax - Born: Massachusetts - Age: 20 - Released on 2 May 1815.

Clark, William - Seaman - Nbr: 2044 - How taken: Gave himself up from HMS North Star - When taken: 16 May 1812 - Date Received: 3 Aug 1814 - From what ship: HMS Lyffey, Chatham Depot - Born: Philadelphia - Age: 33 - Released on 26 Apr 1815.

Clark, William - Seaman - Nbr: 191 - Prize: Star - Ship Type: MV - How taken: HMS Superb - When taken: 9 Feb 1813 - Where taken: Bay of Biscay - Date Received: 2 Apr 1813 - From what ship: Plymouth - Born:

American Prisoners of War Held at Dartmoor during the War of 1812

## Alphabetical listing of names

Newport - Age: 19 - Died on 21 Oct 1813 from small pox.

Clark, William - Seaman - Nbr: 3280 - Prize: Montgomery - Ship Type: P - How taken: HMS Nymphe - When taken: 5 May 1813 - Where taken: off Cape Ann - Date Received: 13 Sep 1814 - From what ship: HMT Niobe, Chatham - Born: Marblehead - Age: 25 - Released on 28 May 1815.

Clarke, Aaron - Marine - Nbr: 5942 - Prize: Harlequin - Ship Type: P - How taken: HMS Bulwark - When taken: 23 Nov 1814 - Where taken: off Halifax - Date Received: 27 Dec 1814 - From what ship: HMT Penelope - Born: Berwick - Age: 19 - Released on 3 Jul 1815.

Clarke, Isaac - Seaman - Nbr: 4490 - Prize: Enterprize - Ship Type: P - How taken: HMS Tenedos - When taken: 21 May 1813 - Where taken: off Cape Cod - Date Received: 8 Oct 1814 - From what ship: HMT Leyden, Chatham - Born: Massachusetts - Age: 25 - Released on 15 Jun 1815.

Clarke, John - Seaman - Nbr: 4651 - How taken: Apprehended at Gosport (England) - When taken: 2 Sep 1813 - Date Received: 9 Oct 1814 - From what ship: HMT Leyden, Chatham - Born: Boston - Age: 29 - Released on 15 Jun 1815.

Clarke, Peter - Seaman - Nbr: 2163 - Prize: Rattlesnake - Ship Type: P - How taken: HMS Hyperion - When taken: 26 Jun 1814 - Where taken: off Cape Finisterre - Date Received: 16 Aug 1814 - From what ship: HMS Dublin, Halifax - Born: New Jersey - Age: 16 - Released on 2 May 1815.

Clarke, Simon - Seaman - Nbr: 2367 - Prize: Snap Dragon - Ship Type: P - How taken: HMS Martin - When taken: 10 Jun 1814 - Where taken: off Halifax - Date Received: 16 Aug 1814 - From what ship: HMT Queen, Halifax - Born: North Carolina - Age: 16 - Race: Negro - Died on 24 Jan 1815 from variola.

Clarke, William - Seaman - Nbr: 6146 - How taken: Gave himself up from HMS Kennerine - Date rec'd: 17 Jan 1815 - What ship: HMT Impregnable - Born: Newbury - Age: 24 - Race: Black - Released on 5 Jul 1815.

Clarke, William - Gunner - Nbr: 4919 - Prize: Starks - Ship Type: P - How taken: HMS Sophia - When taken: 24 Apr 1814 - Where taken: off Bermuda - Date Received: 28 Oct 1814 - From what ship: HMT Alkbar, Halifax - Born: New York - Age: 21 - Released on 21 Jun 1815.

Clausen, John - Seaman - Nbr: 3176 - How taken: Gave himself up from HMS Centaur - When taken: 10 Sep 1813 - Date Received: 11 Sep 1814 - From what ship: HMT Freya, Chatham - Born: Vermont - Age: 19 - Released on 28 May 1815.

Claw, Morris - Seaman - Nbr: 1809 - How taken: Gave himself up from HMS Ville de Paris - When taken: 28 Apr 1813 - Date Received: 20 Jul 1814 - From what ship: HMS Milford, Plymouth - Born: Long Island - Age: 31 - Race: Black - Released on 1 May 1815.

Clay, David - Seaman - Nbr: 5948 - Prize: Harlequin - Ship Type: P - How taken: HMS Bulwark - When taken: 23 Nov 1814 - Where taken: off Halifax - Date Received: 27 Dec 1814 - From what ship: HMT Penelope - Born: Lee - Age: 23 - Released on 3 Jul 1815.

Clay, John - Seaman - Nbr: 2626 - How taken: Gave himself up from HMS Freya - When taken: 13 Dec 1812 - Date Received: 21 Aug 1814 - From what ship: HMT Freya, Chatham - Born: Norfolk - Age: 57 - Released on 26 Apr 1815.

Cleeves, Jonathan - Seaman - Nbr: 6012 - Prize: McDonough - Ship Type: P - How taken: HMS Bacchante - When taken: 1 Nov 1814 - Where taken: Lat 42 Long 67 - Date Received: 28 Dec 1814 - From what ship: HMT Penelope - Born: Norfolk - Age: 57 - Released on 3 Jul 1815.

Clements, John C. - Seaman - Nbr: 196 - Prize: Star - Ship Type: MV - How taken: HMS Superb - When taken: 9 Feb 1813 - Where taken: Bay of Biscay - Date Received: 2 Apr 1813 - From what ship: Plymouth - Born: New Jersey - Age: 23 - Released on 20 Apr 1815.

Clements, Samuel - Seaman - Nbr: 2987 - Prize: Frolic - Ship Type: P - How taken: HMS Heron - When taken: 25 Jan 1814 - Where taken: off St. Thomas - Date Received: 29 Aug 1814 - From what ship: HMT Bittern - Born: Salem - Age: 50 - Released on 19 May 1815.

Clements, William - Seaman - Nbr: 4143 - How taken: Impressed at Shields - When taken: 16 Jun 1813 - Date Received: 6 Oct 1814 - From what ship: HMT Niobe, Chatham - Born: Guildford - Age: 29 - Released on 13 Jun 1815.

Clements, William - Seaman - Nbr: 3126 - How taken: Impressed at London - When taken: 9 Sep 1813 - Date Received: 11 Sep 1814 - From what ship: HMT Freya, Chatham - Born: Salem - Age: 23 - Released on 28 May 1815.

Clepp, Abraham - Seaman - Nbr: 1488 - Prize: Napoleon - Ship Type: LM - How taken: HMS Belle Poule - When taken: 3 Apr 1813 - Where taken: off Cape Ortegal (Spain) - Date Received: 19 Jun 1814 - From what ship: Stapleton - Born: Massachusetts - Age: 22 - Released on 1 May 1815.

Clerk, James - Seaman - Nbr: 6258 - How taken: Impressed at Liverpool - When taken: 21 Dec 1813 - Date

American Prisoners of War Held at Dartmoor during the War of 1812

## Alphabetical listing of names

Received: 4 Feb 1815 - From what ship: HMT Ganges - Born: Bristol - Age: 23 - Released on 5 Jul 1815.

Clerke, George - Master of Arms - Nbr: 559 - Prize: US Brig Argus - Ship Type: MW - How taken: HMS Pelican - When taken: 14 Apr 1813 - Where taken: Irish Channel - Date Received: 8 Sep 1813 - From what ship: Plymouth - Born: New York - Age: 22 - Sent to Dartmouth on 2 Nov 1814.

Cleveland, Davis - Seaman - Nbr: 1971 - Prize: Orbit - Ship Type: MV - How taken: HMS Achates - When taken: 29 Jan 1813 - Where taken: Lat 44N Long 13W - Date Received: 3 Aug 1814 - From what ship: HMS Lyffey, Chatham Depot - Born: Nantucket - Age: 19 - Released on 2 May 1815.

Cleveland, Ebenezer - Seaman - Nbr: 1865 - Prize: Porcupine - Ship Type: LM - How taken: HMS Acasta - When taken: 17 Jun 1813 - Where taken: off Sable Island - Date Received: 29 Jul 1814 - From what ship: HMS Ville de Paris, Chatham Depot - Born: Maryland - Age: 24 - Race: Black - Released on 2 May 1815.

Cleverley, William - Seaman - Nbr: 6081 - Prize: David Porter - Ship Type: LM - How taken: HMS Pylades - When taken: 13 Sep 1814 - Where taken: Georges Bank - Date Received: 28 Dec 1814 - From what ship: HMT Penelope - Born: Massachusetts - Age: 17 - Released on 3 Jul 1815.

Clexton, R. - Seaman - Nbr: 2973 - Prize: Frolic - Ship Type: P - How taken: HMS Heron - When taken: 25 Jan 1814 - Where taken: off St. Thomas - Date Received: 29 Aug 1814 - From what ship: HMT Bittern - Born: Beverly - Age: 19 - Released on 19 May 1815.

Cliff, Peter - Armourer - Nbr: 2282 - Prize: Diomede - Ship Type: P - How taken: HMS Rifleman - When taken: 28 May 1814 - Where taken: off Sable Island - Date Received: 16 Aug 1814 - From what ship: HMS Dublin, Halifax - Born: Ipswich - Age: 23 - Released on 3 May 1815.

Clifford, L. L. - Seaman - Nbr: 2763 - How taken: Gave himself up from HMS Bruin - When taken: 19 Jul 1813 - Date Received: 24 Aug 1814 - From what ship: HMT Liverpool, Chatham - Born: Philadelphia - Age: 33 - Released on 19 May 1815.

Clim, George - Seaman - Nbr: 5610 - How taken: Gave himself up from HMS Ashea - When taken: 13 May 1814 - Date rec'd: 24 Dec 1814 - What ship: HMT Tay - Born: Philadelphia - Age: 21 - Escaped on 1 Jun 1815.

Cline, Lewis - Seaman - Nbr: 4138 - Prize: Bordeaux Packet - Ship Type: LM - How taken: HMS Niemen - When taken: 28 Jun 1814 - Where taken: off Delaware - Date Received: 6 Oct 1814 - From what ship: HMT Chesapeake, Halifax - Born: Baltimore - Age: 26 - Released on 13 Jun 1815.

Clintic, Ralph - Seaman - Nbr: 5518 - Prize: Regent - Ship Type: LM - How taken: HMS Forth - When taken: 19 Sep 1814 - Where taken: off Egg Harbor (New Jersey) - Date Received: 17 Dec 1814 - From what ship: HMT Loire, Halifax - Born: New York - Age: 21 - Released on 1 Jul 1815.

Clothey, Thomas - Seaman - Nbr: 1574 - Prize: Essex - Ship Type: MV - How taken: HMS Pyramus - When taken: 20 Apr 1813 - Where taken: Bay of Biscay - Date Received: 23 Jun 1814 - From what ship: Stapleton - Born: Marblehead - Age: 20 - Released on 1 May 1815.

Clothy, John - Seaman - Nbr: 3292 - Prize: Enterprize - Ship Type: P - How taken: HMS Tenedos - When taken: 21 May 1813 - Where taken: off Cape Cod - Date Received: 13 Sep 1814 - From what ship: HMT Niobe, Chatham - Born: Marblehead - Age: 41 - Released on 28 May 1815.

Clothy, William - Seaman - Nbr: 3296 - Prize: Enterprize - Ship Type: P - How taken: HMS Tenedos - When taken: 21 May 1813 - Where taken: off Cape Cod - Date Received: 13 Sep 1814 - From what ship: HMT Niobe, Chatham - Born: Marblehead - Age: 34 - Released on 28 May 1815.

Clough, David - Seaman - Nbr: 6025 - Prize: McDonough - Ship Type: P - How taken: HMS Bacchante - When taken: 1 Nov 1814 - Where taken: Lat 42 Long 67 - Date Received: 28 Dec 1814 - From what ship: HMT Penelope - Born: Massachusetts - Age: 34 - Released on 3 Jul 1815.

Clough, Enoch - Seaman - Nbr: 6002 - Prize: McDonough - Ship Type: P - How taken: HMS Bacchante - When taken: 1 Nov 1814 - Where taken: Lat 42 Long 67 - Date Received: 28 Dec 1814 - From what ship: HMT Penelope - Born: Massachusetts - Age: 30 - Released on 3 Jul 1815.

Clough, Obed - Seaman - Nbr: 6001 - Prize: McDonough - Ship Type: P - How taken: HMS Bacchante - When taken: 1 Nov 1814 - Where taken: Lat 42 Long 67 - Date Received: 28 Dec 1814 - From what ship: HMT Penelope - Born: Massachusetts - Age: 27 - Released on 11 Jul 1815.

Clough, Shadrick - Seaman - Nbr: 6003 - Prize: McDonough - Ship Type: P - How taken: HMS Bacchante - When taken: 1 Nov 1814 - Where taken: Lat 42 Long 67 - Date Received: 28 Dec 1814 - From what ship: HMT Penelope - Born: Massachusetts - Age: 24 - Released on 3 Jul 1815.

Cloutman, Ephraim - Seaman - Nbr: 6374 - Prize: US Brig Syren - Ship Type: MW - How taken: HMS Medway - When taken: 12 Jul 1814 - Where taken: off Cape of Good Hope - Date Received: 24 Feb 1815 - From what ship: HMT Ganges, Plymouth - Born: Salem - Age: 32 - Released on 11 Jul 1815.

Cloutman, Joseph - Seaman - Nbr: 3288 - Prize: Enterprize - Ship Type: P - How taken: HMS Tenedos - When

American Prisoners of War Held at Dartmoor during the War of 1812

## Alphabetical listing of names

taken: 21 May 1813 - Where taken: off Cape Cod - Date Received: 13 Sep 1814 - From what ship: HMT Niobe, Chatham - Born: Massachusetts - Age: 21 - Released on 28 May 1815.

Cloutman, Robert - Master - Nbr: 4529 - Prize: Wolf Cove - Ship Type: MV - How taken: HMS Briton - When taken: 1 Dec 1812 - Where taken: off Brest - Date Received: 8 Oct 1814 - From what ship: HMT Leyden, Chatham - Born: Massachusetts - Age: 51 - Released on 27 Apr 1815.

Cloutman, Samuel - Carpenter - Nbr: 5467 - Prize: General Putnam - Ship Type: P - How taken: HMS Leander - When taken: 8 Nov 1814 - Where taken: Long 65 Lat 42 - Date Received: 17 Dec 1814 - From what ship: HMT Loire, Halifax - Born: Salem - Age: 27 - Released on 1 Jul 1815.

Cloutman, Thomas - 1st Lieutenant - Nbr: 5461 - Prize: General Putnam - Ship Type: P - How taken: HMS Leander - When taken: 8 Nov 1814 - Where taken: Long 65 Lat 42 - Date Received: 17 Dec 1814 - From what ship: HMT Loire, Halifax - Born: Marblehead - Age: 39 - Released on 1 Jul 1815.

Clover, Louis - Seaman - Nbr: 5392 - Prize: Union - Ship Type: MV - How taken: HMS Malabar - When taken: 17 Jan 1814 - Where taken: off Calcutta - Date Received: 31 Oct 1814 - From what ship: HMT Leyden, Chatham - Born: Jersey - Age: 23 - Released on 1 Jul 1815.

Clowes, Philip - Seaman - Nbr: 6235 - Prize: Prince de Neufchatel - Ship Type: P - How taken: Leander (Newcastle Acasta) - When taken: 20 Dec 1814 - Where taken: Lat 38 Long 56 - Date Received: 30 Jan 1815 - From what ship: HMT Pheasant - Born: Boston - Age: 21 - Released on 5 Jul 1815.

Clyde, John - 2nd Gunner - Nbr: 6370 - Prize: US Brig Syren - Ship Type: MW - How taken: HMS Medway - When taken: 12 Jul 1814 - Where taken: off Cape of Good Hope - Date Received: 24 Feb 1815 - From what ship: HMT Ganges, Plymouth - Born: Oldenburg - Age: 41 - Released on 6 Jul 1815.

Coader, Anthony - Steward - Nbr: 660 - Prize: Marmion - Ship Type: MV - How taken: HMS President - When taken: 14 Aug 1813 - Where taken: off Nantes - Date Received: 8 Sep 1813 - From what ship: Plymouth - Born: New Orleans - Age: 28 - Race: Mulatto - Sent to Ashburton (England) on 20 Sep 1813.

Coas, Samuel - Seaman - Nbr: 1886 - Prize: Watson - Ship Type: Prize - How taken: HMS Poictiers - When taken: 3 Sep 1813 - Where taken: off Sable Island - Date Received: 29 Jul 1814 - From what ship: HMS Ville de Paris, Chatham Depot - Born: Cape Ann - Age: 25 - Released on 2 May 1815.

Coates, Russell - Seaman - Nbr: 5618 - Prize: Mary - Ship Type: P - How taken: Not legible - When taken: 18 Jul 1814 - Date Received: 24 Dec 1814 - From what ship: HMT Tay - Born: Boston - Age: 20 - Released on 3 Jul 1815.

Coates, Samuel M. - Boatswain - Nbr: 6367 - Prize: US Brig Syren - Ship Type: MW - How taken: HMS Medway - When taken: 12 Jul 1814 - Where taken: off Cape of Good Hope - Date Received: 24 Feb 1815 - From what ship: HMT Ganges, Plymouth - Born: Baltimore - Age: 33 - Escaped on 1 Jun 1815.

Cobb, Josiah - Seaman - Nbr: 6234 - Prize: Prince de Neufchatel - Ship Type: P - How taken: Leander (Newcastle Acasta) - When taken: 20 Dec 1814 - Where taken: Lat 38 Long 56 - Date Received: 30 Jan 1815 - From what ship: HMT Pheasant - Born: Massachusetts - Age: 19 - Released on 5 Jul 1815.

Cobbs, James - Seaman - Nbr: 635 - Prize: US Brig Argus - Ship Type: MW - How taken: HMS Pelican - When taken: 14 Aug 1813 - Where taken: Irish Channel - Date Received: 8 Sep 1813 - From what ship: Plymouth - Born: West Easset - Age: 22 - Died on 20 Mar 1814 from diarrhea.

Coburn, Abel - Seaman - Nbr: 1326 - How taken: Impressed - When taken: 7 Jul 1813 - Date Received: 19 Jun 1814 - From what ship: Stapleton - Born: Connecticut - Age: 24 - Released on 28 Apr 1815.

Cochrane, William - 2nd Lieutenant - Nbr: 4929 - Prize: Herald - Ship Type: P - How taken: HMS Endymion - When taken: 15 Aug 1814 - Where taken: off Nantucket - Date Received: 28 Oct 1814 - From what ship: HMT Alkbar, Halifax - Born: New York - Age: 27 - Released on 21 Jun 1815.

Codding, Caleb - Seaman - Nbr: 1819 - How taken: Gave himself up from HMS Swiftsure - When taken: 26 Dec 1813 - Date Received: 20 Jul 1814 - From what ship: HMS Milford, Plymouth - Born: Massachusetts - Age: 48 - Released on 28 May 1815.

Codman, Richard - Seaman - Nbr: 1434 - Prize: Leo - Ship Type: LM - How taken: HMS Magicienne - When taken: 4 Jun 1813 - Where taken: off France - Date Received: 19 Jun 1814 - From what ship: Stapleton - Born: Portland - Age: 20 - Released on 28 Apr 1815.

Codshell, Joseph - Seaman - Nbr: 2812 - Prize: Hope & Anchor - Ship Type: MV - How taken: HMS Nightingale - When taken: 25 Mar 1813 - Where taken: off the Traw - Date Received: 24 Aug 1814 - From what ship: HMT Liverpool, Chatham - Born: Newport - Age: 21 - Released on 19 May 1815.

Cody, James - Seaman - Nbr: 5056 - Prize: Ida - Ship Type: LM - How taken: HMS Newcastle - When taken: 9 Aug 1814 - Where taken: Long 34 - Date Received: 28 Oct 1814 - From what ship: HMT Alkbar, Halifax - Born: Boston - Age: 22 - Released on 29 Jun 1815.

American Prisoners of War Held at Dartmoor during the War of 1812

## Alphabetical listing of names

Coffee, Jacob - Seaman - Nbr: 6284 - How taken: Gave himself up from HMS Growler - Date Received: 17 Feb 1815 - From what ship: HMT Ganges, Plymouth - Born: Boston - Age: 22 - Race: Mulatto - Released on 5 Jul 1815.

Coffee, Ram'l - Seaman - Nbr: 5118 - Prize: Portsmouth - Ship Type: P - How taken: HMS Pylades - When taken: 13 Sep 1814 - Where taken: Long 65 - Date Received: 28 Oct 1814 - From what ship: HMT Alkbar, Halifax - Born: New York - Age: 26 - Race: Negro - Died on 4 Dec 1814 from pneumonia.

Coffee, William - Seaman - Nbr: 968 - Prize: Siro - Ship Type: LM - How taken: HMS Pelican - When taken: 13 Jan 1814 - Where taken: at sea - Date Received: 31 Jan 1814 - From what ship: Plymouth - Born: Georgetown - Age: 24 - Released on 27 Apr 1815.

Coffin, Abel - Seaman - Nbr: 4593 - Prize: Prize to the Diomede - Ship Type: P - How taken: HMS Sapphire - When taken: 24 Feb 1814 - Where taken: off America - Date Received: 9 Oct 1814 - From what ship: HMT Leyden, Chatham - Born: Newbury - Age: 22 - Released on 15 Jun 1815.

Coffin, Alexander - Seaman - Nbr: 910 - Prize: Zephyr - Ship Type: MV - How taken: HMS Pyramus - When taken: 30 Nov 1813 - Where taken: off L'Orient (France) - Date Received: 31 Jan 1814 - From what ship: Plymouth - Born: Nantucket - Age: 20 - Sent to Ashburton (England) on 15 Feb 1814.

Coffin, Caleb - Prize Master - Nbr: 3455 - Prize: Prize to Chasseur - How taken: HMS Tartarus - When taken: 1 Sep 1814 - Where taken: Atlantic - Date Received: 19 Sep 1814 - From what ship: HMT Salvador del Mundo - Born: Nantucket - Age: 34 - Released on 11 Jul 1815.

Coffin, Edward - Seaman - Nbr: 4283 - Prize: Pilot - Ship Type: LM - How taken: Victoria (Privateer) - When taken: 28 Jan 1814 - Where taken: off Bordeaux - Date Received: 7 Oct 1814 - From what ship: HMT Niobe, Chatham - Born: Nantucket - Age: 24 - Released on 14 Jun 1815.

Coffin, Frederick - Seaman - Nbr: 2201 - Prize: Hussar - Ship Type: P - How taken: HMS Saturn - When taken: 25 May 1814 - Where taken: off Sandy Hook - Date Received: 16 Aug 1814 - From what ship: HMS Dublin, Halifax - Born: Nantucket - Age: 41 - Released on 2 May 1815.

Coffin, George - Seaman - Nbr: 2422 - Prize: US Sloop Frolic - Ship Type: MW - How taken: HMS Orpheus - When taken: 20 Apr 1814 - Where taken: off Cuba - Date Received: 16 Aug 1814 - From what ship: HMT Queen, Halifax - Born: Massachusetts - Age: 26 - Released on 3 May 1815.

Coffin, James - Seaman - Nbr: 3231 - Prize: Sampson - Ship Type: MV - How taken: Rebuff - When taken: 12 May 1813 - Where taken: off Cape St. Vincent - Date Received: 11 Sep 1814 - From what ship: HMT Freya, Chatham - Born: Washington - Age: 28 - Released on 28 May 1815.

Coffin, John - Seaman - Nbr: 722 - Prize: Francis and Ann - Ship Type: MV - How taken: HMS Lightning - When taken: 20 Apr 1813 - Where taken: at sea - Date Received: 27 Sep 1813 - From what ship: Plymouth - Born: New York - Age: 23 - Released on 26 Apr 1815.

Coffin, Joseph - Seaman - Nbr: 4687 - How taken: Gave himself up from HMS Invincible - When taken: 14 Jan 1813 - Date Received: 9 Oct 1814 - From what ship: HMT Leyden, Chatham - Born: Rhode Island - Age: 29 - Released on 15 Jun 1815.

Coffin, Samuel - Cook - Nbr: 273 - Prize: Cannoneer - Ship Type: MV - How taken: HMS Warspite - When taken: 14 Mar 1813 - Where taken: Bay of Biscay - Date Received: 28 Jun 1813 - From what ship: Plymouth - Born: Boston - Age: 26 - Released on 20 Apr 1815.

Coffin, Theodore - Prize master - Nbr: 6303 - Prize: John - Ship Type: Prize - How taken: Leander (Newcastle Acasta) - When taken: 30 Jan 1815 - Date Received: 19 Feb 1815 - From what ship: HMT Ganges, Plymouth - Born: Nantucket - Age: 26 - Escaped on 1 Jun 1815.

Coffin, Valentine - 2nd Mate - Nbr: 4762 - Prize: Ocean - Ship Type: MV - How taken: HMS Atalante - When taken: 16 Nov 1812 - Where taken: off Cape of Good Hope - Date Received: 9 Oct 1814 - From what ship: HMT Freya, Chatham - Born: Nantucket - Age: 33 - Released on 27 Apr 1815.

Cogent, Benjamin - Seaman - Nbr: 3947 - Prize: Rolla - Ship Type: P - How taken: HMS Loire - When taken: 10 Dec 1813 - Where taken: off Bull Island (South Carolina) - Date Received: 5 Oct 1814 - From what ship: HMT President, Halifax - Born: Newport - Age: 22 - Released on 9 Jun 1815.

Cogwell, James - Seaman - Nbr: 2261 - Prize: Prize to the Scourge - Ship Type: P - How taken: HMS Martin - When taken: 4 May 1814 - Where taken: off Newfoundland - Date Received: 16 Aug 1814 - From what ship: HMS Dublin, Halifax - Born: Milford - Age: 19 - Released on 3 May 1815.

Cohoun, William - Seaman - Nbr: 2334 - Prize: Snap Dragon - Ship Type: P - How taken: HMS Martin - When taken: 10 Jun 1814 - Where taken: off Halifax - Date Received: 16 Aug 1814 - From what ship: HMS Dublin, Halifax - Born: Philadelphia - Age: 22 - Released on 3 May 1815.

Colborne, John L. - Seaman - Nbr: 3337 - Prize: Prize of the President - Ship Type: MV - How taken: HMS Regulus

American Prisoners of War Held at Dartmoor during the War of 1812

## Alphabetical listing of names

- When taken: 4 Sep 1813 - Where taken: off Newfoundland - Date Received: 13 Sep 1814 - From what ship: HMT Niobe, Chatham - Born: North Carolina - Age: 33 - Released on 28 May 1815.
Colbourn, James - Seaman - Nbr: 4940 - Prize: Herald - Ship Type: P - How taken: HMS Endymion - When taken: 15 Aug 1814 - Where taken: off Nantucket - Date Received: 28 Oct 1814 - From what ship: HMT Alkbar, Halifax - Born: Virginia - Age: 19 - Released on 21 Jun 1815.
Colby, Benjamin - Seaman - Nbr: 6192 - Prize: Prince de Neufchatel - Ship Type: P - How taken: Leander (Newcastle Acasta) - When taken: 20 Dec 1814 - Where taken: Lat 38 Long 56 - Date Received: 30 Jan 1815 - From what ship: HMT Pheasant - Born: New Hampshire - Age: 23 - Released on 5 Jul 1815.
Colcocha, Anthony - Seaman - Nbr: 83 - Prize: Rolla - Ship Type: MV - How taken: HMS Surveillante - When taken: 11 Feb 1813 - Where taken: Bay of Biscay - Date Received: 2 Apr 1813 - From what ship: Plymouth - Born: New Orleans - Age: 26 - Sent to Plymouth on 7 Dec 1813.
Cole, Benjamin - Seaman - Nbr: 3020 - How taken: Gave himself up from HMS Zealous - When taken: 27 Aug 1814 - Date Received: 2 Sep 1814 - From what ship: HMT Centaur - Born: New York - Age: 48 - Released on 28 May 1815.
Cole, Charles - Seaman - Nbr: 6158 - Prize: Lion - Ship Type: P - How taken: HMS Granicus - When taken: 2 Dec 1814 - Where taken: off Lisbon - Date Received: 18 Jan 1815 - From what ship: HMT Impregnable - Born: Bordeaux - Age: 24 - Released on 5 Jul 1815.
Cole, David - Seaman - Nbr: 5983 - Prize: Cossack - Ship Type: Private - How taken: HMS Bulwark - When taken: 1 Nov 1814 - Where taken: Georges Bank - Date Received: 27 Dec 1814 - From what ship: HMT Penelope - Born: Brattleboro - Age: 26 - Released on 3 Jul 1815.
Cole, H. A. - Prize master - Nbr: 3913 - Prize: Prize to Yankee - Ship Type: Prize - How taken: HMS Narcissus - When taken: 4 Oct 1813 - Where taken: Nantucket Shoals - Date Received: 5 Oct 1814 - From what ship: HMT President, Halifax - Born: North Kingstown - Age: 23 - Released on 9 Jun 1815.
Cole, John - Seaman - Nbr: 1253 - Prize: Adeline - Ship Type: MV - How taken: HMS Magicienne - When taken: 16 Mar 1814 - Where taken: off Cape Finisterre - Date Received: 14 Jun 1814 - From what ship: Mill Prison (Plymouth, England) - Born: Baltimore - Age: 38 - Race: Black - Died on 26 Nov 1814 from pneumonia.
Cole, John - Seaman - Nbr: 3759 - Prize: Alfred - Ship Type: P - How taken: HMS Epervier - When taken: 23 Feb 1814 - Where taken: off Newfoundland - Date Received: 30 Sep 1814 - From what ship: HMT President, Halifax - Born: Marblehead - Age: 19 - Released on 4 Jun 1815.
Cole, Nathaniel - Prize master - Nbr: 5606 - Prize: Mentor - Ship Type: Prize - How taken: HMS Maidstone - When taken: 26 Oct 1814 - Where taken: off Cape Sable Island (Canada) - Date Received: 18 Dec 1814 - From what ship: Returned after escaping - Born: Biddeford - Age: 25 - Released on 5 Jul 1815.
Cole, Nathaniel - Seaman - Nbr: 4052 - Prize: US Brig Rattlesnake - Ship Type: MW - How taken: HMS Leander - When taken: 13 Jul 1814 - Where taken: off Shelburne - Date Received: 6 Oct 1814 - From what ship: HMT Chesapeake, Halifax - Born: Massachusetts - Age: 32 - Released on 13 Jun 1815.
Cole, Nathaniel - Seaman - Nbr: 3966 - Prize: Rolla - Ship Type: P - How taken: HMS Loire - When taken: 10 Dec 1813 - Where taken: off Bull Island (South Carolina) - Date Received: 5 Oct 1814 - From what ship: HMT President, Halifax - Born: Charleston - Age: 32 - Released on 9 Jun 1815.
Cole, Perry - Seaman - Nbr: 3104 - Prize: Mary - Ship Type: Prize - How taken: Taken by pilot boat - When taken: 27 Aug 1814 - Where taken: Bristol Channel - Date Received: 9 Sep 1814 - From what ship: HMS Abercrombie - Born: Baltimore - Age: 23 - Released on 28 May 1815.
Cole, William - Seaman - Nbr: 2088 - Prize: True Blooded Yankee - Ship Type: P - How taken: HMS Hope - When taken: 24 Jun 1813 - Where taken: off Brest - Date Received: 3 Aug 1814 - From what ship: HMS Bittern, Chatham Depot - Born: Virginia - Age: 25 - Released on 2 May 1815.
Cole, William - Seaman - Nbr: 2787 - Prize: Prompt - Ship Type: MV - How taken: Chance (Privateer) - When taken: 28 Mar 1813 - Where taken: Bay of Biscay - Date Received: 24 Aug 1814 - From what ship: HMT Liverpool, Chatham - Born: Cape Cod - Age: 23 - Released on 19 May 1815.
Coleby, Ebenezer - Seaman - Nbr: 2569 - Prize: Rattlesnake - Ship Type: P - How taken: HMS Hyperion - When taken: 25 Jun 1814 - Where taken: off Cape Finisterre - Date Received: 21 Aug 1814 - From what ship: HMS Hyperion - Born: Salisbury - Age: 26 - Released on 3 May 1815.
Coleman, Andrew - Seaman - Nbr: 5758 - Prize: Hope - Ship Type: MV - How taken: HMS Nereus - When taken: 14 May 1814 - Where taken: Rio de la Plata - Date Received: 26 Dec 1814 - From what ship: HMT Argo - Born: Hamburg - Age: 39 - Released on 3 Jul 1815.
Coleman, James - Seaman - Nbr: 5732 - How taken: Gave himself up from HMS Ashea - When taken: 12 Apr 1814 - Date Received: 26 Dec 1814 - From what ship: HMT Argo - Born: Boston - Age: 33 - Released on 27 Apr

American Prisoners of War Held at Dartmoor during the War of 1812

## Alphabetical listing of names

    1815.

Coleman, Joseph - Seaman - Nbr: 4984 - Prize: Invincible - Ship Type: LM - How taken: HMS Armide - When taken: 15 Aug 1814 - Where taken: off Nantucket - Date Received: 28 Oct 1814 - From what ship: HMT Alkbar, Halifax - Born: Baltimore - Age: 19 - Released on 21 Jun 1815.

Coleman, William - Seaman - Nbr: 1082 - Prize: Lyon - Ship Type: MV - How taken: HMS Brilliant - When taken: 12 Aug 1813 - Date Received: 10 May 1814 - From what ship: Plymouth - Born: Salem - Age: 23 - Race: Mulatto - Released on 27 Apr 1815.

Coleman, William - Seaman - Nbr: 3547 - Prize: Hawk - Ship Type: P - How taken: HMS Pique - When taken: 26 Apr 1814 - Where taken: off Bermuda - Date Received: 30 Sep 1814 - From what ship: HMT Sybella - Born: North Carolina - Age: 21 - Died on 5 Nov 1814.

Coles, Charles - Seaman - Nbr: 3621 - Prize: Monarch - Ship Type: MV - How taken: HMS Dotterel - When taken: 14 Dec 1813 - Where taken: off Charleston - Date Received: 30 Sep 1814 - From what ship: HMT Sybella - Born: Massachusetts - Age: 27 - Released on 4 Jun 1815.

Colhoun, Joseph - Seaman - Nbr: 2454 - Prize: US Sloop Frolic - Ship Type: MW - How taken: HMS Orpheus - When taken: 20 Apr 1814 - Where taken: off Cuba - Date Received: 16 Aug 1814 - From what ship: HMT Queen, Halifax - Born: Weymouth - Age: 26 - Released on 3 May 1815.

Collett, John - Seaman - Nbr: 2407 - Prize: Hazard - Ship Type: P - How taken: HMS Caledonia - When taken: 21 Feb 1813 - Where taken: off Savannah - Date Received: 16 Aug 1814 - From what ship: HMT Queen, Halifax - Born: Morristown - Age: 30 - Race: Creole - Released on 3 May 1815.

Colley, Thomas - Seaman - Nbr: 5102 - Prize: David Porter - Ship Type: LM - How taken: HMS Pylades - When taken: 12 Sep 1814 - Where taken: Georges Bank - Date Received: 28 Oct 1814 - From what ship: HMT Alkbar, Halifax - Born: Marblehead - Age: 26 - Released on 29 Jun 1815.

Collier, Thomas - Seaman - Nbr: 5387 - How taken: Gave himself up at Chatham - When taken: 9 Sep 1814 - Date rec'd: 31 Oct 1814 - What ship: HMT Leyden, Chatham - Born: Virginia - Age: 32 - Released on 1 Jul 1815.

Collins, Andrew - Seaman - Nbr: 621 - Prize: US Brig Argus - Ship Type: MW - How taken: HMS Pelican - When taken: 14 Aug 1813 - Where taken: Irish Channel - Date Received: 8 Sep 1813 - From what ship: Plymouth - Born: Annapolis - Age: 25 - Sent to Dartmouth on 2 Nov 1814.

Collins, George - Seaman - Nbr: 6038 - Prize: Regent - Ship Type: LM - How taken: HMS Forth - When taken: 19 Sep 1814 - Where taken: off Egg Harbor (New Jersey) - Date Received: 28 Dec 1814 - From what ship: HMT Penelope - Born: Norfolk - Age: 13 - Released on 3 May 1815.

Collins, George - Seaman - Nbr: 3092 - How taken: Gave himself up from HMS Vigo (Spain) - When taken: 8 Dec 1813 - Date Received: 3 Sep 1814 - From what ship: HMT Hydra, Chatham - Born: Boston - Age: 28 - Race: Black - Released on 28 May 1815.

Collins, John - Seaman - Nbr: 6026 - Prize: Regent - Ship Type: LM - How taken: HMS Forth - When taken: 19 Sep 1814 - Where taken: off Egg Harbor (New Jersey) - Date Received: 28 Dec 1814 - From what ship: HMT Penelope - Born: Baltimore - Age: 35 - Released on 3 Jul 1815.

Collins, John - 3rd Lieutenant - Nbr: 5360 - Prize: Argus - Ship Type: MV - How taken: HMS San Domingo - When taken: 1 Mar 1814 - Where taken: off Savannah - Date Received: 31 Oct 1814 - From what ship: HMT Leyden, Chatham - Born: Washington - Age: 32 - Released on 1 Jul 1815.

Collins, John - Seaman - Nbr: 3103 - Prize: Mary - Ship Type: Prize - How taken: Taken by pilot boat - When taken: 27 Aug 1814 - Where taken: Bristol Channel - Date Received: 9 Sep 1814 - From what ship: HMS Abercrombie - Born: Philadelphia - Age: 24 - Died on 8 Oct 1814 from pneumonia.

Collins, Thomas - Seaman - Nbr: 2318 - Prize: Hussar - Ship Type: P - How taken: HMS Saturn - When taken: 25 May 1814 - Where taken: off Sandy Hook - Date Received: 16 Aug 1814 - From what ship: HMS Dublin, Halifax - Born: Stettin, Prussia - Age: 30 - Released on 3 May 1815.

Collison, Elliot - Seaman - Nbr: 3853 - Prize: Dominique - Ship Type: LM - How taken: HMS Dotterel - When taken: 21 May 1814 - Where taken: off Charleston - Date Received: 5 Oct 1814 - From what ship: HMT Orpheus, Halifax - Born: Baltimore - Age: 24 - Released on 9 Jun 1815.

Colman, David - Seaman - Nbr: 1365 - Prize: Paul Jones - Ship Type: P - How taken: HMS Leonidas - When taken: 23 May 1813 - Where taken: Channel - Date Received: 19 Jun 1814 - From what ship: Stapleton - Born: New York - Age: 15 - Released on 28 Apr 1815.

Colquhoun, William - Seaman - Nbr: 2621 - How taken: Gave himself up from HMS Christian 7th - When taken: 21 Dec 1812 - Date Received: 21 Aug 1814 - From what ship: HMT Freya, Chatham - Born: Philadelphia - Age: 37 - Released on 26 Apr 1815.

Colston, John - Seaman - Nbr: 5734 - Prize: US Schooner Somers - Ship Type: MW - How taken: British gunboats -

American Prisoners of War Held at Dartmoor during the War of 1812

## Alphabetical listing of names

When taken: 12 Aug 1814 - Where taken: Fort Erie - Date Received: 26 Dec 1814 - From what ship: HMT Argo - Born: New York - Age: 19 - Released on 12 Jun 1815.

Colton, Charles - Seaman - Nbr: 1208 - Prize: Comet - Ship Type: Prize - How taken: HMS Fairy - When taken: 25 Dec 1812 - Where taken: off St. Antonio - Date Received: 4 Jun 1814 - From what ship: Dartmouth - Born: LA - Age: 23 - Sent to Plymouth on 8 Jul 1814.

Colton, Joseph - Seaman - Nbr: 3004 - Prize: St. Lawrence - How taken: HMS Aquilon - When taken: 9 Aug 1814 - Where taken: off Western Islands (England) - Date Received: 29 Aug 1814 - From what ship: HMT Bittern - Born: Boston - Age: 26 - Released on 28 May 1815.

Colton, Walter - Seaman - Nbr: 1368 - Prize: Paul Jones - Ship Type: P - How taken: HMS Leonidas - When taken: 23 May 1813 - Where taken: Channel - Date Received: 19 Jun 1814 - From what ship: Stapleton - Born: Massachusetts - Age: 30 - Released on 27 Apr 1815.

Columbia, John - Seaman - Nbr: 2254 - Prize: Otario - Ship Type: Prize - How taken: HMS Curlew - When taken: 2 May 1814 - Where taken: off Halifax - Date Received: 16 Aug 1814 - From what ship: HMS Dublin, Halifax - Born: Maryland - Age: 19 - Released on 3 May 1815.

Colvell, John B. - Seaman - Nbr: 5479 - Prize: General Putnam - Ship Type: P - How taken: HMS Leander - When taken: 8 Nov 1814 - Where taken: Long 65 Lat 42 - Date Received: 17 Dec 1814 - From what ship: HMT Loire, Halifax - Born: New Hampshire - Age: 20 - Released on 11 Jul 1815.

Colvelli, John - Seaman - Nbr: 285 - Prize: Cannoneer - Ship Type: MV - How taken: HMS Warspite - When taken: 14 Mar 1813 - Where taken: Bay of Biscay - Date Received: 28 Jun 1813 - From what ship: Plymouth - Born: New York - Age: 23 - Released on 20 Apr 1815.

Commous, John - Seaman - Nbr: 1193 - Prize: Imperatrice Reine - Ship Type: P - How taken: HMS Hotspur - When taken: 13 Jan 1813 - Date Received: 4 Jun 1814 - From what ship: Dartmouth - Born: Salem - Age: 32 - Sent to Mill Prison (Plymouth, England) on 8 Jul 1814.

Complaro, A. A. - Seaman - Nbr: 4815 - Prize: President - Ship Type: P - How taken: HMS Pique - When taken: 7 May 1814 - Where taken: off Porto Rico - Date Received: 9 Oct 1814 - From what ship: HMT Freya, Halifax - Born: Cartagena - Age: 24 - Released on 21 Jun 1815.

Condon, Michael - Soldier - Nbr: 5261 - Prize: 13th US Infantry - Ship Type: Troops - How taken: British Army - When taken: 13 Oct 1812 - Where taken: Canada - Date Received: 31 Oct 1814 - From what ship: HMT Leyden, Chatham - Born: Dublin - Age: 40 - Released on 29 Jun 1815.

Condon, Samuel - Seaman - Nbr: 6205 - Prize: Prince de Neufchatel - Ship Type: P - How taken: Leander (Newcastle Acasta) - When taken: 20 Dec 1814 - Where taken: Lat 38 Long 56 - Date Received: 30 Jan 1815 - From what ship: HMT Pheasant - Born: Charlestown - Age: 19 - Released on 14 Jun 1815.

Conely, Samuel - Seaman - Nbr: 629 - Prize: US Brig Argus - Ship Type: MW - How taken: HMS Pelican - When taken: 14 Aug 1813 - Where taken: Irish Channel - Date Received: 8 Sep 1813 - From what ship: Plymouth - Born: Tolbert - Age: 23 - Sent to Dartmouth on 2 Nov 1814.

Congdon, James - Seaman - Nbr: 4893 - Prize: Goree - Ship Type: LM - How taken: HMS Rolla - When taken: Jul 1814 - Where taken: off Havannah - Date Received: 24 Oct 1814 - From what ship: HMT Salvador del Mundo - Born: Rhode Island - Age: 19 - Died on 11 Nov 1814.

Congleton, A. - Seaman - Nbr: 4312 - How taken: Gave himself up from HMS Leda - When taken: 6 Jul 1814 - Date Received: 7 Oct 1814 - From what ship: HMT Niobe, Chatham - Born: Washington - Age: 20 - Released on 13 Jun 1815.

Congleton, Smith - Seaman - Nbr: 4740 - How taken: Gave himself up from a brig - When taken: 29 Dec 1813 - Date Received: 9 Oct 1814 - From what ship: HMT Freya, Chatham - Born: New York - Age: 21 - Released on 15 Jun 1815.

Conklin, Edward - Seaman - Nbr: 1256 - Prize: Adeline - Ship Type: MV - How taken: HMS Magicienne - When taken: 16 Mar 1814 - Where taken: off Cape Finisterre - Date Received: 14 Jun 1814 - From what ship: Mill Prison (Plymouth, England) - Born: New York - Age: 31 - Released on 28 Apr 1815.

Conklin, Robert - Gunner - Nbr: 1029 - Prize: US Brig Argus - Ship Type: MW - How taken: HMS Pelican - When taken: 16 Aug 1813 - Where taken: Irish Channel - Date Received: 10 May 1814 - From what ship: Plymouth - Born: Kingston - Age: 26 - Sent to Dartmouth on 2 Nov 1814.

Conklin, Vertius - Seaman - Nbr: 5657 - Prize: Herald - Ship Type: P - How taken: HMS Endymion - When taken: 15 Aug 1814 - Where taken: off Nantucket - Date Received: 24 Dec 1814 - From what ship: HMT Impregnable - Born: Nantucket - Age: 20 - Died on 23 Jun 1815 from phthisis pulmonalis.

Conklin, William - Boatswain - Nbr: 336 - Prize: Tiger - Ship Type: MV - How taken: HMS Scylla - When taken: 22 Mar 1813 - Where taken: Bay of Biscay - Date Received: 28 Jun 1813 - From what ship: Plymouth -

American Prisoners of War Held at Dartmoor during the War of 1812

## Alphabetical listing of names

Born: New York - Age: 32 - Released on 20 Apr 1815.

Conlay, Cornelius - Seaman - Nbr: 5159 - Prize: Volante - Ship Type: P - How taken: HMS Curlew - When taken: 25 Nov 1813 - Where taken: off Halifax - Date Received: 31 Oct 1814 - From what ship: HMT Mermaid, Chatham - Born: Boston - Age: 22 - Released on 29 Jun 1815.

Connelly, James - Seaman - Nbr: 2624 - How taken: Gave himself up from HMS Christian 7th - When taken: 21 Dec 1812 - Date Received: 21 Aug 1814 - From what ship: HMT Freya, Chatham - Born: Baltimore - Age: 20 - Released on 26 Apr 1815.

Conner, Edward - Seaman - Nbr: 5619 - How taken: Gave himself up from HMS Lion - When taken: 18 Jun 1813 - Date Received: 24 Dec 1814 - From what ship: HMT Tay - Born: New York - Age: 37 - Race: Black - Released on 3 Jul 1815.

Conner, Edward - Prize Master - Nbr: 1095 - Prize: Bunker Hill - Ship Type: P - How taken: HMS Pomone - When taken: 8 Mar 1814 - Where taken: at sea - Date Received: 10 May 1814 - From what ship: Plymouth - Born: Philadelphia - Age: 23 - Released on 27 Apr 1815.

Conner, Galen - Mate - Nbr: 3605 - Prize: Gothland - Ship Type: MV - How taken: HMS Barbados - When taken: 31 Jan 1814 - Where taken: off St. Bartholomew - Date Received: 30 Sep 1814 - From what ship: HMT Sybella - Born: Maryland - Age: 25 - Released on 4 Jun 1815.

Conner, John - Seaman - Nbr: 3258 - How taken: Gave himself up from HMS Invincible - When taken: 1 Feb 1813 - Date Received: 11 Sep 1814 - From what ship: HMT Freya, Chatham - Born: New York - Age: 25 - Race: Black - Released on 28 May 1815.

Conner, Michael - Seaman - Nbr: 4207 - Prize: Minerva - Ship Type: MV - How taken: HMS Conquistador - When taken: 19 Jan 1814 - Where taken: Bay of Biscay - Date Received: 7 Oct 1814 - From what ship: HMT Niobe, Chatham - Born: Massachusetts - Age: 22 - Released on 13 Jun 1815.

Conner, William - Seaman - Nbr: 5731 - How taken: Gave himself up from MV Warrior - When taken: 12 Nov 1814 - Date rec'd: 26 Dec 1814 - What ship: HMT Argo - Born: Maryland - Age: 27 - Released on 11 Jul 1815.

Conner, William - Passenger - Nbr: 6154 - Prize: Lion - Ship Type: P - How taken: HMS Granicus - When taken: 2 Dec 1814 - Where taken: off Lisbon - Date Received: 18 Jan 1815 - From what ship: HMT Impregnable - Born: Boston - Age: 54 - Released on 11 Jul 1815.

Connor, Fenton - Captain's Clerk - Nbr: 2344 - Prize: Snap Dragon - Ship Type: P - How taken: HMS Martin - When taken: 10 Jun 1814 - Where taken: off Halifax - Date Received: 16 Aug 1814 - From what ship: HMS Dublin, Halifax - Born: Newbury - Age: 21 - Released on 3 May 1815.

Connor, Joseph - Purser - Nbr: 2365 - Prize: Snap Dragon - Ship Type: P - How taken: HMS Martin - When taken: 10 Jun 1814 - Where taken: off Halifax - Date Received: 16 Aug 1814 - From what ship: HMT Queen, Halifax - Born: New Berne - Age: 29 - Released on 3 May 1815.

Conray, William M. - 2nd Mate - Nbr: 884 - Prize: Agnes - Ship Type: LM - How taken: Jane (Cutter) - When taken: 28 Nov 1813 - Where taken: Bay of Biscay - Date Received: 31 Jan 1814 - From what ship: Plymouth - Born: Boston - Age: 54 - Released on 27 Apr 1815.

Constant, William - Seaman - Nbr: 6342 - Prize: Prince de Neufchatel - Ship Type: P - How taken: Leander (Newcastle Acasta) - When taken: 20 Dec 1814 - Where taken: Lat 38 Long 56 - Date Received: 19 Feb 1815 - From what ship: HMT Ganges, Plymouth - Born: Modena - Age: 30 - Released on 11 Jul 1815.

Conton, Philip - Seaman - Nbr: 6282 - How taken: Gave himself up from HMS Adamats - Date Received: 17 Feb 1815 - From what ship: HMT Ganges, Plymouth - Born: New York - Age: 28 - Released on 5 Jul 1815.

Conway, Charles - Seaman - Nbr: 472 - Prize: Hebe - Ship Type: MV - How taken: HMS Stag - When taken: 13 Apr 1813 - Where taken: Bay of Biscay - Date Received: 8 Sep 1813 - From what ship: Plymouth - Born: Baltimore - Age: 43 - Sent to Plymouth on 8 Jul 1814.

Conway, James - Seaman - Nbr: 2110 - How taken: Gave himself up from HMS Mars - When taken: 9 Dec 1812 - Date Received: 4 Aug 1814 - From what ship: Chatham - Born: North Carolina - Age: 54 - Released on 26 Apr 1815.

Conway, Samuel - Seaman - Nbr: 3681 - Prize: Alfred - Ship Type: P - How taken: HMS Epervier - When taken: 23 Feb 1812 - Where taken: off Newfoundland - Date Received: 30 Sep 1814 - From what ship: HMT President - Born: Salem - Age: 31 - Released on 4 Jun 1815.

Conway, William - Seaman - Nbr: 31 - How taken: Apprehended at Gibraltar - When taken: 8 Aug 1813 - Date Received: 2 Apr 1813 - From what ship: Plymouth - Born: Marblehead - Age: 22 - Sent to Dartmouth on 30 Jul 1813.

Cook, Benjamin - Master - Nbr: 3730 - Prize: Alfred - Ship Type: P - How taken: HMS Epervier - When taken: 23

American Prisoners of War Held at Dartmoor during the War of 1812

## Alphabetical listing of names

Feb 1814 - Where taken: off Newfoundland - Date Received: 30 Sep 1814 - From what ship: HMT President, Halifax - Born: Salem - Age: 28 - Released on 4 Jun 1815.

Cook, Benjamin - Seaman - Nbr: 840 - Prize: Chesapeake - Ship Type: LM - How taken: HMS Hotspur & HMS Pyramus - When taken: 26 Oct 1813 - Where taken: off Nantes - Date Received: 29 Nov 1813 - From what ship: Plymouth - Born: Baltimore - Age: 26 - Race: Black - Died on 6 Apr 1814 from pneumonia.

Cook, Charles Howe - Seaman - Nbr: 1352 - Prize: Paul Jones - Ship Type: P - How taken: HMS Leonidas - When taken: 23 May 1813 - Where taken: Channel - Date Received: 19 Jun 1814 - From what ship: Stapleton - Born: SC - Age: 26 - Released on 28 Apr 1815.

Cook, Edward - Seaman - Nbr: 5854 - Prize: Lion - Ship Type: P - How taken: HMS Granicus - When taken: 2 Dec 1814 - Where taken: off Lisbon - Date Received: 26 Dec 1814 - From what ship: HMT Impregnable - Born: Maryland - Age: 31 - Released on 28 May 1815.

Cook, Isaac - Seaman - Nbr: 3344 - How taken: Gave himself up from HMS Prince of Wales - When taken: 28 May 1813 - Date Received: 13 Sep 1814 - From what ship: HMT Niobe, Chatham - Born: Long Island - Age: 24 - Released on 28 May 1815.

Cook, James - Seaman - Nbr: 50 - Prize: Terrible - Ship Type: MV - How taken: HMS Foxhound - When taken: 8 Feb 1813 - Where taken: Channel - Date Received: 2 Apr 1813 - From what ship: Plymouth - Born: Norfolk - Age: 25 - Sent to Dartmouth on 7 Aug 1813.

Cook, Jeremiah - Seaman - Nbr: 5517 - Prize: Robert - When taken: 20 Feb 1814 - Date Received: 17 Dec 1814 - From what ship: HMT Loire, Halifax - Born: New York - Age: 21.

Cook, John - Seaman - Nbr: 3333 - Prize: Polly - Ship Type: P - How taken: HMS Ringdove - When taken: 27 Jul 1813 - Where taken: at sea - Date Received: 13 Sep 1814 - From what ship: HMT Niobe, Chatham - Born: Boston - Age: 34 - Released on 28 May 1815.

Cook, John - Seaman - Nbr: 2875 - How taken: Gave himself up from HMS Scorpion - When taken: 27 May 1813 - Date Received: 24 Aug 1814 - From what ship: HMT Alpheus, Chatham - Born: Frankfort - Age: 25 - Released on 28 Apr 1815.

Cook, Samuel - Seaman - Nbr: 480 - Prize: Price - Ship Type: MV - How taken: HMS Iris - When taken: 13 Apr 1813 - Where taken: Bay of Biscay - Date Received: 8 Sep 1813 - From what ship: Plymouth - Born: Tiverton - Age: 22 - Entered British service on 11 Oct 1813.

Cook, Samuel - Mate - Nbr: 4914 - Prize: Dominique - Ship Type: LM - How taken: HMS Dotterel - When taken: 21 May 1814 - Where taken: off Charleston - Date Received: 28 Oct 1814 - From what ship: HMT Alkbar, Halifax - Born: Salem - Age: 29 - Released on 1 May 1815.

Cook, Silvanus - Carpenter - Nbr: 4421 - Prize: Elbridge Gerry - Ship Type: LM - How taken: HMS Crescent - When taken: 16 Sep 1813 - Where taken: at sea - Date Received: 8 Oct 1814 - From what ship: HMT Leyden, Chatham - Born: New York - Age: 25 - Released on 14 Jun 1815.

Cook, William - Seaman - Nbr: 1348 - Prize: Paul Jones - Ship Type: P - How taken: HMS Leonidas - When taken: 23 May 1813 - Where taken: Channel - Date Received: 19 Jun 1814 - From what ship: Stapleton - Born: New York - Age: 19 - Released on 28 Apr 1815.

Cooke, John - Seaman - Nbr: 2192 - Prize: Fiere Facia - Ship Type: P - How taken: HMS Ramillies - When taken: 27 Feb 1814 - Where taken: off NY - Date Received: 16 Aug 1814 - From what ship: HMS Dublin, Halifax - Born: Massachusetts - Age: 23 - Released on 2 May 1815.

Coombe, Michael - Seaman - Nbr: 4543 - Prize: Growler - Ship Type: P - How taken: HMS Electra - When taken: 7 Jul 1813 - Where taken: at sea - Date Received: 8 Oct 1814 - From what ship: HMT Leyden, Chatham - Born: Marblehead - Age: 21 - Released on 15 Jun 1815.

Coombes, William - Seaman - Nbr: 382 - Prize: Virginia Planter - Ship Type: MV - How taken: HMS Pyramus - When taken: 18 Mar 1813 - Where taken: off Nantes - Date Received: 1 Jul 1813 - From what ship: Plymouth - Born: New York - Age: 19 - Released on 20 Apr 1815.

Coombs, John - Seaman - Nbr: 6469 - Prize: Decatur - Ship Type: P - How taken: HMS Rhin - When taken: 5 Jun 1814 - Where taken: off San Domingo - Date Received: 3 Mar 1815 - From what ship: HMT Ganges, Plymouth - Born: Norfolk - Age: 22 - Race: Mulatto - Released on 11 Jul 1815.

Coombs, John - Seaman - Nbr: 2752 - How taken: Gave himself up from HMS Royal William - When taken: 15 Oct 1812 - Date Received: 24 Aug 1814 - From what ship: HMT Liverpool, Chatham - Born: Pennsylvania - Age: 24 - Released on 26 Apr 1815.

Cooney, Samuel - Seaman - Nbr: 4023 - Prize: US Brig Rattlesnake - Ship Type: MW - How taken: HMS Leander - When taken: 13 Jul 1814 - Where taken: off Shelburne - Date Received: 6 Oct 1814 - From what ship: HMT Chesapeake, Halifax - Born: Wiscasset - Age: 19 - Released on 9 Jun 1815.

American Prisoners of War Held at Dartmoor during the War of 1812

## Alphabetical listing of names

Cooper, Alfred - Seaman - Nbr: 2600 - Prize: Tom Thumb - Ship Type: MV - How taken: Lyon (Privateer) - When taken: 15 Feb 1813 - Where taken: Bay of Biscay - Date Received: 21 Aug 1814 - From what ship: HMT Freya, Chatham - Born: Newbury - Age: 32 - Released on 15 Sep 1814.

Cooper, Andrew A. - Seaman - Nbr: 1663 - Prize: Paul Jones - Ship Type: P - How taken: HMS Leonidas - When taken: 23 May 1813 - Where taken: Channel - Date Received: 23 Jun 1814 - From what ship: Stapleton - Born: Albany - Age: 29 - Released on 1 May 1815.

Cooper, Daniel - Seaman - Nbr: 90 - Prize: Cashier - Ship Type: MV - How taken: HMS Reindeer - When taken: 3 Feb 1813 - Where taken: Bay of Biscay - Date Received: 2 Apr 1813 - From what ship: Plymouth - Born: Baltimore - Age: 30 - Sent to Dartmouth on 30 Jul 1813.

Cooper, Edward - Seaman - Nbr: 3317 - Prize: Wasp - Ship Type: P - How taken: HMS Bream - When taken: 10 Jun 1813 - Where taken: off Halifax - Date Received: 13 Sep 1814 - From what ship: HMT Niobe, Chatham - Born: Boothbay - Age: 20 - Released on 28 May 1815.

Cooper, Ezekiel - Seaman - Nbr: 2293 - Prize: Hussar - Ship Type: P - How taken: HMS Saturn - When taken: 25 May 1814 - Where taken: off Sandy Hook - Date Received: 16 Aug 1814 - From what ship: HMS Dublin, Halifax - Born: New Jersey - Age: 19 - Released on 3 May 1815.

Cooper, James - Seaman - Nbr: 2037 - How taken: Gave himself up from HMS Impeteux - When taken: 2 Dec 1812 - Date Received: 3 Aug 1814 - From what ship: HMS Lyffey, Chatham Depot - Born: Boston - Age: 24 - Released on 26 Apr 1815.

Cooper, James - Seaman - Nbr: 309 - Prize: King George - Ship Type: MV - How taken: HMS Piercer - When taken: 9 Mar 1813 - Where taken: Isle of Bas - Date Received: 28 Jun 1813 - From what ship: Plymouth - Born: Long Island - Age: 25 - Released on 20 Apr 1815.

Cooper, James - Seaman - Nbr: 1032 - Prize: US Brig Argus - Ship Type: MW - How taken: HMS Pelican - When taken: 16 Aug 1813 - Where taken: Irish Channel - Date Received: 10 May 1814 - From what ship: Plymouth - Born: Jamaica, NY - Age: 25 - Sent to Dartmouth on 19 Oct 1814.

Cooper, James - Boy - Nbr: 3797 - Prize: Hannah - Ship Type: Prize - How taken: HMS Martin - When taken: 29 Apr 1814 - Where taken: off Cape Lopez - Date Received: 5 Oct 1814 - From what ship: HMT President, Halifax - Born: Maryland - Age: 19 - Released on 9 Jun 1815.

Cooper, John - Seaman - Nbr: 975 - Prize: Siro - Ship Type: LM - How taken: HMS Pelican - When taken: 13 Jan 1814 - Where taken: at sea - Date Received: 31 Jan 1814 - From what ship: Plymouth - Born: Charlestown - Age: 27 - Released on 27 Apr 1815.

Cooper, John - Seaman - Nbr: 2571 - Prize: Rattlesnake - Ship Type: P - How taken: HMS Hyperion - When taken: 25 Jun 1814 - Where taken: off Cape Finisterre - Date Received: 21 Aug 1814 - From what ship: HMS Hyperion - Born: Baltimore - Age: 33 - Race: Negro - Released on 3 May 1815.

Cooper, John - Seaman - Nbr: 3014 - How taken: Gave himself up from HMS Zealous - When taken: 27 Aug 1814 - Date Received: 2 Sep 1814 - From what ship: HMT Centaur - Born: Jersey - Age: 48 - Race: Black - Released on 28 May 1815.

Cooper, Ludwig - Seaman - Nbr: 3964 - Prize: Rolla - Ship Type: P - How taken: HMS Loire - When taken: 10 Dec 1813 - Where taken: off Bull Island (South Carolina) - Date Received: 5 Oct 1814 - From what ship: HMT President, Halifax - Born: North Kingston - Age: 18 - Released on 9 Jun 1815.

Cooper, Peter - Seaman - Nbr: 6398 - Prize: Nellenville - Ship Type: MV - How taken: HMS Onyx - When taken: 25 Dec 1814 - Where taken: off San Domingo - Date Received: 3 Mar 1815 - From what ship: HMT Ganges, Plymouth - Born: Baltimore - Age: 23 - Race: Black - Released on 11 Jul 1815.

Cooper, Tannick - Seaman - Nbr: 1083 - How taken: Gave himself up - Date Received: 10 May 1814 - From what ship: Plymouth - Born: Baltimore - Age: 34 - Released on 27 Apr 1815.

Cooper, Thomas - 2nd Mate - Nbr: 4549 - Prize: Union - Ship Type: MV - How taken: Julian (East Indian ship) - When taken: 10 Jun 1812 - Date Received: 8 Oct 1814 - From what ship: HMT Leyden, Chatham - Born: Massachusetts - Age: 34 - Died on 8 Nov 1814.

Cooper, Thomas - Carpenter - Nbr: 6262 - Prize: Plutarch - Ship Type: MV - How taken: HMS Helicon - When taken: 5 Feb 1815 - Where taken: off Bordeaux - Date Received: 10 Feb 1815 - From what ship: HMT Ganges, Plymouth - Born: Boston - Age: 26 - Released on 20 Apr 1815.

Cooper, Thomas - Seaman - Nbr: 2724 - How taken: Gave himself up from HMS Nisus - Date Received: 21 Aug 1814 - From what ship: HMT Freya, Chatham - Born: Philadelphia - Age: 22 - Released on 19 May 1815.

Cooper, Thomas - Seaman - Nbr: 4744 - How taken: Gave himself up from HMS Phoenix - When taken: 17 Jul 1814 - Date Received: 9 Oct 1814 - From what ship: HMT Freya, Chatham - Born: Maryland - Age: 26 - Race: Mulatto - Released on 14 Jun 1815.

American Prisoners of War Held at Dartmoor during the War of 1812

## Alphabetical listing of names

Cooper, William - Master's Mate - Nbr: 4555 - How taken: Gave himself up from HMS Apistance - When taken: 26 Aug 1814 - Date Received: 8 Oct 1814 - From what ship: HMT Leyden, Chatham - Born: New York - Age: 34 - Released on 15 Jun 1815.

Cops, Darius - Seaman - Nbr: 5593 - Prize: Albion - Ship Type: Prize - How taken: HMS Jaseur - When taken: 21 Sep 1814 - Where taken: off Halifax - Date Received: 17 Dec 1814 - From what ship: HMT Loire, Halifax - Born: Boston - Age: 22 - Released on 1 Jul 1815.

Corbett, Michael - Seaman - Nbr: 3698 - Prize: Alfred - Ship Type: P - How taken: HMS Epervier - When taken: 23 Feb 1814 - Where taken: off Newfoundland - Date Received: 30 Sep 1814 - From what ship: HMT President, Halifax - Born: Marblehead - Age: 20 - Released on 4 Jun 1815.

Coren, Hugh - Seaman - Nbr: 6256 - How taken: Sent into custody from MV Brothers - When taken: 13 Jan 1813 - Date Received: 4 Feb 1815 - From what ship: HMT Ganges - Born: Maryland - Age: 24 - Released on 5 Jul 1815.

Cornish, Charles - Seaman - Nbr: 838 - Prize: Chesapeake - Ship Type: LM - How taken: HMS Hotspur & HMS Pyramus - When taken: 26 Oct 1813 - Where taken: off Nantes - Date Received: 29 Nov 1813 - From what ship: Plymouth - Born: Maryland - Age: 40 - Race: Black - Died on 10 Jan 1814.

Cornish, John - Seaman - Nbr: 3047 - How taken: Gave himself up from HMS Prince of Wales - When taken: 25 Aug 1814 - Date Received: 2 Sep 1814 - From what ship: HMT Sultan - Born: Pennsylvania - Age: 21 - Race: Negro - Released on 28 Apr 1815.

Cornish, Thomas - Seaman - Nbr: 3511 - How taken: Gave himself up from HMS Warspite - When taken: 24 Sep 1814 - Date Received: 28 Sep 1814 - From what ship: HMT Salvador del Mundo - Born: Salem - Age: 24 - Released on 4 Jun 1815.

Cornivalt, James - Seaman - Nbr: 5887 - Prize: Harlequin - Ship Type: P - How taken: HMS Bulwark - When taken: 23 Nov 1814 - Where taken: off Halifax - Date Received: 27 Dec 1814 - From what ship: HMT Penelope - Born: New York - Age: 39 - Released on 3 Jul 1815.

Cornwall, Arthur - Seaman - Nbr: 1377 - Prize: Grand Napoleon - Ship Type: P - How taken: HMS Goldfinch - When taken: 17 Apr 1813 - Where taken: Bay of Biscay - Date Received: 19 Jun 1814 - From what ship: Stapleton - Born: Philadelphia - Age: 25 - Released on 28 Apr 1815.

Coston, Thomas - Seaman - Nbr: 3939 - Prize: Rolla - Ship Type: P - How taken: HMS Loire - When taken: 10 Dec 1813 - Where taken: off Bull Island (South Carolina) - Date Received: 5 Oct 1814 - From what ship: HMT President, Halifax - Born: New York - Age: 19 - Race: Negro - Released on 9 Jun 1815.

Cotterell, Henry C. - Master - Nbr: 1129 - Prize: Young Dixon - Ship Type: MV - How taken: HMS Fly & HMS Avon - When taken: 3 Apr 1814 - Where taken: at sea - Date Received: 10 May 1814 - From what ship: Plymouth - Born: Baltimore - Age: 28 - Escaped on 8 Aug 1814.

Cottle, John - Prize Master - Nbr: 1696 - Prize: Fanny - Ship Type: Prize - How taken: HMS Sceptre - When taken: 12 May 1814 - Date Received: 2 Jul 1814 - From what ship: Plymouth - Born: Nantucket - Age: 30 - Released on 26 Apr 1815.

Cotton, Edward - Seaman - Nbr: 5313 - Prize: Thomas - Ship Type: P - How taken: HMS Nymphe - When taken: 24 Jun 1813 - Where taken: off Halifax - Date Received: 31 Oct 1814 - From what ship: HMT Leyden, Chatham - Born: Portsmouth - Age: 43 - Released on 29 Jun 1815.

Cotton, Samuel - Seaman - Nbr: 1323 - How taken: Impressed at Bristol - When taken: 30 Jun 1813 - Date Received: 19 Jun 1814 - From what ship: Stapleton - Born: Massachusetts - Age: 25 - Race: Negro - Released on 28 Apr 1815.

Cotton, William - Seaman - Nbr: 5051 - Prize: Ida - Ship Type: LM - How taken: HMS Newcastle - When taken: 9 Aug 1814 - Where taken: Long 34 - Date Received: 28 Oct 1814 - From what ship: HMT Alkbar, Halifax - Born: Massachusetts - Age: 23 - Released on 29 Jun 1815.

Cottrell, Henry - Seaman - Nbr: 3217 - How taken: Gave himself up from HMS Majestic - When taken: 17 Aug 1813 - Date Received: 11 Sep 1814 - From what ship: HMT Freya, Chatham - Born: New London - Age: 53 - Released on 28 May 1815.

Counthon, John - Prize Master - Nbr: 3566 - Prize: Hawk - Ship Type: P - How taken: HMS Pique - When taken: 26 Apr 1814 - Where taken: off Bermuda - Date Received: 30 Sep 1814 - From what ship: HMT Sybella - Born: Beaufort - Age: 25 - Released on 4 Jun 1815.

Couret, Francis - Seaman - Nbr: 607 - Prize: Betsy - Ship Type: MV - How taken: HMS Leonidas - When taken: 12 Aug 1813 - Where taken: Channel - Date Received: 8 Sep 1813 - From what ship: Plymouth - Born: Philadelphia - Age: 19 - Sent to Plymouth on 8 Jul 1814.

Court, Robert - Seaman - Nbr: 224 - Prize: Pert - Ship Type: MV - How taken: HMS Warspite - When taken: 1 Mar

American Prisoners of War Held at Dartmoor during the War of 1812

## Alphabetical listing of names

1813 - Where taken: off Basque Roads (France) - Date Received: 2 Apr 1813 - From what ship: Plymouth - Born: Philadelphia - Age: 19 - Released on 20 Apr 1815.

Courtis, Thomas - Seaman - Nbr: 1909 - How taken: Gave himself up from HMS Mars - When taken: 28 Oct 1812 - Date Received: 3 Aug 1814 - From what ship: HMS Alceste, Chatham Depot - Born: Marblehead - Age: 37 - Released on 26 Apr 1815.

Cousins, Samuel - Seaman - Nbr: 3490 - Prize: US Sloop Frolic - Ship Type: MW - How taken: HMS Orpheus - When taken: 20 Apr 1814 - Where taken: off Cuba - Date Received: 19 Sep 1814 - From what ship: HMT Salvador del Mundo - Born: Massachusetts - Age: 25 - Released on 3 May 1815.

Covall, Ephraim - Seaman - Nbr: 1923 - How taken: Gave himself up from HMS Royal William - When taken: 3 Feb 1813 - Date Received: 3 Aug 1814 - From what ship: HMS Alceste, Chatham Depot - Born: Wellfleet - Age: 40 - Released on 2 May 1815.

Cove, John - Seaman - Nbr: 2680 - Prize: Joseph - Ship Type: MV - How taken: HMS Iris - When taken: 8 Jun 1813 - Where taken: off Spain - Date Received: 21 Aug 1814 - From what ship: HMT Freya, Chatham - Born: Massachusetts - Age: 24 - Released on 19 May 1815.

Covell, Isaac - Seaman - Nbr: 97 - Prize: St. Martin's Planter - Ship Type: P - How taken: HMS Dublin - When taken: 9 Feb 1813 - Where taken: Lat 43 N, Long 33 50 W - Date Received: 2 Apr 1813 - From what ship: Plymouth - Born: Ellington - Age: 29 - Released on 20 Apr 1815.

Coven, Clement - Seaman - Nbr: 5528 - Prize: Sparks - Ship Type: LM - How taken: HMS Maidstone - When taken: 28 Sep 1814 - Where taken: off Nantucket - Date Received: 17 Dec 1814 - From what ship: HMT Loire, Halifax - Born: New Bedford - Age: 20 - Released on 1 Jul 1815.

Coville, Nathaniel - Seaman - Nbr: 6265 - Prize: Plutarch - Ship Type: MV - How taken: HMS Helicon - When taken: 5 Feb 1815 - Where taken: off Bordeaux - Date Received: 10 Feb 1815 - From what ship: HMT Ganges, Plymouth - Born: Chatham - Age: 19 - Released on 5 Jul 1815.

Cowder, James - Prize Master - Nbr: 1103 - Prize: Diamond - Ship Type: P - How taken: HMS Vengeur - When taken: 6 Mar 1814 - Where taken: Lat 47.4 Long 5.60 - Date Received: 10 May 1814 - From what ship: Plymouth - Born: Colry - Age: 49 - Released on 27 Apr 1815.

Cowen, Robert - Seaman - Nbr: 404 - Prize: Young Holkar - Ship Type: MV - How taken: HMS Superb - When taken: 10 Apr 1813 - Where taken: off Belle Isle, France - Date Received: 1 Jul 1813 - From what ship: Plymouth - Born: Philadelphia - Age: 39 - Released on 20 Apr 1815.

Cowen, Thomas - Seaman - Nbr: 3567 - Prize: Hawk - Ship Type: P - How taken: HMS Pique - When taken: 26 Apr 1814 - Where taken: off Bermuda - Date Received: 30 Sep 1814 - From what ship: HMT Sybella - Born: New York - Age: 30 - Released on 4 Jun 1815.

Cowen, William - Seaman - Nbr: 5369 - How taken: Gave himself up from HMS Indefatigable - When taken: 14 Aug 1813 - Date Received: 31 Oct 1814 - From what ship: HMT Leyden, Chatham - Born: Long Island - Age: 56 - Released on 1 Jul 1815.

Cowing, Charles - Seaman - Nbr: 1207 - Prize: Gazelle - Ship Type: P - How taken: HMS Leonidas - When taken: 15 Feb 1812 - Where taken: off Ireland - Date Received: 4 Jun 1814 - From what ship: Dartmouth - Born: America - Age: 40 - Released on 26 Apr 1815.

Cowing, Samuel - Seaman - Nbr: 5903 - Prize: Harlequin - Ship Type: P - How taken: HMS Bulwark - When taken: 23 Nov 1814 - Where taken: off Halifax - Date Received: 27 Dec 1814 - From what ship: HMT Penelope - Born: Bath - Age: 17 - Released on 3 Jul 1815.

Cowley, Samuel - Seaman - Nbr: 2285 - Prize: Diomede - Ship Type: P - How taken: HMS Rifleman - When taken: 28 May 1814 - Where taken: off Sable Island - Date Received: 16 Aug 1814 - From what ship: HMS Dublin, Halifax - Born: Newburyport - Age: 22 - Released on 3 May 1815.

Cox, Abraham - Seaman - Nbr: 2926 - How taken: Gave himself up from HMS Scorpion - Date Received: 24 Aug 1814 - What ship: HMT Alpheus, Chatham - Born: Massachusetts - Age: 25 - Race: Black - Released on 19 May 1815.

Cox, Caleb - Seaman - Nbr: 2362 - Prize: Snap Dragon - Ship Type: P - How taken: HMS Martin - When taken: 10 Jun 1814 - Where taken: off Halifax - Date Received: 16 Aug 1814 - From what ship: HMT Queen, Halifax - Born: Connecticut - Age: 21 - Released on 3 May 1815.

Cox, Daniel - Seaman - Nbr: 3695 - Prize: Alfred - Ship Type: P - How taken: HMS Epervier - When taken: 23 Feb 1814 - Where taken: off Newfoundland - Date Received: 30 Sep 1814 - From what ship: HMT President, Halifax - Born: Delaware - Age: 40 - Race: Negro - Released on 4 Jun 1815.

Cox, John - Seaman - Nbr: 1245 - How taken: Sent into custody from HMS Elizabeth - When taken: 7 Aug 1812 - Date Received: 14 Jun 1814 - From what ship: Mill Prison (Plymouth, England) - Born: Portsmouth - Age:

American Prisoners of War Held at Dartmoor during the War of 1812
## Alphabetical listing of names

44 - Released on 26 Apr 1815.
Cox, John - Seaman - Nbr: 203 - Prize: Star - Ship Type: MV - How taken: HMS Superb - When taken: 9 Feb 1813 - Where taken: Bay of Biscay - Date Received: 2 Apr 1813 - From what ship: Plymouth - Born: Chester - Age: 20 - Sent to Mill Prison (Plymouth, England) on 10 Jul 1813.
Cox, Miles - 2nd Mate - Nbr: 240 - Prize: Orbit - Ship Type: MV - How taken: HMS Achates - When taken: 29 Jul 1813 - Where taken: Lat 44N Long 13W - Date Received: 2 Apr 1813 - From what ship: Plymouth - Born: Philadelphia - Age: 29 - Sent to Dartmouth on 30 Jul 1813.
Cox, Peter J. - Seaman - Nbr: 3714 - Prize: Alfred - Ship Type: P - How taken: HMS Epervier - When taken: 23 Feb 1814 - Where taken: off Newfoundland - Date Received: 30 Sep 1814 - From what ship: HMT President, Halifax - Born: New York - Age: 16 - Released on 4 Jun 1815.
Crafford, Robert - Soldier - Nbr: 1028 - Prize: US Brig Argus - Ship Type: MW - How taken: HMS Pelican - When taken: 16 Aug 1813 - Where taken: Irish Channel - Date Received: 10 May 1814 - From what ship: Plymouth - Born: New York - Age: 41 - Sent to Dartmouth on 19 Oct 1814.
Craft, Richard - Seaman - Nbr: 4683 - Prize: Growler - Ship Type: P - How taken: HMS Electra - When taken: 7 Jul 1813 - Where taken: at sea - Date Received: 9 Oct 1814 - From what ship: HMT Leyden, Chatham - Born: Philadelphia - Age: 21 - Released on 15 Jun 1815.
Craig, John - Seaman - Nbr: 4337 - Prize: Venus - Ship Type: P - How taken: HMS Loire - When taken: 18 Feb 1814 - Where taken: off St. Thomas - Date Received: 7 Oct 1814 - From what ship: HMT Salvador del Mundo, Halifax - Born: New Jersey - Age: 23 - Released on 14 Jun 1815.
Craig, John - Seaman - Nbr: 6083 - Prize: David Porter - Ship Type: LM - How taken: HMS Pylades - When taken: 13 Sep 1814 - Where taken: Georges Bank - Date Received: 28 Dec 1814 - From what ship: HMT Penelope - Born: Virginia - Age: 20 - Released on 3 Jul 1815.
Craig, John E. - Surgeon - Nbr: 4829 - Prize: President - Ship Type: P - How taken: HMS Pique - When taken: 7 May 1814 - Where taken: off Porto Rico - Date Received: 9 Oct 1814 - From what ship: HMT Freya, Halifax - Born: Lynn - Age: 25 - Sent to HMS Impregnable on 27 Dec 1814.
Craig, William - Seaman - Nbr: 1239 - How taken: Sent into custody from HMS Alemine - When taken: 25 Jan 1813 - Date Received: 14 Jun 1814 - From what ship: Mill Prison (Plymouth, England) - Born: Dorset - Age: 29 - Released on 26 Apr 1815.
Craig, William - Seaman - Nbr: 4909 - Prize: Little Belt - Ship Type: MV - How taken: HMS Armide - When taken: 21 Sep 1813 - Where taken: off VA - Date Received: 28 Oct 1814 - From what ship: HMT Alkbar, Halifax - Born: Manchester - Age: 35 - Escaped on 1 Jun 1815.
Cramprey, James - Seaman - Nbr: 5796 - Prize: General Putnam - Ship Type: P - How taken: HMS Leander - When taken: 8 Nov 1814 - Where taken: Long 65 Lat 42 - Date Received: 26 Dec 1814 - From what ship: HMT Argo - Born: Massachusetts - Age: 18 - Released on 3 Jul 1815.
Cramstead, James - Seaman - Nbr: 1343 - Prize: Paul Jones - Ship Type: P - How taken: HMS Leonidas - When taken: 23 May 1813 - Where taken: Channel - Date Received: 19 Jun 1814 - From what ship: Stapleton - Born: New York - Age: 32 - Released on 28 Apr 1815.
Crandell, John - Seaman - Nbr: 1793 - Prize: Ferox - Ship Type: MV - How taken: HMS Medusa & HMS Lyra - When taken: 28 Mar 1813 - Where taken: off Cape Ortegal (Spain) - Date Received: 20 Jul 1814 - From what ship: HMS Milford, Plymouth - Born: Duchess County - Age: 29 - Released on 1 May 1815.
Cranley, James - Boatswain - Nbr: 5908 - Prize: Harlequin - Ship Type: P - How taken: HMS Bulwark - When taken: 23 Nov 1814 - Where taken: off Halifax - Date Received: 27 Dec 1814 - From what ship: HMT Penelope - Born: Boston - Age: 17 - Released on 3 Jul 1815.
Crapeman, Henry - Boy - Nbr: 4091 - Prize: Tyger - Ship Type: Prize - How taken: HMS Bulwark - When taken: 20 Jul 1814 - Where taken: Georges Bank - Date Received: 6 Oct 1814 - From what ship: HMT Chesapeake, Halifax - Born: Rhode Island - Age: 14 - Released on 3 May 1815.
Crapon, George - Seaman - Nbr: 4764 - Prize: Valentine - Ship Type: MV - How taken: HMS Minden - When taken: 16 Nov 1812 - Where taken: Malager Bay - Date Received: 9 Oct 1814 - From what ship: HMT Freya, Chatham - Born: Providence - Age: 19 - Released on 27 Apr 1815.
Crasey, James - Seaman - Nbr: 4184 - Prize: Boros - Ship Type: MV - How taken: HMS Opossum - When taken: 10 Nov 1812 - Where taken: off Barbados - Date Received: 7 Oct 1814 - From what ship: HMT Niobe, Chatham - Born: Kingsbury - Age: 24 - Released on 22 Oct 1814.
Crassey, Benjamin - Marine - Nbr: 5932 - Prize: Harlequin - Ship Type: P - How taken: HMS Bulwark - When taken: 23 Nov 1814 - Where taken: off Halifax - Date Received: 27 Dec 1814 - From what ship: HMT Penelope - Born: Buxton - Age: 26 - Released on 3 Jul 1815.

American Prisoners of War Held at Dartmoor during the War of 1812

## Alphabetical listing of names

Crasus, Richard - Seaman - Nbr: 4603 - Prize: Requin - Ship Type: LM - How taken: HMS Venus - When taken: 6 Mar 1814 - Where taken: off Bordeaux - Date Received: 9 Oct 1814 - From what ship: HMT Leyden, Chatham - Born: Vienna - Age: 23 - Released on 15 Jun 1815.

Crawford, George - Seaman - Nbr: 1720 - How taken: Gave himself up from HMS Rodney - Date Received: 8 Jul 1814 - From what ship: Labrador - Born: Philadelphia - Age: 45 - Released on 1 May 1815.

Crawford, John - Seaman - Nbr: 5848 - Prize: James - Ship Type: MV - How taken: HMS Harpy - When taken: 20 Dec 1812 - Where taken: off Isle of France - Date Received: 26 Dec 1814 - From what ship: HMT Argo - Born: Pennsylvania - Age: 24 - Released on 27 Apr 1815.

Crawford, Nelson - Seaman - Nbr: 1830 - Prize: Lightning - Ship Type: MV - How taken: HMS Medusa - When taken: 2 Apr 1813 - Where taken: Bay of Biscay - Date Received: 21 Jul 1814 - From what ship: HMT Redbeard & Pincher, Chatham Depot - Born: Richmond - Age: 26 - Released on 1 May 1815.

Crawford, William - Seaman - Nbr: 845 - Prize: Chesapeake - Ship Type: LM - How taken: HMS Hotspur & HMS Pyramus - When taken: 26 Oct 1813 - Where taken: off Nantes - Date Received: 29 Nov 1813 - From what ship: Plymouth - Born: Delaware - Age: 27 - Released on 26 Apr 1815.

Creamer, George - Steward - Nbr: 5045 - Prize: Ida - Ship Type: LM - How taken: HMS Newcastle - When taken: 9 Aug 1814 - Where taken: Long 34 - Date Received: 28 Oct 1814 - From what ship: HMT Alkbar, Halifax - Born: Massachusetts - Age: 24 - Released on 29 Jun 1815.

Cree, William - Seaman - Nbr: 1877 - Prize: Elbridge Gerry - Ship Type: P - How taken: HMS Crescent - When taken: 16 Sep 1813 - Where taken: off St. George's - Date Received: 29 Jul 1814 - From what ship: HMS Ville de Paris, Chatham Depot - Born: Portland - Age: 22 - Released on 2 May 1815.

Creek, Frederick - Boy - Nbr: 5532 - Prize: Sparks - Ship Type: LM - How taken: HMS Maidstone - When taken: 28 Sep 1814 - Where taken: off Nantucket - Date Received: 17 Dec 1814 - From what ship: HMT Loire, Halifax - Born: New York - Age: 15 - Released on 2 May 1815.

Creiger, John - Seaman - Nbr: 4225 - Prize: Liberty - Ship Type: MV - How taken: Surrendered - When taken: 30 Dec 1813 - Where taken: Stromess - Date Received: 7 Oct 1814 - From what ship: HMT Niobe, Chatham - Born: New York - Age: 21 - Released on 13 Jun 1815.

Creighton, William - Seaman - Nbr: 6216 - Prize: Prince de Neufchatel - Ship Type: P - How taken: Leander (Newcastle Acasta) - When taken: 20 Dec 1814 - Where taken: Lat 38 Long 56 - Date Received: 30 Jan 1815 - From what ship: HMT Pheasant - Born: Boston - Age: 21 - Released on 5 Jul 1815.

Crellin, William - Seaman - Nbr: 3890 - Prize: Sally - Ship Type: MV - How taken: HMS Acasta - When taken: 6 Jun 1814 - Where taken: off VA - Date Received: 5 Oct 1814 - From what ship: HMT Orpheus, Halifax - Born: New York - Age: 42 - Released on 9 Jun 1815.

Crepo, William - Seaman - Nbr: 853 - Prize: Amiable - Ship Type: LM - How taken: HMS Magnificent - When taken: 30 Oct 1813 - Where taken: off Bordeaux - Date Received: 4 Dec 1813 - From what ship: Plymouth - Born: Massachusetts - Age: 26 - Released on 26 Apr 1815.

Crescoll, Joseph - Boy - Nbr: 5486 - Prize: General Putnam - Ship Type: P - How taken: HMS Leander - When taken: 8 Nov 1814 - Where taken: Long 65 Lat 42 - Date Received: 17 Dec 1814 - From what ship: HMT Loire, Halifax - Born: Bordeaux - Age: 30 - Released on 1 Jul 1815.

Crete, John Isaac - Seaman - Nbr: 176 - Prize: Hope - Ship Type: MV - How taken: Chance (Privateer) - When taken: 15 Feb 1813 - Where taken: off Bordeaux - Date Received: 2 Apr 1813 - From what ship: Plymouth - Born: New Orleans - Age: 24 - Sent to Plymouth on 8 Jul 1814.

Crocker, Silvester - Prize Master - Nbr: 3379 - Prize: Growler - Ship Type: P - How taken: HMS Electra - When taken: 7 Jul 1813 - Where taken: at sea - Date Received: 13 Sep 1814 - From what ship: HMT Niobe, Chatham - Born: Massachusetts - Age: 32 - Released on 28 May 1815.

Crofts, John - Seaman - Nbr: 3927 - Prize: Rolla - Ship Type: P - How taken: HMS Loire - When taken: 10 Dec 1813 - Where taken: off Bull Island (South Carolina) - Date Received: 5 Oct 1814 - From what ship: HMT President, Halifax - Born: Philadelphia - Age: 34 - Released on 9 Jun 1815.

Croilt, Aaron - Seaman - Nbr: 4098 - Prize: Hartford - Ship Type: MV - How taken: HMS Junon - When taken: 6 Jun 1814 - Where taken: off Cape Ann - Date Received: 6 Oct 1814 - From what ship: HMT Chesapeake, Halifax - Born: Salem - Age: 30 - Released on 13 Jun 1815.

Croker, Nathaniel - Seaman - Nbr: 642 - Prize: Betsy - Ship Type: MV - How taken: HMS Leonidas - When taken: 12 Aug 1813 - Where taken: Channel - Date Received: 8 Sep 1813 - From what ship: Plymouth - Born: Boston - Age: 26 - Sent to Dartmouth on 30 Jul 1813.

Cromwell, Glacio - Seaman - Nbr: 2622 - How taken: Gave himself up from HMS Christian 7th - When taken: 21 Dec 1812 - Date Received: 21 Aug 1814 - From what ship: HMT Freya, Chatham - Born: Baltimore - Age:

American Prisoners of War Held at Dartmoor during the War of 1812

## Alphabetical listing of names

23 - Race: Black - Released on 26 Apr 1815.

Crosbeck, Nathaniel - Seaman - Nbr: 2520 - Prize: Snap Dragon - Ship Type: P - How taken: HMS Martin - When taken: 10 Jun 1814 - Where taken: off Halifax - Date Received: 16 Aug 1814 - From what ship: HMT Queen, Halifax - Born: New London - Age: 25 - Race: Black - Released on 3 May 1815.

Crosby, Andrew - Seaman - Nbr: 4271 - Prize: York Town - Ship Type: P - How taken: HMS Maidstone - When taken: 18 Jul 1813 - Where taken: Grand Banks - Date Received: 7 Oct 1814 - From what ship: HMT Niobe, Chatham - Born: New York - Age: 17 - Released on 13 Jun 1815.

Crosby, George - Seaman - Nbr: 856 - Prize: Amiable - Ship Type: LM - How taken: HMS Magnificent - When taken: 30 Oct 1813 - Where taken: off Bordeaux - Date Received: 4 Dec 1813 - From what ship: Plymouth - Born: Fairfield - Age: 23 - Released on 26 Apr 1815.

Crosby, John - Seaman - Nbr: 4152 - Prize: Argus - Ship Type: MV - How taken: HMS San Domingo - When taken: 1 Mar 1814 - Where taken: off Savannah - Date Received: 6 Oct 1814 - From what ship: HMT Niobe, Chatham - Born: New York - Age: 42 - Released on 13 Jun 1815.

Crosby, John - Seaman - Nbr: 1161 - How taken: Gave himself up from HMS Hebrius - Date Received: 10 May 1814 - From what ship: Plymouth - Born: Titting - Age: 25 - Released on 28 Apr 1815.

Cross, Ephraim - Seaman - Nbr: 3312 - Prize: Thomas - Ship Type: P - How taken: HMS Nymphe - When taken: 28 Jun 1813 - Where taken: off Halifax - Date Received: 13 Sep 1814 - From what ship: HMT Niobe, Chatham - Born: Massachusetts - Age: 42 - Released on 28 May 1815.

Cross, Nathaniel - Seaman - Nbr: 4898 - Prize: Atlantic - Ship Type: Prize - How taken: HMS Maidstone - When taken: 10 Oct 1813 - Where taken: off Sable Island - Date Received: 28 Oct 1814 - From what ship: HMT Alkbar, Halifax - Born: Newbury - Age: 21 - Released on 21 Jun 1815.

Cross, Oliver - Seaman - Nbr: 1423 - Prize: Governor Gerry - Ship Type: MV - How taken: HMS Lyra - When taken: 29 May 1813 - Where taken: Bay of Biscay - Date Received: 19 Jun 1814 - From what ship: Stapleton - Born: Virginia - Age: 44 - Race: Negro - Released on 28 Apr 1815.

Cross, Peter - Seaman - Nbr: 919 - Prize: Porcupine - Ship Type: MV - How taken: HMS Acasta - When taken: 15 Jun 1813 - Where taken: at sea - Date Received: 31 Jan 1814 - From what ship: Plymouth - Born: Point Sicily - Age: 23 - Impersonated a Frenchman on the general release.

Cross, Stephen - Seaman - Nbr: 5202 - Prize: Porcupine - Ship Type: LM - How taken: HMS Acasta - When taken: 18 Jul 1813 - Where taken: off Halifax - Date Received: 31 Oct 1814 - From what ship: HMT Mermaid, Chatham - Born: Port au Prince - Age: 20 - Race: Black - Released on 29 Jun 1815.

Crouch, Richard - Seaman - Nbr: 431 - Prize: Viper - Ship Type: MV - How taken: HMS Superb - When taken: 15 Apr 1813 - Where taken: Bay of Biscay - Date Received: 1 Jul 1813 - From what ship: Plymouth - Born: New York - Age: 31 - Released on 20 Apr 1815.

Crow, John - Seaman - Nbr: 1890 - How taken: Gave himself up from HMS William Pitt No. 421 - When taken: 2 Dec 1813 - Date Received: 29 Jul 1814 - From what ship: HMS Ville de Paris, Chatham Depot - Born: Maryland - Age: 23 - Released on 3 May 1815.

Crowel, Mathew - Seaman - Nbr: 4505 - Prize: Enterprize - Ship Type: P - How taken: HMS Tenedos - When taken: 21 May 1813 - Where taken: off Cape Cod - Date Received: 8 Oct 1814 - From what ship: HMT Leyden, Chatham - Born: Massachusetts - Age: 23 - Released on 15 Jun 1815.

Crowell, Elijah - Seaman - Nbr: 5712 - Prize: Swift - Ship Type: MV - How taken: HMS Niemen - When taken: 2 Sep 1814 - Where taken: off Delaware - Date Received: 24 Dec 1814 - From what ship: HMT Penelope - Born: Yarmouth - Age: 28 - Released on 3 Jul 1815.

Crowell, Seth - Seaman - Nbr: 3009 - Prize: Rattlesnake - Ship Type: MV - How taken: HMS Rhin - When taken: 10 Mar 1814 - Where taken: Western ocean - Date Received: 1 Sep 1814 - From what ship: from Bristol - Born: Dennis - Age: 21 - Released on 10 Apr 1814.

Crump, Joseph R. - Seaman - Nbr: 4911 - Prize: Snap Dragon - Ship Type: P - How taken: HMS Martin - When taken: 30 Jun 1814 - Where taken: off Halifax - Date Received: 28 Oct 1814 - From what ship: HMT Alkbar, Halifax - Born: Virginia - Age: 24 - Released on 27 Apr 1815.

Crumpton, William - Seaman - Nbr: 3088 - Prize: Eliza (Whaler) - Ship Type: MV - How taken: HMS Charles - When taken: 20 Jul 1813 - Where taken: off Peter Head - Date Received: 3 Sep 1814 - From what ship: HMT Hydra, Chatham - Born: Philadelphia - Age: 19 - Released on 28 May 1815.

Cudsworth, Henry - Seaman - Nbr: 1420 - Prize: Governor Gerry - Ship Type: MV - How taken: HMS Lyra - When taken: 29 May 1813 - Where taken: Bay of Biscay - Date Received: 19 Jun 1814 - From what ship: Stapleton - Born: Charleston - Age: 19 - Released on 28 Apr 1815.

Cullett, William - Seaman - Nbr: 3366 - How taken: Impressed at London - When taken: 10 Nov 1813 - Date

American Prisoners of War Held at Dartmoor during the War of 1812

## Alphabetical listing of names

Received: 13 Sep 1814 - From what ship: HMT Niobe, Chatham - Born: Philadelphia - Age: 60 - Released on 28 May 1815.

Cummings, James - Seaman - Nbr: 1634 - Prize: Tom - Ship Type: LM - How taken: HMS Surveillante - When taken: 24 Apr 1813 - Where taken: Bay of Biscay - Date Received: 23 Jun 1814 - From what ship: Stapleton - Born: Connecticut - Age: 23 - Released on 1 May 1815.

Cummings, James - Seaman - Nbr: 2923 - How taken: Gave himself up from HMS Union - When taken: 27 May 1813 - Date Received: 24 Aug 1814 - From what ship: HMT Alpheus, Chatham - Born: Pennsylvania - Age: 31 - Released on 19 May 1815.

Cummings, Nathaniel - Seaman - Nbr: 5481 - Prize: General Putnam - Ship Type: P - How taken: HMS Leander - When taken: 8 Nov 1814 - Where taken: Long 65 Lat 42 - Date Received: 17 Dec 1814 - From what ship: HMT Loire, Halifax - Born: Salem - Age: 17 - Released on 1 Jul 1815.

Cummins, Edward - Seaman - Nbr: 2140 - How taken: Gave himself up from HMS Ruby - When taken: 1 Aug 1812 - Date Received: 8 Aug 1814 - From what ship: HMT Raven, Chatham - Born: Philadelphia - Age: 35 - Released on 26 Apr 1815.

Cunningham, John - Fifer - Nbr: 5499 - Prize: 38th US Infantry - Ship Type: Troops - How taken: British Army - When taken: 24 Aug 1814 - Where taken: Washington, DC - Date Received: 17 Dec 1814 - From what ship: HMT Loire, Halifax - Born: New York - Age: 21 - Released on 29 Jun 1815.

Curden, John - Seaman - Nbr: 6041 - Prize: Regent - Ship Type: LM - How taken: HMS Forth - When taken: 19 Sep 1814 - Where taken: off Egg Harbor (New Jersey) - Date Received: 28 Dec 1814 - From what ship: HMT Penelope - Born: Maryland - Age: 28 - Race: Mulatto - Released on 3 Jul 1815.

Curgan, John - Seaman - Nbr: 5802 - Prize: US Frigate Superior (Gig) - Ship Type: MW - How taken: British gunboats - When taken: 26 Aug 1814 - Where taken: Lake Ontario - Date Received: 26 Dec 1814 - From what ship: HMT Argo - Born: Not legible - Age: 34 - Released on 3 Jul 1815.

Curlis, Aaron - 1st Mate - Nbr: 5570 - Prize: Unity - Ship Type: MV - How taken: HMS Asia - When taken: 24 Jul 1814 - Where taken: off Cape VA - Date Received: 17 Dec 1814 - From what ship: HMT Loire, Halifax - Born: Philadelphia - Age: 56 - Released on 1 Jul 1815.

Currant, Michael - Seaman - Nbr: 4115 - Prize: Bordeaux Packet - Ship Type: LM - How taken: HMS Niemen - When taken: 28 Jun 1814 - Where taken: off Delaware - Date Received: 6 Oct 1814 - From what ship: HMT Chesapeake, Halifax - Born: New York - Age: 19 - Released on 13 Jun 1815.

Curren, Nathaniel - Gunner - Nbr: 3733 - Prize: Alfred - Ship Type: P - How taken: HMS Epervier - When taken: 23 Feb 1814 - Where taken: off Newfoundland - Date Received: 30 Sep 1814 - From what ship: HMT President, Halifax - Born: Salem - Age: 22 - Died on 1 Jun 1815 from phthisis pulmonalis.

Currie, James - Prize master - Nbr: 5699 - Prize: David Porter - Ship Type: LM - How taken: HMS Pylades - When taken: 12 Sep 1814 - Where taken: Georges Bank - Date Received: 24 Dec 1814 - From what ship: HMT Penelope - Born: Portsmouth - Age: 27 - Released on 11 Jul 1815.

Currien, Stephen - Seaman - Nbr: 521 - Prize: Friends - Ship Type: Bey of Pool - How taken: HMS Whiting - When taken: 15 Jul 1813 - Where taken: Lat 67 N, Long 8 W - Date Received: 8 Sep 1813 - From what ship: Plymouth - Born: Massachusetts - Age: 28 - Released on 26 Apr 1815.

Curry, Samuel - Seaman - Nbr: 1721 - How taken: Impressed - Date Received: 8 Jul 1814 - From what ship: Labrador - Born: Petersburg - Age: 31 - Released on 1 May 1815.

Curtes, Zebediah - Steward - Nbr: 3637 - Prize: Bordeaux Packet - Ship Type: LM - How taken: HMS Niemen - When taken: 28 Jan 1814 - Where taken: off Delaware - Date Received: 30 Sep 1814 - From what ship: HMT Sybella - Born: Shannon - Age: 25 - Released on 4 Jun 1815.

Curtis, Enoch - Seaman - Nbr: 2939 - How taken: Gave himself up from HMS Dauntless - Date Received: 24 Aug 1814 - From what ship: HMT Alpheus, Chatham - Born: Boston - Age: 37 - Released on 19 May 1815.

Curtis, Ephraim - Seaman - Nbr: 4300 - Prize: Liberty - Ship Type: MV - How taken: Surrendered - When taken: 30 Dec 1813 - Where taken: Stromess - Date Received: 7 Oct 1814 - From what ship: HMT Niobe, Chatham - Born: Freeport - Age: 24 - Released on 14 Jun 1815.

Curtis, Francis - Seaman - Nbr: 523 - Prize: Friends - Ship Type: Bey of Pool - How taken: HMS Whiting - When taken: 15 Jul 1813 - Where taken: Lat 67 N, Long 8 W - Date Received: 8 Sep 1813 - From what ship: Plymouth - Born: Massachusetts - Age: 21 - Released on 26 Apr 1815.

Curtis, Henry - Seaman - Nbr: 524 - Prize: Friends - Ship Type: Bey of Pool - How taken: HMS Whiting - When taken: 15 Jul 1813 - Where taken: Lat 67 N, Long 8 W - Date Received: 8 Sep 1813 - From what ship: Plymouth - Born: New York - Age: 21 - Released on 26 Apr 1815.

Curtis, Henry - Seaman - Nbr: 2572 - Prize: Rattlesnake - Ship Type: P - How taken: HMS Hyperion - When taken:

American Prisoners of War Held at Dartmoor during the War of 1812

## Alphabetical listing of names

25 Jun 1814 - Where taken: off Cape Finisterre - Date Received: 21 Aug 1814 - From what ship: HMS Hyperion - Born: Africa - Age: 27 - Race: Negro - Released on 3 May 1815.

Curtis, John - Seaman - Nbr: 536 - Prize: Union - Ship Type: LM - How taken: HMS Goldfinch - When taken: 17 Jul 1813 - Where taken: Bay of Biscay - Date Received: 8 Sep 1813 - From what ship: Plymouth - Born: Woolwich - Age: 20 - Released on 26 Apr 1815.

Curtis, Joseph - Seaman - Nbr: 567 - Prize: US Brig Argus - Ship Type: MW - How taken: HMS Pelican - When taken: 14 Apr 1813 - Where taken: Irish Channel - Date Received: 8 Sep 1813 - From what ship: Plymouth - Born: Arron in America - Age: 42 - Sent to Plymouth on 8 Jul 1814.

Curtis, Stacey - Sail maker - Nbr: 2404 - Prize: US Sloop Frolic - Ship Type: MW - How taken: HMS Orpheus - When taken: 20 Apr 1814 - Where taken: off Cuba - Date Received: 16 Aug 1814 - From what ship: HMT Queen, Halifax - Born: Marblehead - Age: 23 - Sent to Dartmouth on 19 Oct 1814.

Curtis, William - Boatswain - Nbr: 4018 - Prize: US Brig Rattlesnake - Ship Type: MW - How taken: HMS Leander - When taken: 13 Jul 1814 - Where taken: off Shelburne - Date Received: 6 Oct 1814 - From what ship: HMT Chesapeake, Halifax - Born: Baltimore - Age: 38 - Released on 9 Jun 1815.

Cushman, Orson - Seaman - Nbr: 2000 - How taken: Impressed at Gravesend - When taken: 27 Mar 1813 - Date Received: 3 Aug 1814 - From what ship: HMS Lyffey, Chatham Depot - Born: Stafford - Age: 24 - Released on 2 May 1815.

Cussar, Jacob O. - Seaman - Nbr: 4705 - Prize: Volunteer - Ship Type: MV - How taken: Victoria (Privateer) - When taken: 26 Dec 1813 - Where taken: at sea - Date Received: 9 Oct 1814 - From what ship: HMT Leyden, Chatham - Born: New York - Age: 39 - Died on 7 Dec 1814 from pneumonia.

Cutler, Thomas - Seaman - Nbr: 4735 - How taken: Gave himself up from HMS L'Aigle - When taken: 9 May 1814 - Date Received: 9 Oct 1814 - From what ship: HMT Freya, Chatham - Born: Exeter - Age: 22 - Released on 11 Jul 1815.

Cutts, Ceaser - Seaman - Nbr: 5955 - Prize: Harlequin - Ship Type: P - How taken: HMS Bulwark - When taken: 23 Nov 1814 - Where taken: off Halifax - Date Received: 27 Dec 1814 - From what ship: HMT Penelope - Born: Portsmouth - Age: 49 - Race: Black - Released on 3 Jul 1815.

Daggett, Robert - Prize Master - Nbr: 1151 - Prize: Bunker Hill - Ship Type: P - How taken: HMS Pomone - When taken: 8 Mar 1814 - Where taken: at sea - Date Received: 10 May 1814 - From what ship: Plymouth - Born: Providence - Age: 21 - Released on 28 Apr 1815.

Daigney, Jasper - Seaman - Nbr: 3744 - Prize: Lizard - Ship Type: P - How taken: HMS Prometheus - When taken: 5 May 1814 - Where taken: off Halifax - Date Received: 30 Sep 1814 - From what ship: HMT President, Halifax - Born: Salem - Age: 18 - Race: Creole - Released on 4 Jun 1815.

Dailey, Daniel - Seaman - Nbr: 5968 - Prize: Amazon - Ship Type: Prize - How taken: HMS Bulwark - When taken: 22 Sep 1814 - Where taken: Georges Bank - Date Received: 27 Dec 1814 - From what ship: HMT Penelope - Born: Massachusetts - Age: 28 - Released on 3 Jul 1815.

Dailey, George W. - Seaman - Nbr: 3125 - Prize: Prize of the Blockage - Ship Type: Prize - How taken: HMS Brazen - When taken: 29 Jun 1813 - Where taken: off Scotland - Date Received: 11 Sep 1814 - From what ship: HMT Freya, Chatham - Born: Newport - Age: 25 - Released on 28 May 1815.

Dale, Hercules - Seaman - Nbr: 1195 - Prize: Imperatrice Reine - Ship Type: P - How taken: HMS Hotspur - When taken: 13 Jan 1813 - Date Received: 4 Jun 1814 - From what ship: Dartmouth - Born: Newbury - Age: 33 - Released on 28 Apr 1815.

Dale, John - Seaman - Nbr: 4859 - Prize: Amelia - Ship Type: Prize - How taken: HMS Hardy - When taken: 28 Mar 1814 - Where taken: West Indies - Date Received: 9 Oct 1814 - From what ship: HMT Freya, Halifax - Born: Baltimore - Age: 25 - Released on 21 Jun 1815.

Dallison, James - Seaman - Nbr: 3906 - Prize: Emmeline - Ship Type: MV - How taken: HMS Loire - When taken: 10 Jul 1814 - Where taken: Chesapeake Bay - Date Received: 5 Oct 1814 - From what ship: HMT Orpheus, Halifax - Born: Baltimore - Age: 20 - Race: Negro - Released on 9 Jun 1815.

Dalloway, John - Seaman - Nbr: 726 - Prize: Ned - Ship Type: LM - How taken: HMS Royalist - When taken: 6 Sep 1813 - Where taken: Bay of Biscay - Date Received: 27 Sep 1813 - From what ship: Plymouth - Born: Boston - Age: 26 - Sent to Plymouth on 8 Jul 1814.

Dalmark, William - Seaman - Nbr: 2159 - Prize: Rattlesnake - Ship Type: P - How taken: HMS Hyperion - When taken: 26 Jun 1814 - Where taken: off Cape Finisterre - Date Received: 16 Aug 1814 - From what ship: HMS Dublin, Halifax - Born: Baltimore - Age: 32 - Released on 2 May 1815.

Dalton, Fred W. - Purser steward - Nbr: 2397 - Prize: US Sloop Frolic - Ship Type: MW - How taken: HMS Orpheus - When taken: 20 Apr 1814 - Where taken: off Cuba - Date Received: 16 Aug 1814 - From what

American Prisoners of War Held at Dartmoor during the War of 1812

## Alphabetical listing of names

ship: HMT Queen, Halifax - Born: Connecticut - Age: 25 - Sent to Dartmouth on 19 Oct 1814.

Dalton, John - Soldier - Nbr: 5257 - Prize: 13th US Infantry - Ship Type: Troops - How taken: British Army - When taken: 13 Oct 1812 - Where taken: Canada - Date Received: 31 Oct 1814 - From what ship: HMT Leyden, Chatham - Born: Dublin - Age: 46 - Released on 29 Jun 1815.

Dalton, Samuel - Seaman - Nbr: 1754 - How taken: Gave himself up from HMS Bacchante - When taken: 11 Mar 1813 - Date Received: 20 Jul 1814 - From what ship: HMS Milford, Plymouth - Born: Salem - Age: 23 - Released on 1 May 1815.

Dame, Aesop - Marine - Nbr: 5877 - Prize: Harlequin - Ship Type: P - How taken: HMS Bulwark - When taken: 23 Nov 1814 - Where taken: off Halifax - Date Received: 27 Dec 1814 - From what ship: HMT Penelope - Born: Durham - Age: 25 - Released on 3 Jul 1815.

Dame, John - Boy - Nbr: 5395 - Prize: Hawke - Ship Type: P - How taken: HMS Pique - When taken: 27 Apr 1814 - Where taken: West Indies - Date Received: 31 Oct 1814 - From what ship: HMT Leyden, Chatham - Born: Norfolk - Age: 16 - Released on 3 May 1815.

Damrell, William - Prize master - Nbr: 5110 - Prize: Portsmouth - Ship Type: P - How taken: HMS Pylades - When taken: 13 Sep 1814 - Where taken: Long 65 - Date Received: 28 Oct 1814 - From what ship: HMT Alkbar, Halifax - Born: Portsmouth - Age: 25 - Released on 11 Jul 1815.

Dandy, Hamilton - Seaman - Nbr: 2324 - Prize: Hussar - Ship Type: P - How taken: HMS Saturn - When taken: 25 May 1814 - Where taken: off Sandy Hook - Date Received: 16 Aug 1814 - From what ship: HMS Dublin, Halifax - Born: Philadelphia - Age: 24 - Released on 3 May 1815.

Daniel, John - Seaman - Nbr: 646 - Prize: Betsy - Ship Type: MV - How taken: HMS Leonidas - When taken: 12 Aug 1813 - Where taken: Channel - Date Received: 8 Sep 1813 - From what ship: Plymouth - Born: Baltimore - Age: 19 - Sent to Mill Prison (Plymouth, England) on 10 Jul 1813.

Daniel, Robert - Seaman - Nbr: 1243 - How taken: Sent into custody - When taken: 17 Mar 1813 - Date Received: 14 Jun 1814 - From what ship: Mill Prison (Plymouth, England) - Born: New York - Age: 27 - Released on 28 Apr 1815.

Daniels, Henry - Seaman - Nbr: 6538 - Prize: US Brig Syren - Ship Type: MW - How taken: HMS Medway - When taken: 12 Jul 1814 - Where taken: off Cape of Good Hope - Date Received: 7 Mar 1815 - From what ship: HMT Ganges, Plymouth - Born: New Jersey - Age: 25 - Released on 11 Jul 1815.

Daniels, John - Seaman - Nbr: 5456 - How taken: Gave himself up from HMS York - When taken: 30 Nov 1814 - Date Received: 10 Dec 1814 - From what ship: HMT Impregnable - Born: Charleston - Age: 37 - Released on 1 Jul 1815.

Darran, Daniel - Seaman - Nbr: 2119 - How taken: Gave himself up from HMS Namur - When taken: 1 Nov 1812 - Date Received: 8 Aug 1814 - From what ship: HMT Raven, Chatham - Born: Sanford - Age: 26 - Released on 26 Apr 1815.

Darrow, Aaron - Seaman - Nbr: 3226 - How taken: Gave himself up from HMS Trent - Date Received: 11 Sep 1814 - From what ship: HMT Freya, Chatham - Born: Massachusetts - Age: 24 - Released on 28 May 1815.

Dashells, Richard - Midshipman - Nbr: 4074 - Prize: New Zealander - Ship Type: Prize - How taken: HMS Belviders - When taken: 21 Apr 1814 - Where taken: off Delaware - Date Received: 6 Oct 1814 - What ship: HMT Chesapeake, Halifax - Born: Queen Anne's County - Age: 24 - Sent to Ashburton (England) on 29 Nov 1814.

Dave, David - Seaman - Nbr: 4080 - Prize: New Zealander - Ship Type: Prize - How taken: HMS Belviders - When taken: 21 Apr 1814 - Where taken: off Delaware - Date Received: 6 Oct 1814 - From what ship: HMT Chesapeake, Halifax - Born: Norwich - Age: 29 - Released on 13 Jun 1815.

Davenport, John - Seaman - Nbr: 3522 - Prize: Sabine - Ship Type: P - How taken: HMS Conquistador - When taken: 3 Aug 1814 - Where taken: off the Banks - Date Received: 28 Sep 1814 - From what ship: HMT Salvador del Mundo - Born: East Haven - Age: 21 - Died on 10 Jun 1815 from dysentery.

Davenport, Russell - Seaman - Nbr: 1713 - How taken: Gave himself up from HMS Amphion - Date Received: 8 Jul 1814 - From what ship: Labrador - Born: New London - Age: 33 - Released on 1 May 1815.

Daverille, William - Seaman - Nbr: 5083 - Prize: Ida - Ship Type: LM - How taken: HMS Newcastle - When taken: 9 Aug 1814 - Where taken: Long 34 - Date Received: 28 Oct 1814 - From what ship: HMT Alkbar, Halifax - Born: Boston - Age: 32 - Released on 29 Jun 1815.

Davey, Charles - Seaman - Nbr: 1081 - How taken: Impressed at Liverpool - When taken: 15 Feb 1814 - Date Received: 10 May 1814 - From what ship: Plymouth - Born: New Orleans - Age: 19 - Impersonated a Frenchman on the general release.

David, James - Seaman - Nbr: 6314 - Prize: Prince de Neufchatel - Ship Type: P - How taken: Leander (Newcastle

American Prisoners of War Held at Dartmoor during the War of 1812

## Alphabetical listing of names

Acasta) - When taken: 20 Dec 1814 - Where taken: Lat 38 Long 56 - Date Received: 19 Feb 1815 - From what ship: HMT Ganges, Plymouth - Born: Albany - Age: 18 - Released on 5 Jul 1815.

David, John - Seaman - Nbr: 3322 - Prize: York Town - Ship Type: P - How taken: HMS Nimrod - When taken: 17 Jul 1813 - Where taken: off St. Johns - Date Received: 13 Sep 1814 - From what ship: HMT Niobe, Chatham - Born: Maryland - Age: 29 - Released on 28 May 1815.

David, John M. - Seaman - Nbr: 6283 - How taken: Gave himself up from HMS Phoenix - When taken: 12 Jan 1814 - Date Received: 17 Feb 1815 - From what ship: HMT Ganges, Plymouth - Born: New Orleans - Age: 24 - Race: Mulatto - Released on 5 Jul 1815.

David, William - Seaman - Nbr: 982 - Prize: Amity - Ship Type: P - How taken: HMS Achates - When taken: 22 Dec 1813 - Where taken: Bay of Biscay - Date Received: 31 Jan 1814 - From what ship: Plymouth - Born: New Orleans - Age: 24 - Sent to Plymouth on 27 Apr 1814.

Davidson, Henry - Seaman - Nbr: 353 - Prize: Thrasher - Ship Type: P - How taken: HMS Magicienne - When taken: 17 Jan 1813 - Where taken: off Western Islands (England) - Date Received: 1 Jul 1813 - From what ship: Plymouth - Born: Maryland - Age: 50 - Sent to Dartmouth on 30 Jul 1813.

Davies, John - Seaman - Nbr: 1436 - Prize: Leo - Ship Type: LM - How taken: HMS Magicienne - When taken: 4 Jun 1813 - Where taken: off France - Date Received: 19 Jun 1814 - From what ship: Stapleton - Born: Biddeford - Age: 29 - Released on 20 Mar 1815.

Davies, Nathaniel - Seaman - Nbr: 2945 - Prize: Atalanta - Ship Type: MV - How taken: HMS Barbados - When taken: 19 Jan 1814 - Where taken: off St. Bartholomew - Date Received: 24 Aug 1814 - From what ship: HMT Hannibal - Born: Massachusetts - Age: 20 - Released on 19 May 1815.

Davies, William - Gunner - Nbr: 5091 - Prize: Ida - Ship Type: LM - How taken: HMS Newcastle - When taken: 9 Aug 1814 - Where taken: Long 34 - Date Received: 28 Oct 1814 - From what ship: HMT Alkbar, Halifax - Born: Long Island - Age: 28 - Released on 29 Jun 1815.

Davis, Andrew - Seaman - Nbr: 2660 - How taken: Apprehended at London - When taken: 10 Aug 1813 - Date Received: 21 Aug 1814 - From what ship: HMT Freya, Chatham - Born: Philadelphia - Age: 23 - Released on 19 May 1815.

Davis, Benjamin S. - Prize Master - Nbr: 126 - Prize: Good Intent - Ship Type: P - How taken: HMS Rota - When taken: 26 Jan 1813 - Where taken: Lat 43'30" N, Long 20 W - Date Received: 2 Apr 1813 - From what ship: Plymouth - Born: Gloucester - Age: 26 - Sent to Dartmouth on 30 Jul 1813.

Davis, Charles - Seaman - Nbr: 233 - Prize: Charlotte - Ship Type: MV - How taken: HMS Warspite - When taken: 3 Mar 1813 - Where taken: Bay of Biscay - Date Received: 2 Apr 1813 - From what ship: Plymouth - Born: Norfolk - Age: 27 - Released on 20 Apr 1815.

Davis, Daniel - Seaman - Nbr: 1949 - How taken: Gave himself up from HMS Royal William - When taken: 26 Mar 1813 - Date Received: 3 Aug 1814 - From what ship: HMS Lyffey, Chatham Depot - Born: Carolina - Age: 20 - Released on 2 May 1815.

Davis, Daniel - Seaman - Nbr: 2731 - How taken: Gave himself up from HMS Aboukir - When taken: 28 Oct 1812 - Date Received: 24 Aug 1814 - From what ship: HMT Liverpool, Chatham - Born: Kennebec - Age: 36 - Released on 26 Apr 1815.

Davis, David - Seaman - Nbr: 5112 - Prize: Portsmouth - Ship Type: P - How taken: HMS Pylades - When taken: 13 Sep 1814 - Where taken: Long 65 - Date Received: 28 Oct 1814 - From what ship: HMT Alkbar, Halifax - Born: Leigh - Age: 40 - Released on 29 Jun 1815.

Davis, Elias S. - Seaman - Nbr: 6384 - Prize: US Brig Syren - Ship Type: MW - How taken: HMS Medway - When taken: 12 Jul 1814 - Where taken: off Cape of Good Hope - Date Received: 24 Feb 1815 - From what ship: HMT Ganges, Plymouth - Born: Vermont - Age: 23 - Released on 11 Jul 1815.

Davis, Elisha - Seaman - Nbr: 2506 - Prize: Snap Dragon - Ship Type: P - How taken: HMS Martin - When taken: 10 Jun 1814 - Where taken: off Halifax - Date Received: 16 Aug 1814 - From what ship: HMT Queen, Halifax - Born: North Carolina - Age: 28 - Released on 3 May 1815.

Davis, Frederick - Seaman - Nbr: 6315 - Prize: Prince de Neufchatel - Ship Type: P - How taken: Leander (Newcastle Acasta) - When taken: 20 Dec 1814 - Where taken: Lat 38 Long 56 - Date Received: 19 Feb 1815 - From what ship: HMT Ganges, Plymouth - Born: Boston - Age: 22 - Race: Mulatto - Released on 5 Jul 1815.

Davis, George - Seaman - Nbr: 4840 - Prize: President - Ship Type: P - How taken: HMS Pique - When taken: 7 May 1814 - Where taken: off Porto Rico - Date Received: 9 Oct 1814 - From what ship: HMT Freya, Halifax - Born: Pennsylvania - Age: 19 - Race: Black - Released on 21 Jun 1815.

Davis, George - Seaman - Nbr: 1937 - How taken: Gave himself up from HMS Royal William - When taken: 3 Feb

American Prisoners of War Held at Dartmoor during the War of 1812

## Alphabetical listing of names

1813 - Date Received: 3 Aug 1814 - From what ship: HMS Alceste, Chatham Depot - Born: Albany - Age: 20 - Released on 2 May 1815.

Davis, George - Seaman - Nbr: 753 - How taken: Impressed at Liverpool - Date Received: 3 Nov 1813 - From what ship: Plymouth - Born: Dayfield - Age: 33 - Released on 20 Apr 1815.

Davis, George - Seaman - Nbr: 3635 - Prize: Monarch - Ship Type: MV - How taken: HMS Dotterel - When taken: 14 Dec 1813 - Where taken: off Charleston - Date Received: 30 Sep 1814 - From what ship: HMT Sybella - Born: New York - Age: 32 - Released on 30 Mar 1815.

Davis, George - Seaman - Nbr: 3147 - How taken: Gave himself up from HMS Sabina - When taken: 9 Aug 1813 - Date Received: 11 Sep 1814 - From what ship: HMT Freya, Chatham - Born: New Jersey - Age: 34 - Race: Mulatto - Released on 28 May 1815.

Davis, Hamilton - Seaman - Nbr: 2481 - Prize: US Sloop Frolic - Ship Type: MW - How taken: HMS Orpheus - When taken: 20 Apr 1814 - Where taken: off Cuba - Date Received: 16 Aug 1814 - From what ship: HMT Queen, Halifax - Born: New York - Age: 21 - Released on 3 May 1815.

Davis, Henry - Seaman - Nbr: 4249 - Prize: Requin - Ship Type: LM - How taken: HMS Venus - When taken: 6 Mar 1814 - Where taken: off Bordeaux - Date Received: 7 Oct 1814 - From what ship: HMT Niobe, Chatham - Born: Providence - Age: 23 - Released on 13 Jun 1815.

Davis, Jacob - Seaman - Nbr: 5015 - Prize: Charlotte - Ship Type: Prize - How taken: Censor - When taken: 5 Aug 1814 - Where taken: off Azores - Date Received: 28 Oct 1814 - From what ship: HMT Alkbar, Halifax - Born: Lancaster - Age: 25 - Race: Negro - Released on 21 Jun 1815.

Davis, James - Seaman - Nbr: 6519 - How taken: Gave himself up from HMS Orontes - When taken: Feb 1815 - Date Received: 3 Mar 1815 - From what ship: HMT Ganges, Plymouth - Born: Raymond - Age: 33 - Released on 11 Jul 1815.

Davis, James - Seaman - Nbr: 1701 - Prize: Fanny - Ship Type: Prize - How taken: HMS Sceptre - When taken: 12 May 1814 - Date Received: 2 Jul 1814 - From what ship: Plymouth - Born: New London - Age: 26 - Released on 1 May 1815.

Davis, James - Seaman - Nbr: 266 - Prize: William Bayard - Ship Type: MV - How taken: HMS Warspite - When taken: 3 Mar 1813 - Where taken: Bay of Biscay - Date Received: 28 Jun 1813 - From what ship: Plymouth - Born: Bristol - Age: 34 - Released on 20 Apr 1815.

Davis, James - Seaman - Nbr: 3079 - How taken: Gave himself up from HMS Eliza (Tender) - When taken: 1 Sep 1813 - Date Received: 3 Sep 1814 - From what ship: HMT Hydra, Chatham - Born: New Jersey - Age: 31 - Released on 28 May 1815.

Davis, John - Seaman - Nbr: 4890 - How taken: Gave himself up from HMS Vestal - When taken: Jul 1814 - Date Received: 24 Oct 1814 - From what ship: HMT Salvador del Mundo - Born: Baltimore - Age: 30 - Released on 21 Jun 1815.

Davis, John - Seaman - Nbr: 6088 - Prize: Minerva - Ship Type: MV - How taken: HMS Maidstone - When taken: 28 Mar 1814 - Where taken: off New London - Date Received: 28 Dec 1814 - From what ship: HMT Penelope - Born: St. Bartholomew - Age: 25 - Race: Black - Released on 11 Jul 1815.

Davis, John - Seaman - Nbr: 5240 - Prize: York Town - Ship Type: P - How taken: HMS Racehorse - When taken: 17 Feb 1813 - Where taken: off Cape Sable Island (Canada) - Date Received: 31 Oct 1814 - From what ship: HMT Mermaid, Chatham - Born: Savannah - Age: 25 - Died on 26 Feb 1815 from phthisis pulmonalis.

Davis, John - Seaman - Nbr: 3719 - Prize: Alfred - Ship Type: P - How taken: HMS Epervier - When taken: 23 Feb 1814 - Where taken: off Newfoundland - Date Received: 30 Sep 1814 - From what ship: HMT President, Halifax - Born: Marblehead - Age: 40 - Released on 4 Jun 1815.

Davis, John - Seaman - Nbr: 1631 - Prize: Tom - Ship Type: LM - How taken: HMS Surveillante - When taken: 24 Apr 1813 - Where taken: Bay of Biscay - Date Received: 23 Jun 1814 - From what ship: Stapleton - Born: Massachusetts - Age: 25 - Released on 1 May 1815.

Davis, John - Seaman - Nbr: 1620 - Prize: Tom - Ship Type: LM - How taken: HMS Surveillante - When taken: 24 Apr 1813 - Where taken: Bay of Biscay - Date Received: 23 Jun 1814 - From what ship: Stapleton - Born: New Orleans - Age: 19 - Sent to Mill Prison (Plymouth, England) on 30 Jun 1814.

Davis, John - Seaman - Nbr: 1527 - Prize: Essex - Ship Type: MV - How taken: HMS Pyramus - When taken: 20 Apr 1813 - Where taken: Bay of Biscay - Date Received: 23 Jun 1814 - From what ship: Stapleton - Born: New Brunswick - Age: 26 - Released on 1 May 1815.

Davis, John - Seaman - Nbr: 1143 - Prize: Bunker Hill - Ship Type: P - How taken: HMS Pomone - When taken: 8 Mar 1814 - Where taken: at sea - Date Received: 10 May 1814 - From what ship: Plymouth - Born: Camden - Age: 22 - Released on 28 Apr 1815.

American Prisoners of War Held at Dartmoor during the War of 1812

## Alphabetical listing of names

Davis, John - Seaman - Nbr: 5209 - Prize: Porcupine - Ship Type: LM - How taken: HMS Acasta - When taken: 18 Jul 1813 - Where taken: off Halifax - Date Received: 31 Oct 1814 - From what ship: HMT Mermaid, Chatham - Born: Newbury - Age: 33 - Released on 29 Jun 1815.

Davis, John - Seaman - Nbr: 3970 - Prize: Rolla - Ship Type: P - How taken: HMS Loire - When taken: 10 Dec 1813 - Where taken: off Bull Island (South Carolina) - Date Received: 5 Oct 1814 - From what ship: HMT President, Halifax - Born: Rhode Island - Age: 19 - Race: Negro - Released on 9 Jun 1815.

Davis, John - Seaman - Nbr: 3483 - How taken: Gave himself up from HMS Prince - When taken: 14 Sep 1814 - Date Received: 19 Sep 1814 - From what ship: HMT Salvador del Mundo - Born: Baltimore - Age: 35 - Race: Negro - Released on 4 Jun 1815.

Davis, John - Seaman - Nbr: 4979 - Prize: Invincible - Ship Type: LM - How taken: HMS Armide - When taken: 15 Aug 1814 - Where taken: off Nantucket - Date Received: 28 Oct 1814 - From what ship: HMT Alkbar, Halifax - Born: Baltimore - Age: 32 - Released on 21 Jun 1815.

Davis, John - Boatswain - Nbr: 3562 - Prize: Hawk - Ship Type: P - How taken: HMS Pique - When taken: 26 Apr 1814 - Where taken: off Bermuda - Date Received: 30 Sep 1814 - From what ship: HMT Sybella - Born: Baltimore - Age: 48 - Released on 4 Jun 1815.

Davis, John - Pilot - Nbr: 4426 - Prize: Elbridge Gerry - Ship Type: LM - How taken: HMS Crescent - When taken: 16 Sep 1813 - Where taken: at sea - Date Received: 8 Oct 1814 - From what ship: HMT Leyden, Chatham - Born: Massachusetts - Age: 27 - Released on 14 Jun 1815.

Davis, John - Seaman - Nbr: 5396 - Prize: Gloucester - Ship Type: P - How taken: HMS Pique - When taken: 4 May 1814 - Where taken: West Indies - Date Received: 31 Oct 1814 - From what ship: HMT Leyden, Chatham - Born: New York - Age: 40 - Released on 1 Jul 1815.

Davis, Joseph - Seaman - Nbr: 5916 - Prize: Harlequin - Ship Type: P - How taken: HMS Bulwark - When taken: 23 Nov 1814 - Where taken: off Halifax - Date Received: 27 Dec 1814 - From what ship: HMT Penelope - Born: Massachusetts - Age: 28 - Released on 15 Jul 1815.

Davis, Joseph - Seaman - Nbr: 6252 - How taken: Sent into custody from MV Kulabra - When taken: 25 Nov 1814 - Date Received: 4 Feb 1815 - From what ship: HMT Ganges - Born: Baltimore - Age: 21 - Released on 11 Jul 1815.

Davis, Joseph - Seaman - Nbr: 2813 - How taken: Impressed at London - When taken: 22 Jul 1813 - Date Received: 24 Aug 1814 - From what ship: HMT Liverpool, Chatham - Born: Portland - Age: 20 - Released on 19 May 1815.

Davis, Lott - Seaman - Nbr: 5341 - Prize: Elbridge Gerry - Ship Type: P - How taken: HMS Crescent - When taken: 16 Sep 1813 - Where taken: at sea - Date Received: 31 Oct 1814 - From what ship: HMT Leyden, Chatham - Born: Barnstable - Age: 21 - Released on 1 Jul 1815.

Davis, Moses - Seaman - Nbr: 2673 - Prize: Joseph - Ship Type: MV - How taken: HMS Iris - When taken: 8 Jun 1813 - Where taken: off Spain - Date Received: 21 Aug 1814 - From what ship: HMT Freya, Chatham - Born: Cape Ann - Age: 22 - Released on 19 May 1815.

Davis, Nicholas - Seaman - Nbr: 4692 - Prize: Echo - When taken: 3 Nov 1812 - Date Received: 9 Oct 1814 - From what ship: HMT Leyden, Chatham - Born: Hudson - Age: 31 - Race: Mulatto.

Davis, Osborne - Seaman - Nbr: 3196 - How taken: Gave himself up from HMS America - When taken: 16 Jul 1813 - Date Received: 11 Sep 1814 - From what ship: HMT Freya, Chatham - Born: New Jersey - Age: 29 - Released on 28 May 1815.

Davis, Peter - Seaman - Nbr: 5786 - How taken: Gave himself up from HMS Pelican - When taken: 13 Sep 1814 - Date Received: 26 Dec 1814 - From what ship: HMT Argo - Born: Salem - Age: 32 - Sent to HMS Impregnable on 6 Jan 1815.

Davis, Richard - Prize master - Nbr: 5605 - Prize: General Putnam - Ship Type: P - How taken: HMS Leander - When taken: 8 Nov 1814 - Where taken: Long 65 Lat 42 - Date Received: 18 Dec 1814 - From what ship: Returned after escaping - Born: Salem - Age: 33 - Released on 5 Jul 1815.

Davis, Richman - Seaman - Nbr: 6527 - How taken: Gave himself up from MV Triton - When taken: Nov 1814 - Date Received: 3 Mar 1815 - From what ship: HMT Ganges, Plymouth - Born: Philadelphia - Age: 24 - Released on 11 Jul 1815.

Davis, Solomon - Boy - Nbr: 4767 - Prize: Argus - Ship Type: MV - How taken: HMS San Domingo - When taken: 1 Mar 1814 - Where taken: off Savannah - Date Received: 9 Oct 1814 - From what ship: HMT Freya, Chatham - Born: Boston - Age: 17 - Released on 15 Jun 1815.

Davis, Stephen - Seaman - Nbr: 4569 - How taken: Gave himself up from MV Liberty - When taken: 30 Dec 1813 - Date Received: 9 Oct 1814 - From what ship: HMT Leyden, Chatham - Born: Morington - Age: 37 -

American Prisoners of War Held at Dartmoor during the War of 1812

## Alphabetical listing of names

Released on 15 Jun 1815.

Davis, Thomas - Boy - Nbr: 3994 - Prize: US Brig Rattlesnake - Ship Type: MW - How taken: HMS Leander - When taken: 13 Jul 1814 - Where taken: off Shelburne - Date Received: 6 Oct 1814 - From what ship: HMT Chesapeake, Halifax - Born: Arundel - Age: 15 - Released on 9 Jun 1815.

Davis, William - Seaman - Nbr: 6202 - Prize: Prince de Neufchatel - Ship Type: P - How taken: Leander (Newcastle Acasta) - When taken: 20 Dec 1814 - Where taken: Lat 38 Long 56 - Date Received: 30 Jan 1815 - From what ship: HMT Pheasant - Born: Massachusetts - Age: 36 - Released on 5 Jul 1815.

Davis, William - Seaman - Nbr: 1932 - How taken: Gave himself up from HMS Desiree - When taken: 8 Jan 1814 - Date Received: 3 Aug 1814 - From what ship: HMS Alceste, Chatham Depot - Born: Baltimore - Age: 36 - Released on 26 Apr 1815.

Davis, William - Seaman - Nbr: 1547 - Prize: Zebra - Ship Type: LM - How taken: HMS Pyramus - When taken: 20 Apr 1813 - Where taken: Bay of Biscay - Date Received: 23 Jun 1814 - From what ship: Stapleton - Born: Charlestown - Age: 21 - Released on 30 Mar 1815.

Davis, William - Seaman - Nbr: 625 - Prize: US Brig Argus - Ship Type: MW - How taken: HMS Pelican - When taken: 14 Aug 1813 - Where taken: Irish Channel - Date Received: 8 Sep 1813 - From what ship: Plymouth - Born: Virginia - Age: 26 - Sent to Dartmouth on 2 Nov 1814.

Davis, William - Seaman - Nbr: 2050 - How taken: Gave himself up from HMS Pomona - When taken: 8 Jan 1812 - Date Received: 3 Aug 1814 - From what ship: HMS Lyffey, Chatham Depot - Born: Savannah - Age: 44 - Released on 26 Apr 1815.

Davis, William - Seaman - Nbr: 4371 - Prize: Growler - Ship Type: P - How taken: HMS Electra - When taken: 7 Jul 1813 - Where taken: at sea - Date Received: 8 Oct 1814 - From what ship: HMT Leyden, Chatham - Born: Massachusetts - Age: 44 - Released on 14 Jun 1815.

Davis, William - Seaman - Nbr: 3404 - Prize: Thomas - Ship Type: P - How taken: HMS Nymphe - When taken: 24 Jun 1813 - Where taken: off Halifax - Date Received: 13 Sep 1814 - From what ship: HMT Niobe, Chatham - Born: New York - Age: 29 - Released on 28 May 1815.

Dawes, W. B. - Seaman - Nbr: 2549 - Prize: Caromaned - Ship Type: Prize - How taken: HMS Eridanus - When taken: 13 Aug 1814 - Where taken: Lat 40, Long 16 - Date Received: 16 Aug 1814 - From what ship: HMT Salvador del Mundo - Born: Massachusetts - Age: 16 - Released on 3 May 1815.

Dawson, John - Seaman - Nbr: 6421 - Prize: Farmer's Daughter - Ship Type: MV - How taken: HMS Leviathan - When taken: 29 May 1814 - Date Received: 3 Mar 1815 - From what ship: HMT Ganges, Plymouth - Born: Fredericksburg - Age: 49 - Released on 11 Jul 1815.

Dawson, John - Seaman - Nbr: 2741 - How taken: Gave himself up from HMS Trinculo - When taken: 4 Dec 1812 - Date Received: 24 Aug 1814 - From what ship: HMT Liverpool, Chatham - Born: Philadelphia - Age: 20 - Released on 26 Apr 1815.

Day, Benjamin - Seaman - Nbr: 1147 - Prize: Bunker Hill - Ship Type: P - How taken: HMS Pomone - When taken: 8 Mar 1814 - Where taken: at sea - Date Received: 10 May 1814 - From what ship: Plymouth - Born: Bath - Age: 34 - Released on 28 Apr 1815.

Day, Frederick - Seaman - Nbr: 4375 - Prize: Growler - Ship Type: P - How taken: HMS Electra - When taken: 7 Jul 1813 - Where taken: at sea - Date Received: 8 Oct 1814 - From what ship: HMT Leyden, Chatham - Born: Salem - Age: 19 - Released on 14 Jun 1815.

Day, James - Seaman - Nbr: 1304 - How taken: Sent into custody from HMS Phoenix - Date Received: 14 Jun 1814 - From what ship: Mill Prison (Plymouth, England) - Born: Connecticut - Age: 20 - Released on 28 Apr 1815.

Day, James - Gunner's Mate - Nbr: 2543 - Prize: Polly - Ship Type: P - How taken: HMS Barbados - When taken: 21 Mar 1814 - Where taken: off Hispaniola - Date Received: 16 Aug 1814 - From what ship: HMT Salvador del Mundo - Born: Salem - Age: 46 - Released on 3 May 1815.

Day, John - Seaman - Nbr: 4418 - Prize: Fire Fly - Ship Type: LM - How taken: HMS Revolutionnaire - When taken: 19 Oct 1813 - Where taken: at sea - Date Received: 8 Oct 1814 - From what ship: HMT Leyden, Chatham - Born: Massachusetts - Age: 30 - Released on 14 Jun 1815.

Day, John - Seaman - Nbr: 1241 - How taken: Sent into custody from HMS Guadeloupe - When taken: 19 Dec 1812 - Date Received: 14 Jun 1814 - From what ship: Mill Prison (Plymouth, England) - Born: Springfield - Age: 32 - Released on 26 Apr 1815.

Day, Lyles C. - Sail Maker's Mate - Nbr: 561 - Prize: US Brig Argus - Ship Type: MW - How taken: HMS Pelican - When taken: 14 Apr 1813 - Where taken: Irish Channel - Date Received: 8 Sep 1813 - From what ship: Plymouth - Born: New York - Age: 21 - Sent to Dartmouth on 2 Nov 1814.

American Prisoners of War Held at Dartmoor during the War of 1812

## Alphabetical listing of names

Day, Samuel - Seaman - Nbr: 6474 - Prize: Decatur - Ship Type: P - How taken: HMS Rhin - When taken: 5 Jun 1814 - Where taken: off San Domingo - Date Received: 3 Mar 1815 - From what ship: HMT Ganges, Plymouth - Born: Massachusetts - Age: 22 - Released on 11 Jul 1815.

Day, Thomas - Seaman - Nbr: 2700 - How taken: Gave himself up from HMS Espoir - When taken: 8 Dec 1812 - Date Received: 21 Aug 1814 - From what ship: HMT Freya, Chatham - Born: Virginia - Age: 23 - Released on 19 May 1815.

Day, William - Seaman - Nbr: 3921 - Prize: Rolla - Ship Type: P - How taken: HMS Loire - When taken: 10 Dec 1813 - Where taken: off Bull Island (South Carolina) - Date Received: 5 Oct 1814 - From what ship: HMT President, Halifax - Born: Philadelphia - Age: 36 - Released on 9 Jun 1815.

Days, Thomas - Seaman - Nbr: 3537 - Prize: Hawk - Ship Type: P - How taken: HMS Pique - When taken: 26 Apr 1814 - Where taken: off Bermuda - Date Received: 30 Sep 1814 - From what ship: HMT Sybella - Born: Portsmouth - Age: 17 - Released on 4 Jun 1815.

De La Canto Sepo, I. - Seaman - Nbr: 4823 - Prize: President - Ship Type: P - How taken: HMS Pique - When taken: 7 May 1814 - Where taken: off Porto Rico - Date Received: 9 Oct 1814 - From what ship: HMT Freya, Halifax - Born: Cartagena - Age: 18 - Race: Mulatto - Released on 21 Jun 1815.

Deagle, James - Seaman - Nbr: 3142 - How taken: Impressed at London - When taken: 22 Sep 1813 - Date Received: 11 Sep 1814 - From what ship: HMT Freya, Chatham - Born: Yarmouth - Age: 24 - Released on 28 May 1815.

Deal, John - Seaman - Nbr: 1792 - Prize: Ferox - Ship Type: MV - How taken: HMS Medusa & HMS Lyra - When taken: 28 Mar 1813 - Where taken: off Cape Ortegal (Spain) - Date Received: 20 Jul 1814 - From what ship: HMS Milford, Plymouth - Born: Philadelphia - Age: 40 - Released on 1 May 1815.

Deal, William - Seaman - Nbr: 5 - Prize: Cashier - Ship Type: LM - How taken: HMS Reindeer - When taken: 3 Feb 1813 - Where taken: Bay of Biscay - Date Received: 2 Apr 1813 - From what ship: Plymouth - Born: Bell Haven - Age: 17 - Sent to Dartmouth on 30 Jul 1813.

Dealing, Elisha - Seaman - Nbr: 5222 - Prize: Fly - Ship Type: P - How taken: HMS Dover - When taken: 27 Jun 1813 - Where taken: off Newfoundland - Date Received: 31 Oct 1814 - From what ship: HMT Mermaid, Chatham - Born: Portsmouth - Age: 20 - Released on 29 Jun 1815.

Deamon, Thomas - Seaman - Nbr: 6534 - Prize: US Brig Syren - Ship Type: MW - How taken: HMS Medway - When taken: 12 Jul 1814 - Where taken: off Cape of Good Hope - Date Received: 7 Mar 1815 - From what ship: HMT Ganges, Plymouth - Born: Pembroke - Age: 21 - Released on 11 Jul 1815.

Dean, James B. - Lieutenant Marines - Nbr: 4277 - Prize: Elbridge Gerry - Ship Type: MV - How taken: HMS Crescent - When taken: 16 Feb 1813 - Where taken: at sea - Date Received: 7 Oct 1814 - From what ship: HMT Niobe, Chatham - Born: Boston - Age: 24 - Released on 13 Jun 1815.

Dean, Jonas - Seaman - Nbr: 1585 - Prize: Price - Ship Type: MV - How taken: HMS Pyramus - When taken: 6 Apr 1813 - Where taken: Bay of Biscay - Date Received: 23 Jun 1814 - From what ship: Stapleton - Born: Massachusetts - Age: 29 - Released on 1 May 1815.

Dean, Moses - Seaman - Nbr: 1137 - Prize: Mary - Ship Type: P - How taken: HMS Crocodile - When taken: 6 Aug 1813 - Where taken: off Corsuma - Date Received: 10 May 1814 - From what ship: Plymouth - Born: Boston - Age: 45 - Released on 28 Apr 1815.

Dean, N. B. - Seaman - Nbr: 1637 - Prize: Tom - Ship Type: LM - How taken: HMS Surveillante - When taken: 24 Apr 1813 - Where taken: Bay of Biscay - Date Received: 23 Jun 1814 - From what ship: Stapleton - Born: New Hampshire - Age: 22 - Released on 1 May 1815.

Dean, William - Prize master - Nbr: 4066 - Prize: Leicester - Ship Type: Prize - How taken: HMS Prometheus - When taken: Jun 1814 - Where taken: off Halifax - Date Received: 6 Oct 1814 - From what ship: HMT Chesapeake, Halifax - Born: Salem - Age: 25 - Released on 13 Jun 1815.

Deane, Peter - Seaman - Nbr: 4583 - Prize: Bunker Hill - Ship Type: P - How taken: HMS Pomone - When taken: 2 Mar 1814 - Where taken: Channel - Date Received: 9 Oct 1814 - From what ship: HMT Leyden, Chatham - Born: New Jersey - Age: 30 - Race: Mulatto - Released on 15 Jun 1815.

Dear, Andrew - Seaman - Nbr: 364 - Prize: Two Brothers - Ship Type: MV - How taken: Beetle (LM) - When taken: 18 Mar 1813 - Where taken: off Western Islands (England) - Date Received: 1 Jul 1813 - From what ship: Plymouth - Born: New Orleans - Age: 20 - Sent to Liverpool on 25 Aug 1813.

DeBates, Amos - Seaman - Nbr: 5063 - Prize: Ida - Ship Type: LM - How taken: HMS Newcastle - When taken: 9 Aug 1814 - Where taken: Long 34 - Date Received: 28 Oct 1814 - From what ship: HMT Alkbar, Halifax - Born: Hamburg - Age: 22 - Died on 18 Nov 1814.

DeCoine, John - Seaman - Nbr: 5737 - Prize: US Schooner Somers - Ship Type: MW - How taken: British gunboats

American Prisoners of War Held at Dartmoor during the War of 1812

## Alphabetical listing of names

- When taken: 12 Aug 1814 - Where taken: Fort Erie - Date Received: 26 Dec 1814 - From what ship: HMT Argo - Born: Philadelphia - Age: 18 - Released on 3 Jul 1815.

Deering, M. H. - Carpenter - Nbr: 5909 - Prize: Harlequin - Ship Type: P - How taken: HMS Bulwark - When taken: 23 Nov 1814 - Where taken: off Halifax - Date Received: 27 Dec 1814 - From what ship: HMT Penelope - Born: Portsmouth - Age: 23 - Released on 3 Jul 1815.

Defray, Edward - Seaman - Nbr: 4218 - Prize: Liberty - Ship Type: MV - How taken: Surrendered - When taken: 30 Dec 1813 - Where taken: Stromess - Date Received: 7 Oct 1814 - From what ship: HMT Niobe, Chatham - Born: Nantucket - Age: 26 - Released on 13 Jun 1815.

Degget, Hamel - Seaman - Nbr: 2905 - How taken: Gave himself up from HMS Ministrel - When taken: 5 Jul 1813 - Date Received: 24 Aug 1814 - From what ship: HMT Alpheus, Chatham - Born: Massachusetts - Age: 22 - Released on 19 May 1815.

Delabar, Joseph - Prize Master - Nbr: 2595 - Prize: Rachael - Ship Type: MV - How taken: HMS Bramble - When taken: 2 Dec 1812 - Where taken: at sea - Date Received: 21 Aug 1814 - From what ship: HMT Freya, Chatham - Born: Marblehead - Age: 44 - Released on 26 Apr 1815.

Delaney, Mathew - Seaman - Nbr: 5126 - How taken: Gave himself up from HMS Eridanus - When taken: 27 Oct 1814 - Date Received: 28 Oct 1814 - From what ship: HMT Salvador del Mundo - Born: Philadelphia - Age: 23 - Released on 11 Jul 1815.

Delaware, John - Boy - Nbr: 6318 - Prize: Prince de Neufchatel - Ship Type: P - How taken: Leander (Newcastle Acasta) - When taken: 20 Dec 1814 - Where taken: Lat 38 Long 56 - Date Received: 19 Feb 1815 - From what ship: HMT Ganges, Plymouth - Born: Boston - Age: 17 - Released on 5 Jul 1815.

Dellanon, L. - Seaman - Nbr: 5581 - Prize: Alexander - Ship Type: Prize - How taken: HMS Wasp - When taken: 14 Sep 1814 - Where taken: off Cape Sable Island (Canada) - Date Received: 17 Dec 1814 - From what ship: HMT Loire, Halifax - Born: New Bedford - Age: 25 - Released on 1 Jul 1815.

Dellson, John - Seaman - Nbr: 6125 - Prize: Betsey - Ship Type: Prize - How taken: HMS Bellerophon - When taken: 2 Nov 1814 - Where taken: Long 61 - Date Received: 6 Jan 1814 - From what ship: HMT Impregnable - Born: Salem - Age: 23 - Released on 5 Jul 1815.

Delond, Zaba - Seaman - Nbr: 2439 - Prize: US Sloop Frolic - Ship Type: MW - How taken: HMS Orpheus - When taken: 20 Apr 1814 - Where taken: off Cuba - Date Received: 16 Aug 1814 - From what ship: HMT Queen, Halifax - Born: New Bedford - Age: 26 - Released on 3 May 1815.

Demarlow, Francis - Seaman - Nbr: 4287 - Prize: Pilot - Ship Type: LM - How taken: Victoria (Privateer) - When taken: 28 Jan 1814 - Where taken: off Bordeaux - Date Received: 7 Oct 1814 - From what ship: HMT Niobe, Chatham - Born: Philadelphia - Age: 22 - Race: Mulatto - Released on 14 Jun 1815.

Dematra, George - Seaman - Nbr: 935 - Prize: Fanny - Ship Type: MV - How taken: HMS Eurotas - When taken: 23 Dec 1813 - Where taken: at sea - Date Received: 31 Jan 1814 - From what ship: Plymouth - Born: Virginia - Age: 30 - Sent to Plymouth on 8 Jul 1814.

Demedorff, John - Seaman - Nbr: 6340 - Prize: Prince de Neufchatel - Ship Type: P - How taken: Leander (Newcastle Acasta) - When taken: 20 Dec 1814 - Where taken: Lat 38 Long 56 - Date Received: 19 Feb 1815 - From what ship: HMT Ganges, Plymouth - Born: Petersburg - Age: 42 - Released on 6 Jul 1815.

Dempsey, John - 3rd Lieutenant - Nbr: 2275 - Prize: Diomede - Ship Type: P - How taken: HMS Rifleman - When taken: 23 Jun 1814 - Where taken: off Halifax - Date Received: 16 Aug 1814 - From what ship: HMS Dublin, Halifax - Born: Beverly - Age: 28 - Released on 2 May 1815.

Dempston, Daniel - Seaman - Nbr: 2427 - Prize: US Sloop Frolic - Ship Type: MW - How taken: HMS Orpheus - When taken: 20 Apr 1814 - Where taken: off Cuba - Date Received: 16 Aug 1814 - From what ship: HMT Queen, Halifax - Born: West-Cappel (France) - Age: 23 - Released on 3 May 1815.

Denham, Cornelius - Seaman - Nbr: 530 - How taken: Impressed at Greenock - When taken: 13 Jun 1813 - Date Received: 8 Sep 1813 - From what ship: Plymouth - Born: Massachusetts - Age: 26 - Released on 15 Apr 1815.

Denham, John - Seaman - Nbr: 473 - Prize: Hebe - Ship Type: MV - How taken: HMS Stag - When taken: 13 Apr 1813 - Where taken: Bay of Biscay - Date Received: 8 Sep 1813 - From what ship: Plymouth - Born: Philadelphia - Age: 25 - Released on 20 Apr 1815.

Denham, Silas - Seaman - Nbr: 5074 - Prize: Ida - Ship Type: LM - How taken: HMS Newcastle - When taken: 9 Aug 1814 - Where taken: Long 34 - Date Received: 28 Oct 1814 - From what ship: HMT Alkbar, Halifax - Born: Boston - Age: 21 - Died on 14 Nov 1814.

Denning, Joseph - Seaman - Nbr: 5744 - Prize: US Schooner Ohio - Ship Type: MW - How taken: British gunboats - When taken: 12 Aug 1814 - Where taken: Fort Erie - Date Received: 26 Dec 1814 - From what ship: HMT

American Prisoners of War Held at Dartmoor during the War of 1812

## Alphabetical listing of names

Argo - Born: Massachusetts - Age: 26 - Died on 12 Apr 1815.

Dennis, Francis - Seaman - Nbr: 4942 - Prize: Herald - Ship Type: P - How taken: HMS Endymion - When taken: 15 Aug 1814 - Where taken: off Nantucket - Date Received: 28 Oct 1814 - From what ship: HMT Alkbar, Halifax - Born: Havre de Grace - Age: 22 - Released on 21 Jun 1815.

Dennis, Jonas - Seaman - Nbr: 3728 - Prize: Alfred - Ship Type: P - How taken: HMS Epervier - When taken: 23 Feb 1814 - Where taken: off Newfoundland - Date Received: 30 Sep 1814 - From what ship: HMT President, Halifax - Born: Not listed - Released on 4 Jun 1815.

Dennis, Thomas - Seaman - Nbr: 2802 - How taken: Impressed at London - When taken: 10 Jun 1813 - Date Received: 24 Aug 1814 - From what ship: HMT Liverpool, Chatham - Born: Marblehead - Age: 23 - Released on 19 May 1815.

Dennis, William - Lieutenant - Nbr: 4790 - Prize: Frolic - Ship Type: P - How taken: HMS Heron - When taken: 25 Jun 1814 - Where taken: off St. Thomas - Date Received: 9 Oct 1814 - From what ship: HMT Freya, Halifax - Born: Marblehead - Age: 28 - Released on 11 Jul 1815.

Dennis, William - Seaman - Nbr: 4010 - Prize: US Brig Rattlesnake - Ship Type: MW - How taken: HMS Leander - When taken: 13 Jul 1814 - Where taken: off Shelburne - Date Received: 6 Oct 1814 - From what ship: HMT Chesapeake, Halifax - Born: Marblehead - Age: 40 - Released on 9 Jun 1815.

Dennison, Andrew - Seaman - Nbr: 4406 - Prize: Portsmouth Packet - Ship Type: Prize - How taken: HMS Fantome - When taken: 5 Oct 1813 - Where taken: off Portland - Date Received: 8 Oct 1814 - From what ship: HMT Leyden, Chatham - Born: Massachusetts - Age: 21 - Released on 14 Jun 1815.

Dennison, George - Prize Master - Nbr: 940 - Prize: Siro - Ship Type: LM - How taken: HMS Pelican - When taken: 13 Jan 1814 - Where taken: at sea - Date Received: 31 Jan 1814 - From what ship: Plymouth - Born: Freeport - Age: 26 - Escaped on 13 Mar 1815.

Dennison, Judah - Seaman - Nbr: 183 - Prize: Star - Ship Type: MV - How taken: HMS Superb - When taken: 9 Feb 1813 - Where taken: Bay of Biscay - Date Received: 2 Apr 1813 - From what ship: Plymouth - Born: Saybrook - Age: 25 - Sent to Plymouth on 15 Jun 1813.

Dennison, Laurence - Seaman - Nbr: 2605 - How taken: Gave himself up from HMS Niobe - When taken: 1 Aug 1812 - Date Received: 21 Aug 1814 - From what ship: HMT Freya, Chatham - Born: Vermont - Age: 33 - Released on 26 Apr 1815.

Dennison, Thomas - Seaman - Nbr: 799 - Prize: Collin - Ship Type: MV - How taken: HMS Helicon and HMS Whiting - When taken: 26 Oct 1813 - Where taken: Channel - Date Received: 3 Nov 1813 - From what ship: Plymouth - Born: Denmark - Age: 27 - Sent to Plymouth on 8 Dec 1813.

Denny, James - Seaman - Nbr: 2460 - Prize: US Sloop Frolic - Ship Type: MW - How taken: HMS Orpheus - When taken: 20 Apr 1814 - Where taken: off Cuba - Date Received: 16 Aug 1814 - From what ship: HMT Queen, Halifax - Born: New York - Age: 22 - Released on 3 May 1815.

Densey, Peter - Seaman - Nbr: 4931 - Prize: Herald - Ship Type: P - How taken: HMS Endymion - When taken: 15 Aug 1814 - Where taken: off Nantucket - Date Received: 28 Oct 1814 - From what ship: HMT Alkbar, Halifax - Born: New York - Age: 24 - Race: Negro - Released on 21 Jun 1815.

Denyer, Richard - Seaman - Nbr: 2303 - Prize: Hussar - Ship Type: P - How taken: HMS Saturn - When taken: 25 May 1814 - Where taken: off Sandy Hook - Date Received: 16 Aug 1814 - From what ship: HMS Dublin, Halifax - Born: New York - Age: 16 - Released on 3 May 1815.

Deparvier, John - Seaman - Nbr: 1014 - Prize: Rachel & Ann - Ship Type: P - How taken: HMS Cydnus - When taken: 6 Jan 1814 - Where taken: at sea - Date Received: 31 Jan 1814 - From what ship: Plymouth - Born: Erebough, Sweden - Age: 39 - Sent to Plymouth on 8 Jul 1814.

Depero, Henry - Marine - Nbr: 2497 - Prize: US Sloop Frolic - Ship Type: MW - How taken: HMS Orpheus - When taken: 20 Apr 1814 - Where taken: off Cuba - Date Received: 16 Aug 1814 - From what ship: HMT Queen, Halifax - Born: New York - Age: 21 - Released on 3 May 1815.

Depuyster, Pierre - 1st Officer - Nbr: 6152 - Prize: Lion - Ship Type: P - How taken: HMS Granicus - When taken: 2 Dec 1814 - Where taken: off Lisbon - Date Received: 18 Jan 1815 - From what ship: HMT Impregnable - Born: New Jersey - Age: 33 - Released on 2 May 1815.

DeRichardson, Comte - Seaman - Nbr: 5073 - Prize: Ida - Ship Type: LM - How taken: HMS Newcastle - When taken: 9 Aug 1814 - Where taken: Long 34 - Date Received: 28 Oct 1814 - From what ship: HMT Alkbar, Halifax - Born: St. Bartholomew - Age: 19 - Race: Negro - Released on 29 Jun 1815.

Deroche, John Baptiste - Seaman - Nbr: 5591 - Prize: Albion - Ship Type: Prize - How taken: HMS Jaseur - When taken: 21 Sep 1814 - Where taken: off Halifax - Date Received: 17 Dec 1814 - From what ship: HMT Loire, Halifax - Born: Havre de Grace - Age: 19 - Released on 1 Jul 1815.

American Prisoners of War Held at Dartmoor during the War of 1812

## Alphabetical listing of names

Derrill, Benjamin - Seaman - Nbr: 6020 - Prize: McDonough - Ship Type: P - How taken: HMS Bacchante - When taken: 1 Nov 1814 - Where taken: Lat 42 Long 67 - Date Received: 28 Dec 1814 - From what ship: HMT Penelope - Born: Massachusetts - Age: 21 - Released on 3 Jul 1815.

Derring, William F. - Sail maker - Nbr: 4657 - Prize: Growler - Ship Type: P - How taken: HMS Electra - When taken: 7 Jul 1813 - Where taken: at sea - Date Received: 9 Oct 1814 - From what ship: HMT Leyden, Chatham - Born: Marblehead - Age: 22 - Released on 15 Jun 1815.

Desharn, George - Seaman - Nbr: 2304 - Prize: Hussar - Ship Type: P - How taken: HMS Saturn - When taken: 25 May 1814 - Where taken: off Sandy Hook - Date Received: 16 Aug 1814 - From what ship: HMS Dublin, Halifax - Born: Baltimore - Age: 18 - Released on 3 May 1815.

DeSilvin, Jean - Seaman - Nbr: 4997 - Prize: Invincible - Ship Type: LM - How taken: HMS Armide - When taken: 15 Aug 1814 - Where taken: off Nantucket - Date Received: 28 Oct 1814 - From what ship: HMT Alkbar, Halifax - Born: Lisbon - Age: 27 - Released on 21 Jun 1815.

Desley, Charles - Seaman - Nbr: 6092 - Prize: Black Swan - Ship Type: MV - How taken: HMS Maidstone - When taken: 24 Oct 1814 - Where taken: Georges Bank - Date Received: 28 Dec 1814 - From what ship: HMT Penelope - Born: Massachusetts - Age: 18 - Released on 11 Jul 1815.

Detandes, Joseph - Seaman - Nbr: 6500 - Prize: Dorthea - Ship Type: MV - How taken: HMS North Star - When taken: Dec 1814 - Where taken: St. Jago (Jamaica) - Date Received: 3 Mar 1815 - From what ship: HMT Ganges, Plymouth - Born: Madeira - Age: 35 - Released on 5 May 1815.

Devereux, Benjamin - Cooper - Nbr: 3385 - Prize: Growler - Ship Type: P - How taken: HMS Electra - When taken: 7 Jul 1813 - Where taken: at sea - Date Received: 13 Sep 1814 - From what ship: HMT Niobe, Chatham - Born: Salem - Age: 27 - Released on 28 May 1815.

Deverter, William - Seaman - Nbr: 1950 - How taken: Gave himself up from HMS Niobe - When taken: 26 Mar 1813 - Date Received: 3 Aug 1814 - From what ship: HMS Lyffey, Chatham Depot - Born: Philadelphia - Age: 31 - Released on 2 May 1815.

Devinas, John - Seaman - Nbr: 5743 - Prize: US Schooner Ohio - Ship Type: MW - How taken: British gunboats - When taken: 12 Aug 1814 - Where taken: Fort Erie - Date Received: 26 Dec 1814 - From what ship: HMT Argo - Born: Kentucky - Age: 26 - Died on 12 Apr 1815 from variola.

Devol, Alexander - Seaman - Nbr: 4523 - Prize: Yankee - Ship Type: P - How taken: HMS Shannon - When taken: 20 Aug 1813 - Where taken: at sea - Date Received: 8 Oct 1814 - From what ship: HMT Leyden, Chatham - Born: Rhode Island - Age: 27 - Released on 15 Jun 1815.

Dew, Frederick - Prize master - Nbr: 4383 - Prize: Globe - Ship Type: P - How taken: HMS Spartan - When taken: 13 Jun 1813 - Where taken: off Delaware - Date Received: 8 Oct 1814 - From what ship: HMT Leyden, Chatham - Born: Baltimore - Age: 25 - Released on 14 Jun 1815.

Dewett, John - Boy - Nbr: 147 - Prize: Criterion - Ship Type: MV - How taken: HMS Belle Poule - When taken: 14 Feb 1813 - Where taken: Bay of Biscay - Date Received: 2 Apr 1813 - From what ship: Plymouth - Born: New York - Age: 18 - Released on 20 Apr 1815.

Dexter, Charles - Mate - Nbr: 3262 - Prize: Derby - Ship Type: MV - How taken: HMS Nereus - When taken: 4 Feb 1813 - Where taken: off Cape of Good Hope - Date Received: 11 Sep 1814 - From what ship: HMT Freya, Chatham - Born: Gloucester - Age: 26 - Released on 28 May 1815.

Dexter, David - Seaman - Nbr: 4110 - Prize: Flash - Ship Type: MV - How taken: HMS Niemen - When taken: 7 Jun 1814 - Where taken: off Delaware - Date Received: 6 Oct 1814 - From what ship: HMT Chesapeake, Halifax - Born: Rhode Island - Age: 20 - Released on 13 Jun 1815.

Dexter, George W. - Boy - Nbr: 3263 - Prize: Derby - Ship Type: MV - How taken: HMS Nereus - When taken: 4 Feb 1813 - Where taken: off Cape of Good Hope - Date Received: 11 Sep 1814 - From what ship: HMT Freya, Chatham - Born: Gloucester - Age: 15 - Released on 28 May 1815.

Diamond, William - Seaman - Nbr: 1140 - Prize: Mary - Ship Type: P - How taken: HMS Crocodile - When taken: 6 Aug 1813 - Where taken: off Corsuma - Date Received: 10 May 1814 - From what ship: Plymouth - Born: Rhode Island - Age: 19 - Died on 23 Jan 1815 from variola.

Dias, Joseph - Master - Nbr: 1734 - Prize: Hugh Jones - Ship Type: MV - How taken: HMS Bittern - When taken: 14 Jun 1814 - Where taken: at sea - Date Received: 18 Jul 1814 - From what ship: HMT Salvador del Mundo, Plymouth - Born: Massachusetts - Age: 34 - Released on 1 May 1815.

Dibble, Reuben - Seaman - Nbr: 1354 - Prize: Paul Jones - Ship Type: P - How taken: HMS Leonidas - When taken: 23 May 1813 - Where taken: Channel - Date Received: 19 Jun 1814 - From what ship: Stapleton - Born: Connecticut - Age: 22 - Released on 28 Apr 1815.

Dibble, Zachariah - Seaman - Nbr: 1888 - How taken: Gave himself up from HMS Malta - When taken: 2 Nov 1813

American Prisoners of War Held at Dartmoor during the War of 1812

## Alphabetical listing of names

- Date Received: 29 Jul 1814 - From what ship: HMS Ville de Paris, Chatham Depot - Born: Connecticut - Age: 40 - Released on 3 May 1815.

Dickenson, Chester - Seaman - Nbr: 1511 - Prize: Courier - Ship Type: LM - How taken: HMS Rover - When taken: 14 Mar 1813 - Where taken: Bay of Biscay - Date Received: 23 Jun 1814 - From what ship: Stapleton - Born: Massachusetts - Age: 28 - Released on 1 May 1815.

Dickenson, Henry - Boy - Nbr: 2389 - Prize: US Sloop Frolic - Ship Type: MW - How taken: HMS Orpheus - When taken: 20 Apr 1814 - Where taken: off Cuba - Date Received: 16 Aug 1814 - From what ship: HMT Queen, Halifax - Born: Portsmouth - Age: 14 - Sent to Dartmouth on 19 Oct 1814.

Dickinson, Francis - Seaman - Nbr: 2309 - Prize: Hussar - Ship Type: P - How taken: HMS Saturn - When taken: 25 May 1814 - Where taken: off Sandy Hook - Date Received: 16 Aug 1814 - From what ship: HMS Dublin, Halifax - Born: Philadelphia - Age: 10 - Released on 3 May 1815.

Dickson, Abraham - Seaman - Nbr: 5784 - How taken: Gave himself up from HMS Ariel - When taken: 8 Dec 1812 - Date Received: 26 Dec 1814 - From what ship: HMT Argo - Born: Long Island - Age: 33 - Race: Black - Released on 3 Jul 1815.

Dickson, Charles - Seaman - Nbr: 139 - Prize: Criterion - Ship Type: MV - How taken: HMS Belle Poule - When taken: 14 Feb 1813 - Where taken: Bay of Biscay - Date Received: 2 Apr 1813 - From what ship: Plymouth - Born: New York - Age: 24 - Released on 20 Apr 1815.

Didderas, William - Seaman - Nbr: 5198 - Prize: America - Ship Type: P - How taken: HMS Shannon - When taken: 13 Jul 1813 - Where taken: at sea - Date Received: 31 Oct 1814 - From what ship: HMT Mermaid, Chatham - Born: Salem - Age: 26 - Released on 29 Jun 1815.

Didler, Henry - Seaman - Nbr: 5291 - Prize: Volante - Ship Type: P - How taken: HMS Curlew - When taken: 25 Nov 1813 - Where taken: off Halifax - Date Received: 31 Oct 1814 - From what ship: HMT Leyden, Chatham - Born: Baltimore - Age: 20 - Released on 29 Jun 1815.

Dildure, Samuel - Seaman - Nbr: 1910 - How taken: Gave himself up from HMS Mars - When taken: 9 Dec 1812 - Date Received: 3 Aug 1814 - From what ship: HMS Alceste, Chatham Depot - Born: New Jersey - Age: 22 - Released on 26 Apr 1815.

Dill, Samuel - Seaman - Nbr: 4868 - Prize: US Brig Rattlesnake - Ship Type: MW - How taken: HMS Leander - When taken: 13 Jul 1813 - Where taken: off Shelburne - Date Received: 24 Oct 1814 - From what ship: Royal Hospital, Plymouth - Born: Boston - Age: 25 - Released on 21 Jun 1815.

Dill, William - Seaman - Nbr: 4619 - Prize: Argus - Ship Type: MV - How taken: HMS San Domingo - When taken: 1 Mar 1814 - Where taken: off Savannah - Date Received: 9 Oct 1814 - From what ship: HMT Leyden, Chatham - Born: Boston - Age: 20 - Released on 15 Jun 1815.

Dill, William - Seaman - Nbr: 2345 - Prize: Snap Dragon - Ship Type: P - How taken: HMS Martin - When taken: 10 Jun 1814 - Where taken: off Halifax - Date Received: 16 Aug 1814 - From what ship: HMS Dublin, Halifax - Born: North Carolina - Age: 21 - Released on 3 May 1815.

Dillin, Pierce - Seaman - Nbr: 479 - Prize: Price - Ship Type: MV - How taken: HMS Iris - When taken: 13 Apr 1813 - Where taken: Bay of Biscay - Date Received: 8 Sep 1813 - From what ship: Plymouth - Born: New York - Age: 18 - Released on 20 Apr 1815.

Dillingham, F. A. - Seaman - Nbr: 1186 - Prize: Agile - Ship Type: P - How taken: HMS Sybille - When taken: 19 Jul 1812 - Date Received: 4 Jun 1814 - From what ship: Dartmouth - Born: Newport - Age: 28 - Released on 28 Apr 1815.

Dillon, Richard - Seaman - Nbr: 5440 - How taken: Gave himself up from HMS Lightning - When taken: 31 Oct 1814 - Date Received: 11 Nov 1814 - From what ship: HMT Impregnable - Born: North Carolina - Age: 40 - Released on 1 Jul 1815.

Dillon, William - Marine - Nbr: 599 - Prize: US Brig Argus - Ship Type: MW - How taken: HMS Pelican - When taken: 14 Aug 1813 - Where taken: Irish Channel - Date Received: 8 Sep 1813 - From what ship: Plymouth - Born: New Jersey - Age: 48 - Died on 10 May 1814 from dysentery.

Diluo, Benjamin - Seaman - Nbr: 1522 - Prize: Essex - Ship Type: MV - How taken: HMS Pyramus - When taken: 20 Apr 1813 - Where taken: Bay of Biscay - Date Received: 23 Jun 1814 - From what ship: Stapleton - Born: Massachusetts - Age: 19 - Died on 30 Mar 1814 from phthisis pulmonalis.

Dimond, George - Seaman - Nbr: 5316 - Prize: Thomas - Ship Type: P - How taken: HMS Nymphe - When taken: 24 Jun 1813 - Where taken: off Halifax - Date Received: 31 Oct 1814 - From what ship: HMT Leyden, Chatham - Born: New York - Age: 20 - Race: Black - Released on 29 Jun 1815.

Dine, William - Seaman - Nbr: 6287 - How taken: Gave himself up from HMS Namur - Date Received: 17 Feb 1815 - From what ship: HMT Ganges, Plymouth - Born: Philadelphia - Age: 36 - Released on 28 Apr 1815.

American Prisoners of War Held at Dartmoor during the War of 1812

## Alphabetical listing of names

Dinsmore, John - Seaman - Nbr: 506 - Prize: Tickler - Ship Type: LM - How taken: HMS Magicienne - When taken: 5 Jun 1813 - Where taken: Bay of Biscay - Date Received: 8 Sep 1813 - From what ship: Plymouth - Born: New Castle - Age: 23 - Released on 26 Apr 1815.

Dissmore, Abraham - Mate - Nbr: 3261 - Prize: Rambler - Ship Type: MV - How taken: Morley (Transport) - When taken: 10 Feb 1813 - Where taken: off Isle of France - Date Received: 11 Sep 1814 - From what ship: HMT Freya, Chatham - Born: Boston - Age: 31 - Released on 25 Sep 1814.

Diverall, John - Seaman - Nbr: 6444 - Prize: Jemmell - Ship Type: MV - How taken: HMS Rhin - When taken: 28 May 1814 - Where taken: off Bermuda - Date Received: 3 Mar 1815 - From what ship: HMT Ganges, Plymouth - Born: Massachusetts - Age: 18 - Released on 11 Jul 1815.

Divers, Charles - Seaman - Nbr: 287 - Prize: Cannoneer - Ship Type: MV - How taken: HMS Warspite - When taken: 14 Mar 1813 - Where taken: Bay of Biscay - Date Received: 28 Jun 1813 - From what ship: Plymouth - Born: Lancaster - Age: 18 - Released on 20 Apr 1815.

Dixey, William - Seaman - Nbr: 6115 - Prize: Garland - Ship Type: Prize - How taken: HMS Hamadryad - When taken: 22 Nov 1814 - Where taken: Georges Bank - Date Received: 6 Jan 1814 - From what ship: HMT Impregnable - Born: Marblehead - Age: 21 - Released on 11 Jul 1815.

Dixon, Archibald - Seaman - Nbr: 3914 - Prize: Hornsby - Ship Type: P - How taken: HMS Sceptre - When taken: 2 Feb 1814 - Where taken: off Block Island - Date Received: 5 Oct 1814 - From what ship: HMT President, Halifax - Born: Philadelphia - Age: 18 - Released on 9 Jun 1815.

Dixon, Benjamin - Seaman - Nbr: 915 - How taken: Gave himself up - Date Received: 31 Jan 1814 - From what ship: Plymouth - Born: Baltimore - Age: 30 - Released on 27 Apr 1815.

Dixon, John - Pilot - Nbr: 4608 - Prize: Argus - Ship Type: MV - How taken: HMS San Domingo - When taken: 1 Mar 1814 - Where taken: off Savannah - Date Received: 9 Oct 1814 - From what ship: HMT Leyden, Chatham - Born: Savannah - Age: 26 - Released on 15 Jun 1815.

Dixon, John - Seaman - Nbr: 3969 - Prize: Rolla - Ship Type: P - How taken: HMS Loire - When taken: 10 Dec 1813 - Where taken: off Bull Island (South Carolina) - Date Received: 5 Oct 1814 - From what ship: HMT President, Halifax - Born: Boston - Age: 19 - Released on 9 Jun 1815.

Dixon, Peter - Seaman - Nbr: 4275 - How taken: Gave himself up from HMS Malta - When taken: 4 Feb 1813 - Date Received: 7 Oct 1814 - From what ship: HMT Niobe, Chatham - Born: New York - Age: 26 - Released on 13 Jun 1815.

Dizere, John - Seaman - Nbr: 2335 - Prize: Snap Dragon - Ship Type: P - How taken: HMS Martin - When taken: 10 Jun 1814 - Where taken: off Halifax - Date Received: 16 Aug 1814 - From what ship: HMS Dublin, Halifax - Born: Boston - Age: 21 - Released on 3 May 1815.

Doane, Joshua - Seaman - Nbr: 4948 - Prize: Herald - Ship Type: P - How taken: HMS Endymion - When taken: 15 Aug 1814 - Where taken: off Nantucket - Date Received: 28 Oct 1814 - From what ship: HMT Alkbar, Halifax - Born: Massachusetts - Age: 24 - Released on 21 Jun 1815.

Dobins, John - Seaman - Nbr: 130 - How taken: Sent into custody from William Transport No. 69 - Date Received: 12 Apr 1813 - From what ship: Plymouth - Born: Hartford - Age: 27 - Sent to Dartmouth on 30 Jul 1813.

Dobson, Joseph - Seaman - Nbr: 2476 - Prize: US Sloop Frolic - Ship Type: MW - How taken: HMS Orpheus - When taken: 20 Apr 1814 - Where taken: off Cuba - Date Received: 16 Aug 1814 - From what ship: HMT Queen, Halifax - Born: New York - Age: 51 - Released on 3 May 1815.

Dobson, Robert - Pilot - Nbr: 3865 - Prize: Lewis Warrington - Ship Type: MV - How taken: HMS Loire - When taken: 23 May 1814 - Where taken: off VA - Date Received: 5 Oct 1814 - From what ship: HMT Orpheus, Halifax - Born: Virginia - Age: 27 - Released on 10 May 1815.

Dodd, Samuel - Seaman - Nbr: 62 - Prize: Spitfire - Ship Type: MV - How taken: HMS Achates - When taken: 14 Feb 1813 - Where taken: off Ushant (France) - Date Received: 2 Apr 1813 - From what ship: Plymouth - Born: Marblehead - Age: 17 - Released on 20 Apr 1815.

Doddard, John - Seaman - Nbr: 2487 - Prize: US Sloop Frolic - Ship Type: MW - How taken: HMS Orpheus - When taken: 20 Apr 1814 - Where taken: off Cuba - Date Received: 16 Aug 1814 - From what ship: HMT Queen, Halifax - Born: Massachusetts - Age: 31 - Released on 3 May 1815.

Dodge, David - Seaman - Nbr: 6214 - Prize: Prince de Neufchatel - Ship Type: P - How taken: Leander (Newcastle Acasta) - When taken: 20 Dec 1814 - Where taken: Lat 38 Long 56 - Date Received: 30 Jan 1815 - From what ship: HMT Pheasant - Born: Edgecombe - Age: 23 - Released on 5 Jul 1815.

Dodge, Joseph - Seaman - Nbr: 5400 - Prize: Perfect - Ship Type: P - How taken: HMS Grinder - When taken: 6 Jul 1814 - Where taken: off St. Bartholomew - Date Received: 31 Oct 1814 - From what ship: HMT Leyden, Chatham - Born: Charleston - Age: 19 - Released on 1 Jul 1815.

American Prisoners of War Held at Dartmoor during the War of 1812

## Alphabetical listing of names

Dodge, Joseph - 1st Mate - Nbr: 2001 - Prize: Governor Middleton - Ship Type: MV - How taken: Thetis (Privateer) - When taken: 2 May 1813 - Where taken: Bay of Biscay - Date Received: 3 Aug 1814 - From what ship: HMS Lyffey, Chatham Depot - Born: Lact Isle - Age: 26 - Released on 2 May 1815.

Doer, James - Seaman - Nbr: 2002 - Prize: Governor Middleton - Ship Type: MV - How taken: Thetis (Privateer) - When taken: 2 May 1813 - Where taken: Bay of Biscay - Date Received: 3 Aug 1814 - From what ship: HMS Lyffey, Chatham Depot - Born: Massachusetts - Age: 20 - Released on 2 May 1815.

Doggett, James - Seaman - Nbr: 6199 - Prize: Prince de Neufchatel - Ship Type: P - How taken: Leander (Newcastle Acasta) - When taken: 20 Dec 1814 - Where taken: Lat 38 Long 56 - Date Received: 30 Jan 1815 - From what ship: HMT Pheasant - Born: Massachusetts - Age: 32 - Released on 5 Jul 1815.

Doing, Denis O. - Captain Marines - Nbr: 6414 - Prize: Enterprize - Ship Type: P - How taken: HMS Argo - When taken: 16 Nov 1814 - Where taken: off San Domingo - Date Received: 3 Mar 1815 - From what ship: HMT Ganges, Plymouth - Born: Massachusetts - Age: 23 - Released on 3 May 1815.

Dolabar, John - Seaman - Nbr: 1771 - Prize: Print - Ship Type: MV - How taken: HMS Colossus - When taken: 21 Jan 1813 - Where taken: off Long Island - Date Received: 20 Jul 1814 - From what ship: Chatham - Born: Marblehead - Age: 33 - Released on 1 May 1815.

Dolavice, Henry - Prize master - Nbr: 5767 - Prize: James - Ship Type: Prize - How taken: HMS Galatea - When taken: 10 Sep 1814 - Where taken: Long 13 Lat 29 - Date Received: 26 Dec 1814 - From what ship: HMT Argo - Born: Gloucester - Age: 36 - Released on 1 May 1815.

Dole, Andrew - Soldier - Nbr: 5272 - Prize: 13th US Infantry - Ship Type: Troops - How taken: British Army - When taken: 13 Oct 1812 - Where taken: Canada - Date Received: 31 Oct 1814 - From what ship: HMT Leyden, Chatham - Born: St. David, Upper Canada - Age: 25 - Released on 29 Jun 1815.

Dole, Henry - Mate - Nbr: 3369 - How taken: Impressed at Leith - When taken: 4 Dec 1813 - Date Received: 13 Sep 1814 - What ship: HMT Niobe, Chatham - Born: Newburyport - Age: 24 - Released on 28 May 1815.

Dole, William - Seaman - Nbr: 6185 - Prize: Garland - Ship Type: Prize - How taken: HMS Hamadryad - When taken: 22 Nov 1814 - Where taken: Georges Bank - Date Received: 21 Jan 1815 - From what ship: HMT Impregnable - Born: Newburyport - Age: 28 - Released on 5 Jul 1815.

Doliver, Joseph - Seaman - Nbr: 2677 - Prize: Joseph - Ship Type: MV - How taken: HMS Iris - When taken: 8 Jun 1813 - Where taken: off Spain - Date Received: 21 Aug 1814 - From what ship: HMT Freya, Chatham - Born: Gloucester - Age: 18 - Released on 1 May 1815.

Doliver, William - Seaman - Nbr: 1599 - Prize: Essex - Ship Type: MV - How taken: HMS Pyramus - When taken: 2 Apr 1813 - Where taken: Bay of Biscay - Date Received: 23 Jun 1814 - From what ship: Stapleton - Born: Massachusetts - Age: 17 - Released on 1 May 1815.

D'Olivera, Manuel - Seaman - Nbr: 528 - How taken: Impressed at Liverpool - When taken: 6 Jul 1813 - Date Received: 8 Sep 1813 - What ship: Plymouth - Born: Connecticut - Age: 31 - Race: Negro - Released on 26 Apr 1815.

Dolling, Gannet - Sailing master - Nbr: 6132 - Prize: US Schooner Somers - Ship Type: MW - How taken: British gunboats - When taken: 12 Aug 1814 - Where taken: Lake Erie - Date Received: 17 Jan 1815 - From what ship: HMT Impregnable - Born: New Jersey - Age: 25 - Sent to Ashburton (England) on 22 Jan 1815.

Dolliver, Francis - Seaman - Nbr: 66 - Prize: Spitfire - Ship Type: MV - How taken: HMS Achates - When taken: 14 Feb 1813 - Where taken: off Ushant (France) - Date Received: 2 Apr 1813 - From what ship: Plymouth - Born: Marblehead - Age: 28 - Released on 20 Apr 1815.

Dolliver, Richard - Seaman - Nbr: 890 - Prize: General Kempt - Ship Type: P - How taken: HMS Foxhound - When taken: 18 Dec 1813 - Where taken: Lat 48'60" Long 5'7" - Date Received: 31 Jan 1814 - From what ship: Plymouth - Born: Marblehead - Age: 19 - Released on 27 Apr 1815.

Dolphin, Francis - Seaman - Nbr: 6063 - Prize: Chasseur - Ship Type: P - How taken: HMS Recruit - When taken: 6 Apr 1814 - Where taken: Lat 29 Long 76 - Date Received: 28 Dec 1814 - From what ship: HMT Penelope - Born: Bordeaux - Age: 20 - Released on 3 Jul 1815.

Dolphin, Joseph - Seaman - Nbr: 6407 - Prize: Nancy - Ship Type: P - How taken: Papillion - When taken: 27 Dec 1814 - Where taken: Lat 29 Long 20 - Date Received: 3 Mar 1815 - From what ship: HMT Ganges, Plymouth - Born: Exeter - Age: 22 - Released on 11 Jul 1815.

Domeree, John - Seaman - Nbr: 1893 - Prize: General Kempt - Ship Type: P - How taken: HMS Foxhound - When taken: 18 Dec 1813 - Where taken: Lat 48'60" Long 5'7" - Date Received: 29 Jul 1814 - From what ship: HMS Ville de Paris, Chatham Depot - Born: Marblehead - Age: 21 - Released on 2 May 1815.

Domerell, Edward - Seaman - Nbr: 5298 - Prize: Fox - Ship Type: P - How taken: HMS Shannon - When taken: 7 Nov 1813 - Where taken: off Newfoundland - Date Received: 31 Oct 1814 - From what ship: HMT Leyden,

American Prisoners of War Held at Dartmoor during the War of 1812

## Alphabetical listing of names

Chatham - Born: Portsmouth - Age: 23 - Released on 29 Jun 1815.

Dominico, John - Cook - Nbr: 6440 - Prize: Chance - Ship Type: P - How taken: HMS Statira - When taken: 1 Apr 1814 - Where taken: Lat 38 Long 24 - Date Received: 3 Mar 1815 - From what ship: HMT Ganges, Plymouth - Born: Maryland - Age: 25 - Race: Black - Released on 11 Jul 1815.

Dominigue, Joseph - Seaman - Nbr: 2398 - Prize: US Sloop Frolic - Ship Type: MW - How taken: HMS Orpheus - When taken: 20 Apr 1814 - Where taken: off Cuba - Date Received: 16 Aug 1814 - From what ship: HMT Queen, Halifax - Born: New York - Age: 21 - Sent to Dartmouth on 19 Oct 1814.

Donaldson, Joseph - Seaman - Nbr: 2860 - Prize: Dick - Ship Type: MV - How taken: HMS Dispatch - When taken: 17 Mar 1812 - Where taken: at sea - Date Received: 24 Aug 1814 - From what ship: HMT Alpheus, Chatham - Born: New York - Age: 21 - Released on 19 May 1815.

Done, Elisha - Seaman - Nbr: 3734 - Prize: Alfred - Ship Type: P - How taken: HMS Epervier - When taken: 23 Feb 1814 - Where taken: off Newfoundland - Date Received: 30 Sep 1814 - From what ship: HMT President, Halifax - Born: Salem - Age: 23 - Released on 4 Jun 1815.

Donelly, Anthony - Soldier - Nbr: 5263 - Prize: 13th US Infantry - Ship Type: Troops - How taken: British Army - When taken: 13 Oct 1812 - Where taken: Canada - Date Received: 31 Oct 1814 - From what ship: HMT Leyden, Chatham - Born: Isle of Man - Age: 39 - Released on 29 Jun 1815.

Donham, Ebenezer - Marine - Nbr: 601 - Prize: US Brig Argus - Ship Type: MW - How taken: HMS Pelican - When taken: 14 Aug 1813 - Where taken: Irish Channel - Date Received: 8 Sep 1813 - From what ship: Plymouth - Born: Connecticut - Age: 18 - Sent to Dartmouth on 2 Nov 1814.

Doniner, John - Soldier - Nbr: 5259 - Prize: 13th US Infantry - Ship Type: Troops - How taken: British Army - When taken: 13 Oct 1812 - Where taken: Canada - Date Received: 31 Oct 1814 - From what ship: HMT Leyden, Chatham - Born: Cavern - Age: 41 - Released on 29 Jun 1815.

Donison, William - Seaman - Nbr: 5294 - Prize: Polly - Ship Type: P - How taken: Prize - When taken: 30 Jul 1813 - Where taken: Grand Banks - Date Received: 31 Oct 1814 - From what ship: HMT Leyden, Chatham - Born: Salem - Age: 21 - Released on 29 Jun 1815.

Donnell, Samuel - Seaman - Nbr: 4713 - Prize: Elbridge Gerry - Ship Type: P - How taken: HMS Crescent - When taken: 26 Sep 1813 - Where taken: at sea - Date Received: 9 Oct 1814 - From what ship: HMT Freya, Chatham - Born: Portland - Age: 19 - Released on 20 Mar 1815.

Donnelson, Joseph - Seaman - Nbr: 1003 - Prize: Zephyr - Ship Type: MV - How taken: HMS Surveillante - When taken: 6 Jan 1814 - Where taken: Bay of Biscay - Date Received: 31 Jan 1814 - From what ship: Plymouth - Born: Gradaro - Age: 39 - Sent to Mill Prison (Plymouth, England) on 17 Jun 1814.

Doogood, Abraham - Seaman - Nbr: 5505 - Prize: US Gunboat Nbr 2 - Ship Type: MW - How taken: British forces - When taken: 22 Aug 1814 - Where taken: Chesapeake Bay - Date Received: 17 Dec 1814 - From what ship: HMT Loire, Halifax - Born: Waterford - Age: 60 - Released on 29 Jun 1815.

Doolittle, H. - Seaman - Nbr: 1633 - Prize: Tom - Ship Type: LM - How taken: HMS Surveillante - When taken: 24 Apr 1813 - Where taken: Bay of Biscay - Date Received: 23 Jun 1814 - From what ship: Stapleton - Born: Connecticut - Age: 24 - Released on 1 May 1815.

Door, Ebenezer - 6th Lieutenant - Nbr: 6308 - Prize: Prince de Neufchatel - Ship Type: P - How taken: Leander (Newcastle Acasta) - When taken: 20 Dec 1814 - Where taken: Lat 38 Long 56 - Date Received: 19 Feb 1815 - From what ship: HMT Ganges, Plymouth - Born: Roxborough - Age: 26 - Released on 3 Jul 1815.

Doosenbery, R. - Seaman - Nbr: 4237 - Prize: Prize to the Diomede - Ship Type: Prize - How taken: HMS Sapphire - When taken: 27 Feb 1814 - Where taken: at sea - Date Received: 7 Oct 1814 - From what ship: HMT Niobe, Chatham - Born: New York - Age: 32 - Released on 13 Jun 1815.

Dorchester, Preston - Seaman - Nbr: 1755 - How taken: Gave himself up from HMS Elizabeth - When taken: 31 Jan 1812 - Date Received: 20 Jul 1814 - From what ship: HMS Milford, Plymouth - Born: Connecticut - Age: 31 - Released on 26 Apr 1815.

Dore, Charles - Seaman - Nbr: 456 - Prize: Courier - Ship Type: LM - How taken: HMS Andromache - When taken: 14 Mar 1813 - Where taken: Bay of Biscay - Date Received: 8 Sep 1813 - From what ship: Plymouth - Born: Chapeley - Age: 23 - Released on 26 Apr 1815.

Dorman, Israel - Marine - Nbr: 5693 - Prize: McDonough - Ship Type: P - How taken: HMS Bacchante - When taken: 7 Nov 1814 - Where taken: Lat 42 Long 67 - Date Received: 24 Dec 1814 - From what ship: HMT Penelope - Born: Arundel - Age: 21 - Released on 3 Jul 1815.

Dorman, John - Boatswain - Nbr: 2373 - Prize: US Sloop Frolic - Ship Type: MW - How taken: HMS Orpheus - When taken: 20 Apr 1814 - Where taken: off Cuba - Date Received: 16 Aug 1814 - From what ship: HMT Queen, Halifax - Born: Massachusetts - Age: 40 - Sent to Dartmouth on 19 Oct 1814.

American Prisoners of War Held at Dartmoor during the War of 1812

## Alphabetical listing of names

Dorr, Edward - Seaman - Nbr: 2250 - Prize: Frederick - Ship Type: Prize - How taken: HMS Curlew - When taken: 4 May 1814 - Where taken: off Halifax - Date Received: 16 Aug 1814 - From what ship: HMS Dublin, Halifax - Born: Salisbury - Age: 31 - Released on 3 May 1815.

Dorrell, John - Seaman - Nbr: 267 - Prize: William Bayard - Ship Type: MV - How taken: HMS Warspite - When taken: 3 Mar 1813 - Where taken: Bay of Biscay - Date Received: 28 Jun 1813 - From what ship: Plymouth - Born: Wethersfield - Age: 27 - Released on 20 Apr 1815.

Dorrell, John - Seaman - Nbr: 3867 - Prize: Lewis Warrington - Ship Type: MV - How taken: HMS Loire - When taken: 23 May 1814 - Where taken: off VA - Date Received: 5 Oct 1814 - From what ship: HMT Orpheus, Halifax - Born: Boston - Age: 22 - Released on 10 May 1815.

Dorsey, John - Seaman - Nbr: 3027 - Prize: Prize of the Bunker Hill - Ship Type: Prize - How taken: HMS Fly - When taken: Mar 1814 - Where taken: off St. Malo - Date Received: 2 Sep 1814 - From what ship: Naval Hospital, Plymouth - Born: Philadelphia - Age: 30 - Race: Negro - Released on 28 May 1815.

Dotts, Cornelius - Seaman - Nbr: 5157 - Prize: Volante - Ship Type: P - How taken: HMS Curlew - When taken: 25 Nov 1813 - Where taken: off Halifax - Date Received: 31 Oct 1814 - From what ship: HMT Mermaid, Chatham - Born: Marblehead - Age: 21 - Released on 29 Jun 1815.

Douarte, Angelo - Seaman - Nbr: 322 - Prize: Pallas - Ship Type: MV - How taken: Rebuff - When taken: 23 Dec 1812 - Where taken: off Cadiz - Date Received: 28 Jun 1813 - From what ship: Plymouth - Born: New Orleans - Age: 42 - Sent to Dartmouth on 30 Jul 1813.

Doue, Alexander - Seaman - Nbr: 6065 - Prize: Chasseur - Ship Type: P - How taken: HMS Recruit - When taken: 6 Apr 1814 - Where taken: Lat 29 Long 76 - Date Received: 28 Dec 1814 - From what ship: HMT Penelope - Born: Nantes - Age: 22 - Released on 3 Jul 1815.

Dougall, Thomas - Seaman - Nbr: 1446 - Prize: Tickler - Ship Type: LM - How taken: HMS Magicienne - When taken: 5 Jun 1813 - Where taken: Bay of Biscay - Date Received: 19 Jun 1814 - From what ship: Stapleton - Born: Waterford - Age: 20 - Released on 28 Apr 1815.

Doughty, Jesse - Seaman - Nbr: 1438 - Prize: Leo - Ship Type: LM - How taken: HMS Magicienne - When taken: 4 Jun 1813 - Where taken: off France - Date Received: 19 Jun 1814 - From what ship: Stapleton - Born: Massachusetts - Age: 21 - Released on 28 Apr 1815.

Doughty, Levi - Seaman - Nbr: 1415 - Prize: Orders in Council - Ship Type: LM - How taken: HMS Surveillante - When taken: 1 Jun 1813 - Where taken: off Cape Ortegal (Spain) - Date Received: 19 Jun 1814 - From what ship: Stapleton - Born: Brunswick - Age: 22 - Released on 28 Apr 1815.

Douglas, Dover - Seaman - Nbr: 2255 - Prize: Otario - Ship Type: Prize - How taken: HMS Curlew - When taken: 2 May 1814 - Where taken: off Halifax - Date Received: 16 Aug 1814 - From what ship: HMS Dublin, Halifax - Born: Delaware - Age: 20 - Race: Black - Released on 11 Jul 1815.

Douglas, John - Seaman - Nbr: 3136 - How taken: Gave himself up from HMS Union - When taken: 15 Nov 1812 - Date Received: 11 Sep 1814 - From what ship: HMT Freya, Chatham - Born: New London - Age: 22 - Released on 27 Apr 1815.

Douglas, Samuel - Seaman - Nbr: 5331 - Prize: Elbridge Gerry - Ship Type: P - How taken: HMS Crescent - When taken: 16 Sep 1813 - Where taken: at sea - Date Received: 31 Oct 1814 - From what ship: HMT Leyden, Chatham - Born: North Carolina - Age: 22 - Released on 29 Jun 1815.

Douglass, John - Seaman - Nbr: 2535 - Prize: US Frigate Chesapeake - Ship Type: MW - How taken: HMS Shannon - When taken: 1 Jul 1814 - Where taken: off Boston - Date Received: 16 Aug 1814 - From what ship: HMT Queen, Halifax - Born: New York - Age: 29 - Released on 3 May 1815.

Douglass, Mathew - Seaman - Nbr: 6245 - Prize: Albion - Ship Type: Prize - How taken: HMS Harlequin - When taken: 6 Jan 1814 - Where taken: Long 20 - Date Received: 4 Feb 1815 - From what ship: HMT Ganges - Born: Maryland - Age: 24 - Released on 3 Jul 1815.

Douglass, William - Seaman - Nbr: 4220 - Prize: Liberty - Ship Type: MV - How taken: Surrendered - When taken: 30 Dec 1813 - Where taken: Stromess - Date Received: 7 Oct 1814 - From what ship: HMT Niobe, Chatham - Born: Norway - Age: 18 - Released on 13 Jun 1815.

Dourville, Francis - Boatswain's Mate - Nbr: 5488 - Prize: General Putnam - Ship Type: P - How taken: HMS Leander - When taken: 8 Nov 1814 - Where taken: Long 65 Lat 42 - Date Received: 17 Dec 1814 - From what ship: HMT Loire, Halifax - Born: Granville - Age: 35 - Released on 1 Jul 1815.

Dow, Henry - Seaman - Nbr: 5165 - Prize: Volante - Ship Type: P - How taken: HMS Curlew - When taken: 25 Nov 1813 - Where taken: off Halifax - Date Received: 31 Oct 1814 - From what ship: HMT Mermaid, Chatham - Born: Hampton - Age: 25 - Released on 29 Jun 1815.

Dowell, Isaac - 2nd Mate - Nbr: 346 - Prize: Courier - Ship Type: LM - How taken: HMS Andromache - When

American Prisoners of War Held at Dartmoor during the War of 1812

## Alphabetical listing of names

taken: 14 Mar 1813 - Where taken: Bay of Biscay - Date Received: 28 Jun 1813 - From what ship: Plymouth - Born: Virginia - Age: 25 - Released on 20 Apr 1815.

Dowling, John - Seaman - Nbr: 1764 - Prize: Cygnet - Ship Type: Prize - How taken: HMS Statira - When taken: 8 May 1814 - Where taken: at sea - Date Received: 20 Jul 1814 - From what ship: HMS Milford, Plymouth - Born: New York - Age: 19 - Released on 1 May 1815.

Downe, William - Seaman - Nbr: 1100 - Prize: Fair American - Ship Type: MV - How taken: HMS Andromache - When taken: 19 Jun 1814 - Where taken: Irish Channel - Date Received: 10 May 1814 - From what ship: Plymouth - Born: Boston - Age: 27 - Released on 27 Apr 1815.

Downey, Charles - Seaman - Nbr: 2284 - Prize: Diomede - Ship Type: P - How taken: HMS Rifleman - When taken: 28 May 1814 - Where taken: off Sable Island - Date Received: 16 Aug 1814 - From what ship: HMS Dublin, Halifax - Born: Philadelphia - Age: 25 - Race: Negro - Released on 3 May 1815.

Downey, Richard - 2nd Lieutenant - Nbr: 2274 - Prize: Diomede - Ship Type: P - How taken: HMS Rifleman - When taken: 23 Jun 1814 - Where taken: off Halifax - Date Received: 16 Aug 1814 - From what ship: HMS Dublin, Halifax - Born: Salem - Age: 27 - Released on 3 May 1815.

Downing, Benjamin - Boy - Nbr: 2264 - Prize: Grand Turk - Ship Type: P - How taken: HMS Martin - When taken: 26 May 1814 - Where taken: off Cape Sable Island (Canada) - Date Received: 16 Aug 1814 - From what ship: HMS Dublin, Halifax - Born: Salem - Age: 13 - Released on 3 May 1815.

Downing, George - 1st Mate - Nbr: 3761 - Prize: Grecian - Ship Type: LM - How taken: HMS Jaseur - When taken: 2 May 1814 - Where taken: Chesapeake Bay - Date Received: 5 Oct 1814 - From what ship: HMT Orpheus, Halifax - Born: Newport - Age: 38 - Released on 9 Jun 1815.

Downing, John - Seaman - Nbr: 3007 - How taken: Gave himself up from HMS Portia - When taken: Dec 1814 - Date Received: 29 Aug 1814 - From what ship: HMT Bittern - Born: New London - Age: 38 - Released on 28 May 1815.

Downs, Jesse - Marine - Nbr: 6322 - Prize: Prince de Neufchatel - Ship Type: P - How taken: Leander (Newcastle Acasta) - When taken: 20 Dec 1814 - Where taken: Lat 38 Long 56 - Date Received: 19 Feb 1815 - From what ship: HMT Ganges, Plymouth - Born: Boston - Age: 20 - Released on 11 Jul 1815.

Downs, John - Boy - Nbr: 4281 - Prize: Pilot - Ship Type: LM - How taken: Victoria (Privateer) - When taken: 28 Jan 1814 - Where taken: off Bordeaux - Date Received: 7 Oct 1814 - From what ship: HMT Niobe, Chatham - Born: Philadelphia - Age: 10 - Released on 14 Jun 1815.

Downs, William - Boy - Nbr: 4280 - Prize: Pilot - Ship Type: LM - How taken: Victoria (Privateer) - When taken: 28 Jan 1814 - Where taken: off Bordeaux - Date Received: 7 Oct 1814 - From what ship: HMT Niobe, Chatham - Born: Philadelphia - Age: 16 - Released on 14 Jun 1815.

Doyle, James - Seaman - Nbr: 2321 - Prize: Hussar - Ship Type: P - How taken: HMS Saturn - When taken: 25 May 1814 - Where taken: off Sandy Hook - Date Received: 16 Aug 1814 - From what ship: HMS Dublin, Halifax - Born: Philadelphia - Age: 17 - Released on 3 May 1815.

Doyle, John - Seaman - Nbr: 3532 - Prize: Rover - Ship Type: Prize - How taken: HMS Conquistador - When taken: 22 Aug 1814 - Where taken: Long 19 Lat 107 - Date Received: 28 Sep 1814 - From what ship: HMT Salvador del Mundo - Born: Massachusetts - Age: 22 - Released on 4 Jun 1815.

Drake, Daniel - Seaman - Nbr: 1871 - Prize: Elbridge Gerry - Ship Type: P - How taken: HMS Crescent - When taken: 16 Sep 1813 - Where taken: off St. George's - Date Received: 29 Jul 1814 - From what ship: HMS Ville de Paris, Chatham Depot - Born: Salem - Age: 18 - Released on 2 May 1815.

Drake, Henry - Seaman - Nbr: 15 - Prize: Cashier - Ship Type: LM - How taken: HMS Reindeer - When taken: 3 Feb 1813 - Where taken: Bay of Biscay - Date Received: 2 Apr 1813 - From what ship: Plymouth - Born: New York - Age: 23 - Race: Negro - Sent to Dartmouth on 30 Jul 1813.

Drake, John - Seaman - Nbr: 1124 - Prize: Hope - Ship Type: P - How taken: HMS Sea Horse - When taken: 22 Mar 1814 - Where taken: at sea - Date Received: 10 May 1814 - From what ship: Plymouth - Born: Philadelphia - Age: 33 - Released on 28 Apr 1815.

Drayton, John - Seaman - Nbr: 4784 - How taken: Gave himself up from HMS Queen Charlotte - When taken: 23 Sep 1814 - Date Received: 9 Oct 1814 - From what ship: HMT Freya, Chatham - Born: Baltimore - Age: 23 - Race: Mulatto - Released on 11 Jul 1815.

Drew, Charles - Seaman - Nbr: 1149 - Prize: Bunker Hill - Ship Type: P - How taken: HMS Pomone - When taken: 8 Mar 1814 - Where taken: at sea - Date Received: 10 May 1814 - From what ship: Plymouth - Born: Boston - Age: 31 - Released on 28 Apr 1815.

Drew, Samuel - Cook - Nbr: 4016 - Prize: US Brig Rattlesnake - Ship Type: MW - How taken: HMS Leander - When taken: 13 Jul 1814 - Where taken: off Shelburne - Date Received: 6 Oct 1814 - From what ship: HMT

American Prisoners of War Held at Dartmoor during the War of 1812

## Alphabetical listing of names

Chesapeake, Halifax - Born: New Hampshire - Age: 27 - Released on 9 Jun 1815.

Drew, William - Seaman - Nbr: 535 - Prize: Union - Ship Type: LM - How taken: HMS Goldfinch - When taken: 17 Jul 1813 - Where taken: Bay of Biscay - Date Received: 8 Sep 1813 - From what ship: Plymouth - Born: Massachusetts - Age: 17 - Released on 26 Apr 1815.

Drinkwater, Daniel - Seaman - Nbr: 5922 - Prize: Harlequin - Ship Type: P - How taken: HMS Bulwark - When taken: 23 Nov 1814 - Where taken: off Halifax - Date Received: 27 Dec 1814 - From what ship: HMT Penelope - Born: New York - Age: 28 - Released on 3 Jul 1815.

Drinkwater, Peter - 3rd Lieutenant - Nbr: 937 - Prize: Siro - Ship Type: LM - How taken: HMS Pelican - When taken: 13 Jan 1814 - Where taken: at sea - Date Received: 31 Jan 1814 - From what ship: Plymouth - Born: Poston - Age: 25 - Released on 10 Apr 1815.

Drisco, James - Seaman - Nbr: 5193 - Prize: Governor Plumer - Ship Type: P - How taken: HMS Bold - When taken: 17 May 1813 - Where taken: off Newfoundland - Date Received: 31 Oct 1814 - From what ship: HMT Mermaid, Chatham - Born: Portsmouth - Age: 15 - Released on 29 Jun 1815.

Driscol, Jeremiah - Seaman - Nbr: 4369 - Prize: Thomas - How taken: HMS Nymphe - When taken: 28 Jun 1813 - Where taken: off Halifax - Date Received: 8 Oct 1814 - From what ship: HMT Leyden, Chatham - Born: Portsmouth - Age: 29 - Released on 14 Jun 1815.

Driver, John - Seaman - Nbr: 1247 - How taken: Sent into custody from HMS Edinburgh - When taken: 28 Oct 1812 - Date Received: 14 Jun 1814 - From what ship: Mill Prison (Plymouth, England) - Born: Woodstock - Age: 29 - Released on 26 Apr 1815.

Driver, Thomas - Seaman - Nbr: 2003 - Prize: Governor Middleton - Ship Type: MV - How taken: Thetis (Privateer) - When taken: 2 May 1813 - Where taken: Bay of Biscay - Date Received: 3 Aug 1814 - From what ship: HMS Lyffey, Chatham Depot - Born: Philadelphia - Age: 20 - Released on 2 May 1815.

Dubois, Alexander - Seaman - Nbr: 810 - Prize: Sybille - Ship Type: MV - How taken: HMS Zenobia - When taken: 27 Jun 1813 - Where taken: off Cape St. Mary's - Date Received: 3 Nov 1813 - From what ship: Plymouth - Born: New York - Age: 26 - Released on 26 Apr 1815.

Ducat, William - Seaman - Nbr: 4470 - Prize: Governor Plumer - Ship Type: P - How taken: HMS Shamrock - When taken: 4 Mar 1813 - Where taken: at sea - Date Received: 8 Oct 1814 - From what ship: HMT Leyden, Chatham - Born: New Jersey - Age: 30 - Released on 15 Jun 1815.

Duff, James - Seaman - Nbr: 983 - Prize: Amity - Ship Type: P - How taken: HMS Achates - When taken: 22 Dec 1813 - Where taken: Bay of Biscay - Date Received: 31 Jan 1814 - From what ship: Plymouth - Born: Philadelphia - Age: 26 - Released on 27 Apr 1815.

Duffel, Barnett - Seaman - Nbr: 6064 - Prize: Chasseur - Ship Type: P - How taken: HMS Recruit - When taken: 6 Apr 1814 - Where taken: Lat 29 Long 76 - Date Received: 28 Dec 1814 - From what ship: HMT Penelope - Born: Charlestown - Age: 35 - Released on 19 May 1815.

Duffy, James - Seaman - Nbr: 2553 - Prize: Caromaned - Ship Type: Prize - How taken: HMS Eridanus - When taken: 13 Aug 1814 - Where taken: Lat 40, Long 16 - Date Received: 16 Aug 1814 - From what ship: HMT Salvador del Mundo - Born: Massachusetts - Age: 38 - Released on 3 May 1815.

Dugaretz, Augusta - Seaman - Nbr: 1210 - Prize: Pandour - Ship Type: Prize - How taken: HMS Fairy - When taken: 25 Dec 1812 - Where taken: off St. Antonio - Date Received: 4 Jun 1814 - From what ship: Dartmouth - Born: Jamaica, NY - Age: 20 - Sent to Plymouth on 8 Jul 1814.

Dullivan, James - Seaman - Nbr: 5320 - Prize: Industry - Ship Type: P - How taken: HMS Maidstone - When taken: 17 Jul 1813 - Where taken: Grand Banks - Date Received: 31 Oct 1814 - From what ship: HMT Leyden, Chatham - Born: Marblehead - Age: 16 - Released on 29 Jun 1815.

Dunbar, Luther - Seaman - Nbr: 1176 - Prize: Gazelle - Ship Type: P - How taken: HMS Leonidas - When taken: 15 Feb 1812 - Where taken: off Ireland - Date Received: 4 Jun 1814 - From what ship: Dartmouth - Born: Boston - Age: 33 - Released on 26 Apr 1815.

Dunbar, William - Seaman - Nbr: 2290 - Prize: Hussar - Ship Type: P - How taken: HMS Saturn - When taken: 25 May 1814 - Where taken: off Sandy Hook - Date Received: 16 Aug 1814 - From what ship: HMS Dublin, Halifax - Born: Philadelphia - Age: 21 - Race: Negro - Released on 3 May 1815.

Duncan, Abel - Prize master - Nbr: 5650 - Prize: Prize of the Chasseur - Ship Type: Prize - How taken: HMS Castilian - When taken: 30 Sep 1814 - Where taken: Long 14 Lat 52 - Date Received: 24 Dec 1814 - From what ship: HMT Impregnable - Born: Baltimore - Age: 22 - Escaped on 1 Jun 1815.

Duncan, Edward - Seaman - Nbr: 3182 - How taken: Gave himself up from HMS Hibernia - When taken: 27 Jul 1813 - Date Received: 11 Sep 1814 - From what ship: HMT Freya, Chatham - Born: Rhode Island - Age: 38 - Race: Mulatto - Released on 28 May 1815.

American Prisoners of War Held at Dartmoor during the War of 1812

## Alphabetical listing of names

Duncan, George - Seaman - Nbr: 707 - Prize: Ned - Ship Type: LM - How taken: HMS Royalist - When taken: 6 Sep 1813 - Where taken: Bay of Biscay - Date Received: 27 Sep 1813 - From what ship: Plymouth - Born: New York - Age: 18 - Released on 26 Apr 1815.

Duncan, George - Seaman - Nbr: 1891 - How taken: Impressed at London - When taken: 25 Feb 1814 - Date Received: 29 Jul 1814 - From what ship: HMS Ville de Paris, Chatham Depot - Born: Maryland - Age: 28 - Race: Black - Released on 2 May 1815.

Duncan, Jesse - Seaman - Nbr: 2946 - Prize: Gotley - Ship Type: MV - How taken: HMS Barbados - When taken: 20 Jan 1814 - Where taken: off St. Bartholomew - Date Received: 24 Aug 1814 - From what ship: HMT Hannibal - Born: Virginia - Age: 23 - Released on 19 May 1815.

Duncan, Nathaniel - Seaman - Nbr: 4142 - Prize: US Sloop Growler - Ship Type: MW - How taken: British gunboats - When taken: 3 Jun 1813 - Where taken: Great Lakes - Date Received: 6 Oct 1814 - From what ship: HMT Chesapeake, Halifax - Born: Wilmington - Age: 41 - Released on 13 Jun 1815.

Duncan, Thomas - Seaman - Nbr: 4482 - Prize: Enterprize - Ship Type: P - How taken: HMS Tenedos - When taken: 21 May 1813 - Where taken: off Cape Cod - Date Received: 8 Oct 1814 - From what ship: HMT Leyden, Chatham - Born: Massachusetts - Age: 19 - Released on 15 Jun 1815.

Dunchellier, Isaac - Seaman - Nbr: 1811 - Prize: Eliza - Ship Type: MV - How taken: HMS Surveillante - When taken: 23 Mar 1813 - Where taken: Bay of Biscay - Date Received: 20 Jul 1814 - From what ship: HMS Milford, Plymouth - Born: Boston - Age: 21 - Released on 1 May 1815.

Dunham, Daniel - Seaman - Nbr: 3265 - Prize: Fame (Whaler) - Ship Type: MV - How taken: HMS Cressy - When taken: 20 Jul 1813 - Where taken: at sea - Date Received: 11 Sep 1814 - From what ship: HMT Freya, Chatham - Born: Massachusetts - Age: 27 - Released on 28 May 1815.

Dunham, George - Seaman - Nbr: 5470 - Prize: General Putnam - Ship Type: P - How taken: HMS Leander - When taken: 8 Nov 1814 - Where taken: Long 65 Lat 42 - Date Received: 17 Dec 1814 - From what ship: HMT Loire, Halifax - Born: Salem - Age: 23 - Released on 1 Jul 1815.

Dunham, John - Seaman - Nbr: 2317 - Prize: Hussar - Ship Type: P - How taken: HMS Saturn - When taken: 25 May 1814 - Where taken: off Sandy Hook - Date Received: 16 Aug 1814 - From what ship: HMS Dublin, Halifax - Born: Massachusetts - Age: 29 - Released on 3 May 1815.

Dunham, John - Seaman - Nbr: 3340 - How taken: Gave himself up from HMS Scorpion - When taken: 2 Dec 1812 - Date Received: 13 Sep 1814 - From what ship: HMT Niobe, Chatham - Born: Massachusetts - Age: 20 - Released on 27 Apr 1815.

Dunham, William - Seaman - Nbr: 6376 - Prize: US Brig Syren - Ship Type: MW - How taken: HMS Medway - When taken: 12 Jul 1814 - Where taken: off Cape of Good Hope - Date Received: 24 Feb 1815 - From what ship: HMT Ganges, Plymouth - Born: Lyme - Age: 20 - Released on 11 Jul 1815.

Dunklin, Jesse - Seaman - Nbr: 6361 - Prize: Prince de Neufchatel - Ship Type: P - How taken: Leander (Newcastle Acasta) - When taken: 20 Dec 1814 - Where taken: Lat 38 Long 56 - Date Received: 19 Feb 1815 - From what ship: HMT Ganges, Plymouth - Born: New Hampshire - Age: 31 - Released on 5 Jul 1815.

Dunn, Charles - Seaman - Nbr: 2296 - Prize: Hussar - Ship Type: P - How taken: HMS Saturn - When taken: 25 May 1814 - Where taken: off Sandy Hook - Date Received: 16 Aug 1814 - From what ship: HMS Dublin, Halifax - Born: New York - Age: 22 - Released on 3 May 1815.

Dunn, David - Seaman - Nbr: 2800 - Prize: Weazel - Ship Type: MV - How taken: HMS Foxhound - When taken: 25 Mar 1813 - Where taken: Bay of Biscay - Date Received: 24 Aug 1814 - From what ship: HMT Liverpool, Chatham - Born: New York - Age: 27 - Released on 19 May 1815.

Dunn, Edward - Seaman - Nbr: 3848 - Prize: Dominique - Ship Type: LM - How taken: HMS Dotterel - When taken: 21 May 1814 - Where taken: off Charleston - Date Received: 5 Oct 1814 - From what ship: HMT Orpheus, Halifax - Born: Norfolk - Age: 22 - Released on 9 Jun 1815.

Dunn, Henry - Seaman - Nbr: 2434 - Prize: US Sloop Frolic - Ship Type: MW - How taken: HMS Orpheus - When taken: 20 Apr 1814 - Where taken: off Cuba - Date Received: 16 Aug 1814 - From what ship: HMT Queen, Halifax - Born: Pennsylvania - Age: 34 - Released on 3 May 1815.

Dunn, Henry G. - Seaman - Nbr: 2149 - How taken: Apprehended at London - When taken: 10 Feb 1812 - Date Received: 8 Aug 1814 - From what ship: HMT Raven, Chatham - Born: Tappahannock - Age: 27 - Released on 2 May 1815.

Dunn, James - Seaman - Nbr: 1814 - How taken: Gave himself up from HMS Unicorn - When taken: 17 Jun 1812 - Date Received: 20 Jul 1814 - From what ship: HMS Milford, Plymouth - Born: Boston - Age: 30 - Released on 26 Apr 1815.

Dunn, Robert - Quartermaster - Nbr: 4076 - Prize: New Zealander - Ship Type: Prize - How taken: HMS Belviders -

American Prisoners of War Held at Dartmoor during the War of 1812

## Alphabetical listing of names

When taken: 21 Apr 1814 - Where taken: off Delaware - Date Received: 6 Oct 1814 - From what ship: HMT Chesapeake, Halifax - Born: Pennsylvania - Age: 38 - Released on 13 Jun 1815.

Dunn, William - Seaman - Nbr: 927 - Prize: Growler - Ship Type: MV - How taken: HMS Wolf - When taken: 11 Aug 1813 - Where taken: at sea - Date Received: 31 Jan 1814 - From what ship: Plymouth - Born: Alexandria - Age: 23 - Sent to Dartmouth on 19 Oct 1814.

Dunning, William - Seaman - Nbr: 1710 - How taken: Gave himself up from HMS Forth - Date Received: 8 Jul 1814 - From what ship: Labrador - Born: Massachusetts - Age: 38 - Released on 1 May 1815.

Dunningberg, Henry - Seaman - Nbr: 4521 - Prize: Yankee - Ship Type: P - How taken: HMS Shannon - When taken: 20 Aug 1813 - Where taken: at sea - Date Received: 8 Oct 1814 - From what ship: HMT Leyden, Chatham - Born: Maryland - Age: 25 - Released on 15 Jun 1815.

Dunnings, Charles - Boy - Nbr: 3855 - Prize: Dominique - Ship Type: LM - How taken: HMS Dotterel - When taken: 21 May 1814 - Where taken: off Charleston - Date Received: 5 Oct 1814 - From what ship: HMT Orpheus, Halifax - Born: Baltimore - Age: 12 - Released on 9 Jun 1815.

Durand, John - Seaman - Nbr: 1393 - Prize: Zebra - Ship Type: LM - How taken: HMS Pyramus - When taken: 20 Apr 1813 - Where taken: Bay of Biscay - Date Received: 19 Jun 1814 - From what ship: Stapleton - Born: New Orleans - Age: 22 - Sent to Mill Prison (Plymouth, England) on 26 Jun 1814.

Durgens, John - Seaman - Nbr: 5919 - Prize: Harlequin - Ship Type: P - How taken: HMS Bulwark - When taken: 23 Nov 1814 - Where taken: off Halifax - Date Received: 27 Dec 1814 - From what ship: HMT Penelope - Born: Portsmouth - Age: 17 - Released on 3 Jul 1815.

Durham, Charles - Seaman - Nbr: 1791 - Prize: Ferox - Ship Type: MV - How taken: HMS Medusa & HMS Lyra - When taken: 28 Mar 1813 - Where taken: off Cape Ortegal (Spain) - Date Received: 20 Jul 1814 - From what ship: HMS Milford, Plymouth - Born: New York - Age: 37 - Released on 1 May 1815.

Durham, Samuel - Seaman - Nbr: 3856 - Prize: Leicester - Ship Type: Prize - How taken: HMS Prometheus - When taken: Jun 1814 - Where taken: off Halifax - Date Received: 5 Oct 1814 - From what ship: HMT Orpheus, Halifax - Born: Massachusetts - Age: 18 - Released on 9 Jun 1815.

Durval, N. D. - Seaman - Nbr: 3170 - Prize: Hindortar - Ship Type: MV - How taken: HMS Tenedos - When taken: 25 Jun 1813 - Where taken: off Lisbon - Date Received: 11 Sep 1814 - From what ship: HMT Freya, Chatham - Born: Maryland - Age: 23 - Released on 28 May 1815.

Durvolf, Stephen - Seaman - Nbr: 672 - Prize: Joel Barlow - Ship Type: LM - How taken: HMS Briton - When taken: 3 Jul 1813 - Where taken: off Bordeaux - Date Received: 8 Sep 1813 - From what ship: Plymouth - Born: Connecticut - Age: 24 - Released on 26 Apr 1815.

Dusheels, Arthur - Seaman - Nbr: 834 - Prize: Chesapeake - Ship Type: LM - How taken: HMS Hotspur & HMS Pyramus - When taken: 26 Oct 1813 - Where taken: off Nantes - Date Received: 29 Nov 1813 - From what ship: Plymouth - Born: Maryland - Age: 20 - Released on 26 Apr 1815.

Duston, Peter - Seaman - Nbr: 184 - Prize: Star - Ship Type: MV - How taken: HMS Superb - When taken: 9 Feb 1813 - Where taken: Bay of Biscay - Date Received: 2 Apr 1813 - From what ship: Plymouth - Born: New York - Age: 19 - Released on 20 Apr 1815.

Dutton, Thomas - Seaman - Nbr: 4085 - Prize: New Zealander - Ship Type: Prize - How taken: HMS Belviders - When taken: 21 Apr 1814 - Where taken: off Delaware - Date Received: 6 Oct 1814 - From what ship: HMT Chesapeake, Halifax - Born: Connecticut - Age: 22 - Released on 13 Jun 1815.

Dyer, Benjamin - Seaman - Nbr: 3886 - Prize: US Brig Rattlesnake - Ship Type: MW - How taken: HMS Leander - When taken: 11 Jul 1814 - Where taken: off Shelburne - Date Received: 5 Oct 1814 - From what ship: HMT Orpheus, Halifax - Born: Massachusetts - Age: 29 - Released on 9 Jun 1815.

Dyer, Israel - Seaman - Nbr: 2364 - Prize: Snap Dragon - Ship Type: P - How taken: HMS Martin - When taken: 10 Jun 1814 - Where taken: off Halifax - Date Received: 16 Aug 1814 - From what ship: HMT Queen, Halifax - Born: Boston - Age: 25 - Released on 3 May 1815.

Dyer, Jonathan - Seaman - Nbr: 788 - Prize: Avon - Ship Type: MV - How taken: HMS Eurotas - When taken: 27 Oct 1813 - Where taken: off Ushant (France) - Date Received: 3 Nov 1813 - From what ship: Plymouth - Born: Cape Cod - Age: 40 - Died on 11 Mar 1815 from variola.

Dyer, Joseph - Seaman - Nbr: 6454 - Prize: Jemmell - Ship Type: MV - How taken: HMS Rhin - When taken: 28 May 1814 - Where taken: off Bermuda - Date Received: 3 Mar 1815 - From what ship: HMT Ganges, Plymouth - Born: Not legible - Released on 11 Jul 1815.

Dyer, Samuel - Seaman - Nbr: 2522 - Prize: Snap Dragon - Ship Type: P - How taken: HMS Martin - When taken: 10 Jun 1814 - Where taken: off Halifax - Date Received: 16 Aug 1814 - From what ship: HMT Queen, Halifax - Born: Baltimore - Age: 15 - Released on 3 May 1815.

American Prisoners of War Held at Dartmoor during the War of 1812

## Alphabetical listing of names

Dyer, Thomas - Seaman - Nbr: 3443 - Prize: Derby - Ship Type: MV - How taken: HMS Nereus - When taken: 4 Feb 1813 - Where taken: off Cape of Good Hope - Date Received: 13 Sep 1814 - From what ship: HMT Niobe, Chatham - Born: Boston - Age: 23 - Released on 28 May 1815.

Dyke, Stuart - Seaman - Nbr: 3026 - Prize: US Sloop Frolic - Ship Type: MW - How taken: HMS Orpheus - When taken: 20 Apr 1814 - Where taken: off Cuba - Date Received: 2 Sep 1814 - From what ship: Naval Hospital, Plymouth - Born: Massachusetts - Age: 26 - Released on 3 May 1815.

Dykes, Leven - Seaman - Nbr: 4336 - Prize: Venus - Ship Type: P - How taken: HMS Loire - When taken: 18 Feb 1814 - Where taken: off St. Thomas - Date Received: 7 Oct 1814 - From what ship: HMT Salvador del Mundo, Halifax - Born: Virginia - Age: 19 - Race: Negro - Released on 14 Jun 1815.

Dymoss, Peter - Seaman - Nbr: 517 - Prize: Fox - Ship Type: Packet prize to the Fox - How taken: Superior - When taken: 25 Jun 1813 - Where taken: Lat 50 N, Long 21 W - Date Received: 8 Sep 1813 - From what ship: Plymouth - Born: Bath - Age: 32 - Released on 26 Apr 1815.

Eagerly, Ely - Seaman - Nbr: 5335 - Prize: Elbridge Gerry - Ship Type: P - How taken: HMS Crescent - When taken: 16 Sep 1813 - Where taken: at sea - Date Received: 31 Oct 1814 - From what ship: HMT Leyden, Chatham - Born: Durham - Age: 30 - Released on 29 Jun 1815.

Earle, James - Carpenter - Nbr: 5772 - Prize: William Penn - Ship Type: MV - How taken: HMS Acorn - When taken: 27 Oct 1812 - Where taken: Lat 14 - Date Received: 26 Dec 1814 - From what ship: HMT Argo - Born: Massachusetts - Age: 34 - Released on 27 Apr 1815.

East, John - Seaman - Nbr: 4063 - Prize: US Brig Rattlesnake - Ship Type: MW - How taken: HMS Leander - When taken: 13 Jul 1814 - Where taken: off Shelburne - Date Received: 6 Oct 1814 - From what ship: HMT Chesapeake, Halifax - Born: Dover - Age: 48 - Released on 13 Jun 1815.

Easten, Ephraim - Seaman - Nbr: 4440 - Prize: Growler - Ship Type: P - How taken: HMS Electra - When taken: 7 Jul 1814 - Where taken: at sea - Date Received: 8 Oct 1814 - From what ship: HMT Leyden, Chatham - Born: Massachusetts - Age: 20 - Released on 14 Jun 1815.

Eastlake, James - Seaman - Nbr: 2654 - Prize: Jane - Ship Type: MV - How taken: HMS Crescent - When taken: 28 Jun 1813 - Where taken: off Newfoundland - Date Received: 21 Aug 1814 - From what ship: HMT Freya, Chatham - Born: North Carolina - Age: 21 - Released on 19 May 1815.

Eastland, James - Seaman - Nbr: 469 - Prize: Essex - Ship Type: MV - How taken: HMS Pyramus - When taken: 2 Apr 1813 - Where taken: Bay of Biscay - Date Received: 8 Sep 1813 - From what ship: Plymouth - Born: Marblehead - Age: 35 - Released on 20 Apr 1815.

Eaton, George - Seaman - Nbr: 278 - Prize: Cannoneer - Ship Type: MV - How taken: HMS Warspite - When taken: 14 Mar 1813 - Where taken: Bay of Biscay - Date Received: 28 Jun 1813 - From what ship: Plymouth - Born: Milton - Age: 27 - Released on 20 Apr 1815.

Eaton, Israel - Lieutenant - Nbr: 3421 - Prize: Industry - Ship Type: P - How taken: HMS Heron - When taken: 3 Nov 1813 - Where taken: off Halifax - Date Received: 13 Sep 1814 - From what ship: HMT Niobe, Chatham - Born: Marblehead - Age: 37 - Released on 28 May 1815.

Ebby, Samuel - Seaman - Nbr: 5773 - Prize: William Penn - Ship Type: MV - How taken: HMS Acorn - When taken: 27 Oct 1812 - Where taken: Latitude 14 - Date Received: 26 Dec 1814 - From what ship: HMT Argo - Born: Rhode Island - Age: 28 - Race: Mulatto - Released on 27 Apr 1815.

Ebier, Joseph - Seaman - Nbr: 4809 - Prize: President - Ship Type: P - How taken: HMS Pique - When taken: 7 May 1814 - Where taken: off Porto Rico - Date Received: 9 Oct 1814 - From what ship: HMT Freya, Halifax - Born: Washington - Age: 26 - Released on 21 Jun 1815.

Eburn, Aaron - Clerk - Nbr: 3561 - Prize: Hawk - Ship Type: P - How taken: HMS Pique - When taken: 26 Apr 1814 - Where taken: off Bermuda - Date Received: 30 Sep 1814 - From what ship: HMT Sybella - Born: Washington - Age: 21 - Released on 4 Jun 1815.

Eddey, John - Seaman - Nbr: 168 - How taken: Sent into custody from Flag of Truce PA - Date Received: 2 Apr 1813 - From what ship: Plymouth - Born: Hampton - Age: 18 - Sent to Dartmouth on 30 Jul 1813.

Eddy, Richard - Seaman - Nbr: 3328 - Prize: York Town - Ship Type: P - How taken: HMS Nimrod - When taken: 17 Jul 1813 - Where taken: off St. Johns - Date Received: 13 Sep 1814 - From what ship: HMT Niobe, Chatham - Born: Freetown - Age: 31 - Released on 28 May 1815.

Eden, Mack - Seaman - Nbr: 6022 - Prize: McDonough - Ship Type: P - How taken: HMS Bacchante - When taken: 1 Nov 1814 - Where taken: Lat 42 Long 67 - Date Received: 28 Dec 1814 - From what ship: HMT Penelope - Born: Somersworth - Age: 23 - Race: Black - Released on 3 Jul 1815.

Edgar, William - Seaman - Nbr: 816 - Prize: Hepsa - Ship Type: MV - How taken: HMS Zenobia - When taken: 20 Jun 1813 - Where taken: off Lisbon - Date Received: 3 Nov 1813 - From what ship: Plymouth - Born: New

American Prisoners of War Held at Dartmoor during the War of 1812

## Alphabetical listing of names

Jersey - Age: 36 - Died on 28 Jan 1814.
Edgerley, George - Seaman - Nbr: 317 - Prize: Pallas - Ship Type: MV - How taken: Rebuff - When taken: 23 Dec 1812 - Where taken: off Cadiz - Date Received: 28 Jun 1813 - From what ship: Plymouth - Born: Essex - Age: 19 - Sent to Dartmouth on 30 Jul 1813.
Edmonds, Francis - Seaman - Nbr: 2134 - How taken: Gave himself up from HMS Namur - When taken: 2 Jan 1813 - Date Received: 8 Aug 1814 - From what ship: HMT Raven, Chatham - Born: Virginia - Age: 41 - Released on 2 May 1815.
Edmonds, John - Seaman - Nbr: 5317 - Prize: Wasp - Ship Type: P - How taken: HMS Bream - When taken: 10 Jun 1813 - Where taken: off Halifax - Date Received: 31 Oct 1814 - From what ship: HMT Leyden, Chatham - Born: Newbury - Age: 18 - Released on 29 Jun 1815.
Edsom, John - Seaman - Nbr: 1513 - Prize: Zebra - Ship Type: LM - How taken: HMS Pyramus - When taken: 20 Apr 1813 - Where taken: Bay of Biscay - Date Received: 23 Jun 1814 - From what ship: Stapleton - Born: New Hampshire - Age: 28 - Released on 1 May 1815.
Edwards, David - Seaman - Nbr: 1555 - Prize: Caroline - Ship Type: MV - How taken: HMS Medusa - When taken: 12 Apr 1813 - Where taken: Bay of Biscay - Date Received: 23 Jun 1814 - From what ship: Stapleton - Born: New York - Age: 29 - Race: Colored - Released on 14 Apr 1815.
Edwards, Isaac - Seaman - Nbr: 2720 - How taken: Gave himself up from HMS Mulgrave - When taken: 13 Dec 1812 - Date Received: 21 Aug 1814 - From what ship: HMT Freya, Chatham - Born: Gorham - Age: 34 - Released on 26 Apr 1815.
Edwards, John - Seaman - Nbr: 4623 - Prize: Argus - Ship Type: MV - How taken: HMS San Domingo - When taken: 1 Mar 1814 - Where taken: off Savannah - Date Received: 9 Oct 1814 - From what ship: HMT Leyden, Chatham - Born: New York - Age: 35 - Race: Black - Released on 15 Jun 1815.
Edwards, John - Seaman - Nbr: 1659 - Prize: Paul Jones - Ship Type: P - How taken: HMS Leonidas - When taken: 23 May 1813 - Where taken: Channel - Date Received: 23 Jun 1814 - From what ship: Stapleton - Born: LA - Age: 29 - Sent to Mill Prison (Plymouth, England) on 30 Jun 1814.
Edwards, John - Carpenter - Nbr: 1641 - Prize: Paul Jones - Ship Type: P - How taken: HMS Leonidas - When taken: 23 May 1813 - Where taken: Channel - Date Received: 23 Jun 1814 - From what ship: Stapleton - Born: New York - Age: 43 - Released on 11 Jul 1815.
Edwards, John - Seaman - Nbr: 1007 - How taken: Gave himself up from HMS Crescent - Date Received: 31 Jan 1814 - From what ship: Plymouth - Born: New York - Age: 20 - Race: Black - Released on 27 Apr 1815.
Edwards, John - Seaman - Nbr: 2414 - Prize: US Sloop Frolic - Ship Type: MW - How taken: HMS Orpheus - When taken: 20 Apr 1814 - Where taken: off Cuba - Date Received: 16 Aug 1814 - From what ship: HMT Queen, Halifax - Born: Salem - Age: 30 - Sent to Dartmouth on 19 Oct 1814.
Edwards, John - Seaman - Nbr: 4954 - Prize: Herald - Ship Type: P - How taken: HMS Endymion - When taken: 15 Aug 1814 - Where taken: off Nantucket - Date Received: 28 Oct 1814 - From what ship: HMT Alkbar, Halifax - Born: Massachusetts - Age: 30 - Released on 21 Jun 1815.
Edwards, Paul - Seaman - Nbr: 5504 - Prize: US Gunboat Nbr 2 - Ship Type: MW - How taken: British forces - When taken: 22 Aug 1814 - Where taken: Chesapeake Bay - Date Received: 17 Dec 1814 - From what ship: HMT Loire, Halifax - Born: Harlem - Age: 63 - Released on 29 Jun 1815.
Edwards, Price - Cook - Nbr: 325 - Prize: Paulina - Ship Type: MV - How taken: Rebuff - When taken: 23 Dec 1812 - Where taken: off Cadiz - Date Received: 28 Jun 1813 - From what ship: Plymouth - Born: Staten Island - Age: 40 - Released on 20 Apr 1815.
Edwards, Thomas - Seaman - Nbr: 5359 - How taken: Gave himself up from HMS Eagle - When taken: 9 May 1814 - Date Received: 31 Oct 1814 - From what ship: HMT Leyden, Chatham - Born: Charleston - Age: 47 - Race: Black - Released on 1 Jul 1815.
Edwards, William - Seaman - Nbr: 6422 - Prize: Farmer's Daughter - Ship Type: MV - How taken: HMS Leviathan - When taken: 29 May 1814 - Date Received: 3 Mar 1815 - From what ship: HMT Ganges, Plymouth - Born: Accomack - Age: 30 - Released on 11 Jul 1815.
Edwards, William - Seaman - Nbr: 4297 - Prize: Commodore Perry - Ship Type: Schooner - How taken: Sent into custody from a cutter - When taken: 25 Feb 1814 - Where taken: off Bordeaux - Date Received: 7 Oct 1814 - From what ship: HMT Niobe, Chatham - Born: Sussex County - Age: 23 - Released on 14 Jun 1815.
Egbert, Peter - Seaman - Nbr: 1204 - Prize: Melanie - Ship Type: MV - How taken: HMS Briton - When taken: 15 Sep 1813 - Where taken: off Bordeaux - Date Received: 4 Jun 1814 - From what ship: Dartmouth - Born: Staten Island - Age: 44 - Released on 28 Apr 1815.
Egoss, Joseph - Seaman - Nbr: 1846 - Prize: Polly - Ship Type: MV - How taken: HMS Surveillante - When taken:

American Prisoners of War Held at Dartmoor during the War of 1812

## Alphabetical listing of names

23 Mar 1813 - Where taken: Bay of Biscay - Date Received: 21 Jul 1814 - From what ship: HMT Redbeard & Pincher, Chatham Depot - Born: Marblehead - Age: 14 - Released on 1 May 1815.

Eldon, James - Lieutenant Marines - Nbr: 3560 - Prize: Hawk - Ship Type: P - How taken: HMS Pique - When taken: 26 Apr 1814 - Where taken: off Bermuda - Date Received: 30 Sep 1814 - From what ship: HMT Sybella - Born: Massachusetts - Age: 23 - Released on 4 Jun 1815.

Eldred, Peter - Prize master - Nbr: 5763 - Prize: Young William - Ship Type: Prize - How taken: HMS Plover - When taken: 10 Sep 1814 - Where taken: off Newfoundland - Date Received: 26 Dec 1814 - From what ship: HMT Argo - Born: Rhode Island - Age: 25 - Released on 11 Jul 1815.

Eldridge, Ephraim - Seaman - Nbr: 3820 - Prize: Nimble - Ship Type: Prize - How taken: HMS Arab - When taken: 5 Apr 1814 - Where taken: Lat 37 Long 65 - Date Received: 5 Oct 1814 - From what ship: HMT Orpheus, Halifax - Born: Barnstable - Age: 21 - Released on 9 Jun 1815.

Eldridge, Nathaniel - Seaman - Nbr: 4368 - How taken: Gave himself up from HMS Malta - When taken: 7 May 1813 - Date Received: 8 Oct 1814 - From what ship: HMT Leyden, Chatham - Born: Massachusetts - Age: 32 - Released on 14 Jun 1815.

Eldridge, William - Seaman - Nbr: 3138 - Prize: Juliana Smith - Ship Type: P - How taken: HMS Nymphe - When taken: 11 May 1813 - Where taken: off Cape Sable Island (Canada) - Date Received: 11 Sep 1814 - From what ship: HMT Freya, Chatham - Born: Savannah - Age: 25 - Released on 28 May 1815.

Elfs, James - Boy - Nbr: 3172 - Prize: Hindortar - Ship Type: MV - How taken: HMS Tenedos - When taken: 25 Jun 1813 - Where taken: off Lisbon - Date Received: 11 Sep 1814 - From what ship: HMT Freya, Chatham - Born: Charleston - Age: 15 - Released on 3 May 1815.

Elger, Richard - Seaman - Nbr: 3024 - Prize: Frolic - Ship Type: MV - How taken: HMS Hyperion - When taken: 20 Apr 1814 - Where taken: off Cuba - Date Received: 2 Sep 1814 - From what ship: Naval Hospital, Plymouth - Born: Philadelphia - Age: 24 - Released on 3 May 1815.

Elisha, Thomas - Seaman - Nbr: 2075 - How taken: Impressed at London - When taken: 28 Jun 1813 - Date Received: 3 Aug 1814 - From what ship: HMS Bittern, Chatham Depot - Born: Baltimore - Age: 27 - Released on 2 May 1815.

Elles, William - Seaman - Nbr: 96 - Prize: St. Martin's Planter - Ship Type: P - How taken: HMS Dublin - When taken: 9 Feb 1813 - Where taken: Lat 43 N, Long 33 50 W - Date Received: 2 Apr 1813 - From what ship: Plymouth - Born: Rhode Island - Age: 17 - Released on 20 Apr 1815.

Ellingwood, Joshua - Seaman - Nbr: 5540 - Prize: Lion - Ship Type: MV - How taken: HMS Brilliant - When taken: 4 Jan 1814 - Where taken: Bahamas Banks - Date Received: 17 Dec 1814 - From what ship: HMT Loire, Halifax - Born: Beverly - Age: 35 - Released on 1 Jul 1815.

Elliott, Andrew - Seaman - Nbr: 5299 - Prize: Fox - Ship Type: P - How taken: HMS Shannon - When taken: 7 Nov 1813 - Where taken: off Newfoundland - Date Received: 31 Oct 1814 - From what ship: HMT Leyden, Chatham - Born: Portsmouth - Age: 16 - Released on 29 Jun 1815.

Elliott, Benjamin - Seaman - Nbr: 773 - How taken: Sent into custody from a Russian MV, prize to the French frigate Weser - Date Received: 3 Nov 1813 - From what ship: Plymouth - Born: Beverly - Age: 23 - Released on 26 Apr 1815.

Elliott, Robert - Mate - Nbr: 1076 - Prize: Three Brothers - Ship Type: MV - How taken: HMS Ringdove - When taken: 11 Nov 1813 - Where taken: at sea - Date Received: 10 May 1814 - From what ship: Plymouth - Born: New York - Age: 35 - Released on 27 Apr 1815.

Elliott, Robert - Seaman - Nbr: 2837 - How taken: Gave himself up from HMS Armide - When taken: 8 Jun 1813 - Date Received: 24 Aug 1814 - From what ship: HMT Liverpool, Chatham - Born: Massachusetts - Age: 44 - Released on 19 May 1815.

Elliott, Stephen - Prize master - Nbr: 5705 - Prize: Amazon - Ship Type: Prize - How taken: HMS Bulwark - When taken: 22 Sep 1814 - Where taken: Georges Bank - Date Received: 24 Dec 1814 - From what ship: HMT Penelope - Born: Portsmouth - Age: 25 - Released on 8 May 1815.

Ellis, Cornelius - Seaman - Nbr: 4561 - How taken: Gave himself up from HMS Blenheim - When taken: 28 Aug 1814 - Date Received: 8 Oct 1814 - From what ship: HMT Leyden, Chatham - Born: Virginia - Age: 26 - Released on 15 Jun 1815.

Ellis, George - Seaman - Nbr: 5787 - How taken: HMS Indefatigable - When taken: Aug 1813 - Date Received: 26 Dec 1814 - From what ship: HMT Argo - Born: New York - Age: 40 - Released on 3 Jul 1815.

Ellis, John - Seaman - Nbr: 5021 - Prize: Landrail - Ship Type: Prize - How taken: HMS Wasp - When taken: 27 Jul 1814 - Where taken: Georges Bank - Date Received: 28 Oct 1814 - From what ship: HMT Alkbar, Halifax - Born: New York - Age: 26 - Released on 21 Jun 1815.

American Prisoners of War Held at Dartmoor during the War of 1812

## Alphabetical listing of names

Ellis, John - Seaman - Nbr: 5318 - Prize: Industry - Ship Type: P - How taken: HMS Maidstone - When taken: 17 Jul 1813 - Where taken: Grand Banks - Date Received: 31 Oct 1814 - From what ship: HMT Leyden, Chatham - Born: Topson - Age: 17 - Released on 11 Jul 1815.

Ellis, John - Seaman - Nbr: 6042 - Prize: Regent - Ship Type: LM - How taken: HMS Forth - When taken: 19 Sep 1814 - Where taken: off Egg Harbor (New Jersey) - Date Received: 28 Dec 1814 - From what ship: HMT Penelope - Born: Charlestown - Age: 31 - Released on 3 Jul 1815.

Ellis, John - Seaman - Nbr: 2577 - How taken: Gave himself up from HMS Prince Frederick - Date Received: 21 Aug 1814 - From what ship: HMT Salvador del Mundo - Born: Virginia - Age: 25 - Race: Negro - Released on 3 May 1815.

Ellis, Thomas - Seaman - Nbr: 5519 - Prize: James - Ship Type: Prize - How taken: HMS Galatea - When taken: 7 Sep 1814 - Where taken: Channel - Date Received: 17 Dec 1814 - From what ship: HMT Loire, Halifax - Born: Philadelphia - Age: 22 - Released on 1 Jul 1815.

Ellison, Benjamin - Boy - Nbr: 3962 - Prize: Rolla - Ship Type: P - How taken: HMS Loire - When taken: 10 Dec 1813 - Where taken: off Bull Island (South Carolina) - Date Received: 5 Oct 1814 - From what ship: HMT President, Halifax - Born: New York - Age: 10 - Released on 27 Apr 1815.

Ellison, Joseph - Marine - Nbr: 5944 - Prize: Harlequin - Ship Type: P - How taken: HMS Bulwark - When taken: 23 Nov 1814 - Where taken: off Halifax - Date Received: 27 Dec 1814 - From what ship: HMT Penelope - Born: Barrington - Age: 21 - Released on 3 Jul 1815.

Elm, John - Seaman - Nbr: 271 - Prize: William Bayard - Ship Type: MV - How taken: HMS Warspite - When taken: 3 Mar 1813 - Where taken: Bay of Biscay - Date Received: 28 Jun 1813 - From what ship: Plymouth - Born: New York - Age: 14 - Sent to Plymouth on 8 Jul 1814.

Elmick, Joseph - Seaman - Nbr: 3788 - Prize: James - Ship Type: Prize - How taken: Rebecca (LM) - When taken: 13 Apr 1814 - Where taken: off Delaware - Date Received: 5 Oct 1814 - From what ship: HMT President, Halifax - Born: Philadelphia - Age: 25 - Released on 9 Jun 1815.

Elvell, Robert - Seaman - Nbr: 1678 - Prize: Polly - Ship Type: P - How taken: HMS Barbados - When taken: 2 Apr 1814 - Where taken: off San Domingo - Date Received: 2 Jul 1814 - From what ship: Plymouth - Born: Cape Ann - Age: 59 - Released on 1 May 1815.

Elves, Joseph - Seaman - Nbr: 1122 - Prize: Hope - Ship Type: P - How taken: HMS Sea Horse - When taken: 22 Mar 1814 - Where taken: at sea - Date Received: 10 May 1814 - From what ship: Plymouth - Born: Lisbon - Age: 20 - Released on 28 Apr 1815.

Elves, Manuel - Seaman - Nbr: 1121 - Prize: Hope - Ship Type: P - How taken: HMS Sea Horse - When taken: 22 Mar 1814 - Where taken: at sea - Date Received: 10 May 1814 - From what ship: Plymouth - Born: Lisbon - Age: 23 - Released on 28 Apr 1815.

Elvin, John - Seaman - Nbr: 2458 - Prize: US Sloop Frolic - Ship Type: MW - How taken: HMS Orpheus - When taken: 20 Apr 1814 - Where taken: off Cuba - Date Received: 16 Aug 1814 - From what ship: HMT Queen, Halifax - Born: Maryland - Age: 20 - Released on 3 May 1815.

Elwell, Lewis - Master - Nbr: 4097 - Prize: Harford - Ship Type: MV - How taken: HMS Junon - When taken: 6 Jun 1814 - Where taken: off Cape Ann - Date Received: 6 Oct 1814 - From what ship: HMT Chesapeake, Halifax - Born: Massachusetts - Age: 31 - Released on 13 Jun 1815.

Elwell, Samuel - Seaman - Nbr: 3540 - Prize: Hawk - Ship Type: P - How taken: HMS Pique - When taken: 26 Apr 1814 - Where taken: off Bermuda - Date Received: 30 Sep 1814 - From what ship: HMT Sybella - Born: Gloucester - Age: 29 - Released on 4 Jun 1815.

Ely, Abraham - Seaman - Nbr: 3131 - How taken: Impressed from MV Thames - When taken: 8 Sep 1811 - Date Received: 11 Sep 1814 - From what ship: HMT Freya, Chatham - Born: Baltimore - Age: 29 - Released on 27 Apr 1815.

Ely, Daniel - Seaman - Nbr: 5830 - Prize: US Schooner Tigress - Ship Type: MW - How taken: British gunboats - When taken: 2 Sep 1814 - Where taken: Lake Erie - Date Received: 26 Dec 1814 - From what ship: HMT Argo - Born: Pennsylvania - Age: 34 - Released on 3 Jul 1815.

Emas, Comfort - Seaman - Nbr: 4964 - Prize: Herald - Ship Type: P - How taken: HMS Endymion - When taken: 15 Aug 1814 - Where taken: off Nantucket - Date Received: 28 Oct 1814 - From what ship: HMT Alkbar, Halifax - Born: Connecticut - Age: 67 - Released on 19 May 1815.

Emerson, David - Seaman - Nbr: 4441 - Prize: Growler - Ship Type: P - How taken: HMS Electra - When taken: 7 Jul 1814 - Where taken: at sea - Date Received: 8 Oct 1814 - From what ship: HMT Leyden, Chatham - Born: Massachusetts - Age: 22 - Released on 14 Jun 1815.

Emery, John - Seaman - Nbr: 6017 - Prize: McDonough - Ship Type: P - How taken: HMS Bacchante - When taken:

American Prisoners of War Held at Dartmoor during the War of 1812

## Alphabetical listing of names

1 Nov 1814 - Where taken: Lat 42 Long 67 - Date Received: 28 Dec 1814 - From what ship: HMT Penelope - Born: Biddeford - Age: 23 - Released on 3 Jul 1815.

Emlyn, Isaac - Seaman - Nbr: 5992 - Prize: McDonough - Ship Type: P - How taken: HMS Bacchante - When taken: 1 Nov 1814 - Where taken: Lat 42 Long 67 - Date Received: 27 Dec 1814 - From what ship: HMT Penelope - Born: Massachusetts - Age: 19 - Released on 3 Jul 1815.

Emming, Thomas - Seaman - Nbr: 2839 - How taken: Gave himself up from HMS Armide - When taken: 8 Jun 1813 - Date Received: 24 Aug 1814 - From what ship: HMT Liverpool, Chatham - Born: New York - Age: 42 - Race: Black - Released on 19 May 1815.

Enderson, James - Seaman - Nbr: 1823 - How taken: Gave himself up from Greenock - When taken: 22 Jan 1813 - Date Received: 21 Jul 1814 - From what ship: HMT Redbeard & Pincher, Chatham Depot - Born: New York - Age: 29 - Released on 1 May 1815.

Englis, David - Seaman - Nbr: 1451 - How taken: Impressed from MV Nile - When taken: 15 Mar 1813 - Date Received: 19 Jun 1814 - From what ship: Stapleton - Born: Chester - Age: 30 - Released on 28 Apr 1815.

English, Edward - Seaman - Nbr: 1426 - Prize: Revenge - Ship Type: LM - How taken: HMS Belle Poule - When taken: 10 May 1814 - Where taken: off Cornwall - Date Received: 19 Jun 1814 - From what ship: Stapleton - Born: New York - Age: 25 - Released on 28 Apr 1815.

Enoch, Joseph - Seaman - Nbr: 1758 - How taken: Gave himself up from HMS Badger - When taken: 15 Nov 1812 - Date Received: 20 Jul 1814 - From what ship: HMS Milford, Plymouth - Born: New Jersey - Age: 32 - Released on 26 Apr 1815.

Erighton, Manuel - Seaman - Nbr: 1015 - Prize: Rachel & Ann - Ship Type: P - How taken: HMS Cydnus - When taken: 6 Jan 1814 - Where taken: at sea - Date Received: 31 Jan 1814 - From what ship: Plymouth - Born: Philadelphia - Age: 27 - Released on 27 Apr 1815.

Erlstroom, John - Seaman - Nbr: 283 - Prize: Cannoneer - Ship Type: MV - How taken: HMS Warspite - When taken: 14 Mar 1813 - Where taken: Bay of Biscay - Date Received: 28 Jun 1813 - From what ship: Plymouth - Born: Sweden - Age: 36 - Released on 27 Dec 1813 (Swedish citizen).

Erwin, Elijah - Carpenter - Nbr: 371 - How taken: Impressed at sea - When taken: 29 Mar 1813 - Date Received: 1 Jul 1813 - From what ship: Plymouth - Born: Norwich - Age: 31 - Released on 20 Apr 1815.

Erwin, William - Seaman - Nbr: 182 - Prize: Star - Ship Type: MV - How taken: HMS Superb - When taken: 9 Feb 1813 - Where taken: Bay of Biscay - Date Received: 2 Apr 1813 - From what ship: Plymouth - Born: Cumberland - Age: 33 - Died on 14 Mar 1815 from mania.

Esdale, Thomas - Seaman - Nbr: 735 - Prize: Shadow - Ship Type: LM - How taken: HMS Reindeer - When taken: 6 Apr 1813 - Where taken: Bay of Biscay - Date Received: 3 Nov 1813 - From what ship: Plymouth - Born: Burlington - Age: 47 - Released on 26 Apr 1815.

Eskinson, George - Seaman - Nbr: 677 - Prize: Matilda - Ship Type: MV - How taken: HMS Revolutionnaire - When taken: 27 Jul 1813 - Where taken: off L'Orient (France) - Date Received: 8 Sep 1813 - From what ship: Plymouth - Born: Cape Ann - Age: 19 - Released on 26 Apr 1815.

Esterstone, John - Seaman - Nbr: 3909 - Prize: Thorn - Ship Type: MV - How taken: HMS Bulwark - When taken: 9 Jul 1814 - Where taken: off Nantucket - Date Received: 5 Oct 1814 - From what ship: HMT President, Halifax - Born: Sweden - Age: 32 - Released on 9 Jun 1815.

Evans, Benjamin - Seaman - Nbr: 5032 - Prize: Landrail - Ship Type: Prize - How taken: HMS Wasp - When taken: 27 Jul 1814 - Where taken: Georges Bank - Date Received: 28 Oct 1814 - From what ship: HMT Alkbar, Halifax - Born: New Hampshire - Age: 25 - Released on 21 Jun 1815.

Evans, Edward - Seaman - Nbr: 4853 - Prize: North Star - Ship Type: MV - How taken: HMS Crane - When taken: 10 Jun 1814 - Where taken: off St. Bartholomew - Date Received: 9 Oct 1814 - From what ship: HMT Freya, Halifax - Born: Virginia - Age: 27 - Died on 5 Jan 1815 from peuismeounica.

Evans, Henry - Seaman - Nbr: 17 - Prize: Cashier - Ship Type: LM - How taken: HMS Reindeer - When taken: 3 Feb 1813 - Where taken: Bay of Biscay - Date Received: 2 Apr 1813 - From what ship: Plymouth - Born: Philadelphia - Age: 29 - Sent to Dartmouth on 30 Jul 1813.

Evans, Hezekiah - Seaman - Nbr: 4558 - How taken: Gave himself up from HMS Progich - When taken: 15 Aug 1813 - Date Received: 8 Oct 1814 - From what ship: HMT Leyden, Chatham - Born: Newport - Age: 30 - Released on 15 Jun 1815.

Evans, Jacob - Prize Master - Nbr: 308 - Prize: King George - Ship Type: MV - How taken: HMS Piercer - When taken: 9 Mar 1813 - Where taken: Isle of Bas - Date Received: 28 Jun 1813 - From what ship: Plymouth - Born: Baltimore - Age: 22 - Released on 15 Apr 1815.

Evans, James - Seaman - Nbr: 1807 - How taken: Gave himself up from HMS Foxhound - When taken: 22 Feb 1813

American Prisoners of War Held at Dartmoor during the War of 1812

## Alphabetical listing of names

- Date Received: 20 Jul 1814 - From what ship: HMS Milford, Plymouth - Born: Charlestown - Age: 30 - Released on 28 Apr 1815.

Evans, James - Seaman - Nbr: 111 - How taken: Gave himself up from HMS Foxhound - Date Received: 2 Apr 1813 - From what ship: Plymouth - Born: Charlestown - Age: 30 - Sent to Chatham on 27 May 1813.

Evans, Moses - Seaman - Nbr: 1560 - Prize: Caroline - Ship Type: MV - How taken: HMS Medusa - When taken: 12 Apr 1813 - Where taken: Bay of Biscay - Date Received: 23 Jun 1814 - From what ship: Stapleton - Born: Madison County - Age: 26 - Race: Negro - Released on 1 May 1815.

Evans, Piers - Seaman - Nbr: 5876 - Prize: Lion - Ship Type: P - How taken: HMS Granicus - When taken: 2 Dec 1814 - Where taken: off Lisbon - Date Received: 26 Dec 1814 - From what ship: HMT Impregnable - Born: L'Orient (France) - Age: 24 - Released on 3 Jul 1815.

Evans, Robert - Master - Nbr: 1683 - Prize: Polly - Ship Type: P - How taken: HMS Barbados - When taken: 2 Apr 1814 - Where taken: off San Domingo - Date Received: 2 Jul 1814 - From what ship: Plymouth - Born: Derby - Age: 29 - Released on 1 May 1815.

Evans, Thomas - Seaman - Nbr: 6281 - How taken: Gave himself up from HMS Prince of Wales - Date Received: 17 Feb 1815 - From what ship: HMT Ganges, Plymouth - Born: Virginia - Age: 33 - Released on 1 Jul 1815.

Evans, William - Seaman - Nbr: 5377 - Prize: James - Ship Type: MV - How taken: HMS Harpy - When taken: 18 Dec 1813 - Where taken: off Isle of France - Date Received: 31 Oct 1814 - From what ship: HMT Leyden, Chatham - Born: Baltimore - Age: 40 - Released on 27 Apr 1815.

Evans, William - Seaman - Nbr: 4863 - Prize: Grecian - Ship Type: LM - How taken: HMS Jaseur - When taken: 2 May 1814 - Where taken: Chesapeake Bay - Date Received: 9 Oct 1814 - From what ship: HMT Freya, Halifax - Born: Pennsylvania - Age: 23 - Released on 21 Jun 1815.

Evelish, William - Seaman - Nbr: 1428 - Prize: Revenge - Ship Type: LM - How taken: HMS Belle Poule - When taken: 10 May 1814 - Where taken: off Cornwall - Date Received: 19 Jun 1814 - From what ship: Stapleton - Born: Providence - Age: 17 - Released on 28 Apr 1815.

Everard, Joseph - Seaman - Nbr: 3335 - Prize: York Town - Ship Type: P - How taken: HMS Nimrod - When taken: 17 Jul 1813 - Where taken: off St. Johns - Date Received: 13 Sep 1814 - From what ship: HMT Niobe, Chatham - Born: New York - Age: 36 - Released on 28 May 1815.

Everley, John - Seaman - Nbr: 330 - Prize: Pert - Ship Type: MV - How taken: HMS Warspite - When taken: 1 Mar 1813 - Where taken: off Basque Roads (France) - Date Received: 28 Jun 1813 - From what ship: Plymouth - Born: Pennsylvania - Age: 19 - Released on 20 Apr 1815.

Ewell, Edward - Seaman - Nbr: 4215 - Prize: Devon - Ship Type: Prize - How taken: HMS Fly - When taken: 2 Jan 1814 - Where taken: at sea - Date Received: 7 Oct 1814 - From what ship: HMT Niobe, Chatham - Born: Norfolk - Age: 21 - Released on 20 Mar 1815.

Ewer, David - Seaman - Nbr: 5588 - Prize: Young William - Ship Type: Prize - How taken: HMS Plover - When taken: 8 Sep 1814 - Where taken: off St. Johns - Date Received: 17 Dec 1814 - From what ship: HMT Loire, Halifax - Born: Nantucket - Age: 36 - Released on 1 Jul 1815.

Ewing, Thomas - Seaman - Nbr: 2567 - Prize: Rattlesnake - Ship Type: P - How taken: HMS Hyperion - When taken: 25 Jun 1814 - Where taken: off Cape Finisterre - Date Received: 21 Aug 1814 - From what ship: HMS Hyperion - Born: Philadelphia - Age: 21 - Released on 3 May 1815.

Eyers, John - Seaman - Nbr: 1238 - How taken: Sent into custody from HMS Alemine - When taken: 25 Jan 1813 - Date Received: 14 Jun 1814 - From what ship: Mill Prison (Plymouth, England) - Born: Cambridge - Age: 26 - Released on 26 Apr 1815.

Fabin, James - Seaman - Nbr: 3739 - Prize: Lizard - Ship Type: P - How taken: HMS Prometheus - When taken: 5 May 1814 - Where taken: off Halifax - Date Received: 30 Sep 1814 - From what ship: HMT President, Halifax - Born: Salem - Age: 25 - Released on 4 Jun 1815.

Fackney, John - Seaman - Nbr: 2172 - Prize: Rattlesnake - Ship Type: P - How taken: HMS Hyperion - When taken: 26 Jun 1814 - Where taken: off Cape Finisterre - Date Received: 16 Aug 1814 - From what ship: HMS Dublin, Halifax - Born: Maryland - Age: 25 - Released on 2 May 1815.

Fadden, Charles - Boy - Nbr: 4643 - Prize: Thomas - Ship Type: MV - How taken: HMS Frolic - When taken: 29 Jun 1813 - Where taken: off St. Thomas - Date Received: 9 Oct 1814 - From what ship: HMT Leyden, Chatham - Born: New York - Age: 16 - Released on 15 Jun 1815.

Fadden, John - Seaman - Nbr: 636 - Prize: US Brig Argus - Ship Type: MW - How taken: HMS Pelican - When taken: 14 Aug 1813 - Where taken: Irish Channel - Date Received: 8 Sep 1813 - From what ship: Plymouth - Born: Hampton - Age: 38 - Sent to Dartmouth on 2 Nov 1814.

Fairfield, Asa - Seaman - Nbr: 6000 - Prize: McDonough - Ship Type: P - How taken: HMS Bacchante - When

American Prisoners of War Held at Dartmoor during the War of 1812

## Alphabetical listing of names

taken: 1 Nov 1814 - Where taken: Lat 42 Long 67 - Date Received: 28 Dec 1814 - From what ship: HMT Penelope - Born: Massachusetts - Age: 17 - Released on 28 Apr 1815.

Fairfield, James - Seaman - Nbr: 5678 - Prize: McDonough - Ship Type: P - How taken: HMS Bacchante - When taken: 7 Nov 1814 - Where taken: Lat 42 Long 67 - Date Received: 24 Dec 1814 - From what ship: HMT Penelope - Born: Arundel - Age: 31 - Released on 28 Apr 1815.

Falcon, J. B. - Seaman - Nbr: 4813 - Prize: President - Ship Type: P - How taken: HMS Pique - When taken: 7 May 1814 - Where taken: off Porto Rico - Date Received: 9 Oct 1814 - From what ship: HMT Freya, Halifax - Born: Cartagena - Age: 40 - Race: Black - Released on 21 Jun 1815.

Falkern, W. - Seaman - Nbr: 4331 - Prize: Flash - Ship Type: LM - How taken: HMS Acasta and HMS Loire - When taken: 8 Feb 1814 - Where taken: off NY - Date Received: 7 Oct 1814 - From what ship: HMT Salvador del Mundo, Halifax - Born: New York - Age: 20 - Race: Negro - Released on 14 Jun 1815.

Fall, James - Seaman - Nbr: 4463 - Prize: Governor Plumer - Ship Type: P - How taken: HMS Shamrock - When taken: 4 Mar 1813 - Where taken: at sea - Date Received: 8 Oct 1814 - From what ship: HMT Leyden, Chatham - Born: Massachusetts - Age: 33 - Released on 15 Jun 1815.

Fardell, Thomas - Seaman - Nbr: 738 - Prize: Minerva - Ship Type: MV - How taken: HMS Goldfinch - When taken: 27 Jun 1813 - Where taken: off Nantes - Date Received: 3 Nov 1813 - From what ship: Plymouth - Born: Virginia - Age: 30 - Released on 26 Apr 1815.

Farford, Samuel - Master's mate - Nbr: 5464 - Prize: General Putnam - Ship Type: P - How taken: HMS Leander - When taken: 8 Nov 1814 - Where taken: Long 65 Lat 42 - Date Received: 17 Dec 1814 - From what ship: HMT Loire, Halifax - Born: Salem - Age: 33 - Released on 1 Jul 1815.

Fargo, Elijah - Seaman - Nbr: 2128 - How taken: Gave himself up from HMS Rolus - When taken: 10 Oct 1812 - Date Received: 8 Aug 1814 - From what ship: HMT Raven, Chatham - Born: New London - Age: 25 - Released on 26 Apr 1815.

Farley, John - Seaman - Nbr: 1840 - How taken: Gave himself up from HMS Clarence - When taken: 7 May 1813 - Date Received: 21 Jul 1814 - From what ship: HMT Redbeard & Pincher, Chatham Depot - Born: Baltimore - Age: 33 - Released on 2 May 1815.

Farmer, George - Cook - Nbr: 1300 - Prize: John & Frances - Ship Type: Prize - How taken: HMS Sterling Castle - When taken: 10 May 1814 - Where taken: off Cape Clear - Date Received: 14 Jun 1814 - From what ship: Mill Prison (Plymouth, England) - Born: New Jersey - Age: 30 - Race: Black - Released on 28 Apr 1815.

Farmer, Joseph - Seaman - Nbr: 2867 - How taken: Impressed at London - When taken: 23 Jul 1813 - Date Received: 24 Aug 1814 - From what ship: HMT Alpheus, Chatham - Born: Salem - Age: 34 - Race: Black - Released on 19 May 1815.

Farmer, Thomas - Seaman - Nbr: 6089 - Prize: Black Swan - Ship Type: MV - How taken: HMS Maidstone - When taken: 24 Oct 1814 - Where taken: Georges Bank - Date Received: 28 Dec 1814 - From what ship: HMT Penelope - Born: Plymouth - Age: 20 - Released on 20 Jul 1815.

Farquhar, John - Seaman - Nbr: 5936 - Prize: Harlequin - Ship Type: P - How taken: HMS Bulwark - When taken: 23 Nov 1814 - Where taken: off Halifax - Date Received: 27 Dec 1814 - From what ship: HMT Penelope - Born: Baltimore - Age: 44 - Released on 3 Jul 1815.

Farrell, Andrew - Seaman - Nbr: 4636 - How taken: Gave himself up from HMS Clarence - Date Received: 9 Oct 1814 - From what ship: HMT Leyden, Chatham - Born: Florida - Age: 25 - Released on 15 Jun 1815.

Farrell, James - Seaman - Nbr: 1767 - Prize: Cygnet - Ship Type: Prize - How taken: HMS Statira - When taken: 8 May 1814 - Where taken: at sea - Date Received: 20 Jul 1814 - From what ship: HMS Milford, Plymouth - Born: Philadelphia - Age: 20 - Released on 1 May 1815.

Farrell, John - Boy - Nbr: 6486 - Prize: John - Ship Type: MV - How taken: Variable - When taken: 11 Aug 1814 - Where taken: off Cuba - Date Received: 3 Mar 1815 - From what ship: HMT Ganges, Plymouth - Born: New Orleans - Age: 29 - Race: Mulatto - Released on 11 Jul 1815.

Farrell, John - Seaman - Nbr: 418 - Prize: Viper - Ship Type: MV - How taken: HMS Superb - When taken: 15 Apr 1813 - Where taken: Bay of Biscay - Date Received: 1 Jul 1813 - From what ship: Plymouth - Born: New Orleans - Age: 29 - Released on 20 Apr 1815.

Farrell, Richard - Seaman - Nbr: 5824 - Prize: US Schooner Tigress - Ship Type: MW - How taken: British gunboats - When taken: 2 Sep 1814 - Where taken: Lake Erie - Date Received: 26 Dec 1814 - From what ship: HMT Argo - Born: Baltimore - Age: 23 - Released on 11 Jul 1815.

Farrett, George - Seaman - Nbr: 310 - Prize: King George - Ship Type: MV - How taken: HMS Piercer - When taken: 9 Mar 1813 - Where taken: Isle of Bas - Date Received: 28 Jun 1813 - From what ship: Plymouth - Born: Georgetown - Age: 34 - Released on 20 Apr 1815.

American Prisoners of War Held at Dartmoor during the War of 1812

## Alphabetical listing of names

Farrol, Francis - Seaman - Nbr: 120 - Prize: Governor McKean - Ship Type: LM - How taken: HMS Rover - When taken: 26 Jan 1813 - Where taken: off Bordeaux - Date Received: 2 Apr 1813 - From what ship: Plymouth - Born: Philadelphia - Age: 23 - Sent to Dartmouth on 30 Jul 1813.

Favish, John - Seaman - Nbr: 6480 - Prize: Mary - Ship Type: P - How taken: Lapphin - When taken: 15 Jun 1814 - Where taken: off Saint Marys - Date Received: 3 Mar 1815 - From what ship: HMT Ganges, Plymouth - Born: Virginia - Age: 15 - Released on 11 Jul 1815.

Fawcett, William - Seaman - Nbr: 152 - How taken: Gave himself up from HMS Lavinia - Date Received: 2 Apr 1813 - From what ship: Plymouth - Born: Peterborough - Age: 29 - Released on 20 Apr 1815.

Faye, Samuel - Seaman - Nbr: 1543 - Prize: Zebra - Ship Type: LM - How taken: HMS Pyramus - When taken: 20 Apr 1813 - Where taken: Bay of Biscay - Date Received: 23 Jun 1814 - From what ship: Stapleton - Born: Massachusetts - Age: 30 - Released on 1 May 1815.

Fedie, John - Seaman - Nbr: 653 - Prize: Marmion - Ship Type: MV - How taken: HMS President - When taken: 14 Aug 1813 - Where taken: off Nantes - Date Received: 8 Sep 1813 - From what ship: Plymouth - Born: Charlestown - Age: 23 - Race: Mulatto - Released on 26 Apr 1815.

Fedy, Philip - Seaman - Nbr: 6160 - Prize: Lion - Ship Type: P - How taken: HMS Granicus - When taken: 2 Dec 1814 - Where taken: off Lisbon - Date Received: 18 Jan 1815 - From what ship: HMT Impregnable - Born: New Orleans - Age: 23 - Race: Mulatto - Released on 5 Jul 1815.

Feilman, John - Seaman - Nbr: 4771 - Prize: James - Ship Type: MV - How taken: HMS Harpy - When taken: 18 Dec 1813 - Where taken: off Isle of France - Date Received: 9 Oct 1814 - From what ship: HMT Freya, Chatham - Born: Boston - Age: 26 - Released on 27 Apr 1815.

Fellebrown, William - Seaman - Nbr: 2744 - How taken: Gave himself up from HMS Ceres - When taken: 20 Dec 1812 - Date Received: 24 Aug 1814 - From what ship: HMT Liverpool, Chatham - Born: Boston - Age: 26 - Released on 26 Apr 1815.

Fellows, Isaac - Seaman - Nbr: 1679 - Prize: Christiana - Ship Type: MV - How taken: Found at sea by an English convoy in Gulf of Florida - Date Received: 2 Jul 1814 - From what ship: Plymouth - Born: Ipswich - Age: 43 - Released on 1 May 1815.

Fellows, Nathaniel - Captain's clerk - Nbr: 3942 - Prize: Rolla - Ship Type: P - How taken: HMS Loire - When taken: 10 Dec 1813 - Where taken: off Bull Island (South Carolina) - Date Received: 5 Oct 1814 - From what ship: HMT President, Halifax - Born: Connecticut - Age: 25 - Released on 9 Jun 1815.

Felt, George - 2nd Mate - Nbr: 4768 - Prize: Rambler - Ship Type: MV - How taken: Morley (Transport) - When taken: 10 Feb 1813 - Where taken: off Isle of France - Date Received: 9 Oct 1814 - From what ship: HMT Freya, Chatham - Born: Salem - Age: 32 - Released on 15 Jun 1815.

Felt, John - Seaman - Nbr: 2399 - Prize: US Sloop Frolic - Ship Type: MW - How taken: HMS Orpheus - When taken: 20 Apr 1814 - Where taken: off Cuba - Date Received: 16 Aug 1814 - From what ship: HMT Queen, Halifax - Born: Salem - Age: 40 - Sent to Dartmouth on 19 Oct 1814.

Felton, John - Seaman - Nbr: 5807 - Prize: US Schooner Scorpion - Ship Type: MW - How taken: British gunboats - When taken: 6 Sep 1814 - Where taken: Lake Erie - Date Received: 26 Dec 1814 - From what ship: HMT Argo - Born: Massachusetts - Age: 31 - Released on 3 Jul 1815.

Fenderson, Nathaniel - Seaman - Nbr: 4915 - Prize: US Brig Rattlesnake - Ship Type: MW - How taken: HMS Leander - When taken: 13 Jul 1814 - Where taken: off Shelburne - Date Received: 28 Oct 1814 - From what ship: HMT Alkbar, Halifax - Born: Massachusetts - Age: 28 - Released on 21 Jun 1815.

Fenhouse, John - Seaman - Nbr: 3460 - Prize: Prize to Chasseur - How taken: HMS Tartarus - When taken: 1 Sep 1814 - Where taken: Atlantic - Date Received: 19 Sep 1814 - From what ship: HMT Salvador del Mundo - Born: Breda, Holland - Age: 20 - Released on 4 Jun 1815.

Ferguson, John - Seaman - Nbr: 3145 - Prize: Francis - Ship Type: MV - How taken: HMS Surinam - When taken: 19 Aug 1813 - Where taken: at sea - Date Received: 11 Sep 1814 - From what ship: HMT Freya, Chatham - Born: Carolina - Age: 45 - Released on 28 May 1815.

Ferguson, Thomas - Seaman - Nbr: 3314 - Prize: Globe - Ship Type: P - How taken: HMS Victorious - When taken: 8 Jun 1813 - Where taken: Chesapeake Bay - Date Received: 13 Sep 1814 - From what ship: HMT Niobe, Chatham - Born: Baltimore - Age: 25 - Released on 28 May 1815.

Fernald, John - Seaman - Nbr: 1784 - How taken: Gave himself up from HMS Leyden - When taken: 23 Mar 1813 - Date Received: 20 Jul 1814 - From what ship: HMS Milford, Plymouth - Born: Massachusetts - Age: 27 - Released on 1 May 1815.

Fernald, William - Prize master - Nbr: 6112 - Prize: Garland - Ship Type: Prize - How taken: HMS Hamadryad - When taken: 22 Nov 1814 - Where taken: Georges Bank - Date Received: 6 Jan 1814 - From what ship:

American Prisoners of War Held at Dartmoor during the War of 1812

## Alphabetical listing of names

HMT Impregnable - Born: Kittery - Age: 24 - Died on 23 Jan 1815 from febris.

Fernald, William - Seaman - Nbr: 4471 - Prize: Governor Plumer - Ship Type: P - How taken: HMS Shamrock - When taken: 4 Mar 1813 - Where taken: at sea - Date Received: 8 Oct 1814 - From what ship: HMT Leyden, Chatham - Born: New Hampshire - Age: 20 - Released on 15 Jun 1815.

Fernandez, George - Seaman - Nbr: 6348 - Prize: Prince de Neufchatel - Ship Type: P - How taken: Leander (Newcastle Acasta) - When taken: 20 Dec 1814 - Where taken: Lat 38 Long 56 - Date rec'd: 19 Feb 1815 - What ship: HMT Ganges, Plymouth - Born: Brazil - Age: 24 - Race: Mulatto - Released on 5 Jul 1815.

Ferrin, Daniel - Seaman - Nbr: 6009 - Prize: McDonough - Ship Type: P - How taken: HMS Bacchante - When taken: 1 Nov 1814 - Where taken: Lat 42 Long 67 - Date Received: 28 Dec 1814 - From what ship: HMT Penelope - Born: Massachusetts - Age: 22 - Released on 3 Jul 1815.

Ferris, Jacob - Seaman - Nbr: 1914 - How taken: Gave himself up from HMS Mars - When taken: 9 Dec 1812 - Date Received: 3 Aug 1814 - From what ship: HMS Alceste, Chatham Depot - Born: New York - Age: 33 - Released on 26 Apr 1815.

Ferris, James - Seaman - Nbr: 3093 - How taken: Gave himself up from HMS Invincible - When taken: 14 Jan 1813 - Date Received: 3 Sep 1814 - From what ship: HMT Hydra, Chatham - Born: Cambridge - Age: 25 - Race: Black - Released on 28 May 1815.

Ferris, Jonathan - Seaman - Nbr: 2568 - Prize: Rattlesnake - Ship Type: P - How taken: HMS Hyperion - When taken: 25 Jun 1814 - Where taken: off Cape Finisterre - Date Received: 21 Aug 1814 - From what ship: HMS Hyperion - Born: Barnstable - Age: 29 - Released on 3 May 1815.

Ferris, Joseph - Seaman - Nbr: 1020 - Prize: Rachel & Ann - Ship Type: P - How taken: HMS Cydnus - When taken: 6 Jan 1814 - Where taken: at sea - Date Received: 31 Jan 1814 - From what ship: Plymouth - Born: New York - Age: 25 - Released on 27 Apr 1815.

Ferry, William - Seaman - Nbr: 6128 - Prize: Betsey - Ship Type: Prize - How taken: HMS Bellerophon - When taken: 2 Nov 1814 - Where taken: Long 61 - Date Received: 6 Jan 1814 - From what ship: HMT Impregnable - Born: Long Island - Age: 23 - Released on 11 Jul 1815.

Fetch, Theodore - 3rd Lieutenant - Nbr: 2331 - Prize: Snap Dragon - Ship Type: P - How taken: HMS Martin - When taken: 10 Jun 1814 - Where taken: off Halifax - Date Received: 16 Aug 1814 - From what ship: HMS Dublin, Halifax - Born: Connecticut - Age: 23 - Released on 3 May 1815.

Fetters, Robert - Seaman - Nbr: 6123 - Prize: Betsey - Ship Type: Prize - How taken: HMS Bellerophon - When taken: 2 Nov 1814 - Where taken: Long 61 - Date Received: 6 Jan 1814 - From what ship: HMT Impregnable - Born: Bath - Age: 22 - Race: Mulatto - Released on 5 Jul 1815.

Fields, Alexander - Seaman - Nbr: 1851 - How taken: Gave himself up from HMS Unicorn - When taken: 17 Jun 1812 - Date Received: 21 Jul 1814 - From what ship: HMT Redbeard & Pincher, Chatham Depot - Born: Charlestown - Age: 28 - Released on 26 Apr 1815.

Fields, Charles - Seaman - Nbr: 6215 - Prize: Prince de Neufchatel - Ship Type: P - How taken: Leander (Newcastle Acasta) - When taken: 20 Dec 1814 - Where taken: Lat 38 Long 56 - Date Received: 30 Jan 1815 - From what ship: HMT Pheasant - Born: Quincy - Age: 27 - Released on 5 Jul 1815.

Fields, George - Seaman - Nbr: 2411 - Prize: US Sloop Frolic - Ship Type: MW - How taken: HMS Orpheus - When taken: 20 Apr 1814 - Where taken: off Cuba - Date Received: 16 Aug 1814 - From what ship: HMT Queen, Halifax - Born: Portsmouth - Age: 24 - Sent to Dartmouth on 19 Oct 1814.

Fields, Jacob - Seaman - Nbr: 4270 - Prize: Vivid - Ship Type: P - How taken: HMS Nymphe - When taken: 20 Apr 1813 - Where taken: off Cape Cod - Date Received: 7 Oct 1814 - From what ship: HMT Niobe, Chatham - Born: Quincy - Age: 20 - Released on 13 Jun 1815.

Fields, Michael - Seaman - Nbr: 4983 - Prize: Invincible - Ship Type: LM - How taken: HMS Armide - When taken: 15 Aug 1814 - Where taken: off Nantucket - Date Received: 28 Oct 1814 - From what ship: HMT Alkbar, Halifax - Born: Portsmouth - Age: 27 - Released on 8 May 1815.

Fields, Robert - Seaman - Nbr: 6110 - Prize: Garland - Ship Type: Prize - How taken: HMS Hamadryad - When taken: 22 Nov 1814 - Where taken: Georges Bank - Date Received: 6 Jan 1814 - From what ship: HMT Impregnable - Born: Portsmouth - Age: 27 - Released on 11 Jul 1815.

Fields, Samuel - Seaman - Nbr: 2424 - Prize: US Sloop Frolic - Ship Type: MW - How taken: HMS Orpheus - When taken: 20 Apr 1814 - Where taken: off Cuba - Date Received: 16 Aug 1814 - From what ship: HMT Queen, Halifax - Born: Salem - Age: 30 - Released on 3 May 1815.

Fife, Thomas - Seaman - Nbr: 5378 - Prize: James - Ship Type: MV - How taken: HMS Harpy - When taken: 18 Dec 1813 - Where taken: off Isle of France - Date Received: 31 Oct 1814 - From what ship: HMT Leyden, Chatham - Born: Philadelphia - Age: 25 - Released on 27 Apr 1815.

American Prisoners of War Held at Dartmoor during the War of 1812

## Alphabetical listing of names

Figasson, Henry - Seaman - Nbr: 605 - Prize: US Brig Argus - Ship Type: MW - How taken: HMS Pelican - When taken: 14 Aug 1813 - Where taken: Irish Channel - Date Received: 8 Sep 1813 - From what ship: Plymouth - Born: Philadelphia - Age: 26 - Sent to Dartmouth on 2 Nov 1814.

Filch, John - Seaman - Nbr: 2904 - How taken: Gave himself up from HMS Ministrel - When taken: 5 Jul 1813 - Date Received: 24 Aug 1814 - From what ship: HMT Alpheus, Chatham - Born: New Hampshire - Age: 22 - Released on 19 May 1815.

Finch, Abraham - Seaman - Nbr: 6507 - How taken: Apprehended at Kingston - Date Received: 3 Mar 1815 - From what ship: HMT Ganges, Plymouth - Born: New Hampshire - Age: 26 - Released on 11 Jul 1815.

Finch, Etienne - Seaman - Nbr: 6491 - Prize: Louisa - Ship Type: Prize - How taken: HMS Dasher - When taken: 12 Feb 1814 - Where taken: off America - Date Received: 3 Mar 1815 - From what ship: HMT Ganges, Plymouth - Born: Bordeaux - Age: 18 - Released on 11 Jul 1815.

Finch, William - Seaman - Nbr: 249 - Prize: Union - Ship Type: LM - How taken: HMS Iris - When taken: 17 Jun 1813 - Where taken: Lat 44, Long 7 - Date Received: 2 Apr 1813 - From what ship: Plymouth - Born: Orange County - Age: 25 - Sent to Dartmouth on 30 Jul 1813.

Findley, Thomas - Seaman - Nbr: 4493 - Prize: Enterprize - Ship Type: P - How taken: HMS Tenedos - When taken: 21 May 1813 - Where taken: off Cape Cod - Date Received: 8 Oct 1814 - From what ship: HMT Leyden, Chatham - Born: Massachusetts - Age: 18 - Released on 15 Jun 1815.

Fingall, William - Seaman - Nbr: 2707 - Prize: Dick - Ship Type: MV - How taken: HMS Dispatch - When taken: 15 Mar 1813 - Where taken: near Corcova Lights - Date Received: 21 Aug 1814 - From what ship: HMT Freya, Chatham - Born: Philadelphia - Age: 25 - Released on 19 May 1815.

Fink, Jonas - Seaman - Nbr: 1643 - Prize: Paul Jones - Ship Type: P - How taken: HMS Leonidas - When taken: 23 May 1813 - Where taken: Channel - Date Received: 23 Jun 1814 - From what ship: Stapleton - Born: Hesse-Cassel - Age: 26 - Released on 5 Jul 1814.

Finn, John - Seaman - Nbr: 4148 - Prize: Requin - Ship Type: LM - How taken: HMS Venus - When taken: 6 Mar 1814 - Where taken: off Bordeaux - Date Received: 6 Oct 1814 - From what ship: HMT Niobe, Chatham - Born: Boston - Age: 19 - Released on 13 Jun 1815.

Finney, James - Seaman - Nbr: 6295 - How taken: Impressed at London - Date Received: 17 Feb 1815 - From what ship: HMT Ganges, Plymouth - Born: Sandwich - Age: 31 - Released on 11 Jul 1815.

Fish, Joseph - Seaman - Nbr: 1480 - Prize: Hebe - Ship Type: MV - How taken: HMS Stag - When taken: 18 Apr 1813 - Where taken: Bay of Biscay - Date Received: 19 Jun 1814 - From what ship: Stapleton - Born: Boston - Age: 22 - Race: Mulatto - Released on 1 May 1815.

Fisher, Charles - Seaman - Nbr: 4886 - Prize: Saratoga - Ship Type: P - How taken: HMS Barracouta - When taken: 9 Oct 1814 - Where taken: off Western Islands (England) - Date Received: 24 Oct 1814 - From what ship: HMT Salvador del Mundo - Born: Delaware - Age: 33 - Race: Negro - Died on 6 Apr 1815 from an abscess.

Fisher, James - Seaman - Nbr: 6546 - How taken: Gave himself up from HMS Quebec - When taken: 22 Feb 1813 - Date Received: 7 Mar 1815 - From what ship: HMT Ganges, Plymouth - Born: Maryland - Age: 29 - Escaped on 1 Jun 1815.

Fisher, James - Seaman - Nbr: 2359 - Prize: Snap Dragon - Ship Type: P - How taken: HMS Martin - When taken: 10 Jun 1814 - Where taken: off Halifax - Date Received: 16 Aug 1814 - From what ship: HMT Queen, Halifax - Born: Maryland - Age: 29 - Released on 3 May 1815.

Fisher, James - Mate - Nbr: 1169 - How taken: Sent into custody at Liverpool - Date Received: 10 May 1814 - From what ship: Plymouth - Born: Philadelphia - Age: 25 - Released on 28 Apr 1815.

Fisher, John - Seaman - Nbr: 1146 - Prize: Bunker Hill - Ship Type: P - How taken: HMS Pomone - When taken: 8 Mar 1814 - Where taken: at sea - Date Received: 10 May 1814 - From what ship: Plymouth - Born: Salem - Age: 16 - Released on 28 Apr 1815.

Fisher, John - Seaman - Nbr: 887 - Prize: General Kempt - Ship Type: P - How taken: HMS Foxhound - When taken: 18 Dec 1813 - Where taken: Lat 48'60" Long 5'7" - Date Received: 31 Jan 1814 - From what ship: Plymouth - Born: New York - Age: 22 - Race: Negro - Released on 27 Apr 1815.

Fisher, Lewis - Seaman - Nbr: 1475 - Prize: Essex - Ship Type: MV - How taken: HMS Pyramus - When taken: 2 Apr 1813 - Where taken: Bay of Biscay - Date Received: 19 Jun 1814 - From what ship: Stapleton - Born: Pennsylvania - Age: 25 - Race: Colored - Released on 28 Apr 1815.

Fisher, Robert - Seaman - Nbr: 6297 - Prize: John - Ship Type: Prize - How taken: Leander (Newcastle Acasta) - When taken: 30 Jan 1815 - Date Received: 19 Feb 1815 - From what ship: HMT Ganges, Plymouth - Born: Denmark - Age: 26 - Released on 5 Jul 1815.

Fisher, William - Seaman - Nbr: 5810 - Prize: US Schooner Scorpion - Ship Type: MW - How taken: British

American Prisoners of War Held at Dartmoor during the War of 1812

## Alphabetical listing of names

gunboats - When taken: 6 Sep 1814 - Where taken: Lake Erie - Date Received: 26 Dec 1814 - From what ship: HMT Argo - Born: New York - Age: 24 - Released on 3 Jul 1815.

Fisk, George - Seaman - Nbr: 5994 - Prize: McDonough - Ship Type: P - How taken: HMS Bacchante - When taken: 1 Nov 1814 - Where taken: Lat 42 Long 67 - Date Received: 27 Dec 1814 - From what ship: HMT Penelope - Born: Massachusetts - Age: 28 - Released on 3 Jul 1815.

Fiske, Cyrus - Seaman - Nbr: 4276 - How taken: Sent into custody from Alderney - When taken: 26 Jan 1814 - Date Received: 7 Oct 1814 - From what ship: HMT Niobe, Chatham - Born: Massachusetts - Age: 27 - Released on 14 Jun 1815.

Fitch, William - Seaman - Nbr: 4176 - Prize: Pallas - Ship Type: MV - How taken: Rebuff - When taken: 28 Jan 1813 - Where taken: off Cadiz - Date Received: 7 Oct 1814 - From what ship: HMT Niobe, Chatham - Born: New York - Age: 21 - Released on 27 Mar 1815.

Fitherly, Robert - Seaman - Nbr: 4468 - Prize: Governor Plumer - Ship Type: P - How taken: HMS Shamrock - When taken: 4 Mar 1813 - Where taken: at sea - Date Received: 8 Oct 1814 - From what ship: HMT Leyden, Chatham - Born: Massachusetts - Age: 22 - Released on 15 Jun 1815.

Fitts, Joseph - Seaman - Nbr: 1558 - Prize: Caroline - Ship Type: MV - How taken: HMS Medusa - When taken: 12 Apr 1813 - Where taken: Bay of Biscay - Date Received: 23 Jun 1814 - From what ship: Stapleton - Born: Philadelphia - Age: 42 - Released on 1 May 1815.

Fitzgerald, Edward - Seaman - Nbr: 5522 - Prize: James - Ship Type: Prize - How taken: HMS Galatea - When taken: 7 Sep 1814 - Where taken: Channel - Date Received: 17 Dec 1814 - From what ship: HMT Loire, Halifax - Born: New York - Age: 28 - Released on 1 Jul 1815.

Fitzgerald, James - Soldier - Nbr: 5264 - Prize: 13th US Infantry - Ship Type: Troops - How taken: British Army - When taken: 13 Oct 1812 - Where taken: Canada - Date Received: 31 Oct 1814 - From what ship: HMT Leyden, Chatham - Born: Dublin - Age: 39 - Released on 29 Jun 1815.

Fitzgerald, John - Seaman - Nbr: 2947 - Prize: Hawke - Ship Type: P - How taken: HMS Pique - When taken: 26 Mar 1814 - Where taken: off St. Thomas - Date Received: 24 Aug 1814 - From what ship: HMT Hannibal - Born: North Carolina - Age: 22 - Released on 19 May 1815.

Fitzsmond, Jeremiah - Seaman - Nbr: 4354 - Prize: Fiere Facia - Ship Type: P - How taken: HMS Ramillies - When taken: 26 Feb 1814 - Where taken: off NY - Date Received: 7 Oct 1814 - From what ship: HMT Salvador del Mundo, Halifax - Born: New Hampshire - Age: 23 - Released on 14 Jun 1815.

Flatt, Rob - Seaman - Nbr: 1259 - Prize: Adeline - Ship Type: MV - How taken: HMS Magicienne - When taken: 16 Mar 1814 - Where taken: off Cape Finisterre - Date Received: 14 Jun 1814 - From what ship: Mill Prison (Plymouth, England) - Born: North Carolina - Age: 37 - Released on 28 Apr 1815.

Fleming, Alexander - Seaman - Nbr: 3237 - Prize: Maydock - Ship Type: MV - How taken: Rebuff - When taken: 16 Jun 1813 - Where taken: off St. Marys - Date Received: 11 Sep 1814 - From what ship: HMT Freya, Chatham - Born: Massachusetts - Age: 26 - Released on 30 Mar 1815.

Fleming, John J. - Sail Maker - Nbr: 560 - Prize: US Brig Argus - Ship Type: MW - How taken: HMS Pelican - When taken: 14 Apr 1813 - Where taken: Irish Channel - Date Received: 8 Sep 1813 - From what ship: Plymouth - Born: New York - Age: 20 - Sent to Dartmouth on 2 Nov 1814.

Fletcher, B. - Seaman - Nbr: 2944 - Prize: Atalanta - Ship Type: MV - How taken: HMS Barbados - When taken: 19 Jan 1814 - Where taken: off St. Bartholomew - Date Received: 24 Aug 1814 - From what ship: HMT Hannibal - Born: Massachusetts - Age: 27 - Released on 19 May 1815.

Fletcher, Henry - Seaman - Nbr: 2358 - Prize: Snap Dragon - Ship Type: P - How taken: HMS Martin - When taken: 10 Jun 1814 - Where taken: off Halifax - Date Received: 16 Aug 1814 - From what ship: HMT Queen, Halifax - Born: Manchester - Age: 32 - Released on 3 May 1815.

Fletcher, James - Seaman - Nbr: 6055 - Prize: William - Ship Type: Prize - How taken: HMS Armide - When taken: 11 Oct 1814 - Where taken: off Newport - Date Received: 28 Dec 1814 - From what ship: HMT Penelope - Born: Saint George - Age: 23 - Released on 3 Jul 1815.

Fletcher, John - Boy - Nbr: 847 - Prize: Chesapeake - Ship Type: LM - How taken: HMS Hotspur & HMS Pyramus - When taken: 26 Oct 1813 - Where taken: off Nantes - Date Received: 29 Nov 1813 - From what ship: Plymouth - Born: Norfolk - Age: 18 - Released on 26 Apr 1815.

Fletcher, John - Seaman - Nbr: 5302 - Prize: Growler - Ship Type: P - How taken: HMS Maidstone - When taken: 20 Jul 1813 - Where taken: Grand Banks - Date Received: 31 Oct 1814 - From what ship: HMT Leyden, Chatham - Born: Marblehead - Age: 22 - Released on 29 Jun 1815.

Fletcher, John W. - Seaman - Nbr: 1093 - Prize: Man - Ship Type: Prize - How taken: Recaptured from English merchant vessel - When taken: 18 Dec 1813 - Date Received: 10 May 1814 - From what ship: Plymouth -

American Prisoners of War Held at Dartmoor during the War of 1812

## Alphabetical listing of names

Born: Alexandria - Age: 26 - Escaped on 13 Feb 1815.

Fletcher, P. - Seaman - Nbr: 5634 - How taken: Gave himself up from HMS Semiramis - When taken: 20 Sep 1813 - Date Received: 24 Dec 1814 - What ship: HMT Tay - Born: Boston - Age: 27 - Released on 11 Jul 1815.

Fletcher, Robert - Seaman - Nbr: 5995 - Prize: McDonough - Ship Type: P - How taken: HMS Bacchante - When taken: 1 Nov 1814 - Where taken: Lat 42 Long 67 - Date Received: 27 Dec 1814 - From what ship: HMT Penelope - Born: Arundel - Age: 19 - Released on 3 Jul 1815.

Fletcher, William B. - Seaman - Nbr: 69 - Prize: Spitfire - Ship Type: MV - How taken: HMS Achates - When taken: 14 Feb 1813 - Where taken: off Ushant (France) - Date Received: 2 Apr 1813 - From what ship: Plymouth - Born: Marblehead - Age: 45 - Died on 16 Jul 1813 from pneumonia.

Flinn, Abraham - Seaman - Nbr: 1429 - Prize: Revenge - Ship Type: LM - How taken: HMS Belle Poule - When taken: 10 May 1814 - Where taken: off Cornwall - Date Received: 19 Jun 1814 - From what ship: Stapleton - Born: Boston - Age: 23 - Released on 28 Apr 1815.

Flinn, John F. - Lieutenant - Nbr: 4827 - Prize: President - Ship Type: P - How taken: HMS Pique - When taken: 7 May 1814 - Where taken: off Porto Rico - Date Received: 9 Oct 1814 - From what ship: HMT Freya, Halifax - Born: Ireland - Age: 36 - Sent to HMS Impregnable on 27 Dec 1814.

Flinn, Pearce - Seaman - Nbr: 1291 - Prize: Traveler - Ship Type: Prize - How taken: HM Schooner Canso - When taken: 11 May 1814 - Where taken: off Cape Clear - Date Received: 14 Jun 1814 - From what ship: Mill Prison (Plymouth, England) - Born: Charlestown - Age: 23 - Released on 28 Apr 1815.

Flint, John - Carpenter - Nbr: 2280 - Prize: Diomede - Ship Type: P - How taken: HMS Rifleman - When taken: 28 May 1814 - Where taken: off Sable Island - Date Received: 16 Aug 1814 - From what ship: HMS Dublin, Halifax - Born: Salem - Age: 25 - Released on 19 May 1815.

Flood, David - Seaman - Nbr: 170 - How taken: Impressed at Liverpool - When taken: 10 Feb 1813 - Date Received: 2 Apr 1813 - From what ship: Plymouth - Born: Portland - Age: 25 - Escaped on 14 Mar 1815.

Flood, Francis - Seaman - Nbr: 5132 - How taken: Gave himself up from HMS Crescent - When taken: 20 Oct 1814 - Date Received: 31 Oct 1814 - From what ship: HMT Castillian - Born: Maryland - Age: 24 - Released on 29 Jun 1815.

Flood, John - Seaman - Nbr: 2830 - How taken: Gave himself up from HMS Berwick - When taken: 29 Oct 1812 - Date Received: 24 Aug 1814 - From what ship: HMT Liverpool, Chatham - Born: Boston - Age: 32 - Race: Black - Released on 26 Apr 1815.

Flood, John - Seaman - Nbr: 2713 - How taken: Gave himself up from HMS Sterling Castle - When taken: 12 Jun 1813 - Date Received: 21 Aug 1814 - From what ship: HMT Freya, Chatham - Born: Portland - Age: 22 - Released on 26 Apr 1815.

Florence, Charles - Gunner - Nbr: 3451 - Prize: Growler - Ship Type: P - How taken: HMS Electra - When taken: 7 Jul 1813 - Where taken: at sea - Date Received: 13 Sep 1814 - From what ship: HMT Niobe, Chatham - Born: Marblehead - Age: 56 - Released on 28 May 1815.

Florence, John - Prize master - Nbr: 5007 - Prize: Charlotte - Ship Type: Prize - How taken: HMS Wasp - When taken: 31 Aug 1814 - Where taken: off Cape Sable Island (Canada) - Date Received: 28 Oct 1814 - From what ship: HMT Alkbar, Halifax - Born: Massachusetts - Age: 35 - Released on 1 May 1815.

Florence, John - Seaman - Nbr: 4042 - Prize: US Brig Rattlesnake - Ship Type: MW - How taken: HMS Leander - When taken: 13 Jul 1814 - Where taken: off Shelburne - Date Received: 6 Oct 1814 - From what ship: HMT Chesapeake, Halifax - Born: Marblehead - Age: 48 - Released on 13 Jun 1815.

Flowers, John - Seaman - Nbr: 6169 - Prize: Lion - Ship Type: P - How taken: HMS Granicus - When taken: 2 Dec 1814 - Where taken: off Lisbon - Date Received: 18 Jan 1815 - From what ship: HMT Impregnable - Born: Boston - Age: 56 - Died on 12 May 1815 from bursting blood vessel.

Fluken, William - 1st Lieutenant - Nbr: 6189 - Prize: Prince de Neufchatel - Ship Type: P - How taken: Leander (Newcastle Acasta) - When taken: 20 Dec 1814 - Where taken: Lat 38 Long 56 - Date Received: 30 Jan 1815 - From what ship: HMT Pheasant - Born: Not legible - Sent to Ashburton (England) on 11 Feb 1815.

Flynn, John - Seaman - Nbr: 4937 - Prize: Herald - Ship Type: P - How taken: HMS Endymion - When taken: 15 Aug 1814 - Where taken: off Nantucket - Date Received: 28 Oct 1814 - From what ship: HMT Alkbar, Halifax - Born: Philadelphia - Age: 54 - Released on 21 Jun 1815.

Fogerty, Archibald - Seaman - Nbr: 2027 - How taken: Gave himself up from HMS Horatio - When taken: 16 May 1813 - Date Received: 3 Aug 1814 - From what ship: HMS Lyffey, Chatham Depot - Born: Lincoln - Age: 38 - Died on 18 Mar 1815 from variola.

Fogg, John - Seaman - Nbr: 4842 - Prize: President - Ship Type: P - How taken: HMS Pique - When taken: 7 May 1814 - Where taken: off Porto Rico - Date Received: 9 Oct 1814 - From what ship: HMT Freya, Halifax -

American Prisoners of War Held at Dartmoor during the War of 1812

## Alphabetical listing of names

Born: Exeter - Age: 19 - Race: Mulatto - Released on 14 Jun 1815.

Fogg, Noel - Seaman - Nbr: 2071 - How taken: Impressed at London - When taken: 23 Jun 1813 - Date Received: 3 Aug 1814 - From what ship: HMS Bittern, Chatham Depot - Born: New Hampshire - Age: 28 - Released on 2 May 1815.

Folger, Frederick - Seaman - Nbr: 3190 - How taken: Gave himself up from HMS Swiftsure - When taken: 26 Dec 1812 - Date Received: 11 Sep 1814 - From what ship: HMT Freya, Chatham - Born: Baltimore - Age: 27 - Sent to Dartmouth on 23 Sep 1814.

Follingsby, William - Seaman - Nbr: 2085 - Prize: True Blooded Yankee - Ship Type: P - How taken: HMS Hope - When taken: 24 Jun 1813 - Where taken: off Brest - Date Received: 3 Aug 1814 - From what ship: HMS Bittern, Chatham Depot - Born: Newburyport - Age: 20 - Released on 2 May 1815.

Folwett, Joseph - Seaman - Nbr: 2524 - Prize: Hussar - Ship Type: P - How taken: HMS Saturn - When taken: 25 May 1814 - Where taken: off Sandy Hook - Date Received: 16 Aug 1814 - From what ship: HMT Queen, Halifax - Born: New Jersey - Age: 20 - Released on 3 May 1815.

Fool, Garret - Seaman - Nbr: 3198 - How taken: Gave himself up from HMS America - When taken: 16 Jul 1813 - Date Received: 11 Sep 1814 - From what ship: HMT Freya, Chatham - Born: Virginia - Age: 42 - Released on 28 May 1815.

Foot, Benjamin - Seaman - Nbr: 1879 - Prize: Elbridge Gerry - Ship Type: P - How taken: HMS Crescent - When taken: 16 Sep 1813 - Where taken: off St. George's - Date Received: 29 Jul 1814 - From what ship: HMS Ville de Paris, Chatham Depot - Born: Bath - Age: 18 - Released on 2 May 1815.

Foran, William - Seaman - Nbr: 4004 - Prize: US Brig Rattlesnake - Ship Type: MW - How taken: HMS Leander - When taken: 13 Jul 1814 - Where taken: off Shelburne - Date Received: 6 Oct 1814 - From what ship: HMT Chesapeake, Halifax - Born: Alexandria - Age: 21 - Released on 9 Jun 1815.

Forbes, James - Prize Master - Nbr: 2546 - Prize: Caromaned - Ship Type: Prize - How taken: HMS Eridanus - When taken: 13 Aug 1814 - Where taken: Lat 40, Long 16 - Date Received: 18 Aug 1814 - From what ship: HMT Salvador del Mundo - Born: Philadelphia - Age: 30 - Released on 3 May 1815.

Forbes, John - Seaman - Nbr: 3282 - Prize: Montgomery - Ship Type: P - How taken: HMS Nymphe - When taken: 5 May 1813 - Where taken: off Cape Ann - Date Received: 13 Sep 1814 - From what ship: HMT Niobe, Chatham - Born: Salem - Age: 29 - Released on 28 May 1815.

Forbes, Robert - Seaman - Nbr: 4565 - How taken: Gave himself up from HMS Leviathan - When taken: 1812 - Date Received: 8 Oct 1814 - From what ship: HMT Leyden, Chatham - Born: New Jersey - Age: 32 - Released on 27 Apr 1815.

Forbes, William - Seaman - Nbr: 231 - Prize: Charlotte - Ship Type: MV - How taken: HMS Warspite - When taken: 3 Mar 1813 - Where taken: Bay of Biscay - Date Received: 2 Apr 1813 - From what ship: Plymouth - Born: Boston - Age: 20 - Released on 20 Apr 1815.

Forbes, William - Seaman - Nbr: 4981 - Prize: Invincible - Ship Type: LM - How taken: HMS Armide - When taken: 15 Aug 1814 - Where taken: off Nantucket - Date Received: 28 Oct 1814 - From what ship: HMT Alkbar, Halifax - Born: Massachusetts - Age: 32 - Released on 21 Jun 1815.

Ford, Charles - Seaman - Nbr: 1250 - How taken: Sent into custody from HMS Edinburgh - When taken: 28 Oct 1812 - Date Received: 14 Jun 1814 - From what ship: Mill Prison (Plymouth, England) - Born: North America - Age: 31 - Sent to Exeter Jail on 17 Apr 1814.

Ford, Philip - Seaman - Nbr: 1236 - How taken: Sent into custody from HMS Sultan - When taken: 16 Oct 1812 - Date Received: 14 Jun 1814 - From what ship: Mill Prison (Plymouth, England) - Born: Wilmington - Age: 55 - Released on 28 May 1815.

Ford, Stan - Seaman - Nbr: 4134 - Prize: Bordeaux Packet - Ship Type: LM - How taken: HMS Niemen - When taken: 28 Jun 1814 - Where taken: off Delaware - Date Received: 6 Oct 1814 - From what ship: HMT Chesapeake, Halifax - Born: Pennsylvania - Age: 21 - Released on 13 Jun 1815.

Forester, Francis - Seaman - Nbr: 461 - Prize: Meteor - Ship Type: MV - How taken: HMS Briton - When taken: 12 Mar 1813 - Where taken: Bay of Biscay - Date Received: 8 Sep 1813 - From what ship: Plymouth - Born: Charlestown - Age: 19 - Sent to Plymouth on 7 Dec 1813.

Fornver, Virgil - Seaman - Nbr: 1180 - Prize: Melanie - Ship Type: MV - How taken: HMS Briton - When taken: 15 Sep 1813 - Where taken: off Bordeaux - Date Received: 4 Jun 1814 - From what ship: Dartmouth - Born: New Orleans - Age: 22 - Race: Mulatto - Sent to Plymouth on 8 Jul 1814.

Forrest, James - Seaman - Nbr: 2771 - How taken: Gave himself up from HMS Malta - Date Received: 24 Aug 1814 - What ship: HMT Liverpool, Chatham - Born: Philadelphia - Age: 40 - Released on 19 May 1815.

Forrest, John - Seaman - Nbr: 3773 - Prize: Fame - Ship Type: P - How taken: HMS Thistle - When taken: 10 Apr

American Prisoners of War Held at Dartmoor during the War of 1812

## Alphabetical listing of names

1814 - Where taken: after being cast ashore on a seal island - Date Received: 30 Sep 1814 - From what ship: HMT President, Halifax - Born: East Guilford - Age: 30 - Released on 9 Jun 1815.

Forrest, William - Seaman - Nbr: 2850 - How taken: Gave himself up from HMS Epervier - When taken: 15 Aug 1812 - Date Received: 24 Aug 1814 - From what ship: HMT Alpheus, Chatham - Born: Philadelphia - Age: 32 - Released on 27 Apr 1815.

Forrier, John - Seaman - Nbr: 5730 - How taken: Gave himself up from MV Warner - When taken: 12 Nov 1814 - Date rec'd: 26 Dec 1814 - What ship: HMT Argo - Born: Philadelphia - Age: 23 - Released on 3 Jul 1815.

Forsdick, Joseph H. - Seaman - Nbr: 3109 - Prize: Rattlesnake - Ship Type: LM - How taken: HMS Orion - When taken: 11 Mar 1814 - Where taken: off Bermuda - Date Received: 11 Sep 1814 - From what ship: HMT Salvador del Mundo - Born: Boston - Age: 18 - Released on 28 May 1815.

Forster, George - Seaman - Nbr: 4544 - Prize: Growler - Ship Type: P - How taken: HMS Electra - When taken: 7 Jul 1813 - Where taken: at sea - Date Received: 8 Oct 1814 - From what ship: HMT Leyden, Chatham - Born: Marblehead - Age: 42 - Released on 15 Jun 1815.

Forster, Joseph - Seaman - Nbr: 2074 - How taken: Impressed at London - When taken: 25 Jun 1813 - Date Received: 3 Aug 1814 - From what ship: HMS Bittern, Chatham Depot - Born: Dorchester - Age: 26 - Released on 2 May 1815.

Forster, Samuel - Seaman - Nbr: 5497 - Prize: General Putnam - Ship Type: P - How taken: HMS Leander - When taken: 8 Nov 1814 - Where taken: Long 65 Lat 42 - Date Received: 17 Dec 1814 - From what ship: HMT Loire, Halifax - Born: Beverly - Age: 19 - Released on 1 Jul 1815.

Forster, Thomas - Seaman - Nbr: 1443 - Prize: Tickler - Ship Type: LM - How taken: HMS Magicienne - When taken: 5 Jun 1813 - Where taken: Bay of Biscay - Date Received: 19 Jun 1814 - From what ship: Stapleton - Born: Gloucester - Age: 25 - Released on 28 Apr 1815.

Forster, Thomas - Seaman - Nbr: 2692 - How taken: Gave himself up from HMS Scorpion - Date Received: 21 Aug 1814 - From what ship: HMT Freya, Chatham - Born: Plymouth - Age: 28 - Released on 19 May 1815.

Forstman, John - Seaman - Nbr: 4524 - Prize: Yankee - Ship Type: P - How taken: HMS Shannon - When taken: 20 Aug 1813 - Where taken: at sea - Date Received: 8 Oct 1814 - From what ship: HMT Leyden, Chatham - Born: Stockholm - Age: 20 - Released on 15 Jun 1815.

Forsyth, Robert - Seaman - Nbr: 3316 - Prize: Globe - Ship Type: P - How taken: HMS Victorious - When taken: 8 Jun 1813 - Where taken: Chesapeake Bay - Date Received: 13 Sep 1814 - From what ship: HMT Niobe, Chatham - Born: Baltimore - Age: 23 - Released on 28 May 1815.

Fortune, John - Seaman - Nbr: 4382 - Prize: Teazer - Ship Type: P - How taken: HMS Boyce - When taken: 15 Jun 1813 - Where taken: off Halifax - Date Received: 8 Oct 1814 - From what ship: HMT Leyden, Chatham - Born: New London - Age: 29 - Race: Black - Released on 14 Jun 1815.

Foss, Abijah - Boy - Nbr: 6004 - Prize: McDonough - Ship Type: P - How taken: HMS Bacchante - When taken: 1 Nov 1814 - Where taken: Lat 42 Long 67 - Date Received: 28 Dec 1814 - From what ship: HMT Penelope - Born: Massachusetts - Age: 15 - Released on 3 Jul 1815.

Foss, Edward - Seaman - Nbr: 1439 - Prize: Leo - Ship Type: LM - How taken: HMS Magicienne - When taken: 4 Jun 1813 - Where taken: off France - Date Received: 19 Jun 1814 - From what ship: Stapleton - Born: Lymington - Age: 21 - Released on 28 Apr 1815.

Foss, Joseph - Seaman - Nbr: 1442 - Prize: Leo - Ship Type: LM - How taken: HMS Magicienne - When taken: 4 Jun 1813 - Where taken: off France - Date Received: 19 Jun 1814 - From what ship: Stapleton - Born: Tarbres - Age: 22 - Released on 28 Apr 1815.

Foss, Supply - Seaman - Nbr: 5001 - Prize: Invincible - Ship Type: LM - How taken: HMS Armide - When taken: 15 Aug 1814 - Where taken: off Nantucket - Date Received: 28 Oct 1814 - From what ship: HMT Alkbar, Halifax - Born: Portsmouth - Age: 24 - Released on 8 May 1815.

Fossendor, William - Seaman - Nbr: 153 - How taken: Gave himself up from HMS Lavinia - Date Received: 2 Apr 1813 - From what ship: Plymouth - Born: Homburg - Age: 24 - Sent to Plymouth on 15 Jun 1813.

Fosset, Rodwell - Seaman - Nbr: 5420 - How taken: Gave himself up from HMS Indian - When taken: 20 Aug 1814 - Date Received: 31 Oct 1814 - From what ship: HMT Leyden, Chatham - Born: Burlington - Age: 28 - Race: Black - Released on 1 Jul 1815.

Foster, Cato - Seaman - Nbr: 157 - How taken: Gave himself up from HMS Lavinia - Date Received: 2 Apr 1813 - From what ship: Plymouth - Born: Marblehead - Age: 33 - Race: Negro - Released on 20 Apr 1815.

Foster, David - Seaman - Nbr: 410 - Prize: Viper - Ship Type: MV - How taken: HMS Superb - When taken: 15 Apr 1813 - Where taken: Bay of Biscay - Date Received: 1 Jul 1813 - From what ship: Plymouth - Born: Maryland - Age: 38 - Race: Black - Released on 20 Apr 1815.

American Prisoners of War Held at Dartmoor during the War of 1812

## Alphabetical listing of names

Foster, James - Gunner - Nbr: 2374 - Prize: US Sloop Frolic - Ship Type: MW - How taken: HMS Orpheus - When taken: 20 Apr 1814 - Where taken: off Cuba - Date Received: 16 Aug 1814 - From what ship: HMT Queen, Halifax - Born: New York - Age: 45 - Sent to Dartmouth on 19 Oct 1814.

Foster, James - Seaman - Nbr: 3475 - How taken: Gave himself up from HMS Trident - When taken: 20 Jul 1814 - Date Received: 19 Sep 1814 - From what ship: HMT Salvador del Mundo - Born: Baltimore - Age: 36 - Released on 11 Jul 1815.

Foster, Joseph - Seaman - Nbr: 36 - How taken: Apprehended at Gibraltar - When taken: 8 Aug 1813 - Date Received: 2 Apr 1813 - What ship: Plymouth - Born: Beverly - Age: 23 - Sent to Dartmouth on 30 Jul 1813.

Foster, Joseph - Seaman - Nbr: 252 - Prize: Cashier - Ship Type: LM - How taken: HMS Reindeer - When taken: 3 Feb 1813 - Where taken: Bay of Biscay - Date Received: 28 Jun 1813 - From what ship: Plymouth - Born: Talbot County - Age: 18 - Sent to Dartmouth on 30 Jul 1813.

Foster, Joseph - Seaman - Nbr: 3267 - Prize: Wily Reynard - Ship Type: P - How taken: HMS Shannon - When taken: 2 Oct 1812 - Where taken: off Halifax - Date Received: 11 Sep 1814 - From what ship: HMT Freya, Chatham - Born: Baltimore - Age: 22 - Race: Black - Released on 27 Apr 1815.

Foster, Samuel - Seaman - Nbr: 4374 - Prize: Growler - Ship Type: P - How taken: HMS Electra - When taken: 7 Jul 1813 - Where taken: at sea - Date Received: 8 Oct 1814 - From what ship: HMT Leyden, Chatham - Born: Massachusetts - Age: 19 - Released on 14 Jun 1815.

Foster, Thomas - Seaman - Nbr: 437 - Prize: Magdalen - Ship Type: MV - How taken: HMS Superb - When taken: 15 Apr 1813 - Where taken: off Belle Isle, France - Date Received: 1 Jul 1813 - From what ship: Plymouth - Born: Portsmouth - Age: 24 - Released on 20 Apr 1815.

Fowler, Isaac - Seaman - Nbr: 2491 - Prize: US Sloop Frolic - Ship Type: MW - How taken: HMS Orpheus - When taken: 20 Apr 1814 - Where taken: off Cuba - Date Received: 16 Aug 1814 - From what ship: HMT Queen, Halifax - Born: Massachusetts - Age: 19 - Released on 3 May 1815.

Fowler, Isaac - Seaman - Nbr: 3042 - How taken: Gave himself up from HMS Galatea - When taken: 24 Aug 1814 - Date Received: 2 Sep 1814 - From what ship: HMT Sultan - Born: Baltimore - Age: 30 - Released on 28 Apr 1815.

Fowler, Joshua - Seaman - Nbr: 4730 - How taken: Gave himself up from Garmane - When taken: 3 Sep 1814 - Date Received: 9 Oct 1814 - From what ship: HMT Freya, Chatham - Born: Boston - Age: 30 - Died on 30 Jan 1815 from phthisis pulmonalis.

Fowler, Robert - Seaman - Nbr: 3831 - Prize: Hussar - Ship Type: P - How taken: HMS Saturn - When taken: 24 May 1814 - Where taken: off Sandy Hook - Date Received: 5 Oct 1814 - From what ship: HMT Orpheus, Halifax - Born: Salem - Age: 28 - Released on 9 Jun 1815.

Fowler, William - Seaman - Nbr: 3021 - Prize: Rattlesnake - Ship Type: P - How taken: HMS Hyperion - When taken: 25 Jun 1814 - Where taken: off Cape Finisterre - Date Received: 2 Sep 1814 - From what ship: Naval Hospital, Plymouth - Born: St. Marys - Age: 35 - Released on 28 May 1815.

Fowling, Jeremiah - Seaman - Nbr: 5058 - Prize: Ida - Ship Type: LM - How taken: HMS Newcastle - When taken: 9 Aug 1814 - Where taken: Long 34 - Date Received: 28 Oct 1814 - From what ship: HMT Alkbar, Halifax - Born: Boston - Age: 21 - Released on 29 Jun 1815.

Fox, Edward - Master - Nbr: 6498 - Prize: Fox - Ship Type: MV - How taken: HMS Lacedaemonian - Date Received: 3 Mar 1815 - From what ship: HMT Ganges, Plymouth - Born: Lancaster - Age: 34 - Released on 11 Jul 1815.

Fox, Washington - Seaman - Nbr: 500 - Prize: Margaret - How taken: HMS Nimrod - When taken: 9 Mar 1813 - Where taken: off Morlaix - Date Received: 8 Sep 1813 - From what ship: Plymouth - Born: Alexandria - Age: 38 - Released on 20 Apr 1815.

Fox, Washington - Seaman - Nbr: 163 - Prize: Margaret - Ship Type: P - How taken: HMS Nimrod - When taken: 9 Mar 1813 - Where taken: off Morlaix - Date Received: 2 Apr 1813 - From what ship: Plymouth - Born: Alexandria - Age: 38 - Sent to Plymouth on 6 Jun 1813.

Foxey, Thomas - Seaman - Nbr: 2961 - How taken: Gave himself up from HMS Scout - When taken: 21 May 1814 - Date rec'd: 24 Aug 1814 - What ship: HMT Hannibal - Born: North Carolina - Age: 49 - Released on 19 May 1815.

Frack, Daniel - Seaman - Nbr: 4900 - Prize: Atlantic - Ship Type: Prize - How taken: HMS Maidstone - When taken: 10 Oct 1813 - Where taken: off Sable Island - Date Received: 28 Oct 1814 - From what ship: HMT Alkbar, Halifax - Born: Boston - Age: 17 - Released on 21 Jun 1815.

Frain, Joshua - Cook - Nbr: 6331 - Prize: Prince de Neufchatel - Ship Type: P - How taken: Leander (Newcastle Acasta) - When taken: 20 Dec 1814 - Where taken: Lat 38 Long 56 - Date Received: 19 Feb 1815 - From

American Prisoners of War Held at Dartmoor during the War of 1812

## Alphabetical listing of names

what ship: HMT Ganges, Plymouth - Born: New Orleans - Age: 20 - Race: Black - Released on 5 Jul 1815.

Francis, Abraham - Seaman - Nbr: 3303 - Prize: Porcupine - Ship Type: LM - How taken: British squadron - When taken: 3 Jun 1813 - Where taken: off Cape Sable Island (Canada) - Date Received: 13 Sep 1814 - From what ship: HMT Niobe, Chatham - Born: Marblehead - Age: 34 - Race: Mulatto - Released on 28 May 1815.

Francis, Benjamin - Seaman - Nbr: 262 - Prize: William Bayard - Ship Type: MV - How taken: HMS Warspite - When taken: 3 Mar 1813 - Where taken: Bay of Biscay - Date Received: 28 Jun 1813 - From what ship: Plymouth - Born: Philadelphia - Age: 22 - Released on 20 Apr 1815.

Francis, Hugh - Cook - Nbr: 5493 - Prize: General Putnam - Ship Type: P - How taken: HMS Leander - When taken: 8 Nov 1814 - Where taken: Long 65 Lat 42 - Date Received: 17 Dec 1814 - From what ship: HMT Loire, Halifax - Born: Africa - Age: 21 - Race: Black - Released on 1 Jul 1815.

Francis, John - Seaman - Nbr: 4626 - Prize: Argus - Ship Type: MV - How taken: HMS San Domingo - When taken: 1 Mar 1814 - Where taken: off Savannah - Date Received: 9 Oct 1814 - From what ship: HMT Leyden, Chatham - Born: Newport - Age: 23 - Race: Black - Released on 15 Jun 1815.

Francis, John - Seaman - Nbr: 1582 - Prize: Price - Ship Type: MV - How taken: HMS Pyramus - When taken: 6 Apr 1813 - Where taken: Bay of Biscay - Date Received: 23 Jun 1814 - From what ship: Stapleton - Born: Rhode Island - Age: 24 - Sent to Mill Prison (Plymouth, England) on 30 Jun 1814.

Francis, John - Seaman - Nbr: 2300 - Prize: Hussar - Ship Type: P - How taken: HMS Saturn - When taken: 25 May 1814 - Where taken: off Sandy Hook - Date Received: 16 Aug 1814 - From what ship: HMS Dublin, Halifax - Born: Nantes - Age: 26 - Released on 3 May 1815.

Francis, John - Seaman - Nbr: 2018 - How taken: Gave himself up from HMS Royal William - When taken: 5 May 1813 - Date Received: 3 Aug 1814 - From what ship: HMS Lyffey, Chatham Depot - Born: Portsmouth - Age: 38 - Race: Mulatto - Died on 16 Apr 1815, found dead in prison.

Francis, John - Seaman - Nbr: 4564 - How taken: Gave himself up from MV Liberty - When taken: 30 Dec 1813 - Date Received: 8 Oct 1814 - From what ship: HMT Leyden, Chatham - Born: Boston - Age: 20 - Released on 15 Jun 1815.

Francis, John - Seaman - Nbr: 3091 - How taken: Gave himself up from HMS Bonnie Cetoyenus - When taken: 27 Nov 1813 - Date Received: 3 Sep 1814 - From what ship: HMT Hydra, Chatham - Born: Nantucket - Age: 22 - Race: Black - Released on 28 May 1815.

Francis, Oliver - Seaman - Nbr: 3707 - Prize: Alfred - Ship Type: P - How taken: HMS Epervier - When taken: 23 Feb 1814 - Where taken: off Newfoundland - Date Received: 30 Sep 1814 - From what ship: HMT President, Halifax - Born: Marblehead - Age: 17 - Released on 4 Jun 1815.

Francis, Prince - Seaman - Nbr: 1939 - How taken: Gave himself up from HMS Albacore - When taken: 3 Feb 1813 - Date Received: 3 Aug 1814 - From what ship: HMS Alceste, Chatham Depot - Born: Connecticut - Age: 33 - Released on 2 May 1815.

Francis, Thomas - Boy - Nbr: 5700 - Prize: David Porter - Ship Type: LM - How taken: HMS Pylades - When taken: 12 Sep 1814 - Where taken: Georges Bank - Date Received: 24 Dec 1814 - From what ship: HMT Penelope - Born: Boston - Age: 16 - Released on 3 Jul 1815.

Francisco, Thomas - Seaman - Nbr: 5130 - Prize: Calabria - Ship Type: MV - How taken: HMS Castilian - When taken: 29 Sep 1814 - Where taken: off Ireland - Date Received: 31 Oct 1814 - From what ship: HMT Castillian - Born: Bordeaux - Age: 24 - Released on 29 Jun 1815.

Francois - Seaman - Nbr: 5598 - Prize: Mentor - Ship Type: Prize - How taken: HMS Maidstone - When taken: 26 Oct 1814 - Where taken: off Cape Sable Island (Canada) - Date Received: 17 Dec 1814 - From what ship: HMT Loire, Halifax - Born: Washington - Age: 23 - Race: Negro - Released on 1 Jul 1815.

Franics, John - Seaman - Nbr: 5961 - Prize: Harlequin - Ship Type: P - How taken: HMS Bulwark - When taken: 23 Nov 1814 - Where taken: off Halifax - Date Received: 27 Dec 1814 - From what ship: HMT Penelope - Born: Lisbon - Age: 26 - Race: Black - Released on 3 Jul 1815.

Franics, Michael - Seaman - Nbr: 3640 - Prize: Bordeaux Packet - Ship Type: LM - How taken: HMS Niemen - When taken: 28 Jan 1814 - Where taken: off Delaware - Date Received: 30 Sep 1814 - From what ship: HMT Sybella - Born: Philadelphia - Age: 36 - Released on 4 Jun 1815.

Frank, Joshua - Seaman - Nbr: 5106 - Prize: David Porter - Ship Type: LM - How taken: HMS Pylades - When taken: 12 Sep 1814 - Where taken: Georges Bank - Date Received: 28 Oct 1814 - From what ship: HMT Alkbar, Halifax - Born: Salem - Age: 19 - Released on 29 Jun 1815.

Frank, Louis Ville - Boy - Nbr: 5489 - Prize: General Putnam - Ship Type: P - How taken: HMS Leander - When taken: 8 Nov 1814 - Where taken: Long 65 Lat 42 - Date Received: 17 Dec 1814 - From what ship: HMT Loire, Halifax - Born: Bordeaux - Age: 13 - Released on 1 Jul 1815.

American Prisoners of War Held at Dartmoor during the War of 1812

## Alphabetical listing of names

Franklin, Edward - Seaman - Nbr: 4363 - How taken: Gave himself up from HMS Ulysses - When taken: 6 Oct 1814 - Date Received: 7 Oct 1814 - From what ship: HMT Salvador del Mundo, Halifax - Born: Charleston - Age: 33 - Released on 14 Jun 1815.

Franklin, W. H. - Seaman - Nbr: 4845 - Prize: Hawke - Ship Type: P - How taken: HMS Pique - When taken: 26 Apr 1814 - Where taken: off Porto Rico - Date Received: 9 Oct 1814 - From what ship: HMT Freya, Halifax - Born: Virginia - Age: 21 - Released on 21 Jun 1815.

Franklin, William - Seaman - Nbr: 4315 - How taken: Gave himself up from HMS Daphne - When taken: 20 Aug 1814 - Date Received: 7 Oct 1814 - From what ship: HMT Niobe, Chatham - Born: Philadelphia - Age: 34 - Released on 14 Jun 1815.

Franks, Francis - Steward - Nbr: 823 - Prize: Amiable - Ship Type: LM - How taken: HMS Magnificent - When taken: 30 Oct 1813 - Where taken: off Bordeaux - Date Received: 29 Nov 1813 - From what ship: Plymouth - Born: New York - Age: 37 - Race: Black - Released on 26 Apr 1815.

Frasher, Daniel - Marine - Nbr: 2503 - Prize: US Sloop Frolic - Ship Type: MW - How taken: HMS Orpheus - When taken: 20 Apr 1814 - Where taken: off Cuba - Date Received: 16 Aug 1814 - From what ship: HMT Queen, Halifax - Born: Portland - Age: 17 - Released on 3 May 1815.

Fraste, Samuel - Seaman - Nbr: 2469 - Prize: US Sloop Frolic - Ship Type: MW - How taken: HMS Orpheus - When taken: 20 Apr 1814 - Where taken: off Cuba - Date Received: 16 Aug 1814 - From what ship: HMT Queen, Halifax - Born: Newcastle - Age: 34 - Released on 3 May 1815.

Fray, James - Seaman - Nbr: 5227 - Prize: York Town - Ship Type: P - How taken: HMS Maidstone - When taken: 18 Jul 1813 - Where taken: Grand Banks - Date Received: 31 Oct 1814 - From what ship: HMT Mermaid, Chatham - Born: Baltimore - Age: 23 - Released on 29 Jun 1815.

Frazer, William - Boy - Nbr: 5923 - Prize: Harlequin - Ship Type: P - How taken: HMS Bulwark - When taken: 23 Nov 1814 - Where taken: off Halifax - Date Received: 27 Dec 1814 - From what ship: HMT Penelope - Born: Boston - Age: 13 - Released on 2 May 1815.

Frazier, Charles - 2nd Mate - Nbr: 6275 - Prize: Plutarch - Ship Type: MV - How taken: HMS Helicon - When taken: 5 Feb 1815 - Where taken: off Bordeaux - Date Received: 17 Feb 1815 - From what ship: HMT Ganges, Plymouth - Born: New York - Age: 44 - Released on 5 Jul 1815.

Frazier, James - Seaman - Nbr: 2858 - How taken: Gave himself up from HMS Leonidas - Date Received: 24 Aug 1814 - From what ship: HMT Alpheus, Chatham - Born: New York - Age: 29 - Released on 19 May 1815.

Frazier, William - Seaman - Nbr: 3245 - How taken: Impressed at Cork - When taken: 26 Sep 1813 - Date Received: 11 Sep 1814 - What ship: HMT Freya, Chatham - Born: Pennsylvania - Age: 31 - Released on 10 Apr 1815.

Freddle, John - Seaman - Nbr: 221 - Prize: Pert - Ship Type: MV - How taken: HMS Warspite - When taken: 1 Mar 1813 - Where taken: off Basque Roads (France) - Date Received: 2 Apr 1813 - From what ship: Plymouth - Born: New Orleans - Age: 28 - Released on 12 Apr 1815.

Frederick, Charles - Seaman - Nbr: 1050 - How taken: Sent into custody from Mary (Transport) - Date Received: 10 May 1814 - From what ship: Plymouth - Born: New York - Age: 25 - Released on 27 Apr 1815.

Frederick, Charles - Seaman - Nbr: 2164 - Prize: Rattlesnake - Ship Type: P - How taken: HMS Hyperion - When taken: 26 Jun 1814 - Where taken: off Cape Finisterre - Date Received: 16 Aug 1814 - From what ship: HMS Dublin, Halifax - Born: Portland - Age: 32 - Released on 2 May 1815.

Frederick, John - Seaman - Nbr: 5275 - Prize: Eliza (Whaler) - Ship Type: MV - How taken: HMS Charles - When taken: 20 Jul 1813 - Where taken: off Peter Head - Date Received: 31 Oct 1814 - From what ship: HMT Leyden, Chatham - Born: Virginia - Age: 32 - Released on 29 Jun 1815.

Frederick, John - Seaman - Nbr: 5871 - Prize: Lion - Ship Type: P - How taken: HMS Granicus - When taken: 2 Dec 1814 - Where taken: off Lisbon - Date Received: 26 Dec 1814 - From what ship: HMT Impregnable - Born: Bordeaux - Age: 21 - Released on 3 Jul 1815.

Frederique, John - Seaman - Nbr: 1183 - Prize: Melanie - Ship Type: MV - How taken: HMS Briton - When taken: 15 Sep 1813 - Where taken: off Bordeaux - Date Received: 4 Jun 1814 - From what ship: Dartmouth - Born: New Orleans - Age: 19 - Sent to Plymouth on 8 Jul 1814.

Freeborn, Joseph - Seaman - Nbr: 2137 - How taken: Gave himself up from HMS Alfred - When taken: 26 Jan 1813 - Date Received: 8 Aug 1814 - From what ship: HMT Raven, Chatham - Born: Philadelphia - Age: 23 - Released on 2 May 1815.

Freely, Henry - Seaman - Nbr: 555 - How taken: Gave himself up from HMS Pompeii - When taken: 12 Jul 1813 - Date Received: 8 Sep 1813 - From what ship: Plymouth - Born: Pennsylvania - Age: 29 - Died on 20 Jan 1814.

Freeman, Charles - Seaman - Nbr: 1794 - Prize: Ferox - Ship Type: MV - How taken: HMS Medusa & HMS Lyra -

American Prisoners of War Held at Dartmoor during the War of 1812

## Alphabetical listing of names

When taken: 28 Mar 1813 - Where taken: off Cape Ortegal (Spain) - Date Received: 20 Jul 1814 - From what ship: HMS Milford, Plymouth - Born: Delaware - Age: 23 - Race: Black - Released on 1 May 1815.

Freeman, David - Seaman - Nbr: 1075 - How taken: Impressed at Liverpool - When taken: 2 Feb 1814 - Date Received: 10 May 1814 - From what ship: Plymouth - Born: New Bedford - Age: 24 - Race: Mulatto - Released on 27 Apr 1815.

Freeman, Halkins - Seaman - Nbr: 4945 - Prize: Herald - Ship Type: P - How taken: HMS Endymion - When taken: 15 Aug 1814 - Where taken: off Nantucket - Date Received: 28 Oct 1814 - From what ship: HMT Alkbar, Halifax - Born: Sweden - Age: 29 - Released on 21 Jun 1815.

Freeman, Isaac - Seaman - Nbr: 5172 - Prize: Vivid - Ship Type: P - How taken: HMS Nymphe - When taken: 20 Apr 1813 - Where taken: off Cape Cod - Date Received: 31 Oct 1814 - From what ship: HMT Mermaid, Chatham - Born: Wellfleet - Age: 27 - Released on 29 Jun 1815.

Freeman, John - Seaman - Nbr: 6150 - Prize: William Penn - Ship Type: MV - How taken: HMS Acorn - When taken: 7 Feb 1812 - Where taken: Lat 14 - Date Received: 17 Jan 1815 - From what ship: HMT Impregnable - Born: Barnstable - Age: 18 - Race: Black - Released on 27 Apr 1815.

Freeman, John - 2nd Gunner - Nbr: 687 - Prize: US Brig Argus - Ship Type: MW - How taken: HMS Pelican - When taken: 14 Aug 1813 - Where taken: Irish Channel - Date Received: 27 Sep 1813 - From what ship: Plymouth - Born: Maryland - Age: 35 - Sent to Dartmouth on 2 Nov 1814.

Freeman, John - Seaman - Nbr: 1670 - Prize: Paul Jones - Ship Type: P - How taken: HMS Leonidas - When taken: 23 May 1813 - Where taken: Channel - Date Received: 26 Jun 1814 - From what ship: Exeter - Born: Boston - Age: 33 - Race: Negro - Released on 1 May 1815.

Freeman, Nathaniel - Prize Master - Nbr: 3519 - Prize: Hussar - Ship Type: P - How taken: HMS Saturn - When taken: 25 May 1814 - Where taken: off Sandy Hook - Date Received: 28 Sep 1814 - From what ship: HMT Salvador del Mundo - Born: Massachusetts - Age: 26 - Released on 4 Jun 1815.

Freeman, Plim - Seaman - Nbr: 4627 - Prize: Argus - Ship Type: MV - How taken: HMS San Domingo - When taken: 1 Mar 1814 - Where taken: off Savannah - Date Received: 9 Oct 1814 - From what ship: HMT Leyden, Chatham - Born: Lippstadt - Age: 32 - Race: Black - Released on 15 Jun 1815.

Freeman, Prince - Seaman - Nbr: 52 - How taken: Gave himself up from HMS Boyne - Date Received: 2 Apr 1813 - From what ship: Plymouth - Born: Gloucester - Age: 34 - Sent to Chatham on 27 May 1813.

Freeman, Prince - Seaman - Nbr: 1806 - How taken: Gave himself up from HMS Boyne - When taken: 22 Jun 1813 - Date Received: 20 Jul 1814 - From what ship: HMS Milford, Plymouth - Born: Brewster - Age: 34 - Released on 28 Apr 1815.

Frees, James - Seaman - Nbr: 4175 - Prize: Orbit - Ship Type: MV - How taken: HMS Achates - When taken: 29 Jan 1813 - Where taken: Lat 44N Long 13W - Date Received: 7 Oct 1814 - From what ship: HMT Niobe, Chatham - Born: Lancaster - Age: 40 - Released on 22 Dec 1814.

Frenage, Michael - Marine - Nbr: 4062 - Prize: US Brig Rattlesnake - Ship Type: MW - How taken: HMS Leander - When taken: 13 Jul 1814 - Where taken: off Shelburne - Date Received: 6 Oct 1814 - From what ship: HMT Chesapeake, Halifax - Born: Danzig - Age: 40 - Released on 13 Jun 1815.

Frenair, Domingo - Seaman - Nbr: 2548 - Prize: Caromaned - Ship Type: Prize - How taken: HMS Eridanus - When taken: 13 Aug 1814 - Where taken: Lat 40, Long 16 - Date Received: 16 Aug 1814 - From what ship: HMT Salvador del Mundo - Born: Canary Islands - Age: 25 - Released on 3 May 1815.

Fresher, Henry - Seaman - Nbr: 5515 - Prize: Mary - Ship Type: Prize - How taken: HMS Wasp - When taken: 6 Oct 1814 - Date rec'd: 17 Dec 1814 - From what ship: HMT Loire, Halifax - Born: Portsmouth - Age: 45.

Frewan, Nathaniel - Seaman - Nbr: 5010 - Prize: Charlotte - Ship Type: Prize - How taken: Censor - When taken: 5 Aug 1814 - Where taken: off Azores - Date Received: 28 Oct 1814 - From what ship: HMT Alkbar, Halifax - Born: L'Orient (France) - Age: 23 - Released on 27 Apr 1815.

Friday, John - Steward - Nbr: 1642 - Prize: Paul Jones - Ship Type: P - How taken: HMS Leonidas - When taken: 23 May 1813 - Where taken: Channel - Date Received: 23 Jun 1814 - From what ship: Stapleton - Born: New Orleans - Age: 22 - Race: Colored - Sent to Mill Prison (Plymouth, England) on 30 Jun 1814.

Frizle, John - Seaman - Nbr: 3456 - Prize: Prize to Chasseur - How taken: HMS Tartarus - When taken: 1 Sep 1814 - Where taken: Atlantic - Date Received: 19 Sep 1814 - From what ship: HMT Salvador del Mundo - Born: Baltimore - Age: 27 - Released on 4 Jun 1815.

Frizzle, David - Seaman - Nbr: 1750 - How taken: Gave himself up from HMS Thames - When taken: 30 Dec 1812 - Date Received: 20 Jul 1814 - From what ship: HMS Milford, Plymouth - Born: Charlestown - Age: 33 - Released on 26 Apr 1815.

Frobus, Henry - Seaman - Nbr: 2347 - Prize: Snap Dragon - Ship Type: P - How taken: HMS Martin - When taken:

American Prisoners of War Held at Dartmoor during the War of 1812

## Alphabetical listing of names

10 Jun 1814 - Where taken: off Halifax - Date Received: 16 Aug 1814 - From what ship: HMS Dublin, Halifax - Born: Baltimore - Age: 18 - Released on 3 May 1815.

Frost, Daphne - Seaman - Nbr: 5513 - Prize: Mary - Ship Type: Prize - How taken: HMS Wasp - When taken: 6 Oct 1814 - Date Received: 17 Dec 1814 - From what ship: HMT Loire, Halifax - Born: Massachusetts - Age: 19.

Frost, John - Seaman - Nbr: 2997 - Prize: St. Lawrence - How taken: HMS Aquilon - When taken: 9 Aug 1814 - Where taken: off Western Islands (England) - Date Received: 29 Aug 1814 - From what ship: HMT Bittern - Born: Westchester - Age: 22 - Released on 28 May 1815.

Frost, Thomas Bell - Prize master - Nbr: 6274 - Prize: Nancy - Ship Type: Prize - How taken: HMS Granicus - When taken: 27 Dec 1814 - Where taken: off Cape L'Orient (France) - Date Received: 17 Feb 1815 - From what ship: HMT Ganges, Plymouth - Born: Newcastle - Age: 19 - Released on 11 Jul 1815.

Fry, A. L. - Seaman - Nbr: 4806 - Prize: President - Ship Type: P - How taken: HMS Pique - When taken: 7 May 1814 - Where taken: off Porto Rico - Date Received: 9 Oct 1814 - From what ship: HMT Freya, Halifax - Born: Rhode Island - Age: 38 - Race: Black - Released on 21 Jun 1815.

Fry, John - Carpenter - Nbr: 510 - Prize: Hannah & Eliza - Ship Type: MV - How taken: HMS Lyra - When taken: 29 May 1813 - Where taken: off Bayonne - Date Received: 8 Sep 1813 - From what ship: Plymouth - Born: Andover - Age: 27 - Released on 26 Apr 1815.

Fry, Peter - Seaman - Nbr: 2004 - Prize: Governor Middleton - Ship Type: MV - How taken: Thetis (Privateer) - When taken: 2 May 1813 - Where taken: Bay of Biscay - Date Received: 3 Aug 1814 - From what ship: HMS Lyffey, Chatham Depot - Born: Alexandria - Age: 23 - Released on 2 May 1815.

Fry, Thomas - Seaman - Nbr: 3208 - How taken: Gave himself up from HMS Achilles - When taken: 8 Sep 1813 - Date Received: 11 Sep 1814 - From what ship: HMT Freya, Chatham - Born: Newport - Age: 36 - Released on 28 May 1815.

Fulgar, William - Seaman - Nbr: 4082 - Prize: New Zealander - Ship Type: Prize - How taken: HMS Belviders - When taken: 21 Apr 1814 - Where taken: off Delaware - Date Received: 6 Oct 1814 - From what ship: HMT Chesapeake, Halifax - Born: East Chester - Age: 19 - Released on 13 Jun 1815.

Fuller, Benjamin - Seaman - Nbr: 5480 - Prize: General Putnam - Ship Type: P - How taken: HMS Leander - When taken: 8 Nov 1814 - Where taken: Long 65 Lat 42 - Date Received: 17 Dec 1814 - From what ship: HMT Loire, Halifax - Born: Salem - Age: 18 - Released on 1 Jul 1815.

Fuller, Bernard - Seaman - Nbr: 6541 - How taken: Gave himself up from HMS Transit - When taken: 24 Nov 1814 - Date Received: 7 Mar 1815 - From what ship: HMT Ganges, Plymouth - Born: New Jersey - Age: 22 - Released on 11 Jul 1815.

Fuller, Enoch - Seaman - Nbr: 1155 - How taken: Impressed at Cork - When taken: 20 Mar 1814 - Date Received: 10 May 1814 - From what ship: Plymouth - Born: Norfolk - Age: 22 - Released on 28 Apr 1815.

Fuller, Gideon - Seaman - Nbr: 1766 - Prize: Cygnet - Ship Type: Prize - How taken: HMS Statira - When taken: 8 May 1814 - Where taken: at sea - Date Received: 20 Jul 1814 - From what ship: HMS Milford, Plymouth - Born: Boston - Age: 19 - Race: Mulatto - Released on 1 May 1815.

Fuller, Nathaniel - Seaman - Nbr: 3291 - Prize: Enterprize - Ship Type: P - How taken: HMS Tenedos - When taken: 21 May 1813 - Where taken: off Cape Cod - Date Received: 13 Sep 1814 - From what ship: HMT Niobe, Chatham - Born: Ipswich - Age: 22 - Released on 28 May 1815.

Fuller, Nathaniel - Seaman - Nbr: 3712 - Prize: Alfred - Ship Type: P - How taken: HMS Epervier - When taken: 23 Feb 1814 - Where taken: off Newfoundland - Date Received: 30 Sep 1814 - From what ship: HMT President, Halifax - Born: Ipswich - Age: 54 - Released on 4 Jun 1815.

Fuller, Thomas - Seaman - Nbr: 4330 - Prize: Martha - Ship Type: Prize - How taken: HMS Belviders - When taken: 25 Feb 1814 - Where taken: off New London - Date Received: 7 Oct 1814 - From what ship: HMT Salvador del Mundo, Halifax - Born: Salem - Age: 26 - Released on 14 Jun 1815.

Fulton, James - Seaman - Nbr: 5319 - Prize: Industry - Ship Type: P - How taken: HMS Maidstone - When taken: 17 Jul 1813 - Where taken: Grand Banks - Date Received: 31 Oct 1814 - From what ship: HMT Leyden, Chatham - Born: Marblehead - Age: 15 - Released on 29 Jun 1815.

Fulton, William - Prize master - Nbr: 5017 - Prize: Alexander - Ship Type: Prize - How taken: HMS Wasp - When taken: 6 Sep 1814 - Where taken: off Cape Sable Island (Canada) - Date Received: 28 Oct 1814 - From what ship: HMT Alkbar, Halifax - Born: Baltimore - Age: 24 - Released on 21 Jun 1815.

Furguson, L. - Seaman - Nbr: 5617 - How taken: Gave himself up from HMS Ashea - When taken: 13 May 1814 - Date Received: 24 Dec 1814 - From what ship: HMT Tay - Born: New Hampshire - Age: 28 - Released on 11 Jul 1815.

Furlong, William - 2nd Mate - Nbr: 979 - Prize: Siro - Ship Type: LM - How taken: HMS Pelican - When taken: 13

American Prisoners of War Held at Dartmoor during the War of 1812

## Alphabetical listing of names

Jan 1814 - Where taken: at sea - Date Received: 31 Jan 1814 - From what ship: Plymouth - Born: Baltimore - Age: 19 - Sent to London on 30 May 1814.

Furness, Jesse - Seaman - Nbr: 4263 - Prize: Argus - Ship Type: MV - How taken: HMS San Domingo - When taken: 1 Mar 1814 - Where taken: off Savannah - Date Received: 7 Oct 1814 - From what ship: HMT Niobe, Chatham - Born: New Jersey - Age: 24 - Released on 13 Jun 1815.

Furs, Theodore - Seaman - Nbr: 5066 - Prize: Ida - Ship Type: LM - How taken: HMS Newcastle - When taken: 9 Aug 1814 - Where taken: Long 34 - Date Received: 28 Oct 1814 - From what ship: HMT Alkbar, Halifax - Born: Boston - Age: 24 - Released on 29 Jun 1815.

Gabbett, Henry - Seaman - Nbr: 4055 - Prize: US Brig Rattlesnake - Ship Type: MW - How taken: HMS Leander - When taken: 13 Jul 1814 - Where taken: off Shelburne - Date Received: 6 Oct 1814 - From what ship: HMT Chesapeake, Halifax - Born: Prussia - Age: 34 - Released on 13 Jun 1815.

Gable, John - Seaman - Nbr: 906 - Prize: Squirrel - Ship Type: MV - How taken: HMS Belle Poule - When taken: 14 Dec 1813 - Where taken: at sea - Date Received: 31 Jan 1814 - From what ship: Plymouth - Born: New York - Age: 29 - Race: Mulatto - Sent to Plymouth on 8 Jul 1814.

Gabriel, John - Seaman - Nbr: 1466 - Prize: Governor Gerry - Ship Type: MV - How taken: HMS Lyra - When taken: 29 May 1813 - Where taken: Bay of Biscay - Date Received: 19 Jun 1814 - From what ship: Stapleton - Born: New Orleans - Age: 30 - Race: Negro - Sent to Mill Prison (Plymouth, England) on 26 Jun 1814.

Gage, Isaac - Seaman - Nbr: 1662 - Prize: Governor Gerry - Ship Type: MV - How taken: HMS Royalist - When taken: 31 May 1813 - Where taken: Bay of Biscay - Date Received: 23 Jun 1814 - From what ship: Stapleton - Born: Baltimore - Age: 28 - Released on 1 May 1815.

Gage, Lot - Seaman - Nbr: 4261 - Prize: Argus - Ship Type: MV - How taken: HMS San Domingo - When taken: 1 Mar 1814 - Where taken: off Savannah - Date Received: 7 Oct 1814 - From what ship: HMT Niobe, Chatham - Born: Barnstable - Age: 19 - Released on 13 Jun 1815.

Gair, John - Seaman - Nbr: 5769 - Prize: Rambler - Ship Type: MV - How taken: Morley (Transport) - When taken: 10 Feb 1813 - Where taken: off Isle of France - Date Received: 26 Dec 1814 - From what ship: HMT Argo - Born: Boston - Age: 26 - Released on 11 Jul 1815.

Gair, Thomas - Lieutenant - Nbr: 4794 - Prize: Rattlesnake - Ship Type: MV - How taken: HMS Rhin - When taken: 11 Mar 1814 - Where taken: off San Domingo - Date Received: 9 Oct 1814 - From what ship: HMT Freya, Halifax - Born: Boston - Age: 30 - Released on 28 May 1815.

Gale, Amos - Armourer - Nbr: 3976 - Prize: US Brig Rattlesnake - Ship Type: MW - How taken: HMS Leander - When taken: 13 Jul 1814 - Where taken: off Shelburne - Date Received: 6 Oct 1814 - From what ship: HMT Chesapeake, Halifax - Born: New Haven - Age: 26 - Released on 9 Jun 1815.

Gale, Benjamin G. - 2nd Master - Nbr: 3676 - Prize: Alfred - Ship Type: P - How taken: HMS Epervier - When taken: 23 Feb 1812 - Where taken: off Newfoundland - Date Received: 30 Sep 1814 - From what ship: HMT President - Born: Marblehead - Age: 23 - Released on 4 Jun 1815.

Gale, Edward - Seaman - Nbr: 3993 - Prize: US Brig Rattlesnake - Ship Type: MW - How taken: HMS Leander - When taken: 13 Jul 1814 - Where taken: off Shelburne - Date Received: 6 Oct 1814 - From what ship: HMT Chesapeake, Halifax - Born: Salem - Age: 18 - Released on 9 Jun 1815.

Gale, Russell - Seaman - Nbr: 6438 - Prize: Chance - Ship Type: P - How taken: HMS Statira - When taken: 1 Apr 1814 - Where taken: Lat 38 Long 24 - Date Received: 3 Mar 1815 - From what ship: HMT Ganges, Plymouth - Born: North Carolina - Age: 18 - Released on 14 Jun 1815.

Gale, Samuel - Sailing master - Nbr: 5463 - Prize: General Putnam - Ship Type: P - How taken: HMS Leander - When taken: 8 Nov 1814 - Where taken: Long 65 Lat 42 - Date Received: 17 Dec 1814 - From what ship: HMT Loire, Halifax - Born: Salem - Age: 25 - Released on 1 Jul 1815.

Gale, William - Seaman - Nbr: 4156 - Prize: Argus - Ship Type: MV - How taken: HMS San Domingo - When taken: 1 Mar 1814 - Where taken: off Savannah - Date Received: 6 Oct 1814 - From what ship: HMT Niobe, Chatham - Born: Baltimore - Age: 28 - Race: Black - Released on 13 Jun 1815.

Gall, Michael - Boatswain's Mate - Nbr: 944 - Prize: Siro - Ship Type: LM - How taken: HMS Pelican - When taken: 13 Jan 1814 - Where taken: at sea - Date Received: 31 Jan 1814 - From what ship: Plymouth - Born: Palermo - Age: 25 - Released on 27 Apr 1815.

Galloway, Joseph - Seaman - Nbr: 648 - Prize: Helen and Emeline - Ship Type: MV - How taken: HMS Telegraph - When taken: 13 Aug 1813 - Where taken: off St. Anders - Date Received: 8 Sep 1813 - From what ship: Plymouth - Born: Talbot County - Age: 21 - Sent to Dartmouth on 30 Jul 1813.

Galloway, Joseph - Seaman - Nbr: 9 - Prize: Cashier - Ship Type: LM - How taken: HMS Reindeer - When taken: 3 Feb 1813 - Where taken: Bay of Biscay - Date Received: 2 Apr 1813 - From what ship: Plymouth - Born:

# American Prisoners of War Held at Dartmoor during the War of 1812

## Alphabetical listing of names

Talbot County - Age: 21 - Sent to Dartmouth on 30 Jul 1813.

Gallibrandt, Bernard - Seaman - Nbr: 485 - Prize: Henry Clements - Ship Type: MV - How taken: HMS Orestes - When taken: 13 Apr 1813 - Where taken: Bay of Biscay - Date Received: 8 Sep 1813 - From what ship: Plymouth - Born: Leghorn, Italy - Age: 45 - Sent to Plymouth on 8 Jul 1814.

Gallin, John - Seaman - Nbr: 883 - Prize: Charlotte - Ship Type: MV - How taken: HMS Dwarf - When taken: 4 Nov 1813 - Where taken: off Bordeaux - Date Received: 31 Jan 1814 - From what ship: Plymouth - Born: Africa - Age: 21 - Released on 27 Apr 1815.

Gamble, James - Seaman - Nbr: 5762 - How taken: Impressed at London - When taken: Oct 1814 - Date Received: 26 Dec 1814 - From what ship: HMT Argo - Born: Delaware - Age: 28 - Released on 3 Jul 1815.

Gandell, Epne - Seaman - Nbr: 496 - How taken: Impressed at Belfast - When taken: 16 Jan 1813 - Date Received: 8 Sep 1813 - From what ship: Plymouth - Born: Boston - Age: 24 - Released on 20 Apr 1815.

Ganell, James - Seaman - Nbr: 3764 - Prize: Ann - Ship Type: Prize - How taken: HMS Lacedaemonian - When taken: 10 Jul 1814 - Where taken: off Charleston - Date Received: 30 Sep 1814 - From what ship: HMT President, Halifax - Born: Maryland - Age: 24 - Released on 9 Jun 1815.

Ganganes, Edward - Soldier - Nbr: 5267 - Prize: 6th US Infantry - Ship Type: Troops - How taken: British Army - When taken: 13 Oct 1812 - Where taken: Canada - Date Received: 31 Oct 1814 - From what ship: HMT Leyden, Chatham - Born: Donegal - Age: 30 - Released on 29 Jun 1815.

Gangler, George - Seaman - Nbr: 1274 - How taken: Sent into custody from MV Cygnet - Date Received: 14 Jun 1814 - From what ship: Mill Prison (Plymouth, England) - Born: Boston - Age: 29 - Released on 28 Apr 1815.

Gannett, Mathew - Seaman - Nbr: 2919 - How taken: Impressed at Shields - When taken: 13 Aug 1813 - Date Received: 24 Aug 1814 - From what ship: HMT Alpheus, Chatham - Born: Massachusetts - Age: 26 - Released on 19 May 1815.

Ganster, John - Seaman - Nbr: 6332 - Prize: Prince de Neufchatel - Ship Type: P - How taken: Leander (Newcastle Acasta) - When taken: 20 Dec 1814 - Where taken: Lat 38 Long 56 - Date Received: 19 Feb 1815 - From what ship: HMT Ganges, Plymouth - Born: Boston - Age: 25 - Released on 5 Jul 1815.

Garbonne, John - Mate - Nbr: 6465 - Prize: Decatur - Ship Type: P - How taken: HMS Rhin - When taken: 5 Jun 1814 - Where taken: off San Domingo - Date Received: 3 Mar 1815 - From what ship: HMT Ganges, Plymouth - Born: Charleston - Age: 30 - Race: Mulatto - Released on 19 May 1815.

Garcia, Francis - Seaman - Nbr: 475 - Prize: Zebra - Ship Type: LM - How taken: HMS Pyramus - When taken: 20 Apr 1813 - Where taken: Bay of Biscay - Date Received: 8 Sep 1813 - From what ship: Plymouth - Born: New Orleans - Age: 39 - Sent to Plymouth on 8 Jul 1814.

Garder, George - Seaman - Nbr: 3209 - How taken: Gave himself up from HMS Achilles - When taken: 8 Sep 1813 - Date Received: 11 Sep 1814 - From what ship: HMT Freya, Chatham - Born: New York - Age: 21 - Race: Black - Released on 28 May 1815.

Gardiner, John - Seaman - Nbr: 1465 - Prize: Revenge - Ship Type: LM - How taken: HMS Belle Poule - When taken: 10 Mar 1813 - Where taken: off Cornwall - Date Received: 19 Jun 1814 - From what ship: Stapleton - Born: Boston - Age: 21 - Race: Negro - Released on 28 Apr 1815.

Gardner, Andrew - Seaman - Nbr: 6217 - Prize: Prince de Neufchatel - Ship Type: P - How taken: Leander (Newcastle Acasta) - When taken: 20 Dec 1814 - Where taken: Lat 38 Long 56 - Date Received: 30 Jan 1815 - From what ship: HMT Pheasant - Born: Boston - Age: 18 - Released on 5 Jul 1815.

Gardner, Daniel - Seaman - Nbr: 5441 - How taken: Sent into custody from MV David - When taken: 9 Oct 1814 - Date Received: 11 Nov 1814 - From what ship: HMT Impregnable - Born: Nantucket - Age: 22 - Race: Mulatto - Released on 1 Jul 1815.

Gardner, Edward - Seaman - Nbr: 1024 - Prize: Joseph - Ship Type: MV - How taken: HMS Royalist - When taken: 18 Jan 1814 - Where taken: Bay of Biscay - Date Received: 31 Jan 1814 - From what ship: Plymouth - Born: South Kingston - Age: 23 - Released on 28 May 1815.

Gardner, Edward - 2nd Mate - Nbr: 1059 - Prize: Fair American - Ship Type: MV - How taken: HMS Andromache - When taken: 19 Jan 1814 - Where taken: Bay of Biscay - Date Received: 10 May 1814 - From what ship: Plymouth - Born: Nantucket - Age: 23 - Released on 27 Apr 1815.

Gardner, James - Seaman - Nbr: 3186 - How taken: Gave himself up from HMS Hibernia - When taken: 27 Jul 1813 - Date Received: 11 Sep 1814 - From what ship: HMT Freya, Chatham - Born: Hartford - Age: 31 - Released on 28 May 1815.

Gardner, Jerry - Seaman - Nbr: 4739 - How taken: Gave himself up from a brig - When taken: 10 Dec 1813 - Date Received: 9 Oct 1814 - From what ship: HMT Freya, Chatham - Born: Rhode Island - Age: 27 - Race: Black

American Prisoners of War Held at Dartmoor during the War of 1812

## Alphabetical listing of names

- Died on 1 Mar 1815 from variola.

Gardner, John - Seaman - Nbr: 5028 - Prize: Landrail - Ship Type: Prize - How taken: HMS Wasp - When taken: 27 Jul 1814 - Where taken: Georges Bank - Date Received: 28 Oct 1814 - From what ship: HMT Alkbar, Halifax - Born: New Jersey - Age: 30 - Released on 21 Jun 1815.

Gardner, John - Seaman - Nbr: 3338 - Prize: Prize of the President - Ship Type: MV - How taken: HMS Regulus - When taken: 4 Sep 1813 - Where taken: off Newfoundland - Date Received: 13 Sep 1814 - From what ship: HMT Niobe, Chatham - Born: Salem - Age: 22 - Released on 28 May 1815.

Gardner, Jonathan - Seaman - Nbr: 2770 - How taken: Gave himself up from HMS Moncour - When taken: 27 Apr 1813 - Date Received: 24 Aug 1814 - From what ship: HMT Liverpool, Chatham - Born: Ingram - Age: 43 - Released on 19 May 1815.

Gardner, Joseph - Seaman - Nbr: 1109 - Prize: Bunker Hill - Ship Type: P - How taken: HMS Pomone & HMS Cadmus - When taken: 4 Mar 1814 - Where taken: Bay of Biscay - Date Received: 10 May 1814 - From what ship: Plymouth - Born: Beverly - Age: 25 - Released on 27 Apr 1815.

Gardner, Peter - Seaman - Nbr: 1844 - How taken: Gave himself up from HMS Royal Sovereign - When taken: 19 May 1813 - Date Received: 21 Jul 1814 - From what ship: HMT Redbeard & Pincher, Chatham Depot - Born: New York - Age: 24 - Released on 2 May 1815.

Gardner, Timothy - Seaman - Nbr: 3953 - Prize: Rolla - Ship Type: P - How taken: HMS Loire - When taken: 10 Dec 1813 - Where taken: off Bull Island (South Carolina) - Date Received: 5 Oct 1814 - From what ship: HMT President, Halifax - Born: Rhode Island - Age: 19 - Race: Negro - Died on 15 Jan 1814 from variola.

Garnder, Francis - Seaman - Nbr: 2948 - Prize: Flath - Ship Type: P - How taken: HMS Ramillies - Date Received: 24 Aug 1814 - From what ship: HMT Hannibal - Born: North Carolina - Age: 16 - Race: Negro - Died on 4 Nov 1814.

Garner, James - Seaman - Nbr: 585 - How taken: Impressed at Dublin - When taken: 25 Jul 1813 - Date Received: 8 Sep 1813 - From what ship: Plymouth - Born: Boston - Age: 23 - Released on 26 Apr 1815.

Garner, William - Seaman - Nbr: 2670 - Prize: Orders in Council - Ship Type: MV - How taken: HMS Surveillante - When taken: 1 Jun 1813 - Where taken: off Cape Ortegal (Spain) - Date Received: 21 Aug 1814 - From what ship: HMT Freya, Chatham - Born: Boston - Age: 20 - Released on 10 Apr 1815.

Garney, John - Gunner's mate - Nbr: 3980 - Prize: US Brig Rattlesnake - Ship Type: MW - How taken: HMS Leander - When taken: 13 Jul 1814 - Where taken: off Shelburne - Date Received: 6 Oct 1814 - From what ship: HMT Chesapeake, Halifax - Born: Marblehead - Age: 26 - Released on 9 Jun 1815.

Garreai, Anthony - Seaman - Nbr: 1127 - Prize: Marie Christiana - Ship Type: P - How taken: HMS Pactolus - When taken: 25 Mar 1814 - Where taken: off France - Date Received: 10 May 1814 - From what ship: Plymouth - Born: Malaga - Age: 31 - Released on 28 Apr 1815.

Garret, James - Seaman - Nbr: 304 - Prize: Ducornau - Ship Type: MV - How taken: HMS Pheasant - When taken: 15 Mar 1813 - Where taken: Bay of Biscay - Date Received: 28 Jun 1813 - From what ship: Plymouth - Born: Charlestown - Age: 20 - Released on 30 Mar 1815.

Garrett, L. P. - Seaman - Nbr: 3382 - Prize: Growler - Ship Type: P - How taken: HMS Electra - When taken: 7 Jul 1813 - Where taken: at sea - Date Received: 13 Sep 1814 - From what ship: HMT Niobe, Chatham - Born: Virginia - Age: 21 - Released on 30 Mar 1815.

Garrett, Robert - Seaman - Nbr: 3613 - Prize: Hornsby - Ship Type: P - How taken: HMS Sceptre - When taken: 4 Feb 1814 - Where taken: off Block Island - Date Received: 30 Sep 1814 - From what ship: HMT Sybella - Born: New York - Age: 21 - Released on 4 Jun 1815.

Garrett, William - Seaman - Nbr: 800 - Prize: Collin - Ship Type: MV - How taken: HMS Helicon and HMS Whiting - When taken: 26 Oct 1813 - Where taken: Channel - Date Received: 3 Nov 1813 - From what ship: Plymouth - Born: New Jersey - Age: 33 - Released on 10 Apr 1815.

Garrison, Babet - Seaman - Nbr: 5012 - Prize: Charlotte - Ship Type: Prize - How taken: Censor - When taken: 5 Aug 1814 - Where taken: off Azores - Date Received: 28 Oct 1814 - From what ship: HMT Alkbar, Halifax - Born: Baltimore - Age: 26 - Race: Negro - Released on 21 Jun 1815.

Garrison, Charles - Seaman - Nbr: 4071 - Prize: Hussar - Ship Type: P - How taken: HMS Saturn - When taken: 24 May 1814 - Where taken: off Sandy Hook - Date Received: 6 Oct 1814 - From what ship: HMT Chesapeake, Halifax - Born: New York - Age: 17 - Released on 13 Jun 1815.

Garrison, Charles - Seaman - Nbr: 4366 - How taken: Gave himself up from HMS President - When taken: 27 Apr 1813 - Date Received: 8 Oct 1814 - From what ship: HMT Leyden, Chatham - Born: New Jersey - Age: 27 - Released on 14 Jun 1815.

Garrison, Cornelius - Seaman - Nbr: 5003 - Prize: Invincible - Ship Type: LM - How taken: HMS Armide - When

American Prisoners of War Held at Dartmoor during the War of 1812

## Alphabetical listing of names

taken: 15 Aug 1814 - Where taken: off Nantucket - Date Received: 28 Oct 1814 - From what ship: HMT Alkbar, Halifax - Born: Baltimore - Age: 24 - Released on 21 Jun 1815.

Garrison, James - Seaman - Nbr: 5868 - Prize: Lion - Ship Type: P - How taken: HMS Granicus - When taken: 2 Dec 1814 - Where taken: off Lisbon - Date Received: 26 Dec 1814 - From what ship: HMT Impregnable - Born: Norway - Age: 28 - Released on 3 Jul 1815.

Garrison, Richard - Seaman - Nbr: 1001 - How taken: Impressed at Falmouth - When taken: 1 Nov 1813 - Date Received: 31 Jan 1814 - From what ship: Plymouth - Born: Norfolk - Age: 18 - Released on 27 Apr 1815.

Garthy, James - Seaman - Nbr: 1828 - Prize: Lightning - Ship Type: MV - How taken: HMS Medusa - When taken: 2 Apr 1813 - Where taken: Bay of Biscay - Date Received: 21 Jul 1814 - From what ship: HMT Redbeard & Pincher, Chatham Depot - Born: Philadelphia - Age: 21 - Released on 1 May 1815.

Gasey, Raymond - Seaman - Nbr: 1051 - How taken: Sent into custody from Mary (Transport) - Date Received: 10 May 1814 - From what ship: Plymouth - Born: New Orleans - Age: 57 - Sent to Plymouth on 8 Jul 1814.

Gaskin, William - Seaman - Nbr: 5246 - How taken: Gave himself up from HMS Queen Charlotte - When taken: 6 Sep 1814 - Date Received: 31 Oct 1814 - From what ship: HMT Mermaid, Chatham - Born: Norfolk - Age: 26 - Race: Black - Released on 29 Jun 1815.

Gaston, John - Seaman - Nbr: 5385 - How taken: Gave himself up at Chatham - When taken: 28 Aug 1814 - Date Received: 31 Oct 1814 - From what ship: HMT Leyden, Chatham - Born: North Carolina - Age: 33 - Released on 14 Jun 1815.

Gatchell, Benjamin - Seaman - Nbr: 5991 - Prize: McDonough - Ship Type: P - How taken: HMS Bacchante - When taken: 1 Nov 1814 - Where taken: Lat 42 Long 67 - Date Received: 27 Dec 1814 - From what ship: HMT Penelope - Born: Wales - Age: 35 - Released on 3 Jul 1815.

Gatchell, John G. - Prize master - Nbr: 4438 - Prize: Growler - Ship Type: P - How taken: HMS Electra - When taken: 7 Jul 1814 - Where taken: at sea - Date Received: 8 Oct 1814 - From what ship: HMT Leyden, Chatham - Born: Massachusetts - Age: 26 - Released on 14 Jun 1815.

Gatewood, James - Seaman - Nbr: 1113 - Prize: Bunker Hill - Ship Type: P - How taken: HMS Pomone & HMS Cadmus - When taken: 4 Mar 1814 - Where taken: Bay of Biscay - Date Received: 10 May 1814 - From what ship: Plymouth - Born: Portsmouth - Age: 33 - Race: Mulatto - Died on 17 Feb 1815 from variola.

Gault, William - Seaman - Nbr: 1781 - Prize: William Bayard - Ship Type: MV - How taken: HMS Warspite - When taken: 3 Mar 1813 - Where taken: Bay of Biscay - Date Received: 20 Jul 1814 - From what ship: HMS Milford, Plymouth - Born: New York - Age: 21 - Released on 1 May 1815.

Gavet, James - Gunner - Nbr: 4606 - Prize: Argus - Ship Type: MV - How taken: HMS San Domingo - When taken: 1 Mar 1814 - Where taken: off Savannah - Date Received: 9 Oct 1814 - From what ship: HMT Leyden, Chatham - Born: Salem - Age: 26 - Released on 15 Jun 1815.

Gayer, Joseph - Seaman - Nbr: 1455 - Prize: Revenge - Ship Type: LM - How taken: HMS Belle Poule - When taken: 10 Mar 1813 - Where taken: off Cornwall - Date Received: 19 Jun 1814 - From what ship: Stapleton - Born: Boston - Age: 28 - Sent to Dartmouth on 2 Nov 1813.

Gayler, James - Seaman - Nbr: 2541 - How taken: Gave himself up from HMS America - Date Received: 16 Aug 1814 - From what ship: HMT Salvador del Mundo - Born: North Carolina - Age: 32 - Died on 3 Dec 1814 from phthisis pulmonalis.

Geline, John - Seaman - Nbr: 1769 - How taken: Gave himself up from HMS Trent - Date Received: 20 Jul 1814 - From what ship: Chatham - Born: New York - Age: 30 - Released on 1 May 1815.

Gellens, William - Seaman - Nbr: 223 - Prize: Pert - Ship Type: MV - How taken: HMS Warspite - When taken: 1 Mar 1813 - Where taken: off Basque Roads (France) - Date Received: 2 Apr 1813 - From what ship: Plymouth - Born: Philadelphia - Age: 20 - Released on 20 Apr 1815.

Gennison, Michael - Seaman - Nbr: 5025 - Prize: Landrail - Ship Type: Prize - How taken: HMS Wasp - When taken: 27 Jul 1814 - Where taken: Georges Bank - Date Received: 28 Oct 1814 - From what ship: HMT Alkbar, Halifax - Born: Baltimore - Age: 25 - Died on 12 Nov 1814.

Gennodo, Samuel H. - Prize Master - Nbr: 1044 - Prize: Prince of Wales - Ship Type: P - How taken: Nelson (Transport) - When taken: 4 Feb 1814 - Where taken: at sea - Date Received: 10 May 1814 - From what ship: Plymouth - Born: Rhode Island - Age: 32 - Released on 27 Apr 1815.

Gentile, Joseph - Seaman - Nbr: 11 - Prize: Cashier - Ship Type: LM - How taken: HMS Reindeer - When taken: 3 Feb 1813 - Where taken: Bay of Biscay - Date Received: 2 Apr 1813 - From what ship: Plymouth - Born: New Orleans - Age: 28 - Sent to Dartmouth on 30 Jul 1813.

Gently, Thomas - Seaman - Nbr: 5477 - Prize: General Putnam - Ship Type: P - How taken: HMS Leander - When taken: 8 Nov 1814 - Where taken: Long 65 Lat 42 - Date Received: 17 Dec 1814 - From what ship: HMT

American Prisoners of War Held at Dartmoor during the War of 1812

## Alphabetical listing of names

Loire, Halifax - Born: Massachusetts - Age: 18 - Released on 3 Jul 1815.

George W. W. - Seaman - Nbr: 3273 - Prize: Rachael - Ship Type: MV - How taken: HMS Herring - When taken: 9 Feb 1813 - Where taken: at sea - Date Received: 13 Sep 1814 - From what ship: HMT Niobe, Chatham - Born: Marblehead - Age: 20 - Released on 28 May 1815.

George, John - Seaman - Nbr: 4839 - Prize: President - Ship Type: P - How taken: HMS Pique - When taken: 7 May 1814 - Where taken: off Porto Rico - Date Received: 9 Oct 1814 - From what ship: HMT Freya, Halifax - Born: Boston - Age: 23 - Race: Black - Released on 21 Jun 1815.

George, Thomas - Seaman - Nbr: 4000 - Prize: US Brig Rattlesnake - Ship Type: MW - How taken: HMS Leander - When taken: 13 Jul 1814 - Where taken: off Shelburne - Date Received: 6 Oct 1814 - From what ship: HMT Chesapeake, Halifax - Born: Norfolk - Age: 28 - Released on 9 Jun 1815.

Gerard, Henry - Seaman - Nbr: 2383 - Prize: US Sloop Frolic - Ship Type: MW - How taken: HMS Orpheus - When taken: 20 Apr 1814 - Where taken: off Cuba - Date Received: 16 Aug 1814 - From what ship: HMT Queen, Halifax - Born: Philadelphia - Age: 18 - Sent to Dartmouth on 19 Oct 1814.

Gerard, Pierre - Seaman - Nbr: 5874 - Prize: Lion - Ship Type: P - How taken: HMS Granicus - When taken: 2 Dec 1814 - Where taken: off Lisbon - Date Received: 26 Dec 1814 - From what ship: HMT Impregnable - Born: L'Orient (France) - Age: 18 - Released on 3 Jul 1815.

Gerard, William - Seaman - Nbr: 1070 - Prize: Fair American - Ship Type: MV - How taken: HMS Andromache - When taken: 19 Jan 1814 - Where taken: Bay of Biscay - Date Received: 10 May 1814 - From what ship: Plymouth - Born: Boston - Age: 44 - Released on 27 Apr 1815.

Gibbs, Daniel - Seaman - Nbr: 2790 - Prize: Weazel - Ship Type: MV - How taken: HMS Foxhound - When taken: 25 Mar 1813 - Where taken: Bay of Biscay - Date Received: 24 Aug 1814 - From what ship: HMT Liverpool, Chatham - Born: Newport - Age: 19 - Released on 19 May 1815.

Gibbs, Henry - Seaman - Nbr: 1335 - Prize: Paul Jones - Ship Type: P - How taken: HMS Leonidas - When taken: 23 May 1813 - Where taken: Channel - Date Received: 19 Jun 1814 - From what ship: Stapleton - Born: Massachusetts - Age: 18 - Released on 28 Apr 1815.

Gibbs, Thomas - Seaman - Nbr: 4933 - Prize: Herald - Ship Type: P - How taken: HMS Endymion - When taken: 15 Aug 1814 - Where taken: off Nantucket - Date Received: 28 Oct 1814 - From what ship: HMT Alkbar, Halifax - Born: Providence - Age: 21 - Released on 21 Jun 1815.

Gibbs, Valentine - Seaman - Nbr: 3229 - How taken: Gave himself up from HMS Curacoe - When taken: 3 May 1813 - Date Received: 11 Sep 1814 - From what ship: HMT Freya, Chatham - Born: North Carolina - Age: 29 - Released on 28 May 1815.

Gibbs, William - Seaman - Nbr: 538 - How taken: Impressed at Cork - When taken: 25 Jul 1813 - Date Received: 8 Sep 1813 - From what ship: Plymouth - Born: Springfield - Age: 48 - Released on 26 Apr 1815.

Gibley, Thomas - Boy - Nbr: 2360 - Prize: Snap Dragon - Ship Type: P - How taken: HMS Martin - When taken: 10 Jun 1814 - Where taken: off Halifax - Date Received: 16 Aug 1814 - From what ship: HMT Queen, Halifax - Born: North Carolina - Age: 11 - Released on 3 May 1815.

Gibson, Francis - Seaman - Nbr: 6549 - How taken: Gave himself up from HMS Imogene - When taken: Jun 1814 - Date Received: 7 Mar 1815 - From what ship: HMT Ganges, Plymouth - Born: Pennsylvania - Age: 42 - Race: Black - Released on 11 Jul 1815.

Gibson, Richard - Seaman - Nbr: 4876 - Prize: Hussar - Ship Type: P - How taken: HMS Saturn - When taken: 24 May 1814 - Where taken: off Sandy Hook - Date Received: 24 Oct 1814 - From what ship: Royal Hospital, Plymouth - Born: Long Island - Age: 35 - Released on 21 Jun 1815.

Gibson, Samuel - Seaman - Nbr: 4710 - How taken: Gave himself up from HMS Burisey - When taken: 3 Feb 1814 - Date Received: 9 Oct 1814 - From what ship: HMT Freya, Chatham - Born: Oronoke - Age: 22 - Race: Mulatto - Released on 11 Jul 1815.

Gibson, William - Seaman - Nbr: 3981 - Prize: US Brig Rattlesnake - Ship Type: MW - How taken: HMS Leander - When taken: 13 Jul 1814 - Where taken: off Shelburne - Date Received: 6 Oct 1814 - From what ship: HMT Chesapeake, Halifax - Born: New York - Age: 22 - Died on 22 Oct 1814 from pneumonia.

Giddens, John - Seaman - Nbr: 2186 - Prize: Diomede - Ship Type: P - How taken: HMS Rifleman - When taken: 28 Jun 1813 - Where taken: off Halifax - Date Received: 16 Aug 1814 - From what ship: HMS Dublin, Halifax - Born: Danvers - Age: 19 - Released on 2 May 1815.

Giddings, John - Marine - Nbr: 6321 - Prize: Prince de Neufchatel - Ship Type: P - How taken: Leander (Newcastle Acasta) - When taken: 20 Dec 1814 - Where taken: Lat 38 Long 56 - Date Received: 19 Feb 1815 - From what ship: HMT Ganges, Plymouth - Born: Danvers - Age: 19 - Released on 5 Jul 1815.

Giddins, Hiram - Seaman - Nbr: 4480 - Prize: Enterprize - Ship Type: P - How taken: HMS Tenedos - When taken:

American Prisoners of War Held at Dartmoor during the War of 1812

## Alphabetical listing of names

21 May 1813 - Where taken: off Cape Cod - Date Received: 8 Oct 1814 - From what ship: HMT Leyden, Chatham - Born: Massachusetts - Age: 23 - Released on 15 Jun 1815.

Giddons, Andrew - Seaman - Nbr: 3899 - Prize: Robert - Ship Type: MV - How taken: HMS Loire - When taken: 10 Jul 1814 - Where taken: Chesapeake Bay - Date Received: 5 Oct 1814 - From what ship: HMT Orpheus, Halifax - Born: Baltimore - Age: 39 - Race: Black - Released on 9 Jun 1815.

Gier, Alexander - Seaman - Nbr: 5122 - How taken: Gave himself up from HMS Invincible - When taken: 10 Oct 1814 - Date Received: 28 Oct 1814 - From what ship: HMT Salvador del Mundo - Born: Boston - Age: 23 - Released on 29 Jun 1815.

Gifford, Robert - Seaman - Nbr: 368 - How taken: Impressed at sea - When taken: 4 Apr 1813 - Date Received: 1 Jul 1813 - From what ship: Plymouth - Born: Rhode Island - Age: 29 - Released on 9 Apr 1815.

Gilbert, David - Seaman - Nbr: 3709 - Prize: Alfred - Ship Type: P - How taken: HMS Epervier - When taken: 23 Feb 1814 - Where taken: off Newfoundland - Date Received: 30 Sep 1814 - From what ship: HMT President, Halifax - Born: Marblehead - Age: 18 - Released on 4 Jun 1815.

Gilbert, George - Seaman - Nbr: 2102 - Prize: Ferox - Ship Type: MV - How taken: HMS Medusa & HMS Lyra - When taken: 28 Mar 1813 - Where taken: off Cape Ortegal (Spain) - Date Received: 3 Aug 1814 - From what ship: HMS Bittern, Chatham Depot - Born: Gloucester - Age: 17 - Released on 2 May 1815.

Gilbert, Gurdonel - Seaman - Nbr: 1036 - Prize: US Brig Argus - Ship Type: MW - How taken: HMS Pelican - When taken: 16 Aug 1813 - Where taken: Irish Channel - Date Received: 10 May 1814 - From what ship: Plymouth - Born: Baltimore - Age: 26 - Sent to Dartmouth on 19 Oct 1814.

Gilbert, Henry - Seaman - Nbr: 3541 - Prize: Hawk - Ship Type: P - How taken: HMS Pique - When taken: 26 Apr 1814 - Where taken: off Bermuda - Date Received: 30 Sep 1814 - From what ship: HMT Sybella - Born: North Carolina - Age: 27 - Released on 4 Jun 1815.

Gilbert, Isaac - Seaman - Nbr: 3325 - Prize: York Town - Ship Type: P - How taken: HMS Nimrod - When taken: 17 Jul 1813 - Where taken: off St. Johns - Date Received: 13 Sep 1814 - From what ship: HMT Niobe, Chatham - Born: New York - Age: 23 - Released on 28 May 1815.

Gilbert, John - Landsman - Nbr: 105 - Prize: St. Martin's Planter - Ship Type: P - How taken: HMS Dublin - When taken: 9 Feb 1813 - Where taken: Lat 43 N, Long 33 50 W - Date Received: 2 Apr 1813 - From what ship: Plymouth - Born: New York - Age: 20 - Released on 20 Apr 1815.

Gilbert, Thomas - Seaman - Nbr: 169 - How taken: Impressed at Liverpool - When taken: 15 Jan 1813 - Date Received: 2 Apr 1813 - What ship: Plymouth - Born: Norfolk - Age: 30 - Sent to Dartmouth on 30 Jul 1813.

Gilbert, Thomas - Seaman - Nbr: 14 - Prize: Cashier - Ship Type: LM - How taken: HMS Reindeer - When taken: 3 Feb 1813 - Where taken: Bay of Biscay - Date Received: 2 Apr 1813 - From what ship: Plymouth - Born: Philadelphia - Age: 52 - Sent to Dartmouth on 30 Jul 1813.

Gilbert, Thomas - Seaman - Nbr: 2020 - How taken: Gave himself up from HMS Talbot - When taken: 1 May 1813 - Date Received: 3 Aug 1814 - From what ship: HMS Lyffey, Chatham Depot - Born: Wilmington - Age: 40 - Released on 2 May 1815.

Gilchrist, John - Seaman - Nbr: 899 - Prize: Squirrel - Ship Type: MV - How taken: HMS Belle Poule - When taken: 14 Dec 1813 - Where taken: at sea - Date Received: 31 Jan 1814 - From what ship: Plymouth - Born: New York - Age: 24 - Released on 27 Apr 1815.

Gilchrist, Samuel - Seaman - Nbr: 2943 - Prize: Atalanta - Ship Type: MV - How taken: HMS Barbados - When taken: 19 Jan 1814 - Where taken: off St. Bartholomew - Date Received: 24 Aug 1814 - From what ship: HMT Hannibal - Born: Norfolk - Age: 32 - Race: Black - Released on 19 May 1815.

Giles, Edward - Seaman - Nbr: 525 - Prize: Friends - Ship Type: Bey of Pool - How taken: HMS Whiting - When taken: 15 Jul 1813 - Where taken: Lat 67 N, Long 8 W - Date Received: 8 Sep 1813 - From what ship: Plymouth - Born: Marblehead - Age: 26 - Released on 26 Apr 1815.

Giles, John - Seaman - Nbr: 5309 - Prize: Stark - Ship Type: P - How taken: Surrendered at off Barbados - When taken: 13 Nov 1813 - Date Received: 31 Oct 1814 - From what ship: HMT Leyden, Chatham - Born: Boothbay - Age: 29 - Released on 29 Jun 1815.

Gill, James - Soldier - Nbr: 5255 - Prize: 13th US Infantry - Ship Type: Troops - How taken: British Army - When taken: 13 Oct 1812 - Where taken: Canada - Date Received: 31 Oct 1814 - From what ship: HMT Leyden, Chatham - Born: Sligo - Age: 21 - Released on 29 Jun 1815.

Gill, John - Seaman - Nbr: 3983 - Prize: US Brig Rattlesnake - Ship Type: MW - How taken: HMS Leander - When taken: 13 Jul 1814 - Where taken: off Shelburne - Date Received: 6 Oct 1814 - From what ship: HMT Chesapeake, Halifax - Born: New York - Age: 28 - Released on 9 Jun 1815.

Gillett, Francis - Boy - Nbr: 1120 - Prize: Hope - Ship Type: P - How taken: HMS Sea Horse - When taken: 22 Mar

American Prisoners of War Held at Dartmoor during the War of 1812

## Alphabetical listing of names

1814 - Where taken: at sea - Date Received: 10 May 1814 - From what ship: Plymouth - Born: Brest - Age: 17 - Sent to Mill Prison (Plymouth, England) on 21 Jun 1814.

Gillia, Joseph - Seaman - Nbr: 1173 - Prize: Ville de Milan - Ship Type: MV - How taken: HMS Leander - When taken: 10 Apr 1805 - Where taken: off Bermuda - Date Received: 4 Jun 1814 - From what ship: Dartmouth - Born: New Orleans - Age: 23 - Sent to Plymouth on 8 Jul 1814.

Gilliken, Daniel - Seaman - Nbr: 3834 - Prize: Snap Dragon - Ship Type: P - How taken: HMS Martin - When taken: 31 May 1814 - Where taken: off Charleston - Date Received: 5 Oct 1814 - From what ship: HMT Orpheus, Halifax - Born: North Carolina - Age: 17 - Released on 9 Jun 1815.

Gillstone, George - Seaman - Nbr: 841 - Prize: Chesapeake - Ship Type: LM - How taken: HMS Hotspur & HMS Pyramus - When taken: 26 Oct 1813 - Where taken: off Nantes - Date Received: 29 Nov 1813 - From what ship: Plymouth - Born: New Jersey - Age: 32 - Race: Black - Released on 26 Apr 1815.

Gilmore, William H. - Seaman - Nbr: 484 - Prize: Tom - Ship Type: LM - How taken: HMS Surveillante - When taken: 27 Apr 1813 - Where taken: Bay of Biscay - Date Received: 8 Sep 1813 - From what ship: Plymouth - Born: Pennsylvania - Age: 23 - Released on 20 Apr 1815.

Gilpin, William - Seaman - Nbr: 1275 - How taken: Sent into custody from MV Bittern - Date Received: 14 Jun 1814 - From what ship: Mill Prison (Plymouth, England) - Born: Amboy - Age: 24 - Released on 28 Apr 1815.

Girdler, James - Seaman - Nbr: 5213 - Prize: Porcupine - Ship Type: LM - How taken: HMS Acasta - When taken: 18 Jul 1813 - Where taken: off Halifax - Date Received: 31 Oct 1814 - From what ship: HMT Mermaid, Chatham - Born: Manchester - Age: 18 - Released on 29 Jun 1815.

Givell, Oliver - Seaman - Nbr: 573 - Prize: US Brig Argus - Ship Type: MW - How taken: HMS Pelican - When taken: 14 Apr 1813 - Where taken: Irish Channel - Date Received: 8 Sep 1813 - From what ship: Plymouth - Born: New Orleans - Age: 28 - Sent to Dartmouth on 2 Nov 1814.

Gladding, Joseph - Seaman - Nbr: 2957 - Prize: Rattlesnake - Ship Type: LM - How taken: HMS Rhin - When taken: 10 Mar 1814 - Where taken: West Indies - Date Received: 24 Aug 1814 - From what ship: HMT Hannibal - Born: Rhode Island - Age: 37 - Died on 14 Mar 1815 from febris.

Gladding, William - Quarter Gunner - Nbr: 4011 - Prize: US Brig Rattlesnake - Ship Type: MW - How taken: HMS Leander - When taken: 13 Jul 1814 - Where taken: off Shelburne - Date Received: 6 Oct 1814 - From what ship: HMT Chesapeake, Halifax - Born: New Jersey - Age: 37 - Released on 9 Jun 1815.

Glaridge, Stephen - Boy - Nbr: 5675 - Prize: Harlequin - Ship Type: P - How taken: HMS Bulwark - When taken: 23 Nov 1814 - Where taken: off Halifax - Date Received: 24 Dec 1814 - From what ship: HMT Penelope - Born: Portsmouth - Age: 18 - Released on 3 Jul 1815.

Glashon, Richard - Lieutenant - Nbr: 2563 - Prize: Rattlesnake - Ship Type: P - How taken: HMS Hyperion - When taken: 25 Jun 1814 - Where taken: off Cape Finisterre - Date Received: 21 Aug 1814 - From what ship: HMS Hyperion - Born: New York - Age: 24 - Escaped on 3 Nov 1814.

Glass, John - Prize master - Nbr: 5536 - Prize: Ann - Ship Type: Prize - How taken: HMS Hamadryad - When taken: 10 Sep 1814 - Where taken: Lat 41 Long 23 - Date Received: 17 Dec 1814 - From what ship: HMT Loire, Halifax - Born: Virginia - Age: 29 - Released on 1 Jul 1815.

Glover, Allen - Lieutenant Marines - Nbr: 2512 - Prize: Snap Dragon - Ship Type: P - How taken: HMS Martin - When taken: 10 Jun 1814 - Where taken: off Halifax - Date Received: 16 Aug 1814 - From what ship: HMT Queen, Halifax - Born: North Carolina - Age: 21 - Released on 3 May 1815.

Glover, Benjamin - Seaman - Nbr: 4728 - How taken: Gave himself up from MV John - When taken: 28 Aug 1814 - Date Received: 9 Oct 1814 - From what ship: HMT Freya, Chatham - Born: Massachusetts - Age: 23 - Released on 15 Jun 1815.

Glover, John - Boy - Nbr: 4673 - Prize: Industry - Ship Type: P - How taken: HMS Heron - When taken: 3 Nov 1813 - Where taken: off Halifax - Date Received: 9 Oct 1814 - From what ship: HMT Leyden, Chatham - Born: Marblehead - Age: 18 - Released on 15 Jun 1815.

Glover, John - Seaman - Nbr: 34 - How taken: Apprehended at Gibraltar - When taken: 8 Aug 1813 - Date Received: 2 Apr 1813 - What ship: Plymouth - Born: Massachusetts - Age: 35 - Sent to Dartmouth on 30 Jul 1813.

Glover, Samuel - Seaman - Nbr: 3191 - How taken: Gave himself up from HMS Swiftsure - When taken: 26 Dec 1812 - Date Received: 11 Sep 1814 - From what ship: HMT Freya, Chatham - Born: Carolina - Age: 21 - Released on 27 Apr 1815.

Glover, William - Seaman - Nbr: 6357 - Prize: Prince de Neufchatel - Ship Type: P - How taken: Leander (Newcastle Acasta) - When taken: 20 Dec 1814 - Where taken: Lat 38 Long 56 - Date Received: 19 Feb

American Prisoners of War Held at Dartmoor during the War of 1812

## Alphabetical listing of names

1815 - From what ship: HMT Ganges, Plymouth - Born: Salem - Age: 39 - Released on 6 Jul 1815.

Glynn, Hanson - Officer - Nbr: 3826 - Prize: Quiz - Ship Type: LM - How taken: HMS Niemen - When taken: 22 May 1814 - Where taken: off Egg Harbor (New Jersey) - Date Received: 5 Oct 1814 - From what ship: HMT Orpheus, Halifax - Born: Maryland - Age: 52 - Released on 9 Jun 1815.

Godfrey, Edward - Seaman - Nbr: 637 - Prize: US Brig Argus - Ship Type: MW - How taken: HMS Pelican - When taken: 14 Aug 1813 - Where taken: Irish Channel - Date Received: 8 Sep 1813 - From what ship: Plymouth - Born: Pennsylvania - Age: 21 - Race: Black - Sent to Dartmouth on 2 Nov 1814.

Godfrey, John - Seaman - Nbr: 4860 - Prize: Globe - Ship Type: MV - How taken: Wybrow (Privateer) - When taken: 1 Jan 1814 - Where taken: off St. Bartholomew - Date Received: 9 Oct 1814 - From what ship: HMT Freya, Halifax - Born: Dresden - Age: 23 - Released on 21 Jun 1815.

Godfrey, William - Seaman - Nbr: 1330 - Prize: Paul Jones - Ship Type: P - How taken: HMS Leonidas - When taken: 23 May 1813 - Where taken: Channel - Date Received: 19 Jun 1814 - From what ship: Stapleton - Born: Rhode Island - Age: 19 - Race: Colored - Released on 28 Apr 1815.

Goff, James - Seaman - Nbr: 234 - Prize: Charlotte - Ship Type: MV - How taken: HMS Warspite - When taken: 3 Mar 1813 - Where taken: Bay of Biscay - Date Received: 2 Apr 1813 - From what ship: Plymouth - Born: New York - Age: 20 - Released on 20 Apr 1815.

Golandre, John - Seaman - Nbr: 1005 - Prize: Zephyr - Ship Type: MV - How taken: HMS Surveillante - When taken: 6 Jan 1814 - Where taken: Bay of Biscay - Date Received: 31 Jan 1814 - From what ship: Plymouth - Born: Stockholm - Age: 18 - Released on 8 Nov 1814.

Golden, Edward - Master - Nbr: 1686 - Prize: LA - Ship Type: MV - How taken: Two Tender - When taken: 25 Jan 1813 - Where taken: West Indies - Date Received: 2 Jul 1814 - From what ship: Plymouth - Born: New York - Age: 23 - Released on 1 May 1815.

Golding, Abijah - Seaman - Nbr: 4596 - How taken: Impressed at London - When taken: 4 Apr 1814 - Date Received: 9 Oct 1814 - From what ship: HMT Leyden, Chatham - Born: Lancaster - Age: 27 - Released on 15 Jun 1815.

Golding, William - Seaman - Nbr: 5545 - Prize: Phaeton - Ship Type: MV - How taken: Cast Away - When taken: 7 Feb 1814 - Where taken: Old Heights - Date Received: 17 Dec 1814 - From what ship: HMT Loire, Halifax - Born: North Carolina - Age: 18 - Released on 1 Jul 1815.

Goldsbury, William - Seaman - Nbr: 832 - Prize: Chesapeake - Ship Type: LM - How taken: HMS Hotspur & HMS Pyramus - When taken: 26 Oct 1813 - Where taken: off Nantes - Date Received: 29 Nov 1813 - From what ship: Plymouth - Born: Maryland - Age: 32 - Released on 26 Apr 1815.

Goldsmith, Nathaniel - Seaman - Nbr: 4048 - Prize: US Brig Rattlesnake - Ship Type: MW - How taken: HMS Leander - When taken: 13 Jul 1814 - Where taken: off Shelburne - Date Received: 6 Oct 1814 - From what ship: HMT Chesapeake, Halifax - Born: Marblehead - Age: 34 - Released on 13 Jun 1815.

Goldsmith, Nathaniel - Boatswain's Mate - Nbr: 4013 - Prize: US Brig Rattlesnake - Ship Type: MW - How taken: HMS Leander - When taken: 13 Jul 1814 - Where taken: off Shelburne - Date Received: 6 Oct 1814 - From what ship: HMT Chesapeake, Halifax - Born: New Hampshire - Age: 33 - Released on 9 Jun 1815.

Goldsmith, Thomas - Corporal - Nbr: 5623 - Prize: Rambler - Ship Type: MV - How taken: HMS Lion - When taken: 1 Feb 1813 - Where taken: Cape - Date Received: 24 Dec 1814 - From what ship: HMT Tay - Born: Massachusetts - Age: 38 - Released on 3 Jul 1815.

Golf, John - Seaman - Nbr: 2349 - Prize: Snap Dragon - Ship Type: P - How taken: HMS Martin - When taken: 10 Jun 1814 - Where taken: off Halifax - Date Received: 16 Aug 1814 - From what ship: HMS Dublin, Halifax - Born: North Carolina - Age: 19 - Released on 11 Jul 1815.

Golf, Peter - Boatswain - Nbr: 4392 - Prize: Portsmouth Packet - Ship Type: Prize - How taken: HMS Fantome - When taken: 5 Oct 1813 - Where taken: off Portland - Date Received: 8 Oct 1814 - From what ship: HMT Leyden, Chatham - Born: Connecticut - Age: 28 - Released on 14 Jun 1815.

Golliver, Thomas - Seaman - Nbr: 5490 - Prize: General Putnam - Ship Type: P - How taken: HMS Leander - When taken: 8 Nov 1814 - Where taken: Long 65 Lat 42 - Date Received: 17 Dec 1814 - From what ship: HMT Loire, Halifax - Born: Marblehead - Age: 57 - Released on 1 Jul 1815.

Golliver, William - Seaman - Nbr: 2147 - How taken: Gave himself up from HMS Vigo (Spain) - When taken: 31 Dec 1812 - Date Received: 8 Aug 1814 - From what ship: HMT Raven, Chatham - Born: Boston - Age: 40 - Released on 26 Apr 1815.

Goodall, William - Seaman - Nbr: 2272 - Prize: Success - Ship Type: Prize - How taken: HMS Charybdis - When taken: 29 May 1814 - Where taken: off Cape Logan - Date Received: 16 Aug 1814 - From what ship: HMS Dublin, Halifax - Born: Maryland - Age: 26 - Released on 3 May 1815.

American Prisoners of War Held at Dartmoor during the War of 1812

## Alphabetical listing of names

Goodhall, Joseph - Seaman - Nbr: 6347 - Prize: Prince de Neufchatel - Ship Type: P - How taken: Leander (Newcastle Acasta) - When taken: 20 Dec 1814 - Where taken: Lat 38 Long 56 - Date Received: 19 Feb 1815 - From what ship: HMT Ganges, Plymouth - Born: Bordeaux - Age: 33 - Released on 6 Jul 1815.

Goodings, Thomas - Seaman - Nbr: 5917 - Prize: Harlequin - Ship Type: P - How taken: HMS Bulwark - When taken: 23 Nov 1814 - Where taken: off Halifax - Date Received: 27 Dec 1814 - From what ship: HMT Penelope - Born: Massachusetts - Age: 43 - Released on 20 Mar 1815.

Goodman, Caleb - Seaman - Nbr: 4019 - Prize: US Brig Rattlesnake - Ship Type: MW - How taken: HMS Leander - When taken: 13 Jul 1814 - Where taken: off Shelburne - Date Received: 6 Oct 1814 - From what ship: HMT Chesapeake, Halifax - Born: Baltimore - Age: 22 - Released on 9 Jun 1815.

Goodrich, Edward - Sailing master - Nbr: 6175 - Prize: David Porter - Ship Type: LM - How taken: HMS Pylades - When taken: 12 Sep 1814 - Where taken: Georges Bank - Date Received: 18 Jan 1815 - From what ship: HMT Impregnable - Born: Gloucester - Age: 23 - Escaped on 1 Jun 1815.

Goodwin, John - Seaman - Nbr: 5187 - Prize: Catherine - Ship Type: P - How taken: HMS La Hogue - When taken: 2 May 1813 - Where taken: off Cape Sable Island (Canada) - Date Received: 31 Oct 1814 - From what ship: HMT Mermaid, Chatham - Born: Fort George - Age: 27 - Released on 29 Jun 1815.

Goodwin, Simon - Marine - Nbr: 5896 - Prize: Harlequin - Ship Type: P - How taken: HMS Bulwark - When taken: 23 Nov 1814 - Where taken: off Halifax - Date Received: 27 Dec 1814 - From what ship: HMT Penelope - Born: Arundel - Age: 27 - Released on 3 Jul 1815.

Goodwin, William - Seaman - Nbr: 1413 - Prize: Orders in Council - Ship Type: LM - How taken: HMS Surveillante - When taken: 1 Jun 1813 - Where taken: off Cape Ortegal (Spain) - Date Received: 19 Jun 1814 - From what ship: Stapleton - Born: New York - Age: 29 - Released on 28 Apr 1815.

Gooley, James - Seaman - Nbr: 2772 - How taken: Gave himself up from HMS Leonidas - Date Received: 24 Aug 1814 - From what ship: HMT Liverpool, Chatham - Born: Boston - Age: 31 - Released on 19 May 1815.

Gordan, William - Steward - Nbr: 2857 - How taken: Apprehended at London - When taken: 29 Jun 1813 - Date Received: 24 Aug 1814 - From what ship: HMT Alpheus, Chatham - Born: Portsmouth - Age: 33 - Released on 19 May 1815.

Gordnow, S. B. - Gunner's mate - Nbr: 943 - Prize: Siro - Ship Type: LM - How taken: HMS Pelican - When taken: 13 Jan 1814 - Where taken: at sea - Date Received: 31 Jan 1814 - From what ship: Plymouth - Born: Boston - Age: 25 - Released on 27 Apr 1815.

Gordon, Abraham - Seaman - Nbr: 2924 - How taken: Gave himself up from HMS Union - When taken: 27 May 1813 - Date Received: 24 Aug 1814 - From what ship: HMT Alpheus, Chatham - Born: New York - Age: 26 - Race: Black - Released on 19 May 1815.

Gordon, James - Seaman - Nbr: 2870 - How taken: Gave himself up from HMS Scorpion - When taken: 27 May 1813 - Date Received: 24 Aug 1814 - From what ship: HMT Alpheus, Chatham - Born: Massachusetts - Age: 25 - Released on 19 May 1815.

Gordon, John - Seaman - Nbr: 4645 - How taken: Gave himself up from HMS Bucephalus - When taken: 20 Aug 1813 - Date Received: 9 Oct 1814 - From what ship: HMT Leyden, Chatham - Born: Baltimore - Age: 27 - Released on 15 Jun 1815.

Gordon, John - Seaman - Nbr: 3052 - How taken: Gave himself up from HMS Sultan - When taken: 2 Sep 1814 - Date Received: 2 Sep 1814 - From what ship: HMT Sultan - Born: Philadelphia - Age: 29 - Race: Negro - Released on 28 Apr 1815.

Gordon, Joseph - Master's mate - Nbr: 5665 - Prize: Harlequin - Ship Type: P - How taken: HMS Bulwark - When taken: 23 Nov 1814 - Where taken: off Halifax - Date Received: 24 Dec 1814 - From what ship: HMT Penelope - Born: Massachusetts - Age: 32 - Released on 3 Jul 1815.

Gordon, Philip - Seaman - Nbr: 2182 - Prize: Rattlesnake - Ship Type: P - How taken: HMS Hyperion - When taken: 26 Jun 1814 - Where taken: off Cape Finisterre - Date Received: 16 Aug 1814 - From what ship: HMS Dublin, Halifax - Born: New York - Age: 36 - Released on 2 May 1815.

Gordon, Richard - Seaman - Nbr: 4590 - Prize: Bunker Hill - Ship Type: P - How taken: HMS Pomone - When taken: 2 Mar 1814 - Where taken: Channel - Date Received: 9 Oct 1814 - From what ship: HMT Leyden, Chatham - Born: Boston - Age: 30 - Race: Mulatto - Released on 15 Jun 1815.

Gordon, Thomas - Seaman - Nbr: 188 - Prize: Star - Ship Type: MV - How taken: HMS Superb - When taken: 9 Feb 1813 - Where taken: Bay of Biscay - Date Received: 2 Apr 1813 - From what ship: Plymouth - Born: New Orleans - Age: 22 - Released on 20 Apr 1815.

Gordon, William - Seaman - Nbr: 3643 - Prize: Bordeaux Packet - Ship Type: LM - How taken: HMS Niemen - When taken: 28 Jan 1814 - Where taken: off Delaware - Date Received: 30 Sep 1814 - From what ship:

American Prisoners of War Held at Dartmoor during the War of 1812

## Alphabetical listing of names

HMT Sybella - Born: Maryland - Age: 33 - Released on 30 Mar 1815.

Gore, William - Seaman - Nbr: 1658 - How taken: Apprehended at Bristol - When taken: 5 Mar 1814 - Date Received: 23 Jun 1814 - From what ship: Stapleton - Born: Waterford - Age: 20 - Released on 1 May 1815.

Gorham, Emanuel - Seaman - Nbr: 2183 - Prize: Diomede - Ship Type: P - How taken: HMS Rifleman - When taken: 28 Jun 1813 - Where taken: off Halifax - Date Received: 16 Aug 1814 - From what ship: HMS Dublin, Halifax - Born: St. Michael - Age: 45 - Released on 2 May 1815.

Gorling, George C. - Seaman - Nbr: 250 - Prize: William Bayard - Ship Type: MV - How taken: HMS Warspite - When taken: 3 Mar 1813 - Where taken: Bay of Biscay - Date Received: 2 Apr 1813 - From what ship: Plymouth - Born: Prussia - Age: 36 - Sent to Mill Prison (Plymouth, England) on 17 Jun 1814.

Gornerson, James - Seaman - Nbr: 5283 - Prize: Enterprize - Ship Type: P - How taken: HMS Tenedos - When taken: 21 May 1813 - Where taken: off Cape Ann - Date Received: 31 Oct 1814 - From what ship: HMT Leyden, Chatham - Born: Massachusetts - Age: 25 - Released on 29 Jun 1815.

Gorton, William - Seaman - Nbr: 6140 - How taken: Gave himself up from HMS Iphigenia - When taken: 19 Jun 1814 - Date Received: 17 Jan 1815 - From what ship: HMT Impregnable - Born: Rhode Island - Age: 33 - Released on 5 Jul 1815.

Gosling, Joseph - Seaman - Nbr: 4164 - Prize: Commodore Perry - Ship Type: Schooner - How taken: Sent into custody from a cutter - When taken: 25 Feb 1814 - Where taken: off Bordeaux - Date Received: 6 Oct 1814 - From what ship: HMT Niobe, Chatham - Born: Quebec - Age: 26 - Released on 13 Jun 1815.

Goss, Jesse - Seaman - Nbr: 3295 - Prize: Enterprize - Ship Type: P - How taken: HMS Tenedos - When taken: 21 May 1813 - Where taken: off Cape Cod - Date Received: 13 Sep 1814 - From what ship: HMT Niobe, Chatham - Born: Marblehead - Age: 18 - Released on 28 May 1815.

Goss, Richard - Prize Master - Nbr: 3664 - Prize: Alfred - Ship Type: P - How taken: HMS Epervier - When taken: 23 Feb 1812 - Where taken: off Newfoundland - Date Received: 30 Sep 1814 - From what ship: HMT President - Born: Massachusetts - Age: 31 - Released on 11 Jul 1815.

Gotier, Charles J. - Seaman - Nbr: 4597 - Prize: Pilot - Ship Type: LM - How taken: Victoria (Privateer) - When taken: 28 Jan 1814 - Where taken: off Bordeaux - Date Received: 9 Oct 1814 - From what ship: HMT Leyden, Chatham - Born: Salem - Age: 23 - Released on 15 Jun 1815.

Gould, Henry - Seaman - Nbr: 5386 - How taken: Gave himself up at Chatham - When taken: 28 Aug 1814 - Date Received: 31 Oct 1814 - From what ship: HMT Leyden, Chatham - Born: Nantucket - Age: 27 - Race: Black - Released on 1 Jul 1815.

Gould, John - Seaman - Nbr: 4936 - Prize: Herald - Ship Type: P - How taken: HMS Endymion - When taken: 15 Aug 1814 - Where taken: off Nantucket - Date Received: 28 Oct 1814 - From what ship: HMT Alkbar, Halifax - Born: Africa - Age: 44 - Race: Negro - Released on 21 Jun 1815.

Gould, John - Seaman - Nbr: 2063 - Prize: Dart - Ship Type: LM - How taken: HMS Petrel - When taken: 5 Mar 1812 - Where taken: at sea - Date Received: 3 Aug 1814 - From what ship: HMS Bittern, Chatham Depot - Born: Kettering - Age: 25 - Released on 26 Apr 1815.

Gould, Obadiah - Seaman - Nbr: 5138 - Prize: Calabria - Ship Type: MV - How taken: HMS Castilian - When taken: 29 Sep 1814 - Where taken: off Ireland - Date Received: 31 Oct 1814 - From what ship: HMT Castillian - Born: Vermont - Age: 30 - Released on 29 Jun 1815.

Gould, Samuel - Seaman - Nbr: 2314 - Prize: Hussar - Ship Type: P - How taken: HMS Saturn - When taken: 25 May 1814 - Where taken: off Sandy Hook - Date Received: 16 Aug 1814 - From what ship: HMS Dublin, Halifax - Born: New Jersey - Age: 27 - Race: Mulatto - Released on 3 May 1815.

Gould, Samuel - Seaman - Nbr: 5036 - Prize: Betsey - Ship Type: Prize - How taken: HMS Pylades - When taken: 7 Sep 1814 - Where taken: Canten - Date Received: 28 Oct 1814 - From what ship: HMT Alkbar, Halifax - Born: New Jersey - Age: 27 - Released on 29 Jun 1815.

Goulding, Samuel - Seaman - Nbr: 3324 - Prize: York Town - Ship Type: P - How taken: HMS Nimrod - When taken: 17 Jul 1813 - Where taken: off St. Johns - Date Received: 13 Sep 1814 - From what ship: HMT Niobe, Chatham - Born: New York - Age: 25 - Released on 28 May 1815.

Gowalter, John - Seaman - Nbr: 3425 - Prize: Industry - Ship Type: P - How taken: HMS Heron - When taken: 3 Nov 1813 - Where taken: off Halifax - Date Received: 13 Sep 1814 - From what ship: HMT Niobe, Chatham - Born: Marblehead - Age: 22 - Released on 28 May 1815.

Gowner, Joseph - Seaman - Nbr: 6449 - Prize: Jemmell - Ship Type: MV - How taken: HMS Rhin - When taken: 28 May 1814 - Where taken: off Bermuda - Date Received: 3 Mar 1815 - From what ship: HMT Ganges, Plymouth - Born: Nantucket - Age: 38 - Race: Black - Released on 11 Jul 1815.

Grace, Allen - Seaman - Nbr: 529 - How taken: Impressed at Liverpool - When taken: 6 Jul 1813 - Date Received: 8

American Prisoners of War Held at Dartmoor during the War of 1812

## Alphabetical listing of names

Sep 1813 - From what ship: Plymouth - Born: Maryland - Age: 22 - Race: Negro - Released on 26 Apr 1815.

Gracio, Anthony - Seaman - Nbr: 6119 - Prize: Johannes - Ship Type: Prize - How taken: Lacine - When taken: 26 Oct 1814 - Where taken: Grand Banks - Date Received: 6 Jan 1814 - From what ship: HMT Impregnable - Born: Salem - Age: 18 - Released on 5 Jul 1815.

Grafton, James - Seaman - Nbr: 5127 - Prize: US Sloop-of-War Frolic - Ship Type: MW - How taken: HMS Orpheus - When taken: 20 Apr 1814 - Where taken: off Cuba - Date Received: 28 Oct 1814 - From what ship: HMT Salvador del Mundo - Born: Portsmouth - Age: 21 - Released on 3 May 1815.

Graham, Benjamin - Prize Master - Nbr: 3665 - Prize: Alfred - Ship Type: P - How taken: HMS Epervier - When taken: 23 Feb 1812 - Where taken: off Newfoundland - Date Received: 30 Sep 1814 - From what ship: HMT President - Born: Massachusetts - Age: 28 - Released on 4 Jun 1815.

Graham, David - Seaman - Nbr: 578 - Prize: US Brig Argus - Ship Type: MW - How taken: HMS Pelican - When taken: 14 Apr 1813 - Where taken: Irish Channel - Date Received: 8 Sep 1813 - From what ship: Plymouth - Born: Baltimore - Age: 25 - Sent to Dartmouth on 2 Nov 1814.

Graham, James - Seaman - Nbr: 909 - Prize: Zephyr - Ship Type: MV - How taken: HMS Pyramus - When taken: 30 Nov 1813 - Where taken: off L'Orient (France) - Date Received: 31 Jan 1814 - From what ship: Plymouth - Born: New York - Age: 18 - Released on 27 Apr 1815.

Graham, John - Seaman - Nbr: 1053 - How taken: Sent into custody from Mary (Transport) - Date Received: 10 May 1814 - From what ship: Plymouth - Born: Cape Ann - Age: 59 - Released on 27 Apr 1815.

Graham, Maurice - Seaman - Nbr: 2325 - Prize: Hussar - Ship Type: P - How taken: HMS Saturn - When taken: 25 May 1814 - Where taken: off Sandy Hook - Date Received: 16 Aug 1814 - From what ship: HMS Dublin, Halifax - Born: New York - Age: 21 - Released on 3 May 1815.

Graham, Robert - Seaman - Nbr: 3740 - Prize: Lizard - Ship Type: P - How taken: HMS Prometheus - When taken: 5 May 1814 - Where taken: off Halifax - Date Received: 30 Sep 1814 - From what ship: HMT President, Halifax - Born: Cape Ann - Age: 40 - Released on 4 Jun 1815.

Graham, W. R. - Commander - Nbr: 2368 - Prize: Snap Dragon - Ship Type: P - How taken: HMS Martin - When taken: 10 Jun 1814 - Where taken: off Halifax - Date Received: 16 Aug 1814 - From what ship: HMT Queen, Halifax - Born: Virginia - Age: 29 - Released on 3 Apr 1815.

Grahl, Charles - Seaman - Nbr: 3108 - Prize: Rattlesnake - Ship Type: LM - How taken: HMS Orion - When taken: 11 Mar 1814 - Where taken: off Bermuda - Date Received: 11 Sep 1814 - From what ship: HMT Salvador del Mundo - Born: Boston - Age: 17 - Released on 28 May 1815.

Gramber, Gustoff - Boy - Nbr: 2190 - Prize: Diomede - Ship Type: P - How taken: HMS Rifleman - When taken: 28 Jun 1813 - Where taken: off Halifax - Date Received: 16 Aug 1814 - From what ship: HMS Dublin, Halifax - Born: Sweden - Age: 19 - Released on 2 May 1815.

Grandison, William - Seaman - Nbr: 3013 - How taken: Gave himself up from HMS Zealous - When taken: 27 Aug 1814 - Date Received: 2 Sep 1814 - From what ship: HMT Centaur - Born: Massachusetts - Age: 27 - Race: Black - Released on 11 Jul 1815.

Grant, James - Seaman - Nbr: 1135 - How taken: Discharged from HMS Volontaire - When taken: 2 Feb 1814 - Date Received: 10 May 1814 - From what ship: Plymouth - Born: Salem - Age: 27 - Race: Mulatto - Released on 28 Apr 1815.

Grant, Peter - Seaman - Nbr: 276 - Prize: Cannoneer - Ship Type: MV - How taken: HMS Warspite - When taken: 14 Mar 1813 - Where taken: Bay of Biscay - Date Received: 28 Jun 1813 - From what ship: Plymouth - Born: Maryland - Age: 36 - Released on 20 Apr 1815.

Grant, Samuel - Seaman - Nbr: 5921 - Prize: Harlequin - Ship Type: P - How taken: HMS Bulwark - When taken: 23 Nov 1814 - Where taken: off Halifax - Date Received: 27 Dec 1814 - From what ship: HMT Penelope - Born: Massachusetts - Age: 21 - Released on 3 Jul 1815.

Grant, Samuel - Seaman - Nbr: 4409 - Prize: Portsmouth Packet - Ship Type: Prize - How taken: HMS Fantome - When taken: 5 Oct 1813 - Where taken: off Portland - Date Received: 8 Oct 1814 - From what ship: HMT Leyden, Chatham - Born: Massachusetts - Age: 21 - Released on 14 Jun 1815.

Grant, Thomas - Seaman - Nbr: 5494 - Prize: General Putnam - Ship Type: P - How taken: HMS Leander - When taken: 8 Nov 1814 - Where taken: Long 65 Lat 42 - Date Received: 17 Dec 1814 - From what ship: HMT Loire, Halifax - Born: Marblehead - Age: 22 - Released on 1 Jul 1815.

Grant, William - Seaman - Nbr: 4534 - Prize: Wolf Cove - Ship Type: MV - How taken: HMS Briton - When taken: 1 Dec 1812 - Where taken: off Brest - Date Received: 8 Oct 1814 - From what ship: HMT Leyden, Chatham - Born: Massachusetts - Age: 17 - Released on 27 Apr 1815.

Gratton, Edward - Seaman - Nbr: 4003 - Prize: US Brig Rattlesnake - Ship Type: MW - How taken: HMS Leander -

American Prisoners of War Held at Dartmoor during the War of 1812

## Alphabetical listing of names

When taken: 13 Jul 1814 - Where taken: off Shelburne - Date Received: 6 Oct 1814 - From what ship: HMT Chesapeake, Halifax - Born: Virginia - Age: 21 - Released on 9 Jun 1815.

Graves, Ebenezer - Seaman - Nbr: 1111 - Prize: Bunker Hill - Ship Type: P - How taken: HMS Pomone & HMS Cadmus - When taken: 4 Mar 1814 - Where taken: Bay of Biscay - Date Received: 10 May 1814 - From what ship: Plymouth - Born: Marblehead - Age: 24 - Released on 27 Apr 1815.

Graves, Samuel - Seaman - Nbr: 1775 - Prize: Print - Ship Type: MV - How taken: HMS Colossus - When taken: 21 Jan 1813 - Where taken: off Long Island - Date Received: 20 Jul 1814 - From what ship: HMS Milford, Plymouth - Born: Massachusetts - Age: 43 - Released on 1 May 1815.

Graves, Thomas - Seaman - Nbr: 4785 - How taken: Gave himself up from HMS Port Mahon - When taken: 20 Sep 1814 - Date Received: 9 Oct 1814 - From what ship: HMT Freya, Chatham - Born: Boston - Age: 28 - Race: Mulatto - Died on 23 Feb 1815 from variola.

Gray, Ephraim - Seaman - Nbr: 3510 - How taken: Gave himself up from HMS Warspite - When taken: 24 Sep 1814 - Date Received: 28 Sep 1814 - From what ship: HMT Salvador del Mundo - Born: Salem - Age: 36 - Released on 4 Jun 1815.

Gray, Isaac - Seaman - Nbr: 533 - Prize: Union - Ship Type: LM - How taken: HMS Goldfinch - When taken: 17 Jul 1813 - Where taken: Bay of Biscay - Date Received: 8 Sep 1813 - From what ship: Plymouth - Born: Mason - Age: 18 - Released on 26 Apr 1815.

Gray, James - Seaman - Nbr: 3860 - Prize: Thorn - Ship Type: MV - How taken: HMS Bulwark - When taken: 9 Jul 1814 - Where taken: off Nantucket - Date Received: 5 Oct 1814 - From what ship: HMT Orpheus, Halifax - Born: New Jersey - Age: 26 - Race: Negro - Released on 9 Jun 1815.

Gray, Thomas - Seaman - Nbr: 1177 - Prize: Espadron - Ship Type: P - How taken: HMS Rota - When taken: 25 May 1812 - Where taken: Lat 47N Long 7W - Date Received: 4 Jun 1814 - From what ship: Dartmouth - Born: Fairmouth - Age: 20 - Sent to Plymouth on 8 Jul 1814.

Gray, William - Seaman - Nbr: 4291 - How taken: Impressed at Harwich - When taken: 21 Feb 1814 - Date Received: 7 Oct 1814 - From what ship: HMT Niobe, Chatham - Born: Boston - Age: 29 - Released on 14 Jun 1815.

Gray, William (Robert Elliott) - Seaman - Nbr: 5655 - Prize: Herald - Ship Type: P - How taken: HMS Endymion - When taken: 15 Aug 1814 - Where taken: off Nantucket - Date Received: 24 Dec 1814 - From what ship: HMT Impregnable - Born: Not listed - Released on 11 Jul 1815.

Greaves, John - Seaman - Nbr: 3892 - Prize: Julianne - Ship Type: MV - How taken: HMS Dragon - When taken: 10 Jul 1814 - Where taken: off VA - Date Received: 5 Oct 1814 - From what ship: HMT Orpheus, Halifax - Born: Ipswich - Age: 42 - Released on 9 Jun 1815.

Greaves, John - Seaman - Nbr: 2932 - How taken: Gave himself up from HMS Berwick - Date Received: 24 Aug 1814 - What ship: HMT Alpheus, Chatham - Born: Massachusetts - Age: 34 - Race: Black - Released on 19 May 1815.

Green, Charles - Seaman - Nbr: 3276 - Prize: Cossack - Ship Type: P - How taken: HMS Amelia - When taken: 14 Apr 1813 - Where taken: off St. Johns - Date Received: 13 Sep 1814 - From what ship: HMT Niobe, Chatham - Born: Newbury - Age: 21 - Released on 28 May 1815.

Green, Elijah - Seaman - Nbr: 5031 - Prize: Landrail - Ship Type: Prize - How taken: HMS Wasp - When taken: 27 Jul 1814 - Where taken: Georges Bank - Date Received: 28 Oct 1814 - From what ship: HMT Alkbar, Halifax - Born: Virginia - Age: 22 - Released on 21 Jun 1815.

Green, George - Seaman - Nbr: 3202 - How taken: Gave himself up from HMS America - When taken: 16 Jul 1813 - Date Received: 11 Sep 1814 - From what ship: HMT Freya, Chatham - Born: Baltimore - Age: 44 - Race: Black - Released on 28 May 1815.

Green, James - Seaman - Nbr: 2114 - How taken: Gave himself up from HMS Lender - When taken: 24 Oct 1812 - Date Received: 8 Aug 1814 - From what ship: HMT Raven, Chatham - Born: Dorchester - Age: 36 - Released on 26 Apr 1815.

Green, John - Seaman - Nbr: 4150 - Prize: Requin - Ship Type: LM - How taken: HMS Venus - When taken: 6 Mar 1814 - Where taken: off Bordeaux - Date Received: 6 Oct 1814 - From what ship: HMT Niobe, Chatham - Born: Philadelphia - Age: 18 - Released on 13 Jun 1815.

Green, John - Seaman - Nbr: 5004 - Prize: Invincible - Ship Type: LM - How taken: HMS Armide - When taken: 15 Aug 1814 - Where taken: off Nantucket - Date Received: 28 Oct 1814 - From what ship: HMT Alkbar, Halifax - Born: Salisbury - Age: 29 - Released on 21 Jun 1815.

Green, John - Seaman - Nbr: 1742 - Prize: Hugh Jones - Ship Type: MV - How taken: HMS Bittern - When taken: 14 Jun 1814 - Where taken: at sea - Date Received: 18 Jul 1814 - From what ship: HMT Salvador del

American Prisoners of War Held at Dartmoor during the War of 1812

## Alphabetical listing of names

Mundo, Plymouth - Born: New York - Age: 35 - Race: Black - Released on 1 May 1815.

Green, John - Seaman - Nbr: 1948 - How taken: Gave himself up from HMS Cornwall - When taken: 21 May 1813 - Date Received: 3 Aug 1814 - From what ship: HMS Alceste, Chatham Depot - Born: Baltimore - Age: 24 - Released on 2 May 1815.

Green, John - Seaman - Nbr: 3617 - Prize: Monarch - Ship Type: MV - How taken: HMS Dotterel - When taken: 14 Dec 1813 - Where taken: off Charleston - Date Received: 30 Sep 1814 - From what ship: HMT Sybella - Born: West-Cappel (France) - Age: 37 - Released on 4 Jun 1815.

Green, John - Seaman - Nbr: 5153 - Prize: Fox - Ship Type: P - How taken: HMS Shannon - When taken: 7 Nov 1813 - Where taken: off Newfoundland - Date Received: 31 Oct 1814 - From what ship: HMT Mermaid, Chatham - Born: Norfolk - Age: 25 - Race: Black - Released on 29 Jun 1815.

Green, John - Seaman - Nbr: 5521 - Prize: James - Ship Type: Prize - How taken: HMS Galatea - When taken: 7 Sep 1814 - Where taken: Channel - Date Received: 17 Dec 1814 - From what ship: HMT Loire, Halifax - Born: Baltimore - Age: 22 - Released on 1 Jul 1815.

Green, John M. - Seaman - Nbr: 1717 - How taken: Gave himself up from HMS Rodney - Date Received: 8 Jul 1814 - From what ship: Labrador - Born: Hudson - Age: 24 - Released on 1 May 1815.

Green, Joseph W. - Clerk - Nbr: 3666 - Prize: Alfred - Ship Type: P - How taken: HMS Epervier - When taken: 23 Feb 1812 - Where taken: off Newfoundland - Date Received: 30 Sep 1814 - From what ship: HMT President - Born: Massachusetts - Age: 20 - Released on 4 Jun 1815.

Green, Leveret - Seaman - Nbr: 4227 - Prize: Liberty - Ship Type: MV - How taken: Surrendered - When taken: 30 Dec 1813 - Where taken: Stromess - Date Received: 7 Oct 1814 - From what ship: HMT Niobe, Chatham - Born: New Haven - Age: 21 - Released on 13 Jun 1815.

Green, Moses - Seaman - Nbr: 520 - Prize: Friends - Ship Type: Bey of Pool - How taken: HMS Whiting - When taken: 15 Jul 1813 - Where taken: Lat 67 N, Long 8 W - Date Received: 8 Sep 1813 - From what ship: Plymouth - Born: Massachusetts - Age: 19 - Released on 26 Apr 1815.

Green, Peter - Seaman - Nbr: 6247 - Prize: Albion - Ship Type: Prize - How taken: HMS Harlequin - When taken: 6 Jan 1814 - Where taken: Long 20 - Date Received: 4 Feb 1815 - From what ship: HMT Ganges - Born: Salem - Age: 23 - Race: Black - Released on 5 Jul 1815.

Green, Peter - Seaman - Nbr: 852 - Prize: Amiable - Ship Type: LM - How taken: HMS Magnificent - When taken: 30 Oct 1813 - Where taken: off Bordeaux - Date Received: 4 Dec 1813 - From what ship: Plymouth - Born: Philadelphia - Age: 47 - Released on 6 May 1814.

Green, Reuben - Seaman - Nbr: 781 - Prize: Betsy - Ship Type: MV - How taken: HMS Eurotas - When taken: 26 Oct 1813 - Where taken: off Ushant (France) - Date Received: 3 Nov 1813 - From what ship: Plymouth - Born: Connecticut - Age: 22 - Released on 26 Apr 1815.

Green, Robert - Seaman - Nbr: 5760 - Prize: Hope - Ship Type: MV - How taken: HMS Nereus - When taken: 14 May 1814 - Where taken: Rio de la Plata - Date Received: 26 Dec 1814 - From what ship: HMT Argo - Released on 3 Jul 1815.

Green, Samuel - Seaman - Nbr: 5099 - Prize: David Porter - Ship Type: LM - How taken: HMS Pylades - When taken: 12 Sep 1814 - Where taken: Georges Bank - Date Received: 28 Oct 1814 - From what ship: HMT Alkbar, Halifax - Born: Beverly - Age: 30 - Released on 29 Jun 1815.

Green, Solomon - Seaman - Nbr: 1234 - How taken: Sent into custody from HMS Thames - When taken: 15 Nov 1812 - Date Received: 14 Jun 1814 - From what ship: Mill Prison (Plymouth, England) - Born: Baltimore - Age: 38 - Released on 26 Apr 1815.

Green, Thomas - Seaman - Nbr: 4783 - How taken: Gave himself up from HMS Rolus - When taken: 23 Sep 1814 - Date Received: 9 Oct 1814 - From what ship: HMT Freya, Chatham - Born: Boston - Age: 49 - Race: Mulatto - Released on 15 Jun 1815.

Green, Timothy - Passenger - Nbr: 3814 - Prize: Nimble - Ship Type: Prize - How taken: HMS Arab - When taken: 5 Apr 1814 - Where taken: Lat 37 Long 65 - Date Received: 5 Oct 1814 - From what ship: HMT Orpheus, Halifax - Born: Boston - Age: 27 - Released on 9 Jun 1815.

Green, Tobias - Seaman - Nbr: 3752 - Prize: George - Ship Type: Prize - How taken: HMS Recruit - When taken: 29 Aug 1814 - Where taken: off Cape Sable Island (Canada) - Date Received: 30 Sep 1814 - From what ship: HMT President, Halifax - Born: Rhode Island - Age: 36 - Race: Negro - Released on 4 Jun 1815.

Green, Walter - Seaman - Nbr: 3873 - Prize: Tickler - Ship Type: MV - How taken: HMS Saturn - When taken: 13 Jul 1814 - Where taken: off America - Date Received: 5 Oct 1814 - From what ship: HMT Orpheus, Halifax - Born: Rhode Island - Age: 30 - Released on 9 Jun 1815.

Green, William - Seaman - Nbr: 5469 - Prize: General Putnam - Ship Type: P - How taken: HMS Leander - When

American Prisoners of War Held at Dartmoor during the War of 1812

## Alphabetical listing of names

taken: 8 Nov 1814 - Where taken: Long 65 Lat 42 - Date Received: 17 Dec 1814 - From what ship: HMT Loire, Halifax - Born: Newbury - Age: 23 - Released on 1 Jul 1815.

Green, William - Seaman - Nbr: 1351 - Prize: Paul Jones - Ship Type: P - How taken: HMS Leonidas - When taken: 23 May 1813 - Where taken: Channel - Date Received: 19 Jun 1814 - From what ship: Stapleton - Born: New Jersey - Age: 29 - Released on 28 Apr 1815.

Greenfield, James - Seaman - Nbr: 4935 - Prize: Herald - Ship Type: P - How taken: HMS Endymion - When taken: 15 Aug 1814 - Where taken: off Nantucket - Date Received: 28 Oct 1814 - From what ship: HMT Alkbar, Halifax - Born: Wilmington - Age: 32 - Released on 21 Jun 1815.

Greenland, Stephen - 2nd Mate - Nbr: 1220 - Prize: Monticello - Ship Type: MV - How taken: HMS Racehorse - When taken: 12 Nov 1813 - Where taken: off Cape of Good Hope - Date Received: 14 Jun 1814 - From what ship: Mill Prison (Plymouth, England) - Born: Bellerica - Age: 39 - Released on 26 Apr 1815.

Greenlaw, Jeremiah - 3rd Lieutenant - Nbr: 6305 - Prize: Prince de Neufchatel - Ship Type: P - How taken: Leander (Newcastle Acasta) - When taken: 20 Dec 1814 - Where taken: Lat 38 Long 56 - Date Received: 19 Feb 1815 - From what ship: HMT Ganges, Plymouth - Born: Virginia - Age: 21 - Released on 11 Jul 1815.

Greenleaf, Thomas - Seaman - Nbr: 1975 - Prize: Orbit - Ship Type: MV - How taken: HMS Achates - When taken: 29 Jan 1813 - Where taken: Lat 44N Long 13W - Date Received: 3 Aug 1814 - From what ship: HMS Lyffey, Chatham Depot - Born: Startwater - Age: 17 - Released on 2 May 1815.

Greenleaf, William - Boy - Nbr: 2189 - Prize: Diomede - Ship Type: P - How taken: HMS Rifleman - When taken: 28 Jun 1813 - Where taken: off Halifax - Date Received: 16 Aug 1814 - From what ship: HMS Dublin, Halifax - Born: New Hampshire - Age: 13 - Released on 1 May 1815.

Greenough, James - 2nd Lieutenant - Nbr: 5564 - Prize: Harlequin - Ship Type: P - How taken: HMS Bulwark - When taken: 23 Nov 1814 - Where taken: off Halifax - Date Received: 17 Dec 1814 - From what ship: HMT Loire, Halifax - Born: Massachusetts - Age: 40 - Released on 20 Jun 1815.

Greenwood, Joseph - Third Lieutenant - Nbr: 3152 - Prize: Blockade - Ship Type: P - How taken: HMS Charybdis - When taken: 31 Oct 1812 - Where taken: at sea - Date Received: 11 Sep 1814 - From what ship: HMT Freya, Chatham - Born: Connecticut - Age: 27 - Released on 27 Apr 1815.

Greenwood, Thales - Seaman - Nbr: 277 - Prize: Cannoneer - Ship Type: MV - How taken: HMS Warspite - When taken: 14 Mar 1813 - Where taken: Bay of Biscay - Date Received: 28 Jun 1813 - From what ship: Plymouth - Born: Providence - Age: 23 - Released on 20 Apr 1815.

Gregory, Elijah - Seaman - Nbr: 1313 - How taken: Sent into custody from HMS Cornwallis - Date Received: 14 Jun 1814 - From what ship: Mill Prison (Plymouth, England) - Born: New York - Age: 28 - Released on 28 Apr 1815.

Gregory, George - Seaman - Nbr: 4298 - Prize: Squirrel - Ship Type: MV - How taken: HMS Belle Poule - When taken: 14 Dec 1813 - Where taken: at sea - Date Received: 7 Oct 1814 - From what ship: HMT Niobe, Chatham - Born: Maryland - Age: 23 - Released on 14 Jun 1815.

Gregory, Joseph - Seaman - Nbr: 2468 - Prize: US Sloop Frolic - Ship Type: MW - How taken: HMS Orpheus - When taken: 20 Apr 1814 - Where taken: off Cuba - Date Received: 16 Aug 1814 - From what ship: HMT Queen, Halifax - Born: Lisbon - Age: 35 - Released on 3 May 1815.

Grenaux, Yves - Volunteer - Nbr: 1011 - Prize: Harvest - Ship Type: P - How taken: HMS Orestes - When taken: 21 Jan 1814 - Where taken: Bay of Biscay - Date Received: 31 Jan 1814 - From what ship: Plymouth - Born: Brest - Age: 16 - Sent to Mill Prison (Plymouth, England) on 21 Jun 1814.

Greswold, Truman - Clerk - Nbr: 4867 - Prize: Yankee Lass - Ship Type: P - How taken: HMS Surprize - When taken: 1 May 1814 - Where taken: off Western Islands (England) - Date Received: 9 Oct 1814 - From what ship: HMT Freya, Halifax - Born: Connecticut - Age: 25 - Released on 3 May 1815.

Grey, Francis - Seaman - Nbr: 2981 - Prize: Frolic - Ship Type: P - How taken: HMS Heron - When taken: 25 Jan 1814 - Where taken: off St. Thomas - Date Received: 29 Aug 1814 - From what ship: HMT Bittern - Born: Maryland - Age: 21 - Released on 19 May 1815.

Grey, John - Seaman - Nbr: 94 - Prize: St. Martin's Planter - Ship Type: P - How taken: HMS Dublin - When taken: 9 Feb 1813 - Where taken: Lat 43 N, Long 33 50 W - Date Received: 2 Apr 1813 - From what ship: Plymouth - Born: Richmond - Age: 20 - Died on 26 Apr 1815 from amputated arm.

Grey, John - Seaman - Nbr: 2833 - How taken: Gave himself up from HMS Resistance - When taken: 20 Nov 1812 - Date Received: 24 Aug 1814 - From what ship: HMT Liverpool, Chatham - Born: Snow Hill - Age: 29 - Released on 19 May 1815.

Grey, Morehouse - Seaman - Nbr: 1484 - Prize: Napoleon - Ship Type: LM - How taken: HMS Belle Poule - When taken: 3 Apr 1813 - Where taken: off Cape Ortegal (Spain) - Date Received: 19 Jun 1814 - From what ship:

American Prisoners of War Held at Dartmoor during the War of 1812

## Alphabetical listing of names

Stapleton - Born: Connecticut - Age: 25 - Released on 1 May 1815.

Grey, Thomas - Cook - Nbr: 113 - Prize: Rolla - Ship Type: MV - How taken: HMS Surveillante - When taken: 11 Feb 1813 - Where taken: Bay of Biscay - Date Received: 2 Apr 1813 - From what ship: Plymouth - Born: Baltimore - Age: 28 - Race: Black - Released on 20 Apr 1815.

Grey, Thomas - Seaman - Nbr: 2711 - How taken: Gave himself up from HMS Ringdove - When taken: 29 Mar 1813 - Date Received: 21 Aug 1814 - From what ship: HMT Freya, Chatham - Born: New York - Age: 30 - Released on 19 May 1815.

Gribble, George - Seaman - Nbr: 2346 - Prize: Snap Dragon - Ship Type: P - How taken: HMS Martin - When taken: 10 Jun 1814 - Where taken: off Halifax - Date Received: 16 Aug 1814 - From what ship: HMS Dublin, Halifax - Born: North Carolina - Age: 22 - Released on 3 May 1815.

Griffee, Thomas - Seaman - Nbr: 2545 - Prize: Fanny - Ship Type: Prize - How taken: HMS Sceptre - When taken: 1 Jun 1814 - Date Received: 16 Aug 1814 - From what ship: HMT Salvador del Mundo - Born: New Jersey - Age: 39 - Released on 3 May 1815.

Griffen, Andrew - Seaman - Nbr: 5569 - Prize: Daedalus - Ship Type: MV - How taken: HMS Niemen - When taken: 20 Sep 1814 - Where taken: off Delaware - Date Received: 17 Dec 1814 - From what ship: HMT Loire, Halifax - Born: Newport - Age: 27 - Released on 1 Jul 1815.

Griffin, David - Seaman - Nbr: 3990 - Prize: US Brig Rattlesnake - Ship Type: MW - How taken: HMS Leander - When taken: 13 Jul 1814 - Where taken: off Shelburne - Date Received: 6 Oct 1814 - From what ship: HMT Chesapeake, Halifax - Born: Gloucester - Age: 23 - Released on 9 Jun 1815.

Griffin, Heathcoat - Passenger - Nbr: 6427 - Prize: Farmer's Daughter - Ship Type: MV - How taken: HMS Leviathan - When taken: 29 May 1814 - Date Received: 3 Mar 1815 - From what ship: HMT Ganges, Plymouth - Born: New Haven - Age: 30 - Released on 11 Jul 1815.

Griffin, John - Seaman - Nbr: 4149 - Prize: Requin - Ship Type: LM - How taken: HMS Venus - When taken: 6 Mar 1814 - Where taken: off Bordeaux - Date Received: 6 Oct 1814 - From what ship: HMT Niobe, Chatham - Born: Philadelphia - Age: 18 - Race: Mulatto - Released on 13 Jun 1815.

Griffin, John - Seaman - Nbr: 4782 - How taken: Gave himself up from HMS Rolus - When taken: 23 Sep 1814 - Date Received: 9 Oct 1814 - From what ship: HMT Freya, Chatham - Born: Providence - Age: 19 - Race: Mulatto - Released on 11 Jul 1815.

Griffin, John - Seaman - Nbr: 388 - Prize: John & Frances - Ship Type: MV - How taken: HMS Belle Poule - When taken: 13 Mar 1813 - Where taken: Bay of Biscay - Date Received: 1 Jul 1813 - From what ship: Plymouth - Born: LA - Age: 23 - Sent to Mill Prison (Plymouth, England) on 21 Jun 1814.

Griffin, John - Seaman - Nbr: 2258 - Prize: Prize to the Scourge - Ship Type: P - How taken: HMS Martin - When taken: 4 May 1814 - Where taken: off Newfoundland - Date Received: 16 Aug 1814 - From what ship: HMS Dublin, Halifax - Born: Virginia - Age: 33 - Released on 3 May 1815.

Griffin, William - Seaman - Nbr: 666 - Prize: Joel Barlow - Ship Type: LM - How taken: HMS Briton - When taken: 3 Jul 1813 - Where taken: off Bordeaux - Date Received: 8 Sep 1813 - From what ship: Plymouth - Born: Freeport - Age: 21 - Released on 26 Apr 1815.

Griffin, William - Seaman - Nbr: 373 - Prize: Independence - Ship Type: MV - How taken: HMS Superb - When taken: 16 Mar 1813 - Where taken: Bay of Biscay - Date Received: 1 Jul 1813 - From what ship: Plymouth - Born: Marblehead - Age: 20 - Released on 20 Apr 1815.

Griffin, William - Seaman - Nbr: 2266 - Prize: Scourge - Ship Type: P - How taken: HMS Martin - When taken: 27 May 1814 - Where taken: off Cape Sable Island (Canada) - Date Received: 16 Aug 1814 - From what ship: HMS Dublin, Halifax - Born: Marblehead - Age: 49 - Released on 3 May 1815.

Griffin, William - Seaman - Nbr: 5833 - Prize: US Schooner Tigress - Ship Type: MW - How taken: British gunboats - When taken: 2 Sep 1814 - Where taken: Lake Erie - Date Received: 26 Dec 1814 - From what ship: HMT Argo - Born: New York - Age: 23 - Race: Black - Released on 3 Jul 1815.

Griffin, William - Seaman - Nbr: 3386 - Prize: Mary - Ship Type: Prize - How taken: HMS Bellerophon - When taken: 6 Dec 1813 - Where taken: off Land's End - Date Received: 13 Sep 1814 - From what ship: HMT Niobe, Chatham - Born: Philadelphia - Age: 45 - Released on 30 Mar 1815.

Griffiths, Benjamin - Seaman - Nbr: 632 - Prize: US Brig Argus - Ship Type: MW - How taken: HMS Pelican - When taken: 14 Aug 1813 - Where taken: Irish Channel - Date Received: 8 Sep 1813 - From what ship: Plymouth - Born: Chestertown - Age: 23 - Sent to Dartmouth on 2 Nov 1814.

Griffiths, Henry - Quarter Gunner - Nbr: 5735 - Prize: US Schooner Somers - Ship Type: MW - How taken: British gunboats - When taken: 12 Aug 1814 - Where taken: Fort Erie - Date Received: 26 Dec 1814 - From what ship: HMT Argo - Born: Massachusetts - Age: 40 - Released on 3 Jul 1815.

American Prisoners of War Held at Dartmoor during the War of 1812

## Alphabetical listing of names

Griffiths, Joseph - Mate - Nbr: 4266 - Prize: Argus - Ship Type: MV - How taken: HMS San Domingo - When taken: 1 Mar 1814 - Where taken: off Savannah - Date Received: 7 Oct 1814 - From what ship: HMT Niobe, Chatham - Born: Philadelphia - Age: 30 - Released on 13 Jun 1815.

Griffiths, Thomas - Seaman - Nbr: 138 - Prize: Criterion - Ship Type: MV - How taken: HMS Belle Poule - When taken: 14 Feb 1813 - Where taken: Bay of Biscay - Date Received: 2 Apr 1813 - From what ship: Plymouth - Born: New York - Age: 31 - Released on 9 Apr 1815.

Grimes, George - Seaman - Nbr: 4124 - Prize: Bordeaux Packet - Ship Type: LM - How taken: HMS Niemen - When taken: 28 Jun 1814 - Where taken: off Delaware - Date Received: 6 Oct 1814 - From what ship: HMT Chesapeake, Halifax - Born: Alexandria - Age: 26 - Released on 13 Jun 1815.

Grimes, William - Seaman - Nbr: 2323 - Prize: Hussar - Ship Type: P - How taken: HMS Saturn - When taken: 25 May 1814 - Where taken: off Sandy Hook - Date Received: 16 Aug 1814 - From what ship: HMS Dublin, Halifax - Born: Portsmouth - Age: 35 - Released on 3 May 1815.

Grinard, Bray - Seaman - Nbr: 5870 - Prize: Lion - Ship Type: P - How taken: HMS Granicus - When taken: 2 Dec 1814 - Where taken: off Lisbon - Date Received: 26 Dec 1814 - From what ship: HMT Impregnable - Born: Portsmouth - Age: 28 - Released on 3 Jul 1815.

Grindall, Joseph - Seaman - Nbr: 2959 - Prize: US Sloop Frolic - Ship Type: MW - How taken: HMS Orpheus - When taken: 20 Apr 1814 - Where taken: off Cuba - Date Received: 24 Aug 1814 - From what ship: HMT Hannibal - Born: Massachusetts - Age: 25 - Released on 3 May 1815.

Grist, William - Seaman - Nbr: 3551 - Prize: Hawk - Ship Type: P - How taken: HMS Pique - When taken: 26 Apr 1814 - Where taken: off Bermuda - Date Received: 30 Sep 1814 - From what ship: HMT Sybella - Born: Southampton - Age: 27 - Released on 4 Jun 1815.

Groger, Henry - Seaman - Nbr: 602 - Prize: US Brig Argus - Ship Type: MW - How taken: HMS Pelican - When taken: 14 Aug 1813 - Where taken: Irish Channel - Date Received: 8 Sep 1813 - From what ship: Plymouth - Born: Lubbock, Germany - Age: 20 - Sent to Plymouth on 8 Dec 1814.

Gross, James - Seaman - Nbr: 5149 - Prize: Thorn - Ship Type: P - How taken: HMS Shannon - When taken: 7 Nov 1813 - Where taken: off Newfoundland - Date Received: 31 Oct 1814 - From what ship: HMT Mermaid, Chatham - Born: Marblehead - Age: 36 - Released on 29 Jun 1815.

Gross, William - Seaman - Nbr: 1364 - Prize: Paul Jones - Ship Type: P - How taken: HMS Leonidas - When taken: 23 May 1813 - Where taken: Channel - Date Received: 19 Jun 1814 - From what ship: Stapleton - Born: New York - Age: 14 - Released on 28 Apr 1815.

Grough, Jacob - Seaman - Nbr: 2173 - Prize: Rattlesnake - Ship Type: P - How taken: HMS Hyperion - When taken: 26 Jun 1814 - Where taken: off Cape Finisterre - Date Received: 16 Aug 1814 - From what ship: HMS Dublin, Halifax - Born: Philadelphia - Age: 25 - Released on 2 May 1815.

Groveman, Frederick - Seaman - Nbr: 6530 - Prize: US Brig Syren - Ship Type: MW - How taken: HMS Medway - When taken: 12 Jul 1814 - Where taken: off Cape of Good Hope - Date Received: 7 Mar 1815 - From what ship: HMT Ganges, Plymouth - Born: Winchester - Age: 16 - Released on 11 Jul 1815.

Grover, Edmund - Prize Master - Nbr: 487 - Prize: Miranda - Ship Type: Prize - How taken: HMS Unicorn - When taken: 21 May 1813 - Where taken: off Ushant (France) - Date Received: 8 Sep 1813 - From what ship: Plymouth - Born: Cape Ann - Age: 40 - Released on 25 mar 1815.

Groves, George W. - Seaman - Nbr: 3586 - How taken: Gave himself up from MV Fortune - When taken: 16 Mar 1814 - Date Received: 30 Sep 1814 - From what ship: HMT Sybella - Born: Baltimore - Age: 28 - Released on 4 Jun 1815.

Groves, Pierce - Seaman - Nbr: 1996 - Prize: Kergettos - Ship Type: MV - How taken: HMS Castilian - When taken: 12 Mar 1813 - Where taken: off Cape Ortegal (Spain) - Date Received: 3 Aug 1814 - From what ship: HMS Lyffey, Chatham Depot - Born: Talbot - Age: 23 - Released on 2 May 1815.

Groves, Thomas - Seaman - Nbr: 5383 - Prize: Rattlesnake - Ship Type: LM - How taken: HMS Rhin - When taken: 17 Mar 1814 - Where taken: off Bermuda - Date Received: 31 Oct 1814 - From what ship: HMT Leyden, Chatham - Born: Marblehead - Age: 21 - Released on 1 Jul 1815.

Groward, Peter - Seaman - Nbr: 251 - How taken: Impressed at Belfast - When taken: Aug 1812 - Date Received: 28 Jun 1813 - From what ship: Plymouth - Born: Portsmouth - Age: 46 - Sent to Dartmouth on 30 Jul 1813.

Grubb, Andrew - Seaman - Nbr: 1499 - Prize: Margaret - Ship Type: Prize - How taken: HMS Foxhound - When taken: 27 May 1814 - Where taken: off Isles of Scilly - Date Received: 20 Jun 1814 - From what ship: Mill Prison (Plymouth, England) - Born: Baltimore - Age: 29 - Released on 1 May 1815.

Grundy, Edward - Seaman - Nbr: 4551 - How taken: Impressed at London - When taken: 29 Jan 1814 - Date Received: 8 Oct 1814 - From what ship: HMT Leyden, Chatham - Born: North Carolina - Age: 26 - Escaped.

American Prisoners of War Held at Dartmoor during the War of 1812

## Alphabetical listing of names

Grush, John - Seaman - Nbr: 3654 - Prize: Growler - Ship Type: P - How taken: HMS Electra - When taken: 7 Jul 1813 - Where taken: at sea - Date Received: 30 Sep 1814 - From what ship: HMT President - Born: Marblehead - Age: 30 - Released on 4 Jun 1815.

Grush, Joseph - Seaman - Nbr: 2659 - How taken: Apprehended at London - When taken: 13 Aug 1813 - Date Received: 21 Aug 1814 - From what ship: HMT Freya, Chatham - Born: Marblehead - Age: 21 - Released on 19 May 1815.

Guages, Charles - Seaman - Nbr: 5284 - Prize: Enterprize - Ship Type: P - How taken: HMS Tenedos - When taken: 21 May 1813 - Where taken: off Cape Cod - Date Received: 31 Oct 1814 - From what ship: HMT Leyden, Chatham - Born: Massachusetts - Age: 23 - Released on 29 Jun 1815.

Guard, Caleb - Seaman - Nbr: 3989 - Prize: US Brig Rattlesnake - Ship Type: MW - How taken: HMS Leander - When taken: 13 Jul 1814 - Where taken: off Shelburne - Date Received: 6 Oct 1814 - From what ship: HMT Chesapeake, Halifax - Born: New London - Age: 22 - Released on 9 Jun 1815.

Guiho, Pierre - Pilot - Nbr: 998 - Prize: Apparencen - Ship Type: MV - How taken: HMS Castilian - When taken: 27 Jan 1814 - Where taken: off Ushant (France) - Date Received: 31 Jan 1814 - From what ship: Plymouth - Born: Brest - Age: 57 - Sent to Mill Prison (Plymouth, England) on 21 Jun 1814.

Guillard, Louis - Seaman - Nbr: 1360 - Prize: Paul Jones - Ship Type: P - How taken: HMS Leonidas - When taken: 23 May 1813 - Where taken: Channel - Date Received: 19 Jun 1814 - From what ship: Stapleton - Born: New Orleans - Age: 31 - Sent to Mill Prison (Plymouth, England) on 26 Jun 1814.

Guillard, Peter - Seaman - Nbr: 1506 - Prize: Paul Jones - Ship Type: P - How taken: HMS Leonidas - When taken: 22 May 1813 - Where taken: Channel - Date Received: 23 Jun 1814 - From what ship: Stapleton - Born: Nantes - Age: 20 - Sent to Mill Prison (Plymouth, England) on 26 Jun 1814.

Guillaume, Jean Marie - Seaman - Nbr: 5875 - Prize: Lion - Ship Type: P - How taken: HMS Granicus - When taken: 2 Dec 1814 - Where taken: off Lisbon - Date Received: 26 Dec 1814 - From what ship: HMT Impregnable - Born: L'Orient (France) - Age: 45 - Released on 3 Jul 1815.

Guilmot, Richard - Seaman - Nbr: 976 - Prize: Siro - Ship Type: LM - How taken: HMS Pelican - When taken: 13 Jan 1814 - Where taken: at sea - Date Received: 31 Jan 1814 - From what ship: Plymouth - Born: Portsmouth - Age: 25 - Race: Mulatto - Released on 27 Apr 1815.

Guire, Andrew - Seaman - Nbr: 4640 - How taken: Gave himself up from HMS Leviathan - When taken: 28 Oct 1813 - Date Received: 9 Oct 1814 - From what ship: HMT Leyden, Chatham - Born: Lancaster - Age: 42 - Released on 15 Jun 1815.

Gunby, James - Boy - Nbr: 742 - Prize: Ned - Ship Type: LM - How taken: HMS Royalist - When taken: 6 Sep 1813 - Where taken: Bay of Biscay - Date Received: 3 Nov 1813 - From what ship: Plymouth - Born: Baltimore - Age: 13 - Released on 26 Apr 1815.

Gunn, Charles William - Boy - Nbr: 2022 - Prize: Henrietta - Ship Type: MV - How taken: HMS Castilian - When taken: 12 Mar 1813 - Where taken: off Cape Ortegal (Spain) - Date Received: 3 Aug 1814 - From what ship: HMS Lyffey, Chatham Depot - Born: North Carolina - Age: 15 - Died on 2 Nov 1814.

Gunnell, William - Seaman - Nbr: 2737 - How taken: Gave himself up from HMS Romulus - When taken: 1 Jan 1813 - Date Received: 24 Aug 1814 - From what ship: HMT Liverpool, Chatham - Born: New York - Age: 45 - Released on 26 Apr 1815.

Gurney, John - Boy - Nbr: 6360 - Prize: Prince de Neufchatel - Ship Type: P - How taken: Leander (Newcastle Acasta) - When taken: 20 Dec 1814 - Where taken: Lat 38 Long 56 - Date Received: 19 Feb 1815 - From what ship: HMT Ganges, Plymouth - Born: Charlestown - Age: 16 - Released on 19 May 1815.

Gursh, Nathaniel - Seaman - Nbr: 4546 - Prize: Growler - Ship Type: P - How taken: HMS Electra - When taken: 7 Jul 1813 - Where taken: at sea - Date Received: 8 Oct 1814 - From what ship: HMT Leyden, Chatham - Born: Marblehead - Age: 17 - Released on 15 Jun 1815.

Gustable, Joseph - Boatswain's Mate - Nbr: 5640 - Prize: Lion - Ship Type: P - How taken: HMS Granicus - When taken: 2 Dec 1814 - Where taken: off Lisbon - Date Received: 24 Dec 1814 - From what ship: HMT Hanover, Gibraltar - Born: Leghorn, Italy - Age: 23 - Released on 3 Jul 1815.

Gustave, Peter - Seaman - Nbr: 826 - Prize: Chesapeake - Ship Type: LM - How taken: HMS Hotspur & HMS Pyramus - When taken: 26 Oct 1813 - Where taken: off Nantes - Date Received: 29 Nov 1813 - From what ship: Plymouth - Born: Sweden - Age: 18 - Released on 6 May 1814.

Gustavus, John - Seaman - Nbr: 5805 - Prize: US Frigate Superior (Gig) - Ship Type: MW - How taken: British gunboats - When taken: 26 Aug 1814 - Where taken: Lake Ontario - Date Received: 26 Dec 1814 - From what ship: HMT Argo - Born: Swedesburg - Age: 22 - Released on 3 Jul 1815.

Gwynn, Josiah - Seaman - Nbr: 4930 - Prize: Herald - Ship Type: P - How taken: HMS Endymion - When taken: 15

American Prisoners of War Held at Dartmoor during the War of 1812

## Alphabetical listing of names

Aug 1814 - Where taken: off Nantucket - Date Received: 28 Oct 1814 - From what ship: HMT Alkbar, Halifax - Born: Salem - Age: 18 - Died on 22 Feb 1815 from variols.

Haddart, Robert - Seaman - Nbr: 3319 - Prize: Wasp - Ship Type: P - How taken: HMS Bream - When taken: 10 Jun 1813 - Where taken: off Halifax - Date Received: 13 Sep 1814 - From what ship: HMT Niobe, Chatham - Born: Salem - Age: 29 - Released on 28 May 1815.

Hadley, Andrew - Seaman - Nbr: 3748 - Prize: Pallas - Ship Type: P - How taken: HMS Ringdove - When taken: 28 Apr 1813 - Where taken: off Nantucket - Date Received: 30 Sep 1814 - From what ship: HMT President, Halifax - Born: Reading - Age: 34 - Released on 4 Jun 1815.

Hadley, George - Seaman - Nbr: 3167 - Prize: Falcon - Ship Type: Prize - How taken: Spanish Army - When taken: 2 Jul 1813 - Where taken: Papage Harbor - Date Received: 11 Sep 1814 - From what ship: HMT Freya, Chatham - Born: Chester - Age: 32 - Sent to Dartmouth on 23 Sep 1814.

Hadley, James - Marine - Nbr: 2498 - Prize: US Sloop Frolic - Ship Type: MW - How taken: HMS Orpheus - When taken: 20 Apr 1814 - Where taken: off Cuba - Date Received: 16 Aug 1814 - From what ship: HMT Queen, Halifax - Born: Massachusetts - Age: 34 - Released on 3 May 1815.

Hadlock, Nathaniel - Seaman - Nbr: 5062 - Prize: Ida - Ship Type: LM - How taken: HMS Newcastle - When taken: 9 Aug 1814 - Where taken: Long 34 - Date Received: 28 Oct 1814 - From what ship: HMT Alkbar, Halifax - Born: Cape Ann - Age: 30 - Released on 29 Jun 1815.

Haffer, Robert - Seaman - Nbr: 5057 - Prize: Ida - Ship Type: LM - How taken: HMS Newcastle - When taken: 9 Aug 1814 - Where taken: Long 34 - Date Received: 10 Dec 1814 - From what ship: HMT Impregnable - Born: Richmond - Age: 21 - Released on 1 Jul 1815.

Hagen, Joel - Seaman - Nbr: 5203 - Prize: Porcupine - Ship Type: LM - How taken: HMS Acasta - When taken: 18 Jul 1813 - Where taken: off Halifax - Date Received: 31 Oct 1814 - From what ship: HMT Mermaid, Chatham - Born: Manchester - Age: 27 - Released on 29 Jun 1815.

Haines, Daniel - Seaman - Nbr: 2311 - Prize: Hussar - Ship Type: P - How taken: HMS Saturn - When taken: 25 May 1814 - Where taken: off Sandy Hook - Date Received: 16 Aug 1814 - From what ship: HMS Dublin, Halifax - Born: Pennsylvania - Age: 30 - Released on 3 May 1815.

Halbert, Henry - Seaman - Nbr: 4793 - Prize: Rattlesnake - Ship Type: MV - How taken: HMS Rhin - When taken: 11 Mar 1814 - Where taken: off San Domingo - Date Received: 9 Oct 1814 - From what ship: HMT Freya, Halifax - Born: New Orleans - Age: 20 - Race: Black - Released on 21 Jun 1815.

Hale, John - Seaman - Nbr: 4998 - Prize: Invincible - Ship Type: LM - How taken: HMS Armide - When taken: 15 Aug 1814 - Where taken: off Nantucket - Date Received: 28 Oct 1814 - From what ship: HMT Alkbar, Halifax - Born: Orleans - Age: 17 - Released on 21 Jun 1815.

Hale, Shederick - Seaman - Nbr: 1573 - Prize: Messenger - Ship Type: MV - How taken: HMS Iris - When taken: 10 Mar 1813 - Where taken: Bay of Biscay - Date Received: 23 Jun 1814 - From what ship: Stapleton - Born: Baltimore - Age: 43 - Released on 1 May 1815.

Haley, John - Seaman - Nbr: 4244 - Prize: Requin - Ship Type: LM - How taken: HMS Venus - When taken: 6 Mar 1814 - Where taken: off Bordeaux - Date Received: 7 Oct 1814 - From what ship: HMT Niobe, Chatham - Born: Bedford - Age: 34 - Released on 13 Jun 1815.

Haley, John - Seaman - Nbr: 2234 - Prize: General Hart - Ship Type: P - How taken: HMS Sophie - When taken: 24 Apr 1814 - Where taken: off Bermuda - Date Received: 16 Aug 1814 - From what ship: HMS Dublin, Halifax - Born: Boston - Age: 20 - Released on 2 May 1815.

Haley, Samuel - Seaman - Nbr: 6326 - Prize: Prince de Neufchatel - Ship Type: P - How taken: Leander (Newcastle Acasta) - When taken: 20 Dec 1814 - Where taken: Lat 38 Long 56 - Date Received: 19 Feb 1815 - From what ship: HMT Ganges, Plymouth - Born: Boston - Age: 18 - Released on 5 Jul 1815.

Haley, Samuel - Seaman - Nbr: 5920 - Prize: Harlequin - Ship Type: P - How taken: HMS Bulwark - When taken: 23 Nov 1814 - Where taken: off Halifax - Date Received: 27 Dec 1814 - From what ship: HMT Penelope - Born: Massachusetts - Age: 23 - Released on 3 Jul 1815.

Halfpenny, Robert - Seaman - Nbr: 6 - Prize: Cashier - Ship Type: LM - How taken: HMS Reindeer - When taken: 3 Feb 1813 - Where taken: Bay of Biscay - Date Received: 2 Apr 1813 - From what ship: Plymouth - Born: Baltimore - Age: 19 - Sent to Dartmouth on 30 Jul 1813.

Halfpenny, William - Pilot - Nbr: 3823 - Prize: Bordeaux Packet - Ship Type: LM - How taken: HMS Niemen - When taken: 28 Jan 1814 - Where taken: off Delaware - Date Received: 5 Oct 1814 - From what ship: HMT Orpheus, Halifax - Born: Baltimore - Age: 50 - Released on 9 Jun 1815.

Hall, Charles - Seaman - Nbr: 2477 - Prize: US Sloop Frolic - Ship Type: MW - How taken: HMS Orpheus - When taken: 20 Apr 1814 - Where taken: off Cuba - Date Received: 16 Aug 1814 - From what ship: HMT Queen,

American Prisoners of War Held at Dartmoor during the War of 1812

## Alphabetical listing of names

Halifax - Born: Burlington - Age: 21 - Released on 3 May 1815.

Hall, Daniel - Seaman - Nbr: 3098 - How taken: Gave himself up from HMS Norge - Date Received: 3 Sep 1814 - From what ship: HMT Bristol - Born: New Haven - Age: 25 - Race: Black - Released on 28 May 1815.

Hall, David - Seaman - Nbr: 347 - Prize: Courier - Ship Type: LM - How taken: HMS Andromache - When taken: 14 Mar 1813 - Where taken: Bay of Biscay - Date Received: 28 Jun 1813 - From what ship: Plymouth - Born: Maryland - Age: 22 - Released on 20 Apr 1815.

Hall, David - Seaman - Nbr: 3118 - Prize: Harmony - Ship Type: Prize - How taken: HMS Brisk - When taken: 24 Aug 1814 - Where taken: Bristol Channel - Date Received: 11 Sep 1814 - From what ship: HMT Salvador del Mundo - Born: Weston - Age: 28 - Released on 28 May 1815.

Hall, David - Seaman - Nbr: 4224 - Prize: Liberty - Ship Type: MV - How taken: Surrendered - When taken: 30 Dec 1813 - Where taken: Stromess - Date Received: 7 Oct 1814 - From what ship: HMT Niobe, Chatham - Born: Delaware - Age: 26 - Race: Black - Released on 13 Jun 1815.

Hall, Hammond - Prize Master - Nbr: 3099 - Prize: Mary - Ship Type: Prize - How taken: Taken by pilot boat - When taken: 27 Aug 1814 - Where taken: Bristol Channel - Date Received: 9 Sep 1814 - From what ship: HMS Abercrombie - Born: Massachusetts - Age: 28 - Released on 28 May 1815.

Hall, James - Seaman - Nbr: 5728 - Prize: Enterprize (Saratoga) - Ship Type: Prize - How taken: HMS Barracouta - When taken: 10 Oct 1814 - Where taken: off Western Islands (England) - Date Received: 26 Dec 1814 - From what ship: HMT Argo - Born: Rhode Island - Age: 52 - Released on 3 Jul 1815.

Hall, James - Seaman - Nbr: 4517 - Prize: Grand Turk - Ship Type: P - How taken: HMS Tenedos - When taken: 26 May 1813 - Where taken: off Cape Cod - Date Received: 8 Oct 1814 - From what ship: HMT Leyden, Chatham - Born: Massachusetts - Age: 23 - Released on 28 Apr 1815.

Hall, James - Seaman - Nbr: 674 - Prize: US Brig Argus - Ship Type: MW - How taken: HMS Pelican - When taken: 14 Aug 1813 - Where taken: Irish Channel - Date Received: 8 Sep 1813 - From what ship: Plymouth - Born: Cambridge - Age: 26 - Sent to Dartmouth on 2 Nov 1814.

Hall, James - Seaman - Nbr: 541 - Prize: Orders in Council - Ship Type: MV - How taken: HMS Surveillante - When taken: 1 Jun 1813 - Where taken: off Cape Ortegal (Spain) - Date Received: 8 Sep 1813 - From what ship: Plymouth - Born: Virginia - Age: 29 - Released on 26 Apr 1815.

Hall, James - Seaman - Nbr: 2930 - How taken: Gave himself up from HMS Berwick - Date Received: 24 Aug 1814 - From what ship: HMT Alpheus, Chatham - Born: Connecticut - Age: 29 - Released on 19 May 1815.

Hall, John - Marine - Nbr: 6339 - Prize: Prince de Neufchatel - Ship Type: P - How taken: Leander (Newcastle Acasta) - When taken: 20 Dec 1814 - Where taken: Lat 38 Long 56 - Date Received: 19 Feb 1815 - From what ship: HMT Ganges, Plymouth - Born: Methuen - Age: 20 - Released on 6 Jul 1815.

Hall, John - Seaman - Nbr: 4685 - Prize: Yankee - Ship Type: P - How taken: HMS Shannon - When taken: 20 Aug 1813 - Where taken: at sea - Date Received: 9 Oct 1814 - From what ship: HMT Leyden, Chatham - Born: Maryland - Age: 22 - Released on 15 Jun 1815.

Hall, John - Seaman - Nbr: 2686 - How taken: Gave himself up from HMS Royal George - When taken: 29 Oct 1812 - Date Received: 21 Aug 1814 - From what ship: HMT Freya, Chatham - Born: Baltimore - Age: 31 - Released on 26 Apr 1815.

Hall, Joseph - Seaman - Nbr: 6147 - How taken: Gave himself up from HMS Tremendous - When taken: 7 Sep 1812 - Date Received: 17 Jan 1815 - From what ship: HMT Impregnable - Born: Philadelphia - Age: 25 - Released on 11 Jul 1815.

Hall, Lewis - Seaman - Nbr: 2509 - Prize: Snap Dragon - Ship Type: P - How taken: HMS Martin - When taken: 10 Jun 1814 - Where taken: off Halifax - Date Received: 16 Aug 1814 - From what ship: HMT Queen, Halifax - Born: Pennsylvania - Age: 24 - Race: Black - Released on 3 May 1815.

Hall, Perry - Seaman - Nbr: 3252 - Prize: Wiley Reynard - Ship Type: P - How taken: HMS Shannon - When taken: 15 Aug 1812 - Where taken: off Halifax - Date Received: 11 Sep 1814 - From what ship: HMT Freya, Chatham - Born: Baltimore - Age: 25 - Race: Black - Released on 27 Apr 1815.

Hall, Reuben - Seaman - Nbr: 5599 - Prize: Mentor - Ship Type: Prize - How taken: HMS Maidstone - When taken: 26 Oct 1814 - Where taken: off Cape Sable Island (Canada) - Date Received: 17 Dec 1814 - From what ship: HMT Loire, Halifax - Born: Boston - Age: 17 - Race: Negro - Released on 5 Jul 1815.

Hall, Richard - Passenger - Nbr: 4573 - How taken: Taken off a Swedish ship bound for America - Date Received: 9 Oct 1814 - From what ship: HMT Leyden, Chatham - Born: Boston - Age: 35 - Race: Black - Released on 15 Jun 1815.

Hall, Richard - Seaman - Nbr: 348 - Prize: Courier - Ship Type: LM - How taken: HMS Andromache - When taken: 14 Mar 1813 - Where taken: Bay of Biscay - Date Received: 28 Jun 1813 - From what ship: Plymouth -

American Prisoners of War Held at Dartmoor during the War of 1812

## Alphabetical listing of names

Born: Snow Hill - Age: 25 - Released on 20 Apr 1815.

Hall, Robert - Seaman - Nbr: 5864 - Prize: Lion - Ship Type: P - How taken: HMS Granicus - When taken: 2 Dec 1814 - Where taken: off Lisbon - Date Received: 26 Dec 1814 - From what ship: HMT Impregnable - Born: Philadelphia - Age: 24 - Released on 3 Jul 1815.

Hall, Silvester - Seaman - Nbr: 4525 - Prize: Yankee - Ship Type: P - How taken: HMS Shannon - When taken: 20 Aug 1813 - Where taken: at sea - Date Received: 8 Oct 1814 - From what ship: HMT Leyden, Chatham - Born: Rhode Island - Age: 20 - Released on 15 Jun 1815.

Hall, Stephen - Prize master - Nbr: 6237 - Prize: Albion - Ship Type: Prize - How taken: HMS Harlequin - When taken: 6 Jan 1814 - Where taken: Long 20 - Date Received: 4 Feb 1815 - From what ship: HMT Ganges - Born: Delaware - Age: 26 - Released on 11 Jul 1815.

Hall, Stephen - Seaman - Nbr: 3467 - How taken: Gave himself up from HMS Prince - When taken: 1 Sep 1814 - Date Received: 19 Sep 1814 - From what ship: HMT Salvador del Mundo - Born: Delaware - Age: 26 - Race: Black - Released on 4 Jun 1815.

Hall, Thomas - Prize Master - Nbr: 1708 - Prize: Traveler - Ship Type: Prize - How taken: HM Schooner Canso - When taken: 11 May 1814 - Where taken: off Cape Clear - Date Received: 8 Jul 1814 - From what ship: Labrador - Born: Maryland - Age: 36 - Died on 18 Apr 1815 from phthisis pulmonalis.

Hall, William - Seaman - Nbr: 2768 - How taken: Gave himself up from HMS Franchise - When taken: 20 Dec 1812 - Date Received: 24 Aug 1814 - From what ship: HMT Liverpool, Chatham - Born: Charlestown - Age: 25 - Released on 19 May 1815.

Hall, Zachariah - Seaman - Nbr: 4951 - Prize: Herald - Ship Type: P - How taken: HMS Endymion - When taken: 15 Aug 1814 - Where taken: off Nantucket - Date Received: 28 Oct 1814 - From what ship: HMT Alkbar, Halifax - Born: New York - Age: 28 - Race: Mulatto - Released on 21 Jun 1815.

Hallet, William - Seaman - Nbr: 4615 - Prize: Argus - Ship Type: MV - How taken: HMS San Domingo - When taken: 1 Mar 1814 - Where taken: off Savannah - Date Received: 9 Oct 1814 - From what ship: HMT Leyden, Chatham - Born: Barnstable - Age: 24 - Released on 15 Jun 1815.

Halton, William - Farmer - Nbr: 3897 - How taken: HMS Albion - When taken: 31 May 1814 - Where taken: on shore at St. Marys, MD - Date Received: 5 Oct 1814 - From what ship: HMT Orpheus, Halifax - Born: St. Marys - Age: 21 - Released on 23 Nov 1814.

Ham, John - Seaman - Nbr: 1045 - Prize: Prince of Wales - Ship Type: P - How taken: Nelson (Transport) - When taken: 4 Feb 1814 - Where taken: at sea - Date Received: 10 May 1814 - From what ship: Plymouth - Born: Portsmouth - Age: 24 - Released on 27 Apr 1815.

Ham, Robert - Marine - Nbr: 4061 - Prize: US Brig Rattlesnake - Ship Type: MW - How taken: HMS Leander - When taken: 13 Jul 1814 - Where taken: off Shelburne - Date Received: 6 Oct 1814 - From what ship: HMT Chesapeake, Halifax - Born: Portsmouth - Age: 32 - Released on 29 Jun 1815.

Hambleton, John - Seaman - Nbr: 5055 - Prize: Ida - Ship Type: LM - How taken: HMS Newcastle - When taken: 9 Aug 1814 - Where taken: Long 34 - Date Received: 28 Oct 1814 - From what ship: HMT Alkbar, Halifax - Born: Boston - Age: 27 - Released on 29 Jun 1815.

Hamcomb, Thomas - Seaman - Nbr: 2912 - Prize: Matilda - Ship Type: MV - How taken: HMS Revolutionnaire - When taken: 25 Jul 1813 - Where taken: off L'Orient (France) - Date Received: 24 Aug 1814 - From what ship: HMT Alpheus, Chatham - Born: Massachusetts - Age: 21 - Released on 2 Nov 1814.

Hamel, William - Seaman - Nbr: 1810 - How taken: Gave himself up from HMS Clarence - When taken: 7 May 1813 - Date Received: 20 Jul 1814 - From what ship: HMS Milford, Plymouth - Born: New York - Age: 22 - Released on 1 May 1815.

Hamilton, Alexander M. - Seaman - Nbr: 1362 - Prize: Paul Jones - Ship Type: P - How taken: HMS Leonidas - When taken: 23 May 1813 - Where taken: Channel - Date Received: 19 Jun 1814 - From what ship: Stapleton - Born: Boston - Age: 14 - Released on 28 Apr 1815.

Hamilton, Clayton - Seaman - Nbr: 5710 - Prize: Regent - Ship Type: LM - How taken: HMS Forth - When taken: 19 Sep 1814 - Where taken: off Egg Harbor (New Jersey) - Date Received: 24 Dec 1814 - From what ship: HMT Penelope - Born: New Jersey - Age: 27 - Released on 3 Jul 1815.

Hamilton, Elijah - Seaman - Nbr: 2271 - Prize: Success - Ship Type: Prize - How taken: HMS Charybdis - When taken: 29 May 1814 - Where taken: off Cape Logan - Date Received: 16 Aug 1814 - From what ship: HMS Dublin, Halifax - Born: Chatham - Age: 24 - Released on 3 May 1815.

Hamilton, George W. - Seaman - Nbr: 3320 - Prize: Wasp - Ship Type: P - How taken: HMS Bream - When taken: 10 Jun 1813 - Where taken: off Halifax - Date Received: 13 Sep 1814 - From what ship: HMT Niobe, Chatham - Born: New York - Age: 29 - Released on 28 May 1815.

American Prisoners of War Held at Dartmoor during the War of 1812

## Alphabetical listing of names

Hamilton, John - Seaman - Nbr: 2247 - Prize: Carbineer - Ship Type: Prize - How taken: HMS Ringdove - When taken: 24 Apr 1814 - Where taken: off Bermuda - Date Received: 16 Aug 1814 - From what ship: HMS Dublin, Halifax - Born: Massachusetts - Age: 25 - Released on 3 May 1815.

Hamilton, John - Seaman - Nbr: 4223 - Prize: Liberty - Ship Type: MV - How taken: Surrendered - When taken: 30 Dec 1813 - Where taken: Stromess - Date Received: 7 Oct 1814 - From what ship: HMT Niobe, Chatham - Born: Salem - Age: 39 - Race: Black - Released on 13 Jun 1815.

Hamilton, Richard - Prize Master - Nbr: 1821 - Prize: True Blooded Yankee - Ship Type: P - How taken: HMS Hamadryad - When taken: 24 Jul 1813 - Where taken: off Norway - Date Received: 20 Jul 1814 - From what ship: HMS Milford, Plymouth - Born: New London - Age: 24 - Released on 11 Jul 1815.

Hamilton, Robert - Seaman - Nbr: 4189 - Prize: Commodore Perry - Ship Type: Schooner - How taken: Sent into custody from a cutter - When taken: 25 Feb 1814 - Where taken: off Bordeaux - Date Received: 7 Oct 1814 - From what ship: HMT Niobe, Chatham - Born: Philadelphia - Age: 31 - Released on 13 Jun 1815.

Hamilton, William - Seaman - Nbr: 5131 - How taken: Gave himself up from HMS Crescent - When taken: 20 Oct 1814 - Date Received: 31 Oct 1814 - From what ship: HMT Castillian - Born: Philadelphia - Age: 23 - Released on 29 Jun 1815.

Hammond, Abner - Seaman - Nbr: 5980 - Prize: Cossack - Ship Type: Private - How taken: HMS Bulwark - When taken: 1 Nov 1814 - Where taken: Georges Bank - Date Received: 27 Dec 1814 - From what ship: HMT Penelope - Born: Plymouth - Age: 19 - Released on 3 Jul 1815.

Hammond, William - Seaman - Nbr: 5290 - Prize: Fox - Ship Type: P - How taken: HMS Shannon - When taken: 7 Nov 1813 - Where taken: off Newfoundland - Date Received: 31 Oct 1814 - From what ship: HMT Leyden, Chatham - Born: New York - Age: 25 - Released on 29 Jun 1815.

Hammond, Benjamin - Seaman - Nbr: 2401 - Prize: US Sloop Frolic - Ship Type: MW - How taken: HMS Orpheus - When taken: 20 Apr 1814 - Where taken: off Cuba - Date Received: 16 Aug 1814 - From what ship: HMT Queen, Halifax - Born: Marblehead - Age: 23 - Sent to Dartmouth on 19 Oct 1814.

Hammond, Edward - Seaman - Nbr: 3760 - Prize: Alfred - Ship Type: P - How taken: HMS Epervier - When taken: 23 Feb 1814 - Where taken: off Newfoundland - Date Received: 30 Sep 1814 - From what ship: HMT President, Halifax - Born: Marblehead - Age: 19 - Released on 4 Jun 1815.

Hammond, Edward - Seaman - Nbr: 3690 - Prize: Alfred - Ship Type: P - How taken: HMS Epervier - When taken: 23 Feb 1812 - Where taken: off Newfoundland - Date Received: 30 Sep 1814 - From what ship: HMT President - Born: Marblehead - Age: 18 - Released on 4 Jun 1815.

Hammond, Isaac - Seaman - Nbr: 4165 - How taken: Gave himself up from HMS L'Aigle - When taken: 9 May 1814 - Date Received: 6 Oct 1814 - From what ship: HMT Niobe, Chatham - Born: Rochester - Age: 34 - Race: Black - Released on 13 Jun 1815.

Hammond, John - Seaman - Nbr: 5148 - Prize: Thorn - Ship Type: P - How taken: HMS Shannon - When taken: 7 Nov 1813 - Where taken: off Newfoundland - Date Received: 31 Oct 1814 - From what ship: HMT Mermaid, Chatham - Born: Marblehead - Age: 19 - Released on 2 May 1815.

Hammond, Joseph - Seaman - Nbr: 1593 - Prize: Eliza - Ship Type: MV - How taken: HMS Surveillante - When taken: 27 Mar 1813 - Where taken: Bay of Biscay - Date Received: 23 Jun 1814 - From what ship: Stapleton - Born: Marblehead - Age: 22 - Released on 1 May 1815.

Hammond, Samuel - Seaman - Nbr: 3950 - Prize: Rolla - Ship Type: P - How taken: HMS Loire - When taken: 10 Dec 1813 - Where taken: off Bull Island (South Carolina) - Date Received: 5 Oct 1814 - From what ship: HMT President, Halifax - Born: North Kingston - Age: 19 - Released on 9 Jun 1815.

Hammond, William - Seaman - Nbr: 2108 - Prize: Thetis - Ship Type: MV - How taken: HMS Flora - When taken: 11 Jun 1814 - Where taken: Lat 30N Long 40W - Date Received: 3 Aug 1814 - From what ship: Plymouth - Born: Marblehead - Age: 26 - Released on 2 May 1815.

Hammond, William - Seaman - Nbr: 1023 - Prize: Joseph - Ship Type: MV - How taken: HMS Royalist - When taken: 18 Jan 1814 - Where taken: Bay of Biscay - Date Received: 31 Jan 1814 - From what ship: Plymouth - Born: Marblehead - Age: 22 - Released on 27 Apr 1815.

Hampstead, Cambridge - Seaman - Nbr: 3803 - Prize: Margetits - Ship Type: Prize - How taken: HMS Maidstone - When taken: 20 Mar 1814 - Where taken: off NY - Date Received: 5 Oct 1814 - From what ship: HMT Orpheus, Halifax - Born: Washington - Age: 26 - Race: Negro - Released on 9 Jun 1815.

Hampton, Thomas - Seaman - Nbr: 2998 - Prize: St. Lawrence - How taken: HMS Aquilon - When taken: 9 Aug 1814 - Where taken: off Western Islands (England) - Date Received: 29 Aug 1814 - From what ship: HMT Bittern - Born: Philadelphia - Age: 38 - Race: Black - Released on 28 May 1815.

Hamson, John - Seaman - Nbr: 4483 - Prize: Enterprize - Ship Type: P - How taken: HMS Tenedos - When taken:

American Prisoners of War Held at Dartmoor during the War of 1812

## Alphabetical listing of names

21 May 1813 - Where taken: off Cape Cod - Date Received: 8 Oct 1814 - From what ship: HMT Leyden, Chatham - Born: Massachusetts - Age: 23 - Released on 15 Jun 1815.

Hamson, William - Seaman - Nbr: 4476 - Prize: Enterprize - Ship Type: P - How taken: HMS Tenedos - When taken: 21 May 1813 - Where taken: off Cape Cod - Date Received: 8 Oct 1814 - From what ship: HMT Leyden, Chatham - Born: Massachusetts - Age: 21 - Released on 15 Jun 1815.

Hancock, James - Seaman - Nbr: 3591 - Prize: Atlas - Ship Type: MV - How taken: Leephe - When taken: 12 Jul 1813 - Where taken: off NC - Date Received: 30 Sep 1814 - From what ship: HMT Sybella - Born: Newbury - Age: 30 - Released on 4 Jun 1815.

Handell, Henry - Seaman - Nbr: 2471 - Prize: US Sloop Frolic - Ship Type: MW - How taken: HMS Orpheus - When taken: 20 Apr 1814 - Where taken: off Cuba - Date Received: 16 Aug 1814 - From what ship: HMT Queen, Halifax - Born: Philadelphia - Age: 24 - Released on 3 May 1815.

Handfield, Robert - Seaman - Nbr: 1088 - How taken: Impressed from a merchant vessel - When taken: 2 Mar 1814 - Date Received: 10 May 1814 - From what ship: Plymouth - Born: Havana - Age: 30 - Released on 27 Apr 1815.

Handley, John - Seaman - Nbr: 2844 - How taken: Gave himself up from HMS Repulse - When taken: 27 May 1813 - Date Received: 24 Aug 1814 - From what ship: HMT Liverpool, Chatham - Born: Massachusetts - Age: 32 - Released on 19 May 1815.

Hane, W. B. - 2nd Mate - Nbr: 1601 - Prize: Fox - Ship Type: LM - How taken: HMS Pheasant - When taken: 22 Apr 1813 - Where taken: Bay of Biscay - Date Received: 23 Jun 1814 - From what ship: Stapleton - Born: New Jersey - Age: 29 - Released on 1 May 1815.

Hanely, Thomas - Seaman - Nbr: 5199 - Prize: Porcupine - Ship Type: LM - How taken: HMS Acasta - When taken: 18 Jul 1813 - Where taken: off Halifax - Date Received: 31 Oct 1814 - From what ship: HMT Mermaid, Chatham - Born: Bristol - Age: 18 - Race: Black - Released on 29 Jun 1815.

Hanes, Simon - Seaman - Nbr: 4199 - Prize: Zephyr - Ship Type: MV - How taken: HMS Surveillante - When taken: 4 Jan 1814 - Where taken: Bay of Biscay - Date Received: 7 Oct 1814 - From what ship: HMT Niobe, Chatham - Born: Virginia - Age: 40 - Race: Black - Released on 13 Jun 1815.

Haney, Edward - Seaman - Nbr: 1657 - How taken: Apprehended at Bristol - When taken: 5 Mar 1814 - Date Received: 23 Jun 1814 - What ship: Stapleton - Born: Philadelphia - Age: 36 - Released on 1 May 1815.

Hanfield, Enos - Seaman - Nbr: 3279 - Prize: Montgomery - Ship Type: P - How taken: HMS Nymphe - When taken: 5 May 1813 - Where taken: off Cape Ann - Date Received: 13 Sep 1814 - From what ship: HMT Niobe, Chatham - Born: Salem - Age: 23 - Released on 28 May 1815.

Hanford, William - Seaman - Nbr: 1551 - Prize: Zebra - Ship Type: LM - How taken: HMS Pyramus - When taken: 20 Apr 1813 - Where taken: Bay of Biscay - Date Received: 23 Jun 1814 - From what ship: Stapleton - Born: Connecticut - Age: 24 - Released on 1 May 1815.

Hankey, Frederick - Prize Master - Nbr: 3038 - Prize: Commerce - How taken: HMS Superb - When taken: 23 Jun 1814 - Where taken: off Sandy Hook - Date Received: 2 Sep 1814 - From what ship: HMT Sultan - Born: Berlin - Age: 30 - Released on 28 May 1815.

Hannah, Thomas - Seaman - Nbr: 5111 - Prize: Portsmouth - Ship Type: P - How taken: HMS Pylades - When taken: 13 Sep 1814 - Where taken: Long 65 - Date Received: 28 Oct 1814 - From what ship: HMT Alkbar, Halifax - Born: Baltimore - Age: 32 - Released on 29 Jun 1815.

Hanner, Daniel - Seaman - Nbr: 4192 - Prize: Commodore Perry - Ship Type: Schooner - How taken: Sent into custody from a cutter - When taken: 25 Feb 1814 - Where taken: off Bordeaux - Date Received: 7 Oct 1814 - From what ship: HMT Niobe, Chatham - Born: Charleston - Age: 29 - Released on 13 Jun 1815.

Hannon, Alexander - Seaman - Nbr: 2511 - Prize: Snap Dragon - Ship Type: P - How taken: HMS Martin - When taken: 10 Jun 1814 - Where taken: off Halifax - Date Received: 16 Aug 1814 - From what ship: HMT Queen, Halifax - Born: New York - Age: 22 - Released on 3 May 1815.

Hanscom, Moses - Seaman - Nbr: 630 - Prize: US Brig Argus - Ship Type: MW - How taken: HMS Pelican - When taken: 14 Aug 1813 - Where taken: Irish Channel - Date Received: 8 Sep 1813 - From what ship: Plymouth - Born: Kittery - Age: 27 - Sent to Dartmouth on 2 Nov 1814.

Hansey, Peter - Seaman - Nbr: 2645 - How taken: Gave himself up from HMS Caledonia - When taken: 5 Dec 1812 - Date Received: 21 Aug 1814 - From what ship: HMT Freya, Chatham - Born: New York - Age: 55 - Race: Black - Released on 26 Apr 1815.

Hanson, Henry - Seaman - Nbr: 2229 - Prize: General Hart - Ship Type: P - How taken: HMS Sophie - When taken: 24 Apr 1814 - Where taken: off Bermuda - Date Received: 16 Aug 1814 - From what ship: HMS Dublin, Halifax - Born: Marblehead - Age: 34 - Released on 2 May 1815.

American Prisoners of War Held at Dartmoor during the War of 1812

## Alphabetical listing of names

Hanson, J. A. - Seaman - Nbr: 1097 - Prize: Lyon - Ship Type: P - How taken: Brilliant (Privateer) - When taken: 12 Aug 1814 - Where taken: off Charlestown - Date Received: 10 May 1814 - From what ship: Plymouth - Born: New York - Age: 42 - Released on 27 Apr 1815.

Hanson, John - Seaman - Nbr: 5034 - Prize: Betsey - Ship Type: Prize - How taken: HMS Pylades - When taken: 7 Sep 1814 - Where taken: Canten - Date Received: 28 Oct 1814 - From what ship: HMT Alkbar, Halifax - Born: New York - Age: 38 - Released on 29 Jun 1815.

Hanson, John - Seaman - Nbr: 3535 - Prize: Rover - Ship Type: Prize - How taken: HMS Conquistador - When taken: 22 Aug 1814 - Where taken: Long 19 Lat 107 - Date Received: 28 Sep 1814 - From what ship: HMT Salvador del Mundo - Born: Denmark - Age: 18 - Released on 4 Jun 1815.

Hanson, Peter - Seaman - Nbr: 2910 - Prize: Matilda - Ship Type: MV - How taken: HMS Revolutionnaire - When taken: 25 Jul 1813 - Where taken: off L'Orient (France) - Date Received: 24 Aug 1814 - From what ship: HMT Alpheus, Chatham - Born: Gothenburg - Age: 19 - Released on 11 Jul 1815.

Hanson, Stephen - Seaman - Nbr: 3526 - Prize: Sabine - Ship Type: P - How taken: HMS Conquistador - When taken: 3 Aug 1814 - Where taken: off the Banks - Date Received: 28 Sep 1814 - From what ship: HMT Salvador del Mundo - Born: Gothenburg - Age: 46 - Released on 4 Jun 1815.

Hanson, William - Seaman - Nbr: 4397 - Prize: Portsmouth Packet - Ship Type: Prize - How taken: HMS Fantome - When taken: 5 Oct 1813 - Where taken: off Portland - Date Received: 8 Oct 1814 - From what ship: HMT Leyden, Chatham - Born: Hampshire - Age: 20 - Escaped on 1 Jun 1815.

Harbrook, Richard - Seaman - Nbr: 3411 - Prize: Thomas - Ship Type: P - How taken: HMS Nymphe - When taken: 24 Jun 1813 - Where taken: off Halifax - Date Received: 13 Sep 1814 - From what ship: HMT Niobe, Chatham - Born: New York - Age: 18 - Released on 28 May 1815.

Harburn, Thomas - Seaman - Nbr: 1197 - Prize: Imperatrice Reine - Ship Type: P - How taken: HMS Hotspur - When taken: 13 Jan 1813 - Date Received: 4 Jun 1814 - From what ship: Dartmouth - Born: Baltimore - Age: 30 - Released on 28 Apr 1815.

Harding, John - Seaman - Nbr: 4154 - Prize: Argus - Ship Type: MV - How taken: HMS San Domingo - When taken: 1 Mar 1814 - Where taken: off Savannah - Date Received: 6 Oct 1814 - From what ship: HMT Niobe, Chatham - Born: Boston - Age: 18 - Released on 13 Jun 1815.

Harding, John - Seaman - Nbr: 383 - Prize: VA Planter - Ship Type: MV - How taken: HMS Pyramus - When taken: 18 Mar 1813 - Where taken: off Nantes - Date Received: 1 Jul 1813 - From what ship: Plymouth - Born: Newport - Age: 19 - Released on 20 Apr 1815.

Harding, John - Seaman - Nbr: 3786 - Prize: James - Ship Type: Prize - How taken: Rebecca (LM) - When taken: 13 Apr 1814 - Where taken: off Delaware - Date Received: 5 Oct 1814 - From what ship: HMT President, Halifax - Born: Mammet - Age: 21 - Released on 9 Jun 1815.

Harding, Jonathan - Seaman - Nbr: 5988 - Prize: McDonough - Ship Type: P - How taken: HMS Bacchante - When taken: 1 Nov 1814 - Where taken: Lat 42 Long 67 - Date Received: 27 Dec 1814 - From what ship: HMT Penelope - Born: Massachusetts - Age: 35 - Released on 3 Jul 1815.

Harding, Joseph - Prize master - Nbr: 5190 - Prize: Juliana Smith - Ship Type: P - How taken: HMS Nymphe - When taken: 11 May 1813 - Where taken: off Cape Sable Island (Canada) - Date Received: 31 Oct 1814 - From what ship: HMT Mermaid, Chatham - Born: Chatham - Age: 33 - Released on 11 Jul 1815.

Harding, William - Seaman - Nbr: 4235 - Prize: Pilot - Ship Type: LM - How taken: Victoria (Privateer) - When taken: 28 Jan 1814 - Where taken: off Bordeaux - Date Received: 7 Oct 1814 - From what ship: HMT Niobe, Chatham - Born: SC - Age: 28 - Race: Black - Released on 13 Jun 1815.

Hardingbrook, Theop. - Seaman - Nbr: 1539 - Prize: Zebra - Ship Type: LM - How taken: HMS Pyramus - When taken: 20 Apr 1813 - Where taken: Bay of Biscay - Date Received: 23 Jun 1814 - From what ship: Stapleton - Born: New Jersey - Age: 22 - Released on 1 May 1815.

Hardman, John - Seaman - Nbr: 5041 - Prize: Betsey - Ship Type: Prize - How taken: HMS Pylades - When taken: 7 Sep 1814 - Where taken: Canten - Date Received: 28 Oct 1814 - From what ship: HMT Alkbar, Halifax - Born: Portsmouth - Age: 20 - Released on 29 Jun 1815.

Hardy, Andrew H. - Soldier - Nbr: 5841 - Prize: 25th US Infantry - Ship Type: Troops - How taken: British Army - When taken: 17 Sep 1814 - Where taken: Niagara - Date Received: 26 Dec 1814 - From what ship: HMT Argo - Born: GA - Age: 36 - Released on 29 Jun 1815.

Hardy, Robert - Seaman - Nbr: 1692 - Prize: Nonsuch - Ship Type: MV - How taken: HMS Dotterel - When taken: 14 Dec 1813 - Where taken: off Charlestown - Date Received: 2 Jul 1814 - From what ship: Plymouth - Born: Philadelphia - Age: 16 - Released on 1 May 1815.

Harker, John - Boy - Nbr: 1294 - Prize: Traveler - Ship Type: Prize - How taken: HM Schooner Canso - When

American Prisoners of War Held at Dartmoor during the War of 1812

## Alphabetical listing of names

taken: 11 May 1814 - Where taken: off Cape Clear - Date Received: 14 Jun 1814 - From what ship: Mill Prison (Plymouth, England) - Born: Maryland - Age: 17 - Released on 28 Apr 1815.
Harker, William - Seaman - Nbr: 5585 - Prize: Young William - Ship Type: Prize - How taken: HMS Plover - When taken: 8 Sep 1814 - Where taken: off St. Johns - Date Received: 17 Dec 1814 - From what ship: HMT Loire, Halifax - Born: North Carolina - Age: 25 - Released on 14 Jun 1815.
Harley, Edward - Seaman - Nbr: 4130 - Prize: Bordeaux Packet - Ship Type: LM - How taken: HMS Niemen - When taken: 28 Jun 1814 - Where taken: off Delaware - Date Received: 6 Oct 1814 - From what ship: HMT Chesapeake, Halifax - Born: Baltimore - Age: 20 - Released on 13 Jun 1815.
Harley, Joshua - Seaman - Nbr: 3845 - Prize: Dominique - Ship Type: LM - How taken: HMS Dotterel - When taken: 21 May 1814 - Where taken: off Charleston - Date Received: 5 Oct 1814 - From what ship: HMT Orpheus, Halifax - Born: Biddeford - Age: 46 - Released on 9 Jun 1815.
Harman, Isaac - 2nd Mate - Nbr: 4423 - Prize: Elbridge Gerry - Ship Type: LM - How taken: HMS Crescent - When taken: 16 Sep 1813 - Where taken: at sea - Date Received: 8 Oct 1814 - From what ship: HMT Leyden, Chatham - Born: Massachusetts - Age: 24 - Died on 9 Nov 1814.
Harman, John - Seaman - Nbr: 4838 - Prize: President - Ship Type: P - How taken: HMS Pique - When taken: 7 May 1814 - Where taken: off Porto Rico - Date Received: 9 Oct 1814 - From what ship: HMT Freya, Halifax - Born: Minden - Age: 42 - Released on 21 Jun 1815.
Harman, Mathew - Seaman - Nbr: 4096 - Prize: Tyger - Ship Type: Prize - How taken: HMS Bulwark - When taken: 20 Jul 1814 - Where taken: Georges Bank - Date Received: 6 Oct 1814 - From what ship: HMT Chesapeake, Halifax - Born: Finland - Age: 29 - Released on 13 Jun 1815.
Harman, Peter - Seaman - Nbr: 4093 - Prize: Tyger - Ship Type: Prize - How taken: HMS Bulwark - When taken: 20 Jul 1814 - Where taken: Georges Bank - Date Received: 6 Oct 1814 - From what ship: HMT Chesapeake, Halifax - Born: New Orleans - Age: 23 - Released on 13 Jun 1815.
Haro, John - Seaman - Nbr: 4119 - Prize: Bordeaux Packet - Ship Type: LM - How taken: HMS Niemen - When taken: 28 Jun 1814 - Where taken: off Delaware - Date Received: 6 Oct 1814 - From what ship: HMT Chesapeake, Halifax - Born: Philadelphia - Age: 24 - Released on 13 Jun 1815.
Harrington, John - Seaman - Nbr: 782 - Prize: Betsy - Ship Type: MV - How taken: HMS Eurotas - When taken: 26 Oct 1813 - Where taken: off Ushant (France) - Date Received: 3 Nov 1813 - From what ship: Plymouth - Born: New York - Age: 23 - Released on 26 Apr 1815.
Harrington, Simon - Seaman - Nbr: 1535 - Prize: Zebra - Ship Type: LM - How taken: HMS Pyramus - When taken: 20 Apr 1813 - Where taken: Bay of Biscay - Date Received: 23 Jun 1814 - From what ship: Stapleton - Born: Massachusetts - Age: 24 - Released on 1 May 1815.
Harrington, William - Seaman - Nbr: 686 - Prize: US Brig Argus - Ship Type: MW - How taken: HMS Pelican - When taken: 14 Aug 1813 - Where taken: Irish Channel - Date Received: 27 Sep 1813 - From what ship: Plymouth - Born: Delaware - Age: 25 - Race: Black - Sent to Dartmouth on 2 Nov 1814.
Harris, Benjamin - Seaman - Nbr: 3008 - Prize: Perseverance - Ship Type: MV - How taken: HMS Barbados - When taken: 20 Jan 1814 - Where taken: off St. Bartholomew - Date Received: 1 Sep 1814 - From what ship: from Bristol - Born: Boothbay - Age: 25 - Released on 28 May 1815.
Harris, Bradley - Boy - Nbr: 950 - Prize: Siro - Ship Type: LM - How taken: HMS Pelican - When taken: 13 Jan 1814 - Where taken: at sea - Date Received: 31 Jan 1814 - From what ship: Plymouth - Born: Boston - Age: 18 - Released on 27 Apr 1815.
Harris, David - Seaman - Nbr: 2421 - Prize: US Sloop Frolic - Ship Type: MW - How taken: HMS Orpheus - When taken: 20 Apr 1814 - Where taken: off Cuba - Date Received: 16 Aug 1814 - From what ship: HMT Queen, Halifax - Born: Norwich - Age: 22 - Released on 3 May 1815.
Harris, Ebenezer - Seaman - Nbr: 4205 - Prize: Minerva - Ship Type: MV - How taken: HMS Conquistador - When taken: 19 Jan 1814 - Where taken: Bay of Biscay - Date Received: 7 Oct 1814 - From what ship: HMT Niobe, Chatham - Born: Yarmouth - Age: 25 - Released on 13 Jun 1815.
Harris, Alpheus - Seaman - Nbr: 4265 - Prize: Argus - Ship Type: MV - How taken: HMS San Domingo - When taken: 1 Mar 1814 - Where taken: off Savannah - Date Received: 7 Oct 1814 - From what ship: HMT Niobe, Chatham - Born: Chesterfield - Age: 22 - Released on 13 Jun 1815.
Harris, George - Seaman - Nbr: 5366 - Prize: Penn - Ship Type: Whaler - How taken: HMS Acorn - When taken: 27 Oct 1813 - Where taken: South Sea - Date Received: 31 Oct 1814 - From what ship: HMT Leyden, Chatham - Born: Nantucket - Age: 17 - Released on 1 Jul 1815.
Harris, George - Seaman - Nbr: 344 - Prize: Good Friends - Ship Type: MV - How taken: HMS Andromache - When taken: 2 Apr 1813 - Where taken: Bay of Biscay - Date Received: 28 Jun 1813 - From what ship:

American Prisoners of War Held at Dartmoor during the War of 1812

## Alphabetical listing of names

Plymouth - Born: Portsmouth - Age: 19 - Released on 20 Apr 1815.
Harris, James - Seaman - Nbr: 1468 - Prize: Governor Gerry - Ship Type: MV - How taken: HMS Lyra - When taken: 29 May 1813 - Where taken: Bay of Biscay - Date Received: 19 Jun 1814 - From what ship: Stapleton - Born: New Orleans - Age: 19 - Race: Negro - Released on 28 Apr 1815.
Harris, James - Seaman - Nbr: 2715 - How taken: Gave himself up from HMS Sterling Castle - When taken: 12 Jun 1813 - Date Received: 21 Aug 1814 - From what ship: HMT Freya, Chatham - Born: Pennsylvania - Age: 24 - Race: Black - Released on 26 Apr 1815.
Harris, James - Seaman - Nbr: 3495 - Prize: Hannibal - Ship Type: MV - How taken: HMS Pique - When taken: Apr 1814 - Date Received: 24 Sep 1814 - From what ship: HMT Salvador del Mundo - Born: Philadelphia - Age: 29 - Released on 4 Jun 1815.
Harris, John - Seaman - Nbr: 4741 - How taken: Gave himself up from HMS Salvador - When taken: 1 Feb 1814 - Date Received: 9 Oct 1814 - From what ship: HMT Freya, Chatham - Born: Virginia - Age: 26 - Race: Mulatto - Released on 15 Jun 1815.
Harris, John - Seaman - Nbr: 87 - Prize: Cashier - Ship Type: MV - How taken: HMS Reindeer - When taken: 3 Feb 1813 - Where taken: Bay of Biscay - Date Received: 2 Apr 1813 - From what ship: Plymouth - Born: Maryland - Age: 25 - Sent to Dartmouth on 30 Jul 1813.
Harris, John - Seaman - Nbr: 1318 - How taken: Gave himself up from HMS Venus - When taken: 2 Jan 1813 - Date Received: 19 Jun 1814 - From what ship: Stapleton - Born: Abbott County - Age: 20 - Race: Negro - Released on 28 Apr 1815.
Harris, John - Marine - Nbr: 2504 - Prize: US Sloop Frolic - Ship Type: MW - How taken: HMS Orpheus - When taken: 20 Apr 1814 - Where taken: off Cuba - Date Received: 16 Aug 1814 - From what ship: HMT Queen, Halifax - Born: Homer - Age: 22 - Released on 3 May 1815.
Harris, John - Seaman - Nbr: 1605 - Prize: Fox - Ship Type: LM - How taken: HMS Pheasant - When taken: 23 Apr 1813 - Where taken: Bay of Biscay - Date Received: 23 Jun 1814 - From what ship: Stapleton - Born: Pennsylvania - Age: 27 - Race: Negro - Released on 1 May 1815.
Harris, Joseph - Lieutenant - Nbr: 5282 - Prize: Enterprize - Ship Type: P - How taken: HMS Tenedos - When taken: 21 May 1813 - Where taken: off Cape Ann - Date Received: 31 Oct 1814 - From what ship: HMT Leyden, Chatham - Born: Ipswich - Age: 32 - Released on 4 Jun 1815.
Harris, Moses - 2nd Mate - Nbr: 5771 - Prize: William Penn - Ship Type: MV - How taken: HMS Acorn - When taken: 27 Oct 1812 - Where taken: Lat 14 - Date Received: 26 Dec 1814 - From what ship: HMT Argo - Born: Massachusetts - Age: 29 - Released on 27 Apr 1815.
Harris, Simeon - Seaman - Nbr: 443 - Prize: Magdalen - Ship Type: MV - How taken: HMS Superb - When taken: 15 Apr 1813 - Where taken: off Belle Isle, France - Date Received: 1 Jul 1813 - From what ship: Plymouth - Born: Massachusetts - Age: 20 - Died on 5 Mar 1814 from dysentery and dropsy.
Harris, William - Seaman - Nbr: 1833 - Prize: Lightning - Ship Type: MV - How taken: HMS Medusa - When taken: 2 Apr 1813 - Where taken: Bay of Biscay - Date Received: 21 Jul 1814 - From what ship: HMT Redbeard & Pincher, Chatham Depot - Born: Marblehead - Age: 43 - Released on 1 May 1815.
Harris, William - Seaman - Nbr: 2095 - How taken: Gave himself up from HMS Malta - When taken: 1 Jan 1813 - Date Received: 3 Aug 1814 - From what ship: HMS Bittern, Chatham Depot - Born: New York - Age: 37 - Released on 2 May 1815.
Harris, William - Seaman - Nbr: 5116 - Prize: Portsmouth - Ship Type: P - How taken: HMS Pylades - When taken: 13 Sep 1814 - Where taken: Long 65 - Date Received: 28 Oct 1814 - From what ship: HMT Alkbar, Halifax - Born: Portsmouth - Age: 16 - Race: Negro - Died on 24 Nov 1814.
Harrison, George - Seaman - Nbr: 6194 - Prize: Prince de Neufchatel - Ship Type: P - How taken: Leander (Newcastle Acasta) - When taken: 20 Dec 1814 - Where taken: Lat 38 Long 56 - Date Received: 30 Jan 1815 - From what ship: HMT Pheasant - Born: New York - Age: 17 - Released on 5 Jul 1815.
Harrison, Henry - Seaman - Nbr: 4185 - How taken: Gave himself up from HMS Fame - When taken: 4 May 1813 - Date Received: 7 Oct 1814 - From what ship: HMT Niobe, Chatham - Born: New York - Age: 28 - Released on 13 Jun 1815.
Harrison, James - Seaman - Nbr: 2283 - Prize: Diomede - Ship Type: P - How taken: HMS Rifleman - When taken: 28 May 1814 - Where taken: off Sable Island - Date Received: 16 Aug 1814 - From what ship: HMS Dublin, Halifax - Born: Salem - Age: 26 - Released on 3 May 1815.
Harrison, John - Second Lieutenant - Nbr: 2515 - Prize: Snap Dragon - Ship Type: P - How taken: HMS Martin - When taken: 10 Jun 1814 - Where taken: off Halifax - Date Received: 16 Aug 1814 - From what ship: HMT Queen, Halifax - Born: Maryland - Age: 29 - Released on 3 May 1815.

American Prisoners of War Held at Dartmoor during the War of 1812

## Alphabetical listing of names

Harrison, Silas - Seaman - Nbr: 3571 - Prize: Hawk - Ship Type: P - How taken: HMS Pique - When taken: 26 Apr 1814 - Where taken: off Bermuda - Date Received: 30 Sep 1814 - From what ship: HMT Sybella - Born: North Carolina - Age: 21 - Died on 6 Jan 1815 from variola.

Harry, John - Seaman - Nbr: 151 - How taken: Gave himself up from HMS Abercrombie - Date Received: 2 Apr 1813 - From what ship: Plymouth - Born: New Orleans - Age: 45 - Sent to Chatham on 27 May 1813.

Hart, Bartholomew - Seaman - Nbr: 664 - Prize: Joel Barlow - Ship Type: LM - How taken: HMS Briton - When taken: 3 Jul 1813 - Where taken: off Bordeaux - Date Received: 8 Sep 1813 - From what ship: Plymouth - Born: New York - Age: 42 - Released on 26 Apr 1815.

Hart, Emuis - Seaman - Nbr: 5628 - Prize: William Penn - Ship Type: MV - How taken: HMS Acorn - When taken: 27 Oct 1812 - Where taken: Lat 14 - Date Received: 24 Dec 1814 - From what ship: HMT Tay - Born: Dartmouth - Age: 33 - Race: Mulatto - Released on 3 Jul 1815.

Hart, George - Seaman - Nbr: 2473 - Prize: US Sloop Frolic - Ship Type: MW - How taken: HMS Orpheus - When taken: 20 Apr 1814 - Where taken: off Cuba - Date Received: 16 Aug 1814 - From what ship: HMT Queen, Halifax - Born: New York - Age: 23 - Released on 3 May 1815.

Hart, James - Seaman - Nbr: 1508 - Prize: Courier - Ship Type: LM - How taken: HMS Rover - When taken: 14 Mar 1813 - Where taken: Bay of Biscay - Date Received: 23 Jun 1814 - From what ship: Stapleton - Born: New London - Age: 28 - Died on 8 Jul 1814 from pneumonia.

Hart, Samuel - 2nd Mate - Nbr: 5714 - Prize: Tickler - Ship Type: MV - How taken: HMS Niemen - When taken: 2 Oct 1814 - Where taken: off Delaware - Date Received: 24 Dec 1814 - From what ship: HMT Penelope - Born: Philadelphia - Age: 23 - Released on 3 Jul 1815.

Hartar, Henry - Seaman - Nbr: 384 - Prize: VA Planter - Ship Type: MV - How taken: HMS Pyramus - When taken: 18 Mar 1813 - Where taken: off Nantes - Date Received: 1 Jul 1813 - From what ship: Plymouth - Born: Charlestown - Age: 20 - Sent to Mill Prison (Plymouth, England) on 3 Aug 1813.

Harter, Henry - Seaman - Nbr: 540 - Prize: Matilda - Ship Type: MV - How taken: HMS Revolutionnaire - When taken: 27 Jul 1813 - Where taken: off L'Orient (France) - Date Received: 8 Sep 1813 - From what ship: Plymouth - Born: Charlestown - Age: 20 - Released on 26 Apr 1815.

Hartfield, James - Seaman - Nbr: 549 - How taken: Gave himself up from HMS Scorpion - When taken: 4 Nov 1812 - Date Received: 8 Sep 1813 - From what ship: Plymouth - Born: Havan County - Age: 36 - Released on 26 Apr 1815.

Harts, William - Seaman - Nbr: 446 - Prize: Magdalen - Ship Type: MV - How taken: HMS Superb - When taken: 15 Apr 1813 - Where taken: off Belle Isle, France - Date Received: 1 Jul 1813 - From what ship: Plymouth - Born: New York - Age: 24 - Released on 20 Apr 1815.

Hartsoln, John - Seaman - Nbr: 2980 - Prize: Frolic - Ship Type: P - How taken: HMS Heron - When taken: 25 Jan 1814 - Where taken: off St. Thomas - Date Received: 29 Aug 1814 - From what ship: HMT Bittern - Born: Massachusetts - Age: 23 - Released on 19 May 1815.

Hartwell, Berry - Seaman - Nbr: 1235 - How taken: Sent into custody from HMS Thames - When taken: 28 Dec 1812 - Date Received: 14 Jun 1814 - From what ship: Mill Prison (Plymouth, England) - Born: Philadelphia - Age: 37 - Released on 26 Apr 1815.

Harvey, Anthony - Seaman - Nbr: 4693 - Prize: Growler - Ship Type: P - How taken: HMS Maidstone - When taken: 20 Jul 1813 - Where taken: Grand Banks - Date Received: 9 Oct 1814 - From what ship: HMT Leyden, Chatham - Born: Boston - Age: 26 - Released on 29 Jun 1815.

Harvey, Dover - Seaman - Nbr: 5498 - Prize: General Putnam - Ship Type: P - How taken: HMS Leander - When taken: 8 Nov 1814 - Where taken: Long 65 Lat 42 - Date Received: 17 Dec 1814 - From what ship: HMT Loire, Halifax - Born: Salem - Age: 50 - Race: Black - Released on 1 Jul 1815.

Harvey, Edward - Boy - Nbr: 2390 - Prize: US Sloop Frolic - Ship Type: MW - How taken: HMS Orpheus - When taken: 20 Apr 1814 - Where taken: off Cuba - Date Received: 16 Aug 1814 - From what ship: HMT Queen, Halifax - Born: North Carolina - Age: 16 - Sent to Dartmouth on 19 Oct 1814.

Harvey, James - Seaman - Nbr: 6392 - Prize: Nellenville - Ship Type: MV - How taken: HMS Onyx - When taken: 25 Dec 1814 - Where taken: off San Domingo - Date Received: 3 Mar 1815 - From what ship: HMT Ganges, Plymouth - Born: North Carolina - Age: 32 - Released on 11 Jul 1815.

Harvey, John - Seaman - Nbr: 1843 - How taken: Gave himself up from HMS Foxhound - When taken: 22 Feb 1813 - Date Received: 21 Jul 1814 - From what ship: HMT Redbeard & Pincher, Chatham Depot - Born: New Orleans - Age: 45 - Released on 26 Apr 1815.

Harvey, Joseph - Seaman - Nbr: 377 - Prize: Independence - Ship Type: MV - How taken: HMS Superb - When taken: 16 Mar 1813 - Where taken: Bay of Biscay - Date Received: 1 Jul 1813 - From what ship: Plymouth -

American Prisoners of War Held at Dartmoor during the War of 1812

## Alphabetical listing of names

Born: Salem - Age: 19 - Released on 20 Apr 1815.

Harvey, Joseph - Seaman - Nbr: 4202 - Prize: Hannah - Ship Type: MV - How taken: HMS Conquistador - When taken: 15 Jan 1814 - Where taken: at sea - Date Received: 7 Oct 1814 - From what ship: HMT Niobe, Chatham - Born: Beverly - Age: 33 - Released on 13 Jun 1815.

Harvey, Joseph - Boy - Nbr: 3419 - Prize: Fox - Ship Type: P - How taken: Manley - When taken: 1 Aug 1813 - Where taken: off Halifax - Date Received: 13 Sep 1814 - From what ship: HMT Niobe, Chatham - Born: Bath - Age: 15 - Released on 28 May 1815.

Harwood, William - Seaman - Nbr: 3238 - Prize: Maydock - Ship Type: MV - How taken: Rebuff - When taken: 16 Jun 1813 - Where taken: off St. Marys - Date Received: 11 Sep 1814 - From what ship: HMT Freya, Chatham - Born: Maryland - Age: 31 - Released on 28 May 1815.

Haskell, Thomas - Seaman - Nbr: 822 - Prize: Fire Fly - Ship Type: LM - How taken: HMS Revolutionnaire - When taken: 19 Oct 1813 - Where taken: at sea - Date Received: 29 Nov 1813 - From what ship: Plymouth - Born: Cape Ann - Age: 38 - Released on 26 Apr 1815.

Haskett, Luther - Seaman - Nbr: 6226 - Prize: Prince de Neufchatel - Ship Type: P - How taken: Leander (Newcastle Acasta) - When taken: 20 Dec 1814 - Where taken: Lat 38 Long 56 - Date Received: 30 Jan 1815 - From what ship: HMT Pheasant - Born: Hampden - Age: 28 - Released on 5 Jul 1815.

Haskett, Robert - Seaman - Nbr: 1416 - Prize: Orders in Council - Ship Type: LM - How taken: HMS Surveillante - When taken: 1 Jun 1813 - Where taken: off Cape Ortegal (Spain) - Date Received: 19 Jun 1814 - From what ship: Stapleton - Born: Massachusetts - Age: 30 - Released on 28 Apr 1815.

Haskins, Joseph - Seaman - Nbr: 2463 - Prize: US Sloop Frolic - Ship Type: MW - How taken: HMS Orpheus - When taken: 20 Apr 1814 - Where taken: off Cuba - Date Received: 16 Aug 1814 - From what ship: HMT Queen, Halifax - Born: New Hampton - Age: 25 - Released on 3 May 1815.

Haster, John - Boatman - Nbr: 4316 - Prize: Union - Ship Type: MV - How taken: HMS Malabar - When taken: 17 Jan 1813 - Where taken: off Calcutta - Date Received: 7 Oct 1814 - From what ship: HMT Niobe, Chatham - Born: Not legible - Age: 32 - Released on 14 Jun 1815.

Hastings, Ephraim - 2nd Mate - Nbr: 5543 - Prize: Black Swan - Ship Type: MV - How taken: HMS Maidstone - When taken: 24 Oct 1814 - Where taken: Georges Bank - Date Received: 17 Dec 1814 - From what ship: HMT Loire, Halifax - Born: Plymouth - Age: 27 - Released on 27 Apr 1815.

Hatch, William - Seaman - Nbr: 2908 - Prize: Matilda - Ship Type: MV - How taken: HMS Revolutionnaire - When taken: 25 Jul 1813 - Where taken: off L'Orient (France) - Date Received: 24 Aug 1814 - From what ship: HMT Alpheus, Chatham - Born: Connecticut - Age: 32 - Released on 2 Nov 1814.

Hatheway, Philip - Seaman - Nbr: 4686 - Prize: Yankee - Ship Type: P - How taken: HMS Shannon - When taken: 20 Aug 1813 - Where taken: at sea - Date Received: 9 Oct 1814 - From what ship: HMT Leyden, Chatham - Born: Massachusetts - Age: 23 - Released on 15 Jun 1815.

Hatkins, Henry - Seaman - Nbr: 3018 - How taken: Gave himself up from HMS Zealous - When taken: 27 Aug 1814 - Date Received: 2 Sep 1814 - From what ship: HMT Centaur - Born: New Haven - Age: 24 - Released on 28 May 1815.

Hatton, Peter - Seaman - Nbr: 5288 - Prize: Fly - Ship Type: Prize - How taken: HMS Melpomene - When taken: 4 Sep 1813 - Where taken: at sea - Date Received: 31 Oct 1814 - From what ship: HMT Leyden, Chatham - Born: New Hampshire - Age: 40 - Released on 29 Jun 1815.

Hautell, Stephen - Seaman - Nbr: 6266 - Prize: Plutarch - Ship Type: MV - How taken: HMS Helicon - When taken: 5 Feb 1815 - Where taken: off Bordeaux - Date Received: 10 Feb 1815 - From what ship: HMT Ganges, Plymouth - Born: Not legible - Age: 23 - Released on 5 Jul 1815.

Hawes, Caleb - Seaman - Nbr: 6323 - Prize: Prince de Neufchatel - Ship Type: P - How taken: Leander (Newcastle Acasta) - When taken: 20 Dec 1814 - Where taken: Lat 38 Long 56 - Date Received: 19 Feb 1815 - From what ship: HMT Ganges, Plymouth - Born: Boston - Age: 22 - Released on 5 Jul 1815.

Hawker, Edward - Seaman - Nbr: 2841 - How taken: Gave himself up from HMS Armide - When taken: 8 Jun 1813 - Date Received: 24 Aug 1814 - From what ship: HMT Liverpool, Chatham - Born: Massachusetts - Age: 24 - Released on 19 May 1815.

Hawkins, Isaac - Seaman - Nbr: 3334 - Prize: Columbia - Ship Type: MV - How taken: Sent into custody from a privateer - When taken: 14 May 1813 - Where taken: at sea - Date Received: 13 Sep 1814 - From what ship: HMT Niobe, Chatham - Born: Maryland - Age: 31 - Race: Mulatto - Released on 28 May 1815.

Hawkins, John - Seaman - Nbr: 6301 - Prize: John - Ship Type: Prize - How taken: Leander (Newcastle Acasta) - When taken: 30 Jan 1815 - Date Received: 19 Feb 1815 - From what ship: HMT Ganges, Plymouth - Born: Rhode Island - Age: 25 - Race: Black - Released on 5 Jul 1815.

American Prisoners of War Held at Dartmoor during the War of 1812

## Alphabetical listing of names

Hawkins, John - Seaman - Nbr: 1782 - How taken: Gave himself up from HMS Leyden - When taken: 23 Mar 1813 - Date Received: 20 Jul 1814 - From what ship: HMS Milford, Plymouth - Born: Philadelphia - Age: 32 - Released on 1 May 1815.

Hawkins, Parker - Seaman - Nbr: 5733 - How taken: Impressed at London - When taken: 18 Oct 1814 - Date Received: 26 Dec 1814 - From what ship: HMT Argo - Born: GA - Age: 28 - Released on 3 Jul 1815.

Hawks, Benjamin - Seaman - Nbr: 3711 - Prize: Alfred - Ship Type: P - How taken: HMS Epervier - When taken: 23 Feb 1814 - Where taken: off Newfoundland - Date Received: 30 Sep 1814 - From what ship: HMT President, Halifax - Born: Marblehead - Age: 19 - Released on 4 Jun 1815.

Hawley, Frederick - Seaman - Nbr: 2848 - How taken: Gave himself up from HMS Royal William - When taken: 3 Feb 1813 - Date Received: 24 Aug 1814 - From what ship: HMT Alpheus, Chatham - Born: Wilmington - Age: 23 - Died on 5 Feb 1815 from variola.

Hay, John - 2nd Lieutenant - Nbr: 5189 - Prize: Juliana Smith - Ship Type: P - How taken: HMS Nymphe - When taken: 11 May 1813 - Where taken: off Cape Sable Island (Canada) - Date Received: 31 Oct 1814 - From what ship: HMT Mermaid, Chatham - Born: New Jersey - Age: 34 - Released on 7 Jun 1815.

Haycock, John - Seaman - Nbr: 4978 - Prize: Invincible - Ship Type: LM - How taken: HMS Armide - When taken: 15 Aug 1814 - Where taken: off Nantucket - Date Received: 28 Oct 1814 - From what ship: HMT Alkbar, Halifax - Born: New York - Age: 24 - Released on 21 Jun 1815.

Haycock, Joseph - 2nd Gunner - Nbr: 6371 - Prize: US Brig Syren - Ship Type: MW - How taken: HMS Medway - When taken: 12 Jul 1814 - Where taken: off Cape of Good Hope - Date Received: 24 Feb 1815 - From what ship: HMT Ganges, Plymouth - Born: Portland - Age: 55 - Died on 20 Mar 1815 from variola.

Hayday, Larsay - Seaman - Nbr: 3356 - How taken: Gave himself up from HMS Union - When taken: 9 Dec 1812 - Date Received: 13 Sep 1814 - From what ship: HMT Niobe, Chatham - Born: New Jersey - Age: 38 - Released on 27 Apr 1815.

Haydon, William - Seaman - Nbr: 1959 - How taken: Gave himself up from HMS Bellerophon - When taken: 17 Oct 1812 - Date Received: 3 Aug 1814 - From what ship: HMS Lyffey, Chatham Depot - Born: Boston - Age: 32 - Released on 26 Apr 1815.

Haye, Moses - Seaman - Nbr: 1589 - Prize: Price - Ship Type: MV - How taken: HMS Pyramus - When taken: 6 Apr 1813 - Where taken: Bay of Biscay - Date Received: 23 Jun 1814 - From what ship: Stapleton - Born: Savannah - Age: 17 - Released on 1 May 1815.

Hayes, Benjamin - Seaman - Nbr: 931 - Prize: Fanny - Ship Type: MV - How taken: HMS Eurotas - When taken: 23 Dec 1813 - Where taken: at sea - Date Received: 31 Jan 1814 - From what ship: Plymouth - Born: Maryland - Age: 24 - Released on 27 Apr 1815.

Hayes, Edward - Quarter Gunner - Nbr: 3977 - Prize: US Brig Rattlesnake - Ship Type: MW - How taken: HMS Leander - When taken: 13 Jul 1814 - Where taken: off Shelburne - Date Received: 6 Oct 1814 - From what ship: HMT Chesapeake, Halifax - Born: Salem - Age: 38 - Released on 9 Jun 1815.

Hayes, John - Gunner - Nbr: 5435 - Prize: Invincible - Ship Type: LM - How taken: HMS Armide - When taken: 15 Aug 1814 - Where taken: off Nantucket - Date Received: 11 Nov 1814 - From what ship: HMT Impregnable - Born: Baltimore - Age: 31 - Released on 1 Jul 1815.

Hayes, Simon - Seaman - Nbr: 121 - Prize: Governor McKean - Ship Type: LM - How taken: HMS Rover - When taken: 26 Jan 1813 - Where taken: off Bordeaux - Date Received: 2 Apr 1813 - From what ship: Plymouth - Born: Virginia - Age: 19 - Sent to Dartmouth on 30 Jul 1813.

Hayes, Simon - Seaman - Nbr: 1671 - Prize: Vivid - Ship Type: Prize - How taken: HMS Ceres - When taken: 5 Jun 1814 - Where taken: Channel - Date Received: 2 Jul 1814 - From what ship: Plymouth - Born: Dumfries - Age: 21 - Released on 1 May 1815.

Haynes, Stephen - Seaman - Nbr: 3812 - Prize: Nimble - Ship Type: Prize - How taken: HMS Arab - When taken: 5 Apr 1814 - Where taken: Lat 37 Long 65 - Date Received: 5 Oct 1814 - From what ship: HMT Orpheus, Halifax - Born: Massachusetts - Age: 35 - Released on 9 Jun 1815.

Haywood, John - Seaman - Nbr: 2719 - How taken: Gave himself up from HMS Mulgrave - When taken: 13 Dec 1812 - Date Received: 21 Aug 1814 - From what ship: HMT Freya, Chatham - Born: Baltimore - Age: 25 - Race: Mulatto - Released on 26 Apr 1815.

Haywood, John - Seaman - Nbr: 3134 - How taken: Gave himself up from HMS Scorpion - When taken: 20 Dec 1812 - Date Received: 11 Sep 1814 - From what ship: HMT Freya, Chatham - Born: Maryland - Age: 25 - Race: Black - Died on 6 Apr 1815 from gunshot wound in prison.

Haywood, Samuel - Seaman - Nbr: 6336 - Prize: Prince de Neufchatel - Ship Type: P - How taken: Leander (Newcastle Acasta) - When taken: 20 Dec 1814 - Where taken: Lat 38 Long 56 - Date Received: 19 Feb

American Prisoners of War Held at Dartmoor during the War of 1812

## Alphabetical listing of names

1815 - From what ship: HMT Ganges, Plymouth - Born: Roxborough - Age: 26 - Released on 6 Jul 1815.

Hazard, Charles - Seaman - Nbr: 4578 - Prize: Bunker Hill - Ship Type: P - How taken: HMS Pomone - When taken: 2 Mar 1814 - Where taken: Channel - Date Received: 9 Oct 1814 - From what ship: HMT Leyden, Chatham - Born: Providence - Age: 24 - Race: Mulatto - Released on 15 Jun 1815.

Hazard, D. - Seaman - Nbr: 4907 - Prize: Rolla - Ship Type: P - How taken: HMS Loire - When taken: 10 Dec 1813 - Where taken: off Bull Island (South Carolina) - Date Received: 28 Oct 1814 - From what ship: HMT Alkbar, Halifax - Born: Rhode Island - Age: 22 - Race: Mulatto - Released on 21 Jun 1815.

Hazard, Prince - Seaman - Nbr: 4679 - Prize: Blockade - Ship Type: P - How taken: HMS Recruit - When taken: 17 Aug 1813 - Where taken: off America - Date Received: 9 Oct 1814 - From what ship: HMT Leyden, Chatham - Born: Rhode Island - Age: 22 - Race: Black - Released on 15 Jun 1815.

Hazard, Robert - Seaman - Nbr: 4734 - How taken: Gave himself up from HMS Stork - When taken: 24 Feb 1814 - Date Received: 9 Oct 1814 - From what ship: HMT Freya, Chatham - Born: Providence - Age: 23 - Race: Mulatto - Released on 15 Jun 1815.

Headen, John - Seaman - Nbr: 4120 - Prize: Bordeaux Packet - Ship Type: LM - How taken: HMS Niemen - When taken: 28 Jun 1814 - Where taken: off Delaware - Date Received: 6 Oct 1814 - From what ship: HMT Chesapeake, Halifax - Born: Boston - Age: 29 - Released on 13 Jun 1815.

Headley, John - Seaman - Nbr: 2582 - How taken: Gave himself up from HMS Royal William - When taken: 5 Feb 1813 - Date Received: 21 Aug 1814 - From what ship: HMT Freya, Chatham - Born: New York - Age: 29 - Released on 3 May 1815.

Hearns, Patrick - Soldier - Nbr: 5270 - Prize: 1st US Artillery - Ship Type: Troops - How taken: British Army - When taken: 13 Oct 1812 - Where taken: Canada - Date Received: 31 Oct 1814 - From what ship: HMT Leyden, Chatham - Born: King's County - Age: 38 - Released on 29 Jun 1815.

Heasey, John - Seaman - Nbr: 6544 - How taken: Gave himself up from HMS Orlando - When taken: 7 May 1813 - Date Received: 7 Mar 1815 - From what ship: HMT Ganges, Plymouth - Born: Martha's Vineyard - Age: 23 - Released on 11 Jul 1815.

Heath, Henry - Seaman - Nbr: 289 - Prize: Cannoneer - Ship Type: MV - How taken: HMS Warspite - When taken: 14 Mar 1813 - Where taken: Bay of Biscay - Date Received: 28 Jun 1813 - From what ship: Plymouth - Born: New York - Age: 18 - Released on 20 Apr 1815.

Heath, James - Seaman - Nbr: 3793 - Prize: James - Ship Type: Prize - How taken: Rebecca (LM) - When taken: 13 Apr 1814 - Where taken: off Delaware - Date Received: 5 Oct 1814 - From what ship: HMT President, Halifax - Born: Philadelphia - Age: 49 - Released on 9 Jun 1815.

Heaton, Henry - Seaman - Nbr: 3197 - How taken: Gave himself up from HMS America - When taken: 16 Jul 1813 - Date Received: 11 Sep 1814 - From what ship: HMT Freya, Chatham - Born: Lancaster - Age: 33 - Released on 28 May 1815.

Heaton, William - Seaman - Nbr: 2698 - How taken: Gave himself up from HMS Carlotte - Date Received: 21 Aug 1814 - From what ship: HMT Freya, Chatham - Born: Hampshire - Age: 26 - Released on 28 May 1815.

Hebron, John - Seaman - Nbr: 724 - Prize: Francis and Ann - Ship Type: MV - How taken: HMS Lightning - When taken: 20 Apr 1813 - Where taken: at sea - Date Received: 27 Sep 1813 - From what ship: Plymouth - Born: Philadelphia - Age: 19 - Released on 26 Apr 1815.

Hecock, William - Seaman - Nbr: 4899 - Prize: Atlantic - Ship Type: Prize - How taken: HMS Maidstone - When taken: 10 Oct 1813 - Where taken: off Sable Island - Date Received: 28 Oct 1814 - From what ship: HMT Alkbar, Halifax - Born: New Haven - Age: 36 - Released on 21 Jun 1815.

Hedenburg, Jacob - Seaman - Nbr: 1108 - Prize: Bunker Hill - Ship Type: P - How taken: HMS Pomone & HMS Cadmus - When taken: 4 Mar 1814 - Where taken: Bay of Biscay - Date Received: 10 May 1814 - From what ship: Plymouth - Born: Philadelphia - Age: 23 - Sent to Mill Prison (Plymouth, England) on 17 Jun 1814.

Hedley, John - Seaman - Nbr: 1704 - Prize: Fanny - Ship Type: Prize - How taken: HMS Sceptre - When taken: 12 May 1814 - Date Received: 2 Jul 1814 - From what ship: Plymouth - Born: Virginia - Age: 28 - Released on 1 May 1815.

Helair, Gasper - Seaman - Nbr: 499 - Prize: Courier - Ship Type: LM - How taken: HMS Rover - When taken: 14 Mar 1813 - Where taken: Bay of Biscay - Date Received: 8 Sep 1813 - From what ship: Plymouth - Born: Philadelphia - Age: 52 - Released on 26 Apr 1815.

Hellin, John P. - Seaman - Nbr: 5654 - Prize: Valentine - Ship Type: MV - How taken: British forces - When taken: 17 Nov 1812 - Where taken: off Cape of Good Hope - Date Received: 24 Dec 1814 - From what ship: HMT Impregnable - Born: Norwich - Age: 25 - Released on 27 Apr 1815.

American Prisoners of War Held at Dartmoor during the War of 1812

## Alphabetical listing of names

Hely, John - Seaman - Nbr: 1460 - Prize: Revenge - Ship Type: LM - How taken: HMS Belle Poule - When taken: 10 Mar 1813 - Where taken: off Cornwall - Date Received: 19 Jun 1814 - From what ship: Stapleton - Born: Boston - Age: 20 - Released on 28 Apr 1815.

Hencock, William - Seaman - Nbr: 908 - Prize: Zephyr - Ship Type: MV - How taken: HMS Pyramus - When taken: 30 Nov 1813 - Where taken: off L'Orient (France) - Date Received: 31 Jan 1814 - From what ship: Plymouth - Born: New York - Age: 40 - Released on 27 Apr 1815.

Henderson, Alexander - Seaman - Nbr: 1780 - Prize: Criterion - Ship Type: MV - How taken: HMS Belle Poule - When taken: 14 Feb 1813 - Where taken: Bay of Biscay - Date Received: 20 Jul 1814 - From what ship: HMS Milford, Plymouth - Born: Connecticut - Age: 26 - Died on 27 Dec 1814 from phthisis pulmonalis.

Henderson, Benjamin - Seaman - Nbr: 5715 - Prize: Theodore - Ship Type: P - How taken: HMS Saturn - When taken: 5 Nov 1814 - Where taken: off Brazil - Date Received: 24 Dec 1814 - From what ship: HMT Penelope - Born: Not listed - Released on 3 Jul 1815.

Henderson, George - Soldier - Nbr: 5838 - Prize: 4th US Rifles - Ship Type: Troops - How taken: British Army - When taken: 17 Sep 1814 - Where taken: Fort Erie - Date Received: 26 Dec 1814 - From what ship: HMT Argo - Born: Maryland - Age: 46 - Released on 29 Jun 1815.

Henderson, James - Seaman - Nbr: 2265 - Prize: Grand Turk - Ship Type: P - How taken: HMS Martin - When taken: 26 May 1814 - Where taken: off Cape Sable Island (Canada) - Date Received: 16 Aug 1814 - From what ship: HMS Dublin, Halifax - Born: Salem - Age: 46 - Released on 3 May 1815.

Henderson, John - Seaman - Nbr: 5104 - Prize: David Porter - Ship Type: LM - How taken: HMS Pylades - When taken: 12 Sep 1814 - Where taken: Georges Bank - Date Received: 28 Oct 1814 - From what ship: HMT Alkbar, Halifax - Born: Salem - Age: 23 - Released on 29 Jun 1815.

Henderson, Joseph - Seaman - Nbr: 6484 - Prize: John - Ship Type: MV - How taken: Variable - When taken: 11 Aug 1814 - Where taken: off Cuba - Date Received: 3 Mar 1815 - From what ship: HMT Ganges, Plymouth - Born: New London - Age: 21 - Released on 11 Jul 1815.

Hendrickson, John - Seaman - Nbr: 4122 - Prize: Bordeaux Packet - Ship Type: LM - How taken: HMS Niemen - When taken: 28 Jun 1814 - Where taken: off Delaware - Date Received: 6 Oct 1814 - From what ship: HMT Chesapeake, Halifax - Born: Maryland - Age: 30 - Released on 13 Jun 1815.

Hendrickson, T. - Seaman - Nbr: 2510 - Prize: Snap Dragon - Ship Type: P - How taken: HMS Martin - When taken: 10 Jun 1814 - Where taken: off Halifax - Date Received: 16 Aug 1814 - From what ship: HMT Queen, Halifax - Born: New York - Age: 24 - Released on 3 May 1815.

Hendry, William - Seaman - Nbr: 6505 - How taken: Gave himself up from HMS Shark - Date Received: 3 Mar 1815 - From what ship: HMT Ganges, Plymouth - Born: SC - Age: 49 - Released on 28 Apr 1815.

Henley, John - Seaman - Nbr: 5185 - Prize: Montgomery - Ship Type: P - How taken: HMS Nymphe - When taken: 1 May 1813 - Where taken: off Cape Cod - Date Received: 31 Oct 1814 - From what ship: HMT Mermaid, Chatham - Born: Marblehead - Age: 53 - Released on 29 Jun 1815.

Henman, Elisha - Seaman - Nbr: 2112 - How taken: Gave himself up from HMS Cornwallis - When taken: 26 Jul 1814 - Date Received: 8 Aug 1814 - From what ship: HMS Sterling Castle - Born: Massachusetts - Age: 33 - Released on 2 May 1815.

Henory, James - Seaman - Nbr: 6540 - Prize: US Brig Syren - Ship Type: MW - How taken: HMS Medway - When taken: 12 Jul 1814 - Where taken: off Cape of Good Hope - Date Received: 7 Mar 1815 - From what ship: HMT Ganges, Plymouth - Born: Portland - Age: 33 - Released on 11 Jul 1815.

Henry, Henry - Seaman - Nbr: 3166 - Prize: Falcon - Ship Type: Prize - How taken: Spanish Army - When taken: 2 Jul 1813 - Where taken: Papage Harbor - Date Received: 11 Sep 1814 - From what ship: HMT Freya, Chatham - Born: Massachusetts - Age: 25 - Sent to Dartmouth on 23 Sep 1814.

Henry, Isaac - Seaman - Nbr: 4550 - How taken: Gave himself up from HMS Bruiser - When taken: 3 Feb 1814 - Date Received: 8 Oct 1814 - From what ship: HMT Leyden, Chatham - Born: Frederick - Age: 25 - Released on 15 Jun 1815.

Henry, James - Seaman - Nbr: 572 - Prize: US Brig Argus - Ship Type: MW - How taken: HMS Pelican - When taken: 14 Apr 1813 - Where taken: Irish Channel - Date Received: 8 Sep 1813 - From what ship: Plymouth - Born: New York - Age: 18 - Died on 3 Jul 1814, killed fighting with Thomas Hill.

Henry, John - Seaman - Nbr: 633 - Prize: US Brig Argus - Ship Type: MW - How taken: HMS Pelican - When taken: 14 Aug 1813 - Where taken: Irish Channel - Date Received: 8 Sep 1813 - From what ship: Plymouth - Born: Norfolk - Age: 24 - Sent to Dartmouth on 2 Nov 1814.

Henry, Nathanial - Seaman - Nbr: 4113 - Prize: Thorn - Ship Type: MV - How taken: HMS Bulwark - When taken: 9 Jul 1814 - Where taken: off Nantucket - Date Received: 6 Oct 1814 - From what ship: HMT Chesapeake,

American Prisoners of War Held at Dartmoor during the War of 1812

## Alphabetical listing of names

Halifax - Born: Portsmouth - Age: 27 - Race: Negro - Released on 13 Jun 1815.
Henry, Robert - Seaman - Nbr: 3095 - How taken: Gave himself up from HMS Norge - Date Received: 3 Sep 1814 - From what ship: HMT Bristol - Born: Delaware - Age: 35 - Released on 28 May 1815.
Henry, William - Seaman - Nbr: 6163 - Prize: Lion - Ship Type: P - How taken: HMS Granicus - When taken: 2 Dec 1814 - Where taken: off Lisbon - Date Received: 18 Jan 1815 - From what ship: HMT Impregnable - Born: Albany - Age: 21 - Released on 5 Jul 1815.
Hensell, John - Seaman - Nbr: 1597 - Prize: Shadow - Ship Type: LM - How taken: HMS Reindeer - When taken: 6 Apr 1813 - Where taken: Bay of Biscay - Date Received: 23 Jun 1814 - From what ship: Stapleton - Born: Philadelphia - Age: 28 - Released on 1 May 1815.
Henshaw, George - Seaman - Nbr: 2036 - How taken: Gave himself up from HMS Impeteux - When taken: 2 Dec 1812 - Date Received: 3 Aug 1814 - From what ship: HMS Lyffey, Chatham Depot - Born: Salem - Age: 34 - Released on 26 Apr 1815.
Henson, Peter - Boatswain - Nbr: 6431 - Prize: Chance - Ship Type: P - How taken: HMS Statira - When taken: 1 Apr 1814 - Where taken: Lat 38 Long 24 - Date Received: 3 Mar 1815 - From what ship: HMT Ganges, Plymouth - Born: Not legible - Age: 50 - Released on 11 Jul 1815.
Hensow, John - Seaman - Nbr: 3783 - Prize: James - Ship Type: Prize - How taken: Rebecca (LM) - When taken: 13 Apr 1814 - Where taken: off Delaware - Date Received: 5 Oct 1814 - From what ship: HMT President, Halifax - Born: Baltimore - Age: 27 - Released on 9 Jun 1815.
Henltey, Jacob - Seaman - Nbr: 6451 - Prize: Jemmell - Ship Type: MV - How taken: HMS Rhin - When taken: 28 May 1814 - Where taken: off Bermuda - Date Received: 3 Mar 1815 - From what ship: HMT Ganges, Plymouth - Born: Salem - Age: 17 - Died on 16 Apr 1815 from phthisis pulmonalis.
Heny, Daniel - Prize Master - Nbr: 3595 - Prize: Frolic - Ship Type: P - How taken: HMS Heron - When taken: 25 Jan 1814 - Where taken: off St. Thomas - Date Received: 30 Sep 1814 - From what ship: HMT Sybella - Born: Salem - Age: 22 - Died on 25 Jan 1815 from febris.
Henzeman, Christopher - Seaman - Nbr: 4188 - Prize: Commodore Perry - Ship Type: Schooner - How taken: Sent into custody from a cutter - When taken: 25 Feb 1814 - Where taken: off Bordeaux - Date Received: 7 Oct 1814 - From what ship: HMT Niobe, Chatham - Born: Prussia - Age: 21 - Released on 22 Dec 1814.
Hepburn, James - Seaman - Nbr: 6276 - Prize: Plutarch - Ship Type: MV - How taken: HMS Helicon - When taken: 5 Feb 1815 - Where taken: off Bordeaux - Date Received: 17 Feb 1815 - From what ship: HMT Ganges, Plymouth - Born: Charleston - Age: 30 - Race: Mulatto - Released on 14 Jun 1815.
Hermanden, Peter - Seaman - Nbr: 4240 - Prize: Sally - Ship Type: MV - How taken: HMS Derwent - When taken: 21 Jan 1814 - Where taken: Grand Banks - Date Received: 7 Oct 1814 - From what ship: HMT Niobe, Chatham - Born: Weissbach - Age: 20 - Released on 13 Jun 1815.
Herrimond, Hezekiel - Seaman - Nbr: 4923 - Prize: Polly - Ship Type: MV - How taken: Wolverine (Privateer) - When taken: 9 Dec 1813 - Where taken: off Cape Ann - Date Received: 28 Oct 1814 - From what ship: HMT Alkbar, Halifax - Born: Castine - Age: 21 - Released on 21 Jun 1815.
Herring, William - 2nd Mate - Nbr: 4864 - Prize: Grecian - Ship Type: LM - How taken: HMS Jaseur - When taken: 2 May 1814 - Where taken: Chesapeake Bay - Date Received: 9 Oct 1814 - From what ship: HMT Freya, Halifax - Born: Philadelphia - Age: 24 - Released on 21 Jun 1815.
Hesty, John - Seaman - Nbr: 3515 - Prize: Aeolus - Ship Type: Prize - How taken: HMS Warspite - When taken: 2 Sep 1814 - Where taken: off Newfoundland - Date Received: 28 Sep 1814 - From what ship: HMT Salvador del Mundo - Born: Delaware - Age: 25 - Released on 4 Jun 1815.
Heydon, Elie - Seaman - Nbr: 302 - Prize: Ducornau - Ship Type: MV - How taken: HMS Pheasant - When taken: 15 Mar 1813 - Where taken: Bay of Biscay - Date Received: 28 Jun 1813 - From what ship: Plymouth - Born: Hartford - Age: 35 - Released on 9 Apr 1815.
Hibben, Jeremiah - Seaman - Nbr: 2064 - How taken: Apprehended at London - When taken: 17 Jun 1813 - Date Received: 3 Aug 1814 - From what ship: HMS Bittern, Chatham Depot - Born: Virginia - Age: 29 - Released on 2 May 1815.
Hickman, Joseph - Seaman - Nbr: 1492 - Prize: Price - Ship Type: MV - How taken: HMS Iris - When taken: 13 Apr 1813 - Where taken: Bay of Biscay - Date Received: 19 Jun 1814 - From what ship: Stapleton - Born: New Jersey - Age: 22 - Released on 1 May 1815.
Hicks, Benjamin - Seaman - Nbr: 6476 - Prize: Mary - Ship Type: P - How taken: Lapphin - When taken: 15 Jun 1814 - Where taken: off Saint Marys - Date Received: 3 Mar 1815 - From what ship: HMT Ganges, Plymouth - Born: Rhode Island - Age: 28 - Escaped on 1 Jun 1815.
Hicks, James - Seaman - Nbr: 5422 - How taken: Gave himself up from HMS Indian - When taken: 20 Aug 1814 -

American Prisoners of War Held at Dartmoor during the War of 1812

## Alphabetical listing of names

Date Received: 31 Oct 1814 - From what ship: HMT Leyden, Chatham - Born: Providence - Age: 32 - Race: Black - Released on 11 Jul 1815.

Hicks, Ogershill - Seaman - Nbr: 5414 - Prize: John & Mary - Ship Type: Prize - How taken: HMS Epervier - When taken: 20 Oct 1812 - Where taken: off Long Island - Date Received: 31 Oct 1814 - From what ship: HMT Leyden, Chatham - Born: Tiverton - Age: 26 - Released on 27 Apr 1815.

Hickson, John - Seaman - Nbr: 5024 - Prize: Landrail - Ship Type: Prize - How taken: HMS Wasp - When taken: 27 Jul 1814 - Where taken: Georges Bank - Date Received: 28 Oct 1814 - From what ship: HMT Alkbar, Halifax - Born: Massachusetts - Age: 21 - Released on 21 Jun 1815.

Hidalgo, Valentine - Seaman - Nbr: 4814 - Prize: President - Ship Type: P - How taken: HMS Pique - When taken: 7 May 1814 - Where taken: off Porto Rico - Date Received: 9 Oct 1814 - From what ship: HMT Freya, Halifax - Born: Cartagena - Age: 15 - Race: Mulatto - Released on 21 Jun 1815.

Hier, Andrew - 2nd Officer - Nbr: 5533 - Prize: Sparks - Ship Type: LM - How taken: HMS Maidstone - When taken: 28 Sep 1814 - Where taken: off Nantucket - Date Received: 17 Dec 1814 - From what ship: HMT Loire, Halifax - Born: New York - Age: 33 - Released on 1 Jul 1815.

Higby, James - Seaman - Nbr: 4308 - Prize: Alligator - Ship Type: MV - How taken: HMS San Domingo - When taken: 16 May 1813 - Where taken: off Bengal - Date Received: 7 Oct 1814 - From what ship: HMT Niobe, Chatham - Born: Bedford - Age: 17 - Released on 4 Jun 1815.

Higginbotham, William - Seaman - Nbr: 3824 - Prize: Bordeaux Packet - Ship Type: LM - How taken: HMS Niemen - When taken: 28 Jan 1814 - Where taken: off Delaware - Date Received: 5 Oct 1814 - From what ship: HMT Orpheus, Halifax - Born: Baltimore - Age: 36 - Released on 9 Jun 1815.

Higgins, Asa - Seaman - Nbr: 3285 - Prize: Juliana Smith - Ship Type: P - How taken: HMS Nymphe - When taken: 11 May 1813 - Where taken: off Cape Sable Island (Canada) - Date Received: 13 Sep 1814 - From what ship: HMT Niobe, Chatham - Born: Orleans - Age: 21 - Released on 28 May 1815.

Higgins, Francis - Seaman - Nbr: 3575 - Prize: William - Ship Type: MV - How taken: HMS Hasty - When taken: 14 May 1814 - Where taken: off Charleston - Date Received: 30 Sep 1814 - From what ship: HMT Sybella - Born: Massachusetts - Age: 22 - Released on 4 Jun 1815.

Higgins, George - Seaman - Nbr: 32 - How taken: Apprehended at Gibraltar - When taken: 8 Aug 1813 - Date Received: 2 Apr 1813 - From what ship: Plymouth - Born: Boston - Age: 26 - Race: Negro - Sent to Dartmouth on 30 Jul 1813.

Higgins, Isaac - Seaman - Nbr: 4946 - Prize: Herald - Ship Type: P - How taken: HMS Endymion - When taken: 15 Aug 1814 - Where taken: off Nantucket - Date Received: 28 Oct 1814 - From what ship: HMT Alkbar, Halifax - Born: Massachusetts - Age: 26 - Released on 21 Jun 1815.

Higgins, James - Seaman - Nbr: 392 - Prize: John & Frances - Ship Type: MV - How taken: HMS Belle Poule - When taken: 13 Mar 1813 - Where taken: Bay of Biscay - Date Received: 1 Jul 1813 - From what ship: Plymouth - Born: Philadelphia - Age: 27 - Released on 20 Apr 1815.

Higgins, John - Seaman - Nbr: 5178 - Prize: Vivid - Ship Type: P - How taken: HMS Nymphe - When taken: 20 Apr 1813 - Where taken: off Cape Cod - Date Received: 31 Oct 1814 - From what ship: HMT Mermaid, Chatham - Born: Wellfleet - Age: 19 - Released on 29 Jun 1815.

Higginson, David - Seaman - Nbr: 3050 - How taken: Gave himself up from HMS Sultan - When taken: 2 Sep 1814 - Date Received: 2 Sep 1814 - From what ship: HMT Sultan - Born: Virginia - Age: 36 - Race: Negro - Released on 28 Apr 1815.

Highby, William - Seaman - Nbr: 5107 - Prize: David Porter - Ship Type: LM - How taken: HMS Pylades - When taken: 12 Sep 1814 - Where taken: Georges Bank - Date Received: 28 Oct 1814 - From what ship: HMT Alkbar, Halifax - Born: New Jersey - Age: 19 - Released on 29 Jun 1815.

Higley, John - Seaman - Nbr: 3801 - Prize: Margetits - Ship Type: Prize - How taken: HMS Maidstone - When taken: 20 Mar 1814 - Where taken: off NY - Date Received: 5 Oct 1814 - From what ship: HMT Orpheus, Halifax - Born: Long Island - Age: 19 - Released on 9 Jun 1815.

Hilbert, Henry - Seaman - Nbr: 1112 - Prize: Bunker Hill - Ship Type: P - How taken: HMS Pomone & HMS Cadmus - When taken: 4 Mar 1814 - Where taken: Bay of Biscay - Date Received: 10 May 1814 - From what ship: Plymouth - Born: Philadelphia - Age: 17 - Released on 27 Apr 1815.

Hilby, Thomas - Mate - Nbr: 4217 - Prize: Liberty - Ship Type: MV - How taken: Surrendered - When taken: 30 Dec 1813 - Where taken: Stromess - Date Received: 7 Oct 1814 - From what ship: HMT Niobe, Chatham - Born: Stratford - Age: 24 - Released on 13 Jun 1815.

Hill, Benjamin - Seaman - Nbr: 3278 - Prize: Cossack - Ship Type: P - How taken: HMS Amelia - When taken: 14 Apr 1813 - Where taken: off St. Johns - Date Received: 13 Sep 1814 - From what ship: HMT Niobe,

American Prisoners of War Held at Dartmoor during the War of 1812

## Alphabetical listing of names

Chatham - Born: Salem - Age: 22 - Released on 28 May 1815.
Hill, Daniel - Seaman - Nbr: 4208 - Prize: Minerva - Ship Type: MV - How taken: HMS Conquistador - When taken: 19 Jan 1814 - Where taken: Bay of Biscay - Date Received: 7 Oct 1814 - From what ship: HMT Niobe, Chatham - Born: Charleston - Age: 27 - Race: Black - Released on 13 Jun 1815.
Hill, Ephraim - Seaman - Nbr: 2785 - Prize: Tiger - Ship Type: MV - How taken: HMS Scylla - When taken: 22 May 1813 - Where taken: Bay of Biscay - Date Received: 24 Aug 1814 - From what ship: HMT Liverpool, Chatham - Born: Hatford - Age: 24 - Released on 19 May 1815.
Hill, Francis - Seaman - Nbr: 6439 - Prize: Chance - Ship Type: P - How taken: HMS Statira - When taken: 1 Apr 1814 - Where taken: Lat 38 Long 24 - Date Received: 3 Mar 1815 - From what ship: HMT Ganges, Plymouth - Born: Virginia - Age: 18 - Released on 11 Jul 1815.
Hill, Hannel - Seaman - Nbr: 5212 - Prize: Porcupine - Ship Type: LM - How taken: HMS Acasta - When taken: 18 Jul 1813 - Where taken: off Halifax - Date Received: 31 Oct 1814 - From what ship: HMT Mermaid, Chatham - Born: Andover - Age: 19 - Released on 29 Jun 1815.
Hill, James - Seaman - Nbr: 4526 - How taken: Gave himself up from HMS Invincible - When taken: 14 Jan 1813 - Date Received: 8 Oct 1814 - From what ship: HMT Leyden, Chatham - Born: Boston - Age: 31 - Released on 28 May 1815.
Hill, James A. - Seaman - Nbr: 2378 - Prize: US Sloop Frolic - Ship Type: MW - How taken: HMS Orpheus - When taken: 20 Apr 1814 - Where taken: off Cuba - Date Received: 16 Aug 1814 - From what ship: HMT Queen, Halifax - Born: Boston - Age: 25 - Sent to Dartmouth on 19 Oct 1814.
Hill, Jeremiah - Seaman - Nbr: 4708 - Prize: Watson - Ship Type: MV - How taken: HMS Briton - When taken: 13 Dec 1813 - Where taken: off France - Date Received: 9 Oct 1814 - From what ship: HMT Leyden, Chatham - Born: Baltimore - Age: 27 - Released on 15 Jun 1815.
Hill, John - Seaman - Nbr: 1517 - Prize: Governor Gerry - Ship Type: LM - How taken: HMS Royalist - When taken: 31 May 1813 - Where taken: Bay of Biscay - Date Received: 23 Jun 1814 - From what ship: Stapleton - Born: Philadelphia - Age: 21 - Race: Colored - Released on 1 May 1815.
Hill, John - Boatswain - Nbr: 4434 - Prize: Taken off an English whaler - Ship Type: MV - How taken: HMS Illustrious - When taken: 22 Oct 1813 - Where taken: at sea - Date Received: 8 Oct 1814 - From what ship: HMT Leyden, Chatham - Born: Salem - Age: 20 - Escaped on 28 May 1815.
Hill, Joseph - Seaman - Nbr: 2781 - Prize: Tiger - Ship Type: MV - How taken: HMS Scylla - When taken: 22 May 1813 - Where taken: Bay of Biscay - Date Received: 24 Aug 1814 - From what ship: HMT Liverpool, Chatham - Born: New Orleans - Age: 20 - Race: Black - Released on 19 May 1815.
Hill, Justice - Seaman - Nbr: 4228 - Prize: Liberty - Ship Type: MV - How taken: Surrendered - When taken: 30 Dec 1813 - Where taken: Stromess - Date Received: 7 Oct 1814 - From what ship: HMT Niobe, Chatham - Born: Stockbridge - Age: 27 - Released on 13 Jun 1815.
Hill, Leonard - Seaman - Nbr: 465 - Prize: Meteor - Ship Type: MV - How taken: HMS Briton - When taken: 12 Mar 1813 - Where taken: Bay of Biscay - Date Received: 8 Sep 1813 - From what ship: Plymouth - Born: Alexandria - Age: 19 - Released on 20 Apr 1815.
Hill, Pompey - Seaman - Nbr: 3938 - Prize: Rolla - Ship Type: P - How taken: HMS Loire - When taken: 10 Dec 1813 - Where taken: off Bull Island (South Carolina) - Date Received: 5 Oct 1814 - From what ship: HMT President, Halifax - Born: Hutton County - Age: 38 - Race: Negro - Released on 9 Jun 1815.
Hill, Thomas - Seaman - Nbr: 1040 - Prize: US Brig Argus - Ship Type: MW - How taken: HMS Pelican - When taken: 16 Aug 1813 - Where taken: Irish Channel - Date Received: 10 May 1814 - What ship: Plymouth - Born: Brooklyn - Age: 27 - Sent to Exeter to be tried for the murder of James Henry on 15 Jul 1814.
Hill, Thomas - Seaman - Nbr: 2109 - Prize: US Brig Argus - Ship Type: MW - How taken: HMS Pelican - When taken: 14 Aug 1813 - Where taken: Irish Channel - Date Received: 3 Aug 1814 - From what ship: from Exeter jail - Born: Long Island - Age: 27 - Sent to Dartmouth on 19 Oct 1814.
Hill, Timothy - Marine - Nbr: 5881 - Prize: Harlequin - Ship Type: P - How taken: HMS Bulwark - When taken: 23 Nov 1814 - Where taken: off Halifax - Date Received: 27 Dec 1814 - From what ship: HMT Penelope - Born: Nottingham - Age: 25 - Released on 3 Jul 1815.
Hill, Timothy - Seaman - Nbr: 2832 - How taken: Gave himself up from HMS Resistance - When taken: 20 Nov 1812 - Date Received: 24 Aug 1814 - From what ship: HMT Liverpool, Chatham - Born: Nottingham - Age: 33 - Released on 26 Apr 1815.
Himes, Walter - Sergeant - Nbr: 594 - Prize: US Brig Argus - Ship Type: MW - How taken: HMS Pelican - When taken: 14 Aug 1813 - Where taken: Irish Channel - Date Received: 8 Sep 1813 - From what ship: Plymouth - Born: New Hampshire - Age: 21 - Sent to Dartmouth on 2 Nov 1814.

American Prisoners of War Held at Dartmoor during the War of 1812

## Alphabetical listing of names

Hinchman, George - Captain - Nbr: 6488 - Prize: John - Ship Type: MV - How taken: Variable - When taken: 11 Aug 1814 - Where taken: off Cuba - Date Received: 3 Mar 1815 - From what ship: HMT Ganges, Plymouth - Born: Boston - Age: 21 - Released on 11 Jul 1815.

Hindman, John - Seaman - Nbr: 6393 - Prize: Nellenville - Ship Type: MV - How taken: HMS Onyx - When taken: 25 Dec 1814 - Where taken: off San Domingo - Date Received: 3 Mar 1815 - From what ship: HMT Ganges, Plymouth - Born: Annapolis - Age: 41 - Race: Black - Released on 11 Jul 1815.

Hingston, Richard - Mate - Nbr: 6238 - Prize: Albion - Ship Type: Prize - How taken: HMS Harlequin - When taken: 6 Jan 1814 - Where taken: Long 20 - Date Received: 4 Feb 1815 - From what ship: HMT Ganges - Born: Vienna - Age: 21 - Released on 5 Jul 1815.

Hinkley, Aaron - Seaman - Nbr: 416 - Prize: Viper - Ship Type: MV - How taken: HMS Superb - When taken: 15 Apr 1813 - Where taken: Bay of Biscay - Date Received: 1 Jul 1813 - From what ship: Plymouth - Born: Massachusetts - Age: 25 - Sent to Plymouth on 29 Aug 1813.

Hinton, John - Seaman - Nbr: 3075 - How taken: Gave himself up from HMS Tigre - When taken: 10 Feb 1813 - Date Received: 3 Sep 1814 - From what ship: HMT Hydra, Chatham - Born: North Carolina - Age: 22 - Released on 28 May 1815.

Hitchcock, Edward - Seaman - Nbr: 3121 - How taken: Gave himself up from HMS Blake - When taken: 10 Dec 1812 - Date Received: 11 Sep 1814 - From what ship: HMT Freya, Chatham - Born: New York - Age: 46 - Race: Black - Released on 27 Apr 1815.

Hitchcock, Moses - Seaman - Nbr: 2658 - How taken: Apprehended at London - When taken: 13 Aug 1813 - Date Received: 21 Aug 1814 - From what ship: HMT Freya, Chatham - Born: New York - Age: 30 - Race: Black - Released on 19 May 1815.

Hitchell, George - Seaman - Nbr: 6477 - Prize: Mary - Ship Type: P - How taken: Lapphin - When taken: 15 Jun 1814 - Where taken: off Saint Marys - Date Received: 3 Mar 1815 - From what ship: HMT Ganges, Plymouth - Born: New York - Age: 19 - Released on 11 Jul 1815.

Hitchins, William - Prize master - Nbr: 4681 - Prize: Elbridge Gerry - Ship Type: P - How taken: HMS Crescent - When taken: 16 Sep 1813 - Where taken: at sea - Date Received: 9 Oct 1814 - From what ship: HMT Leyden, Chatham - Born: Massachusetts - Age: 23 - Released on 4 Jun 1815.

Hobart, Samuel B. - Prize master - Nbr: 5704 - Prize: Halifax Packet - Ship Type: Prize - How taken: HMS Bulwark - When taken: 20 Sep 1814 - Where taken: Georges Bank - Date Received: 24 Dec 1814 - From what ship: HMT Penelope - Born: Portsmouth - Age: 24 - Released on 20 Mar 1815.

Hobart, William - Prize Master - Nbr: 980 - Prize: Amity - Ship Type: P - How taken: HMS Achates - When taken: 22 Dec 1813 - Where taken: Bay of Biscay - Date Received: 31 Jan 1814 - From what ship: Plymouth - Born: Connecticut - Age: 26 - Released on 27 Apr 1815.

Hobday, William - Seaman - Nbr: 3924 - Prize: Rolla - Ship Type: P - How taken: HMS Loire - When taken: 10 Dec 1813 - Where taken: off Bull Island (South Carolina) - Date Received: 5 Oct 1814 - From what ship: HMT President, Halifax - Born: Norfolk - Age: 23 - Released on 9 Jun 1815.

Hobert, James - Seaman - Nbr: 4961 - Prize: Herald - Ship Type: P - How taken: HMS Endymion - When taken: 15 Aug 1814 - Where taken: off Nantucket - Date Received: 28 Oct 1814 - From what ship: HMT Alkbar, Halifax - Born: Maryland - Age: 49 - Released on 21 Jun 1815.

Hobson, Abraham - Seaman - Nbr: 1487 - Prize: Napoleon - Ship Type: LM - How taken: HMS Belle Poule - When taken: 3 Apr 1813 - Where taken: off Cape Ortegal (Spain) - Date Received: 19 Jun 1814 - From what ship: Stapleton - Born: Connecticut - Age: 21 - Released on 10 Apr 1815.

Hock, J. N. - Seaman - Nbr: 279 - Prize: Cannoneer - Ship Type: MV - How taken: HMS Warspite - When taken: 14 Mar 1813 - Where taken: Bay of Biscay - Date Received: 28 Jun 1813 - From what ship: Plymouth - Born: Denmark - Age: 23 - Sent to Plymouth on 3 Dec 1813.

Hockman, William - Seaman - Nbr: 324 - Prize: Pallas - Ship Type: MV - How taken: Rebuff - When taken: 23 Dec 1812 - Where taken: off Cadiz - Date Received: 28 Jun 1813 - From what ship: Plymouth - Born: Prussia - Age: 28 - Sent to Dartmouth on 30 Jul 1813.

Hodge, Edward - Seaman - Nbr: 6378 - Prize: US Brig Syren - Ship Type: MW - How taken: HMS Medway - When taken: 12 Jul 1814 - Where taken: off Cape of Good Hope - Date Received: 24 Feb 1815 - From what ship: HMT Ganges, Plymouth - Born: Boston - Age: 26 - Released on 11 Jul 1815.

Hodge, Rufus - Seaman - Nbr: 16 - Prize: Cashier - Ship Type: LM - How taken: HMS Reindeer - When taken: 3 Feb 1813 - Where taken: Bay of Biscay - Date Received: 2 Apr 1813 - From what ship: Plymouth - Born: Wiscasset - Age: 22 - Sent to Dartmouth on 30 Jul 1813.

Hodges, Hercules - Seaman - Nbr: 4618 - Prize: Argus - Ship Type: MV - How taken: HMS San Domingo - When

American Prisoners of War Held at Dartmoor during the War of 1812

## Alphabetical listing of names

taken: 1 Mar 1814 - Where taken: off Savannah - Date Received: 9 Oct 1814 - From what ship: HMT Leyden, Chatham - Born: Barnstable - Age: 18 - Released on 15 Jun 1815.

Hodgkins, Daniel - Seaman - Nbr: 376 - Prize: Independence - Ship Type: MV - How taken: HMS Superb - When taken: 16 Mar 1813 - Where taken: Bay of Biscay - Date Received: 1 Jul 1813 - From what ship: Plymouth - Born: Ipswich - Age: 24 - Released on 20 Apr 1815.

Hodsdale, Daniel - Seaman - Nbr: 6021 - Prize: McDonough - Ship Type: P - How taken: HMS Bacchante - When taken: 1 Nov 1814 - Where taken: Lat 42 Long 67 - Date Received: 28 Dec 1814 - From what ship: HMT Penelope - Born: New Hampshire - Age: 38 - Released on 3 Jul 1815.

Hoff, Abraham - Seaman - Nbr: 5941 - Prize: Harlequin - Ship Type: P - How taken: HMS Bulwark - When taken: 23 Nov 1814 - Where taken: off Halifax - Date Received: 27 Dec 1814 - From what ship: HMT Penelope - Born: Arundel - Age: 24 - Released on 3 Jul 1815.

Hoff, Charles - Seaman - Nbr: 2420 - Prize: US Sloop Frolic - Ship Type: MW - How taken: HMS Orpheus - When taken: 20 Apr 1814 - Where taken: off Cuba - Date Received: 16 Aug 1814 - From what ship: HMT Queen, Halifax - Born: Homer's Brush - Age: 22 - Released on 11 Jul 1815.

Hoff, Nicholas - Seaman - Nbr: 5993 - Prize: McDonough - Ship Type: P - How taken: HMS Bacchante - When taken: 1 Nov 1814 - Where taken: Lat 42 Long 67 - Date Received: 27 Dec 1814 - From what ship: HMT Penelope - Born: Arundel - Age: 31 - Released on 3 Jul 1815.

Hoffman, James - Seaman - Nbr: 2123 - How taken: Gave himself up from HMS Monmouth - When taken: 18 Sep 1812 - Date Received: 8 Aug 1814 - From what ship: HMT Raven, Chatham - Born: Boston - Age: 28 - Released on 26 Apr 1815.

Hofslider, Jessie - Seaman - Nbr: 2070 - How taken: Impressed at Hull - When taken: 18 May 1813 - Date Received: 3 Aug 1814 - From what ship: HMS Bittern, Chatham Depot - Born: Pennsylvania - Age: 33 - Released on 2 May 1815.

Hogabets, John - Seaman - Nbr: 6550 - Prize: Good Friends - Ship Type: MV - How taken: Rosemache - When taken: 2 Apr 1813 - Where taken: Bay of Biscay - Date Received: 26 Mar 1815 - From what ship: from Exeter jail - Born: Philadelphia - Age: 25 - Released on 9 Jun 1815.

Hogabets, John - 2nd Mate - Nbr: 338 - Prize: Good Friends - Ship Type: MV - How taken: HMS Andromache - When taken: 2 Apr 1813 - Where taken: Bay of Biscay - Date Received: 28 Jun 1813 - From what ship: Plymouth - Born: Philadelphia - Age: 25 - Sent to Exeter Jail on 9 Mar 1815.

Hogan, John - Seaman - Nbr: 5139 - Prize: Calabria - Ship Type: MV - How taken: HMS Castilian - When taken: 29 Sep 1814 - Where taken: off Ireland - Date Received: 31 Oct 1814 - From what ship: HMT Castillian - Born: Danzig - Age: 28 - Released on 29 Jun 1815.

Hogan, William - Seaman - Nbr: 2150 - How taken: Apprehended at London - When taken: 6 Nov 1812 - Date Received: 8 Aug 1814 - From what ship: HMT Raven, Chatham - Born: Portland - Age: 43 - Released on 26 Apr 1815.

Holbery, Emil - Seaman - Nbr: 1246 - How taken: Sent into custody from HMS Edinburgh - When taken: 28 Oct 1812 - Date Received: 14 Jun 1814 - From what ship: Mill Prison (Plymouth, England) - Born: New York - Age: 28 - Released on 26 Apr 1815.

Holbrook, Benjamin - Seaman - Nbr: 3482 - How taken: Gave himself up from HMS Zealous - When taken: 29 Aug 1814 - Date Received: 19 Sep 1814 - From what ship: HMT Salvador del Mundo - Born: Bath - Age: 28 - Released on 4 Jun 1815.

Holbrook, David - Seaman - Nbr: 2882 - How taken: Gave himself up from HMS Bombay - When taken: 27 May 1813 - Date Received: 24 Aug 1814 - From what ship: HMT Alpheus, Chatham - Born: Connecticut - Age: 31 - Released on 19 May 1815.

Holbrook, Ebenezer - Seaman - Nbr: 5793 - Prize: Derby - Ship Type: MV - How taken: HMS Nereus - When taken: 4 Feb 1813 - Where taken: off Cape of Good Hope - Date Received: 26 Dec 1814 - From what ship: HMT Argo - Born: Weymouth - Age: 24 - Died on 9 Mar 1815 from variola.

Holbrook, Elias - Seaman - Nbr: 1261 - Prize: Adeline - Ship Type: MV - How taken: HMS Magicienne - When taken: 16 Mar 1814 - Where taken: off Cape Finisterre - Date Received: 14 Jun 1814 - From what ship: Mill Prison (Plymouth, England) - Born: Freeport - Age: 30 - Released on 28 Apr 1815.

Holbrook, Richard - Seaman - Nbr: 4347 - Prize: Union - Ship Type: LM - How taken: HMS Curlew - When taken: 1 Apr 1814 - Where taken: off Halifax - Date Received: 7 Oct 1814 - From what ship: HMT Salvador del Mundo, Halifax - Born: Wellfleet - Age: 18 - Released on 14 Jun 1815.

Holbrook, Robert - Seaman - Nbr: 3308 - Prize: Thomas - Ship Type: P - How taken: HMS Nymphe - When taken: 28 Jun 1813 - Where taken: off Halifax - Date Received: 13 Sep 1814 - From what ship: HMT Niobe,

American Prisoners of War Held at Dartmoor during the War of 1812

## Alphabetical listing of names

Chatham - Born: Portsmouth - Age: 23 - Released on 28 May 1815.

Holden, Andrew - Seaman - Nbr: 1008 - Prize: Harvest - Ship Type: P - How taken: HMS Orestes - When taken: 21 Jan 1814 - Where taken: Bay of Biscay - Date Received: 31 Jan 1814 - From what ship: Plymouth - Born: New York - Age: 20 - Released on 27 Apr 1815.

Holden, Charles - Boatswain - Nbr: 2985 - Prize: Frolic - Ship Type: P - How taken: HMS Heron - When taken: 25 Jan 1814 - Where taken: off St. Thomas - Date Received: 29 Aug 1814 - From what ship: HMT Bittern - Born: Salem - Age: 33 - Released on 19 May 1815.

Holden, George - Seaman - Nbr: 802 - Prize: Collin - Ship Type: MV - How taken: HMS Helicon and HMS Whiting - When taken: 26 Oct 1813 - Where taken: Channel - Date Received: 3 Nov 1813 - From what ship: Plymouth - Born: Connecticut - Age: 20 - Released on 26 Apr 1815.

Holden, James - Seaman - Nbr: 3920 - Prize: Rolla - Ship Type: P - How taken: HMS Loire - When taken: 10 Dec 1813 - Where taken: off Bull Island (South Carolina) - Date Received: 5 Oct 1814 - From what ship: HMT President, Halifax - Born: Rhode Island - Age: 24 - Released on 9 Jun 1815.

Holden, John - Seaman - Nbr: 5150 - Prize: Thorn - Ship Type: P - How taken: HMS Shannon - When taken: 7 Nov 1813 - Where taken: off Newfoundland - Date Received: 31 Oct 1814 - From what ship: HMT Mermaid, Chatham - Born: Marblehead - Age: 25 - Released on 29 Jun 1815.

Holden, John - Seaman - Nbr: 2406 - Prize: US Sloop Frolic - Ship Type: MW - How taken: HMS Orpheus - When taken: 20 Apr 1814 - Where taken: off Cuba - Date Received: 16 Aug 1814 - From what ship: HMT Queen, Halifax - Born: Salem - Age: 35 - Sent to Dartmouth on 19 Oct 1814.

Holding, Henry - Seaman - Nbr: 3054 - How taken: Gave himself up from HMS Sultan - When taken: 2 Sep 1814 - Date Received: 2 Sep 1814 - From what ship: HMT Sultan - Born: Boston - Age: 27 - Race: Black - Died on 6 Apr 1815 from variola.

Holding, Nathaniel - Seaman - Nbr: 3259 - Prize: Wiley Reynard - Ship Type: P - How taken: HMS Shannon - When taken: 15 Aug 1812 - Where taken: off Halifax - Date Received: 11 Sep 1814 - From what ship: HMT Freya, Chatham - Born: Charleston - Age: 18 - Released on 27 Apr 1815.

Holdridge, Hector - Seaman - Nbr: 3269 - Prize: Eliza (Whaler) - Ship Type: MV - How taken: HMS Charles - When taken: 20 Jul 1813 - Where taken: off Peter Head - Date Received: 11 Sep 1814 - From what ship: HMT Freya, Chatham - Born: Chatham - Age: 21 - Released on 28 May 1815.

Holford, Elisha - Seaman - Nbr: 5853 - How taken: Gave himself up from HMS Barfleur - Date Received: 26 Dec 1814 - From what ship: HMT Argo - Born: New York - Age: 19 - Died on 5 Jan 1815 from febris.

Holiday, Francis - Marine - Nbr: 4041 - Prize: US Brig Rattlesnake - Ship Type: MW - How taken: HMS Leander - When taken: 13 Jul 1814 - Where taken: off Shelburne - Date Received: 6 Oct 1814 - From what ship: HMT Chesapeake, Halifax - Born: Gloucester - Age: 25 - Died on 24 Feb 1815 from variola.

Holiday, John H. - Seaman - Nbr: 5947 - Prize: Harlequin - Ship Type: P - How taken: HMS Bulwark - When taken: 23 Nov 1814 - Where taken: off Halifax - Date Received: 27 Dec 1814 - From what ship: HMT Penelope - Born: Not legible - Released on 3 Jul 1815.

Holland, James - Seaman - Nbr: 1285 - Prize: Indian Lass - Ship Type: Prize - How taken: Not listed - When taken: 29 Apr 1814 - Date Received: 14 Jun 1814 - From what ship: Mill Prison (Plymouth, England) - Born: Cape Ann - Age: 27 - Released on 28 Apr 1815.

Holland, Richard - Seaman - Nbr: 1432 - Prize: Revenge - Ship Type: LM - How taken: HMS Belle Poule - When taken: 10 May 1814 - Where taken: off Cornwall - Date Received: 19 Jun 1814 - From what ship: Stapleton - Born: Maryland - Age: 25 - Released on 28 May 1815.

Hollinger, William - Seaman - Nbr: 1618 - Prize: Tom - Ship Type: LM - How taken: HMS Surveillante - When taken: 24 Apr 1813 - Where taken: Bay of Biscay - Date Received: 23 Jun 1814 - From what ship: Stapleton - Born: Virginia - Age: 30 - Released on 1 May 1815.

Hollis, Nathaniel - Seaman - Nbr: 1699 - Prize: Fanny - Ship Type: Prize - How taken: HMS Sceptre - When taken: 12 May 1814 - Date Received: 2 Jul 1814 - From what ship: Plymouth - Born: Massachusetts - Age: 27 - Released on 1 May 1815.

Holmes, Almoran - Seaman - Nbr: 4453 - Prize: Juliana Smith - Ship Type: P - How taken: HMS Nymphe - When taken: 11 May 1813 - Where taken: off Cape Sable Island (Canada) - Date Received: 8 Oct 1814 - From what ship: HMT Leyden, Chatham - Born: Massachusetts - Age: 19 - Released on 15 Jun 1815.

Holmes, Andrew - Seaman - Nbr: 229 - Prize: Charlotte - Ship Type: MV - How taken: HMS Warspite - When taken: 3 Mar 1813 - Where taken: Bay of Biscay - Date Received: 2 Apr 1813 - From what ship: Plymouth - Born: New York - Age: 21 - Released on 20 Apr 1815.

Holmes, Caleb R. - Seaman - Nbr: 282 - Prize: Cannoneer - Ship Type: MV - How taken: HMS Warspite - When

American Prisoners of War Held at Dartmoor during the War of 1812

## Alphabetical listing of names

taken: 14 Mar 1813 - Where taken: Bay of Biscay - Date Received: 28 Jun 1813 - From what ship: Plymouth - Born: Connecticut - Age: 25 - Escaped on 3 Feb 1815.

Holmes, Charles - Seaman - Nbr: 769 - How taken: Sent into custody from HMS Pallas - Date Received: 3 Nov 1813 - From what ship: Plymouth - Born: Thomastown - Age: 20 - Released on 26 Apr 1815.

Holmes, Christopher - Seaman - Nbr: 2467 - Prize: US Sloop Frolic - Ship Type: MW - How taken: HMS Orpheus - When taken: 20 Apr 1814 - Where taken: off Cuba - Date Received: 16 Aug 1814 - From what ship: HMT Queen, Halifax - Born: Gothenburg - Age: 20 - Released on 3 May 1815.

Holmes, Elisha - Seaman - Nbr: 4665 - Prize: Thomas - Ship Type: P - How taken: HMS Nymphe - When taken: 28 Jun 1813 - Where taken: off Halifax - Date Received: 9 Oct 1814 - From what ship: HMT Leyden, Chatham - Born: Massachusetts - Age: 18 - Released on 15 Jun 1815.

Holmes, Isaac - Seaman - Nbr: 6446 - Prize: Jemmell - Ship Type: MV - How taken: HMS Rhin - When taken: 28 May 1814 - Where taken: off Bermuda - Date Received: 3 Mar 1815 - From what ship: HMT Ganges, Plymouth - Born: Cape Cod - Age: 23 - Race: Black - Released on 11 Jul 1815.

Holmes, James - Seaman - Nbr: 440 - Prize: Magdalen - Ship Type: MV - How taken: HMS Superb - When taken: 15 Apr 1813 - Where taken: off Belle Isle, France - Date Received: 1 Jul 1813 - From what ship: Plymouth - Born: Portsmouth - Age: 23 - Sent to Plymouth on 29 Aug 1813.

Holmes, John - Seaman - Nbr: 6073 - Prize: Daedalus - Ship Type: MV - How taken: HMS Niemen - When taken: 20 Sep 1814 - Where taken: off Delaware - Date Received: 28 Dec 1814 - From what ship: HMT Penelope - Born: New York - Age: 20 - Released on 3 Jul 1815.

Holmes, John - Marine - Nbr: 4038 - Prize: US Brig Rattlesnake - Ship Type: MW - How taken: HMS Leander - When taken: 13 Jul 1814 - Where taken: off Shelburne - Date Received: 6 Oct 1814 - From what ship: HMT Chesapeake, Halifax - Born: Pennsylvania - Age: 28 - Released on 9 Jun 1815.

Holmes, John - Seaman - Nbr: 3636 - Prize: Monarch - Ship Type: MV - How taken: HMS Dotterel - When taken: 14 Dec 1813 - Where taken: off Charleston - Date Received: 30 Sep 1814 - From what ship: HMT Sybella - Born: Exbridge - Age: 24 - Released on 4 Jun 1815.

Holmes, John - Seaman - Nbr: 2803 - How taken: Gave himself up from HMS York - Date Received: 24 Aug 1814 - From what ship: HMT Liverpool, Chatham - Born: New York - Age: 29 - Released on 19 May 1815.

Holmes, Robert - Seaman - Nbr: 2213 - Prize: Hussar - Ship Type: P - How taken: HMS Saturn - When taken: 25 May 1814 - Where taken: off Sandy Hook - Date Received: 16 Aug 1814 - From what ship: HMS Dublin, Halifax - Born: Salem - Age: 40 - Released on 2 May 1815.

Holmes, Samuel - Seaman - Nbr: 6230 - Prize: Prince de Neufchatel - Ship Type: P - How taken: Leander (Newcastle Acasta) - When taken: 20 Dec 1814 - Where taken: Lat 38 Long 56 - Date Received: 30 Jan 1815 - From what ship: HMT Pheasant - Born: Plymouth - Age: 23 - Released on 5 Jul 1815.

Holsein, Peter - Seaman - Nbr: 1222 - How taken: Impressed on the Atlantic - When taken: 28 Aug 1813 - Date Received: 14 Jun 1814 - From what ship: Mill Prison (Plymouth, England) - Born: Norway - Age: 30 - Released on 27 Jun 1814.

Holsley, John - Seaman - Nbr: 1690 - Prize: Circe - Ship Type: MV - How taken: HMS Acteon - When taken: 22 Oct 1813 - Where taken: off Cape Hatteras - Date Received: 2 Jul 1814 - From what ship: Plymouth - Born: Raintree - Age: 29 - Released on 1 May 1815.

Holstade, Joseph - Seaman - Nbr: 1224 - How taken: Sent into custody from MV Havana - Date Received: 14 Jun 1814 - From what ship: Mill Prison (Plymouth, England) - Born: New York - Age: 48 - Released on 27 Apr 1815.

Holstein, Richard - Seaman - Nbr: 5144 - Prize: Baroness Longueville - Ship Type: MV - How taken: HMS Illustrious - When taken: 5 Aug 1813 - Where taken: off St. Helena - Date Received: 31 Oct 1814 - From what ship: HMT Mermaid, Chatham - Born: Virginia - Age: 33 - Died on 25 May 1815 from venereal disease.

Holts, Daniel - Boatswain - Nbr: 178 - Prize: Star - Ship Type: MV - How taken: HMS Superb - When taken: 9 Feb 1813 - Where taken: Bay of Biscay - Date Received: 2 Apr 1813 - From what ship: Plymouth - Born: New London - Age: 26 - Sent to Plymouth on 15 Jun 1813.

Homan, Edward - Seaman - Nbr: 4353 - Prize: Plutus - Ship Type: P - How taken: HMS Curlew - When taken: 1 Apr 1814 - Where taken: off Halifax - Date Received: 7 Oct 1814 - From what ship: HMT Salvador del Mundo, Halifax - Born: Marblehead - Age: 29 - Released on 14 Jun 1815.

Homell, Christopher - Seaman - Nbr: 842 - Prize: Chesapeake - Ship Type: LM - How taken: HMS Hotspur & HMS Pyramus - When taken: 26 Oct 1813 - Where taken: off Nantes - Date Received: 29 Nov 1813 - From what ship: Plymouth - Born: SC - Age: 43 - Released on 26 Apr 1815.

American Prisoners of War Held at Dartmoor during the War of 1812

## Alphabetical listing of names

Homer, Henry - Seaman - Nbr: 1873 - Prize: Elbridge Gerry - Ship Type: P - How taken: HMS Crescent - When taken: 16 Sep 1813 - Where taken: off St. George's - Date Received: 29 Jul 1814 - From what ship: HMS Ville de Paris, Chatham Depot - Born: Amsterdam - Age: 22 - Released on 2 May 1815.

Homer, John - Seaman - Nbr: 3168 - Prize: Falcon - Ship Type: Prize - How taken: Spanish Army - When taken: 2 Jul 1813 - Where taken: Papage Harbor - Date Received: 11 Sep 1814 - From what ship: HMT Freya, Chatham - Born: Boston - Age: 27 - Sent to Dartmouth on 23 Sep 1814.

Homer, Michael - Seaman - Nbr: 1068 - Prize: Fair American - Ship Type: MV - How taken: HMS Andromache - When taken: 19 Jan 1814 - Where taken: Bay of Biscay - Date Received: 10 May 1814 - From what ship: Plymouth - Born: Boston - Age: 19 - Released on 27 Apr 1815.

Homes, Ensign E. - Seaman - Nbr: 1753 - How taken: Gave himself up from HMS Alemene - When taken: 24 Jan 1813 - Date Received: 20 Jul 1814 - From what ship: HMS Milford, Plymouth - Born: Massachusetts - Age: 26 - Released on 1 May 1815.

Honner, John - Seaman - Nbr: 3843 - Prize: Dominique - Ship Type: LM - How taken: HMS Dotterel - When taken: 21 May 1814 - Where taken: off Charleston - Date Received: 5 Oct 1814 - From what ship: HMT Orpheus, Halifax - Born: New Jersey - Age: 22 - Released on 9 Jun 1815.

Hook, John - Seaman - Nbr: 119 - Prize: Governor McKean - Ship Type: LM - How taken: HMS Rover - When taken: 26 Jan 1813 - Where taken: off Bordeaux - Date Received: 2 Apr 1813 - From what ship: Plymouth - Born: Pennsylvania - Age: 26 - Sent to Dartmouth on 30 Jul 1813.

Hooke, George - Boy - Nbr: 2256 - Prize: Otario - Ship Type: Prize - How taken: HMS Curlew - When taken: 2 May 1814 - Where taken: off Halifax - Date Received: 16 Aug 1814 - From what ship: HMS Dublin, Halifax - Born: Pennsylvania - Age: 18 - Released on 3 May 1815.

Hooper, John - Prize Master - Nbr: 3609 - Prize: Martha - Ship Type: Prize - How taken: HMS Belviders - When taken: 6 Feb 1814 - Where taken: off Long Island - Date Received: 30 Sep 1814 - From what ship: HMT Sybella - Born: Marblehead - Age: 22 - Released on 4 Jun 1815.

Hooper, Samuel - Seaman - Nbr: 5328 - Prize: Portsmouth Packet - Ship Type: P - How taken: HMS Maidstone - When taken: 17 Jul 1813 - Where taken: Grand Banks - Date Received: 31 Oct 1814 - From what ship: HMT Leyden, Chatham - Born: Portsmouth - Age: 17 - Released on 29 Jun 1815.

Hooper, William - Seaman - Nbr: 1244 - How taken: Sent into custody from HMS Elizabeth - When taken: 7 Aug 1812 - Date Received: 14 Jun 1814 - From what ship: Mill Prison (Plymouth, England) - Born: Marblehead - Age: 33 - Released on 26 Apr 1815.

Hoper, Joseph - Mate - Nbr: 3433 - Prize: Growler - Ship Type: P - How taken: HMS Electra - When taken: 7 Jul 1813 - Where taken: at sea - Date Received: 13 Sep 1814 - From what ship: HMT Niobe, Chatham - Born: Marblehead - Age: 23 - Released on 28 May 1815.

Hopkins, Daniel - Seaman - Nbr: 1391 - Prize: Courier - Ship Type: P - How taken: HMS Rover - When taken: 14 May 1813 - Where taken: Bay of Biscay - Date Received: 19 Jun 1814 - From what ship: Stapleton - Born: Maryland - Age: 28 - Released on 28 Apr 1815.

Hopkins, John - Prize master - Nbr: 4912 - Prize: William - Ship Type: Prize - How taken: HMS Warspite - When taken: 19 Jun 1814 - Where taken: Georges Bank - Date Received: 28 Oct 1814 - From what ship: HMT Alkbar, Halifax - Born: Cape Cod - Age: 32 - Released on 21 Jun 1815.

Hopkins, Samuel - Seaman - Nbr: 3077 - How taken: Gave himself up from HMS Gloucester - When taken: 26 Dec 1812 - Date Received: 3 Sep 1814 - From what ship: HMT Hydra, Chatham - Born: Rhode Island - Age: 42 - Released on 27 Apr 1815.

Hopkins, William - Seaman - Nbr: 569 - Prize: US Brig Argus - Ship Type: MW - How taken: HMS Pelican - When taken: 14 Apr 1813 - Where taken: Irish Channel - Date Received: 8 Sep 1813 - From what ship: Plymouth - Born: Boston - Age: 49 - Sent to Plymouth on 8 Jul 1814.

Hoppins, William - Seaman - Nbr: 874 - Prize: Agnes - Ship Type: LM - How taken: Jane (Cutter) - When taken: 29 Nov 1813 - Where taken: Bay of Biscay - Date Received: 31 Jan 1814 - From what ship: Plymouth - Born: Newburyport - Age: 23 - Released on 27 Apr 1815.

Horman, Edward - Seaman - Nbr: 5966 - Prize: Amazon - Ship Type: Prize - How taken: HMS Bulwark - When taken: 22 Sep 1814 - Where taken: Georges Bank - Date Received: 27 Dec 1814 - From what ship: HMT Penelope - Born: Marblehead - Age: 51 - Released on 3 Jul 1815.

Horn, Abner - Seaman - Nbr: 813 - How taken: Impressed at Cork - When taken: 25 Oct 1813 - Date Received: 3 Nov 1813 - From what ship: Plymouth - Born: Massachusetts - Age: 33 - Released on 26 Apr 1815.

Horne, E. L. - Seaman - Nbr: 5900 - Prize: Harlequin - Ship Type: P - How taken: HMS Bulwark - When taken: 23 Nov 1814 - Where taken: off Halifax - Date Received: 27 Dec 1814 - From what ship: HMT Penelope -

American Prisoners of War Held at Dartmoor during the War of 1812

## Alphabetical listing of names

Born: Dover - Age: 23 - Released on 3 Jul 1815.
Horne, John - Seaman - Nbr: 4481 - Prize: Enterprize - Ship Type: P - How taken: HMS Tenedos - When taken: 21 May 1813 - Where taken: off Cape Cod - Date Received: 8 Oct 1814 - From what ship: HMT Leyden, Chatham - Born: Massachusetts - Age: 42 - Released on 15 Jun 1815.
Horney, Charles - Boy - Nbr: 5674 - Prize: Harlequin - Ship Type: P - How taken: HMS Bulwark - When taken: 23 Nov 1814 - Where taken: off Halifax - Date Received: 24 Dec 1814 - From what ship: HMT Penelope - Born: Portsmouth - Age: 14 - Released on 3 Jul 1815.
Horney, Gilbert - Sergeant Marines - Nbr: 5672 - Prize: Harlequin - Ship Type: P - How taken: HMS Bulwark - When taken: 23 Nov 1814 - Where taken: off Halifax - Date Received: 24 Dec 1814 - From what ship: HMT Penelope - Born: Portsmouth - Age: 25 - Released on 3 Jul 1815.
Horrow, Ansel - Seaman - Nbr: 6094 - Prize: Black Swan - Ship Type: MV - How taken: HMS Maidstone - When taken: 24 Oct 1814 - Where taken: Georges Bank - Date Received: 28 Dec 1814 - From what ship: HMT Penelope - Born: Plymouth - Age: 32 - Released on 3 Jul 1815.
Horsey, J. H. - Seaman - Nbr: 1999 - How taken: Impressed - When taken: 20 Mar 1813 - Date Received: 3 Aug 1814 - From what ship: HMS Lyffey, Chatham Depot - Born: Boston - Age: 28 - Released on 2 May 1815.
Horsfalt, William - Seaman - Nbr: 3361 - How taken: Impressed at London - When taken: 15 Oct 1813 - Date rec'd: 13 Sep 1814 - What ship: HMT Niobe, Chatham - Born: New York - Age: 26 - Released on 28 May 1815.
Horton, G. A. - Seaman - Nbr: 2402 - Prize: US Sloop Frolic - Ship Type: MW - How taken: HMS Orpheus - When taken: 20 Apr 1814 - Where taken: off Cuba - Date Received: 16 Aug 1814 - From what ship: HMT Queen, Halifax - Born: Marblehead - Age: 38 - Sent to Dartmouth on 19 Oct 1814.
Horwell, John - Seaman - Nbr: 3937 - Prize: Rolla - Ship Type: P - How taken: HMS Loire - When taken: 10 Dec 1813 - Where taken: off Bull Island (South Carolina) - Date Received: 5 Oct 1814 - From what ship: HMT President, Halifax - Born: New York - Age: 22 - Race: Negro - Released on 9 Jun 1815.
Hose, Richard - Carpenter - Nbr: 6350 - Prize: Prince de Neufchatel - Ship Type: P - How taken: Leander (Newcastle Acasta) - When taken: 20 Dec 1814 - Where taken: Lat 38 Long 56 - Date Received: 19 Feb 1815 - From what ship: HMT Ganges, Plymouth - Born: Boston - Age: 23 - Released on 11 Jul 1815.
Hoselquist, John - Seaman - Nbr: 930 - Prize: Fanny - Ship Type: MV - How taken: HMS Eurotas - When taken: 23 Dec 1813 - Where taken: at sea - Date Received: 31 Jan 1814 - From what ship: Plymouth - Born: Wexchy, Sweden - Age: 38 - Released on 6 May 1814.
Hosgood, B. - Seaman - Nbr: 5915 - Prize: Harlequin - Ship Type: P - How taken: HMS Bulwark - When taken: 23 Nov 1814 - Where taken: off Halifax - Date Received: 27 Dec 1814 - From what ship: HMT Penelope - Born: Not legible - Age: 21 - Released on 15 Jul 1815.
Hoskins, James - Seaman - Nbr: 261 - Prize: William Bayard - Ship Type: MV - How taken: HMS Warspite - When taken: 3 Mar 1813 - Where taken: Bay of Biscay - Date Received: 28 Jun 1813 - From what ship: Plymouth - Born: Massachusetts - Age: 20 - Released on 7 Dec 1813.
Hoskins, John - Seaman - Nbr: 3225 - How taken: Gave himself up from HMS Trent - Date Received: 11 Sep 1814 - From what ship: HMT Freya, Chatham - Born: Bedford - Age: 20 - Released on 28 May 1815.
Hosson, John - Seaman - Nbr: 2337 - Prize: Snap Dragon - Ship Type: P - How taken: HMS Martin - When taken: 10 Jun 1814 - Where taken: off Halifax - Date Received: 16 Aug 1814 - From what ship: HMS Dublin, Halifax - Born: North Carolina - Age: 23 - Died on 14 Mar 1815 from variola.
Hoston, Cato - Seaman - Nbr: 3968 - Prize: Rolla - Ship Type: P - How taken: HMS Loire - When taken: 10 Dec 1813 - Where taken: off Bull Island (South Carolina) - Date Received: 5 Oct 1814 - From what ship: HMT President, Halifax - Born: New Haven - Age: 48 - Race: Negro - Released on 9 Jun 1815.
Hough, Ebenezer - Seaman - Nbr: 27 - How taken: Apprehended at Gibraltar - When taken: 8 Aug 1813 - Date rec'd: 2 Apr 1813 - What ship: Plymouth - Born: Avondale - Age: 32 - Sent to Dartmouth on 30 Jul 1813.
Hough, Samuel - Seaman - Nbr: 2576 - Prize: Rattlesnake - Ship Type: P - How taken: HMS Hyperion - When taken: 25 Jun 1814 - Where taken: off Cape Finisterre - Date Received: 21 Aug 1814 - From what ship: HMS Hyperion - Born: Philadelphia - Age: 29 - Released on 3 May 1815.
Houghman, John - Seaman - Nbr: 4891 - How taken: Gave himself up at Plymouth - When taken: Oct 1814 - Date Received: 24 Oct 1814 - From what ship: HMT Salvador del Mundo - Born: New York - Age: 34 - Released on 21 Jun 1815.
Houghton, Timothy - Seaman - Nbr: 3859 - Prize: Thorn - Ship Type: MV - How taken: HMS Bulwark - When taken: 9 Jul 1814 - Where taken: off Nantucket - Date Received: 5 Oct 1814 - From what ship: HMT Orpheus, Halifax - Born: Baltimore - Age: 41 - Released on 9 Jun 1815.
Houseman, John - Seaman - Nbr: 1924 - How taken: Gave himself up from HMS Royal William - When taken: 3

American Prisoners of War Held at Dartmoor during the War of 1812

## Alphabetical listing of names

Feb 1813 - Date Received: 3 Aug 1814 - From what ship: HMS Alceste, Chatham Depot - Born: Maryland - Age: 23 - Released on 2 May 1815.

Housewife, Mathew - Seaman - Nbr: 2969 - How taken: Gave himself up from HMS Teazer - Date Received: 24 Aug 1814 - From what ship: HMT Hannibal - Born: New York - Age: 27 - Released on 19 May 1815.

Hovey, Joseph - Seaman - Nbr: 35 - How taken: Apprehended at Gibraltar - When taken: 8 Aug 1813 - Date Received: 2 Apr 1813 - What ship: Plymouth - Born: Massachusetts - Age: 32 - Sent to Dartmouth on 30 Jul 1813.

How, William - Seaman - Nbr: 1861 - Prize: Vivid - Ship Type: P - How taken: HMS Nymphe - When taken: 20 Apr 1813 - Where taken: off Cape Cod - Date Received: 29 Jul 1814 - From what ship: HMS Ville de Paris, Chatham Depot - Born: Potersham - Age: 27 - Released on 2 May 1815.

Howard, Benjamin - Lieutenant - Nbr: 1695 - Prize: Fiere Facia - Ship Type: P - How taken: HMS Ramillies - When taken: 25 Feb 1814 - Where taken: off NY - Date Received: 2 Jul 1814 - From what ship: Plymouth - Born: West Springfield - Age: 25 - Escaped on 8 Dec 1814.

Howard, Bethnel - Seaman - Nbr: 5538 - Prize: Ann - Ship Type: Prize - How taken: HMS Hamadryad - When taken: 10 Sep 1814 - Where taken: Lat 41 Long 23 - Date Received: 17 Dec 1814 - From what ship: HMT Loire, Halifax - Born: New York - Age: 21 - Released on 1 Jul 1815.

Howard, Frederick - Seaman - Nbr: 4109 - Prize: Flash - Ship Type: MV - How taken: HMS Niemen - When taken: 7 Jun 1814 - Where taken: off Delaware - Date Received: 6 Oct 1814 - From what ship: HMT Chesapeake, Halifax - Born: Bridgewater - Age: 19 - Released on 13 Jun 1815.

Howard, Samuel - Seaman - Nbr: 393 - Prize: John & Frances - Ship Type: MV - How taken: HMS Belle Poule - When taken: 13 Mar 1813 - Where taken: Bay of Biscay - Date Received: 1 Jul 1813 - From what ship: Plymouth - Born: Reed Hark - Age: 31 - Released on 20 Apr 1815.

Howdy, Joseph - Seaman - Nbr: 6526 - How taken: Gave himself up from MV Triton - When taken: Nov 1814 - Date Received: 3 Mar 1815 - From what ship: HMT Ganges, Plymouth - Born: Maryland - Age: 24 - Released on 11 Jul 1815.

Howe, Jacob - Seaman - Nbr: 4612 - Prize: Argus - Ship Type: MV - How taken: HMS San Domingo - When taken: 1 Mar 1814 - Where taken: off Savannah - Date Received: 9 Oct 1814 - From what ship: HMT Leyden, Chatham - Born: Bridgetown - Age: 24 - Released on 15 Jun 1815.

Howe, John - Seaman - Nbr: 438 - Prize: Magdalen - Ship Type: MV - How taken: HMS Superb - When taken: 15 Apr 1813 - Where taken: off Belle Isle, France - Date Received: 1 Jul 1813 - From what ship: Plymouth - Born: Blockfield - Age: 23 - Released on 20 Apr 1815.

Howe, Phineas - Seaman - Nbr: 1502 - How taken: Sent into custody from HMS Dublin - Date Received: 20 Jun 1814 - From what ship: Mill Prison (Plymouth, England) - Born: New Bedford - Age: 26 - Released on 1 Jul 1815.

Howell, John - Seaman - Nbr: 2721 - How taken: Gave himself up from HMS Mulgrave - When taken: 13 Dec 1812 - Date Received: 21 Aug 1814 - From what ship: HMT Freya, Chatham - Born: Philadelphia - Age: 24 - Released on 26 Apr 1815.

Howell, John - Seaman - Nbr: 3135 - How taken: Gave himself up from HMS Paulina - When taken: 29 May 1813 - Date Received: 11 Sep 1814 - From what ship: HMT Freya, Chatham - Born: Baltimore - Age: 46 - Released on 28 May 1815.

Howell, Thomas - Seaman - Nbr: 374 - Prize: Independence - Ship Type: MV - How taken: HMS Superb - When taken: 16 Mar 1813 - Where taken: Bay of Biscay - Date Received: 1 Jul 1813 - From what ship: Plymouth - Born: Beverly - Age: 24 - Sent to Plymouth on 7 Dec 1813.

Howell, Thomas - Seaman - Nbr: 3468 - How taken: Gave himself up from HMS Norge - When taken: 16 Aug 1814 - Date Received: 19 Sep 1814 - From what ship: HMT Salvador del Mundo - Born: Beverly - Age: 27 - Released on 4 Jun 1815.

Howitt, William - Pilot - Nbr: 4279 - Prize: Pilot - Ship Type: LM - How taken: Victoria (Privateer) - When taken: 28 Jan 1814 - Where taken: off Bordeaux - Date Received: 7 Oct 1814 - From what ship: HMT Niobe, Chatham - Born: New York - Age: 23 - Released on 14 Jun 1815.

Howland, J. W. - Prize master - Nbr: 4908 - Prize: Rolla - Ship Type: P - How taken: HMS Loire - When taken: 10 Dec 1813 - Where taken: off Bull Island (South Carolina) - Date Received: 28 Oct 1814 - From what ship: HMT Alkbar, Halifax - Born: Newport - Age: 35 - Released on 21 Jun 1815.

Howland, Solomon - Seaman - Nbr: 2348 - Prize: Snap Dragon - Ship Type: P - How taken: HMS Martin - When taken: 10 Jun 1814 - Where taken: off Halifax - Date Received: 16 Aug 1814 - From what ship: HMS Dublin, Halifax - Born: North Carolina - Age: 21 - Released on 3 May 1815.

American Prisoners of War Held at Dartmoor during the War of 1812

## Alphabetical listing of names

Howland, William - Carpenter - Nbr: 2843 - How taken: Gave himself up from HMS Scorpion - When taken: 27 May 1813 - Date Received: 24 Aug 1814 - From what ship: HMT Liverpool, Chatham - Born: Massachusetts - Age: 40 - Released on 28 Apr 1815.

Howman, John - Seaman - Nbr: 2596 - Prize: Rachael - Ship Type: MV - How taken: HMS Bramble - When taken: 2 Dec 1812 - Where taken: at sea - Date Received: 21 Aug 1814 - From what ship: HMT Freya, Chatham - Born: Marblehead - Age: 27 - Released on 26 Apr 1815.

Howman, Jonas - Seaman - Nbr: 2618 - Prize: Rachael - Ship Type: MV - How taken: Hening - When taken: 9 Feb 1813 - Where taken: at sea - Date Received: 21 Aug 1814 - From what ship: HMT Freya, Chatham - Born: Marblehead - Age: 27 - Released on 3 May 1815.

Hoye, Cornelius - Boatswain - Nbr: 429 - Prize: Viper - Ship Type: MV - How taken: HMS Superb - When taken: 15 Apr 1813 - Where taken: Bay of Biscay - Date Received: 1 Jul 1813 - From what ship: Plymouth - Born: Chester County - Age: 33 - Released on 20 Apr 1815.

Hoyt, David - Seaman - Nbr: 4112 - Prize: Plutus - Ship Type: P - How taken: HMS Curlew - When taken: 1 Apr 1814 - Where taken: off Halifax - Date Received: 6 Oct 1814 - From what ship: HMT Chesapeake, Halifax - Born: Salisbury - Age: 23 - Released on 13 Jun 1815.

Hoyt, David - Seaman - Nbr: 3796 - Prize: Hannah - Ship Type: Prize - How taken: HMS Martin - When taken: 29 Apr 1814 - Where taken: off Cape Lopez - Date Received: 5 Oct 1814 - From what ship: HMT President, Halifax - Born: Virginia - Age: 24 - Released on 9 Jun 1815.

Hoyt, James - Seaman - Nbr: 1280 - How taken: Sent into custody from HMS Nerus - Date Received: 14 Jun 1814 - From what ship: Mill Prison (Plymouth, England) - Born: Norfolk - Age: 46 - Race: Black - Released on 28 Apr 1815.

Huane, Benjamin - Seaman - Nbr: 6100 - Prize: Harlequin - Ship Type: P - How taken: HMS Bulwark - When taken: 23 Nov 1814 - Where taken: off Halifax - Date Received: 28 Dec 1814 - From what ship: HMT Penelope - Born: Salem - Age: 31 - Released on 20 Mar 1815.

Hubbard, Alfred - Prize Master - Nbr: 2786 - Prize: Prompt - Ship Type: MV - How taken: Chance (Privateer) - When taken: 28 Mar 1813 - Where taken: Bay of Biscay - Date Received: 24 Aug 1814 - From what ship: HMT Liverpool, Chatham - Born: New Haven - Age: 28 - Released on 19 May 1815.

Hubbard, Daniel - Seaman - Nbr: 6096 - Prize: Black Swan - Ship Type: MV - How taken: HMS Maidstone - When taken: 24 Oct 1814 - Where taken: Georges Bank - Date Received: 28 Dec 1814 - From what ship: HMT Penelope - Born: Plymouth - Age: 26 - Race: Black - Released on 11 Jul 1815.

Hubbard, John - Seaman - Nbr: 24 - How taken: Apprehended at Gibraltar - When taken: 8 Aug 1813 - Date Received: 2 Apr 1813 - From what ship: Plymouth - Born: Philadelphia - Age: 29 - Sent to Plymouth on 15 Jun 1813.

Hubbard, John - Seaman - Nbr: 551 - How taken: Gave himself up from HMS Bombay - When taken: 30 Nov 1812 - Date Received: 8 Sep 1813 - From what ship: Plymouth - Born: Norfolk - Age: 27 - Released on 26 Apr 1815.

Hubbard, John G. - Seaman - Nbr: 4786 - How taken: Gave himself up from HMS Fame - When taken: 28 Sep 1814 - Date Received: 9 Oct 1814 - From what ship: HMT Freya, Chatham - Born: Marblehead - Age: 25 - Released on 11 Jul 1815.

Hubbard, Joseph - Seaman - Nbr: 5739 - Prize: US Schooner Somers - Ship Type: MW - How taken: British gunboats - When taken: 12 Aug 1814 - Where taken: Fort Erie - Date Received: 26 Dec 1814 - From what ship: HMT Argo - Born: Philadelphia - Age: 24 - Released on 3 Jul 1815.

Hubbard, William - Seaman - Nbr: 4511 - Prize: America - Ship Type: P - How taken: HMS Shannon - When taken: 23 May 1813 - Where taken: off Cape Cod - Date Received: 8 Oct 1814 - From what ship: HMT Leyden, Chatham - Born: Rhode Island - Age: 30 - Released on 15 Jun 1815.

Hubble, James - Seaman - Nbr: 5231 - Prize: Devon - Ship Type: Prize - How taken: HMS Fly - When taken: 2 Jan 1814 - Where taken: at sea - Date Received: 31 Oct 1814 - From what ship: HMT Mermaid, Chatham - Born: Fairfield - Age: 29 - Released on 29 Jun 1815.

Huckber, Venus - Boy - Nbr: 4006 - Prize: US Brig Rattlesnake - Ship Type: MW - How taken: HMS Leander - When taken: 13 Jul 1814 - Where taken: off Shelburne - Date Received: 6 Oct 1814 - From what ship: HMT Chesapeake, Halifax - Born: Marseille - Age: 13 - Released on 9 Jun 1815.

Hudson, Daniel - Seaman - Nbr: 1404 - Prize: Eliza - Ship Type: MV - How taken: HMS Surveillante - When taken: 27 Mar 1813 - Where taken: Bay of Biscay - Date Received: 19 Jun 1814 - From what ship: Stapleton - Born: Rhode Island - Age: 28 - Released on 28 Apr 1815.

Hudson, James - Seaman - Nbr: 1680 - How taken: Gave himself up from HMS Eclipse - Date Received: 2 Jul 1814

American Prisoners of War Held at Dartmoor during the War of 1812

## Alphabetical listing of names

- From what ship: Plymouth - Born: Providence - Age: 57 - Released on 1 May 1815.
Hudson, James - Seaman - Nbr: 1049 - How taken: Sent into custody from Mary (Transport) - Date Received: 10 May 1814 - From what ship: Plymouth - Born: Baltimore - Age: 24 - Released on 27 Apr 1815.
Hudson, Thomas - Boy - Nbr: 1424 - Prize: Revenge - Ship Type: LM - How taken: HMS Belle Poule - When taken: 10 May 1814 - Where taken: off Cornwall - Date Received: 19 Jun 1814 - From what ship: Stapleton - Born: Richmond - Age: 13 - Released on 28 Apr 1815.
Hudson, William - Seaman - Nbr: 4139 - Prize: Bordeaux Packet - Ship Type: LM - How taken: HMS Niemen - When taken: 28 Jun 1814 - Where taken: off Delaware - Date Received: 6 Oct 1814 - From what ship: HMT Chesapeake, Halifax - Born: Tiverton - Age: 39 - Race: Black - Released on 13 Jun 1815.
Hugans, Daniel - Carpenter - Nbr: 481 - Prize: Shadow - Ship Type: LM - How taken: HMS Reindeer - When taken: 6 Apr 1813 - Where taken: Bay of Biscay - Date Received: 8 Sep 1813 - From what ship: Plymouth - Born: Hamburg - Age: 39 - Sent to Mill Prison (Plymouth, England) on 17 Jun 1814.
Hugg, Jacob - Seaman - Nbr: 4292 - Prize: Commodore Perry - Ship Type: Schooner - How taken: Sent into custody from a cutter - When taken: 25 Feb 1814 - Where taken: off Bordeaux - Date Received: 7 Oct 1814 - From what ship: HMT Niobe, Chatham - Born: New Jersey - Age: 40 - Released on 14 Jun 1815.
Hughes, Abijah - Seaman - Nbr: 3839 - Prize: Dominique - Ship Type: LM - How taken: HMS Dotterel - When taken: 21 May 1814 - Where taken: off Charleston - Date Received: 5 Oct 1814 - From what ship: HMT Orpheus, Halifax - Born: New Jersey - Age: 36 - Released on 9 Jun 1815.
Hughes, Ebenezer - Seaman - Nbr: 5975 - Prize: Halifax Packet - Ship Type: Prize - How taken: HMS Bulwark - When taken: 22 Sep 1814 - Where taken: Georges Bank - Date Received: 27 Dec 1814 - From what ship: HMT Penelope - Born: Newbury - Age: 40 - Released on 3 Jul 1815.
Hughes, John - Seaman - Nbr: 5362 - Prize: Adeline - Ship Type: LM - How taken: HMS Magicienne - When taken: 14 Mar 1814 - Where taken: off Cape Ortegal (Spain) - Date Received: 31 Oct 1814 - From what ship: HMT Leyden, Chatham - Born: Philadelphia - Age: 20 - Released on 1 Jul 1815.
Hughes, John - Seaman - Nbr: 2051 - How taken: Impressed at London - When taken: May 1812 - Date Received: 3 Aug 1814 - From what ship: HMS Lyffey, Chatham Depot - Born: Baltimore - Age: 24 - Released on 26 Apr 1815.
Hughes, John - Seaman - Nbr: 180 - Prize: Star - Ship Type: MV - How taken: HMS Superb - When taken: 9 Feb 1813 - Where taken: Bay of Biscay - Date Received: 2 Apr 1813 - From what ship: Plymouth - Born: Philadelphia - Age: 26 - Sent to Mill Prison (Plymouth, England) on 10 Jul 1813.
Hughes, John - Seaman - Nbr: 897 - Prize: Squirrel - Ship Type: MV - How taken: HMS Belle Poule - When taken: 14 Dec 1813 - Where taken: at sea - Date Received: 31 Jan 1814 - From what ship: Plymouth - Born: New Jersey - Age: 29 - Released on 27 Apr 1815.
Hughes, Peter - Seaman - Nbr: 3999 - Prize: US Brig Rattlesnake - Ship Type: MW - How taken: HMS Leander - When taken: 13 Jul 1814 - Where taken: off Shelburne - Date Received: 6 Oct 1814 - From what ship: HMT Chesapeake, Halifax - Born: New Jersey - Age: 20 - Released on 9 Jun 1815.
Hughes, Thomas - Seaman - Nbr: 2956 - How taken: Apprehended at Curacao - When taken: 12 Dec 1813 - Date Received: 24 Aug 1814 - From what ship: HMT Hannibal - Born: Queens County, Ireland - Age: 40 - Released on 19 May 1815.
Hughes, William - Seaman - Nbr: 3633 - Prize: Monarch - Ship Type: MV - How taken: HMS Dotterel - When taken: 14 Dec 1813 - Where taken: off Charleston - Date Received: 30 Sep 1814 - From what ship: HMT Sybella - Born: Baltimore - Age: 34 - Released on 4 Jun 1815.
Huish, John - Seaman - Nbr: 6099 - Prize: Harlequin - Ship Type: P - How taken: HMS Bulwark - When taken: 23 Nov 1814 - Where taken: off Halifax - Date Received: 28 Dec 1814 - From what ship: HMT Penelope - Born: Not legible - Released on 11 Jul 1815.
Hulen, Edward - Seaman - Nbr: 6174 - Prize: Lion - Ship Type: P - How taken: HMS Granicus - When taken: 2 Dec 1814 - Where taken: off Lisbon - Date Received: 18 Jan 1815 - From what ship: HMT Impregnable - Born: Salem - Age: 37 - Released on 5 Jul 1815.
Hulet, Michael - Landsman - Nbr: 101 - Prize: St. Martin's Planter - Ship Type: P - How taken: HMS Dublin - When taken: 9 Feb 1813 - Where taken: Lat 43 N, Long 33 50 W - Date Received: 2 Apr 1813 - From what ship: Plymouth - Born: Shrewsbury - Age: 21 - Released on 20 Apr 1815.
Hulin, Abraham - Prize master - Nbr: 3732 - Prize: Alfred - Ship Type: P - How taken: HMS Epervier - When taken: 23 Feb 1814 - Where taken: off Newfoundland - Date Received: 30 Sep 1814 - From what ship: HMT President, Halifax - Born: Salem - Age: 30 - Released on 4 Jun 1815.
Hull, Edward - Seaman - Nbr: 2791 - Prize: Weazel - Ship Type: MV - How taken: HMS Foxhound - When taken:

American Prisoners of War Held at Dartmoor during the War of 1812

## Alphabetical listing of names

25 Mar 1813 - Where taken: Bay of Biscay - Date Received: 24 Aug 1814 - From what ship: HMT Liverpool, Chatham - Born: New York - Age: 19 - Released on 19 May 1815.

Hull, George - Lieutenant Marines - Nbr: 5563 - Prize: Harlequin - Ship Type: P - How taken: HMS Bulwark - When taken: 23 Oct 1814 - Where taken: off Halifax - Date Received: 17 Dec 1814 - From what ship: HMT Loire, Halifax - Born: Durham - Age: 35 - Released on 1 Jul 1815.

Hull, Homer - Mate - Nbr: 3112 - Prize: Harmony - Ship Type: Prize - How taken: HMS Brisk - When taken: 24 Aug 1814 - Where taken: Bristol Channel - Date Received: 11 Sep 1814 - From what ship: HMT Salvador del Mundo - Born: Connecticut - Age: 24 - Released on 26 Apr 1815.

Hull, William - Seaman - Nbr: 225 - Prize: Pert - Ship Type: MV - How taken: HMS Warspite - When taken: 1 Mar 1813 - Where taken: off Basque Roads (France) - Date Received: 2 Apr 1813 - From what ship: Plymouth - Born: New York - Age: 31 - Released on 20 Apr 1815.

Hult, Thomas - Seaman - Nbr: 4650 - Prize: Thomas - Ship Type: P - How taken: HMS Nymphe - When taken: 28 Jun 1813 - Where taken: off Halifax - Date Received: 9 Oct 1814 - From what ship: HMT Leyden, Chatham - Born: Portland - Age: 26 - Race: Mulatto - Released on 15 Jun 1815.

Hume, George - Seaman - Nbr: 505 - Prize: Leo - Ship Type: LM - How taken: HMS Magicienne - When taken: 4 Jun 1813 - Where taken: off France - Date Received: 8 Sep 1813 - From what ship: Plymouth - Born: Winslow - Age: 35 - Sent to Plymouth on 7 Dec 1813.

Humphrey, Asa - Seaman - Nbr: 5339 - Prize: Elbridge Gerry - Ship Type: P - How taken: HMS Crescent - When taken: 16 Sep 1813 - Where taken: at sea - Date Received: 31 Oct 1814 - From what ship: HMT Leyden, Chatham - Born: Grey - Age: 21 - Released on 1 Jul 1815.

Humphries, Jacob - Seaman - Nbr: 124 - Prize: Governor McKean - Ship Type: LM - How taken: HMS Rover - When taken: 26 Jan 1813 - Where taken: off Bordeaux - Date Received: 2 Apr 1813 - From what ship: Plymouth - Born: Philadelphia - Age: 28 - Sent to Dartmouth on 30 Jul 1813.

Humphries, Warren - Seaman - Nbr: 801 - Prize: Collin - Ship Type: MV - How taken: HMS Helicon and HMS Whiting - When taken: 26 Oct 1813 - Where taken: Channel - Date Received: 3 Nov 1813 - From what ship: Plymouth - Born: Connecticut - Age: 25 - Released on 26 Apr 1815.

Hunberly, Elisha - Seaman - Nbr: 1897 - Prize: Quebec - Ship Type: Prize - How taken: HMS Derwent - When taken: 29 Jan 1813 - Where taken: off Lisbon - Date Received: 3 Aug 1814 - From what ship: HMS Alceste, Chatham Depot - Born: New Haven - Age: 28 - Released on 2 May 1815.

Hunn, John - Seaman - Nbr: 3438 - How taken: Gave himself up from HMS Clorinda - When taken: 18 Dec 1813 - Date Received: 13 Sep 1814 - From what ship: HMT Niobe, Chatham - Born: Delaware - Age: 36 - Race: Black - Released on 28 May 1815.

Hunt, Charles - Marine - Nbr: 591 - Prize: US Brig Argus - Ship Type: MW - How taken: HMS Pelican - When taken: 14 Aug 1813 - Where taken: Irish Channel - Date Received: 8 Sep 1813 - From what ship: Plymouth - Born: New York - Age: 22 - Sent to Dartmouth on 2 Nov 1814.

Hunt, James - Seaman - Nbr: 6419 - Prize: Farmer's Daughter - Ship Type: MV - How taken: HMS Leviathan - When taken: 29 May 1814 - Date Received: 3 Mar 1815 - From what ship: HMT Ganges, Plymouth - Born: Baltimore - Age: 24 - Released on 3 Jul 1815.

Hunt, Job - Seaman - Nbr: 3678 - Prize: Alfred - Ship Type: P - How taken: HMS Epervier - When taken: 23 Feb 1812 - Where taken: off Newfoundland - Date Received: 30 Sep 1814 - From what ship: HMT President - Born: Marblehead - Age: 25 - Released on 4 Jun 1815.

Hunt, Samuel - Seaman - Nbr: 5351 - Prize: Volunteer - Ship Type: MV - How taken: Victoria (Privateer) - When taken: 26 Dec 1813 - Where taken: at sea - Date Received: 31 Oct 1814 - From what ship: HMT Leyden, Chatham - Born: New York - Age: 29 - Released on 1 Jul 1815.

Hunter, Alexander - Seaman - Nbr: 2953 - Prize: Ellen and Elizabeth - Ship Type: MV - How taken: Rose & Thistle - When taken: 24 Feb 1814 - Where taken: off St. Bartholomew - Date Received: 24 Aug 1814 - From what ship: HMT Hannibal - Born: Baltimore - Age: 25 - Released on 19 May 1815.

Hunter, G. S. - Prize master - Nbr: 6482 - Prize: Adamant - Ship Type: Prize - How taken: HMS Leviathan - When taken: 29 May 1813 - Where taken: off Bermuda - Date Received: 3 Mar 1815 - From what ship: HMT Ganges, Plymouth - Born: New York - Age: 27 - Released on 11 Jul 1815.

Hunter, George - Seaman - Nbr: 424 - Prize: Viper - Ship Type: MV - How taken: HMS Superb - When taken: 15 Apr 1813 - Where taken: Bay of Biscay - Date Received: 1 Jul 1813 - From what ship: Plymouth - Born: Fredericktown - Age: 27 - Released on 20 Apr 1815.

Hunter, James - Seaman - Nbr: 3315 - Prize: Globe - Ship Type: P - How taken: HMS Victorious - When taken: 8 Jun 1813 - Where taken: Chesapeake Bay - Date Received: 13 Sep 1814 - From what ship: HMT Niobe,

American Prisoners of War Held at Dartmoor during the War of 1812

## Alphabetical listing of names

Chatham - Born: Maryland - Age: 28 - Released on 28 May 1815.

Hunter, William - Seaman - Nbr: 1583 - Prize: Price - Ship Type: MV - How taken: HMS Pyramus - When taken: 6 Apr 1813 - Where taken: Bay of Biscay - Date Received: 23 Jun 1814 - From what ship: Stapleton - Born: New York - Age: 19 - Released on 1 May 1815.

Hunter, William - Seaman - Nbr: 4932 - Prize: Herald - Ship Type: P - How taken: HMS Endymion - When taken: 15 Aug 1814 - Where taken: off Nantucket - Date Received: 28 Oct 1814 - From what ship: HMT Alkbar, Halifax - Born: Philadelphia - Age: 43 - Released on 21 Jun 1815.

Huntingdon, Edward - Seaman - Nbr: 3925 - Prize: Rolla - Ship Type: P - How taken: HMS Loire - When taken: 10 Dec 1813 - Where taken: off Bull Island (South Carolina) - Date Received: 5 Oct 1814 - From what ship: HMT President, Halifax - Born: Newport - Age: 17 - Released on 9 Jun 1815.

Huntley, Charles - Seaman - Nbr: 2087 - Prize: True Blooded Yankee - Ship Type: P - How taken: HMS Hope - When taken: 24 Jun 1813 - Where taken: off Brest - Date Received: 3 Aug 1814 - From what ship: HMS Bittern, Chatham Depot - Born: Hartford - Age: 26 - Released on 2 May 1815.

Huntress, Robert - Boatswain - Nbr: 1966 - Prize: Orbit - Ship Type: MV - How taken: HMS Achates - When taken: 29 Jan 1813 - Where taken: Lat 44N Long 13W - Date Received: 3 Aug 1814 - From what ship: HMS Lyffey, Chatham Depot - Born: Portsmouth - Age: 37 - Released on 2 May 1815.

Hupton, Henry - Lieutenant - Nbr: 2238 - Prize: General Hart - Ship Type: P - How taken: HMS Sophie - When taken: 24 Apr 1814 - Where taken: off Bermuda - Date Received: 16 Aug 1814 - From what ship: HMS Dublin, Halifax - Born: Salem - Age: 21 - Released on 2 May 1815.

Hurd, Abel - Seaman - Nbr: 3224 - How taken: Gave himself up from HMS Trent - Date Received: 11 Sep 1814 - From what ship: HMT Freya, Chatham - Born: Boston - Age: 35 - Released on 28 May 1815.

Hurd, William - Seaman - Nbr: 300 - Prize: Ducornau - Ship Type: MV - How taken: HMS Pheasant - When taken: 15 Mar 1813 - Where taken: Bay of Biscay - Date Received: 28 Jun 1813 - From what ship: Plymouth - Born: New York - Age: 24 - Released on 20 Apr 1815.

Hurt, John A. - Seaman - Nbr: 3885 - Prize: US Brig Rattlesnake - Ship Type: MW - How taken: HMS Leander - When taken: 11 Jul 1814 - Where taken: off Shelburne - Date Received: 5 Oct 1814 - From what ship: HMT Orpheus, Halifax - Born: Brandenburg - Age: 23 - Released on 9 Jun 1815.

Huse, Ebenezer - Seaman - Nbr: 3149 - Prize: Kitty - Ship Type: Prize - How taken: Dart of Guernsey (Privateer) - When taken: 20 Jun 1813 - Where taken: off the Western Isles (England) - Date Received: 11 Sep 1814 - From what ship: HMT Freya, Chatham - Born: Massachusetts - Age: 24 - Sent to Dartmouth on 23 Sep 1814.

Hussey, Edward - Seaman - Nbr: 3898 - Prize: Recovery - Ship Type: MV - How taken: HMS Thistle - When taken: 12 Jul 1814 - Where taken: off VA - Date Received: 5 Oct 1814 - From what ship: HMT Orpheus, Halifax - Born: Nantucket - Age: 19 - Released on 9 Jun 1815.

Hussey, Joseph - Seaman - Nbr: 5779 - Prize: William Penn - Ship Type: MV - How taken: HMS Acorn - When taken: 27 Oct 1812 - Where taken: Lat 14 - Date Received: 26 Dec 1814 - From what ship: HMT Argo - Born: Westminster - Age: 30 - Race: Black - Released on 27 Apr 1815.

Hussey, Thomas - Seaman - Nbr: 2801 - How taken: Impressed at London - When taken: 5 Jul 1813 - Date Received: 24 Aug 1814 - From what ship: HMT Liverpool, Chatham - Born: New York - Age: 34 - Released on 19 May 1815.

Huston, James - Seaman - Nbr: 3084 - Prize: Eliza (Whaler) - Ship Type: MV - How taken: HMS Charles - When taken: 20 Jul 1813 - Where taken: off Peter Head - Date Received: 3 Sep 1814 - From what ship: HMT Hydra, Chatham - Born: New York - Age: 29 - Released on 28 May 1815.

Hutchins, Edward - Seaman - Nbr: 1380 - Prize: Courier - Ship Type: P - How taken: HMS Rover - When taken: 14 May 1813 - Where taken: Bay of Biscay - Date Received: 19 Jun 1814 - From what ship: Stapleton - Born: Maryland - Age: 22 - Race: Negro - Released on 28 Apr 1815.

Hutchins, Henry - Seaman - Nbr: 1664 - Prize: Grand Napoleon - Ship Type: P - How taken: HMS Goldfinch - When taken: 17 Apr 1813 - Where taken: Bay of Biscay - Date Received: 23 Jun 1814 - From what ship: Stapleton - Born: New York - Age: 21 - Released on 1 May 1815.

Hutchins, John - Seaman - Nbr: 5882 - Prize: Harlequin - Ship Type: P - How taken: HMS Bulwark - When taken: 23 Nov 1814 - Where taken: off Halifax - Date Received: 27 Dec 1814 - From what ship: HMT Penelope - Born: Arundel - Age: 19 - Released on 3 Jul 1815.

Hutchins, John - Seaman - Nbr: 3581 - Prize: Greyhound - Ship Type: MV - How taken: Amith (Privateer) - When taken: 5 Jun 1814 - Where taken: off Bermuda - Date Received: 30 Sep 1814 - From what ship: HMT Sybella - Born: South Kingston - Age: 52 - Released on 4 Jun 1815.

Hutchins, Joseph - Seaman - Nbr: 5906 - Prize: Harlequin - Ship Type: P - How taken: HMS Bulwark - When taken:

American Prisoners of War Held at Dartmoor during the War of 1812

## Alphabetical listing of names

23 Nov 1814 - Where taken: off Halifax - Date Received: 27 Dec 1814 - From what ship: HMT Penelope - Born: Alfred - Age: 39 - Released on 3 Jul 1815.

Hutchins, Nathaniel - Seaman - Nbr: 5998 - Prize: McDonough - Ship Type: P - How taken: HMS Bacchante - When taken: 1 Nov 1814 - Where taken: Lat 42 Long 67 - Date Received: 28 Dec 1814 - From what ship: HMT Penelope - Born: Massachusetts - Age: 20 - Released on 3 Jul 1815.

Hutchins, Samuel - Seaman - Nbr: 5651 - How taken: Sent into custody at Cork - When taken: 28 Nov 1814 - Date Received: 24 Dec 1814 - From what ship: HMT Impregnable - Born: Alford - Age: 22 - Released on 3 Jul 1815.

Hutchins, Theodore - Seaman - Nbr: 5930 - Prize: Harlequin - Ship Type: P - How taken: HMS Bulwark - When taken: 23 Nov 1814 - Where taken: off Halifax - Date Received: 27 Dec 1814 - From what ship: HMT Penelope - Born: Alfred - Age: 33 - Released on 3 Jul 1815.

Hutchins, William - Armourer - Nbr: 942 - Prize: Siro - Ship Type: LM - How taken: HMS Pelican - When taken: 13 Jan 1814 - Where taken: at sea - Date Received: 31 Jan 1814 - From what ship: Plymouth - Born: Philadelphia - Age: 32 - Released on 27 Apr 1815.

Hutchinson, Lamore - Seaman - Nbr: 3881 - Prize: Patriot - Ship Type: MV - How taken: HMS Niemen - When taken: 21 May 1814 - Where taken: off Cape Hatteras - Date Received: 5 Oct 1814 - From what ship: HMT Orpheus, Halifax - Born: Maryland - Age: 19 - Race: Negro - Released on 9 Jun 1815.

Hutchinson, Townsend - Seaman - Nbr: 1956 - How taken: Gave himself up from HMS Bellerophon - When taken: 17 Oct 1812 - Date Received: 3 Aug 1814 - From what ship: HMS Lyffey, Chatham Depot - Born: Long Island - Age: 52 - Released on 26 Apr 1815.

Hutes, Cyrus - Seaman - Nbr: 3435 - How taken: Impressed at London - When taken: 8 Nov 1813 - Date Received: 13 Sep 1814 - From what ship: HMT Niobe, Chatham - Born: Boston - Age: 22 - Released on 28 May 1815.

Hutson, John - Seaman - Nbr: 2894 - How taken: Gave himself up from HMS Union - When taken: 27 May 1813 - Date Received: 24 Aug 1814 - From what ship: HMT Alpheus, Chatham - Born: Copenhagen - Age: 28 - Released on 19 May 1815.

Hutson, Peter - Cook - Nbr: 2866 - Prize: Tickler - Ship Type: MVLM - How taken: HMS Magicienne - When taken: 5 Jun 1813 - Where taken: Bay of Biscay - Date Received: 24 Aug 1814 - From what ship: HMT Alpheus, Chatham - Born: New York - Age: 23 - Race: Black - Released on 19 May 1815.

Huxtable, William - Seaman - Nbr: 6501 - How taken: Gave himself up - Date Received: 3 Mar 1815 - From what ship: HMT Ganges, Plymouth - Born: Concord - Age: 21 - Released on 11 Jul 1815.

Huzzey, John - Seaman - Nbr: 5002 - Prize: Invincible - Ship Type: LM - How taken: HMS Armide - When taken: 15 Aug 1814 - Where taken: off Nantucket - Date Received: 28 Oct 1814 - From what ship: HMT Alkbar, Halifax - Born: Boston - Age: 26 - Released on 21 Jun 1815.

Hydalgo, Vincent - Seaman - Nbr: 4834 - Prize: President - Ship Type: P - How taken: HMS Pique - When taken: 7 May 1814 - Where taken: off Porto Rico - Date Received: 9 Oct 1814 - From what ship: HMT Freya, Halifax - Born: Cartagena - Age: 30 - Race: Black - Released on 21 Jun 1815.

Hydra, Dempsey - Seaman - Nbr: 491 - Prize: Paul Jones - Ship Type: P - How taken: HMS Leonidas - When taken: 23 May 1813 - Where taken: Channel - Date Received: 8 Sep 1813 - From what ship: Plymouth - Born: North Carolina - Age: 25 - Died on 23 Dec 1814.

Inbritson, Nicholas - Seaman - Nbr: 88 - Prize: Cashier - Ship Type: MV - How taken: HMS Reindeer - When taken: 3 Feb 1813 - Where taken: Bay of Biscay - Date Received: 2 Apr 1813 - From what ship: Plymouth - Born: Sweden - Age: 22 - Sent to Dartmouth on 30 Jul 1813.

Ingersole, John - Boy - Nbr: 4766 - Prize: Alligator - Ship Type: MV - How taken: HMS San Domingo - When taken: 16 May 1813 - Where taken: off Bengal - Date Received: 9 Oct 1814 - From what ship: HMT Freya, Chatham - Born: Richmond - Age: 18 - Released on 10 Apr 1815.

Ingersoll, Abraham - Seaman - Nbr: 2076 - Prize: True Blooded Yankee - Ship Type: P - How taken: HMS Hope - When taken: 24 Jun 1813 - Where taken: off Brest - Date Received: 3 Aug 1814 - From what ship: HMS Bittern, Chatham Depot - Born: Boston - Age: 33 - Released on 2 May 1815.

Ingerson, James B. - Seaman - Nbr: 2809 - How taken: Impressed at London - When taken: 9 Jul 1813 - Date Received: 24 Aug 1814 - From what ship: HMT Liverpool, Chatham - Born: Portland - Age: 19 - Released on 19 May 1815.

Ingerson, Michael - Seaman - Nbr: 1581 - Prize: Price - Ship Type: MV - How taken: HMS Pyramus - When taken: 6 Apr 1813 - Where taken: Bay of Biscay - Date Received: 23 Jun 1814 - From what ship: Stapleton - Born: New Jersey - Age: 22 - Released on 1 May 1815.

Ingerson, Nathaniel - Boy - Nbr: 5484 - Prize: General Putnam - Ship Type: P - How taken: HMS Leander - When

American Prisoners of War Held at Dartmoor during the War of 1812

## Alphabetical listing of names

taken: 8 Nov 1814 - Where taken: Long 65 Lat 42 - Date Received: 17 Dec 1814 - From what ship: HMT Loire, Halifax - Born: Richmond - Age: 17 - Released on 10 Apr 1815.

Ingles, John - Seaman - Nbr: 3424 - Prize: Industry - Ship Type: P - How taken: HMS Heron - When taken: 3 Nov 1813 - Where taken: off Halifax - Date Received: 13 Sep 1814 - From what ship: HMT Niobe, Chatham - Born: Marblehead - Age: 25 - Released on 28 May 1815.

Ingles, John - Seaman - Nbr: 1569 - Prize: Messenger - Ship Type: MV - How taken: HMS Iris - When taken: 10 Mar 1813 - Where taken: Bay of Biscay - Date Received: 23 Jun 1814 - From what ship: Stapleton - Born: Baltimore - Age: 27 - Released on 1 May 1815.

Ingles, Daniel - Steward - Nbr: 875 - Prize: Zephyr - Ship Type: MV - How taken: HMS Pyramus - When taken: 30 Nov 1813 - Where taken: off L'Orient (France) - Date Received: 31 Jan 1814 - From what ship: Plymouth - Born: Dover - Age: 39 - Race: Black - Sent to Liverpool on 25 Aug 1813.

Ingram, John - Seaman - Nbr: 2012 - Prize: a recaptured MV - How taken: HMS Revolutionnaire - When taken: 10 Apr 1813 - Where taken: off Western Islands (England) - Date Received: 3 Aug 1814 - From what ship: HMS Lyffey, Chatham Depot - Born: New York - Age: 20 - Released on 2 May 1815.

Innes, J. - Seaman - Nbr: 1614 - Prize: Shadow - Ship Type: LM - How taken: HMS Reindeer - When taken: 6 Apr 1813 - Where taken: Bay of Biscay - Date Received: 23 Jun 1814 - From what ship: Stapleton - Born: Guadeloupe - Age: 21 - Race: Negro - Released on 11 Jul 1815.

Ireland, James - Seaman - Nbr: 75 - Prize: Rolla - Ship Type: MV - How taken: HMS Surveillante - When taken: 11 Feb 1813 - Where taken: Bay of Biscay - Date Received: 2 Apr 1813 - From what ship: Plymouth - Born: Egg Harbor (New Jersey) - Age: 28 - Released on 20 Apr 1815.

Ireson, Robert B. - Seaman - Nbr: 1774 - Prize: Print - Ship Type: MV - How taken: HMS Colossus - When taken: 21 Jan 1813 - Where taken: off Long Island - Date Received: 20 Jul 1814 - From what ship: HMS Milford, Plymouth - Born: Marblehead - Age: 15 - Released on 1 May 1815.

Irvin, Arthur - Seaman - Nbr: 1363 - Prize: Paul Jones - Ship Type: P - How taken: HMS Leonidas - When taken: 23 May 1813 - Where taken: Channel - Date Received: 19 Jun 1814 - From what ship: Stapleton - Born: Philadelphia - Age: 16 - Released on 28 Apr 1815.

Irvine, Andrew - Seaman - Nbr: 2538 - How taken: Gave himself up from HMS President - When taken: 1 May 1813 - Date Received: 16 Aug 1814 - From what ship: HMT Salvador del Mundo - Born: Pennsylvania - Age: 29 - Released on 3 May 1815.

Irwin, Andrew - Seaman - Nbr: 1991 - How taken: Gave himself up from HMS President - Date Received: 3 Aug 1814 - From what ship: HMS Lyffey, Chatham Depot - Born: Yorkinton - Age: 29 - Released on 2 May 1815.

Irwin, Mathew - Boy - Nbr: 483 - Prize: Fox - Ship Type: LM - How taken: HMS Pheasant - When taken: 23 Apr 1813 - Where taken: Bay of Biscay - Date Received: 8 Sep 1813 - From what ship: Plymouth - Born: Philadelphia - Age: 14 - Released on 20 Apr 1815.

Isaac, Moses - Seaman - Nbr: 190 - Prize: Star - Ship Type: MV - How taken: HMS Superb - When taken: 9 Feb 1813 - Where taken: Bay of Biscay - Date Received: 2 Apr 1813 - From what ship: Plymouth - Born: New York - Age: 21 - Released on 20 Apr 1815.

Isdale, James - Seaman - Nbr: 2013 - Prize: a recaptured MV - How taken: HMS Revolutionnaire - When taken: 10 Apr 1813 - Where taken: off Western Islands (England) - Date Received: 3 Aug 1814 - From what ship: HMS Lyffey, Chatham Depot - Born: Philadelphia - Age: 25 - Released on 2 May 1815.

Isles, Robert - Seaman - Nbr: 1448 - Prize: Tickler - Ship Type: LM - How taken: HMS Magicienne - When taken: 5 Jun 1813 - Where taken: Bay of Biscay - Date Received: 19 Jun 1814 - From what ship: Stapleton - Born: Rhode Island - Age: 37 - Released on 28 Apr 1815.

Ivy, John - Seaman - Nbr: 4833 - Prize: President - Ship Type: P - How taken: HMS Pique - When taken: 7 May 1814 - Where taken: off Porto Rico - Date Received: 9 Oct 1814 - From what ship: HMT Freya, Halifax - Born: Brest - Age: 30 - Released on 3 May 1815.

Jack, John - Seaman - Nbr: 6418 - Prize: Enterprize - Ship Type: P - How taken: HMS Argo - When taken: 16 Nov 1814 - Where taken: off San Domingo - Date Received: 3 Mar 1815 - From what ship: HMT Ganges, Plymouth - Born: New York - Age: 24 - Race: Black - Released on 11 Jul 1815.

Jack, John - Seaman - Nbr: 6514 - How taken: Gave himself up from HMS Orontes - When taken: Feb 1815 - Date Received: 3 Mar 1815 - From what ship: HMT Ganges, Plymouth - Born: Baltimore - Age: 36 - Race: Black - Died in Mar 1815 from pneumonia.

Jack, John - Seaman - Nbr: 4904 - How taken: Sent into custody from an English schooner - Date Received: 28 Oct 1814 - From what ship: HMT Alkbar, Halifax - Born: St. Thomas - Age: 20 - Race: Negro - Released on 21

American Prisoners of War Held at Dartmoor during the War of 1812

## Alphabetical listing of names

Jun 1815.

Jack, John - Seaman - Nbr: 1741 - Prize: Hugh Jones - Ship Type: MV - How taken: HMS Bittern - When taken: 14 Jun 1814 - Where taken: at sea - Date Received: 18 Jul 1814 - From what ship: HMT Salvador del Mundo, Plymouth - Born: New York - Age: 21 - Race: Black - Released on 1 May 1815.

Jackson, Benjamin - Seaman - Nbr: 5723 - How taken: Gave himself up from HMS Pelican - When taken: 13 Sep 1813 - Date Received: 26 Dec 1814 - From what ship: HMT Argo - Born: Alexandria - Age: 33 - Race: Black - Released on 9 Jan 1815.

Jackson, Charles - Seaman - Nbr: 4229 - Prize: Liberty - Ship Type: MV - How taken: Surrendered - When taken: 30 Dec 1813 - Where taken: Stromess - Date Received: 7 Oct 1814 - From what ship: HMT Niobe, Chatham - Born: Troy - Age: 17 - Released on 13 Jun 1815.

Jackson, Curtis - Seaman - Nbr: 872 - Prize: Agnes - Ship Type: LM - How taken: Jane (Cutter) - When taken: 29 Nov 1813 - Where taken: Bay of Biscay - Date Received: 31 Jan 1814 - From what ship: Plymouth - Born: Boston - Age: 17 - Released on 27 Apr 1815.

Jackson, Daniel - Boy - Nbr: 366 - Prize: Two Brothers - Ship Type: MV - How taken: Beetle (LM) - When taken: 18 Mar 1813 - Where taken: off Western Islands (England) - Date Received: 1 Jul 1813 - From what ship: Plymouth - Born: Boston - Age: 15 - Sent to Liverpool on 25 Aug 1813.

Jackson, Daniel - Seaman - Nbr: 1799 - How taken: Gave himself up from HMS Ajax - When taken: 14 Oct 1813 - Date Received: 20 Jul 1814 - From what ship: HMS Milford, Plymouth - Born: Connecticut - Age: 37 - Released on 1 May 1815.

Jackson, David - Cook - Nbr: 697 - Prize: Ann - Ship Type: MV - How taken: HMS Tenedos - When taken: 5 May 1813 - Where taken: Boston Bay - Date Received: 27 Sep 1813 - From what ship: Plymouth - Born: Albany - Age: 20 - Race: Black - Released on 11 Jul 1815.

Jackson, David - Steward - Nbr: 3601 - Prize: Frolic - Ship Type: P - How taken: HMS Heron - When taken: 25 Jan 1814 - Where taken: off St. Thomas - Date Received: 30 Sep 1814 - From what ship: HMT Sybella - Born: Long Island - Age: 34 - Race: Negro - Released on 4 Jun 1815.

Jackson, Ebenezer - Boy - Nbr: 6359 - Prize: Prince de Neufchatel - Ship Type: P - How taken: Leander (Newcastle Acasta) - When taken: 20 Dec 1814 - Where taken: Lat 38 Long 56 - Date Received: 19 Feb 1815 - From what ship: HMT Ganges, Plymouth - Born: Charlestown - Age: 17 - Released on 11 Jul 1815.

Jackson, Elias - Seaman - Nbr: 3072 - How taken: Gave himself up from HMS Hotspur - When taken: 12 Oct 1812 - Date Received: 3 Sep 1814 - From what ship: HMT Hydra, Chatham - Born: Virginia - Age: 26 - Race: Mulatto - Released on 27 Apr 1815.

Jackson, Frederick - Seaman - Nbr: 4289 - Prize: Pilot - Ship Type: LM - How taken: Victoria (Privateer) - When taken: 28 Jan 1814 - Where taken: off Bordeaux - Date Received: 7 Oct 1814 - From what ship: HMT Niobe, Chatham - Born: Maryland - Age: 22 - Race: Black - Released on 14 Jun 1815.

Jackson, George - Gunner - Nbr: 2984 - Prize: Frolic - Ship Type: P - How taken: HMS Heron - When taken: 25 Jan 1814 - Where taken: off St. Thomas - Date Received: 29 Aug 1814 - From what ship: HMT Bittern - Born: Marblehead - Age: 23 - Released on 19 May 1815.

Jackson, George - Seaman - Nbr: 2954 - Prize: Sally & Betsey - Ship Type: MV - How taken: HMS Pique - When taken: 5 Feb 1814 - Where taken: off St. Johns - Date Received: 24 Aug 1814 - From what ship: HMT Hannibal - Born: Delaware - Age: 25 - Race: Black - Released on 19 May 1815.

Jackson, Henry - Seaman - Nbr: 5886 - Prize: Harlequin - Ship Type: P - How taken: HMS Bulwark - When taken: 23 Nov 1814 - Where taken: off Halifax - Date Received: 27 Dec 1814 - From what ship: HMT Penelope - Born: New York - Age: 39 - Race: Black - Released on 3 Jul 1815.

Jackson, Henry - Seaman - Nbr: 4570 - How taken: Gave himself up from MV Liberty - When taken: 30 Dec 1813 - Date Received: 9 Oct 1814 - From what ship: HMT Leyden, Chatham - Born: New York - Age: 46 - Race: Black - Released on 15 Jun 1815.

Jackson, Henry - Seaman - Nbr: 3840 - Prize: Dominique - Ship Type: LM - How taken: HMS Dotterel - When taken: 21 May 1814 - Where taken: off Charleston - Date Received: 5 Oct 1814 - From what ship: HMT Orpheus, Halifax - Born: New Jersey - Age: 36 - Released on 9 Jun 1815.

Jackson, Isaac - Seaman - Nbr: 3146 - How taken: Gave himself up from HMS Cadmus - When taken: 13 Aug 1813 - Date Received: 11 Sep 1814 - From what ship: HMT Freya, Chatham - Born: New Castle - Age: 25 - Race: Black - Released on 28 May 1815.

Jackson, James - Seaman - Nbr: 5327 - Prize: Enterprize - Ship Type: P - How taken: HMS Tenedos - When taken: 21 May 1813 - Where taken: off Cape Cod - Date Received: 31 Oct 1814 - From what ship: HMT Leyden, Chatham - Born: Salem - Age: 24 - Released on 29 Jun 1815.

American Prisoners of War Held at Dartmoor during the War of 1812

## Alphabetical listing of names

Jackson, James - Seaman - Nbr: 5785 - How taken: Gave himself up from HMS Ariel - When taken: 8 Dec 1812 - Date Received: 26 Dec 1814 - From what ship: HMT Argo - Born: New York - Age: 22 - Race: Black - Released on 3 Jul 1815.

Jackson, James - Seaman - Nbr: 679 - Prize: Paul Jones - Ship Type: P - How taken: HMS Leonidas - When taken: 23 May 1813 - Where taken: Channel - Date Received: 14 Sep 1813 - From what ship: Plymouth - Born: New York - Age: 27 - Sent to Plymouth on 7 Dec 1813.

Jackson, James - Seaman - Nbr: 451 - Prize: Paul Jones - Ship Type: P - How taken: HMS Leonidas - When taken: 23 May 1813 - Where taken: Channel - Date Received: 22 Jul 1813 - From what ship: Berry Head Barracks by account of 28th Regiment of Foot - Born: New York - Age: 27 - Sent to Mill Prison (Plymouth, England) on 13 Sep 1813.

Jackson, James - Seaman - Nbr: 1098 - Prize: Lyon - Ship Type: P - How taken: Brilliant (Privateer) - When taken: 12 Aug 1814 - Where taken: off Charlestown - Date Received: 10 May 1814 - From what ship: Plymouth - Born: Salem - Age: 40 - Released on 27 Apr 1815.

Jackson, James - Seaman - Nbr: 5722 - How taken: Gave himself up from HMS Pelican - When taken: 13 Sep 1813 - Date Received: 26 Dec 1814 - From what ship: HMT Argo - Born: Cape Elizabeth - Age: 34 - Released on 9 Jan 1815.

Jackson, John - Seaman - Nbr: 1724 - How taken: Gave himself up from HMS Hebrius - When taken: May 1814 - Date Received: 8 Jul 1814 - From what ship: Labrador - Born: New York - Age: 28 - Race: Negro - Released on 1 May 1815.

Jackson, John - Seaman - Nbr: 258 - Prize: William Bayard - Ship Type: MV - How taken: HMS Warspite - When taken: 3 Mar 1813 - Where taken: Bay of Biscay - Date Received: 28 Jun 1813 - From what ship: Plymouth - Born: Dover - Age: 38 - Released on 20 Apr 1815.

Jackson, John - Seaman - Nbr: 3247 - Prize: Wiley Reynard - Ship Type: P - How taken: HMS Shannon - When taken: 15 Aug 1812 - Where taken: off Halifax - Date Received: 11 Sep 1814 - From what ship: HMT Freya, Chatham - Born: New York - Age: 52 - Race: Black - Released on 27 Apr 1815.

Jackson, Joseph - Seaman - Nbr: 6552 - Prize: Zebra - Ship Type: MV - How taken: HMS Pyramus - When taken: 20 Apr 1813 - Where taken: Bay of Biscay - Date Received: 26 Mar 1815 - From what ship: from Exeter jail - Born: New Brunswick - Age: 22 - Released on 1 May 1815.

Jackson, Joseph - Seaman - Nbr: 1536 - Prize: Zebra - Ship Type: LM - How taken: HMS Pyramus - When taken: 20 Apr 1813 - Where taken: Bay of Biscay - Date Received: 23 Jun 1814 - From what ship: Stapleton - Born: New Brunswick - Age: 26 - Sent to Exeter Jail on 9 Mar 1814.

Jackson, Joseph - Seaman - Nbr: 4849 - Prize: Melville - Ship Type: MV - How taken: HMS Albion - When taken: 1 Jan 1814 - Where taken: off St. Martins - Date Received: 9 Oct 1814 - From what ship: HMT Freya, Halifax - Born: Bath - Age: 24 - Released on 21 Jun 1815.

Jackson, Samuel - Seaman - Nbr: 978 - Prize: Siro - Ship Type: LM - How taken: HMS Pelican - When taken: 13 Jan 1814 - Where taken: at sea - Date Received: 31 Jan 1814 - From what ship: Plymouth - Born: New York - Age: 29 - Race: Mulatto - Released on 27 Apr 1815.

Jackson, Sidney - Seaman - Nbr: 2838 - How taken: Gave himself up from HMS Armide - When taken: 8 Jun 1813 - Date Received: 24 Aug 1814 - From what ship: HMT Liverpool, Chatham - Born: Virginia - Age: 23 - Race: Black - Released on 19 May 1815.

Jackson, Thomas - Seaman - Nbr: 6520 - How taken: Gave himself up from HMS Orontes - When taken: Feb 1815 - Date Received: 3 Mar 1815 - From what ship: HMT Ganges, Plymouth - Born: New York - Age: 14 - Race: Black - Died on 7 Apr 1815 from gunshot wound in prison.

Jackson, Thomas - Seaman - Nbr: 442 - Prize: Magdalen - Ship Type: MV - How taken: HMS Superb - When taken: 15 Apr 1813 - Where taken: off Belle Isle, France - Date Received: 1 Jul 1813 - From what ship: Plymouth - Born: Charlestown - Age: 19 - Released on 20 Apr 1815.

Jackson, Thomas - Cook - Nbr: 1162 - How taken: Gave himself up from HMS Hebrius - Date Received: 10 May 1814 - From what ship: Plymouth - Born: New York - Age: 24 - Race: Black - Died on 6 Jun 1814 from pneumonia.

Jackson, Thomas - Seaman - Nbr: 2542 - How taken: Sent from Exeter jail to Plymouth - Date Received: 16 Aug 1814 - What ship: HMT Salvador del Mundo - Born: New York - Age: 33 - Race: Black - Released on 3 May 1815.

Jackson, William - Seaman - Nbr: 868 - Prize: Amiable - Ship Type: LM - How taken: HMS Magnificent - When taken: 30 Oct 1813 - Where taken: off Bordeaux - Date Received: 4 Dec 1813 - From what ship: Plymouth - Born: Philadelphia - Age: 21 - Released on 27 Apr 1815.

American Prisoners of War Held at Dartmoor during the War of 1812

## Alphabetical listing of names

Jackson, William - Seaman - Nbr: 516 - Prize: Fox - Ship Type: Packet prize to the Fox - How taken: Superior - When taken: 25 Jun 1813 - Where taken: Lat 50 N, Long 21 W - Date Received: 8 Sep 1813 - From what ship: Plymouth - Born: Dover - Age: 23 - Released on 26 Apr 1815.

Jackson, William - Sail maker - Nbr: 1270 - How taken: Sent into custody from HMS Pyramus - Date Received: 14 Jun 1814 - What ship: Mill Prison (Plymouth, England) - Born: Cambridge - Age: 32 - Race: Black - Released on 28 Apr 1815.

Jackson, William - Seaman - Nbr: 4428 - Prize: Baroness Longueville - Ship Type: MV - How taken: HMS Illustrious - When taken: 5 Aug 1813 - Where taken: off St. Helena - Date Received: 8 Oct 1814 - From what ship: HMT Leyden, Chatham - Born: New York - Age: 28 - Race: Black - Released on 29 Dec 1814.

Jackson, William - Seaman - Nbr: 2845 - How taken: Gave himself up from HMS Charlotte - When taken: 28 May 1813 - Date Received: 24 Aug 1814 - From what ship: HMT Liverpool, Chatham - Born: Charleston - Age: 53 - Released on 19 May 1815.

Jackson, William - Seaman - Nbr: 2687 - How taken: Gave himself up from HMS Fawn - When taken: 11 Dec 1812 - Date Received: 21 Aug 1814 - From what ship: HMT Freya, Chatham - Born: Virginia - Age: 27 - Released on 1 Jul 1815.

Jackson, William - Seaman - Nbr: 2714 - How taken: Gave himself up from HMS Sterling Castle - When taken: 12 Jun 1813 - Date Received: 21 Aug 1814 - From what ship: HMT Freya, Chatham - Born: New London - Age: 38 - Race: Black - Released on 26 Apr 1815.

Jackson, William - Seaman - Nbr: 3645 - Prize: Stockholm - Ship Type: MV - How taken: HMS Niemen - When taken: Jan 1814 - Where taken: off Canada - Date Received: 30 Sep 1814 - From what ship: HMT Sybella - Born: Scarborough - Age: 32 - Race: Negro - Released on 4 Jun 1815.

Jacob, Louis - Seaman - Nbr: 5163 - Prize: Volante - Ship Type: P - How taken: HMS Curlew - When taken: 25 Nov 1813 - Where taken: off Halifax - Date Received: 31 Oct 1814 - From what ship: HMT Mermaid, Chatham - Born: New Orleans - Age: 30 - Race: Black - Released on 29 Jun 1815.

Jacob, Peter - Seaman - Nbr: 4962 - Prize: Herald - Ship Type: P - How taken: HMS Endymion - When taken: 15 Aug 1814 - Where taken: off Nantucket - Date Received: 28 Oct 1814 - From what ship: HMT Alkbar, Halifax - Born: France - Age: 50 - Released on 21 Jun 1815.

Jacobs, Hans - Seaman - Nbr: 5457 - How taken: Gave himself up - When taken: 30 Nov 1814 - Date Received: 10 Dec 1814 - From what ship: HMT Impregnable - Born: Copenhagen - Age: 40 - Released on 1 Jul 1815.

Jacobs, William - Seaman - Nbr: 5292 - Prize: Vivid - Ship Type: P - How taken: HMS Nymphe - When taken: 20 Apr 1813 - Where taken: off Cope Cod - Date Received: 31 Oct 1814 - From what ship: HMT Leyden, Chatham - Born: Norway - Age: 15 - Released on 29 Jun 1815.

Jacobson, Mathew - Seaman - Nbr: 5869 - Prize: Lion - Ship Type: P - How taken: HMS Granicus - When taken: 2 Dec 1814 - Where taken: off Lisbon - Date Received: 26 Dec 1814 - From what ship: HMT Impregnable - Born: Stockholm - Age: 28 - Released on 3 Jul 1815.

Jacobson, William - Seaman - Nbr: 2130 - Prize: Joseph Ricketson - Ship Type: MV - How taken: HMS Rifleman - When taken: 22 Aug 1812 - Where taken: off Jutland - Date Received: 8 Aug 1814 - From what ship: HMT Raven, Chatham - Born: Oldenburg - Age: 39 - Released on 28 Apr 1815.

James, Benjamin - Seaman - Nbr: 2554 - Prize: Caromaned - Ship Type: Prize - How taken: HMS Eridanus - When taken: 13 Aug 1814 - Where taken: Lat 40, Long 16 - Date Received: 16 Aug 1814 - From what ship: HMT Salvador del Mundo - Born: Maryland - Age: 31 - Race: Mulatto - Released on 5 Jul 1815.

James, Daniel - Seaman - Nbr: 1607 - Prize: Fox - Ship Type: LM - How taken: HMS Pheasant - When taken: 23 Apr 1813 - Where taken: Bay of Biscay - Date Received: 23 Jun 1814 - From what ship: Stapleton - Born: Boston - Age: 27 - Released on 1 May 1815.

James, George - Seaman - Nbr: 5162 - Prize: Volante - Ship Type: P - How taken: HMS Curlew - When taken: 25 Nov 1813 - Where taken: off Halifax - Date Received: 31 Oct 1814 - From what ship: HMT Mermaid, Chatham - Born: Rhode Island - Age: 35 - Released on 29 Jun 1815.

James, Jeremiah - Seaman - Nbr: 6118 - Prize: Johannes - Ship Type: Prize - How taken: Lacine - When taken: 26 Oct 1814 - Where taken: Grand Banks - Date Received: 6 Jan 1814 - From what ship: HMT Impregnable - Born: New Jersey - Age: 17.

James, John - Seaman - Nbr: 4161 - How taken: Gave himself up from HMS Minos - When taken: 28 Dec 1813 - Date Received: 6 Oct 1814 - From what ship: HMT Niobe, Chatham - Born: Philadelphia - Age: 26 - Race: Mulatto - Released on 13 Jun 1815.

James, John - Seaman - Nbr: 3343 - How taken: Gave himself up from HMS Prince of Wales - When taken: 28 May 1813 - Date Received: 13 Sep 1814 - From what ship: HMT Niobe, Chatham - Born: SC - Age: 26 - Race:

American Prisoners of War Held at Dartmoor during the War of 1812

## Alphabetical listing of names

Black - Released on 11 Jul 1815.

James, John - Seaman - Nbr: 3616 - Prize: Monarch - Ship Type: MV - How taken: HMS Dotterel - When taken: 14 Dec 1813 - Where taken: off Charleston - Date Received: 30 Sep 1814 - From what ship: HMT Sybella - Born: Baltimore - Age: 27 - Released on 4 Jun 1815.

James, John - Seaman - Nbr: 3497 - Prize: Old Friend - Ship Type: Prize - How taken: Viper (Privateer) - When taken: 1 Jan 1814 - Where taken: at sea - Date Received: 24 Sep 1814 - From what ship: HMT Salvador del Mundo - Born: Boston - Age: 19 - Released on 4 Jun 1815.

James, Joseph - Seaman - Nbr: 3032 - Prize: Achille - Ship Type: MV - How taken: HMS Belvidera - When taken: 23 Jul 1813 - Where taken: off Sandy Hook - Date Received: 2 Sep 1814 - From what ship: HMT Sultan - Born: Portland - Age: 21 - Released on 10 Apr 1815.

James, Thomas - Seaman - Nbr: 398 - Prize: Young Holkar - Ship Type: MV - How taken: HMS Superb - When taken: 10 Apr 1813 - Where taken: off Belle Isle, France - Date Received: 1 Jul 1813 - From what ship: Plymouth - Born: New Orleans - Age: 29 - Impersonated a Frenchman on the general release.

James, William - Seaman - Nbr: 6259 - Prize: Plutarch - Ship Type: MV - How taken: HMS Helicon - When taken: 5 Feb 1815 - Where taken: off Bordeaux - Date Received: 10 Feb 1815 - From what ship: HMT Ganges, Plymouth - Born: New York - Age: 27 - Released on 5 Jul 1815.

Jameson, George - Seaman - Nbr: 3181 - How taken: Gave himself up from HMS Hibernia - When taken: 27 Jul 1813 - Date Received: 11 Sep 1814 - From what ship: HMT Freya, Chatham - Born: Philadelphia - Age: 53 - Race: Black - Released on 28 May 1815.

Jameson, Robert - Seaman - Nbr: 850 - Prize: US Brig Argus - Ship Type: MW - How taken: HMS Pelican - When taken: 14 Aug 1813 - Where taken: Irish Channel - Date Received: 29 Nov 1813 - From what ship: Plymouth - Born: Delaware - Age: 54 - Sent to Dartmouth on 2 Nov 1814.

Jamieson, Daniel - Seaman - Nbr: 2384 - Prize: US Sloop Frolic - Ship Type: MW - How taken: HMS Orpheus - When taken: 20 Apr 1814 - Where taken: off Cuba - Date Received: 16 Aug 1814 - From what ship: HMT Queen, Halifax - Born: Baltimore - Age: 28 - Sent to Dartmouth on 19 Oct 1814.

Jamison, Peter - Seaman - Nbr: 495 - How taken: Impressed at Belfast - When taken: 16 Jan 1813 - Date Received: 8 Sep 1813 - From what ship: Plymouth - Born: Northumberland - Age: 33 - Released on 20 Apr 1815.

Jane, William - Seaman - Nbr: 1740 - Prize: Hugh Jones - Ship Type: MV - How taken: HMS Bittern - When taken: 14 Jun 1814 - Where taken: at sea - Date Received: 18 Jul 1814 - From what ship: HMT Salvador del Mundo, Plymouth - Born: Rhode Island - Age: 28 - Released on 1 May 1815.

Jansen, Knute - Seaman - Nbr: 761 - How taken: Impressed at Liverpool - Date Received: 3 Nov 1813 - From what ship: Plymouth - Born: New York - Age: 21 - Released on 6 May 1814.

Jardine, Samuel - Seaman - Nbr: 3071 - How taken: Gave himself up from HMS Ceylon - When taken: 20 Jan 1814 - Date Received: 2 Sep 1814 - From what ship: HMT Hydra, Chatham - Born: Massachusetts - Age: 25 - Released on 28 May 1815.

Jarey, Bartholomew - Seaman - Nbr: 2550 - Prize: Caromaned - Ship Type: Prize - How taken: HMS Eridanus - When taken: 13 Aug 1814 - Where taken: Lat 40, Long 16 - Date Received: 16 Aug 1814 - From what ship: HMT Salvador del Mundo - Born: Minorca - Age: 21 - Released on 3 May 1815.

Jarvis, George - Seaman - Nbr: 3056 - How taken: Gave himself up from HMS Andromache - Date Received: 2 Sep 1814 - From what ship: HMT Hydra, Chatham - Born: Albion - Age: 46 - Race: Black - Released on 28 Apr 1815.

Jarvis, James - Seaman - Nbr: 1226 - How taken: Sent into custody from HMS Haughty - When taken: 8 Aug 1812 - Date Received: 14 Jun 1814 - From what ship: Mill Prison (Plymouth, England) - Born: Long Island - Age: 27 - Race: Black - Released on 26 Apr 1815.

Jarvis, Peter - Seaman - Nbr: 6312 - Prize: Prince de Neufchatel - Ship Type: P - How taken: Leander (Newcastle Acasta) - When taken: 20 Dec 1814 - Where taken: Lat 38 Long 56 - Date Received: 19 Feb 1815 - From what ship: HMT Ganges, Plymouth - Born: Saint Malo - Age: 25 - Released on 5 Jul 1815.

Jarvis, Thomas - Seaman - Nbr: 5321 - Prize: Industry - Ship Type: P - How taken: HMS Maidstone - When taken: 17 Jul 1813 - Where taken: Grand Banks - Date Received: 31 Oct 1814 - From what ship: HMT Leyden, Chatham - Born: Marblehead - Age: 18 - Died on 25 Jan 1815 from phthisis pulmonalis.

Jasmine, Paul - Seaman - Nbr: 2913 - Prize: Matilda - Ship Type: MV - How taken: HMS Revolutionnaire - When taken: 25 Jul 1813 - Where taken: off L'Orient (France) - Date Received: 24 Aug 1814 - From what ship: HMT Alpheus, Chatham - Born: Boston - Age: 18 - Race: Black - Released on 2 Nov 1814.

Jefferson, Edward - Seaman - Nbr: 1652 - How taken: Apprehended at Bristol - When taken: 14 Sep 1813 - Date Received: 23 Jun 1814 - From what ship: Stapleton - Born: Virginia - Age: 21 - Released on 1 May 1815.

American Prisoners of War Held at Dartmoor during the War of 1812

## Alphabetical listing of names

Jefferson, Jacob - Seaman - Nbr: 3639 - Prize: Bordeaux Packet - Ship Type: LM - How taken: HMS Niemen - When taken: 28 Jan 1814 - Where taken: off Delaware - Date Received: 30 Sep 1814 - From what ship: HMT Sybella - Born: Delaware - Age: 30 - Released on 25 Mar 1815.

Jefferson, Richard - Seaman - Nbr: 4985 - Prize: Invincible - Ship Type: LM - How taken: HMS Armide - When taken: 15 Aug 1814 - Where taken: off Nantucket - Date Received: 28 Oct 1814 - From what ship: HMT Alkbar, Halifax - Born: New York - Age: 22 - Released on 21 Jun 1815.

Jeffery, Robert - Boy - Nbr: 688 - Prize: Joel Barlow - Ship Type: LM - How taken: HMS Briton - When taken: 3 Jul 1813 - Where taken: off Bordeaux - Date Received: 27 Sep 1813 - From what ship: Plymouth - Born: New London - Age: 12 - Released on 26 Apr 1815.

Jeffrey, Francis - Seaman - Nbr: 1616 - Prize: Essex - Ship Type: MV - How taken: HMS Pyramus - When taken: 2 Apr 1813 - Where taken: Bay of Biscay - Date Received: 23 Jun 1814 - From what ship: Stapleton - Born: Marblehead - Age: 17 - Released on 1 May 1815.

Jeffreys, Henry - Seaman - Nbr: 4420 - Prize: Fire Fly - Ship Type: LM - How taken: HMS Revolutionnaire - When taken: 19 Oct 1813 - Where taken: at sea - Date Received: 8 Oct 1814 - From what ship: HMT Leyden, Chatham - Born: Delaware - Age: 23 - Released on 14 Jun 1815.

Jeffreys, Henry - Seaman - Nbr: 2709 - Prize: Dick - Ship Type: MV - How taken: HMS Dispatch - When taken: 15 Mar 1813 - Where taken: near Corcova Lights - Date Received: 21 Aug 1814 - From what ship: HMT Freya, Chatham - Born: Elizabeth - Age: 23 - Released on 19 May 1815.

Jeffs, Joseph - Seaman - Nbr: 2797 - Prize: Weazel - Ship Type: MV - How taken: HMS Foxhound - When taken: 25 Mar 1813 - Where taken: Bay of Biscay - Date Received: 24 Aug 1814 - From what ship: HMT Liverpool, Chatham - Born: Gloucester - Age: 22 - Released on 19 May 1815.

Jenkins, Elijah - Seaman - Nbr: 6098 - Prize: Harlequin - Ship Type: P - How taken: HMS Bulwark - When taken: 23 Nov 1814 - Where taken: off Halifax - Date Received: 28 Dec 1814 - From what ship: HMT Penelope - Born: Old York - Age: 24 - Released on 3 Jul 1815.

Jenkins, James - Seaman - Nbr: 6034 - Prize: Regent - Ship Type: LM - How taken: HMS Forth - When taken: 19 Sep 1814 - Where taken: off Egg Harbor (New Jersey) - Date Received: 28 Dec 1814 - From what ship: HMT Penelope - Born: New Bedford - Age: 26 - Released on 3 Jul 1815.

Jenkins, Joseph - Prize Master - Nbr: 515 - Prize: Fox - Ship Type: Packet prize to the Fox - How taken: Superior - When taken: 25 Jun 1813 - Where taken: Lat 50 N, Long 21 W - Date Received: 8 Sep 1813 - From what ship: Plymouth - Born: Massachusetts - Age: 39 - Released on 26 Apr 1815.

Jenkins, Nathaniel - Seaman - Nbr: 1636 - Prize: Tom (Thumb) - Ship Type: LM - How taken: HMS Surveillante - When taken: 24 Apr 1813 - Where taken: Bay of Biscay - Date Received: 23 Jun 1814 - From what ship: Stapleton - Born: Baltimore - Age: 20 - Race: Negro - Died on 21 Feb 1815 from variola.

Jenkins, Samuel - Seaman - Nbr: 4073 - Prize: Fame - Ship Type: MV - How taken: HMS Niemen - When taken: 31 May 1814 - Where taken: off Cape Hatteras - Date Received: 6 Oct 1814 - From what ship: HMT Chesapeake, Halifax - Born: Wiscasset - Age: 25 - Released on 13 Jun 1815.

Jenkins, Samuel - Seaman - Nbr: 2544 - How taken: Gave himself up from HMS Salvador - Date Received: 16 Aug 1814 - What ship: HMT Salvador del Mundo - Born: Delaware - Age: 39 - Race: Black - Released on 3 May 1815.

Jenkins, Thomas - 2nd Mate - Nbr: 6391 - Prize: Nellenville - Ship Type: MV - How taken: HMS Onyx - When taken: 25 Dec 1814 - Where taken: off San Domingo - Date Received: 3 Mar 1815 - From what ship: HMT Ganges, Plymouth - Born: Virginia - Age: 23 - Released on 11 Jul 1815.

Jenkins, William - Seaman - Nbr: 628 - Prize: US Brig Argus - Ship Type: MW - How taken: HMS Pelican - When taken: 14 Aug 1813 - Where taken: Irish Channel - Date Received: 8 Sep 1813 - From what ship: Plymouth - Born: Pennsylvania - Age: 26 - Sent to Dartmouth on 2 Nov 1814.

Jennings, Francis - Seaman - Nbr: 6417 - Prize: Enterprize - Ship Type: P - How taken: HMS Argo - When taken: 16 Nov 1814 - Where taken: off San Domingo - Date Received: 3 Mar 1815 - From what ship: HMT Ganges, Plymouth - Born: New York - Age: 30 - Released on 11 Jul 1815.

Jennings, Henry - Seaman - Nbr: 3775 - Prize: Fame - Ship Type: P - How taken: HMS Thistle - When taken: 10 Apr 1814 - Where taken: after being cast ashore on a seal island - Date Received: 30 Sep 1814 - From what ship: HMT President, Halifax - Born: New York - Age: 30 - Race: Negro - Released on 9 Jun 1815.

Jennings, John - Seaman - Nbr: 4630 - How taken: Gave himself up from HMS Inane - When taken: 9 Aug 1812 - Date Received: 9 Oct 1814 - From what ship: HMT Leyden, Chatham - Born: Annapolis - Age: 29 - Released on 27 Apr 1815.

Jennings, John - Seaman - Nbr: 4846 - Prize: Hawk - Ship Type: P - How taken: HMS Pique - When taken: 26 Apr

American Prisoners of War Held at Dartmoor during the War of 1812

## Alphabetical listing of names

1814 - Where taken: off Porto Rico - Date Received: 9 Oct 1814 - From what ship: HMT Freya, Halifax - Born: Martha's Vineyard - Age: 18 - Race: Mulatto - Died on 22 Feb 1815 from variola.

Jennings, Luther - Seaman - Nbr: 2663 - Prize: Thomas - Ship Type: MV - How taken: HMS Frolic - When taken: 29 Jun 1813 - Where taken: off St. Thomas - Date Received: 21 Aug 1814 - From what ship: HMT Freya, Chatham - Born: Massachusetts - Age: 24 - Released on 19 May 1815.

Jennings, Nathaniel - Seaman - Nbr: 272 - Prize: Cannoneer - Ship Type: MV - How taken: HMS Warspite - When taken: 14 Mar 1813 - Where taken: Bay of Biscay - Date Received: 28 Jun 1813 - From what ship: Plymouth - Born: Connecticut - Age: 26 - Released on 20 Apr 1815.

Jennings, Samuel - Seaman - Nbr: 5951 - Prize: Harlequin - Ship Type: P - How taken: HMS Bulwark - When taken: 23 Nov 1814 - Where taken: off Halifax - Date Received: 27 Dec 1814 - From what ship: HMT Penelope - Born: Baltimore - Age: 41 - Race: Black - Released on 3 Jul 1815.

Jerez, Duly - Seaman - Nbr: 4348 - Prize: Union - Ship Type: LM - How taken: HMS Curlew - When taken: 1 Apr 1814 - Where taken: off Halifax - Date Received: 7 Oct 1814 - From what ship: HMT Salvador del Mundo, Halifax - Born: Yarmouth - Age: 22 - Released on 14 Jun 1815.

Jerald, Abraham - Seaman - Nbr: 2007 - Prize: Darby - Ship Type: MV - How taken: HMS Narcissus - When taken: 4 Feb 1813 - Where taken: off St. Helena - Date Received: 3 Aug 1814 - From what ship: HMS Lyffey, Chatham Depot - Born: Dinis - Age: 22 - Released on 2 May 1815.

Jervis, Isaac - Seaman - Nbr: 3554 - Prize: Hawk - Ship Type: P - How taken: HMS Pique - When taken: 26 Apr 1814 - Where taken: off Bermuda - Date Received: 30 Sep 1814 - From what ship: HMT Sybella - Born: Long Island - Age: 30 - Released on 4 Jun 1815.

Jervis, John - Seaman - Nbr: 60 - Prize: Spitfire - Ship Type: MV - How taken: HMS Achates - When taken: 14 Feb 1813 - Where taken: off Ushant (France) - Date Received: 2 Apr 1813 - From what ship: Plymouth - Born: Marblehead - Age: 21 - Released on 20 Apr 1815.

Jessamine, John - Seaman - Nbr: 3321 - Prize: York Town - Ship Type: P - How taken: HMS Nimrod - When taken: 17 Jul 1813 - Where taken: off St. Johns - Date Received: 13 Sep 1814 - From what ship: HMT Niobe, Chatham - Born: Bremen - Age: 22 - Released on 28 May 1815.

Jessamine, John - Seaman - Nbr: 5277 - Prize: York Town - Ship Type: P - How taken: HMS Maidstone - When taken: 18 Jul 1813 - Where taken: Grand Banks - Date Received: 31 Oct 1814 - From what ship: HMT Leyden, Chatham - Born: Bremen - Age: 24 - Released on 29 Jun 1815.

Jetson, Samuel - Seaman - Nbr: 2885 - How taken: Gave himself up from HMS Shearwater - When taken: 28 May 1813 - Date Received: 24 Aug 1814 - From what ship: HMT Alpheus, Chatham - Born: Rhode Island - Age: 39 - Released on 19 May 1815.

Jeuvmeson, John - Seaman - Nbr: 2824 - How taken: Gave himself up from HMS Philomel - When taken: 28 Dec 1812 - Date Received: 24 Aug 1814 - From what ship: HMT Liverpool, Chatham - Born: Delaware - Age: 28 - Race: Black - Released on 26 Apr 1815.

Jewitt, Jasper - Seaman - Nbr: 4727 - How taken: Gave himself up from MV John - When taken: 28 Aug 1814 - Date Received: 9 Oct 1814 - From what ship: HMT Freya, Chatham - Born: New Hampshire - Age: 23 - Released on 15 Jun 1815.

Jewitt, Samuel - Seaman - Nbr: 2184 - Prize: Diomede - Ship Type: P - How taken: HMS Rifleman - When taken: 28 Jun 1813 - Where taken: off Halifax - Date Received: 16 Aug 1814 - From what ship: HMS Dublin, Halifax - Born: Massachusetts - Age: 28 - Released on 2 May 1815.

Jiett, Benjamin - Seaman - Nbr: 65 - Prize: Spitfire - Ship Type: MV - How taken: HMS Achates - When taken: 14 Feb 1813 - Where taken: off Ushant (France) - Date Received: 2 Apr 1813 - From what ship: Plymouth - Born: Marblehead - Age: 22 - Released on 20 Apr 1815.

Jocelyn, Robert - Seaman - Nbr: 254 - Prize: Star - Ship Type: MV - How taken: HMS Superb - When taken: 9 Feb 1813 - Where taken: Bay of Biscay - Date Received: 28 Jun 1813 - From what ship: Plymouth - Born: Connecticut - Age: 27 - Released on 20 Apr 1815.

Johannes, John - Seaman - Nbr: 4836 - Prize: President - Ship Type: P - How taken: HMS Pique - When taken: 7 May 1814 - Where taken: off Porto Rico - Date Received: 9 Oct 1814 - From what ship: HMT Freya, Halifax - Born: St. Thomas - Age: 55 - Race: Black - Died on 8 Jan 1815 from perissneuoniria.

Johannes, Logan - Seaman - Nbr: 6402 - Prize: Nellenville - Ship Type: MV - How taken: HMS Onyx - When taken: 25 Dec 1814 - Where taken: off San Domingo - Date Received: 3 Mar 1815 - From what ship: HMT Ganges, Plymouth - Born: Calcutta - Age: 26 - Race: Black - Released on 11 Jul 1815.

Johansen, Johan - Sergeant - Nbr: 744 - Prize: 14th US Infantry - Ship Type: Troops - How taken: British Army - When taken: 24 Jun 1813 - Where taken: Forty Mile Creek - Date Received: 3 Nov 1813 - From what ship:

American Prisoners of War Held at Dartmoor during the War of 1812

## Alphabetical listing of names

Plymouth - Born: Amsterdam - Age: 27 - Sent to Dartmouth on 29 Jun 1814.

Johns, Thomas - Seaman - Nbr: 624 - Prize: US Brig Argus - Ship Type: MW - How taken: HMS Pelican - When taken: 14 Aug 1813 - Where taken: Irish Channel - Date Received: 8 Sep 1813 - From what ship: Plymouth - Born: Norfolk - Age: 23 - Sent to Dartmouth on 2 Nov 1814.

Johnson, Abraham - Seaman - Nbr: 6502 - How taken: Apprehended at Almigion - Date Received: 3 Mar 1815 - From what ship: HMT Ganges, Plymouth - Born: New York - Age: 42 - Released on 11 Jul 1815.

Johnson, Albert - Seaman - Nbr: 6027 - Prize: Regent - Ship Type: LM - How taken: HMS Forth - When taken: 19 Sep 1814 - Where taken: off Egg Harbor (New Jersey) - Date Received: 28 Dec 1814 - From what ship: HMT Penelope - Born: Not legible - Released on 3 Jul 1815.

Johnson, Alexander - Seaman - Nbr: 4171 - Prize: Quebec - Ship Type: Prize - How taken: HMS Derwent - When taken: 29 Jan 1813 - Where taken: off Lisbon - Date Received: 7 Oct 1814 - From what ship: HMT Niobe, Chatham - Born: New York - Age: 23 - Released on 13 Jun 1815.

Johnson, Andrew - Seaman - Nbr: 1237 - How taken: Sent into custody from HMS Ajax - When taken: 22 Jan 1813 - Date Received: 14 Jun 1814 - From what ship: Mill Prison (Plymouth, England) - Born: Portland - Age: 24 - Released on 26 Apr 1815.

Johnson, Charles - Seaman - Nbr: 6467 - Prize: Decatur - Ship Type: P - How taken: HMS Rhin - When taken: 5 Jun 1814 - Where taken: off San Domingo - Date Received: 3 Mar 1815 - From what ship: HMT Ganges, Plymouth - Born: New York - Age: 42 - Race: Black - Released on 11 Jul 1815.

Johnson, Edward - Seaman - Nbr: 2760 - How taken: Gave himself up from HMS Bruin - When taken: 19 Jul 1813 - Date Received: 24 Aug 1814 - From what ship: HMT Liverpool, Chatham - Born: Bennington - Age: 26 - Released on 19 May 1815.

Johnson, Elias - Seaman - Nbr: 792 - Prize: Avon - Ship Type: MV - How taken: HMS Eurotas - When taken: 27 Oct 1813 - Where taken: off Ushant (France) - Date Received: 3 Nov 1813 - From what ship: Plymouth - Born: Sweden - Age: 59 - Released on 6 May 1814.

Johnson, Frederick - Seaman - Nbr: 1929 - How taken: Gave himself up from HMS Antelope - When taken: 25 Dec 1812 - Date Received: 3 Aug 1814 - From what ship: HMS Alceste, Chatham Depot - Born: Connecticut - Age: 53 - Released on 26 Apr 1815.

Johnson, Frederick - Seaman - Nbr: 2320 - Prize: Hussar - Ship Type: P - How taken: HMS Saturn - When taken: 25 May 1814 - Where taken: off Sandy Hook - Date Received: 16 Aug 1814 - From what ship: HMS Dublin, Halifax - Born: Alexandria - Age: 21 - Race: Mulatto - Released on 3 May 1815.

Johnson, George - Soldier - Nbr: 5269 - Prize: 6th US Infantry - Ship Type: Troops - How taken: British Army - When taken: 13 Oct 1812 - Where taken: Canada - Date Received: 31 Oct 1814 - From what ship: HMT Leyden, Chatham - Born: Armagh - Age: 45 - Released on 19 Mar 1815.

Johnson, George - Seaman - Nbr: 4302 - How taken: Gave himself up from HMS Acteon - When taken: 9 May 1814 - Date Received: 7 Oct 1814 - From what ship: HMT Niobe, Chatham - Born: Philadelphia - Age: 31 - Released on 14 Jun 1815.

Johnson, George - Seaman - Nbr: 4502 - Prize: Enterprize - Ship Type: P - How taken: HMS Tenedos - When taken: 21 May 1813 - Where taken: off Cape Cod - Date Received: 8 Oct 1814 - From what ship: HMT Leyden, Chatham - Born: Massachusetts - Age: 19 - Released on 11 Jul 1815.

Johnson, George - Seaman - Nbr: 3514 - Prize: Aeolus - Ship Type: Prize - How taken: HMS Warspite - When taken: 2 Sep 1814 - Where taken: off Newfoundland - Date Received: 28 Sep 1814 - From what ship: HMT Salvador del Mundo - Born: New York - Age: 24 - Race: Negro - Released on 4 Jun 1815.

Johnson, Henry - Seaman - Nbr: 4720 - How taken: Gave himself up from HMS Phoenix - When taken: 17 Jul 1813 - Date Received: 9 Oct 1814 - From what ship: HMT Freya, Chatham - Born: New York - Age: 26 - Race: Black - Released on 15 Jun 1815.

Johnson, Henry - Boy - Nbr: 1301 - Prize: John & Frances - Ship Type: Prize - How taken: HMS Sterling Castle - When taken: 10 May 1814 - Where taken: off Cape Clear - Date Received: 14 Jun 1814 - From what ship: Mill Prison (Plymouth, England) - Born: Baltimore - Age: 15 - Race: Black - Released on 28 Apr 1815.

Johnson, Henry - Seaman - Nbr: 3244 - How taken: Impressed at Cork - When taken: 26 Sep 1813 - Date Received: 11 Sep 1814 - From what ship: HMT Freya, Chatham - Born: Danvers - Age: 20 - Released on 10 Apr 1815.

Johnson, Jacob - Seaman - Nbr: 1977 - Prize: Orbit - Ship Type: MV - How taken: HMS Achates - When taken: 29 Jan 1813 - Where taken: Lat 44N Long 13W - Date Received: 3 Aug 1814 - From what ship: HMS Lyffey, Chatham Depot - Born: Long Island - Age: 33 - Released on 2 May 1815.

Johnson, Jacob - Steward - Nbr: 4117 - Prize: Bordeaux Packet - Ship Type: LM - How taken: HMS Niemen - When taken: 28 Jun 1814 - Where taken: off Delaware - Date Received: 6 Oct 1814 - From what ship: HMT

American Prisoners of War Held at Dartmoor during the War of 1812

## Alphabetical listing of names

Chesapeake, Halifax - Born: Snow Hill - Age: 22 - Race: Negro - Released on 13 Jun 1815.
Johnson, Jacob - Seaman - Nbr: 2690 - Prize: Porcupine - Ship Type: LM - How taken: British squadron - When taken: 3 Jun 1813 - Where taken: off Cape Sable Island (Canada) - Date Received: 21 Aug 1814 - From what ship: HMT Freya, Chatham - Born: Marblehead - Age: 22 - Race: Black - Released on 19 May 1815.
Johnson, James - Seaman - Nbr: 4604 - Prize: Requin - Ship Type: LM - How taken: HMS Venus - When taken: 6 Mar 1814 - Where taken: off Bordeaux - Date Received: 9 Oct 1814 - From what ship: HMT Leyden, Chatham - Born: Providence - Age: 27 - Released on 15 Jun 1815.
Johnson, James - Seaman - Nbr: 2385 - Prize: US Sloop Frolic - Ship Type: MW - How taken: HMS Orpheus - When taken: 20 Apr 1814 - Where taken: off Cuba - Date Received: 16 Aug 1814 - From what ship: HMT Queen, Halifax - Born: New Hampshire - Age: 23 - Sent to Dartmouth on 19 Oct 1814.
Johnson, James - Seaman - Nbr: 1976 - Prize: Orbit - Ship Type: MV - How taken: HMS Achates - When taken: 29 Jan 1813 - Where taken: Lat 44N Long 13W - Date Received: 3 Aug 1814 - From what ship: HMS Lyffey, Chatham Depot - Born: Northumberland - Age: 34 - Released on 2 May 1815.
Johnson, James - Seaman - Nbr: 4078 - Prize: New Zealander - Ship Type: Prize - How taken: HMS Belviders - When taken: 21 Apr 1814 - Where taken: off Delaware - Date Received: 6 Oct 1814 - From what ship: HMT Chesapeake, Halifax - Born: Georgetown - Age: 19 - Released on 13 Jun 1815.
Johnson, James - Seaman - Nbr: 3931 - Prize: Rolla - Ship Type: P - How taken: HMS Loire - When taken: 10 Dec 1813 - Where taken: off Bull Island (South Carolina) - Date Received: 5 Oct 1814 - From what ship: HMT President, Halifax - Born: Rhode Island - Age: 17 - Released on 9 Jun 1815.
Johnson, James - Boy - Nbr: 3608 - Prize: Fiere Facia - Ship Type: P - How taken: HMS Ramillies - When taken: 26 Feb 1814 - Where taken: off NY - Date Received: 30 Sep 1814 - From what ship: HMT Sybella - Born: New York - Age: 13 - Race: Black - Released on 4 Jun 1815.
Johnson, Jesse - Seaman - Nbr: 6246 - Prize: Albion - Ship Type: Prize - How taken: HMS Harlequin - When taken: 6 Jan 1814 - Where taken: Long 20 - Date Received: 4 Feb 1815 - From what ship: HMT Ganges - Born: New York - Age: 20 - Race: Black - Released on 5 Jul 1815.
Johnson, John - Seaman - Nbr: 4807 - Prize: President - Ship Type: P - How taken: HMS Pique - When taken: 7 May 1814 - Where taken: off Porto Rico - Date Received: 9 Oct 1814 - From what ship: HMT Freya, Halifax - Born: New York - Age: 22 - Race: Black - Released on 21 Jun 1815.
Johnson, John - Boatswain - Nbr: 132 - Prize: Criterion - Ship Type: MV - How taken: HMS Belle Poule - When taken: 14 Feb 1813 - Where taken: Bay of Biscay - Date Received: 12 Apr 1813 - From what ship: Plymouth - Born: Rhode Island - Age: 23 - Died on 1 Feb 1815 from phthisis pulmonalis.
Johnson, John - Seaman - Nbr: 833 - Prize: Chesapeake - Ship Type: LM - How taken: HMS Hotspur & HMS Pyramus - When taken: 26 Oct 1813 - Where taken: off Nantes - Date Received: 29 Nov 1813 - From what ship: Plymouth - Born: New York - Age: 29 - Released on 26 Apr 1815.
Johnson, John - Seaman - Nbr: 2357 - Prize: Snap Dragon - Ship Type: P - How taken: HMS Martin - When taken: 10 Jun 1814 - Where taken: off Halifax - Date Received: 16 Aug 1814 - From what ship: HMT Queen, Halifax - Born: Maryland - Age: 17 - Released on 3 May 1815.
Johnson, John - Seaman - Nbr: 2379 - Prize: US Sloop Frolic - Ship Type: MW - How taken: HMS Orpheus - When taken: 20 Apr 1814 - Where taken: off Cuba - Date Received: 16 Aug 1814 - From what ship: HMT Queen, Halifax - Born: North Carolina - Age: 30 - Sent to Dartmouth on 19 Oct 1814.
Johnson, John - Seaman - Nbr: 2006 - Prize: Governor Middleton - Ship Type: MV - How taken: Thetis (Privateer) - When taken: 2 May 1813 - Where taken: Bay of Biscay - Date Received: 3 Aug 1814 - From what ship: HMS Lyffey, Chatham Depot - Born: Connecticut - Age: 18 - Released on 2 May 1815.
Johnson, John - Seaman - Nbr: 72 - Prize: Rolla - Ship Type: MV - How taken: HMS Surveillante - When taken: 11 Feb 1813 - Where taken: Bay of Biscay - Date Received: 2 Apr 1813 - From what ship: Plymouth - Born: New Castle - Age: 28 - Released in May 1813.
Johnson, John - Seaman - Nbr: 2462 - Prize: US Sloop Frolic - Ship Type: MW - How taken: HMS Orpheus - When taken: 20 Apr 1814 - Where taken: off Cuba - Date Received: 16 Aug 1814 - From what ship: HMT Queen, Halifax - Born: Haiti - Age: 28 - Race: Mulatto - Released on 3 May 1815.
Johnson, John - Seaman - Nbr: 3051 - How taken: Gave himself up from HMS Sultan - When taken: 2 Sep 1814 - Date Received: 2 Sep 1814 - From what ship: HMT Sultan - Born: Baltimore - Age: 29 - Race: Negro - Released on 28 Apr 1815.
Johnson, John - Seaman - Nbr: 2590 - How taken: Gave himself up from HMS Electra - When taken: 20 Sep 1812 - Date Received: 21 Aug 1814 - From what ship: HMT Freya, Chatham - Born: Boston - Age: 27 - Released on 26 Apr 1815.

American Prisoners of War Held at Dartmoor during the War of 1812

## Alphabetical listing of names

Johnson, John - Seaman - Nbr: 4870 - Prize: Jane - Ship Type: MV - How taken: Not listed - Date Received: 24 Oct 1814 - From what ship: Royal Hospital, Plymouth - Born: Boston - Age: 27 - Released on 21 Jun 1815.

Johnson, John - Seaman - Nbr: 2440 - Prize: US Sloop Frolic - Ship Type: MW - How taken: HMS Orpheus - When taken: 20 Apr 1814 - Where taken: off Cuba - Date Received: 16 Aug 1814 - From what ship: HMT Queen, Halifax - Born: Philadelphia - Age: 27 - Race: Negro - Released on 3 May 1815.

Johnson, John - Seaman - Nbr: 3903 - Prize: Henry Guilder - Ship Type: MV - How taken: HMS Niemen - When taken: 14 Jul 1814 - Where taken: at sea - Date Received: 5 Oct 1814 - From what ship: HMT Orpheus, Halifax - Born: New York - Age: 25 - Race: Negro - Released on 9 Jun 1815.

Johnson, Joseph - Seaman - Nbr: 6090 - Prize: Black Swan - Ship Type: MV - How taken: HMS Maidstone - When taken: 24 Oct 1814 - Where taken: Georges Bank - Date Received: 28 Dec 1814 - From what ship: HMT Penelope - Born: Plymouth - Age: 19 - Released on 20 Jul 1815.

Johnson, Joseph - Seaman - Nbr: 1565 - Prize: Messenger - Ship Type: MV - How taken: HMS Iris - When taken: 10 Mar 1813 - Where taken: Bay of Biscay - Date Received: 23 Jun 1814 - From what ship: Stapleton - Born: Maryland - Age: 20 - Released on 1 May 1815.

Johnson, Joseph - Seaman - Nbr: 2117 - How taken: Gave himself up from HMS Prince William - When taken: 13 Oct 1812 - Date Received: 8 Aug 1814 - From what ship: HMT Raven, Chatham - Born: Philadelphia - Age: 30 - Released on 26 Apr 1815.

Johnson, Joseph Toker - Seaman - Nbr: 1347 - Prize: Paul Jones - Ship Type: P - How taken: HMS Leonidas - When taken: 23 May 1813 - Where taken: Channel - Date Received: 19 Jun 1814 - From what ship: Stapleton - Born: Connecticut - Age: 19 - Died on 6 Apr 1815 from gunshot wound in prison.

Johnson, Luke - Seaman - Nbr: 4852 - Prize: North Star - Ship Type: MV - How taken: HMS Crane - When taken: 10 Jun 1814 - Where taken: off St. Bartholomew - Date Received: 9 Oct 1814 - From what ship: HMT Freya, Halifax - Born: Delaware - Age: 42 - Race: Black - Released on 21 Jun 1815.

Johnson, Mark - Gunner - Nbr: 5749 - Prize: US Schooner Ohio - Ship Type: MW - How taken: British gunboats - When taken: 12 Aug 1814 - Where taken: Fort Erie - Date Received: 26 Dec 1814 - From what ship: HMT Argo - Born: Albany - Age: 29 - Released on 3 Jul 1815.

Johnson, Mathew - Seaman - Nbr: 2820 - How taken: Gave himself up from HMS Philomel - When taken: 28 Dec 1812 - Date Received: 24 Aug 1814 - From what ship: HMT Liverpool, Chatham - Born: Charleston - Age: 35 - Released on 26 Apr 1815.

Johnson, Oliver - Seaman - Nbr: 2613 - How taken: Gave himself up from HMS Romulus - When taken: 14 Aug 1812 - Date Received: 21 Aug 1814 - From what ship: HMT Freya, Chatham - Born: Connecticut - Age: 32 - Released on 26 Apr 1815.

Johnson, Peter - Seaman - Nbr: 3611 - Prize: Hornsby - Ship Type: P - How taken: HMS Sceptre - When taken: 4 Feb 1814 - Where taken: off Block Island - Date Received: 30 Sep 1814 - From what ship: HMT Sybella - Born: Dartmouth - Age: 27 - Released on 4 Jun 1815.

Johnson, Peter - Seaman - Nbr: 4878 - Prize: Saratoga - Ship Type: P - How taken: HMS Barracouta - When taken: 9 Oct 1814 - Where taken: off Western Islands (England) - Date Received: 24 Oct 1814 - From what ship: HMT Salvador del Mundo - Born: Boston - Age: 20 - Released on 21 Jun 1815.

Johnson, Richard - Seaman - Nbr: 3339 - Prize: Fame (Whaler) - Ship Type: MV - How taken: HMS Cressy - When taken: 20 Jul 1813 - Where taken: at sea - Date Received: 13 Sep 1814 - From what ship: HMT Niobe, Chatham - Born: New York - Age: 42 - Race: Mulatto - Released on 28 May 1815.

Johnson, Richard - Seaman - Nbr: 2938 - How taken: Gave himself up from HMS Ocean - Date Received: 24 Aug 1814 - From what ship: HMT Alpheus, Chatham - Born: Norfolk - Age: 24 - Race: Black - Released on 19 May 1815.

Johnson, Robert - Seaman - Nbr: 4433 - Prize: Taken off an English whaler - Ship Type: MV - How taken: HMS Illustrious - When taken: 22 Oct 1813 - Where taken: at sea - Date Received: 8 Oct 1814 - From what ship: HMT Leyden, Chatham - Born: Baltimore - Age: 24 - Released on 14 Jun 1815.

Johnson, Samuel - Seaman - Nbr: 550 - How taken: Gave himself up from HMS Scorpion - When taken: 1 Dec 1812 - Date Received: 8 Sep 1813 - From what ship: Plymouth - Born: Staten Island - Age: 28 - Race: Negro - Released on 26 Apr 1815.

Johnson, Samuel - Seaman - Nbr: 343 - Prize: Good Friends - Ship Type: MV - How taken: HMS Andromache - When taken: 2 Apr 1813 - Where taken: Bay of Biscay - Date Received: 28 Jun 1813 - From what ship: Plymouth - Born: Suffolk - Age: 26 - Released on 12 Apr 1815.

Johnson, Samuel - Seaman - Nbr: 3065 - How taken: Gave himself up from HMS Eridanus - Date Received: 2 Sep 1814 - From what ship: HMT Hydra, Chatham - Born: Salem - Age: 39 - Released on 28 May 1815.

American Prisoners of War Held at Dartmoor during the War of 1812

## Alphabetical listing of names

Johnson, Samuel - Seaman - Nbr: 2847 - How taken: Gave himself up from HMS Vigo (Spain) - When taken: 31 Dec 1812 - Date Received: 24 Aug 1814 - From what ship: HMT Alpheus, Chatham - Born: Providence - Age: 59 - Released on 27 Apr 1815.

Johnson, Samuel - Seaman - Nbr: 5103 - Prize: David Porter - Ship Type: LM - How taken: HMS Pylades - When taken: 12 Sep 1814 - Where taken: Georges Bank - Date Received: 28 Oct 1814 - From what ship: HMT Alkbar, Halifax - Born: Salem - Age: 33 - Released on 29 Jun 1815.

Johnson, Samuel - Seaman - Nbr: 3502 - How taken: Impressed at Liverpool - When taken: 9 Sep 1814 - Date Received: 24 Sep 1814 - From what ship: HMT Salvador del Mundo - Born: New York - Age: 27 - Race: Negro - Released on 11 Jul 1815.

Johnson, Stephen - Seaman - Nbr: 5013 - Prize: Charlotte - Ship Type: Prize - How taken: Censor - When taken: 5 Aug 1814 - Where taken: off Azores - Date Received: 28 Oct 1814 - From what ship: HMT Alkbar, Halifax - Born: New York - Age: 21 - Race: Negro - Released on 21 Jun 1815.

Johnson, Thomas - Seaman - Nbr: 6069 - Prize: Emperor Napoleon - Ship Type: MV - How taken: HMS Loire - When taken: 26 Sep 1814 - Where taken: off Delaware - Date Received: 28 Dec 1814 - From what ship: HMT Penelope - Born: New London - Age: 18 - Released on 3 Jul 1815.

Johnson, Thomas - Seaman - Nbr: 6424 - Prize: Farmer's Daughter - Ship Type: MV - How taken: HMS Leviathan - When taken: 29 May 1814 - Date Received: 3 Mar 1815 - From what ship: HMT Ganges, Plymouth - Born: Taunton - Age: 25 - Released on 11 Jul 1815.

Johnson, Thomas - Seaman - Nbr: 2097 - How taken: Gave himself up from HMS Malta - When taken: 1 Jan 1813 - Date Received: 3 Aug 1814 - From what ship: HMS Bittern, Chatham Depot - Born: Baltimore - Age: 27 - Released on 2 May 1815.

Johnson, Thomas - Boy - Nbr: 2338 - Prize: Snap Dragon - Ship Type: P - How taken: HMS Martin - When taken: 10 Jun 1814 - Where taken: off Halifax - Date Received: 16 Aug 1814 - From what ship: HMS Dublin, Halifax - Born: Virginia - Age: 17 - Released on 3 May 1815.

Johnson, Thomas - Seaman - Nbr: 3904 - Prize: Henry Guilder - Ship Type: MV - How taken: HMS Niemen - When taken: 14 Jul 1814 - Where taken: at sea - Date Received: 5 Oct 1814 - From what ship: HMT Orpheus, Halifax - Born: Albany - Age: 21 - Race: Negro - Released on 9 Jun 1815.

Johnson, Thomas - Seaman - Nbr: 2993 - Prize: Frolic - Ship Type: P - How taken: HMS Heron - When taken: 25 Jan 1814 - Where taken: off St. Thomas - Date Received: 29 Aug 1814 - From what ship: HMT Bittern - Born: Martinsburg - Age: 19 - Released on 28 May 1815.

Johnson, Thomas - Seaman - Nbr: 3406 - Prize: Thomas - Ship Type: P - How taken: HMS Nymphe - When taken: 24 Jun 1813 - Where taken: off Halifax - Date Received: 13 Sep 1814 - From what ship: HMT Niobe, Chatham - Born: New York - Age: 38 - Released on 28 May 1815.

Johnson, William - Seaman - Nbr: 5059 - Prize: Ida - Ship Type: LM - How taken: HMS Newcastle - When taken: 9 Aug 1814 - Where taken: Long 34 - Date Received: 28 Oct 1814 - From what ship: HMT Alkbar, Halifax - Born: New York - Age: 28 - Released on 29 Jun 1815.

Johnson, William - Seaman - Nbr: 6468 - Prize: Decatur - Ship Type: P - How taken: HMS Rhin - When taken: 5 Jun 1814 - Where taken: off San Domingo - Date Received: 3 Mar 1815 - From what ship: HMT Ganges, Plymouth - Born: New York - Age: 23 - Race: Black - Released on 11 Jul 1815.

Johnson, William - Seaman - Nbr: 1920 - How taken: Gave himself up from HMS Antelope - When taken: 3 Feb 1813 - Date Received: 3 Aug 1814 - From what ship: HMS Alceste, Chatham Depot - Born: Philadelphia - Age: 29 - Died on 9 Mar 1815 from variola.

Johnson, William - Seaman - Nbr: 670 - Prize: Joel Barlow - Ship Type: LM - How taken: HMS Briton - When taken: 3 Jul 1813 - Where taken: off Bordeaux - Date Received: 8 Sep 1813 - From what ship: Plymouth - Born: Connecticut - Age: 23 - Released on 26 Apr 1815.

Johnson, William - Seaman - Nbr: 159 - How taken: Gave himself up - Date Received: 2 Apr 1813 - From what ship: Plymouth - Born: New York - Age: 26 - Released on 20 Apr 1815.

Johnson, William - Seaman - Nbr: 3574 - Prize: William - Ship Type: MV - How taken: HMS Hasty - When taken: 14 May 1814 - Where taken: off Charleston - Date Received: 30 Sep 1814 - From what ship: HMT Sybella - Born: Charleston - Age: 25 - Race: Negro - Died on 2 Nov 1814.

Johnson, William - Seaman - Nbr: 3487 - How taken: Gave himself up from HMS Prince - When taken: 14 Sep 1814 - Date Received: 19 Sep 1814 - From what ship: HMT Salvador del Mundo - Born: Virginia - Age: 45 - Released on 4 Jun 1815.

Johnston, Charles - Seaman - Nbr: 4269 - Prize: York Town - Ship Type: P - How taken: HMS Maidstone - When taken: 18 Jul 1813 - Where taken: Grand Banks - Date Received: 7 Oct 1814 - From what ship: HMT Niobe,

American Prisoners of War Held at Dartmoor during the War of 1812

## Alphabetical listing of names

Chatham - Born: Gothenburg - Age: 32 - Released on 13 Jun 1815.

Johnston, Edward - Seaman - Nbr: 81 - Prize: Rolla - Ship Type: MV - How taken: HMS Surveillante - When taken: 11 Feb 1813 - Where taken: Bay of Biscay - Date Received: 2 Apr 1813 - From what ship: Plymouth - Born: Darlington, England - Age: 68 - Released on 20 Apr 1815.

Johnston, Edward - Seaman - Nbr: 1252 - Prize: Adeline - Ship Type: MV - How taken: HMS Magicienne - When taken: 16 Mar 1814 - Where taken: off Cape Finisterre - Date Received: 14 Jun 1814 - From what ship: Mill Prison (Plymouth, England) - Born: Virginia - Age: 33 - Released on 28 Apr 1815.

Johnston, Michael - Seaman - Nbr: 3805 - Prize: Stark - Ship Type: P - How taken: HMS Sophie - When taken: 20 Apr 1814 - Where taken: off Bermuda - Date Received: 5 Oct 1814 - From what ship: HMT Orpheus, Halifax - Born: Salem - Age: 25 - Released on 9 Jun 1815.

Johnston, Perry - Seaman - Nbr: 6240 - Prize: Albion - Ship Type: Prize - How taken: HMS Harlequin - When taken: 6 Jan 1814 - Where taken: Long 20 - Date Received: 4 Feb 1815 - From what ship: HMT Ganges - Born: Maryland - Age: 26 - Race: Black - Released on 5 Jul 1815.

Johnston, Thomas - Mate - Nbr: 5142 - Prize: Revenge - Ship Type: MV - How taken: HMS Monkey - When taken: 20 May 1813 - Where taken: Bay of Biscay - Date Received: 31 Oct 1814 - From what ship: HMT Mermaid, Chatham - Born: Wilmington - Age: 25 - Escaped on 4 Apr 1815.

Johnston, William - Seaman - Nbr: 5544 - Prize: Leander - Ship Type: MV - How taken: HMS Rifleman - When taken: 16 Jan 1813 - Where taken: off America - Date Received: 17 Dec 1814 - From what ship: HMT Loire, Halifax - Born: Boston - Age: 26 - Released on 1 Jul 1815.

Johnstone, Gresham - Seaman - Nbr: 186 - Prize: Star - Ship Type: MV - How taken: HMS Superb - When taken: 9 Feb 1813 - Where taken: Bay of Biscay - Date Received: 2 Apr 1813 - From what ship: Plymouth - Born: New York - Age: 26 - Released on 20 Apr 1815.

Johnstone, John - Seaman - Nbr: 448 - Prize: Magdalen - Ship Type: MV - How taken: HMS Superb - When taken: 15 Apr 1813 - Where taken: off Belle Isle, France - Date Received: 1 Jul 1813 - From what ship: Plymouth - Born: Norfolk - Age: 36 - Race: Black - Released on 20 Apr 1815.

Johnstone, Peter - Seaman - Nbr: 449 - Prize: Magdalen - Ship Type: MV - How taken: HMS Superb - When taken: 15 Apr 1813 - Where taken: off Belle Isle, France - Date Received: 1 Jul 1813 - From what ship: Plymouth - Born: New Orleans - Age: 32 - Sent to Mill Prison (Plymouth, England) on 21 Jun 1814.

Johnstone, Thomas - Seaman - Nbr: 1779 - Prize: Criterion - Ship Type: MV - How taken: HMS Belle Poule - When taken: 14 Feb 1813 - Where taken: Bay of Biscay - Date Received: 20 Jul 1814 - From what ship: HMS Milford, Plymouth - Born: Albany - Age: 22 - Released on 1 May 1815.

Jonathan, Jonathan - Seaman - Nbr: 4749 - How taken: Gave himself up from HMS Minden - When taken: 4 Oct 1813 - Date Received: 9 Oct 1814 - From what ship: HMT Freya, Chatham - Born: Baltimore - Age: 26 - Released on 15 Jun 1815.

Jones, Abraham - Seaman - Nbr: 6157 - Prize: Lion - Ship Type: P - How taken: HMS Granicus - When taken: 2 Dec 1814 - Where taken: off Lisbon - Date Received: 18 Jan 1815 - From what ship: HMT Impregnable - Born: New Jersey - Age: 31 - Released on 5 Jul 1815.

Jones, Anthony - Seaman - Nbr: 4268 - How taken: Gave himself up from HMS Hope - When taken: 3 Oct 1813 - Date Received: 7 Oct 1814 - From what ship: HMT Niobe, Chatham - Born: Orleans - Age: 32 - Released on 13 Jun 1815.

Jones, Benjamin - Cook - Nbr: 211 - Prize: Mars - Ship Type: MV - How taken: HMS Warspite - When taken: 26 Feb 1813 - Where taken: off Basque Roads (France) - Date Received: 2 Apr 1813 - From what ship: Plymouth - Born: Washington - Age: 25 - Released on 20 Apr 1815.

Jones, Benjamin - Seaman - Nbr: 1825 - Prize: William Bayard - Ship Type: MV - How taken: HMS Warspite - When taken: 3 Mar 1813 - Where taken: Bay of Biscay - Date Received: 21 Jul 1814 - From what ship: HMT Redbeard & Pincher, Chatham Depot - Born: Milford - Age: 27 - Released on 1 May 1815.

Jones, Benjamin - Seaman - Nbr: 248 - Prize: William Bayard - Ship Type: MV - How taken: HMS Warspite - When taken: 3 Mar 1813 - Where taken: Bay of Biscay - Date Received: 2 Apr 1813 - From what ship: Plymouth - Born: Philadelphia - Age: 27 - Released on 20 Apr 1815.

Jones, Calvin - Seaman - Nbr: 6410 - Prize: Nancy - Ship Type: P - How taken: Papillion - When taken: 27 Dec 1814 - Where taken: Lat 29 Long 20 - Date Received: 3 Mar 1815 - From what ship: HMT Ganges, Plymouth - Born: Allentown - Age: 18 - Released on 11 Jul 1815.

Jones, Charles - Seaman - Nbr: 584 - How taken: Impressed at Dublin - When taken: Jun 1813 - Date Received: 8 Sep 1813 - From what ship: Plymouth - Born: Connecticut - Age: 19 - Released on 26 Apr 1815.

Jones, Charles - Seaman - Nbr: 476 - Prize: Zebra - Ship Type: LM - How taken: HMS Pyramus - When taken: 20

American Prisoners of War Held at Dartmoor during the War of 1812

## Alphabetical listing of names

Apr 1813 - Where taken: Bay of Biscay - Date Received: 8 Sep 1813 - From what ship: Plymouth - Born: New York - Age: 22 - Released on 20 Apr 1815.

Jones, Charles - Seaman - Nbr: 3644 - Prize: Rising States - Ship Type: MV - How taken: HMS Niemen - When taken: 17 Dec 1813 - Where taken: off Cape Sable Island (Canada) - Date Received: 30 Sep 1814 - From what ship: HMT Sybella - Born: New York - Age: 26 - Race: Negro - Released on 4 Jun 1815.

Jones, David - Seaman - Nbr: 3577 - Prize: William - Ship Type: MV - How taken: HMS Hasty - When taken: 14 May 1814 - Where taken: off Charleston - Date Received: 30 Sep 1814 - From what ship: HMT Sybella - Born: New York - Age: 20 - Released on 4 Jun 1815.

Jones, Edward - Seaman - Nbr: 3945 - Prize: Rolla - Ship Type: P - How taken: HMS Loire - When taken: 10 Dec 1813 - Where taken: off Bull Island (South Carolina) - Date Received: 5 Oct 1814 - From what ship: HMT President, Halifax - Born: Charleston - Age: 27 - Released on 9 Jun 1815.

Jones, Edward - Marine - Nbr: 6319 - Prize: Prince de Neufchatel - Ship Type: P - How taken: Leander (Newcastle Acasta) - When taken: 20 Dec 1814 - Where taken: Lat 38 Long 56 - Date Received: 19 Feb 1815 - From what ship: HMT Ganges, Plymouth - Born: Roxborough - Age: 18 - Released on 5 Jul 1815.

Jones, Edward - Seaman - Nbr: 5523 - Prize: James - Ship Type: Prize - How taken: HMS Galatea - When taken: 7 Sep 1814 - Where taken: Channel - Date Received: 17 Dec 1814 - From what ship: HMT Loire, Halifax - Born: Philadelphia - Age: 18 - Released on 1 Jul 1815.

Jones, Ezekiel - Seaman - Nbr: 3534 - Prize: Rover - Ship Type: Prize - How taken: HMS Conquistador - When taken: 22 Aug 1814 - Where taken: Long 19 Lat 107 - Date Received: 28 Sep 1814 - From what ship: HMT Salvador del Mundo - Born: Boston - Age: 30 - Race: Black - Released on 4 Jun 1815.

Jones, F. V. - Seaman - Nbr: 59 - Prize: Spitfire - Ship Type: MV - How taken: HMS Achates - When taken: 14 Feb 1813 - Where taken: off Ushant (France) - Date Received: 2 Apr 1813 - From what ship: Plymouth - Born: Marblehead - Age: 21 - Released on 20 Apr 1815.

Jones, Francis - Seaman - Nbr: 58 - Prize: Spitfire - Ship Type: MV - How taken: HMS Achates - When taken: 14 Feb 1813 - Where taken: off Ushant (France) - Date Received: 2 Apr 1813 - From what ship: Plymouth - Born: Marblehead - Age: 38 - Released on 20 Apr 1815.

Jones, George - Seaman - Nbr: 430 - Prize: Viper - Ship Type: MV - How taken: HMS Superb - When taken: 15 Apr 1813 - Where taken: Bay of Biscay - Date Received: 1 Jul 1813 - From what ship: Plymouth - Born: New Orleans - Age: 24 - Race: Black - Died on 30 Apr 1814 from phthisis.

Jones, Henry - Seaman - Nbr: 6492 - Prize: Louisa - Ship Type: Prize - How taken: HMS Dasher - When taken: 12 Feb 1814 - Where taken: off America - Date Received: 3 Mar 1815 - From what ship: HMT Ganges, Plymouth - Born: Boston - Age: 19 - Released on 11 Jul 1815.

Jones, Henry - Seaman - Nbr: 5590 - Prize: Young William - Ship Type: Prize - How taken: HMS Plover - When taken: 8 Sep 1814 - Where taken: off St. Johns - Date Received: 17 Dec 1814 - From what ship: HMT Loire, Halifax - Born: Maryland - Age: 25 - Race: Colored - Released on 1 Jul 1815.

Jones, Isaac - Seaman - Nbr: 4556 - How taken: Gave himself up from HMS Hussar - When taken: 6 Jul 1813 - Date Received: 8 Oct 1814 - From what ship: HMT Leyden, Chatham - Born: Boston - Age: 22 - Died on 23 Jan 1815 from perissneuoniria.

Jones, James - Seaman - Nbr: 6400 - Prize: Nellenville - Ship Type: MV - How taken: HMS Onyx - When taken: 25 Dec 1814 - Where taken: off San Domingo - Date Received: 3 Mar 1815 - From what ship: HMT Ganges, Plymouth - Born: Baltimore - Age: 28 - Released on 11 Jul 1815.

Jones, James - Seaman - Nbr: 4718 - How taken: Gave himself up from HMS Hussar - When taken: 5 Jul 1813 - Date Received: 9 Oct 1814 - From what ship: HMT Freya, Chatham - Born: New York - Age: 27 - Died on 27 May 1815 from variola.

Jones, James - Boy - Nbr: 2513 - Prize: Snap Dragon - Ship Type: P - How taken: HMS Martin - When taken: 10 Jun 1814 - Where taken: off Halifax - Date Received: 16 Aug 1814 - From what ship: HMT Queen, Halifax - Born: New Berne - Age: 16 - Released on 3 May 1815.

Jones, James - Seaman - Nbr: 3408 - Prize: Thomas - Ship Type: P - How taken: HMS Nymphe - When taken: 24 Jun 1813 - Where taken: off Halifax - Date Received: 13 Sep 1814 - From what ship: HMT Niobe, Chatham - Born: Massachusetts - Age: 32 - Race: Black - Released on 28 May 1815.

Jones, Jeremiah - Carpenter - Nbr: 4012 - Prize: US Brig Rattlesnake - Ship Type: MW - How taken: HMS Leander - When taken: 13 Jul 1814 - Where taken: off Shelburne - Date Received: 6 Oct 1814 - From what ship: HMT Chesapeake, Halifax - Born: Boston - Age: 39 - Released on 9 Jun 1815.

Jones, John - Seaman - Nbr: 4847 - Prize: Hawke - Ship Type: P - How taken: HMS Pique - When taken: 26 Apr 1814 - Where taken: off Porto Rico - Date Received: 9 Oct 1814 - From what ship: HMT Freya, Halifax -

American Prisoners of War Held at Dartmoor during the War of 1812

## Alphabetical listing of names

Born: North Carolina - Age: 21 - Released on 14 Jun 1815.
Jones, John - Seaman - Nbr: 6409 - Prize: Nancy - Ship Type: P - How taken: Papillion - When taken: 27 Dec 1814 - Where taken: Lat 29 Long 20 - Date Received: 3 Mar 1815 - From what ship: HMT Ganges, Plymouth - Born: Portsmouth - Age: 26 - Released on 11 Jul 1815.
Jones, John - Seaman - Nbr: 6043 - Prize: Tickler - Ship Type: MV - How taken: HMS Niemen - When taken: 14 Sep 1814 - Where taken: off Delaware - Date Received: 28 Dec 1814 - From what ship: HMT Penelope - Born: Norfolk - Age: 32 - Race: Black - Released on 3 Jul 1815.
Jones, John - Seaman - Nbr: 269 - Prize: William Bayard - Ship Type: MV - How taken: HMS Warspite - When taken: 3 Mar 1813 - Where taken: Bay of Biscay - Date Received: 28 Jun 1813 - From what ship: Plymouth - Born: Newcastle - Age: 54 - Released on 20 Apr 1815.
Jones, John - Seaman - Nbr: 4758 - How taken: Gave himself up from HMS Venus - When taken: 4 Sep 1813 - Date Received: 9 Oct 1814 - From what ship: HMT Freya, Chatham - Born: New York - Age: 32 - Released on 15 Jun 1815.
Jones, John - Seaman - Nbr: 3493 - How taken: Gave himself up from HMS Dublin - When taken: 25 Aug 1814 - Date Received: 19 Sep 1814 - From what ship: HMT Salvador del Mundo - Born: Connecticut - Age: 27 - Released on 4 Jun 1815.
Jones, John - 2nd Officer - Nbr: 5042 - Prize: Ida - Ship Type: LM - How taken: HMS Newcastle - When taken: 9 Aug 1814 - Where taken: Long 34 - Date Received: 28 Oct 1814 - From what ship: HMT Alkbar, Halifax - Born: Massachusetts - Age: 33 - Released on 29 Jun 1815.
Jones, Kenny - Seaman - Nbr: 1501 - Prize: Margaret - Ship Type: Prize - How taken: HMS Foxhound - When taken: 27 May 1814 - Where taken: off Isles of Scilly - Date Received: 20 Jun 1814 - From what ship: Mill Prison (Plymouth, England) - Born: Clement Parish - Age: 20 - Released on 1 May 1815.
Jones, Lawrence - Seaman - Nbr: 3838 - Prize: Dominique - Ship Type: LM - How taken: HMS Dotterel - When taken: 21 May 1814 - Where taken: off Charleston - Date Received: 5 Oct 1814 - From what ship: HMT Orpheus, Halifax - Born: Port au Prince - Age: 21 - Released on 9 Jun 1815.
Jones, Lewis - Seaman - Nbr: 4595 - Prize: Caroline - Ship Type: MV - How taken: HMS Moselle - When taken: 12 Aug 1813 - Where taken: off Charleston - Date Received: 9 Oct 1814 - From what ship: HMT Leyden, Chatham - Born: Maryland - Age: 25 - Released on 15 Jun 1815.
Jones, Paul - Seaman - Nbr: 3538 - Prize: Hawk - Ship Type: P - How taken: HMS Pique - When taken: 26 Apr 1814 - Where taken: off Bermuda - Date Received: 30 Sep 1814 - From what ship: HMT Sybella - Born: Lisbon - Age: 48 - Race: Mulatto - Released on 4 Jun 1815.
Jones, Peter - Seaman - Nbr: 1801 - How taken: Gave himself up from HMS Magnificent - When taken: 2 May 1813 - Date Received: 20 Jul 1814 - From what ship: HMS Milford, Plymouth - Born: Maryland - Age: 24 - Race: Black - Released on 1 May 1815.
Jones, Reuben - Boy - Nbr: 2558 - Prize: Caromaned - Ship Type: Prize - How taken: HMS Eridanus - When taken: 13 Aug 1814 - Where taken: Lat 40, Long 16 - Date Received: 16 Aug 1814 - From what ship: HMT Salvador del Mundo - Born: Norway - Age: 12 - Released on 3 May 1815.
Jones, Richard - Seaman - Nbr: 5448 - Prize: Prize to the Lawrence - Ship Type: P - How taken: HMS Glasgow - When taken: 2 Nov 1814 - Where taken: Channel - Date Received: 10 Dec 1814 - From what ship: HMT Impregnable - Born: Hartford - Age: 54 - Released on 1 Jul 1815.
Jones, Richard - Seaman - Nbr: 680 - Prize: Lightning - Ship Type: MV - How taken: HMS Medusa - When taken: 2 Apr 1813 - Where taken: Bay of Biscay - Date Received: 27 Sep 1813 - From what ship: Plymouth - Born: Baltimore - Age: 25 - Released on 26 Apr 1815.
Jones, Samuel - Seaman - Nbr: 5851 - Prize: Hornet - Ship Type: MV - How taken: HMS Surprize - When taken: 19 Aug 1814 - Where taken: Lat 35 Long 24 - Date Received: 26 Dec 1814 - From what ship: HMT Argo - Born: Wilmington - Age: 21 - Released on 3 Jul 1815.
Jones, Samuel B. - Seaman - Nbr: 4789 - Prize: Valentine - Ship Type: MV - How taken: HMS Minden - When taken: 16 Nov 1812 - Where taken: Malager Bay - Date Received: 9 Oct 1814 - From what ship: HMT Freya, Chatham - Born: Providence - Age: 22 - Released on 27 Jul 1815.
Jones, Saul - Seaman - Nbr: 3473 - How taken: Gave himself up from HMS Boyne - When taken: 14 Sep 1814 - Date Received: 19 Sep 1814 - From what ship: HMT Salvador del Mundo - Born: Sandwich - Age: 38 - Released on 4 Jun 1815.
Jones, Stephen - Seaman - Nbr: 4416 - Prize: Fire Fly - Ship Type: LM - How taken: HMS Revolutionnaire - When taken: 19 Oct 1813 - Where taken: at sea - Date Received: 8 Oct 1814 - From what ship: HMT Leyden, Chatham - Born: Massachusetts - Age: 18 - Released on 14 Jun 1815.

American Prisoners of War Held at Dartmoor during the War of 1812

## Alphabetical listing of names

Jones, Stephen - Seaman - Nbr: 4707 - Prize: Volunteer - Ship Type: MV - How taken: Victoria (Privateer) - When taken: 26 Dec 1813 - Where taken: at sea - Date Received: 9 Oct 1814 - From what ship: HMT Leyden, Chatham - Born: New York - Age: 27 - Died on 4 Nov 1814.

Jones, Theodore - Seaman - Nbr: 4158 - How taken: Gave himself up from HMS Bulwark - When taken: 28 Dec 1813 - Date Received: 6 Oct 1814 - From what ship: HMT Niobe, Chatham - Born: Maryland - Age: 26 - Released on 13 Jun 1815.

Jones, Thomas - Seaman - Nbr: 4780 - How taken: Gave himself up from HMS Blenheim - When taken: 2 Sep 1814 - Date Received: 9 Oct 1814 - From what ship: HMT Freya, Chatham - Born: Boston - Age: 26 - Released on 1 May 1815.

Jones, Thomas - Seaman - Nbr: 1494 - How taken: Impressed at Liverpool - Date Received: 19 Jun 1814 - From what ship: Stapleton - Born: Baltimore - Age: 22 - Released on 1 May 1815.

Jones, Thomas - Seaman - Nbr: 913 - Prize: Zephyr - Ship Type: MV - How taken: HMS Pyramus - When taken: 30 Nov 1813 - Where taken: off L'Orient (France) - Date Received: 31 Jan 1814 - From what ship: Plymouth - Born: Philadelphia - Age: 48 - Released on 27 Apr 1815.

Jones, Thomas - Seaman - Nbr: 1474 - Prize: Governor Gerry - Ship Type: MV - How taken: HMS Lyra - When taken: 29 May 1813 - Where taken: Bay of Biscay - Date Received: 19 Jun 1814 - From what ship: Stapleton - Born: New York - Age: 28 - Released on 28 Apr 1815.

Jones, Thomas - Seaman - Nbr: 4812 - Prize: President - Ship Type: P - How taken: HMS Pique - When taken: 7 May 1814 - Where taken: off Porto Rico - Date Received: 9 Oct 1814 - From what ship: HMT Freya, Halifax - Born: St. Martins - Age: 29 - Released on 21 Jun 1815.

Jones, Thomas - Cook - Nbr: 3434 - Prize: Growler - Ship Type: P - How taken: HMS Electra - When taken: 7 Jul 1813 - Where taken: at sea - Date Received: 13 Sep 1814 - From what ship: HMT Niobe, Chatham - Born: Baltimore - Age: 38 - Race: Black - Died on 23 Feb 1815 from phthisis pulmonalis.

Jones, Thomas - Seaman - Nbr: 4688 - How taken: Gave himself up from HMS Buzzard - When taken: 14 Jan 1813 - Date Received: 9 Oct 1814 - From what ship: HMT Leyden, Chatham - Born: Baltimore - Age: 25 - Released on 15 Jun 1815.

Jones, Walter - Seaman - Nbr: 797 - Prize: Collin - Ship Type: MV - How taken: HMS Helicon and HMS Whiting - When taken: 26 Oct 1813 - Where taken: Channel - Date Received: 3 Nov 1813 - From what ship: Plymouth - Born: Litchfield - Age: 28 - Released on 26 Apr 1815.

Jones, William - Seaman - Nbr: 2915 - How taken: Impressed in Brazil - When taken: 16 May 1813 - Date Received: 24 Aug 1814 - From what ship: HMT Alpheus, Chatham - Born: New York - Age: 45 - Released on 28 May 1815.

Jones, William - Boy - Nbr: 5135 - Prize: Calabria - Ship Type: MV - How taken: HMS Castilian - When taken: 29 Sep 1814 - Where taken: off Ireland - Date Received: 31 Oct 1814 - From what ship: HMT Castillian - Born: Baltimore - Age: 15 - Released on 29 Jun 1815.

Jones, William - Boatswain's Mate - Nbr: 4151 - Prize: Argus - Ship Type: MV - How taken: HMS San Domingo - When taken: 1 Mar 1814 - Where taken: off Savannah - Date Received: 6 Oct 1814 - From what ship: HMT Niobe, Chatham - Born: Philadelphia - Age: 27 - Released on 13 Jun 1815.

Jones, William - Seaman - Nbr: 2307 - Prize: Hussar - Ship Type: P - How taken: HMS Saturn - When taken: 25 May 1814 - Where taken: off Sandy Hook - Date Received: 16 Aug 1814 - From what ship: HMS Dublin, Halifax - Born: New Jersey - Age: 21 - Released on 3 May 1815.

Jones, William P. - Prize Master - Nbr: 93 - Prize: St. Martin's Planter - Ship Type: P - How taken: HMS Dublin - When taken: 9 Feb 1813 - Where taken: Lat 43 N, Long 33 50 W - Date Received: 2 Apr 1813 - From what ship: Plymouth - Born: Haverhill - Age: 35 - Sent to Dartmouth on 2 Aug 1813.

Jones, William P. - Seaman - Nbr: 280 - Prize: Cannoneer - Ship Type: MV - How taken: HMS Warspite - When taken: 14 Mar 1813 - Where taken: Bay of Biscay - Date Received: 28 Jun 1813 - From what ship: Plymouth - Born: Philadelphia - Age: 18 - Released on 20 Apr 1815.

Jonson, Edward - Seaman - Nbr: 1164 - How taken: Gave himself up from HMS Hebrius - Date Received: 10 May 1814 - From what ship: Plymouth - Born: Philadelphia - Age: 20 - Race: Black - Released on 28 Apr 1815.

Jordan, Artemus - Seaman - Nbr: 3351 - How taken: Gave himself up from HMS Gordon - When taken: 1 Nov 1812 - Date Received: 13 Sep 1814 - From what ship: HMT Niobe, Chatham - Born: Plymouth - Age: 28 - Released on 27 Apr 1815.

Jordan, Christopher - Carpenter - Nbr: 2372 - Prize: US Sloop Frolic - Ship Type: MW - How taken: HMS Orpheus - When taken: 20 Apr 1814 - Where taken: off Cuba - Date Received: 16 Aug 1814 - From what ship: HMT Queen, Halifax - Born: Massachusetts - Age: 33 - Sent to Dartmouth on 19 Oct 1814.

American Prisoners of War Held at Dartmoor during the War of 1812

## Alphabetical listing of names

Jordan, Peter - Seaman - Nbr: 1827 - Prize: Lightning - Ship Type: MV - How taken: HMS Medusa - When taken: 2 Apr 1813 - Where taken: Bay of Biscay - Date Received: 21 Jul 1814 - From what ship: HMT Redbeard & Pincher, Chatham Depot - Born: Massina - Age: 31 - Released on 1 May 1815.

Jordan, Richard - Seaman - Nbr: 545 - Prize: Orders in Council - Ship Type: MV - How taken: HMS Surveillante - When taken: 1 Jun 1813 - Where taken: off Cape Ortegal (Spain) - Date Received: 8 Sep 1813 - From what ship: Plymouth - Born: New York - Age: 17 - Released on 26 Apr 1815.

Jordan, Samuel - Seaman - Nbr: 5950 - Prize: Harlequin - Ship Type: P - How taken: HMS Bulwark - When taken: 23 Nov 1814 - Where taken: off Halifax - Date Received: 27 Dec 1814 - From what ship: HMT Penelope - Born: Sacho - Age: 25 - Released on 20 Mar 1815.

Jordan, Simon - 3rd Lieutenant - Nbr: 3559 - Prize: Hawk - Ship Type: P - How taken: HMS Pique - When taken: 26 Apr 1814 - Where taken: off Bermuda - Date Received: 30 Sep 1814 - From what ship: HMT Sybella - Born: Massachusetts - Age: 30 - Released on 4 Jun 1815.

Joscelyn, Ambrose - Seaman - Nbr: 5597 - Prize: Mentor - Ship Type: Prize - How taken: HMS Maidstone - When taken: 26 Oct 1814 - Where taken: off Cape Sable Island (Canada) - Date Received: 17 Dec 1814 - From what ship: HMT Loire, Halifax - Born: Massachusetts - Age: 23 - Released on 1 Jul 1815.

Jose, Emanuel - Seaman - Nbr: 5095 - Prize: David Porter - Ship Type: LM - How taken: HMS Pylades - When taken: 12 Sep 1814 - Where taken: Georges Bank - Date Received: 28 Oct 1814 - From what ship: HMT Alkbar, Halifax - Born: Portugal - Age: 20 - Died on 25 Nov 1814.

Joseph, Fois - Seaman - Nbr: 197 - Prize: Star - Ship Type: MV - How taken: HMS Superb - When taken: 9 Feb 1813 - Where taken: Bay of Biscay - Date Received: 2 Apr 1813 - From what ship: Plymouth - Born: New Orleans - Age: 20 - Sent to Mill Prison (Plymouth, England) on 10 Jul 1813.

Joseph, Francis - 2nd Lieutenant - Nbr: 1150 - Prize: Bunker Hill - Ship Type: P - How taken: HMS Pomone - When taken: 8 Mar 1814 - Where taken: at sea - Date Received: 10 May 1814 - From what ship: Plymouth - Born: Salem - Age: 38 - Released on 27 Apr 1815.

Joseph, John - Stevedore - Nbr: 3854 - Prize: Dominique - Ship Type: LM - How taken: HMS Dotterel - When taken: 21 May 1814 - Where taken: off Charleston - Date Received: 5 Oct 1814 - From what ship: HMT Orpheus, Halifax - Born: Marseille - Age: 23 - Released on 9 Jun 1815.

Joseph, John - Seaman - Nbr: 3461 - Prize: Prize to Chasseur - How taken: HMS Tartarus - When taken: 1 Sep 1814 - Where taken: Atlantic - Date Received: 19 Sep 1814 - From what ship: HMT Salvador del Mundo - Born: Vigo (Spain) - Age: 25 - Released on 4 Jun 1815.

Joseph, Michael - Seaman - Nbr: 4325 - Prize: Fiere Facia - Ship Type: P - How taken: HMS Ramillies - When taken: 26 Feb 1814 - Where taken: off NY - Date Received: 7 Oct 1814 - From what ship: HMT Salvador del Mundo, Halifax - Born: Canary Islands - Age: 27 - Race: Negro - Released on 14 Jun 1815.

Joseph, Nicholas - Boy - Nbr: 5278 - Prize: Growler - Ship Type: P - How taken: HMS Electra - When taken: 7 Jul 1813 - Where taken: at sea - Date Received: 31 Oct 1814 - From what ship: HMT Leyden, Chatham - Born: Marblehead - Age: 14 - Released on 29 Jun 1815.

Joseph, Pedro - Seaman - Nbr: 4810 - Prize: President - Ship Type: P - How taken: HMS Pique - When taken: 7 May 1814 - Where taken: off Porto Rico - Date Received: 9 Oct 1814 - From what ship: HMT Freya, Halifax - Born: Guadeloupe - Age: 26 - Race: Black - Died on 25 Feb 1815 from phthisis pulmonalis.

Joseph, Thomas - Seaman - Nbr: 386 - Prize: John & Frances - Ship Type: MV - How taken: HMS Belle Poule - When taken: 13 Mar 1813 - Where taken: Bay of Biscay - Date Received: 1 Jul 1813 - From what ship: Plymouth - Born: Charlestown - Age: 27 - Released on 20 Apr 1815.

Jouranne, Louis - Seaman - Nbr: 4999 - Prize: Invincible - Ship Type: LM - How taken: HMS Armide - When taken: 15 Aug 1814 - Where taken: off Nantucket - Date Received: 28 Oct 1814 - From what ship: HMT Alkbar, Halifax - Born: Brest - Age: 30 - Released on 21 Jun 1815.

Judah, David - Master - Nbr: 6291 - How taken: Sent into custody from a West Indiaman - When taken: 26 Jun 1813 - Where taken: L - Date Received: 17 Feb 1815 - From what ship: HMT Ganges, Plymouth - Born: Connecticut - Age: 27 - Sent to Ashburton (England) on 2 Mar 1815.

Judson, Obadiah - Seaman - Nbr: 1320 - How taken: Gave himself up from HMS Harmony - When taken: 26 Mar 1813 - Date Received: 19 Jun 1814 - From what ship: Stapleton - Born: New Haven - Age: 40 - Race: Negro - Released on 28 Apr 1815.

Jupiter, James - Seaman - Nbr: 4704 - Prize: Volunteer - Ship Type: MV - How taken: Victoria (Privateer) - When taken: 26 Dec 1813 - Where taken: at sea - Date Received: 9 Oct 1814 - From what ship: HMT Leyden, Chatham - Born: Philadelphia - Age: 24 - Race: Black - Released on 15 Jun 1815.

Justice, John - Seaman - Nbr: 2426 - Prize: US Sloop Frolic - Ship Type: MW - How taken: HMS Orpheus - When

American Prisoners of War Held at Dartmoor during the War of 1812

## Alphabetical listing of names

taken: 20 Apr 1814 - Where taken: off Cuba - Date Received: 16 Aug 1814 - From what ship: HMT Queen, Halifax - Born: Virginia - Age: 26 - Released on 3 May 1815.

Justin, William - Seaman - Nbr: 3965 - Prize: Rolla - Ship Type: P - How taken: HMS Loire - When taken: 10 Dec 1813 - Where taken: off Bull Island (South Carolina) - Date Received: 5 Oct 1814 - From what ship: HMT President, Halifax - Born: Newport - Age: 17 - Released on 9 Jun 1815.

Kales, John - Seaman - Nbr: 2443 - Prize: US Sloop Frolic - Ship Type: MW - How taken: HMS Orpheus - When taken: 20 Apr 1814 - Where taken: off Cuba - Date Received: 16 Aug 1814 - From what ship: HMT Queen, Halifax - Born: New Bedford - Age: 18 - Released on 3 May 1815.

Kaller, Joseph - Seaman - Nbr: 1993 - Prize: Kergettos - Ship Type: MV - How taken: HMS Castilian - When taken: 12 Mar 1813 - Where taken: off Cape Ortegal (Spain) - Date Received: 3 Aug 1814 - From what ship: HMS Lyffey, Chatham Depot - Born: Connecticut - Age: 24 - Released on 2 May 1815.

Kanada, Joseph - Seaman - Nbr: 2602 - How taken: Gave himself up from HMS Hamadryad - When taken: 12 Sep 1812 - Date Received: 21 Aug 1814 - From what ship: HMT Freya, Chatham - Born: Baltimore - Age: 20 - Released on 26 Apr 1815.

Kanes, William - Seaman - Nbr: 4568 - How taken: Gave himself up from MV Liberty - When taken: 30 Dec 1813 - Date Received: 9 Oct 1814 - From what ship: HMT Leyden, Chatham - Born: New York - Age: 24 - Released on 19 May 1815.

Karnes, James - Seaman - Nbr: 2217 - Prize: Hussar - Ship Type: P - How taken: HMS Saturn - When taken: 25 May 1814 - Where taken: off Sandy Hook - Date Received: 16 Aug 1814 - From what ship: HMS Dublin, Halifax - Born: New York - Age: 21 - Released on 2 May 1815.

Kean, Robert - Marine - Nbr: 4037 - Prize: US Brig Rattlesnake - Ship Type: MW - How taken: HMS Leander - When taken: 13 Jul 1814 - Where taken: off Shelburne - Date Received: 6 Oct 1814 - From what ship: HMT Chesapeake, Halifax - Born: Salem - Age: 25 - Released on 9 Jun 1815.

Keane, Daniel - Seaman - Nbr: 4132 - Prize: Bordeaux Packet - Ship Type: LM - How taken: HMS Niemen - When taken: 28 Jun 1814 - Where taken: off Delaware - Date Received: 6 Oct 1814 - From what ship: HMT Chesapeake, Halifax - Born: New Jersey - Age: 22 - Released on 13 Jun 1815.

Keeffe, Alexander - Seaman - Nbr: 1746 - How taken: Gave himself up from HMS Malta - When taken: 26 Dec 1812 - Date Received: 20 Jul 1814 - From what ship: HMS Milford, Plymouth - Born: Massachusetts - Age: 28 - Released on 26 Apr 1815.

Keen, Benjamin - Seaman - Nbr: 4464 - Prize: Governor Plumer - Ship Type: P - How taken: HMS Shamrock - When taken: 4 Mar 1813 - Where taken: at sea - Date Received: 8 Oct 1814 - From what ship: HMT Leyden, Chatham - Born: Massachusetts - Age: 24 - Released on 15 Jun 1815.

Keen, Joseph - Seaman - Nbr: 4198 - Prize: Zephyr - Ship Type: MV - How taken: HMS Surveillante - When taken: 4 Jan 1814 - Where taken: Bay of Biscay - Date Received: 7 Oct 1814 - From what ship: HMT Niobe, Chatham - Born: Massachusetts - Age: 26 - Released on 13 Jun 1815.

Keen, Nathaniel - Boatswain - Nbr: 455 - Prize: Courier - Ship Type: LM - How taken: HMS Andromache - When taken: 14 Mar 1813 - Where taken: Bay of Biscay - Date Received: 8 Sep 1813 - From what ship: Plymouth - Born: Kittery - Age: 40 - Released on 26 Apr 1815.

Keeting, John - Seaman - Nbr: 2409 - Prize: US Sloop Frolic - Ship Type: MW - How taken: HMS Orpheus - When taken: 20 Apr 1814 - Where taken: off Cuba - Date Received: 16 Aug 1814 - From what ship: HMT Queen, Halifax - Born: Marblehead - Age: 35 - Sent to Dartmouth on 19 Oct 1814.

Kegs, Philip - Seaman - Nbr: 1648 - How taken: Apprehended at Bristol - When taken: 16 Aug 1813 - Date Received: 23 Jun 1814 - From what ship: Stapleton - Born: Pennsylvania - Age: 37 - Released on 1 May 1815.

Kegs, Zenas - Seaman - Nbr: 4442 - Prize: Growler - Ship Type: P - How taken: HMS Electra - When taken: 7 Jul 1814 - Where taken: at sea - Date Received: 8 Oct 1814 - From what ship: HMT Leyden, Chatham - Born: Massachusetts - Age: 22 - Released on 14 Jun 1815.

Keitch, Jonah - Seaman - Nbr: 2294 - Prize: Hussar - Ship Type: P - How taken: HMS Saturn - When taken: 25 May 1814 - Where taken: off Sandy Hook - Date Received: 16 Aug 1814 - From what ship: HMS Dublin, Halifax - Born: New York - Age: 17 - Released on 3 May 1815.

Keith, James - Seaman - Nbr: 6209 - Prize: Prince de Neufchatel - Ship Type: P - How taken: Leander (Newcastle Acasta) - When taken: 20 Dec 1814 - Where taken: Lat 38 Long 56 - Date Received: 30 Jan 1815 - From what ship: HMT Pheasant - Born: Massachusetts - Age: 22 - Released on 5 Jul 1815.

Kell, Francis - Seaman - Nbr: 6395 - Prize: Nellenville - Ship Type: MV - How taken: HMS Onyx - When taken: 25 Dec 1814 - Where taken: off San Domingo - Date Received: 3 Mar 1815 - From what ship: HMT Ganges,

American Prisoners of War Held at Dartmoor during the War of 1812

## Alphabetical listing of names

Plymouth - Born: Baltimore - Age: 22 - Released on 11 Jul 1815.

Kellem, Daniel - Seaman - Nbr: 5474 - Prize: General Putnam - Ship Type: P - How taken: HMS Leander - When taken: 8 Nov 1814 - Where taken: Long 65 Lat 42 - Date Received: 17 Dec 1814 - From what ship: HMT Loire, Halifax - Born: Salem - Age: 27 - Released on 1 Jul 1815.

Kellem, Frederick - Seaman - Nbr: 5072 - Prize: Ida - Ship Type: LM - How taken: HMS Newcastle - When taken: 9 Aug 1814 - Where taken: Long 34 - Date Received: 28 Oct 1814 - From what ship: HMT Alkbar, Halifax - Born: Cape Ann - Age: 17 - Released on 2 May 1815.

Kellem, John C. - Seaman - Nbr: 260 - Prize: William Bayard - Ship Type: MV - How taken: HMS Warspite - When taken: 3 Mar 1813 - Where taken: Bay of Biscay - Date Received: 28 Jun 1813 - From what ship: Plymouth - Born: Virginia - Age: 23 - Released on 20 Apr 1815.

Kelley, Henry - Soldier - Nbr: 5266 - Prize: 13th US Infantry - Ship Type: Troops - How taken: British Army - When taken: 13 Oct 1812 - Where taken: Canada - Date Received: 31 Oct 1814 - From what ship: HMT Leyden, Chatham - Born: Antrim - Age: 30 - Released on 29 Jun 1815.

Kelley, James - Boatswain - Nbr: 3772 - Prize: Fame - Ship Type: P - How taken: HMS Thistle - When taken: 10 Apr 1814 - Where taken: after being cast ashore on a seal island - Date Received: 30 Sep 1814 - From what ship: HMT President, Halifax - Born: Salem - Age: 52 - Released on 9 Jun 1815.

Kelley, James - Cooper - Nbr: 4902 - Prize: Alfred - Ship Type: P - How taken: HMS Epervier - When taken: 23 Feb 1814 - Where taken: off Newfoundland - Date Received: 28 Oct 1814 - From what ship: HMT Alkbar, Halifax - Born: Marblehead - Age: 23 - Released on 21 Jun 1815.

Kelley, John - Seaman - Nbr: 3756 - Prize: Alfred - Ship Type: P - How taken: HMS Epervier - When taken: 23 Feb 1814 - Where taken: off Newfoundland - Date Received: 30 Sep 1814 - From what ship: HMT President, Halifax - Born: Marblehead - Age: 62 - Died on 29 Mar 1815 from Ana area.

Kelley, John - Seaman - Nbr: 4492 - Prize: Enterprize - Ship Type: P - How taken: HMS Tenedos - When taken: 21 May 1813 - Where taken: off Cape Cod - Date Received: 8 Oct 1814 - From what ship: HMT Leyden, Chatham - Born: Delaware - Age: 33 - Released on 15 Jun 1815.

Kelley, John - Seaman - Nbr: 2822 - How taken: Gave himself up from HMS Philomel - When taken: 28 Dec 1812 - Date Received: 24 Aug 1814 - From what ship: HMT Liverpool, Chatham - Born: New York - Age: 19 - Released on 26 Apr 1815.

Kellick, Stephen - Seaman - Nbr: 6531 - Prize: US Brig Syren - Ship Type: MW - How taken: HMS Medway - When taken: 12 Jul 1814 - Where taken: off Cape of Good Hope - Date Received: 7 Mar 1815 - From what ship: HMT Ganges, Plymouth - Born: Wilmington - Age: 33 - Released on 11 Jul 1815.

Kellogg, Amos - Seaman - Nbr: 2089 - How taken: Gave himself up from HMS Hyperion - When taken: 18 Jun 1812 - Date Received: 3 Aug 1814 - From what ship: HMS Bittern, Chatham Depot - Born: Sheffield - Age: 30 - Released on 2 May 1815.

Kellum, Smith - Seaman - Nbr: 2876 - How taken: Gave himself up from HMS Bellerophon - When taken: 27 May 1813 - Date Received: 24 Aug 1814 - From what ship: HMT Alpheus, Chatham - Born: Virginia - Age: 31 - Released on 19 May 1815.

Kelly, Henry - Seaman - Nbr: 2556 - Prize: Caromaned - Ship Type: Prize - How taken: HMS Eridanus - When taken: 13 Aug 1814 - Where taken: Lat 40, Long 16 - Date Received: 16 Aug 1814 - From what ship: HMT Salvador del Mundo - Born: Virginia - Age: 26 - Race: Black - Released on 3 May 1815.

Kelly, Samuel - Seaman - Nbr: 4580 - Prize: Bunker Hill - Ship Type: P - How taken: HMS Pomone - When taken: 2 Mar 1814 - Where taken: Channel - Date Received: 9 Oct 1814 - From what ship: HMT Leyden, Chatham - Born: New York - Age: 33 - Race: Mulatto - Released on 20 Mar 1815.

Kemble, John - Sail maker - Nbr: 5429 - Prize: Rambler - Ship Type: MV - How taken: Morley (Transport) - When taken: 10 Feb 1813 - Where taken: off Isle of France - Date Received: 31 Oct 1814 - From what ship: HMT Leyden, Chatham - Born: Ipswich - Age: 23 - Released on 1 Jul 1815.

Kembourn, John - Seaman - Nbr: 5022 - Prize: Landrail - Ship Type: Prize - How taken: HMS Wasp - When taken: 27 Jul 1814 - Where taken: Georges Bank - Date Received: 28 Oct 1814 - From what ship: HMT Alkbar, Halifax - Born: Philadelphia - Age: 32 - Released on 21 Jun 1815.

Kemp, James - Seaman - Nbr: 2101 - How taken: Gave himself up from HMS Leonidas - When taken: 13 Jun 1813 - Date Received: 3 Aug 1814 - From what ship: HMS Bittern, Chatham Depot - Born: New York - Age: 27 - Released on 2 May 1815.

Kempt, Eric - Carpenter - Nbr: 4970 - Prize: Invincible - Ship Type: LM - How taken: HMS Armide - When taken: 15 Aug 1814 - Where taken: off Nantucket - Date Received: 28 Oct 1814 - From what ship: HMT Alkbar, Halifax - Born: Massachusetts - Age: 26 - Released on 21 Jun 1815.

American Prisoners of War Held at Dartmoor during the War of 1812

## Alphabetical listing of names

Kempt, George - Seaman - Nbr: 4990 - Prize: Invincible - Ship Type: LM - How taken: HMS Armide - When taken: 15 Aug 1814 - Where taken: off Nantucket - Date Received: 28 Oct 1814 - From what ship: HMT Alkbar, Halifax - Born: New York - Age: 18 - Released on 21 Jun 1815.

Kendal, Franklin - Boy - Nbr: 6248 - Prize: Albion - Ship Type: Prize - How taken: HMS Harlequin - When taken: 6 Jan 1814 - Where taken: Long 20 - Date Received: 4 Feb 1815 - From what ship: HMT Ganges - Born: Leominster - Age: 16 - Released on 5 Jul 1815.

Kendrick, Benjamin - Prize master - Nbr: 4345 - Prize: Union - Ship Type: LM - How taken: HMS Curlew - When taken: 1 Apr 1814 - Where taken: off Halifax - Date Received: 7 Oct 1814 - From what ship: HMT Salvador del Mundo, Halifax - Born: Wareham - Age: 31 - Released on 14 Jun 1815.

Kennard, N. - Seaman - Nbr: 5663 - Prize: Harlequin - Ship Type: P - How taken: HMS Bulwark - When taken: 23 Nov 1814 - Where taken: off Halifax - Date Received: 24 Dec 1814 - From what ship: HMT Penelope - Born: Portsmouth - Age: 22 - Released on 20 Mar 1815.

Kenney, Jacob - Seaman - Nbr: 803 - Prize: Collin - Ship Type: MV - How taken: HMS Helicon and HMS Whiting - When taken: 26 Oct 1813 - Where taken: Channel - Date Received: 3 Nov 1813 - From what ship: Plymouth - Born: Prussia - Age: 28 - Released on 6 May 1814.

Kennedy, Dennis - Seaman - Nbr: 4699 - Prize: Volunteer - Ship Type: MV - How taken: Victoria (Privateer) - When taken: 26 Dec 1813 - Where taken: at sea - Date Received: 9 Oct 1814 - From what ship: HMT Leyden, Chatham - Born: Carolina - Age: 31 - Released on 14 Jun 1815.

Kennedy, James - Seaman - Nbr: 5419 - How taken: Gave himself up from HMS Blenheim - When taken: 27 Aug 1814 - Date Received: 31 Oct 1814 - From what ship: HMT Leyden, Chatham - Born: Philadelphia - Age: 36 - Released on 11 Jul 1815.

Kennedy, John - Seaman - Nbr: 464 - Prize: Meteor - Ship Type: MV - How taken: HMS Briton - When taken: 12 Mar 1813 - Where taken: Bay of Biscay - Date Received: 8 Sep 1813 - From what ship: Plymouth - Born: Hudson Island - Age: 24 - Released on 14 Apr 1815.

Kennedy, John - Seaman - Nbr: 439 - Prize: Magdalen - Ship Type: MV - How taken: HMS Superb - When taken: 15 Apr 1813 - Where taken: off Belle Isle, France - Date Received: 1 Jul 1813 - From what ship: Plymouth - Born: New Jersey - Age: 25 - Released on 20 Apr 1815.

Kennedy, Peter - Seaman - Nbr: 1970 - Prize: Orbit - Ship Type: MV - How taken: HMS Achates - When taken: 29 Jan 1813 - Where taken: Lat 44N Long 13W - Date Received: 3 Aug 1814 - From what ship: HMS Lyffey, Chatham Depot - Born: New Jersey - Age: 20 - Released on 2 May 1815.

Kennedy, Richard - Seaman - Nbr: 1041 - Prize: US Brig Argus - Ship Type: MW - How taken: HMS Pelican - When taken: 16 Aug 1813 - Where taken: Irish Channel - Date Received: 10 May 1814 - From what ship: Plymouth - Born: Brooklyn - Age: 21 - Sent to Dartmouth on 19 Oct 1814.

Kennedy, William - Seaman - Nbr: 1800 - How taken: Gave himself up from HMS Ajax - When taken: 14 Oct 1813 - Date Received: 20 Jul 1814 - From what ship: HMS Milford, Plymouth - Born: Boston - Age: 29 - Released on 1 May 1815.

Kennison, William - Seaman - Nbr: 876 - How taken: Impressed at Falmouth - When taken: 30 Nov 1813 - Date Received: 31 Jan 1814 - From what ship: Plymouth - Born: Deerfield - Age: 23 - Released on 10 Apr 1815.

Kenny, George - Boy - Nbr: 3770 - Prize: Fame - Ship Type: P - How taken: HMS Thistle - When taken: 10 Apr 1814 - Where taken: after being cast ashore on a seal island - Date Received: 30 Sep 1814 - From what ship: HMT President, Halifax - Born: Salem - Age: 15 - Released on 9 Jun 1815.

Kenny, Harris - Pilot - Nbr: 502 - Prize: Orders in Council - Ship Type: MV - How taken: HMS Surveillante - When taken: 1 Jun 1813 - Where taken: off Cape Ortegal (Spain) - Date Received: 8 Sep 1813 - From what ship: Plymouth - Born: New London - Age: 30 - Released on 26 Apr 1815.

Kenny, William - Seaman - Nbr: 4247 - Prize: Requin - Ship Type: LM - How taken: HMS Venus - When taken: 6 Mar 1814 - Where taken: off Bordeaux - Date Received: 7 Oct 1814 - From what ship: HMT Niobe, Chatham - Born: New London - Age: 22 - Race: Black - Released on 13 Jun 1815.

Kent, James - Seaman - Nbr: 2753 - How taken: Impressed at Cork - When taken: 19 May 1813 - Date Received: 24 Aug 1814 - From what ship: HMT Liverpool, Chatham - Born: Washington - Age: 24 - Released on 19 May 1815.

Kent, William - Seaman - Nbr: 4513 - Prize: America - Ship Type: P - How taken: HMS Shannon - When taken: 23 May 1813 - Where taken: off Cape Cod - Date Received: 8 Oct 1814 - From what ship: HMT Leyden, Chatham - Born: Rhode Island - Age: 36 - Released on 15 Jun 1815.

Keogh, Edward - Seaman - Nbr: 5085 - Prize: Ida - Ship Type: LM - How taken: HMS Newcastle - When taken: 9 Aug 1814 - Where taken: Long 34 - Date Received: 28 Oct 1814 - From what ship: HMT Alkbar, Halifax -

American Prisoners of War Held at Dartmoor during the War of 1812

## Alphabetical listing of names

Born: Bristol - Age: 28 - Released on 29 Jun 1815.

Kevel, Alexander - Seaman - Nbr: 5765 - Prize: Young William - Ship Type: Prize - How taken: HMS Plover - When taken: 10 Sep 1814 - Where taken: off Newfoundland - Date Received: 26 Dec 1814 - From what ship: HMT Argo - Born: France - Age: 28 - Released on 3 Jul 1815.

Keen, Edward - Gunner - Nbr: 489 - Prize: Paul Jones - Ship Type: P - How taken: HMS Leonidas - When taken: 23 May 1813 - Where taken: Channel - Date Received: 8 Sep 1813 - From what ship: Plymouth - Born: Petersburg - Age: 25 - Released on 14 Apr 1815.

Kilda, John - Seaman - Nbr: 4724 - How taken: Gave himself up from HMS Leda - When taken: 4 Jul 1813 - Date Received: 9 Oct 1814 - From what ship: HMT Freya, Chatham - Born: Salem - Age: 30 - Released on 15 Jun 1815.

Killam, James - Seaman - Nbr: 748 - Prize: US Brig Argus - Ship Type: MW - How taken: HMS Pelican - When taken: 14 Aug 1813 - Where taken: Irish Channel - Date Received: 3 Nov 1813 - From what ship: Plymouth - Born: Virginia - Age: 39 - Sent to Dartmouth on 2 Nov 1814.

Kiplinger, John - Seaman - Nbr: 1458 - Prize: Revenge - Ship Type: LM - How taken: HMS Belle Poule - When taken: 10 Mar 1813 - Where taken: off Cornwall - Date Received: 19 Jun 1814 - From what ship: Stapleton - Born: Maryland - Age: 32 - Released on 28 Apr 1815.

Killer, John - Seaman - Nbr: 1467 - Prize: Governor Gerry - Ship Type: MV - How taken: HMS Lyra - When taken: 29 May 1813 - Where taken: Bay of Biscay - Date Received: 19 Jun 1814 - From what ship: Stapleton - Born: Boston - Age: 40 - Released on 28 Apr 1815.

Killerman, M. - Seaman - Nbr: 5295 - Prize: York Town - Ship Type: P - How taken: HMS Maidstone - When taken: 18 Jul 1813 - Where taken: Grand Banks - Date Received: 31 Oct 1814 - From what ship: HMT Leyden, Chatham - Born: Leghorn, Italy - Age: 17 - Released on 29 Jun 1815.

Killett, William - Boy - Nbr: 5485 - Prize: General Putnam - Ship Type: P - How taken: HMS Leander - When taken: 8 Nov 1814 - Where taken: Long 65 Lat 42 - Date Received: 17 Dec 1814 - From what ship: HMT Loire, Halifax - Born: Marblehead - Age: 17 - Released on 1 Jul 1815.

Killingworth, John - Seaman - Nbr: 4295 - Prize: Commodore Perry - Ship Type: Schooner - How taken: Sent into custody from a cutter - When taken: 25 Feb 1814 - Where taken: off Bordeaux - Date Received: 7 Oct 1814 - From what ship: HMT Niobe, Chatham - Born: Sussex County - Age: 20 - Released on 14 Jun 1815.

Kimball, Nathaniel - Seaman - Nbr: 1303 - How taken: Sent into custody from HMS Phoenix - Date Received: 14 Jun 1814 - From what ship: Mill Prison (Plymouth, England) - Born: East Harford - Age: 19 - Released on 28 Apr 1815.

Kimball, Samuel - Seaman - Nbr: 5363 - How taken: Impressed at Shields - When taken: 20 Mar 1814 - Date Received: 31 Oct 1814 - From what ship: HMT Leyden, Chatham - Born: Nantucket - Age: 28 - Released on 1 Jul 1815.

Kimmins, John - Seaman - Nbr: 19 - Prize: Cashier - Ship Type: LM - How taken: HMS Reindeer - When taken: 3 Feb 1813 - Where taken: Bay of Biscay - Date Received: 2 Apr 1813 - From what ship: Plymouth - Born: New York - Age: 25 - Sent to Dartmouth on 30 Jul 1813.

Kindan, Ephraim - Seaman - Nbr: 4496 - Prize: Enterprize - Ship Type: P - How taken: HMS Tenedos - When taken: 21 May 1813 - Where taken: off Cape Cod - Date Received: 8 Oct 1814 - From what ship: HMT Leyden, Chatham - Born: Massachusetts - Age: 25 - Released on 15 Jun 1815.

King, John - Seaman - Nbr: 3114 - Prize: Harmony - Ship Type: Prize - How taken: HMS Brisk - When taken: 24 Aug 1814 - Where taken: Bristol Channel - Date Received: 11 Sep 1814 - From what ship: HMT Salvador del Mundo - Born: Cherbourg - Age: 26 - Released on 28 May 1815.

King, John - Seaman - Nbr: 2795 - Prize: Weazel - Ship Type: MV - How taken: HMS Foxhound - When taken: 25 Mar 1813 - Where taken: Bay of Biscay - Date Received: 24 Aug 1814 - From what ship: HMT Liverpool, Chatham - Born: New York - Age: 20 - Released on 19 May 1815.

King, Joseph - Seaman - Nbr: 4861 - Prize: Two Sisters - Ship Type: MV - How taken: HMS Bermuda - When taken: 26 May 1814 - Where taken: off Maderia - Date Received: 9 Oct 1814 - From what ship: HMT Freya, Halifax - Born: New Orleans - Age: 32 - Race: Black - Released on 14 Jun 1815.

King, Peter - Seaman - Nbr: 5349 - How taken: Sent into custody from Alderney - When taken: 26 Jan 1814 - Date Received: 31 Oct 1814 - From what ship: HMT Leyden, Chatham - Born: Boston - Age: 40 - Released on 1 Jul 1815.

King, Peter - Seaman - Nbr: 5009 - Prize: Charlotte - Ship Type: Prize - How taken: HMS Wasp - When taken: 31 Aug 1814 - Where taken: off Cape Sable Island (Canada) - Date Received: 28 Oct 1814 - From what ship: HMT Alkbar, Halifax - Born: Marseille - Age: 36 - Released on 21 Jun 1815.

American Prisoners of War Held at Dartmoor during the War of 1812

## Alphabetical listing of names

King, Seth - Seaman - Nbr: 6227 - Prize: Prince de Neufchatel - Ship Type: P - How taken: Leander (Newcastle Acasta) - When taken: 20 Dec 1814 - Where taken: Lat 38 Long 56 - Date Received: 30 Jan 1815 - From what ship: HMT Pheasant - Born: Connecticut - Age: 21 - Released on 5 Jul 1815.

King, Solomon - Prize master - Nbr: 4395 - Prize: Portsmouth Packet - Ship Type: Prize - How taken: HMS Fantome - When taken: 5 Oct 1813 - Where taken: off Portland - Date Received: 8 Oct 1814 - From what ship: HMT Leyden, Chatham - Born: New York - Age: 27 - Released on 14 Jun 1815.

King, Uriel - Seaman - Nbr: 3847 - Prize: Dominique - Ship Type: LM - How taken: HMS Dotterel - When taken: 21 May 1814 - Where taken: off Charleston - Date Received: 5 Oct 1814 - From what ship: HMT Orpheus, Halifax - Born: Massachusetts - Age: 22 - Died on 3 Feb 1815 from pneumonitis.

Kingbutton, John - Seaman - Nbr: 4429 - Prize: Taken off an English whaler - Ship Type: MV - How taken: HMS Illustrious - When taken: 22 Oct 1813 - Where taken: at sea - Date Received: 8 Oct 1814 - From what ship: HMT Leyden, Chatham - Born: Providence Town - Age: 26 - Released on 14 Jun 1815.

Kingdom, John - Seaman - Nbr: 6423 - Prize: Farmer's Daughter - Ship Type: MV - How taken: HMS Leviathan - When taken: 29 May 1814 - Date Received: 3 Mar 1815 - From what ship: HMT Ganges, Plymouth - Born: Maryland - Age: 38 - Released on 11 Jul 1815.

Kingman, Charles - Seaman - Nbr: 3260 - Prize: Wiley Reynard - Ship Type: P - How taken: HMS Shannon - When taken: 15 Aug 1812 - Where taken: off Halifax - Date Received: 11 Sep 1814 - From what ship: HMT Freya, Chatham - Born: Boston - Age: 77 - Released on 27 Apr 1815.

Kinsley, Benjamin - Seaman - Nbr: 1964 - How taken: Gave himself up from HMS Andromache - When taken: 24 Dec 1812 - Date Received: 3 Aug 1814 - From what ship: HMS Lyffey, Chatham Depot - Born: Nobleboro - Age: 35 - Released on 26 Apr 1815.

Kirbly, Anthony - Seaman - Nbr: 3828 - Prize: Modelle - Ship Type: MV - How taken: HMS Niemen - When taken: 22 May 1814 - Where taken: off Egg Harbor (New Jersey) - Date Received: 5 Oct 1814 - From what ship: HMT Orpheus, Halifax - Born: Baltimore - Age: 20 - Released on 9 Jun 1815.

Kirby, Robert - Boy - Nbr: 6386 - Prize: US Brig Syren - Ship Type: MW - How taken: HMS Medway - When taken: 12 Jul 1814 - Where taken: off Cape of Good Hope - Date Received: 24 Feb 1815 - From what ship: HMT Ganges, Plymouth - Born: Boston - Age: 17 - Released on 11 Jul 1815.

Kirk, William - Seaman - Nbr: 1869 - Prize: Fox - Ship Type: P - How taken: HMS Merlin - When taken: 6 Aug 1813 - Where taken: off Sable Island - Date Received: 29 Jul 1814 - From what ship: HMS Ville de Paris, Chatham Depot - Born: Charlestown - Age: 21 - Released on 2 May 1815.

Kirkpatrick, William - Seaman - Nbr: 3251 - Prize: Wiley Reynard - Ship Type: P - How taken: HMS Shannon - When taken: 15 Aug 1812 - Where taken: off Halifax - Date Received: 11 Sep 1814 - From what ship: HMT Freya, Chatham - Born: Wilmington - Age: 28 - Released on 27 Apr 1815.

Kister, John - Seaman - Nbr: 1356 - Prize: Paul Jones - Ship Type: P - How taken: HMS Leonidas - When taken: 23 May 1813 - Where taken: Channel - Date Received: 19 Jun 1814 - From what ship: Stapleton - Born: New York - Age: 20 - Released on 28 Apr 1815.

Kitchen, Daniel - Mate - Nbr: 4251 - Prize: Argus - Ship Type: MV - How taken: HMS San Domingo - When taken: 1 Mar 1814 - Where taken: off Savannah - Date Received: 7 Oct 1814 - From what ship: HMT Niobe, Chatham - Born: Westchester - Age: 31 - Released on 13 Jun 1815.

Kitchen, George - Seaman - Nbr: 6380 - Prize: US Brig Syren - Ship Type: MW - How taken: HMS Medway - When taken: 12 Jul 1814 - Where taken: off Cape of Good Hope - Date Received: 24 Feb 1815 - From what ship: HMT Ganges, Plymouth - Born: Boston - Age: 23 - Released on 11 Jul 1815.

Kitchen, Richard - Seaman - Nbr: 864 - Prize: Amiable - Ship Type: LM - How taken: HMS Magnificent - When taken: 30 Oct 1813 - Where taken: off Bordeaux - Date Received: 4 Dec 1813 - From what ship: Plymouth - Born: New York - Age: 23 - Released on 27 Apr 1815.

Kitchen, Robert - Seaman - Nbr: 4140 - Prize: Bordeaux Packet - Ship Type: LM - How taken: HMS Niemen - When taken: 28 Jun 1814 - Where taken: off Delaware - Date Received: 6 Oct 1814 - From what ship: HMT Chesapeake, Halifax - Born: Philadelphia - Age: 34 - Released on 13 Jun 1815.

Kitton, Abraham - Boy - Nbr: 6396 - Prize: Nellenville - Ship Type: MV - How taken: HMS Onyx - When taken: 25 Dec 1814 - Where taken: off San Domingo - Date Received: 3 Mar 1815 - From what ship: HMT Ganges, Plymouth - Born: Baltimore - Age: 14 - Released on 11 May 1815.

Knabbs, James - Seaman - Nbr: 4798 - Prize: President - Ship Type: P - How taken: HMS Pique - When taken: 7 May 1814 - Where taken: off Porto Rico - Date Received: 9 Oct 1814 - From what ship: HMT Freya, Halifax - Born: Baltimore - Age: 22 - Died on 26 Feb 1815 from perissneuoniria.

Knap, Samuel - Seaman - Nbr: 4538 - Prize: Wolf Cove - Ship Type: MV - How taken: HMS Briton - When taken:

American Prisoners of War Held at Dartmoor during the War of 1812

## Alphabetical listing of names

1 Dec 1812 - Where taken: off Brest - Date Received: 8 Oct 1814 - From what ship: HMT Leyden, Chatham - Born: Massachusetts - Age: 45 - Released on 27 Apr 1815.

Knapp, Joseph - Prize master - Nbr: 4897 - Prize: Atlantic - Ship Type: Prize - How taken: HMS Maidstone - When taken: 10 Oct 1813 - Where taken: off Sable Island - Date Received: 28 Oct 1814 - From what ship: HMT Alkbar, Halifax - Born: New Jersey - Age: 25 - Released on 21 Jun 1815.

Knight, Daniel - Seaman - Nbr: 3405 - Prize: Thomas - Ship Type: P - How taken: HMS Nymphe - When taken: 24 Jun 1813 - Where taken: off Halifax - Date Received: 13 Sep 1814 - From what ship: HMT Niobe, Chatham - Born: Massachusetts - Age: 28 - Released on 28 May 1815.

Knight, Elisha - Seaman - Nbr: 2144 - How taken: Gave himself up from HMS Victory - When taken: 5 Dec 1812 - Date Received: 8 Aug 1814 - From what ship: HMT Raven, Chatham - Born: Portland - Age: 26 - Released on 26 Apr 1815.

Knight, George - Seaman - Nbr: 1770 - Prize: Print - Ship Type: MV - How taken: HMS Colossus - When taken: 21 Jan 1813 - Where taken: off Long Island - Date Received: 20 Jul 1814 - From what ship: Chatham - Born: Marblehead - Age: 25 - Released on 1 May 1815.

Knight, Samuel - Seaman - Nbr: 4350 - Prize: Plutus - Ship Type: P - How taken: HMS Curlew - When taken: 1 Apr 1814 - Where taken: off Halifax - Date Received: 7 Oct 1814 - From what ship: HMT Salvador del Mundo, Halifax - Born: West-Cappel (France) - Age: 23 - Released on 14 Jun 1815.

Knight, William - Seaman - Nbr: 1772 - Prize: Print - Ship Type: MV - How taken: HMS Colossus - When taken: 21 Jan 1813 - Where taken: off Long Island - Date Received: 20 Jul 1814 - From what ship: Chatham - Born: Marblehead - Age: 19 - Released on 1 May 1815.

Knowlton, Enos - Seaman - Nbr: 5069 - Prize: Ida - Ship Type: LM - How taken: HMS Newcastle - When taken: 9 Aug 1814 - Where taken: Long 34 - Date Received: 28 Oct 1814 - From what ship: HMT Alkbar, Halifax - Born: Hambleton - Age: 24 - Released on 29 Jun 1815.

Knox, John - Seaman - Nbr: 1278 - How taken: Sent into custody from MV Nereus - Date Received: 14 Jun 1814 - From what ship: Mill Prison (Plymouth, England) - Born: New York - Age: 47 - Released on 28 Apr 1815.

Kromkout, Barney - Steward - Nbr: 44 - Prize: Terrible - Ship Type: MV - How taken: HMS Foxhound - When taken: 8 Feb 1813 - Where taken: Channel - Date Received: 2 Apr 1813 - From what ship: Plymouth - Born: Baltimore - Age: 23 - Sent to Dartmouth on 30 Jul 1813.

Kyler, John - Seaman - Nbr: 5416 - How taken: Impressed at London - When taken: 13 Oct 1814 - Date Received: 31 Oct 1814 - What ship: HMT Leyden, Chatham - Born: Boddington - Age: 36 - Released on 1 Jul 1815.

La Roche, Jean - Seaman - Nbr: 5280 - Prize: Pomona - Ship Type: Prize - How taken: HMS Ethalion - When taken: 14 Dec 1813 - Where taken: at sea - Date Received: 31 Oct 1814 - From what ship: HMT Leyden, Chatham - Born: New Hampshire - Age: 24 - Released on 29 Jun 1815.

Laborde, Peter - Seaman - Nbr: 918 - Prize: Volante - Ship Type: MV - How taken: Calalou - When taken: 10 Nov 1813 - Where taken: at sea - Date Received: 31 Jan 1814 - From what ship: Plymouth - Born: Genoa - Age: 20 - Sent to Plymouth on 8 Jul 1814.

Lacey, Henry - Seaman - Nbr: 5404 - How taken: Impressed at St. Johns - When taken: 1 May 1814 - Date Received: 31 Oct 1814 - What ship: HMT Leyden, Chatham - Born: New York - Age: 25 - Released on 1 Jul 1815.

Lacey, Zachariah - Pilot - Nbr: 6464 - Prize: Decatur - Ship Type: P - How taken: HMS Rhin - When taken: 5 Jun 1814 - Where taken: off San Domingo - Date Received: 3 Mar 1815 - From what ship: HMT Ganges, Plymouth - Born: Charleston - Age: 31 - Released on 11 Jul 1815.

Lachame, Thomas - Seaman - Nbr: 3619 - Prize: Monarch - Ship Type: MV - How taken: HMS Dotterel - When taken: 14 Dec 1813 - Where taken: off Charleston - Date Received: 30 Sep 1814 - From what ship: HMT Sybella - Born: Philadelphia - Age: 24 - Released on 4 Jun 1815.

Lackey, Joseph - Seaman - Nbr: 4486 - Prize: Enterprize - Ship Type: P - How taken: HMS Tenedos - When taken: 21 May 1813 - Where taken: off Cape Cod - Date Received: 8 Oct 1814 - From what ship: HMT Leyden, Chatham - Born: Massachusetts - Age: 30 - Died on 4 Feb 1815 from phthisis pulmonalis.

Lackman, Isaac - Seaman - Nbr: 2231 - Prize: General Hart - Ship Type: P - How taken: HMS Sophie - When taken: 24 Apr 1814 - Where taken: off Bermuda - Date Received: 16 Aug 1814 - From what ship: HMS Dublin, Halifax - Born: Beverly - Age: 26 - Released on 2 May 1815.

Lacour, John Baptist - Seaman - Nbr: 1646 - Prize: Paul Jones - Ship Type: P - How taken: HMS Leonidas - When taken: 23 May 1813 - Where taken: Channel - Date Received: 23 Jun 1814 - From what ship: Stapleton - Born: Naples - Age: 52 - Sent to Mill Prison (Plymouth, England) on 30 Jun 1814.

Lad, Daniel - Marine - Nbr: 3025 - Prize: US Sloop Frolic - Ship Type: MW - How taken: HMS Orpheus - When

American Prisoners of War Held at Dartmoor during the War of 1812

## Alphabetical listing of names

taken: 20 Apr 1814 - Where taken: off Cuba - Date Received: 2 Sep 1814 - From what ship: Naval Hospital, Plymouth - Born: Massachusetts - Age: 21 - Released on 3 May 1815.

Laingan, Daniel - Seaman - Nbr: 4777 - How taken: Gave himself up from HMS Revenge - When taken: 23 Sep 1814 - Date Received: 9 Oct 1814 - From what ship: HMT Freya, Chatham - Born: Boston - Age: 47 - Released on 11 Jul 1815.

Lake, Charles - Seaman - Nbr: 1654 - How taken: Apprehended at Bristol - When taken: 9 Nov 1813 - Date Received: 23 Jun 1814 - From what ship: Stapleton - Born: Philadelphia - Age: 23 - Released on 1 May 1815.Lake, Daniel - Seaman - Nbr: 1715 - How taken: Gave himself up from HMS Amphion - Date Received: 8 Jul 1814 - From what ship: Labrador - Born: Staten Island - Age: 35 - Released on 1 May 1815.

Lake, George - Seaman - Nbr: 1884 - Prize: Thomas - Ship Type: P - How taken: HMS Nymphe - When taken: 24 Jun 1813 - Where taken: off Halifax - Date Received: 29 Jul 1814 - From what ship: HMS Ville de Paris, Chatham Depot - Born: Cumberland County - Age: 23 - Race: Black - Released on 2 May 1815.

Lamar, Edward - Seaman - Nbr: 1181 - Prize: Melanie - Ship Type: MV - How taken: HMS Briton - When taken: 15 Sep 1813 - Where taken: off Bordeaux - Date Received: 4 Jun 1814 - From what ship: Dartmouth - Born: New Orleans - Age: 22 - Race: Mulatto - Sent to Plymouth on 8 Jul 1814.

Lamataille, John - Seaman - Nbr: 3849 - Prize: Dominique - Ship Type: LM - How taken: HMS Dotterel - When taken: 21 May 1814 - Where taken: off Charleston - Date Received: 5 Oct 1814 - From what ship: HMT Orpheus, Halifax - Born: Pennsylvania - Age: 24 - Race: Mulatto - Released on 9 Jun 1815.

Lamb, Anthony - Seaman - Nbr: 5008 - Prize: Charlotte - Ship Type: Prize - How taken: HMS Wasp - When taken: 31 Aug 1814 - Where taken: off Cape Sable Island (Canada) - Date Received: 28 Oct 1814 - From what ship: HMT Alkbar, Halifax - Born: Connecticut - Age: 19 - Race: Negro - Died on 22 Nov 1814.

Lamb, Jack - Cook - Nbr: 3171 - Prize: Hindustan - Ship Type: MV - How taken: HMS Tenedos - When taken: 25 Jun 1813 - Where taken: off Lisbon - Date Received: 11 Sep 1814 - From what ship: HMT Freya, Chatham - Born: Africa - Age: 20 - Race: Black - Released on 28 May 1815.

Lamb, Joseph - Marine - Nbr: 6198 - Prize: Prince de Neufchatel - Ship Type: P - How taken: Leander (Newcastle Acasta) - When taken: 20 Dec 1814 - Where taken: Lat 38 Long 56 - Date Received: 30 Jan 1815 - From what ship: HMT Pheasant - Born: Boston - Age: 18 - Released on 5 Jul 1815.

Lambert, Andres - Seaman - Nbr: 6263 - Prize: Plutarch - Ship Type: MV - How taken: HMS Helicon - When taken: 5 Feb 1815 - Where taken: off Bordeaux - Date Received: 10 Feb 1815 - From what ship: HMT Ganges, Plymouth - Born: Hanover - Age: 26 - Released on 5 Jul 1815.

Lambert, Calvin - Seaman - Nbr: 5346 - Prize: Sally - Ship Type: MV - How taken: HMS Maidstone - When taken: 17 Jul 1813 - Where taken: Grand Banks - Date Received: 31 Oct 1814 - From what ship: HMT Leyden, Chatham - Born: Barnstable - Age: 18 - Released on 1 Jul 1815.

Lambert, Ephraim - Seaman - Nbr: 5176 - Prize: Vivid - Ship Type: P - How taken: HMS Nymphe - When taken: 20 Apr 1813 - Where taken: off Cape Cod - Date Received: 31 Oct 1814 - From what ship: HMT Mermaid, Chatham - Born: Truro - Age: 19 - Released on 11 Jul 1815.

Lambert, Samuel - Seaman - Nbr: 1762 - Prize: Cygnet - Ship Type: Prize - How taken: HMS Statira - When taken: 8 May 1814 - Where taken: at sea - Date Received: 20 Jul 1814 - From what ship: HMS Milford, Plymouth - Born: Salem - Age: 19 - Released on 1 May 1815.

Lambert, Thomas - Seaman - Nbr: 4644 - How taken: Gave himself up from HMS Ville de Paris - When taken: 1 Jul 1813 - Date Received: 9 Oct 1814 - From what ship: HMT Leyden, Chatham - Born: Massachusetts - Age: 39 - Race: Black - Released on 15 Jun 1815.

Lambert, William - Prize Master - Nbr: 928 - Prize: Fanny - Ship Type: MV - How taken: HMS Eurotas - When taken: 23 Dec 1813 - Where taken: at sea - Date Received: 31 Jan 1814 - From what ship: Plymouth - Born: New York - Age: 42 - Impersonated a Frenchman on the general release.

Lambert, William - Seaman - Nbr: 2552 - Prize: Caromaned - Ship Type: Prize - How taken: HMS Eridanus - When taken: 13 Aug 1814 - Where taken: Lat 40, Long 16 - Date Received: 16 Aug 1814 - From what ship: HMT Salvador del Mundo - Born: New York - Age: 26 - Race: Black - Released on 3 May 1815.

Lamond, John - Seaman - Nbr: 1425 - Prize: Revenge - Ship Type: LM - How taken: HMS Belle Poule - When taken: 10 May 1814 - Where taken: off Cornwall - Date Received: 19 Jun 1814 - From what ship: Stapleton - Born: Philadelphia - Age: 37 - Released on 28 Apr 1815.

Lampriere, David - Seaman - Nbr: 3988 - Prize: US Brig Rattlesnake - Ship Type: MW - How taken: HMS Leander - When taken: 13 Jul 1814 - Where taken: off Shelburne - Date Received: 6 Oct 1814 - From what ship: HMT Chesapeake, Halifax - Born: Beaumont - Age: 24 - Released on 9 Jun 1815.

Lamson, Amos - Seaman - Nbr: 4484 - Prize: Enterprize - Ship Type: P - How taken: HMS Tenedos - When taken:

American Prisoners of War Held at Dartmoor during the War of 1812

## Alphabetical listing of names

21 May 1813 - Where taken: off Cape Cod - Date Received: 8 Oct 1814 - From what ship: HMT Leyden, Chatham - Born: Massachusetts - Age: 20 - Released on 15 Jun 1815.

Lamson, Charles - 2nd Mate - Nbr: 53 - Prize: Spitfire - Ship Type: MV - How taken: HMS Achates - When taken: 14 Feb 1813 - Where taken: off Ushant (France) - Date Received: 2 Apr 1813 - From what ship: Plymouth - Born: Beverly - Age: 23 - Released on 29 Apr 1815.

Lamson, Noah - Seaman - Nbr: 5169 - Prize: Volante - Ship Type: P - How taken: HMS Curlew - When taken: 25 Nov 1813 - Where taken: off Halifax - Date Received: 31 Oct 1814 - From what ship: HMT Mermaid, Chatham - Born: Plymouth - Age: 17 - Released on 29 Jun 1815.

Lamson, William - Boy - Nbr: 1866 - Prize: Porcupine - Ship Type: LM - How taken: HMS Acasta - When taken: 17 Jun 1813 - Where taken: off Sable Island - Date Received: 29 Jul 1814 - From what ship: HMS Ville de Paris, Chatham Depot - Born: Charlestown - Age: 18 - Released on 2 May 1815.

Lander, John - Seaman - Nbr: 1744 - How taken: Gave himself up from HMS Royalist - When taken: 16 Jun 1814 - Date Received: 18 Jul 1814 - From what ship: HMT Salvador del Mundo, Plymouth - Born: Massachusetts - Age: 35 - Released on 1 May 1815.

Lander, Warren - Seaman - Nbr: 5795 - Prize: General Putnam - Ship Type: P - How taken: HMS Leander - When taken: 8 Nov 1814 - Where taken: Long 65 Lat 42 - Date Received: 26 Dec 1814 - From what ship: HMT Argo - Born: Massachusetts - Age: 19 - Released on 3 Jul 1815.

Landerkin, Daniel - Seaman - Nbr: 733 - Prize: Ned - Ship Type: LM - How taken: HMS Royalist - When taken: 6 Sep 1813 - Where taken: Bay of Biscay - Date Received: 27 Sep 1813 - From what ship: Plymouth - Born: Marblehead - Age: 33 - Released on 26 Apr 1815.

Landerman, Joseph - Steward - Nbr: 274 - Prize: Cannoneer - Ship Type: MV - How taken: HMS Warspite - When taken: 14 Mar 1813 - Where taken: Bay of Biscay - Date Received: 28 Jun 1813 - From what ship: Plymouth - Born: Havana - Age: 23 - Released on 20 Apr 1815.

Landford, John - Prize Master - Nbr: 774 - Prize: Betsy - Ship Type: MV - How taken: HMS Eurotas - When taken: 26 Oct 1813 - Where taken: off Ushant (France) - Date Received: 3 Nov 1813 - From what ship: Plymouth - Born: Somerset County - Age: 25 - Escaped on 3 Feb 1815.

Landsbury, John - Master - Nbr: 5576 - How taken: Sent into custody from a fishing boat on the Baltimore River - Date Received: 17 Dec 1814 - From what ship: HMT Loire, Halifax - Born: Annapolis - Age: 33 - Released on 1 Jul 1815.

Lane, Henry - Seaman - Nbr: 4905 - Prize: Diomede - Ship Type: P - How taken: HMS Maidstone - When taken: 26 Mar 1814 - Where taken: off NY - Date Received: 28 Oct 1814 - From what ship: HMT Alkbar, Halifax - Born: New York - Age: 32 - Released on 21 Jun 1815.

Lane, Henry - Seaman - Nbr: 3701 - Prize: Alfred - Ship Type: P - How taken: HMS Epervier - When taken: 23 Feb 1814 - Where taken: off Newfoundland - Date Received: 30 Sep 1814 - From what ship: HMT President, Halifax - Born: Marblehead - Age: 47 - Released on 4 Jun 1815.

Lane, James - Seaman - Nbr: 1371 - Prize: Grand Napoleon - Ship Type: P - How taken: HMS Goldfinch - When taken: 17 Apr 1813 - Where taken: Bay of Biscay - Date Received: 19 Jun 1814 - From what ship: Stapleton - Born: New York - Age: 24 - Released on 28 Apr 1815.

Lane, James - Prize master - Nbr: 4452 - Prize: Prize to Prince - Ship Type: Prize - How taken: HMS Ethalion - When taken: 14 Dec 1813 - Where taken: at sea - Date Received: 8 Oct 1814 - From what ship: HMT Leyden, Chatham - Born: Massachusetts - Age: 32 - Released on 14 Jun 1815.

Lane, John - Boy - Nbr: 1899 - Prize: Quebec - Ship Type: Prize - How taken: HMS Derwent - When taken: 29 Jan 1813 - Where taken: off Lisbon - Date Received: 3 Aug 1814 - From what ship: HMS Alceste, Chatham Depot - Born: New York - Age: 18 - Released on 2 May 1815.

Lane, John - Seaman - Nbr: 1745 - How taken: Gave himself up from HMS Sparrow Hawk - When taken: 8 Aug 1812 - Date Received: 20 Jul 1814 - From what ship: HMS Milford, Plymouth - Born: Stillwater - Age: 24 - Released on 26 Apr 1815.

Lane, Titus - Seaman - Nbr: 3721 - Prize: Alfred - Ship Type: P - How taken: HMS Epervier - When taken: 23 Feb 1814 - Where taken: off Newfoundland - Date Received: 30 Sep 1814 - From what ship: HMT President, Halifax - Born: Salem - Age: 30 - Race: Negro - Released on 4 Jun 1815.

Lane, W. S. - Prize Master - Nbr: 1760 - Prize: Cygnet - Ship Type: Prize - How taken: HMS Statira - When taken: 8 May 1814 - Where taken: at sea - Date Received: 20 Jul 1814 - From what ship: HMS Milford, Plymouth - Born: Gloucester - Age: 22 - Released on 1 May 1815.

Lane, William - Seaman - Nbr: 1887 - How taken: Gave himself up from HMS Comet - When taken: 1813 - Date Received: 29 Jul 1814 - From what ship: HMS Ville de Paris, Chatham Depot - Born: New York - Age: 33 -

American Prisoners of War Held at Dartmoor during the War of 1812

## Alphabetical listing of names

Released on 20 Apr 1815.
Lane, William - Commander - Nbr: 4684 - Prize: Prize to the Wiley Reynan - Ship Type: P - How taken: HMS Shannon - When taken: 11 Oct 1812 - Where taken: off Newfoundland - Date Received: 9 Oct 1814 - From what ship: HMT Leyden, Chatham - Born: Boston - Age: 33 - Race: Mulatto - Released on 27 Apr 1815.
Lang, William - Seaman - Nbr: 1202 - Prize: Lyon - Ship Type: P - How taken: HMS Pheasant - When taken: 8 Jun 1813 - Where taken: off St. Antonio - Date Received: 4 Jun 1814 - From what ship: Dartmouth - Born: New York - Age: 23 - Released on 28 Apr 1815.
Langford, Samuel - Seaman - Nbr: 3400 - Prize: Thomas - Ship Type: P - How taken: HMS Nymphe - When taken: 24 Jun 1813 - Where taken: off Halifax - Date Received: 13 Sep 1814 - From what ship: HMT Niobe, Chatham -Born: Massachusetts - Age: 18 - Released on 28 May 1815.
Langhane, Bell'm - Seaman - Nbr: 5453 - Prize: Prize to the Lawrence - Ship Type: P - How taken: HMS Glasgow - When taken: 2 Nov 1814 - Where taken: Channel - Date Received: 10 Dec 1814 - From what ship: HMT Impregnable - Born: Virginia - Age: 22 - Released on 3 Jul 1815.
Langley, Richard - Seaman - Nbr: 4332 - Prize: Flash - Ship Type: LM - How taken: HMS Acasta and HMS Loire - When taken: 8 Feb 1814 - Where taken: off NY - Date Received: 7 Oct 1814 - From what ship: HMT Salvador del Mundo, Halifax - Born: Virginia - Age: 42 - Race: Negro - Released on 14 Jun 1815.
Langley, Richard - Prize Master - Nbr: 3113 - Prize: Harmony - Ship Type: Prize - How taken: HMS Brisk - When taken: 24 Aug 1814 - Where taken: Bristol Channel - Date Received: 11 Sep 1814 - From what ship: HMT Salvador del Mundo - Born: Portland - Age: 35 - Released on 28 May 1815.
Langroth, Francis - Seaman - Nbr: 4174 - How taken: Gave himself up from HMS Machal - Date Received: 7 Oct 1814 - From what ship: HMT Niobe, Chatham - Born: Hungary - Age: 33 - Released on 10 Oct 1814.
Lanstone, Peter - Seaman - Nbr: 3908 - Prize: Thorn - Ship Type: MV - How taken: HMS Bulwark - When taken: 9 Jul 1814 - Where taken: off Nantucket - Date Received: 5 Oct 1814 - From what ship: HMT Orpheus, Halifax - Born: Gothenburg - Age: 37 - Released on 9 Jun 1815.
Lapham, Cushman - Prize Master - Nbr: 2604 - How taken: Gave himself up from HMS Quebec - When taken: 29 Jan 1813 - Date Received: 21 Aug 1814 - From what ship: HMT Freya, Chatham - Born: Rhode Islandng - Age: 32 - Released on 27 Apr 1815.
Lapish, Andrew - Sergeant - Nbr: 4674 - Prize: Portsmouth Packet - Ship Type: Prize - How taken: HMS Fantome - When taken: 5 Oct 1813 - Where taken: off Portland - Date Received: 9 Oct 1814 - From what ship: HMT Leyden, Chatham - Born: Hampshire - Age: 26 - Released on 15 Jun 1815.
Larabee, Thomas - Seaman - Nbr: 4404 - Prize: Portsmouth Packet - Ship Type: Prize - How taken: HMS Fantome - When taken: 5 Oct 1813 - Where taken: off Portland - Date Received: 8 Oct 1814 - From what ship: HMT Leyden, Chatham - Born: Massachusetts - Age: 21 - Released on 14 Jun 1815.
Larkin, Amos - Seaman - Nbr: 1768 - How taken: Gave himself up from HMS Reynard - When taken: 4 Jul 1814 - Date Received: 20 Jul 1814 - From what ship: HMS Milford, Plymouth - Born: Beverly - Age: 30 - Race: Black - Died on 29 Jan 1815 from phthisis pulmonalis.
Larkin, Lewis - Seaman - Nbr: 2240 - Prize: Rolla - Ship Type: Prize - How taken: HMS Loire - When taken: 9 Dec 1813 - Where taken: off Newport - Date Received: 16 Aug 1814 - From what ship: HMS Dublin, Halifax - Born: Connecticut - Age: 2 - Died on 30 Sep 1814.
Larrabee, William - Seaman - Nbr: 1978 - Prize: Orbit - Ship Type: MV - How taken: HMS Achates - When taken: 29 Jan 1813 - Where taken: Lat 44N Long 13W - Date Received: 3 Aug 1814 - From what ship: HMS Lyffey, Chatham Depot - Born: Portland - Age: 44 - Released on 2 May 1815.
Laskey, Benjamin - Seaman - Nbr: 3448 - Prize: Industry - Ship Type: P - How taken: HMS Heron - When taken: 3 Nov 1813 - Where taken: off Halifax - Date Received: 13 Sep 1814 - From what ship: HMT Niobe, Chatham - Born: Marblehead - Age: 18 - Released on 28 May 1815.
Latham, Emil - Seaman - Nbr: 6087 - Prize: Foria - Ship Type: MV - How taken: HMS Loire - When taken: 26 Sep 1814 - Where taken: off Bahamas - Date Received: 28 Dec 1814 - From what ship: HMT Penelope - Born: SC - Age: 27 - Released on 14 Jun 1815.
Latham, Giles - Prize master - Nbr: 4862 - Prize: Chasseur - Ship Type: P - How taken: HMS Recruit - When taken: 6 Apr 1814 - Where taken: Lat 29 Long 76 - Date Received: 9 Oct 1814 - From what ship: HMT Freya, Halifax - Born: Connecticut - Age: 34 - Released on 21 Jun 1815.
Latham, John - Seaman - Nbr: 2888 - How taken: Gave himself up from HMS Shearwater - When taken: 28 May 1813 - Date Received: 24 Aug 1814 - From what ship: HMT Alpheus, Chatham - Born: New York - Age: 25 - Released on 19 May 1815.
Latimer, John - Seaman - Nbr: 5852 - Prize: Prize of the Enterprize - Ship Type: P - How taken: HMS Barracouta -

American Prisoners of War Held at Dartmoor during the War of 1812

## Alphabetical listing of names

When taken: 10 Oct 1814 - Where taken: off Western Islands (England) - Date Received: 26 Dec 1814 - From what ship: HMT Argo - Born: Belgium - Age: 20 - Released on 3 Jul 1815.

Latimore, Mathew - Seaman - Nbr: 460 - Prize: Meteor - Ship Type: MV - How taken: HMS Briton - When taken: 12 Mar 1813 - Where taken: Bay of Biscay - Date Received: 8 Sep 1813 - From what ship: Plymouth - Born: New York - Age: 19 - Sent to Plymouth on 7 Dec 1813.

Lattimore, John - Seaman - Nbr: 5508 - Prize: Ann Dorothy - Ship Type: Prize - How taken: HMS Maidstone - When taken: 30 Oct 1814 - Where taken: off Cape Sable Island (Canada) - Date Received: 17 Dec 1814 - From what ship: HMT Loire, Halifax - Born: Philadelphia - Age: 19 - Released on 29 Jun 1815.

Laurence, Jean - Seaman - Nbr: 1542 - Prize: Zebra - Ship Type: LM - How taken: HMS Pyramus - When taken: 20 Apr 1813 - Where taken: Bay of Biscay - Date Received: 23 Jun 1814 - From what ship: Stapleton - Born: Isle of France - Age: 27 - Race: Colored - Sent to Mill Prison (Plymouth, England) on 26 Jun 1814.

Laurence, John - Seaman - Nbr: 5577 - Prize: Sarah - Ship Type: Prize - How taken: HMS Maidstone - When taken: 28 Sep 1814 - Date Received: 17 Dec 1814 - From what ship: HMT Loire, Halifax - Born: Buckston - Age: 26 - Released on 1 Jul 1815.

Laurenceau, John - Boy - Nbr: 501 - Prize: Hannah & Eliza - Ship Type: MV - How taken: HMS Lyra - When taken: 29 Mar 1813 - Where taken: off Bayonne - Date Received: 8 Sep 1813 - From what ship: Plymouth - Born: Bordeaux - Age: 9 - Sent to Mill Prison (Plymouth, England) on 21 Jun 1814.

Laurent, Jonathan - Seaman - Nbr: 3539 - Prize: Hawk - Ship Type: P - How taken: HMS Pique - When taken: 26 Apr 1814 - Where taken: off Bermuda - Date Received: 30 Sep 1814 - From what ship: HMT Sybella - Born: Not listed - Released on 4 Jun 1815.

Lavasseur, Rene - Seaman - Nbr: 1184 - Prize: Melanie - Ship Type: MV - How taken: HMS Briton - When taken: 15 Sep 1813 - Where taken: off Bordeaux - Date Received: 4 Jun 1814 - From what ship: Dartmouth - Born: New Orleans - Age: 20 - Race: Mulatto - Sent to Plymouth on 8 Jul 1814.

Law, Thomas - Seaman - Nbr: 4135 - Prize: Bordeaux Packet - Ship Type: LM - How taken: HMS Niemen - When taken: 28 Jun 1814 - Where taken: off Delaware - Date Received: 6 Oct 1814 - From what ship: HMT Chesapeake, Halifax - Born: Delaware - Age: 20 - Released on 13 Jun 1815.

Lawrence, Francis - Seaman - Nbr: 6499 - Prize: Dorthea - Ship Type: MV - How taken: HMS North Star - When taken: Dec 1814 - Where taken: St. Jago (Jamaica) - Date Received: 3 Mar 1815 - From what ship: HMT Ganges, Plymouth - Born: Denmark - Age: 21 - Released on 11 Jul 1815.

Lawrence, George - Seaman - Nbr: 3286 - Prize: Juliana Smith - Ship Type: P - How taken: HMS Nymphe - When taken: 11 May 1813 - Where taken: off Cape Sable Island (Canada) - Date Received: 13 Sep 1814 - From what ship: HMT Niobe, Chatham - Born: Alexandria - Age: 20 - Released on 28 May 1815.

Lawrence, Robert - Seaman - Nbr: 1848 - How taken: Gave himself up from HMS Decoy - When taken: 28 May 1813 - Date Received: 21 Jul 1814 - From what ship: HMT Redbeard & Pincher, Chatham Depot - Born: Philadelphia - Age: 37 - Released on 2 May 1815.

Laws, Peter - Seaman - Nbr: 3129 - How taken: Impressed at London - When taken: 3 Sep 1813 - Date Received: 11 Sep 1814 - From what ship: HMT Freya, Chatham - Born: New York - Age: 30 - Race: Black - Released on 27 Apr 1815.

Lawson, James - Seaman - Nbr: 210 - Prize: Mars - Ship Type: MV - How taken: HMS Warspite - When taken: 26 Feb 1813 - Where taken: off Basque Roads (France) - Date Received: 2 Apr 1813 - From what ship: Plymouth - Born: Baltimore - Age: 27 - Race: Black - Died on 5 Jan 1814 from small pox.

Lawson, Thomas - Seaman - Nbr: 4666 - Prize: Thomas - Ship Type: P - How taken: HMS Nymphe - When taken: 28 Jun 1813 - Where taken: off Halifax - Date Received: 9 Oct 1814 - From what ship: HMT Leyden, Chatham - Born: Massachusetts - Age: 27 - Released on 15 Jun 1815.

Lawson, Thomas - Seaman - Nbr: 222 - Prize: Pert - Ship Type: MV - How taken: HMS Warspite - When taken: 1 Mar 1813 - Where taken: off Basque Roads (France) - Date Received: 2 Apr 1813 - From what ship: Plymouth - Born: New York - Age: 36 - Released on 2 Apr 1815.

Lawton, William - Seaman - Nbr: 4589 - Prize: Bunker Hill - Ship Type: P - How taken: HMS Pomone - When taken: 2 Mar 1814 - Where taken: Channel - Date Received: 9 Oct 1814 - From what ship: HMT Leyden, Chatham - Born: Portsmouth - Age: 28 - Released on 15 Jun 1815.

Lay, Lee - Seaman - Nbr: 5652 - How taken: Sent into custody from MV Purssan - Date Received: 24 Dec 1814 - From what ship: HMT Impregnable - Born: Connecticut - Age: 33 - Released on 3 Jul 1815.

Layfield, Littleton - Seaman - Nbr: 1622 - Prize: Tom - Ship Type: LM - How taken: HMS Surveillante - When taken: 24 Apr 1813 - Where taken: Bay of Biscay - Date Received: 23 Jun 1814 - From what ship: Stapleton - Born: Maryland - Age: 21 - Released on 1 May 1815.

American Prisoners of War Held at Dartmoor during the War of 1812

## Alphabetical listing of names

Layton, William - Boy - Nbr: 1786 - Prize: Gleaner - Ship Type: MV - How taken: Brothers (Privateer) - When taken: 16 Mar 1813 - Where taken: Bay of Biscay - Date Received: 20 Jul 1814 - From what ship: HMS Milford, Plymouth - Born: Castine - Age: 18 - Released on 1 May 1815.

Le More, John - Seaman - Nbr: 4387 - Prize: Globe - Ship Type: P - How taken: HMS Spartan - When taken: 13 Jun 1813 - Where taken: off Delaware - Date Received: 8 Oct 1814 - From what ship: HMT Leyden, Chatham - Born: New York - Age: 23 - Race: Black - Released on 14 Jun 1815.

Leach, Charles - Seaman - Nbr: 5415 - Prize: Prize to the Diomede - How taken: HMS Sapphire - When taken: 24 Feb 1814 - Where taken: off America - Date Received: 31 Oct 1814 - From what ship: HMT Leyden, Chatham - Born: Salem - Age: 27 - Released on 1 Jul 1815.

Leach, Daniel - Seaman - Nbr: 4403 - Prize: Portsmouth Packet - Ship Type: Prize - How taken: HMS Fantome - When taken: 5 Oct 1813 - Where taken: off Portland - Date Received: 8 Oct 1814 - From what ship: HMT Leyden, Chatham - Born: Massachusetts - Age: 51 - Released on 14 Jun 1815.

Leach, George - Seaman - Nbr: 2988 - Prize: Frolic - Ship Type: P - How taken: HMS Heron - When taken: 25 Jan 1814 - Where taken: off St. Thomas - Date Received: 29 Aug 1814 - From what ship: HMT Bittern - Born: Salem - Age: 60 - Released on 19 May 1815.

Leach, Josem - Seaman - Nbr: 619 - Prize: US Brig Argus - Ship Type: MW - How taken: HMS Pelican - When taken: 14 Aug 1813 - Where taken: Irish Channel - Date Received: 8 Sep 1813 - From what ship: Plymouth - Born: Onondaga - Age: 34 - Sent to Dartmouth on 2 Nov 1814.

Leach, William - Seaman - Nbr: 264 - Prize: William Bayard - Ship Type: MV - How taken: HMS Warspite - When taken: 3 Mar 1813 - Where taken: Bay of Biscay - Date Received: 28 Jun 1813 - From what ship: Plymouth - Born: Newburgh - Age: 19 - Released on 20 Apr 1815.

Leach, William - Seaman - Nbr: 48 - Prize: Terrible - Ship Type: MV - How taken: HMS Foxhound - When taken: 8 Feb 1813 - Where taken: Channel - Date Received: 2 Apr 1813 - From what ship: Plymouth - Born: Debry - Age: 20 - Sent to Dartmouth on 7 Aug 1813.

Lear, Alexander - Seaman - Nbr: 2416 - Prize: US Sloop Frolic - Ship Type: MW - How taken: HMS Orpheus - When taken: 20 Apr 1814 - Where taken: off Cuba - Date Received: 16 Aug 1814 - From what ship: HMT Queen, Halifax - Born: Portsmouth - Age: 38 - Released on 3 May 1815.

Lear, George - Seaman - Nbr: 5904 - Prize: Harlequin - Ship Type: P - How taken: HMS Bulwark - When taken: 23 Nov 1814 - Where taken: off Halifax - Date Received: 27 Dec 1814 - From what ship: HMT Penelope - Born: Portsmouth - Age: 24 - Released on 3 Jul 1815.

Leary, Daniel - Seaman - Nbr: 1035 - Prize: US Brig Argus - Ship Type: MW - How taken: HMS Pelican - When taken: 16 Aug 1813 - Where taken: Irish Channel - Date Received: 10 May 1814 - From what ship: Plymouth - Born: New York - Age: 24 - Sent to Dartmouth on 19 Oct 1814.

Leas, Anthony - Seaman - Nbr: 358 - Prize: Two Brothers - Ship Type: MV - How taken: Beetle (LM) - When taken: 18 Mar 1813 - Where taken: off Western Islands (England) - Date Received: 1 Jul 1813 - From what ship: Plymouth - Born: New Orleans - Age: 30 - Sent to Mill Prison (Plymouth, England) on 21 Jun 1814.

Lebaith, John P. - Seaman - Nbr: 396 - Prize: Young Holkar - Ship Type: MV - How taken: HMS Superb - When taken: 10 Apr 1813 - Where taken: off Belle Isle, France - Date Received: 1 Jul 1813 - From what ship: Plymouth - Born: New Orleans - Age: 20 - Sent to Plymouth on 8 Jul 1814.

LeBaron, Peter - Seaman - Nbr: 2640 - Prize: True Blooded Yankee - Ship Type: P - How taken: HMS Hamadryad - When taken: 24 Jul 1813 - Where taken: off Norway - Date Received: 21 Aug 1814 - From what ship: HMT Freya, Chatham - Born: Cape Ann - Age: 28 - Released on 10 Apr 1815.

Leckler, Anthony - Seaman - Nbr: 1311 - How taken: Sent into custody from HMS Cornwallis - Date Received: 14 Jun 1814 - From what ship: Mill Prison (Plymouth, England) - Born: Lancaster - Age: 26 - Released on 28 Apr 1815.

LeCore, Peter M. - Pilot - Nbr: 989 - Prize: Harvest - Ship Type: P - How taken: HMS Orestes - When taken: 20 Jan 1814 - Where taken: Bay of Biscay - Date Received: 31 Jan 1814 - From what ship: Plymouth - Born: Roscoff - Age: 63 - Sent to Mill Prison (Plymouth, England) on 21 Jun 1814.

Lee, Abraham - Seaman - Nbr: 6241 - Prize: Albion - Ship Type: Prize - How taken: HMS Harlequin - When taken: 6 Jan 1814 - Where taken: Long 20 - Date Received: 4 Feb 1815 - From what ship: HMT Ganges - Born: New York - Age: 16 - Released on 5 Jul 1815.

Lee, Charles - Seaman - Nbr: 6121 - Prize: Betsey - Ship Type: Prize - How taken: HMS Bellerophon - When taken: 2 Nov 1814 - Where taken: Long 61 - Date Received: 6 Jan 1814 - From what ship: HMT Impregnable - Born: Manchester - Age: 24 - Released on 5 Jul 1815.

Lee, Edward - Seaman - Nbr: 2718 - How taken: Gave himself up from HMS Mulgrave - When taken: 13 Dec 1812

American Prisoners of War Held at Dartmoor during the War of 1812

## Alphabetical listing of names

- Date Received: 21 Aug 1814 - From what ship: HMT Freya, Chatham - Born: Maryland - Age: 25 - Race: Mulatto - Released on 26 Apr 1815.
Lee, Ely - Seaman - Nbr: 571 - Prize: US Brig Argus - Ship Type: MW - How taken: HMS Pelican - When taken: 14 Apr 1813 - Where taken: Irish Channel - Date Received: 8 Sep 1813 - From what ship: Plymouth - Born: Connecticut - Age: 34 - Sent to Dartmouth on 2 Nov 1814.
Lee, George - Seaman - Nbr: 1789 - Prize: Ferox - Ship Type: MV - How taken: HMS Medusa & HMS Lyra - When taken: 28 Mar 1813 - Where taken: off Cape Ortegal (Spain) - Date Received: 20 Jul 1814 - From what ship: HMS Milford, Plymouth - Born: New York - Age: 33 - Released on 1 May 1815.
Lee, Isaac - Seaman - Nbr: 3089 - Prize: Eliza (Whaler) - Ship Type: MV - How taken: HMS Charles - When taken: 20 Jul 1813 - Where taken: off Peter Head - Date Received: 3 Sep 1814 - From what ship: HMT Hydra, Chatham - Born: New York - Age: 19 - Released on 28 May 1815.
Lee, John - Seaman - Nbr: 6031 - Prize: Regent - Ship Type: LM - How taken: HMS Forth - When taken: 19 Sep 1814 - Where taken: off Egg Harbor (New Jersey) - Date Received: 28 Dec 1814 - From what ship: HMT Penelope - Born: New York - Age: 21 - Released on 11 Jul 1815.
Lee, John - Seaman - Nbr: 2695 - How taken: Gave himself up from HMS Spartan - When taken: 29 Oct 1813 - Date Received: 21 Aug 1814 - From what ship: HMT Freya, Chatham - Born: New Orleans - Age: 40 - Released on 19 May 1815.
Lee, Joseph - Seaman - Nbr: 2052 - How taken: Gave himself up from Chatham - When taken: 21 Jun 1813 - Date Received: 3 Aug 1814 - From what ship: HMS Lyffey, Chatham Depot - Born: New York - Age: 35 - Released on 2 May 1815.
Lee, Michael - Seaman - Nbr: 1875 - Prize: Elbridge Gerry - Ship Type: P - How taken: HMS Crescent - When taken: 16 Sep 1813 - Where taken: off St. George's - Date Received: 29 Jul 1814 - From what ship: HMS Ville de Paris, Chatham Depot - Born: Portland - Age: 19 - Released on 2 May 1815.
Lee, Nathaniel - Boy - Nbr: 2657 - Prize: Growler - Ship Type: MV - How taken: HMS Electra - When taken: 7 Jul 1813 - Where taken: off Newfoundland - Date Received: 21 Aug 1814 - From what ship: HMT Freya, Chatham - Born: Marblehead - Age: 12 - Released on 19 May 1815.
Lee, Richard - Seaman - Nbr: 6126 - Prize: Betsey - Ship Type: Prize - How taken: HMS Bellerophon - When taken: 2 Nov 1814 - Where taken: Long 61 - Date Received: 6 Jan 1814 - From what ship: HMT Impregnable - Born: Marblehead - Age: 25 - Died on 19 Jun 1815 from phthisis pulmonalis.
Lee, Richard - Seaman - Nbr: 3758 - Prize: Alfred - Ship Type: P - How taken: HMS Epervier - When taken: 23 Feb 1814 - Where taken: off Newfoundland - Date Received: 30 Sep 1814 - From what ship: HMT President, Halifax - Born: Marblehead - Age: 52 - Released on 4 Jun 1815.
Lee, Richard Robert - Seaman - Nbr: 5537 - Prize: Ann - Ship Type: Prize - How taken: HMS Hamadryad - When taken: 10 Sep 1814 - Where taken: Lat 41 Long 43 - Date Received: 17 Dec 1814 - From what ship: HMT Loire, Halifax - Born: Massachusetts - Age: 23 - Died on 20 Jan 1815 from phthisis pulmonalis.
Lee, Samuel - Seaman - Nbr: 4874 - Prize: US Brig Rattlesnake - Ship Type: MW - How taken: HMS Leander - When taken: 13 Jul 1814 - Where taken: off Shelburne - Date Received: 24 Oct 1814 - From what ship: Royal Hospital, Plymouth - Born: Not listed - Released on 11 Jul 1815.
Lee. John - Seaman - Nbr: 4610 - Prize: Argus - Ship Type: MV - How taken: HMS San Domingo - When taken: 1 Mar 1814 - Where taken: off Savannah - Date Received: 9 Oct 1814 - From what ship: HMT Leyden, Chatham - Born: Providence - Age: 25 - Released on 15 Jun 1815.
Leech, Ezekiel - Seaman - Nbr: 3708 - Prize: Alfred - Ship Type: P - How taken: HMS Epervier - When taken: 23 Feb 1814 - Where taken: off Newfoundland - Date Received: 30 Sep 1814 - From what ship: HMT President, Halifax - Born: Marblehead - Age: 18 - Released on 4 Jun 1815.
Leech, William - Seaman - Nbr: 2207 - Prize: Hussar - Ship Type: P - How taken: HMS Saturn - When taken: 25 May 1814 - Where taken: off Sandy Hook - Date Received: 16 Aug 1814 - From what ship: HMS Dublin, Halifax - Born: Connecticut - Age: 23 - Released on 2 May 1815.
Leeds, Leon - Seaman - Nbr: 25 - How taken: Apprehended at Gibraltar - When taken: 8 Aug 1813 - Date Received: 2 Apr 1813 - From what ship: Plymouth - Born: Boston - Age: 18 - Sent to Dartmouth on 30 Jul 1813.
Legatt, Charles - Seaman - Nbr: 4094 - Prize: Tyger - Ship Type: Prize - How taken: HMS Bulwark - When taken: 20 Jul 1814 - Where taken: Georges Bank - Date Received: 6 Oct 1814 - From what ship: HMT Chesapeake, Halifax - Born: New Orleans - Age: 23 - Released on 13 Jun 1815.
Leger, Andrew - Seaman - Nbr: 985 - Prize: Amity - Ship Type: P - How taken: HMS Achates - When taken: 22 Dec 1813 - Where taken: Bay of Biscay - Date Received: 31 Jan 1814 - From what ship: Plymouth - Born: New Orleans - Age: 25 - Sent to Plymouth on 8 Jul 1814.

American Prisoners of War Held at Dartmoor during the War of 1812

## Alphabetical listing of names

Leggatt, Charles - Seaman - Nbr: 5718 - Prize: Tyger - Ship Type: Prize - How taken: HMS Bulwark - When taken: 20 Jul 1814 - Where taken: Georges Bank - Date Received: 24 Dec 1814 - From what ship: HMT Penelope - Born: Finland - Age: 40 - Released on 3 Jul 1815.

Leggett, William - Seaman - Nbr: 1107 - Prize: Bunker Hill - Ship Type: P - How taken: HMS Pomone & HMS Cadmus - When taken: 4 Mar 1814 - Where taken: Bay of Biscay - Date Received: 10 May 1814 - From what ship: Plymouth - Born: Boston - Age: 21 - Released on 27 Apr 1815.

Leggins, Ebenezer - Seaman - Nbr: 4081 - Prize: New Zealander - Ship Type: Prize - How taken: HMS Belviders - When taken: 21 Apr 1814 - Where taken: off Delaware - Date Received: 6 Oct 1814 - From what ship: HMT Chesapeake, Halifax - Born: Massachusetts - Age: 23 - Released on 13 Jun 1815.

Legos, Philip - Seaman - Nbr: 2909 - Prize: Matilda - Ship Type: MV - How taken: HMS Revolutionnaire - When taken: 25 Jul 1813 - Where taken: off L'Orient (France) - Date Received: 24 Aug 1814 - From what ship: HMT Alpheus, Chatham - Born: Massachusetts - Age: 31 - Released on 2 Nov 1814.

Legur, Ephraim - Seaman - Nbr: 3512 - Prize: Aeolus - Ship Type: Prize - How taken: HMS Warspite - When taken: 2 Sep 1814 - Where taken: off Newfoundland - Date Received: 28 Sep 1814 - From what ship: HMT Salvador del Mundo - Born: New York - Age: 28 - Released on 4 Jun 1815.

Lehart, Jacob - Seaman - Nbr: 871 - Prize: Agnes - Ship Type: LM - How taken: Jane (Cutter) - When taken: 29 Nov 1813 - Where taken: Bay of Biscay - Date Received: 31 Jan 1814 - From what ship: Plymouth - Born: Cape Cod - Age: 22 - Released on 27 Apr 1815.

Lehens, Andrew - Seaman - Nbr: 116 - Prize: Governor McKean - Ship Type: LM - How taken: HMS Rover - When taken: 26 Jan 1813 - Where taken: off Bordeaux - Date Received: 2 Apr 1813 - From what ship: Plymouth - Born: Medford - Age: 25 - Sent to Dartmouth on 30 Jul 1813.

Lehman, Francis - 2nd Lieutenant - Nbr: 4436 - Prize: Growler - Ship Type: P - How taken: HMS Electra - When taken: 7 Jul 1814 - Where taken: at sea - Date Received: 8 Oct 1814 - From what ship: HMT Leyden, Chatham - Born: Marblehead - Age: 28 - Released on 27 Apr 1815.

Lemercier, Peter - Seaman - Nbr: 297 - Prize: Ducornau - Ship Type: MV - How taken: HMS Pheasant - When taken: 15 Mar 1813 - Where taken: Bay of Biscay - Date Received: 28 Jun 1813 - From what ship: Plymouth - Born: Boston - Age: 25 - Released on 20 Apr 1815.

Lemon, James - Seaman - Nbr: 3489 - Prize: US Sloop Frolic - Ship Type: MW - How taken: HMS Orpheus - When taken: 20 Apr 1814 - Where taken: off Cuba - Date Received: 19 Sep 1814 - From what ship: HMT Salvador del Mundo - Born: Delaware - Age: 35 - Released on 3 May 1815.

Lemon, John - Seaman - Nbr: 6496 - Prize: Herald - Ship Type: P - How taken: HMS Endymion - When taken: 15 Aug 1814 - Where taken: off Nantucket - Date Received: 3 Mar 1815 - From what ship: HMT Ganges, Plymouth - Born: New York - Age: 26 - Race: Black - Released on 11 Jul 1815.

Lemon, N. C. - 2nd Lieutenant - Nbr: 2723 - Prize: John of Salem - Ship Type: P - How taken: HMS Peruvian - When taken: 6 Feb 1813 - Where taken: at sea - Date Received: 21 Aug 1814 - From what ship: HMT Freya, Chatham - Born: Marblehead - Age: 33 - Escaped on 23 Oct 1815.

Lemorrett, John P. - Seaman - Nbr: 5642 - Prize: Lion - Ship Type: P - How taken: HMS Granicus - When taken: 2 Dec 1814 - Where taken: off Lisbon - Date Received: 24 Dec 1814 - From what ship: HMT Hanover, Gibraltar - Born: France - Age: 24 - Released on 3 Jul 1815.

Lenham, John - Seaman - Nbr: 5531 - Prize: Sparks - Ship Type: LM - How taken: HMS Maidstone - When taken: 28 Sep 1814 - Where taken: off Nantucket - Date Received: 17 Dec 1814 - From what ship: HMT Loire, Halifax - Born: Elsingford - Age: 24 - Released on 1 Jul 1815.

Leno, Frederick - Seaman - Nbr: 4147 - Prize: Bunker Hill - Ship Type: P - How taken: HMS Pomone - When taken: 2 Mar 1814 - Where taken: Channel - Date Received: 6 Oct 1814 - From what ship: HMT Niobe, Chatham - Born: Prussia - Age: 29 - Released on 13 Jun 1815.

Lent, Joseph - Seaman - Nbr: 1951 - How taken: Gave himself up from HMS Niobe - When taken: 26 Mar 1813 - Date Received: 3 Aug 1814 - From what ship: HMS Lyffey, Chatham Depot - Born: Salem - Age: 39 - Released on 2 May 1815.

Lent, Samuel - Seaman - Nbr: 5399 - Prize: Perverance - Ship Type: P - How taken: HMS Barbados - When taken: 29 Jan 1814 - Where taken: off St. Bartholomew - Date Received: 31 Oct 1814 - From what ship: HMT Leyden, Chatham - Born: Cumberland - Age: 27 - Released on 1 Jul 1815.

Leonard, John - Seaman - Nbr: 2916 - How taken: Impressed at London - When taken: 4 Aug 1813 - Date Received: 24 Aug 1814 - From what ship: HMT Alpheus, Chatham - Born: Sandwich - Age: 30 - Race: Mulatto - Released on 19 May 1815.

Leonard, Robert - Seaman - Nbr: 3230 - Prize: Sampson - Ship Type: MV - How taken: Rebuff - When taken: 12

American Prisoners of War Held at Dartmoor during the War of 1812

## Alphabetical listing of names

May 1813 - Where taken: off Cape St. Vincent - Date Received: 11 Sep 1814 - From what ship: HMT Freya, Chatham - Born: New York - Age: 29 - Released on 28 May 1815.

Leonard, Thomas - Seaman - Nbr: 6341 - Prize: Prince de Neufchatel - Ship Type: P - How taken: Leander (Newcastle Acasta) - When taken: 20 Dec 1814 - Where taken: Lat 38 Long 56 - Date Received: 19 Feb 1815 - From what ship: HMT Ganges, Plymouth - Born: Belgium - Age: 24 - Released on 6 Jul 1815.

Lepberg, John P. - Seaman - Nbr: 1006 - Prize: Minerva - Ship Type: P - How taken: HMS Conquistador - When taken: 19 Jan 1814 - Where taken: Bay of Biscay - Date Received: 31 Jan 1814 - From what ship: Plymouth - Born: Gothenburg - Age: 20 - Released on 8 Nov 1814.

Lerna, John - Seaman - Nbr: 1403 - Prize: Good Friends - Ship Type: MV - How taken: HMS Andromache - When taken: 2 Apr 1813 - Where taken: Bay of Biscay - Date Received: 19 Jun 1814 - From what ship: Stapleton - Born: Leghorn, Italy - Age: 18 - Released on 25 Mar 1815.

Leroy, Alexander - Seaman - Nbr: 290 - Prize: Cannoneer - Ship Type: MV - How taken: HMS Warspite - When taken: 14 Mar 1813 - Where taken: Bay of Biscay - Date Received: 28 Jun 1813 - From what ship: Plymouth - Born: Rhode Island - Age: 18 - Sent to Plymouth on 8 Jul 1814.

Lerwich, John - Seaman - Nbr: 1136 - Prize: Indostan - Ship Type: MV - How taken: HMS Zenobia - When taken: 25 Jun 1813 - Where taken: off Lisbon - Date Received: 10 May 1814 - From what ship: Plymouth - Born: Virginia - Age: 23 - Released on 28 Apr 1815.

Lesle, Richard - Seaman - Nbr: 2026 - How taken: Gave himself up from HMS Cornwall - When taken: 14 Mar 1813 - Date Received: 3 Aug 1814 - From what ship: HMS Lyffey, Chatham Depot - Born: New York - Age: 36 - Released on 2 May 1815.

Lessall, Francis - Seaman - Nbr: 82 - Prize: Rolla - Ship Type: MV - How taken: HMS Surveillante - When taken: 11 Feb 1813 - Where taken: Bay of Biscay - Date Received: 2 Apr 1813 - From what ship: Plymouth - Born: San Sebastian - Age: 22 - Sent to Mill Prison (Plymouth, England) on 10 Jul 1813.

Lester, James - Seaman - Nbr: 4855 - Prize: Fairy - Ship Type: MV - How taken: HMS Hardy - When taken: 28 Mar 1814 - Where taken: West Indies - Date Received: 9 Oct 1814 - From what ship: HMT Freya, Halifax - Born: Pennsylvania - Age: 40 - Released on 28 May 1815.

Lester, Nathaniel R. - Prize Master - Nbr: 2211 - Prize: Hussar - Ship Type: P - How taken: HMS Saturn - When taken: 25 May 1814 - Where taken: off Sandy Hook - Date Received: 16 Aug 1814 - From what ship: HMS Dublin, Halifax - Born: Connecticut - Age: 33 - Released on 2 May 1815.

Lethorpe, James - Seaman - Nbr: 1431 - Prize: Revenge - Ship Type: LM - How taken: HMS Belle Poule - When taken: 10 May 1814 - Where taken: off Cornwall - Date Received: 19 Jun 1814 - From what ship: Stapleton - Born: Boston - Age: 26 - Released on 28 Apr 1815.

Lethorpe, James - Seaman - Nbr: 1942 - How taken: Gave himself up from HMS Royal William - When taken: 3 Feb 1813 - Date Received: 3 Aug 1814 - From what ship: HMS Alceste, Chatham Depot - Born: Boston - Age: 29 - Released on 2 May 1815.

Lettington, James - Seaman - Nbr: 723 - Prize: Francis and Ann - Ship Type: MV - How taken: HMS Lightning - When taken: 20 Apr 1813 - Where taken: at sea - Date Received: 27 Sep 1813 - From what ship: Plymouth - Born: Jersey - Age: 25 - Released on 26 Apr 1815.

Leurand, Ambrose - Seaman - Nbr: 4824 - Prize: President - Ship Type: P - How taken: HMS Pique - When taken: 7 May 1814 - Where taken: off Porto Rico - Date Received: 9 Oct 1814 - From what ship: HMT Freya, Halifax - Born: Cartagena - Age: 19 - Race: Mulatto - Died on 24 Oct 1814 from pneumonia.

Leverage, Zachariah - Seaman - Nbr: 4944 - Prize: Herald - Ship Type: P - How taken: HMS Endymion - When taken: 15 Aug 1814 - Where taken: off Nantucket - Date Received: 28 Oct 1814 - From what ship: HMT Alkbar, Halifax - Born: Long Island - Age: 16 - Released on 1 May 1815.

Leverage, William - Seaman - Nbr: 4884 - Prize: Saratoga - Ship Type: P - How taken: HMS Barracouta - When taken: 9 Oct 1814 - Where taken: off Western Islands (England) - Date Received: 24 Oct 1814 - From what ship: HMT Salvador del Mundo - Born: New York - Age: 18 - Died on 6 Apr 1815 from gunshot wound in prison.

Leversage, William - Seaman - Nbr: 436 - Prize: Magdalen - Ship Type: MV - How taken: HMS Superb - When taken: 15 Apr 1813 - Where taken: off Belle Isle, France - Date Received: 1 Jul 1813 - From what ship: Plymouth - Born: Pennsylvania - Age: 19 - Released on 28 May 1815.

Levin, Alexander - Seaman - Nbr: 2629 - How taken: Gave himself up from HMS Clarence - When taken: 7 May 1813 - Date Received: 21 Aug 1814 - From what ship: HMT Freya, Chatham - Born: Boston - Age: 20 - Released on 3 May 1815.

Levit, Caleb - Seaman - Nbr: 6325 - Prize: Prince de Neufchatel - Ship Type: P - How taken: Leander (Newcastle

American Prisoners of War Held at Dartmoor during the War of 1812

## Alphabetical listing of names

Acasta) - When taken: 20 Dec 1814 - Where taken: Lat 38 Long 56 - Date Received: 19 Feb 1815 - From what ship: HMT Ganges, Plymouth - Born: Boston - Age: 18 - Released on 5 Jul 1815.

Levy, Charles - Seaman - Nbr: 5348 - Prize: Yankee - Ship Type: P - How taken: HMS Maidstone - When taken: 4 Nov 1814 - Where taken: off Nantucket - Date Received: 31 Oct 1814 - From what ship: HMT Leyden, Chatham - Born: Vineyard - Age: 21 - Released on 20 Mar 1815.

Lewes, John - Seaman - Nbr: 246 - Prize: William Bayard - Ship Type: MV - How taken: HMS Warspite - When taken: 3 Mar 1813 - Where taken: Bay of Biscay - Date Received: 2 Apr 1813 - From what ship: Plymouth - Born: New York - Age: 20 - Race: Black - Released on 20 Apr 1815.

Lewes, John - Seaman - Nbr: 205 - Prize: Star - Ship Type: MV - How taken: HMS Superb - When taken: 9 Feb 1813 - Where taken: Bay of Biscay - Date Received: 2 Apr 1813 - From what ship: Plymouth - Born: Staten Island - Age: 18 - Released on 20 Apr 1815.

Lewis, Benjamin - Seaman - Nbr: 6011 - Prize: McDonough - Ship Type: P - How taken: HMS Bacchante - When taken: 1 Nov 1814 - Where taken: Lat 42 Long 67 - Date Received: 28 Dec 1814 - From what ship: HMT Penelope - Born: Massachusetts - Age: 27 - Released on 3 Jul 1815.

Lewis, Daniel - Seaman - Nbr: 3956 - Prize: Rolla - Ship Type: P - How taken: HMS Loire - When taken: 10 Dec 1813 - Where taken: off Bull Island (South Carolina) - Date Received: 5 Oct 1814 - From what ship: HMT President, Halifax - Born: Newport - Age: 21 - Released on 9 Jun 1815.

Lewis, Edward - Steward - Nbr: 327 - Prize: Paulina - Ship Type: MV - How taken: Rebuff - When taken: 23 Dec 1812 - Where taken: off Cadiz - Date Received: 28 Jun 1813 - From what ship: Plymouth - Born: Philadelphia - Age: 30 - Released on 20 Apr 1815.

Lewis, Francis B. - Seaman - Nbr: 441 - Prize: Magdalen - Ship Type: MV - How taken: HMS Superb - When taken: 15 Apr 1813 - Where taken: off Belle Isle, France - Date Received: 1 Jul 1813 - From what ship: Plymouth - Born: New York - Age: 36 - Released on 20 Apr 1815.

Lewis, Gabriel - Seaman - Nbr: 2010 - Prize: Dictator - Ship Type: MV - How taken: HMS Derwent - When taken: 7 May 1813 - Where taken: off Nantes - Date Received: 3 Aug 1814 - From what ship: HMS Lyffey, Chatham Depot - Born: Virginia - Age: 23 - Released on 2 May 1815.

Lewis, George - Seaman - Nbr: 2315 - Prize: Hussar - Ship Type: P - How taken: HMS Saturn - When taken: 25 May 1814 - Where taken: off Sandy Hook - Date Received: 16 Aug 1814 - From what ship: HMS Dublin, Halifax - Born: Manila - Age: 30 - Race: Creole - Released on 3 May 1815.

Lewis, Gitt - Seaman - Nbr: 5078 - Prize: Ida - Ship Type: LM - How taken: HMS Newcastle - When taken: 9 Aug 1814 - Where taken: Long 34 - Date Received: 28 Oct 1814 - From what ship: HMT Alkbar, Halifax - Born: Boston - Age: 19 - Released on 29 Jun 1815.

Lewis, Henry - Seaman - Nbr: 4256 - Prize: Argus - Ship Type: MV - How taken: HMS San Domingo - When taken: 1 Mar 1814 - Where taken: off Savannah - Date Received: 7 Oct 1814 - From what ship: HMT Niobe, Chatham - Born: Barnstable - Age: 17 - Released on 13 Jun 1815.

Lewis, Henry - Passenger - Nbr: 3880 - Prize: Fame - Ship Type: MV - How taken: HMS Niemen - When taken: 21 May 1814 - Where taken: off Cape Hatteras - Date Received: 5 Oct 1814 - From what ship: HMT Orpheus, Halifax - Born: Nantucket - Age: 20 - Released on 9 Jun 1815.

Lewis, James - Quartermaster - Nbr: 1126 - Prize: Marie Christiana - Ship Type: P - How taken: HMS Pactolus - When taken: 25 Mar 1814 - Where taken: off France - Date Received: 10 May 1814 - From what ship: Plymouth - Born: Southey Town - Age: 30 - Released on 2 Apr 1815.

Lewis, Jesse - Marine - Nbr: 6328 - Prize: Prince de Neufchatel - Ship Type: P - How taken: Leander (Newcastle Acasta) - When taken: 20 Dec 1814 - Where taken: Lat 38 Long 56 - Date Received: 19 Feb 1815 - From what ship: HMT Ganges, Plymouth - Born: Connecticut - Age: 22 - Released on 5 Jul 1815.

Lewis, John - Seaman - Nbr: 6171 - Prize: Lion - Ship Type: P - How taken: HMS Granicus - When taken: 2 Dec 1814 - Where taken: off Lisbon - Date Received: 18 Jan 1815 - From what ship: HMT Impregnable - Born: Massachusetts - Age: 24 - Released on 5 Jul 1815.

Lewis, John - Seaman - Nbr: 4756 - How taken: Gave himself up from HMS Theban - When taken: 21 Jul 1813 - Date Received: 9 Oct 1814 - From what ship: HMT Freya, Chatham - Born: Annapolis - Age: 28 - Race: Black - Released on 15 Jun 1815.

Lewis, John - Seaman - Nbr: 1716 - How taken: Gave himself up from HMS Amphion - Date Received: 8 Jul 1814 - From what ship: Labrador - Born: Pennsylvania - Age: 21 - Released on 1 May 1815.

Lewis, John - Seaman - Nbr: 1086 - How taken: Impressed at Falmouth - When taken: 20 Mar 1814 - Date Received: 10 May 1814 - From what ship: Plymouth - Born: Rhode Island - Age: 30 - Race: Mulatto - Released on 27 Apr 1815.

American Prisoners of War Held at Dartmoor during the War of 1812

## Alphabetical listing of names

Lewis, John - Seaman - Nbr: 5322 - Prize: Juliana Smith - Ship Type: P - How taken: HMS Nymphe - When taken: 11 May 1813 - Where taken: off Cape Sable Island (Canada) - Date Received: 31 Oct 1814 - From what ship: HMT Leyden, Chatham - Born: Boston - Age: 19 - Released on 29 Jun 1815.

Lewis, John - Boy - Nbr: 3655 - Prize: Elbridge Gerry - Ship Type: P - How taken: HMS Crescent - When taken: Jun 1813 - Where taken: off St. Johns - Date Received: 30 Sep 1814 - From what ship: HMT President - Born: New York - Age: 13 - Released on 28 Apr 1815.

Lewis, John - Seaman - Nbr: 5251 - Prize: Dart - Ship Type: P - How taken: HMS Peterel - When taken: 5 May 1813 - Where taken: at sea - Date Received: 31 Oct 1814 - From what ship: HMT Leyden, Chatham - Born: Nantucket - Age: 27 - Race: Black - Released on 29 Jun 1815.

Lewis, Peter - Seaman - Nbr: 3387 - Prize: Mary (True Blooded Yankee) - Ship Type: Prize - How taken: HMS Bellerophon - When taken: 6 Dec 1813 - Where taken: off Land's End - Date Received: 13 Sep 1814 - From what ship: HMT Niobe, Chatham - Born: Charleston - Age: 34 - Released on 28 May 1815.

Lewis, Solomon - Seaman - Nbr: 2072 - How taken: Impressed at London - When taken: 23 Jun 1813 - Date Received: 3 Aug 1814 - From what ship: HMS Bittern, Chatham Depot - Born: Stratford - Age: 22 - Released on 2 May 1815.

Lewis, Thomas - Seaman - Nbr: 5403 - How taken: Impressed at Trinidad - When taken: 1 Mar 1814 - Date Received: 31 Oct 1814 - From what ship: HMT Leyden, Chatham - Born: Calais - Age: 31 - Race: Black - Released on 1 Jul 1815.

Lewis, Thomas - Seaman - Nbr: 708 - Prize: Ned - Ship Type: LM - How taken: HMS Royalist - When taken: 6 Sep 1813 - Where taken: Bay of Biscay - Date Received: 27 Sep 1813 - From what ship: Plymouth - Born: Norfolk - Age: 35 - Released on 26 Apr 1815.

Lewis, Thomas - Seaman - Nbr: 2537 - Prize: Hussar - Ship Type: P - How taken: HMS Saturn - When taken: 25 May 1814 - Where taken: off Sandy Hook - Date Received: 16 Aug 1814 - From what ship: HMT Queen, Halifax - Born: Prussia - Age: 30 - Released on 3 May 1815.

Lewis, William - Seaman - Nbr: 5228 - Prize: York Town - Ship Type: P - How taken: HMS Maidstone - When taken: 18 Jul 1813 - Where taken: Grand Banks - Date Received: 31 Oct 1814 - From what ship: HMT Mermaid, Chatham - Born: Boston - Age: 28 - Released on 29 Jun 1815.

Lewis, William - Seaman - Nbr: 5047 - Prize: Ida - Ship Type: LM - How taken: HMS Newcastle - When taken: 9 Aug 1814 - Where taken: Long 34 - Date Received: 28 Oct 1814 - From what ship: HMT Alkbar, Halifax - Born: Boston - Age: 22 - Released on 29 Jun 1815.

Lewis, Winslow - Seaman - Nbr: 6382 - Prize: US Brig Syren - Ship Type: MW - How taken: HMS Medway - When taken: 12 Jul 1814 - Where taken: off Cape of Good Hope - Date Received: 24 Feb 1815 - From what ship: HMT Ganges, Plymouth - Born: Boston - Age: 25 - Released on 11 Jul 1815.

Ley, John - Seaman - Nbr: 1218 - Prize: James - Ship Type: MV - How taken: HMS Harpy - When taken: 19 Dec 1812 - Where taken: off Isle of France - Date Received: 14 Jun 1814 - From what ship: Mill Prison (Plymouth, England) - Born: Philadelphia - Age: 19 - Released on 26 Apr 1815.

Libbey, John - Marine - Nbr: 5943 - Prize: Harlequin - Ship Type: P - How taken: HMS Bulwark - When taken: 23 Nov 1814 - Where taken: off Halifax - Date Received: 27 Dec 1814 - From what ship: HMT Penelope - Born: New Hampshire - Age: 22 - Released on 3 Jul 1815.

Libley, Moses - Seaman - Nbr: 1531 - Prize: Essex - Ship Type: MV - How taken: HMS Pyramus - When taken: 20 Apr 1813 - Where taken: Bay of Biscay - Date Received: 23 Jun 1814 - From what ship: Stapleton - Born: New Hampshire - Age: 28 - Released on 1 May 1815.

Lilley, Samuel - Seaman - Nbr: 4030 - Prize: US Brig Rattlesnake - Ship Type: MW - How taken: HMS Leander - When taken: 13 Jul 1814 - Where taken: off Shelburne - Date Received: 6 Oct 1814 - From what ship: HMT Chesapeake, Halifax - Born: Boston - Age: 19 - Died on 16 May 1815 from debility.

Lilley, Simon - Seaman - Nbr: 1600 - Prize: Shadow - Ship Type: LM - How taken: HMS Reindeer - When taken: 6 Apr 1813 - Where taken: Bay of Biscay - Date Received: 23 Jun 1814 - From what ship: Stapleton - Born: Massachusetts - Age: 20 - Released on 1 May 1815.

Lilliford, Jacob - Cook - Nbr: 329 - Prize: Dick - Ship Type: MV - How taken: HMS Dispatch - When taken: 17 Dec 1813 - Where taken: off Bordeaux - Date Received: 28 Jun 1813 - From what ship: Plymouth - Born: Pennsylvania - Age: 35 - Race: Black - Released on 20 Apr 1815.

Limbourg, Charles - Seaman - Nbr: 6255 - How taken: Sent into custody from Liverpool - Date Received: 4 Feb 1815 - From what ship: HMT Ganges - Born: Savannah - Age: 37 - Released on 5 Jul 1815.

Lincoln, Ephraim - Seaman - Nbr: 4153 - Prize: Argus - Ship Type: MV - How taken: HMS San Domingo - When taken: 1 Mar 1814 - Where taken: off Savannah - Date Received: 6 Oct 1814 - From what ship: HMT Niobe,

American Prisoners of War Held at Dartmoor during the War of 1812

## Alphabetical listing of names

Chatham - Born: Boston - Age: 18 - Released on 13 Jun 1815.

Lincoln, Solomon - Seaman - Nbr: 3058 - How taken: Gave himself up from HMS Andromache - Date Received: 2 Sep 1814 - What ship: HMT Hydra, Chatham - Born: New Jersey - Age: 35 - Race: Black - Released on 28 Apr 1815.

Lindburgh, Charles - Seaman - Nbr: 795 - Prize: Collin - Ship Type: MV - How taken: HMS Helicon and HMS Whiting - When taken: 26 Oct 1813 - Where taken: Channel - Date Received: 3 Nov 1813 - From what ship: Plymouth - Born: Sweden - Age: 36 - Released on 6 May 1814.

Lindsay, Nathaniel - Captain - Nbr: 2656 - Prize: Growler - Ship Type: MV - How taken: HMS Electra - When taken: 7 Jul 1813 - Where taken: off Newfoundland - Date Received: 21 Aug 1814 - From what ship: HMT Freya, Chatham - Born: Salem - Age: 42 - Released on 19 May 1815.

Lindsay, Samuel - Seaman - Nbr: 2881 - How taken: Gave himself up from HMS Bombay - When taken: 27 May 1813 - Date Received: 24 Aug 1814 - From what ship: HMT Alpheus, Chatham - Born: Virginia - Age: 33 - Released on 19 May 1815.

Lindsey, James - Seaman - Nbr: 3476 - How taken: Gave himself up from HMS Revenge - When taken: 26 May 1814 - Date Received: 19 Sep 1814 - From what ship: HMT Salvador del Mundo - Born: Norwich - Age: 38 - Released on 4 Jun 1815.

Lindsey, William - Seaman - Nbr: 3253 - Prize: Wiley Reynard - Ship Type: P - How taken: HMS Shannon - When taken: 15 Aug 1812 - Where taken: off Halifax - Date Received: 11 Sep 1814 - From what ship: HMT Freya, Chatham - Born: Philadelphia - Age: 21 - Released on 27 Apr 1815.

Lingall, George - Seaman - Nbr: 167 - How taken: Sent into custody from Flag of Truce PA - Date Received: 2 Apr 1813 - From what ship: Plymouth - Born: Maryland - Age: 29 - Race: Black - Sent to Dartmouth on 30 Jul 1813.

Lingrin, Peter - Seaman - Nbr: 922 - Prize: US Frigate Chesapeake - Ship Type: MW - How taken: HMS Shannon - When taken: 1 Jun 1813 - Where taken: off Boston - Date Received: 31 Jan 1814 - From what ship: Plymouth - Born: Lisbon - Age: 33 - Sent to Dartmouth on 19 Oct 1814.

Linnett, William - Seaman - Nbr: 5310 - Prize: Yankee - Ship Type: P - How taken: HMS Shannon - When taken: 20 Aug 1813 - Where taken: at sea - Date Received: 31 Oct 1814 - From what ship: HMT Leyden, Chatham - Born: Barnstable - Age: 23 - Released on 11 Jul 1815.

Linnott, John - Seaman - Nbr: 247 - Prize: William Bayard - Ship Type: MV - How taken: HMS Warspite - When taken: 3 Mar 1813 - Where taken: Bay of Biscay - Date Received: 2 Apr 1813 - From what ship: Plymouth - Born: Philadelphia - Age: 34 - Entered British service on 15 Jun 1813.

Linscott, Josiah - Seaman - Nbr: 6015 - Prize: McDonough - Ship Type: P - How taken: HMS Bacchante - When taken: 1 Nov 1814 - Where taken: Lat 42 Long 67 - Date Received: 28 Dec 1814 - From what ship: HMT Penelope - Born: Old York - Age: 49 - Released on 3 Jul 1815.

Linsey, Alexander - Seaman - Nbr: 1471 - Prize: Governor Gerry - Ship Type: MV - How taken: HMS Lyra - When taken: 29 May 1813 - Where taken: Bay of Biscay - Date Received: 19 Jun 1814 - From what ship: Stapleton - Born: Baltimore - Age: 16 - Released on 28 Apr 1815.

Linton, Joseph - Prize master - Nbr: 3782 - Prize: James - Ship Type: Prize - How taken: Rebecca (LM) - When taken: 13 Apr 1814 - Where taken: off Delaware - Date Received: 5 Oct 1814 - From what ship: HMT President, Halifax - Born: Pennsylvania - Age: 35 - Released on 9 Jun 1815.

Lippart, Thomas D. - Seaman - Nbr: 488 - Prize: Paul Jones - Ship Type: P - How taken: HMS Leonidas - When taken: 23 May 1813 - Where taken: Channel - Date Received: 8 Sep 1813 - From what ship: Plymouth - Born: Pennsylvania - Age: 49 - Died on 9 Mar 1815 from apoplexy.

Liscomb, William - Seaman - Nbr: 4569 - How taken: Gave himself up from MV Liberty - When taken: 30 Dec 1813 - Date Received: 9 Oct 1814 - From what ship: HMT Leyden, Chatham - Born: New York - Age: 25 - Released on 15 Jun 1815.

Lister, Louis - Seaman - Nbr: 2586 - How taken: Gave himself up from HMS Ulysses - When taken: 13 Jan 1813 - Date Received: 21 Aug 1814 - From what ship: HMT Freya, Chatham - Born: New York - Age: 26 - Released on 3 May 1815.

Listrades, Joseph - Seaman - Nbr: 2167 - Prize: Rattlesnake - Ship Type: P - How taken: HMS Hyperion - When taken: 26 Jun 1814 - Where taken: off Cape Finisterre - Date Received: 16 Aug 1814 - From what ship: HMS Dublin, Halifax - Born: New York - Age: 25 - Released on 2 May 1815.

Litchfield, Erick - Seaman - Nbr: 2030 - How taken: Apprehended at London - When taken: 24 May 1813 - Date Received: 3 Aug 1814 - From what ship: HMS Lyffey, Chatham Depot - Born: Cohasset - Age: 31 - Released on 2 May 1815.

American Prisoners of War Held at Dartmoor during the War of 1812

## Alphabetical listing of names

Little, Charles - Seaman - Nbr: 6509 - How taken: Apprehended at Trinidad - Date Received: 3 Mar 1815 - From what ship: HMT Ganges, Plymouth - Born: Rhode Island - Age: 28 - Released on 11 Jul 1815.

Little, George - Seaman - Nbr: 1367 - Prize: Paul Jones - Ship Type: P - How taken: HMS Leonidas - When taken: 23 May 1813 - Where taken: Channel - Date Received: 19 Jun 1814 - From what ship: Stapleton - Born: Massachusetts - Age: 25 - Released on 25 Mar 1815.

Little, John - Seaman - Nbr: 537 - Prize: Minerva - Ship Type: MV - How taken: HMS Goldfinch - When taken: 27 Jul 1813 - Where taken: off Nantes - Date Received: 8 Sep 1813 - From what ship: Plymouth - Born: Philadelphia - Age: 29 - Entered British service on 11 Oct 1813.

Little, Thomas - Seaman - Nbr: 4747 - How taken: Gave himself up from HMS Treban - When taken: 21 Jul 1814 - Date Received: 9 Oct 1814 - From what ship: HMT Freya, Chatham - Born: Philadelphia - Age: 26 - Released on 15 Jun 1815.

Little, William - Seaman - Nbr: 3929 - Prize: Rolla - Ship Type: P - How taken: HMS Loire - When taken: 10 Dec 1813 - Where taken: off Bull Island (South Carolina) - Date Received: 5 Oct 1814 - From what ship: HMT President, Halifax - Born: Philadelphia - Age: 31 - Released on 9 Jun 1815.

Little, William - Seaman - Nbr: 6433 - Prize: Chance - Ship Type: P - How taken: HMS Statira - When taken: 1 Apr 1814 - Where taken: Lat 38 Long 24 - Date Received: 3 Mar 1815 - From what ship: HMT Ganges, Plymouth - Born: New York - Age: 24 - Released on 11 Jul 1815.

Littlefield, Israel - Seaman - Nbr: 6018 - Prize: McDonough - Ship Type: P - How taken: HMS Bacchante - When taken: 1 Nov 1814 - Where taken: Lat 42 Long 67 - Date Received: 28 Dec 1814 - From what ship: HMT Penelope - Born: Massachusetts - Age: 23 - Released on 3 Jul 1815.

Littlefield, James - Seaman - Nbr: 1660 - Prize: Paul Jones - Ship Type: P - How taken: HMS Leonidas - When taken: 23 May 1813 - Where taken: Channel - Date Received: 23 Jun 1814 - From what ship: Stapleton - Born: Massachusetts - Age: 28 - Released on 1 May 1815.

Littlefield, Nicholas - Passenger - Nbr: 6045 - Prize: Little Belt - Ship Type: MV - How taken: HMS Armide - When taken: 21 Sep 1813 - Where taken: off VA - Date Received: 28 Dec 1814 - From what ship: HMT Penelope - Born: New York - Age: 33 - Released on 11 Jul 1815.

Littlefield, Rufus - Seaman - Nbr: 1411 - Prize: Miranda - Ship Type: Prize - How taken: HMS Unicorn - When taken: 21 May 1813 - Where taken: off Ushant (France) - Date Received: 19 Jun 1814 - From what ship: Stapleton - Born: Massachusetts - Age: 17 - Released on 28 Apr 1815.

Littlefield, Oliver - Seaman - Nbr: 6013 - Prize: McDonough - Ship Type: P - How taken: HMS Bacchante - When taken: 1 Nov 1814 - Where taken: Lat 42 Long 67 - Date Received: 28 Dec 1814 - From what ship: HMT Penelope - Born: Massachusetts - Age: 31 - Escaped on 1 Jun 1815.

Littleford, L. - Seaman - Nbr: 5427 - Prize: Rattlesnake - Ship Type: LM - How taken: HMS Rhin - When taken: 12 May 1814 - Where taken: off Bermuda - Date Received: 31 Oct 1814 - From what ship: HMT Leyden, Chatham - Born: Frankford - Age: 21 - Released on 1 Jul 1815.

Livermore, Arthur - Seaman - Nbr: 796 - Prize: Collin - Ship Type: MV - How taken: HMS Helicon and HMS Whiting - When taken: 26 Oct 1813 - Where taken: Channel - Date Received: 3 Nov 1813 - From what ship: Plymouth - Born: Rockingham - Age: 22 - Released on 26 Apr 1815.

Livesley, Thomas - Seaman - Nbr: 4584 - Prize: Bunker Hill - Ship Type: P - How taken: HMS Pomone - When taken: 2 Mar 1814 - Where taken: Channel - Date Received: 9 Oct 1814 - From what ship: HMT Leyden, Chatham - Born: New York - Age: 34 - Released on 20 Mar 1815.

Livingston, Henry - Seaman - Nbr: 4282 - Prize: Pilot - Ship Type: LM - How taken: Victoria (Privateer) - When taken: 28 Jan 1814 - Where taken: off Bordeaux - Date Received: 7 Oct 1814 - From what ship: HMT Niobe, Chatham - Born: New York - Age: 28 - Race: Black - Released on 14 Jun 1815.

Lloyd, David - Seaman - Nbr: 1200 - Prize: Ocean - Ship Type: P - How taken: HMS Achates - When taken: 14 Jun 1810 - Where taken: Channel - Date Received: 4 Jun 1814 - From what ship: Dartmouth - Born: Newport - Age: 23 - Released on 28 Apr 1815.

Lock, James - Seaman - Nbr: 2555 - Prize: Caromaned - Ship Type: Prize - How taken: HMS Eridanus - When taken: 13 Aug 1814 - Where taken: Lat 40, Long 16 - Date Received: 16 Aug 1814 - From what ship: HMT Salvador del Mundo - Born: Massachusetts - Age: 18 - Released on 3 May 1815.

Locker, Michael - Seaman - Nbr: 4717 - How taken: Gave himself up at London - When taken: 26 Aug 1814 - Date Received: 9 Oct 1814 - From what ship: HMT Freya, Chatham - Born: St. Marys - Age: 21 - Released on 15 Jun 1815.

Lockwood, Benjamin - Seaman - Nbr: 5761 - Prize: Valentine - Ship Type: MV - How taken: HMS Racehorse - When taken: Oct 1812 - Where taken: off Cape of Good Hope - Date Received: 26 Dec 1814 - From what

American Prisoners of War Held at Dartmoor during the War of 1812

## Alphabetical listing of names

ship: HMT Argo - Born: Rhode Island - Age: 26 - Released on 27 Apr 1815.
Lockwood, Caleb - Seaman - Nbr: 1900 - Prize: Quebec - Ship Type: Prize - How taken: HMS Derwent - When taken: 29 Jan 1813 - Where taken: off Lisbon - Date Received: 3 Aug 1814 - From what ship: HMS Alceste, Chatham Depot - Born: Rhode Island - Age: 21 - Released on 2 May 1815.
Lockwood, William - Seaman - Nbr: 4966 - Prize: Herald - Ship Type: P - How taken: HMS Endymion - When taken: 15 Aug 1814 - Where taken: off Nantucket - Date Received: 28 Oct 1814 - From what ship: HMT Alkbar, Halifax - Born: Philadelphia - Age: 27 - Race: Negro - Released on 21 Jun 1815.
Logan, Charles - Seaman - Nbr: 1541 - Prize: Zebra - Ship Type: LM - How taken: HMS Pyramus - When taken: 20 Apr 1813 - Where taken: Bay of Biscay - Date Received: 23 Jun 1814 - From what ship: Stapleton - Born: New Jersey - Age: 22 - Released on 1 May 1815.
Logan, James - Seaman - Nbr: 2405 - Prize: Fiere Facia - Ship Type: P - How taken: HMS Ramillies - When taken: 27 Feb 1814 - Where taken: off NY - Date Received: 16 Aug 1814 - From what ship: HMT Queen, Halifax - Born: Philadelphia - Age: 49 - Race: Black - Released on 3 May 1815.
Logan, William - Seaman - Nbr: 5956 - Prize: Harlequin - Ship Type: P - How taken: HMS Bulwark - When taken: 23 Nov 1814 - Where taken: off Halifax - Date Received: 27 Dec 1814 - From what ship: HMT Penelope - Born: Baltimore - Age: 23 - Released on 3 Jul 1815.
Logan, William - Seaman - Nbr: 1387 - Prize: Courier - Ship Type: P - How taken: HMS Rover - When taken: 14 May 1813 - Where taken: Bay of Biscay - Date Received: 19 Jun 1814 - From what ship: Stapleton - Born: Baltimore - Age: 23 - Released on 28 Apr 1815.
Logere, H. C. - Sergeant Marines - Nbr: 4014 - Prize: US Brig Rattlesnake - Ship Type: MW - How taken: HMS Leander - When taken: 13 Jul 1814 - Where taken: off Shelburne - Date Received: 6 Oct 1814 - From what ship: HMT Chesapeake, Halifax - Born: Baltimore - Age: 25 - Released on 9 Jun 1815.
Long, Charles - Seaman - Nbr: 904 - Prize: Squirrel - Ship Type: MV - How taken: HMS Belle Poule - When taken: 14 Dec 1813 - Where taken: at sea - Date Received: 31 Jan 1814 - From what ship: Plymouth - Born: Swinehunde - Age: 21 - Sent to Plymouth on 28 Feb 1814.
Long, Charles - Seaman - Nbr: 2210 - Prize: Hussar - Ship Type: P - How taken: HMS Saturn - When taken: 25 May 1814 - Where taken: off Sandy Hook - Date Received: 16 Aug 1814 - From what ship: HMS Dublin, Halifax - Born: New York - Age: 25 - Released on 2 May 1815.
Long, David - Seaman - Nbr: 5627 - Prize: William Penn - Ship Type: MV - How taken: HMS Acorn - When taken: 27 Oct 1812 - Where taken: Lat 14 - Date Received: 24 Dec 1814 - From what ship: HMT Tay - Born: Nantucket - Age: 23 - Released on 3 Jul 1815.
Long, John - Seaman - Nbr: 4896 - Prize: Atlas - Ship Type: LM - How taken: HMS Aelous - When taken: 2 Jun 1814 - Where taken: West Indies - Date Received: 28 Oct 1814 - From what ship: HMT Alkbar, Halifax - Born: Guadeloupe - Age: 28 - Race: Negro - Released on 21 Jun 1815.
Long, John - Seaman - Nbr: 5970 - Prize: Halifax Packet - Ship Type: Prize - How taken: HMS Bulwark - When taken: 22 Sep 1814 - Where taken: Georges Bank - Date Received: 27 Dec 1814 - From what ship: HMT Penelope - Born: Guadeloupe - Age: 28 - Race: Black - Released on 3 Jul 1815.
Long, Joseph - Seaman - Nbr: 4460 - Prize: Fame - Ship Type: P - How taken: HMS Pratteo - When taken: 3 May 1813 - Where taken: Bay of Biscay - Date Received: 8 Oct 1814 - From what ship: HMT Leyden, Chatham - Born: Massachusetts - Age: 21 - Died on 29 May 1815 from phthisis pulmonalis.
Long, Joseph - Seaman - Nbr: 1293 - Prize: Traveler - Ship Type: Prize - How taken: HM Schooner Canso - When taken: 11 May 1814 - Where taken: off Cape Clear - Date Received: 14 Jun 1814 - From what ship: Mill Prison (Plymouth, England) - Born: Baltimore - Age: 36 - Released on 28 Apr 1815.
Long, R. S. - Prize Master - Nbr: 494 - Prize: Paul Jones - Ship Type: P - How taken: HMS Leonidas - When taken: 23 May 1813 - Where taken: Channel - Date Received: 8 Sep 1813 - From what ship: Plymouth - Born: Charlestown - Age: 30 - Released on 26 Apr 1815.
Longford, Samuel - Seaman - Nbr: 1592 - Prize: Eliza - Ship Type: MV - How taken: HMS Surveillante - When taken: 27 Mar 1813 - Where taken: Bay of Biscay - Date Received: 23 Jun 1814 - From what ship: Stapleton - Born: Maryland - Age: 24 - Sent to Mill Prison (Plymouth, England) on 30 Jun 1814.
Longworthy, John - Seaman - Nbr: 86 - Prize: Cashier - Ship Type: MV - How taken: HMS Reindeer - When taken: 3 Feb 1813 - Where taken: Bay of Biscay - Date Received: 2 Apr 1813 - From what ship: Plymouth - Born: New Berne - Age: 41 - Sent to Dartmouth on 30 Jul 1813.
Lonie, James - Seaman - Nbr: 4045 - Prize: US Brig Rattlesnake - Ship Type: MW - How taken: HMS Leander - When taken: 13 Jul 1814 - Where taken: off Shelburne - Date Received: 6 Oct 1814 - From what ship: HMT Chesapeake, Halifax - Born: Salem - Age: 48 - Released on 13 Jun 1815.

American Prisoners of War Held at Dartmoor during the War of 1812

## Alphabetical listing of names

Lopans, William - Seaman - Nbr: 2862 - Prize: Prompt - Ship Type: MV - How taken: Change (Privateer) - When taken: 28 Mar 1813 - Where taken: Bay of Biscay - Date Received: 24 Aug 1814 - From what ship: HMT Alpheus, Chatham - Born: Boston - Age: 21 - Released on 19 May 1815.

Lopez, B. Eldridge - Seaman - Nbr: 1472 - Prize: Governor Gerry - Ship Type: MV - How taken: HMS Lyra - When taken: 29 May 1813 - Where taken: Bay of Biscay - Date Received: 19 Jun 1814 - From what ship: Stapleton - Born: New York - Age: 23 - Released on 28 Apr 1815.

Lopez, Domingo - Seaman - Nbr: 4817 - Prize: President - Ship Type: P - How taken: HMS Pique - When taken: 7 May 1814 - Where taken: off Porto Rico - Date Received: 9 Oct 1814 - From what ship: HMT Freya, Halifax - Born: Cartagena - Age: 20 - Race: Mulatto - Released on 21 Jun 1815.

Lopez, Joseph - Seaman - Nbr: 1128 - Prize: Marie Christiana - Ship Type: P - How taken: HMS Pactolus - When taken: 25 Mar 1814 - Where taken: off France - Date Received: 10 May 1814 - From what ship: Plymouth - Born: Malaga - Age: 20 - Released on 28 Apr 1815.

Lorang, M. - Seaman - Nbr: 4804 - Prize: President - Ship Type: P - How taken: HMS Pique - When taken: 7 May 1814 - Where taken: off Porto Rico - Date Received: 9 Oct 1814 - From what ship: HMT Freya, Halifax - Born: Guadeloupe - Age: 25 - Race: Black - Released on 21 Jun 1815.

Lord, Benjamin - Boy - Nbr: 5683 - Prize: McDonough - Ship Type: P - How taken: HMS Bacchante - When taken: 7 Nov 1814 - Where taken: Lat 42 Long 67 - Date Received: 24 Dec 1814 - From what ship: HMT Penelope - Born: Arundel - Age: 18 - Released on 3 Jul 1815.

Lord, Dormer - Master's mate - Nbr: 5681 - Prize: McDonough - Ship Type: P - How taken: HMS Bacchante - When taken: 7 Nov 1814 - Where taken: Lat 42 Long 67 - Date Received: 24 Dec 1814 - From what ship: HMT Penelope - Born: Arundel - Age: 27 - Released on 3 Jul 1815.

Lord, John - Seaman - Nbr: 5682 - Prize: McDonough - Ship Type: P - How taken: HMS Bacchante - When taken: 7 Nov 1814 - Where taken: Lat 42 Long 67 - Date Received: 24 Dec 1814 - From what ship: HMT Penelope - Born: Arundel - Age: 25 - Released on 3 Jul 1815.

Lord, Joseph - Prize master - Nbr: 5680 - Prize: McDonough - Ship Type: P - How taken: HMS Bacchante - When taken: 7 Nov 1814 - Where taken: Lat 42 Long 67 - Date Received: 24 Dec 1814 - From what ship: HMT Penelope - Born: Arundel - Age: 38 - Released on 28 Apr 1815.

Lord, Samuel - Seaman - Nbr: 5818 - Prize: US Schooner Scorpion - Ship Type: MW - How taken: British gunboats - When taken: 6 Sep 1814 - Where taken: Lake Erie - Date Received: 26 Dec 1814 - From what ship: HMT Argo - Born: Pennsylvania - Age: 17 - Released on 29 Jun 1815.

Lorenze, Thomas - Seaman - Nbr: 1257 - Prize: Adeline - Ship Type: MV - How taken: HMS Magicienne - When taken: 16 Mar 1814 - Where taken: off Cape Finisterre - Date Received: 14 Jun 1814 - From what ship: Mill Prison (Plymouth, England) - Born: Lippstadt - Age: 17 - Released on 28 Apr 1815.

Loring, Caleb - Seaman - Nbr: 6224 - Prize: Prince de Neufchatel - Ship Type: P - How taken: Leander (Newcastle Acasta) - When taken: 20 Dec 1814 - Where taken: Lat 38 Long 56 - Date Received: 30 Jan 1815 - From what ship: HMT Pheasant - Born: Boston - Age: 19 - Released on 5 Jul 1815.

Loring, George - Seaman - Nbr: 3957 - Prize: Rolla - Ship Type: P - How taken: HMS Loire - When taken: 10 Dec 1813 - Where taken: off Bull Island (South Carolina) - Date Received: 5 Oct 1814 - From what ship: HMT President, Halifax - Born: Newport - Age: 32 - Released on 9 Jun 1815.

Loring, Rufus - Boy - Nbr: 5552 - Prize: Minerva - Ship Type: MV - How taken: Lunenburg (Privateer) - When taken: 5 Sep 1814 - Where taken: Irish Shoals - Date Received: 17 Dec 1814 - From what ship: HMT Loire, Halifax - Born: Massachusetts - Age: 15 - Released on 1 Jul 1815.

Loring, Samuel - Seaman - Nbr: 4307 - Prize: Argus - Ship Type: MV - How taken: HMS San Domingo - When taken: 1 Mar 1814 - Where taken: off Savannah - Date Received: 7 Oct 1814 - From what ship: HMT Niobe, Chatham - Born: Boston - Age: 18 - Released on 4 Jun 1815.

Loring, William - Seaman - Nbr: 5011 - Prize: Charlotte - Ship Type: Prize - How taken: Censor - When taken: 5 Aug 1814 - Where taken: off Azores - Date Received: 28 Oct 1814 - From what ship: HMT Alkbar, Halifax - Born: Portland - Age: 28 - Released on 21 Jun 1815.

Louis, John - Seaman - Nbr: 1739 - Prize: Hugh Jones - Ship Type: MV - How taken: HMS Bittern - When taken: 14 Jun 1814 - Where taken: at sea - Date Received: 18 Jul 1814 - From what ship: HMT Salvador del Mundo, Plymouth - Born: New Orleans - Age: 24 - Race: Black - Died on 5 Aug 1814 from pneumonia.

Louis, Nicholas - Seaman - Nbr: 1331 - Prize: Paul Jones - Ship Type: P - How taken: HMS Leonidas - When taken: 23 May 1813 - Where taken: Channel - Date Received: 19 Jun 1814 - From what ship: Stapleton - Born: Hampshire - Age: 44 - Sent to Mill Prison (Plymouth, England) on 26 Jun 1814.

Lounsbury, Benjamin F. - Servant - Nbr: 588 - Prize: US Brig Argus - Ship Type: MW - How taken: HMS Pelican -

American Prisoners of War Held at Dartmoor during the War of 1812

## Alphabetical listing of names

When taken: 14 Aug 1813 - Where taken: Irish Channel - Date Received: 8 Sep 1813 - From what ship: Plymouth - Born: West Chester - Age: 16 - Sent to Dartmouth on 2 Nov 1814.

Lourie, Solomon - Seaman - Nbr: 3806 - Prize: Stark - Ship Type: P - How taken: HMS Sophie - When taken: 20 Apr 1814 - Where taken: off Bermuda - Date Received: 5 Oct 1814 - From what ship: HMT Orpheus, Halifax - Born: North Yarmouth - Age: 22 - Released on 9 Jun 1815.

Louring, Henry - Seaman - Nbr: 199 - Prize: Star - Ship Type: MV - How taken: HMS Superb - When taken: 9 Feb 1813 - Where taken: Bay of Biscay - Date Received: 2 Apr 1813 - From what ship: Plymouth - Born: Norfolk - Age: 35 - Released on 20 Apr 1815.

Lovel, John - 2nd Mate - Nbr: 41 - Prize: Terrible - Ship Type: MV - How taken: HMS Foxhound - When taken: 8 Feb 1813 - Where taken: Channel - Date Received: 2 Apr 1813 - From what ship: Plymouth - Born: Newbury - Age: 40 - Sent to Dartmouth on 30 Jul 1813.

Loveland, Daniel - Seaman - Nbr: 2203 - Prize: Hussar - Ship Type: P - How taken: HMS Saturn - When taken: 25 May 1814 - Where taken: off Sandy Hook - Date Received: 16 Aug 1814 - From what ship: HMS Dublin, Halifax - Born: Connecticut - Age: 27 - Released on 2 May 1815.

Loveland, Leonard - Seaman - Nbr: 3811 - Prize: Nimble - Ship Type: Prize - How taken: HMS Arab - When taken: 5 Apr 1814 - Where taken: Lat 37 Long 65 - Date Received: 5 Oct 1814 - From what ship: HMT Orpheus, Halifax - Born: Chatham - Age: 22 - Released on 9 Jun 1815.

Lovell, Robert - Seaman - Nbr: 3122 - Prize: Prize of the Blockage - Ship Type: Prize - How taken: HMS Brazen - When taken: 29 Jun 1813 - Where taken: off Scotland - Date Received: 11 Sep 1814 - From what ship: HMT Freya, Chatham - Born: Rhode Island - Age: 21 - Race: Mulatto - Released on 28 May 1815.

Lovell, Samuel - Boy - Nbr: 4342 - Prize: Union - Ship Type: LM - How taken: HMS Curlew - When taken: 1 Apr 1814 - Where taken: off Halifax - Date Received: 7 Oct 1814 - From what ship: HMT Salvador del Mundo, Halifax - Born: Boston - Age: 17 - Released on 19 May 1815.

Lovell, William - Seaman - Nbr: 4752 - How taken: Gave himself up from HMS Owen Glendower - When taken: 28 Jun 1813 - Date Received: 9 Oct 1814 - From what ship: HMT Freya, Chatham - Born: Not legible - Age: 36 - Released on 15 Jun 1815.

Lovely, Placid - Seaman - Nbr: 3544 - Prize: Hawk - Ship Type: P - How taken: HMS Pique - When taken: 26 Apr 1814 - Where taken: off Bermuda - Date Received: 30 Sep 1814 - From what ship: HMT Sybella - Born: New Orleans - Age: 29 - Race: Mulatto - Died on 1 Nov 1814.

Lovett, Henry - Servant - Nbr: 589 - Prize: US Brig Argus - Ship Type: MW - How taken: HMS Pelican - When taken: 14 Aug 1813 - Where taken: Irish Channel - Date Received: 8 Sep 1813 - From what ship: Plymouth - Born: Providence - Age: 20 - Race: Negro - Sent to Dartmouth on 2 Nov 1814.

Lovett, William - Seaman - Nbr: 67 - Prize: Spitfire - Ship Type: MV - How taken: HMS Achates - When taken: 14 Feb 1813 - Where taken: off Ushant (France) - Date Received: 2 Apr 1813 - From what ship: Plymouth - Born: Marblehead - Age: 21 - Escaped on 10 Feb 1815.

Lowdie, Samuel - Seaman - Nbr: 2104 - How taken: Gave himself up from HMS Hyperion - Date Received: 3 Aug 1814 - From what ship: HMS Bittern, Chatham Depot - Born: New York - Age: 25 - Race: Mulatto - Released on 2 May 1815.

Lowe, Frederick - 3rd Officer - Nbr: 5090 - Prize: Ida - Ship Type: LM - How taken: HMS Newcastle - When taken: 9 Aug 1814 - Where taken: Long 34 - Date Received: 28 Oct 1814 - From what ship: HMT Alkbar, Halifax - Born: Gloucester - Age: 24 - Released on 29 Jun 1815.

Lowe, George - Seaman - Nbr: 5249 - Prize: Thomas - Ship Type: P - How taken: HMS Nymphe - When taken: 24 Jun 1813 - Where taken: off Halifax - Date Received: 31 Oct 1814 - From what ship: HMT Mermaid, Chatham - Born: Portsmouth - Age: 23 - Released on 29 Jun 1815.

Lowe, Isaac - Seaman - Nbr: 6239 - Prize: Albion - Ship Type: Prize - How taken: HMS Harlequin - When taken: 6 Jan 1814 - Where taken: Long 20 - Date Received: 4 Feb 1815 - From what ship: HMT Ganges - Born: Maryland - Age: 34 - Race: Black - Released on 5 Jul 1815.

Lowe, John - Seaman - Nbr: 1533 - Prize: Courier - Ship Type: LM - How taken: HMS Rover - When taken: 14 Mar 1813 - Where taken: Bay of Biscay - Date Received: 23 Jun 1814 - From what ship: Stapleton - Born: Baltimore - Age: 18 - Released on 1 May 1815.

Lowe, Thomas - Seaman - Nbr: 3349 - How taken: Gave himself up from HMS Gordon - When taken: 1 Nov 1812 - Date Received: 13 Sep 1814 - From what ship: HMT Niobe, Chatham - Born: Massachusetts - Age: 28 - Released on 27 Apr 1815.

Lowman, David - Seaman - Nbr: 3660 - Prize: Peggy - Ship Type: MV - How taken: HMS Epervier - When taken: 3 Nov 1813 - Where taken: off Portsmouth - Date Received: 30 Sep 1814 - From what ship: HMT President -

American Prisoners of War Held at Dartmoor during the War of 1812

## Alphabetical listing of names

Born: Massachusetts - Age: 29 - Released on 4 Jun 1815.

Lowrie, William - Seaman - Nbr: 3001 - Prize: St. Lawrence - How taken: HMS Aquilon - When taken: 9 Aug 1814 - Where taken: off Western Islands (England) - Date Received: 29 Aug 1814 - From what ship: HMT Bittern - Born: New York - Age: 23 - Released on 28 May 1815.

Lowton, John - Seaman - Nbr: 5808 - Prize: US Schooner Scorpion - Ship Type: MW - How taken: British gunboats - When taken: 6 Sep 1814 - Where taken: Lake Erie - Date Received: 26 Dec 1814 - From what ship: HMT Argo - Born: Newport - Age: 33 - Released on 11 Jul 1815.

Lubery, John C. - Seaman - Nbr: 4187 - Prize: Commodore Perry - Ship Type: Schooner - How taken: Sent into custody from a cutter - When taken: 25 Feb 1814 - Where taken: off Bordeaux - Date Received: 7 Oct 1814 - From what ship: HMT Niobe, Chatham - Born: Prussia - Age: 28 - Released on 13 Jun 1815.

Lucas, Daniel - Seaman - Nbr: 3304 - Prize: Porcupine - Ship Type: LM - How taken: British squadron - When taken: 3 Jun 1813 - Where taken: off Cape Sable Island (Canada) - Date Received: 13 Sep 1814 - From what ship: HMT Niobe, Chatham - Born: Virginia - Age: 23 - Race: Black - Released on 28 May 1815.

Lucas, Martin - Seaman - Nbr: 3188 - How taken: Gave himself up from HMS Swiftsure - When taken: 26 Dec 1812 - Date Received: 11 Sep 1814 - From what ship: HMT Freya, Chatham - Born: New York - Age: 50 - Race: Black - Released on 27 Apr 1815.

Lucas, Robert - Seaman - Nbr: 3437 - How taken: Gave himself up from HMS Acorn - When taken: 27 Jul 1813 - Date Received: 13 Sep 1814 - From what ship: HMT Niobe, Chatham - Born: Philadelphia - Age: 22 - Released on 26 Apr 1815.

Luce, Prosper - Seaman - Nbr: 798 - Prize: Collin - Ship Type: MV - How taken: HMS Helicon and HMS Whiting - When taken: 26 Oct 1813 - Where taken: Channel - Date Received: 3 Nov 1813 - From what ship: Plymouth - Born: Massachusetts - Age: 21 - Released on 20 Mar 1814.

Ludlow, Charles - Boy - Nbr: 150 - Prize: Criterion - Ship Type: MV - How taken: HMS Belle Poule - When taken: 14 Feb 1813 - Where taken: Bay of Biscay - Date Received: 2 Apr 1813 - From what ship: Plymouth - Born: New Jersey - Age: 17 - Released on 20 Apr 1815.

Luffie, Warren - Seaman - Nbr: 1878 - Prize: Elbridge Gerry - Ship Type: P - How taken: HMS Crescent - When taken: 16 Sep 1813 - Where taken: off St. George's - Date Received: 29 Jul 1814 - From what ship: HMS Ville de Paris, Chatham Depot - Born: New York - Age: 42 - Race: Black - Released on 2 May 1815.

Lufkin, William - Seaman - Nbr: 4497 - Prize: Enterprize - Ship Type: P - How taken: HMS Tenedos - When taken: 21 May 1813 - Where taken: off Cape Cod - Date Received: 8 Oct 1814 - From what ship: HMT Leyden, Chatham - Born: Massachusetts - Age: 22 - Released on 15 Jun 1815.

Lumsby, Frederick - Seaman - Nbr: 6478 - Prize: Mary - Ship Type: P - How taken: Lapphin - When taken: 15 Jun 1814 - Where taken: off Saint Marys - Date Received: 3 Mar 1815 - From what ship: HMT Ganges, Plymouth - Born: New York - Age: 20 - Released on 11 Jul 1815.

Lunberg, Berg - Seaman - Nbr: 1052 - How taken: Sent into custody from Mary (Transport) - Date Received: 10 May 1814 - From what ship: Plymouth - Born: Gothenburg - Age: 22 - Released on 8 Nov 1814.

Lunt, Daniel - Seaman - Nbr: 5361 - Prize: Argus - Ship Type: MV - How taken: HMS San Domingo - When taken: 1 Mar 1814 - Where taken: off Savannah - Date Received: 31 Oct 1814 - From what ship: HMT Leyden, Chatham - Born: Newbury - Age: 23 - Released on 1 Jul 1815.

Lupton, John - Seaman - Nbr: 2996 - Prize: St. Lawrence - How taken: HMS Aquilon - When taken: 9 Aug 1814 - Where taken: off Western Islands (England) - Date Received: 29 Aug 1814 - From what ship: HMT Bittern - Born: Long Island - Age: 24 - Released on 28 May 1815.

Lush, John G. - Seaman - Nbr: 5912 - Prize: Harlequin - Ship Type: P - How taken: HMS Bulwark - When taken: 23 Nov 1814 - Where taken: off Halifax - Date Received: 27 Dec 1814 - From what ship: HMT Penelope - Born: Portsmouth - Age: 26 - Released on 15 Jul 1815.

Luther, Cromwell - Seaman - Nbr: 4763 - Prize: Valentine - Ship Type: MV - How taken: HMS Minden - When taken: 16 Nov 1812 - Where taken: Malager Bay - Date Received: 9 Oct 1814 - From what ship: HMT Freya, Chatham - Born: Warren - Age: 23 - Released on 27 Apr 1815.

Lyman, Saul - Seaman - Nbr: 2453 - Prize: US Sloop Frolic - Ship Type: MW - How taken: HMS Orpheus - When taken: 20 Apr 1814 - Where taken: off Cuba - Date Received: 16 Aug 1814 - From what ship: HMT Queen, Halifax - Born: Norfolk - Age: 24 - Released on 3 May 1815.

Lynch, Elisha - Seaman - Nbr: 2039 - How taken: Gave himself up from HMS Impeteux - When taken: 2 Dec 1812 - Date Received: 3 Aug 1814 - From what ship: HMS Lyffey, Chatham Depot - Born: Dorchester - Age: 27 - Released on 26 Apr 1815.

Lynch, James - Seaman - Nbr: 3833 - Prize: Snap Dragon - Ship Type: P - How taken: HMS Martin - When taken:

American Prisoners of War Held at Dartmoor during the War of 1812

## Alphabetical listing of names

31 May 1814 - Where taken: off Charleston - Date Received: 5 Oct 1814 - From what ship: HMT Orpheus, Halifax - Born: Beauford - Age: 17 - Released on 9 Jun 1815.

Lynch, William - Seaman - Nbr: 4190 - Prize: Commodore Perry - Ship Type: Schooner - How taken: Sent into custody from a cutter - When taken: 25 Feb 1814 - Where taken: off Bordeaux - Date Received: 7 Oct 1814 - From what ship: HMT Niobe, Chatham - Born: Philadelphia - Age: 35 - Released on 13 Jun 1815.

Lynch, William - Seaman - Nbr: 4857 - Prize: Fairy - Ship Type: MV - How taken: HMS Hardy - When taken: 28 Mar 1814 - Where taken: West Indies - Date Received: 9 Oct 1814 - From what ship: HMT Freya, Halifax - Born: Maryland - Age: 23 - Released on 21 Jun 1815.

Lynes, Amos - Seaman - Nbr: 6032 - Prize: Regent - Ship Type: LM - How taken: HMS Forth - When taken: 19 Sep 1814 - Where taken: off Egg Harbor (New Jersey) - Date Received: 28 Dec 1814 - From what ship: HMT Penelope - Born: New Haven - Age: 21 - Released on 11 Jul 1815.

Lyons, Charles - Seaman - Nbr: 4533 - Prize: Wolf Cove - Ship Type: MV - How taken: HMS Briton - When taken: 1 Dec 1812 - Where taken: off Brest - Date Received: 8 Oct 1814 - From what ship: HMT Leyden, Chatham - Born: Massachusetts - Age: 20 - Released on 27 Apr 1815.

Lyons, Henry - Seaman - Nbr: 3915 - Prize: Hornsby - Ship Type: P - How taken: HMS Sceptre - When taken: 2 Feb 1814 - Where taken: off Block Island - Date Received: 5 Oct 1814 - From what ship: HMT President, Halifax - Born: New York - Age: 20 - Released on 9 Jun 1815.

Lyons, John - Carpenter's mate - Nbr: 6351 - Prize: Prince de Neufchatel - Ship Type: P - How taken: Leander (Newcastle Acasta) - When taken: 20 Dec 1814 - Where taken: Lat 38 Long 56 - Date Received: 19 Feb 1815 - From what ship: HMT Ganges, Plymouth - Born: Virginia - Age: 34 - Released on 30 Apr 1815.

Maceman, Thomas - Seaman - Nbr: 6436 - Prize: Chance - Ship Type: P - How taken: HMS Statira - When taken: 1 Apr 1814 - Where taken: Lat 38 Long 24 - Date Received: 3 Mar 1815 - From what ship: HMT Ganges, Plymouth - Born: New York - Age: 30 - Released on 11 Jul 1815.

Mack, John - Seaman - Nbr: 3165 - How taken: Impressed from MV Alfred - When taken: 2 Sep 1813 - Date Received: 11 Sep 1814 - From what ship: HMT Freya, Chatham - Born: Salem - Age: 17 - Released on 28 May 1815.

Mack, Thereon - Seaman - Nbr: 1706 - Prize: Fox - Ship Type: LM - How taken: HMS Pheasant - When taken: 23 Apr 1813 - Where taken: Bay of Biscay - Date Received: 6 Jul 1814 - From what ship: Bristol - Born: New York - Age: 24 - Released on 1 May 1815.

Mackay, Charles - Seaman - Nbr: 5301 - Prize: Centurion - Ship Type: Prize - How taken: HMS Maidstone - When taken: 20 Jul 1813 - Where taken: Grand Banks - Date Received: 31 Oct 1814 - From what ship: HMT Leyden, Chatham - Born: Boston - Age: 26 - Released on 29 Jun 1815.

Mackay, Joseph - Seaman - Nbr: 5027 - Prize: Landrail - Ship Type: Prize - How taken: HMS Wasp - When taken: 27 Jul 1814 - Where taken: Georges Bank - Date Received: 28 Oct 1814 - From what ship: HMT Alkbar, Halifax - Born: Baltimore - Age: 21 - Race: Negro - Released on 21 Jun 1815.

Mackay, Thomas - Seaman - Nbr: 5850 - How taken: Impressed at London - When taken: Nov 1814 - Date Received: 26 Dec 1814 - From what ship: HMT Argo - Born: Easton - Age: 24 - Race: Negro - Released on 3 Jul 1815.

Mackenzie, George - Seaman - Nbr: 2922 - How taken: Gave himself up from HMS Berwick - When taken: 29 Oct 1812 - Date Received: 24 Aug 1814 - From what ship: HMT Alpheus, Chatham - Born: Georgetown - Age: 27 - Released on 27 Apr 1815.

Mackey, John - Seaman - Nbr: 3329 - Prize: York Town - Ship Type: P - How taken: HMS Nimrod - When taken: 17 Jul 1813 - Where taken: off St. Johns - Date Received: 13 Sep 1814 - From what ship: HMT Niobe, Chatham - Born: New York - Age: 36 - Released on 28 May 1815.

Mackey, Richard - 1st Lieutenant - Nbr: 4924 - Prize: Herald - Ship Type: P - How taken: HMS Endymion - When taken: 15 Aug 1814 - Where taken: off Nantucket - Date Received: 28 Oct 1814 - From what ship: HMT Alkbar, Halifax - Born: New York - Age: 40 - Released on 21 Jun 1815.

Macklin, John - Seaman - Nbr: 6437 - Prize: Chance - Ship Type: P - How taken: HMS Statira - When taken: 1 Apr 1814 - Where taken: Lat 38 Long 24 - Date Received: 3 Mar 1815 - From what ship: HMT Ganges, Plymouth - Born: Georgetown - Age: 23 - Released on 11 Jul 1815.

Macure, Angelo - Seaman - Nbr: 2138 - How taken: Gave himself up from HMS Alexander - When taken: 11 Feb 1813 - Date Received: 8 Aug 1814 - From what ship: HMT Raven, Chatham - Born: New Orleans - Age: 22 - Released on 2 May 1815.

Madden, Frederick - Seaman - Nbr: 20 - Prize: Cashier - Ship Type: LM - How taken: HMS Reindeer - When taken: 3 Feb 1813 - Where taken: Bay of Biscay - Date Received: 2 Apr 1813 - From what ship: Plymouth - Born:

American Prisoners of War Held at Dartmoor during the War of 1812

## Alphabetical listing of names

Alexandria - Age: 19 - Sent to Dartmouth on 30 Jul 1813.

Madden, John - Seaman - Nbr: 6399 - Prize: Nellenville - Ship Type: MV - How taken: HMS Onyx - When taken: 25 Dec 1814 - Where taken: off San Domingo - Date Received: 3 Mar 1815 - From what ship: HMT Ganges, Plymouth - Born: Talbot County - Age: 26 - Race: Black - Released on 11 Jul 1815.

Maddison, Alexander - Boy - Nbr: 3600 - Prize: Frolic - Ship Type: P - How taken: HMS Heron - When taken: 25 Jan 1814 - Where taken: off St. Thomas - Date Received: 30 Sep 1814 - From what ship: HMT Sybella - Born: Salem - Age: 12 - Released on 1 May 1815.

Magrath, Samuel - Seaman - Nbr: 2142 - How taken: Gave himself up from HMS Minerva - When taken: 5 Dec 1812 - Date Received: 8 Aug 1814 - From what ship: HMT Raven, Chatham - Born: Philadelphia - Age: 30 - Released on 26 Apr 1815.

Maigot, Abner - Seaman - Nbr: 3959 - Prize: Rolla - Ship Type: P - How taken: HMS Loire - When taken: 10 Dec 1813 - Where taken: off Bull Island (South Carolina) - Date Received: 5 Oct 1814 - From what ship: HMT President, Halifax - Born: North Carolina - Age: 18 - Released on 9 Jun 1815.

Main, Charles - Seaman - Nbr: 2155 - Prize: Rattlesnake - Ship Type: P - How taken: HMS Hyperion - When taken: 26 Jun 1814 - Where taken: off Cape Finisterre - Date Received: 16 Aug 1814 - From what ship: HMS Dublin, Halifax - Born: Philadelphia - Age: 37 - Released on 2 May 1815.

Maine, John - Seaman - Nbr: 4311 - How taken: Gave himself up from HMS Whiting - When taken: 14 Feb 1814 - Date Received: 7 Oct 1814 - From what ship: HMT Niobe, Chatham - Born: Shrewsbury - Age: 39 - Released on 4 Jun 1815.

Maine, William - Seaman - Nbr: 1476 - Prize: Essex - Ship Type: MV - How taken: HMS Pyramus - When taken: 2 Apr 1813 - Where taken: Bay of Biscay - Date Received: 19 Jun 1814 - From what ship: Stapleton - Born: Marblehead - Age: 20 - Released on 28 Apr 1815.

Mainey, John - Prize master - Nbr: 4334 - Prize: Venus - Ship Type: P - How taken: HMS Loire - When taken: 18 Feb 1814 - Where taken: off St. Thomas - Date Received: 7 Oct 1814 - From what ship: HMT Salvador del Mundo, Halifax - Born: Bremen - Age: 30 - Released on 14 Jun 1815.

Mains, Darius - Seaman - Nbr: 99 - Prize: St. Martin's Planter - Ship Type: P - How taken: HMS Dublin - When taken: 9 Feb 1813 - Where taken: Lat 43 N, Long 33 50 W - Date Received: 2 Apr 1813 - From what ship: Plymouth - Born: Georgetown - Age: 24 - Released on 20 Apr 1815.

Mains, Henry - Seaman - Nbr: 4616 - Prize: Argus - Ship Type: MV - How taken: HMS San Domingo - When taken: 1 Mar 1814 - Where taken: off Savannah - Date Received: 9 Oct 1814 - From what ship: HMT Leyden, Chatham - Born: Boston - Age: 19 - Race: Black - Released on 15 Jun 1815.

Mainwarring, John - Seaman - Nbr: 5721 - How taken: Gave himself up from HMS Pelican - When taken: 13 Sep 1813 - Date Received: 26 Dec 1814 - From what ship: HMT Argo - Born: Newport - Age: 34 - Released on 9 Jan 1815.

Major, James - Seaman - Nbr: 1099 - Prize: Fair American - Ship Type: MV - How taken: HMS Andromache - When taken: 19 Jun 1814 - Where taken: Irish Channel - Date Received: 10 May 1814 - From what ship: Plymouth - Born: Salem - Age: 32 - Sent to Plymouth on 8 Jul 1814.

Malcomb, Alexander - Seaman - Nbr: 4444 - Prize: Growler - Ship Type: P - How taken: HMS Electra - When taken: 7 Jul 1814 - Where taken: at sea - Date Received: 8 Oct 1814 - From what ship: HMT Leyden, Chatham - Born: Massachusetts - Age: 23 - Released on 14 Jun 1815.

Malis, John - Seaman - Nbr: 1928 - How taken: Gave himself up from HMS Royal William - When taken: 3 Feb 1813 - Date Received: 3 Aug 1814 - From what ship: HMS Alceste, Chatham Depot - Born: New Jersey - Age: 27 - Released on 2 May 1815.

Mallard, James - Seaman - Nbr: 3416 - Prize: Thomas - Ship Type: P - How taken: HMS Nymphe - When taken: 24 Jun 1813 - Where taken: off Halifax - Date Received: 13 Sep 1814 - From what ship: HMT Niobe, Chatham - Born: Canterbury - Age: 24 - Released on 28 May 1815.

Mallery, William - Gunner - Nbr: 179 - Prize: Star - Ship Type: MV - How taken: HMS Superb - When taken: 9 Feb 1813 - Where taken: Bay of Biscay - Date Received: 2 Apr 1813 - From what ship: Plymouth - Born: New Orleans - Age: 32 - Sent to Dartmouth on 7 Aug 1813.

Malloy, William - Seaman - Nbr: 4516 - Prize: Grand Turk - Ship Type: P - How taken: HMS Tenedos - When taken: 26 May 1813 - Where taken: off Cape Cod - Date Received: 8 Oct 1814 - From what ship: HMT Leyden, Chatham - Born: Massachusetts - Age: 30 - Released on 13 Jun 1815.

Malone, James - Seaman - Nbr: 2762 - How taken: Gave himself up from HMS Bruin - When taken: 19 Jul 1813 - Date Received: 24 Aug 1814 - From what ship: HMT Liverpool, Chatham - Born: New Jersey - Age: 37 - Released on 19 May 1815.

American Prisoners of War Held at Dartmoor during the War of 1812

## Alphabetical listing of names

Maloon, Bryant - Seaman - Nbr: 3373 - How taken: Impressed at London - When taken: 7 Jul 1813 - Date Received: 13 Sep 1814 - From what ship: HMT Niobe, Chatham - Born: Portland - Age: 22 - Released on 28 May 1815.

Mangin, John B. - Marine - Nbr: 2500 - Prize: US Sloop Frolic - Ship Type: MW - How taken: HMS Orpheus - When taken: 20 Apr 1814 - Where taken: off Cuba - Date Received: 16 Aug 1814 - From what ship: HMT Queen, Halifax - Born: Rennes - Age: 29 - Released on 3 May 1815.

Manley, David - Seaman - Nbr: 4234 - Prize: Pilot - Ship Type: LM - How taken: Victoria (Privateer) - When taken: 28 Jan 1814 - Where taken: off Bordeaux - Date Received: 7 Oct 1814 - From what ship: HMT Niobe, Chatham - Born: SC - Age: 23 - Race: Mulatto - Released on 13 Jun 1815.

Mann, Charles - Seaman - Nbr: 6133 - Prize: US Frigate Superior - Ship Type: MW - How taken: British gunboats - When taken: 26 Aug 1814 - Where taken: Lake Ontario - Date Received: 17 Jan 1815 - From what ship: HMT Impregnable - Born: Germany - Age: 30 - Released on 10 Apr 1815.

Mann, James - Seaman - Nbr: 970 - Prize: Siro - Ship Type: LM - How taken: HMS Pelican - When taken: 13 Jan 1814 - Where taken: at sea - Date Received: 31 Jan 1814 - From what ship: Plymouth - Born: Boston - Age: 30 - Died on 6 Apr 1815 from gunshot wound in prison.

Mann, Richard - Quarter Gunner - Nbr: 4027 - Prize: US Brig Rattlesnake - Ship Type: MW - How taken: HMS Leander - When taken: 13 Jul 1814 - Where taken: off Shelburne - Date Received: 6 Oct 1814 - From what ship: HMT Chesapeake, Halifax - Born: Virginia - Age: 31 - Released on 9 Jun 1815.

Mann, Samuel - Seaman - Nbr: 2941 - Prize: Old Friend - How taken: HMS Viper - When taken: 1 Jan 1814 - Where taken: at sea - Date Received: 24 Aug 1814 - From what ship: HMT Hannibal - Born: Philadelphia - Age: 27 - Released on 19 May 1815.

Mannett, Richard - Seaman - Nbr: 143 - Prize: Criterion - Ship Type: MV - How taken: HMS Belle Poule - When taken: 14 Feb 1813 - Where taken: Bay of Biscay - Date Received: 2 Apr 1813 - From what ship: Plymouth - Born: New York - Age: 26 - Released on 20 Apr 1815.

Manning, George - Seaman - Nbr: 5244 - How taken: Gave himself up from HMS Africanus - When taken: 4 Oct 1813 - Date Received: 31 Oct 1814 - From what ship: HMT Mermaid, Chatham - Born: New Brunswick - Age: 36 - Released on 29 Jun 1815.

Manning, Samuel - Seaman - Nbr: 5496 - Prize: General Putnam - Ship Type: P - How taken: HMS Leander - When taken: 8 Nov 1814 - Where taken: Long 65 Lat 42 - Date Received: 17 Dec 1814 - From what ship: HMT Loire, Halifax - Born: Balboa - Age: 23 - Released on 1 Jul 1815.

Mansfield, George - Seaman - Nbr: 5560 - Prize: Levant - Ship Type: MV - How taken: HMS Forester - When taken: 4 Jan 1814 - Where taken: Bahamas Banks - Date Received: 17 Dec 1814 - From what ship: HMT Loire, Halifax - Born: Lynn - Age: 17 - Released on 1 Jul 1815.

Mansfield, James - Seaman - Nbr: 1629 - Prize: Tom - Ship Type: LM - How taken: HMS Surveillante - When taken: 24 Apr 1813 - Where taken: Bay of Biscay - Date Received: 23 Jun 1814 - From what ship: Stapleton - Born: Boston - Age: 21 - Released on 1 May 1815.

Manson, Jeremiah - Seaman - Nbr: 1666 - Prize: Zebra - Ship Type: LM - How taken: HMS Pyramus - When taken: 20 Apr 1813 - Where taken: Bay of Biscay - Date Received: 26 Jun 1814 - From what ship: Exeter - Born: Philadelphia - Age: 26 - Released on 1 May 1815.

Manson, William 1st - Seaman - Nbr: 1509 - Prize: Courier - Ship Type: LM - How taken: HMS Rover - When taken: 14 Mar 1813 - Where taken: Bay of Biscay - Date Received: 23 Jun 1814 - From what ship: Stapleton - Born: Massachusetts - Age: 25 - Released on 1 May 1815.

Manson, William 2nd - Seaman - Nbr: 1510 - Prize: Courier - Ship Type: LM - How taken: HMS Rover - When taken: 14 Mar 1813 - Where taken: Bay of Biscay - Date Received: 23 Jun 1814 - From what ship: Stapleton - Born: Massachusetts - Age: 16 - Released on 1 May 1815.

Manuel, Diego - Seaman - Nbr: 547 - Prize: Jane Barns - Ship Type: MV - How taken: HMS Comus - When taken: 14 Mar 1813 - Where taken: off Lisbon - Date Received: 8 Sep 1813 - From what ship: Plymouth - Born: New Orleans - Age: 23 - Released on 26 Apr 1815.

Manuel, John - Seaman - Nbr: 5745 - Prize: US Schooner Ohio - Ship Type: MW - How taken: British gunboats - When taken: 12 Aug 1814 - Where taken: Fort Erie - Date Received: 26 Dec 1814 - From what ship: HMT Argo - Born: New Orleans - Age: 21 - Released on 3 Jul 1815.

Marble, Jabez - Seaman - Nbr: 137 - Prize: Criterion - Ship Type: MV - How taken: HMS Belle Poule - When taken: 14 Feb 1813 - Where taken: Bay of Biscay - Date Received: 2 Apr 1813 - From what ship: Plymouth - Born: Massachusetts - Age: 23 - Released on 20 Apr 1815.

Marble, Samuel - Seaman - Nbr: 3413 - Prize: Thomas - Ship Type: P - How taken: HMS Nymphe - When taken: 24

American Prisoners of War Held at Dartmoor during the War of 1812

## Alphabetical listing of names

Jun 1813 - Where taken: off Halifax - Date Received: 13 Sep 1814 - From what ship: HMT Niobe, Chatham - Born: Hampshire - Age: 46 - Released on 28 May 1815.

March, Beverley - Seaman - Nbr: 752 - How taken: Impressed at Liverpool - Date Received: 3 Nov 1813 - From what ship: Plymouth - Born: New Jersey - Age: 20 - Released on 20 Apr 1815.

March, Jesse - Seaman - Nbr: 5691 - Prize: McDonough - Ship Type: P - How taken: HMS Bacchante - When taken: 7 Nov 1814 - Where taken: Lat 42 Long 67 - Date Received: 24 Dec 1814 - From what ship: HMT Penelope - Born: Massachusetts - Age: 28 - Died on 5 Feb 1815 from pneumonia.

Marchand, Angelo - Seaman - Nbr: 6056 - Prize: William - Ship Type: Prize - How taken: HMS Armide - When taken: 11 Oct 1814 - Where taken: off Newport - Date Received: 28 Dec 1814 - From what ship: HMT Penelope - Born: Genoa - Age: 26 - Released on 3 Jul 1815.

Marchant, Isaac - Seaman - Nbr: 5443 - How taken: Sent into custody from MV Eagle - When taken: 20 Oct 1814 - Date Received: 11 Nov 1814 - From what ship: HMT Impregnable - Born: Rhode Island - Age: 22 - Race: Negro - Released on 1 Jul 1815.

Maret, Ebenezer - Seaman - Nbr: 1763 - Prize: Cygnet - Ship Type: Prize - How taken: HMS Statira - When taken: 8 May 1814 - Where taken: at sea - Date Received: 20 Jul 1814 - From what ship: HMS Milford, Plymouth - Born: Salem - Age: 24 - Released on 1 May 1815.

Maria, Hosea - Seaman - Nbr: 4703 - Prize: Volunteer - Ship Type: MV - How taken: Victoria (Privateer) - When taken: 26 Dec 1813 - Where taken: at sea - Date Received: 9 Oct 1814 - From what ship: HMT Leyden, Chatham - Born: Cartagena - Age: 22 - Released on 15 Jun 1815.

Maria, Josea - Seaman - Nbr: 4709 - How taken: Impressed at Dublin - When taken: 13 Aug 1813 - Date Received: 9 Oct 1814 - From what ship: HMT Leyden, Chatham - Born: Frederick - Age: 25 - Race: Black - Released on 15 Jun 1815.

Marie, John - Seaman - Nbr: 4803 - Prize: President - Ship Type: P - How taken: HMS Pique - When taken: 7 May 1814 - Where taken: off Porto Rico - Date Received: 9 Oct 1814 - From what ship: HMT Freya, Halifax - Born: Nantes - Age: 32 - Released on 21 Jun 1815.

Mariner, Joseph - Seaman - Nbr: 2382 - Prize: US Sloop Frolic - Ship Type: MW - How taken: HMS Orpheus - When taken: 20 Apr 1814 - Where taken: off Cuba - Date Received: 16 Aug 1814 - From what ship: HMT Queen, Halifax - Born: Oporto, Portugal - Age: 43 - Sent to Dartmouth on 19 Oct 1814.

Marion, John - Seaman - Nbr: 6067 - Prize: Chasseur - Ship Type: P - How taken: HMS Recruit - When taken: 6 Apr 1814 - Where taken: Lat 29 Long 76 - Date Received: 28 Dec 1814 - From what ship: HMT Penelope - Born: Baltimore - Age: 25 - Race: Black - Released on 3 Jul 1815.

Mark, James - Seaman - Nbr: 4571 - How taken: Gave himself up from MV Liberty - When taken: 30 Dec 1813 - Date Received: 9 Oct 1814 - From what ship: HMT Leyden, Chatham - Born: New York - Age: 17 - Released on 15 Jun 1815.

Markins, John - Seaman - Nbr: 2259 - Prize: Prize to the Scourge - Ship Type: P - How taken: HMS Martin - When taken: 4 May 1814 - Where taken: off Newfoundland - Date Received: 16 Aug 1814 - From what ship: HMS Dublin, Halifax - Born: New York - Age: 20 - Released on 3 May 1815.

Marks, Peter - Seaman - Nbr: 2616 - How taken: Gave himself up from HMS Tigre - When taken: 15 Aug 1812 - Date Received: 21 Aug 1814 - From what ship: HMT Freya, Chatham - Born: New Orleans - Age: 31 - Released on 26 Apr 1815.

Marlborough, Francis - Seaman - Nbr: 2726 - How taken: Gave himself up from HMS Denmark - When taken: 4 Jan 1814 - Date Received: 21 Aug 1814 - From what ship: HMT Freya, Chatham - Born: Hartford - Age: 47 - Race: Black - Released on 19 May 1815.

Marlow, Owen - Seaman - Nbr: 4700 - Prize: Volunteer - Ship Type: MV - How taken: Victoria (Privateer) - When taken: 26 Dec 1813 - Where taken: at sea - Date Received: 9 Oct 1814 - From what ship: HMT Leyden, Chatham - Born: Massachusetts - Age: 22 - Released on 15 Jun 1815.

Marral, Ephraim - Seaman - Nbr: 4340 - Prize: Lizard - Ship Type: P - How taken: HMS Barbados - When taken: 1 Mar 1814 - Where taken: off Halifax - Date Received: 7 Oct 1814 - From what ship: HMT Salvador del Mundo, Halifax - Born: Massachusetts - Age: 24 - Released on 14 Jun 1815.

Mars, George - Seaman - Nbr: 2780 - Prize: Tiger - Ship Type: MV - How taken: HMS Scylla - When taken: 22 May 1813 - Where taken: Bay of Biscay - Date Received: 24 Aug 1814 - From what ship: HMT Liverpool, Chatham - Born: Massachusetts - Age: 39 - Released on 19 May 1815.

Marsh, Hercules - Seaman - Nbr: 3157 - Prize: Fame (Whaler) - Ship Type: MV - How taken: HMS Cressy - When taken: 20 Jul 1813 - Where taken: at sea - Date Received: 11 Sep 1814 - From what ship: HMT Freya, Chatham - Born: Rhode Island - Age: 58 - Race: Black - Released on 28 May 1815.

American Prisoners of War Held at Dartmoor during the War of 1812

## Alphabetical listing of names

Marsh, Jesse - Boy - Nbr: 378 - Prize: Independence - Ship Type: MV - How taken: HMS Superb - When taken: 16 Mar 1813 - Where taken: Bay of Biscay - Date Received: 1 Jul 1813 - From what ship: Plymouth - Born: Boston - Age: 14 - Released on 20 Apr 1815.

Marshall, Alexander - Seaman - Nbr: 1596 - Prize: Shadow - Ship Type: LM - How taken: HMS Reindeer - When taken: 6 Apr 1813 - Where taken: Bay of Biscay - Date Received: 23 Jun 1814 - From what ship: Stapleton - Born: Philadelphia - Age: 24 - Released on 1 May 1815.

Marshall, Anthony - Seaman - Nbr: 6268 - How taken: Impressed at Plymouth - When taken: 19 Jan 1815 - Date Received: 10 Feb 1815 - From what ship: HMT Ganges, Plymouth - Born: New Orleans - Age: 28 - Race: Mulatto - Released on 5 Jul 1815.

Marshall, B. - Seaman - Nbr: 5632 - Prize: Mary - Ship Type: Prize - How taken: Not legible - When taken: 18 Jun 1814 - Date Received: 24 Dec 1814 - From what ship: HMT Tay - Born: Islesboro - Age: 23 - Released on 5 Jul 1815.

Marshall, Benjamin - Seaman - Nbr: 5245 - How taken: Gave himself up from HMS Minden - When taken: 17 Jul 1813 - Date Received: 31 Oct 1814 - From what ship: HMT Mermaid, Chatham - Born: Hillsborough - Age: 23 - Died on 27 Mar 1814 from variola.

Marshall, Emil - Seaman - Nbr: 844 - Prize: Chesapeake - Ship Type: LM - How taken: HMS Hotspur & HMS Pyramus - When taken: 26 Oct 1813 - Where taken: off Nantes - Date Received: 29 Nov 1813 - From what ship: Plymouth - Born: Providence - Age: 40 - Released on 26 Apr 1815.

Marshall, Francis - Seaman - Nbr: 2587 - How taken: Gave himself up from HMS Ulysses - When taken: 13 Jan 1813 - Date Received: 21 Aug 1814 - From what ship: HMT Freya, Chatham - Born: Virginia - Age: 27 - Released on 3 May 1815.

Marshall, George - Seaman - Nbr: 3620 - Prize: Monarch - Ship Type: MV - How taken: HMS Dotterel - When taken: 14 Dec 1813 - Where taken: off Charleston - Date Received: 30 Sep 1814 - From what ship: HMT Sybella - Born: Philadelphia - Age: 29 - Released on 4 Jun 1815.

Marshall, John - Seaman - Nbr: 3907 - Prize: Thorn - Ship Type: MV - How taken: HMS Bulwark - When taken: 9 Jul 1814 - Where taken: off Nantucket - Date Received: 5 Oct 1814 - From what ship: HMT Orpheus, Halifax - Born: Massachusetts - Age: 23 - Released on 9 Jun 1815.

Marshall, John - Seaman - Nbr: 1752 - How taken: Gave himself up from HMS Alemene - When taken: 24 Jan 1813 - Date Received: 20 Jul 1814 - From what ship: HMS Milford, Plymouth - Born: New Bedford - Age: 41 - Died on 8 Apr 1815 from erysipelatous inflammation.

Marshall, John - Mate - Nbr: 817 - How taken: Impressed at Liverpool - When taken: 10 Sep 1813 - Date Received: 3 Nov 1813 - From what ship: Plymouth - Born: Nantucket - Age: 31 - Released on 26 Apr 1815.

Marshall, John - Seaman - Nbr: 3310 - Prize: Thomas - Ship Type: P - How taken: HMS Nymphe - When taken: 28 Jun 1813 - Where taken: off Halifax - Date Received: 13 Sep 1814 - From what ship: HMT Niobe, Chatham - Born: New York - Age: 21 - Released on 28 May 1815.

Marshall, Joseph - Seaman - Nbr: 5902 - Prize: Harlequin - Ship Type: P - How taken: HMS Bulwark - When taken: 23 Nov 1814 - Where taken: off Halifax - Date Received: 27 Dec 1814 - From what ship: HMT Penelope - Born: Boston - Age: 24 - Released on 3 Jul 1815.

Marshall, Levy - Seaman - Nbr: 3546 - Prize: Hawk - Ship Type: P - How taken: HMS Pique - When taken: 26 Apr 1814 - Where taken: off Bermuda - Date Received: 30 Sep 1814 - From what ship: HMT Sybella - Born: Baltimore - Age: 29 - Released on 4 Jun 1815.

Marshall, M. - Boy - Nbr: 6007 - Prize: McDonough - Ship Type: P - How taken: HMS Bacchante - When taken: 1 Nov 1814 - Where taken: Lat 42 Long 67 - Date Received: 28 Dec 1814 - From what ship: HMT Penelope - Born: Massachusetts - Age: 16 - Released on 3 Jul 1815.

Marshall, Solomon - Seaman - Nbr: 5019 - Prize: Alexander - Ship Type: Prize - How taken: HMS Wasp - When taken: 6 Sep 1814 - Where taken: off Cape Sable Island (Canada) - Date Received: 28 Oct 1814 - From what ship: HMT Alkbar, Halifax - Born: Massachusetts - Age: 27 - Died on 20 Nov 1814.

Marshall, Thomas - Seaman - Nbr: 6008 - Prize: McDonough - Ship Type: P - How taken: HMS Bacchante - When taken: 1 Nov 1814 - Where taken: Lat 42 Long 67 - Date Received: 28 Dec 1814 - From what ship: HMT Penelope - Born: Berwick - Age: 47 - Released on 3 Jul 1815.

Marshall, Thomas - Seaman - Nbr: 553 - How taken: Gave himself up from HMS Pompeii - When taken: 31 Oct 1812 - Date Received: 8 Sep 1813 - From what ship: Plymouth - Born: Drummond Town - Age: 26 - Released on 26 Apr 1815.

Marshall, William - Boy - Nbr: 6366 - Prize: St. Johanna - Ship Type: Prize - How taken: HMS Sabine - When taken: 25 Oct 1814 - Where taken: off Newfoundland Banks - Date Received: 24 Feb 1815 - From what ship:

American Prisoners of War Held at Dartmoor during the War of 1812

## Alphabetical listing of names

HMT Ganges, Plymouth - Born: Bath - Age: 18 - Released on 11 Jul 1815.

Marshall, William - Seaman - Nbr: 5450 - Prize: Prize to the Lawrence - Ship Type: P - How taken: HMS Glasgow - When taken: 2 Nov 1814 - Where taken: Channel - Date Received: 10 Dec 1814 - From what ship: HMT Impregnable - Born: New York - Age: 18 - Released on 2 May 1815.

Marthoup, John - Seaman - Nbr: 2459 - Prize: US Sloop Frolic - Ship Type: MW - How taken: HMS Orpheus - When taken: 20 Apr 1814 - Where taken: off Cuba - Date Received: 16 Aug 1814 - From what ship: HMT Queen, Halifax - Born: New Haven - Age: 29 - Released on 3 May 1815.

Martin, Andrew - Seaman - Nbr: 3480 - How taken: Gave himself up from HMS Queen Charlotte - When taken: Aug 1814 - Date Received: 19 Sep 1814 - From what ship: HMT Salvador del Mundo - Born: Lancaster - Age: 23 - Race: Black - Released on 4 Jun 1815.

Martin, Anthony - Seaman - Nbr: 1556 - Prize: Caroline - Ship Type: MV - How taken: HMS Medusa - When taken: 12 Apr 1813 - Where taken: Bay of Biscay - Date Received: 23 Jun 1814 - From what ship: Stapleton - Born: New Orleans - Age: 33 - Sent to Mill Prison (Plymouth, England) on 30 Jun 1814.

Martin, Daniel - Seaman - Nbr: 1346 - Prize: Paul Jones - Ship Type: P - How taken: HMS Leonidas - When taken: 23 May 1813 - Where taken: Channel - Date Received: 19 Jun 1814 - From what ship: Stapleton - Born: New Orleans - Age: 17 - Died on 22 Sep 1814 from pneumonia.

Martin, Francis - Seaman - Nbr: 4706 - Prize: Volunteer - Ship Type: MV - How taken: Victoria (Privateer) - When taken: 26 Dec 1813 - Where taken: at sea - Date Received: 9 Oct 1814 - From what ship: HMT Leyden, Chatham - Born: Galatean - Age: 24 - Released on 15 Jun 1815.

Martin, Henry - Seaman - Nbr: 1401 - Prize: Good Friends - Ship Type: MV - How taken: HMS Andromache - When taken: 2 Apr 1813 - Where taken: Bay of Biscay - Date Received: 19 Jun 1814 - From what ship: Stapleton - Born: Prussia - Age: 25 - Released on 3 Jul 1814.

Martin, Henry - Seaman - Nbr: 1954 - How taken: Gave himself up from HMS Christian 7th - When taken: 19 May 1813 - Date Received: 3 Aug 1814 - From what ship: HMS Lyffey, Chatham Depot - Born: Albany - Age: 28 - Released on 2 May 1815.

Martin, Henry - Prize Master - Nbr: 2241 - Prize: Rolla - Ship Type: Prize - How taken: HMS Loire - When taken: 9 Dec 1813 - Where taken: off Newport - Date Received: 16 Aug 1814 - From what ship: HMS Dublin, Halifax - Born: Philadelphia - Age: 32 - Released on 2 May 1815.

Martin, Isaac - Seaman - Nbr: 1361 - Prize: Paul Jones - Ship Type: P - How taken: HMS Leonidas - When taken: 23 May 1813 - Where taken: Channel - Date Received: 19 Jun 1814 - From what ship: Stapleton - Born: Baltimore - Age: 16 - Released on 28 Apr 1815.

Martin, James - Seaman - Nbr: 1388 - Prize: Paul Jones - Ship Type: P - How taken: HMS Leonidas - When taken: 23 May 1813 - Where taken: Channel - Date Received: 19 Jun 1814 - From what ship: Stapleton - Born: Salem - Age: 37 - Released on 28 Apr 1815.

Martin, John - Seaman - Nbr: 6179 - Prize: Lion - Ship Type: P - How taken: HMS Granicus - When taken: 2 Dec 1814 - Where taken: off Lisbon - Date Received: 21 Jan 1815 - From what ship: HMT Impregnable - Born: New Jersey - Age: 25 - Released on 5 Jul 1815.

Martin, John - 2nd Lieutenant - Nbr: 6304 - Prize: Prince de Neufchatel - Ship Type: P - How taken: Leander (Newcastle Acasta) - When taken: 20 Dec 1814 - Where taken: Lat 38 Long 56 - Date Received: 19 Feb 1815 - What ship: HMT Ganges, Plymouth - Born: Boston - Age: 25 - Sent to Ashburton (England) on 24 Feb 1815.

Martin, John - Seaman - Nbr: 902 - Prize: Squirrel - Ship Type: MV - How taken: HMS Belle Poule - When taken: 14 Dec 1813 - Where taken: at sea - Date Received: 31 Jan 1814 - From what ship: Plymouth - Born: Doliver - Age: 27 - Released on 27 Apr 1815.

Martin, John - Seaman - Nbr: 1319 - How taken: Gave himself up from HMS Venus - When taken: 2 Jan 1813 - Date Received: 19 Jun 1814 - From what ship: Stapleton - Born: Abbott County - Age: 31 - Race: Negro - Released on 28 Apr 1815.

Martin, Peter - Steward - Nbr: 43 - Prize: Terrible - Ship Type: MV - How taken: HMS Foxhound - When taken: 8 Feb 1813 - Where taken: Channel - Date Received: 2 Apr 1813 - From what ship: Plymouth - Born: Burlington - Age: 24 - Sent to Dartmouth on 30 Jul 1813.

Martin, Peter - Seaman - Nbr: 3513 - Prize: Aeolus - Ship Type: Prize - How taken: HMS Warspite - When taken: 2 Sep 1814 - Where taken: off Newfoundland - Date Received: 28 Sep 1814 - From what ship: HMT Salvador del Mundo - Born: Virginia - Age: 24 - Race: Mulatto - Released on 4 Jun 1815.

Martin, Philip - Seaman - Nbr: 3523 - Prize: Sabine - Ship Type: P - How taken: HMS Conquistador - When taken: 3 Aug 1814 - Where taken: off the Banks - Date Received: 28 Sep 1814 - From what ship: HMT Salvador del

American Prisoners of War Held at Dartmoor during the War of 1812

## Alphabetical listing of names

Mundo - Born: Baltimore - Age: 25 - Released on 4 Jun 1815.

Martin, Stephen - Boy - Nbr: 846 - Prize: Chesapeake - Ship Type: LM - How taken: HMS Hotspur & HMS Pyramus - When taken: 26 Oct 1813 - Where taken: off Nantes - Date Received: 29 Nov 1813 - From what ship: Plymouth - Born: New York - Age: 13 - Sent to Plymouth on 8 Jul 1814.

Martin, Thomas - Seaman - Nbr: 683 - How taken: Impressed at Belfast - When taken: 29 Jun 1813 - Date Received: 27 Sep 1813 - From what ship: Plymouth - Born: New York - Age: 22 - Released on 26 Apr 1815.

Martin, William - Seaman - Nbr: 2669 - Prize: Orders in Council - Ship Type: MV - How taken: HMS Surveillante - When taken: 1 Jun 1813 - Where taken: off Cape Ortegal (Spain) - Date Received: 21 Aug 1814 - From what ship: HMT Freya, Chatham - Born: Norfolk - Age: 25 - Race: Black - Released on 19 May 1815.

Martling, Abraham - Seaman - Nbr: 1697 - Prize: Fanny - Ship Type: Prize - How taken: HMS Sceptre - When taken: 12 May 1814 - Date Received: 2 Jul 1814 - From what ship: Plymouth - Born: New York - Age: 22 - Released on 1 May 1815.

Martyn, John - Seaman - Nbr: 6493 - Prize: Harlequin - Ship Type: P - How taken: HMS Bulwark - When taken: 23 Nov 1814 - Where taken: off Halifax - Date Received: 3 Mar 1815 - From what ship: HMT Ganges, Plymouth - Born: Holland - Age: 30 - Released on 11 Jul 1815.

Martyn, John - Seaman - Nbr: 6363 - Prize: Prince de Neufchatel - Ship Type: P - How taken: Leander (Newcastle Acasta) - When taken: 20 Dec 1814 - Where taken: Lat 38 Long 56 - Date Received: 19 Feb 1815 - From what ship: HMT Ganges, Plymouth - Born: New Orleans - Age: 35 - Released on 6 Jul 1815.

Marvel, David - Seaman - Nbr: 4725 - How taken: Gave himself up from HMS Hecate - When taken: 4 Jul 1813 - Date Received: 9 Oct 1814 - From what ship: HMT Freya, Chatham - Born: Rhode Island - Age: 36 - Released on 15 Jun 1815.

Masick, Joseph - Seaman - Nbr: 1228 - How taken: Sent into custody from HMS Furieuse - When taken: 23 Sep 1812 - Date Received: 14 Jun 1814 - From what ship: Mill Prison (Plymouth, England) - Born: Charlestown - Age: 31 - Race: Mulatto - Released on 9 Jun 1815.

Mason, Aaron - Seaman - Nbr: 5229 - Prize: Polly - Ship Type: MV - How taken: HMS Maidstone - When taken: 17 Jul 1813 - Where taken: off Cape Sable Island (Canada) - Date Received: 31 Oct 1814 - From what ship: HMT Mermaid, Chatham - Born: Salem - Age: 16 - Released on 29 Jun 1815.

Mason, Daniel - Seaman - Nbr: 1867 - Prize: General Starks - Ship Type: P - How taken: HMS Martin - When taken: 9 Oct 1813 - Where taken: at sea - Date Received: 29 Jul 1814 - From what ship: HMS Ville de Paris, Chatham Depot - Born: New Hampshire - Age: 20 - Released on 2 May 1815.

Mason, Daniel - Sailing master - Nbr: 5575 - Prize: McDonough - Ship Type: P - How taken: HMS Bacchante - When taken: 1 Nov 1814 - Where taken: Lat 42 Long 67 - Date Received: 17 Dec 1814 - From what ship: HMT Loire, Halifax - Born: Arundel - Age: 34 - Released on 1 Jul 1815.

Mason, J. Bude - Seaman - Nbr: 768 - How taken: Sent into custody from HMS Pallas - Date Received: 3 Nov 1813 - From what ship: Plymouth - Born: Marblehead - Age: 22 - Released on 26 Apr 1815.

Mason, James - Seaman - Nbr: 5297 - Prize: Fox - Ship Type: P - How taken: HMS Shannon - When taken: 7 Nov 1813 - Where taken: off Newfoundland - Date Received: 31 Oct 1814 - From what ship: HMT Leyden, Chatham - Born: Boswick - Age: 27 - Released on 29 Jun 1815.

Mason, John - Seaman - Nbr: 2593 - How taken: Gave himself up from HMS Ariel - When taken: 21 Oct 1813 - Date Received: 21 Aug 1814 - From what ship: HMT Freya, Chatham - Born: New Haven - Age: 32 - Released on 26 Apr 1815.

Mason, John - Seaman - Nbr: 2664 - Prize: Eliza - Ship Type: MV - How taken: HMS Tenedos - Where taken: off America - Date Received: 21 Aug 1814 - From what ship: HMT Freya, Chatham - Born: Jacobstown - Age: 34 - Released on 19 May 1815.

Mason, Joseph J. - Master - Nbr: 3153 - Prize: John of Salem - Ship Type: P - How taken: HMS Peruvian - When taken: 6 Feb 1813 - Where taken: at sea - Date Received: 11 Sep 1814 - From what ship: HMT Freya, Chatham - Born: Marblehead - Age: 26 - Released on 28 May 1815.

Mason, Joshua - Seaman - Nbr: 5879 - Prize: Harlequin - Ship Type: P - How taken: HMS Bulwark - When taken: 23 Nov 1814 - Where taken: off Halifax - Date Received: 27 Dec 1814 - From what ship: HMT Penelope - Born: Arundel - Age: 23 - Released on 3 Jul 1815.

Mason, Moses - Seaman - Nbr: 5016 - Prize: Charlotte - Ship Type: Prize - How taken: Censor - When taken: 5 Aug 1814 - Where taken: off Azores - Date Received: 28 Oct 1814 - From what ship: HMT Alkbar, Halifax - Born: Lymington - Age: 13 - Released on 21 Jun 1815.

Mason, Nathaniel - Seaman - Nbr: 1435 - Prize: Leo - Ship Type: LM - How taken: HMS Magicienne - When taken: 4 Jun 1813 - Where taken: off France - Date Received: 19 Jun 1814 - From what ship: Stapleton - Born:

American Prisoners of War Held at Dartmoor during the War of 1812

## Alphabetical listing of names

Lymington - Age: 22 - Released on 28 Apr 1815.
Mason, Richard - Seaman - Nbr: 2433 - Prize: US Sloop Frolic - Ship Type: MW - How taken: HMS Orpheus - When taken: 20 Apr 1814 - Where taken: off Cuba - Date Received: 16 Aug 1814 - From what ship: HMT Queen, Halifax - Born: New York - Age: 24 - Released on 3 May 1815.
Mason, Richard - Seaman - Nbr: 5413 - Prize: Franklin - Ship Type: MV - How taken: HMS Weazle - When taken: 8 Jul 1813 - Where taken: York River - Date Received: 31 Oct 1814 - From what ship: HMT Leyden, Chatham - Born: Caiosso, Africa - Age: 47 - Race: Mulatto - Released on 1 Jul 1815.
Mason, Thomas - Seaman - Nbr: 4125 - Prize: Bordeaux Packet - Ship Type: LM - How taken: HMS Niemen - When taken: 28 Jun 1814 - Where taken: off Delaware - Date Received: 6 Oct 1814 - From what ship: HMT Chesapeake, Halifax - Born: Philadelphia - Age: 17 - Released on 13 Jun 1815.
Mason, William - Prize Master - Nbr: 2523 - Prize: Aeolus - Ship Type: Prize - How taken: HMS Cyane - When taken: 22 May 1814 - Where taken: off Newfoundland - Date Received: 16 Aug 1814 - From what ship: HMT Queen, Halifax - Born: Baltimore - Age: 25 - Released on 3 May 1815.
Mason, William - Seaman - Nbr: 966 - Prize: Siro - Ship Type: LM - How taken: HMS Pelican - When taken: 13 Jan 1814 - Where taken: at sea - Date Received: 31 Jan 1814 - From what ship: Plymouth - Born: New York - Age: 40 - Released on 27 Apr 1815.
Masser, Enoch - Marine - Nbr: 4057 - Prize: US Brig Rattlesnake - Ship Type: MW - How taken: HMS Leander - When taken: 13 Jul 1814 - Where taken: off Shelburne - Date Received: 6 Oct 1814 - From what ship: HMT Chesapeake, Halifax - Born: Methuen - Age: 23 - Released on 13 Jun 1815.
Mastin, James - Seaman - Nbr: 5827 - Prize: US Schooner Tigress - Ship Type: MW - How taken: British gunboats - When taken: 2 Sep 1814 - Where taken: Lake Erie - Date Received: 26 Dec 1814 - From what ship: HMT Argo - Born: Delaware - Age: 45 - Released on 3 Jul 1815.
Mathais, Louis - Seaman - Nbr: 2078 - Prize: True Blooded Yankee - Ship Type: P - How taken: HMS Hope - When taken: 24 Jun 1813 - Where taken: off Brest - Date Received: 3 Aug 1814 - From what ship: HMS Bittern, Chatham Depot - Born: Delaware - Age: 35 - Released on 2 May 1815.
Mather, John - Seaman - Nbr: 244 - Prize: William Bayard - Ship Type: MV - How taken: HMS Warspite - When taken: 3 Mar 1813 - Where taken: Bay of Biscay - Date Received: 2 Apr 1813 - From what ship: Plymouth - Born: Philadelphia - Age: 27 - Released on 20 Apr 1815.
Mathews, Cornelius - Seaman - Nbr: 5381 - Prize: Harriett - Ship Type: MV - How taken: HMS Thistle - When taken: 24 Feb 1813 - Where taken: off St. Bartholomew - Date Received: 31 Oct 1814 - From what ship: HMT Leyden, Chatham - Born: Baltimore - Age: 25 - Released on 1 Jul 1815.
Mathews, Henry - Seaman - Nbr: 916 - How taken: Gave himself up from a French frigate - Date Received: 31 Jan 1814 - From what ship: Plymouth - Born: Virginia - Age: 32 - Released on 27 Apr 1815.
Mathews, Joseph - Seaman - Nbr: 911 - Prize: Zephyr - Ship Type: MV - How taken: HMS Pyramus - When taken: 30 Nov 1813 - Where taken: off L'Orient (France) - Date Received: 31 Jan 1814 - From what ship: Plymouth - Born: Providence - Age: 32 - Released on 27 Apr 1815.
Mathews, Joseph - Seaman - Nbr: 156 - How taken: Gave himself up from HMS Lavinia - Date Received: 2 Apr 1813 - From what ship: Plymouth - Born: Rhode Island - Age: 30 - Released on 20 Apr 1815.
Mathews, P. - Seaman - Nbr: 3863 - Prize: Buzi - Ship Type: MV - How taken: HMS Dragon - When taken: 20 Jul 1814 - Where taken: off VA - Date Received: 5 Oct 1814 - From what ship: HMT Orpheus, Halifax - Born: North Carolina - Age: 18 - Released on 10 May 1815.
Mathews, Pedro - Seaman - Nbr: 2301 - Prize: Hussar - Ship Type: P - How taken: HMS Saturn - When taken: 25 May 1814 - Where taken: off Sandy Hook - Date Received: 16 Aug 1814 - From what ship: HMS Dublin, Halifax - Born: Bordeaux - Age: 24 - Released on 3 May 1815.
Mathews, Richard - Seaman - Nbr: 1534 - How taken: Impressed at Bristol - When taken: 1 May 1813 - Date Received: 23 Jun 1814 - From what ship: Stapleton - Born: New Jersey - Age: 34 - Released on 1 May 1815.
Mathews, William - Seaman - Nbr: 1743 - Prize: Hugh Jones - Ship Type: MV - How taken: HMS Bittern - When taken: 14 Jun 1814 - Where taken: at sea - Date Received: 18 Jul 1814 - From what ship: HMT Salvador del Mundo, Plymouth - Born: Rhode Island - Age: 24 - Released on 1 May 1815.
Mathews, Williams - Seaman - Nbr: 1067 - Prize: Fair American - Ship Type: MV - How taken: HMS Andromache - When taken: 19 Jan 1814 - Where taken: Bay of Biscay - Date Received: 10 May 1814 - From what ship: Plymouth - Born: Alsace - Age: 25 - Released on 27 Apr 1815.
Mathias, Henry - Seaman - Nbr: 709 - Prize: Ned - Ship Type: LM - How taken: HMS Royalist - When taken: 6 Sep 1813 - Where taken: Bay of Biscay - Date Received: 27 Sep 1813 - From what ship: Plymouth - Born: New York - Age: 18 - Sent to Plymouth on 8 Jul 1814.

American Prisoners of War Held at Dartmoor during the War of 1812

## Alphabetical listing of names

Mathy, James - Captain - Nbr: 4831 - Prize: President - Ship Type: P - How taken: HMS Pique - When taken: 7 May 1814 - Where taken: off Porto Rico - Date Received: 9 Oct 1814 - From what ship: HMT Freya, Halifax - Born: Paris - Age: 25 - Released on 3 May 1815.

Maxen, Joseph - Seaman - Nbr: 1182 - Prize: Melanie - Ship Type: MV - How taken: HMS Briton - When taken: 15 Sep 1813 - Where taken: off Bordeaux - Date Received: 4 Jun 1814 - From what ship: Dartmouth - Born: New Orleans - Age: 21 - Sent to Plymouth on 8 Jul 1814.

Maxine, Joseph - Seaman - Nbr: 6156 - Prize: Lion - Ship Type: P - How taken: HMS Granicus - When taken: 2 Dec 1814 - Where taken: off Lisbon - Date Received: 18 Jan 1815 - From what ship: HMT Impregnable - Born: New Orleans - Age: 22 - Released on 5 Jul 1815.

May, Henry - Seaman - Nbr: 695 - Prize: Montgomery - Ship Type: P - How taken: HMS Nymphe - When taken: 5 May 1813 - Where taken: Boston Bay - Date Received: 27 Sep 1813 - From what ship: Plymouth - Born: Staten Island - Age: 25 - Sent to Mill Prison (Plymouth, England) on 17 Jun 1814.

May, John - Seaman - Nbr: 6223 - Prize: Prince de Neufchatel - Ship Type: P - How taken: Leander (Newcastle Acasta) - When taken: 20 Dec 1814 - Where taken: Lat 38 Long 56 - Date Received: 30 Jan 1815 - From what ship: HMT Pheasant - Born: Boston - Age: 35 - Released on 5 Jul 1815.

May, Joseph - Seaman - Nbr: 3741 - Prize: Lizard - Ship Type: P - How taken: HMS Prometheus - When taken: 5 May 1814 - Where taken: off Halifax - Date Received: 30 Sep 1814 - From what ship: HMT President, Halifax - Born: Boston - Age: 30 - Released on 4 Jun 1815.

May, Walter - Seaman - Nbr: 1577 - Prize: Price - Ship Type: MV - How taken: HMS Pyramus - When taken: 6 Apr 1813 - Where taken: Bay of Biscay - Date Received: 23 Jun 1814 - From what ship: Stapleton - Born: Norfolk - Age: 29 - Released on 1 May 1815.

Mayeau, Morris - Seaman - Nbr: 4367 - Prize: Decatur - Ship Type: MV - How taken: Desire - When taken: 7 May 1813 - Where taken: off Nantes - Date Received: 8 Oct 1814 - From what ship: HMT Leyden, Chatham - Born: Not legible - Age: 19 - Released on 14 Jun 1815.

Mayers, James - Seaman - Nbr: 3206 - How taken: Gave himself up from HMS Ganymede - When taken: 16 Jul 1813 - Date Received: 11 Sep 1814 - From what ship: HMT Freya, Chatham - Born: Philadelphia - Age: 23 - Released on 28 May 1815.

Maynard, Humphrey - Cooper - Nbr: 5753 - Prize: Hope - Ship Type: MV - How taken: HMS Nereus - When taken: 14 May 1814 - Where taken: Rio de la Plata - Date Received: 26 Dec 1814 - From what ship: HMT Argo - Born: New Bedford - Age: 27 - Released on 3 Jul 1815.

Mayo, Nathaniel - Seaman - Nbr: 2641 - Prize: True Blooded Yankee - Ship Type: P - How taken: HMS Hamadryad - When taken: 24 Jul 1813 - Where taken: off Norway - Date Received: 21 Aug 1814 - From what ship: HMT Freya, Chatham - Born: Cape Cod - Age: 26 - Released on 3 May 1815.

Mazal, James - Seaman - Nbr: 1836 - How taken: Gave himself up from HMS Malta - When taken: 1 Jan 1813 - Date Received: 21 Jul 1814 - From what ship: HMT Redbeard & Pincher, Chatham Depot - Born: Morristown - Age: 20 - Released on 1 May 1815.

Mazely, William - Seaman - Nbr: 2336 - Prize: Snap Dragon - Ship Type: P - How taken: HMS Martin - When taken: 10 Jun 1814 - Where taken: off Halifax - Date Received: 16 Aug 1814 - From what ship: HMS Dublin, Halifax - Born: Virginia - Age: 22 - Released on 3 May 1815.

McAlpin, Charles - Seaman - Nbr: 3227 - How taken: Gave himself up from HMS Hope - When taken: 18 Oct 1813 - Date Received: 11 Sep 1814 - From what ship: HMT Freya, Chatham - Born: Rhode Island - Age: 28 - Released on 28 May 1815.

McBride, James - Seaman - Nbr: 1841 - How taken: Gave himself up from HMS Clarence - When taken: 7 May 1813 - Date Received: 21 Jul 1814 - From what ship: HMT Redbeard & Pincher, Chatham Depot - Born: Baltimore - Age: 27 - Released on 2 May 1815.

McBuchey, Patrick - Soldier - Nbr: 5254 - Prize: 13th US Infantry - Ship Type: Troops - How taken: British Army - When taken: 13 Oct 1812 - Where taken: Canada - Date Received: 31 Oct 1814 - From what ship: HMT Leyden, Chatham - Born: Donegal - Age: 40 - Released on 29 Jun 1815.

McCalla, David - Seaman - Nbr: 4950 - Prize: Herald - Ship Type: P - How taken: HMS Endymion - When taken: 15 Aug 1814 - Where taken: off Nantucket - Date Received: 28 Oct 1814 - From what ship: HMT Alkbar, Halifax - Born: SC - Age: 25 - Released on 21 Jun 1815.

McCannon, Joseph - Soldier - Nbr: 5252 - Prize: 13th US Infantry - Ship Type: Troops - How taken: British Army - When taken: 13 Oct 1812 - Where taken: Canada - Date Received: 31 Oct 1814 - From what ship: HMT Leyden, Chatham - Born: Ireland - Age: 39 - Released on 29 Jun 1815.

McCanon, John - Seaman - Nbr: 5117 - Prize: Portsmouth - Ship Type: P - How taken: HMS Pylades - When taken:

American Prisoners of War Held at Dartmoor during the War of 1812

## Alphabetical listing of names

13 Sep 1814 - Where taken: Long 65 - Date Received: 28 Oct 1814 - From what ship: HMT Alkbar, Halifax - Born: New York - Age: 14 - Race: Negro - Released on 29 Jun 1815.

McCarthy, Henry - Seaman - Nbr: 2983 - Prize: Moranda - Ship Type: MV - How taken: Circa (Privateer) - When taken: 14 Feb 1814 - Where taken: off Barbados - Date Received: 29 Aug 1814 - From what ship: HMT Bittern - Born: Savannah - Age: 42 - Released on 19 May 1815.

McCarthy, John - Seaman - Nbr: 1233 - How taken: Sent into custody from HMS Nautilus - When taken: 20 Dec 1812 - Date Received: 14 Jun 1814 - From what ship: Mill Prison (Plymouth, England) - Born: Philadelphia - Age: 35 - Released on 26 Apr 1815.

McCarthy, Samuel - Seaman - Nbr: 6490 - Prize: Louisa - Ship Type: Prize - How taken: HMS Dasher - When taken: 12 Feb 1814 - Where taken: off America - Date Received: 3 Mar 1815 - From what ship: HMT Ganges, Plymouth - Born: Baltimore - Age: 18 - Released on 11 Jul 1815.

McCauley, George - Boy - Nbr: 84 - Prize: Rolla - Ship Type: MV - How taken: HMS Surveillante - When taken: 11 Feb 1813 - Where taken: Bay of Biscay - Date Received: 2 Apr 1813 - From what ship: Plymouth - Born: Philadelphia - Age: 14 - Released on 20 Apr 1815.

McConnell, John - Seaman - Nbr: 1089 - How taken: Impressed at Friend - When taken: 5 Mar 1815 - Date Received: 10 May 1814 - From what ship: Plymouth - Born: Wilmington - Age: 25 - Released on 27 Apr 1815.

McCormick, Daniel - Seaman - Nbr: 2166 - Prize: Rattlesnake - Ship Type: P - How taken: HMS Hyperion - When taken: 26 Jun 1814 - Where taken: off Cape Finisterre - Date Received: 16 Aug 1814 - From what ship: HMS Dublin, Halifax - Born: New Jersey - Age: 23 - Released on 2 May 1815.

McCormick, James - Seaman - Nbr: 6456 - Prize: Nonsuch - How taken: HMS Dotterel - When taken: 14 Dec 1815 - Date Received: 3 Mar 1815 - From what ship: HMT Ganges, Plymouth - Born: New York - Age: 27 - Released on 11 Jul 1815.

McCormick, Simon - Seaman - Nbr: 6091 - Prize: Black Swan - Ship Type: MV - How taken: HMS Maidstone - When taken: 24 Oct 1814 - Where taken: Georges Bank - Date Received: 28 Dec 1814 - From what ship: HMT Penelope - Born: Boston - Age: 54 - Released on 20 Jul 1815.

McCormick, Simon - Passenger - Nbr: 312 - Prize: Ducornau - Ship Type: MV - How taken: HMS Pheasant - When taken: 15 Mar 1813 - Where taken: Bay of Biscay - Date Received: 28 Jun 1813 - From what ship: Plymouth - Born: Boston - Age: 54 - Sent to Dartmouth on 16 Jun 1814.

McCormick, William - Carpenter - Nbr: 2127 - How taken: Gave himself up from HMS Dublin - When taken: 1 Nov 1812 - Date Received: 8 Aug 1814 - From what ship: HMT Raven, Chatham - Born: Philadelphia - Age: 45 - Released on 26 Apr 1815.

McDaniel, John - Mate - Nbr: 3530 - Prize: Rover - Ship Type: Prize - How taken: HMS Conquistador - When taken: 22 Aug 1814 - Where taken: Long 19 Lat 107 - Date Received: 28 Sep 1814 - From what ship: HMT Salvador del Mundo - Born: Baltimore - Age: 22 - Released on 12 Jun 1815.

McDonald, John - Seaman - Nbr: 622 - Prize: US Brig Argus - Ship Type: MW - How taken: HMS Pelican - When taken: 14 Aug 1813 - Where taken: Irish Channel - Date Received: 8 Sep 1813 - From what ship: Plymouth - Born: Baltimore - Age: 25 - Sent to Dartmouth on 2 Nov 1814.

McDonald, John - Seaman - Nbr: 835 - Prize: Chesapeake - Ship Type: LM - How taken: HMS Hotspur & HMS Pyramus - When taken: 26 Oct 1813 - Where taken: off Nantes - Date Received: 29 Nov 1813 - From what ship: Plymouth - Born: Norfolk - Age: 27 - Released on 26 Apr 1815.

McDonald, John - Seaman - Nbr: 2252 - Prize: Otario - Ship Type: Prize - How taken: HMS Curlew - When taken: 2 May 1814 - Where taken: off Halifax - Date Received: 16 Aug 1814 - From what ship: HMS Dublin, Halifax - Born: New Jersey - Age: 24 - Released on 3 May 1815.

McDonald, John - Seaman - Nbr: 1918 - How taken: Gave himself up from HMS Tweed - When taken: 20 Dec 1812 - Date Received: 3 Aug 1814 - From what ship: HMS Alceste, Chatham Depot - Born: New York - Age: 45 - Released on 26 Apr 1815.

McDougall, Hugh - Seaman - Nbr: 6061 - Prize: Chasseur - Ship Type: P - How taken: HMS Recruit - When taken: 6 Apr 1814 - Where taken: Lat 29 Long 76 - Date Received: 28 Dec 1814 - From what ship: HMT Penelope - Born: New York - Age: 23 - Released on 3 Jul 1815.

McDowell, John - Seaman - Nbr: 790 - Prize: Avon - Ship Type: MV - How taken: HMS Eurotas - When taken: 27 Oct 1813 - Where taken: off Ushant (France) - Date Received: 3 Nov 1813 - From what ship: Plymouth - Born: Pennsylvania - Age: 26 - Released on 26 Apr 1815.

McEvoy, John - Seaman - Nbr: 2228 - Prize: US Schooner Growler - Ship Type: MW - How taken: HMS Melville - When taken: 11 Aug 1813 - Where taken: Lakes, Upper Canada - Date Received: 16 Aug 1814 - From what

American Prisoners of War Held at Dartmoor during the War of 1812

## Alphabetical listing of names

ship: HMS Dublin, Halifax - Born: New York - Age: 38 - Sent to Dartmouth on 19 Oct 1814.

McFadden, John - Seaman - Nbr: 4118 - Prize: Bordeaux Packet - Ship Type: LM - How taken: HMS Niemen - When taken: 28 Jun 1814 - Where taken: off Delaware - Date Received: 6 Oct 1814 - From what ship: HMT Chesapeake, Halifax - Born: Baltimore - Age: 23 - Released on 13 Jun 1815.

McFadon, James - 2nd Prize Master - Nbr: 929 - Prize: Fanny - Ship Type: MV - How taken: HMS Eurotas - When taken: 23 Dec 1813 - Where taken: at sea - Date Received: 31 Jan 1814 - From what ship: Plymouth - Born: Baltimore - Age: 26 - Escaped on 3 Jun 1814.

McFall, William - Seaman - Nbr: 691 - How taken: Impressed at Liverpool - When taken: 10 Sep 1813 - Date Received: 27 Sep 1813 - From what ship: Plymouth - Born: New York - Age: 33 - Released on 26 Apr 1815.

McFarlane, Daniel - Seaman - Nbr: 4581 - Prize: Bunker Hill - Ship Type: P - How taken: HMS Pomone - When taken: 2 Mar 1814 - Where taken: Channel - Date Received: 9 Oct 1814 - From what ship: HMT Leyden, Chatham - Born: Philadelphia - Age: 17 - Race: Mulatto - Released on 15 Jun 1815.

McFarlane, John - Seaman - Nbr: 903 - Prize: Squirrel - Ship Type: MV - How taken: HMS Belle Poule - When taken: 14 Dec 1813 - Where taken: at sea - Date Received: 31 Jan 1814 - From what ship: Plymouth - Born: Massachusetts - Age: 21 - Released on 27 Apr 1815.

McFarlane, Robert - Seaman - Nbr: 2355 - Prize: Snap Dragon - Ship Type: P - How taken: HMS Martin - When taken: 10 Jun 1814 - Where taken: off Halifax - Date Received: 16 Aug 1814 - From what ship: HMT Queen, Halifax - Born: Virginia - Age: 18 - Released on 3 May 1815.

McFree, John - Seaman - Nbr: 1983 - How taken: Gave himself up from HMS Barham - When taken: Apr 1813 - Date Received: 3 Aug 1814 - From what ship: HMS Lyffey, Chatham Depot - Born: Alexandria - Age: 21 - Released on 2 May 1815.

McGee, John - Seaman - Nbr: 3958 - Prize: Rolla - Ship Type: P - How taken: HMS Loire - When taken: 10 Dec 1813 - Where taken: off Bull Island (South Carolina) - Date Received: 5 Oct 1814 - From what ship: HMT President, Halifax - Born: Baltimore - Age: 32 - Released on 9 Jun 1815.

McGee, Robert - Seaman - Nbr: 4639 - How taken: Gave himself up from HMS Swiftsure - When taken: 26 Dec 1812 - Date Received: 9 Oct 1814 - From what ship: HMT Leyden, Chatham - Born: Pennsylvania - Age: 36 - Released on 27 Apr 1815.

McGeorge, William - Seaman - Nbr: 1960 - How taken: Gave himself up from HMS Colossus - When taken: 18 Jan 1813 - Date Received: 3 Aug 1814 - From what ship: HMS Lyffey, Chatham Depot - Born: Watertown - Age: 23 - Released on 2 May 1815.

McGill, Robert - Seaman - Nbr: 3414 - Prize: Thomas - Ship Type: P - How taken: HMS Nymphe - When taken: 24 Jun 1813 - Where taken: off Halifax - Date Received: 13 Sep 1814 - From what ship: HMT Niobe, Chatham - Born: Portsmouth - Age: 32 - Released on 28 May 1815.

McGilmore, John - Seaman - Nbr: 3866 - Prize: Lewis Warrington - Ship Type: MV - How taken: HMS Loire - When taken: 23 May 1814 - Where taken: off VA - Date Received: 5 Oct 1814 - From what ship: HMT Orpheus, Halifax - Born: Petersburg - Age: 18 - Released on 10 May 1815.

McGowen, Joseph - Soldier - Nbr: 5256 - Prize: 13th US Infantry - Ship Type: Troops - How taken: British Army - When taken: 13 Oct 1812 - Where taken: Canada - Date Received: 31 Oct 1814 - From what ship: HMT Leyden, Chatham - Born: Ireland - Age: 49 - Released on 29 Jun 1815.

McGowen, Patrick - Soldier - Nbr: 5843 - Prize: General Patch's Volunteers - Ship Type: Troops - How taken: British Army - When taken: 17 Sep 1814 - Where taken: Fort Erie - Date Received: 26 Dec 1814 - From what ship: HMT Argo - Born: Pennsylvania - Age: 33 - Released on 29 Jun 1815.

McGuire, John - Seaman - Nbr: 5506 - Prize: Ann Dorothy - Ship Type: Prize - How taken: HMS Maidstone - When taken: 30 Oct 1814 - Where taken: off Cape Sable Island (Canada) - Date Received: 17 Dec 1814 - From what ship: HMT Loire, Halifax - Born: Baltimore - Age: 24 - Released on 29 Jun 1815.

McInley, James - Seaman - Nbr: 2704 - How taken: Gave himself up from HMS Denmark - When taken: 12 Dec 1813 - Date Received: 21 Aug 1814 - From what ship: HMT Freya, Chatham - Born: Yorkshire - Age: 33 - Released on 19 May 1815.

McIntire, A. - Prize Master - Nbr: 1694 - Prize: Hornsby - Ship Type: P - How taken: HMS Sceptre - When taken: 4 Feb 1814 - Where taken: off Block Island - Date Received: 2 Jul 1814 - From what ship: Plymouth - Born: New York - Age: 28 - Released on 1 May 1815.

McIntire, John - Seaman - Nbr: 4495 - Prize: Enterprize - Ship Type: P - How taken: HMS Tenedos - When taken: 21 May 1813 - Where taken: off Cape Cod - Date Received: 8 Oct 1814 - From what ship: HMT Leyden, Chatham - Born: Massachusetts - Age: 22 - Released on 15 Jun 1815.

McIntire, Petty - Seaman - Nbr: 3311 - Prize: Thomas - Ship Type: P - How taken: HMS Nymphe - When taken: 28

American Prisoners of War Held at Dartmoor during the War of 1812

## Alphabetical listing of names

Jun 1813 - Where taken: off Halifax - Date Received: 13 Sep 1814 - From what ship: HMT Niobe, Chatham - Born: New York - Age: 29 - Released on 28 May 1815.

McIntire, Samuel - Seaman - Nbr: 1623 - Prize: Tom - Ship Type: LM - How taken: HMS Surveillante - When taken: 24 Apr 1813 - Where taken: Bay of Biscay - Date Received: 23 Jun 1814 - From what ship: Stapleton - Born: Massachusetts - Age: 24 - Released on 1 May 1815.

McIntire, William - Seaman - Nbr: 2655 - Prize: Jane - Ship Type: MV - How taken: HMS Crescent - When taken: 28 Jun 1813 - Where taken: off Newfoundland - Date Received: 21 Aug 1814 - From what ship: HMT Freya, Chatham - Born: Ireland - Age: 26 - Released on 19 May 1815.

McIver, John - Seaman - Nbr: 2646 - How taken: Gave himself up from HMS Kent - When taken: 26 Oct 1812 - Date Received: 21 Aug 1814 - From what ship: HMT Freya, Chatham - Born: Massachusetts - Age: 29 - Released on 26 Apr 1815.

McJugen, Robert - Seaman - Nbr: 5736 - Prize: US Schooner Somers - Ship Type: MW - How taken: British gunboats - When taken: 12 Aug 1814 - Where taken: Fort Erie - Date Received: 26 Dec 1814 - From what ship: HMT Argo - Born: Pennsylvania - Age: 16 - Released on 3 Jul 1815.

McKeige, Denis - Seaman - Nbr: 3015 - How taken: Gave himself up from HMS Zealous - When taken: 27 Aug 1814 - Date Received: 2 Sep 1814 - From what ship: HMT Centaur - Born: Maryland - Age: 34 - Released on 28 May 1815.

McKennie, Barney - Seaman - Nbr: 3212 - How taken: Gave himself up from HMS Crebenis - When taken: 8 Sep 1813 - Date Received: 11 Sep 1814 - From what ship: HMT Freya, Chatham - Born: Pennsylvania - Age: 22 - Released on 11 Jul 1815.

McKensie, William - Seaman - Nbr: 3241 - Prize: Hepsey - Ship Type: MV - How taken: HMS Tenedos - When taken: 22 Jun 1813 - Where taken: off Lisbon - Date Received: 11 Sep 1814 - From what ship: HMT Freya, Chatham - Born: New York - Age: 23 - Released on 28 May 1815.

McKenzie, Alexander - Seaman - Nbr: 457 - Prize: Courier - Ship Type: LM - How taken: HMS Andromache - When taken: 14 Mar 1813 - Where taken: Bay of Biscay - Date Received: 8 Sep 1813 - From what ship: Plymouth - Born: New York - Age: 30 - Released on 26 Apr 1815.

McKenzie, John - Seaman - Nbr: 2879 - How taken: Gave himself up from HMS Barfleur - When taken: 27 May 1813 - Date Received: 24 Aug 1814 - From what ship: HMT Alpheus, Chatham - Born: Baltimore - Age: 36 - Race: Black - Released on 19 May 1815.

McKertre, Abraham - Prize Master - Nbr: 1295 - Prize: John & Frances - Ship Type: Prize - How taken: HMS Sterling Castle - When taken: 10 May 1814 - Where taken: off Cape Clear - Date Received: 14 Jun 1814 - From what ship: Mill Prison (Plymouth, England) - Born: Baltimore - Age: 28 - Released on 28 Apr 1815.

McKey, James Abercomby - Seaman - Nbr: 1610 - Prize: Fox - Ship Type: LM - How taken: HMS Pheasant - When taken: 23 Apr 1813 - Where taken: Bay of Biscay - Date Received: 23 Jun 1814 - From what ship: Stapleton - Born: Philadelphia - Age: 22 - Released on 1 May 1815.

McKinney, Isaac - Seaman - Nbr: 3299 - Prize: Governor Plumer - Ship Type: P - How taken: Sent into custody from a privateer - When taken: 1 Jun 1813 - Where taken: off Cape Ann - Date Received: 13 Sep 1814 - From what ship: HMT Niobe, Chatham - Born: Massachusetts - Age: 20 - Released on 28 May 1815.

McKinney, John - Seaman - Nbr: 1427 - Prize: Revenge - Ship Type: LM - How taken: HMS Belle Poule - When taken: 10 May 1814 - Where taken: off Cornwall - Date Received: 19 Jun 1814 - From what ship: Stapleton - Born: Georgetown - Age: 35 - Released on 28 Apr 1815.

McKinnon, John - Seaman - Nbr: 3250 - Prize: Wiley Reynard - Ship Type: P - How taken: HMS Shannon - When taken: 15 Aug 1812 - Where taken: off Halifax - Date Received: 11 Sep 1814 - From what ship: HMT Freya, Chatham - Born: Chester - Age: 23 - Released on 27 Apr 1815.

McKinnon, Nathaniel - Seaman - Nbr: 1445 - Prize: Tickler - Ship Type: LM - How taken: HMS Magicienne - When taken: 5 Jun 1813 - Where taken: Bay of Biscay - Date Received: 19 Jun 1814 - From what ship: Stapleton - Born: Baltimore - Age: 32 - Race: Negro - Released on 28 Apr 1815.

McKinnon, Noel - Captain - Nbr: 3023 - Prize: Hussar - Ship Type: P - How taken: HMS Saturn - When taken: 25 May 1814 - Where taken: off Sandy Hook - Date Received: 2 Sep 1814 - From what ship: Naval Hospital, Plymouth - Born: New York - Age: 28 - Released on 27 Apr 1815.

McKinzie, Daniel - Seaman - Nbr: 5774 - Prize: William Penn - Ship Type: MV - How taken: HMS Acorn - When taken: 27 Oct 1812 - Where taken: Lat 14 - Date Received: 26 Dec 1814 - From what ship: HMT Argo - Born: Massachusetts - Age: 21 - Released on 27 Apr 1815.

McKray, John W. - Seaman - Nbr: 2220 - Prize: Hussar - Ship Type: P - How taken: HMS Saturn - When taken: 25 May 1814 - Where taken: off Sandy Hook - Date Received: 16 Aug 1814 - From what ship: HMS Dublin,

American Prisoners of War Held at Dartmoor during the War of 1812

## Alphabetical listing of names

Halifax - Born: Virginia - Age: 22 - Released on 2 May 1815.
McLane, George - Seaman - Nbr: 5410 - Prize: Frolic - Ship Type: P - How taken: HMS Maidstone - When taken: 18 Jul 1813 - Where taken: Grand Banks - Date Received: 31 Oct 1814 - From what ship: HMT Leyden, Chatham - Born: Amsterdam - Age: 47 - Released on 19 May 1815.
McLane, John - Seaman - Nbr: 5161 - Prize: Volante - Ship Type: P - How taken: HMS Curlew - When taken: 25 Nov 1813 - Where taken: off Halifax - Date Received: 31 Oct 1814 - From what ship: HMT Mermaid, Chatham - Born: Providence - Age: 28 - Released on 29 Jun 1815.
McLaughlan, M. - Seaman - Nbr: 3888 - Prize: US Brig Rattlesnake - Ship Type: MW - How taken: HMS Leander - When taken: 11 Jul 1814 - Where taken: off Shelburne - Date Received: 5 Oct 1814 - From what ship: HMT Orpheus, Halifax - Born: Lancaster - Age: 40 - Released on 9 Jun 1815.
McLean, Thomas - Boy - Nbr: 2257 - Prize: Otario - Ship Type: Prize - How taken: HMS Curlew - When taken: 2 May 1814 - Where taken: off Halifax - Date Received: 16 Aug 1814 - From what ship: HMS Dublin, Halifax - Born: Maryland - Age: 18 - Released on 3 May 1815.
McLelland, William - Seaman - Nbr: 154 - How taken: Gave himself up from HMS Lavinia - Date Received: 2 Apr 1813 - From what ship: Plymouth - Born: Fredericksburg - Age: 30 - Released on 9 Apr 1815.
McLeod, Colin - Boatswain - Nbr: 1675 - Prize: US Brig Argus - Ship Type: MW - How taken: HMS Pelican - When taken: 14 Aug 1813 - Where taken: Irish Channel - Date Received: 2 Jul 1814 - From what ship: Plymouth - Born: Philadelphia - Age: 30 - Sent to Dartmouth on 2 Nov 1814.
McLeod, M. - Boatswain - Nbr: 849 - Prize: US Brig Argus - Ship Type: MW - How taken: HMS Pelican - When taken: 14 Aug 1813 - Where taken: Irish Channel - Date Received: 29 Nov 1813 - From what ship: Plymouth - Born: Philadelphia - Age: 30 - Sent to Plymouth on 13 Feb 1814.
McManus, Michael - Seaman - Nbr: 2490 - Prize: US Sloop Frolic - Ship Type: MW - How taken: HMS Orpheus - When taken: 20 Apr 1814 - Where taken: off Cuba - Date Received: 16 Aug 1814 - From what ship: HMT Queen, Halifax - Born: New York - Age: 28 - Released on 3 May 1815.
McMiller, Andrew - Seaman - Nbr: 2816 - Prize: Kitty - Ship Type: Prize - How taken: Dart of Guernsey (Privateer) - When taken: 20 Jun 1813 - Where taken: off the Western Isles (England) - Date Received: 24 Aug 1814 - From what ship: HMT Liverpool, Chatham - Born: Salem - Age: 45 - Released on 2 Nov 1814.
McNab, John - Seaman - Nbr: 731 - Prize: Ned - Ship Type: LM - How taken: HMS Royalist - When taken: 6 Sep 1813 - Where taken: Bay of Biscay - Date Received: 27 Sep 1813 - From what ship: Plymouth - Born: New York - Age: 19 - Released on 26 Apr 1815.
McNeel, Philip - Seaman - Nbr: 3829 - Prize: Hussar - Ship Type: P - How taken: HMS Saturn - When taken: 24 May 1814 - Where taken: off Sandy Hook - Date Received: 5 Oct 1814 - From what ship: HMT Orpheus, Halifax - Born: Boston - Age: 19 - Released on 9 Jun 1815.
McNeil, Dennis - Seaman - Nbr: 6462 - Prize: Decatur - Ship Type: P - How taken: HMS Rhin - When taken: 5 Jun 1814 - Where taken: off San Domingo - Date Received: 3 Mar 1815 - From what ship: HMT Ganges, Plymouth - Born: Boston - Age: 32 - Released on 11 Jul 1815.
McNeil, John - Seaman - Nbr: 1002 - How taken: Impressed at Cork - When taken: 15 Dec 1813 - Date Received: 31 Jan 1814 - From what ship: Plymouth - Born: Boston - Age: 20 - Released on 27 Apr 1815.
McNelly, Thomas - Carpenter - Nbr: 420 - Prize: Viper - Ship Type: MV - How taken: HMS Superb - When taken: 15 Apr 1813 - Where taken: Bay of Biscay - Date Received: 1 Jul 1813 - From what ship: Plymouth - Born: Maryland - Age: 57 - Released on 20 Apr 1815.
McQueen, Charles - Seaman - Nbr: 3573 - Prize: Hawk - Ship Type: P - How taken: HMS Pique - When taken: 26 Apr 1814 - Where taken: off Bermuda - Date Received: 30 Sep 1814 - From what ship: HMT Sybella - Born: Virginia - Age: 21 - Released on 4 Jun 1815.
McQuillan, James - Lieutenant - Nbr: 1309 - Prize: Margaret - Ship Type: Prize - How taken: HMS Foxhound - When taken: 27 Mar 1814 - Where taken: off Isles of Scilly - Date Received: 14 Jun 1814 - From what ship: Mill Prison (Plymouth, England) - Born: SC - Age: 31 - Released on 28 Apr 1815.
McStarbuck, Jordan - Purser - Nbr: 5823 - Prize: US Schooner Ohio - Ship Type: MW - How taken: British gunboats - When taken: 12 Jun 1814 - Where taken: Fort Erie - Date Received: 26 Dec 1814 - From what ship: HMT Argo - Born: New York - Age: 24 - Released on 3 Jul 1815.
McUmber, Jacob - Seaman - Nbr: 2688 - How taken: Gave himself up from HMS Orpheus - When taken: 11 Dec 1812 - Date Received: 21 Aug 1814 - From what ship: HMT Freya, Chatham - Born: Dartmouth - Age: 27 - Released on 26 Apr 1815.
Mead, Ezel - Seaman - Nbr: 3067 - Prize: Liberty - Ship Type: MV - How taken: Surrendered - When taken: 30 Dec 1813 - Where taken: Stromess - Date Received: 2 Sep 1814 - From what ship: HMT Hydra, Chatham - Born:

American Prisoners of War Held at Dartmoor during the War of 1812

## Alphabetical listing of names

New York - Age: 26 - Released on 28 May 1815.
Mead, William - Seaman - Nbr: 3271 - Prize: US Sloop Frolic - Ship Type: MW - How taken: HMS Orpheus - When taken: 20 Apr 1814 - Where taken: off Cuba - Date Received: 13 Sep 1814 - From what ship: Naval Hospital, Plymouth - Born: North Carolina - Age: 20 - Died on 24 Jul 1815 from anasarca.
Meadows, Timothy - Seaman - Nbr: 1203 - Prize: Renommee - Ship Type: P - How taken: Rebecca - When taken: 18 Jun 1813 - Where taken: at sea - Date Received: 4 Jun 1814 - From what ship: Dartmouth - Born: Nantucket - Age: 21 - Released on 28 Apr 1815.
Meath, Samuel - Seaman - Nbr: 4689 - How taken: Gave himself up from HMS Invincible - When taken: 14 Jan 1813 - Date Received: 9 Oct 1814 - From what ship: HMT Leyden, Chatham - Born: Maryland - Age: 25 - Race: Black - Released on 11 Jul 1815.
Medker, Charles D. - Seaman - Nbr: 3933 - Prize: Rolla - Ship Type: P - How taken: HMS Loire - When taken: 10 Dec 1813 - Where taken: off Bull Island (South Carolina) - Date Received: 5 Oct 1814 - From what ship: HMT President, Halifax - Born: Fenton - Age: 26 - Released on 9 Jun 1815.
Meech, David - Seaman - Nbr: 1101 - Prize: Mary - Ship Type: MV - How taken: Recaptured - When taken: 18 Dec 1813 - Date Received: 10 May 1814 - From what ship: Plymouth - Born: New London - Age: 24 - Released on 27 Apr 1815.
Meeden, William - Seaman - Nbr: 785 - Prize: Betsy - Ship Type: MV - How taken: HMS Eurotas - When taken: 26 Oct 1813 - Where taken: off Ushant (France) - Date Received: 3 Nov 1813 - From what ship: Plymouth - Born: Philadelphia - Age: 24 - Released on 26 Apr 1815.
Meeds, Joseph - Prize Master - Nbr: 3529 - Prize: Rover - Ship Type: Prize - How taken: HMS Conquistador - When taken: 22 Aug 1814 - Where taken: Long 19 Lat 107 - Date Received: 28 Sep 1814 - From what ship: HMT Salvador del Mundo - Born: Rhode Island - Age: 24 - Sent to Ashburton (England) on 17 Nov 1814.
Meek, Thomas - Seaman - Nbr: 2429 - Prize: US Sloop Frolic - Ship Type: MW - How taken: HMS Orpheus - When taken: 20 Apr 1814 - Where taken: off Cuba - Date Received: 16 Aug 1814 - From what ship: HMT Queen, Halifax - Born: Salem - Age: 21 - Released on 3 May 1815.
Meigs, John - 2nd Mate - Nbr: 328 - Prize: Gold Coiner - Ship Type: P - How taken: HMS Lyra - When taken: 29 Mar 1813 - Where taken: Lat 44, Long 20 - Date Received: 28 Jun 1813 - From what ship: Plymouth - Born: Thomastown - Age: 27 - Released on 20 Apr 1815.
Meinier, Benjamin - Seaman - Nbr: 2619 - Prize: Union - Ship Type: MV - How taken: HMS Iris - When taken: 17 Jan 1813 - Where taken: at sea - Date Received: 21 Aug 1814 - From what ship: HMT Freya, Chatham - Born: New London - Age: 22 - Released on 3 May 1815.
Meker, James - Seaman - Nbr: 2066 - How taken: Apprehended at London - When taken: 12 Jun 1813 - Date Received: 3 Aug 1814 - From what ship: HMS Bittern, Chatham Depot - Born: New Jersey - Age: 30 - Race: Mulatto - Released on 2 May 1815.
Melbourne, William - Seaman - Nbr: 4239 - Prize: Sally - Ship Type: MV - How taken: HMS Derwent - When taken: 21 Jan 1814 - Where taken: Grand Banks - Date Received: 7 Oct 1814 - From what ship: HMT Niobe, Chatham - Born: Salem - Age: 39 - Escaped on 1 Jun 1815.
Melcher, John - Seaman - Nbr: 5154 - Prize: Revenge - Ship Type: P - How taken: HMS Shannon - When taken: 5 Nov 1813 - Where taken: off Halifax - Date Received: 31 Oct 1814 - From what ship: HMT Mermaid, Chatham - Born: Wilmington - Age: 24 - Released on 29 Jun 1815.
Mellett, James - Seaman - Nbr: 128 - Prize: Good Intent - Ship Type: P - How taken: HMS Rota - When taken: 26 Jan 1813 - Where taken: Lat 43'30" N, Long 20 W - Date Received: 12 Apr 1813 - From what ship: Plymouth - Born: Cape Ann - Age: 21 - Sent to Dartmouth on 30 Jul 1813.
Mellim, George - Seaman - Nbr: 2730 - How taken: Gave himself up from HMS Cherub - When taken: 5 Dec 1812 - Date Received: 24 Aug 1814 - From what ship: HMT Liverpool, Chatham - Born: New York - Age: 26 - Released on 26 Apr 1815.
Melvin, John - Seaman - Nbr: 2697 - How taken: Gave himself up from HMS Ministrel - When taken: 28 Jul 1813 - Date Received: 21 Aug 1814 - From what ship: HMT Freya, Chatham - Born: Boston - Age: 22 - Released on 19 May 1815.
Melzard, George - Seaman - Nbr: 4670 - Prize: Industry - Ship Type: P - How taken: HMS Heron - When taken: 3 Nov 1813 - Where taken: off Halifax - Date Received: 9 Oct 1814 - From what ship: HMT Leyden, Chatham - Born: Marblehead - Age: 42 - Released on 15 Jun 1815.
Melzard, Peter - Seaman - Nbr: 3290 - Prize: Enterprize - Ship Type: P - How taken: HMS Tenedos - When taken: 21 May 1813 - Where taken: off Cape Cod - Date Received: 13 Sep 1814 - From what ship: HMT Niobe, Chatham - Born: Marblehead - Age: 20 - Released on 28 May 1815.

American Prisoners of War Held at Dartmoor during the War of 1812

## Alphabetical listing of names

Menard, Augustus - Boy - Nbr: 659 - Prize: Marmion - Ship Type: MV - How taken: HMS President - When taken: 14 Aug 1813 - Where taken: off Nantes - Date Received: 8 Sep 1813 - From what ship: Plymouth - Born: Niort, France - Age: 12 - Sent to Ashburton (England) on 20 Sep 1813.

Mendez, Joseph - Seaman - Nbr: 6078 - Prize: Daedalus - Ship Type: MV - How taken: HMS Niemen - When taken: 20 Sep 1814 - Where taken: off Delaware - Date Received: 28 Dec 1814 - From what ship: HMT Penelope - Born: Spain - Age: 23 - Released on 3 Jul 1815.

Mendoza, Caesar - Seaman - Nbr: 4811 - Prize: President - Ship Type: P - How taken: HMS Pique - When taken: 7 May 1814 - Where taken: off Porto Rico - Date Received: 9 Oct 1814 - From what ship: HMT Freya, Halifax - Born: Cartagena - Age: 28 - Race: Mulatto - Died on 27 Oct 1814 from enteritis.

Menillo, John - Seaman - Nbr: 4917 - Prize: US Brig Rattlesnake - Ship Type: MW - How taken: HMS Leander - When taken: 13 Jul 1814 - Where taken: off Shelburne - Date Received: 28 Oct 1814 - From what ship: HMT Alkbar, Halifax - Born: Alexandria - Age: 21 - Died on 18 Nov 1814.

Mercer, Benjamin - Seaman - Nbr: 477 - Prize: Zebra - Ship Type: LM - How taken: HMS Pyramus - When taken: 20 Apr 1813 - Where taken: Bay of Biscay - Date Received: 8 Sep 1813 - From what ship: Plymouth - Born: New York - Age: 21 - Released on 20 Apr 1815.

Mercey, Thomas - Seaman - Nbr: 3010 - How taken: Gave himself up from HMS Tigre - Date Received: 2 Sep 1814 - From what ship: HMT Salvador del Mundo - Born: Swansea - Age: 27 - Race: Negro - Released on 28 May 1815.

Merchant, Elijah - Seaman - Nbr: 4963 - Prize: Herald - Ship Type: P - How taken: HMS Endymion - When taken: 15 Aug 1814 - Where taken: off Nantucket - Date Received: 28 Oct 1814 - From what ship: HMT Alkbar, Halifax - Born: Virginia - Age: 50 - Released on 21 Jun 1815.

Merchant, John - Seaman - Nbr: 5524 - Prize: Theodore - Ship Type: Prize - How taken: HMS Galatea - When taken: 7 Sep 1814 - Where taken: Channel - Date Received: 17 Dec 1814 - From what ship: HMT Loire, Halifax - Born: Bordeaux - Age: 19 - Released on 1 Jul 1815.

Merchant, William - Seaman - Nbr: 1242 - How taken: Sent into custody from HMS Guadeloupe - When taken: 19 Dec 1812 - Date Received: 14 Jun 1814 - From what ship: Mill Prison (Plymouth, England) - Born: Martha's Vineyard - Age: 23 - Released on 26 Apr 1815.

Merckett, John - Seaman - Nbr: 2627 - How taken: Gave himself up from HMS Royal William - Date Received: 21 Aug 1814 - From what ship: HMT Freya, Chatham - Born: Boston - Age: 23 - Released on 3 May 1815.

Merle, John - Seaman - Nbr: 1953 - How taken: Gave himself up from HMS Christian 7th - When taken: 19 May 1813 - Date Received: 3 Aug 1814 - From what ship: HMS Lyffey, Chatham Depot - Born: New York - Age: 24 - Released on 2 May 1815.

Merlo, Christopher - Seaman - Nbr: 5000 - Prize: Invincible - Ship Type: LM - How taken: HMS Armide - When taken: 15 Aug 1814 - Where taken: off Nantucket - Date Received: 28 Oct 1814 - From what ship: HMT Alkbar, Halifax - Born: Rochelle - Age: 32 - Released on 21 Jun 1815.

Merrick, William - Seaman - Nbr: 6036 - Prize: Regent - Ship Type: LM - How taken: HMS Forth - When taken: 19 Sep 1814 - Where taken: off Egg Harbor (New Jersey) - Date Received: 28 Dec 1814 - From what ship: HMT Penelope - Born: Newbury - Age: 26 - Released on 3 Jul 1815.

Merrill, Enoch - Seaman - Nbr: 4772 - How taken: Impressed at Shields - When taken: 14 Mar 1814 - Date Received: 9 Oct 1814 - From what ship: HMT Freya, Chatham - Born: Salisbury - Age: 22 - Released on 15 Jun 1815.

Merrill, Abraham - Seaman - Nbr: 5582 - Prize: Alexander - Ship Type: Prize - How taken: HMS Wasp - When taken: 14 Sep 1814 - Where taken: off Cape Sable Island (Canada) - Date Received: 17 Dec 1814 - From what ship: HMT Loire, Halifax - Born: Massachusetts - Age: 21 - Released on 1 Jul 1815.

Merrill, Jacob - Seaman - Nbr: 5987 - Prize: McDonough - Ship Type: P - How taken: HMS Bacchante - When taken: 1 Nov 1814 - Where taken: Lat 42 Long 67 - Date Received: 27 Dec 1814 - From what ship: HMT Penelope - Born: Arundel - Age: 55 - Released on 3 Jul 1815.

Merrill, John - Seaman - Nbr: 1287 - Prize: Columbia - Ship Type: MV - How taken: HMS Sir John Sherbrook - When taken: 13 Apr 1814 - Where taken: off Boston - Date Received: 14 Jun 1814 - From what ship: Mill Prison (Plymouth, England) - Born: Yarmouth - Age: 33 - Released on 28 Apr 1815.

Merrish, Joseph - Seaman - Nbr: 5234 - Prize: Teazer - Ship Type: P - How taken: HMS Maidstone - When taken: 18 Jul 1813 - Where taken: at sea - Date Received: 31 Oct 1814 - From what ship: HMT Mermaid, Chatham - Born: New Orleans - Age: 38 - Released on 29 Jun 1815.

Merritt, Almon - Landsman - Nbr: 103 - Prize: St. Martin's Planter - Ship Type: P - How taken: HMS Dublin - When taken: 9 Feb 1813 - Where taken: Lat 43 N, Long 33 50 W - Date Received: 2 Apr 1813 - From what

American Prisoners of War Held at Dartmoor during the War of 1812

## Alphabetical listing of names

ship: Plymouth - Born: Massachusetts - Age: 21 - Released on 20 Apr 1815.

Merritt, Enoch - Seaman - Nbr: 4204 - Prize: Minerva - Ship Type: MV - How taken: HMS Conquistador - When taken: 19 Jan 1814 - Where taken: Bay of Biscay - Date Received: 7 Oct 1814 - From what ship: HMT Niobe, Chatham - Born: Falmouth - Age: 19 - Released on 13 Jun 1815.

Merritt, Jonah - Seaman - Nbr: 1462 - Prize: Revenge - Ship Type: LM - How taken: HMS Belle Poule - When taken: 10 Mar 1813 - Where taken: off Cornwall - Date Received: 19 Jun 1814 - From what ship: Stapleton - Born: New York - Age: 19 - Released on 28 Apr 1815.

Merritt, Robert - Seaman - Nbr: 1419 - Prize: Governor Gerry - Ship Type: MV - How taken: HMS Lyra - When taken: 29 May 1813 - Where taken: Bay of Biscay - Date Received: 19 Jun 1814 - From what ship: Stapleton - Born: New York - Age: 34 - Released on 28 Apr 1815.

Merritt, Thomas - Seaman - Nbr: 1549 - Prize: Essex - Ship Type: MV - How taken: HMS Pyramus - When taken: 2 Apr 1813 - Where taken: Bay of Biscay - Date Received: 23 Jun 1814 - From what ship: Stapleton - Born: Boston - Age: 20 - Released on 1 May 1815.

Metley, Thomas - Seaman - Nbr: 1566 - Prize: Messenger - Ship Type: MV - How taken: HMS Iris - When taken: 10 Mar 1813 - Where taken: Bay of Biscay - Date Received: 23 Jun 1814 - From what ship: Stapleton - Born: Philadelphia - Age: 26 - Released on 1 May 1815.

Meurinose, John - Seaman - Nbr: 1449 - Prize: Tickler - Ship Type: LM - How taken: HMS Magicienne - When taken: 5 Jun 1813 - Where taken: Bay of Biscay - Date Received: 19 Jun 1814 - From what ship: Stapleton - Born: New York - Age: 30 - Released on 28 Apr 1815.

Meyers, David - Seaman - Nbr: 1790 - Prize: Ferox - Ship Type: MV - How taken: HMS Medusa & HMS Lyra - When taken: 28 Mar 1813 - Where taken: off Cape Ortegal (Spain) - Date Received: 20 Jul 1814 - From what ship: HMS Milford, Plymouth - Born: Torrington - Age: 28 - Released on 1 May 1815.

Mezich, Elisha - Seaman - Nbr: 1459 - Prize: Revenge - Ship Type: LM - How taken: HMS Belle Poule - When taken: 10 Mar 1813 - Where taken: off Cornwall - Date Received: 19 Jun 1814 - From what ship: Stapleton - Born: Massachusetts - Age: 30 - Released on 28 Apr 1815.

Michaels, Henry - Seaman - Nbr: 5659 - How taken: Taken off a Swedish vessel at Halifax - Date Received: 24 Dec 1814 - From what ship: HMT Impregnable - Born: Gorham - Age: 25 - Released on 3 Jul 1815.

Michaels, John - Seaman - Nbr: 3816 - Prize: Nimble - Ship Type: Prize - How taken: HMS Arab - When taken: 5 Apr 1814 - Where taken: Lat 37 Long 65 - Date Received: 5 Oct 1814 - From what ship: HMT Orpheus, Halifax - Born: Boston - Age: 20 - Released on 9 Jun 1815.

Michaels, John - Seaman - Nbr: 3587 - How taken: Sent into custody at Barbados - Date Received: 30 Sep 1814 - From what ship: HMT Sybella - Born: SC - Age: 28 - Released on 4 Jun 1815.

Michell, Charles - Seaman - Nbr: 4993 - Prize: Invincible - Ship Type: LM - How taken: HMS Armide - When taken: 15 Aug 1814 - Where taken: off Nantucket - Date Received: 28 Oct 1814 - From what ship: HMT Alkbar, Halifax - Born: Paris - Age: 31 - Released on 21 Jun 1815.

Middlefield, Abraham - Seaman - Nbr: 5990 - Prize: McDonough - Ship Type: P - How taken: HMS Bacchante - When taken: 1 Nov 1814 - Where taken: Lat 42 Long 67 - Date Received: 27 Dec 1814 - From what ship: HMT Penelope - Born: Massachusetts - Age: 39 - Released on 11 Jul 1815.

Middleton, John W. - Mate - Nbr: 2547 - Prize: Caromaned - Ship Type: Prize - How taken: HMS Eridanus - When taken: 13 Aug 1814 - Where taken: Lat 40, Long 16 - Date Received: 16 Aug 1814 - From what ship: HMT Salvador del Mundo - Born: Virginia - Age: 26 - Released on 3 May 1815.

Middleton, L. - Seaman - Nbr: 4732 - How taken: Gave himself up from HMS Rosamond - When taken: 28 Dec 1813 - Date Received: 9 Oct 1814 - From what ship: HMT Freya, Chatham - Born: Maryland - Age: 24 - Race: Mulatto - Released on 15 Jun 1815.

Middleton, Reuben - Seaman - Nbr: 3082 - Prize: Sword Fish - Ship Type: P - How taken: HMS Elephant - When taken: 8 Dec 1812 - Where taken: off Ireland - Date Received: 3 Sep 1814 - From what ship: HMT Hydra, Chatham - Born: Salem - Age: 22 - Released on 27 Apr 1815.

Mids, Michael - Seaman - Nbr: 4286 - Prize: Pilot - Ship Type: LM - How taken: Victoria (Privateer) - When taken: 28 Jan 1814 - Where taken: off Bordeaux - Date Received: 7 Oct 1814 - From what ship: HMT Niobe, Chatham - Born: Baltimore - Age: 26 - Released on 14 Jun 1815.

Miflin, Richard - Seaman - Nbr: 5635 - Prize: Monticello - Ship Type: MV - How taken: HMS Racehorse - When taken: Dec 1812 - Date Received: 24 Dec 1814 - From what ship: HMT Tay - Born: Baltimore - Age: 26 - Race: Negro - Released on 27 Apr 1815.

Milborne, William - Seaman - Nbr: 4837 - Prize: President - Ship Type: P - How taken: HMS Pique - When taken: 7 May 1814 - Where taken: off Porto Rico - Date Received: 9 Oct 1814 - From what ship: HMT Freya, Halifax

American Prisoners of War Held at Dartmoor during the War of 1812

## Alphabetical listing of names

- Born: GA Age: 21 - Released on 21 Jun 1815.
Miles, Thomas - Seaman - Nbr: 3651 - Prize: Ulysses - Ship Type: MV - How taken: HMS Majestic - When taken: 29 Jun 1813 - Where taken: off Western Islands (England) - Date Received: 30 Sep 1814 - From what ship: HMT President - Born: Massachusetts - Age: 19 - Released on 4 Jun 1815.
Miles, William - Seaman - Nbr: 1170 - Prize: Bunker Hill - Ship Type: P - How taken: HMS Pomone & HMS Cadmus - When taken: 4 Mar 1814 - Where taken: Bay of Biscay - Date Received: 10 May 1814 - From what ship: Plymouth - Born: Pennsylvania - Age: 24 - Released on 28 Apr 1815.
Miller, Edward - Seaman - Nbr: 5014 - Prize: Charlotte - Ship Type: Prize - How taken: Censor - When taken: 5 Aug 1814 - Where taken: off Azores - Date Received: 28 Oct 1814 - From what ship: HMT Alkbar, Halifax - Born: New Jersey - Age: 26 - Race: Negro - Died on 23 Feb 1815 from phthisis pulmonalis.
Miller, George - Seaman - Nbr: 5076 - Prize: Ida - Ship Type: LM - How taken: HMS Newcastle - When taken: 9 Aug 1814 - Where taken: Long 34 - Date Received: 28 Oct 1814 - From what ship: HMT Alkbar, Halifax - Born: Belgium - Age: 28 - Released on 29 Jun 1815.
Miller, George - Seaman - Nbr: 2914 - Prize: Matilda - Ship Type: MV - How taken: HMS Revolutionnaire - When taken: 25 Jul 1813 - Where taken: off L'Orient (France) - Date Received: 24 Aug 1814 - From what ship: HMT Alpheus, Chatham - Born: Rhode Island - Age: 24 - Race: Black - Released on 2 Nov 1814.
Miller, Henry - Seaman - Nbr: 70 - Prize: Spitfire - Ship Type: MV - How taken: HMS Achates - When taken: 14 Feb 1813 - Where taken: off Ushant (France) - Date Received: 2 Apr 1813 - From what ship: Plymouth - Born: Germantown - Age: 29 - Released on 20 Apr 1815.
Miller, Henry - Seaman - Nbr: 2475 - Prize: US Sloop Frolic - Ship Type: MW - How taken: HMS Orpheus - When taken: 20 Apr 1814 - Where taken: off Cuba - Date Received: 16 Aug 1814 - From what ship: HMT Queen, Halifax - Born: New London - Age: 23 - Released on 3 May 1815.
Miller, Isaac - Seaman - Nbr: 2328 - Prize: Hussar - Ship Type: P - How taken: HMS Saturn - When taken: 25 May 1814 - Where taken: off Sandy Hook - Date Received: 16 Aug 1814 - From what ship: HMS Dublin, Halifax - Born: Philadelphia - Age: 20 - Race: Creole - Released on 3 May 1815.
Miller, James - Seaman - Nbr: 4448 - Prize: Dart - Ship Type: LM - How taken: HMS Niger - When taken: 13 Nov 1815 - Where taken: at sea - Date Received: 8 Oct 1814 - From what ship: HMT Leyden, Chatham - Born: Charleston - Age: 27 - Released on 14 Jun 1815.
Miller, James - Boy - Nbr: 5307 - Prize: Elbridge Gerry - Ship Type: P - How taken: HMS Crescent - When taken: 25 Aug 1813 - Where taken: off St. Johns - Date Received: 31 Oct 1814 - From what ship: HMT Leyden, Chatham - Born: Cape Elizabeth - Age: 15 - Released on 29 Jun 1815.
Miller, Jeremiah - Seaman - Nbr: 6292 - How taken: Gave himself up from HMS Montague - Date Received: 17 Feb 1815 - From what ship: HMT Ganges, Plymouth - Born: Kennebec - Age: 50 - Released on 4 Jun 1815.
Miller, John - Prize master - Nbr: 5592 - Prize: Albion - Ship Type: Prize - How taken: HMS Jaseur - When taken: 21 Sep 1814 - Where taken: off Halifax - Date Received: 17 Dec 1814 - From what ship: HMT Loire, Halifax - Born: Alexandria - Age: 28 - Released on 1 Jul 1815.
Miller, John - Seaman - Nbr: 6356 - Prize: Prince de Neufchatel - Ship Type: P - How taken: Leander (Newcastle Acasta) - When taken: 20 Dec 1814 - Where taken: Lat 38 Long 56 - Date Received: 19 Feb 1815 - From what ship: HMT Ganges, Plymouth - Born: Baltimore - Age: 30 - Released on 6 Jul 1815.
Miller, John - Cook - Nbr: 134 - Prize: Criterion - Ship Type: MV - How taken: HMS Belle Poule - When taken: 14 Feb 1813 - Where taken: Bay of Biscay - Date Received: 12 Apr 1813 - From what ship: Plymouth - Born: Queen Anne's County - Age: 25 - Race: Black - Released on 20 Apr 1815.
Miller, John - Seaman - Nbr: 1674 - Prize: Vivid - Ship Type: Prize - How taken: HMS Ceres - When taken: 5 Jun 1814 - Where taken: Channel - Date Received: 2 Jul 1814 - From what ship: Plymouth - Born: Philadelphia - Age: 38 - Released on 1 May 1815.
Miller, John - Seaman - Nbr: 2963 - Prize: True Blooded Yankee - Ship Type: P - When taken: 15 Feb 1813 - Date Received: 24 Aug 1814 - What ship: HMT Hannibal - Born: Troy - Age: 29 - Released on 19 May 1815.
Miller, John - Seaman - Nbr: 3646 - How taken: Gave himself up from HMS Barfleur - Date Received: 30 Sep 1814 - From what ship: HMT Sybella - Born: Baltimore - Age: 30 - Released on 4 Jun 1815.
Miller, John - Seaman - Nbr: 3132 - How taken: Gave himself up from HMS Royal George - When taken: 29 Oct 1812 - Date Received: 11 Sep 1814 - From what ship: HMT Freya, Chatham - Born: New Jersey - Age: 37 - Released on 27 Apr 1815.
Miller, Jonathan - Seaman - Nbr: 2529 - Prize: Hussar - Ship Type: P - How taken: HMS Saturn - When taken: 25 May 1814 - Where taken: off Sandy Hook - Date Received: 16 Aug 1814 - From what ship: HMT Queen, Halifax - Born: Delaware - Age: 26 - Race: Black - Released on 3 May 1815.

American Prisoners of War Held at Dartmoor during the War of 1812

## Alphabetical listing of names

Miller, Peter - Seaman - Nbr: 4108 - Prize: Resolution - Ship Type: MV - How taken: HMS Junon - When taken: 6 Jun 1814 - Where taken: off Cape Ann - Date Received: 6 Oct 1814 - From what ship: HMT Chesapeake, Halifax - Born: Friendship - Age: 28 - Released on 13 Jun 1815.

Miller, Richard - Seaman - Nbr: 2351 - Prize: Snap Dragon - Ship Type: P - How taken: HMS Martin - When taken: 10 Jun 1814 - Where taken: off Halifax - Date Received: 16 Aug 1814 - From what ship: HMS Dublin, Halifax - Born: Pennsylvania - Age: 26 - Died on 20 Nov 1814.

Miller, Stephen - Seaman - Nbr: 387 - Prize: John & Frances - Ship Type: MV - How taken: HMS Belle Poule - When taken: 13 Mar 1813 - Where taken: Bay of Biscay - Date Received: 1 Jul 1813 - From what ship: Plymouth - Born: New York - Age: 28 - Sent to Plymouth on 8 Jul 1814.

Miller, Thomas - Seaman - Nbr: 1705 - Prize: Paul Jones - Ship Type: P - How taken: HMS Leonidas - When taken: 23 May 1813 - Where taken: Channel - Date Received: 6 Jul 1814 - From what ship: Bristol - Born: New York - Age: 21 - Race: Mulatto - Released on 1 May 1815.

Miller, Thomas - Seaman - Nbr: 2339 - Prize: Snap Dragon - Ship Type: P - How taken: HMS Martin - When taken: 10 Jun 1814 - Where taken: off Halifax - Date Received: 16 Aug 1814 - From what ship: HMS Dublin, Halifax - Born: Craven County - Age: 19 - Released on 3 May 1815.

Miller, Thomas - Seaman - Nbr: 3642 - Prize: Bordeaux Packet - Ship Type: LM - How taken: HMS Niemen - When taken: 28 Jan 1814 - Where taken: off Delaware - Date Received: 30 Sep 1814 - From what ship: HMT Sybella - Born: Charleston - Age: 22 - Released on 4 Jun 1815.

Miller, William - Seaman - Nbr: 6101 - Prize: Circe - Ship Type: MV - How taken: HMS Acteon - When taken: 22 Oct 1813 - Where taken: off Cape Hatteras - Date Received: 28 Dec 1814 - From what ship: HMT Penelope - Born: New York - Age: 18 - Released on 5 Jul 1815.

Miller, William - Seaman - Nbr: 5434 - How taken: Gave himself up from HMS Ruby - When taken: 5 Nov 1814 - Date Received: 11 Nov 1814 - From what ship: HMT Impregnable - Born: Baltimore - Age: 20 - Released on 1 Jul 1815.

Miller, William - Seaman - Nbr: 972 - Prize: Siro - Ship Type: LM - How taken: HMS Pelican - When taken: 13 Jan 1814 - Where taken: at sea - Date Received: 31 Jan 1814 - From what ship: Plymouth - Born: Portsmouth - Age: 19 - Released on 27 Apr 1815.

Miller, William - Seaman - Nbr: 2628 - How taken: Gave himself up from HMS Hyperion - Date Received: 21 Aug 1814 - From what ship: HMT Freya, Chatham - Born: Boston - Age: 22 - Released on 3 May 1815.

Millet, Joseph - Steward - Nbr: 2281 - Prize: Diomede - Ship Type: P - How taken: HMS Rifleman - When taken: 28 May 1814 - Where taken: off Sable Island - Date Received: 16 Aug 1814 - From what ship: HMS Dublin, Halifax - Born: Salem - Age: 18 - Released on 2 May 1815.

Millett, John - Seaman - Nbr: 4500 - Prize: Enterprize - Ship Type: P - How taken: HMS Tenedos - When taken: 21 May 1813 - Where taken: off Cape Cod - Date Received: 8 Oct 1814 - From what ship: HMT Leyden, Chatham - Born: Massachusetts - Age: 19 - Released on 15 Jun 1815.

Millett, Joseph - Seaman - Nbr: 4415 - Prize: Fire Fly - Ship Type: LM - How taken: HMS Revolutionnaire - When taken: 19 Oct 1813 - Where taken: at sea - Date Received: 8 Oct 1814 - From what ship: HMT Leyden, Chatham - Born: Massachusetts - Age: 39 - Released on 14 Jun 1815.

Millow, John - Seaman - Nbr: 6379 - Prize: US Brig Syren - Ship Type: MW - How taken: HMS Medway - When taken: 12 Jul 1814 - Where taken: off Cape of Good Hope - Date Received: 24 Feb 1815 - From what ship: HMT Ganges, Plymouth - Born: Alexandria - Age: 25 - Released on 11 Jul 1815.

Mills, John - Seaman - Nbr: 5433 - How taken: Gave himself up from HMS Belle Poule - Date Received: 11 Nov 1814 - From what ship: HMT Impregnable - Born: Liverpool - Age: 47 - Released on 1 Jul 1815.

Mills, Joseph - Seaman - Nbr: 2557 - Prize: Caromaned - Ship Type: Prize - How taken: HMS Eridanus - When taken: 13 Aug 1814 - Where taken: Lat 40, Long 16 - Date Received: 16 Aug 1814 - From what ship: HMT Salvador del Mundo - Born: Havana - Age: 26 - Released on 3 May 1815.

Mills, William - Seaman - Nbr: 1390 - Prize: Courier - Ship Type: P - How taken: HMS Rover - When taken: 14 May 1813 - Where taken: Bay of Biscay - Date Received: 19 Jun 1814 - From what ship: Stapleton - Born: Maryland - Age: 30 - Released on 28 Apr 1815.

Mills, William - Seaman - Nbr: 1538 - Prize: Zebra - Ship Type: LM - How taken: HMS Pyramus - When taken: 20 Apr 1813 - Where taken: Bay of Biscay - Date Received: 23 Jun 1814 - From what ship: Stapleton - Born: New Jersey - Age: 21 - Died on 24 Mar 1815 from erysipelas.

Miln, Nicholas - Captain - Nbr: 6188 - Prize: Prince de Neufchatel - Ship Type: P - How taken: Leander (Newcastle Acasta) - When taken: 20 Dec 1814 - Where taken: Lat 38 Long 56 - Date Received: 30 Jan 1815 - From what ship: HMT Pheasant - Born: New Orleans - Age: 28 - Sent to Ashburton (England) on 11 Feb 1815.

American Prisoners of War Held at Dartmoor during the War of 1812

## Alphabetical listing of names

Milner, Benjamin - Seaman - Nbr: 6053 - Prize: William - Ship Type: Prize - How taken: HMS Armide - When taken: 11 Oct 1814 - Where taken: off Newport - Date Received: 28 Dec 1814 - From what ship: HMT Penelope - Born: Philadelphia - Age: 26 - Released on 3 Jul 1815.

Mines, Artemas - Seaman - Nbr: 719 - Prize: Ned - Ship Type: LM - How taken: HMS Royalist - When taken: 6 Sep 1813 - Where taken: Bay of Biscay - Date Received: 27 Sep 1813 - From what ship: Plymouth - Born: Maryland - Age: 32 - Race: Black - Released on 26 Apr 1815.

Mingalls, Robert - Seaman - Nbr: 1714 - How taken: Gave himself up from HMS Amphion - Date Received: 8 Jul 1814 - From what ship: Labrador - Born: Massachusetts - Age: 25 - Race: Mulatto - Released on 1 May 1815.

Mingle, Thomas - Seaman - Nbr: 1603 - Prize: Fox - Ship Type: LM - How taken: HMS Pheasant - When taken: 23 Apr 1813 - Where taken: Bay of Biscay - Date Received: 23 Jun 1814 - From what ship: Stapleton - Born: Africa - Age: 55 - Race: Negro - Released on 1 May 1815.

Mingo, Albert - Passenger - Nbr: 3827 - Prize: Quiz - Ship Type: LM - How taken: HMS Niemen - When taken: 22 May 1814 - Where taken: off Egg Harbor (New Jersey) - Date Received: 5 Oct 1814 - From what ship: HMT Orpheus, Halifax - Born: New Orleans - Age: 29 - Race: Negro - Died on 25 Oct 1814 from pneumonia.

Minifee, Charles - Seaman - Nbr: 721 - Prize: Ned - Ship Type: LM - How taken: HMS Royalist - When taken: 6 Sep 1813 - Where taken: Bay of Biscay - Date Received: 27 Sep 1813 - From what ship: Plymouth - Born: Baltimore - Age: 21 - Sent to Plymouth on 7 Dec 1813.

Minks, Jacob - Seaman - Nbr: 2204 - Prize: Hussar - Ship Type: P - How taken: HMS Saturn - When taken: 25 May 1814 - Where taken: off Sandy Hook - Date Received: 16 Aug 1814 - From what ship: HMS Dublin, Halifax - Born: Lancaster - Age: 24 - Released on 2 May 1815.

Minois, Pierre - Seaman - Nbr: 1160 - Prize: Caroline - Ship Type: Prize - How taken: HMS Vengeur - When taken: 24 Feb 1813 - Where taken: off Belle Isles - Date Received: 10 May 1814 - From what ship: Plymouth - Born: Belle Isle - Age: 16 - Sent to Mill Prison (Plymouth, England) on 8 Jul 1814.

Minor, David - Seaman - Nbr: 4179 - How taken: Gave himself up from HMS Quebec - When taken: 20 Mar 1813 - Date Received: 7 Oct 1814 - From what ship: HMT Niobe, Chatham - Born: New London - Age: 27 - Released on 13 Jun 1815.

Minor, John - Seaman - Nbr: 5124 - Prize: Liberty - Ship Type: MV - How taken: Surrendered - When taken: 30 Dec 1813 - Where taken: Stromess - Date Received: 28 Oct 1814 - From what ship: HMT Salvador del Mundo - Born: New London - Age: 15 - Released on 27 Apr 1815.

Mires, John - Seaman - Nbr: 33 - How taken: Apprehended at Gibraltar - When taken: 8 Aug 1813 - Date Received: 2 Apr 1813 - What ship: Plymouth - Born: Prussia - Age: 24 - Sent to Dartmouth on 30 Jul 1813.

Mista, Sullivan - Seaman - Nbr: 5247 - Prize: Alatanta - Ship Type: MV - How taken: HMS Barbados - When taken: 12 Jan 1814 - Where taken: off St. Bartholomew - Date Received: 31 Oct 1814 - From what ship: HMT Mermaid, Chatham - Born: Virginia - Age: 36 - Died on 13 Feb 1815 from variola.

Mitchell, Carr - Seaman - Nbr: 2681 - Prize: Joseph - Ship Type: MV - How taken: HMS Iris - When taken: 8 Jun 1813 - Where taken: off Spain - Date Received: 21 Aug 1814 - From what ship: HMT Freya, Chatham - Born: Virginia - Age: 18 - Released on 19 May 1815.

Mitchell, Ezekiel - Seaman - Nbr: 237 - Prize: Charlotte - Ship Type: MV - How taken: HMS Warspite - When taken: 3 Mar 1813 - Where taken: Bay of Biscay - Date Received: 2 Apr 1813 - From what ship: Plymouth - Born: Massachusetts - Age: 23 - Died on 12 Jan 1815 from variola.

Mitchell, Faisly - Seaman - Nbr: 811 - Prize: Sybelle - Ship Type: MV - How taken: HMS Zenobia - When taken: 27 Jun 1813 - Where taken: off Cape St. Mary's - Date Received: 3 Nov 1813 - From what ship: Plymouth - Born: New York - Age: 19 - Released on 26 Apr 1815.

Mitchell, Henry - Seaman - Nbr: 3509 - How taken: Gave himself up from HMS Warspite - When taken: 24 Sep 1814 - Date Received: 28 Sep 1814 - From what ship: HMT Salvador del Mundo - Born: Philadelphia - Age: 26 - Released on 4 Jun 1815.

Mitchell, James - Seaman - Nbr: 5688 - Prize: McDonough - Ship Type: P - How taken: HMS Bacchante - When taken: 7 Nov 1814 - Where taken: Lat 42 Long 67 - Date Received: 24 Dec 1814 - From what ship: HMT Penelope - Born: Arundel - Age: 23 - Released on 3 Jul 1815.

Mitchell, James M. - Seaman - Nbr: 2597 - Prize: Tom Thumb - Ship Type: MV - How taken: Lyon (Privateer) - When taken: 15 Feb 1813 - Where taken: Bay of Biscay - Date Received: 21 Aug 1814 - From what ship: HMT Freya, Chatham - Born: New York - Age: 19 - Released on 3 May 1815.

Mitchell, John - Seaman - Nbr: 6105 - How taken: Gave himself up from HMS Tyrien - When taken: 31 Dec 1814 - Date Received: 6 Jan 1814 - From what ship: HMT Impregnable - Born: Massachusetts - Age: 27 - Released

American Prisoners of War Held at Dartmoor during the War of 1812

## Alphabetical listing of names

on 13 Jun 1815.

Mitchell, John - Seaman - Nbr: 1123 - Prize: Hope - Ship Type: P - How taken: HMS Sea Horse - When taken: 22 Mar 1814 - Where taken: at sea - Date Received: 10 May 1814 - From what ship: Plymouth - Born: Maryland - Age: 22 - Race: Negro - Released on 28 Apr 1815.

Mitchell, John - Seaman - Nbr: 5215 - Prize: Teazer - Ship Type: P - How taken: HMS La Hogue - When taken: 26 Jul 1813 - Where taken: at sea - Date Received: 31 Oct 1814 - From what ship: HMT Mermaid, Chatham - Born: Bordeaux - Age: 41 - Released on 29 Jun 1815.

Mitchell, John - Seaman - Nbr: 3841 - Prize: Dominique - Ship Type: LM - How taken: HMS Dotterel - When taken: 21 May 1814 - Where taken: off Charleston - Date Received: 5 Oct 1814 - From what ship: HMT Orpheus, Halifax - Born: Guadeloupe - Age: 36 - Race: Mulatto - Released on 9 Jun 1815.

Mitchell, Reuben - Gunner - Nbr: 5500 - Prize: US Gunboat Nbr 2 - Ship Type: MW - How taken: British forces - When taken: 22 Aug 1814 - Where taken: Chesapeake Bay - Date Received: 17 Dec 1814 - From what ship: HMT Loire, Halifax - Born: Maryland - Age: 29 - Died on 11 May 1815 from variola.

Mitchell, Thomas - Seaman - Nbr: 1935 - How taken: Gave himself up from HMS Royal William - When taken: 3 Feb 1813 - Date Received: 3 Aug 1814 - From what ship: HMS Alceste, Chatham Depot - Born: Marblehead - Age: 35 - Released on 2 May 1815.

Mitchell, William - Seaman - Nbr: 3450 - Prize: Ann Packet - Ship Type: Prize - How taken: HMS La Hogue - When taken: 1813 - Where taken: off America - Date Received: 13 Sep 1814 - From what ship: HMT Niobe, Chatham - Born: Salem - Age: 27 - Released on 28 May 1815.

Mix, William A. - Seaman - Nbr: 201 - Prize: Star - Ship Type: MV - How taken: HMS Superb - When taken: 9 Feb 1813 - Where taken: Bay of Biscay - Date Received: 2 Apr 1813 - From what ship: Plymouth - Born: New Haven - Age: 18 - Sent to Plymouth on 8 Jul 1814.

Mode, David - Cook - Nbr: 1477 - Prize: Hebe - Ship Type: MV - How taken: HMS Stag - When taken: 18 Apr 1813 - Where taken: Bay of Biscay - Date Received: 19 Jun 1814 - From what ship: Stapleton - Born: Delaware - Age: 21 - Race: Mulatto - Released on 1 May 1815.

Modge, Daniel - Marine - Nbr: 5883 - Prize: Harlequin - Ship Type: P - How taken: HMS Bulwark - When taken: 23 Nov 1814 - Where taken: off Halifax - Date Received: 27 Dec 1814 - From what ship: HMT Penelope - Born: Lee - Age: 13 - Died on 16 Jan 1815 from pneumonia.

Modre, John - Seaman - Nbr: 4196 - Prize: Zephyr - Ship Type: MV - How taken: HMS Surveillante - When taken: 4 Jan 1814 - Where taken: Bay of Biscay - Date Received: 7 Oct 1814 - From what ship: HMT Niobe, Chatham - Born: Portugal - Age: 26 - Released on 13 Jun 1815.

Moffatt, David - Commander - Nbr: 2561 - Prize: Rattlesnake - Ship Type: P - How taken: HMS Hyperion - When taken: 25 Jun 1814 - Where taken: off Cape Finisterre - Date Received: 21 Aug 1814 - From what ship: HMS Hyperion - Born: Ireland - Age: 50 - Sent to Ashburton (England) on 27 Aug 1814.

Moffatt, Hugh - Seaman - Nbr: 5237 - Prize: Teazer - Ship Type: P - How taken: HMS Maidstone - When taken: 18 Jul 1813 - Where taken: at sea - Date Received: 31 Oct 1814 - From what ship: HMT Mermaid, Chatham - Born: Philadelphia - Age: 38 - Released on 29 Jun 1815.

Moffett, John - Seaman - Nbr: 1822 - Prize: Napoleon - Ship Type: MV - How taken: HMS Belle Poule - When taken: 3 Apr 1813 - Where taken: off Cape Ortegal (Spain) - Date Received: 21 Jul 1814 - From what ship: HMT Redbeard & Pincher, Chatham Depot - Born: New York - Age: 38 - Released on 1 May 1815.

Mold, John - Seaman - Nbr: 2738 - How taken: Gave himself up from HMS Pomona - When taken: 9 Jan 1813 - Date Received: 24 Aug 1814 - From what ship: HMT Liverpool, Chatham - Born: New York - Age: 20 - Released on 26 Apr 1815.

Molley, William - Seaman - Nbr: 5196 - Prize: Enterprize - Ship Type: P - How taken: HMS Tenedos - When taken: 21 May 1813 - Where taken: off Cape Cod - Date Received: 31 Oct 1814 - From what ship: HMT Mermaid, Chatham - Born: Salem - Age: 16 - Released on 29 Jun 1815.

Molloy, Peter - Seaman - Nbr: 4159 - How taken: Gave himself up from HMS Minos - When taken: 28 Dec 1813 - Date Received: 6 Oct 1814 - From what ship: HMT Niobe, Chatham - Born: New York - Age: 44 - Released on 13 Jun 1815.

Money, Henry - Seaman - Nbr: 435 - Prize: Magdalen - Ship Type: MV - How taken: HMS Superb - When taken: 15 Apr 1813 - Where taken: off Belle Isle, France - Date Received: 1 Jul 1813 - From what ship: Plymouth - Born: New Orleans - Age: 18 - Released on 20 Apr 1815.

Monk, Joseph - Seaman - Nbr: 4770 - Prize: James - Ship Type: MV - How taken: HMS Harpy - When taken: 18 Dec 1813 - Where taken: off Isle of France - Date Received: 9 Oct 1814 - From what ship: HMT Freya, Chatham - Born: Philadelphia - Age: 18 - Released on 27 Apr 1815.

American Prisoners of War Held at Dartmoor during the War of 1812

## Alphabetical listing of names

Monk, Philip - Seaman - Nbr: 6300 - Prize: John - Ship Type: Prize - How taken: Leander (Newcastle Acasta) - When taken: 30 Jan 1815 - Date Received: 19 Feb 1815 - From what ship: HMT Ganges, Plymouth - Born: Baltimore - Age: 22 - Race: Black - Released on 5 Jul 1815.

Monks, John - Seaman - Nbr: 207 - Prize: Criterion - Ship Type: MV - How taken: HMS Belle Poule - When taken: 14 Feb 1813 - Where taken: Bay of Biscay - Date Received: 2 Apr 1813 - From what ship: Plymouth - Born: Massachusetts - Age: 29 - Released on 20 Apr 1815.

Monroe, James - Seaman - Nbr: 3445 - How taken: Gave himself up from HMS Freya - When taken: 21 Aug 1813 - Date Received: 13 Sep 1814 - From what ship: HMT Niobe, Chatham - Born: New York - Age: 34 - Released on 28 May 1815.

Monroe, John - Seaman - Nbr: 5644 - Prize: Lion - Ship Type: P - How taken: HMS Granicus - When taken: 2 Dec 1814 - Where taken: off Lisbon - Date Received: 24 Dec 1814 - From what ship: HMT Tay - Born: New York - Age: 35 - Released on 3 Jul 1815.

Monsieur, Peter - Seaman - Nbr: 1209 - Prize: Comet - Ship Type: Prize - How taken: HMS Fairy - When taken: 25 Dec 1812 - Where taken: off St. Antonio - Date Received: 4 Jun 1814 - From what ship: Dartmouth - Born: New Orleans - Age: 26 - Sent to Plymouth on 8 Jul 1814.

Montbelly, William - Seaman - Nbr: 1544 - Prize: Zebra - Ship Type: LM - How taken: HMS Pyramus - When taken: 20 Apr 1813 - Where taken: Bay of Biscay - Date Received: 23 Jun 1814 - From what ship: Stapleton - Born: Savannah - Age: 25 - Sent to Mill Prison (Plymouth, England) on 30 Jun 1814.

Montcalm, Henry - Seaman - Nbr: 3614 - Prize: Hornsby - Ship Type: P - How taken: HMS Sceptre - When taken: 4 Feb 1814 - Where taken: off Block Island - Date Received: 30 Sep 1814 - From what ship: HMT Sybella - Born: Boston - Age: 21 - Released on 4 Jun 1815.

Monte, Charles - Seaman - Nbr: 3879 - Prize: Fame - Ship Type: MV - How taken: HMS Niemen - When taken: 21 May 1814 - Where taken: off Cape Hatteras - Date Received: 5 Oct 1814 - From what ship: HMT Orpheus, Halifax - Born: St. Antonio - Age: 22 - Race: Black - Died on 21 Feb 1815 from variola.

Montgomery, John - Seaman - Nbr: 754 - How taken: Impressed at Liverpool - Date Received: 3 Nov 1813 - From what ship: Plymouth - Born: New York - Age: 21 - Died on 24 Feb 1814 from pneumonia.

Montgomery, William - Seaman - Nbr: 4830 - Prize: President - Ship Type: P - How taken: HMS Pique - When taken: 7 May 1814 - Where taken: off Porto Rico - Date Received: 9 Oct 1814 - From what ship: HMT Freya, Halifax - Born: New York - Age: 20 - Released on 3 May 1815.

Montgomery, William - Seaman - Nbr: 5365 - Prize: Vengeance - Ship Type: LM - How taken: HMS Herald - When taken: 26 Jun 1814 - Where taken: Mississippi - Date Received: 31 Oct 1814 - From what ship: HMT Leyden, Chatham - Born: New York - Age: 41 - Released on 1 Jul 1815.

Monturn, Francis - Seaman - Nbr: 394 - Prize: John & Frances - Ship Type: MV - How taken: HMS Belle Poule - When taken: 13 Mar 1813 - Where taken: Bay of Biscay - Date Received: 1 Jul 1813 - From what ship: Plymouth - Born: Sardinia - Age: 23 - Sent to Plymouth on 27 Apr 1814.

Monturne, Jais - Seaman - Nbr: 1159 - Prize: John & Frances - Ship Type: MV - How taken: HMS Belle Poule - When taken: 13 Mar 1813 - Where taken: Bay of Biscay - Date Received: 10 May 1814 - From what ship: Plymouth - Born: Sardinia - Age: 24 - Sent to Mill Prison (Plymouth, England) on 8 Jul 1814.

Moody, John - Seaman - Nbr: 6033 - Prize: Regent - Ship Type: LM - How taken: HMS Forth - When taken: 19 Sep 1814 - Where taken: off Egg Harbor (New Jersey) - Date Received: 28 Dec 1814 - From what ship: HMT Penelope - Born: Woodbridge - Age: 28 - Released on 11 Jul 1815.

Mooney, Mathew - Soldier - Nbr: 5258 - Prize: 13th US Infantry - Ship Type: Troops - How taken: British Army - When taken: 13 Oct 1812 - Where taken: Canada - Date Received: 31 Oct 1814 - From what ship: HMT Leyden, Chatham - Born: Londonderry - Age: 43 - Released on 29 Jun 1815.

Mooney, Peter - Seaman - Nbr: 1359 - Prize: Paul Jones - Ship Type: P - How taken: HMS Leonidas - When taken: 23 May 1813 - Where taken: Channel - Date Received: 19 Jun 1814 - From what ship: Stapleton - Born: New Orleans - Age: 18 - Sent to Mill Prison (Plymouth, England) on 26 Jun 1814.

Moor, Abraham - Seaman - Nbr: 5230 - How taken: Impressed at Gravesend - When taken: 29 Jul 1813 - Date Received: 31 Oct 1814 - From what ship: HMT Mermaid, Chatham - Born: Chester - Age: 26 - Released on 29 Jun 1815.

Moor, Joshua - Marine - Nbr: 6324 - Prize: Prince de Neufchatel - Ship Type: P - How taken: Leander (Newcastle Acasta) - When taken: 20 Dec 1814 - Where taken: Lat 38 Long 56 - Date Received: 19 Feb 1815 - From what ship: HMT Ganges, Plymouth - Born: Portsmouth - Age: 21 - Released on 5 Jul 1815.

Moor, Samuel - Steward's mate - Nbr: 5666 - Prize: Harlequin - Ship Type: P - How taken: HMS Bulwark - When taken: 23 Nov 1814 - Where taken: off Halifax - Date Received: 24 Dec 1814 - From what ship: HMT

American Prisoners of War Held at Dartmoor during the War of 1812

## Alphabetical listing of names

Penelope - Born: Portsmouth - Age: 17 - Released on 3 Jul 1815.
Moor, William - Seaman - Nbr: 5568 - Prize: Daedalus - Ship Type: MV - How taken: HMS Niemen - When taken: 20 Sep 1814 - Where taken: off Delaware - Date Received: 17 Dec 1814 - From what ship: HMT Loire, Halifax - Born: Londonderry - Age: 29 - Released on 1 Jul 1815.
Moore, Alexander - Seaman - Nbr: 973 - Prize: Siro - Ship Type: LM - How taken: HMS Pelican - When taken: 13 Jan 1814 - Where taken: at sea - Date Received: 31 Jan 1814 - From what ship: Plymouth - Born: New Jersey - Age: 21 - Released on 27 Apr 1815.
Moore, Edward - Seaman - Nbr: 4582 - Prize: Bunker Hill - Ship Type: P - How taken: HMS Pomone - When taken: 2 Mar 1814 - Where taken: Channel - Date Received: 9 Oct 1814 - From what ship: HMT Leyden, Chatham - Born: New York - Age: 22 - Race: Mulatto - Released on 15 Jun 1815.
Moore, George - Seaman - Nbr: 6068 - Prize: Chasseur - Ship Type: P - How taken: HMS Recruit - When taken: 6 Apr 1814 - Where taken: Lat 29 Long 76 - Date Received: 28 Dec 1814 - From what ship: HMT Penelope - Born: Boston - Age: 35 - Race: Black - Died on 29 Mar 1815 from pneumonia.
Moore, George - Seaman - Nbr: 805 - Prize: Collin - Ship Type: MV - How taken: HMS Helicon and HMS Whiting - When taken: 26 Oct 1813 - Where taken: Channel - Date Received: 3 Nov 1813 - From what ship: Plymouth - Born: New York - Age: 22 - Released on 26 Apr 1815.
Moore, Henry - Seaman - Nbr: 654 - Prize: Marmion - Ship Type: MV - How taken: HMS President - When taken: 14 Aug 1813 - Where taken: off Nantes - Date Received: 8 Sep 1813 - From what ship: Plymouth - Born: New York - Age: 30 - Race: Negro - Died on 4 Jan 1814 from dropsy.
Moore, J. B. - 2nd Lieutenant - Nbr: 4969 - Prize: Invincible - Ship Type: LM - How taken: HMS Armide - When taken: 15 Aug 1814 - Where taken: off Nantucket - Date Received: 28 Oct 1814 - From what ship: HMT Alkbar, Halifax - Born: Portland - Age: 23 - Released on 21 Jun 1815.
Moore, James - Seaman - Nbr: 5935 - Prize: Harlequin - Ship Type: P - How taken: HMS Bulwark - When taken: 23 Nov 1814 - Where taken: off Halifax - Date Received: 27 Dec 1814 - From what ship: HMT Penelope - Born: Philadelphia - Age: 26 - Released on 4 Jun 1815.
Moore, James - Seaman - Nbr: 1091 - How taken: Impressed from Cartel St. Francis - When taken: 26 May 1814 - Date Received: 10 May 1814 - From what ship: Plymouth - Born: North Carolina - Age: 40 - Released on 27 Apr 1815.
Moore, James - Seaman - Nbr: 1751 - How taken: Gave himself up from HMS Alemene - When taken: 24 Jan 1813 - Date Received: 20 Jul 1814 - From what ship: HMS Milford, Plymouth - Born: North Carolina - Age: 27 - Race: Mulatto - Released on 1 May 1815.
Moore, John - Seaman - Nbr: 837 - Prize: Chesapeake - Ship Type: LM - How taken: HMS Hotspur & HMS Pyramus - When taken: 26 Oct 1813 - Where taken: off Nantes - Date Received: 29 Nov 1813 - From what ship: Plymouth - Born: Maryland - Age: 29 - Race: Black - Released on 26 Apr 1815.
Moore, John - Seaman - Nbr: 2081 - Prize: True Blooded Yankee - Ship Type: P - How taken: HMS Hope - When taken: 24 Jun 1813 - Where taken: off Brest - Date Received: 3 Aug 1814 - From what ship: HMS Bittern, Chatham Depot - Born: Norwich - Age: 26 - Released on 2 May 1815.
Moore, John - Seaman - Nbr: 984 - Prize: Amity - Ship Type: P - How taken: HMS Achates - When taken: 22 Dec 1813 - Where taken: Bay of Biscay - Date Received: 31 Jan 1814 - From what ship: Plymouth - Born: Messina - Age: 56 - Released on 27 Apr 1815.
Moore, John - Seaman - Nbr: 2999 - Prize: St. Lawrence - How taken: HMS Aquilon - When taken: 9 Aug 1814 - Where taken: off Western Islands (England) - Date Received: 29 Aug 1814 - From what ship: HMT Bittern - Born: New Jersey - Age: 24 - Race: Black - Released on 28 May 1815.
Moore, John - Seaman - Nbr: 2814 - How taken: Impressed at London - When taken: 26 Jul 1813 - Date rec'd: 24 Aug 1814 - What ship: HMT Liverpool, Chatham - Born: Baltimore - Age: 30 - Released on 19 May 1815.
Moore, Laurence - Seaman - Nbr: 326 - Prize: Paulina - Ship Type: MV - How taken: Rebuff - When taken: 23 Dec 1812 - Where taken: off Cadiz - Date Received: 28 Jun 1813 - From what ship: Plymouth - Born: New York - Age: 47 - Released on 20 Apr 1815.
Moore, Richard - Seaman - Nbr: 1604 - Prize: Fox - Ship Type: LM - How taken: HMS Pheasant - When taken: 23 Apr 1813 - Where taken: Bay of Biscay - Date Received: 23 Jun 1814 - From what ship: Stapleton - Born: Pennsylvania - Age: 36 - Race: Negro - Released on 1 May 1815.
Moore, Robert - Seaman - Nbr: 6411 - Prize: Nancy - Ship Type: P - How taken: Papillion - When taken: 27 Dec 1814 - Where taken: Lat 29 Long 20 - Date Received: 3 Mar 1815 - From what ship: HMT Ganges, Plymouth - Born: Kittery - Age: 20 - Released on 11 Jul 1815.
Moore, Samuel - Seaman - Nbr: 3301 - Prize: Governor Plumer - Ship Type: P - How taken: Sent into custody from

American Prisoners of War Held at Dartmoor during the War of 1812

## Alphabetical listing of names

a privateer - When taken: 1 Jun 1813 - Where taken: off Cape Ann - Date Received: 13 Sep 1814 - From what ship: HMT Niobe, Chatham - Born: New York - Age: 24 - Released on 28 May 1815.

Moore, T. T. - Seaman - Nbr: 6074 - Prize: Daedalus - Ship Type: MV - How taken: HMS Niemen - When taken: 20 Sep 1814 - Where taken: off Delaware - Date Received: 28 Dec 1814 - From what ship: HMT Penelope - Born: Newbury - Age: 25 - Released on 3 Jul 1815.

Moore, Warren - Seaman - Nbr: 6345 - Prize: Prince de Neufchatel - Ship Type: P - How taken: Leander (Newcastle Acasta) - When taken: 20 Dec 1814 - Where taken: Lat 38 Long 56 - Date Received: 19 Feb 1815 - From what ship: HMT Ganges, Plymouth - Born: Massachusetts - Age: 22 - Released on 6 Jul 1815.

Moore, William - Seaman - Nbr: 6475 - Prize: Mary - Ship Type: P - How taken: Lapphin - When taken: 15 Jun 1814 - Where taken: off Saint Marys - Date Received: 3 Mar 1815 - From what ship: HMT Ganges, Plymouth - Born: Londonderry - Age: 34 - Released on 11 Jul 1815.

Moore, William - Seaman - Nbr: 704 - Prize: Ned - Ship Type: LM - How taken: HMS Royalist - When taken: 6 Sep 1813 - Where taken: Bay of Biscay - Date Received: 27 Sep 1813 - From what ship: Plymouth - Born: Portsmouth - Age: 40 - Released on 26 Apr 1815.

Mordaunt, John - Seaman - Nbr: 2158 - Prize: Rattlesnake - Ship Type: P - How taken: HMS Hyperion - When taken: 26 Jun 1814 - Where taken: off Cape Finisterre - Date Received: 16 Aug 1814 - From what ship: HMS Dublin, Halifax - Born: New Orleans - Age: 32 - Released on 2 May 1815.

Morel, Thomas - Seaman - Nbr: 18 - Prize: Cashier - Ship Type: LM - How taken: HMS Reindeer - When taken: 3 Feb 1813 - Where taken: Bay of Biscay - Date Received: 2 Apr 1813 - From what ship: Plymouth - Born: New York - Age: 32 - Sent to Dartmouth on 30 Jul 1813.

Morell, Benjamin - Seaman - Nbr: 667 - Prize: Joel Barlow - Ship Type: LM - How taken: HMS Briton - When taken: 3 Jul 1813 - Where taken: off Bordeaux - Date Received: 8 Sep 1813 - From what ship: Plymouth - Born: Connecticut - Age: 18 - Released on 26 Apr 1815.

Morell, John - Seaman - Nbr: 2764 - How taken: Gave himself up from HMS Mermaid - When taken: 30 Nov 1812 - Date Received: 24 Aug 1814 - From what ship: HMT Liverpool, Chatham - Born: New Haven - Age: 31 - Released on 19 May 1815.

Morell, Samuel J. - Seaman - Nbr: 5661 - Prize: Harlequin - Ship Type: P - How taken: HMS Bulwark - When taken: 23 Nov 1814 - Where taken: off Halifax - Date Received: 24 Dec 1814 - From what ship: HMT Penelope - Born: Saco - Age: 33 - Released on 20 Mar 1815.

Morey, Ezekiel - Seaman - Nbr: 5836 - Prize: US Sloop-of-War Frolic - Ship Type: MW - How taken: HMS Orpheus - When taken: 20 Apr 1814 - Where taken: off Cuba - Date Received: 26 Dec 1814 - From what ship: HMT Argo - Born: Boston - Age: 20 - Released on 3 May 1815.

Morgan, Henry - Seaman - Nbr: 4988 - Prize: Invincible - Ship Type: LM - How taken: HMS Armide - When taken: 15 Aug 1814 - Where taken: off Nantucket - Date Received: 28 Oct 1814 - From what ship: HMT Alkbar, Halifax - Born: Beverly - Age: 22 - Released on 21 Jun 1815.

Morgan, Jacob - Seaman - Nbr: 2326 - Prize: Hussar - Ship Type: P - How taken: HMS Saturn - When taken: 25 May 1814 - Where taken: off Sandy Hook - Date Received: 16 Aug 1814 - From what ship: HMS Dublin, Halifax - Born: Connecticut - Age: 20 - Released on 3 May 1815.

Morgan, James - Seaman - Nbr: 4702 - Prize: Volunteer - Ship Type: MV - How taken: Victoria (Privateer) - When taken: 26 Dec 1813 - Where taken: at sea - Date Received: 9 Oct 1814 - From what ship: HMT Leyden, Chatham - Born: New York - Age: 19 - Released on 15 Jun 1815.

Morgan, John - Seaman - Nbr: 1571 - Prize: Messenger - Ship Type: MV - How taken: HMS Iris - When taken: 10 Mar 1813 - Where taken: Bay of Biscay - Date Received: 23 Jun 1814 - From what ship: Stapleton - Born: North Carolina - Age: 22 - Released on 1 May 1815.

Morgan, John - Seaman - Nbr: 1018 - Prize: Rachel & Ann - Ship Type: P - How taken: HMS Cydnus - When taken: 6 Jan 1814 - Where taken: at sea - Date Received: 31 Jan 1814 - From what ship: Plymouth - Born: Boston - Age: 32 - Released on 27 Apr 1815.

Morgan, John - Seaman - Nbr: 1072 - Prize: Fair American - Ship Type: MV - How taken: HMS Andromache - When taken: 19 Jan 1814 - Where taken: Bay of Biscay - Date Received: 10 May 1814 - From what ship: Plymouth - Born: Manchester - Age: 21 - Released on 10 Apr 1815.

Morgan, Joseph - Seaman - Nbr: 405 - Prize: Viper - Ship Type: MV - How taken: HMS Superb - When taken: 15 Apr 1813 - Where taken: Bay of Biscay - Date Received: 1 Jul 1813 - From what ship: Plymouth - Born: New Jersey - Age: 26 - Released on 20 Apr 1815.

Morgan, William - Seaman - Nbr: 5812 - Prize: US Schooner Scorpion - Ship Type: MW - How taken: British gunboats - When taken: 6 Sep 1814 - Where taken: Lake Erie - Date Received: 26 Dec 1814 - From what

American Prisoners of War Held at Dartmoor during the War of 1812

## Alphabetical listing of names

ship: HMT Argo - Born: Virginia - Age: 28 - Released on 3 Jul 1815.
Morgan, William - Seaman - Nbr: 4049 - Prize: US Brig Rattlesnake - Ship Type: MW - How taken: HMS Leander - When taken: 13 Jul 1814 - Where taken: off Shelburne - Date Received: 6 Oct 1814 - From what ship: HMT Chesapeake, Halifax - Born: New York - Age: 24 - Released on 13 Jun 1815.
Morie, Denis - Seaman - Nbr: 888 - Prize: General Kempt - Ship Type: P - How taken: HMS Foxhound - When taken: 18 Dec 1813 - Where taken: Lat 48'60" Long 5'7" - Date Received: 31 Jan 1814 - From what ship: Plymouth - Born: Salem - Age: 26 - Released on 27 Apr 1815.
Morier, John - Seaman - Nbr: 4991 - Prize: Invincible - Ship Type: LM - How taken: HMS Armide - When taken: 15 Aug 1814 - Where taken: off Nantucket - Date Received: 28 Oct 1814 - From what ship: HMT Alkbar, Halifax - Born: L'Orient (France) - Age: 30 - Released on 21 Jun 1815.
Morrell, Jacob - Seaman - Nbr: 4871 - Prize: Fox - Ship Type: Prize - How taken: HMS Dover - When taken: Apr 1813 - Where taken: Newfoundland Banks - Date Received: 24 Oct 1814 - From what ship: Royal Hospital, Plymouth - Born: Massachusetts - Age: 22 - Died on 27 Apr 1815 from phthisis pulmonalis.
Morris, Benjamin - Seaman - Nbr: 5509 - Prize: Ann Dorothy - Ship Type: Prize - How taken: HMS Maidstone - When taken: 30 Oct 1814 - Where taken: off Cape Sable Island (Canada) - Date Received: 17 Dec 1814 - From what ship: HMT Loire, Halifax - Born: Massachusetts - Age: 25 - Released on 29 Jun 1815.
Morris, George - Seaman - Nbr: 5783 - How taken: Gave himself up from HMS Ariel - When taken: 8 Dec 1812 - Date Received: 26 Dec 1814 - From what ship: HMT Argo - Born: Philadelphia - Age: 22 - Race: Mulatto - Released on 11 Jul 1815.
Morris, Isaac - Seaman - Nbr: 1251 - Prize: Adeline - Ship Type: MV - How taken: HMS Magicienne - When taken: 16 Mar 1814 - Where taken: off Cape Finisterre - Date Received: 14 Jun 1814 - From what ship: Mill Prison (Plymouth, England) - Born: Virginia - Age: 21 - Race: Black - Released on 28 Apr 1815.
Morris, Jacob - Seaman - Nbr: 3428 - Prize: Industry - Ship Type: P - How taken: HMS Heron - When taken: 3 Nov 1813 - Where taken: off Halifax - Date Received: 13 Sep 1814 - From what ship: HMT Niobe, Chatham - Born: Pennsylvania - Age: 29 - Race: Black - Released on 28 May 1815.
Morris, John - Boatswain's mate - Nbr: 6368 - Prize: US Brig Syren - Ship Type: MW - How taken: HMS Medway - When taken: 12 Jul 1814 - Where taken: off Cape of Good Hope - Date Received: 24 Feb 1815 - From what ship: HMT Ganges, Plymouth - Born: Marblehead - Age: 27 - Released on 6 Jul 1815.
Morris, John - Seaman - Nbr: 6085 - Prize: David Porter - Ship Type: LM - How taken: HMS Pylades - When taken: 13 Sep 1814 - Where taken: Georges Bank - Date Received: 28 Dec 1814 - From what ship: HMT Penelope - Born: New York - Age: 20 - Race: Creole - Released on 11 Jul 1815.
Morris, John - Seaman - Nbr: 786 - Prize: Avon - Ship Type: MV - How taken: HMS Eurotas - When taken: 27 Oct 1813 - Where taken: off Ushant (France) - Date Received: 3 Nov 1813 - From what ship: Plymouth - Born: Virginia - Age: 23 - Released on 26 Apr 1815.
Morris, John - Seaman - Nbr: 73 - Prize: Rolla - Ship Type: MV - How taken: HMS Surveillante - When taken: 11 Feb 1813 - Where taken: Bay of Biscay - Date Received: 2 Apr 1813 - From what ship: Plymouth - Born: New York - Age: 20 - Released on 20 Apr 1815.
Morris, John - Seaman - Nbr: 341 - Prize: Good Friends - Ship Type: MV - How taken: HMS Andromache - When taken: 2 Apr 1813 - Where taken: Bay of Biscay - Date Received: 28 Jun 1813 - From what ship: Plymouth - Born: New Orleans - Age: 48 - Sent to Plymouth on 8 Jul 1814.
Morris, John - Seaman - Nbr: 4084 - Prize: New Zealander - Ship Type: Prize - How taken: HMS Belviders - When taken: 21 Apr 1814 - Where taken: off Delaware - Date Received: 6 Oct 1814 - From what ship: HMT Chesapeake, Halifax - Born: Boston - Age: 29 - Race: Negro - Released on 13 Jun 1815.
Morris, Joseph - Seaman - Nbr: 4128 - Prize: Bordeaux Packet - Ship Type: LM - How taken: HMS Niemen - When taken: 28 Jun 1814 - Where taken: off Delaware - Date Received: 6 Oct 1814 - From what ship: HMT Chesapeake, Halifax - Born: Newcastle - Age: 24 - Released on 13 Jun 1815.
Morris, Louis - Seaman - Nbr: 1952 - How taken: Gave himself up from HMS Christian 7th - When taken: 19 May 1813 - Date Received: 3 Aug 1814 - From what ship: HMS Lyffey, Chatham Depot - Born: New Haven - Age: 23 - Released on 2 May 1815.
Morris, Manuel - Seaman - Nbr: 5535 - Prize: Theodore - Ship Type: Prize - How taken: HMS Galatea - When taken: 7 Sep 1814 - Where taken: Channel - Date Received: 17 Dec 1814 - From what ship: HMT Loire, Halifax - Born: Corunna - Age: 40 - Released on 1 Jul 1815.
Morris, Manuel - Seaman - Nbr: 2448 - Prize: US Sloop Frolic - Ship Type: MW - How taken: HMS Orpheus - When taken: 20 Apr 1814 - Where taken: off Cuba - Date Received: 16 Aug 1814 - From what ship: HMT Queen, Halifax - Born: Coast of Peru - Age: 29 - Race: Negro - Released on 3 May 1815.

American Prisoners of War Held at Dartmoor during the War of 1812

## Alphabetical listing of names

Morris, Peter - Seaman - Nbr: 5725 - How taken: HMS Ringdove - When taken: 2 Oct 1814 - Where taken: Not legible - Date Received: 26 Dec 1814 - From what ship: HMT Argo - Born: Bayonne - Age: 30 - Released on 3 Jul 1815.

Morris, Richard - Seaman - Nbr: 6172 - Prize: Lion - Ship Type: P - How taken: HMS Granicus - When taken: 2 Dec 1814 - Where taken: off Lisbon - Date Received: 18 Jan 1815 - From what ship: HMT Impregnable - Born: Pennsylvania - Age: 42 - Race: Black - Released on 5 Jul 1815.

Morris, Samuel - Seaman - Nbr: 4757 - How taken: Gave himself up from HMS Africa - When taken: 4 Oct 1813 - Date Received: 9 Oct 1814 - From what ship: HMT Freya, Chatham - Born: New York - Age: 25 - Race: Black - Released on 15 Jun 1815.

Morrison, D. - Seaman - Nbr: 5609 - How taken: Gave himself up from HMS Ashea - When taken: 13 May 1814 - Date Received: 24 Dec 1814 - From what ship: HMT Tay - Born: Portsmouth - Age: 24 - Released on 27 Mar 1815.

Morrison, Davis - Landsman - Nbr: 106 - Prize: St. Martin's Planter - Ship Type: P - How taken: HMS Dublin - When taken: 9 Feb 1813 - Where taken: Lat 43 N, Long 33 50 W - Date Received: 2 Apr 1813 - From what ship: Plymouth - Born: Northumberland - Age: 21 - Released on 20 Apr 1815.

Morrison, James - Seaman - Nbr: 4953 - Prize: Herald - Ship Type: P - How taken: HMS Endymion - When taken: 15 Aug 1814 - Where taken: off Nantucket - Date Received: 28 Oct 1814 - From what ship: HMT Alkbar, Halifax - Born: New York - Age: 23 - Released on 28 Apr 1815.

Morrison, John Baptist - Seaman - Nbr: 1009 - Prize: Harvest - Ship Type: P - How taken: HMS Orestes - When taken: 21 Jan 1814 - Where taken: Bay of Biscay - Date Received: 31 Jan 1814 - From what ship: Plymouth - Born: Cape Francis - Age: 18 - Sent to Mill Prison (Plymouth, England) on 21 Jun 1814.

Morrison, Joseph - Seaman - Nbr: 5446 - Prize: Prize to the Lawrence - Ship Type: P - How taken: HMS Glasgow - When taken: 2 Nov 1814 - Where taken: Channel - Date Received: 10 Dec 1814 - From what ship: HMT Impregnable - Born: New Jersey - Age: 19 - Released on 28 May 1815.

Morrison, Michael - Seaman - Nbr: 5442 - How taken: Impressed at Dublin - When taken: 7 Oct 1814 - Date Received: 11 Nov 1814 - What ship: HMT Impregnable - Born: Troy - Age: 23 - Released on 1 Jul 1815.

Morrison, Samuel - Seaman - Nbr: 2531 - Prize: Hussar - Ship Type: P - How taken: HMS Saturn - When taken: 25 May 1814 - Where taken: off Sandy Hook - Date Received: 16 Aug 1814 - From what ship: HMT Queen, Halifax - Born: New York - Age: 32 - Released on 3 May 1815.

Morrison, Thomas - Seaman - Nbr: 4575 - Prize: Bunker Hill - Ship Type: P - How taken: HMS Pomone - When taken: 2 Mar 1814 - Where taken: Channel - Date Received: 9 Oct 1814 - From what ship: HMT Leyden, Chatham - Born: Galway - Age: 24 - Released on 15 Jun 1815.

Morrison, William - Boy - Nbr: 2637 - How taken: Gave himself up from HMS Leviathan - When taken: 22 Oct 1813 - Date Received: 21 Aug 1814 - From what ship: HMT Freya, Chatham - Born: New York - Age: 17 - Released on 3 May 1815.

Morriss, Andrew - Seaman - Nbr: 4624 - Prize: Argus - Ship Type: MV - How taken: HMS San Domingo - When taken: 1 Mar 1814 - Where taken: off Savannah - Date Received: 9 Oct 1814 - From what ship: HMT Leyden, Chatham - Born: Baltimore - Age: 18 - Race: Black - Released on 15 Jun 1815.

Morse, Henry - Seaman - Nbr: 345 - Prize: Good Friends - Ship Type: MV - How taken: HMS Andromache - When taken: 2 Apr 1813 - Where taken: Bay of Biscay - Date Received: 28 Jun 1813 - From what ship: Plymouth - Born: Vermont - Age: 19 - Released on 25 Mar 1815.

Mortice, William - Seaman - Nbr: 1895 - How taken: Impressed at Plymouth - When taken: Jul 1813 - Date Received: 29 Jul 1814 - From what ship: HMS Ville de Paris, Chatham Depot - Born: Massachusetts - Age: 33 - Race: Negro - Released on 2 May 1815.

Morton, Samuel - Seaman - Nbr: 3517 - Prize: Aeolus - Ship Type: Prize - How taken: HMS Warspite - When taken: 2 Sep 1814 - Where taken: off Newfoundland - Date Received: 28 Sep 1814 - From what ship: HMT Salvador del Mundo - Born: Milford - Age: 22 - Released on 4 Jun 1815.

Morton, William - Seaman - Nbr: 3120 - How taken: Gave himself up from HMS Derwent - When taken: Sep 1814 - Date Received: 11 Sep 1814 - From what ship: HMT Salvador del Mundo - Born: Philadelphia - Age: 28 - Released on 28 May 1815.

Midweek, Andrew - Seaman - Nbr: 720 - Prize: Ned - Ship Type: LM - How taken: HMS Royalist - When taken: 6 Sep 1813 - Where taken: Bay of Biscay - Date Received: 27 Sep 1813 - From what ship: Plymouth - Born: Portland - Age: 21 - Released on 26 Apr 1815.

Moses, James - Seaman - Nbr: 2810 - How taken: Impressed at London - When taken: 9 Jul 1813 - Date Received: 24 Aug 1814 - What ship: HMT Liverpool, Chatham - Born: Standish - Age: 21 - Released on 19 May 1815.

American Prisoners of War Held at Dartmoor during the War of 1812

## Alphabetical listing of names

Moses, John - Seaman - Nbr: 2235 - Prize: General Hart - Ship Type: P - How taken: HMS Sophie - When taken: 24 Apr 1814 - Where taken: off Bermuda - Date Received: 16 Aug 1814 - From what ship: HMS Dublin, Halifax - Born: Portsmouth - Age: 24 - Released on 2 May 1815.

Moss, John - Seaman - Nbr: 342 - Prize: Good Friends - Ship Type: MV - How taken: HMS Andromache - When taken: 2 Apr 1813 - Where taken: Bay of Biscay - Date Received: 28 Jun 1813 - From what ship: Plymouth - Born: Stralsund - Age: 34 - Released on 6 May 1814.

Moss, N. J. - Seaman - Nbr: 5060 - Prize: Ida - Ship Type: LM - How taken: HMS Newcastle - When taken: 9 Aug 1814 - Where taken: Long 34 - Date Received: 28 Oct 1814 - From what ship: HMT Alkbar, Halifax - Born: Cape Ann - Age: 25 - Released on 29 Jun 1815.

Moss, Richard - Seaman - Nbr: 6510 - How taken: Apprehended at Queen - When taken: 30 Nov 1813 - Date Received: 3 Mar 1815 - From what ship: HMT Ganges, Plymouth - Born: Bridgewater - Age: 24 - Released on 11 Jul 1815.

Moss, Thomas - Seaman - Nbr: 1529 - Prize: Essex - Ship Type: MV - How taken: HMS Pyramus - When taken: 20 Apr 1813 - Where taken: Bay of Biscay - Date Received: 23 Jun 1814 - From what ship: Stapleton - Born: Marblehead - Age: 44 - Released on 1 May 1815.

Mott, Thomas - Seaman - Nbr: 5123 - Prize: York Town - Ship Type: P - How taken: HMS Maidstone - When taken: 18 Jul 1813 - Where taken: Grand Banks - Date Received: 28 Oct 1814 - From what ship: HMT Salvador del Mundo - Born: Philadelphia - Age: 17 - Released on 27 Apr 1815.

Malden, William - Seaman - Nbr: 2758 - How taken: Gave himself up from HMS Blake - When taken: 10 Dec 1812 - Date Received: 24 Aug 1814 - From what ship: HMT Liverpool, Chatham - Born: Andover - Age: 41 - Released on 26 Apr 1815.

Moulds, Benjamin - Seaman - Nbr: 736 - Prize: Orders in Council - Ship Type: LM - How taken: HMS Surveillante - When taken: 1 Jun 1813 - Where taken: off Cape Ortegal (Spain) - Date Received: 3 Nov 1813 - From what ship: Plymouth - Born: Virginia - Age: 38 - Released on 26 Apr 1815.

Mauston, Nathaniel - Seaman - Nbr: 6513 - How taken: Gave himself up from HMS Orontes - When taken: Feb 1815 - Date Received: 3 Mar 1815 - From what ship: HMT Ganges, Plymouth - Born: Newburyport - Age: 18 - Released on 11 Jul 1815.

Mount, James - Soldier - Nbr: 5840 - Prize: 22nd US Infantry - Ship Type: Troops - How taken: British Army - When taken: 17 Sep 1814 - Where taken: Niagara - Date Received: 26 Dec 1814 - From what ship: HMT Argo - Born: Pennsylvania - Age: 21 - Released on 29 Jun 1815.

Mounts, John - Seaman - Nbr: 6242 - Prize: Albion - Ship Type: Prize - How taken: HMS Harlequin - When taken: 6 Jan 1814 - Where taken: Long 20 - Date Received: 4 Feb 1815 - From what ship: HMT Ganges - Born: New Jersey - Age: 24 - Released on 5 Jul 1815.

Muckleroy, Samuel - Seaman - Nbr: 2047 - How taken: Gave himself up from HMS Menelaus - When taken: 6 Jun 1812 - Date Received: 3 Aug 1814 - From what ship: HMS Lyffey, Chatham Depot - Born: Philadelphia - Age: 40 - Released on 26 Apr 1815.

Mugford, William - Seaman - Nbr: 5622 - Prize: Rambler - Ship Type: MV - How taken: HMS Lion - When taken: 1 Feb 1813 - Where taken: Cape - Date Received: 24 Dec 1814 - From what ship: HMT Tay - Born: Salem - Age: 43 - Released on 3 Jul 1815.

Muggin, John - Seaman - Nbr: 766 - How taken: Sent into custody from HMS Pallas - Date Received: 3 Nov 1813 - From what ship: Plymouth - Born: Warren - Age: 39 - Released on 26 Apr 1815.

Muley, Etienne - Seaman - Nbr: 1329 - Prize: Paul Jones - Ship Type: P - How taken: HMS Leonidas - When taken: 23 May 1813 - Where taken: Channel - Date Received: 19 Jun 1814 - From what ship: Stapleton - Born: L'Orient (France) - Age: 23 - Sent to Mill Prison (Plymouth, England) on 21 Jun 1814.

Mullan, John - Seaman - Nbr: 643 - Prize: US Brig Argus - Ship Type: MW - How taken: HMS Pelican - When taken: 14 Aug 1813 - Where taken: Irish Channel - Date Received: 8 Sep 1813 - From what ship: Plymouth - Born: New York - Age: 23 - Sent to Dartmouth on 30 Jul 1813.

Mullan, John - Seaman - Nbr: 4 - Prize: Cashier - Ship Type: LM - How taken: HMS Reindeer - When taken: 3 Feb 1813 - Where taken: Bay of Biscay - Date Received: 2 Apr 1813 - From what ship: Plymouth - Born: New York - Age: 25 - Sent to Dartmouth on 30 Jul 1813.

Muller, Edward - Seaman - Nbr: 1615 - Prize: Shadow - Ship Type: LM - How taken: HMS Reindeer - When taken: 6 Apr 1813 - Where taken: Bay of Biscay - Date Received: 23 Jun 1814 - From what ship: Stapleton - Born: Pennsylvania - Age: 40 - Released on 1 May 1815.

Muller, John - Seaman - Nbr: 4879 - Prize: Saratoga - Ship Type: P - How taken: HMS Barracouta - When taken: 9 Oct 1814 - Where taken: off Western Islands (England) - Date Received: 24 Oct 1814 - From what ship:

American Prisoners of War Held at Dartmoor during the War of 1812

## Alphabetical listing of names

HMT Salvador del Mundo - Born: Newport - Age: 36 - Released on 21 Jun 1815.

Mullett, Joseph - Seaman - Nbr: 5194 - Prize: Enterprize - Ship Type: P - How taken: HMS Tenedos - When taken: 21 May 1813 - Where taken: off Cape Cod - Date Received: 31 Oct 1814 - From what ship: HMT Mermaid, Chatham - Born: Marblehead - Age: 19 - Released on 29 Jun 1815.

Mullin, Robert - Seaman - Nbr: 3650 - Prize: Ulysses - Ship Type: MV - How taken: HMS Majestic - When taken: 29 Jun 1813 - Where taken: off Western Islands (England) - Date Received: 30 Sep 1814 - From what ship: HMT President - Born: Albany - Age: 25 - Released on 4 Jun 1815.

Mullins, James - Seaman - Nbr: 1384 - Prize: Courier - Ship Type: P - How taken: HMS Rover - When taken: 14 May 1813 - Where taken: Bay of Biscay - Date Received: 19 Jun 1814 - From what ship: Stapleton - Born: North Carolina - Age: 40 - Released on 28 Apr 1815.

Mumford, Thomas - 2nd Mate - Nbr: 390 - Prize: John & Frances - Ship Type: MV - How taken: HMS Belle Poule - When taken: 13 Mar 1813 - Where taken: Bay of Biscay - Date Received: 1 Jul 1813 - From what ship: Plymouth - Born: Newport - Age: 41 - Released on 20 Apr 1815.

Munro, Harry - Seaman - Nbr: 5250 - Prize: King George - Ship Type: Prize - How taken: HMS Recruit - When taken: 29 Aug 1814 - Where taken: off Cape Sable Island (Canada) - Date Received: 31 Oct 1814 - From what ship: HMT Mermaid, Chatham - Born: Rhode Island - Age: 19 - Race: Negro - Released on 29 Jun 1815.

Munsey, Daniel - Seaman - Nbr: 4920 - Prize: Bordeaux Packet - Ship Type: LM - How taken: HMS Niemen - When taken: 28 Jan 1814 - Where taken: off Delaware - Date Received: 28 Oct 1814 - From what ship: HMT Alkbar, Halifax - Born: Cumberland County - Age: 26 - Released on 21 Jun 1815.

Munster, Isaac - Seaman - Nbr: 2804 - How taken: Gave himself up from HMS York - Date Received: 24 Aug 1814 - From what ship: HMT Liverpool, Chatham - Born: Marblehead - Age: 24 - Race: Black - Released on 19 May 1815.

Murphy, Samuel - Seaman - Nbr: 6010 - Prize: McDonough - Ship Type: P - How taken: HMS Bacchante - When taken: 1 Nov 1814 - Where taken: Lat 42 Long 67 - Date Received: 28 Dec 1814 - From what ship: HMT Penelope - Born: Massachusetts - Age: 26 - Released on 3 Jul 1815.

Murphy, William - Seaman - Nbr: 1171 - Prize: Harvest - Ship Type: P - How taken: HMS Orestes - When taken: 21 Jan 1814 - Where taken: Bay of Biscay - Date Received: 10 May 1814 - From what ship: Plymouth - Born: Philadelphia - Age: 26 - Released on 28 Apr 1815.

Murray, Alexander - Seaman - Nbr: 2395 - Prize: US Sloop Frolic - Ship Type: MW - How taken: HMS Orpheus - When taken: 20 Apr 1814 - Where taken: off Cuba - Date Received: 16 Aug 1814 - From what ship: HMT Queen, Halifax - Born: Virginia - Age: 30 - Sent to Dartmouth on 19 Oct 1814.

Murray, Colton - Seaman - Nbr: 4987 - Prize: Invincible - Ship Type: LM - How taken: HMS Armide - When taken: 15 Aug 1814 - Where taken: off Nantucket - Date Received: 28 Oct 1814 - From what ship: HMT Alkbar, Halifax - Born: Portland - Age: 23 - Released on 21 Jun 1815.

Murray, David - Seaman - Nbr: 4310 - Prize: Vivid - Ship Type: MV - How taken: HMS Nymphe - When taken: 20 Apr 1813 - Where taken: off Cape Cod - Date Received: 7 Oct 1814 - From what ship: HMT Niobe, Chatham - Born: Boston - Age: 20 - Released on 4 Jun 1815.

Murray, George - Seaman - Nbr: 3746 - Prize: Mariner - Ship Type: Prize - How taken: HMS Poictiers - When taken: 30 Aug 1813 - Where taken: off Newfoundland - Date Received: 30 Sep 1814 - From what ship: HMT President, Halifax - Born: Bordeaux - Age: 36 - Released on 4 Jun 1815.

Murray, Jacob - Seaman - Nbr: 1591 - Prize: Eliza - Ship Type: MV - How taken: HMS Surveillante - When taken: 27 Mar 1813 - Where taken: Bay of Biscay - Date Received: 23 Jun 1814 - From what ship: Stapleton - Born: SC - Age: 26 - Race: Colored - Released on 1 May 1815.

Murray, James - Seaman - Nbr: 676 - Prize: Messenger - Ship Type: MV - How taken: HMS Iris - When taken: 10 Mar 1813 - Where taken: Bay of Biscay - Date Received: 8 Sep 1813 - From what ship: Plymouth - Born: Kent County - Age: 24 - Died on 17 Oct 1813 from pneumonia.

Murray, James - Seaman - Nbr: 2874 - How taken: Gave himself up from HMS Scorpion - When taken: 27 May 1813 - Date Received: 24 Aug 1814 - From what ship: HMT Alpheus, Chatham - Born: Salem - Age: 38 - Released on 19 May 1815.

Murray, John - Seaman - Nbr: 6102 - How taken: Gave himself up from HMS Tigre - When taken: 3 Nov 1814 - Date Received: 6 Jan 1814 - What ship: HMT Impregnable - Born: New York - Age: 21 - Released on 5 Jul 1815.

Murray, M. - Boy - Nbr: 2978 - Prize: Frolic - Ship Type: P - How taken: HMS Heron - When taken: 25 Jan 1814 - Where taken: off St. Thomas - Date Received: 29 Aug 1814 - From what ship: HMT Bittern - Born: Santiago

American Prisoners of War Held at Dartmoor during the War of 1812

## Alphabetical listing of names

- Age: 19 - Race: Black - Released on 19 May 1815.
Murray, Nathaniel - Seaman - Nbr: 1216 - Prize: James - Ship Type: MV - How taken: HMS Harpy - When taken: 19 Dec 1812 - Where taken: off Isle of France - Date Received: 14 Jun 1814 - From what ship: Mill Prison (Plymouth, England) - Born: Lancaster - Age: 23 - Race: Black - Released on 26 Apr 1815.
Murray, Oliver - Seaman - Nbr: 5631 - Prize: Mary - Ship Type: Prize - How taken: Not legible - When taken: 18 Jun 1814 - Date Received: 24 Dec 1814 - From what ship: HMT Tay - Born: Massachusetts - Age: 21 - Released on 5 Jul 1815.
Murray, Peter - Seaman - Nbr: 2898 - How taken: Gave himself up from HMS Ocean - When taken: 28 May 1813 - Date Received: 24 Aug 1814 - From what ship: HMT Alpheus, Chatham - Born: Harlem - Age: 28 - Released on 19 May 1815.
Murray, Richard - Steward - Nbr: 5186 - Prize: Montgomery - Ship Type: P - How taken: HMS Nymphe - When taken: 1 May 1813 - Where taken: off Cape Cod - Date Received: 31 Oct 1814 - From what ship: HMT Mermaid, Chatham - Born: Salem - Age: 16 - Released on 29 Jun 1815.
Murray, Richard - Seaman - Nbr: 4209 - Prize: Commodore Perry - Ship Type: Schooner - How taken: Sent into custody from a cutter - When taken: 25 Feb 1814 - Where taken: off Bordeaux - Date Received: 7 Oct 1814 - From what ship: HMT Niobe, Chatham - Born: Maryland - Age: 40 - Race: Black - Released on 13 Jun 1815.
Murray, William - Seaman - Nbr: 3036 - Prize: Commerce - How taken: HMS Superb - When taken: 23 Jun 1814 - Where taken: off Sandy Hook - Date Received: 2 Sep 1814 - From what ship: HMT Sultan - Born: Philadelphia - Age: 18 - Released on 28 May 1815.
Murrell, Mark - Cook - Nbr: 1667 - Prize: Shadow - Ship Type: LM - How taken: HMS Reindeer - When taken: 6 Apr 1813 - Where taken: Bay of Biscay - Date Received: 26 Jun 1814 - From what ship: Exeter - Born: Marblehead - Age: 29 - Released on 1 May 1815.
Muss, Joseph - Seaman - Nbr: 5638 - Prize: Lion - Ship Type: P - How taken: HMS Granicus - When taken: 2 Dec 1814 - Where taken: off Lisbon - Date Received: 24 Dec 1814 - From what ship: HMT Hanover, Gibraltar - Born: L'Orient (France) - Age: 34 - Released on 3 Jul 1815.
Mutch, James - Seaman - Nbr: 2886 - How taken: Gave himself up from HMS Shearwater - When taken: 28 May 1813 - Date Received: 24 Aug 1814 - From what ship: HMT Alpheus, Chatham - Born: Philadelphia - Age: 35 - Released on 19 May 1815.
Myers, Daniel - Seaman - Nbr: 1063 - Prize: Fair American - Ship Type: MV - How taken: HMS Andromache - When taken: 19 Jan 1814 - Where taken: Bay of Biscay - Date Received: 10 May 1814 - From what ship: Plymouth - Born: Philadelphia - Age: 24 - Released on 27 Apr 1815.
Myers, Edward - Seaman - Nbr: 2292 - Prize: Hussar - Ship Type: P - How taken: HMS Saturn - When taken: 25 May 1814 - Where taken: off Sandy Hook - Date Received: 16 Aug 1814 - From what ship: HMS Dublin, Halifax - Born: Maryland - Age: 22 - Released on 3 May 1815.
Myers, George - Seaman - Nbr: 1189 - Prize: Vice Admiral Martin - Ship Type: P - How taken: HMS Fortune - When taken: 8 Oct 1811 - Date Received: 4 Jun 1814 - From what ship: Dartmouth - Born: Long Island - Age: 30 - Sent to Mill Prison (Plymouth, England) on 17 Jun 1814.
Myers, John - Seaman - Nbr: 3857 - Prize: Leicester - Ship Type: Prize - How taken: HMS Prometheus - When taken: Jun 1814 - Where taken: off Halifax - Date Received: 5 Oct 1814 - From what ship: HMT Orpheus, Halifax - Born: Boston - Age: 21 - Race: Negro - Released on 9 Jun 1815.
Myers, John - Seaman - Nbr: 1584 - Prize: Price - Ship Type: MV - How taken: HMS Pyramus - When taken: 6 Apr 1813 - Where taken: Bay of Biscay - Date Received: 23 Jun 1814 - From what ship: Stapleton - Born: New York - Age: 19 - Released on 1 May 1815.
Myers, Joseph - Seaman - Nbr: 5033 - Prize: Landrail - Ship Type: Prize - How taken: HMS Wasp - When taken: 27 Jul 1814 - Where taken: Georges Bank - Date Received: 28 Oct 1814 - From what ship: HMT Alkbar, Halifax - Born: Pennsylvania - Age: 23 - Released on 21 Jun 1815.
Naborne, Nicholas - Seaman - Nbr: 3123 - Prize: Prize of the Blockage - Ship Type: Prize - How taken: HMS Brazen - When taken: 29 Jun 1813 - Where taken: off Scotland - Date Received: 11 Sep 1814 - From what ship: HMT Freya, Chatham - Born: Rhode Island - Age: 17 - Released on 28 May 1815.
Nagle, George - Seaman - Nbr: 576 - Prize: US Brig Argus - Ship Type: MW - How taken: HMS Pelican - When taken: 14 Apr 1813 - Where taken: Irish Channel - Date Received: 8 Sep 1813 - From what ship: Plymouth - Born: Maryland - Age: 32 - Sent to Dartmouth on 2 Nov 1814.
Nanson, Henry - Seaman - Nbr: 960 - Prize: Siro - Ship Type: LM - How taken: HMS Pelican - When taken: 13 Jan 1814 - Where taken: at sea - Date Received: 31 Jan 1814 - From what ship: Plymouth - Born: Boston - Age: 42 - Sent to Dartmouth on 2 Nov 1814.

American Prisoners of War Held at Dartmoor during the War of 1812

## Alphabetical listing of names

Nanson, Peter - Seaman - Nbr: 961 - Prize: Siro - Ship Type: LM - How taken: HMS Pelican - When taken: 13 Jan 1814 - Where taken: at sea - Date Received: 31 Jan 1814 - From what ship: Plymouth - Born: Gothenburg - Age: 27 - Released on 6 May 1814.

Nash, Alexander - Seaman - Nbr: 6349 - Prize: Prince de Neufchatel - Ship Type: P - How taken: Leander (Newcastle Acasta) - When taken: 20 Dec 1814 - Where taken: Lat 38 Long 56 - Date Received: 19 Feb 1815 - From what ship: HMT Ganges, Plymouth - Born: Boston - Age: 36 - Released on 6 Jul 1815.

Nash, Daniel - Seaman - Nbr: 3485 - How taken: Gave himself up from HMS Prince - When taken: 14 Sep 1814 - Date Received: 19 Sep 1814 - From what ship: HMT Salvador del Mundo - Born: Dorset - Age: 31 - Died on 14 Feb 1815 from variola.

Nash, Manuel - Seaman - Nbr: 608 - Prize: US Brig Argus - Ship Type: MW - How taken: HMS Pelican - When taken: 14 Aug 1813 - Where taken: Irish Channel - Date Received: 8 Sep 1813 - From what ship: Plymouth - Born: New Orleans - Age: 18 - Sent to Dartmouth on 2 Nov 1814.

Nasson, Joseph - Boy - Nbr: 2806 - Prize: Leo - Ship Type: LM - How taken: HMS Magicienne - When taken: 4 Jun 1813 - Where taken: off France - Date Received: 24 Aug 1814 - From what ship: HMT Liverpool, Chatham - Born: Limerick - Age: 18 - Released on 19 May 1815.

Navar, Francis - Seaman - Nbr: 3117 - Prize: Harmony - Ship Type: Prize - How taken: HMS Brisk - When taken: 24 Aug 1814 - Where taken: Bristol Channel - Date Received: 11 Sep 1814 - From what ship: HMT Salvador del Mundo - Born: Bordeaux - Age: 38 - Released on 28 May 1815.

Neal, Daniel - Seaman - Nbr: 993 - Prize: Apparencen - Ship Type: MV - How taken: HMS Castilian - When taken: 27 Jan 1814 - Where taken: off Ushant (France) - Date Received: 31 Jan 1814 - From what ship: Plymouth - Born: Portsmouth - Age: 21 - Released on 27 Apr 1815.

Neal, John - Gunner - Nbr: 939 - Prize: Siro - Ship Type: LM - How taken: HMS Pelican - When taken: 13 Jan 1814 - Where taken: at sea - Date Received: 31 Jan 1814 - From what ship: Plymouth - Born: Baltimore - Age: 36 - Released on 27 Apr 1815.

Neale, Dennis - Seaman - Nbr: 1407 - Prize: Courier - Ship Type: P - How taken: HMS Rover - When taken: 14 May 1813 - Where taken: Bay of Biscay - Date Received: 19 Jun 1814 - From what ship: Stapleton - Born: Maryland - Age: 25 - Released on 28 Apr 1815.

Ned, Deaf - Seaman - Nbr: 4226 - Prize: Liberty - Ship Type: MV - How taken: Surrendered - When taken: 30 Dec 1813 - Where taken: Stromess - Date Received: 7 Oct 1814 - From what ship: HMT Niobe, Chatham - Born: Norfolk - Age: 28 - Released on 13 Jun 1815.

Neel, David - Seaman - Nbr: 141 - Prize: Criterion - Ship Type: MV - How taken: HMS Belle Poule - When taken: 14 Feb 1813 - Where taken: Bay of Biscay - Date Received: 2 Apr 1813 - From what ship: Plymouth - Born: Philadelphia - Age: 27 - Released on 20 Apr 1815.

Neel, William - Seaman - Nbr: 1138 - Prize: Mary - Ship Type: P - How taken: HMS Crocodile - When taken: 6 Aug 1813 - Where taken: off Corsuma - Date Received: 10 May 1814 - From what ship: Plymouth - Born: Fredericktown - Age: 23 - Released on 28 Apr 1815.

Neil, David A. - Clerk - Nbr: 3911 - Prize: Diomede - Ship Type: Prize - How taken: HMS Maidstone - When taken: 20 Mar 1814 - Where taken: off NY - Date Received: 5 Oct 1814 - From what ship: HMT President, Halifax - Born: Salem - Age: 21 - Released on 2 Apr 1815.

Neil, John - Seaman - Nbr: 3066 - How taken: Gave himself up from HMS Saturn - When taken: 3 Dec 1813 - Date Received: 2 Sep 1814 - From what ship: HMT Hydra, Chatham - Born: Gloucester - Age: 30 - Released on 29 Jun 1815.

Nelly, Richard J. - Seaman - Nbr: 1625 - Prize: Tom - Ship Type: LM - How taken: HMS Surveillante - When taken: 24 Apr 1813 - Where taken: Bay of Biscay - Date Received: 23 Jun 1814 - From what ship: Stapleton - Born: Pennsylvania - Age: 34 - Released on 1 May 1815.

Nelson, David - Seaman - Nbr: 6195 - Prize: Prince de Neufchatel - Ship Type: P - How taken: Leander (Newcastle Acasta) - When taken: 20 Dec 1814 - Where taken: Lat 38 Long 56 - Date Received: 30 Jan 1815 - From what ship: HMT Pheasant - Born: Boston - Age: 24 - Released on 5 Jul 1815.

Nelson, Richard - Seaman - Nbr: 1838 - How taken: Gave himself up from HMS Ajax - When taken: 2 May 1813 - Date Received: 21 Jul 1814 - From what ship: HMT Redbeard & Pincher, Chatham Depot - Born: New York - Age: 28 - Released on 1 May 1815.

Nelson, Thomas - Seaman - Nbr: 5173 - Prize: Vivid - Ship Type: P - How taken: HMS Nymphe - When taken: 20 Apr 1813 - Where taken: off Cape Cod - Date Received: 31 Oct 1814 - From what ship: HMT Mermaid, Chatham - Born: Truro - Age: 17 - Released on 29 Jun 1815.

Nelson, Thomas - Seaman - Nbr: 858 - Prize: Amiable - Ship Type: LM - How taken: HMS Magnificent - When

American Prisoners of War Held at Dartmoor during the War of 1812

## Alphabetical listing of names

taken: 30 Oct 1813 - Where taken: off Bordeaux - Date Received: 4 Dec 1813 - From what ship: Plymouth - Born: Delaware - Age: 18 - Released on 26 Apr 1815.

Nelson, William - Seaman - Nbr: 5858 - Prize: Lion - Ship Type: P - How taken: HMS Granicus - When taken: 2 Dec 1814 - Where taken: off Lisbon - Date Received: 26 Dec 1814 - From what ship: HMT Impregnable - Born: Pennsylvania - Age: 34 - Released on 3 Jul 1815.

Neptune, Daniel - Seaman - Nbr: 2249 - Prize: Carbineer - Ship Type: Prize - How taken: HMS Ringdove - When taken: 24 Apr 1814 - Where taken: off Bermuda - Date Received: 16 Aug 1814 - From what ship: HMS Dublin, Halifax - Born: New Jersey - Age: 24 - Released on 3 May 1815.

Nesbit, Richard - Seaman - Nbr: 1156 - Prize: Dorothea - Ship Type: MV - How taken: HMS Teazer - When taken: 26 Mar 1814 - Where taken: off Cork - Date Received: 10 May 1814 - From what ship: Plymouth - Born: Philadelphia - Age: 27 - Released on 28 Apr 1815.

Neumann, Gustavus - Seaman - Nbr: 5048 - Prize: Ida - Ship Type: LM - How taken: HMS Newcastle - When taken: 9 Aug 1814 - Where taken: Long 34 - Date Received: 28 Oct 1814 - From what ship: HMT Alkbar, Halifax - Born: Finland - Age: 26 - Released on 29 Jun 1815.

Newall, J. B. L. - Marine - Nbr: 6329 - Prize: Prince de Neufchatel - Ship Type: P - How taken: Leander (Newcastle Acasta) - When taken: 20 Dec 1814 - Where taken: Lat 38 Long 56 - Date Received: 19 Feb 1815 - From what ship: HMT Ganges, Plymouth - Born: Lynn - Age: 20 - Released on 5 Jul 1815.

Newall, John - Seaman - Nbr: 6236 - Prize: Prince de Neufchatel - Ship Type: P - How taken: Leander (Newcastle Acasta) - When taken: 20 Dec 1814 - Where taken: Lat 38 Long 56 - Date Received: 30 Jan 1815 - From what ship: HMT Pheasant - Born: Reading - Age: 21 - Released on 5 Jul 1815.

Newberry, John - Seaman - Nbr: 2992 - Prize: Frolic - Ship Type: P - How taken: HMS Heron - When taken: 25 Jan 1814 - Where taken: off St. Thomas - Date Received: 29 Aug 1814 - From what ship: HMT Bittern - Born: Philadelphia - Age: 23 - Race: Black - Released on 28 May 1815.

Newby, Lambert - Seaman - Nbr: 693 - How taken: Impressed at Liverpool - When taken: 28 Aug 1813 - Date Received: 27 Sep 1813 - From what ship: Plymouth - Born: New York - Age: 40 - Released on 26 Apr 1815.

Newcomb, John - Seaman - Nbr: 6232 - Prize: Prince de Neufchatel - Ship Type: P - How taken: Leander (Newcastle Acasta) - When taken: 20 Dec 1814 - Where taken: Lat 38 Long 56 - Date Received: 30 Jan 1815 - From what ship: HMT Pheasant - Born: Hampden - Age: 19 - Released on 5 Jul 1815.

Newcomb, Tilton - Lieutenant - Nbr: 4877 - Prize: Saratoga - Ship Type: P - How taken: HMS Barracouta - When taken: 9 Oct 1814 - Where taken: off Western Islands (England) - Date Received: 24 Oct 1814 - From what ship: HMT Salvador del Mundo - Born: Maryland - Age: 31 - Released on 21 Jun 1815.

Newell, Charles - Seaman - Nbr: 4854 - Prize: North Star - Ship Type: MV - How taken: HMS Crane - When taken: 10 Jun 1814 - Where taken: off St. Bartholomew - Date Received: 9 Oct 1814 - From what ship: HMT Freya, Halifax - Born: Boston - Age: 43 - Race: Black - Released on 11 Jul 1815.

Newell, George - Seaman - Nbr: 2008 - Prize: Dictator - Ship Type: MV - How taken: HMS Derwent - When taken: 7 May 1813 - Where taken: off Nantes - Date Received: 3 Aug 1814 - From what ship: HMS Lyffey, Chatham Depot - Born: Newburyport - Age: 23 - Released on 2 May 1815.

Newell, Isaac - Seaman - Nbr: 1339 - Prize: Paul Jones - Ship Type: P - How taken: HMS Leonidas - When taken: 23 May 1813 - Where taken: Channel - Date Received: 19 Jun 1814 - From what ship: Stapleton - Born: Massachusetts - Age: 30 - Released on 28 Apr 1815.

Newell, John - Cook - Nbr: 22 - How taken: Apprehended at Gibraltar - When taken: 8 Aug 1813 - Date Received: 2 Apr 1813 - From what ship: Plymouth - Born: Africa - Age: 32 - Race: Negro - Released on 20 Apr 1815.

Newell, Joseph - Seaman - Nbr: 6221 - Prize: Prince de Neufchatel - Ship Type: P - How taken: Leander (Newcastle Acasta) - When taken: 20 Dec 1814 - Where taken: Lat 38 Long 56 - Date Received: 30 Jan 1815 - From what ship: HMT Pheasant - Born: Reading - Age: 18 - Released on 5 Jul 1815.

Newell, Joseph - Seaman - Nbr: 6204 - Prize: Prince de Neufchatel - Ship Type: P - How taken: Leander (Newcastle Acasta) - When taken: 20 Dec 1814 - Where taken: Lat 38 Long 56 - Date Received: 30 Jan 1815 - From what ship: HMT Pheasant - Born: Boston - Age: 17 - Released on 5 Jul 1815.

Newell, Nathaniel - Carpenter - Nbr: 867 - Prize: Amiable - Ship Type: LM - How taken: HMS Magnificent - When taken: 30 Oct 1813 - Where taken: off Bordeaux - Date Received: 4 Dec 1813 - From what ship: Plymouth - Born: Providence - Age: 30 - Released on 27 Apr 1815.

Newell, Paul - Seaman - Nbr: 3436 - How taken: Gave himself up from HMS Ashea - When taken: 18 Nov 1813 - Date Received: 13 Sep 1814 - From what ship: HMT Niobe, Chatham - Born: Marblehead - Age: 22 - Released on 28 May 1815.

Newman, Daniel - Seaman - Nbr: 1254 - Prize: Adeline - Ship Type: MV - How taken: HMS Magicienne - When

American Prisoners of War Held at Dartmoor during the War of 1812

## Alphabetical listing of names

taken: 16 Mar 1814 - Where taken: off Cape Finisterre - Date Received: 14 Jun 1814 - From what ship: Mill Prison (Plymouth, England) - Born: Newbury - Age: 26 - Released on 28 Apr 1815.

Newman, Henry - Seaman - Nbr: 3444 - How taken: Gave himself up from HMS Clorinda - When taken: 24 Aug 1813 - Date Received: 13 Sep 1814 - From what ship: HMT Niobe, Chatham - Born: Washington - Age: 30 - Released on 28 May 1815.

Newman, John - Seaman - Nbr: 6111 - Prize: Garland - Ship Type: Prize - How taken: HMS Hamadryad - When taken: 22 Nov 1814 - Where taken: Georges Bank - Date Received: 6 Jan 1814 - From what ship: HMT Impregnable - Born: Cape Ann - Age: 25 - Released on 5 Jul 1815.

Newman, John - Seaman - Nbr: 5740 - Prize: US Schooner Somers - Ship Type: MW - How taken: British gunboats - When taken: 12 Aug 1814 - Where taken: Fort Erie - Date Received: 26 Dec 1814 - From what ship: HMT Argo - Born: New Jersey - Age: 39 - Released on 3 Jul 1815.

Newons, Francis - Seaman - Nbr: 3851 - Prize: Dominique - Ship Type: LM - How taken: HMS Dotterel - When taken: 21 May 1814 - Where taken: off Charleston - Date Received: 5 Oct 1814 - From what ship: HMT Orpheus, Halifax - Born: Maderia - Age: 23 - Released on 9 Jun 1815.

Newport, Mathew - Quartermaster - Nbr: 4008 - Prize: US Brig Rattlesnake - Ship Type: MW - How taken: HMS Leander - When taken: 13 Jul 1814 - Where taken: off Shelburne - Date Received: 6 Oct 1814 - From what ship: HMT Chesapeake, Halifax - Born: Manchester - Age: 29 - Released on 9 Jun 1815.

Newton, John - Pilot - Nbr: 4211 - Prize: Commodore Perry - Ship Type: Schooner - How taken: Sent into custody from a cutter - When taken: 25 Feb 1814 - Where taken: off Bordeaux - Date Received: 7 Oct 1814 - From what ship: HMT Niobe, Chatham - Born: New Jersey - Age: 25 - Released on 13 Jun 1815.

Neyren, John - Seaman - Nbr: 1185 - Prize: Melanie - Ship Type: MV - How taken: HMS Briton - When taken: 15 Sep 1813 - Where taken: off Bordeaux - Date Received: 4 Jun 1814 - From what ship: Dartmouth - Born: New Orleans - Age: 25 - Race: Mulatto - Sent to Plymouth on 8 Jul 1814.

Nichelle, Pierre - Passenger - Nbr: 6273 - Prize: Lion - Ship Type: P - How taken: HMS Granicus - When taken: 2 Dec 1814 - Where taken: off Lisbon - Date Received: 17 Feb 1815 - From what ship: HMT Ganges, Plymouth - Born: L'Orient (France) - Age: 34 - Escaped on 1 Jun 1815.

Nickerson, Joseph - Seaman - Nbr: 1418 - Prize: Orders in Council - Ship Type: LM - How taken: HMS Surveillante - When taken: 1 Jun 1813 - Where taken: off Cape Ortegal (Spain) - Date Received: 19 Jun 1814 - From what ship: Stapleton - Born: Boston - Age: 19 - Released on 28 Apr 1815.

Nicholas, Jacob - Seaman - Nbr: 6533 - Prize: US Brig Syren - Ship Type: MW - How taken: HMS Medway - When taken: 12 Jul 1814 - Where taken: off Cape of Good Hope - Date Received: 7 Mar 1815 - From what ship: HMT Ganges, Plymouth - Born: Not legible - Released on 11 Jul 1815.

Nicholas, John - Seaman - Nbr: 2106 - How taken: Gave himself up from HMS Ocean - When taken: 1 Aug 1814 - Date Received: 3 Aug 1814 - What ship: Plymouth - Born: Philadelphia - Age: 25 - Released on 2 May 1815.

Nicholas, John Buess - Seaman - Nbr: 869 - Prize: Dart - Ship Type: MV - How taken: HMS Niger - When taken: 12 Nov 1813 - Where taken: off Cape Finisterre - Date Received: 31 Jan 1814 - From what ship: Plymouth - Born: Massachusetts - Age: 24 - Released on 27 Apr 1815.

Nicholls, John - Seaman - Nbr: 1074 - How taken: Impressed at Liverpool - When taken: 2 Feb 1814 - Date Received: 10 May 1814 - What ship: Plymouth - Born: New Orleans - Age: 23 - Released on 27 Apr 1815.

Nicholls, John - Seaman - Nbr: 2783 - Prize: Tiger - Ship Type: MV - How taken: HMS Scylla - When taken: 22 May 1813 - Where taken: Bay of Biscay - Date Received: 24 Aug 1814 - From what ship: HMT Liverpool, Chatham - Born: New York - Age: 18 - Released on 19 May 1815.

Nicholls, Thomas - Seaman - Nbr: 3155 - Prize: Fame (Whaler) - Ship Type: MV - How taken: HMS Cressy - When taken: 20 Jul 1813 - Where taken: at sea - Date Received: 11 Sep 1814 - From what ship: HMT Freya, Chatham - Born: Rhode Island - Age: 25 - Race: Black - Released on 11 Jul 1815.

Nicholls, William - Seaman - Nbr: 1854 - Prize: Jason - Ship Type: MV - How taken: HMS Venerable - When taken: 1 Jun 1814 - Where taken: off Tenerife - Date Received: 21 Jul 1814 - From what ship: HMT Redbeard, Chatham Depot - Born: Nashville - Age: 19 - Released on 2 May 1815.

Nichols, John - Seaman - Nbr: 4222 - Prize: Liberty - Ship Type: MV - How taken: Surrendered - When taken: 30 Dec 1813 - Where taken: Stromess - Date Received: 7 Oct 1814 - From what ship: HMT Niobe, Chatham - Born: Plymouth - Age: 23 - Released on 13 Jun 1815.

Nichols, John - Seaman - Nbr: 6066 - Prize: Chasseur - Ship Type: P - How taken: HMS Recruit - When taken: 6 Apr 1814 - Where taken: Lat 29 Long 76 - Date Received: 28 Dec 1814 - From what ship: HMT Penelope - Born: Boston - Age: 23 - Race: Black - Released on 3 Jul 1815.

Nichols, John - Seaman - Nbr: 2146 - How taken: Gave himself up from HMS Aboukir - When taken: 28 Oct 1812 -

American Prisoners of War Held at Dartmoor during the War of 1812

## Alphabetical listing of names

Date Received: 8 Aug 1814 - From what ship: HMT Raven, Chatham - Born: Durham - Age: 22 - Released on 10 Apr 1815.

Nichols, Silas - Seaman - Nbr: 3992 - Prize: US Brig Rattlesnake - Ship Type: MW - How taken: HMS Leander - When taken: 13 Jul 1814 - Where taken: off Shelburne - Date Received: 6 Oct 1814 - From what ship: HMT Chesapeake, Halifax - Born: Connecticut - Age: 35 - Released on 9 Jun 1815.

Nichols, William - Seaman - Nbr: 4858 - Prize: Fairy - Ship Type: MV - How taken: HMS Hardy - When taken: 28 Mar 1814 - Where taken: West Indies - Date Received: 9 Oct 1814 - From what ship: HMT Freya, Halifax - Born: Philadelphia - Age: 19 - Released on 21 Jun 1815.

Nicholson, James - Seaman - Nbr: 4443 - Prize: Growler - Ship Type: P - How taken: HMS Electra - When taken: 7 Jul 1814 - Where taken: at sea - Date Received: 8 Oct 1814 - From what ship: HMT Leyden, Chatham - Born: Maryland - Age: 21 - Released on 14 Jun 1815.

Nicholson, James - Seaman - Nbr: 762 - How taken: Impressed at Liverpool - Date Received: 3 Nov 1813 - From what ship: Plymouth - Born: Jersey - Age: 24 - Released on 26 Apr 1815.

Nicholson, John - Seaman - Nbr: 4656 - Prize: Growler - Ship Type: P - How taken: HMS Electra - When taken: 7 Jul 1813 - Where taken: at sea - Date Received: 9 Oct 1814 - From what ship: HMT Leyden, Chatham - Born: Marblehead - Age: 23 - Released on 15 Jun 1815.

Nicholson, Thomas - Seaman - Nbr: 4172 - Prize: Sea Nymphe - Ship Type: MV - How taken: HMS Thraster - When taken: 4 Mar 1813 - Where taken: North Sea - Date Received: 7 Oct 1814 - From what ship: HMT Niobe, Chatham - Born: Prussia - Age: 39 - Released on 20 Dec 1814.

Nickerson, James - 2nd Mate - Nbr: 772 - How taken: Sent into custody from a Russian MV, prize to the French frigate Weser - Date Received: 3 Nov 1813 - From what ship: Plymouth - Born: Chatham - Age: 22 - Released on 26 Apr 1815.

Nickerson, Warren - Seaman - Nbr: 5711 - Prize: Regent - Ship Type: LM - How taken: HMS Forth - When taken: 19 Sep 1814 - Where taken: off Egg Harbor (New Jersey) - Date Received: 24 Dec 1814 - From what ship: HMT Penelope - Born: Chatham - Age: 23 - Released on 3 Jul 1815.

Niel, Henry - Seaman - Nbr: 705 - Prize: Ned - Ship Type: LM - How taken: HMS Royalist - When taken: 6 Sep 1813 - Where taken: Bay of Biscay - Date Received: 27 Sep 1813 - From what ship: Plymouth - Born: New York - Age: 25 - Sent to Plymouth on 7 Dec 1813.

Niles, George - Carpenter - Nbr: 945 - Prize: Siro - Ship Type: LM - How taken: HMS Pelican - When taken: 13 Jan 1814 - Where taken: at sea - Date Received: 31 Jan 1814 - From what ship: Plymouth - Born: Boston - Age: 27 - Released on 27 Apr 1815.

Nisbett, James - Seaman - Nbr: 4934 - Prize: Herald - Ship Type: P - How taken: HMS Endymion - When taken: 15 Aug 1814 - Where taken: off Nantucket - Date Received: 28 Oct 1814 - From what ship: HMT Alkbar, Halifax - Born: New York - Age: 37 - Released on 21 Jun 1815.

Nivers, James - Seaman - Nbr: 2482 - Prize: US Sloop Frolic - Ship Type: MW - How taken: HMS Orpheus - When taken: 20 Apr 1814 - Where taken: off Cuba - Date Received: 16 Aug 1814 - From what ship: HMT Queen, Halifax - Born: Massachusetts - Age: 21 - Released on 3 May 1815.

Nixon, Charles - Seaman - Nbr: 2928 - How taken: Gave himself up from HMS Bombay - Date Received: 24 Aug 1814 - From what ship: HMT Alpheus, Chatham - Born: Boston - Age: 43 - Race: Black - Released on 19 May 1815.

Nixon, Primos - Seaman - Nbr: 3504 - How taken: Impressed at Liverpool - When taken: 11 Sep 1814 - Date Received: 24 Sep 1814 - From what ship: HMT Salvador del Mundo - Born: Philadelphia - Age: 34 - Race: Mulatto - Released on 4 Jun 1815.

Noble, Charles - Seaman - Nbr: 3342 - How taken: Gave himself up from HMS Scorpion - When taken: 2 Dec 1812 - Date Received: 13 Sep 1814 - From what ship: HMT Niobe, Chatham - Born: Cape Ann - Age: 22 - Released on 27 Apr 1815.

Noble, Daniel - Seaman - Nbr: 5401 - How taken: Gave himself up from MV Martha - When taken: 2 Apr 1814 - Date Received: 31 Oct 1814 - From what ship: HMT Leyden, Chatham - Born: Calais - Age: 24 - Released on 1 Jul 1815.

Noble, Joseph - Seaman - Nbr: 6006 - Prize: McDonough - Ship Type: P - How taken: HMS Bacchante - When taken: 1 Nov 1814 - Where taken: Lat 42 Long 67 - Date Received: 28 Dec 1814 - From what ship: HMT Penelope - Born: Wales - Age: 22 - Released on 3 Jul 1815.

Noel, Stephen C. - Seaman - Nbr: 3243 - Prize: Hepsey - Ship Type: MV - How taken: HMS Tenedos - When taken: 22 Jun 1813 - Where taken: off Lisbon - Date Received: 11 Sep 1814 - From what ship: HMT Freya, Chatham - Born: Newbury - Age: 23 - Released on 28 May 1815.

American Prisoners of War Held at Dartmoor during the War of 1812

## Alphabetical listing of names

Nolen, William - Seaman - Nbr: 5866 - Prize: Lion - Ship Type: P - How taken: HMS Granicus - When taken: 2 Dec 1814 - Where taken: off Lisbon - Date Received: 26 Dec 1814 - From what ship: HMT Impregnable - Born: Delaware - Age: 38 - Released on 3 Jul 1815.

Nolton, John - Seaman - Nbr: 1962 - How taken: Impressed at Sunderland - When taken: 6 Mar 1814 - Date Received: 3 Aug 1814 - From what ship: HMS Lyffey, Chatham Depot - Born: Boston - Age: 22 - Released on 2 May 1815.

Noney, Peter - Boy - Nbr: 638 - Prize: US Brig Argus - Ship Type: MW - How taken: HMS Pelican - When taken: 14 Aug 1813 - Where taken: Irish Channel - Date Received: 8 Sep 1813 - From what ship: Plymouth - Born: Bordeaux - Age: 16 - Impersonated a Frenchman on the general release.

Nooney, William - Seaman - Nbr: 1118 - Prize: Bunker Hill - Ship Type: P - How taken: HMS Pomone & HMS Cadmus - When taken: 4 Mar 1814 - Where taken: Bay of Biscay - Date Received: 10 May 1814 - From what ship: Plymouth - Born: Philadelphia - Age: 25 - Released on 20 Mar 1815.

Norbury, Joseph - Seaman - Nbr: 2248 - Prize: Carbineer - Ship Type: Prize - How taken: HMS Ringdove - When taken: 24 Apr 1814 - Where taken: off Bermuda - Date Received: 16 Aug 1814 - From what ship: HMS Dublin, Halifax - Born: Pennsylvania - Age: 24 - Released on 3 May 1815.

Norcross, Abel - Seaman - Nbr: 1892 - How taken: Impressed at London - When taken: 11 Feb 1814 - Date Received: 29 Jul 1814 - From what ship: HMS Ville de Paris, Chatham Depot - Born: Connecticut - Age: 23 - Race: Black - Released on 2 May 1815.

Norcross, Archibald - Seaman - Nbr: 4401 - Prize: Portsmouth Packet - Ship Type: Prize - How taken: HMS Fantome - When taken: 5 Oct 1813 - Where taken: off Portland - Date Received: 8 Oct 1814 - From what ship: HMT Leyden, Chatham - Born: Massachusetts - Age: 35 - Released on 14 Jun 1815.

Norcross, Phillipe - Seaman - Nbr: 1056 - How taken: Impressed at Cork - Date Received: 10 May 1814 - From what ship: Plymouth - Born: Boston - Age: 27 - Released on 27 Apr 1815.

Norcross, Thomas - Seaman - Nbr: 4258 - Prize: Argus - Ship Type: MV - How taken: HMS San Domingo - When taken: 1 Mar 1814 - Where taken: off Savannah - Date Received: 7 Oct 1814 - From what ship: HMT Niobe, Chatham - Born: Boston - Age: 19 - Released on 13 Jun 1815.

Norman, Michael - Seaman - Nbr: 1310 - How taken: Sent into custody from HMS Cornwallis - Date Received: 14 Jun 1814 - From what ship: Mill Prison (Plymouth, England) - Born: Baltimore - Age: 25 - Released on 28 Apr 1815.

Norman, Peter - Seaman - Nbr: 227 - Prize: Charlotte - Ship Type: MV - How taken: HMS Warspite - When taken: 3 Mar 1813 - Where taken: Bay of Biscay - Date Received: 2 Apr 1813 - From what ship: Plymouth - Born: Falmouth - Age: 36 - Released on 6 May 1814.

Norman, William - Seaman - Nbr: 6517 - How taken: Gave himself up from HMS Orontes - When taken: Feb 1815 - Date Received: 3 Mar 1815 - From what ship: HMT Ganges, Plymouth - Born: Boston - Age: 26 - Released on 13 Jun 1815.

Norris, August G. - Seaman - Nbr: 1453 - How taken: Impressed at Liverpool - When taken: 1 May 1813 - Date Received: 19 Jun 1814 - From what ship: Stapleton - Born: Rhode Island - Age: 25 - Released on 28 Apr 1815.

Norris, Benjamin - Seaman - Nbr: 6405 - Prize: Nancy - Ship Type: P - How taken: Papillion - When taken: 27 Dec 1814 - Where taken: Lat 29 Long 20 - Date Received: 3 Mar 1815 - From what ship: HMT Ganges, Plymouth - Born: Not legible - Age: 30 - Released on 11 Jul 1815.

Norris, Henry - Seaman - Nbr: 4949 - Prize: Herald - Ship Type: P - How taken: HMS Endymion - When taken: 15 Aug 1814 - Where taken: off Nantucket - Date Received: 28 Oct 1814 - From what ship: HMT Alkbar, Halifax - Born: Connecticut - Age: 21 - Released on 21 Jun 1815.

Norris, Philip - Seaman - Nbr: 4141 - Prize: Bordeaux Packet - Ship Type: LM - How taken: HMS Niemen - When taken: 28 Jun 1814 - Where taken: off Delaware - Date Received: 6 Oct 1814 - From what ship: HMT Chesapeake, Halifax - Born: New York - Age: 23 - Released on 13 Jun 1815.

Norris, Robert - Seaman - Nbr: 714 - Prize: Ned - Ship Type: LM - How taken: HMS Royalist - When taken: 6 Sep 1813 - Where taken: Bay of Biscay - Date Received: 27 Sep 1813 - From what ship: Plymouth - Born: New York - Age: 17 - Released on 26 Apr 1815.

North, Thomas - Seaman - Nbr: 1653 - How taken: Apprehended at Bristol - When taken: 9 Nov 1813 - Date Received: 23 Jun 1814 - From what ship: Stapleton - Born: Maryland - Age: 39 - Released on 1 May 1815.

Northcote, George - Seaman - Nbr: 2666 - How taken: Impressed at London - When taken: 27 Aug 1813 - Date Received: 21 Aug 1814 - From what ship: HMT Freya, Chatham - Born: Berkshire - Age: 20 - Released on 19 May 1815.

American Prisoners of War Held at Dartmoor during the War of 1812

## Alphabetical listing of names

Northey, Joseph - Seaman - Nbr: 4722 - How taken: Gave himself up from HMS Hussar - When taken: 17 Jul 1813 - Date Received: 9 Oct 1814 - From what ship: HMT Freya, Chatham - Born: Massachusetts - Age: 25 - Released on 15 Jun 1815.

Northey, Joseph - Seaman - Nbr: 3659 - Prize: Industry - Ship Type: P - How taken: HMS Arab - When taken: Sep 1813 - Where taken: off Halifax - Date Received: 30 Sep 1814 - From what ship: HMT President - Born: Marblehead - Age: 19 - Released on 4 Jun 1815.

Norton, Andrew - Seaman - Nbr: 5832 - Prize: US Schooner Tigress - Ship Type: MW - How taken: British gunboats - When taken: 2 Sep 1814 - Where taken: Lake Erie - Date Received: 26 Dec 1814 - From what ship: HMT Argo - Born: Virginia - Age: 22 - Race: Black - Released on 3 Jul 1815.

Norton, Edward - Seaman - Nbr: 2495 - Prize: US Sloop Frolic - Ship Type: MW - How taken: HMS Orpheus - When taken: 20 Apr 1814 - Where taken: off Cuba - Date Received: 16 Aug 1814 - From what ship: HMT Queen, Halifax - Born: Massachusetts - Age: 21 - Died on 29 Sep 1814.

Norton, George - Seaman - Nbr: 296 - Prize: Ducornau - Ship Type: MV - How taken: HMS Pheasant - When taken: 15 Mar 1813 - Where taken: Bay of Biscay - Date Received: 28 Jun 1813 - From what ship: Plymouth - Born: New Bedford - Age: 44 - Released on 20 Apr 1815.

Norton, Joseph - Seaman - Nbr: 3403 - Prize: Thomas - Ship Type: P - How taken: HMS Nymphe - When taken: 24 Jun 1813 - Where taken: off Halifax - Date Received: 13 Sep 1814 - From what ship: HMT Niobe, Chatham - Born: Massachusetts - Age: 17 - Released on 28 May 1815.

Norton, Richard - Seaman - Nbr: 2667 - How taken: Impressed at London - When taken: 21 Aug 1813 - Date Received: 21 Aug 1814 - From what ship: HMT Freya, Chatham - Born: Massachusetts - Age: 27 - Released on 19 May 1815.

Norton, Solomon - Seaman - Nbr: 4458 - Prize: Juliana Smith - Ship Type: P - How taken: HMS Nymphe - When taken: 11 May 1813 - Where taken: off Cape Sable Island (Canada) - Date Received: 8 Oct 1814 - From what ship: HMT Leyden, Chatham - Born: Massachusetts - Age: 42 - Released on 15 Jun 1815.

Nowland, Andrew - Seaman - Nbr: 3704 - Prize: Alfred - Ship Type: P - How taken: HMS Epervier - When taken: 23 Feb 1814 - Where taken: off Newfoundland - Date Received: 30 Sep 1814 - From what ship: HMT President, Halifax - Born: Marblehead - Age: 17 - Released on 4 Jun 1815.

Nowland, Edward - Gunner's Mate - Nbr: 4973 - Prize: Invincible - Ship Type: LM - How taken: HMS Armide - When taken: 15 Aug 1814 - Where taken: off Nantucket - Date Received: 28 Oct 1814 - From what ship: HMT Alkbar, Halifax - Born: Barnstable - Age: 37 - Released on 21 Jun 1815.

Nowland, Thomas - Seaman - Nbr: 6208 - Prize: Prince de Neufchatel - Ship Type: P - How taken: Leander (Newcastle Acasta) - When taken: 20 Dec 1814 - Where taken: Lat 38 Long 56 - Date Received: 30 Jan 1815 - From what ship: HMT Pheasant - Born: Marblehead - Age: 22 - Released on 5 Jul 1815.

Noyes, Charles - Seaman - Nbr: 616 - Prize: Betsy - Ship Type: MV - How taken: HMS Leonidas - When taken: 12 Aug 1813 - Where taken: Channel - Date Received: 8 Sep 1813 - From what ship: Plymouth - Born: Andover - Age: 28 - Sent to Dartmouth on 2 Nov 1814.

Nugent, John - Seaman - Nbr: 1725 - Prize: US Brig Argus - Ship Type: MW - How taken: HMS Pelican - When taken: 14 Aug 1813 - Where taken: Irish Channel - Date Received: 8 Jul 1814 - From what ship: Mill Prison (Plymouth, England) - Born: Philadelphia - Age: 27 - Sent to Dartmouth on 2 Nov 1814.

Nugent, John - Seaman - Nbr: 778 - Prize: Betsy - Ship Type: MV - How taken: HMS Eurotas - When taken: 26 Oct 1813 - Where taken: off Ushant (France) - Date Received: 3 Nov 1813 - From what ship: Plymouth - Born: New York - Age: 18 - Sent to Plymouth on 7 Dec 1813.

Nunns, William - Seaman - Nbr: 1907 - How taken: Gave himself up from HMS Royal William - When taken: 25 Jan 1813 - Date Received: 3 Aug 1814 - From what ship: HMS Alceste, Chatham Depot - Born: Philadelphia - Age: 34 - Released on 2 May 1815.

Nye, Cornelius - Seaman - Nbr: 959 - Prize: Siro - Ship Type: LM - How taken: HMS Pelican - When taken: 13 Jan 1814 - Where taken: at sea - Date Received: 31 Jan 1814 - From what ship: Plymouth - Born: Boston - Age: 21 - Released on 27 Apr 1815.

Nye, Stephen - Seaman - Nbr: 2388 - Prize: US Sloop Frolic - Ship Type: MW - How taken: HMS Orpheus - When taken: 20 Apr 1814 - Where taken: off Cuba - Date Received: 16 Aug 1814 - From what ship: HMT Queen, Halifax - Born: Massachusetts - Age: 23 - Sent to Dartmouth on 19 Oct 1814.

Nye, William - Seaman - Nbr: 4738 - How taken: British Army - When taken: 20 Mar 1814 - Where taken: off Bordeaux - Date Received: 9 Oct 1814 - From what ship: HMT Freya, Chatham - Born: Boston - Age: 20 - Released on 15 Jun 1815.

Nye, William - Seaman - Nbr: 2525 - Prize: Hussar - Ship Type: P - How taken: HMS Saturn - When taken: 25 May

American Prisoners of War Held at Dartmoor during the War of 1812

## Alphabetical listing of names

1814 - Where taken: off Sandy Hook - Date Received: 16 Aug 1814 - From what ship: HMT Queen, Halifax - Born: Holyoake - Age: 21 - Released on 3 May 1815.

Oakes, George - Seaman - Nbr: 4501 - Prize: Enterprize - Ship Type: P - How taken: HMS Tenedos - When taken: 21 May 1813 - Where taken: off Cape Cod - Date Received: 8 Oct 1814 - From what ship: HMT Leyden, Chatham - Born: Massachusetts - Age: 19 - Released on 11 Jul 1815.

Oakley, Jacob - Seaman - Nbr: 746 - How taken: Gave himself up from HMS Amide - Date Received: 3 Nov 1813 - From what ship: Plymouth - Born: New York - Age: 25 - Released on 26 Apr 1815.

Oakley, Joseph - Seaman - Nbr: 839 - Prize: Chesapeake - Ship Type: LM - How taken: HMS Hotspur & HMS Pyramus - When taken: 26 Oct 1813 - Where taken: off Nantes - Date Received: 29 Nov 1813 - From what ship: Plymouth - Born: Delaware - Age: 27 - Released on 26 Apr 1815.

Obrey, Mathew - Seaman - Nbr: 6313 - Prize: Prince de Neufchatel - Ship Type: P - How taken: Leander (Newcastle Acasta) - When taken: 20 Dec 1814 - Where taken: Lat 38 Long 56 - Date Received: 19 Feb 1815 - From what ship: HMT Ganges, Plymouth - Born: Calais - Age: 25 - Released on 5 Jul 1815.

Odeen, John - Seaman - Nbr: 28 - How taken: Apprehended at Gibraltar - When taken: 8 Aug 1813 - Date Received: 2 Apr 1813 - From what ship: Plymouth - Born: Alexandria - Age: 21 - Sent to Dartmouth on 30 Jul 1813.

Odgen, James - Seaman - Nbr: 750 - How taken: Impressed at Liverpool - Date Received: 3 Nov 1813 - From what ship: Plymouth - Born: Eastport - Age: 27 - Released on 20 Apr 1815.

Odiam, Joseph H. - Seaman - Nbr: 4439 - Prize: Growler - Ship Type: P - How taken: HMS Electra - When taken: 7 Jul 1814 - Where taken: at sea - Date Received: 8 Oct 1814 - From what ship: HMT Leyden, Chatham - Born: New Hampshire - Age: 21 - Released on 14 Jun 1815.

Odiorne, John - Captain - Nbr: 3592 - Prize: Frolic - Ship Type: P - How taken: HMS Heron - When taken: 25 Jan 1814 - Where taken: off St. Thomas - Date Received: 30 Sep 1814 - From what ship: HMT Sybella - Born: Salisbury - Age: 30 - Released on 28 Apr 1815.

Ogle, Charles - Seaman - Nbr: 5452 - Prize: Prize to the Lawrence - Ship Type: P - How taken: HMS Glasgow - When taken: 2 Nov 1814 - Where taken: Channel - Date Received: 10 Dec 1814 - From what ship: HMT Impregnable - Born: North Carolina - Age: 20 - Race: Black - Released on 1 Jul 1815.

O'Hara, Terrence - Marine - Nbr: 4060 - Prize: US Brig Rattlesnake - Ship Type: MW - How taken: HMS Leander - When taken: 13 Jul 1814 - Where taken: off Shelburne - Date Received: 6 Oct 1814 - From what ship: HMT Chesapeake, Halifax - Born: Harlem - Age: 33 - Released on 13 Jun 1815.

Oliver, Anthony - Seaman - Nbr: 5430 - Prize: York Town - Ship Type: P - How taken: HMS Maidstone - When taken: 18 Jul 1813 - Where taken: Grand Banks - Date Received: 31 Oct 1814 - From what ship: HMT Leyden, Chatham - Born: Baltimore - Age: 47 - Race: Negro - Released on 1 Jul 1815.

Oliver, Griffith - Seaman - Nbr: 5800 - Prize: US Frigate Superior (Gig) - Ship Type: MW - How taken: British gunboats - When taken: 26 Aug 1814 - Where taken: Lake Ontario - Date Received: 26 Dec 1814 - From what ship: HMT Argo - Born: Virginia - Age: 21 - Released on 12 Jun 1815.

Oliver, John - Seaman - Nbr: 3895 - Prize: Governor Shelby - Ship Type: MV - How taken: HMS Saturn - When taken: 9 Jul 1814 - Where taken: off Long Island - Date Received: 5 Oct 1814 - From what ship: HMT Orpheus, Halifax - Born: Virginia - Age: 26 - Released on 9 Jun 1815.

Oliver, Joseph - Seaman - Nbr: 4591 - Prize: Bunker Hill - Ship Type: P - How taken: HMS Pomone - When taken: 2 Mar 1814 - Where taken: Channel - Date Received: 9 Oct 1814 - From what ship: HMT Leyden, Chatham - Born: Portugal - Age: 33 - Released on 15 Jun 1815.

Oliver, Mathew - Seaman - Nbr: 6251 - How taken: Sent into custody from MV Philaer - When taken: 25 Nov 1814 - Date Received: 4 Feb 1815 - From what ship: HMT Ganges - Born: Bath - Age: 26 - Released on 3 Jul 1815.

Oliver, Samuel C. - Seaman - Nbr: 2329 - Prize: Hussar - Ship Type: P - How taken: HMS Saturn - When taken: 25 May 1814 - Where taken: off Sandy Hook - Date Received: 16 Aug 1814 - From what ship: HMS Dublin, Halifax - Born: Milford - Age: 21 - Released on 3 May 1815.

Oliver, Stephen - Boy - Nbr: 2295 - Prize: Hussar - Ship Type: P - How taken: HMS Saturn - When taken: 25 May 1814 - Where taken: off Sandy Hook - Date Received: 16 Aug 1814 - From what ship: HMS Dublin, Halifax - Born: New York - Age: 16 - Released on 3 May 1815.

Olliver, John - Seaman - Nbr: 4077 - Prize: New Zealander - Ship Type: Prize - How taken: HMS Belviders - When taken: 21 Apr 1814 - Where taken: off Delaware - Date Received: 6 Oct 1814 - From what ship: HMT Chesapeake, Halifax - Born: Massachusetts - Age: 19 - Released on 13 Jun 1815.

Olseen, Elias - Seaman - Nbr: 791 - Prize: Avon - Ship Type: MV - How taken: HMS Eurotas - When taken: 27 Oct 1813 - Where taken: off Ushant (France) - Date Received: 3 Nov 1813 - From what ship: Plymouth - Born:

American Prisoners of War Held at Dartmoor during the War of 1812

## Alphabetical listing of names

Sweden - Age: 23 - Released on 6 May 1814.

O'Neil, Joseph - Seaman - Nbr: 6196 - Prize: Prince de Neufchatel - Ship Type: P - How taken: Leander (Newcastle Acasta) - When taken: 20 Dec 1814 - Where taken: Lat 38 Long 56 - Date Received: 30 Jan 1815 - From what ship: HMT Pheasant - Born: Methuen - Age: 23 - Released on 5 Jul 1815.

Orcroft, Lewis - Seaman - Nbr: 5503 - Prize: US Scorpion (Commodore Barney) - Ship Type: MW - How taken: British forces - When taken: 22 Aug 1814 - Where taken: Chesapeake Bay - Date Received: 17 Dec 1814 - From what ship: HMT Loire, Halifax - Born: Marblehead - Age: 34 - Released on 29 Jun 1815.

Orkitt, Horea - Seaman - Nbr: 6233 - Prize: Prince de Neufchatel - Ship Type: P - How taken: Leander (Newcastle Acasta) - When taken: 20 Dec 1814 - Where taken: Lat 38 Long 56 - Date Received: 30 Jan 1815 - From what ship: HMT Pheasant - Born: Massachusetts - Age: 19 - Released on 5 Jul 1815.

Orne, Israel - Seaman - Nbr: 3380 - Prize: Growler - Ship Type: P - How taken: HMS Electra - When taken: 7 Jul 1813 - Where taken: at sea - Date Received: 13 Sep 1814 - From what ship: HMT Niobe, Chatham - Born: Salem - Age: 25 - Released on 28 May 1815.

Orne, W. B. - 3rd Lieutenant - Nbr: 1106 - Prize: Bunker Hill - Ship Type: P - How taken: HMS Pomone & HMS Cadmus - When taken: 4 Mar 1814 - Where taken: Bay of Biscay - Date Received: 10 May 1814 - From what ship: Plymouth - Born: Marblehead - Age: 29 - Released on 27 Apr 1815.

Orwick, Thomas C. - Prize Master - Nbr: 1213 - Prize: Diamond - Ship Type: MV - How taken: HMS Vengeur - When taken: 6 Mar 1814 - Where taken: Lat 47.4 Long 5.60 - Date Received: 14 Jun 1814 - From what ship: Mill Prison (Plymouth, England) - Born: Portland - Age: 35 - Released on 26 Apr 1815.

Osborn, Stephen - Seaman - Nbr: 1883 - Prize: Margaret - Ship Type: Prize - How taken: HMS Martin - When taken: 9 Oct 1813 - Where taken: at sea - Date Received: 29 Jul 1814 - From what ship: HMS Ville de Paris, Chatham Depot - Born: Pembroke - Age: 21 - Released on 2 May 1815.

Osborne, Henry - Seaman - Nbr: 2419 - Prize: US Sloop Frolic - Ship Type: MW - How taken: HMS Orpheus - When taken: 20 Apr 1814 - Where taken: off Cuba - Date Received: 16 Aug 1814 - From what ship: HMT Queen, Halifax - Born: Beverly - Age: 22 - Released on 3 May 1815.

Osborne, John - Boy - Nbr: 3572 - Prize: Hawk - Ship Type: P - How taken: HMS Pique - When taken: 26 Apr 1814 - Where taken: off Bermuda - Date Received: 30 Sep 1814 - From what ship: HMT Sybella - Born: Washington - Age: 15 - Released on 4 Jun 1815.

Osborne, John L. - Seaman - Nbr: 6406 - Prize: Nancy - Ship Type: P - How taken: Papillion - When taken: 27 Dec 1814 - Where taken: Lat 39' 40" - Date Received: 3 Mar 1815 - From what ship: HMT Ganges, Plymouth - Born: Newburyport - Age: 18 - Died on 24 May 1815 from variola.

Osborne, Louis - Seaman - Nbr: 3358 - How taken: Gave himself up from HMS Scorpion - When taken: 10 Dec 1812 - Date Received: 13 Sep 1814 - From what ship: HMT Niobe, Chatham - Born: New York - Age: 30 - Released on 27 Apr 1815.

Osborne, Samuel - Seaman - Nbr: 4587 - Prize: Bunker Hill - Ship Type: P - How taken: HMS Pomone - When taken: 2 Mar 1814 - Where taken: Channel - Date Received: 9 Oct 1814 - From what ship: HMT Leyden, Chatham - Born: Sussex - Age: 30 - Race: Black - Released on 15 Jun 1815.

Osbourne, Archibald - Seaman - Nbr: 3049 - How taken: Gave himself up from HMS Pembroke - When taken: 16 Aug 1814 - Date Received: 2 Sep 1814 - From what ship: HMT Sultan - Born: Albany - Age: 26 - Released on 28 Apr 1815.

Osgood, David - Seaman - Nbr: 2584 - How taken: Impressed at Cowes - Date Received: 21 Aug 1814 - From what ship: HMT Freya, Chatham - Born: Baltimore - Age: 26 - Released on 3 May 1815.

O'Sheffey, Jacobus - Seaman - Nbr: 4241 - Prize: Bunker Hill - Ship Type: MV - How taken: HMS Pomone - When taken: 4 Mar 1814 - Where taken: Channel - Date Received: 7 Oct 1814 - From what ship: HMT Niobe, Chatham - Born: Danzig - Age: 42 - Released on 13 Jun 1815.

Ostand, Michael - Seaman - Nbr: 1378 - Prize: Grand Napoleon - Ship Type: P - How taken: HMS Goldfinch - When taken: 17 Apr 1813 - Where taken: Bay of Biscay - Date Received: 19 Jun 1814 - From what ship: Stapleton - Born: Paris - Age: 32 - Sent to Mill Prison (Plymouth, England) on 21 Jun 1814.

Osten, John - Seaman - Nbr: 109 - How taken: Gave himself up from HMS Foxhound - Date Received: 2 Apr 1813 - From what ship: Plymouth - Born: Boston - Age: 23 - Released on 20 Apr 1815.

Osten, John - Boy - Nbr: 5487 - Prize: General Putnam - Ship Type: P - How taken: HMS Leander - When taken: 8 Nov 1814 - Where taken: Long 65 Lat 42 - Date Received: 17 Dec 1814 - From what ship: HMT Loire, Halifax - Born: Bordeaux - Age: 29 - Released on 1 Jul 1815.

Ostman, David - Seaman - Nbr: 3159 - Prize: Fame (Whaler) - Ship Type: MV - How taken: HMS Cressy - When taken: 20 Jul 1813 - Where taken: at sea - Date Received: 11 Sep 1814 - From what ship: HMT Freya,

American Prisoners of War Held at Dartmoor during the War of 1812

## Alphabetical listing of names

Chatham - Born: Connecticut - Age: 20 - Released on 28 May 1815.
Otiel, Charles - Seaman - Nbr: 5742 - Prize: US Schooner Somers - Ship Type: MW - How taken: British gunboats - When taken: 12 Aug 1814 - Where taken: Fort Erie - Date Received: 26 Dec 1814 - From what ship: HMT Argo - Born: Albany - Age: 34 - Released on 3 Jul 1815.
Owen, Burden - Seaman - Nbr: 3345 - How taken: Gave himself up from HMS Ocean - When taken: 29 Oct 1812 - Date Received: 13 Sep 1814 - From what ship: HMT Niobe, Chatham - Born: New York - Age: 47 - Released on 27 Apr 1815.
Owen, Zachariah - Seaman - Nbr: 4647 - Prize: Prize to the Hunter - Ship Type: P - How taken: Not legible - When taken: 14 Jan 1813 - Where taken: off Halifax - Date Received: 9 Oct 1814 - From what ship: HMT Leyden, Chatham - Born: Massachusetts - Age: 19 - Released on 15 Jun 1815.
Oytiel, John - Seaman - Nbr: 1105 - Prize: Diamond - Ship Type: P - How taken: HMS Vengeur - When taken: 6 Mar 1814 - Where taken: Lat 47.4 Long 5.60 - Date Received: 10 May 1814 - From what ship: Plymouth - Born: Brest - Age: 60 - Sent to Mill Prison (Plymouth, England) on 21 Jun 1814.
Pacher, John - Seaman - Nbr: 5857 - Prize: Lion - Ship Type: P - How taken: HMS Granicus - When taken: 2 Dec 1814 - Where taken: off Lisbon - Date Received: 26 Dec 1814 - From what ship: HMT Impregnable - Born: Connecticut - Age: 25 - Released on 3 Jul 1815.
Pack, Abraham - Seaman - Nbr: 91 - Prize: Cashier - Ship Type: MV - How taken: HMS Reindeer - When taken: 3 Feb 1813 - Where taken: Bay of Biscay - Date Received: 2 Apr 1813 - From what ship: Plymouth - Born: Harford - Age: 26 - Race: Negro - Sent to Dartmouth on 30 Jul 1813.
Packett, John - Seaman - Nbr: 5828 - Prize: US Schooner Tigress - Ship Type: MW - How taken: British gunboats - When taken: 2 Sep 1814 - Where taken: Lake Erie - Date Received: 26 Dec 1814 - From what ship: HMT Argo - Born: L'Orient (France) - Age: 22 - Released on 3 Jul 1815.
Packhouse, William - Seaman - Nbr: 3111 - How taken: Apprehended at Bristol - When taken: 23 May 1814 - Date Received: 11 Sep 1814 - From what ship: HMT Salvador del Mundo - Born: Long Island - Age: 22 - Race: Mulatto - Released on 28 May 1815.
Page, Cato - Seaman - Nbr: 2958 - How taken: Picked up in a boat from the Fox - Date Received: 24 Aug 1814 - From what ship: HMT Hannibal - Born: Baltimore - Age: 23 - Race: Black - Released on 19 May 1815.
Page, John - Seaman - Nbr: 513 - Prize: Revenge - Ship Type: LM - How taken: HMS Belle Poule - When taken: 10 May 1813 - Where taken: off Cornwall - Date Received: 8 Sep 1813 - From what ship: Plymouth - Born: Massachusetts - Age: 25 - Released on 26 Apr 1815.
Page, Joseph - Seaman - Nbr: 3889 - Prize: Sally - Ship Type: MV - How taken: HMS Acasta - When taken: 6 Jun 1814 - Where taken: off VA - Date Received: 5 Oct 1814 - From what ship: HMT Orpheus, Halifax - Born: New York - Age: 23 - Sent to Ashburton (England) on 4 Jan 1815.
Page, Reuben - Master's Mate - Nbr: 2277 - Prize: Diomede - Ship Type: P - How taken: HMS Rifleman - When taken: 23 Jun 1814 - Where taken: off Halifax - Date Received: 16 Aug 1814 - From what ship: HMS Dublin, Halifax - Born: Bridgton - Age: 23 - Released on 3 May 1815.
Page, Thomas - Seaman - Nbr: 1479 - Prize: Hebe - Ship Type: MV - How taken: HMS Stag - When taken: 18 Apr 1813 - Where taken: Bay of Biscay - Date Received: 19 Jun 1814 - From what ship: Stapleton - Born: Massachusetts - Age: 20 - Released on 1 May 1815.
Pagechin, Henry - Seaman - Nbr: 905 - Prize: Squirrel - Ship Type: MV - How taken: HMS Belle Poule - When taken: 14 Dec 1813 - Where taken: at sea - Date Received: 31 Jan 1814 - From what ship: Plymouth - Born: Bremen - Age: 21 - Sent to Mill Prison (Plymouth, England) on 17 Jun 1814.
Paine, Charles - Seaman - Nbr: 3612 - Prize: Hornsby - Ship Type: P - How taken: HMS Sceptre - When taken: 4 Feb 1814 - Where taken: off Block Island - Date Received: 30 Sep 1814 - From what ship: HMT Sybella - Born: New York - Age: 19 - Released on 4 Jun 1815.
Paine, George - Seaman - Nbr: 957 - Prize: Siro - Ship Type: LM - How taken: HMS Pelican - When taken: 13 Jan 1814 - Where taken: at sea - Date Received: 31 Jan 1814 - From what ship: Plymouth - Born: Boston - Age: 24 - Released on 27 Apr 1815.
Paine, Joseph - Seaman - Nbr: 428 - Prize: Viper - Ship Type: MV - How taken: HMS Superb - When taken: 15 Apr 1813 - Where taken: Bay of Biscay - Date Received: 1 Jul 1813 - From what ship: Plymouth - Born: New Orleans - Age: 33 - Released on 20 Apr 1815.
Paine, Mark - Seaman - Nbr: 3040 - Prize: Commerce - How taken: HMS Superb - When taken: 23 Jun 1814 - Where taken: off Sandy Hook - Date Received: 2 Sep 1814 - From what ship: HMT Sultan - Born: Philadelphia - Age: 20 - Race: Creole - Released on 28 May 1815.
Paine, R. B. - Seaman - Nbr: 1990 - How taken: Sent into custody from West Indiaman - Date Received: 3 Aug

American Prisoners of War Held at Dartmoor during the War of 1812

## Alphabetical listing of names

1814 - From what ship: HMS Lyffey, Chatham Depot - Born: New York - Age: 31 - Released on 2 May 1815.

Palfrey, John - Seaman - Nbr: 4328 - Prize: Martha - Ship Type: Prize - How taken: HMS Belviders - When taken: 25 Feb 1814 - Where taken: off New London - Date Received: 7 Oct 1814 - From what ship: HMT Salvador del Mundo, Halifax - Born: Salem - Age: 19 - Released on 14 Jun 1815.

Palmer, Benjamin - Seaman - Nbr: 3944 - Prize: Rolla - Ship Type: P - How taken: HMS Loire - When taken: 10 Dec 1813 - Where taken: off Bull Island (South Carolina) - Date Received: 5 Oct 1814 - From what ship: HMT President, Halifax - Born: Connecticut - Age: 21 - Released on 9 Jun 1815.

Palmer, George H. - Seaman - Nbr: 3372 - How taken: Impressed at Hull - When taken: 3 Oct 1813 - Date Received: 13 Sep 1814 - From what ship: HMT Niobe, Chatham - Born: Massachusetts - Age: 24 - Released on 28 May 1815.

Palmer, John - Seaman - Nbr: 836 - Prize: Chesapeake - Ship Type: LM - How taken: HMS Hotspur & HMS Pyramus - When taken: 26 Oct 1813 - Where taken: off Nantes - Date Received: 29 Nov 1813 - From what ship: Plymouth - Born: Maryland - Age: 37 - Released on 26 Apr 1815.

Palmer, John - Seaman - Nbr: 284 - Prize: Cannoneer - Ship Type: MV - How taken: HMS Warspite - When taken: 14 Mar 1813 - Where taken: Bay of Biscay - Date Received: 28 Jun 1813 - From what ship: Plymouth - Born: Wilmington - Age: 27 - Released on 20 Apr 1815.

Palmer, Joseph - Seaman - Nbr: 5054 - Prize: Ida - Ship Type: LM - How taken: HMS Newcastle - When taken: 9 Aug 1814 - Where taken: Long 34 - Date Received: 28 Oct 1814 - From what ship: HMT Alkbar, Halifax - Born: Portsmouth - Age: 18 - Race: Negro - Released on 29 Jun 1815.

Palmer, Pero - Seaman - Nbr: 665 - Prize: Joel Barlow - Ship Type: LM - How taken: HMS Briton - When taken: 3 Jul 1813 - Where taken: off Bordeaux - Date Received: 8 Sep 1813 - From what ship: Plymouth - Born: Connecticut - Age: 22 - Released on 26 Apr 1815.

Palmer, Peter - Seaman - Nbr: 6170 - Prize: Lion - Ship Type: P - How taken: HMS Granicus - When taken: 2 Dec 1814 - Where taken: off Lisbon - Date Received: 18 Jan 1815 - From what ship: HMT Impregnable - Born: Genoa - Age: 25 - Released on 5 Jul 1815.

Palmer, Peter - Seaman - Nbr: 3769 - Prize: Fame - Ship Type: P - How taken: HMS Thistle - When taken: 10 Apr 1814 - Where taken: after being cast ashore on a seal island - Date Received: 30 Sep 1814 - From what ship: HMT President, Halifax - Born: Salem - Age: 16 - Released on 9 Jun 1815.

Palmer, Peter - Seaman - Nbr: 2599 - Prize: Tom Thumb - Ship Type: MV - How taken: Lyon (Privateer) - When taken: 15 Feb 1813 - Where taken: Bay of Biscay - Date Received: 21 Aug 1814 - From what ship: HMT Freya, Chatham - Born: Bunt ford - Age: 22 - Released on 3 May 1815.

Palmer, Robert - Seaman - Nbr: 3825 - Prize: Bordeaux Packet - Ship Type: LM - How taken: HMS Niemen - When taken: 28 Jan 1814 - Where taken: off Delaware - Date Received: 5 Oct 1814 - From what ship: HMT Orpheus, Halifax - Born: New Jersey - Age: 21 - Released on 9 Jun 1815.

Palmer, Thomas - Seaman - Nbr: 5829 - Prize: US Schooner Tigress - Ship Type: MW - How taken: British gunboats - When taken: 2 Sep 1814 - Where taken: Lake Erie - Date Received: 26 Dec 1814 - From what ship: HMT Argo - Born: New York - Age: 22 - Race: Black - Released on 3 Jul 1815.

Palmer, William - Seaman - Nbr: 2092 - How taken: Gave himself up from HMS Horatio - When taken: 11 Jun 1813 - Date Received: 3 Aug 1814 - From what ship: HMS Bittern, Chatham Depot - Born: Portsmouth - Age: 21 - Released on 2 May 1815.

Palma, Vincent - Seaman - Nbr: 4821 - Prize: President - Ship Type: P - How taken: HMS Pique - When taken: 7 May 1814 - Where taken: off Porto Rico - Date Received: 9 Oct 1814 - From what ship: HMT Freya, Halifax - Born: Cartagena - Age: 14 - Race: Black - Released on 21 Jun 1815.

Pane, William - Seaman - Nbr: 2570 - Prize: Rattlesnake - Ship Type: P - How taken: HMS Hyperion - When taken: 25 Jun 1814 - Where taken: off Cape Finisterre - Date Received: 21 Aug 1814 - From what ship: HMS Hyperion - Born: Philadelphia - Age: 44 - Released on 3 May 1815.

Pinion, Henry - Cook - Nbr: 3569 - Prize: Hawk - Ship Type: P - How taken: HMS Pique - When taken: 26 Apr 1814 - Where taken: off Bermuda - Date Received: 30 Sep 1814 - From what ship: HMT Sybella - Born: Maryland - Age: 34 - Race: Negro - Released on 4 Jun 1815.

Pardett, Charles - Seaman - Nbr: 2846 - How taken: Gave himself up from HMS Ocean - When taken: 15 Aug 1813 - Date Received: 24 Aug 1814 - From what ship: HMT Alpheus, Chatham - Born: New Orleans - Age: 28 - Released on 10 Mar 1815.

Parnell, Benjamin - Seaman - Nbr: 1661 - Prize: Miranda - Ship Type: Prize - How taken: HMS Unicorn - When taken: 21 May 1813 - Where taken: off Ushant (France) - Date Received: 23 Jun 1814 - From what ship:

American Prisoners of War Held at Dartmoor during the War of 1812

## Alphabetical listing of names

Stapleton - Born: Connecticut - Age: 29 - Released on 1 May 1815.

Paris, John - Seaman - Nbr: 3080 - How taken: Gave himself up from HMS Tigre - When taken: 1 Sep 1813 - Date Received: 3 Sep 1814 - From what ship: HMT Hydra, Chatham - Born: North Yarmouth - Age: 28 - Released on 28 May 1815.

Paris, Peter - Seaman - Nbr: 4721 - How taken: Gave himself up from HMS Hussar - When taken: 17 Jul 1813 - Date Received: 9 Oct 1814 - From what ship: HMT Freya, Chatham - Born: Portland - Age: 23 - Released on 15 Jun 1815.

Paris, William - Seaman - Nbr: 623 - Prize: Betsy - Ship Type: MV - How taken: HMS Leonidas - When taken: 12 Aug 1813 - Where taken: Channel - Date Received: 8 Sep 1813 - From what ship: Plymouth - Born: South Hanistead - Age: 21 - Sent to Dartmouth on 2 Nov 1814.

Parish, Samuel - Seaman - Nbr: 1507 - Prize: Grand Napoleon - Ship Type: P - How taken: HMS Goldfinch - When taken: 17 Apr 1813 - Where taken: Bay of Biscay - Date Received: 23 Jun 1814 - From what ship: Stapleton - Born: Norfolk - Age: 31 - Died on 1 Apr 1815 from phthisis pulmonalis.

Parker, Colby - Seaman - Nbr: 2180 - Prize: Rattlesnake - Ship Type: P - How taken: HMS Hyperion - When taken: 26 Jun 1814 - Where taken: off Cape Finisterre - Date Received: 16 Aug 1814 - From what ship: HMS Dublin, Halifax - Born: Philadelphia - Age: 30 - Released on 2 May 1815.

Parker, Edward - Seaman - Nbr: 1312 - How taken: Sent into custody from HMS Cornwallis - Date Received: 14 Jun 1814 - From what ship: Mill Prison (Plymouth, England) - Born: Elizabethtown - Age: 25 - Released on 28 Apr 1815.

Parker, George - Seaman - Nbr: 212 - Prize: Mars - Ship Type: MV - How taken: HMS Warspite - When taken: 26 Feb 1813 - Where taken: off Basque Roads (France) - Date Received: 2 Apr 1813 - From what ship: Plymouth - Born: New York - Age: 25 - Released on 20 Apr 1815.

Parker, George - Seaman - Nbr: 5143 - How taken: Gave himself up from HMS Menelaus - Date Received: 31 Oct 1814 - From what ship: HMT Mermaid, Chatham - Born: Cambridge - Age: 26 - Released on 19 May 1815.

Parker, J. A. - Seaman - Nbr: 1524 - Prize: Essex - Ship Type: MV - How taken: HMS Pyramus - When taken: 20 Apr 1813 - Where taken: Bay of Biscay - Date Received: 23 Jun 1814 - From what ship: Stapleton - Born: Boston - Age: 16 - Released on 1 May 1815.

Parker, John - Seaman - Nbr: 2470 - Prize: US Sloop Frolic - Ship Type: MW - How taken: HMS Orpheus - When taken: 20 Apr 1814 - Where taken: off Cuba - Date Received: 16 Aug 1814 - From what ship: HMT Queen, Halifax - Born: Boston - Age: 23 - Released on 3 May 1815.

Parker, Peter - Seaman - Nbr: 1133 - Prize: Vengeance - Ship Type: MV - How taken: Hazel - When taken: 26 Jun 1813 - Where taken: off Charleston (South Carolina) - Date Received: 10 May 1814 - From what ship: Plymouth - Born: Boston - Age: 24 - Released on 28 Apr 1815.

Parker, Richard - Seaman - Nbr: 1756 - How taken: Gave himself up from HMS Edinburgh - When taken: 3 Dec 1812 - Date Received: 20 Jul 1814 - From what ship: HMS Milford, Plymouth - Born: North Carolina - Age: 34 - Released on 26 Apr 1815.

Parker, Robert - Seaman - Nbr: 3266 - Prize: Wily Reynard - Ship Type: P - How taken: HMS Shannon - When taken: 2 Oct 1812 - Where taken: off Halifax - Date Received: 11 Sep 1814 - From what ship: HMT Freya, Chatham - Born: Boston - Age: 51 - Race: Black - Released on 27 Apr 1815.

Parker, Samuel - Seaman - Nbr: 6104 - How taken: Gave himself up from HMS Glasgow - When taken: 31 Dec 1814 - Date Received: 6 Jan 1814 - From what ship: HMT Impregnable - Born: Bristol - Age: 46 - Released on 5 Jul 1815.

Parker, Thomas - Seaman - Nbr: 2244 - Prize: Carbineer - Ship Type: Prize - How taken: HMS Ringdove - When taken: 24 Apr 1814 - Where taken: off Bermuda - Date Received: 16 Aug 1814 - From what ship: HMS Dublin, Halifax - Born: Providence - Age: 30 - Released on 3 May 1815.

Parker, Thomas - Seaman - Nbr: 3842 - Prize: Dominique - Ship Type: LM - How taken: HMS Dotterel - When taken: 21 May 1814 - Where taken: off Charleston - Date Received: 5 Oct 1814 - From what ship: HMT Orpheus, Halifax - Born: Delaware - Age: 22 - Died on 5 Nov 1814.

Parker, William - Seaman - Nbr: 5375 - Prize: Derby - Ship Type: MV - How taken: HMS Nereus - When taken: 4 Feb 1813 - Where taken: off Cape of Good Hope - Date Received: 31 Oct 1814 - From what ship: HMT Leyden, Chatham - Born: Barnstable - Age: 20 - Died on 28 Nov 1814 from rubeula.

Parker, William - Seaman - Nbr: 3097 - How taken: Gave himself up from HMS Norge - Date Received: 3 Sep 1814 - From what ship: HMT Bristol - Born: Philadelphia - Age: 25 - Released on 28 May 1815.

Parkinson, Stephen - Prize master - Nbr: 3799 - Prize: Star - Ship Type: Prize - How taken: HMS Shannon - When taken: 12 Nov 1814 - Where taken: off St. Martins - Date Received: 5 Oct 1814 - From what ship: HMT

American Prisoners of War Held at Dartmoor during the War of 1812

## Alphabetical listing of names

Orpheus, Halifax - Born: Connecticut - Age: 24 - Released on 3 May 1815.

Parks, A. D. La Enoh - Seaman - Nbr: 4818 - Prize: President - Ship Type: P - How taken: HMS Pique - When taken: 7 May 1814 - Where taken: off Porto Rico - Date Received: 9 Oct 1814 - From what ship: HMT Freya, Halifax - Born: Cartagena - Age: 18 - Race: Black - Released on 21 Jun 1815.

Parley, William - Seaman - Nbr: 889 - Prize: General Kempt - Ship Type: P - How taken: HMS Foxhound - When taken: 18 Dec 1813 - Where taken: Lat 48'60" Long 5'7" - Date Received: 31 Jan 1814 - From what ship: Plymouth - Born: Exeter - Age: 20 - Released on 27 Apr 1815.

Parnell, Hugh - Seaman - Nbr: 4121 - Prize: Bordeaux Packet - Ship Type: LM - How taken: HMS Niemen - When taken: 28 Jun 1814 - Where taken: off Delaware - Date Received: 6 Oct 1814 - From what ship: HMT Chesapeake, Halifax - Born: Baltimore - Age: 28 - Released on 13 Jun 1815.

Parr, James - Seaman - Nbr: 1687 - Prize: William Wilson - Ship Type: MV - How taken: HMS San Domingo - When taken: 19 Sep 1813 - Where taken: West Indies - Date Received: 2 Jul 1814 - From what ship: Plymouth - Born: Philadelphia - Age: 28 - Released on 1 May 1815.

Parr, John - Boy - Nbr: 2975 - Prize: Frolic - Ship Type: P - How taken: HMS Heron - When taken: 25 Jan 1814 - Where taken: off St. Thomas - Date Received: 29 Aug 1814 - From what ship: HMT Bittern - Born: Massachusetts - Age: 16 - Released on 19 May 1815.

Parr, Samuel - Seaman - Nbr: 3106 - How taken: Gave himself up from HMS L'Aigle - When taken: 19 Jul 1814 - Date Received: 11 Sep 1814 - From what ship: HMT Salvador del Mundo - Born: Suffolk - Age: 40 - Released on 28 May 1815.

Parrish, William - Seaman - Nbr: 5567 - Prize: Daedalus - Ship Type: MV - How taken: HMS Niemen - When taken: 20 Sep 1814 - Where taken: off Delaware - Date Received: 17 Dec 1814 - From what ship: HMT Loire, Halifax - Born: Princess Ann County - Age: 27 - Released on 14 Jun 1815.

Parrott, Ebenezer - Seaman - Nbr: 5332 - Prize: Elbridge Gerry - Ship Type: P - How taken: HMS Crescent - When taken: 16 Sep 1813 - Where taken: at sea - Date Received: 31 Oct 1814 - From what ship: HMT Leyden, Chatham - Born: Cape Elizabeth - Age: 22 - Released on 29 Jun 1815.

Parsons, Andrew - Seaman - Nbr: 2415 - Prize: US Sloop Frolic - Ship Type: MW - How taken: HMS Orpheus - When taken: 20 Apr 1814 - Where taken: off Cuba - Date Received: 16 Aug 1814 - From what ship: HMT Queen, Halifax - Born: Salem - Age: 30 - Sent to Dartmouth on 19 Oct 1814.

Parsons, Daniel - Seaman - Nbr: 2675 - Prize: Joseph - Ship Type: MV - How taken: HMS Iris - When taken: 8 Jun 1813 - Where taken: off Spain - Date Received: 21 Aug 1814 - From what ship: HMT Freya, Chatham - Born: Gloucester - Age: 28 - Released on 19 May 1815.

Parsons, Eugene - Seaman - Nbr: 1808 - How taken: Gave himself up from HMS Foxhound - When taken: 22 Feb 1813 - Date Received: 20 Jul 1814 - From what ship: HMS Milford, Plymouth - Born: Gloucester - Age: 24 - Released on 28 Apr 1815.

Parsons, Ignatius - Seaman - Nbr: 112 - How taken: Gave himself up from HMS Foxhound - Date Received: 2 Apr 1813 - From what ship: Plymouth - Born: Gloucester - Age: 24 - Sent to Chatham on 27 May 1813.

Parsons, James - Seaman - Nbr: 6508 - How taken: Apprehended at Kingston - Date Received: 3 Mar 1815 - From what ship: HMT Ganges, Plymouth - Born: Long Island - Age: 27 - Released on 11 Jul 1815.

Parsons, John - Armourer - Nbr: 4422 - Prize: Elbridge Gerry - Ship Type: LM - How taken: HMS Crescent - When taken: 16 Sep 1813 - Where taken: at sea - Date Received: 8 Oct 1814 - From what ship: HMT Leyden, Chatham - Born: Massachusetts - Age: 25 - Released on 14 Jun 1815.

Parsons, Junius - Seaman - Nbr: 6024 - Prize: McDonough - Ship Type: P - How taken: HMS Bacchante - When taken: 1 Nov 1814 - Where taken: Lat 42 Long 67 - Date Received: 28 Dec 1814 - From what ship: HMT Penelope - Born: Old York - Age: 34 - Released on 3 Jul 1815.

Parsons, Rufus - Seaman - Nbr: 2061 - Prize: Dart - Ship Type: LM - How taken: HMS Peterel - When taken: 5 Mar 1812 - Where taken: at sea - Date Received: 3 Aug 1814 - From what ship: HMS Bittern, Chatham Depot - Born: York - Age: 22 - Released on 2 May 1815.

Parsons, Samuel D. - Seaman - Nbr: 492 - Prize: Paul Jones - Ship Type: P - How taken: HMS Leonidas - When taken: 23 May 1813 - Where taken: Channel - Date Received: 8 Sep 1813 - From what ship: Plymouth - Born: Springfield - Age: 29 - Released on 20 Apr 1815.

Parsons, Simon - Seaman - Nbr: 4090 - Prize: Tyger - Ship Type: Prize - How taken: HMS Bulwark - When taken: 20 Jul 1814 - Where taken: Georges Bank - Date Received: 6 Oct 1814 - From what ship: HMT Chesapeake, Halifax - Born: Connecticut - Age: 21 - Released on 13 Jun 1815.

Parsons, William - Seaman - Nbr: 6218 - Prize: Prince de Neufchatel - Ship Type: P - How taken: Leander (Newcastle Acasta) - When taken: 20 Dec 1814 - Where taken: Lat 38 Long 56 - Date Received: 30 Jan 1815

American Prisoners of War Held at Dartmoor during the War of 1812

## Alphabetical listing of names

- From what ship: HMT Pheasant - Born: Boston - Age: 19 - Released on 5 Jul 1815.
Pass, Benjamin - Seaman - Nbr: 6136 - How taken: Gave himself up from HMS Freya - When taken: 3 Jan 1814 - Date Received: 17 Jan 1815 - What ship: HMT Impregnable - Born: Rhode Island - Age: 31 - Released on 5 Jul 1815.
Patch, George - Seaman - Nbr: 5898 - Prize: Harlequin - Ship Type: P - How taken: HMS Bulwark - When taken: 23 Nov 1814 - Where taken: off Halifax - Date Received: 27 Dec 1814 - From what ship: HMT Penelope - Born: Massachusetts - Age: 24 - Released on 3 Jul 1815.
Paterson, Hanse - Seaman - Nbr: 38 - How taken: Apprehended at Gibraltar - When taken: 8 Aug 1813 - Date Received: 2 Apr 1813 - What ship: Plymouth - Born: Salem - Age: 17 - Sent to Dartmouth on 30 Jul 1813.
Patten, James - Seaman - Nbr: 546 - Prize: Orders in Council - Ship Type: MV - How taken: HMS Surveillante - When taken: 1 Jun 1813 - Where taken: off Cape Ortegal (Spain) - Date Received: 8 Sep 1813 - From what ship: Plymouth - Born: New York - Age: 21 - Released on 26 Apr 1815.
Patten, John - Seaman - Nbr: 2060 - Prize: Dart - Ship Type: LM - How taken: HMS Peterel - When taken: 5 Mar 1812 - Where taken: at sea - Date Received: 3 Aug 1814 - From what ship: HMS Bittern, Chatham Depot - Born: Durham - Age: 23 - Released on 2 May 1815.
Patten, John U. - Seaman - Nbr: 3674 - Prize: Alfred - Ship Type: P - How taken: HMS Epervier - When taken: 23 Feb 1812 - Where taken: off Newfoundland - Date Received: 30 Sep 1814 - From what ship: HMT President - Born: Marblehead - Age: 21 - Released on 4 Jun 1815.
Patten, Nathaniel - Marine - Nbr: 5695 - Prize: McDonough - Ship Type: P - How taken: HMS Bacchante - When taken: 7 Nov 1814 - Where taken: Lat 42 Long 67 - Date Received: 24 Dec 1814 - From what ship: HMT Penelope - Born: Biddeford - Age: 19 - Released on 3 Jul 1815.
Patten, Robert - Master's mate - Nbr: 5684 - Prize: McDonough - Ship Type: P - How taken: HMS Bacchante - When taken: 7 Nov 1814 - Where taken: Lat 42 Long 67 - Date Received: 24 Dec 1814 - From what ship: HMT Penelope - Born: Arundel - Age: 37 - Released on 3 Jul 1815.
Patterson, Andrew - Seaman - Nbr: 948 - Prize: Siro - Ship Type: LM - How taken: HMS Pelican - When taken: 13 Jan 1814 - Where taken: at sea - Date Received: 31 Jan 1814 - From what ship: Plymouth - Born: Pillau, Prussia - Age: 42 - Sent to Plymouth on 28 Feb 1814.
Patterson, John - Boatswain's Yeoman - Nbr: 3979 - Prize: US Brig Rattlesnake - Ship Type: MW - How taken: HMS Leander - When taken: 13 Jul 1814 - Where taken: off Shelburne - Date Received: 6 Oct 1814 - From what ship: HMT Chesapeake, Halifax - Born: Salem - Age: 24 - Released on 9 Jun 1815.
Patterson, John - Seaman - Nbr: 963 - Prize: Siro - Ship Type: LM - How taken: HMS Pelican - When taken: 13 Jan 1814 - Where taken: at sea - Date Received: 31 Jan 1814 - From what ship: Plymouth - Born: Stettin, Prussia - Age: 32 - Sent to Plymouth on 28 Feb 1814.
Patterson, John - Seaman - Nbr: 4072 - Prize: Hussar - Ship Type: P - How taken: HMS Saturn - When taken: 24 May 1814 - Where taken: off Sandy Hook - Date Received: 6 Oct 1814 - From what ship: HMT Chesapeake, Halifax - Born: New York - Age: 29 - Race: Negro - Released on 13 Jun 1815.
Patterson, Joseph - Seaman - Nbr: 6080 - Prize: Daedalus - Ship Type: MV - How taken: HMS Niemen - When taken: 20 Sep 1814 - Where taken: off Delaware - Date Received: 28 Dec 1814 - From what ship: HMT Penelope - Born: Baltimore - Age: 17 - Race: Mulatto - Released on 3 Jul 1815.
Patterson, Peter - Seaman - Nbr: 6039 - Prize: Regent - Ship Type: LM - How taken: HMS Forth - When taken: 19 Sep 1814 - Where taken: off Egg Harbor (New Jersey) - Date Received: 28 Dec 1814 - From what ship: HMT Penelope - Born: Philadelphia - Age: 40 - Race: Black - Released on 3 Jul 1815.
Patterson, Peter - Seaman - Nbr: 1919 - How taken: Gave himself up from HMS Tweed - When taken: 20 Dec 1812 - Date Received: 3 Aug 1814 - From what ship: HMS Alceste, Chatham Depot - Born: Philadelphia - Age: 40 - Released on 26 Apr 1815.
Patterson, William - Seaman - Nbr: 5145 - How taken: Gave himself up from HMS Racehorse - When taken: 25 Oct 1813 - Date Received: 31 Oct 1814 - From what ship: HMT Mermaid, Chatham - Born: Newburyport - Age: 26 - Released on 29 Jun 1815.
Patton, Robert - Seaman - Nbr: 1839 - How taken: Gave himself up from HMS Dublin - When taken: 2 May 1813 - Date Received: 21 Jul 1814 - From what ship: HMT Redbeard & Pincher, Chatham Depot - Born: Charlestown - Age: 34 - Released on 2 May 1815.
Paul, Daniel - Seaman - Nbr: 3552 - Prize: Hawk - Ship Type: P - How taken: HMS Pique - When taken: 26 Apr 1814 - Where taken: off Bermuda - Date Received: 30 Sep 1814 - From what ship: HMT Sybella - Born: North Carolina - Age: 21 - Released on 4 Jun 1815.
Paul, Jacob - Seaman - Nbr: 2749 - Prize: Dart - Ship Type: MV - How taken: HMS Peterel - When taken: 15 Mar

American Prisoners of War Held at Dartmoor during the War of 1812

## Alphabetical listing of names

1813 - Where taken: at sea - Date Received: 24 Aug 1814 - From what ship: HMT Liverpool, Chatham - Born: Elliott - Age: 21 - Released on 19 May 1815.

Paul, John - Boy - Nbr: 5890 - Prize: Harlequin - Ship Type: P - How taken: HMS Bulwark - When taken: 23 Nov 1814 - Where taken: off Halifax - Date Received: 27 Dec 1814 - From what ship: HMT Penelope - Born: Dover - Age: 16 - Released on 3 Jul 1815.

Paul, Jonathan - Seaman - Nbr: 1998 - How taken: Impressed from HMS Hind - When taken: 17 Mar 1813 - Date Received: 3 Aug 1814 - From what ship: HMS Lyffey, Chatham Depot - Born: Charlestown - Age: 30 - Died on 9 Mar 1815 from wounds made by another prisoner.

Paulin, Nicholas - Seaman - Nbr: 1158 - Prize: Two Brothers - Ship Type: MV - How taken: Beetle (LM) - When taken: 18 Mar 1813 - Where taken: off Western Islands (England) - Date Received: 10 May 1814 - From what ship: Plymouth - Born: Sardinia - Age: 46 - Released on 28 Apr 1815.

Paulm, Nas. - Seaman - Nbr: 360 - Prize: Two Brothers - Ship Type: MV - How taken: Beetle (LM) - When taken: 18 Mar 1813 - Where taken: off Western Islands (England) - Date Received: 1 Jul 1813 - From what ship: Plymouth - Born: New Orleans - Age: 36 - Sent to Plymouth on 27 Apr 1814.

Payer, Walter - Seaman - Nbr: 1537 - Prize: Zebra - Ship Type: LM - How taken: HMS Pyramus - When taken: 20 Apr 1813 - Where taken: Bay of Biscay - Date Received: 23 Jun 1814 - From what ship: Stapleton - Born: New Orleans - Age: 35 - Sent to Mill Prison (Plymouth, England) on 30 Jun 1814.

Payne, Alexander - Seaman - Nbr: 5037 - Prize: Betsey - Ship Type: Prize - How taken: HMS Pylades - When taken: 7 Sep 1814 - Where taken: Canten - Date Received: 28 Oct 1814 - From what ship: HMT Alkbar, Halifax - Born: Portland - Age: 26 - Released on 29 Jun 1815.

Payne, James - Seaman - Nbr: 1961 - How taken: Gave himself up from HMS Sterling Castle - When taken: 14 Mar 1813 - Date Received: 3 Aug 1814 - From what ship: HMS Lyffey, Chatham Depot - Born: Huntingdon - Age: 21 - Released on 2 May 1815.

Payne, John - Seaman - Nbr: 4873 - Prize: US Brig Rattlesnake - Ship Type: MW - How taken: HMS Leander - When taken: 13 Jul 1814 - Where taken: off Shelburne - Date Received: 24 Oct 1814 - From what ship: Royal Hospital, Plymouth - Born: Not listed - Released on 21 Jun 1815.

Payne, Joseph - Seaman - Nbr: 4050 - Prize: US Brig Rattlesnake - Ship Type: MW - How taken: HMS Leander - When taken: 13 Jul 1814 - Where taken: off Shelburne - Date Received: 6 Oct 1814 - From what ship: HMT Chesapeake, Halifax - Born: Bath - Age: 37 - Released on 13 Jun 1815.

Payne, Josiah Smith - 3rd Mate - Nbr: 1669 - Prize: Tom - Ship Type: LM - How taken: HMS Surveillante - When taken: 24 Apr 1813 - Where taken: Bay of Biscay - Date Received: 26 Jun 1814 - From what ship: Exeter - Born: Charleston - Age: 18 - Sent to Ashburton (England) on 23 Dec 1814.

Payne, William - Seaman - Nbr: 5286 - Prize: Yankee - Ship Type: P - How taken: HMS Shannon - When taken: 20 Au 1813 - Where taken: at sea - Date Received: 31 Oct 1814 - From what ship: HMT Leyden, Chatham - Born: Massachusetts - Age: 22 - Released on 4 Jun 1815.

Payton, James - Prize master - Nbr: 6362 - Prize: Prince de Neufchatel - Ship Type: P - How taken: Leander (Newcastle Acasta) - When taken: 20 Dec 1814 - Where taken: Lat 38 Long 56 - Date Received: 19 Feb 1815 - From what ship: HMT Ganges, Plymouth - Born: Charleston - Age: 26 - Released on 20 Mar 1815.

Peabody, John - Prize master - Nbr: 3771 - Prize: Fame - Ship Type: P - How taken: HMS Thistle - When taken: 10 Apr 1814 - Where taken: after being cast ashore on a seal island - Date Received: 30 Sep 1814 - From what ship: HMT President, Halifax - Born: Massachusetts - Age: 56 - Released on 9 Jun 1815.

Peach, John - Seaman - Nbr: 5326 - Prize: Enterprize - Ship Type: P - How taken: HMS Tenedos - When taken: 21 May 1813 - Where taken: off Cape Cod - Date Received: 31 Oct 1814 - From what ship: HMT Leyden, Chatham - Born: Marblehead - Age: 17 - Released on 29 Jun 1815.

Peadon, William - Seaman - Nbr: 4690 - How taken: HMS Dictator - When taken: 3 Nov 1813 - Date Received: 9 Oct 1814 - From what ship: HMT Leyden, Chatham - Born: Philadelphia - Age: 34.

Peak, James - Seaman - Nbr: 4349 - Prize: Plutus - Ship Type: P - How taken: HMS Curlew - When taken: 1 Apr 1814 - Where taken: off Halifax - Date Received: 7 Oct 1814 - From what ship: HMT Salvador del Mundo, Halifax - Born: New Hampshire - Age: 23 - Released on 14 Jun 1815.

Peake, J. W. - Seaman - Nbr: 3447 - How taken: Gave himself up from HMS Christian - When taken: 29 Mar 1813 - Date Received: 13 Sep 1814 - From what ship: HMT Niobe, Chatham - Born: Albany - Age: 26 - Released on 28 May 1815.

Peal, Andrew - Seaman - Nbr: 1711 - How taken: Gave himself up from HMS Urgent - Date Received: 8 Jul 1814 - From what ship: Labrador - Born: Marblehead - Age: 22 - Released on 1 May 1815.

Peane, Henry - Seaman - Nbr: 862 - Prize: Amiable - Ship Type: LM - How taken: HMS Magnificent - When taken:

American Prisoners of War Held at Dartmoor during the War of 1812

## Alphabetical listing of names

30 Oct 1813 - Where taken: off Bordeaux - Date Received: 4 Dec 1813 - From what ship: Plymouth - Born: Philadelphia - Age: 26 - Released on 27 Apr 1815.

Pearce, Alexander - Seaman - Nbr: 824 - Prize: Dart - Ship Type: MV - How taken: HMS Niger - When taken: 12 Nov 1813 - Where taken: off Cape Finisterre - Date Received: 29 Nov 1813 - From what ship: Plymouth - Born: Philadelphia - Age: 17 - Released on 26 Apr 1815.

Pearce, Charles - Seaman - Nbr: 5957 - Prize: Harlequin - Ship Type: P - How taken: HMS Bulwark - When taken: 23 Nov 1814 - Where taken: off Halifax - Date Received: 27 Dec 1814 - From what ship: HMT Penelope - Born: Baltimore - Age: 35 - Released on 3 Jul 1815.

Pearce, David - Seaman - Nbr: 4537 - Prize: Wolf Cove - Ship Type: MV - How taken: HMS Briton - When taken: 1 Dec 1812 - Where taken: off Brest - Date Received: 8 Oct 1814 - From what ship: HMT Leyden, Chatham - Born: Massachusetts - Age: 19 - Released on 27 Apr 1815.

Pearce, Joseph - Chief Mate - Nbr: 870 - Prize: Agnes - Ship Type: LM - How taken: Jane (Cutter) - When taken: 29 Nov 1813 - Where taken: Bay of Biscay - Date Received: 31 Jan 1814 - From what ship: Plymouth - Born: Boston - Age: 21 - Released on 28 Apr 1815.

Pearce, Thomas - Seaman - Nbr: 2048 - How taken: Gave himself up from HMS Amelia - When taken: 1 Jun 1812 - Date Received: 3 Aug 1814 - From what ship: HMS Lyffey, Chatham Depot - Born: Boston - Age: 36 - Released on 26 Apr 1815.

Pearce, William - Seaman - Nbr: 2603 - How taken: Gave himself up from HMS Muros - When taken: 12 Sep 1812 - Date Received: 21 Aug 1814 - From what ship: HMT Freya, Chatham - Born: Providence - Age: 18 - Released on 26 Apr 1815.

Pearcey, John - Boy - Nbr: 5136 - Prize: Calabria - Ship Type: MV - How taken: HMS Castilian - When taken: 29 Sep 1814 - Where taken: off Ireland - Date Received: 31 Oct 1814 - From what ship: HMT Castillian - Born: New York - Age: 15 - Released on 28 May 1815.

Pearer, John - Seaman - Nbr: 831 - Prize: Chesapeake - Ship Type: LM - How taken: HMS Hotspur & HMS Pyramus - When taken: 26 Oct 1813 - Where taken: off Nantes - Date Received: 29 Nov 1813 - From what ship: Plymouth - Born: Pennsylvania - Age: 36 - Released on 26 Apr 1815.

Pearse, John - Seaman - Nbr: 6165 - Prize: Lion - Ship Type: P - How taken: HMS Granicus - When taken: 2 Dec 1814 - Where taken: off Lisbon - Date Received: 18 Jan 1815 - From what ship: HMT Impregnable - Born: New York - Age: 24 - Released on 5 Jul 1815.

Pearse, Prince - Seaman - Nbr: 4540 - Prize: Wolf Cove - Ship Type: MV - How taken: HMS Briton - When taken: 1 Dec 1812 - Where taken: off Brest - Date Received: 8 Oct 1814 - From what ship: HMT Leyden, Chatham - Born: Massachusetts - Age: 24 - Released on 27 Apr 1815.

Pearson, Benjamin - Seaman - Nbr: 2125 - How taken: Gave himself up from HMS Mulgrave - When taken: 17 Nov 1812 - Date Received: 8 Aug 1814 - From what ship: HMT Raven, Chatham - Born: Baltimore - Age: 35 - Released on 26 Apr 1815.

Pearson, Charles - Seaman - Nbr: 3452 - How taken: Gave himself up from HMS Lightning - When taken: 14 Sep 1814 - Date Received: 19 Sep 1814 - From what ship: HMT Salvador del Mundo - Born: Baltimore - Age: 35 - Released on 11 Jun 1815.

Pearson, Samuel - Seaman - Nbr: 4413 - Prize: Fire Fly - Ship Type: LM - How taken: HMS Revolutionnaire - When taken: 19 Oct 1813 - Where taken: at sea - Date Received: 8 Oct 1814 - From what ship: HMT Leyden, Chatham - Born: Massachusetts - Age: 24 - Released on 14 Jun 1815.

Peck, Benjamin - Seaman - Nbr: 3750 - Prize: Enterprize - Ship Type: P - How taken: HMS Tenedos - When taken: 21 May 1813 - Where taken: off Cape Cod - Date Received: 30 Sep 1814 - From what ship: HMT President, Halifax - Born: New Jersey - Age: 50 - Released on 11 Jul 1815.

Peck, Thomas - Seaman - Nbr: 2175 - Prize: Rattlesnake - Ship Type: P - How taken: HMS Hyperion - When taken: 26 Jun 1814 - Where taken: off Cape Finisterre - Date Received: 16 Aug 1814 - From what ship: HMS Dublin, Halifax - Born: Philadelphia - Age: 16 - Released on 2 May 1815.

Peck, Thomas - Seaman - Nbr: 1332 - Prize: Paul Jones - Ship Type: P - How taken: HMS Leonidas - When taken: 23 May 1813 - Where taken: Channel - Date Received: 19 Jun 1814 - From what ship: Stapleton - Born: New London - Age: 39 - Race: Negro - Died on 15 Mar 1815 from variola.

Peckham, Ezekiel - Seaman - Nbr: 4751 - How taken: Gave himself up from HMS Thisbe - When taken: 12 Aug 1813 - Date Received: 9 Oct 1814 - From what ship: HMT Freya, Chatham - Born: Newburyport - Age: 31 - Released on 15 Jun 1815.

Peckham, H. - Seaman - Nbr: 2082 - Prize: True Blooded Yankee - Ship Type: P - How taken: HMS Hope - When taken: 24 Jun 1813 - Where taken: off Brest - Date Received: 3 Aug 1814 - From what ship: HMS Bittern,

American Prisoners of War Held at Dartmoor during the War of 1812

## Alphabetical listing of names

Chatham Depot - Born: Newburyport - Age: 25 - Released on 2 May 1815.

Peckham, Henry - Seaman - Nbr: 557 - Prize: Godfrey and Mary - Ship Type: MV - How taken: Robert Todd - When taken: 23 Jun 1813 - Where taken: Lat 21, 30 Long 53 - Date Received: 8 Sep 1813 - From what ship: Plymouth - Born: Stonington - Age: 20 - Released on 26 Apr 1815.

Peckham, Isaac - Seaman - Nbr: 3235 - Prize: Maydock - Ship Type: MV - How taken: Rebuff - When taken: 16 Jun 1813 - Where taken: off St. Marys - Date Received: 11 Sep 1814 - From what ship: HMT Freya, Chatham - Born: Portsmouth - Age: 23 - Released on 28 May 1815.

Pedang, John - Seaman - Nbr: 3743 - Prize: Lizard - Ship Type: P - How taken: HMS Prometheus - When taken: 5 May 1814 - Where taken: off Halifax - Date Received: 30 Sep 1814 - From what ship: HMT President, Halifax - Born: Sumatra - Age: 21 - Race: Creole - Released on 4 Jun 1815.

Pedrick, George - Seaman - Nbr: 2400 - Prize: US Sloop Frolic - Ship Type: MW - How taken: HMS Orpheus - When taken: 20 Apr 1814 - Where taken: off Cuba - Date Received: 16 Aug 1814 - From what ship: HMT Queen, Halifax - Born: Marblehead - Age: 24 - Sent to Dartmouth on 19 Oct 1814.

Pedro, Francisco - Seaman - Nbr: 6138 - Prize: Jane - Ship Type: Prize - How taken: HMS Phoebe - When taken: 26 Feb 1814 - Where taken: off Porto Rico - Date Received: 17 Jan 1815 - From what ship: HMT Impregnable - Born: Portugal - Age: 33 - Released on 5 Jul 1815.

Pelhem, Robert - Prize Master - Nbr: 991 - Prize: Apparencen - Ship Type: MV - How taken: HMS Castilian - When taken: 27 Jan 1814 - Where taken: off Ushant (France) - Date Received: 31 Jan 1814 - From what ship: Plymouth - Born: Boston - Age: 29 - Sent to Plymouth on 8 Jul 1814.

Pellow, William - Seaman - Nbr: 2263 - Prize: Grand Turk - Ship Type: P - How taken: HMS Martin - When taken: 26 May 1814 - Where taken: off Cape Sable Island (Canada) - Date Received: 16 Aug 1814 - From what ship: HMS Dublin, Halifax - Born: Portsmouth - Age: 18 - Released on 3 May 1815.

Pembroke, James - Seaman - Nbr: 4881 - Prize: Saratoga - Ship Type: P - How taken: HMS Barracouta - When taken: 9 Oct 1814 - Where taken: off Western Islands (England) - Date Received: 24 Oct 1814 - From what ship: HMT Salvador del Mundo - Born: New Marlborough - Age: 19 - Released on 21 Jun 1815.

Pender, John - Seaman - Nbr: 6518 - How taken: Gave himself up from HMS Orontes - When taken: Feb 1815 - Date Received: 3 Mar 1815 - From what ship: HMT Ganges, Plymouth - Born: Philadelphia - Age: 25 - Race: Man of color - Released on 4 Jun 1815.

Pendergrast, Morris - Seaman - Nbr: 2055 - How taken: Apprehended at Gravesend - When taken: 16 May 1813 - Date Received: 3 Aug 1814 - From what ship: HMS Lyffey, Chatham Depot - Born: Boston - Age: 27 - Released on 2 May 1815.

Pendess, Lewis - Seaman - Nbr: 5979 - Prize: Cossack - Ship Type: Private - How taken: HMS Bulwark - When taken: 1 Nov 1814 - Where taken: Georges Bank - Date Received: 27 Dec 1814 - From what ship: HMT Penelope - Born: Salem - Age: 19 - Released on 11 Jul 1815.

Pendleton, Charles - Seaman - Nbr: 2237 - Prize: General Hart - Ship Type: P - How taken: HMS Sophie - When taken: 24 Apr 1814 - Where taken: off Bermuda - Date Received: 16 Aug 1814 - From what ship: HMS Dublin, Halifax - Born: Connecticut - Age: 20 - Released on 2 May 1815.

Pendleton, Evan - Seaman - Nbr: 2732 - How taken: Gave himself up from HMS Aboukir - When taken: 28 Oct 1812 - Date Received: 24 Aug 1814 - From what ship: HMT Liverpool, Chatham - Born: Ellisburg - Age: 21 - Released on 26 Apr 1815.

Pendleton, George - Seaman - Nbr: 4099 - Prize: Harford - Ship Type: MV - How taken: HMS Junon - When taken: 6 Jun 1814 - Where taken: off Cape Ann - Date Received: 6 Oct 1814 - From what ship: HMT Chesapeake, Halifax - Born: Massachusetts - Age: 21 - Released on 13 Jun 1815.

Pendleton, Jonathan - Seaman - Nbr: 5865 - Prize: Lion - Ship Type: P - How taken: HMS Granicus - When taken: 2 Dec 1814 - Where taken: off Lisbon - Date Received: 26 Dec 1814 - From what ship: HMT Impregnable - Born: Connecticut - Age: 20 - Released on 19 May 1815.

Pendleton, Samuel - Prize Master - Nbr: 2517 - Prize: Snap Dragon - Ship Type: P - How taken: HMS Martin - When taken: 10 Jun 1814 - Where taken: off Halifax - Date Received: 16 Aug 1814 - From what ship: HMT Queen, Halifax - Born: New Berne - Age: 27 - Released on 3 May 1815.

Penfold, John - Seaman - Nbr: 4163 - How taken: Gave himself up from HMS Minos - When taken: 2 Feb 1814 - Date Received: 6 Oct 1814 - From what ship: HMT Niobe, Chatham - Born: Baltimore - Age: 26 - Released on 13 Jun 1815.

Penman, Richard - Seaman - Nbr: 1707 - Prize: Tom - Ship Type: LM - How taken: HMS Surveillante - When taken: 24 Apr 1813 - Where taken: Bay of Biscay - Date Received: 6 Jul 1814 - From what ship: Bristol - Born: New York - Age: 25 - Released on 1 May 1815.

American Prisoners of War Held at Dartmoor during the War of 1812

## Alphabetical listing of names

Penn, John - Pilot - Nbr: 2202 - Prize: Hussar - Ship Type: P - How taken: HMS Saturn - When taken: 25 May 1814 - Where taken: off Sandy Hook - Date Received: 16 Aug 1814 - From what ship: HMS Dublin, Halifax - Born: New York - Age: 22 - Released on 2 May 1815.

Penn, William - Seaman - Nbr: 1722 - How taken: Impressed at London - When taken: Nov 1812 - Date Received: 8 Jul 1814 - From what ship: Labrador - Born: Lancaster - Age: 26 - Race: Black - Released on 1 May 1815.

Penn, William - Seaman - Nbr: 3395 - Prize: Thomas - Ship Type: P - How taken: HMS Nymphe - When taken: 24 Jun 1813 - Where taken: off Halifax - Date Received: 13 Sep 1814 - From what ship: HMT Niobe, Chatham - Born: Eastport - Age: 28 - Released on 28 May 1815.

Penny, James - Seaman - Nbr: 3449 - Prize: Jane - Ship Type: Prize - How taken: Current - When taken: 27 Jun 1813 - Where taken: Newfoundland Banks - Date Received: 13 Sep 1814 - From what ship: HMT Niobe, Chatham - Born: Newburgh - Age: 20 - Released on 28 May 1815.

Penny, Richard - Seaman - Nbr: 2869 - How taken: Gave himself up from HMS Scorpion - When taken: 27 May 1813 - Date Received: 24 Aug 1814 - From what ship: HMT Alpheus, Chatham - Born: New York - Age: 36 - Released on 19 May 1815.

Penrose, Abraham - Seaman - Nbr: 2902 - How taken: Gave himself up from HMS Ocean - When taken: 28 May 1813 - Date Received: 24 Aug 1814 - From what ship: HMT Alpheus, Chatham - Born: Albany - Age: 32 - Race: Black - Released on 19 May 1815.

Percival, John - Seaman - Nbr: 5397 - Prize: Hawke - Ship Type: P - How taken: HMS Pique - When taken: 27 Apr 1814 - Where taken: West Indies - Date Received: 31 Oct 1814 - From what ship: HMT Leyden, Chatham - Born: Sandwich - Age: 25 - Released on 1 Jul 1815.

Peregrine, Taggert - Seaman - Nbr: 1564 - Prize: Caroline - Ship Type: MV - How taken: HMS Medusa - When taken: 12 Apr 1813 - Where taken: Bay of Biscay - Date Received: 23 Jun 1814 - From what ship: Stapleton - Born: Maryland - Age: 29 - Released on 1 May 1815.

Perigo, Joel - Seaman - Nbr: 5064 - Prize: Ida - Ship Type: LM - How taken: HMS Newcastle - When taken: 9 Aug 1814 - Where taken: Long 34 - Date Received: 28 Oct 1814 - From what ship: HMT Alkbar, Halifax - Born: Connecticut - Age: 20 - Died on 24 Nov 1814.

Perish, William - Seaman - Nbr: 5337 - Prize: Elbridge Gerry - Ship Type: P - How taken: HMS Crescent - When taken: 16 Sep 1813 - Where taken: at sea - Date Received: 31 Oct 1814 - From what ship: HMT Leyden, Chatham - Born: Richmond - Age: 21 - Race: Black - Released on 29 Jun 1815.

Perkham, Peter - Seaman - Nbr: 3878 - Prize: Fame - Ship Type: MV - How taken: HMS Niemen - When taken: 21 May 1814 - Where taken: off Cape Hatteras - Date Received: 5 Oct 1814 - From what ship: HMT Orpheus, Halifax - Born: Tiverton - Age: 26 - Released on 9 Jun 1815.

Perkington, John - Seaman - Nbr: 3469 - How taken: Gave himself up from HMS Prince - When taken: 14 Sep 1814 - Date Received: 19 Sep 1814 - From what ship: HMT Salvador del Mundo - Born: Virginia - Age: 36 - Released on 4 Jun 1815.

Perkins, Benjamin - Seaman - Nbr: 4405 - Prize: Portsmouth Packet - Ship Type: Prize - How taken: HMS Fantome - When taken: 5 Oct 1813 - Where taken: off Portland - Date Received: 8 Oct 1814 - From what ship: HMT Leyden, Chatham - Born: Massachusetts - Age: 22 - Released on 14 Jun 1815.

Perkins, Clement - Seaman - Nbr: 6408 - Prize: Nancy - Ship Type: P - How taken: Papillion - When taken: 27 Dec 1814 - Where taken: Lat 29 Long 20 - Date Received: 3 Mar 1815 - From what ship: HMT Ganges, Plymouth - Born: Kennebunkport - Age: 21 - Released on 11 Jul 1815.

Perkins, Elijah - Seaman - Nbr: 894 - Prize: General Kempt - Ship Type: P - How taken: HMS Foxhound - When taken: 18 Dec 1813 - Where taken: Lat 48'60" Long 5'7" - Date Received: 31 Jan 1814 - From what ship: Plymouth - Born: Salem - Age: 23 - Released on 27 Apr 1815.

Perkins, George - Seaman - Nbr: 5690 - Prize: McDonough - Ship Type: P - How taken: HMS Bacchante - When taken: 7 Nov 1814 - Where taken: Lat 42 Long 67 - Date Received: 24 Dec 1814 - From what ship: HMT Penelope - Born: Arundel - Age: 18 - Released on 3 Jul 1815.

Perkins, Henry - Seaman - Nbr: 2743 - How taken: Gave himself up from HMS Ceres - When taken: 20 Dec 1812 - Date Received: 24 Aug 1814 - From what ship: HMT Liverpool, Chatham - Born: Boston - Age: 26 - Released on 26 Apr 1815.

Perkins, James - Seaman - Nbr: 4408 - Prize: Portsmouth Packet - Ship Type: Prize - How taken: HMS Fantome - When taken: 5 Oct 1813 - Where taken: off Portland - Date Received: 8 Oct 1814 - From what ship: HMT Leyden, Chatham - Born: Massachusetts - Age: 19 - Released on 14 Jun 1815.

Perkins, John - Carpenter's mate - Nbr: 946 - Prize: Siro - Ship Type: LM - How taken: HMS Pelican - When taken: 13 Jan 1814 - Where taken: at sea - Date Received: 31 Jan 1814 - From what ship: Plymouth - Born: New

American Prisoners of War Held at Dartmoor during the War of 1812

## Alphabetical listing of names

Hampton - Age: 25 - Died on 3 Nov 1814.

Perkins, John - Commander - Nbr: 4399 - Prize: Portsmouth Packet - Ship Type: Prize - How taken: HMS Fantome - When taken: 5 Oct 1813 - Where taken: off Portland - Date Received: 8 Oct 1814 - From what ship: HMT Leyden, Chatham - Born: Massachusetts - Age: 30 - Released on 28 Apr 1815.

Perkins, John - Seaman - Nbr: 3396 - Prize: Thomas - Ship Type: P - How taken: HMS Nymphe - When taken: 24 Jun 1813 - Where taken: off Halifax - Date Received: 13 Sep 1814 - From what ship: HMT Niobe, Chatham - Born: Massachusetts - Age: 21 - Released on 28 May 1815.

Perkins, Jonathan - Prize Master - Nbr: 3597 - Prize: Frolic - Ship Type: P - How taken: HMS Heron - When taken: 25 Jan 1814 - Where taken: off St. Thomas - Date Received: 30 Sep 1814 - From what ship: HMT Sybella - Born: Salem - Age: 24 - Released on 4 Jun 1815.

Perkins, Joseph - Marine - Nbr: 4025 - Prize: US Brig Rattlesnake - Ship Type: MW - How taken: HMS Leander - When taken: 13 Jul 1814 - Where taken: off Shelburne - Date Received: 6 Oct 1814 - From what ship: HMT Chesapeake, Halifax - Born: Massachusetts - Age: 24 - Released on 9 Jun 1815.

Perkins, Joseph - Boy - Nbr: 5553 - Prize: Lucy - Ship Type: Fishing boat - How taken: Lunenburg (Privateer) - When taken: 12 Sep 1814 - Where taken: off Cape Ann - Date Received: 17 Dec 1814 - From what ship: HMT Loire, Halifax - Born: Massachusetts - Age: 18 - Died on 20 Apr 1815 from dysentery.

Perkins, Joseph - Seaman - Nbr: 5677 - Prize: McDonough - Ship Type: P - How taken: HMS Bacchante - When taken: 7 Nov 1814 - Where taken: Lat 42 Long 67 - Date Received: 24 Dec 1814 - From what ship: HMT Penelope - Born: Arundel - Age: 29 - Released on 3 Jul 1815.

Perkins, Samuel - Master's mate - Nbr: 5685 - Prize: McDonough - Ship Type: P - How taken: HMS Bacchante - When taken: 7 Nov 1814 - Where taken: Lat 42 Long 67 - Date Received: 24 Dec 1814 - From what ship: HMT Penelope - Born: Arundel - Age: 25 - Released on 3 Jul 1815.

Perkins, Thomas - Seaman - Nbr: 6452 - Prize: Jemmell - Ship Type: MV - How taken: HMS Rhin - When taken: 28 May 1814 - Where taken: off Bermuda - Date Received: 3 Mar 1815 - From what ship: HMT Ganges, Plymouth - Born: New Bedford - Age: 19 - Released on 11 Jul 1815.

Perkins, Thomas - Seaman - Nbr: 857 - Prize: Amiable - Ship Type: LM - How taken: HMS Magnificent - When taken: 30 Oct 1813 - Where taken: off Bordeaux - Date Received: 4 Dec 1813 - From what ship: Plymouth - Born: Massachusetts - Age: 28 - Released on 26 Apr 1815.

Perkins, William - Seaman - Nbr: 3074 - How taken: Gave himself up from HMS Pembroke - When taken: 15 Jan 1813 - Date Received: 3 Sep 1814 - From what ship: HMT Hydra, Chatham - Born: New Hampshire - Age: 33 - Released on 28 May 1815.

Perkinson, James - Seaman - Nbr: 4402 - Prize: Portsmouth Packet - Ship Type: Prize - How taken: HMS Fantome - When taken: 5 Oct 1813 - Where taken: off Portland - Date Received: 8 Oct 1814 - From what ship: HMT Leyden, Chatham - Born: Massachusetts - Age: 20 - Released on 14 Jun 1815.

Perlie, Abraham - Surgeon - Nbr: 1733 - Prize: Rattlesnake - Ship Type: P - How taken: HMS Hyperion - When taken: 26 Jun 1814 - Where taken: off Cape Finisterre - Date Received: 18 Jul 1814 - From what ship: HMT Salvador del Mundo, Plymouth - Born: New Jersey - Age: 26 - Sent to Ashburton (England) on 5 Aug 1814.

Perne, George - Seaman - Nbr: 468 - Prize: Meteor - Ship Type: MV - How taken: HMS Briton - When taken: 12 Mar 1813 - Where taken: Bay of Biscay - Date Received: 8 Sep 1813 - From what ship: Plymouth - Born: New Orleans - Age: 30 - Sent to Mill Prison (Plymouth, England) on 21 Jun 1814.

Perney, William - Seaman - Nbr: 462 - Prize: Meteor - Ship Type: MV - How taken: HMS Briton - When taken: 12 Mar 1813 - Where taken: Bay of Biscay - Date Received: 8 Sep 1813 - From what ship: Plymouth - Born: New York - Age: 19 - Released on 20 Apr 1815.

Perry, Charles - Seaman - Nbr: 1490 - Prize: Fox - Ship Type: Packet - How taken: Superior - When taken: 25 Jun 1813 - Date Received: 19 Jun 1814 - From what ship: Stapleton - Born: Norfolk - Age: 29 - Released on 1 May 1815.

Perry, Ebenezer - Seaman - Nbr: 6316 - Prize: Prince de Neufchatel - Ship Type: P - How taken: Leander (Newcastle Acasta) - When taken: 20 Dec 1814 - Where taken: Lat 38 Long 56 - Date Received: 19 Feb 1815 - From what ship: HMT Ganges, Plymouth - Born: Cambridge - Age: 19 - Released on 5 Jul 1815.

Perry, John - Seaman - Nbr: 2534 - Prize: Prize of the Young Wasp - Ship Type: Prize - How taken: Not listed - Date Received: 16 Aug 1814 - From what ship: HMT Queen, Halifax - Born: Philadelphia - Age: 22 - Released on 3 May 1815.

Perry, Samuel - Seaman - Nbr: 3205 - How taken: Gave himself up from HMS America - When taken: 16 Jul 1813 - Date Received: 11 Sep 1814 - From what ship: HMT Freya, Chatham - Born: Salem - Age: 25 - Race: Mulatto - Released on 28 May 1815.

American Prisoners of War Held at Dartmoor during the War of 1812

## Alphabetical listing of names

Perry, William - Seaman - Nbr: 3223 - How taken: Gave himself up from HMS Trent - Date Received: 11 Sep 1814 - From what ship: HMT Freya, Chatham - Born: Massachusetts - Age: 31 - Released on 28 May 1815.

Peters, Aaron - Seaman - Nbr: 661 - Prize: Joel Barlow - Ship Type: LM - How taken: HMS Briton - When taken: 3 Jul 1813 - Where taken: off Bordeaux - Date Received: 8 Sep 1813 - From what ship: Plymouth - Born: Rhode Island - Age: 21 - Died on 14 Jan 1815 from variola.

Peters, Benjamin - Boatswain - Nbr: 4255 - Prize: Argus - Ship Type: MV - How taken: HMS San Domingo - When taken: 1 Mar 1814 - Where taken: off Savannah - Date Received: 7 Oct 1814 - From what ship: HMT Niobe, Chatham - Born: Providence - Age: 28 - Released on 13 Jun 1815.

Peters, John - Seaman - Nbr: 5738 - Prize: US Schooner Somers - Ship Type: MW - How taken: British gunboats - When taken: 12 Aug 1814 - Where taken: Fort Erie - Date Received: 26 Dec 1814 - From what ship: HMT Argo - Born: Pennsylvania - Age: 19 - Race: Negro - Released on 3 Jul 1815.

Peters, John - Seaman - Nbr: 4641 - How taken: Gave himself up from HMS Leviathan - When taken: 28 Oct 1813 - Date Received: 9 Oct 1814 - From what ship: HMT Leyden, Chatham - Born: Pennsylvania - Age: 29 - Released on 15 Jun 1815.

Peters, John - Seaman - Nbr: 5200 - Prize: Porcupine - Ship Type: LM - How taken: HMS Acasta - When taken: 18 Jul 1813 - Where taken: off Halifax - Date Received: 31 Oct 1814 - From what ship: HMT Mermaid, Chatham - Born: Boston - Age: 22 - Released on 29 Jun 1815.

Peters, John - Seaman - Nbr: 5358 - Prize: Liberty - Ship Type: MV - How taken: Surrendered - When taken: 30 Dec 1813 - Where taken: Stromess - Date Received: 31 Oct 1814 - From what ship: HMT Leyden, Chatham - Born: Tennessee - Age: 46 - Released on 1 Jul 1815.

Peters, Peter - Seaman - Nbr: 758 - How taken: Impressed at Liverpool - Date Received: 3 Nov 1813 - From what ship: Plymouth - Born: Exeter - Age: 20 - Released on 26 Apr 1815.

Peters, Thomas - Seaman - Nbr: 4646 - How taken: Gave himself up from HMS Berwick - When taken: 25 Nov 1813 - Date Received: 9 Oct 1814 - From what ship: HMT Leyden, Chatham - Born: Baltimore - Age: 39 - Race: Black - Released on 15 Jun 1815.

Peterson, Alexander - Seaman - Nbr: 6270 - How taken: Impressed at Greenock - Date Received: 10 Feb 1815 - From what ship: HMT Ganges, Plymouth - Born: Maryland - Age: 25 - Race: Black - Released on 5 Jul 1815.

Peterson, Alexander - Seaman - Nbr: 173 - How taken: Impressed at Liverpool - When taken: 23 Feb 1813 - Date Received: 2 Apr 1813 - From what ship: Plymouth - Born: New York - Age: 29 - Released on 20 Apr 1815.

Peterson, Christopher - Marine - Nbr: 4040 - Prize: US Brig Rattlesnake - Ship Type: MW - How taken: HMS Leander - When taken: 13 Jul 1814 - Where taken: off Shelburne - Date Received: 6 Oct 1814 - From what ship: HMT Chesapeake, Halifax - Born: Philadelphia - Age: 23 - Released on 13 Jun 1815.

Peterson, David - Seaman - Nbr: 6344 - Prize: Prince de Neufchatel - Ship Type: P - How taken: Leander (Newcastle Acasta) - When taken: 20 Dec 1814 - Where taken: Lat 38 Long 56 - Date Received: 19 Feb 1815 - From what ship: HMT Ganges, Plymouth - Born: Wiscasset - Age: 20 - Released on 6 Jul 1815.

Peterson, Jacob - Seaman - Nbr: 6107 - Prize: US Schooner Somers - Ship Type: MW - How taken: British gunboats - When taken: 12 Aug 1814 - Where taken: Fort Erie - Date Received: 6 Jan 1814 - From what ship: HMT Impregnable - Born: Philadelphia - Age: 20 - Released on 5 Jul 1815.

Peterson, Jacob - Seaman - Nbr: 3588 - How taken: Gave himself up from MV John - When taken: Apr 1814 - Date Received: 30 Sep 1814 - From what ship: HMT Sybella - Born: Rhode Island - Age: 22 - Race: Negro - Died on 4 Nov 1814.

Peterson, James - Seaman - Nbr: 2456 - Prize: US Sloop Frolic - Ship Type: MW - How taken: HMS Orpheus - When taken: 20 Apr 1814 - Where taken: off Cuba - Date Received: 16 Aug 1814 - From what ship: HMT Queen, Halifax - Born: New Jersey - Age: 28 - Released on 3 May 1815.

Peterson, James - Seaman - Nbr: 3625 - Prize: Monarch - Ship Type: MV - How taken: HMS Dotterel - When taken: 14 Dec 1813 - Where taken: off Charleston - Date Received: 30 Sep 1814 - From what ship: HMT Sybella - Born: Maryland - Age: 46 - Released on 4 Jun 1815.

Peterson, John - Seaman - Nbr: 6515 - How taken: Gave himself up from HMS Orontes - When taken: Feb 1815 - Date Received: 3 Mar 1815 - From what ship: HMT Ganges, Plymouth - Born: Albany - Age: 31 - Race: Mulatto - Died on 1 Jun 1815 from phthisis pulmonalis.

Peterson, John - Seaman - Nbr: 5447 - Prize: Prize to the Lawrence - Ship Type: P - How taken: HMS Glasgow - When taken: 2 Nov 1814 - Where taken: Channel - Date Received: 10 Dec 1814 - From what ship: HMT Impregnable - Born: New York - Age: 24 - Race: Negro - Released on 1 Jul 1815.

Peterson, John - Seaman - Nbr: 6269 - Prize: El Patrick - Ship Type: MV - How taken: Superior - When taken: Sep

American Prisoners of War Held at Dartmoor during the War of 1812

## Alphabetical listing of names

1814 - Where taken: off NY - Date Received: 10 Feb 1815 - From what ship: HMT Ganges, Plymouth - Born: Liverpool - Age: 21 - Race: Mulatto - Released on 5 Jul 1815.

Peterson, John - Seaman - Nbr: 3479 - How taken: Gave himself up from HMS Clarence - When taken: 12 Sep 1814 - Date Received: 19 Sep 1814 - From what ship: HMT Salvador del Mundo - Born: New Orleans - Age: 27 - Race: Black - Released on 17 Jul 1815.

Peterson, John - Seaman - Nbr: 2965 - How taken: Gave himself up from MV Spencer - When taken: 29 May 1813 - Date Received: 24 Aug 1814 - From what ship: HMT Hannibal - Born: Baltimore - Age: 34 - Released on 19 May 1815.

Peterson, Laurence - Seaman - Nbr: 860 - Prize: Amiable - Ship Type: LM - How taken: HMS Magnificent - When taken: 30 Oct 1813 - Where taken: off Bordeaux - Date Received: 4 Dec 1813 - From what ship: Plymouth - Born: Norrkoping, Sweden - Age: 20 - Released on 6 May 1814.

Peterson, Lawrence - Seaman - Nbr: 3629 - Prize: Monarch - Ship Type: MV - How taken: HMS Dotterel - When taken: 14 Dec 1813 - Where taken: off Charleston - Date Received: 30 Sep 1814 - From what ship: HMT Sybella - Born: Sweden - Age: 20 - Died on 8 Jan 1815 from phthisis pulmonalis.

Peterson, P. - 3rd Lieutenant - Nbr: 4339 - Prize: Nonsuch - Ship Type: LM - How taken: HMS Dotterel - When taken: 14 Dec 1813 - Where taken: off Block Island - Date Received: 7 Oct 1814 - From what ship: HMT Salvador del Mundo, Halifax - Born: Virginia - Age: 33 - Escaped on 1 Jun 1815.

Peterson, Samuel - Seaman - Nbr: 4976 - Prize: Invincible - Ship Type: LM - How taken: HMS Armide - When taken: 15 Aug 1814 - Where taken: off Nantucket - Date Received: 28 Oct 1814 - From what ship: HMT Alkbar, Halifax - Born: Hamburg - Age: 36 - Released on 21 Jun 1815.

Peterson, Thomas - Seaman - Nbr: 6522 - How taken: Gave himself up from HMS Ashea - Date Received: 3 Mar 1815 - From what ship: HMT Ganges, Plymouth - Born: Norfolk - Age: 25 - Race: Man of color - Released on 13 Jun 1815.

Peterson, William - Seaman - Nbr: 2196 - Prize: Olio - Ship Type: Prize - How taken: HMS Cyane - When taken: 22 May 1814 - Where taken: off NY - Date Received: 16 Aug 1814 - From what ship: HMS Dublin, Halifax - Born: Delaware - Age: 32 - Race: Black - Released on 2 May 1815.

Peterson, William - Seaman - Nbr: 2900 - How taken: Gave himself up from HMS Ocean - When taken: 28 May 1813 - Date Received: 24 Aug 1814 - From what ship: HMT Alpheus, Chatham - Born: Long Island - Age: 28 - Race: Black - Released on 19 May 1815.

Petion, Samuel - Seaman - Nbr: 2366 - Prize: Snap Dragon - Ship Type: P - How taken: HMS Martin - When taken: 10 Jun 1814 - Where taken: off Halifax - Date Received: 16 Aug 1814 - From what ship: HMT Queen, Halifax - Born: Connecticut - Age: 24 - Race: Negro - Released on 3 May 1815.

Petterson, Frederick - Seaman - Nbr: 5510 - Prize: Ann Dorothy - Ship Type: Prize - How taken: HMS Maidstone - When taken: 30 Oct 1814 - Where taken: off Cape Sable Island (Canada) - Date Received: 17 Dec 1814 - From what ship: HMT Loire, Halifax - Born: Prussia - Age: 26 - Released on 29 Jun 1815.

Pettigrene, William - Seaman - Nbr: 3398 - Prize: Thomas - Ship Type: P - How taken: HMS Nymphe - When taken: 24 Jun 1813 - Where taken: off Halifax - Date Received: 13 Sep 1814 - From what ship: HMT Niobe, Chatham - Born: Massachusetts - Age: 16 - Released on 28 May 1815.

Pettingell, James - Seaman - Nbr: 2991 - Prize: Frolic - Ship Type: P - How taken: HMS Heron - When taken: 25 Jan 1814 - Where taken: off St. Thomas - Date Received: 29 Aug 1814 - From what ship: HMT Bittern - Born: Ipswich - Age: 33 - Released on 28 May 1815.

Pettingall, Joseph - Seaman - Nbr: 3297 - Prize: Enterprize - Ship Type: P - How taken: HMS Tenedos - When taken: 21 May 1813 - Where taken: off Cape Cod - Date Received: 13 Sep 1814 - From what ship: HMT Niobe, Chatham - Born: Salem - Age: 18 - Died on 7 Oct 1814 from phthisis.

Petton, Alexander - Seaman - Nbr: 1626 - Prize: Tom - Ship Type: LM - How taken: HMS Surveillante - When taken: 24 Apr 1813 - Where taken: Bay of Biscay - Date Received: 23 Jun 1814 - From what ship: Stapleton - Born: Massachusetts - Age: 22 - Released on 1 May 1815.

Phelps, Elijah - Seaman - Nbr: 3943 - Prize: Rolla - Ship Type: P - How taken: HMS Loire - When taken: 10 Dec 1813 - Where taken: off Bull Island (South Carolina) - Date Received: 5 Oct 1814 - From what ship: HMT President, Halifax - Born: Connecticut - Age: 30 - Released on 9 Jun 1815.

Phenny, John - Seaman - Nbr: 3284 - Prize: Juliana Smith - Ship Type: P - How taken: HMS Nymphe - When taken: 11 May 1813 - Where taken: off Cape Sable Island (Canada) - Date Received: 13 Sep 1814 - From what ship: HMT Niobe, Chatham - Born: Sandwich - Age: 25 - Released on 28 May 1815.

Philbrook, Cyrus - Seaman - Nbr: 3649 - Prize: Ulysses - Ship Type: MV - How taken: HMS Majestic - When taken: 29 Jun 1813 - Where taken: off Western Islands (England) - Date Received: 30 Sep 1814 - From what

American Prisoners of War Held at Dartmoor during the War of 1812

## Alphabetical listing of names

ship: HMT President - Born: Massachusetts - Age: 23 - Released on 4 Jun 1815.
Philen, Richard - 2nd Mate - Nbr: 293 - Prize: Ducornau - Ship Type: MV - How taken: HMS Pheasant - When taken: 15 Mar 1813 - Where taken: Bay of Biscay - Date Received: 28 Jun 1813 - From what ship: Plymouth - Born: Wilmington - Age: 38 - Released on 20 Apr 1815.
Philips, Timothy - Seaman - Nbr: 4514 - Prize: Grand Turk - Ship Type: P - How taken: HMS Tenedos - When taken: 26 May 1813 - Where taken: off Cape Cod - Date Received: 8 Oct 1814 - From what ship: HMT Leyden, Chatham - Born: Massachusetts - Age: 20 - Released on 15 Jun 1815.
Philling, Nathaniel - Seaman - Nbr: 1077 - Prize: Three Brothers - Ship Type: MV - How taken: HMS Ringdove - When taken: 11 Nov 1813 - Where taken: at sea - Date Received: 10 May 1814 - From what ship: Plymouth - Born: Salem - Age: 23 - Released on 27 Apr 1815.
Phillips, Augusta - Seaman - Nbr: 5835 - Prize: US Schooner Tigress - Ship Type: MW - How taken: British gunboats - When taken: 2 Sep 1814 - Where taken: Lake Erie - Date Received: 26 Dec 1814 - From what ship: HMT Argo - Born: Russia - Age: 50 - Released on 3 Jul 1815.
Phillips, Benjamin - Seaman - Nbr: 568 - Prize: US Brig Argus - Ship Type: MW - How taken: HMS Pelican - When taken: 14 Apr 1813 - Where taken: Irish Channel - Date Received: 8 Sep 1813 - From what ship: Plymouth - Born: Massachusetts - Age: 56 - Sent to Plymouth on 8 Jul 1814.
Phillips, Benjamin - Carpenter - Nbr: 3214 - How taken: Gave himself up from HMS Castor - When taken: 13 Dec 1812 - Date Received: 11 Sep 1814 - From what ship: HMT Freya, Chatham - Born: Charleston - Age: 29 - Released on 27 Apr 1815.
Phillips, Burton - Seaman - Nbr: 3481 - How taken: Gave himself up from HMS Prince - When taken: 12 Sep 1814 - Date Received: 19 Sep 1814 - From what ship: HMT Salvador del Mundo - Born: Connecticut - Age: 44 - Race: Black - Released on 4 Jun 1815.
Phillips, George - Seaman - Nbr: 4769 - Prize: Rambler - Ship Type: MV - How taken: Morley (Transport) - When taken: 10 Feb 1813 - Where taken: off Isle of France - Date Received: 9 Oct 1814 - From what ship: HMT Freya, Chatham - Born: Salem - Age: 28 - Released on 15 Jun 1815.
Phillips, George W. - Prize Master - Nbr: 3124 - Prize: Prize of the Blockage - Ship Type: Prize - How taken: HMS Brazen - When taken: 29 Jun 1813 - Where taken: off Scotland - Date Received: 11 Sep 1814 - From what ship: HMT Freya, Chatham - Born: Providence - Age: 27 - Released on 28 May 1815.
Phillips, James - Seaman - Nbr: 1092 - How taken: Impressed on shore - When taken: 23 Mar 1814 - Date Received: 10 May 1814 - From what ship: Plymouth - Born: Harwich - Age: 33 - Released on 27 Apr 1815.
Phillips, John - Seaman - Nbr: 6503 - How taken: Gave himself up from MV Edward - Date Received: 3 Mar 1815 - From what ship: HMT Ganges, Plymouth - Born: Connecticut - Age: 22 - Released on 11 Jul 1815.
Phillips, John - Seaman - Nbr: 4745 - How taken: Gave himself up from HMS Hussar - When taken: 26 Jul 1814 - Date Received: 9 Oct 1814 - From what ship: HMT Freya, Chatham - Born: New York - Age: 40 - Race: Black - Released on 14 Jun 1815.
Phillips, John - Seaman - Nbr: 1595 - How taken: Said he belonged to the Shadow - Date Received: 23 Jun 1814 - From what ship: Stapleton - Born: Pennsylvania - Age: 25 - Released on 1 May 1815.
Phillips, John - Seaman - Nbr: 1297 - Prize: John & Frances - Ship Type: Prize - How taken: HMS Sterling Castle - When taken: 10 May 1814 - Where taken: off Cape Clear - Date Received: 14 Jun 1814 - From what ship: Mill Prison (Plymouth, England) - Born: Philadelphia - Age: 22 - Race: Black - Released on 28 Apr 1815.
Phillips, John - Seaman - Nbr: 5201 - Prize: Porcupine - Ship Type: LM - How taken: HMS Acasta - When taken: 18 Jul 1813 - Where taken: off Halifax - Date Received: 31 Oct 1814 - From what ship: HMT Mermaid, Chatham - Born: New Orleans - Age: 42 - Race: Black - Released on 29 Jun 1815.
Phillips, Peter - Seaman - Nbr: 1188 - How taken: Impressed at Cork - When taken: 24 Nov 1812 - Date Received: 4 Jun 1814 - From what ship: Dartmouth - Born: Washington - Age: 27 - Sent to Plymouth on 8 Jul 1814.
Phillips, Thomas - Seaman - Nbr: 2062 - Prize: Dart - Ship Type: LM - How taken: HMS Peterel - When taken: 5 Mar 1812 - Where taken: at sea - Date Received: 3 Aug 1814 - From what ship: HMS Bittern, Chatham Depot - Born: York - Age: 29 - Released on 26 Apr 1815.
Phillips, William - Pilot - Nbr: 6463 - Prize: Decatur - Ship Type: P - How taken: HMS Rhin - When taken: 5 Jun 1814 - Where taken: off San Domingo - Date Received: 3 Mar 1815 - From what ship: HMT Ganges, Plymouth - Born: Philadelphia - Age: 21 - Released on 11 Jul 1815.
Phillips, William - Seaman - Nbr: 3133 - How taken: Gave himself up from HMS Scorpion - When taken: 20 Dec 1812 - Date Received: 11 Sep 1814 - From what ship: HMT Freya, Chatham - Born: Baltimore - Age: 29 - Race: Black - Released on 27 Apr 1815.
Philman, Adam - Sailing master - Nbr: 2566 - Prize: Rattlesnake - Ship Type: P - How taken: HMS Hyperion -

American Prisoners of War Held at Dartmoor during the War of 1812

## Alphabetical listing of names

When taken: 25 Jun 1814 - Where taken: off Cape Finisterre - Date Received: 21 Aug 1814 - From what ship: HMS Hyperion - Born: Philadelphia - Age: 21 - Released on 6 Apr 1815.

Phip, William - Prize Master - Nbr: 3662 - Prize: Alfred - Ship Type: P - How taken: HMS Epervier - When taken: 23 Feb 1812 - Where taken: off Newfoundland - Date Received: 30 Sep 1814 - From what ship: HMT President - Born: Salem - Age: 28 - Released on 4 Jun 1815.

Phippen, Israel - Seaman - Nbr: 4372 - Prize: Growler - Ship Type: P - How taken: HMS Electra - When taken: 7 Jul 1813 - Where taken: at sea - Date Received: 8 Oct 1814 - From what ship: HMT Leyden, Chatham - Born: Massachusetts - Age: 31 - Released on 14 Jun 1815.

Phippen, John - Seaman - Nbr: 4530 - Prize: Wolf Cove - Ship Type: MV - How taken: HMS Briton - When taken: 1 Dec 1812 - Where taken: off Brest - Date Received: 8 Oct 1814 - From what ship: HMT Leyden, Chatham - Born: Massachusetts - Age: 27 - Released on 27 Apr 1815.

Phipps, Richard - Seaman - Nbr: 4449 - Prize: Dart - Ship Type: LM - How taken: HMS Niger - When taken: 13 Nov 1815 - Where taken: at sea - Date Received: 8 Oct 1814 - From what ship: HMT Leyden, Chatham - Born: New Orleans - Age: 19 - Released on 14 Jun 1815.

Phipps, Stephen - Seaman - Nbr: 1546 - Prize: Zebra - Ship Type: LM - How taken: HMS Pyramus - When taken: 20 Apr 1813 - Where taken: Bay of Biscay - Date Received: 23 Jun 1814 - From what ship: Stapleton - Born: Massachusetts - Age: 27 - Released on 1 May 1815.

Phoenix, John - Master - Nbr: 5534 - Prize: Eagle - Ship Type: MV - How taken: HMS Havana - When taken: 29 Aug 1814 - Where taken: off Cape VA - Date Received: 17 Dec 1814 - From what ship: HMT Loire, Halifax - Born: Monmouth - Age: 30 - Released on 1 Jul 1815.

Pickett, Josiah - Seaman - Nbr: 5982 - Prize: Cossack - Ship Type: Private - How taken: HMS Bulwark - When taken: 1 Nov 1814 - Where taken: Georges Bank - Date Received: 27 Dec 1814 - From what ship: HMT Penelope - Born: Beverly - Age: 19 - Released on 3 Jul 1815.

Picking, George - Cook - Nbr: 365 - Prize: Two Brothers - Ship Type: MV - How taken: Beetle (LM) - When taken: 18 Mar 1813 - Where taken: off Western Islands (England) - Date Received: 1 Jul 1813 - From what ship: Plymouth - Born: Boston - Age: 20 - Sent to Liverpool on 25 Aug 1813.

Pickrage, George - Seaman - Nbr: 507 - Prize: Tickler - Ship Type: LM - How taken: HMS Magicienne - When taken: 5 Jun 1813 - Where taken: Bay of Biscay - Date Received: 8 Sep 1813 - From what ship: Plymouth - Born: Messina - Age: 24 - Released on 26 Apr 1815.

Pickrase, Daniel - Carpenter's mate - Nbr: 5958 - Prize: Harlequin - Ship Type: P - How taken: HMS Bulwark - When taken: 23 Nov 1814 - Taken: off Halifax - Born: Portsmouth - Age: 25 - Released on 3 Jul 1815.

Pidgeon, Jacques - Seaman - Nbr: 2299 - Prize: Hussar - Ship Type: P - How taken: HMS Saturn - When taken: 25 May 1814 - Where taken: off Sandy Hook - Date Received: 16 Aug 1814 - From what ship: HMS Dublin, Halifax - Born: Grumprie - Age: 21 - Released on 3 May 1815.

Pier, Henry - Soldier - Nbr: 5846 - Prize: 1st Militia Regiment - Ship Type: Troops - How taken: British Army - When taken: 17 Sep 1814 - Where taken: Fort Erie - Date Received: 26 Dec 1814 - From what ship: HMT Argo - Born: New Jersey - Age: 25 - Released on 29 Jun 1815.

Pierce, Amos - Seaman - Nbr: 158 - How taken: Gave himself up from HMS Lavinia - Date Received: 2 Apr 1813 - From what ship: Plymouth - Born: New Hampshire - Age: 29 - Released on 20 Apr 1815.

Pierce, Edward - 1st Lieutenant - Nbr: 3765 - Prize: Fame - Ship Type: P - How taken: HMS Thistle - When taken: 10 Apr 1814 - Where taken: after being cast ashore on a seal island - Date Received: 30 Sep 1814 - From what ship: HMT President, Halifax - Born: Salem - Age: 31 - Released on 9 Jun 1815.

Pierce, Edward - Seaman - Nbr: 2747 - How taken: Gave himself up from HMS Circe - When taken: 5 Nov 1812 - Date Received: 24 Aug 1814 - From what ship: HMT Liverpool, Chatham - Born: Baltimore - Age: 26 - Released on 26 Apr 1815.

Pierce, John - Seaman - Nbr: 3790 - Prize: James - Ship Type: Prize - How taken: Rebecca (LM) - When taken: 13 Apr 1814 - Where taken: off Delaware - Date Received: 5 Oct 1814 - From what ship: HMT President, Halifax - Born: New York - Age: 25 - Released on 9 Jun 1815.

Pierce, John - Seaman - Nbr: 3683 - Prize: Alfred - Ship Type: P - How taken: HMS Epervier - When taken: 23 Feb 1812 - Where taken: off Newfoundland - Date Received: 30 Sep 1814 - From what ship: HMT President - Born: Marblehead - Age: 16 - Released on 4 Jun 1815.

Pierce, Joseph - Seaman - Nbr: 4510 - Prize: America - Ship Type: P - How taken: HMS Shannon - When taken: 23 May 1813 - Where taken: off Cape Cod - Date Received: 8 Oct 1814 - From what ship: HMT Leyden, Chatham - Born: Massachusetts - Age: 22 - Released on 15 Jun 1815.

Pierce, Moses - Seaman - Nbr: 2288 - Prize: Diomede - Ship Type: P - How taken: HMS Rifleman - When taken: 28

American Prisoners of War Held at Dartmoor during the War of 1812

## Alphabetical listing of names

May 1814 - Where taken: off Sable Island - Date Received: 16 Aug 1814 - From what ship: HMS Dublin, Halifax - Born: West-Cappel (France) - Age: 23 - Released on 3 May 1815.

Pierce, Samuel - Seaman - Nbr: 6294 - How taken: Gave himself up from HMS Sabine - Date Received: 17 Feb 1815 - From what ship: HMT Ganges, Plymouth - Born: Boston - Age: 23 - Released on 5 Jul 1815.

Pierce, Samuel - Seaman - Nbr: 892 - Prize: Dart - Ship Type: MV - How taken: HMS Niger - When taken: 12 Nov 1813 - Where taken: off Cape Finisterre - Date Received: 31 Jan 1814 - From what ship: Plymouth - Born: Not listed - Died on 12 Mar 1814 from diarrhea and pneumonia.

Pigger, Stephen - Seaman - Nbr: 4355 - Prize: Fiere Facia - Ship Type: P - How taken: HMS Ramillies - When taken: 26 Feb 1814 - Where taken: off NY - Date Received: 7 Oct 1814 - From what ship: HMT Salvador del Mundo, Halifax - Born: Paris - Age: 29 - Released on 14 Jun 1815.

Pike, Benjamin - Seaman - Nbr: 6113 - Prize: Garland - Ship Type: Prize - How taken: HMS Hamadryad - When taken: 22 Nov 1814 - Where taken: Georges Bank - Date Received: 6 Jan 1814 - From what ship: HMT Impregnable - Born: Newbury - Age: 26 - Released on 5 Jul 1815.

Pike, Benjamin - Seaman - Nbr: 5475 - Prize: General Putnam - Ship Type: P - How taken: HMS Leander - When taken: 8 Nov 1814 - Where taken: Long 65 Lat 42 - Date Received: 17 Dec 1814 - From what ship: HMT Loire, Halifax - Born: Salem - Age: 27 - Released on 1 Jul 1815.

Pike, James - Seaman - Nbr: 3836 - Prize: Dominique - Ship Type: LM - How taken: HMS Dotterel - When taken: 21 May 1814 - Where taken: off Charleston - Date Received: 5 Oct 1814 - From what ship: HMT Orpheus, Halifax - Born: Massachusetts - Age: 22 - Released on 9 Jun 1815.

Pike, Jeremiah - Boatswain's Mate - Nbr: 4393 - Prize: Portsmouth Packet - Ship Type: Prize - How taken: HMS Fantome - When taken: 5 Oct 1813 - Where taken: off Portland - Date Received: 8 Oct 1814 - From what ship: HMT Leyden, Chatham - Born: Portsmouth - Age: 26 - Released on 14 Jun 1815.

Pike, John - Seaman - Nbr: 235 - Prize: Charlotte - Ship Type: MV - How taken: HMS Warspite - When taken: 3 Mar 1813 - Where taken: Bay of Biscay - Date Received: 2 Apr 1813 - From what ship: Plymouth - Born: Washington - Age: 18 - Released on 20 Apr 1815.

Piles, James - Seaman - Nbr: 5390 - How taken: Gave himself up at Chatham - When taken: 10 Sep 1814 - Date Received: 31 Oct 1814 - From what ship: HMT Leyden, Chatham - Born: Philadelphia - Age: 37 - Race: Black - Released on 1 Jul 1815.

Piles, John - Seaman - Nbr: 4662 - Prize: Teazer - Ship Type: P - How taken: HMS Boyce - When taken: 15 Jun 1813 - Where taken: off Halifax - Date Received: 9 Oct 1814 - From what ship: HMT Leyden, Chatham - Born: New York - Age: 27 - Released on 13 Jun 1815.

Pilot, Samuel G. - Mate - Nbr: 4791 - Prize: Sarah & Ann - Ship Type: MV - How taken: Dashu - When taken: 8 Dec 1813 - Where taken: off Barbados - Date Received: 9 Oct 1814 - From what ship: HMT Freya, Halifax - Born: Savannah - Age: 25 - Sent to Ashburton (England) on 18 Oct 1814.

Pinckham, Amos - Seaman - Nbr: 956 - Prize: Siro - Ship Type: LM - How taken: HMS Pelican - When taken: 13 Jan 1814 - Where taken: at sea - Date Received: 31 Jan 1814 - From what ship: Plymouth - Born: Bristol - Age: 30 - Released on 27 Apr 1815.

Pinder, George - Seaman - Nbr: 3274 - Prize: Cossack - Ship Type: P - How taken: HMS Amelia - When taken: 14 Apr 1813 - Where taken: off St. Johns - Date Received: 13 Sep 1814 - From what ship: HMT Niobe, Chatham - Born: Ipswich - Age: 21 - Released on 28 May 1815.

Piner, Thomas - Seaman - Nbr: 2333 - Prize: Snap Dragon - Ship Type: P - How taken: HMS Martin - When taken: 10 Jun 1814 - Where taken: off Halifax - Date Received: 16 Aug 1814 - From what ship: HMS Dublin, Halifax - Born: North Carolina - Age: 23 - Released on 3 May 1815.

Pines, Isaac - Seaman - Nbr: 4566 - How taken: Gave himself up from MV Liberty - When taken: 30 Dec 1813 - Date Received: 9 Oct 1814 - From what ship: HMT Leyden, Chatham - Born: Swedesboro - Age: 35 - Released on 15 Jun 1815.

Pinkham, Allen - Seaman - Nbr: 2769 - How taken: Gave himself up from HMS Franchise - When taken: 20 Dec 1812 - Date Received: 24 Aug 1814 - From what ship: HMT Liverpool, Chatham - Born: Bristol - Age: 25 - Released on 19 May 1815.

Pinkham, Jacob - Seaman - Nbr: 3100 - Prize: Mary - Ship Type: Prize - How taken: Taken by pilot boat - When taken: 27 Aug 1814 - Where taken: Bristol Channel - Date Received: 9 Sep 1814 - From what ship: HMS Abercrombie - Born: Massachusetts - Age: 22 - Died on 25 Sep 1814.

Pinne, John - Seaman - Nbr: 3210 - How taken: Gave himself up from HMS Achille - When taken: 8 Sep 1813 - Date Received: 11 Sep 1814 - From what ship: HMT Freya, Chatham - Born: Long Island - Age: 26 - Race: Black - Released on 11 Jul 1815.

American Prisoners of War Held at Dartmoor during the War of 1812

## Alphabetical listing of names

Pipe, Henry - Seaman - Nbr: 1735 - Prize: Hugh Jones - Ship Type: MV - How taken: HMS Bittern - When taken: 14 Jun 1814 - Where taken: at sea - Date Received: 18 Jul 1814 - From what ship: HMT Salvador del Mundo, Plymouth - Born: Philadelphia - Age: 24 - Released on 1 May 1815.

Pith, William - Seaman - Nbr: 3257 - How taken: Gave himself up from HMS Invincible - When taken: 1 Feb 1813 - Date Received: 11 Sep 1814 - From what ship: HMT Freya, Chatham - Born: Salem - Age: 26 - Race: Mulatto - Released on 28 May 1815.

Pitman, Henry - Seaman - Nbr: 3309 - Prize: Thomas - Ship Type: P - How taken: HMS Nymphe - When taken: 28 Jun 1813 - Where taken: off Halifax - Date Received: 13 Sep 1814 - From what ship: HMT Niobe, Chatham - Born: Portsmouth - Age: 23 - Released on 28 May 1815.

Pitman, John - Seaman - Nbr: 2412 - Prize: US Sloop Frolic - Ship Type: MW - How taken: HMS Orpheus - When taken: 20 Apr 1814 - Where taken: off Cuba - Date Received: 16 Aug 1814 - From what ship: HMT Queen, Halifax - Born: Portsmouth - Age: 28 - Sent to Dartmouth on 19 Oct 1814.

Pittman, Benjamin - Seaman - Nbr: 4542 - Prize: Growler - Ship Type: P - How taken: HMS Electra - When taken: 7 Jul 1813 - Where taken: at sea - Date Received: 8 Oct 1814 - From what ship: HMT Leyden, Chatham - Born: Marblehead - Age: 21 - Released on 15 Jun 1815.

Pittman, Benjamin - Seaman - Nbr: 3754 - Prize: Alfred - Ship Type: P - How taken: HMS Epervier - When taken: 23 Feb 1814 - Where taken: off Newfoundland - Date Received: 30 Sep 1814 - From what ship: HMT President, Halifax - Born: Marblehead - Age: 49 - Released on 4 Jun 1815.

Pittman, Samuel - Seaman - Nbr: 2982 - Prize: Frolic - Ship Type: P - How taken: HMS Heron - When taken: 25 Jan 1814 - Where taken: off St. Thomas - Date Received: 29 Aug 1814 - From what ship: HMT Bittern - Born: Salem - Age: 41 - Released on 19 May 1815.

Pitts, Charles - Seaman - Nbr: 3234 - Prize: Maydock - Ship Type: MV - How taken: Rebuff - When taken: 16 Jun 1813 - Where taken: off St. Marys - Date Received: 11 Sep 1814 - From what ship: HMT Freya, Chatham - Born: Massachusetts - Age: 20 - Released on 28 May 1815.

Pitts, Francis - Seaman - Nbr: 6249 - Prize: Steven Getard - Ship Type: MV - How taken: HMS Mars - When taken: 18 Dec 1813 - Date Received: 4 Feb 1815 - From what ship: HMT Ganges - Born: Philadelphia - Age: 24 - Race: Black - Released on 11 Jul 1815.

Pitts, George - Seaman - Nbr: 6261 - Prize: Plutarch - Ship Type: MV - How taken: HMS Helicon - When taken: 5 Feb 1815 - Where taken: off Bordeaux - Date Received: 10 Feb 1815 - From what ship: HMT Ganges, Plymouth - Born: Newport - Age: 35 - Released on 5 Jul 1815.

Pitts, George - Lieutenant - Nbr: 4519 - Prize: Yankee - Ship Type: P - How taken: HMS Shannon - When taken: 20 Aug 1813 - Where taken: at sea - Date Received: 8 Oct 1814 - From what ship: HMT Leyden, Chatham - Born: Massachusetts - Age: 31 - Released on 15 Jun 1815.

Pitts, Thomas - Prize master - Nbr: 3745 - Prize: Mariner - Ship Type: Prize - How taken: HMS Poictiers - When taken: 30 Aug 1813 - Where taken: off Newfoundland - Date Received: 30 Sep 1814 - From what ship: HMT President, Halifax - Born: Massachusetts - Age: 28 - Released on 11 Jul 1815.

Pitts, William - Seaman - Nbr: 1399 - Prize: Zebra - Ship Type: LM - How taken: HMS Pyramus - When taken: 20 Apr 1813 - Where taken: Bay of Biscay - Date Received: 19 Jun 1814 - From what ship: Stapleton - Born: Massachusetts - Age: 29 - Released on 28 Apr 1815.

Place, Aaron - Seaman - Nbr: 2369 - Prize: Snap Dragon - Ship Type: P - How taken: HMS Martin - When taken: 10 Jun 1814 - Where taken: off Halifax - Date Received: 16 Aug 1814 - From what ship: HMT Queen, Halifax - Born: Long Island - Age: 23 - Released on 3 Apr 1815.

Place, James - Prize master - Nbr: 5702 - Prize: Halifax Packet - Ship Type: Prize - How taken: HMS Bulwark - When taken: 20 Sep 1814 - Where taken: Georges Bank - Date Received: 24 Dec 1814 - From what ship: HMT Penelope - Born: Boston - Age: 33 - Released on 28 May 1815.

Place, John - Armourer - Nbr: 564 - Prize: US Brig Argus - Ship Type: MW - How taken: HMS Pelican - When taken: 14 Apr 1813 - Where taken: Irish Channel - Date Received: 8 Sep 1813 - From what ship: Plymouth - Born: New York - Age: 25 - Sent to Dartmouth on 2 Nov 1814.

Place, Thomas - Seaman - Nbr: 4467 - Prize: Governor Plumer - Ship Type: P - How taken: HMS Shamrock - When taken: 4 Mar 1813 - Where taken: at sea - Date Received: 8 Oct 1814 - From what ship: HMT Leyden, Chatham - Born: Massachusetts - Age: 34 - Released on 15 Jun 1815.

Platt, Daniel - Seaman - Nbr: 1889 - Prize: Pearl - Ship Type: LM - How taken: Guernsey (Privateer) - When taken: Mar 1813 - Where taken: at sea - Date Received: 29 Jul 1814 - From what ship: HMS Ville de Paris, Chatham Depot - Born: New York - Age: 19 - Race: Black - Released on 3 May 1815.

Platt, John - Seaman - Nbr: 6095 - Prize: Black Swan - Ship Type: MV - How taken: HMS Maidstone - When taken:

American Prisoners of War Held at Dartmoor during the War of 1812

## Alphabetical listing of names

24 Oct 1814 - Where taken: Georges Bank - Date Received: 28 Dec 1814 - From what ship: HMT Penelope - Born: Not legible - Age: 18 - Released on 3 Jul 1815.

Plowman, Joseph - Seaman - Nbr: 2775 - How taken: Gave himself up from HMS Vigo (Spain) - Date Received: 24 Aug 1814 - From what ship: HMT Liverpool, Chatham - Born: Connecticut - Age: 33 - Released on 19 May 1815.

Plumber, James - Seaman - Nbr: 6412 - Prize: Nancy - Ship Type: P - How taken: Papillion - When taken: 27 Dec 1814 - Where taken: Lat 29 Long 20 - Date Received: 3 Mar 1815 - From what ship: HMT Ganges, Plymouth - Born: Bristol - Age: 26 - Released on 11 Jul 1815.

Plumer, John - Seaman - Nbr: 1553 - Prize: Caroline - Ship Type: MV - How taken: HMS Medusa - When taken: 12 Apr 1813 - Where taken: Bay of Biscay - Date Received: 23 Jun 1814 - From what ship: Stapleton - Born: Virginia - Age: 24 - Released on 1 May 1815.

Plumer, William - Seaman - Nbr: 4195 - Prize: Zephyr - Ship Type: MV - How taken: HMS Surveillante - When taken: 4 Jan 1814 - Where taken: Bay of Biscay - Date Received: 7 Oct 1814 - From what ship: HMT Niobe, Chatham - Born: Connecticut - Age: 32 - Released on 13 Jun 1815.

Plummer, John - Seaman - Nbr: 5967 - Prize: Amazon - Ship Type: Prize - How taken: HMS Bulwark - When taken: 22 Sep 1814 - Where taken: Georges Bank - Date Received: 27 Dec 1814 - From what ship: HMT Penelope - Born: Connecticut - Age: 32 - Released on 3 Jul 1815.

Plummer, William - Seaman - Nbr: 1880 - Prize: Elbridge Gerry - Ship Type: P - How taken: HMS Crescent - When taken: 16 Sep 1813 - Where taken: off St. George's - Date Received: 29 Jul 1814 - From what ship: HMS Ville de Paris, Chatham Depot - Born: Portland - Age: 21 - Released on 2 May 1815.

Pockmitt, Jackness - Seaman - Nbr: 6443 - Prize: Jemmell - Ship Type: MV - How taken: HMS Rhin - When taken: 28 May 1814 - Where taken: off Bermuda - Date Received: 3 Mar 1815 - From what ship: HMT Ganges, Plymouth - Born: Cape Cod - Age: 19 - Race: Creole - Released on 11 Jul 1815.

Poland, John - Seaman - Nbr: 1374 - Prize: Grand Napoleon - Ship Type: P - How taken: HMS Goldfinch - When taken: 17 Apr 1813 - Where taken: Bay of Biscay - Date Received: 19 Jun 1814 - From what ship: Stapleton - Born: Massachusetts - Age: 22 - Released on 28 Apr 1815.

Poland, Thomas - Seaman - Nbr: 988 - How taken: Impressed at Bristol - When taken: 13 Oct 1813 - Date Received: 31 Jan 1814 - From what ship: Plymouth - Born: Selans - Age: 27 - Released on 28 Apr 1815.

Polland, John - Seaman - Nbr: 5052 - Prize: Ida - Ship Type: LM - How taken: HMS Newcastle - When taken: 9 Aug 1814 - Where taken: Long 34 - Date Received: 28 Oct 1814 - From what ship: HMT Alkbar, Halifax - Born: Brazil - Age: 27 - Race: Mulatto - Died on 20 Nov 1814.

Pomp, William - Seaman - Nbr: 1550 - Prize: Zebra - Ship Type: LM - How taken: HMS Pyramus - When taken: 20 Apr 1813 - Where taken: Bay of Biscay - Date Received: 23 Jun 1814 - From what ship: Stapleton - Born: Leighland - Age: 21 - Race: Colored - Released on 1 May 1815.

Pool, John - Seaman - Nbr: 2842 - How taken: Gave himself up from HMS Armide - When taken: 8 Jun 1813 - Date Received: 24 Aug 1814 - From what ship: HMT Liverpool, Chatham - Born: Maryland - Age: 22 - Released on 19 May 1815.

Poole, John - Seaman - Nbr: 1917 - How taken: Gave himself up from HMS Royal William - When taken: 3 Feb 1813 - Date Received: 3 Aug 1814 - From what ship: HMS Alceste, Chatham Depot - Born: Baltimore - Age: 28 - Released on 2 May 1815.

Poor, David - Seaman - Nbr: 4327 - Prize: Martha - Ship Type: Prize - How taken: HMS Belviders - When taken: 25 Feb 1814 - Where taken: off New London - Date Received: 7 Oct 1814 - From what ship: HMT Salvador del Mundo, Halifax - Born: Massachusetts - Age: 23 - Released on 14 Jun 1815.

Popal, Richard - Seaman - Nbr: 981 - Prize: Amity - Ship Type: P - How taken: HMS Achates - When taken: 22 Dec 1813 - Where taken: Bay of Biscay - Date Received: 31 Jan 1814 - From what ship: Plymouth - Born: Philadelphia - Age: 57 - Released on 27 Apr 1815.

Pope, E. - Seaman - Nbr: 5507 - Prize: Ann Dorothy - Ship Type: Prize - How taken: HMS Maidstone - When taken: 30 Oct 1814 - Where taken: off Cape Sable Island (Canada) - Date Received: 17 Dec 1814 - From what ship: HMT Loire, Halifax - Born: Middleburgh - Age: 23 - Released on 29 Jun 1815.

Pope, Oliver - Seaman - Nbr: 6381 - Prize: US Brig Syren - Ship Type: MW - How taken: HMS Medway - When taken: 12 Jul 1814 - Where taken: off Cape of Good Hope - Date Received: 24 Feb 1815 - From what ship: HMT Ganges, Plymouth - Born: Reading - Age: 21 - Released on 6 Jul 1815.

Porston, Joseph - Sailing master - Nbr: 2276 - Prize: Diomede - Ship Type: P - How taken: HMS Rifleman - When taken: 23 Jun 1814 - Where taken: off Halifax - Date Received: 16 Aug 1814 - From what ship: HMS Dublin, Halifax - Born: Salem - Age: 32 - Released on 3 May 1815.

American Prisoners of War Held at Dartmoor during the War of 1812

## Alphabetical listing of names

Port, John - Seaman - Nbr: 4216 - Prize: Harvest - Ship Type: Prize - How taken: HMS Orestes - When taken: 24 Jan 1814 - Where taken: Bay of Biscay - Date Received: 7 Oct 1814 - From what ship: HMT Niobe, Chatham - Born: Rhode Island - Age: 21 - Race: Black - Released on 13 Jun 1815.

Porter, Calvin - Seaman - Nbr: 4157 - Prize: Argus - Ship Type: MV - How taken: HMS San Domingo - When taken: 1 Mar 1814 - Where taken: off Savannah - Date Received: 6 Oct 1814 - From what ship: HMT Niobe, Chatham - Born: Boston - Age: 16 - Race: Black - Released on 3 May 1815.

Porter, Charles - Seaman - Nbr: 4431 - Prize: Taken off an English whaler - Ship Type: MV - How taken: HMS Illustrious - When taken: 22 Oct 1813 - Where taken: at sea - Date Received: 8 Oct 1814 - From what ship: HMT Leyden, Chatham - Born: Massachusetts - Age: 28 - Released on 14 Jun 1815.

Porter, Edward - Seaman - Nbr: 4461 - Prize: Fame - Ship Type: P - How taken: HMS Pratteo - When taken: 3 May 1813 - Where taken: Bay of Biscay - Date Received: 8 Oct 1814 - From what ship: HMT Leyden, Chatham - Born: Salem - Age: 27 - Released on 15 Jun 1815.

Porter, Ephraim - Seaman - Nbr: 4313 - How taken: Gave himself up from HMS Theban - When taken: 21 Jul 1814 - Date Received: 7 Oct 1814 - From what ship: HMT Niobe, Chatham - Born: Bethlehem - Age: 25 - Released on 14 Jun 1815.

Porter, Frederick - Seaman - Nbr: 5613 - How taken: Gave himself up from HMS Trachman - When taken: 16 May 1812 - Date Received: 24 Dec 1814 - From what ship: HMT Tay - Born: Massachusetts - Age: 34 - Released on 27 Mar 1815.

Porter, Gideon - Seaman - Nbr: 4737 - How taken: Gave himself up from HMS Action - When taken: 9 May 1814 - Date Received: 9 Oct 1814 - From what ship: HMT Freya, Chatham - Born: Newport - Age: 32 - Died on 22 Mar 1815 from variola.

Porter, Joseph - Seaman - Nbr: 2518 - Prize: Snap Dragon - Ship Type: P - How taken: HMS Martin - When taken: 10 Jun 1814 - Where taken: off Halifax - Date Received: 16 Aug 1814 - From what ship: HMT Queen, Halifax - Born: Boston - Age: 22 - Released on 3 May 1815.

Porter, Joseph - Seaman - Nbr: 2890 - How taken: Gave himself up from HMS Tremendous - When taken: 15 Dec 1812 - Date Received: 24 Aug 1814 - From what ship: HMT Alpheus, Chatham - Born: Boston - Age: 25 - Released on 27 Apr 1815.

Porter, Louis - Seaman - Nbr: 4520 - Prize: Yankee - Ship Type: P - How taken: HMS Shannon - When taken: 20 Aug 1813 - Where taken: at sea - Date Received: 8 Oct 1814 - From what ship: HMT Leyden, Chatham - Born: North Carolina - Age: 20 - Released on 28 Apr 1815.

Porter, Samuel - Marine - Nbr: 2505 - Prize: US Sloop Frolic - Ship Type: MW - How taken: HMS Orpheus - When taken: 20 Apr 1814 - Where taken: off Cuba - Date Received: 16 Aug 1814 - From what ship: HMT Queen, Halifax - Born: Salem - Age: 27 - Released on 3 May 1815.

Porter, Samuel - Seaman - Nbr: 2635 - How taken: Gave himself up from HMS Leviathan - When taken: 22 Oct 1813 - Date Received: 21 Aug 1814 - From what ship: HMT Freya, Chatham - Born: Boston - Age: 23 - Released on 3 May 1815.

Porter, William - Seaman - Nbr: 208 - Prize: Criterion - Ship Type: MV - How taken: HMS Belle Poule - When taken: 14 Feb 1813 - Where taken: Bay of Biscay - Date Received: 2 Apr 1813 - From what ship: Plymouth - Born: Rhode Island - Age: 25 - Released on 20 Apr 1815.

Portlock, William - Passenger - Nbr: 5641 - Prize: Lion - Ship Type: P - How taken: HMS Granicus - When taken: 2 Dec 1814 - Where taken: off Lisbon - Date Received: 24 Dec 1814 - From what ship: HMT Hanover, Gibraltar - Born: Rochelle - Age: 32 - Released on 3 Jul 1815.

Posey, Valentine - Passenger - Nbr: 2015 - Prize: Messenger - Ship Type: MV - How taken: HMS Iris - When taken: 10 Mar 1813 - Where taken: Bay of Biscay - Date Received: 3 Aug 1814 - From what ship: HMS Lyffey, Chatham Depot - Born: Charlestown - Age: 24 - Released on 2 May 1815.

Postlock, William - 2nd Mate - Nbr: 907 - Prize: Zephyr - Ship Type: MV - How taken: HMS Pyramus - When taken: 30 Nov 1813 - Where taken: off L'Orient (France) - Date Received: 31 Jan 1814 - From what ship: Plymouth - Born: Charlestown - Age: 32 - Sent to Plymouth on 8 Jul 1814.

Poswell, John - Boy - Nbr: 1047 - Prize: Prince of Wales - Ship Type: P - How taken: Nelson (Transport) - When taken: 4 Feb 1814 - Where taken: at sea - Date Received: 10 May 1814 - From what ship: Plymouth - Born: Long Island - Age: 16 - Released on 27 Apr 1815.

Pote, Jeremiah - Seaman - Nbr: 1373 - Prize: Grand Napoleon - Ship Type: P - How taken: HMS Goldfinch - When taken: 17 Apr 1813 - Where taken: Bay of Biscay - Date Received: 19 Jun 1814 - From what ship: Stapleton - Born: Massachusetts - Age: 28 - Released on 28 Apr 1815.

Potter, Jacob - Seaman - Nbr: 3073 - How taken: Gave himself up from HMS Antelope - When taken: 12 Oct 1812 -

American Prisoners of War Held at Dartmoor during the War of 1812

## Alphabetical listing of names

Date Received: 3 Sep 1814 - From what ship: HMT Hydra, Chatham - Born: Lewistown - Age: 30 - Race: Mulatto - Released on 27 Apr 1815.

Potter, John - Seaman - Nbr: 1102 - How taken: Impressed at Bristol - When taken: 25 Feb 1814 - Date Received: 10 May 1814 - From what ship: Plymouth - Born: Nantucket - Age: 26 - Released on 27 Apr 1815.

Potter, John - Seaman - Nbr: 3178 - How taken: Gave himself up from HMS Pompeii - When taken: 27 May 1813 - Date Received: 11 Sep 1814 - From what ship: HMT Freya, Chatham - Born: New York - Age: 27 - Released on 28 May 1815.

Potter, John - Seaman - Nbr: 2854 - How taken: Impressed at Gravesend - When taken: 13 Mar 1813 - Date Received: 24 Aug 1814 - From what ship: HMT Alpheus, Chatham - Born: Philadelphia - Age: 32 - Died on 5 Oct 1814 from pneumonia.

Potter, Richard H. - 2nd Mate - Nbr: 1221 - Prize: Valentine - Ship Type: MV - How taken: HMS Racehorse - When taken: 16 Nov 1812 - Where taken: off Cape of Good Hope - Date Received: 14 Jun 1814 - From what ship: Mill Prison (Plymouth, England) - Born: New London - Age: 30 - Released on 26 Apr 1815.

Potter, Titus - Seaman - Nbr: 3817 - Prize: Nimble - Ship Type: Prize - How taken: HMS Arab - When taken: 5 Apr 1814 - Where taken: Lat 37 Long 65 - Date Received: 5 Oct 1814 - From what ship: HMT Orpheus, Halifax - Born: Massachusetts - Age: 19 - Released on 9 Jun 1815.

Pottigal, John - Seaman - Nbr: 1255 - Prize: Adeline - Ship Type: MV - How taken: HMS Magicienne - When taken: 16 Mar 1814 - Where taken: off Cape Finisterre - Date Received: 14 Jun 1814 - From what ship: Mill Prison (Plymouth, England) - Born: Boston - Age: 46 - Released on 28 Apr 1815.

Poverley, Henry - Seaman - Nbr: 4638 - How taken: Gave himself up from HMS Bombay - When taken: 27 May 1813 - Date Received: 9 Oct 1814 - From what ship: HMT Leyden, Chatham - Born: Hampshire - Age: 33 - Released on 11 Jul 1815.

Powell, Joseph - Seaman - Nbr: 3352 - How taken: Gave himself up from HMS Gordon - When taken: 1 Nov 1812 - Date Received: 13 Sep 1814 - From what ship: HMT Niobe, Chatham - Born: Philadelphia - Age: 32 - Race: Black - Released on 27 Apr 1815.

Powers, James - Seaman - Nbr: 349 - Prize: Courier - Ship Type: LM - How taken: HMS Andromache - When taken: 14 Mar 1813 - Where taken: Bay of Biscay - Date Received: 28 Jun 1813 - From what ship: Plymouth - Born: Philadelphia - Age: 26 - Released on 20 Apr 1815.

Pratt, Lester - Seaman - Nbr: 3085 - Prize: Eliza (Whaler) - Ship Type: MV - How taken: HMS Charles - When taken: 20 Jul 1813 - Where taken: off Peterhead (Scotland) - Date Received: 3 Sep 1814 - From what ship: HMT Hydra, Chatham - Born: Connecticut - Age: 22 - Released on 28 May 1815.

Pratt, Paris - Seaman - Nbr: 558 - Prize: Godfrey and Mary - Ship Type: MV - How taken: Robert Todd - When taken: 23 Jun 1813 - Where taken: Lat 21, 30 Long 53 - Date Received: 8 Sep 1813 - From what ship: Plymouth - Born: New London - Age: 24 - Released on 26 Apr 1815.

Pratt, Philip - Seaman - Nbr: 4664 - Prize: Wasp - Ship Type: P - How taken: HMS Bream - When taken: 10 Jun 1813 - Where taken: off Halifax - Date Received: 9 Oct 1814 - From what ship: HMT Leyden, Chatham - Born: Massachusetts - Age: 20 - Released on 15 Jun 1815.

Pray, Samuel - Prize master - Nbr: 6403 - Prize: Nancy - Ship Type: P - How taken: Papillion - When taken: 27 Dec 1814 - Where taken: Lat 29 Long 20 - Date Received: 3 Mar 1815 - From what ship: HMT Ganges, Plymouth - Born: Kittery - Age: 25 - Released on 11 Jul 1815.

Prely, Charles - Boy - Nbr: 5911 - Prize: Harlequin - Ship Type: P - How taken: HMS Bulwark - When taken: 23 Nov 1814 - Where taken: off Halifax - Date Received: 27 Dec 1814 - From what ship: HMT Penelope - Born: Portsmouth - Age: 18 - Released on 9 Jun 1815.

Preniergath, Nicholas - Seaman - Nbr: 5938 - Prize: Harlequin - Ship Type: P - How taken: HMS Bulwark - When taken: 23 Nov 1814 - Where taken: off Halifax - Date Received: 27 Dec 1814 - From what ship: HMT Penelope - Born: Durham - Age: 26 - Released on 3 Jul 1815.

Prentis, James - Seaman - Nbr: 1896 - How taken: Gave himself up at London - When taken: 15 Feb 1814 - Date Received: 3 Aug 1814 - From what ship: HMS Alceste, Chatham Depot - Born: Boston - Age: 27 - Released on 2 May 1815.

Prescott, Abraham - Seaman - Nbr: 6130 - Prize: Garland - Ship Type: Prize - How taken: Kamiderad - When taken: 22 Nov 1814 - Where taken: Georges Bank - Date Received: 6 Jan 1814 - From what ship: HMT Impregnable - Born: Exeter - Age: 19 - Released on 5 Jul 1815.

Prestley, James - Prize Master - Nbr: 941 - Prize: Siro - Ship Type: LM - How taken: HMS Pelican - When taken: 13 Jan 1814 - Where taken: at sea - Date Received: 31 Jan 1814 - From what ship: Plymouth - Born: Massachusetts - Age: 21 - Released on 27 Apr 1815.

American Prisoners of War Held at Dartmoor during the War of 1812

## Alphabetical listing of names

Preston, John - Seaman - Nbr: 1179 - Prize: Imperatrice Reine - Ship Type: P - How taken: HMS Hotspur - When taken: 13 May 1813 - Where taken: off Grochite - Date Received: 4 Jun 1814 - From what ship: Dartmouth - Born: Hereford County - Age: 30 - Released on 28 Apr 1815.

Preston, John - Prize Master - Nbr: 3390 - Prize: Wasp - Ship Type: P - How taken: HMS Bream - When taken: 10 Jun 1813 - Where taken: off Halifax - Date Received: 13 Sep 1814 - From what ship: HMT Niobe, Chatham - Born: Salem - Age: 26 - Released on 28 May 1815.

Preston, Jonas - Seaman - Nbr: 604 - Prize: Betsy - Ship Type: MV - How taken: HMS Leonidas - When taken: 12 Aug 1813 - Where taken: Channel - Date Received: 8 Sep 1813 - From what ship: Plymouth - Born: Wilmington - Age: 20 - Sent to Dartmouth on 2 Nov 1814.

Preston, Samuel - Seaman - Nbr: 3694 - Prize: Alfred - Ship Type: P - How taken: HMS Epervier - When taken: 23 Feb 1814 - Where taken: off Newfoundland - Date Received: 30 Sep 1814 - From what ship: HMT President, Halifax - Born: Salem - Age: 20 - Released on 4 Jun 1815.

Preston, Thomas - Seaman - Nbr: 5724 - How taken: Gave himself up from HMS Drake - When taken: 6 Oct 1814 - Date Received: 26 Dec 1814 - From what ship: HMT Argo - Born: Philadelphia - Age: 31 - Released on 3 Jul 1815.

Prett, Benjamin - Seaman - Nbr: 1486 - Prize: Napoleon - Ship Type: LM - How taken: HMS Belle Poule - When taken: 3 Apr 1813 - Where taken: off Cape Ortegal (Spain) - Date Received: 19 Jun 1814 - From what ship: Stapleton - Born: Connecticut - Age: 24 - Released on 1 May 1815.

Price, Charlton - Seaman - Nbr: 3228 - How taken: Gave himself up from HMS Lairel - When taken: 18 Oct 1813 - Date Received: 11 Sep 1814 - From what ship: HMT Freya, Chatham - Born: Concord - Age: 21 - Released on 28 May 1815.

Price, Francis - Seaman - Nbr: 5752 - Prize: Hope - Ship Type: MV - How taken: HMS Nereus - When taken: 14 May 1814 - Where taken: Rio de la Plata - Date Received: 26 Dec 1814 - From what ship: HMT Argo - Born: Kingston - Age: 24 - Race: Mulatto - Released on 3 Jul 1815.

Price, Jacob - Seaman - Nbr: 1579 - Prize: Price - Ship Type: MV - How taken: HMS Pyramus - When taken: 6 Apr 1813 - Where taken: Bay of Biscay - Date Received: 23 Jun 1814 - From what ship: Stapleton - Born: Connecticut - Age: 24 - Released on 1 May 1815.

Price, John - Seaman - Nbr: 4474 - Prize: Enterprize - Ship Type: P - How taken: HMS Tenedos - When taken: 21 May 1813 - Where taken: off Cape Cod - Date Received: 8 Oct 1814 - From what ship: HMT Leyden, Chatham - Born: Massachusetts - Age: 33 - Released on 15 Jun 1815.

Price, John - Seaman - Nbr: 5437 - Prize: Fox - Ship Type: MV - How taken: HMS Pheasant - When taken: Apr 1812 - Where taken: off Bordeaux - Date Received: 11 Nov 1814 - From what ship: HMT Impregnable - Born: Bristol - Age: 25 - Released on 1 Jul 1815.

Price, John - Boy - Nbr: 2392 - Prize: US Sloop Frolic - Ship Type: MW - How taken: HMS Orpheus - When taken: 20 Apr 1814 - Where taken: off Cuba - Date Received: 16 Aug 1814 - From what ship: HMT Queen, Halifax - Born: New York - Age: 16 - Sent to Dartmouth on 19 Oct 1814.

Price, John - Seaman - Nbr: 2897 - How taken: Gave himself up from HMS Ocean - When taken: 28 May 1813 - Date Received: 24 Aug 1814 - From what ship: HMT Alpheus, Chatham - Born: New York - Age: 31 - Released on 19 May 1815.

Price, Peter - Seaman - Nbr: 6543 - How taken: Gave himself up from HMS Ramus - When taken: 22 Feb 1813 - Date Received: 7 Mar 1815 - From what ship: HMT Ganges, Plymouth - Born: Old Chester - Age: 28 - Race: Mulatto - Released on 11 Jul 1815.

Price, Samuel - Seaman - Nbr: 1430 - Prize: Revenge - Ship Type: LM - How taken: HMS Belle Poule - When taken: 10 May 1814 - Where taken: off Cornwall - Date Received: 19 Jun 1814 - From what ship: Stapleton - Born: Boston - Age: 21 - Released on 28 Apr 1815.

Price, William - Seaman - Nbr: 1033 - Prize: US Brig Argus - Ship Type: MW - How taken: HMS Pelican - When taken: 16 Aug 1813 - Where taken: Irish Channel - Date Received: 10 May 1814 - From what ship: Plymouth - Born: Philadelphia - Age: 39 - Sent to Dartmouth on 19 Oct 1814.

Prichard, Robert - Seaman - Nbr: 3700 - Prize: Alfred - Ship Type: P - How taken: HMS Epervier - When taken: 23 Feb 1814 - Where taken: off Newfoundland - Date Received: 30 Sep 1814 - From what ship: HMT President, Halifax - Born: Marblehead - Age: 21 - Released on 4 Jun 1815.

Prichard, William - Seaman - Nbr: 2734 - How taken: Gave himself up from HMS Belle Poule - When taken: 5 Sep 1812 - Date Received: 24 Aug 1814 - From what ship: HMT Liverpool, Chatham - Born: New York - Age: 23 - Released on 26 Apr 1815.

Primas, James - Seaman - Nbr: 4427 - Prize: Baroness Longueville - Ship Type: MV - How taken: HMS Illustrious -

American Prisoners of War Held at Dartmoor during the War of 1812

## Alphabetical listing of names

When taken: 5 Aug 1813 - Where taken: off St. Helena - Date Received: 8 Oct 1814 - From what ship: HMT Leyden, Chatham - Born: New York - Age: 28 - Race: Black - Released on 11 Jul 1815.

Prime, James - Seaman - Nbr: 4959 - Prize: Herald - Ship Type: P - How taken: HMS Endymion - When taken: 15 Aug 1814 - Where taken: off Nantucket - Date Received: 28 Oct 1814 - From what ship: HMT Alkbar, Halifax - Born: New York - Age: 30 - Released on 21 Jun 1815.

Prince, Benjamin - Seaman - Nbr: 447 - Prize: Magdalen - Ship Type: MV - How taken: HMS Superb - When taken: 15 Apr 1813 - Where taken: off Belle Isle, France - Date Received: 1 Jul 1813 - From what ship: Plymouth - Born: Massachusetts - Age: 27 - Escaped on 20 Sep 1814.

Prince, George - Seaman - Nbr: 1344 - Prize: Paul Jones - Ship Type: P - How taken: HMS Leonidas - When taken: 23 May 1813 - Where taken: Channel - Date Received: 19 Jun 1814 - From what ship: Stapleton - Born: Massachusetts - Age: 33 - Released on 28 Apr 1815.

Prince, Jeffery - Seaman - Nbr: 2126 - How taken: Gave himself up from HMS Dublin - When taken: 24 Oct 1812 - Date Received: 8 Aug 1814 - From what ship: HMT Raven, Chatham - Born: South Hold - Age: 30 - Released on 26 Apr 1815.

Prince, Lubin - Seaman - Nbr: 3463 - How taken: Gave himself up from HMS Cyane - When taken: 14 Sep 1814 - Date Received: 19 Sep 1814 - From what ship: HMT Salvador del Mundo - Born: Massachusetts - Age: 39 - Race: Mulatto - Released on 3 Jul 1814.

Prince, Prince - Seaman - Nbr: 3706 - Prize: Alfred - Ship Type: P - How taken: HMS Epervier - When taken: 23 Feb 1814 - Where taken: off Newfoundland - Date Received: 30 Sep 1814 - From what ship: HMT President, Halifax - Born: Marblehead - Age: 24 - Race: Negro - Released on 4 Jun 1815.

Prince, Sylvester - Seaman - Nbr: 5589 - Prize: Young William - Ship Type: Prize - How taken: HMS Plover - When taken: 8 Sep 1814 - Where taken: off St. Johns - Date Received: 17 Dec 1814 - From what ship: HMT Loire, Halifax - Born: Massachusetts - Age: 35 - Race: Mulatto - Released on 1 Jul 1815.

Prince, William - Seaman - Nbr: 2023 - How taken: Gave himself up from HMS Cherub - Date Received: 3 Aug 1814 - From what ship: HMS Lyffey, Chatham Depot - Born: New York - Age: 26 - Released on 2 May 1815.

Prindall, Samuel - Seaman - Nbr: 3618 - Prize: Monarch - Ship Type: MV - How taken: HMS Dotterel - When taken: 14 Dec 1813 - Where taken: off Charleston - Date Received: 30 Sep 1814 - From what ship: HMT Sybella - Born: Connecticut - Age: 25 - Released on 4 Jun 1815.

Prindell, Daniel - Corporal Marines - Nbr: 4032 - Prize: US Brig Rattlesnake - Ship Type: MW - How taken: HMS Leander - When taken: 13 Jul 1814 - Where taken: off Shelburne - Date Received: 6 Oct 1814 - From what ship: HMT Chesapeake, Halifax - Born: Gloucester - Age: 28 - Released on 9 Jun 1815.

Prior, Asa - 2nd Prize Master - Nbr: 807 - Prize: Collin - Ship Type: MV - How taken: HMS Helicon and HMS Whiting - When taken: 26 Oct 1813 - Where taken: Channel - Date Received: 3 Nov 1813 - From what ship: Plymouth - Born: Connecticut - Age: 25 - Released on 26 Apr 1815.

Prissey, John - Seaman - Nbr: 5204 - Prize: Porcupine - Ship Type: LM - How taken: HMS Acasta - When taken: 18 Jul 1813 - Where taken: off Halifax - Date Received: 31 Oct 1814 - From what ship: HMT Mermaid, Chatham - Born: Salem - Age: 19 - Released on 29 Jun 1815.

Pritchard, George - Seaman - Nbr: 6210 - Prize: Prince de Neufchatel - Ship Type: P - How taken: Leander (Newcastle Acasta) - When taken: 20 Dec 1814 - Where taken: Lat 38 Long 56 - Date Received: 30 Jan 1815 - From what ship: HMT Pheasant - Born: Dresden - Age: 21 - Released on 5 Jul 1815.

Pritchard, Israel - Seaman - Nbr: 4200 - Prize: Hannah - Ship Type: MV - How taken: HMS Conquistador - When taken: 15 Jan 1814 - Where taken: at sea - Date Received: 7 Oct 1814 - From what ship: HMT Niobe, Chatham - Born: Marblehead - Age: 22 - Released on 13 Jun 1815.

Pritchard, John - Seaman - Nbr: 3713 - Prize: Alfred - Ship Type: P - How taken: HMS Epervier - When taken: 23 Feb 1814 - Where taken: off Newfoundland - Date Received: 30 Sep 1814 - From what ship: HMT President, Halifax - Born: Marblehead - Age: 24 - Race: Creole - Released on 4 Jun 1815.

Pritchard, T. G. - Master's Mate - Nbr: 5892 - Prize: Harlequin - Ship Type: P - How taken: HMS Bulwark - When taken: 23 Nov 1814 - Where taken: off Halifax - Date Received: 27 Dec 1814 - From what ship: HMT Penelope - Born: Newbury - Age: 41 - Released on 3 Jul 1815.

Proctor, Henry - Prize Master - Nbr: 3527 - Prize: Sabine - Ship Type: P - How taken: HMS Conquistador - When taken: 3 Aug 1814 - Where taken: off the Banks - Date Received: 28 Sep 1814 - From what ship: HMT Salvador del Mundo - Born: Boston - Age: 32 - Released on 12 Apr 1815.

Proctor, John - Seaman - Nbr: 3699 - Prize: Alfred - Ship Type: P - How taken: HMS Epervier - When taken: 23 Feb 1814 - Where taken: off Newfoundland - Date Received: 30 Sep 1814 - From what ship: HMT President,

American Prisoners of War Held at Dartmoor during the War of 1812

## Alphabetical listing of names

Halifax - Born: Marblehead - Age: 23 - Released on 4 Jun 1815.

Prout, Henry - Seaman - Nbr: 6537 - Prize: US Brig Syren - Ship Type: MW - How taken: HMS Medway - When taken: 12 Jul 1814 - Where taken: off Cape of Good Hope - Date Received: 7 Mar 1815 - From what ship: HMT Ganges, Plymouth - Born: Boston - Age: 30 - Released on 11 Jul 1815.

Prymvoy, Richard - Seaman - Nbr: 865 - Prize: Amiable - Ship Type: LM - How taken: HMS Magnificent - When taken: 30 Oct 1813 - Where taken: off Bordeaux - Date Received: 4 Dec 1813 - From what ship: Plymouth - Born: SC - Age: 31 - Released on 27 Apr 1815.

Puffer, John - Seaman - Nbr: 3431 - Prize: York Town - Ship Type: P - How taken: British squadron - When taken: 26 Jul 1813 - Where taken: off Halifax - Date Received: 13 Sep 1814 - From what ship: HMT Niobe, Chatham - Born: Canton, China - Age: 20 - Released on 28 May 1815.

Pukiam, E. L. - Marine - Nbr: 5931 - Prize: Harlequin - Ship Type: P - How taken: HMS Bulwark - When taken: 23 Nov 1814 - Where taken: off Halifax - Date Received: 27 Dec 1814 - From what ship: HMT Penelope - Born: New Hampshire - Age: 18 - Released on 3 Jul 1815.

Puss, Nathaniel - Mate - Nbr: 5977 - Prize: Cossack - Ship Type: Private - How taken: HMS Bulwark - When taken: 1 Nov 1814 - Where taken: Georges Bank - Date Received: 27 Dec 1814 - From what ship: HMT Penelope - Born: Newbury - Age: 19 - Released on 3 Jul 1815.

Putman, Charles - Seaman - Nbr: 1456 - Prize: Revenge - Ship Type: LM - How taken: HMS Belle Poule - When taken: 10 Mar 1813 - Where taken: off Cornwall - Date Received: 19 Jun 1814 - From what ship: Stapleton - Born: Massachusetts - Age: 30 - Released on 28 Apr 1815.

Putnam, A. - Seaman - Nbr: 3060 - How taken: Apprehended at Cork - When taken: 28 Sep 1813 - Date Received: 2 Sep 1814 - From what ship: HMT Hydra, Chatham - Born: Danvers - Age: 20 - Released on 10 Apr 1815.

Pyne, Joseph - Seaman - Nbr: 5077 - Prize: Ida - Ship Type: LM - How taken: HMS Newcastle - When taken: 9 Aug 1814 - Where taken: Long 34 - Date Received: 28 Oct 1814 - From what ship: HMT Alkbar, Halifax - Born: Philadelphia - Age: 17 - Released on 29 Jun 1815.

Pyneo, John Rogers - Seaman - Nbr: 770 - How taken: Impressed at Liverpool - Date Received: 3 Nov 1813 - From what ship: Plymouth - Born: Cheesnen - Age: 26 - Released on 26 Apr 1815.

Quarterman, William - Seaman - Nbr: 3179 - How taken: Gave himself up from HMS Ocean - When taken: 27 May 1813 - Date Received: 11 Sep 1814 - From what ship: HMT Freya, Chatham - Born: Charleston - Age: 23 - Released on 28 May 1815.

Queen, Daniel - Seaman - Nbr: 5418 - How taken: Gave himself up from HMS Blenheim - When taken: 27 Aug 1814 - Date Received: 31 Oct 1814 - From what ship: HMT Leyden, Chatham - Born: George County - Age: 31 - Race: Black - Released on 1 Jul 1815.

Queenwell, Peter - Seaman - Nbr: 5629 - Prize: Walker - Ship Type: MV - How taken: HMS Nimrod - When taken: 20 May 1813 - Where taken: off Cape Horn - Date Received: 24 Dec 1814 - From what ship: HMT Tay - Born: Dartmouth - Age: 33 - Race: Mulatto - Died on 27 Feb 1815 from variola.

Quince, Peter - Seaman - Nbr: 1857 - Prize: Thorn - Ship Type: P - How taken: HMS Shannon - When taken: 31 Oct 1812 - Where taken: off Western Islands (England) - Date Received: 29 Jul 1814 - From what ship: HMS Ville de Paris, Chatham Depot - Born: New Orleans - Age: 18 - Released on 26 Apr 1815.

Quincy, Benjamin - Seaman - Nbr: 2232 - Prize: General Hart - Ship Type: P - How taken: HMS Sophie - When taken: 24 Apr 1814 - Where taken: off Bermuda - Date Received: 16 Aug 1814 - From what ship: HMS Dublin, Halifax - Born: Massachusetts - Age: 31 - Released on 2 May 1815.

Quiner, Stephen - Seaman - Nbr: 4539 - Prize: Wolf Cove - Ship Type: MV - How taken: HMS Briton - When taken: 1 Dec 1812 - Where taken: off Brest - Date Received: 8 Oct 1814 - From what ship: HMT Leyden, Chatham - Born: Massachusetts - Age: 18 - Released on 27 Apr 1815.

Quinn, William - Seaman - Nbr: 3037 - Prize: Commerce - How taken: HMS Superb - When taken: 23 Jun 1814 - Where taken: off Sandy Hook - Date Received: 2 Sep 1814 - From what ship: HMT Sultan - Born: Charleston - Age: 25 - Released on 28 May 1815.

Raddick, Ebenezer - Seaman - Nbr: 5354 - How taken: Impressed at Harwich - When taken: 21 Feb 1814 - Date Received: 31 Oct 1814 - From what ship: HMT Leyden, Chatham - Born: Cambridge - Age: 36 - Released on 1 Jul 1815.

Rae, William Iron - Seaman - Nbr: 2056 - How taken: Apprehended at London - When taken: 13 Jun 1813 - Date Received: 3 Aug 1814 - From what ship: HMS Lyffey, Chatham Depot - Born: Boston - Age: 46 - Released on 2 May 1815.

Rake, Martin - Seaman - Nbr: 6495 - Prize: Corinthian - Ship Type: MV - How taken: HMS Forth - Where taken: off NY - Date Received: 3 Mar 1815 - From what ship: HMT Ganges, Plymouth - Born: Prussia - Age: 30 -

American Prisoners of War Held at Dartmoor during the War of 1812

## Alphabetical listing of names

Released on 11 Jul 1815.

Ralph, David - Boy - Nbr: 4320 - Prize: Lizard - Ship Type: P - How taken: HMS Barbados - When taken: 1 Mar 1814 - Where taken: off Halifax - Date Received: 7 Oct 1814 - From what ship: HMT Salvador del Mundo, Halifax - Born: Salem - Age: 16 - Race: Negro - Released on 14 Jun 1815.

Rameson, Thomas - Seaman - Nbr: 3484 - How taken: Gave himself up from HMS Prince - When taken: 14 Sep 1814 - Date Received: 19 Sep 1814 - From what ship: HMT Salvador del Mundo - Born: New York - Age: 28 - Race: Negro - Released on 11 Jul 1815.

Ramsdell, Charles - Prize master - Nbr: 6489 - Prize: Louisa - Ship Type: Prize - How taken: HMS Dasher - When taken: 12 Feb 1814 - Where taken: off America - Date Received: 3 Mar 1815 - From what ship: HMT Ganges, Plymouth - Born: Massachusetts - Age: 33 - Released on 5 Jul 1815.

Ramsden, John - Boy - Nbr: 986 - Prize: Hannah - Ship Type: MV - How taken: HMS Conquistador - When taken: 15 Jan 1814 - Where taken: at sea - Date Received: 31 Jan 1814 - From what ship: Plymouth - Born: Marblehead - Age: 14 - Released on 27 Apr 1815.

Ramsden, William - Seaman - Nbr: 5600 - Prize: Alexander - Ship Type: Prize - How taken: HMS Wasp - When taken: 14 Sep 1814 - Where taken: off Cape Sable Island (Canada) - Date Received: 17 Dec 1814 - From what ship: HMT Loire, Halifax - Born: Baltimore - Age: 21 - Released on 5 Jul 1815.

Ramsey, Samuel - Seaman - Nbr: 2450 - Prize: US Sloop Frolic - Ship Type: MW - How taken: HMS Orpheus - When taken: 20 Apr 1814 - Where taken: off Cuba - Date Received: 16 Aug 1814 - From what ship: HMT Queen, Halifax - Born: Middlesex - Age: 29 - Released on 3 May 1815.

Rand, Charles - Seaman - Nbr: 6201 - Prize: Prince de Neufchatel - Ship Type: P - How taken: Leander (Newcastle Acasta) - When taken: 20 Dec 1814 - Where taken: Lat 38 Long 56 - Date Received: 30 Jan 1815 - From what ship: HMT Pheasant - Born: Boston - Age: 19 - Released on 5 Jul 1815.

Rand, Robert - Captain's clerk - Nbr: 5046 - Prize: Ida - Ship Type: LM - How taken: HMS Newcastle - When taken: 9 Aug 1814 - Where taken: Long 34 - Date Received: 28 Oct 1814 - From what ship: HMT Alkbar, Halifax - Born: Boston - Age: 34 - Released on 29 Jun 1815.

Rand, Thomas - Seaman - Nbr: 6200 - Prize: Prince de Neufchatel - Ship Type: P - How taken: Leander (Newcastle Acasta) - When taken: 20 Dec 1814 - Where taken: Lat 38 Long 56 - Date Received: 30 Jan 1815 - From what ship: HMT Pheasant - Born: Boston - Age: 20 - Released on 5 Jul 1815.

Randall, Charles - Seaman - Nbr: 3802 - Prize: Margetits - Ship Type: Prize - How taken: HMS Maidstone - When taken: 20 Mar 1814 - Where taken: off NY - Date Received: 5 Oct 1814 - From what ship: HMT Orpheus, Halifax - Born: Boston - Age: 26 - Race: Negro - Released on 9 Jun 1815.

Randall, Henry - Seaman - Nbr: 4469 - Prize: Governor Plumer - Ship Type: P - How taken: HMS Shamrock - When taken: 4 Mar 1813 - Where taken: at sea - Date Received: 8 Oct 1814 - From what ship: HMT Leyden, Chatham - Born: New Hampshire - Age: 24 - Released on 15 Jun 1815.

Randall, Jacob - Seaman - Nbr: 4620 - Prize: Argus - Ship Type: MV - How taken: HMS San Domingo - When taken: 1 Mar 1814 - Where taken: off Savannah - Date Received: 9 Oct 1814 - From what ship: HMT Leyden, Chatham - Born: Norfolk - Age: 29 - Released on 15 Jun 1815.

Randall, James - Seaman - Nbr: 6186 - Prize: Garland - Ship Type: Prize - How taken: HMS Hamadryad - When taken: 22 Nov 1814 - Where taken: Georges Bank - Date Received: 21 Jan 1815 - From what ship: HMT Impregnable - Born: Newcastle - Age: 20 - Released on 5 Jul 1815.

Randell, Thomas - Seaman - Nbr: 6413 - Prize: Enterprize - Ship Type: P - How taken: HMS Argo - When taken: 16 Nov 1814 - Where taken: off San Domingo - Date Received: 3 Mar 1815 - From what ship: HMT Ganges, Plymouth - Born: Rochester - Age: 23 - Released on 11 Jul 1815.

Randolph, George - Seaman - Nbr: 2492 - Prize: US Sloop Frolic - Ship Type: MW - How taken: HMS Orpheus - When taken: 20 Apr 1814 - Where taken: off Cuba - Date Received: 16 Aug 1814 - From what ship: HMT Queen, Halifax - Born: Pennsylvania - Age: 25 - Released on 3 May 1815.

Ranford, Henry - Boy - Nbr: 4026 - Prize: US Brig Rattlesnake - Ship Type: MW - How taken: HMS Leander - When taken: 13 Jul 1814 - Where taken: off Shelburne - Date Received: 6 Oct 1814 - From what ship: HMT Chesapeake, Halifax - Born: Boston - Age: 16 - Released on 9 Jun 1815.

Ranger, Desire - Volunteer - Nbr: 1010 - Prize: Harvest - Ship Type: P - How taken: HMS Orestes - When taken: 21 Jan 1814 - Where taken: Bay of Biscay - Date Received: 31 Jan 1814 - From what ship: Plymouth - Born: Vangere - Age: 17 - Sent to Mill Prison (Plymouth, England) on 21 Jun 1814.

Rankin, William - Caulker - Nbr: 3215 - How taken: Gave himself up from HMS Castor - When taken: 13 Dec 1812 - Date Received: 11 Sep 1814 - From what ship: HMT Freya, Chatham - Born: Delaware - Age: 32 - Released on 27 Apr 1815.

American Prisoners of War Held at Dartmoor during the War of 1812

## Alphabetical listing of names

Ranlot, John - Seaman - Nbr: 3219 - How taken: Gave himself up from HMS Orpheus - When taken: 10 Dec 1812 - Date Received: 11 Sep 1814 - From what ship: HMT Freya, Chatham - Born: Newport - Age: 27 - Released on 27 Apr 1815.

Ranson, Joseph - Seaman - Nbr: 700 - Prize: Ned - Ship Type: LM - How taken: HMS Royalist - When taken: 6 Sep 1813 - Where taken: Bay of Biscay - Date Received: 27 Sep 1813 - From what ship: Plymouth - Born: Philadelphia - Age: 23 - Died on 1 Mar 1815 from variola.

Rape, Nicholas - Seaman - Nbr: 1969 - Prize: Orbit - Ship Type: MV - How taken: HMS Achates - When taken: 29 Jan 1813 - Where taken: Lat 44N Long 13W - Date Received: 3 Aug 1814 - From what ship: HMS Lyffey, Chatham Depot - Born: Philadelphia - Age: 40 - Released on 2 May 1815.

Raquinis, Francis - Boy - Nbr: 997 - Prize: Apparencen - Ship Type: MV - How taken: HMS Castilian - When taken: 27 Jan 1814 - Where taken: off Ushant (France) - Date Received: 31 Jan 1814 - From what ship: Plymouth - Born: Brest - Age: 16 - Sent to Mill Prison (Plymouth, England) on 21 Jun 1814.

Ratcliffe, Joseph - Seaman - Nbr: 5578 - Prize: Dash - Ship Type: LM - How taken: HMS Rainbow - When taken: Aug 1814 - Where taken: off Savannah - Date Received: 17 Dec 1814 - From what ship: HMT Loire, Halifax - Born: New York - Age: 30 - Race: Black - Released on 1 Jul 1815.

Raton, Peter - Seaman - Nbr: 3789 - Prize: James - Ship Type: Prize - How taken: Rebecca (LM) - When taken: 13 Apr 1814 - Where taken: off Delaware - Date Received: 5 Oct 1814 - From what ship: HMT President, Halifax - Born: Philadelphia - Age: 22 - Released on 9 Jun 1815.

Ratoon, Thomas - Seaman - Nbr: 315 - Prize: Pallas - Ship Type: MV - How taken: Rebuff - When taken: 23 Dec 1812 - Where taken: off Cadiz - Date Received: 28 Jun 1813 - From what ship: Plymouth - Born: New York - Age: 20 - Sent to Dartmouth on 30 Jul 1813.

Rauzelin, Jacob - Seaman - Nbr: 2273 - Prize: Success - Ship Type: Prize - How taken: HMS Charybdis - When taken: 29 May 1814 - Where taken: off Cape Logan - Date Received: 16 Aug 1814 - From what ship: HMS Dublin, Halifax - Born: Albany - Age: 18 - Race: Negro - Released on 3 May 1815.

Raven, James - Cooper - Nbr: 2219 - Prize: Hussar - Ship Type: P - How taken: HMS Saturn - When taken: 25 May 1814 - Where taken: off Sandy Hook - Date Received: 16 Aug 1814 - From what ship: HMS Dublin, Halifax - Born: Baltimore - Age: 25 - Released on 2 May 1815.

Ravett, William - Seaman - Nbr: 1201 - Prize: Lyon - Ship Type: P - How taken: HMS Pheasant - When taken: 8 Jun 1813 - Where taken: off St. Antonio - Date Received: 4 Jun 1814 - From what ship: Dartmouth - Born: Pinchry - Age: 19 - Released on 28 Apr 1815.

Rawl, Elias B. - Seaman - Nbr: 2493 - Prize: US Sloop Frolic - Ship Type: MW - How taken: HMS Orpheus - When taken: 20 Apr 1814 - Where taken: off Cuba - Date Received: 16 Aug 1814 - From what ship: HMT Queen, Halifax - Born: New York - Age: 28 - Released on 3 May 1815.

Rawlins, Nicholas - Seaman - Nbr: 5601 - Prize: Alexander - Ship Type: Prize - How taken: HMS Wasp - When taken: 14 Sep 1814 - Where taken: off Cape Sable Island (Canada) - Date Received: 17 Dec 1814 - From what ship: HMT Loire, Halifax - Born: Massachusetts - Age: 19 - Released on 5 Jul 1815.

Rawlinson, Thomas - Seaman - Nbr: 5137 - Prize: Calabria - Ship Type: MV - How taken: HMS Castilian - When taken: 29 Sep 1814 - Where taken: off Ireland - Date Received: 31 Oct 1814 - From what ship: HMT Castillian - Born: Virginia - Age: 22 - Race: Mulatto - Died on 26 Nov 1814 from enteritis.

Ray, Charles - Seaman - Nbr: 2578 - How taken: Gave himself up from HMS Chatham - When taken: 1 May 1814 - Date Received: 21 Aug 1814 - From what ship: HMT Salvador del Mundo - Born: Newland - Age: 30 - Released on 3 May 1815.

Ray, David - Seaman - Nbr: 1139 - Prize: Mary - Ship Type: P - How taken: HMS Crocodile - When taken: 6 Aug 1813 - Where taken: off Corsuma - Date Received: 10 May 1814 - From what ship: Plymouth - Born: Adam - Age: 23 - Released on 28 Apr 1815.

Ray, James - Seaman - Nbr: 4183 - How taken: Impressed at Gravesend - When taken: 7 Jul 1813 - Date Received: 7 Oct 1814 - From what ship: HMT Niobe, Chatham - Born: New Jersey - Age: 27 - Released on 28 May 1815.

Ray, Richard - Seaman - Nbr: 6260 - Prize: Plutarch - Ship Type: MV - How taken: HMS Helicon - When taken: 5 Feb 1815 - Where taken: off Bordeaux - Date Received: 10 Feb 1815 - From what ship: HMT Ganges, Plymouth - Born: Newland - Age: 30 - Released on 5 Jul 1815.

Ray, William - Seaman - Nbr: 3584 - How taken: Gave himself up from HMS Crestis - When taken: 2 Jul 1814 - Date Received: 30 Sep 1814 - From what ship: HMT Sybella - Born: Maryfield - Age: 20 - Released on 4 Jun 1815.

Raymond, George - Seaman - Nbr: 2739 - How taken: Gave himself up from HMS Arrow - When taken: 4 Jan 1814

American Prisoners of War Held at Dartmoor during the War of 1812

## Alphabetical listing of names

- Date Received: 24 Aug 1814 - From what ship: HMT Liverpool, Chatham - Born: Newburgh - Age: 26 - Released on 26 Apr 1815.

Raymond, John - Cook - Nbr: 6320 - Prize: Prince de Neufchatel - Ship Type: P - How taken: Leander (Newcastle Acasta) - When taken: 20 Dec 1814 - Where taken: Lat 38 Long 56 - Date Received: 19 Feb 1815 - From what ship: HMT Ganges, Plymouth - Born: New Orleans - Age: 24 - Race: Black - Released on 5 Jul 1815.

Raysden, John - Seaman - Nbr: 3795 - Prize: Hannah - Ship Type: Prize - How taken: HMS Martin - When taken: 29 Apr 1814 - Where taken: off Cape Lopez - Date Received: 5 Oct 1814 - From what ship: HMT President, Halifax - Born: New York - Age: 32 - Died on 14 Feb 1815 from enteritis.

Rea, Charles - Seaman - Nbr: 5815 - Prize: US Schooner Scorpion - Ship Type: MW - How taken: British gunboats - When taken: 6 Sep 1814 - Where taken: Lake Erie - Date Received: 26 Dec 1814 - From what ship: HMT Argo - Born: New London - Age: 29 - Released on 3 Jul 1815.

Read, Charles - Prize master - Nbr: 5380 - Prize: Harriett - Ship Type: MV - How taken: HMS Thistle - When taken: 24 Feb 1813 - Where taken: off St. Bartholomew - Date Received: 31 Oct 1814 - From what ship: HMT Leyden, Chatham - Born: Reading - Age: 32 - Released on 1 Jul 1815.

Read, David - Seaman - Nbr: 4068 - Prize: Leicester - Ship Type: Prize - How taken: HMS Prometheus - When taken: Jun 1814 - Where taken: off Halifax - Date Received: 6 Oct 1814 - From what ship: HMT Chesapeake, Halifax - Born: Wiscasset - Age: 21 - Died on 14 Nov 1814.

Read, David - Seaman - Nbr: 4069 - Prize: Leicester - Ship Type: Prize - How taken: HMS Prometheus - When taken: Jun 1814 - Where taken: off Halifax - Date Received: 6 Oct 1814 - From what ship: HMT Chesapeake, Halifax - Born: Boothbay - Age: 30 - Released on 13 Jun 1815.

Read, John - Marine - Nbr: 4036 - Prize: US Brig Rattlesnake - Ship Type: MW - How taken: HMS Leander - When taken: 13 Jul 1814 - Where taken: off Shelburne - Date Received: 6 Oct 1814 - From what ship: HMT Chesapeake, Halifax - Born: Marblehead - Age: 22 - Released on 9 Jun 1815.

Read, Thomas - Seaman - Nbr: 5816 - Prize: US Schooner Scorpion - Ship Type: MW - How taken: British gunboats - When taken: 6 Sep 1814 - Where taken: Lake Erie - Date Received: 26 Dec 1814 - From what ship: HMT Argo - Born: Boston - Age: 24 - Released on 3 Jul 1815.

Read, Thomas - Seaman - Nbr: 3583 - How taken: Gave himself up at Port Mahan - When taken: Jun 1814 - Date Received: 30 Sep 1814 - From what ship: HMT Sybella - Born: Maryland - Age: 36 - Race: Negro - Released on 4 Jun 1815.

Read, William - Seaman - Nbr: 2703 - How taken: Gave himself up from HMS Denmark - When taken: 12 Dec 1813 - Date Received: 21 Aug 1814 - From what ship: HMT Freya, Chatham - Born: Philadelphia - Age: 44 - Race: Mulatto - Released on 19 May 1815.

Read, William - Seaman - Nbr: 2696 - How taken: Gave himself up from HMS Racehorse - When taken: 25 Oct 1813 - Date Received: 21 Aug 1814 - From what ship: HMT Freya, Chatham - Born: Portsmouth - Age: 25 - Died on 3 Jun 1815 from phthisis pulmonalis.

Reardon, Andrew - Seaman - Nbr: 320 - Prize: Pallas - Ship Type: MV - How taken: Rebuff - When taken: 23 Dec 1812 - Where taken: off Cadiz - Date Received: 28 Jun 1813 - From what ship: Plymouth - Born: Trieste - Age: 27 - Sent to Dartmouth on 30 Jul 1813.

Redding, John - Seaman - Nbr: 1375 - Prize: Grand Napoleon - Ship Type: P - How taken: HMS Goldfinch - When taken: 17 Apr 1813 - Where taken: Bay of Biscay - Date Received: 19 Jun 1814 - From what ship: Stapleton - Born: Massachusetts - Age: 23 - Released on 28 Apr 1815.

Redman, David - Seaman - Nbr: 3218 - How taken: Gave himself up from HMS Sceptre - When taken: 26 Jun 1813 - Date Received: 11 Sep 1814 - From what ship: HMT Freya, Chatham - Born: Maryland - Age: 21 - Released on 28 May 1815.

Redman, William - Seaman - Nbr: 880 - How taken: Impressed at Liverpool - When taken: 28 Nov 1813 - Date Received: 31 Jan 1814 - What ship: Plymouth - Born: Philadelphia - Age: 27 - Released on 27 Apr 1815.

Reed, Abraham - Seaman - Nbr: 4191 - Prize: Commodore Perry - Ship Type: Schooner - How taken: Sent into custody from a cutter - When taken: 25 Feb 1814 - Where taken: off Bordeaux - Date Received: 7 Oct 1814 - From what ship: HMT Niobe, Chatham - Born: New York - Age: 19 - Released on 13 Jun 1815.

Reed, George - Seaman - Nbr: 4512 - Prize: America - Ship Type: P - How taken: HMS Shannon - When taken: 23 May 1813 - Where taken: off Cape Cod - Date Received: 8 Oct 1814 - From what ship: HMT Leyden, Chatham - Born: Rhode Island - Age: 22 - Released on 15 Jun 1815.

Reed, James - Seaman - Nbr: 2049 - How taken: Gave himself up from HMS Arethusa - When taken: Apr 1812 - Date Received: 3 Aug 1814 - From what ship: HMS Lyffey, Chatham Depot - Born: New York - Age: 40 - Released on 26 Apr 1815.

American Prisoners of War Held at Dartmoor during the War of 1812

## Alphabetical listing of names

Reed, John - Seaman - Nbr: 5708 - Prize: Daedalus - Ship Type: MV - How taken: HMS Niemen - When taken: 20 Sep 1814 - Where taken: off Delaware - Date Received: 24 Dec 1814 - From what ship: HMT Penelope - Born: New York - Age: 20 - Released on 24 May 1815.

Reed, John - Seaman - Nbr: 4377 - How taken: Gave himself up from HMS Charlotte - When taken: 15 Oct 1812 - Date Received: 8 Oct 1814 - From what ship: HMT Leyden, Chatham - Born: Philadelphia - Age: 24 - Race: Black - Released on 27 Apr 1815.

Reed, John - Seaman - Nbr: 4654 - How taken: Gave himself up from HMS Undaunted - When taken: 15 Oct 1812 - Date Received: 9 Oct 1814 - From what ship: HMT Leyden, Chatham - Born: Virginia - Age: 49 - Released on 27 Apr 1815.

Reed, Joseph - Seaman - Nbr: 4642 - Prize: Jane - Ship Type: Prize - How taken: HMS Crescent - When taken: 28 Jun 1813 - Where taken: off Newfoundland - Date Received: 9 Oct 1814 - From what ship: HMT Leyden, Chatham - Born: Plymouth - Age: 27 - Released on 15 Jun 1815.

Reed, Robert - Gunner - Nbr: 5466 - Prize: General Putnam - Ship Type: P - How taken: HMS Leander - When taken: 8 Nov 1814 - Where taken: Long 65 Lat 42 - Date Received: 17 Dec 1814 - From what ship: HMT Loire, Halifax - Born: Newbury - Age: 32 - Released on 1 Jul 1815.

Reed, William - Seaman - Nbr: 4351 - Prize: Plutus - Ship Type: P - How taken: HMS Curlew - When taken: 1 Apr 1814 - Where taken: off Halifax - Date Received: 7 Oct 1814 - From what ship: HMT Salvador del Mundo, Halifax - Born: West-Cappel (France) - Age: 20 - Released on 14 Jun 1815.

Refry, Peter - Seaman - Nbr: 1635 - Prize: Essex - Ship Type: MV - How taken: HMS Pyramus - When taken: 2 Apr 1813 - Where taken: Bay of Biscay - Date Received: 23 Jun 1814 - From what ship: Stapleton - Born: Marblehead - Age: 26 - Released on 1 May 1815.

Reeves, Asa - Seaman - Nbr: 4294 - Prize: Commodore Perry - Ship Type: Schooner - How taken: Sent into custody from a cutter - When taken: 25 Feb 1814 - Where taken: off Bordeaux - Date Received: 7 Oct 1814 - From what ship: HMT Niobe, Chatham - Born: Salem - Age: 45 - Released on 14 Jun 1815.

Reeves, James - Seaman - Nbr: 79 - Prize: Rolla - Ship Type: MV - How taken: HMS Surveillante - When taken: 11 Feb 1813 - Where taken: Bay of Biscay - Date Received: 2 Apr 1813 - From what ship: Plymouth - Born: Salem - Age: 25 - Sent to Plymouth on 15 Jun 1813.

Reeves, James - Seaman - Nbr: 682 - Prize: Rolla - Ship Type: MV - How taken: HMS Surveillante - When taken: 11 Feb 1813 - Where taken: Bay of Biscay - Date Received: 27 Sep 1813 - From what ship: Plymouth - Born: Salem - Age: 25 - Released on 26 Apr 1815.

Reeves, William - Boy - Nbr: 148 - Prize: Criterion - Ship Type: MV - How taken: HMS Belle Poule - When taken: 14 Feb 1813 - Where taken: Bay of Biscay - Date Received: 2 Apr 1813 - From what ship: Plymouth - Born: West Chester - Age: 16 - Released on 20 Apr 1815.

Regens, Jonathan - Seaman - Nbr: 118 - Prize: Governor McKean - Ship Type: LM - How taken: HMS Rover - When taken: 26 Jan 1813 - Where taken: off Bordeaux - Date Received: 2 Apr 1813 - From what ship: Plymouth - Born: Salem - Age: 26 - Sent to Dartmouth on 30 Jul 1813.

Reid, Thomas - Seaman - Nbr: 2021 - How taken: Impressed at Gravesend - When taken: 10 Mar 1813 - Date Received: 3 Aug 1814 - From what ship: HMS Lyffey, Chatham Depot - Born: Philadelphia - Age: 46 - Released on 2 May 1815.

Reid, Thomas - Seaman - Nbr: 2193 - Prize: Olio - Ship Type: Prize - How taken: HMS Cyane - When taken: 22 May 1814 - Where taken: off NY - Date Received: 16 Aug 1814 - From what ship: HMS Dublin, Halifax - Born: Philadelphia - Age: 35 - Released on 2 May 1815.

Reily, Lewis - Seaman - Nbr: 5616 - How taken: Gave himself up from HMS Lion - When taken: 18 Jun 1813 - Date rec'd: 24 Dec 1814 - What ship: HMT Tay - Born: Little York - Age: 38 - Released on 27 Apr 1815.

Reley, Michael - Seaman - Nbr: 265 - Prize: William Bayard - Ship Type: MV - How taken: HMS Warspite - When taken: 3 Mar 1813 - Where taken: Bay of Biscay - Date Received: 28 Jun 1813 - From what ship: Plymouth - Born: Eugenia - Age: 25 - Released on 20 Apr 1815.

Reninel, William - Seaman - Nbr: 6145 - How taken: Gave himself up from HMS Iris - Date Received: 17 Jan 1815 - From what ship: HMT Impregnable - Born: New Haven - Age: 20 - Released on 5 Jul 1815.

Rennard, George - Prize master - Nbr: 5660 - Prize: Harlequin - Ship Type: P - How taken: HMS Bulwark - When taken: 23 Nov 1814 - Where taken: off Halifax - Date Received: 24 Dec 1814 - From what ship: HMT Penelope - Born: Portsmouth - Age: 28 - Released on 20 Mar 1815.

Rennel, States - Seaman - Nbr: 4629 - How taken: Gave himself up from HMS Freyer - When taken: 2 Aug 1812 - Date Received: 9 Oct 1814 - From what ship: HMT Leyden, Chatham - Born: Charleston - Age: 30 - Released on 28 Apr 1815.

American Prisoners of War Held at Dartmoor during the War of 1812

## Alphabetical listing of names

Resmabin, Benjamin - Seaman - Nbr: 482 - Prize: Fox - Ship Type: LM - How taken: HMS Pheasant - When taken: 23 Apr 1813 - Where taken: Bay of Biscay - Date Received: 8 Sep 1813 - From what ship: Plymouth - Born: Guadeloupe - Age: 28 - Race: Negro - Died on 16 Nov 1813 from pneumonia.

Restrum, Andrew - Seaman - Nbr: 5094 - Prize: David Porter - Ship Type: LM - How taken: HMS Pylades - When taken: 12 Sep 1814 - Where taken: Georges Bank - Date Received: 28 Oct 1814 - From what ship: HMT Alkbar, Halifax - Born: Not legible - Age: 43 - Released on 29 Jun 1815.

Reynard, John - Master - Nbr: 5526 - Prize: Rover - Ship Type: MV - How taken: HMS Vesta - When taken: 23 Sep 1814 - Where taken: off Delaware - Date Received: 17 Dec 1814 - From what ship: HMT Loire, Halifax - Born: New Bedford - Age: 33 - Escaped on 1 Jun 1815.

Reynolds, Amos - Seaman - Nbr: 1984 - How taken: Gave himself up from HMS Colossus - When taken: Apr 1813 - Date Received: 3 Aug 1814 - From what ship: HMS Lyffey, Chatham Depot - Born: Connecticut - Age: 24 - Released on 2 May 1815.

Reynolds, Frederick - Boy - Nbr: 1796 - Prize: Napoleon - Ship Type: MV - How taken: HMS Belle Poule - When taken: 3 Apr 1813 - Where taken: off Cape Ortegal (Spain) - Date Received: 20 Jul 1814 - From what ship: HMS Milford, Plymouth - Born: New York - Age: 18 - Released on 1 May 1815.

Reynolds, James - Seaman - Nbr: 1400 - Prize: Zebra - Ship Type: LM - How taken: HMS Pyramus - When taken: 20 Apr 1813 - Where taken: Bay of Biscay - Date Received: 19 Jun 1814 - From what ship: Stapleton - Born: Virginia - Age: 28 - Released on 28 Apr 1815.

Reynolds, Owen - Seaman - Nbr: 445 - Prize: Magdalen - Ship Type: MV - How taken: HMS Superb - When taken: 15 Apr 1813 - Where taken: off Belle Isle, France - Date Received: 1 Jul 1813 - From what ship: Plymouth - Born: Newburyport - Age: 32 - Released on 20 Apr 1815.

Reynolds, Stephen - Seaman - Nbr: 2661 - How taken: Apprehended at London - When taken: 17 Aug 1813 - Date Received: 21 Aug 1814 - From what ship: HMT Freya, Chatham - Born: New York - Age: 24 - Released on 19 May 1815.

Reynolds, W. B. - Seaman - Nbr: 2387 - Prize: US Sloop Frolic - Ship Type: MW - How taken: HMS Orpheus - When taken: 20 Apr 1814 - Where taken: off Cuba - Date Received: 16 Aug 1814 - From what ship: HMT Queen, Halifax - Born: Newark - Age: 21 - Sent to Dartmouth on 19 Oct 1814.

Rhodes, Daniel - Seaman - Nbr: 6014 - Prize: McDonough - Ship Type: P - How taken: HMS Bacchante - When taken: 1 Nov 1814 - Where taken: Lat 42 Long 67 - Date Received: 28 Dec 1814 - From what ship: HMT Penelope - Born: Massachusetts - Age: 27 - Released on 3 Jul 1815.

Rhodrick, Joseph - Seaman - Nbr: 174 - Prize: Hope - Ship Type: MV - How taken: Chance (Privateer) - When taken: 15 Feb 1813 - Where taken: off Bordeaux - Date Received: 2 Apr 1813 - From what ship: Plymouth - Born: St. Marys - Age: 39 - Released on 20 Apr 1815.

Rice, Francis - Boy - Nbr: 379 - Prize: VA Planter - Ship Type: MV - How taken: HMS Pyramus - When taken: 18 Mar 1813 - Where taken: off Nantes - Date Received: 1 Jul 1813 - From what ship: Plymouth - Born: Brookfield - Age: 19 - Entered British service on 13 Sep 1813.

Rice, James - Seaman - Nbr: 3507 - Prize: Chasseur - Ship Type: P - How taken: HMS Whiting - When taken: Aug 1814 - Where taken: off the Western Isles (England) - Date Received: 24 Sep 1814 - From what ship: HMT Salvador del Mundo - Born: St. Marys - Age: 20 - Released on 4 Jun 1815.

Rice, John - 2nd Mate - Nbr: 1215 - Prize: James - Ship Type: MV - How taken: HMS Harpy - When taken: 19 Dec 1812 - Where taken: off Isle of France - Date Received: 14 Jun 1814 - From what ship: Mill Prison (Plymouth, England) - Born: Philadelphia - Age: 28 - Released on 26 Apr 1815.

Rice, M. - Mate - Nbr: 4792 - Prize: Rattlesnake - Ship Type: MV - How taken: HMS Rhin - When taken: 11 Mar 1814 - Where taken: off San Domingo - Date Received: 9 Oct 1814 - From what ship: HMT Freya, Halifax - Born: Savannah - Age: 42 - Sent to Ashburton (England) on 18 Oct 1814.

Rice, Thomas - Commander - Nbr: 3420 - Prize: Industry - Ship Type: P - How taken: HMS Heron - When taken: 3 Nov 1813 - Where taken: off Halifax - Date Received: 13 Sep 1814 - From what ship: HMT Niobe, Chatham - Born: Boston - Age: 26 - Released on 28 May 1815.

Rice, William - Seaman - Nbr: 1748 - How taken: Gave himself up from HMS Thames - When taken: 9 Nov 1812 - Date Received: 20 Jul 1814 - From what ship: HMS Milford, Plymouth - Born: Boston - Age: 37 - Released on 26 Apr 1815.

Riceo, Joachim - Seaman - Nbr: 12 - Prize: Cashier - Ship Type: LM - How taken: HMS Reindeer - When taken: 3 Feb 1813 - Where taken: Bay of Biscay - Date Received: 2 Apr 1813 - From what ship: Plymouth - Born: Lima, South America - Age: 28 - Sent to Dartmouth on 30 Jul 1813.

Rich, Elisha - Seaman - Nbr: 1849 - How taken: Gave himself up from HMS Rosario - When taken: 1 May 1813 -

American Prisoners of War Held at Dartmoor during the War of 1812

## Alphabetical listing of names

Date Received: 21 Jul 1814 - From what ship: HMT Redbeard & Pincher, Chatham Depot - Born: Cape Cod - Age: 37 - Released on 2 May 1815.

Rich, Francis - Seaman - Nbr: 5797 - How taken: Gave himself up from MV Norfolk - When taken: Jan 1814 - Date Received: 26 Dec 1814 - From what ship: HMT Argo - Born: Massachusetts - Age: 26 - Released on 3 Jul 1815.

Richards, Benjamin - Seaman - Nbr: 3755 - Prize: Alfred - Ship Type: P - How taken: HMS Epervier - When taken: 23 Feb 1814 - Where taken: off Newfoundland - Date Received: 30 Sep 1814 - From what ship: HMT President, Halifax - Born: Marblehead - Age: 18 - Released on 4 Jun 1815.

Richards, Edward - Seaman - Nbr: 6184 - Prize: Lion - Ship Type: P - How taken: HMS Granicus - When taken: 2 Dec 1814 - Where taken: off Lisbon - Date Received: 21 Jan 1815 - From what ship: HMT Impregnable - Born: Virginia - Age: 19 - Released on 5 Jul 1815.

Richards, George - Seaman - Nbr: 3742 - Prize: Lizard - Ship Type: P - How taken: HMS Prometheus - When taken: 5 May 1814 - Where taken: off Halifax - Date Received: 30 Sep 1814 - From what ship: HMT President, Halifax - Born: Salem - Age: 26 - Released on 4 Jun 1815.

Richards, George - Marine - Nbr: 597 - Prize: US Brig Argus - Ship Type: MW - How taken: HMS Pelican - When taken: 14 Aug 1813 - Where taken: Irish Channel - Date Received: 8 Sep 1813 - From what ship: Plymouth - Born: New York - Age: 26 - Sent to Dartmouth on 2 Nov 1814.

Richards, Henry - Seaman - Nbr: 2896 - How taken: Gave himself up from HMS Ocean - When taken: 28 May 1813 - Date Received: 24 Aug 1814 - From what ship: HMT Alpheus, Chatham - Born: Massachusetts - Age: 32 - Released on 19 May 1815.

Richards, Isaac - Steward - Nbr: 5468 - Prize: General Putnam - Ship Type: P - How taken: HMS Leander - When taken: 8 Nov 1814 - Where taken: Long 65 Lat 42 - Date Received: 17 Dec 1814 - From what ship: HMT Loire, Halifax - Born: Salem - Age: 18 - Released on 1 Jul 1815.

Richards, James - Seaman - Nbr: 4576 - Prize: Bunker Hill - Ship Type: P - How taken: HMS Pomone - When taken: 2 Mar 1814 - Where taken: Channel - Date Received: 9 Oct 1814 - From what ship: HMT Leyden, Chatham - Born: Charleston - Age: 34 - Race: Black - Released on 15 Jun 1815.

Richards, James - Seaman - Nbr: 1842 - How taken: Gave himself up from HMS Trent - When taken: 8 May 1813 - Date Received: 21 Jul 1814 - From what ship: HMT Redbeard & Pincher, Chatham Depot - Born: Newbury - Age: 44 - Released on 2 May 1815.

Richards, John - Seaman - Nbr: 2084 - Prize: True Blooded Yankee - Ship Type: P - How taken: HMS Hope - When taken: 24 Jun 1813 - Where taken: off Brest - Date Received: 3 Aug 1814 - From what ship: HMS Bittern, Chatham Depot - Born: Boston - Age: 22 - Released on 2 May 1815.

Richards, Lawrence - Seaman - Nbr: 5620 - How taken: Gave himself up from HMS Leminanis - When taken: 20 Sep 1813 - Date Received: 24 Dec 1814 - From what ship: HMT Tay - Born: Philadelphia - Age: 24 - Released on 11 Jul 1815.

Richards, Richard - Seaman - Nbr: 6005 - Prize: McDonough - Ship Type: P - How taken: HMS Bacchante - When taken: 1 Nov 1814 - Where taken: Lat 42 Long 67 - Date Received: 28 Dec 1814 - From what ship: HMT Penelope - Born: Not legible - Released on 3 Jul 1815.

Richards, Sandy - Seaman - Nbr: 5308 - Prize: Elbridge Gerry - Ship Type: P - How taken: HMS Crescent - When taken: 25 Aug 1813 - Where taken: off St. Johns - Date Received: 31 Oct 1814 - From what ship: HMT Leyden, Chatham - Born: Baltimore - Age: 15 - Race: Black - Released on 29 Jun 1815.

Richards, Thomas - Seaman - Nbr: 5621 - How taken: Gave himself up from HMS Harpy - When taken: 7 Mar 1813 - Date Received: 24 Dec 1814 - From what ship: HMT Tay - Born: Marblehead - Age: 31 - Released on 3 Jul 1815.

Richards, Willis - Seaman - Nbr: 5476 - Prize: General Putnam - Ship Type: P - How taken: HMS Leander - When taken: 8 Nov 1814 - Where taken: Long 65 Lat 42 - Date Received: 17 Dec 1814 - From what ship: HMT Loire, Halifax - Born: Norwich - Age: 23 - Released on 1 Jul 1815.

Richardson, Andrew - Carpenter - Nbr: 2986 - Prize: Frolic - Ship Type: P - How taken: HMS Heron - When taken: 25 Jan 1814 - Where taken: off St. Thomas - Date Received: 29 Aug 1814 - From what ship: HMT Bittern - Born: Salem - Age: 47 - Released on 19 May 1815.

Richardson, Cheney - Seaman - Nbr: 1061 - Prize: Fair American - Ship Type: MV - How taken: HMS Andromache - When taken: 19 Jan 1814 - Where taken: Bay of Biscay - Date Received: 10 May 1814 - From what ship: Plymouth - Born: Brookfield - Age: 24 - Released on 27 Apr 1815.

Richardson, Edward - Seaman - Nbr: 2098 - How taken: Gave himself up from HMS Strombolo - When taken: 25 Apr 1813 - Date Received: 3 Aug 1814 - From what ship: HMS Bittern, Chatham Depot - Born: Portland -

American Prisoners of War Held at Dartmoor during the War of 1812

## Alphabetical listing of names

Age: 46 - Released on 2 May 1815.

Richardson, Francis - Seaman - Nbr: 270 - Prize: William Bayard - Ship Type: MV - How taken: HMS Warspite - When taken: 3 Mar 1813 - Where taken: Bay of Biscay - Date Received: 28 Jun 1813 - From what ship: Plymouth - Born: Newtown - Age: 22 - Released on 20 Apr 1815.

Richardson, James - Seaman - Nbr: 1904 - How taken: Gave himself up from HMS Denmark - When taken: 21 Oct 1812 - Date Received: 3 Aug 1814 - From what ship: HMS Alceste, Chatham Depot - Born: Massachusetts - Age: 30 - Released on 26 Apr 1815.

Richardson, James - Seaman - Nbr: 4649 - Prize: Wasp - Ship Type: P - How taken: HMS Bream - When taken: 10 Jun 1813 - Where taken: off Halifax - Date Received: 9 Oct 1814 - From what ship: HMT Leyden, Chatham - Born: Salem - Age: 60 - Released on 15 Jun 1815.

Richardson, John - Marine - Nbr: 2496 - Prize: US Sloop Frolic - Ship Type: MW - How taken: HMS Orpheus - When taken: 20 Apr 1814 - Where taken: off Cuba - Date Received: 16 Aug 1814 - From what ship: HMT Queen, Halifax - Born: Middlesex - Age: 39 - Released on 3 May 1815.

Richardson, John - Seaman - Nbr: 1719 - How taken: Gave himself up from HMS Rodney - Date Received: 8 Jul 1814 - From what ship: Labrador - Born: New York - Age: 23 - Released on 1 May 1815.

Richardson, John - Seaman - Nbr: 2964 - How taken: Gave himself up from HMS Andromache - When taken: 25 Feb 1813 - Date Received: 24 Aug 1814 - From what ship: HMT Hannibal - Born: Kennebec - Age: 29 - Race: Black - Released on 19 May 1815.

Richardson, John - Seaman - Nbr: 2631 - How taken: Gave himself up from HMS Leviathan - When taken: 22 Oct 1813 - Date Received: 21 Aug 1814 - From what ship: HMT Freya, Chatham - Born: Jersey - Age: 42 - Race: Black - Released on 3 May 1815.

Richardson, Nathan - Seaman - Nbr: 6222 - Prize: Prince de Neufchatel - Ship Type: P - How taken: Leander (Newcastle Acasta) - When taken: 20 Dec 1814 - Where taken: Lat 38 Long 56 - Date Received: 30 Jan 1815 - From what ship: HMT Pheasant - Born: Woburn - Age: 19 - Released on 5 Jul 1815.

Richardson, Perry - Seaman - Nbr: 3940 - Prize: Rolla - Ship Type: P - How taken: HMS Loire - When taken: 10 Dec 1813 - Where taken: off Bull Island (South Carolina) - Date Received: 5 Oct 1814 - From what ship: HMT President, Halifax - Born: Maryland - Age: 25 - Race: Negro - Released on 9 Jun 1815.

Richardson, Robert - Seaman - Nbr: 3193 - How taken: Gave himself up from HMS Swiftsure - When taken: 26 Dec 1812 - Date Received: 11 Sep 1814 - From what ship: HMT Freya, Chatham - Born: Philadelphia - Age: 43 - Race: Mulatto - Released on 27 Apr 1815.

Richardson, Samuel - Seaman - Nbr: 2918 - How taken: Impressed at Shields - When taken: 3 Aug 1812 - Date Received: 24 Aug 1814 - From what ship: HMT Alpheus, Chatham - Born: Lewistown - Age: 39 - Race: Black - Released on 27 Apr 1815.

Richardson, William - Seaman - Nbr: 4475 - Prize: Enterprize - Ship Type: P - How taken: HMS Tenedos - When taken: 21 May 1813 - Where taken: off Cape Cod - Date Received: 8 Oct 1814 - From what ship: HMT Leyden, Chatham - Born: Massachusetts - Age: 27 - Released on 15 Jun 1815.

Richardson, William - Seaman - Nbr: 291 - Prize: Cannoneer - Ship Type: MV - How taken: HMS Warspite - When taken: 14 Mar 1813 - Where taken: Bay of Biscay - Date Received: 28 Jun 1813 - From what ship: Plymouth - Born: Boston - Age: 18 - Released on 20 Apr 1815.

Richardson, William - Seaman - Nbr: 4509 - Prize: America - Ship Type: P - How taken: HMS Shannon - When taken: 23 May 1813 - Where taken: off Cape Cod - Date Received: 8 Oct 1814 - From what ship: HMT Leyden, Chatham - Born: Massachusetts - Age: 18 - Released on 15 Jun 1815.

Richardson, William - Steward - Nbr: 3667 - Prize: Alfred - Ship Type: P - How taken: HMS Epervier - When taken: 23 Feb 1812 - Where taken: off Newfoundland - Date Received: 30 Sep 1814 - From what ship: HMT President - Born: Salem - Age: 20 - Released on 4 Jun 1815.

Richman, Joshua - Seaman - Nbr: 189 - Prize: Star - Ship Type: MV - How taken: HMS Superb - When taken: 9 Feb 1813 - Where taken: Bay of Biscay - Date Received: 2 Apr 1813 - From what ship: Plymouth - Born: Maryland - Age: 21 - Released on 20 Apr 1815.

Richmond, Caleb - Seaman - Nbr: 6523 - How taken: Gave himself up from HMS Ashea - Date Received: 3 Mar 1815 - From what ship: HMT Ganges, Plymouth - Born: Philadelphia - Age: 34 - Race: Black - Escaped on 1 Jun 1815.

Rick, Francis - Seaman - Nbr: 5860 - Prize: Lion - Ship Type: P - How taken: HMS Granicus - When taken: 2 Dec 1814 - Where taken: off Lisbon - Date Received: 26 Dec 1814 - From what ship: HMT Impregnable - Born: Messina - Age: 27 - Released on 3 Jul 1815.

Rickard, Elijah - Master's mate - Nbr: 5670 - Prize: Harlequin - Ship Type: P - How taken: HMS Bulwark - When

American Prisoners of War Held at Dartmoor during the War of 1812

## Alphabetical listing of names

taken: 23 Nov 1814 - Where taken: off Halifax - Date Received: 24 Dec 1814 - From what ship: HMT Penelope - Born: Massachusetts - Age: 29 - Released on 8 May 1815.

Ricker, Edward - Seaman - Nbr: 2413 - Prize: US Sloop Frolic - Ship Type: MW - How taken: HMS Orpheus - When taken: 20 Apr 1814 - Where taken: off Cuba - Date Received: 16 Aug 1814 - From what ship: HMT Queen, Halifax - Born: Massachusetts - Age: 24 - Sent to Dartmouth on 19 Oct 1814.

Ricker, James - Seaman - Nbr: 1672 - Prize: Vivid - Ship Type: Prize - How taken: HMS Ceres - When taken: 5 Jun 1814 - Where taken: Channel - Date Received: 2 Jul 1814 - From what ship: Plymouth - Born: Baltimore - Age: 22 - Released on 1 May 1815.

Ricker, Thomas - Seaman - Nbr: 4364 - How taken: Gave himself up from HMS Abuneance - When taken: 6 Oct 1814 - Date Received: 7 Oct 1814 - From what ship: HMT Salvador del Mundo, Halifax - Born: New Hampshire - Age: 24 - Released on 10 Apr 1815.

Ricker, Wentworth - Seaman - Nbr: 5901 - Prize: Harlequin - Ship Type: P - How taken: HMS Bulwark - When taken: 23 Nov 1814 - Where taken: off Halifax - Date Received: 27 Dec 1814 - From what ship: HMT Penelope - Born: Lebanon - Age: 24 - Released on 3 Jul 1815.

Ricks, Thomas - Seaman - Nbr: 1649 - How taken: Apprehended at Bristol - When taken: 16 Aug 1813 - Date Received: 23 Jun 1814 - From what ship: Stapleton - Born: New York - Age: 26 - Race: Negro - Died on 22 Jan 1815 from rheumatismus.

Ridden, John - Seaman - Nbr: 5803 - Prize: US Frigate Superior (Gig) - Ship Type: MW - How taken: British gunboats - When taken: 26 Aug 1814 - Where taken: Lake Ontario - Date Received: 26 Dec 1814 - From what ship: HMT Argo - Born: New York - Age: 25 - Released on 12 Jun 1815.

Rideout, J. J. - Seaman - Nbr: 1815 - How taken: Gave himself up from HMS Unicorn - When taken: 17 Jun 1812 - Date Received: 20 Jul 1814 - From what ship: HMS Milford, Plymouth - Born: Baltimore - Age: 28 - Released on 26 Apr 1815.

Rideout, James - Seaman - Nbr: 1142 - How taken: Impressed in Scotland - When taken: 25 Jan 1814 - Date Received: 10 May 1814 - From what ship: Plymouth - Born: Annapolis - Age: 37 - Race: Negro - Released on 28 Apr 1815.

Rider, John - Seaman - Nbr: 5981 - Prize: Cossack - Ship Type: Private - How taken: HMS Bulwark - When taken: 1 Nov 1814 - Where taken: Georges Bank - Date Received: 27 Dec 1814 - From what ship: HMT Penelope - Born: Prussia - Age: 26 - Released on 3 Jul 1815.

Ridland, Ephraim - Seaman - Nbr: 5602 - Prize: Alexander - Ship Type: Prize - How taken: HMS Wasp - When taken: 14 Sep 1814 - Where taken: off Cape Sable Island (Canada) - Date Received: 17 Dec 1814 - From what ship: HMT Loire, Halifax - Born: Massachusetts - Age: 17 - Released on 5 Jul 1815.

Ridley, Nathaniel - Seaman - Nbr: 5959 - Prize: Harlequin - Ship Type: P - How taken: HMS Bulwark - When taken: 23 Nov 1814 - Where taken: off Halifax - Date Received: 27 Dec 1814 - From what ship: HMT Penelope - Born: Sacho - Age: 41 - Released on 20 Mar 1815.

Riggs, Robert - Seaman - Nbr: 6141 - How taken: Gave himself up from HMS Trent - When taken: 19 Jun 1814 - Date Received: 17 Jan 1815 - From what ship: HMT Impregnable - Born: Lancaster - Age: 24 - Released on 5 Jul 1815.

Right, Joshua - Seaman - Nbr: 181 - Prize: Star - Ship Type: MV - How taken: HMS Superb - When taken: 9 Feb 1813 - Where taken: Bay of Biscay - Date Received: 2 Apr 1813 - From what ship: Plymouth - Born: Saybrook - Age: 24 - Sent to Plymouth on 7 Dec 1813.

Rightman, John - Seaman - Nbr: 2817 - Prize: Kitty - Ship Type: Prize - How taken: Dart of Guernsey (Privateer) - When taken: 20 Jun 1813 - Where taken: off the Western Isles (England) - Date Received: 24 Aug 1814 - From what ship: HMT Liverpool, Chatham - Born: London - Age: 24 - Released on 2 Nov 1814.

Ring, Andrew - Seaman - Nbr: 5338 - Prize: Elbridge Gerry - Ship Type: P - How taken: HMS Crescent - When taken: 16 Sep 1813 - Where taken: at sea - Date Received: 31 Oct 1814 - From what ship: HMT Leyden, Chatham - Born: Yarmouth - Age: 17 - Released on 1 Jul 1815.

Ringe, John - Seaman - Nbr: 2185 - Prize: Diomede - Ship Type: P - How taken: HMS Rifleman - When taken: 28 Jun 1813 - Where taken: off Halifax - Date Received: 16 Aug 1814 - From what ship: HMS Dublin, Halifax - Born: Ipswich - Age: 26 - Released on 2 May 1815.

Ringgold, Thomas - Seaman - Nbr: 4607 - Prize: Argus - Ship Type: MV - How taken: HMS San Domingo - When taken: 1 Mar 1814 - Where taken: off Savannah - Date Received: 9 Oct 1814 - From what ship: HMT Leyden, Chatham - Born: Baltimore - Age: 32 - Race: Black - Released on 15 Jun 1815.

Ringold, Peregrine - Seaman - Nbr: 971 - Prize: Siro - Ship Type: LM - How taken: HMS Pelican - When taken: 13 Jan 1814 - Where taken: at sea - Date Received: 31 Jan 1814 - From what ship: Plymouth - Born: Baltimore -

American Prisoners of War Held at Dartmoor during the War of 1812
## Alphabetical listing of names

Age: 36 - Released on 27 Apr 1815.
Ringrove, William - Boy - Nbr: 3555 - Prize: Hawk - Ship Type: P - How taken: HMS Pique - When taken: 26 Apr 1814 - Where taken: off Bermuda - Date Received: 30 Sep 1814 - From what ship: HMT Sybella - Born: Maryland - Age: 12 - Released on 4 Jun 1815.
Ripley, Eden - Seaman - Nbr: 1521 - Prize: Essex - Ship Type: MV - How taken: HMS Pyramus - When taken: 20 Apr 1813 - Where taken: Bay of Biscay - Date Received: 23 Jun 1814 - From what ship: Stapleton - Born: Massachusetts - Age: 19 - Released on 1 May 1815.
Ripton, Solomon - Seaman - Nbr: 1973 - Prize: Orbit - Ship Type: MV - How taken: HMS Achates - When taken: 29 Jan 1813 - Where taken: Lat 44N Long 13W - Date Received: 3 Aug 1814 - From what ship: HMS Lyffey, Chatham Depot - Born: Baltimore - Age: 53 - Released on 2 May 1815.
Risings, John - Seaman - Nbr: 1505 - Prize: Paul Jones - Ship Type: P - How taken: HMS Leonidas - When taken: 22 May 1813 - Where taken: Channel - Date Received: 23 Jun 1814 - From what ship: Stapleton - Born: Albany - Age: 29 - Race: Colored - Sent to Mill Prison (Plymouth, England) on 30 Jun 1814.
Riswell, Palmer - Seaman - Nbr: 5184 - Prize: Montgomery - Ship Type: P - How taken: HMS Nymphe - When taken: 1 May 1813 - Where taken: off Cape Cod - Date Received: 31 Oct 1814 - From what ship: HMT Mermaid, Chatham - Born: Old George - Age: 35 - Released on 29 Jun 1815.
Ritchie, James - Seaman - Nbr: 3986 - Prize: US Brig Rattlesnake - Ship Type: MW - How taken: HMS Leander - When taken: 13 Jul 1814 - Where taken: off Shelburne - Date Received: 6 Oct 1814 - From what ship: HMT Chesapeake, Halifax - Born: Baltimore - Age: 23 - Released on 9 Jun 1815.
Rivers, Thomas - Seaman - Nbr: 6097 - Prize: Black Swan - Ship Type: MV - How taken: HMS Maidstone - When taken: 24 Oct 1814 - Where taken: Georges Bank - Date Received: 28 Dec 1814 - From what ship: HMT Penelope - Born: Bengal, India - Age: 38 - Race: Black - Released on 3 Jul 1815.
Riviere, Toussaint - Seaman - Nbr: 3837 - Prize: Dominique - Ship Type: LM - How taken: HMS Dotterel - When taken: 21 May 1814 - Where taken: off Charleston - Date Received: 5 Oct 1814 - From what ship: HMT Orpheus, Halifax - Born: Toulon - Age: 23 - Released on 9 Jun 1815.
Roach, James - Seaman - Nbr: 5113 - Prize: Portsmouth - Ship Type: P - How taken: HMS Pylades - When taken: 13 Sep 1814 - Where taken: Long 65 - Date Received: 28 Oct 1814 - From what ship: HMT Alkbar, Halifax - Born: Portsmouth - Age: 17 - Released on 29 Jun 1815.
Roach, Pedro - Seaman - Nbr: 3505 - Prize: Maria - Ship Type: MV - How taken: Not listed - When taken: 11 Sep 1814 - Where taken: off Liverpool - Date Received: 24 Sep 1814 - From what ship: HMT Salvador del Mundo - Born: Bilbao - Age: 35 - Released on 4 Jun 1815.
Roach, Reuben - Seaman - Nbr: 1804 - How taken: Gave himself up from HMS Dublin - Date Received: 20 Jul 1814 - From what ship: HMS Milford, Plymouth - Born: Dichery - Age: 25 - Race: Black - Released on 1 May 1815.
Road, John - Seaman - Nbr: 901 - Prize: Squirrel - Ship Type: MV - How taken: HMS Belle Poule - When taken: 14 Dec 1813 - Where taken: at sea - Date Received: 31 Jan 1814 - From what ship: Plymouth - Born: Hamburg - Age: 24 - Sent to Mill Prison (Plymouth, England) on 17 Jun 1814.
Roath, Stephen - Seaman - Nbr: 698 - Prize: Frederick Augusta - Ship Type: MV - How taken: HMS Sir John Sherbrook - When taken: 12 Apr 1813 - Where taken: off Newport - Date Received: 27 Sep 1813 - From what ship: Plymouth - Born: Connecticut - Age: 24 - Released on 9 Apr 1815.
Roberson, Henry - Seaman - Nbr: 656 - Prize: Marmion - Ship Type: MV - How taken: HMS President - When taken: 14 Aug 1813 - Where taken: off Nantes - Date Received: 8 Sep 1813 - From what ship: Plymouth - Born: Richmond - Age: 17 - Race: Negro - Released on 26 Apr 1815.
Roberts, David - Seaman - Nbr: 2038 - How taken: Gave himself up from HMS Impeteux - When taken: 2 Dec 1812 - Date Received: 3 Aug 1814 - From what ship: HMS Lyffey, Chatham Depot - Born: Virginia - Age: 24 - Race: Black - Released on 26 Apr 1815.
Roberts, Francis - Seaman - Nbr: 829 - Prize: Chesapeake - Ship Type: LM - How taken: HMS Hotspur & HMS Pyramus - When taken: 26 Oct 1813 - Where taken: off Nantes - Date Received: 29 Nov 1813 - From what ship: Plymouth - Born: San Sebastian - Age: 26 - Died on 7 Feb 1815 from variola.
Roberts, George - Seaman - Nbr: 1192 - Prize: Miquelonnaiss - Ship Type: P - How taken: Crete - When taken: 25 Sep 1813 - Date Received: 4 Jun 1814 - From what ship: Dartmouth - Born: Baltimore - Age: 30 - Released on 6 Apr 1815.
Roberts, Hugh - Seaman - Nbr: 532 - Prize: Union - Ship Type: LM - How taken: HMS Goldfinch - When taken: 17 Jul 1813 - Where taken: Bay of Biscay - Date Received: 8 Sep 1813 - From what ship: Plymouth - Born: Pennsylvania - Age: 39 - Released on 26 Apr 1815.

American Prisoners of War Held at Dartmoor during the War of 1812

## Alphabetical listing of names

Roberts, James - Seaman - Nbr: 6511 - How taken: Gave himself up from HMS Orontes - When taken: Feb 1815 - Date Received: 3 Mar 1815 - From what ship: HMT Ganges, Plymouth - Born: Wilmington - Age: 34 - Race: Black - Released on 8 Jun 1815.

Roberts, Joel - Seaman - Nbr: 925 - Prize: Growler - Ship Type: MV - How taken: HMS Wolf - When taken: 11 Aug 1813 - Where taken: at sea - Date Received: 31 Jan 1814 - From what ship: Plymouth - Born: Littletown - Age: 20 - Sent to Dartmouth on 19 Oct 1814.

Roberts, John - Seaman - Nbr: 4822 - Prize: President - Ship Type: P - How taken: HMS Pique - When taken: 7 May 1814 - Where taken: off Porto Rico - Date Received: 9 Oct 1814 - From what ship: HMT Freya, Halifax - Born: Africa - Age: 50 - Race: Black - Released on 8 Jun 1815.

Roberts, John - Seaman - Nbr: 486 - How taken: Gave himself up at Cove of Cork - Date Received: 8 Sep 1813 - From what ship: Plymouth - Born: Baltimore - Age: 23 - Died on 12 May 1815 from gunshot wound.

Roberts, Moses - Seaman - Nbr: 1268 - Prize: Adeline - Ship Type: MV - How taken: HMS Magicienne - When taken: 16 Mar 1814 - Where taken: off Cape Finisterre - Date Received: 14 Jun 1814 - From what ship: Mill Prison (Plymouth, England) - Born: Falmouth, Jamaica - Age: 30 - Race: Black - Released on 28 Apr 1815.

Roberts, Nathaniel - Seaman - Nbr: 6187 - Prize: Garland - Ship Type: Prize - How taken: HMS Hamadryad - When taken: 22 Nov 1814 - Where taken: Georges Bank - Date Received: 21 Jan 1815 - From what ship: HMT Impregnable - Born: Kennebec - Age: 18 - Released on 5 Jul 1815.

Roberts, Nathaniel - Seaman - Nbr: 4761 - Prize: Adeline - Ship Type: LM - How taken: HMS Magicienne - When taken: 14 Mar 1814 - Where taken: off Cape Ortegal (Spain) - Date Received: 9 Oct 1814 - From what ship: HMT Freya, Chatham - Born: Beverly - Age: 19 - Released on 15 Jun 1815.

Roberts, Nathaniel - Steward - Nbr: 5694 - Prize: McDonough - Ship Type: P - How taken: HMS Bacchante - When taken: 7 Nov 1814 - Where taken: Lat 42 Long 67 - Date Received: 24 Dec 1814 - From what ship: HMT Penelope - Born: Massachusetts - Age: 27 - Released on 11 Jul 1815.

Roberts, Robert - Seaman - Nbr: 2610 - How taken: Gave himself up from HMS Zephyr - When taken: 12 Sep 1812 - Date Received: 21 Aug 1814 - From what ship: HMT Freya, Chatham - Born: New York - Age: 27 - Released on 26 Apr 1815.

Robertson, James - Seaman - Nbr: 1578 - Prize: Price - Ship Type: MV - How taken: HMS Pyramus - When taken: 6 Apr 1813 - Where taken: Bay of Biscay - Date Received: 23 Jun 1814 - From what ship: Stapleton - Born: Massachusetts - Age: 21 - Died on 1 Apr 1815 from variola.

Robertson, Robert - Seaman - Nbr: 1409 - Prize: Miranda - Ship Type: Prize - How taken: HMS Unicorn - When taken: 21 May 1813 - Where taken: off Ushant (France) - Date Received: 19 Jun 1814 - From what ship: Stapleton - Born: SC - Age: 23 - Released on 28 Apr 1815.

Robertson, Thomas - Seaman - Nbr: 5382 - Prize: Rattlesnake - Ship Type: LM - How taken: HMS Rhin - When taken: 17 Mar 1814 - Where taken: off Bermuda - Date Received: 31 Oct 1814 - From what ship: HMT Leyden, Chatham - Born: Boston - Age: 21 - Released on 1 Jul 1815.

Robertson, William - Seaman - Nbr: 4430 - Prize: Taken off an English whaler - Ship Type: MV - How taken: HMS Illustrious - When taken: 22 Oct 1813 - Where taken: at sea - Date Received: 8 Oct 1814 - From what ship: HMT Leyden, Chatham - Born: Philadelphia - Age: 26 - Escaped on 1 Jun 1815.

Robes, Edward - Boy - Nbr: 4672 - Prize: Industry - Ship Type: P - How taken: HMS Heron - When taken: 3 Nov 1813 - Where taken: off Halifax - Date Received: 9 Oct 1814 - From what ship: HMT Leyden, Chatham - Born: Marblehead - Age: 15 - Released on 15 Jun 1815.

Robinet, Samuel - Seaman - Nbr: 2113 - How taken: Gave himself up from HMS Sarpedon - When taken: 2 Oct 1812 - Date Received: 8 Aug 1814 - From what ship: HMT Raven, Chatham - Born: Philadelphia - Age: 46 - Sent to Exeter Jail on 9 Mar 1815.

Robinett, Samuel - Seaman - Nbr: 6553 - How taken: Gave himself up from HMS Laspidon - When taken: 2 Oct 1812 - Date Received: 26 Mar 1815 - From what ship: from Exeter jail - Born: Philadelphia - Age: 45 - Released on 27 Apr 1815.

Robins, John - Seaman - Nbr: 1012 - Prize: Rachel & Ann - Ship Type: P - How taken: HMS Cydnus - When taken: 6 Jan 1814 - Where taken: at sea - Date Received: 31 Jan 1814 - From what ship: Plymouth - Born: New Orleans - Age: 30 - Sent to Plymouth on 8 Jul 1814.

Robins, Philip - Marine - Nbr: 2499 - Prize: US Sloop Frolic - Ship Type: MW - How taken: HMS Orpheus - When taken: 20 Apr 1814 - Where taken: off Cuba - Date Received: 16 Aug 1814 - From what ship: HMT Queen, Halifax - Born: Massachusetts - Age: 22 - Released on 3 May 1815.

Robins, William - Steward - Nbr: 4210 - Prize: Commodore Perry - Ship Type: Schooner - How taken: Sent into custody from a cutter - When taken: 25 Feb 1814 - Where taken: off Bordeaux - Date Received: 7 Oct 1814 -

American Prisoners of War Held at Dartmoor during the War of 1812

## Alphabetical listing of names

From what ship: HMT Niobe, Chatham - Born: Eddington - Age: 27 - Race: Mulatto - Released on 13 Jun 1815.

Robinson, Benjamin - Seaman - Nbr: 1936 - How taken: Gave himself up from HMS Royal William - When taken: 3 Feb 1813 - Date Received: 3 Aug 1814 - From what ship: HMS Alceste, Chatham Depot - Born: Boston - Age: 32 - Released on 2 May 1815.

Robinson, Charles - Seaman - Nbr: 426 - Prize: Viper - Ship Type: MV - How taken: HMS Superb - When taken: 15 Apr 1813 - Where taken: Bay of Biscay - Date Received: 1 Jul 1813 - From what ship: Plymouth - Born: Sweden - Age: 22 - Released on 20 Apr 1815.

Robinson, Edward - Seaman - Nbr: 1908 - How taken: Gave himself up from HMS Elephant - When taken: 25 Jan 1813 - Date Received: 3 Aug 1814 - From what ship: HMS Alceste, Chatham Depot - Born: Bath - Age: 27 - Released on 2 May 1815.

Robinson, Edward - Seaman - Nbr: 2967 - How taken: Gave himself up from HMS Elephant - When taken: 13 Jan 1812 - Date Received: 24 Aug 1814 - From what ship: HMT Hannibal - Born: Massachusetts - Age: 28 - Released on 27 Apr 1815.

Robinson, Elias - Seaman - Nbr: 1417 - Prize: Orders in Council - Ship Type: LM - How taken: HMS Surveillante - When taken: 1 Jun 1813 - Where taken: off Cape Ortegal (Spain) - Date Received: 19 Jun 1814 - From what ship: Stapleton - Born: Connecticut - Age: 19 - Released on 28 Apr 1815.

Robinson, Garner - Seaman - Nbr: 5778 - Prize: William Penn - Ship Type: MV - How taken: HMS Acorn - When taken: 27 Oct 1812 - Where taken: Lat 14 - Date Received: 26 Dec 1814 - From what ship: HMT Argo - Born: Trinidad - Age: 23 - Race: Black - Released on 27 Apr 1815.

Robinson, George - Seaman - Nbr: 5782 - How taken: Gave himself up from HMS Ariel - When taken: 8 Dec 1812 - Date Received: 26 Dec 1814 - From what ship: HMT Argo - Born: Talbot County - Age: 25 - Race: Black - Released on 3 Jul 1815.

Robinson, Henry - Seaman - Nbr: 2931 - How taken: Gave himself up from HMS Berwick - Date Received: 24 Aug 1814 - From what ship: HMT Alpheus, Chatham - Born: Philadelphia - Age: 33 - Race: Black - Released on 19 May 1815.

Robinson, James - Mate - Nbr: 5455 - How taken: Impressed at Liverpool - When taken: Nov 1814 - Date Received: 10 Dec 1814 - From what ship: HMT Impregnable - Born: Connecticut - Age: 30 - Sent to Ashburton (England) on 16 Dec 1814.

Robinson, James - Seaman - Nbr: 1034 - Prize: US Brig Argus - Ship Type: MW - How taken: HMS Pelican - When taken: 16 Aug 1813 - Where taken: Irish Channel - Date Received: 10 May 1814 - From what ship: Plymouth - Born: New York - Age: 19 - Sent to Dartmouth on 19 Oct 1814.

Robinson, James - Seaman - Nbr: 3140 - How taken: Impressed from MV Rebecca - When taken: 23 Sep 1813 - Date Received: 11 Sep 1814 - From what ship: HMT Freya, Chatham - Born: Rhode Island - Age: 22 - Race: Black - Released on 28 May 1815.

Robinson, John - Chief Mate and Prize Master - Nbr: 531 - Prize: Union - Ship Type: LM - How taken: HMS Goldfinch - When taken: 17 Jul 1813 - Where taken: Bay of Biscay - Date Received: 8 Sep 1813 - From what ship: Plymouth - Born: New London - Age: 30 - Released on 6 Apr 1815.

Robinson, John - Seaman - Nbr: 1644 - Prize: Paul Jones - Ship Type: P - How taken: HMS Leonidas - When taken: 23 May 1813 - Where taken: Channel - Date Received: 23 Jun 1814 - From what ship: Stapleton - Born: Connecticut - Age: 52 - Released on 1 May 1815.

Robinson, John - Seaman - Nbr: 3199 - How taken: Gave himself up from HMS America - When taken: 16 Jul 1813 - Date Received: 11 Sep 1814 - From what ship: HMT Freya, Chatham - Born: Massachusetts - Age: 26 - Released on 28 May 1815.

Robinson, John - Seaman - Nbr: 3268 - How taken: Gave himself up from HMS Cloride - Date Received: 11 Sep 1814 - What ship: HMT Freya, Chatham - Born: Pennsylvania - Age: 32 - Race: Black - Released on 28 May 1815.

Robinson, John - Seaman - Nbr: 3389 - Prize: Wasp - Ship Type: P - How taken: HMS Bream - When taken: 10 Jun 1813 - Where taken: off Halifax - Date Received: 13 Sep 1814 - From what ship: HMT Niobe, Chatham - Born: Nantucket - Age: 28 - Released on 28 May 1815.

Robinson, Joseph - Seaman - Nbr: 5547 - Prize: Phaeton - Ship Type: MV - How taken: Cast Away - When taken: 7 Feb 1814 - Where taken: Old Heights - Date Received: 17 Dec 1814 - From what ship: HMT Loire, Halifax - Born: Falmouth - Age: 22 - Released on 11 Jul 1815.

Robinson, Joshua - 2nd Lieutenant - Nbr: 5574 - Prize: McDonough - Ship Type: P - How taken: HMS Bacchante - When taken: 1 Nov 1814 - Where taken: Lat 42 Long 67 - Date Received: 17 Dec 1814 - From what ship:

American Prisoners of War Held at Dartmoor during the War of 1812

## Alphabetical listing of names

HMT Loire, Halifax - Born: Portland - Age: 29 - Released on 7 Apr 1815.

Robinson, Josiah - Seaman - Nbr: 375 - Prize: Independence - Ship Type: MV - How taken: HMS Superb - When taken: 16 Mar 1813 - Where taken: Bay of Biscay - Date Received: 1 Jul 1813 - From what ship: Plymouth - Born: Beverly - Age: 21 - Released on 20 Apr 1815.

Robinson, Michael - Seaman - Nbr: 6177 - How taken: Apprehended at Capeland - When taken: 7 Oct 1814 - Date Received: 21 Jan 1815 - What ship: HMT Impregnable - Born: Bristol - Age: 21 - Released on 5 Jul 1815.

Robinson, Nicholas - Seaman - Nbr: 703 - Prize: Ned - Ship Type: LM - How taken: HMS Royalist - When taken: 6 Sep 1813 - Where taken: Bay of Biscay - Date Received: 27 Sep 1813 - From what ship: Plymouth - Born: Stockholm - Age: 34 - Released on 7 May 1814.

Robinson, Robert - Seaman - Nbr: 4319 - How taken: Gave himself up from HMS Sanagora - When taken: 10 Mar 1814 - Date Received: 7 Oct 1814 - From what ship: HMT Niobe, Chatham - Born: Philadelphia - Age: 32 - Race: Mulatto - Released on 14 Jun 1815.

Robinson, Robert - Seaman - Nbr: 5300 - Prize: Polly - Ship Type: P - How taken: Prize - When taken: 20 Jul 1813 - Where taken: Grand Banks - Date Received: 31 Oct 1814 - From what ship: HMT Leyden, Chatham - Born: Portsmouth - Age: 35 - Released on 29 Jun 1815.

Robinson, Samuel - Seaman - Nbr: 298 - Prize: Ducornau - Ship Type: MV - How taken: HMS Pheasant - When taken: 15 Mar 1813 - Where taken: Bay of Biscay - Date Received: 28 Jun 1813 - From what ship: Plymouth - Born: Boston - Age: 33 - Died on 15 Feb 1815 while serving as a cook in the hospital.

Robinson, Stephen - Seaman - Nbr: 4621 - Prize: Argus - Ship Type: MV - How taken: HMS San Domingo - When taken: 1 Mar 1814 - Where taken: off Savannah - Date Received: 9 Oct 1814 - From what ship: HMT Leyden, Chatham - Born: Boston - Age: 25 - Race: Black - Released on 15 Jun 1815.

Robinson, Thomas - Seaman - Nbr: 4396 - Prize: Portsmouth Packet - Ship Type: Prize - How taken: HMS Fantome - When taken: 5 Oct 1813 - Where taken: off Portland - Date Received: 8 Oct 1814 - From what ship: HMT Leyden, Chatham - Born: Pennsylvania - Age: 27 - Released on 14 Jun 1815.

Robinson, Thomas - Seaman - Nbr: 5038 - Prize: Betsey - Ship Type: Prize - How taken: HMS Pylades - When taken: 7 Sep 1814 - Where taken: Canten - Date Received: 28 Oct 1814 - From what ship: HMT Alkbar, Halifax - Born: Philadelphia - Age: 20 - Released on 29 Jun 1815.

Robinson, Tony - Seaman - Nbr: 2949 - Prize: Rattlesnake - Ship Type: LM - How taken: HMS Rhin - When taken: 10 Mar 1814 - Where taken: West Indies - Date Received: 24 Aug 1814 - From what ship: HMT Hannibal - Born: Falmouth - Age: 27 - Released on 10 Apr 1814.

Robinson, William - Seaman - Nbr: 2533 - Prize: Diomede - Ship Type: P - How taken: HMS Rifleman - When taken: 28 Jul 1814 - Where taken: off Halifax - Date Received: 16 Aug 1814 - From what ship: HMT Queen, Halifax - Born: Philadelphia - Age: 40 - Released on 3 May 1815.

Robinson, William - Seaman - Nbr: 4778 - How taken: Gave himself up from HMS Revenge - When taken: 23 Sep 1814 - Date Received: 9 Oct 1814 - From what ship: HMT Freya, Chatham - Born: Baltimore - Age: 21 - Race: Mulatto - Released on 15 Jun 1815.

Robinson, William - Seaman - Nbr: 4498 - Prize: Enterprize - Ship Type: P - How taken: HMS Tenedos - When taken: 21 May 1813 - Where taken: off Cape Cod - Date Received: 8 Oct 1814 - From what ship: HMT Leyden, Chatham - Born: Massachusetts - Age: 27 - Released on 15 Jun 1815.

Robinson, William - Seaman - Nbr: 6264 - Prize: Plutarch - Ship Type: MV - How taken: HMS Helicon - When taken: 5 Feb 1815 - Where taken: off Bordeaux - Date Received: 10 Feb 1815 - From what ship: HMT Ganges, Plymouth - Born: Philadelphia - Age: 40 - Died on 18 Apr 1815 from articulo mortis.

Robinson, William - Seaman - Nbr: 318 - Prize: Pallas - Ship Type: MV - How taken: Rebuff - When taken: 23 Dec 1812 - Where taken: off Cadiz - Date Received: 28 Jun 1813 - From what ship: Plymouth - Born: New York - Age: 29 - Sent to Dartmouth on 30 Jul 1813.

Rock, Oliver - Seaman - Nbr: 164 - Prize: Margaret - Ship Type: P - How taken: HMS Nimrod - When taken: 9 Mar 1813 - Where taken: off Morlaix - Date Received: 2 Apr 1813 - From what ship: Plymouth - Born: Morlaix - Age: 34 - Sent to Dartmouth on 30 Jul 1813.

Rockhill, John - Seaman - Nbr: 4123 - Prize: Bordeaux Packet - Ship Type: LM - How taken: HMS Niemen - When taken: 28 Jun 1814 - Where taken: off Delaware - Date Received: 6 Oct 1814 - From what ship: HMT Chesapeake, Halifax - Born: Philadelphia - Age: 25 - Released on 13 Jun 1815.

Roddy, W. B. - Fifer - Nbr: 3272 - Prize: US Sloop Frolic - Ship Type: MW - How taken: HMS Orpheus - When taken: 20 Apr 1814 - Where taken: off Cuba - Date Received: 13 Sep 1814 - From what ship: Naval Hospital, Plymouth - Born: Maryland - Age: 22 - Released on 28 May 1815.

Roderick, John - Seaman - Nbr: 5831 - Prize: US Schooner Tigress - Ship Type: MW - How taken: British gunboats

American Prisoners of War Held at Dartmoor during the War of 1812

## Alphabetical listing of names

- When taken: 2 Sep 1814 - Where taken: Lake Erie - Date Received: 26 Dec 1814 - From what ship: HMT Argo - Born: Havana - Age: 23 - Released on 3 Jul 1815.

Roderigo, Joseph - Seaman - Nbr: 5873 - Prize: Lion - Ship Type: P - How taken: HMS Granicus - When taken: 2 Dec 1814 - Where taken: off Lisbon - Date Received: 26 Dec 1814 - From what ship: HMT Impregnable - Born: Lisbon - Age: 18 - Released on 3 Jul 1815.

Roderique, Joseph - Seaman - Nbr: 6124 - Prize: Betsey - Ship Type: Prize - How taken: HMS Bellerophon - When taken: 2 Nov 1814 - Where taken: Long 61 - Date Received: 6 Jan 1814 - From what ship: HMT Impregnable - Born: Lisbon - Age: 18 - Released on 5 Jul 1815.

Rodgers, Abraham - Seaman - Nbr: 4435 - Prize: Dart - Ship Type: LM - How taken: HMS Niger - When taken: 13 Nov 1815 - Where taken: at sea - Date Received: 8 Oct 1814 - From what ship: HMT Leyden, Chatham - Born: Delaware - Age: 21 - Released on 14 Jun 1815.

Rodgers, Edward - Seaman - Nbr: 3163 - How taken: Impressed at Gravesend - Date Received: 11 Sep 1814 - From what ship: HMT Freya, Chatham - Born: Boston - Age: 29 - Released on 28 May 1815.

Rodgers, Ephraim - Seaman - Nbr: 2651 - Prize: Jane - Ship Type: MV - How taken: HMS Crescent - When taken: 28 Jun 1813 - Where taken: off Newfoundland - Date Received: 21 Aug 1814 - From what ship: HMT Freya, Chatham - Born: Long Island - Age: 22 - Released on 3 May 1815.

Rodgers, George - Seaman - Nbr: 2679 - Prize: Joseph - Ship Type: MV - How taken: HMS Iris - When taken: 8 Jun 1813 - Where taken: off Spain - Date Received: 21 Aug 1814 - From what ship: HMT Freya, Chatham - Born: Gloucester - Age: 22 - Released on 19 May 1815.

Rodgers, John - Seaman - Nbr: 2033 - How taken: Gave himself up from HMS Vega - When taken: 12 Mar 1813 - Date Received: 3 Aug 1814 - From what ship: HMS Lyffey, Chatham Depot - Born: Accomack - Age: 35 - Released on 2 May 1815.

Rodgers, Samuel - Prize Master - Nbr: 2639 - Prize: True Blooded Yankee - Ship Type: P - How taken: HMS Hamadryad - When taken: 24 Jul 1813 - Where taken: off Norway - Date Received: 21 Aug 1814 - From what ship: HMT Freya, Chatham - Born: New York - Age: 20 - Escaped on 16 Feb 1815.

Rodgers, Samuel - Seaman - Nbr: 3150 - Prize: Kitty - Ship Type: Prize - How taken: Dart of Guernsey (Privateer) - When taken: 20 Jun 1813 - Where taken: off the Western Isles (England) - Date Received: 11 Sep 1814 - From what ship: HMT Freya, Chatham - Born: Boston - Age: 18 - Sent to Dartmouth on 23 Sep 1814.

Rodgers, William - Seaman - Nbr: 3327 - Prize: York Town - Ship Type: P - How taken: HMS Nimrod - When taken: 17 Jul 1813 - Where taken: off St. Johns - Date Received: 13 Sep 1814 - From what ship: HMT Niobe, Chatham - Born: Chester - Age: 27 - Released on 28 May 1815.

Rodwin, Walter - Seaman - Nbr: 650 - Prize: Marmion - Ship Type: MV - How taken: HMS President - When taken: 14 Aug 1813 - Where taken: off Nantes - Date Received: 8 Sep 1813 - From what ship: Plymouth - Born: Newport - Age: 27 - Released on 26 Apr 1815.

Rodyman, John - Seaman - Nbr: 863 - Prize: Amiable - Ship Type: LM - How taken: HMS Magnificent - When taken: 30 Oct 1813 - Where taken: off Bordeaux - Date Received: 4 Dec 1813 - From what ship: Plymouth - Born: New York - Age: 21 - Released on 27 Apr 1815.

Roe, John - Seaman - Nbr: 242 - Prize: William Bayard - Ship Type: MV - How taken: HMS Warspite - When taken: 3 Mar 1813 - Where taken: Bay of Biscay - Date Received: 2 Apr 1813 - From what ship: Plymouth - Born: New Orleans - Age: 27 - Sent to Mill Prison (Plymouth, England) on 21 Jun 1814.

Roff, Samuel - Gunner - Nbr: 3670 - Prize: Alfred - Ship Type: P - How taken: HMS Epervier - When taken: 23 Feb 1812 - Where taken: off Newfoundland - Date Received: 30 Sep 1814 - From what ship: HMT President - Born: Marblehead - Age: 53 - Released on 4 Jun 1815.

Rogers, Ambrose - Seaman - Nbr: 3715 - Prize: Alfred - Ship Type: P - How taken: HMS Epervier - When taken: 23 Feb 1814 - Where taken: off Newfoundland - Date Received: 30 Sep 1814 - From what ship: HMT President, Halifax - Born: Marblehead - Age: 20 - Released on 4 Jun 1815.

Rogers, Asa - Seaman - Nbr: 2483 - Prize: US Sloop Frolic - Ship Type: MW - How taken: HMS Orpheus - When taken: 20 Apr 1814 - Where taken: off Cuba - Date Received: 16 Aug 1814 - From what ship: HMT Queen, Halifax - Born: Newbury - Age: 22 - Released on 3 May 1815.

Rogers, Daniel - Seaman - Nbr: 6048 - Prize: Sally - Ship Type: MV - How taken: HMS Majestic - When taken: 18 Oct 1814 - Where taken: off Block Island - Date Received: 28 Dec 1814 - From what ship: HMT Penelope - Born: Worcester - Age: 29 - Released on 3 Jul 1815.

Rogers, Francis - Seaman - Nbr: 2393 - Prize: US Sloop Frolic - Ship Type: MW - How taken: HMS Orpheus - When taken: 20 Apr 1814 - Where taken: off Cuba - Date Received: 16 Aug 1814 - From what ship: HMT Queen, Halifax - Born: Rhode Island - Age: 23 - Sent to Dartmouth on 19 Oct 1814.

American Prisoners of War Held at Dartmoor during the War of 1812

## Alphabetical listing of names

Rogers, Gaff - Seaman - Nbr: 4968 - Prize: Herald - Ship Type: P - How taken: HMS Endymion - When taken: 15 Aug 1814 - Where taken: off Nantucket - Date Received: 28 Oct 1814 - From what ship: HMT Alkbar, Halifax - Born: New York - Age: 27 - Released on 21 Jun 1815.

Rogers, George - Marine - Nbr: 4035 - Prize: US Brig Rattlesnake - Ship Type: MW - How taken: HMS Leander - When taken: 13 Jul 1814 - Where taken: off Shelburne - Date Received: 6 Oct 1814 - From what ship: HMT Chesapeake, Halifax - Born: Boston - Age: 43 - Released on 9 Jun 1815.

Rogers, Henry - Seaman - Nbr: 1016 - Prize: Rachel & Ann - Ship Type: P - How taken: HMS Cydnus - When taken: 6 Jan 1814 - Where taken: at sea - Date Received: 31 Jan 1814 - From what ship: Plymouth - Born: New Orleans - Age: 25 - Sent to Plymouth on 8 Jul 1814.

Rogers, Henry - Seaman - Nbr: 4133 - Prize: Bordeaux Packet - Ship Type: LM - How taken: HMS Niemen - When taken: 28 Jun 1814 - Where taken: off Delaware - Date Received: 6 Oct 1814 - From what ship: HMT Chesapeake, Halifax - Born: New Jersey - Age: 28 - Released on 13 Jun 1815.

Rogers, James - Seaman - Nbr: 389 - Prize: John & Frances - Ship Type: MV - How taken: HMS Belle Poule - When taken: 13 Mar 1813 - Where taken: Bay of Biscay - Date Received: 1 Jul 1813 - From what ship: Plymouth - Born: Philadelphia - Age: 25 - Released on 20 Apr 1815.

Rogers, Luke - Seaman - Nbr: 4856 - Prize: Fairy - Ship Type: MV - How taken: HMS Hardy - When taken: 28 Mar 1814 - Where taken: West Indies - Date Received: 9 Oct 1814 - From what ship: HMT Freya, Halifax - Born: Carolina - Age: 24 - Died on 13 Nov 1814.

Rogers, Nathaniel - Seaman - Nbr: 1798 - Prize: Polly - Ship Type: MV - How taken: HMS Surveillante - When taken: 23 Mar 1813 - Where taken: Bay of Biscay - Date Received: 20 Jul 1814 - From what ship: HMS Milford, Plymouth - Born: Marblehead - Age: 23 - Released on 1 May 1815.

Rogers, Nathaniel - Boy - Nbr: 2403 - Prize: US Sloop Frolic - Ship Type: MW - How taken: HMS Orpheus - When taken: 20 Apr 1814 - Where taken: off Cuba - Date Received: 16 Aug 1814 - From what ship: HMT Queen, Halifax - Born: New Bedford - Age: 14 - Sent to Dartmouth on 19 Oct 1814.

Rogers, Pely - Seaman - Nbr: 5342 - Prize: John & Mary - Ship Type: Prize - How taken: HMS Borer - When taken: 20 Oct 1813 - Where taken: off Long Island - Date Received: 31 Oct 1814 - From what ship: HMT Leyden, Chatham - Born: Warren - Age: 22 - Released on 1 Jul 1815.

Rogers, Robert - Seaman - Nbr: 6114 - Prize: Garland - Ship Type: Prize - How taken: HMS Hamadryad - When taken: 22 Nov 1814 - Where taken: Georges Bank - Date Received: 6 Jan 1814 - From what ship: HMT Impregnable - Born: Newbury - Age: 26 - Released on 5 Jul 1815.

Rogers, Robert - Seaman - Nbr: 893 - Prize: General Kempt - Ship Type: P - How taken: HMS Foxhound - When taken: 18 Dec 1813 - Where taken: Lat 48'60" Long 5'7" - Date Received: 31 Jan 1814 - From what ship: Plymouth - Born: Virginia - Age: 23 - Released on 27 Apr 1815.

Roille, Oliver - Seaman - Nbr: 5872 - Prize: Lion - Ship Type: P - How taken: HMS Granicus - When taken: 2 Dec 1814 - Where taken: off Lisbon - Date Received: 26 Dec 1814 - From what ship: HMT Impregnable - Born: L'Orient (France) - Age: 17 - Released on 3 Jul 1815.

Roles, John - Seaman - Nbr: 1386 - Prize: Courier - Ship Type: P - How taken: HMS Rover - When taken: 14 May 1813 - Where taken: Bay of Biscay - Date Received: 19 Jun 1814 - From what ship: Stapleton - Born: Maryland - Age: 23 - Released on 28 Apr 1815.

Roley, John - Seaman - Nbr: 1355 - Prize: Paul Jones - Ship Type: P - How taken: HMS Leonidas - When taken: 23 May 1813 - Where taken: Channel - Date Received: 19 Jun 1814 - From what ship: Stapleton - Born: Alexandria - Age: 36 - Sent to Mill Prison (Plymouth, England) on 21 Jun 1814.

Rolla, William - Seaman - Nbr: 3446 - How taken: Gave himself up from HMS Royal William - When taken: 22 Nov 1813 - Date Received: 13 Sep 1814 - From what ship: HMT Niobe, Chatham - Born: Gloucester - Age: 33 - Released on 28 May 1815.

Rollands, Joseph - Seaman - Nbr: 268 - Prize: William Bayard - Ship Type: MV - How taken: HMS Warspite - When taken: 3 Mar 1813 - Where taken: Bay of Biscay - Date Received: 28 Jun 1813 - From what ship: Plymouth - Born: Pennsylvania - Age: 25 - Released on 20 Apr 1815.

Romain, Samuel - Seaman - Nbr: 185 - Prize: Star - Ship Type: MV - How taken: HMS Superb - When taken: 9 Feb 1813 - Where taken: Bay of Biscay - Date Received: 2 Apr 1813 - From what ship: Plymouth - Born: New York - Age: 30 - Released on 20 Apr 1815.

Roman, John - Seaman - Nbr: 511 - Prize: Hannah & Eliza - Ship Type: MV - How taken: HMS Lyra - When taken: 29 May 1813 - Where taken: off Bayonne - Date Received: 8 Sep 1813 - From what ship: Plymouth - Born: New York - Age: 35 - Released on 26 Apr 1815.

Ronder, Stephen C. - Seaman - Nbr: 3680 - Prize: Alfred - Ship Type: P - How taken: HMS Epervier - When taken:

American Prisoners of War Held at Dartmoor during the War of 1812

## Alphabetical listing of names

23 Feb 1812 - Where taken: off Newfoundland - Date Received: 30 Sep 1814 - From what ship: HMT President - Born: Marblehead - Age: 19 - Released on 4 Jun 1815.

Ronning, Mathew - Seaman - Nbr: 2286 - Prize: Diomede - Ship Type: P - How taken: HMS Rifleman - When taken: 28 May 1814 - Where taken: off Sable Island - Date Received: 16 Aug 1814 - From what ship: HMS Dublin, Halifax - Born: MI - Age: 24 - Released on 3 May 1815.

Ronse, Joseph - Seaman - Nbr: 2474 - Prize: US Sloop Frolic - Ship Type: MW - How taken: HMS Orpheus - When taken: 20 Apr 1814 - Where taken: off Cuba - Date Received: 16 Aug 1814 - From what ship: HMT Queen, Halifax - Born: Gloucester - Age: 24 - Released on 3 May 1815.

Root, Eleazer F. - Steward - Nbr: 45 - Prize: Terrible - Ship Type: MV - How taken: HMS Foxhound - When taken: 8 Feb 1813 - Where taken: Channel - Date Received: 2 Apr 1813 - From what ship: Plymouth - Born: Hartford - Age: 20 - Sent to Dartmouth on 30 Jul 1813.

Ropes, David - Seaman - Nbr: 5168 - Prize: Volante - Ship Type: P - How taken: HMS Curlew - When taken: 25 Nov 1813 - Where taken: off Halifax - Date Received: 31 Oct 1814 - From what ship: HMT Mermaid, Chatham - Born: Salem - Age: 16 - Released on 29 Jun 1815.

Rosah, Samuel - Seaman - Nbr: 613 - Prize: US Brig Argus - Ship Type: MW - How taken: HMS Pelican - When taken: 14 Aug 1813 - Where taken: Irish Channel - Date Received: 8 Sep 1813 - From what ship: Plymouth - Born: New York - Age: 22 - Sent to Dartmouth on 2 Nov 1814.

Rose, William - Seaman - Nbr: 1230 - How taken: Sent into custody from HMS Nautilus - When taken: 20 Dec 1812 - Date Received: 14 Jun 1814 - From what ship: Mill Prison (Plymouth, England) - Born: Vienna - Age: 30 - Released on 26 Apr 1815.

Roseberry, Charles - Seaman - Nbr: 6542 - How taken: Gave himself up from HMS Ramus - When taken: 22 Feb 1813 - Date Received: 7 Mar 1815 - From what ship: HMT Ganges, Plymouth - Born: Washington - Age: 28 - Released on 28 Jun 1815.

Rosett, Samuel - Seaman - Nbr: 172 - How taken: Impressed at Liverpool - When taken: 23 Feb 1813 - Date Received: 2 Apr 1813 - From what ship: Plymouth - Born: New York - Age: 31 - Released on 20 Apr 1815.

Ross, David - Seaman - Nbr: 4671 - Prize: Industry - Ship Type: P - How taken: HMS Heron - When taken: 3 Nov 1813 - Where taken: off Halifax - Date Received: 9 Oct 1814 - From what ship: HMT Leyden, Chatham - Born: Marblehead - Age: 27 - Released on 15 Jun 1815.

Ross, George - Seaman - Nbr: 4787 - How taken: Gave himself up from HMS Fame - When taken: 28 Sep 1814 - Date Received: 9 Oct 1814 - From what ship: HMT Freya, Chatham - Born: Hanover - Age: 33 - Released on 11 Jul 1815.

Ross, Isaac - Seaman - Nbr: 2725 - Prize: Zebra - Ship Type: LM - How taken: HMS Andromache - When taken: 20 Apr 1813 - Where taken: off Corcova Lights - Date Received: 21 Aug 1814 - From what ship: HMT Freya, Chatham - Born: New Jersey - Age: 23 - Released on 19 May 1815.

Ross, John - Seaman - Nbr: 3216 - How taken: Gave himself up from HMS Majestic - When taken: 17 Aug 1813 - Date Received: 11 Sep 1814 - From what ship: HMT Freya, Chatham - Born: New York - Age: 51 - Released on 28 May 1815.

Ross, Philip - Seaman - Nbr: 809 - Prize: Sybelle - Ship Type: MV - How taken: HMS Zenobia - When taken: 27 Jun 1813 - Where taken: off Cape St. Mary's - Date Received: 3 Nov 1813 - From what ship: Plymouth - Born: LA - Age: 24 - Released on 26 Apr 1815.

Ross, Richard - Seaman - Nbr: 3727 - Prize: Alfred - Ship Type: P - How taken: HMS Epervier - When taken: 23 Feb 1814 - Where taken: off Newfoundland - Date Received: 30 Sep 1814 - From what ship: HMT President, Halifax - Born: Savannah - Age: 18 - Race: Creole - Released on 4 Jun 1815.

Ross, William - Carpenter's Mate - Nbr: 3973 - Prize: US Brig Rattlesnake - Ship Type: MW - How taken: HMS Leander - When taken: 13 Jul 1814 - Where taken: off Shelburne - Date Received: 6 Oct 1814 - From what ship: HMT Chesapeake, Halifax - Born: Massachusetts - Age: 30 - Released on 9 Jun 1815.

Ross, William - Seaman - Nbr: 4797 - Prize: Martin - Ship Type: MV - How taken: HMS Swaggerer - When taken: 13 Mar 1814 - Where taken: off St. Thomas - Date Received: 9 Oct 1814 - From what ship: HMT Freya, Halifax - Born: Pennsylvania - Age: 33 - Released on 21 Jun 1815.

Rossuros, Philip - Seaman - Nbr: 2024 - Prize: Henrietta - Ship Type: MV - How taken: HMS Castilian - When taken: 12 Mar 1813 - Where taken: off Cape Ortegal (Spain) - Date Received: 3 Aug 1814 - From what ship: HMS Lyffey, Chatham Depot - Born: New York - Age: 27 - Released on 2 May 1815.

Roth, James - Seaman - Nbr: 5232 - Prize: Mary - Ship Type: MV - How taken: Raker - When taken: 17 Dec 1813 - Where taken: off Cape Haida - Date Received: 31 Oct 1814 - From what ship: HMT Mermaid, Chatham - Born: Norwich - Age: 25 - Died on 29 Dec 1814 from phthisis pulmonalis.

American Prisoners of War Held at Dartmoor during the War of 1812

## Alphabetical listing of names

Rotner, James W. - Seaman - Nbr: 78 - Prize: Rolla - Ship Type: MV - How taken: HMS Surveillante - When taken: 11 Feb 1813 - Where taken: Bay of Biscay - Date Received: 2 Apr 1813 - From what ship: Plymouth - Born: Philadelphia - Age: 25 - Released on 20 Apr 1815.

Roundy, Francis - Seaman - Nbr: 3677 - Prize: Alfred - Ship Type: P - How taken: HMS Epervier - When taken: 23 Feb 1812 - Where taken: off Newfoundland - Date Received: 30 Sep 1814 - From what ship: HMT President - Born: Marblehead - Age: 21 - Released on 4 Jun 1815.

Roundy, Jeremiah - Boy - Nbr: 4694 - Prize: Growler - Ship Type: P - How taken: HMS Electra - When taken: 7 Jul 1814 - Where taken: at sea - Date Received: 9 Oct 1814 - From what ship: HMT Leyden, Chatham - Born: Marblehead - Age: 21.

Roundy, Samuel - Seaman - Nbr: 3697 - Prize: Alfred - Ship Type: P - How taken: HMS Epervier - When taken: 23 Feb 1814 - Where taken: off Newfoundland - Date Received: 30 Sep 1814 - From what ship: HMT President, Halifax - Born: Marblehead - Age: 18 - Released on 4 Jun 1815.

Roundy, Thomas - Seaman - Nbr: 3384 - Prize: Growler - Ship Type: P - How taken: HMS Electra - When taken: 7 Jul 1813 - Where taken: at sea - Date Received: 13 Sep 1814 - From what ship: HMT Niobe, Chatham - Born: Marblehead - Age: 26 - Released on 28 May 1815.

Routeman, Benedict - Seaman - Nbr: 936 - Prize: Fanny - Ship Type: MV - How taken: HMS Eurotas - When taken: 23 Dec 1813 - Where taken: at sea - Date Received: 31 Jan 1814 - From what ship: Plymouth - Born: Oldenburg - Age: 27 - Sent to Mill Prison (Plymouth, England) on 17 Jun 1814.

Rowe, Abraham - Gunner - Nbr: 4928 - Prize: Herald - Ship Type: P - How taken: HMS Endymion - When taken: 15 Aug 1814 - Where taken: off Nantucket - Date Received: 28 Oct 1814 - From what ship: HMT Alkbar, Halifax - Born: Cape Ann - Age: 39 - Released on 21 Jun 1815.

Rowe, Cornelius - Seaman - Nbr: 2224 - Prize: Hussar - Ship Type: P - How taken: HMS Saturn - When taken: 25 May 1814 - Where taken: off Sandy Hook - Date Received: 16 Aug 1814 - From what ship: HMS Dublin, Halifax - Born: Philadelphia - Age: 25 - Released on 2 May 1815.

Rowe, John - Seaman - Nbr: 1632 - Prize: Tom - Ship Type: LM - How taken: HMS Surveillante - When taken: 24 Apr 1813 - Where taken: Bay of Biscay - Date Received: 23 Jun 1814 - From what ship: Stapleton - Born: Connecticut - Age: 27 - Released on 1 May 1815.

Rowe, Richard - Seaman - Nbr: 2046 - How taken: Gave himself up from HMS North Star - When taken: 16 May 1812 - Date Received: 3 Aug 1814 - From what ship: HMS Lyffey, Chatham Depot - Born: Philadelphia - Age: 43 - Released on 26 Apr 1815.

Rowe, Seth - Lieutenant - Nbr: 3388 - Prize: Wasp - Ship Type: P - How taken: HMS Bream - When taken: 10 Jun 1813 - Where taken: off Halifax - Date Received: 13 Sep 1814 - From what ship: HMT Niobe, Chatham - Born: Massachusetts - Age: 41 - Released on 28 May 1815.

Rowe, William - Seaman - Nbr: 4411 - Prize: Fire Fly - Ship Type: LM - How taken: HMS Revolutionnaire - When taken: 19 Oct 1813 - Where taken: at sea - Date Received: 8 Oct 1814 - From what ship: HMT Leyden, Chatham - Born: Boston - Age: 27 - Released on 14 Jun 1815.

Rowe, William - Seaman - Nbr: 5813 - Prize: US Schooner Scorpion - Ship Type: MW - How taken: British gunboats - When taken: 6 Sep 1814 - Where taken: Lake Erie - Date Received: 26 Dec 1814 - From what ship: HMT Argo - Born: New York - Age: 26 - Released on 3 Jul 1815.

Roweth, William - Seaman - Nbr: 4691 - How taken: HMS Racehorse - Date Received: 9 Oct 1814 - From what ship: HMT Leyden, Chatham - Born: New Jersey - Age: 37.

Rowland, Abner - Seaman - Nbr: 4462 - Prize: Lark - Ship Type: MV - How taken: HMS Brearn - When taken: 12 Apr 1813 - Where taken: off Cape Sable Island (Canada) - Date Received: 8 Oct 1814 - From what ship: HMT Leyden, Chatham - Born: Salem - Age: 27 - Released on 15 Jun 1815.

Rowland, William - Seaman - Nbr: 2990 - Prize: Frolic - Ship Type: P - How taken: HMS Heron - When taken: 25 Jan 1814 - Where taken: off St. Thomas - Date Received: 29 Aug 1814 - From what ship: HMT Bittern - Born: Beverly - Age: 45 - Died on 8 Jun 1815 from mania.

Rowle, Benjamin - Corporal - Nbr: 1602 - Prize: Fox - Ship Type: LM - How taken: HMS Pheasant - When taken: 22 Apr 1813 - Where taken: Bay of Biscay - Date Received: 23 Jun 1814 - From what ship: Stapleton - Born: Philadelphia - Age: 49 - Released on 1 May 1815.

Rowley, Henry - Seaman - Nbr: 1619 - Prize: Tom - Ship Type: LM - How taken: HMS Surveillante - When taken: 24 Apr 1813 - Where taken: Bay of Biscay - Date Received: 23 Jun 1814 - From what ship: Stapleton - Born: Pennsylvania - Age: 37 - Released on 1 May 1815.

Rowley, James - Seaman - Nbr: 3891 - Prize: Julianne - Ship Type: MV - How taken: HMS Dragon - When taken: 10 Jul 1814 - Where taken: off VA - Date Received: 5 Oct 1814 - From what ship: HMT Orpheus, Halifax -

American Prisoners of War Held at Dartmoor during the War of 1812

## Alphabetical listing of names

Born: Richmond - Age: 22 - Race: Mulatto - Released on 9 Jun 1815.

Royal, John - Seaman - Nbr: 136 - Prize: Criterion - Ship Type: MV - How taken: HMS Belle Poule - When taken: 14 Feb 1813 - Where taken: Bay of Biscay - Date Received: 2 Apr 1813 - From what ship: Plymouth - Born: Peterborough - Age: 28 - Released on 20 Apr 1815.

Rozier, William - Seaman - Nbr: 497 - Prize: Grand Napoleon - Ship Type: MV - How taken: HMS Goldfinch - When taken: 17 Apr 1813 - Where taken: Bay of Biscay - Date Received: 8 Sep 1813 - From what ship: Plymouth - Born: New Orleans - Age: 24 - Impersonated a Frenchman on the general release.

Ruby, Isaiah - Passenger - Nbr: 5579 - Prize: Adams - Ship Type: MV - How taken: HMS Lacedaemonian - When taken: 30 Aug 1814 - Where taken: off St. Bartholomew - Date Received: 17 Dec 1814 - From what ship: HMT Loire, Halifax - Born: St. Bartholomew - Age: 21 - Race: Mulatto - Released on 1 Jul 1815.

Ruddick, William - 2nd Mate - Nbr: 1826 - Prize: Lightning - Ship Type: MV - How taken: HMS Medusa - When taken: 2 Apr 1813 - Where taken: Bay of Biscay - Date Received: 21 Jul 1814 - From what ship: HMT Redbeard & Pincher, Chatham Depot - Born: Philadelphia - Age: 25 - Released on 1 May 1815.

Rue, William - 2nd Mate - Nbr: 3974 - Prize: US Brig Rattlesnake - Ship Type: MW - How taken: HMS Leander - When taken: 13 Jul 1814 - Where taken: off Shelburne - Date Received: 6 Oct 1814 - From what ship: HMT Chesapeake, Halifax - Born: Salem - Age: 27 - Released on 9 Jun 1815.

Ruffield, Samuel - Seaman - Nbr: 1349 - Prize: Paul Jones - Ship Type: P - How taken: HMS Leonidas - When taken: 23 May 1813 - Where taken: Channel - Date Received: 19 Jun 1814 - From what ship: Stapleton - Born: Connecticut - Age: 20 - Released on 28 Apr 1815.

Ruling, Arnold - Seaman - Nbr: 2951 - Prize: Rattlesnake - Ship Type: LM - How taken: HMS Rhin - When taken: 10 Mar 1814 - Where taken: West Indies - Date Received: 24 Aug 1814 - From what ship: HMT Hannibal - Born: Africa - Age: 24 - Race: Negro - Released on 19 May 1815.

Rumlet, Ebenezer - Seaman - Nbr: 2043 - How taken: Gave himself up from HMS Juniper - When taken: 22 May 1812 - Date Received: 3 Aug 1814 - From what ship: HMS Lyffey, Chatham Depot - Born: Wiscasset - Age: 30 - Released on 26 Apr 1815.

Rush, Samuel - Seaman - Nbr: 5777 - Prize: William Penn - Ship Type: MV - How taken: HMS Acorn - When taken: 27 Oct 1812 - Where taken: Lat 14 - Date Received: 26 Dec 1814 - From what ship: HMT Argo - Born: Baltimore - Age: 24 - Released on 27 Apr 1815.

Russell, Benjamin - Seaman - Nbr: 4901 - Prize: Alfred - Ship Type: P - How taken: HMS Epervier - When taken: 23 Feb 1814 - Where taken: off Newfoundland - Date Received: 28 Oct 1814 - From what ship: HMT Alkbar, Halifax - Born: Marblehead - Age: 18 - Released on 21 Jun 1815.

Russell, Frederick - Seaman - Nbr: 606 - Prize: US Brig Argus - Ship Type: MW - How taken: HMS Pelican - When taken: 14 Aug 1813 - Where taken: Irish Channel - Date Received: 8 Sep 1813 - From what ship: Plymouth - Born: Norway - Age: 30 - Sent to Dartmouth on 2 Nov 1814.

Russell, Isaac - Landsman - Nbr: 100 - Prize: St. Martin's Planter - Ship Type: P - How taken: HMS Dublin - When taken: 9 Feb 1813 - Where taken: Lat 43 N, Long 33 50 W - Date Received: 2 Apr 1813 - From what ship: Plymouth - Born: New Bedford - Age: 24 - Escaped on 29 Sep 1814.

Russell, James - Seaman - Nbr: 4100 - Prize: Scourge - Ship Type: Prize - How taken: HMS Curlew - When taken: 25 May 1814 - Where taken: off Newport - Date Received: 6 Oct 1814 - From what ship: HMT Chesapeake, Halifax - Born: New Orleans - Age: 24 - Released on 13 Jun 1815.

Russell, James - Seaman - Nbr: 542 - Prize: Orders in Council - Ship Type: MV - How taken: HMS Surveillante - When taken: 1 Jun 1813 - Where taken: off Cape Ortegal (Spain) - Date Received: 8 Sep 1813 - From what ship: Plymouth - Born: New York - Age: 35 - Released on 26 Apr 1815.

Russell, John - Seaman - Nbr: 5079 - Prize: Ida - Ship Type: LM - How taken: HMS Newcastle - When taken: 9 Aug 1814 - Where taken: Long 34 - Date Received: 28 Oct 1814 - From what ship: HMT Alkbar, Halifax - Born: Portland - Age: 30 - Escaped on 27 Jun 1815.

Russell, Louis - Seaman - Nbr: 3427 - Prize: Industry - Ship Type: P - How taken: HMS Heron - When taken: 3 Nov 1813 - Where taken: off Halifax - Date Received: 13 Sep 1814 - From what ship: HMT Niobe, Chatham - Born: Marblehead - Age: 21 - Released on 28 May 1815.

Russell, Moses - Quartermaster - Nbr: 4394 - Prize: Portsmouth Packet - Ship Type: Prize - How taken: HMS Fantome - When taken: 5 Oct 1813 - Where taken: off Portland - Date Received: 8 Oct 1814 - From what ship: HMT Leyden, Chatham - Born: Virginia - Age: 38 - Released on 14 Jun 1815.

Russell, Patton - Seaman - Nbr: 1525 - Prize: Essex - Ship Type: MV - How taken: HMS Pyramus - When taken: 20 Apr 1813 - Where taken: Bay of Biscay - Date Received: 23 Jun 1814 - From what ship: Stapleton - Born: Cambridge - Age: 19 - Released on 1 May 1815.

American Prisoners of War Held at Dartmoor during the War of 1812

## Alphabetical listing of names

Russell, Richard - Surgeon - Nbr: 1677 - Prize: Polly - Ship Type: P - How taken: HMS Barbados - When taken: 2 Apr 1814 - Where taken: off San Domingo - Date Received: 2 Jul 1814 - From what ship: Plymouth - Born: Surry - Age: 27 - Sent to Ashburton (England) on 15 Aug 1814.

Russell, Robert - Seaman - Nbr: 3293 - Prize: Enterprize - Ship Type: P - How taken: HMS Tenedos - When taken: 21 May 1813 - Where taken: off Cape Cod - Date Received: 13 Sep 1814 - From what ship: HMT Niobe, Chatham - Born: Marblehead - Age: 38 - Released on 28 May 1815.

Russell, Thomas - Sailing Master - Nbr: 3762 - Prize: Yankee Lass - Ship Type: P - How taken: HMS Surprize - When taken: 1 May 1814 - Where taken: off Western Islands (England) - Date Received: 5 Oct 1814 - From what ship: HMT Orpheus, Halifax - Born: Rhode Island - Age: 20 - Released on 9 Jun 1815.

Russell, William - Seaman - Nbr: 3377 - Prize: Growler - Ship Type: P - How taken: HMS Electra - When taken: 7 Jul 1813 - Where taken: at sea - Date Received: 13 Sep 1814 - From what ship: HMT Niobe, Chatham - Born: Marblehead - Age: 19 - Released on 28 May 1815.

Rust, John - Mate - Nbr: 4716 - Prize: Sally - Ship Type: MV - How taken: HMS Derwent - When taken: 21 Jan 1814 - Where taken: Grand Banks - Date Received: 9 Oct 1814 - From what ship: HMT Freya, Chatham - Born: Salem - Age: 27 - Released on 17 Apr 1815.

Rust, John - Seaman - Nbr: 4695 - Prize: Growler - Ship Type: P - How taken: HMS Electra - When taken: 7 Jul 1814 - Where taken: at sea - Date Received: 9 Oct 1814 - From what ship: HMT Leyden, Chatham - Born: Salem - Age: 51.

Rust, John - 1st Lieutenant - Nbr: 4659 - Prize: Growler - Ship Type: P - How taken: HMS Electra - When taken: 7 Jul 1813 - Where taken: at sea - Date Received: 9 Oct 1814 - From what ship: HMT Leyden, Chatham - Born: Salem - Age: 51 - Released on 10 Apr 1815.

Rust, Zebulon - Seaman - Nbr: 6383 - Prize: US Brig Syren - Ship Type: MW - How taken: HMS Medway - When taken: 12 Jul 1814 - Where taken: off Cape of Good Hope - Date Received: 24 Feb 1815 - From what ship: HMT Ganges, Plymouth - Born: Gloucester - Age: 29 - Released on 11 Jul 1815.

Rusto, Peter - Boy - Nbr: 2230 - Prize: General Hart - Ship Type: P - How taken: HMS Sophie - When taken: 24 Apr 1814 - Where taken: off Bermuda - Date Received: 16 Aug 1814 - From what ship: HMS Dublin, Halifax - Born: Nantes - Age: 17 - Released on 2 May 1815.

Rutter, Thomas - Sailing master - Nbr: 6131 - Prize: US Schooner Scorpion - Ship Type: MW - How taken: British gunboats - When taken: 7 Sep 1814 - Where taken: Lake Erie - Date Received: 17 Jan 1815 - From what ship: HMT Impregnable - Born: Baltimore - Age: 27 - Sent to Ashburton (England) on 22 Jan 1815.

Rutter, William - Seaman - Nbr: 5814 - Prize: US Schooner Scorpion - Ship Type: MW - How taken: British gunboats - When taken: 6 Sep 1814 - Where taken: Lake Erie - Date Received: 26 Dec 1814 - From what ship: HMT Argo - Born: Pennsylvania - Age: 18 - Released on 11 Jul 1815.

Ryan, Patrick - Master's Mate - Nbr: 3661 - Prize: Alfred - Ship Type: P - How taken: HMS Epervier - When taken: 23 Feb 1812 - Where taken: off Newfoundland - Date Received: 30 Sep 1814 - From what ship: HMT President - Born: Wexford - Age: 41 - Released on 4 Jun 1815.

Ryan, William - Seaman - Nbr: 74 - Prize: Rolla - Ship Type: MV - How taken: HMS Surveillante - When taken: 11 Feb 1813 - Where taken: Bay of Biscay - Date Received: 2 Apr 1813 - From what ship: Plymouth - Born: Philadelphia - Age: 19 - Released on 20 Apr 1815.

Ryan, William - Sailing Master - Nbr: 3673 - Prize: Alfred - Ship Type: P - How taken: HMS Epervier - When taken: 23 Feb 1812 - Where taken: off Newfoundland - Date Received: 30 Sep 1814 - From what ship: HMT President - Born: Marblehead - Age: 22 - Released on 4 Jun 1815.

Ryder, Amos - Seaman - Nbr: 3815 - Prize: Nimble - Ship Type: Prize - How taken: HMS Arab - When taken: 5 Apr 1814 - Where taken: Lat 37 Long 65 - Date Received: 5 Oct 1814 - From what ship: HMT Orpheus, Halifax - Born: Chatham - Age: 20 - Released on 9 Jun 1815.

Ryder, George - Seaman - Nbr: 2194 - Prize: Olio - Ship Type: Prize - How taken: HMS Cyane - When taken: 22 May 1814 - Where taken: off NY - Date Received: 16 Aug 1814 - From what ship: HMS Dublin, Halifax - Born: Philadelphia - Age: 35 - Released on 2 May 1815.

Rye, William - Seaman - Nbr: 4317 - Prize: Ocean - Ship Type: MV - How taken: HMS Atalante - When taken: 16 Nov 1812 - Where taken: off Cape of Good Hope - Date Received: 7 Oct 1814 - From what ship: HMT Niobe, Chatham - Born: Baltimore - Age: 21 - Released on 27 Apr 1815.

Ryley, William - Seaman - Nbr: 253 - How taken: Sent into custody from Flag of Truce PA - Date Received: 28 Jun 1813 - From what ship: Plymouth - Born: Pennsylvania - Age: 40 - Sent to Dartmouth on 30 Jul 1813.

Sacalogos, Nicholas - Seaman - Nbr: 6054 - Prize: William - Ship Type: Prize - How taken: HMS Armide - When taken: 11 Oct 1814 - Where taken: off Newport - Date Received: 28 Dec 1814 - From what ship: HMT

American Prisoners of War Held at Dartmoor during the War of 1812

## Alphabetical listing of names

Penelope - Born: Genoa - Age: 35 - Released on 3 Jul 1815.

Sage, Moses - Seaman - Nbr: 146 - Prize: Criterion - Ship Type: MV - How taken: HMS Belle Poule - When taken: 14 Feb 1813 - Where taken: Bay of Biscay - Date Received: 2 Apr 1813 - From what ship: Plymouth - Born: Middletown - Age: 29 - Released on 20 Apr 1815.

Sale, Thomas - Seaman - Nbr: 6144 - How taken: Gave himself up from HMS Feameyestion - Date Received: 17 Jan 1815 - From what ship: HMT Impregnable - Born: Petersburg - Age: 34 - Race: Black - Released on 29 Jun 1815.

Salisbury, John - Seaman - Nbr: 1737 - Prize: Hugh Jones - Ship Type: MV - How taken: HMS Bittern - When taken: 14 Jun 1814 - Where taken: at sea - Date Received: 18 Jul 1814 - From what ship: HMT Salvador del Mundo, Plymouth - Born: Philadelphia - Age: 24 - Released on 1 May 1815.

Salisbury, Joseph - Passenger - Nbr: 6442 - Prize: Jemmell - Ship Type: MV - How taken: HMS Rhin - When taken: 28 May 1814 - Where taken: off Bermuda - Date Received: 3 Mar 1815 - From what ship: HMT Ganges, Plymouth - Born: Newport - Age: 22 - Died on 15 Mar 1815 from febris.

Salley, James - Quartermaster - Nbr: 563 - Prize: US Brig Argus - Ship Type: MW - How taken: HMS Pelican - When taken: 14 Apr 1813 - Where taken: Irish Channel - Date Received: 8 Sep 1813 - From what ship: Plymouth - Born: Massachusetts - Age: 27 - Sent to Dartmouth on 2 Nov 1814.

Salmon, Archibald - Seaman - Nbr: 4507 - Prize: America - Ship Type: P - How taken: HMS Shannon - When taken: 23 May 1813 - Where taken: off Cape Cod - Date Received: 8 Oct 1814 - From what ship: HMT Leyden, Chatham - Born: Massachusetts - Age: 21 - Released on 15 Jun 1815.

Samerton, George - Seaman - Nbr: 4380 - Prize: Teazer - Ship Type: P - How taken: HMS Boyce - When taken: 15 Jun 1813 - Where taken: off Halifax - Date Received: 8 Oct 1814 - From what ship: HMT Leyden, Chatham - Born: Salem - Age: 27 - Released on 14 Jun 1815.

Sammers, Henry - Seaman - Nbr: 230 - Prize: Charlotte - Ship Type: MV - How taken: HMS Warspite - When taken: 3 Mar 1813 - Where taken: Bay of Biscay - Date Received: 2 Apr 1813 - From what ship: Plymouth - Born: New York - Age: 23 - Released on 20 Apr 1815.

Sampson, Jacob - Seaman - Nbr: 2784 - Prize: Tiger - Ship Type: MV - How taken: HMS Scylla - When taken: 22 May 1813 - Where taken: Bay of Biscay - Date Received: 24 Aug 1814 - From what ship: HMT Liverpool, Chatham - Born: New York - Age: 24 - Race: Negro - Released on 19 May 1815.

Samuel, Nathaniel - Seaman - Nbr: 964 - Prize: Siro - Ship Type: LM - How taken: HMS Pelican - When taken: 13 Jan 1814 - Where taken: at sea - Date Received: 31 Jan 1814 - From what ship: Plymouth - Born: Philadelphia - Age: 37 - Released on 27 Apr 1815.

Samuels, Samuel - Seaman - Nbr: 6299 - Prize: John - Ship Type: Prize - How taken: Leander (Newcastle Acasta) - When taken: 30 Jan 1815 - Date Received: 19 Feb 1815 - From what ship: HMT Ganges, Plymouth - Born: Norfolk - Age: 26 - Race: Black - Released on 5 Jul 1815.

Sanburn, James - Seaman - Nbr: 1864 - Prize: Enterprize - Ship Type: P - How taken: HMS Tenedos - When taken: 21 May 1813 - Where taken: off Cape Cod - Date Received: 29 Jul 1814 - From what ship: HMS Ville de Paris, Chatham Depot - Born: Petersburg, PA - Age: 25 - Released on 2 May 1815.

Sandburn, John - Seaman - Nbr: 765 - How taken: Sent into custody from HMS Pallas - Date Received: 3 Nov 1813 - From what ship: Plymouth - Born: Hampshire - Age: 23 - Released on 26 Apr 1815.

Sanders, Gabriel - Seaman - Nbr: 3486 - How taken: Gave himself up from HMS Prince - When taken: 14 Sep 1814 - Date Received: 19 Sep 1814 - From what ship: HMT Salvador del Mundo - Born: Boston - Age: 23 - Race: Black - Released on 4 Jun 1815.

Sanderson, William - Seaman - Nbr: 2748 - How taken: Gave himself up from HMS Royal William - When taken: 3 Feb 1813 - Date Received: 24 Aug 1814 - From what ship: HMT Liverpool, Chatham - Born: Baltimore - Age: 25 - Released on 19 May 1815.

Sandford, John William - Seaman - Nbr: 490 - Prize: Paul Jones - Ship Type: P - How taken: HMS Leonidas - When taken: 23 May 1813 - Where taken: Channel - Date Received: 8 Sep 1813 - From what ship: Plymouth - Born: New York - Age: 25 - Released on 20 Apr 1815.

Sandford, William - Seaman - Nbr: 1647 - Prize: Paul Jones - Ship Type: P - How taken: HMS Leonidas - When taken: 23 May 1813 - Where taken: Channel - Date Received: 23 Jun 1814 - From what ship: Stapleton - Born: New York - Age: 21 - Released on 1 May 1815.

Sandor, James - Mate - Nbr: 2652 - Prize: Jane - Ship Type: MV - How taken: HMS Crescent - When taken: 28 Jun 1813 - Where taken: off Newfoundland - Date Received: 21 Aug 1814 - From what ship: HMT Freya, Chatham - Born: North Carolina - Age: 23 - Released on 3 May 1815.

Sands, Henry - Seaman - Nbr: 4887 - Prize: Saratoga - Ship Type: P - How taken: HMS Barracouta - When taken: 9

American Prisoners of War Held at Dartmoor during the War of 1812

## Alphabetical listing of names

Oct 1814 - Where taken: off Western Islands (England) - Date Received: 24 Oct 1814 - From what ship: HMT Salvador del Mundo - Born: Rhode Island - Age: 26 - Race: Negro - Released on 21 Jun 1815.

Sands, John - Seaman - Nbr: 5071 - Prize: Ida - Ship Type: LM - How taken: HMS Newcastle - When taken: 9 Aug 1814 - Where taken: Long 34 - Date Received: 28 Oct 1814 - From what ship: HMT Alkbar, Halifax - Born: New York - Age: 36 - Released on 29 Jun 1815.

Sandy, W. P. - Seaman - Nbr: 3531 - Prize: Rover - Ship Type: Prize - How taken: HMS Conquistador - When taken: 22 Aug 1814 - Where taken: Long 19 Lat 107 - Date Received: 28 Sep 1814 - From what ship: HMT Salvador del Mundo - Born: North Carolina - Age: 30 - Released on 4 Jun 1815.

Sanis, Lewis - Seaman - Nbr: 657 - Prize: Marmion - Ship Type: MV - How taken: HMS President - When taken: 14 Aug 1813 - Where taken: off Nantes - Date Received: 8 Sep 1813 - From what ship: Plymouth - Born: New Orleans - Age: 21 - Race: Mulatto - Sent to Plymouth on 8 Jul 1814.

Sankey, Caesar - Seaman - Nbr: 5371 - How taken: Gave himself up from HMS Africanus - When taken: 4 Oct 1813 - Date Received: 31 Oct 1814 - From what ship: HMT Leyden, Chatham - Born: New Hampshire - Age: 32 - Race: Black - Released on 13 Jun 1815.

Sares, Charles - Passenger - Nbr: 771 - How taken: Sent into custody from a Russian MV, prize to the French frigate Weser - Date Received: 3 Nov 1813 - From what ship: Plymouth - Born: Boston - Age: 17 - Released on 13 Nov 1813.

Sargeant, Samuel H. - 2nd Prize Master - Nbr: 127 - Prize: Good Intent - Ship Type: P - How taken: HMS Rota - When taken: 26 Jan 1813 - Where taken: Lat 43'30" N, Long 20 W - Date Received: 12 Apr 1813 - From what ship: Plymouth - Born: Cape Ann - Age: 20 - Sent to Dartmouth on 30 Jul 1813.

Sargeant, Charles - Seaman - Nbr: 2455 - Prize: US Sloop Frolic - Ship Type: MW - How taken: HMS Orpheus - When taken: 20 Apr 1814 - Where taken: off Cuba - Date Received: 16 Aug 1814 - From what ship: HMT Queen, Halifax - Born: New Hampton - Age: 27 - Released on 3 May 1815.

Saul, Francis - Seaman - Nbr: 2135 - How taken: Gave himself up from HMS Mercurious - When taken: 4 Dec 1812 - Date Received: 8 Aug 1814 - From what ship: HMT Raven, Chatham - Born: Westport - Age: 32 - Died on 20 Oct 1814 from pneumonia.

Sault, Manuel - Seaman - Nbr: 5128 - Prize: Calabria - Ship Type: MV - How taken: HMS Castilian - When taken: 29 Sep 1814 - Where taken: off Ireland - Date Received: 31 Oct 1814 - From what ship: HMT Castillian - Born: Lisbon - Age: 44 - Released on 29 Jun 1815.

Saunders, Cornelius - Seaman - Nbr: 1738 - Prize: Hugh Jones - Ship Type: MV - How taken: HMS Bittern - When taken: 14 Jun 1814 - Where taken: at sea - Date Received: 18 Jul 1814 - From what ship: HMT Salvador del Mundo, Plymouth - Born: Antwerp - Age: 25 - Sent to Exeter Jail on 9 Mar 1815.

Saunders, James - Seaman - Nbr: 5114 - Prize: Portsmouth - Ship Type: P - How taken: HMS Pylades - When taken: 13 Sep 1814 - Where taken: Long 65 - Date Received: 28 Oct 1814 - From what ship: HMT Alkbar, Halifax - Born: Portsmouth - Age: 21 - Released on 29 Jun 1815.

Saunders, John - Seaman - Nbr: 5096 - Prize: David Porter - Ship Type: LM - How taken: HMS Pylades - When taken: 12 Sep 1814 - Where taken: Georges Bank - Date Received: 28 Oct 1814 - From what ship: HMT Alkbar, Halifax - Born: Finland - Age: 30 - Released on 29 Jun 1815.

Saunders, Joseph - Seaman - Nbr: 2327 - Prize: Hussar - Ship Type: P - How taken: HMS Saturn - When taken: 25 May 1814 - Where taken: off Sandy Hook - Date Received: 16 Aug 1814 - From what ship: HMS Dublin, Halifax - Born: Delaware - Age: 36 - Released on 3 May 1815.

Saunders, Peter - Seaman - Nbr: 2925 - How taken: Gave himself up from HMS Union - When taken: 27 May 1813 - Date Received: 24 Aug 1814 - From what ship: HMT Alpheus, Chatham - Born: Salem - Age: 32 - Released on 19 May 1815.

Saunders, Richard - Seaman - Nbr: 5152 - Prize: Thorn - Ship Type: P - How taken: HMS Shannon - When taken: 7 Nov 1813 - Where taken: off Newfoundland - Date Received: 31 Oct 1814 - From what ship: HMT Mermaid, Chatham - Born: Boston - Age: 17 - Released on 29 Jun 1815.

Saunders, Thomas - Seaman - Nbr: 1912 - How taken: Gave himself up from HMS Mars - When taken: 9 Dec 1812 - Date Received: 3 Aug 1814 - From what ship: HMS Alceste, Chatham Depot - Born: Norfolk - Age: 29 - Released on 26 Apr 1815.

Saunders, William - Seaman - Nbr: 218 - Prize: Mars - Ship Type: MV - How taken: HMS Warspite - When taken: 26 Feb 1813 - Where taken: off Basque Roads (France) - Date Received: 2 Apr 1813 - From what ship: Plymouth - Born: Massachusetts - Age: 19 - Died on 16 Jan 1814.

Saunderson, Jacob - Seaman - Nbr: 5561 - Prize: Levant - Ship Type: MV - How taken: HMS Forester - When taken: 4 Jan 1814 - Where taken: Bahamas Banks - Date Received: 17 Dec 1814 - From what ship: HMT

American Prisoners of War Held at Dartmoor during the War of 1812

## Alphabetical listing of names

Loire, Halifax - Born: Salem - Age: 18 - Released on 3 May 1815.
Saundes, Cornelius - Seaman - Nbr: 6551 - Prize: Hugh Jones - Ship Type: MV - How taken: HMS Bittern - When taken: 14 Jun 1814 - Where taken: at sea - Date Received: 26 Mar 1815 - From what ship: from Exeter jail - Born: Antwerp - Age: 32 - Released on 1 May 1815.
Saundry, Nathaniel - Seaman - Nbr: 5195 - Prize: Enterprize - Ship Type: P - How taken: HMS Tenedos - When taken: 21 May 1813 - Where taken: off Cape Cod - Date Received: 31 Oct 1814 - From what ship: HMT Mermaid, Chatham - Born: Boston - Age: 19 - Released on 29 Jun 1815.
Savage, James - Mate - Nbr: 3602 - Prize: Atalante - Ship Type: MV - How taken: HMS Barbados - When taken: 19 Jan 1814 - Where taken: off St. Bartholomew - Date Received: 30 Sep 1814 - From what ship: HMT Sybella - Born: Virginia - Age: 29 - Released on 4 Jun 1815.
Savage, John H. - Seaman - Nbr: 3110 - Prize: Perseverance - Ship Type: MV - How taken: HMS Barbados - When taken: 21 Jan 1814 - Where taken: off St. Bartholomew - Date Received: 11 Sep 1814 - From what ship: HMT Salvador del Mundo - Born: West-Cappel (France) - Age: 18 - Released on 10 Apr 1815.
Savage, Robert - Seaman - Nbr: 3543 - Prize: Hawk - Ship Type: P - How taken: HMS Pique - When taken: 26 Apr 1814 - Where taken: off Bermuda - Date Received: 30 Sep 1814 - From what ship: HMT Sybella - Born: North Carolina - Age: 18 - Released on 4 Jun 1815.
Savary, Peter - Seaman - Nbr: 3685 - Prize: Alfred - Ship Type: P - How taken: HMS Epervier - When taken: 23 Feb 1812 - Where taken: off Newfoundland - Date Received: 30 Sep 1814 - From what ship: HMT President - Born: Marblehead - Age: 20 - Released on 4 Jun 1815.
Saverio, John - Seaman - Nbr: 6225 - Prize: Prince de Neufchatel - Ship Type: P - How taken: Leander (Newcastle Acasta) - When taken: 20 Dec 1814 - Where taken: Lat 38 Long 56 - Date Received: 30 Jan 1815 - From what ship: HMT Pheasant - Born: Portugal - Age: 28 - Released on 5 Jul 1815.
Sawyer, Jacob - Seaman - Nbr: 4788 - How taken: Impressed at London - When taken: 4 Sep 1813 - Date Received: 9 Oct 1814 - From what ship: HMT Freya, Chatham - Born: Providence - Age: 27 - Race: Mulatto - Died on 25 Oct 1814 from enteritis.
Sawyer, James - Seaman - Nbr: 4381 - Prize: Teazer - Ship Type: P - How taken: HMS Boyce - When taken: 15 Jun 1813 - Where taken: off Halifax - Date Received: 8 Oct 1814 - From what ship: HMT Leyden, Chatham - Born: Portland - Age: 50 - Released on 14 Jun 1815.
Sawyer, Peter - Seaman - Nbr: 3418 - Prize: Fox - Ship Type: P - How taken: Manley - When taken: 1 Aug 1813 - Where taken: off Halifax - Date Received: 13 Sep 1814 - From what ship: HMT Niobe, Chatham - Born: Portland - Age: 31 - Released on 19 May 1815.
Saywood, John - Seaman - Nbr: 4022 - Prize: US Brig Rattlesnake - Ship Type: MW - How taken: HMS Leander - When taken: 13 Jul 1814 - Where taken: off Shelburne - Date Received: 6 Oct 1814 - From what ship: HMT Chesapeake, Halifax - Born: Cape Ann - Age: 21 - Released on 9 Jun 1815.
Scary, Joseph - Seaman - Nbr: 1175 - Prize: Creole - Ship Type: P - How taken: HMS Surveillante - When taken: 2 May 1811 - Where taken: off Cape Clyde - Date Received: 4 Jun 1814 - From what ship: Dartmouth - Born: Charlestown - Age: 24 - Released on 26 Apr 1815.
Schaeman, Frederick - Seaman - Nbr: 7 - Prize: Cashier - Ship Type: LM - How taken: HMS Reindeer - When taken: 3 Feb 1813 - Where taken: Bay of Biscay - Date Received: 2 Apr 1813 - From what ship: Plymouth - Born: Baltimore - Age: 19 - Sent to Dartmouth on 10 Jul 1813.
Schanck, William - Seaman - Nbr: 2880 - How taken: Gave himself up from HMS Bombay - When taken: 27 May 1813 - Date Received: 24 Aug 1814 - From what ship: HMT Alpheus, Chatham - Born: Philadelphia - Age: 33 - Released on 19 May 1815.
Schew, Richard - Seaman - Nbr: 851 - Prize: Amiable - Ship Type: LM - How taken: HMS Magnificent - When taken: 30 Oct 1813 - Where taken: off Bordeaux - Date Received: 4 Dec 1813 - From what ship: Plymouth - Born: New York - Age: 32 - Died on 8 Feb 1814 from pneumonia.
Schneider, John A. - Seaman - Nbr: 3506 - Prize: Maria - Ship Type: MV - How taken: Not listed - When taken: 11 Sep 1814 - Where taken: off Liverpool - Date Received: 24 Sep 1814 - From what ship: HMT Salvador del Mundo - Born: Finland - Age: 35 - Released on 8 Nov 1814.
Schole, John - Seaman - Nbr: 463 - Prize: Meteor - Ship Type: MV - How taken: HMS Briton - When taken: 12 Mar 1813 - Where taken: Bay of Biscay - Date Received: 8 Sep 1813 - From what ship: Plymouth - Born: Long Island - Age: 30 - Sent to Plymouth on 6 Dec 1813.
Schultz, John - Seaman - Nbr: 142 - Prize: Criterion - Ship Type: MV - How taken: HMS Belle Poule - When taken: 14 Feb 1813 - Where taken: Bay of Biscay - Date Received: 2 Apr 1813 - From what ship: Plymouth - Born: Stralsund - Age: 20 - Sent to Plymouth on 15 Jun 1813.

American Prisoners of War Held at Dartmoor during the War of 1812

## Alphabetical listing of names

Scobe, William - Boy - Nbr: 5525 - Prize: Sally - Ship Type: Prize - How taken: HMS Galatea - When taken: 3 Aug 1814 - Where taken: at sea - Date Received: 17 Dec 1814 - From what ship: HMT Loire, Halifax - Born: Marblehead - Age: 16 - Released on 1 Jul 1815.

Scott, Alexander - Seaman - Nbr: 1982 - How taken: Gave himself up from HMS Barham - When taken: Apr 1813 - Date Received: 3 Aug 1814 - From what ship: HMS Lyffey, Chatham Depot - Born: Philadelphia - Age: 29 - Released on 2 May 1815.

Scott, Anthony - Seaman - Nbr: 4748 - How taken: Gave himself up from HMS Treban - When taken: 21 Jul 1814 - Date Received: 9 Oct 1814 - From what ship: HMT Freya, Chatham - Born: New Orleans - Age: 31 - Released on 15 Jun 1815.

Scott, Henry - Seaman - Nbr: 4232 - Prize: Volante - Ship Type: P - How taken: HMS Curlew - When taken: 25 Nov 1813 - Where taken: off Halifax - Date Received: 7 Oct 1814 - From what ship: HMT Niobe, Chatham - Born: Boston - Age: 43 - Released on 13 Jun 1815.

Scott, John - Seaman - Nbr: 1187 - Prize: Gazelle - Ship Type: P - How taken: HMS Leonidas - When taken: 15 Feb 1812 - Where taken: off Ireland - Date Received: 4 Jun 1814 - From what ship: Dartmouth - Born: Charlestown - Age: 31 - Sent to Plymouth on 8 Jul 1814.

Scott, John - Seaman - Nbr: 741 - Prize: US Brig Argus - Ship Type: MW - How taken: HMS Pelican - When taken: 14 Aug 1813 - Where taken: Irish Channel - Date Received: 3 Nov 1813 - From what ship: Plymouth - Born: Philadelphia - Age: 25 - Sent to Dartmouth on 2 Nov 1814.

Scott, John - Seaman - Nbr: 1454 - How taken: Impressed at Liverpool - When taken: 1 May 1813 - Date Received: 19 Jun 1814 - From what ship: Stapleton - Born: Rhode Island - Age: 48 - Released on 28 Apr 1815.

Scott, John - Seaman - Nbr: 1837 - How taken: Gave himself up from HMS Ajax - When taken: 14 Oct 1813 - Date Received: 21 Jul 1814 - From what ship: HMT Redbeard & Pincher, Chatham Depot - Born: Essex - Age: 35 - Released on 1 May 1815.

Scott, Robert - Seaman - Nbr: 577 - Prize: US Brig Argus - Ship Type: MW - How taken: HMS Pelican - When taken: 14 Apr 1813 - Where taken: Irish Channel - Date Received: 8 Sep 1813 - From what ship: Plymouth - Born: New York - Age: 20 - Sent to Dartmouth on 2 Nov 1814.

Scott, Samuel - Seaman - Nbr: 5141 - How taken: Gave himself up from HMS Caroline - When taken: 26 Mar 1813 - Date Received: 31 Oct 1814 - From what ship: HMT Mermaid, Chatham - Born: Virginia - Age: 20 - Released on 29 Jun 1815.

Scott, William - Seaman - Nbr: 5454 - How taken: Impressed at Liverpool - When taken: Oct 1814 - Date Received: 10 Dec 1814 - From what ship: HMT Impregnable - Born: Philadelphia - Age: 21 - Race: Creole - Released on 1 Jul 1815.

Scott, William - Carpenter - Nbr: 2889 - How taken: Gave himself up from HMS Tremendous - When taken: 15 Dec 1812 - Date Received: 24 Aug 1814 - From what ship: HMT Alpheus, Chatham - Born: Carolina - Age: 32 - Released on 27 Apr 1815.

Scott, William - Sail Maker - Nbr: 4015 - Prize: US Brig Rattlesnake - Ship Type: MW - How taken: HMS Leander - When taken: 13 Jul 1814 - Where taken: off Shelburne - Date Received: 6 Oct 1814 - From what ship: HMT Chesapeake, Halifax - Born: Boston - Age: 24 - Released on 9 Jun 1815.

Scovel, John - Seaman - Nbr: 1493 - Prize: Essex - Ship Type: MV - How taken: HMS Pyramus - When taken: 2 Apr 1813 - Where taken: Bay of Biscay - Date Received: 19 Jun 1814 - From what ship: Stapleton - Born: Boston - Age: 19 - Released on 1 May 1815.

Scribble, Henry - Seaman - Nbr: 2431 - Prize: US Sloop Frolic - Ship Type: MW - How taken: HMS Orpheus - When taken: 20 Apr 1814 - Where taken: off Cuba - Date Received: 16 Aug 1814 - From what ship: HMT Queen, Halifax - Born: Baltimore - Age: 25 - Released on 3 May 1815.

Scribner, William - Seaman - Nbr: 1931 - How taken: Gave himself up from HMS Desiree - When taken: 8 Jan 1814 - Date Received: 3 Aug 1814 - From what ship: HMS Alceste, Chatham Depot - Born: New Haven - Age: 36 - Released on 26 Apr 1815.

Seaberry, John - Seaman - Nbr: 2297 - Prize: Hussar - Ship Type: P - How taken: HMS Saturn - When taken: 25 May 1814 - Where taken: off Sandy Hook - Date Received: 16 Aug 1814 - From what ship: HMS Dublin, Halifax - Born: New York - Age: 19 - Released on 3 May 1815.

Seagell, William - Seaman - Nbr: 3738 - Prize: Lizard - Ship Type: P - How taken: HMS Prometheus - When taken: 5 May 1814 - Where taken: off Halifax - Date Received: 30 Sep 1814 - From what ship: HMT President, Halifax - Born: Salem - Age: 17 - Released on 4 Jun 1815.

Seaman, Francis - Seaman - Nbr: 2270 - Prize: Success - Ship Type: Prize - How taken: HMS Charybdis - When taken: 29 May 1814 - Where taken: off Cape Logan - Date Received: 16 Aug 1814 - From what ship: HMS

American Prisoners of War Held at Dartmoor during the War of 1812

## Alphabetical listing of names

Dublin, Halifax - Born: Portsmouth - Age: 24 - Released on 3 May 1815.

Seaman, M. C. - Seaman - Nbr: 6075 - Prize: Daedalus - Ship Type: MV - How taken: HMS Niemen - When taken: 20 Sep 1814 - Where taken: off Delaware - Date Received: 28 Dec 1814 - From what ship: HMT Penelope - Born: New York - Age: 18 - Released on 24 May 1815.

Seapatch, John - Boy - Nbr: 5889 - Prize: Harlequin - Ship Type: P - How taken: HMS Bulwark - When taken: 23 Nov 1814 - Where taken: off Halifax - Date Received: 27 Dec 1814 - From what ship: HMT Penelope - Born: Massachusetts - Age: 12 - Died on 7 Feb 1815 from pneumonitis.

Sears, Abraham - Steward - Nbr: 566 - Prize: US Brig Argus - Ship Type: MW - How taken: HMS Pelican - When taken: 14 Apr 1813 - Where taken: Irish Channel - Date Received: 8 Sep 1813 - From what ship: Plymouth - Born: New York - Age: 26 - Sent to Dartmouth on 2 Nov 1814.

Secundie, James - Seaman - Nbr: 2363 - Prize: Snap Dragon - Ship Type: P - How taken: HMS Martin - When taken: 10 Jun 1814 - Where taken: off Halifax - Date Received: 16 Aug 1814 - From what ship: HMT Queen, Halifax - Born: North Carolina - Age: 35 - Race: Negro - Released on 3 May 1815.

Segeant, John - Seaman - Nbr: 3864 - Prize: Buzi - Ship Type: MV - How taken: HMS Dragon - When taken: 20 Jul 1814 - Where taken: off VA - Date Received: 5 Oct 1814 - From what ship: HMT Orpheus, Halifax - Born: Norfolk - Age: 18 - Released on 10 May 1815.

Seith, John - Seaman - Nbr: 4386 - Prize: Globe - Ship Type: P - How taken: HMS Spartan - When taken: 13 Jun 1813 - Where taken: off Delaware - Date Received: 8 Oct 1814 - From what ship: HMT Leyden, Chatham - Born: Not legible - Age: 25 - Released on 14 Jun 1815.

Self, Thomas - Seaman - Nbr: 5233 - How taken: Impressed at London - When taken: 2 Mar 1814 - Date Received: 31 Oct 1814 - From what ship: HMT Mermaid, Chatham - Born: Charleston - Age: 20 - Released on 14 Jun 1815.

Selley, Miles - Boy - Nbr: 6071 - Prize: Daedalus - Ship Type: MV - How taken: HMS Niemen - When taken: 20 Sep 1814 - Where taken: off Delaware - Date Received: 28 Dec 1814 - From what ship: HMT Penelope - Born: Philadelphia - Age: 15 - Released on 2 May 1815.

Selman, Edward - Seaman - Nbr: 4083 - Prize: New Zealander - Ship Type: Prize - How taken: HMS Belviders - When taken: 21 Apr 1814 - Where taken: off Delaware - Date Received: 6 Oct 1814 - From what ship: HMT Chesapeake, Halifax - Born: Marblehead - Age: 28 - Released on 13 Jun 1815.

Selman, John - Seaman - Nbr: 4201 - Prize: Hannah - Ship Type: MV - How taken: HMS Conquistador - When taken: 15 Jan 1814 - Where taken: at sea - Date Received: 7 Oct 1814 - From what ship: HMT Niobe, Chatham - Born: Marblehead - Age: 19 - Released on 13 Jun 1815.

Selston, Isaac - Seaman - Nbr: 779 - Prize: Betsy - Ship Type: MV - How taken: HMS Eurotas - When taken: 26 Oct 1813 - Where taken: off Ushant (France) - Date Received: 3 Nov 1813 - From what ship: Plymouth - Born: Sweden - Age: 28 - Released on 6 May 1814.

Senter, Noah - Seaman - Nbr: 4760 - Prize: Adeline - Ship Type: LM - How taken: HMS Magicienne - When taken: 14 Mar 1814 - Where taken: off Cape Ortegal (Spain) - Date Received: 9 Oct 1814 - From what ship: HMT Freya, Chatham - Born: New Hampshire - Age: 27 - Released on 15 Jun 1815.

Sentille, Francis - Seaman - Nbr: 1345 - Prize: Paul Jones - Ship Type: P - How taken: HMS Leonidas - When taken: 23 May 1813 - Where taken: Channel - Date Received: 19 Jun 1814 - From what ship: Stapleton - Born: Charlestown - Age: 19 - Sent to Mill Prison (Plymouth, England) on 26 Jun 1814.

Sereder, Martin - Seaman - Nbr: 934 - Prize: Fanny - Ship Type: MV - How taken: HMS Eurotas - When taken: 23 Dec 1813 - Where taken: at sea - Date Received: 31 Jan 1814 - From what ship: Plymouth - Born: Menzel, Prussia - Age: 22 - Sent to Plymouth on 28 Feb 1814.

Sergeant, Daniel - Seaman - Nbr: 5412 - Prize: Thomas - Ship Type: LM - How taken: HMS Nymphe - When taken: 28 Jun 1813 - Where taken: off Halifax - Date Received: 31 Oct 1814 - From what ship: HMT Leyden, Chatham - Born: Old York - Age: 53 - Released on 1 Jul 1815.

Setchell, Jonathan - Seaman - Nbr: 3922 - Prize: Rolla - Ship Type: P - How taken: HMS Loire - When taken: 10 Dec 1813 - Where taken: off Bull Island (South Carolina) - Date Received: 5 Oct 1814 - From what ship: HMT President, Halifax - Born: Ipswich - Age: 30 - Released on 9 Jun 1815.

Sevear, William - Seaman - Nbr: 6076 - Prize: Daedalus - Ship Type: MV - How taken: HMS Niemen - When taken: 20 Sep 1814 - Where taken: off Delaware - Date Received: 28 Dec 1814 - From what ship: HMT Penelope - Born: Maryland - Age: 20 - Released on 3 Jul 1815.

Sewell, John - Seaman - Nbr: 6545 - How taken: Gave himself up from HMS Orlando - When taken: 7 May 1813 - Date Received: 7 Mar 1815 - From what ship: HMT Ganges, Plymouth - Born: Baltimore - Age: 26 - Released on 11 Jul 1815.

American Prisoners of War Held at Dartmoor during the War of 1812

## Alphabetical listing of names

Sewell, Lot - Seaman - Nbr: 3491 - How taken: Gave himself up from HMS Prospero - When taken: 12 Sep 1814 - Date Received: 19 Sep 1814 - From what ship: HMT Salvador del Mundo - Born: Baltimore - Age: 26 - Released on 4 Jun 1815.

Seyeant, Henry - Seaman - Nbr: 5907 - Prize: Harlequin - Ship Type: P - How taken: HMS Bulwark - When taken: 23 Nov 1814 - Where taken: off Halifax - Date Received: 27 Dec 1814 - From what ship: HMT Penelope - Born: York - Age: 28 - Released on 3 Jul 1815.

Seymond, Abner - Seaman - Nbr: 3046 - How taken: Gave himself up from HMS Galatea - When taken: 24 Aug 1814 - Date Received: 2 Sep 1814 - From what ship: HMT Sultan - Born: Boston - Age: 26 - Race: Black - Released on 28 Apr 1815.

Seywood, James - Seaman - Nbr: 4986 - Prize: Invincible - Ship Type: LM - How taken: HMS Armide - When taken: 15 Aug 1814 - Where taken: off Nantucket - Date Received: 28 Oct 1814 - From what ship: HMT Alkbar, Halifax - Born: Bath - Age: 26 - Released on 21 Jun 1815.

Shackley, John - Seaman - Nbr: 6137 - Prize: Robert - Ship Type: MV - How taken: HMS Phoebe - When taken: 26 Feb 1814 - Where taken: off Porto Rico - Date Received: 17 Jan 1815 - From what ship: HMT Impregnable - Born: New York - Age: 17 - Released on 5 Jul 1815.

Shade, Nathaniel - Marine - Nbr: 5482 - Prize: General Putnam - Ship Type: P - How taken: HMS Leander - When taken: 8 Nov 1814 - Where taken: Long 65 Lat 42 - Date Received: 17 Dec 1814 - From what ship: HMT Loire, Halifax - Born: Massachusetts - Age: 50 - Released on 1 Jul 1815.

Shadwick, John - Seaman - Nbr: 2153 - Prize: Rattlesnake - Ship Type: P - How taken: HMS Hyperion - When taken: 26 Jun 1814 - Where taken: off Cape Finisterre - Date Received: 16 Aug 1814 - From what ship: HMS Dublin, Halifax - Born: Salem - Age: 29 - Released on 2 May 1815.

Shairs, Samuel - 2nd Mate - Nbr: 458 - Prize: Amphitrite - Ship Type: MV - How taken: HMS Gleaner - When taken: 27 Feb 1813 - Where taken: Bay of Biscay - Date Received: 8 Sep 1813 - From what ship: Plymouth - Born: Stratford - Age: 25 - Released on 26 Apr 1815.

Shamtan, Samuel - Seaman - Nbr: 92 - Prize: Cashier - Ship Type: MV - How taken: HMS Reindeer - When taken: 3 Feb 1813 - Where taken: Bay of Biscay - Date Received: 2 Apr 1813 - From what ship: Plymouth - Born: Baltimore - Age: 35 - Sent to Dartmouth on 2 Aug 1813.

Shanks, William - Mate - Nbr: 3893 - Prize: Governor Shelby - Ship Type: MV - How taken: HMS Saturn - When taken: 9 Jul 1814 - Where taken: off Long Island - Date Received: 5 Oct 1814 - From what ship: HMT Orpheus, Halifax - Born: Maryland - Age: 36 - Released on 9 Jun 1815.

Sharp, John - Prize Master - Nbr: 2564 - Prize: Rattlesnake - Ship Type: P - How taken: HMS Hyperion - When taken: 25 Jun 1814 - Where taken: off Cape Finisterre - Date Received: 21 Aug 1814 - From what ship: HMS Hyperion - Born: New Jersey - Age: 29 - Released on 6 Apr 1815.

Sharpless, Robert - Seaman - Nbr: 5039 - Prize: Betsey - Ship Type: Prize - How taken: HMS Pylades - When taken: 7 Sep 1814 - Where taken: Canten - Date Received: 28 Oct 1814 - From what ship: HMT Alkbar, Halifax - Born: Maryland - Age: 28 - Released on 11 Jul 1815.

Shaw, Andrew - Seaman - Nbr: 3370 - How taken: Impressed at Leith - When taken: 4 Dec 1813 - Date Received: 13 Sep 1814 - What ship: HMT Niobe, Chatham - Born: Newburyport - Age: 24 - Released on 28 May 1815.

Shaw, Benjamin - Cook - Nbr: 3784 - Prize: James - Ship Type: Prize - How taken: Rebecca (LM) - When taken: 13 Apr 1814 - Where taken: off Delaware - Date Received: 5 Oct 1814 - From what ship: HMT President, Halifax - Born: Alexandria - Age: 27 - Race: Negro - Released on 9 Jun 1815.

Shaw, Edward - 1st Mate - Nbr: 2728 - Prize: Chesapeake - Ship Type: LM - How taken: HMS Pyramus - When taken: 26 Oct 1813 - Where taken: off Nantes - Date Received: 24 Aug 1814 - From what ship: HMT Liverpool, Chatham - Born: Baltimore - Age: 29 - Released on 11 May 1815.

Shaw, Henry - Cook - Nbr: 4419 - Prize: Fire Fly - Ship Type: LM - How taken: HMS Revolutionnaire - When taken: 19 Oct 1813 - Where taken: at sea - Date Received: 8 Oct 1814 - From what ship: HMT Leyden, Chatham - Born: Boston - Age: 23 - Race: Black - Released on 14 Jun 1815.

Shaw, John - Seaman - Nbr: 2856 - Prize: Governor Middleton - Ship Type: MV - How taken: HMS Thetis - When taken: 2 May 1813 - Where taken: Bay of Biscay - Date Received: 24 Aug 1814 - From what ship: HMT Alpheus, Chatham - Born: Bath - Age: 20 - Released on 19 May 1815.

Shaw, John - Seaman - Nbr: 2966 - Prize: Governor Middleton - Ship Type: MV - How taken: Thetis (Privateer) - When taken: 2 May 1813 - Where taken: Bay of Biscay - Date Received: 24 Aug 1814 - From what ship: HMT Hannibal - Born: Bath - Age: 20 - Released on 19 May 1815.

Shaw, Joseph - Prize master - Nbr: 4906 - Prize: Rolla - Ship Type: P - How taken: HMS Loire - When taken: 10 Dec 1813 - Where taken: off Bull Island (South Carolina) - Date Received: 28 Oct 1814 - From what ship:

American Prisoners of War Held at Dartmoor during the War of 1812

## Alphabetical listing of names

HMT Alkbar, Halifax - Born: Stonington - Age: 33 - Released on 21 Jun 1815.

Shaw, Richard - Seaman - Nbr: 1382 - Prize: Courier - Ship Type: P - How taken: HMS Rover - When taken: 14 May 1813 - Where taken: Bay of Biscay - Date Received: 19 Jun 1814 - From what ship: Stapleton - Born: Maryland - Age: 26 - Released on 28 Apr 1815.

Shaw, Robert - Seaman - Nbr: 3946 - Prize: Rolla - Ship Type: P - How taken: HMS Loire - When taken: 10 Dec 1813 - Where taken: off Bull Island (South Carolina) - Date Received: 5 Oct 1814 - From what ship: HMT President, Halifax - Born: Virginia - Age: 22 - Released on 9 Jun 1815.

Shaw, Samuel - Prize Master - Nbr: 4697 - Prize: Growler - Ship Type: P - How taken: HMS Electra - When taken: 7 Jul 1813 - Where taken: at sea - Date Received: 9 Oct 1814 - From what ship: HMT Leyden, Chatham - Born: Massachusetts - Age: 48.

Shaw, Thomas - Seaman - Nbr: 2174 - Prize: Rattlesnake - Ship Type: P - How taken: HMS Hyperion - When taken: 26 Jun 1814 - Where taken: off Cape Finisterre - Date Received: 16 Aug 1814 - From what ship: HMS Dublin, Halifax - Born: Philadelphia - Age: 24 - Released on 2 May 1815.

Shaw, William - Seaman - Nbr: 1038 - Prize: US Brig Argus - Ship Type: MW - How taken: HMS Pelican - When taken: 16 Aug 1813 - Where taken: Irish Channel - Date Received: 10 May 1814 - From what ship: Plymouth - Born: Philadelphia - Age: 23 - Died on 17 Oct 1814 from dropsy.

Shaw, William - Seaman - Nbr: 4577 - Prize: Bunker Hill - Ship Type: P - How taken: HMS Pomone - When taken: 2 Mar 1814 - Where taken: Channel - Date Received: 9 Oct 1814 - From what ship: HMT Leyden, Chatham - Born: Beverly - Age: 19 - Released on 15 Jun 1815.

Shean, John - Seaman - Nbr: 3830 - Prize: Hussar - Ship Type: P - How taken: HMS Saturn - When taken: 24 May 1814 - Where taken: off Sandy Hook - Date Received: 5 Oct 1814 - From what ship: HMT Orpheus, Halifax - Born: Philadelphia - Age: 28 - Released on 9 Jun 1815.

Sheardon, Ellison - Seaman - Nbr: 949 - Prize: Siro - Ship Type: LM - How taken: HMS Pelican - When taken: 13 Jan 1814 - Where taken: at sea - Date Received: 31 Jan 1814 - From what ship: Plymouth - Born: Rhode Island - Age: 18 - Released on 27 Apr 1815.

Sheilds, Michael - Soldier - Nbr: 5271 - Prize: 13th US Infantry - Ship Type: Troops - How taken: British Army - When taken: 13 Oct 1812 - Where taken: Canada - Date Received: 31 Oct 1814 - From what ship: HMT Leyden, Chatham - Born: Donegal - Age: 21 - Released on 29 Jun 1815.

Shelton, Samuel - Seaman - Nbr: 4385 - Prize: Globe - Ship Type: P - How taken: HMS Spartan - When taken: 13 Jun 1813 - Where taken: off Delaware - Date Received: 8 Oct 1814 - From what ship: HMT Leyden, Chatham - Born: New York - Age: 27 - Released on 14 Jun 1815.

Shelton, Smith - Soldier - Nbr: 5847 - Prize: NY Militia - Ship Type: Troops - How taken: British Army - When taken: 17 Sep 1814 - Where taken: Niagara - Date Received: 26 Dec 1814 - From what ship: HMT Argo - Born: Rhode Island - Age: 25 - Died on 19 Jan 1815 from variola.

Shephard, James - Seaman - Nbr: 4586 - Prize: Bunker Hill - Ship Type: P - How taken: HMS Pomone - When taken: 2 Mar 1814 - Where taken: Channel - Date Received: 9 Oct 1814 - From what ship: HMT Leyden, Chatham - Born: New York - Age: 35 - Released on 15 Jun 1815.

Shepherd, Henry - Seaman - Nbr: 5374 - Prize: James - Ship Type: MV - How taken: HMS Harpy - When taken: 16 Dec 1812 - Where taken: off Isle of France - Date Received: 31 Oct 1814 - From what ship: HMT Leyden, Chatham - Born: Maryland - Age: 28 - Released on 27 Apr 1815.

Shepherd, Isaac D. - Boy - Nbr: 5512 - Prize: Mary - Ship Type: Prize - How taken: HMS Wasp - When taken: 6 Oct 1814 - Date Received: 17 Dec 1814 - From what ship: HMT Loire, Halifax - Born: New Hampshire - Age: 17.

Shepherd, James - Seaman - Nbr: 2685 - How taken: Gave himself up from HMS Royal George - When taken: 29 Oct 1812 - Date Received: 21 Aug 1814 - From what ship: HMT Freya, Chatham - Born: New Hampshire - Age: 29 - Released on 26 Apr 1815.

Shepherd, Samuel - Gunner - Nbr: 4254 - Prize: Argus - Ship Type: MV - How taken: HMS San Domingo - When taken: 1 Mar 1814 - Where taken: off Savannah - Date Received: 7 Oct 1814 - From what ship: HMT Niobe, Chatham - Born: Salem - Age: 29 - Released on 13 Jun 1815.

Shepherd, William - Seaman - Nbr: 1166 - How taken: Gave himself up from HMS Hebrius - Date Received: 10 May 1814 - From what ship: Plymouth - Born: Lancaster - Age: 28 - Released on 28 Apr 1815.

Sheppard, William - Seaman - Nbr: 359 - Prize: Two Brothers - Ship Type: MV - How taken: Beetle (LM) - When taken: 18 Mar 1813 - Where taken: off Western Islands (England) - Date Received: 1 Jul 1813 - From what ship: Plymouth - Born: New York - Age: 41 - Released on 20 Apr 1815.

Sherbon, Henry - Prize master - Nbr: 5703 - Prize: Amazon - Ship Type: Prize - How taken: HMS Bulwark - When

American Prisoners of War Held at Dartmoor during the War of 1812

## Alphabetical listing of names

taken: 22 Sep 1814 - Where taken: Georges Bank - Date Received: 24 Dec 1814 - From what ship: HMT Penelope - Born: Portsmouth - Age: 24 - Released on 3 Jul 1815.

Sherdon, Joseph - Seaman - Nbr: 5305 - Prize: Yankee - Ship Type: P - How taken: HMS Shannon - When taken: 20 Aug 1813 - Where taken: at sea - Date Received: 31 Oct 1814 - From what ship: HMT Leyden, Chatham - Born: Somerset - Age: 18 - Released on 29 Jun 1815.

Sherlock, Charles - Seaman - Nbr: 2461 - Prize: US Sloop Frolic - Ship Type: MW - How taken: HMS Orpheus - When taken: 20 Apr 1814 - Where taken: off Cuba - Date Received: 16 Aug 1814 - From what ship: HMT Queen, Halifax - Born: New York - Age: 25 - Released on 3 May 1815.

Sherman, Reuben - 2nd Mate - Nbr: 334 - Prize: Criterion - Ship Type: MV - How taken: HMS Belle Poule - When taken: 14 Feb 1813 - Where taken: Bay of Biscay - Date Received: 28 Jun 1813 - From what ship: Plymouth - Born: New Bedford - Age: 25 - Released on 20 Apr 1815.

Sherridan, Henry - Seaman - Nbr: 3187 - How taken: Gave himself up from HMS Scorpion - When taken: 16 Jun 1813 - Date Received: 11 Sep 1814 - From what ship: HMT Freya, Chatham - Born: New York - Age: 22 - Race: Mulatto - Died on 24 Jan 1815 in prison.

Sherriff, Benjamin P. - Seaman - Nbr: 3174 - How taken: Gave himself up from HMS Impeccable - When taken: 10 Sep 1813 - Date Received: 11 Sep 1814 - From what ship: HMT Freya, Chatham - Born: Exeter - Age: 22 - Released on 28 May 1815.

Sherry, John - Prize master - Nbr: 6116 - Prize: Johannes - Ship Type: Prize - How taken: Lacine - When taken: 26 Oct 1814 - Where taken: Grand Banks - Date Received: 6 Jan 1814 - From what ship: HMT Impregnable - Born: Salem - Age: 39 - Released on 11 Jul 1815.

Sherry, Peter - Boy - Nbr: 4977 - Prize: Invincible - Ship Type: LM - How taken: HMS Armide - When taken: 15 Aug 1814 - Where taken: off Nantucket - Date Received: 28 Oct 1814 - From what ship: HMT Alkbar, Halifax - Born: Bordeaux - Age: 14 - Released on 21 Jun 1815.

Sherwood, John - Seaman - Nbr: 259 - Prize: William Bayard - Ship Type: MV - How taken: HMS Warspite - When taken: 3 Mar 1813 - Where taken: Bay of Biscay - Date Received: 28 Jun 1813 - From what ship: Plymouth - Born: Poughkeepsie - Age: 23 - Entered British service on 11 Oct 1813.

Shields, John - Gunner's Mate - Nbr: 3653 - Prize: Thomas - Ship Type: P - How taken: HMS Nymphe - When taken: 28 Jun 1813 - Where taken: off Halifax - Date Received: 30 Sep 1814 - From what ship: HMT President - Born: Portsmouth - Age: 25 - Released on 4 Jun 1815.

Shiers, Vincent - Seaman - Nbr: 5789 - How taken: Gave himself up from HMS Clorinda - Date Received: 26 Dec 1814 - From what ship: HMT Argo - Born: Connecticut - Age: 23 - Released on 3 Jul 1815.

Shilling, James - Seaman - Nbr: 5340 - Prize: Elbridge Gerry - Ship Type: P - How taken: HMS Crescent - When taken: 16 Sep 1813 - Where taken: at sea - Date Received: 31 Oct 1814 - From what ship: HMT Leyden, Chatham - Born: Grey - Age: 22 - Released on 1 Jul 1815.

Shilling, Morris - Seaman - Nbr: 2699 - How taken: Gave himself up from HMS Echo - When taken: 3 Nov 1812 - Date Received: 21 Aug 1814 - From what ship: HMT Freya, Chatham - Born: Connecticut - Age: 24 - Released on 26 Apr 1815.

Shipley, Charles - Seaman - Nbr: 2588 - How taken: Gave himself up from Gladiator - When taken: 7 Feb 1813 - Date Received: 21 Aug 1814 - From what ship: HMT Freya, Chatham - Born: Boston - Age: 32 - Released on 3 May 1815.

Shipley, Daniel - Seaman - Nbr: 755 - How taken: Impressed at Liverpool - Date Received: 3 Nov 1813 - From what ship: Plymouth - Born: New Castle - Age: 24 - Released on 26 Apr 1815.

Shipling, Peter - Seaman - Nbr: 5478 - Prize: General Putnam - Ship Type: P - How taken: HMS Leander - When taken: 8 Nov 1814 - Where taken: Long 65 Lat 42 - Date Received: 17 Dec 1814 - From what ship: HMT Loire, Halifax - Born: Danzig - Age: 32 - Released on 1 Jul 1815.

Shoals, Christopher - Carpenter - Nbr: 5643 - Prize: Lion - Ship Type: P - How taken: HMS Granicus - When taken: 2 Dec 1814 - Where taken: off Lisbon - Date Received: 24 Dec 1814 - From what ship: HMT Hanover, Gibraltar - Born: Not legible - Age: 29 - Released on 3 Jul 1815.

Shoe, Bernard - Sail Maker - Nbr: 4146 - Prize: Prize to the Diomede - Ship Type: Prize - How taken: HMS Sapphire - When taken: 27 Feb 1814 - Where taken: at sea - Date Received: 6 Oct 1814 - From what ship: HMT Niobe, Chatham - Born: Holland - Age: 53 - Released on 13 Jun 1815.

Sholtz, Charles - Seaman - Nbr: 5098 - Prize: David Porter - Ship Type: LM - How taken: HMS Pylades - When taken: 12 Sep 1814 - Where taken: Georges Bank - Date Received: 28 Oct 1814 - From what ship: HMT Alkbar, Halifax - Born: Russia - Age: 25 - Released on 29 Jun 1815.

Shomodia, John - Seaman - Nbr: 3130 - How taken: Impressed from MV Thames - When taken: 8 Sep 1811 - Date

American Prisoners of War Held at Dartmoor during the War of 1812

## Alphabetical listing of names

Received: 11 Sep 1814 - From what ship: HMT Freya, Chatham - Born: New London - Age: 25 - Released on 27 Apr 1815.

Shoot, William - Seaman - Nbr: 4955 - Prize: Herald - Ship Type: P - How taken: HMS Endymion - When taken: 15 Aug 1814 - Where taken: off Nantucket - Date Received: 28 Oct 1814 - From what ship: HMT Alkbar, Halifax - Born: Newbury - Age: 27 - Released on 21 Jun 1815.

Shore, Caesar - Seaman - Nbr: 5775 - Prize: William Penn - Ship Type: MV - How taken: HMS Acorn - When taken: 27 Oct 1812 - Where taken: Lat 14 - Date Received: 26 Dec 1814 - From what ship: HMT Argo - Born: New London - Age: 23 - Released on 27 Apr 1815.

Shorts, Clement - Steward - Nbr: 1217 - Prize: James - Ship Type: MV - How taken: HMS Harpy - When taken: 19 Dec 1812 - Where taken: off Isle of France - Date Received: 14 Jun 1814 - From what ship: Mill Prison (Plymouth, England) - Born: New York - Age: 27 - Race: Black - Released on 26 Apr 1815.

Shorts, Richard - Seaman - Nbr: 2165 - Prize: Rattlesnake - Ship Type: P - How taken: HMS Hyperion - When taken: 26 Jun 1814 - Where taken: off Cape Finisterre - Date Received: 16 Aug 1814 - From what ship: HMS Dublin, Halifax - Born: Delaware - Age: 28 - Released on 2 May 1815.

Shorts, Samuel - Seaman - Nbr: 3336 - Prize: Prize of the President - Ship Type: MV - How taken: HMS Regulus - When taken: 4 Sep 1813 - Where taken: off Newfoundland - Date Received: 13 Sep 1814 - From what ship: HMT Niobe, Chatham - Born: Rhode Island - Age: 32 - Released on 28 May 1815.

Shotsenberg, Manuel - Seaman - Nbr: 5241 - How taken: British Army - When taken: 26 Mar 1814 - Where taken: off Bordeaux - Date Received: 31 Oct 1814 - From what ship: HMT Mermaid, Chatham - Born: Boston - Age: 30 - Released on 29 Jun 1815.

Shott, John - Boy - Nbr: 2976 - Prize: Frolic - Ship Type: P - How taken: HMS Heron - When taken: 25 Jan 1814 - Where taken: off St. Thomas - Date Received: 29 Aug 1814 - From what ship: HMT Bittern - Born: Salem - Age: 15 - Released on 19 May 1815.

Shrousy, William - Seaman - Nbr: 4303 - How taken: Gave himself up from HMS Action - When taken: 9 May 1814 - Date Received: 7 Oct 1814 - From what ship: HMT Niobe, Chatham - Born: Philadelphia - Age: 30 - Released on 14 Jun 1815.

Shurman, Thomas - Gunner - Nbr: 2278 - Prize: Diomede - Ship Type: P - How taken: HMS Rifleman - When taken: 28 May 1814 - Where taken: off Sable Island - Date Received: 16 Aug 1814 - From what ship: HMS Dublin, Halifax - Born: Salem - Age: 23 - Released on 3 May 1815.

Shutcliffe, John - Seaman - Nbr: 3916 - Prize: Rolla - Ship Type: P - How taken: HMS Loire - When taken: 10 Dec 1813 - Where taken: off Bull Island (South Carolina) - Date Received: 5 Oct 1814 - From what ship: HMT President, Halifax - Born: Philadelphia - Age: 23 - Released on 9 Jun 1815.

Shute, Thomas - Seaman - Nbr: 2962 - How taken: Gave himself up from HMS Scout - When taken: 21 May 1814 - Date Received: 24 Aug 1814 - From what ship: HMT Hannibal - Born: Philadelphia - Age: 37 - Released on 19 May 1815.

Sickless, J. M. - Seaman - Nbr: 712 - Prize: Ned - Ship Type: LM - How taken: HMS Royalist - When taken: 6 Sep 1813 - Where taken: Bay of Biscay - Date Received: 27 Sep 1813 - From what ship: Plymouth - Born: New York - Age: 22 - Released on 26 Apr 1815.

Sidebottom, John - Seaman - Nbr: 1916 - How taken: Gave himself up from HMS Royal William - When taken: 3 Feb 1813 - Date Received: 3 Aug 1814 - From what ship: HMS Alceste, Chatham Depot - Born: Baltimore - Age: 28 - Released on 2 May 1815.

Sides, Samuel - Seaman - Nbr: 4472 - Prize: Governor Plumer - Ship Type: P - How taken: HMS Shamrock - When taken: 4 Mar 1813 - Where taken: at sea - Date Received: 8 Oct 1814 - From what ship: HMT Leyden, Chatham - Born: Portsmouth - Age: 34 - Released on 15 Jun 1815.

Siffers, Steuben - Seaman - Nbr: 1630 - Prize: Tom - Ship Type: LM - How taken: HMS Surveillante - When taken: 24 Apr 1813 - Where taken: Bay of Biscay - Date Received: 23 Jun 1814 - From what ship: Stapleton - Born: Connecticut - Age: 20 - Released on 1 May 1815.

Silkes, Peter - Seaman - Nbr: 6168 - Prize: Lion - Ship Type: P - How taken: HMS Granicus - When taken: 2 Dec 1814 - Where taken: off Lisbon - Date Received: 18 Jan 1815 - From what ship: HMT Impregnable - Born: Baltimore - Age: 39 - Released on 5 Jul 1815.

Sillick, Thomas - Seaman - Nbr: 4445 - Prize: York Town - Ship Type: P - How taken: HMS Maidstone - When taken: 18 Jul 1813 - Where taken: Grand Banks - Date Received: 8 Oct 1814 - From what ship: HMT Leyden, Chatham - Born: Connecticut - Age: 33 - Released on 14 Jun 1815.

Silsby, Nathaniel - Seaman - Nbr: 1915 - How taken: Gave himself up from HMS Dapper - When taken: 9 Dec 1812 - Date Received: 3 Aug 1814 - From what ship: HMS Alceste, Chatham Depot - Born: Salem - Age: 20 -

American Prisoners of War Held at Dartmoor during the War of 1812

## Alphabetical listing of names

Released on 26 Apr 1815.
Silva, Francis - Seaman - Nbr: 651 - Prize: Marmion - Ship Type: MV - How taken: HMS President - When taken: 14 Aug 1813 - Where taken: off Nantes - Date Received: 8 Sep 1813 - From what ship: Plymouth - Born: New York - Age: 22 - Released on 26 Apr 1815.
Silverlock, John - Seaman - Nbr: 1749 - How taken: Gave himself up from HMS Thames - When taken: 9 Nov 1812 - Date Received: 20 Jul 1814 - From what ship: HMS Milford, Plymouth - Born: Virginia - Age: 21 - Released on 26 Apr 1815.
Silverthorn, James - Seaman - Nbr: 4410 - Prize: Blockade - Ship Type: P - How taken: HMS Recruit - When taken: 17 Aug 1813 - Where taken: off America - Date Received: 8 Oct 1814 - From what ship: HMT Leyden, Chatham - Born: Virginia - Age: 19 - Released on 14 Jun 1815.
Silvey, John - Seaman - Nbr: 5367 - Prize: Penn - Ship Type: Whaler - How taken: HMS Acorn - When taken: 27 Oct 1813 - Where taken: South Sea - Date Received: 31 Oct 1814 - From what ship: HMT Leyden, Chatham - Born: Nantucket - Age: 20 - Released on 1 Jul 1815.
Simes, Henry - Seaman - Nbr: 3500 - How taken: Impressed from an East Indiaman - Date rec'd: 24 Sep 1814 - What ship: HMT Salvador del Mundo - Born: Philadelphia - Age: 22 - Race: Negro - Released on 4 Jun 1815.
Simes, Stephen - Seaman - Nbr: 3501 - Prize: Abel - Ship Type: MV - How taken: Matchless (Privateer) - Where taken: off Antigua - Date Received: 24 Sep 1814 - From what ship: HMT Salvador del Mundo - Born: Philadelphia - Age: 28 - Race: Negro - Released on 4 Jun 1815.
Simmonds, Alexander - Carpenter - Nbr: 5044 - Prize: Ida - Ship Type: LM - How taken: HMS Newcastle - When taken: 9 Aug 1814 - Where taken: Long 34 - Date Received: 28 Oct 1814 - From what ship: HMT Alkbar, Halifax - Born: Boston - Age: 25 - Released on 29 Jun 1815.
Simmonds, Charles - Seaman - Nbr: 4260 - Prize: Argus - Ship Type: MV - How taken: HMS San Domingo - When taken: 1 Mar 1814 - Where taken: off Savannah - Date Received: 7 Oct 1814 - From what ship: HMT Niobe, Chatham - Born: Barnstable - Age: 21 - Released on 13 Jun 1815.
Simmonds, Henry - Seaman - Nbr: 4285 - Prize: Pilot - Ship Type: LM - How taken: Victoria (Privateer) - When taken: 28 Jan 1814 - Where taken: off Bordeaux - Date Received: 7 Oct 1814 - From what ship: HMT Niobe, Chatham - Born: Prussia - Age: 23 - Race: Black - Released on 14 Jun 1815.
Simmonds, Joel - Seaman - Nbr: 4262 - Prize: Argus - Ship Type: MV - How taken: HMS San Domingo - When taken: 1 Mar 1814 - Where taken: off Savannah - Date Received: 7 Oct 1814 - From what ship: HMT Niobe, Chatham - Born: Barnstable - Age: 19 - Released on 13 Jun 1815.
Simmons, David - Seaman - Nbr: 4479 - Prize: Enterprize - Ship Type: P - How taken: HMS Tenedos - When taken: 21 May 1813 - Where taken: off Cape Cod - Date Received: 8 Oct 1814 - From what ship: HMT Leyden, Chatham - Born: Massachusetts - Age: 18 - Died on 22 Jan 1815 from phthisis pulmonalis.
Simmons, Ebenezer - Seaman - Nbr: 4850 - How taken: Gave himself up from HMS Barbadoes - When taken: 1 Feb 1814 - Date Received: 9 Oct 1814 - From what ship: HMT Freya, Halifax - Born: Newburyport - Age: 20 - Died on 12 Jan 1815 from variola.
Simmons, Richard - Seaman - Nbr: 3524 - Prize: Sabine - Ship Type: P - How taken: HMS Conquistador - When taken: 3 Aug 1814 - Where taken: off the Banks - Date Received: 28 Sep 1814 - From what ship: HMT Salvador del Mundo - Born: Philadelphia - Age: 40 - Race: Black - Released on 4 Jun 1815.
Simmons, Robert - Seaman - Nbr: 1503 - How taken: Impressed at Bristol - When taken: 10 Jun 1813 - Date Received: 23 Jun 1814 - From what ship: Stapleton - Born: Charleston - Age: 35 - Race: Negro - Released on 1 May 1815.
Simmons, Thomas - Seaman - Nbr: 5571 - Prize: Saratoga - Ship Type: Prize - How taken: HMS Asia - When taken: 24 Jul 1814 - Where taken: off Cape VA - Date Received: 17 Dec 1814 - From what ship: HMT Loire, Halifax - Born: New Bedford - Age: 55 - Race: Mulatto - Died on 29 Jan 1814 from phthisis pulmonalis.
Simon, Mark - Prize Master - Nbr: 3596 - Prize: Frolic - Ship Type: P - How taken: HMS Heron - When taken: 25 Jan 1814 - Where taken: off St. Thomas - Date Received: 30 Sep 1814 - From what ship: HMT Sybella - Born: Ipswich - Age: 23 - Released on 4 Jun 1815.
Simondson, Isaac - Seaman - Nbr: 4995 - Prize: Invincible - Ship Type: LM - How taken: HMS Armide - When taken: 15 Aug 1814 - Where taken: off Nantucket - Date Received: 28 Oct 1814 - From what ship: HMT Alkbar, Halifax - Born: New York - Age: 20 - Died on 20 Nov 1814.
Simons, John - Seaman - Nbr: 1288 - Prize: Traveler - Ship Type: Prize - How taken: HM Schooner Canso - When taken: 11 May 1814 - Where taken: off Cape Clear - Date Received: 14 Jun 1814 - From what ship: Mill Prison (Plymouth, England) - Born: Long Island - Age: 18 - Released on 28 Apr 1815.

American Prisoners of War Held at Dartmoor during the War of 1812

## Alphabetical listing of names

Simons, John - Seaman - Nbr: 1997 - Prize: Kergettos - Ship Type: MV - How taken: HMS Castilian - When taken: 12 Mar 1813 - Where taken: off Cape Ortegal (Spain) - Date Received: 3 Aug 1814 - From what ship: HMS Lyffey, Chatham Depot - Born: Washington - Age: 18 - Released on 2 May 1815.

Simons, John C. - Seaman - Nbr: 3735 - Prize: Alfred - Ship Type: P - How taken: HMS Epervier - When taken: 23 Feb 1814 - Where taken: off Newfoundland - Date Received: 30 Sep 1814 - From what ship: HMT President, Halifax - Born: Salem - Age: 20 - Released on 4 Jun 1815.

Simons, William - Seaman - Nbr: 1084 - How taken: Impressed at Falmouth - When taken: 20 Mar 1814 - Date Received: 10 May 1814 - From what ship: Plymouth - Born: Nantucket - Age: 24 - Race: Negro - Released on 27 Apr 1815.

Simpson, John - Boy - Nbr: 5668 - Prize: Harlequin - Ship Type: P - How taken: HMS Bulwark - When taken: 23 Nov 1814 - Where taken: off Halifax - Date Received: 24 Dec 1814 - From what ship: HMT Penelope - Born: Portsmouth - Age: 13 - Released on 11 Jul 1815.

Simpson, Joseph - Seaman - Nbr: 5880 - Prize: Harlequin - Ship Type: P - How taken: HMS Bulwark - When taken: 23 Nov 1814 - Where taken: off Halifax - Date Received: 27 Dec 1814 - From what ship: HMT Penelope - Born: New Hampshire - Age: 21 - Released on 28 May 1815.

Simpson, Smith - Seaman - Nbr: 4779 - How taken: Gave himself up from HMS Blenheim - When taken: 2 Sep 1814 - Date Received: 9 Oct 1814 - From what ship: HMT Freya, Chatham - Born: North Carolina - Age: 26 - Released on 15 Jun 1815.

Simpson, Thomas - Seaman - Nbr: 1572 - Prize: Messenger - Ship Type: MV - How taken: HMS Iris - When taken: 10 Mar 1813 - Where taken: Bay of Biscay - Date Received: 23 Jun 1814 - From what ship: Stapleton - Born: Maryland - Age: 37 - Released on 1 May 1815.

Simpson, William - Seaman - Nbr: 1298 - Prize: John & Frances - Ship Type: Prize - How taken: HMS Sterling Castle - When taken: 10 May 1814 - Where taken: off Cape Clear - Date Received: 14 Jun 1814 - From what ship: Mill Prison (Plymouth, England) - Born: Philadelphia - Age: 27 - Released on 28 Apr 1815.

Simpson, William - Seaman - Nbr: 3777 - Prize: Fame - Ship Type: P - How taken: HMS Thistle - When taken: 10 Apr 1814 - Where taken: after being cast ashore on a seal island - Date Received: 30 Sep 1814 - From what ship: HMT President, Halifax - Born: New Jersey - Age: 30 - Released on 9 Jun 1815.

Sims, Clement - Seaman - Nbr: 2765 - How taken: Gave himself up from HMS Mermaid - When taken: 30 Nov 1812 - Date Received: 24 Aug 1814 - From what ship: HMT Liverpool, Chatham - Born: Baltimore - Age: 30 - Released on 19 May 1815.

Sims, Joseph - Seaman - Nbr: 3086 - Prize: Eliza (Whaler) - Ship Type: MV - How taken: HMS Charles - When taken: 20 Jul 1813 - Where taken: off Peter Head - Date Received: 3 Sep 1814 - From what ship: HMT Hydra, Chatham - Born: Africa - Age: 30 - Race: Black - Released on 28 May 1815.

Sims, Oliver - Seaman - Nbr: 1824 - How taken: Impressed at Dublin - When taken: 12 Feb 1813 - Date Received: 21 Jul 1814 - From what ship: HMT Redbeard & Pincher, Chatham Depot - Born: Rhode Island - Age: 10 - Released on 1 May 1815.

Sims, William - Seaman - Nbr: 2933 - How taken: Gave himself up from HMS Berwick - Date Received: 24 Aug 1814 - From what ship: HMT Alpheus, Chatham - Born: Norfolk - Age: 21 - Released on 19 May 1815.

Simson, John - Seaman - Nbr: 5973 - Prize: Halifax Packet - Ship Type: Prize - How taken: HMS Bulwark - When taken: 22 Sep 1814 - Where taken: Georges Bank - Date Received: 27 Dec 1814 - From what ship: HMT Penelope - Born: Westerly - Age: 21 - Released on 3 Jul 1815.

Simson, John - Seaman - Nbr: 3917 - Prize: Rolla - Ship Type: P - How taken: HMS Loire - When taken: 10 Dec 1813 - Where taken: off Bull Island (South Carolina) - Date Received: 5 Oct 1814 - From what ship: HMT President, Halifax - Born: Long Island - Age: 32 - Race: Negro - Released on 9 Jun 1815.

Simson, John - Seaman - Nbr: 2649 - How taken: Gave himself up from HMS Shallow - When taken: 25 May 1813 - Date Received: 21 Aug 1814 - From what ship: HMT Freya, Chatham - Born: New York - Age: 23 - Released on 3 May 1815.

Sinclair, Thomas - Prize master - Nbr: 6190 - Prize: Prince de Neufchatel - Ship Type: P - How taken: Leander (Newcastle Acasta) - When taken: 20 Dec 1814 - Where taken: Lat 38 Long 56 - Date Received: 30 Jan 1815 - From what ship: HMT Pheasant - Born: Boston - Age: 39 - Released on 5 Jul 1815.

Sissons, Pedro - Seaman - Nbr: 4805 - Prize: President - Ship Type: P - How taken: HMS Pique - When taken: 7 May 1814 - Where taken: off Porto Rico - Date Received: 9 Oct 1814 - From what ship: HMT Freya, Halifax - Born: Brazil - Age: 34 - Race: Black - Released on 21 Jun 1815.

Skelder, Samuel - Seaman - Nbr: 780 - Prize: Betsy - Ship Type: MV - How taken: HMS Eurotas - When taken: 26 Oct 1813 - Where taken: off Ushant (France) - Date Received: 3 Nov 1813 - From what ship: Plymouth -

American Prisoners of War Held at Dartmoor during the War of 1812

## Alphabetical listing of names

Born: Connecticut - Age: 22 - Released on 26 Apr 1815.
Skenny, Francis - Seaman - Nbr: 4903 - Prize: Lizard - Ship Type: P - How taken: HMS Prometheus - When taken: 5 May 1814 - Where taken: off Halifax - Date Received: 28 Oct 1814 - From what ship: HMT Alkbar, Halifax - Born: Salem - Age: 22 - Released on 21 Jun 1815.
Skinner, Tileman - Seaman - Nbr: 1302 - Prize: John & Frances - Ship Type: Prize - How taken: HMS Sterling Castle - When taken: 10 May 1814 - Where taken: off Cape Clear - Date Received: 14 Jun 1814 - From what ship: Mill Prison (Plymouth, England) - Born: Philadelphia - Age: 22 - Race: Black - Released on 28 Apr 1815.
Skipper, John - Seaman - Nbr: 6448 - Prize: Jemmell - Ship Type: MV - How taken: HMS Rhin - When taken: 28 May 1814 - Where taken: off Bermuda - Date Received: 3 Mar 1815 - From what ship: HMT Ganges, Plymouth - Born: Cape Cod - Age: 17 - Race: Mulatto - Released on 11 Jul 1815.
Skudder, Alexander - Prize master - Nbr: 4253 - Prize: Argus - Ship Type: MV - How taken: HMS San Domingo - When taken: 1 Mar 1814 - Where taken: off Savannah - Date Received: 7 Oct 1814 - From what ship: HMT Niobe, Chatham - Born: Philadelphia - Age: 30 - Released on 14 Apr 1815.
Slater, Ludowick - Seaman - Nbr: 2191 - Prize: Diomede - Ship Type: P - How taken: HMS Rifleman - When taken: 28 Jun 1813 - Where taken: off Halifax - Date Received: 16 Aug 1814 - From what ship: HMS Dublin, Halifax - Born: New London - Age: 50 - Released on 2 May 1815.
Slebar, Samuel - Seaman - Nbr: 2016 - How taken: Apprehended at Cork - When taken: 7 Apr 1813 - Date Received: 3 Aug 1814 - From what ship: HMS Lyffey, Chatham Depot - Born: Christian Bridge - Age: 22 - Released on 2 May 1815.
Sline, Samuel - Seaman - Nbr: 2952 - Prize: Ellen and Elizabeth - Ship Type: MV - How taken: Rose & Thistle - When taken: 24 Feb 1814 - Where taken: off St. Bartholomew - Date Received: 24 Aug 1814 - From what ship: HMT Hannibal - Born: Philadelphia - Age: 25 - Released on 19 May 1815.
Sloane, William - Seaman - Nbr: 3064 - How taken: Impressed at London - When taken: 5 Mar 1814 - Date Received: 2 Sep 1814 - From what ship: HMT Hydra, Chatham - Born: Stamford - Age: 28 - Released on 28 May 1815.
Slocombe, Abraham - Seaman - Nbr: 1606 - Prize: Fox - Ship Type: LM - How taken: HMS Pheasant - When taken: 23 Apr 1813 - Where taken: Bay of Biscay - Date Received: 23 Jun 1814 - From what ship: Stapleton - Born: New Orleans - Age: 32 - Sent to Mill Prison (Plymouth, England) on 30 Jun 1814.
Slocum, William - Seaman - Nbr: 2668 - Prize: Orders in Council - Ship Type: MV - How taken: HMS Surveillante - When taken: 1 Jun 1813 - Where taken: off Cape Ortegal (Spain) - Date Received: 21 Aug 1814 - From what ship: HMT Freya, Chatham - Born: Rhode Island - Age: 23 - Released on 19 May 1815.
Small, Enoch - Seaman - Nbr: 808 - Prize: Sybelle - Ship Type: MV - How taken: HMS Zenobia - When taken: 27 Jun 1813 - Where taken: off Cape St. Mary's - Date Received: 3 Nov 1813 - From what ship: Plymouth - Born: Massachusetts - Age: 21 - Released on 26 Apr 1815.
Small, George D. - Seaman - Nbr: 4180 - Prize: Tigre - Ship Type: MV - How taken: HMS Scylla - When taken: 22 Mar 1813 - Where taken: Bay of Biscay - Date Received: 7 Oct 1814 - From what ship: HMT Niobe, Chatham - Born: New York - Age: 24 - Released on 1 May 1815.
Small, Joseph - Seaman - Nbr: 3503 - How taken: Impressed at Liverpool - When taken: 10 Sep 1814 - Date Received: 24 Sep 1814 - From what ship: HMT Salvador del Mundo - Born: Massachusetts - Age: 28 - Released on 4 Jun 1815.
Small, Richard - Seaman - Nbr: 427 - Prize: Viper - Ship Type: MV - How taken: HMS Superb - When taken: 15 Apr 1813 - Where taken: Bay of Biscay - Date Received: 1 Jul 1813 - From what ship: Plymouth - Born: Baltimore - Age: 18 - Released on 20 Apr 1815.
Small, William - Seaman - Nbr: 2464 - Prize: US Sloop Frolic - Ship Type: MW - How taken: HMS Orpheus - When taken: 20 Apr 1814 - Where taken: off Cuba - Date Received: 16 Aug 1814 - From what ship: HMT Queen, Halifax - Born: Massachusetts - Age: 31 - Released on 3 May 1815.
Smalley, Thomas - Seaman - Nbr: 415 - Prize: Viper - Ship Type: MV - How taken: HMS Superb - When taken: 15 Apr 1813 - Where taken: Bay of Biscay - Date Received: 1 Jul 1813 - From what ship: Plymouth - Born: Philadelphia - Age: 56 - Released on 20 Apr 1815.
Smallpiece, John - Mate - Nbr: 3031 - Prize: Hazard - Ship Type: MV - How taken: HMS Belvidera - When taken: 27 Jun 1814 - Where taken: off Delaware - Date Received: 2 Sep 1814 - From what ship: HMT Sultan - Born: Boston - Age: 27 - Released on 28 May 1815.
Smart, Richard - Master's mate - Nbr: 5925 - Prize: Harlequin - Ship Type: P - How taken: HMS Bulwark - When taken: 23 Nov 1814 - Where taken: off Halifax - Date Received: 27 Dec 1814 - From what ship: HMT

American Prisoners of War Held at Dartmoor during the War of 1812

## Alphabetical listing of names

Penelope - Born: Portsmouth - Age: 39 - Released on 11 Jul 1815.

Smiles, John - Seaman - Nbr: 977 - Prize: Siro - Ship Type: LM - How taken: HMS Pelican - When taken: 13 Jan 1814 - Where taken: at sea - Date Received: 31 Jan 1814 - From what ship: Plymouth - Born: Batavia - Age: 29 - Race: Mulatto - Released on 16 Jul 1814.

Smith, Abraham - Seaman - Nbr: 6139 - How taken: Gave himself up from HMS Ariel - Date Received: 17 Jan 1815 - From what ship: HMT Impregnable - Born: Not legible - Race: Black - Released on 5 Jul 1815.

Smith, Adam - Seaman - Nbr: 615 - Prize: US Brig Argus - Ship Type: MW - How taken: HMS Pelican - When taken: 14 Aug 1813 - Where taken: Irish Channel - Date Received: 8 Sep 1813 - From what ship: Plymouth - Born: Amsterdam - Age: 24 - Sent to Dartmouth on 2 Nov 1814.

Smith, Aesop - Seaman - Nbr: 2911 - Prize: Matilda - Ship Type: MV - How taken: HMS Revolutionnaire - When taken: 25 Jul 1813 - Where taken: off L'Orient (France) - Date Received: 24 Aug 1814 - From what ship: HMT Alpheus, Chatham - Born: Cape Cod - Age: 30 - Released on 2 Nov 1814.

Smith, Andrew - Seaman - Nbr: 1624 - Prize: Tom - Ship Type: LM - How taken: HMS Surveillante - When taken: 24 Apr 1813 - Where taken: Bay of Biscay - Date Received: 23 Jun 1814 - From what ship: Stapleton - Born: Maryland - Age: 24 - Died on 5 Mar 1815 from variola.

Smith, Benjamin - Seaman - Nbr: 3378 - Prize: Growler - Ship Type: P - How taken: HMS Electra - When taken: 7 Jul 1813 - Where taken: at sea - Date Received: 13 Sep 1814 - From what ship: HMT Niobe, Chatham - Born: Salem - Age: 19 - Released on 28 May 1815.

Smith, Caesar - Seaman - Nbr: 1930 - How taken: Gave himself up from HMS Desiree - When taken: 8 Jan 1814 - Date Received: 3 Aug 1814 - From what ship: HMS Alceste, Chatham Depot - Born: Long Island - Age: 33 - Released on 26 Apr 1815.

Smith, Charles - Seaman - Nbr: 2313 - Prize: Hussar - Ship Type: P - How taken: HMS Saturn - When taken: 25 May 1814 - Where taken: off Sandy Hook - Date Received: 16 Aug 1814 - From what ship: HMS Dublin, Halifax - Born: Delaware - Age: 25 - Released on 3 May 1815.

Smith, Charles - Seaman - Nbr: 177 - How taken: Gave himself up from HMS Mackerel - Date Received: 2 Apr 1813 - From what ship: Plymouth - Born: Albany - Age: 25 - Released on 20 Apr 1815.

Smith, Charles - Boatswain - Nbr: 5639 - Prize: Lion - Ship Type: P - How taken: HMS Granicus - When taken: 2 Dec 1814 - Where taken: off Lisbon - Date Received: 24 Dec 1814 - From what ship: HMT Hanover, Gibraltar - Born: Rhode Island - Age: 29 - Released on 3 Jul 1815.

Smith, Charles - Seaman - Nbr: 5207 - Prize: Porcupine - Ship Type: LM - How taken: HMS Acasta - When taken: 18 Jul 1813 - Where taken: off Halifax - Date Received: 31 Oct 1814 - From what ship: HMT Mermaid, Chatham - Born: Baltimore - Age: 34 - Race: Black - Released on 29 Jun 1815.

Smith, Charles - Boy - Nbr: 2972 - Prize: Frolic - Ship Type: P - How taken: HMS Heron - When taken: 25 Jan 1814 - Where taken: off St. Thomas - Date Received: 29 Aug 1814 - From what ship: HMT Bittern - Born: Northampton - Age: 16 - Released on 28 Apr 1815.

Smith, Charles P. - Seaman - Nbr: 5530 - Prize: Sparks - Ship Type: LM - How taken: HMS Maidstone - When taken: 28 Sep 1814 - Where taken: off Nantucket - Date Received: 17 Dec 1814 - From what ship: HMT Loire, Halifax - Born: New York - Age: 22 - Released on 1 Jul 1815.

Smith, Chester - Seaman - Nbr: 1514 - Prize: Eliza - Ship Type: MV - How taken: HMS Surveillante - When taken: 27 Mar 1813 - Where taken: Bay of Biscay - Date Received: 23 Jun 1814 - From what ship: Stapleton - Born: New Jersey - Age: 26 - Released on 1 May 1815.

Smith, Christopher - Seaman - Nbr: 2156 - Prize: Rattlesnake - Ship Type: P - How taken: HMS Hyperion - When taken: 26 Jun 1814 - Where taken: off Cape Finisterre - Date Received: 16 Aug 1814 - From what ship: HMS Dublin, Halifax - Born: Memel - Age: 32 - Released on 2 May 1815.

Smith, Cornelius - Mate - Nbr: 6044 - Prize: Tickler - Ship Type: MV - How taken: HMS Niemen - When taken: 14 Sep 1814 - Where taken: off Delaware - Date Received: 28 Dec 1814 - From what ship: HMT Penelope - Born: New York - Age: 24 - Released on 3 Jul 1815.

Smith, Daniel - Seaman - Nbr: 4989 - Prize: Invincible - Ship Type: LM - How taken: HMS Armide - When taken: 15 Aug 1814 - Where taken: off Nantucket - Date Received: 28 Oct 1814 - From what ship: HMT Alkbar, Halifax - Born: Newburyport - Age: 35 - Released on 21 Jun 1815.

Smith, David - Seaman - Nbr: 4322 - Prize: Flash - Ship Type: LM - How taken: HMS Acasta and HMS Loire - When taken: 8 Feb 1814 - Where taken: off NY - Date Received: 7 Oct 1814 - From what ship: HMT Salvador del Mundo, Halifax - Born: New York - Age: 16 - Race: Negro - Released on 14 Jun 1815.

Smith, Eldridge - Seaman - Nbr: 46 - Prize: Terrible - Ship Type: MV - How taken: HMS Foxhound - When taken: 8 Feb 1813 - Where taken: Channel - Date Received: 2 Apr 1813 - From what ship: Plymouth - Born:

American Prisoners of War Held at Dartmoor during the War of 1812

## Alphabetical listing of names

Poundstown - Age: 28 - Sent to Dartmouth on 7 Aug 1813.

Smith, Elisha - Seaman - Nbr: 702 - Prize: Ned - Ship Type: LM - How taken: HMS Royalist - When taken: 6 Sep 1813 - Where taken: Bay of Biscay - Date Received: 27 Sep 1813 - From what ship: Plymouth - Born: East Haven - Age: 25 - Released on 26 Apr 1815.

Smith, Elisha - Seaman - Nbr: 4459 - Prize: Juliana Smith - Ship Type: P - How taken: HMS Nymphe - When taken: 11 May 1813 - Where taken: off Cape Sable Island (Canada) - Date Received: 8 Oct 1814 - From what ship: HMT Leyden, Chatham - Born: Massachusetts - Age: 30 - Released on 15 Jun 1815.

Smith, Ezekiel - Prize master - Nbr: 6120 - Prize: Betsey - Ship Type: Prize - How taken: HMS Bellerophon - When taken: 2 Nov 1814 - Where taken: Long 61 - Date Received: 6 Jan 1814 - From what ship: HMT Impregnable - Born: Beverly - Age: 40 - Released on 11 Jul 1815.

Smith, George - Seaman - Nbr: 5557 - Prize: Levant - Ship Type: MV - How taken: HMS Forester - When taken: 4 Jan 1814 - Where taken: Bahamas Banks - Date Received: 17 Dec 1814 - From what ship: HMT Loire, Halifax - Born: Philadelphia - Age: 27 - Released on 1 Jul 1815.

Smith, George - Seaman - Nbr: 323 - Prize: Pallas - Ship Type: MV - How taken: Rebuff - When taken: 23 Dec 1812 - Where taken: off Cadiz - Date Received: 28 Jun 1813 - From what ship: Plymouth - Born: Pennsylvania - Age: 30 - Sent to Dartmouth on 30 Jul 1813.

Smith, George - Seaman - Nbr: 3498 - Ship Type: P - How taken: Taken in a boat of the Fox - Date Received: 24 Sep 1814 - From what ship: HMT Salvador del Mundo - Born: Maryland - Age: 35 - Released on 20 Mar 1815.

Smith, George M. - Seaman - Nbr: 5242 - Prize: Argus - Ship Type: MV - How taken: HMS San Domingo - When taken: 1 Mar 1814 - Where taken: off Savannah - Date Received: 31 Oct 1814 - From what ship: HMT Mermaid, Chatham - Born: South Caroline - Age: 20 - Released on 29 Jun 1815.

Smith, Gerard - Seaman - Nbr: 5608 - How taken: Gave himself up from HMS Ashea - When taken: 13 May 1814 - Date Received: 24 Dec 1814 - From what ship: HMT Tay - Born: Connecticut - Age: 28 - Released on 27 Mar 1815.

Smith, Henry - Seaman - Nbr: 5334 - Prize: Elbridge Gerry - Ship Type: P - How taken: HMS Crescent - When taken: 16 Sep 1813 - Where taken: at sea - Date Received: 31 Oct 1814 - From what ship: HMT Leyden, Chatham - Born: Providence - Age: 25 - Released on 29 Jun 1815.

Smith, Henry - Seaman - Nbr: 3412 - Prize: Thomas - Ship Type: P - How taken: HMS Nymphe - When taken: 24 Jun 1813 - Where taken: off Halifax - Date Received: 13 Sep 1814 - From what ship: HMT Niobe, Chatham - Born: Portsmouth - Age: 21 - Released on 28 May 1815.

Smith, Henry - Seaman - Nbr: 3194 - How taken: Gave himself up from HMS Swiftsure - When taken: 26 Dec 1812 - Date Received: 11 Sep 1814 - From what ship: HMT Freya, Chatham - Born: New Hampshire - Age: 22 - Released on 27 Apr 1815.

Smith, Henry - Seaman - Nbr: 3474 - How taken: Gave himself up from HMS Prince - When taken: 12 Sep 1814 - Date Received: 19 Sep 1814 - From what ship: HMT Salvador del Mundo - Born: New York - Age: 26 - Race: Black - Released on 4 Jun 1815.

Smith, Isaac - Seaman - Nbr: 2457 - Prize: US Sloop Frolic - Ship Type: MW - How taken: HMS Orpheus - When taken: 20 Apr 1814 - Where taken: off Cuba - Date Received: 16 Aug 1814 - From what ship: HMT Queen, Halifax - Born: Baltimore - Age: 26 - Released on 3 May 1815.

Smith, J. W. - Seaman - Nbr: 4843 - Prize: President - Ship Type: P - How taken: HMS Pique - When taken: 7 May 1814 - Where taken: off Porto Rico - Date Received: 9 Oct 1814 - From what ship: HMT Freya, Halifax - Born: Rhode Island - Age: 27 - Released on 21 Jun 1815.

Smith, James - Seaman - Nbr: 878 - How taken: Impressed at Liverpool - When taken: 1 Nov 1813 - Date Received: 31 Jan 1814 - From what ship: Plymouth - Born: New York - Age: 22 - Released on 27 Apr 1815.

Smith, James - Seaman - Nbr: 1469 - Prize: Governor Gerry - Ship Type: MV - How taken: HMS Lyra - When taken: 29 May 1813 - Where taken: Bay of Biscay - Date Received: 19 Jun 1814 - From what ship: Stapleton - Born: Boston - Age: 25 - Race: Negro - Released on 28 Apr 1815.

Smith, James - Seaman - Nbr: 2122 - How taken: Gave himself up from HMS Dublin - When taken: 10 Nov 1812 - Date Received: 8 Aug 1814 - From what ship: HMT Raven, Chatham - Born: Connecticut - Age: 39 - Released on 26 Apr 1815.

Smith, James - 1st Lieutenant - Nbr: 2330 - Prize: Snap Dragon - Ship Type: P - How taken: HMS Martin - When taken: 10 Jun 1814 - Where taken: off Halifax - Date Received: 16 Aug 1814 - From what ship: HMS Dublin, Halifax - Born: Boston - Age: 29 - Released on 28 Apr 1815.

Smith, James - Seaman - Nbr: 582 - Prize: US Brig Argus - Ship Type: MW - How taken: HMS Pelican - When

## Alphabetical listing of names

taken: 14 Apr 1813 - Where taken: Irish Channel - Date Received: 8 Sep 1813 - From what ship: Plymouth - Born: Connecticut - Age: 21 - Sent to Plymouth on 8 Jul 1814.

Smith, James - Seaman - Nbr: 5727 - How taken: Gave himself up from HMS Kamel - Date Received: 26 Dec 1814 - From what ship: HMT Argo - Born: Long Island - Age: 33 - Race: Mulatto - Released on 3 Jul 1815.

Smith, James - Seaman - Nbr: 3492 - How taken: Gave himself up from HMS Prospero - When taken: 12 Sep 1814 - Date Received: 19 Sep 1814 - From what ship: HMT Salvador del Mundo - Born: Massachusetts - Age: 38 - Released on 4 Jun 1815.

Smith, Jeremiah - Prize master - Nbr: 4658 - Prize: Growler - Ship Type: P - How taken: HMS Electra - When taken: 7 Jul 1813 - Where taken: at sea - Date Received: 9 Oct 1814 - From what ship: HMT Leyden, Chatham - Born: Marblehead - Age: 35 - Released on 15 Jun 1815.

Smith, John - Seaman - Nbr: 6109 - Prize: Garland - Ship Type: Prize - How taken: HMS Hamadryad - When taken: 22 Nov 1814 - Where taken: Georges Bank - Date Received: 6 Jan 1814 - From what ship: HMT Impregnable - Born: Portsmouth - Age: 30 - Released on 11 Jul 1815.

Smith, John - Seaman - Nbr: 6285 - How taken: Gave himself up from HMS Cornwallis - Date Received: 17 Feb 1815 - What ship: HMT Ganges, Plymouth - Born: New York - Age: 21 - Race: Black - Released on 5 Jul 1815.

Smith, John - Seaman - Nbr: 3822 - Prize: Bordeaux Packet - Ship Type: LM - How taken: HMS Niemen - When taken: 28 Jan 1814 - Where taken: off Delaware - Date Received: 5 Oct 1814 - From what ship: HMT Orpheus, Halifax - Born: Maryland - Age: 29 - Released on 9 Jun 1815.

Smith, John - Mate - Nbr: 4698 - Ship Type: MV - When taken: 13 Jan 1814 - Date Received: 9 Oct 1814 - From what ship: HMT Leyden, Chatham - Born: New York - Age: 44.

Smith, John - Seaman - Nbr: 3991 - Prize: US Brig Rattlesnake - Ship Type: MW - How taken: HMS Leander - When taken: 13 Jul 1814 - Where taken: off Shelburne - Date Received: 6 Oct 1814 - From what ship: HMT Chesapeake, Halifax - Born: Wilmington - Age: 23 - Released on 9 Jun 1815.

Smith, John - Seaman - Nbr: 6429 - How taken: Sent into custody from Prize Patter - When taken: 12 Apr 1814 - Date Received: 3 Mar 1815 - From what ship: HMT Ganges, Plymouth - Born: Baltimore - Age: 19 - Released on 19 May 1815.

Smith, John - Carpenter - Nbr: 42 - Prize: Terrible - Ship Type: MV - How taken: HMS Foxhound - When taken: 8 Feb 1813 - Where taken: Channel - Date Received: 2 Apr 1813 - From what ship: Plymouth - Born: Groton - Age: 23 - Sent to Dartmouth on 30 Jul 1813.

Smith, John - Seaman - Nbr: 1334 - Prize: Paul Jones - Ship Type: P - How taken: HMS Leonidas - When taken: 23 May 1813 - Where taken: Channel - Date Received: 19 Jun 1814 - From what ship: Stapleton - Born: Rhode Island - Age: 20 - Released on 28 Apr 1815.

Smith, John - Seaman - Nbr: 1132 - How taken: Impressed at Belfast - When taken: 18 Dec 1813 - Date Received: 10 May 1814 - From what ship: Plymouth - Born: New York - Age: 36 - Released on 28 Apr 1815.

Smith, John - 2nd Gunner - Nbr: 1031 - Prize: US Brig Argus - Ship Type: MW - How taken: HMS Pelican - When taken: 16 Aug 1813 - Where taken: Irish Channel - Date Received: 10 May 1814 - From what ship: Plymouth - Born: Middletown - Age: 30 - Sent to Dartmouth on 19 Oct 1814.

Smith, John - Seaman - Nbr: 855 - Prize: Amiable - Ship Type: LM - How taken: HMS Magnificent - When taken: 30 Oct 1813 - Where taken: off Bordeaux - Date Received: 4 Dec 1813 - From what ship: Plymouth - Born: New York - Age: 20 - Released on 26 Apr 1815.

Smith, John - Seaman - Nbr: 2178 - Prize: Rattlesnake - Ship Type: P - How taken: HMS Hyperion - When taken: 26 Jun 1814 - Where taken: off Cape Finisterre - Date Received: 16 Aug 1814 - From what ship: HMS Dublin, Halifax - Born: Boston - Age: 35 - Released on 2 May 1815.

Smith, John - Seaman - Nbr: 1422 - Prize: Governor Gerry - Ship Type: MV - How taken: HMS Lyra - When taken: 29 May 1813 - Where taken: Bay of Biscay - Date Received: 19 Jun 1814 - From what ship: Stapleton - Born: Virginia - Age: 27 - Released on 28 Apr 1815.

Smith, John - Seaman - Nbr: 351 - How taken: Impressed at Exeter - When taken: 12 Mar 1813 - Date Received: 1 Jul 1813 - From what ship: Plymouth - Born: Philadelphia - Age: 30 - Released on 20 Apr 1815.

Smith, John - Quartermaster - Nbr: 2381 - Prize: US Sloop Frolic - Ship Type: MW - How taken: HMS Orpheus - When taken: 20 Apr 1814 - Where taken: off Cuba - Date Received: 16 Aug 1814 - From what ship: HMT Queen, Halifax - Born: Philadelphia - Age: 30 - Sent to Dartmouth on 19 Oct 1814.

Smith, John - Seaman - Nbr: 789 - Prize: Avon - Ship Type: MV - How taken: HMS Eurotas - When taken: 27 Oct 1813 - Where taken: off Ushant (France) - Date Received: 3 Nov 1813 - From what ship: Plymouth - Born: Massachusetts - Age: 34 - Released on 26 Apr 1815.

American Prisoners of War Held at Dartmoor during the War of 1812

## Alphabetical listing of names

Smith, John - Seaman - Nbr: 2105 - Prize: Sea Nymphe - Ship Type: MV - How taken: HMS Thraster - When taken: 4 Mar 1813 - Where taken: North Sea - Date Received: 3 Aug 1814 - From what ship: HMS Bittern, Chatham Depot - Born: New York - Age: 16 - Released on 2 May 1815.

Smith, John - Seaman - Nbr: 1017 - Prize: Rachel & Ann - Ship Type: P - How taken: HMS Cydnus - When taken: 6 Jan 1814 - Where taken: at sea - Date Received: 31 Jan 1814 - From what ship: Plymouth - Born: New York - Age: 37 - Released on 27 Apr 1815.

Smith, John - Seaman - Nbr: 4895 - Prize: Julia - Ship Type: MV - How taken: Sir James Yeo's squadron - When taken: 11 Aug 1813 - Where taken: Lako Balaris - Date Received: 28 Oct 1814 - From what ship: HMT Alkbar, Halifax - Born: Rhode Island - Age: 21 - Released on 3 May 1815.

Smith, John - Seaman - Nbr: 4370 - Prize: Growler - Ship Type: P - How taken: HMS Electra - When taken: 7 Jul 1813 - Where taken: at sea - Date Received: 8 Oct 1814 - From what ship: HMT Leyden, Chatham - Born: Marblehead - Age: 25 - Released on 14 Jun 1815.

Smith, John - Seaman - Nbr: 3081 - How taken: Gave himself up from HMS Ville de Paris - When taken: 28 Apr 1813 - Date Received: 3 Sep 1814 - From what ship: HMT Hydra, Chatham - Born: Massachusetts - Age: 28 - Released on 28 May 1815.

Smith, John - Seaman - Nbr: 3185 - How taken: Gave himself up from HMS Hibernia - When taken: 27 Jul 1813 - Date Received: 11 Sep 1814 - From what ship: HMT Freya, Chatham - Born: Boston - Age: 43 - Released on 28 May 1815.

Smith, John - Seaman - Nbr: 2815 - How taken: Impressed at London - When taken: 24 Jul 1813 - Date Received: 24 Aug 1814 - What ship: HMT Liverpool, Chatham - Born: Salem - Age: 37 - Released on 19 May 1815.

Smith, John - Boy - Nbr: 2849 - Prize: Sea Nymphe - Ship Type: MV - How taken: HMS Thraster - When taken: 4 Mar 1813 - Where taken: North Sea - Date Received: 24 Aug 1814 - From what ship: HMT Alpheus, Chatham - Born: New York - Age: 18 - Race: Black - Released on 19 May 1815.

Smith, John - Seaman - Nbr: 2630 - How taken: Gave himself up from HMS Ville de Paris - When taken: 22 Apr 1813 - Date Received: 21 Aug 1814 - From what ship: HMT Freya, Chatham - Born: Massachusetts - Age: 28 - Released on 3 May 1815.

Smith, John - Seaman - Nbr: 3044 - How taken: Gave himself up from HMS Galatea - When taken: 24 Aug 1814 - Date Received: 2 Sep 1814 - From what ship: HMT Sultan - Born: Staten Island - Age: 40 - Released on 28 Apr 1815.

Smith, John - Seaman - Nbr: 2891 - How taken: Gave himself up from HMS Union - When taken: 27 May 1813 - Date Received: 24 Aug 1814 - From what ship: HMT Alpheus, Chatham - Born: Connecticut - Age: 34 - Released on 19 May 1815.

Smith, John - Seaman - Nbr: 3457 - Prize: Prize to Chasseur - How taken: HMS Tartarus - When taken: 1 Sep 1814 - Where taken: Atlantic - Date Received: 19 Sep 1814 - From what ship: HMT Salvador del Mundo - Born: Staten Island - Age: 22 - Released on 4 Jun 1815.

Smith, John - Seaman - Nbr: 3439 - How taken: Gave himself up from HMS Clorinda - When taken: 18 Dec 1813 - Date Received: 13 Sep 1814 - From what ship: HMT Niobe, Chatham - Born: Philadelphia - Age: 30 - Race: Black - Released on 28 May 1815.

Smith, John - Seaman - Nbr: 3006 - How taken: Gave himself up from HMS Portia - When taken: Dec 1814 - Date Received: 29 Aug 1814 - What ship: HMT Bittern - Born: Charleston - Age: 29 - Released on 28 May 1815.

Smith, John D. - Prize Master - Nbr: 1119 - Prize: Hope - Ship Type: P - How taken: HMS Sea Horse - When taken: 22 Mar 1814 - Where taken: at sea - Date Received: 10 May 1814 - From what ship: Plymouth - Born: Stonington - Age: 27 - Released on 25 Apr 1815.

Smith, Joseph - Seaman - Nbr: 5658 - How taken: Gave himself up from HMS Impregnable - Date Received: 24 Dec 1814 - From what ship: HMT Impregnable - Born: Virginia - Age: 35 - Released on 3 Jul 1815.

Smith, Joseph - Seaman - Nbr: 3689 - Prize: Alfred - Ship Type: P - How taken: HMS Epervier - When taken: 23 Feb 1812 - Where taken: off Newfoundland - Date Received: 30 Sep 1814 - From what ship: HMT President - Born: Newbury - Age: 24 - Released on 4 Jun 1815.

Smith, Kilby - 2nd Mate - Nbr: 3264 - Prize: Derby - Ship Type: MV - How taken: HMS Nereus - When taken: 4 Feb 1813 - Where taken: off Cape of Good Hope - Date Received: 11 Sep 1814 - From what ship: HMT Freya, Chatham - Born: Boston - Age: 27 - Released on 25 Sep 1814.

Smith, Monday - Seaman - Nbr: 5336 - Prize: Elbridge Gerry - Ship Type: P - How taken: HMS Crescent - When taken: 16 Sep 1813 - Where taken: at sea - Date Received: 31 Oct 1814 - From what ship: HMT Leyden, Chatham - Born: Georgetown - Age: 31 - Race: Black - Released on 29 Jun 1815.

Smith, Nicholas - Seaman - Nbr: 4958 - Prize: Herald - Ship Type: P - How taken: HMS Endymion - When taken:

American Prisoners of War Held at Dartmoor during the War of 1812

## Alphabetical listing of names

15 Aug 1814 - Where taken: off Nantucket - Date Received: 28 Oct 1814 - From what ship: HMT Alkbar, Halifax - Born: Richmond - Age: 25 - Died on 9 Jan 1815 from variola.

Smith, Oakley - Seaman - Nbr: 6049 - Prize: Cato - Ship Type: MV - How taken: HMS Forth - When taken: 2 Sep 1814 - Where taken: off Sandy Hook - Date Received: 28 Dec 1814 - From what ship: HMT Penelope - Born: Long Island - Age: 28 - Released on 3 Jul 1815.

Smith, Paul - Seaman - Nbr: 2840 - How taken: Gave himself up from HMS Armide - When taken: 8 Jun 1813 - Date Received: 24 Aug 1814 - From what ship: HMT Liverpool, Chatham - Born: Charleston - Age: 28 - Race: Black - Released on 19 May 1815.

Smith, Peter - Seaman - Nbr: 2620 - How taken: Gave himself up from HMS Christian 7th - When taken: 21 Dec 1812 - Date Received: 21 Aug 1814 - From what ship: HMT Freya, Chatham - Born: New York - Age: 25 - Race: Black - Released on 26 Apr 1815.

Smith, Richard - Seaman - Nbr: 4194 - Prize: General Kempt - Ship Type: P - How taken: HMS Foxhound - When taken: 18 Dec 1813 - Where taken: Lat 48'60" Long 5'7" - Date Received: 7 Oct 1814 - From what ship: HMT Niobe, Chatham - Born: Salem - Age: 24 - Died on 14 Apr 1815 from phthisis pulmonalis.

Smith, Richard - Gunner - Nbr: 2 - Prize: Cashier - Ship Type: LM - How taken: HMS Reindeer - When taken: 3 Feb 1813 - Where taken: Bay of Biscay - Date Received: 2 Apr 1813 - From what ship: Plymouth - Born: New York - Age: 23 - Released on 20 Apr 1815.

Smith, Richard - Gunner - Nbr: 641 - Prize: US Brig Argus - Ship Type: MW - How taken: HMS Pelican - When taken: 14 Aug 1813 - Where taken: Irish Channel - Date Received: 8 Sep 1813 - From what ship: Plymouth - Born: New York - Age: 23 - Released on 20 Apr 1815.

Smith, Richard - Seaman - Nbr: 4365 - How taken: Gave himself up from HMS Abuneance - When taken: 6 Oct 1814 - Date Received: 7 Oct 1814 - From what ship: HMT Salvador del Mundo, Halifax - Born: New York - Age: 27 - Released on 13 Jun 1815.

Smith, Robert - Seaman - Nbr: 5115 - Prize: Portsmouth - Ship Type: P - How taken: HMS Pylades - When taken: 13 Sep 1814 - Where taken: Long 65 - Date Received: 28 Oct 1814 - From what ship: HMT Alkbar, Halifax - Born: New York - Age: 21 - Race: Black - Released on 29 Jun 1815.

Smith, Samuel - Seaman - Nbr: 6435 - Prize: Chance - Ship Type: P - How taken: HMS Statira - When taken: 1 Apr 1814 - Where taken: Lat 38 Long 24 - Date Received: 3 Mar 1815 - From what ship: HMT Ganges, Plymouth - Born: Norfolk - Age: 20 - Released on 11 Jul 1815.

Smith, Samuel - Seaman - Nbr: 2432 - Prize: US Sloop Frolic - Ship Type: MW - How taken: HMS Orpheus - When taken: 20 Apr 1814 - Where taken: off Cuba - Date Received: 16 Aug 1814 - From what ship: HMT Queen, Halifax - Born: Boston - Age: 21 - Released on 3 May 1815.

Smith, Samuel - Seaman - Nbr: 3787 - Prize: James - Ship Type: Prize - How taken: Rebecca (LM) - When taken: 13 Apr 1814 - Where taken: off Delaware - Date Received: 5 Oct 1814 - From what ship: HMT President, Halifax - Born: New Jersey - Age: 13 - Released on 9 Jun 1815.

Smith, Shapley - Passenger - Nbr: 1198 - Prize: Ocean - Ship Type: P - How taken: HMS Achates - When taken: 14 Jun 1810 - Where taken: Channel - Date Received: 4 Jun 1814 - From what ship: Dartmouth - Born: Baltimore - Age: 29 - Escaped on 8 Aug 1814.

Smith, Thomas - Seaman - Nbr: 5583 - Prize: Alexander - Ship Type: Prize - How taken: HMS Wasp - When taken: 14 Sep 1814 - Where taken: off Cape Sable Island (Canada) - Date Received: 17 Dec 1814 - From what ship: HMT Loire, Halifax - Born: Gothenburg - Age: 24 - Released on 1 Jul 1815.

Smith, Thomas - Seaman - Nbr: 4473 - Prize: Enterprize - Ship Type: P - How taken: HMS Tenedos - When taken: 21 May 1813 - Where taken: off Cape Cod - Date Received: 8 Oct 1814 - From what ship: HMT Leyden, Chatham - Born: Massachusetts - Age: 37 - Released on 15 Jun 1815.

Smith, Thomas - Seaman - Nbr: 4735 - How taken: Gave himself up from HMS Rota - When taken: 24 Feb 1814 - Date Received: 9 Oct 1814 - From what ship: HMT Freya, Chatham - Born: Virginia - Age: 26 - Race: Mulatto - Released on 15 Jun 1815.

Smith, Thomas - Seaman - Nbr: 1639 - Prize: Henry Clements - Ship Type: MV - How taken: HMS Orestes - When taken: 15 Apr 1813 - Where taken: Bay of Biscay - Date Received: 23 Jun 1814 - From what ship: Stapleton - Born: Massachusetts - Age: 18 - Released on 1 May 1815.

Smith, Thomas - Boatswain - Nbr: 1640 - Prize: Paul Jones - Ship Type: P - How taken: HMS Leonidas - When taken: 23 May 1813 - Where taken: Channel - Date Received: 23 Jun 1814 - From what ship: Stapleton - Born: New York - Age: 30 - Released on 11 Jul 1815.

Smith, Thomas - Seaman - Nbr: 4450 - Prize: Dart - Ship Type: LM - How taken: HMS Niger - When taken: 13 Nov 1815 - Where taken: at sea - Date Received: 8 Oct 1814 - From what ship: HMT Leyden, Chatham - Born:

American Prisoners of War Held at Dartmoor during the War of 1812

## Alphabetical listing of names

Massachusetts - Age: 25 - Released on 14 Jun 1815.

Smith, Thomas - Seaman - Nbr: 2852 - How taken: Gave himself up from HMS Blake - When taken: 10 Dec 1812 - Date Received: 24 Aug 1814 - From what ship: HMT Alpheus, Chatham - Born: New York - Age: 28 - Race: Black - Released on 27 Apr 1815.

Smith, Thomas - Seaman - Nbr: 2968 - How taken: Sent into custody from a West Indiaman - When taken: 13 Oct 1813 - Date Received: 24 Aug 1814 - From what ship: HMT Hannibal - Born: New York - Age: 25 - Released on 19 May 1815.

Smith, Thomas - Seaman - Nbr: 4633 - How taken: Gave himself up from HMS Raleigh - When taken: 7 Jan 1813 - Date Received: 9 Oct 1814 - From what ship: HMT Leyden, Chatham - Born: Somerset - Age: 23 - Released on 15 Jun 1815.

Smith, Titus - Seaman - Nbr: 5353 - Prize: Pilot - Ship Type: LM - How taken: Victoria (Privateer) - When taken: 28 Jan 1814 - Where taken: off Bordeaux - Date Received: 31 Oct 1814 - From what ship: HMT Leyden, Chatham - Born: New York - Age: 23 - Race: Black - Released on 1 Jul 1815.

Smith, Turpin - Seaman - Nbr: 1761 - Prize: Cygnet - Ship Type: Prize - How taken: HMS Statira - When taken: 8 May 1814 - Where taken: at sea - Date Received: 20 Jul 1814 - From what ship: HMS Milford, Plymouth - Born: Rhode Island - Age: 25 - Released on 1 May 1815.

Smith, W. B. - Cook - Nbr: 4602 - Prize: Sally - Ship Type: MV - How taken: HMS Derwent - When taken: 21 Jan 1814 - Where taken: Grand Banks - Date Received: 9 Oct 1814 - From what ship: HMT Leyden, Chatham - Born: Salem - Age: 17 - Released on 15 Jun 1815.

Smith, W. L. - Mate - Nbr: 2722 - Prize: Maria - Ship Type: MV - How taken: HMS Thais - When taken: 29 Jan 1813 - Date Received: 21 Aug 1814 - From what ship: HMT Freya, Chatham - Born: New York - Age: 49 - Released on 2 Nov 1814.

Smith, William - Seaman - Nbr: 1039 - Prize: US Brig Argus - Ship Type: MW - How taken: HMS Pelican - When taken: 16 Aug 1813 - Where taken: Irish Channel - Date Received: 10 May 1814 - From what ship: Plymouth - Born: Charlestown - Age: 30 - Sent to Dartmouth on 19 Oct 1814.

Smith, William - Seaman - Nbr: 1232 - How taken: Sent into custody from HMS Nautilus - When taken: 20 Dec 1812 - Date Received: 14 Jun 1814 - What ship: Mill Prison (Plymouth, England) - Born: Rhode Island - Age: 33 - Released on 26 Apr 1815.

Smith, William - Seaman - Nbr: 740 - Prize: US Brig Argus - Ship Type: MW - How taken: HMS Pelican - When taken: 14 Aug 1813 - Where taken: Irish Channel - Date Received: 3 Nov 1813 - From what ship: Plymouth - Born: Charlestown - Age: 25 - Sent to Dartmouth on 2 Nov 1814.

Smith, William - Cook - Nbr: 219 - Prize: Pert - Ship Type: MV - How taken: HMS Warspite - When taken: 1 Mar 1813 - Where taken: off Basque Roads (France) - Date Received: 2 Apr 1813 - From what ship: Plymouth - Born: New York - Age: 26 - Released on 20 Apr 1815.

Smith, William - Seaman - Nbr: 49 - Prize: Terrible - Ship Type: MV - How taken: HMS Foxhound - When taken: 8 Feb 1813 - Where taken: Channel - Date Received: 2 Apr 1813 - From what ship: Plymouth - Born: New York - Age: 18 - Sent to Dartmouth on 15 Jun 1813.

Smith, William - Seaman - Nbr: 459 - How taken: Apprehended at Gibraltar - When taken: Oct 1812 - Date Received: 8 Sep 1813 - From what ship: Plymouth - Born: Baltimore - Age: 30 - Released on 26 Apr 1815.

Smith, William - Seaman - Nbr: 3516 - Prize: Aeolus - Ship Type: Prize - How taken: HMS Warspite - When taken: 2 Sep 1814 - Where taken: off Newfoundland - Date Received: 28 Sep 1814 - From what ship: HMT Salvador del Mundo - Born: Connecticut - Age: 25 - Released on 4 Jun 1815.

Smith, William - Seaman - Nbr: 3579 - Prize: William - Ship Type: MV - How taken: HMS Hasty - When taken: 14 May 1814 - Where taken: off Charleston - Date Received: 30 Sep 1814 - From what ship: HMT Sybella - Born: New York - Age: 31 - Released on 4 Jun 1815.

Smithers, Richard - Seaman - Nbr: 4333 - Prize: Flash - Ship Type: LM - How taken: HMS Acasta and HMS Loire - When taken: 8 Feb 1814 - Where taken: off NY - Date Received: 7 Oct 1814 - From what ship: HMT Salvador del Mundo, Halifax - Born: New York - Age: 17 - Race: Negro - Died on 6 Mar 1815 from variola.

Snade, William - Seaman - Nbr: 2245 - Prize: Carbineer - Ship Type: Prize - How taken: HMS Ringdove - When taken: 24 Apr 1814 - Where taken: off Bermuda - Date Received: 16 Aug 1814 - From what ship: HMS Dublin, Halifax - Born: Philadelphia - Age: 23 - Released on 3 May 1815.

Snalt, George - Seaman - Nbr: 6167 - Prize: Lion - Ship Type: P - How taken: HMS Granicus - When taken: 2 Dec 1814 - Where taken: off Lisbon - Date Received: 18 Jan 1815 - From what ship: HMT Impregnable - Born: Nantes - Age: 22 - Released on 5 Jul 1815.

Snate, George - Seaman - Nbr: 466 - Prize: Meteor - Ship Type: MV - How taken: HMS Briton - When taken: 12

# American Prisoners of War Held at Dartmoor during the War of 1812

## Alphabetical listing of names

Mar 1813 - Where taken: Bay of Biscay - Date Received: 8 Sep 1813 - From what ship: Plymouth - Born: New York - Age: 21 - Sent to Plymouth on 8 Jul 1814.

Snell, Shaderick - Fifer - Nbr: 5842 - Prize: 1st US Rifles - Ship Type: Troops - How taken: British Army - When taken: 17 Sep 1814 - Where taken: Sackets Harbor - Date Received: 26 Dec 1814 - From what ship: HMT Argo - Born: Rhode Island - Age: 19 - Died on 16 Mar 1815 from variola.

Sniffon, John - Carpenter's Mate - Nbr: 1726 - Prize: US Brig Argus - Ship Type: MW - How taken: HMS Pelican - When taken: 14 Aug 1813 - Where taken: Irish Channel - Date Received: 8 Jul 1814 - From what ship: Mill Prison (Plymouth, England) - Born: New York - Age: 30 - Sent to Dartmouth on 18 Oct 1814.

Snow, Colger - Seaman - Nbr: 4562 - How taken: Gave himself up from HMS Shannon - When taken: 28 Aug 1814 - Date Received: 8 Oct 1814 - From what ship: HMT Leyden, Chatham - Born: Massachusetts - Age: 23 - Released on 15 Jun 1815.

Snow, Daniel - Seaman - Nbr: 5174 - Prize: Vivid - Ship Type: P - How taken: HMS Nymphe - When taken: 20 Apr 1813 - Where taken: off Cape Cod - Date Received: 31 Oct 1814 - From what ship: HMT Mermaid, Chatham - Born: Haiti - Age: 19 - Released on 29 Jun 1815.

Snow, Daniel - Seaman - Nbr: 6330 - Prize: Prince de Neufchatel - Ship Type: P - How taken: Leander (Newcastle Acasta) - When taken: 20 Dec 1814 - Where taken: Lat 38 Long 56 - Date Received: 19 Feb 1815 - From what ship: HMT Ganges, Plymouth - Born: Litchfield - Age: 19 - Released on 5 Jul 1815.

Snow, James - Seaman - Nbr: 3429 - Prize: Industry - Ship Type: P - How taken: HMS Heron - When taken: 3 Nov 1813 - Where taken: off Halifax - Date Received: 13 Sep 1814 - From what ship: HMT Niobe, Chatham - Born: Salem - Age: 24 - Released on 28 May 1815.

Snow, Thomas - Seaman - Nbr: 3287 - Prize: Juliana Smith - Ship Type: P - How taken: HMS Nymphe - When taken: 11 May 1813 - Where taken: off Cape Sable Island (Canada) - Date Received: 13 Sep 1814 - From what ship: HMT Niobe, Chatham - Born: Massachusetts - Age: 25 - Released on 28 May 1815.

Snowdon, Jacob - Seaman - Nbr: 713 - Prize: Ned - Ship Type: LM - How taken: HMS Royalist - When taken: 6 Sep 1813 - Where taken: Bay of Biscay - Date Received: 27 Sep 1813 - From what ship: Plymouth - Born: Jersey - Age: 34 - Race: Black - Released on 26 Apr 1815.

Snowdon, John - Seaman - Nbr: 2472 - Prize: US Sloop Frolic - Ship Type: MW - How taken: HMS Orpheus - When taken: 20 Apr 1814 - Where taken: off Cuba - Date Received: 16 Aug 1814 - From what ship: HMT Queen, Halifax - Born: Salem - Age: 24 - Released on 3 May 1815.

Snyder, William - Seaman - Nbr: 1315 - Prize: Barracuda - Ship Type: MV - How taken: HMS Barracouta - Date Received: 14 Jun 1814 - From what ship: Mill Prison (Plymouth, England) - Born: Maryland - Age: 32 - Released on 28 Apr 1815.

Sobier, Bernard - Seaman - Nbr: 4994 - Prize: Invincible - Ship Type: LM - How taken: HMS Armide - When taken: 15 Aug 1814 - Where taken: off Nantucket - Date Received: 28 Oct 1814 - From what ship: HMT Alkbar, Halifax - Born: Bayonne - Age: 34 - Released on 21 Jun 1815.

Soderback, Andrew - Seaman - Nbr: 962 - Prize: Siro - Ship Type: LM - How taken: HMS Pelican - When taken: 13 Jan 1814 - Where taken: at sea - Date Received: 31 Jan 1814 - From what ship: Plymouth - Born: Gothenburg - Age: 24 - Released on 6 May 1814.

Soderbury, John - Boy - Nbr: 2188 - Prize: Diomede - Ship Type: P - How taken: HMS Rifleman - When taken: 28 Jun 1813 - Where taken: off Halifax - Date Received: 16 Aug 1814 - From what ship: HMS Dublin, Halifax - Born: Boston - Age: 14 - Released on 2 May 1815.

Soie, Henry - Seaman - Nbr: 1765 - Prize: Cygnet - Ship Type: Prize - How taken: HMS Statira - When taken: 8 May 1814 - Where taken: at sea - Date Received: 20 Jul 1814 - From what ship: HMS Milford, Plymouth - Born: Moravia - Age: 25 - Released on 21 Oct 1814.

Solden, Charles - Lieutenant - Nbr: 2562 - Prize: Rattlesnake - Ship Type: P - How taken: HMS Hyperion - When taken: 25 Jun 1814 - Where taken: off Cape Finisterre - Date Received: 21 Aug 1814 - From what ship: HMS Hyperion - Born: Connecticut - Age: 30 - Sent to Ashburton (England) on 8 Dec 1814.

Sole, Edward - Seaman - Nbr: 1260 - Prize: Adeline - Ship Type: MV - How taken: HMS Magicienne - When taken: 16 Mar 1814 - Where taken: off Cape Finisterre - Date Received: 14 Jun 1814 - From what ship: Mill Prison (Plymouth, England) - Born: Freeport - Age: 30 - Released on 28 Apr 1815.

Sole, Elias - Seaman - Nbr: 1262 - Prize: Adeline - Ship Type: MV - How taken: HMS Magicienne - When taken: 16 Mar 1814 - Where taken: off Cape Finisterre - Date Received: 14 Jun 1814 - From what ship: Mill Prison (Plymouth, England) - Born: Freeport - Age: 22 - Released on 28 Apr 1815.

Sole, John - Prize Master - Nbr: 3565 - Prize: Hawk - Ship Type: P - How taken: HMS Pique - When taken: 26 Apr 1814 - Where taken: off Bermuda - Date Received: 30 Sep 1814 - From what ship: HMT Sybella - Born:

American Prisoners of War Held at Dartmoor during the War of 1812

## Alphabetical listing of names

Massachusetts - Age: 24 - Released on 4 Jun 1815.
Sole, Thomas - Seaman - Nbr: 1265 - Prize: Adeline - Ship Type: MV - How taken: HMS Magicienne - When taken: 16 Mar 1814 - Where taken: off Cape Finisterre - Date Received: 14 Jun 1814 - From what ship: Mill Prison (Plymouth, England) - Born: Freeport - Age: 34 - Released on 28 Apr 1815.
Soley, Nathaniel - Seaman - Nbr: 5223 - Prize: Grand Turk - Ship Type: P - How taken: HMS Dover - When taken: 27 Jun 1813 - Where taken: off Newfoundland - Date Received: 31 Oct 1814 - From what ship: HMT Mermaid, Chatham - Born: Hardwick - Age: 27 - Released on 29 Jun 1815.
Solomon, John - Seaman - Nbr: 6035 - Prize: Regent - Ship Type: LM - How taken: HMS Forth - When taken: 19 Sep 1814 - Where taken: off Egg Harbor (New Jersey) - Date Received: 28 Dec 1814 - From what ship: HMT Penelope - Born: India - Age: 50 - Race: Mulatto - Released on 3 Jul 1815.
Somerville, Charles - Seaman - Nbr: 354 - Prize: Thrasher - Ship Type: P - How taken: HMS Magicienne - When taken: 17 Jan 1813 - Where taken: off Western Islands (England) - Date Received: 1 Jul 1813 - From what ship: Plymouth - Born: Baltimore - Age: 17 - Sent to Dartmouth on 30 Jul 1813.
Sone, John - Seaman - Nbr: 1299 - Prize: John & Frances - Ship Type: Prize - How taken: HMS Sterling Castle - When taken: 10 May 1814 - Where taken: off Cape Clear - Date Received: 14 Jun 1814 - From what ship: Mill Prison (Plymouth, England) - Born: Trieste - Age: 27 - Released on 28 Apr 1815.
Sonnett, James - Seaman - Nbr: 4465 - Prize: Governor Plumer - Ship Type: P - How taken: HMS Shamrock - When taken: 4 Mar 1813 - Where taken: at sea - Date Received: 8 Oct 1814 - From what ship: HMT Leyden, Chatham - Born: Massachusetts - Age: 24 - Released on 15 Jun 1815.
Soreby, Robert - Seaman - Nbr: 1157 - Prize: Dorothea - Ship Type: MV - How taken: HMS Teazer - When taken: 26 Mar 1814 - Where taken: off Cork - Date Received: 10 May 1814 - From what ship: Plymouth - Born: Philadelphia - Age: 21 - Released on 28 Apr 1815.
Southcomb, Peter - 1st Mate - Nbr: 6390 - Prize: Nellenville - Ship Type: MV - How taken: HMS Onyx - When taken: 25 Dec 1814 - Where taken: off San Domingo - Date Received: 3 Mar 1815 - From what ship: HMT Ganges, Plymouth - Born: Philadelphia - Age: 27 - Sent to Ashburton (England) on 10 Mar 1815.
Southcombe, Kemp - Seaman - Nbr: 6062 - Prize: Chasseur - Ship Type: P - How taken: HMS Recruit - When taken: 6 Apr 1814 - Where taken: Lat 29 Long 76 - Date Received: 28 Dec 1814 - From what ship: HMT Penelope - Born: Virginia - Age: 21 - Released on 3 Jul 1815.
Southwich, Israel - Seaman - Nbr: 1797 - How taken: Gave himself up from HMS Swiftsure - When taken: 26 Dec 1813 - Date Received: 20 Jul 1814 - From what ship: HMS Milford, Plymouth - Born: Boston - Age: 22 - Released on 1 May 1815.
Southwick, George - Seaman - Nbr: 2989 - Prize: Frolic - Ship Type: P - How taken: HMS Heron - When taken: 25 Jan 1814 - Where taken: off St. Thomas - Date Received: 29 Aug 1814 - From what ship: HMT Bittern - Born: Salem - Age: 54 - Released on 28 May 1815.
Spangle, Frederick - Seaman - Nbr: 734 - Prize: Ned - Ship Type: LM - How taken: HMS Royalist - When taken: 6 Sep 1813 - Where taken: Bay of Biscay - Date Received: 27 Sep 1813 - From what ship: Plymouth - Born: Carolina - Age: 17 - Released on 26 Apr 1815.
Spanow, W. T. - Seaman - Nbr: 3521 - Prize: Sabine - Ship Type: P - How taken: HMS Conquistador - When taken: 3 Aug 1814 - Where taken: off the Banks - Date Received: 28 Sep 1814 - From what ship: HMT Salvador del Mundo - Born: North Carolina - Age: 33 - Released on 4 Jun 1815.
Spark, Samuel - Seaman - Nbr: 1141 - Prize: Mary - Ship Type: P - How taken: HMS Crocodile - When taken: 6 Aug 1813 - Where taken: off Corsuma - Date Received: 10 May 1814 - From what ship: Plymouth - Born: Philadelphia - Age: 19 - Released on 28 Apr 1815.
Sparkes, John - Gunner - Nbr: 6369 - Prize: US Brig Syren - Ship Type: MW - How taken: HMS Medway - When taken: 12 Jul 1814 - Where taken: off Cape of Good Hope - Date Received: 24 Feb 1815 - From what ship: HMT Ganges, Plymouth - Born: Philadelphia - Age: 36 - Released on 6 Jul 1815.
Sparkes, Thomas - Seaman - Nbr: 3283 - Prize: Montgomery - Ship Type: P - How taken: HMS Nymphe - When taken: 5 May 1813 - Where taken: off Cape Ann - Date Received: 13 Sep 1814 - From what ship: HMT Niobe, Chatham - Born: Salem - Age: 38 - Released on 28 May 1815.
Sparks, Robert - Seaman - Nbr: 1071 - Prize: Fair American - Ship Type: MV - How taken: HMS Andromache - When taken: 19 Jan 1814 - Where taken: Bay of Biscay - Date Received: 10 May 1814 - From what ship: Plymouth - Born: New York - Age: 36 - Race: Black - Released on 27 Apr 1815.
Sparrow, James - Seaman - Nbr: 1588 - Prize: Price - Ship Type: MV - How taken: HMS Pyramus - When taken: 6 Apr 1813 - Where taken: Bay of Biscay - Date Received: 23 Jun 1814 - From what ship: Stapleton - Born: Virginia - Age: 36 - Released on 1 May 1815.

American Prisoners of War Held at Dartmoor during the War of 1812

## Alphabetical listing of names

Sparrow, John - Seaman - Nbr: 1955 - How taken: Gave himself up from HMS Christian 7th - When taken: 19 May 1813 - Date Received: 3 Aug 1814 - From what ship: HMS Lyffey, Chatham Depot - Born: Norfolk - Age: 22 - Released on 2 May 1815.

Sparrow, Joseph - Cook - Nbr: 2906 - How taken: Gave himself up from HMS Ministrel - When taken: 5 Jul 1813 - Date Received: 24 Aug 1814 - From what ship: HMT Alpheus, Chatham - Born: Virginia - Age: 43 - Race: Black - Released on 19 May 1815.

Spear, Joseph - Seaman - Nbr: 6539 - Prize: US Brig Syren - Ship Type: MW - How taken: HMS Medway - When taken: 12 Jul 1814 - Where taken: off Cape of Good Hope - Date Received: 7 Mar 1815 - From what ship: HMT Ganges, Plymouth - Born: Boston - Age: 22 - Released on 11 Jul 1815.

Spear, Joseph - Mate - Nbr: 2103 - How taken: Impressed at London - When taken: 12 May 1813 - Date Received: 3 Aug 1814 - What ship: HMS Bittern, Chatham Depot - Born: Boston - Age: 23 - Released on 2 May 1815.

Spears, Samuel - Carpenter - Nbr: 3 - Prize: Cashier - Ship Type: LM - How taken: HMS Reindeer - When taken: 3 Feb 1813 - Where taken: Bay of Biscay - Date Received: 2 Apr 1813 - From what ship: Plymouth - Born: Boston - Age: 25 - Sent to Dartmouth on 30 Jul 1813.

Spellard, Robert - Marine - Nbr: 2502 - Prize: US Sloop Frolic - Ship Type: MW - How taken: HMS Orpheus - When taken: 20 Apr 1814 - Where taken: off Cuba - Date Received: 16 Aug 1814 - From what ship: HMT Queen, Halifax - Born: Pennsylvania - Age: 30 - Released on 3 May 1815.

Spencer, Job - Seaman - Nbr: 3753 - Prize: Prize of the Yankee - Ship Type: Prize - How taken: HMS Poictiers - When taken: 30 Sep 1813 - Where taken: off Cape Sable Island (Canada) - Date Received: 30 Sep 1814 - From what ship: HMT President, Halifax - Born: Rhode Island - Age: 18 - Released on 4 Jun 1815.

Spencer, John - Seaman - Nbr: 4653 - How taken: Impressed from MV Sampson - When taken: 1 Oct 1813 - Date Received: 9 Oct 1814 - From what ship: HMT Leyden, Chatham - Born: Exeter - Age: 24 - Released on 15 Jun 1815.

Spencer, Leonard - Seaman - Nbr: 2776 - Prize: Tiger - Ship Type: MV - How taken: HMS Scylla - When taken: 22 May 1813 - Where taken: Bay of Biscay - Date Received: 24 Aug 1814 - From what ship: HMT Liverpool, Chatham - Born: Connecticut - Age: 32 - Released on 19 May 1815.

Spencer, Robert - Gunner's Mate - Nbr: 2380 - Prize: US Sloop Frolic - Ship Type: MW - How taken: HMS Orpheus - When taken: 20 Apr 1814 - Where taken: off Cuba - Date Received: 16 Aug 1814 - From what ship: HMT Queen, Halifax - Born: Maryland - Age: 42 - Sent to Dartmouth on 19 Oct 1814.

Spenny, Nathaniel - Seaman - Nbr: 4677 - Prize: Portsmouth Packet - Ship Type: Prize - How taken: HMS Fantome - When taken: 5 Oct 1813 - Where taken: off Portland - Date Received: 9 Oct 1814 - From what ship: HMT Leyden, Chatham - Born: Massachusetts - Age: 19 - Escaped on 1 Jun 1815.

Spern, Elijah - Seaman - Nbr: 4193 - Prize: General Kempt - Ship Type: P - How taken: HMS Foxhound - When taken: 18 Dec 1813 - Where taken: Lat 48'60" Long 5'7" - Date Received: 7 Oct 1814 - From what ship: HMT Niobe, Chatham - Born: Boston - Age: 18 - Released on 13 Jun 1815.

Spicer, Alexander - Seaman - Nbr: 4554 - How taken: Gave himself up from HMS L'Aigle - Date Received: 8 Oct 1814 - From what ship: HMT Leyden, Chatham - Born: Dover - Age: 35 - Race: Mulatto - Released on 15 Jun 1815.

Spick, John - Seaman - Nbr: 5218 - Prize: Rolla - Ship Type: P - How taken: HMS Victorious - When taken: 2 Jun 1813 - Where taken: off Halifax - Date Received: 31 Oct 1814 - From what ship: HMT Mermaid, Chatham - Born: Baltimore - Age: 18 - Released on 29 Jun 1815.

Spiers, Nathaniel - Seaman - Nbr: 4574 - Prize: Bunker Hill - Ship Type: P - How taken: HMS Pomone - When taken: 2 Mar 1814 - Where taken: Channel - Date Received: 9 Oct 1814 - From what ship: HMT Leyden, Chatham - Born: Providence - Age: 23 - Race: Black - Released on 11 Jul 1815.

Spiller, Moses - Prize master - Nbr: 6358 - Prize: Prince de Neufchatel - Ship Type: P - How taken: Leander (Newcastle Acasta) - When taken: 20 Dec 1814 - Where taken: Lat 38 Long 56 - Date Received: 19 Feb 1815 - From what ship: HMT Ganges, Plymouth - Born: Salisbury - Age: 33 - Released on 11 Jul 1815.

Spince, William - Cook - Nbr: 1191 - Prize: L'Aigle - How taken: HMS Phoenix - When taken: 20 May 1809 - Date Received: 4 Jun 1814 - From what ship: Dartmouth - Born: New York - Age: 28 - Race: Black - Released on 11 Jul 1815.

Spires, Thomas - 2nd Mate - Nbr: 725 - Prize: Ned - Ship Type: LM - How taken: HMS Royalist - When taken: 6 Sep 1813 - Where taken: Bay of Biscay - Date Received: 27 Sep 1813 - From what ship: Plymouth - Born: Virginia - Age: 39 - Released on 26 Apr 1815.

Spouding, Joseph - Seaman - Nbr: 3318 - Prize: Wasp - Ship Type: P - How taken: HMS Bream - When taken: 10 Jun 1813 - Where taken: off Halifax - Date Received: 13 Sep 1814 - From what ship: HMT Niobe, Chatham

American Prisoners of War Held at Dartmoor during the War of 1812

## Alphabetical listing of names

- Born: Massachusetts - Age: 21 - Released on 28 May 1815.
Spragg, Henry - Seaman - Nbr: 4065 - Prize: Leicester - Ship Type: Prize - How taken: HMS Prometheus - When taken: Jun 1814 - Where taken: off Halifax - Date Received: 6 Oct 1814 - From what ship: HMT Chesapeake, Halifax - Born: Massachusetts - Age: 27 - Released on 13 Jun 1815.
Sprague, John - Seaman - Nbr: 5402 - How taken: Gave himself up from MV Martha - When taken: 2 Apr 1814 - Date Received: 31 Oct 1814 - From what ship: HMT Leyden, Chatham - Born: Calais - Age: 21 - Released on 1 Jul 1815.
Sprague, Stephen - Seaman - Nbr: 4795 - Prize: Martha - Ship Type: MV - How taken: HMS Eclipse - When taken: 24 Jan 1812 - Where taken: off Barbados - Date Received: 9 Oct 1814 - From what ship: HMT Freya, Halifax - Born: Machias - Age: 24 - Released on 11 Jul 1815.
Spratt, Thomas - Seaman - Nbr: 2583 - How taken: Gave himself up from HMS Ethlion - Date Received: 21 Aug 1814 - From what ship: HMT Freya, Chatham - Born: Virginia - Age: 32 - Released on 3 May 1815.
Spring, Seth - Seaman - Nbr: 5927 - Prize: Harlequin - Ship Type: P - How taken: HMS Bulwark - When taken: 23 Nov 1814 - Where taken: off Halifax - Date Received: 27 Dec 1814 - From what ship: HMT Penelope - Born: Brownfield - Age: 23 - Released on 20 Mar 1815.
Springle, James - Seaman - Nbr: 2233 - Prize: General Hart - Ship Type: P - How taken: HMS Sophie - When taken: 24 Apr 1814 - Where taken: off Bermuda - Date Received: 16 Aug 1814 - From what ship: HMS Dublin, Halifax - Born: Bath - Age: 22 - Released on 2 May 1815.
Springstern, Abraham - Seaman - Nbr: 1178 - Prize: Imperatrice Reine - Ship Type: P - How taken: HMS Hotspur - When taken: 13 May 1813 - Where taken: off Grochite - Date Received: 4 Jun 1814 - From what ship: Dartmouth - Born: New York - Age: 25 - Released on 28 Apr 1815.
Sproson, James - Carpenter - Nbr: 337 - Prize: Tiger - Ship Type: MV - How taken: HMS Scylla - When taken: 22 Mar 1813 - Where taken: Bay of Biscay - Date Received: 28 Jun 1813 - From what ship: Plymouth - Born: New York - Age: 26 - Released on 20 Apr 1815.
Sprust, Anthony - Seaman - Nbr: 6052 - Prize: William - Ship Type: Prize - How taken: HMS Armide - When taken: 11 Oct 1814 - Where taken: off Newport - Date Received: 28 Dec 1814 - From what ship: HMT Penelope - Born: Utica - Age: 34 - Released on 3 Jul 1815.
Squibb, Silas - Seaman - Nbr: 5976 - Prize: Halifax Packet - Ship Type: Prize - How taken: HMS Bulwark - When taken: 22 Sep 1814 - Where taken: Georges Bank - Date Received: 27 Dec 1814 - From what ship: HMT Penelope - Born: New London - Age: 21 - Died on 18 Mar 1815 from variola.
Squires, Prince - Seaman - Nbr: 1989 - How taken: Impressed at London - When taken: 22 Mar 1813 - Date Received: 3 Aug 1814 - From what ship: HMS Lyffey, Chatham Depot - Born: Montfort - Age: 34 - Released on 2 May 1815.
St. Martin, Joseph - Seaman - Nbr: 3835 - Prize: Dominique - Ship Type: LM - How taken: HMS Dotterel - When taken: 21 May 1814 - Where taken: off Charleston - Date Received: 5 Oct 1814 - From what ship: HMT Orpheus, Halifax - Born: Haiti - Age: 37 - Released on 9 Jun 1815.
St. Vincent, Stephen - Seaman - Nbr: 4309 - Prize: Vivid - Ship Type: MV - How taken: HMS Nymphe - When taken: 20 Apr 1813 - Where taken: off Cape Cod - Date Received: 7 Oct 1814 - From what ship: HMT Niobe, Chatham - Born: Boston - Age: 20 - Released on 4 Jun 1815.
Stacey, Benjamin - Seaman - Nbr: 4531 - Prize: Wolf Cove - Ship Type: MV - How taken: HMS Briton - When taken: 1 Dec 1812 - Where taken: off Brest - Date Received: 8 Oct 1814 - From what ship: HMT Leyden, Chatham - Born: Massachusetts - Age: 19 - Released on 27 Apr 1815.
Stacey, John - Seaman - Nbr: 3430 - Prize: Industry - Ship Type: P - How taken: HMS Heron - When taken: 3 Nov 1813 - Where taken: off Halifax - Date Received: 13 Sep 1814 - From what ship: HMT Niobe, Chatham - Born: Marblehead - Age: 37 - Released on 28 May 1815.
Stacey, Osmond C. - Seaman - Nbr: 3687 - Prize: Alfred - Ship Type: P - How taken: HMS Epervier - When taken: 23 Feb 1812 - Where taken: off Newfoundland - Date Received: 30 Sep 1814 - From what ship: HMT President - Born: Marblehead - Age: 19 - Released on 4 Jun 1815.
Stacey, Perry - Seaman - Nbr: 2755 - How taken: Gave himself up from HMS Blake - When taken: 10 Dec 1812 - Date Received: 24 Aug 1814 - From what ship: HMT Liverpool, Chatham - Born: Baltimore - Age: 23 - Released on 26 Apr 1815.
Stacey, Samuel - Seaman - Nbr: 4532 - Prize: Wolf Cove - Ship Type: MV - How taken: HMS Briton - When taken: 1 Dec 1812 - Where taken: off Brest - Date Received: 8 Oct 1814 - From what ship: HMT Leyden, Chatham - Born: Massachusetts - Age: 24 - Released on 27 Apr 1815.
Stacey, Stephen - Seaman - Nbr: 5750 - Prize: US Schooner Ohio - Ship Type: MW - How taken: British gunboats -

American Prisoners of War Held at Dartmoor during the War of 1812

## Alphabetical listing of names

When taken: 12 Aug 1814 - Where taken: Fort Erie - Date Received: 26 Dec 1814 - From what ship: HMT Argo - Born: Marblehead - Age: 35 - Died on 16 Mar 1815 from variola.

Stacey, William - Boy - Nbr: 3725 - Prize: Alfred - Ship Type: P - How taken: HMS Epervier - When taken: 23 Feb 1814 - Where taken: off Newfoundland - Date Received: 30 Sep 1814 - From what ship: HMT President, Halifax - Born: Marblehead - Age: 17 - Released on 4 Jun 1815.

Stacey, William - Seaman - Nbr: 1025 - Prize: Joseph - Ship Type: MV - How taken: HMS Royalist - When taken: 18 Jan 1814 - Where taken: Bay of Biscay - Date Received: 31 Jan 1814 - From what ship: Plymouth - Born: Marblehead - Age: 21 - Released on 27 Apr 1815.

Stackpole, Andrew - Seaman - Nbr: 6023 - Prize: McDonough - Ship Type: P - How taken: HMS Bacchante - When taken: 1 Nov 1814 - Where taken: Lat 42 Long 67 - Date Received: 28 Dec 1814 - From what ship: HMT Penelope - Born: Biddeford - Age: 30 - Released on 3 Jul 1815.

Stacy, Samuel S. - Boy - Nbr: 5978 - Prize: Cossack - Ship Type: Private - How taken: HMS Bulwark - When taken: 1 Nov 1814 - Where taken: Georges Bank - Date Received: 27 Dec 1814 - From what ship: HMT Penelope - Born: Marblehead - Age: 18 - Released on 3 Jul 1815.

Stafford, John - 2nd Mate - Nbr: 3975 - Prize: US Brig Rattlesnake - Ship Type: MW - How taken: HMS Leander - When taken: 13 Jul 1814 - Where taken: off Shelburne - Date Received: 6 Oct 1814 - From what ship: HMT Chesapeake, Halifax - Born: New York - Age: 54 - Released on 9 Jun 1815.

Stafford, John - Seaman - Nbr: 1058 - How taken: Impressed at Cork - Date Received: 10 May 1814 - From what ship: Plymouth - Born: Boston - Age: 29 - Released on 10 Apr 1815.

Stafford, John - Prentice - Nbr: 4288 - Prize: Pilot - Ship Type: LM - How taken: Victoria (Privateer) - When taken: 28 Jan 1814 - Where taken: off Bordeaux - Date Received: 7 Oct 1814 - From what ship: HMT Niobe, Chatham - Born: Virginia - Age: 30 - Race: Black - Released on 14 Jun 1815.

Stafford, Thomas - Seaman - Nbr: 4007 - Prize: US Brig Rattlesnake - Ship Type: MW - How taken: HMS Leander - When taken: 13 Jul 1814 - Where taken: off Shelburne - Date Received: 6 Oct 1814 - From what ship: HMT Chesapeake, Halifax - Born: Massachusetts - Age: 21 - Released on 9 Jun 1815.

Stage, William - Seaman - Nbr: 3169 - Prize: Hindustan - Ship Type: MV - How taken: HMS Tenedos - When taken: 25 Jun 1813 - Where taken: off Lisbon - Date Received: 11 Sep 1814 - From what ship: HMT Freya, Chatham - Born: Charleston - Age: 26 - Released on 28 May 1815.

Stagg, J. Bolton - Seaman - Nbr: 1548 - Prize: Zebra - Ship Type: LM - How taken: HMS Pyramus - When taken: 20 Apr 1813 - Where taken: Bay of Biscay - Date Received: 23 Jun 1814 - From what ship: Stapleton - Born: New York - Age: 24 - Released on 1 May 1815.

Staggs, Henry - Seaman - Nbr: 2614 - How taken: Gave himself up from HMS Romulus - When taken: 14 Aug 1812 - Date Received: 21 Aug 1814 - From what ship: HMT Freya, Chatham - Born: Harford - Age: 32 - Released on 26 Apr 1815.

Staines, Samuel - Seaman - Nbr: 5549 - Prize: Hornet - Ship Type: MV - How taken: HMS Surprize - When taken: 19 Aug 1814 - Where taken: Lat 35 Long 24 - Date Received: 17 Dec 1814 - From what ship: HMT Loire, Halifax - Born: Gloucester - Age: 28 - Released on 1 Jul 1815.

Staitmand, James - Seaman - Nbr: 155 - How taken: Gave himself up from HMS Lavinia - Date Received: 2 Apr 1813 - From what ship: Plymouth - Born: Salisbury - Age: 25 - Released on 20 Apr 1815.

Stamford, James - Seaman - Nbr: 3270 - How taken: Gave himself up from HMS Berwick - Date Received: 11 Sep 1814 - From what ship: HMT Freya, Chatham - Born: Philadelphia - Age: 31 - Released on 28 May 1815.

Stanbeck, Jacob - Seaman - Nbr: 2516 - Prize: Snap Dragon - Ship Type: P - How taken: HMS Martin - When taken: 10 Jun 1814 - Where taken: off Halifax - Date Received: 16 Aug 1814 - From what ship: HMT Queen, Halifax - Born: Sweden - Age: 24 - Released on 3 May 1815.

Stanfield, George - Seaman - Nbr: 4384 - Prize: Globe - Ship Type: P - How taken: HMS Spartan - When taken: 13 Jun 1813 - Where taken: off Delaware - Date Received: 8 Oct 1814 - From what ship: HMT Leyden, Chatham - Born: Philadelphia - Age: 22 - Released on 14 Jun 1815.

Stanfield, Thomas - Seaman - Nbr: 6030 - Prize: Regent - Ship Type: LM - How taken: HMS Forth - When taken: 19 Sep 1814 - Where taken: off Egg Harbor (New Jersey) - Date Received: 28 Dec 1814 - From what ship: HMT Penelope - Born: Philadelphia - Age: 28 - Released on 3 Jul 1815.

Stanford, Nicholas - Seaman - Nbr: 5825 - Prize: US Schooner Tigress - Ship Type: MW - How taken: British gunboats - When taken: 2 Sep 1814 - Where taken: Lake Erie - Date Received: 26 Dec 1814 - From what ship: HMT Argo - Born: Massachusetts - Age: 20 - Released on 3 Jul 1815.

Stanley, Redmond - Seaman - Nbr: 2508 - Prize: Snap Dragon - Ship Type: P - How taken: HMS Martin - When taken: 10 Jun 1814 - Where taken: off Halifax - Date Received: 16 Aug 1814 - From what ship: HMT Queen,

American Prisoners of War Held at Dartmoor during the War of 1812

## Alphabetical listing of names

Halifax - Born: North Carolina - Age: 22 - Released on 3 May 1815.

Stanley, Samuel - Seaman - Nbr: 3578 - Prize: William - Ship Type: MV - How taken: HMS Hasty - When taken: 14 May 1814 - Where taken: off Charleston - Date Received: 30 Sep 1814 - From what ship: HMT Sybella - Born: Rhode Island - Age: 26 - Released on 4 Jun 1815.

Stanton, Henry - 2nd Mate - Nbr: 355 - Prize: Two Brothers - Ship Type: MV - How taken: Beetle (LM) - When taken: 18 Mar 1813 - Where taken: off Western Islands (England) - Date Received: 1 Jul 1813 - From what ship: Plymouth - Born: Connecticut - Age: 23 - Released on 20 Apr 1815.

Stanwood, James - Seaman - Nbr: 3034 - Prize: Commerce - How taken: HMS Superb - When taken: 23 Jun 1814 - Where taken: off Sandy Hook - Date Received: 2 Sep 1814 - From what ship: HMT Sultan - Born: New York - Age: 19 - Released on 28 May 1815.

Stanwood, Timothy - Seaman - Nbr: 2145 - How taken: Gave himself up from HMS Aboukir - When taken: 28 Oct 1812 - Date Received: 8 Aug 1814 - From what ship: HMT Raven, Chatham - Born: Newburyport - Age: 23 - Died on 20 Mar 1815 from variola.

Staples, James - Seaman - Nbr: 4432 - Prize: Taken off an English whaler - Ship Type: MV - How taken: HMS Illustrious - When taken: 22 Oct 1813 - Where taken: at sea - Date Received: 8 Oct 1814 - From what ship: HMT Leyden, Chatham - Born: Boston - Age: 21 - Released on 14 Jun 1815.

Staples, William - Seaman - Nbr: 3648 - Prize: Ulysses - Ship Type: MV - How taken: HMS Majestic - When taken: 29 Jun 1813 - Where taken: off Western Islands (England) - Date Received: 30 Sep 1814 - From what ship: HMT President - Born: Massachusetts - Age: 23 - Released on 4 Jun 1815.

Starberne, John - Seaman - Nbr: 3794 - Prize: Flash - Ship Type: LM - How taken: HMS Acasta - When taken: 27 Feb 1814 - Where taken: off Long Island - Date Received: 5 Oct 1814 - From what ship: HMT President, Halifax - Born: New York - Age: 30 - Released on 9 Jun 1815.

Starboard, William - Seaman - Nbr: 2031 - How taken: Apprehended at London - When taken: 24 May 1813 - Date Received: 3 Aug 1814 - From what ship: HMS Lyffey, Chatham Depot - Born: Portland - Age: 23 - Released on 2 May 1815.

Starbuck, George - Seaman - Nbr: 1043 - Prize: US Brig Argus - Ship Type: MW - How taken: HMS Pelican - When taken: 16 Aug 1813 - Where taken: Irish Channel - Date Received: 10 May 1814 - From what ship: Plymouth - Born: Nantucket - Age: 25 - Sent to Dartmouth on 19 Oct 1814.

Starbuck, Thaddeus B. - Seaman - Nbr: 160 - How taken: Impressed at Dublin - When taken: 7 Jan 1813 - Date Received: 2 Apr 1813 - From what ship: Plymouth - Born: Nantucket - Age: 24 - Sent to Dartmouth on 30 Jul 1813.

Starkweather, James - Seaman - Nbr: 4723 - How taken: Gave himself up from HMS Phoenix - When taken: 17 Jul 1813 - Date Received: 9 Oct 1814 - From what ship: HMT Freya, Chatham - Born: Connecticut - Age: 33 - Released on 15 Jun 1815.

Start, Joshua - Seaman - Nbr: 655 - Prize: Marmion - Ship Type: MV - How taken: HMS President - When taken: 14 Aug 1813 - Where taken: off Nantes - Date Received: 8 Sep 1813 - From what ship: Plymouth - Born: Baltimore - Age: 22 - Race: Negro - Released on 26 Apr 1815.

Steel, John - Seaman - Nbr: 6460 - Prize: Decatur - Ship Type: P - How taken: HMS Rhin - When taken: 5 Jun 1814 - Where taken: off San Domingo - Date Received: 3 Mar 1815 - From what ship: HMT Ganges, Plymouth - Born: Hertford - Age: 32 - Released on 5 Apr 1815.

Steel, John - Seaman - Nbr: 5863 - Prize: Lion - Ship Type: P - How taken: HMS Granicus - When taken: 2 Dec 1814 - Where taken: off Lisbon - Date Received: 26 Dec 1814 - From what ship: HMT Impregnable - Born: Maryland - Age: 39 - Released on 3 Jul 1815.

Steel, John - Seaman - Nbr: 263 - Prize: William Bayard - Ship Type: MV - How taken: HMS Warspite - When taken: 3 Mar 1813 - Where taken: Bay of Biscay - Date Received: 28 Jun 1813 - From what ship: Plymouth - Born: Maryland - Age: 26 - Died on 15 Dec 1814 from pneumonia.

Steel, Mark - Cook - Nbr: 4960 - Prize: Herald - Ship Type: P - How taken: HMS Endymion - When taken: 15 Aug 1814 - Where taken: off Nantucket - Date Received: 28 Oct 1814 - From what ship: HMT Alkbar, Halifax - Born: New Jersey - Age: 44 - Race: Mulatto - Released on 21 Jun 1815.

Steele, Mathew - Prize master - Nbr: 3781 - Prize: James - Ship Type: Prize - How taken: Rebecca (LM) - When taken: 13 Apr 1814 - Where taken: off Delaware - Date Received: 5 Oct 1814 - From what ship: HMT President, Halifax - Born: Lancaster - Age: 33 - Released on 9 Jun 1815.

Steele, Thomas - Seaman - Nbr: 5370 - Prize: Rattlesnake - Ship Type: LM - How taken: HMS Rhin - When taken: 12 Mar 1814 - Where taken: off Bermuda - Date Received: 31 Oct 1814 - From what ship: HMT Leyden, Chatham - Born: Boston - Age: 21 - Released on 1 Jul 1815.

American Prisoners of War Held at Dartmoor during the War of 1812

## Alphabetical listing of names

Steele, William - Seaman - Nbr: 3785 - Prize: James - Ship Type: Prize - How taken: Rebecca (LM) - When taken: 13 Apr 1814 - Where taken: off Delaware - Date Received: 5 Oct 1814 - From what ship: HMT President, Halifax - Born: Lancaster - Age: 25 - Released on 9 Jun 1815.

Stenson, David - Seaman - Nbr: 6250 - How taken: Sent into custody from MV Kulabra - When taken: 25 Nov 1814 - Date Received: 4 Feb 1815 - From what ship: HMT Ganges - Born: Georgetown - Age: 24 -Released on 11 Jul 1815.

Stephens, Amos - Seaman - Nbr: 1115 - Prize: Bunker Hill - Ship Type: P - How taken: HMS Pomone & HMS Cadmus - When taken: 4 Mar 1814 - Where taken: Bay of Biscay - Date Received: 10 May 1814 - From what ship: Plymouth - Born: Boston - Age: 36 - Released on 10 Apr 1815.

Stephens, John - Seaman - Nbr: 1688 - Prize: Vixen - Ship Type: MV - How taken: HMS Southampton - When taken: 22 Nov 1812 - Where taken: off St. Marys - Date Received: 2 Jul 1814 - From what ship: Plymouth - Born: SC - Age: 27 - Sent to Dartmouth on 19 Oct 1814.

Stephens, Obadiah - Seaman - Nbr: 1240 - How taken: Sent into custody from HMS Alemine - When taken: 25 Jan 1813 - Date Received: 14 Jun 1814 - From what ship: Mill Prison (Plymouth, England) - Born: Philadelphia - Age: 27 - Released on 26 Apr 1815.

Stephens, William - Seaman - Nbr: 4661 - How taken: Impressed at London - Date Received: 9 Oct 1814 - From what ship: HMT Leyden, Chatham - Born: Boston - Age: 22 - Released on 13 Jun 1815.

Stephens, William - Seaman - Nbr: 4799 - Prize: President - Ship Type: P - How taken: HMS Pique - When taken: 7 May 1814 - Where taken: off Porto Rico - Date Received: 9 Oct 1814 - From what ship: HMT Freya, Halifax - Born: Not listed - Released on 21 Jun 1815.

Stephenson, John - Seaman - Nbr: 1153 - How taken: Impressed at Cork - When taken: 27 Mar 1814 - Date Received: 10 May 1814 - From what ship: Plymouth - Born: Salem - Age: 40 - Released on 28 Apr 1815.

Sterns, Joseph - Seaman - Nbr: 4599 - Prize: Elbridge Gerry - Ship Type: P - How taken: HMS Crescent - When taken: 13 Nov 1813 - Where taken: off St. Johns - Date Received: 9 Oct 1814 - From what ship: HMT Leyden, Chatham - Born: Portland - Age: 24 - Released on 28 Apr 1815.

Stetson, Hiram - Seaman - Nbr: 4257 - Prize: Argus - Ship Type: MV - How taken: HMS San Domingo - When taken: 1 Mar 1814 - Where taken: off Savannah - Date Received: 7 Oct 1814 - From what ship: HMT Niobe, Chatham - Born: Boston - Age: 17 - Released on 13 Jun 1815.

Stetson, Mathew - Seaman - Nbr: 2447 - Prize: US Sloop Frolic - Ship Type: MW - How taken: HMS Orpheus - When taken: 20 Apr 1814 - Where taken: off Cuba - Date Received: 16 Aug 1814 - From what ship: HMT Queen, Halifax - Born: New Bedford - Age: 18 - Released on 3 May 1815.

Stevens, Charles - Seaman - Nbr: 4017 - Prize: US Brig Rattlesnake - Ship Type: MW - How taken: HMS Leander - When taken: 13 Jul 1814 - Where taken: off Shelburne - Date Received: 6 Oct 1814 - From what ship: HMT Chesapeake, Halifax - Born: Boston - Age: 19 - Released on 9 Jun 1815.

Stevens, Hugh - Seaman - Nbr: 663 - Prize: Joel Barlow - Ship Type: LM - How taken: HMS Briton - When taken: 3 Jul 1813 - Where taken: off Bordeaux - Date Received: 8 Sep 1813 - From what ship: Plymouth - Born: Philadelphia - Age: 27 - Released on 26 Apr 1815.

Stevens, Ira - Seaman - Nbr: 3053 - How taken: Gave himself up from HMS Sultan - When taken: 2 Sep 1814 - Date Received: 2 Sep 1814 - From what ship: HMT Sultan - Born: Boston - Age: 20 - Race: Negro - Released on 28 Apr 1815.

Stevens, James - Seaman - Nbr: 3367 - Prize: Rambler - Ship Type: MV - How taken: HMS Thais - When taken: 30 Mar 1813 - Where taken: off Africa - Date Received: 13 Sep 1814 - From what ship: HMT Niobe, Chatham - Born: Boston - Age: 32 - Race: Black - Released on 28 May 1815.

Stevens, John - Seaman - Nbr: 6529 - Prize: US Brig Syren - Ship Type: MW - How taken: HMS Medway - When taken: 12 Jul 1814 - Where taken: off Cape of Good Hope - Date Received: 7 Mar 1815 - From what ship: HMT Ganges, Plymouth - Born: Ipswich - Age: 25 - Released on 11 Jul 1815.

Stevens, John - Sergeant - Nbr: 593 - Prize: US Brig Argus - Ship Type: MW - How taken: HMS Pelican - When taken: 14 Aug 1813 - Where taken: Irish Channel - Date Received: 8 Sep 1813 - From what ship: Plymouth - Born: Delaware - Age: 38 - Sent to Dartmouth on 2 Nov 1814.

Stevens, Joseph - Seaman - Nbr: 2040 - How taken: Gave himself up from HMS Decoy - When taken: Nov 1812 - Date Received: 3 Aug 1814 - From what ship: HMS Lyffey, Chatham Depot - Born: Philadelphia - Age: 38 - Released on 26 Apr 1815.

Stevens, Leman - Seaman - Nbr: 2446 - Prize: US Sloop Frolic - Ship Type: MW - How taken: HMS Orpheus - When taken: 20 Apr 1814 - Where taken: off Cuba - Date Received: 16 Aug 1814 - From what ship: HMT Queen, Halifax - Born: Massachusetts - Age: 26 - Released on 3 May 1815.

American Prisoners of War Held at Dartmoor during the War of 1812

## Alphabetical listing of names

Stevens, Thomas - Master - Nbr: 3594 - Prize: Frolic - Ship Type: P - How taken: HMS Heron - When taken: 25 Jan 1814 - Where taken: off St. Thomas - Date Received: 30 Sep 1814 - From what ship: HMT Sybella - Born: Marblehead - Age: 28 - Released on 4 Jun 1815.

Stevens, William - Seaman - Nbr: 6254 - How taken: Impressed at Liverpool - When taken: 23 Nov 1814 - Date Received: 4 Feb 1815 - What ship: HMT Ganges - Born: Newburyport - Age: 22 - Released on 5 Jul 1815.

Stevens, William - Seaman - Nbr: 1366 - Prize: Paul Jones - Ship Type: P - How taken: HMS Leonidas - When taken: 23 May 1813 - Where taken: Channel - Date Received: 19 Jun 1814 - From what ship: Stapleton - Born: Connecticut - Age: 14 - Released on 28 Apr 1815.

Stevenson, George - Prize Master - Nbr: 2011 - Prize: a recaptured MV - How taken: HMS Revolutionnaire - When taken: 10 Apr 1813 - Where taken: off Western Islands (England) - Date Received: 3 Aug 1814 - From what ship: HMS Lyffey, Chatham Depot - Born: New York - Age: 34 - Released on 10 Apr 1815.

Stevenson, Levi - Seaman - Nbr: 2895 - How taken: Gave himself up from HMS Prince of Wales - When taken: 28 May 1813 - Date Received: 24 Aug 1814 - From what ship: HMT Alpheus, Chatham - Born: Virginia - Age: 27 - Race: Black - Released on 19 May 1815.

Steward, Isaac - Seaman - Nbr: 804 - Prize: Collin - Ship Type: MV - How taken: HMS Helicon and HMS Whiting - When taken: 26 Oct 1813 - Where taken: Channel - Date Received: 3 Nov 1813 - From what ship: Plymouth - Born: Baltimore - Age: 29 - Released on 26 Apr 1815.

Steward, John - Seaman - Nbr: 2093 - How taken: Gave himself up from HMS Malta - When taken: 1 Jan 1813 - Date Received: 3 Aug 1814 - From what ship: HMS Bittern, Chatham Depot - Born: Virginia - Age: 20 - Released on 2 May 1815.

Steward, William - Seaman - Nbr: 4894 - How taken: Gave himself up from HMS Casillian - When taken: Sep 1814 - Date Received: 24 Oct 1814 - From what ship: HMT Salvador del Mundo - Born: West River - Age: 27 - Released on 11 May 1815.

Stewart, Adam - Seaman - Nbr: 1317 - Prize: Siro - Ship Type: LM - How taken: HMS Pelican - When taken: 13 Jan 1814 - Where taken: at sea - Date Received: 14 Jun 1814 - From what ship: Mill Prison (Plymouth, England) - Born: New York - Age: 27 - Released on 28 Apr 1815.

Stewart, James - Steward - Nbr: 2131 - How taken: Apprehended at London - When taken: 21 Dec 1812 - Date Received: 8 Aug 1814 - What ship: HMT Raven, Chatham - Born: New York - Age: 28 - Released on 26 Apr 1815.

Stewart, John - Seaman - Nbr: 2643 - How taken: Gave himself up from HMS Caledonia - When taken: 5 Dec 1812 - Date Received: 21 Aug 1814 - From what ship: HMT Freya, Chatham - Born: Philadelphia - Age: 35 - Released on 26 Apr 1815.

Stewart, Scipio - Seaman - Nbr: 1757 - How taken: Gave himself up from HMS Edinburgh - When taken: 3 Dec 1812 - Date Received: 20 Jul 1814 - From what ship: HMS Milford, Plymouth - Born: Charlestown - Age: 25 - Race: Black - Released on 26 Apr 1815.

Stewart, William - Seaman - Nbr: 1421 - Prize: Paul Jones - Ship Type: P - How taken: HMS Leonidas - When taken: 23 May 1813 - Where taken: Channel - Date Received: 19 Jun 1814 - From what ship: Stapleton - Born: New York - Age: 19 - Race: Negro - Released on 28 Apr 1815.

Stewart, William - Seaman - Nbr: 512 - Prize: Paul Jones - Ship Type: P - How taken: HMS Leonidas - When taken: 23 May 1813 - Where taken: Channel - Date Received: 8 Sep 1813 - From what ship: Plymouth - Born: Manchester - Age: 45 - Released on 26 Apr 1815.

Stibbens, Stephen - Seaman - Nbr: 4892 - Prize: Goree - Ship Type: LM - How taken: HMS Rolla - When taken: Jul 1814 - Where taken: off Havana - Date Received: 24 Oct 1814 - From what ship: HMT Salvador del Mundo - Born: New Jersey - Age: 30 - Released on 21 Jun 1815.

Stibbins, Thomas - Steward - Nbr: 2818 - How taken: Impressed at Dover - When taken: 10 Jul 1813 - Date Received: 24 Aug 1814 - From what ship: HMT Liverpool, Chatham - Born: Chatham - Age: 27 - Released on 19 May 1815.

Stickney, Abraham - Carpenter - Nbr: 4605 - Prize: Argus - Ship Type: MV - How taken: HMS San Domingo - When taken: 1 Mar 1814 - Where taken: off Savannah - Date Received: 9 Oct 1814 - From what ship: HMT Leyden, Chatham - Born: Newburyport - Age: 24 - Released on 15 Jun 1815.

Stickney, Amos - Seaman - Nbr: 5554 - Prize: Levant - Ship Type: MV - How taken: HMS Forester - When taken: 4 Jan 1814 - Where taken: Bahamas Banks - Date Received: 17 Dec 1814 - From what ship: HMT Loire, Halifax - Born: Beverly - Age: 26 - Released on 1 Jul 1815.

Stiles, Jodick - Seaman - Nbr: 5206 - Prize: Porcupine - Ship Type: LM - How taken: HMS Acasta - When taken: 18 Jul 1813 - Where taken: off Halifax - Date Received: 31 Oct 1814 - From what ship: HMT Mermaid,

American Prisoners of War Held at Dartmoor during the War of 1812

## Alphabetical listing of names

Chatham - Born: Boxford - Age: 28 - Released on 11 Jul 1815.
Still, David - Seaman - Nbr: 1069 - Prize: Fair American - Ship Type: MV - How taken: HMS Andromache - When taken: 19 Jan 1814 - Where taken: Bay of Biscay - Date Received: 10 May 1814 - From what ship: Plymouth - Born: Wiscasset - Age: 42 - Released on 27 Apr 1815.
Stilman, John - Boy - Nbr: 3063 - Prize: Argner - How taken: Jane (Cutter) - When taken: 29 Nov 1813 - Where taken: Bay of Biscay - Date Received: 2 Sep 1814 - From what ship: HMT Hydra, Chatham - Born: Boston - Age: 17 - Released on 1 May 1815.
Stilwell, William - Seaman - Nbr: 4776 - How taken: Gave himself up from HMS Revenge - When taken: 23 Sep 1814 - Date Received: 9 Oct 1814 - From what ship: HMT Freya, Chatham - Born: Middleton - Age: 25 - Escaped on 1 Jun 1815.
Stinchcombe, George - Mate - Nbr: 2014 - Prize: Price - Ship Type: LM - How taken: HMS Iris - When taken: 13 Apr 1813 - Where taken: Bay of Biscay - Date Received: 3 Aug 1814 - From what ship: HMS Lyffey, Chatham Depot - Born: Baltimore - Age: 24 - Released on 2 May 1815.
Stinton, James - Seaman - Nbr: 3211 - How taken: Gave himself up from HMS Achille - When taken: 8 Sep 1813 - Date Received: 11 Sep 1814 - From what ship: HMT Freya, Chatham - Born: Baltimore - Age: 24 - Released on 28 May 1815.
Stirrell, George - Seaman - Nbr: 3000 - Prize: St. Lawrence - How taken: HMS Aquilon - When taken: 9 Aug 1814 - Where taken: off Western Islands (England) - Date Received: 29 Aug 1814 - From what ship: HMT Bittern - Born: Baltimore - Age: 21 - Race: Black - Released on 28 May 1815.
Stitman, A. W. - Mate - Nbr: 3070 - Prize: Argus - Ship Type: MV - How taken: HMS San Domingo - When taken: 1 Mar 1814 - Where taken: off Savannah - Date Received: 2 Sep 1814 - From what ship: HMT Hydra, Chatham - Born: Barnstable - Age: 21 - Released on 2 Apr 1815.
Stockham, W. B. - Seaman - Nbr: 1530 - Prize: Essex - Ship Type: MV - How taken: HMS Pyramus - When taken: 20 Apr 1813 - Where taken: Bay of Biscay - Date Received: 23 Jun 1814 - From what ship: Stapleton - Born: Massachusetts - Age: 29 - Released on 1 May 1815.
Stoddard, Conrad - Seaman - Nbr: 5235 - Prize: Teazer - Ship Type: P - How taken: HMS Maidstone - When taken: 18 Jul 1813 - Where taken: at sea - Date Received: 31 Oct 1814 - From what ship: HMT Mermaid, Chatham - Born: Bremen - Age: 25 - Released on 29 Jun 1815.
Stoddard, Isaac - Seaman - Nbr: 3955 - Prize: Rolla - Ship Type: P - How taken: HMS Loire - When taken: 10 Dec 1813 - Where taken: off Bull Island (South Carolina) - Date Received: 5 Oct 1814 - From what ship: HMT President, Halifax - Born: Rhode Island - Age: 20 - Released on 9 Jun 1815.
Stoddard, Reuben - Seaman - Nbr: 226 - Prize: Charlotte - Ship Type: MV - How taken: HMS Warspite - When taken: 3 Mar 1813 - Where taken: Bay of Biscay - Date Received: 2 Apr 1813 - From what ship: Plymouth - Born: Massachusetts - Age: 17 - Released on 20 Apr 1815.
Stoddard, Samuel - Seaman - Nbr: 5558 - Prize: Levant - Ship Type: MV - How taken: HMS Forester - When taken: 4 Jan 1814 - Where taken: Bahamas Banks - Date Received: 17 Dec 1814 - From what ship: HMT Loire, Halifax - Born: Salem - Age: 25 - Released on 1 Jul 1815.
Stoddard, Thomas - Boy - Nbr: 3780 - Prize: Fame - Ship Type: P - How taken: HMS Thistle - When taken: 10 Apr 1814 - Where taken: after being cast ashore on a seal island - Date Received: 30 Sep 1814 - From what ship: HMT President, Halifax - Born: Salem - Age: 15 - Released on 3 May 1815.
Stoddart, Robert - Seaman - Nbr: 286 - Prize: Cannoneer - Ship Type: MV - How taken: HMS Warspite - When taken: 14 Mar 1813 - Where taken: Bay of Biscay - Date Received: 28 Jun 1813 - From what ship: Plymouth - Born: Long Island - Age: 19 - Released on 20 Apr 1815.
Stone, Abner - Prize master - Nbr: 5676 - Prize: McDonough - Ship Type: P - How taken: HMS Bacchante - When taken: 7 Nov 1814 - Where taken: Lat 42 Long 67 - Date Received: 24 Dec 1814 - From what ship: HMT Penelope - Born: Massachusetts - Age: 26 - Released on 3 Jul 1815.
Stone, Adam - Seaman - Nbr: 6019 - Prize: McDonough - Ship Type: P - How taken: HMS Bacchante - When taken: 1 Nov 1814 - Where taken: Lat 42 Long 67 - Date Received: 28 Dec 1814 - From what ship: HMT Penelope - Born: Arundel - Age: 23 - Released on 3 Jul 1815.
Stone, Benjamin - Seaman - Nbr: 3599 - Prize: Frolic - Ship Type: P - How taken: HMS Heron - When taken: 25 Jan 1814 - Where taken: off St. Thomas - Date Received: 30 Sep 1814 - From what ship: HMT Sybella - Born: Newbury - Age: 18 - Released on 4 Jun 1815.
Stone, Edward - Seaman - Nbr: 818 - Prize: Alexander - Ship Type: MV - How taken: HMS Andromache - When taken: 23 Oct 1813 - Where taken: off Brest - Date Received: 3 Nov 1813 - From what ship: Plymouth - Born: Beverly - Age: 23 - Released on 26 Apr 1815.

American Prisoners of War Held at Dartmoor during the War of 1812

## Alphabetical listing of names

Stone, Henry - Seaman - Nbr: 3348 - How taken: Gave himself up from HMS Gordon - When taken: 1 Nov 1812 - Date Received: 13 Sep 1814 - From what ship: HMT Niobe, Chatham - Born: Connecticut - Age: 26 - Released on 27 Apr 1815.

Stone, Isaac - Passenger - Nbr: 4572 - How taken: Taken off a Swedish ship bound for America - Date Received: 9 Oct 1814 - From what ship: HMT Leyden, Chatham - Born: Danvers - Age: 28 - Race: Mulatto - Released on 11 Jul 1815.

Stone, Jerry - Seaman - Nbr: 6243 - Prize: Albion - Ship Type: Prize - How taken: HMS Harlequin - When taken: 6 Jan 1814 - Where taken: Long 20 - Date Received: 4 Feb 1815 - From what ship: HMT Ganges - Born: New Orleans - Age: 19 - Race: Mulatto - Released on 5 Jul 1815.

Stone, John - Seaman - Nbr: 6532 - Prize: US Brig Syren - Ship Type: MW - How taken: HMS Medway - When taken: 12 Jul 1814 - Where taken: off Cape of Good Hope - Date Received: 7 Mar 1815 - From what ship: HMT Ganges, Plymouth - Born: Gloucester - Age: 33 - Released on 11 Jul 1815.

Stone, John - Seaman - Nbr: 5888 - Prize: Harlequin - Ship Type: P - How taken: HMS Bulwark - When taken: 23 Nov 1814 - Where taken: off Halifax - Date Received: 27 Dec 1814 - From what ship: HMT Penelope - Born: Arundel - Age: 44 - Died on 5 Jan 1815 from pneumonia.

Stone, Jonathan - Marine - Nbr: 5692 - Prize: McDonough - Ship Type: P - How taken: HMS Bacchante - When taken: 7 Nov 1814 - Where taken: Lat 42 Long 67 - Date Received: 24 Dec 1814 - From what ship: HMT Penelope - Born: Arundel - Age: 23 - Released on 3 Jul 1815.

Stone, Robert - Seaman - Nbr: 4400 - Prize: Portsmouth Packet - Ship Type: Prize - How taken: HMS Fantome - When taken: 5 Oct 1813 - Where taken: off Portland - Date Received: 8 Oct 1814 - From what ship: HMT Leyden, Chatham - Born: Massachusetts - Age: 32 - Released on 10 Apr 1815.

Stone, Samuel - Seaman - Nbr: 1286 - Prize: Indian Lass - Ship Type: Prize - How taken: Not listed - When taken: 29 Apr 1814 - Date Received: 14 Jun 1814 - From what ship: Mill Prison (Plymouth, England) - Born: Ipswich - Age: 19 - Released on 28 Apr 1815.

Stone, Samuel - Seaman - Nbr: 2168 - Prize: Rattlesnake - Ship Type: P - How taken: HMS Hyperion - When taken: 26 Jun 1814 - Where taken: off Cape Finisterre - Date Received: 16 Aug 1814 - From what ship: HMS Dublin, Halifax - Born: Philadelphia - Age: 24 - Released on 2 May 1815.

Stone, William - Seaman - Nbr: 2340 - Prize: Snap Dragon - Ship Type: P - How taken: HMS Martin - When taken: 10 Jun 1814 - Where taken: off Halifax - Date Received: 16 Aug 1814 - From what ship: HMS Dublin, Halifax - Born: New Jersey - Age: 20 - Released on 3 May 1815.

Storey, George - Seaman - Nbr: 3723 - Prize: Alfred - Ship Type: P - How taken: HMS Epervier - When taken: 23 Feb 1814 - Where taken: off Newfoundland - Date Received: 30 Sep 1814 - From what ship: HMT President, Halifax - Born: Sumatra - Age: 21 - Race: Mulatto - Released on 4 Jun 1815.

Storey, John - Seaman - Nbr: 3710 - Prize: Alfred - Ship Type: P - How taken: HMS Epervier - When taken: 23 Feb 1814 - Where taken: off Newfoundland - Date Received: 30 Sep 1814 - From what ship: HMT President, Halifax - Born: Cape Ann - Age: 18 - Released on 4 Jun 1815.

Storey, William - Seaman - Nbr: 5436 - Prize: Herald - Ship Type: P - How taken: HMS Endymion - When taken: 15 Aug 1814 - Where taken: off Nantucket - Date Received: 11 Nov 1814 - From what ship: HMT Impregnable - Born: Boston - Age: 26 - Released on 11 Jul 1815.

Storm, Daniel - Seaman - Nbr: 3331 - Prize: York Town - Ship Type: P - How taken: HMS Nimrod - When taken: 17 Jul 1813 - Where taken: off St. Johns - Date Received: 13 Sep 1814 - From what ship: HMT Niobe, Chatham - Born: New York - Age: 28 - Released on 28 May 1815.

Story, William - Prize Master - Nbr: 2638 - Prize: True Blooded Yankee - Ship Type: P - How taken: HMS Hamadryad - When taken: 24 Jul 1813 - Where taken: off Norway - Date Received: 21 Aug 1814 - From what ship: HMT Freya, Chatham - Born: Philadelphia - Age: 40 - Released on 3 May 1815.

Stout, John - Seaman - Nbr: 951 - Prize: Siro - Ship Type: LM - How taken: HMS Pelican - When taken: 13 Jan 1814 - Where taken: at sea - Date Received: 31 Jan 1814 - From what ship: Plymouth - Born: Portland - Age: 16 - Died on 20 Jan 1815 from variola.

Stove, Lewis - Seaman - Nbr: 3874 - Prize: Tickler - Ship Type: MV - How taken: HMS Saturn - When taken: 13 Jul 1814 - Where taken: off America - Date Received: 5 Oct 1814 - From what ship: HMT Orpheus, Halifax - Born: Connecticut - Age: 22 - Died on 21 Nov 1814.

Stover, David - Seaman - Nbr: 5899 - Prize: Harlequin - Ship Type: P - How taken: HMS Bulwark - When taken: 23 Nov 1814 - Where taken: off Halifax - Date Received: 27 Dec 1814 - From what ship: HMT Penelope - Born: York - Age: 28 - Released on 3 Jul 1815.

Stoves, Daniel - Seaman - Nbr: 5984 - Prize: McDonough - Ship Type: P - How taken: HMS Bacchante - When

American Prisoners of War Held at Dartmoor during the War of 1812

## Alphabetical listing of names

taken: 1 Nov 1814 - Where taken: Lat 42 Long 67 - Date Received: 27 Dec 1814 - From what ship: HMT Penelope - Born: York - Age: 18 - Released on 3 Jul 1815.

Stow, Jeremiah - Seaman - Nbr: 1783 - How taken: Gave himself up from HMS Leyden - When taken: 23 Mar 1813 - Date Received: 20 Jul 1814 - From what ship: HMS Milford, Plymouth - Born: Philadelphia - Age: 44 - Released on 1 May 1815.

Strand, Peter - Seaman - Nbr: 5421 - How taken: Gave himself up from HMS Indian - When taken: 20 Aug 1814 - Date Received: 31 Oct 1814 - From what ship: HMT Leyden, Chatham - Born: Long Island - Age: 48 - Race: Black - Released on 1 Jul 1815.

Straye, George L. - Seaman - Nbr: 5388 - Prize: Sister - Ship Type: MV - How taken: HMS Unicorn - When taken: 3 Jul 1814 - Where taken: off Christian Land - Date Received: 31 Oct 1814 - From what ship: HMT Leyden, Chatham - Born: Baltimore - Age: 30 - Released on 1 Jul 1815.

Strong, William - Seaman - Nbr: 5274 - Prize: Sword Fish - Ship Type: P - How taken: HMS Elephant - When taken: 28 Dec 1813 - Where taken: off Newfoundland - Date Received: 31 Oct 1814 - From what ship: HMT Leyden, Chatham - Born: Marblehead - Age: 22 - Released on 29 Jun 1815.

Stroud, Joshua - Captain's clerk - Nbr: 5472 - Prize: General Putnam - Ship Type: P - How taken: HMS Leander - When taken: 8 Nov 1814 - Where taken: Long 65 Lat 42 - Date Received: 17 Dec 1814 - From what ship: HMT Loire, Halifax - Born: Salem - Age: 20 - Released on 1 Jul 1815.

Strouse, Martin - Seaman - Nbr: 2170 - Prize: Rattlesnake - Ship Type: P - How taken: HMS Hyperion - When taken: 26 Jun 1814 - Where taken: off Cape Finisterre - Date Received: 16 Aug 1814 - From what ship: HMS Dublin, Halifax - Born: Alexandria - Age: 23 - Released on 2 May 1815.

Strouts, Joseph - Seaman - Nbr: 3800 - Prize: Margetits - Ship Type: Prize - How taken: HMS Maidstone - When taken: 20 Mar 1814 - Where taken: off NY - Date Received: 5 Oct 1814 - From what ship: HMT Orpheus, Halifax - Born: Salem - Age: 28 - Released on 17 Apr 1815.

Strube, John - Seaman - Nbr: 1594 - Prize: Eliza - Ship Type: MV - How taken: HMS Surveillante - When taken: 27 Mar 1813 - Where taken: Bay of Biscay - Date Received: 23 Jun 1814 - From what ship: Stapleton - Born: Massachusetts - Age: 23 - Released on 1 May 1815.

Studdy, Richard - Seaman - Nbr: 3035 - Prize: Commerce - How taken: HMS Superb - When taken: 23 Jun 1814 - Where taken: off Sandy Hook - Date Received: 2 Sep 1814 - From what ship: HMT Sultan - Born: Virginia - Age: 30 - Died on 3 Nov 1814.

Studley, John - Seaman - Nbr: 3846 - Prize: Dominique - Ship Type: LM - How taken: HMS Dotterel - When taken: 21 May 1814 - Where taken: off Charleston - Date Received: 5 Oct 1814 - From what ship: HMT Orpheus, Halifax - Born: SC - Age: 25 - Released on 9 Jun 1815.

Studley, Simon - Seaman - Nbr: 2361 - Prize: Snap Dragon - Ship Type: P - How taken: HMS Martin - When taken: 10 Jun 1814 - Where taken: off Halifax - Date Received: 16 Aug 1814 - From what ship: HMT Queen, Halifax - Born: Boston - Age: 25 - Released on 3 May 1815.

Studley, Warren - Corporal - Nbr: 6311 - Prize: Prince de Neufchatel - Ship Type: P - How taken: Leander (Newcastle Acasta) - When taken: 20 Dec 1814 - Where taken: Lat 38 Long 56 - Date Received: 19 Feb 1815 - From what ship: HMT Ganges, Plymouth - Born: Not legible - Age: 22 - Released on 5 Jul 1815.

Stuff, Francis - Seaman - Nbr: 1570 - Prize: Messenger - Ship Type: MV - How taken: HMS Iris - When taken: 10 Mar 1813 - Where taken: Bay of Biscay - Date Received: 23 Jun 1814 - From what ship: Stapleton - Born: New York - Age: 21 - Race: Colored - Sent to Mill Prison (Plymouth, England) on 39 Jun 1814.

Sturdwent, David - Seaman - Nbr: 2032 - How taken: Apprehended at London - When taken: 24 May 1813 - Date Received: 3 Aug 1814 - From what ship: HMS Lyffey, Chatham Depot - Born: Summer - Age: 23 - Released on 2 May 1815.

Sturges, Bradley - 2nd Mate - Nbr: 395 - Prize: Young Holkar - Ship Type: MV - How taken: HMS Superb - When taken: 10 Apr 1813 - Where taken: off Belle Isle, France - Date Received: 1 Jul 1813 - From what ship: Plymouth - Born: Connecticut - Age: 19 - Released on 20 Apr 1815.

Sturgess, Major - Seaman - Nbr: 5049 - Prize: Ida - Ship Type: LM - How taken: HMS Newcastle - When taken: 9 Aug 1814 - Where taken: Long 34 - Date Received: 28 Oct 1814 - From what ship: HMT Alkbar, Halifax - Born: Maryland - Age: 30 - Released on 29 Jun 1815.

Suckley, Richard - Seaman - Nbr: 3162 - How taken: Impressed at Gravesend - When taken: 7 Jul 1813 - Date rec'd: 11 Sep 1814 - What ship: HMT Freya, Chatham - Born: Massachusetts - Age: 26 - Released on 28 May 1815.

Sullivan, P. - Boy - Nbr: 2514 - Prize: Snap Dragon - Ship Type: P - How taken: HMS Martin - When taken: 10 Jun 1814 - Where taken: off Halifax - Date Received: 16 Aug 1814 - From what ship: HMT Queen, Halifax -

American Prisoners of War Held at Dartmoor during the War of 1812

## Alphabetical listing of names

Born: North Carolina - Age: 14 - Released on 3 May 1815.
Sullivan, Florence - Seaman - Nbr: 4114 - Prize: Bordeaux Packet - Ship Type: LM - How taken: HMS Niemen - When taken: 28 Jun 1814 - Where taken: off Delaware - Date Received: 6 Oct 1814 - From what ship: HMT Chesapeake, Halifax - Born: Baltimore - Age: 36 - Released on 13 Jun 1815.
Sullivan, Hampton - Seaman - Nbr: 194 - Prize: Star - Ship Type: MV - How taken: HMS Superb - When taken: 9 Feb 1813 - Where taken: Bay of Biscay - Date Received: 2 Apr 1813 - From what ship: Plymouth - Born: Wilmington - Age: 19 - Released on 20 Apr 1815.
Sullivan, John - Seaman - Nbr: 5133 - Prize: Calabria - Ship Type: MV - How taken: HMS Castilian - When taken: 29 Sep 1814 - Where taken: off Ireland - Date Received: 31 Oct 1814 - From what ship: HMT Castillian - Born: Annapolis - Age: 18 - Released on 29 Jun 1815.
Sulton, John - Seaman - Nbr: 2779 - Prize: Tiger - Ship Type: MV - How taken: HMS Scylla - When taken: 22 May 1813 - Where taken: Bay of Biscay - Date Received: 24 Aug 1814 - From what ship: HMT Liverpool, Chatham - Born: Nantucket - Age: 26 - Released on 19 May 1815.
Sumerg, Benjamin - Seaman - Nbr: 5208 - Prize: Porcupine - Ship Type: LM - How taken: HMS Acasta - When taken: 18 Jul 1813 - Where taken: off Halifax - Date Received: 31 Oct 1814 - From what ship: HMT Mermaid, Chatham - Born: New Haven - Age: 33 - Race: Black - Released on 29 Jun 1815.
Sutherland, John - Seaman - Nbr: 4136 - Prize: Bordeaux Packet - Ship Type: LM - How taken: HMS Niemen - When taken: 28 Jun 1814 - Where taken: off Delaware - Date Received: 6 Oct 1814 - From what ship: HMT Chesapeake, Halifax - Born: New Jersey - Age: 28 - Released on 13 Jun 1815.
Sutter, John G. - Seaman - Nbr: 548 - How taken: Gave himself up from HMS Scorpion - When taken: 4 Nov 1812 - Date rec'd: 8 Sep 1813 - What ship: Plymouth - Born: Philadelphia - Age: 33 - Released on 9 Apr 1815.
Sutton, Benjamin - Prize Master - Nbr: 2267 - Prize: Grand Turk - Ship Type: P - How taken: HMS Martin - When taken: 26 May 1814 - Where taken: off Cape Sable Island (Canada) - Date Received: 16 Aug 1814 - From what ship: HMS Dublin, Halifax - Born: North Carolina - Age: 32 - Released on 3 May 1815.
Sutton, John - Seaman - Nbr: 2701 - How taken: Gave himself up from HMS Leopard - When taken: 25 Dec 1813 - Date Received: 21 Aug 1814 - From what ship: HMT Freya, Chatham - Born: Washington - Age: 39 - Released on 19 May 1815.
Sutton, Martin - Seaman - Nbr: 5647 - Prize: Lion - Ship Type: P - How taken: HMS Granicus - When taken: 2 Dec 1814 - Where taken: off Lisbon - Date Received: 24 Dec 1814 - From what ship: HMT Tay - Born: New Bedford - Age: 26 - Race: Black - Died on 22 Feb 1815 from variola.
Sutton, Prince - Seaman - Nbr: 1958 - How taken: Gave himself up from HMS Bellerophon - When taken: 17 Oct 1812 - Date Received: 3 Aug 1814 - From what ship: HMS Lyffey, Chatham Depot - Born: Rhode Island - Age: 45 - Race: Mulatto - Released on 26 Apr 1815.
Sutton, Richard - Seaman - Nbr: 2200 - Prize: Hussar - Ship Type: P - How taken: HMS Saturn - When taken: 25 May 1814 - Where taken: off Sandy Hook - Date Received: 16 Aug 1814 - From what ship: HMS Dublin, Halifax - Born: Philadelphia - Age: 40 - Released on 2 Apr 1815.
Swaby, Jonathan - Seaman - Nbr: 4335 - Prize: Venus - Ship Type: P - How taken: HMS Loire - When taken: 18 Feb 1814 - Where taken: off St. Thomas - Date Received: 7 Oct 1814 - From what ship: HMT Salvador del Mundo, Halifax - Born: Boston - Age: 25 - Released on 14 Jun 1815.
Swadey, Samuel - Boy - Nbr: 4321 - Prize: Lizard - Ship Type: P - How taken: HMS Barbados - When taken: 1 Mar 1814 - Where taken: off Halifax - Date Received: 7 Oct 1814 - From what ship: HMT Salvador del Mundo, Halifax - Born: Salem - Age: 17 - Released on 14 Jun 1815.
Swain, Darius - Seaman - Nbr: 339 - Prize: Good Friends - Ship Type: MV - How taken: HMS Andromache - When taken: 2 Apr 1813 - Where taken: Bay of Biscay - Date Received: 28 Jun 1813 - From what ship: Plymouth - Born: Carolina - Age: 23 - Released on 20 Apr 1815.
Swain, David - Seaman - Nbr: 3368 - Prize: Walker - Ship Type: MV - How taken: HMS Nimrod - When taken: 19 Mar 1813 - Where taken: at sea - Date Received: 13 Sep 1814 - From what ship: HMT Niobe, Chatham - Born: Nantucket - Age: 27 - Released on 23 Apr 1815.
Swain, Edward - Prize Master - Nbr: 3518 - Prize: Aeolus - Ship Type: Prize - How taken: HMS Warspite - When taken: 2 Sep 1814 - Where taken: off Newfoundland - Date Received: 28 Sep 1814 - From what ship: HMT Salvador del Mundo - Born: Newburyport - Age: 38 - Released on 4 Jun 1815.
Swain, James - Seaman - Nbr: 3154 - Prize: Fame (Whaler) - Ship Type: MV - How taken: HMS Cressy - When taken: 20 Jul 1813 - Where taken: at sea - Date Received: 11 Sep 1814 - From what ship: HMT Freya, Chatham - Born: Nantucket - Age: 19 - Released on 1 May 1815.
Swain, Luke - Passenger - Nbr: 3751 - Prize: Dolphin - Ship Type: P - How taken: HMS Lacedaemonian - When

# American Prisoners of War Held at Dartmoor during the War of 1812

## Alphabetical listing of names

taken: 7 Sep 1813 - Where taken: Chesapeake Bay - Date Received: 30 Sep 1814 - From what ship: HMT President, Halifax - Born: Charleston - Age: 30 - Released on 4 Jun 1815.

Swain, Obadiah - Seaman - Nbr: 4773 - Prize: William Penn - Ship Type: MV - How taken: HMS Acorn - When taken: 27 Oct 1812 - Where taken: Lat 14 - Date Received: 9 Oct 1814 - From what ship: HMT Freya, Chatham - Born: Nantucket - Age: 21 - Released on 15 Jun 1815.

Swaine, Joseph - Master - Nbr: 4918 - Prize: Unity - Ship Type: MV - How taken: HMS Asia - When taken: 24 Jul 1814 - Where taken: off Cape VA - Date Received: 28 Oct 1814 - From what ship: HMT Alkbar, Halifax - Born: Nantucket - Age: 23 - Released on 1 May 1815.

Swaine, Thomas - 2nd Lieutenant - Nbr: 3593 - Prize: Frolic - Ship Type: P - How taken: HMS Heron - When taken: 25 Jan 1814 - Where taken: off St. Thomas - Date Received: 30 Sep 1814 - From what ship: HMT Sybella - Born: Salisbury - Age: 30 - Released on 4 Jun 1815.

Swaine, Thomas - Lieutenant - Nbr: 2971 - Prize: Wily Reynard - Ship Type: P - How taken: HMS Shannon - When taken: 11 Oct 1812 - Where taken: off Halifax - Date Received: 24 Aug 1814 - From what ship: HMT Queen - Born: Massachusetts - Age: 29 - Escaped on 2 Nov 1814.

Swaney, William - Seaman - Nbr: 6244 - Prize: Albion - Ship Type: Prize - How taken: HMS Harlequin - When taken: 6 Jan 1814 - Where taken: Long 20 - Date Received: 4 Feb 1815 - From what ship: HMT Ganges - Born: Salem - Age: 19 - Released on 5 Jul 1815.

Swasey, John - Seaman - Nbr: 1110 - Prize: Bunker Hill - Ship Type: P - How taken: HMS Pomone & HMS Cadmus - When taken: 4 Mar 1814 - Where taken: Bay of Biscay - Date Received: 10 May 1814 - From what ship: Plymouth - Born: Salem - Age: 20 - Released on 27 Apr 1815.

Sweet, Francis - Seaman - Nbr: 1773 - Prize: Print - Ship Type: MV - How taken: HMS Colossus - When taken: 21 Jan 1813 - Where taken: off Long Island - Date Received: 20 Jul 1814 - From what ship: Chatham - Born: Marblehead - Age: 24 - Released on 1 May 1815.

Sweetland, William - Seaman - Nbr: 4975 - Prize: Invincible - Ship Type: LM - How taken: HMS Armide - When taken: 15 Aug 1814 - Where taken: off Nantucket - Date Received: 28 Oct 1814 - From what ship: HMT Alkbar, Halifax - Born: Massachusetts - Age: 26 - Released on 21 Jun 1815.

Sweetman, Samuel - Seaman - Nbr: 1979 - Prize: Orbit - Ship Type: MV - How taken: HMS Achates - When taken: 29 Jan 1813 - Where taken: Lat 44N Long 13W - Date Received: 3 Aug 1814 - From what ship: HMS Lyffey, Chatham Depot - Born: New York - Age: 25 - Released on 2 May 1815.

Swinbourne, George - Seaman - Nbr: 4362 - How taken: Gave himself up from HMS Caledonia - When taken: 2 Oct 1814 - Date Received: 7 Oct 1814 - From what ship: HMT Salvador del Mundo, Halifax - Born: Maryland - Age: 22 - Released on 10 Apr 1815.

Swinley, John - Seaman - Nbr: 4980 - Prize: Invincible - Ship Type: LM - How taken: HMS Armide - When taken: 15 Aug 1814 - Where taken: off Nantucket - Date Received: 28 Oct 1814 - From what ship: HMT Alkbar, Halifax - Born: Pennsylvania - Age: 42 - Released on 21 Jun 1815.

Syeway, Peter - Seaman - Nbr: 2589 - How taken: Gave himself up from HMS Electra - When taken: 20 Sep 1812 - Date Received: 21 Aug 1814 - From what ship: HMT Freya, Chatham - Born: Boston - Age: 28 - Released on 26 Apr 1815.

Symes, John - Seaman - Nbr: 1718 - How taken: Gave himself up from HMS Rodney - Date Received: 8 Jul 1814 - From what ship: Labrador - Born: Virginia - Age: 36 - Released on 1 May 1815.

Symonds, Israel - Seaman - Nbr: 1776 - Prize: Print - Ship Type: MV - How taken: HMS Colossus - When taken: 21 Jan 1813 - Where taken: off Long Island - Date Received: 20 Jul 1814 - From what ship: HMS Milford, Plymouth - Born: Massachusetts - Age: 21 - Released on 1 May 1815.

Symonds, M. John - Prize Master - Nbr: 885 - Prize: General Kempt - Ship Type: P - How taken: HMS Foxhound - When taken: 18 Dec 1813 - Where taken: Lat 48'60" Long 5'7" - Date Received: 31 Jan 1814 - From what ship: Plymouth - Born: Salem - Age: 23 - Released on 27 Apr 1815.

Symonds, Moses - Seaman - Nbr: 6302 - Prize: John - Ship Type: Prize - How taken: Leander (Newcastle Acasta) - When taken: 30 Jan 1815 - Date Received: 19 Feb 1815 - From what ship: HMT Ganges, Plymouth - Born: New York - Age: 26 - Race: Black - Released on 5 Jul 1815.

Symons, William - Seaman - Nbr: 819 - Prize: Alexander - Ship Type: MV - How taken: HMS Andromache - When taken: 23 Oct 1813 - Where taken: off Brest - Date Received: 3 Nov 1813 - From what ship: Plymouth - Born: Bedford - Age: 19 - Released on 26 Apr 1815.

Tafe, William - Boy - Nbr: 631 - Prize: US Brig Argus - Ship Type: MW - How taken: HMS Pelican - When taken: 14 Aug 1813 - Where taken: Irish Channel - Date Received: 8 Sep 1813 - From what ship: Plymouth - Born: New York - Age: 13 - Sent to Dartmouth on 2 Nov 1814.

American Prisoners of War Held at Dartmoor during the War of 1812

## Alphabetical listing of names

Taft, Henry - Boy - Nbr: 5673 - Prize: Harlequin - Ship Type: P - How taken: HMS Bulwark - When taken: 23 Nov 1814 - Where taken: off Halifax - Date Received: 24 Dec 1814 - From what ship: HMT Penelope - Born: Portsmouth - Age: 14 - Released on 3 May 1815.

Taggart, John - Seaman - Nbr: 2129 - How taken: Gave himself up from HMS Inconstant - When taken: 1 Dec 1812 - Date Received: 8 Aug 1814 - From what ship: HMT Raven, Chatham - Born: Philadelphia - Age: 25 - Released on 28 Apr 1815.

Talgart, Shepherd - Seaman - Nbr: 333 - Prize: William Bayard - Ship Type: P - How taken: HMS Warspite - When taken: 12 Dec 1813 - Date Received: 28 Jun 1813 - From what ship: Plymouth - Born: Nantucket - Age: 28 - Released on 20 Apr 1815.

Tamer, John - Seaman - Nbr: 749 - Prize: US Brig Argus - Ship Type: MW - How taken: HMS Pelican - When taken: 14 Aug 1813 - Where taken: Irish Channel - Date Received: 3 Nov 1813 - From what ship: Plymouth - Born: Holland - Age: 32 - Sent to Mill Prison (Plymouth, England) on 17 Jun 1814.

Tapley, Isaac - Seaman - Nbr: 5211 - Prize: Porcupine - Ship Type: LM - How taken: HMS Acasta - When taken: 18 Jul 1813 - Where taken: off Halifax - Date Received: 31 Oct 1814 - From what ship: HMT Mermaid, Chatham - Born: Cambridge - Age: 16 - Released on 29 Jun 1815.

Tar, Caleb - Seaman - Nbr: 4412 - Prize: Fire Fly - Ship Type: LM - How taken: HMS Revolutionnaire - When taken: 19 Oct 1813 - Where taken: at sea - Date Received: 8 Oct 1814 - From what ship: HMT Leyden, Chatham - Born: Massachusetts - Age: 29 - Released on 14 Jun 1815.

Tar, Zedediah - Seaman - Nbr: 2222 - Prize: Hussar - Ship Type: P - How taken: HMS Saturn - When taken: 25 May 1814 - Where taken: off Sandy Hook - Date Received: 16 Aug 1814 - From what ship: HMS Dublin, Halifax - Born: Charlestown - Age: 26 - Released on 2 May 1815.

Tarbor, Samuel - Seaman - Nbr: 6327 - Prize: Prince de Neufchatel - Ship Type: P - How taken: Leander (Newcastle Acasta) - When taken: 20 Dec 1814 - Where taken: Lat 38 Long 56 - Date Received: 19 Feb 1815 - From what ship: HMT Ganges, Plymouth - Born: Lynn - Age: 20 - Released on 5 Jul 1815.

Tardy, Andy - Seaman - Nbr: 1336 - Prize: Paul Jones - Ship Type: P - How taken: HMS Leonidas - When taken: 23 May 1813 - Where taken: Channel - Date Received: 19 Jun 1814 - From what ship: Stapleton - Born: Palermo - Age: 32 - Released on 28 Apr 1815.

Tarlton, George - Seaman - Nbr: 4398 - Prize: Portsmouth Packet - Ship Type: Prize - How taken: HMS Fantome - When taken: 5 Oct 1813 - Where taken: off Portland - Date Received: 8 Oct 1814 - From what ship: HMT Leyden, Chatham - Born: Massachusetts - Age: 30 - Released on 14 Jun 1815.

Tarlton, J. P. - Seaman - Nbr: 2751 - Prize: Dart - Ship Type: MV - How taken: HMS Peterel - When taken: 15 Mar 1813 - Where taken: at sea - Date Received: 24 Aug 1814 - From what ship: HMT Liverpool, Chatham - Born: Greenland - Age: 21 - Released on 19 May 1815.

Tarlton, James - Seaman - Nbr: 4528 - Prize: Portsmouth Packet - Ship Type: Prize - How taken: HMS Fantome - When taken: 4 Oct 1813 - Where taken: off Portland - Date Received: 8 Oct 1814 - From what ship: HMT Leyden, Chatham - Born: Massachusetts - Age: 20 - Released on 15 Jun 1815.

Tarlton, John - Seaman - Nbr: 3298 - Prize: Governor Plumer - Ship Type: P - How taken: Sent into custody from a privateer - When taken: 1 Jun 1813 - Where taken: off Cape Ann - Date Received: 13 Sep 1814 - From what ship: HMT Niobe, Chatham - Born: New Castle - Age: 29 - Released on 28 May 1815.

Tarr, James - Seaman - Nbr: 4775 - How taken: Gave himself up from West Indiaman John - When taken: 22 Sep 1814 - Date Received: 9 Oct 1814 - From what ship: HMT Freya, Chatham - Born: Virginia - Age: 22 - Released on 15 Jun 1815.

Tarr, William - Seaman - Nbr: 6338 - Prize: Prince de Neufchatel - Ship Type: P - How taken: Leander (Newcastle Acasta) - When taken: 20 Dec 1814 - Where taken: Lat 38 Long 56 - Date Received: 19 Feb 1815 - From what ship: HMT Ganges, Plymouth - Born: Camden - Age: 24 - Released on 6 Jul 1815.

Tassier, Louis - Seaman - Nbr: 1340 - Prize: Paul Jones - Ship Type: P - How taken: HMS Leonidas - When taken: 23 May 1813 - Where taken: Channel - Date Received: 19 Jun 1814 - From what ship: Stapleton - Born: New Orleans - Age: 18 - Sent to Mill Prison (Plymouth, England) on 26 Jun 1814.

Taylor, Alexander - Seaman - Nbr: 2356 - Prize: Snap Dragon - Ship Type: P - How taken: HMS Martin - When taken: 10 Jun 1814 - Where taken: off Halifax - Date Received: 16 Aug 1814 - From what ship: HMT Queen, Halifax - Born: North Carolina - Age: 17 - Released on 3 May 1815.

Taylor, David - Seaman - Nbr: 5729 - Prize: Enterprize (Saratoga) - Ship Type: Prize - How taken: HMS Barracouta - When taken: 10 Oct 1814 - Where taken: off Western Islands (England) - Date Received: 26 Dec 1814 - From what ship: HMT Argo - Born: Massachusetts - Age: 22 - Released on 3 Jul 1815.

Taylor, David - Seaman - Nbr: 5105 - Prize: David Porter - Ship Type: LM - How taken: HMS Pylades - When

American Prisoners of War Held at Dartmoor during the War of 1812

## Alphabetical listing of names

taken: 12 Sep 1814 - Where taken: Georges Bank - Date Received: 28 Oct 1814 - From what ship: HMT Alkbar, Halifax - Born: Philadelphia - Age: 19 - Race: Negro - Died on 18 Jun 1815 from hemoptysis.

Taylor, David A. - Passenger - Nbr: 6155 - Prize: Lion - Ship Type: P - How taken: HMS Granicus - When taken: 2 Dec 1814 - Where taken: off Lisbon - Date Received: 18 Jan 1815 - From what ship: HMT Impregnable - Born: New York - Age: 27.

Taylor, George - Landsman - Nbr: 104 - Prize: St. Martin's Planter - Ship Type: P - How taken: HMS Dublin - When taken: 9 Feb 1813 - Where taken: Lat 43 N, Long 33 50 W - Date Received: 2 Apr 1813 - From what ship: Plymouth - Born: New Jersey - Age: 27 - Released on 20 Apr 1815.

Taylor, George - Seaman - Nbr: 4290 - Prize: Pilot - Ship Type: LM - How taken: Victoria (Privateer) - When taken: 28 Jan 1814 - Where taken: off Bordeaux - Date Received: 7 Oct 1814 - From what ship: HMT Niobe, Chatham - Born: Rhode Island - Age: 24 - Race: Mulatto - Released on 14 Jun 1815.

Taylor, George - Seaman - Nbr: 2435 - Prize: US Sloop Frolic - Ship Type: MW - How taken: HMS Orpheus - When taken: 20 Apr 1814 - Where taken: off Cuba - Date Received: 16 Aug 1814 - From what ship: HMT Queen, Halifax - Born: Salem - Age: 45 - Released on 3 May 1815.

Taylor, Jacob - Seaman - Nbr: 236 - Prize: Charlotte - Ship Type: MV - How taken: HMS Warspite - When taken: 3 Mar 1813 - Where taken: Bay of Biscay - Date Received: 2 Apr 1813 - From what ship: Plymouth - Born: Philadelphia - Age: 24 - Released on 20 Apr 1815.

Taylor, James - Seaman - Nbr: 5759 - Prize: Hope - Ship Type: MV - How taken: HMS Nereus - When taken: 14 May 1814 - Where taken: Rio de la Plata - Date Received: 26 Dec 1814 - From what ship: HMT Argo - Born: New York - Age: 19 - Released on 3 Jul 1815.

Taylor, James - Seaman - Nbr: 5586 - Prize: Young William - Ship Type: Prize - How taken: HMS Plover - When taken: 8 Sep 1814 - Where taken: off St. Johns - Date Received: 17 Dec 1814 - From what ship: HMT Loire, Halifax - Born: Amsterdam - Age: 21 - Released on 1 Jul 1815.

Taylor, James - Seaman - Nbr: 879 - How taken: Impressed at Liverpool - When taken: 1 Nov 1813 - Date Received: 31 Jan 1814 - From what ship: Plymouth - Born: Maryland - Age: 30 - Released on 27 Apr 1815.

Taylor, James - Seaman - Nbr: 2090 - How taken: Gave himself up from HMS Hyperion - When taken: 28 Mar 1813 - Date Received: 3 Aug 1814 - From what ship: HMS Bittern, Chatham Depot - Born: Philadelphia - Age: 27 - Released on 2 May 1815.

Taylor, James - Seaman - Nbr: 924 - How taken: Sent into custody from HMS Sparrow - Date Received: 31 Jan 1814 - From what ship: Plymouth - Born: Philadelphia - Age: 21 - Race: Negro - Released on 27 Apr 1815.

Taylor, John - Seaman - Nbr: 3997 - Prize: US Brig Rattlesnake - Ship Type: MW - How taken: HMS Leander - When taken: 13 Jul 1814 - Where taken: off Shelburne - Date Received: 6 Oct 1814 - From what ship: HMT Chesapeake, Halifax - Born: New Jersey - Age: 21 - Released on 9 Jun 1815.

Taylor, John - Seaman - Nbr: 6387 - How taken: Impressed at Plymouth - When taken: 7 Feb 1815 - Date Received: 24 Feb 1815 - From what ship: HMT Ganges, Plymouth - Born: Annapolis - Age: 29 - Released on 11 Jul 1815.

Taylor, Joseph - Seaman - Nbr: 2648 - How taken: Gave himself up from HMS Fame - When taken: 31 Dec 1812 - Date Received: 21 Aug 1814 - From what ship: HMT Freya, Chatham - Born: Philadelphia - Age: 24 - Released on 26 Apr 1815.

Taylor, Paul - Seaman - Nbr: 5905 - Prize: Harlequin - Ship Type: P - How taken: HMS Bulwark - When taken: 23 Nov 1814 - Where taken: off Halifax - Date Received: 27 Dec 1814 - From what ship: HMT Penelope - Born: New Hampshire - Age: 20 - Released on 11 Jul 1815.

Taylor, Peter - Seaman - Nbr: 2080 - Prize: True Blooded Yankee - Ship Type: P - How taken: HMS Hope - When taken: 24 Jun 1813 - Where taken: off Brest - Date Received: 3 Aug 1814 - From what ship: HMS Bittern, Chatham Depot - Born: New York - Age: 22 - Released on 2 May 1815.

Taylor, Samuel - Seaman - Nbr: 4841 - Prize: President - Ship Type: P - How taken: HMS Pique - When taken: 7 May 1814 - Where taken: off Porto Rico - Date Received: 9 Oct 1814 - From what ship: HMT Freya, Halifax - Born: Washington - Age: 27 - Released on 21 Jun 1815.

Taylor, Samuel E. - Cook - Nbr: 1668 - Prize: Tom - Ship Type: LM - How taken: HMS Surveillante - When taken: 24 Apr 1813 - Where taken: Bay of Biscay - Date Received: 26 Jun 1814 - From what ship: Exeter - Born: Charleston - Age: 23 - Released on 1 May 1815.

Taylor, Thomas - Seaman - Nbr: 508 - How taken: Impressed at Liverpool - When taken: 4 May 1813 - Date Received: 8 Sep 1813 - From what ship: Plymouth - Born: Boston - Age: 26 - Released on 10 Apr 1815.

Taylor, Thomas - Seaman - Nbr: 3143 - Prize: Volante - Ship Type: MV - How taken: HMS Curlew - When taken: 25 Nov 1813 - Where taken: off Halifax - Date Received: 11 Sep 1814 - From what ship: HMT Freya,

American Prisoners of War Held at Dartmoor during the War of 1812

## Alphabetical listing of names

Chatham - Born: New York - Age: 21 - Released on 27 Apr 1815.
Taylor, William - Seaman - Nbr: 6461 - Prize: Decatur - Ship Type: P - How taken: HMS Rhin - When taken: 5 Jun 1814 - Where taken: off San Domingo - Date Received: 3 Mar 1815 - From what ship: HMT Ganges, Plymouth - Born: Charleston - Age: 23 - Released on 11 Jul 1815.
Taylor, William - Seaman - Nbr: 1231 - How taken: Sent into custody from HMS Nautilus - When taken: 20 Dec 1812 - Date Received: 14 Jun 1814 - From what ship: Mill Prison (Plymouth, England) - Born: Wilmington - Age: 42 - Released on 26 Apr 1815.
Taylor, William - Seaman - Nbr: 2929 - How taken: Gave himself up from HMS Bombay - Date Received: 24 Aug 1814 - From what ship: HMT Alpheus, Chatham - Born: Long Island - Age: 39 - Released on 27 Apr 1815.
Taylor, William - Seaman - Nbr: 3658 - Prize: Globe - Ship Type: P - How taken: HMS Spartan - When taken: 13 Jun 1813 - Where taken: off Delaware - Date Received: 30 Sep 1814 - From what ship: HMT President - Born: Maryland - Age: 22 - Released on 4 Jun 1815.
Techand, John - Seaman - Nbr: 3682 - Prize: Alfred - Ship Type: P - How taken: HMS Epervier - When taken: 23 Feb 1812 - Where taken: off Newfoundland - Date Received: 30 Sep 1814 - From what ship: HMT President - Born: Marblehead - Age: 63 - Released on 4 Jun 1815.
Teggett, John - 2nd Mate - Nbr: 1264 - Prize: Adeline - Ship Type: MV - How taken: HMS Magicienne - When taken: 16 Mar 1814 - Where taken: off Cape Finisterre - Date Received: 14 Jun 1814 - From what ship: Mill Prison (Plymouth, England) - Born: Philadelphia - Age: 32 - Released on 28 Apr 1815.
Tendreth, Robert - Seaman - Nbr: 6151 - Prize: Alexander - Ship Type: Prize - How taken: HMS Wasp - When taken: 6 Sep 1814 - Where taken: off Cape Sable Island (Canada) - Date Received: 17 Jan 1815 - From what ship: HMT Impregnable - Born: Massachusetts - Age: 23 - Released on 5 Jul 1815.
Terry, Daniel - Seaman - Nbr: 5081 - Prize: Ida - Ship Type: LM - How taken: HMS Newcastle - When taken: 9 Aug 1814 - Where taken: Long 34 - Date Received: 28 Oct 1814 - From what ship: HMT Alkbar, Halifax - Born: Lebanon - Age: 22 - Released on 29 Jun 1815.
Terry, Joseph - Seaman - Nbr: 5214 - Prize: Porcupine - Ship Type: LM - How taken: HMS Acasta - When taken: 18 Jul 1813 - Where taken: off Halifax - Date Received: 31 Oct 1814 - From what ship: HMT Mermaid, Chatham - Born: Dorchester - Age: 18 - Race: Mulatto - Released on 11 Jul 1815.
Thaley, William - Seaman - Nbr: 3180 - How taken: Gave himself up from HMS Hibernia - When taken: 27 Jul 1813 - Date Received: 11 Sep 1814 - From what ship: HMT Freya, Chatham - Born: New York - Age: 32 - Race: Black - Released on 28 May 1815.
Thaya, Lubin - Prize master - Nbr: 5352 - Prize: Elbridge Gerry - Ship Type: P - How taken: HMS Crescent - When taken: 16 Sep 1813 - Where taken: at sea - Date Received: 31 Oct 1814 - From what ship: HMT Leyden, Chatham - Born: Taunton - Age: 30 - Released on 1 Jul 1815.
Thayer, Ebenezer - Seaman - Nbr: 3076 - How taken: Gave himself up from HMS Gloucester - When taken: 26 Dec 1812 - Date Received: 3 Sep 1814 - From what ship: HMT Hydra, Chatham - Born: Boston - Age: 21 - Released on 27 Apr 1815.
Theal, Bristo - Seaman - Nbr: 3158 - Prize: Fame (Whaler) - Ship Type: MV - How taken: HMS Cressy - When taken: 20 Jul 1813 - Where taken: at sea - Date Received: 11 Sep 1814 - From what ship: HMT Freya, Chatham - Born: Nantucket - Age: 38 - Race: Black - Released on 28 May 1815.
Theyer, James - Seaman - Nbr: 3350 - How taken: Gave himself up from HMS Gordon - When taken: 1 Nov 1812 - Date Received: 13 Sep 1814 - From what ship: HMT Niobe, Chatham - Born: Massachusetts - Age: 23 - Released on 27 Apr 1815.
Thimonier, Pr. - Seaman - Nbr: 2151 - How taken: Gave himself up from HMS Adaimant - When taken: 11 Oct 1812 - Date Received: 8 Aug 1814 - From what ship: HMT Raven, Chatham - Born: New Orleans - Age: 25 - Released on 26 Apr 1815.
Thinney, Elvin - Seaman - Nbr: 4611 - Prize: Argus - Ship Type: MV - How taken: HMS San Domingo - When taken: 1 Mar 1814 - Where taken: off Savannah - Date Received: 9 Oct 1814 - From what ship: HMT Leyden, Chatham - Born: Massachusetts - Age: 20 - Released on 15 Jun 1815.
Thomas, Allan - Seaman - Nbr: 2341 - Prize: Snap Dragon - Ship Type: P - How taken: HMS Martin - When taken: 10 Jun 1814 - Where taken: off Halifax - Date Received: 16 Aug 1814 - From what ship: HMS Dublin, Halifax - Born: North Carolina - Age: 17 - Released on 3 May 1815.
Thomas, Andrew - Seaman - Nbr: 3016 - How taken: Gave himself up from HMS Zealous - When taken: 27 Aug 1814 - Date Received: 2 Sep 1814 - From what ship: HMT Centaur - Born: New Jersey - Age: 21 - Released on 28 May 1815.
Thomas, Archibald - Carpenter - Nbr: 133 - Prize: Criterion - Ship Type: MV - How taken: HMS Belle Poule -

American Prisoners of War Held at Dartmoor during the War of 1812

## Alphabetical listing of names

When taken: 14 Feb 1813 - Where taken: Bay of Biscay - Date Received: 12 Apr 1813 - From what ship: Plymouth - Born: New York - Age: 25 - Released on 20 Apr 1815.

Thomas, Charles - Prize master - Nbr: 6352 - Prize: Prince de Neufchatel - Ship Type: P - How taken: Leander (Newcastle Acasta) - When taken: 20 Dec 1814 - Where taken: Lat 38 Long 56 - Date Received: 19 Feb 1815 - From what ship: HMT Ganges, Plymouth - Born: Massachusetts - Age: 30 - Released on 3 Jul 1815.

Thomas, Charles - Seaman - Nbr: 2766 - How taken: Gave himself up from HMS Franchise - When taken: 20 Dec 1812 - Date Received: 24 Aug 1814 - From what ship: HMT Liverpool, Chatham - Born: Boston - Age: 28 - Released on 19 May 1815.

Thomas, David - Seaman - Nbr: 1030 - Prize: US Brig Argus - Ship Type: MW - How taken: HMS Pelican - When taken: 16 Aug 1813 - Where taken: Irish Channel - Date Received: 10 May 1814 - From what ship: Plymouth - Born: Newport - Age: 32 - Sent to Dartmouth on 19 Oct 1814.

Thomas, Elisha - Seaman - Nbr: 4407 - Prize: Portsmouth Packet - Ship Type: Prize - How taken: HMS Fantome - When taken: 5 Oct 1813 - Where taken: off Portland - Date Received: 8 Oct 1814 - From what ship: HMT Leyden, Chatham - Born: New Market - Age: 21 - Released on 14 Jun 1815.

Thomas, Francis - Seaman - Nbr: 1957 - How taken: Gave himself up from HMS Bellerophon - When taken: 17 Oct 1812 - Date Received: 3 Aug 1814 - From what ship: HMS Lyffey, Chatham Depot - Born: Salem - Age: 32 - Released on 26 Apr 1815.

Thomas, George - Seaman - Nbr: 5885 - Prize: Harlequin - Ship Type: P - How taken: HMS Bulwark - When taken: 23 Nov 1814 - Where taken: off Halifax - Date Received: 27 Dec 1814 - From what ship: HMT Penelope - Born: Massachusetts - Age: 36 - Race: Black - Released on 3 Jul 1815.

Thomas, Henry - Seaman - Nbr: 2057 - How taken: Gave himself up from HMS Cornwall - When taken: 21 Mar 1812 - Date Received: 3 Aug 1814 - From what ship: HMS Lyffey, Chatham Depot - Born: Bloklic - Age: 21 - Race: Black - Released on 26 Apr 1815.

Thomas, Isaac - Seaman - Nbr: 4913 - Prize: William - Ship Type: Prize - How taken: HMS Warspite - When taken: 19 Jun 1814 - Where taken: Georges Bank - Date Received: 28 Oct 1814 - From what ship: HMT Alkbar, Halifax - Born: Boston - Age: 25 - Released on 21 Jun 1815.

Thomas, James - Seaman - Nbr: 5709 - Prize: Regent - Ship Type: LM - How taken: HMS Forth - When taken: 19 Sep 1814 - Where taken: off Egg Harbor (New Jersey) - Date Received: 24 Dec 1814 - From what ship: HMT Penelope - Born: Philadelphia - Age: 33 - Released on 3 Jul 1815.

Thomas, James - Prize master - Nbr: 4866 - Prize: Yankee Lass - Ship Type: P - How taken: HMS Surprze - When taken: 1 May 1814 - Where taken: off Western Islands (England) - Date Received: 9 Oct 1814 - From what ship: HMT Freya, Halifax - Born: Salem - Age: 36 - Released on 21 Jun 1815.

Thomas, James - Seaman - Nbr: 4318 - Prize: Marengo - Ship Type: MV - How taken: Banashow (Guernsey) - When taken: Feb 1814 - Date Received: 7 Oct 1814 - From what ship: HMT Niobe, Chatham - Born: Connecticut - Age: 21 - Race: Negro - Released on 14 Jun 1815.

Thomas, John - Master at Arms - Nbr: 6372 - Prize: US Brig Syren - Ship Type: MW - How taken: HMS Medway - When taken: 12 Jul 1814 - Where taken: off Cape of Good Hope - Date Received: 24 Feb 1815 - From what ship: HMT Ganges, Plymouth - Born: SC - Age: 42 - Released on 6 Jul 1815.

Thomas, John - Seaman - Nbr: 4186 - Prize: Pilot - Ship Type: LM - How taken: Victoria (Privateer) - When taken: 28 Jan 1814 - Where taken: off Bordeaux - Date Received: 7 Oct 1814 - From what ship: HMT Niobe, Chatham - Born: Maryland - Age: 29 - Race: Black - Released on 11 Jul 1815.

Thomas, John - Seaman - Nbr: 362 - Prize: Two Brothers - Ship Type: MV - How taken: Beetle (LM) - When taken: 18 Mar 1813 - Where taken: off Western Islands (England) - Date Received: 1 Jul 1813 - From what ship: Plymouth - Born: Boston - Age: 21 - Sent to Liverpool on 25 Aug 1813.

Thomas, John - Seaman - Nbr: 403 - Prize: Young Holkar - Ship Type: MV - How taken: HMS Superb - When taken: 10 Apr 1813 - Where taken: off Belle Isle, France - Date Received: 1 Jul 1813 - From what ship: Plymouth - Born: New Orleans - Age: 21 - Sent to Plymouth on 8 Jul 1814.

Thomas, John - Seaman - Nbr: 2872 - How taken: Gave himself up from HMS Scorpion - When taken: 27 May 1813 - Date Received: 24 Aug 1814 - From what ship: HMT Alpheus, Chatham - Born: New York - Age: 24 - Race: Black - Released on 19 May 1815.

Thomas, John - Seaman - Nbr: 2873 - How taken: Gave himself up from HMS Scorpion - When taken: 27 May 1813 - Date Received: 24 Aug 1814 - From what ship: HMT Alpheus, Chatham - Born: Bristol - Age: 26 - Released on 19 May 1815.

Thomas, John - Seaman - Nbr: 5754 - Prize: Hope - Ship Type: MV - How taken: HMS Nereus - When taken: 14 May 1814 - Where taken: Rio de la Plata - Date Received: 26 Dec 1814 - From what ship: HMT Argo -

American Prisoners of War Held at Dartmoor during the War of 1812

## Alphabetical listing of names

Born: New York - Age: 22 - Race: Black - Released on 3 Jul 1815.
Thomas, John - Seaman - Nbr: 4116 - Prize: Bordeaux Packet - Ship Type: LM - How taken: HMS Niemen - When taken: 28 Jun 1814 - Where taken: off Delaware - Date Received: 6 Oct 1814 - From what ship: HMT Chesapeake, Halifax - Born: Havre de Grace - Age: 19 - Released on 13 Jun 1815.
Thomas, John - Seaman - Nbr: 5806 - Prize: US Frigate Superior (Gig) - Ship Type: MW - How taken: British gunboats - When taken: 26 Aug 1814 - Where taken: Lake Ontario - Date Received: 26 Dec 1814 - From what ship: HMT Argo - Born: Maryland - Age: 19 - Released on 3 Jul 1815.
Thomas, John - Seaman - Nbr: 2955 - Prize: Rattlesnake - Ship Type: LM - How taken: HMS Rhin - When taken: 10 Mar 1814 - Where taken: West Indies - Date Received: 24 Aug 1814 - From what ship: HMT Hannibal - Born: Virginia - Age: 24 - Released on 19 May 1815.
Thomas, John - Seaman - Nbr: 5224 - Prize: Wig - Ship Type: P - How taken: HMS Dover - When taken: 27 Jun 1813 - Where taken: off Newfoundland - Date Received: 31 Oct 1814 - From what ship: HMT Mermaid, Chatham - Born: Baltimore - Age: 20 - Race: Black - Released on 14 Jun 1815.
Thomas, John - Seaman - Nbr: 4284 - Prize: Pilot - Ship Type: LM - How taken: Victoria (Privateer) - When taken: 28 Jan 1814 - Where taken: off Bordeaux - Date Received: 7 Oct 1814 - From what ship: HMT Niobe, Chatham - Born: Maryland - Age: 29 - Race: Mulatto - Released on 14 Jun 1815.
Thomas, John - Seaman - Nbr: 3656 - Prize: Elbridge Gerry - Ship Type: P - How taken: HMS Crescent - When taken: Jun 1813 - Where taken: off St. Johns - Date Received: 30 Sep 1814 - From what ship: HMT President - Born: New York - Age: 23 - Died on 25 Oct 1814 from enteritis.
Thomas, John - Boy - Nbr: 5181 - Prize: Montgomery - Ship Type: P - How taken: HMS Nymphe - When taken: 1 May 1813 - Where taken: off Cape Cod - Date Received: 31 Oct 1814 - From what ship: HMT Mermaid, Chatham - Born: Boston - Age: 19 - Race: Black - Released on 11 Jul 1815.
Thomas, Jonah - Gunner's Mate - Nbr: 3672 - Prize: Alfred - Ship Type: P - How taken: HMS Epervier - When taken: 23 Feb 1812 - Where taken: off Newfoundland - Date Received: 30 Sep 1814 - From what ship: HMT President - Born: Marblehead - Age: 24 - Released on 4 Jun 1815.
Thomas, Joseph - Passenger - Nbr: 5637 - Prize: Lion - Ship Type: P - How taken: HMS Granicus - When taken: 2 Dec 1814 - Where taken: off Lisbon - Date Received: 24 Dec 1814 - From what ship: HMT Hanover, Gibraltar - Born: L'Orient (France) - Age: 30 - Released on 3 Jul 1815.
Thomas, Moses - Seaman - Nbr: 3189 - How taken: Gave himself up from HMS Swiftsure - When taken: 26 Dec 1812 - Date Received: 11 Sep 1814 - From what ship: HMT Freya, Chatham - Born: Norfolk - Age: 21 - Race: Black - Released on 27 Apr 1815.
Thomas, Richard - Boatswain - Nbr: 4971 - Prize: Invincible - Ship Type: LM - How taken: HMS Armide - When taken: 15 Aug 1814 - Where taken: off Nantucket - Date Received: 28 Oct 1814 - From what ship: HMT Alkbar, Halifax - Born: Paris - Age: 32 - Released on 21 Jun 1815.
Thomas, Spencer - Seaman - Nbr: 2678 - Prize: Joseph - Ship Type: MV - How taken: HMS Iris - When taken: 8 Jun 1813 - Where taken: off Spain - Date Received: 21 Aug 1814 - From what ship: HMT Freya, Chatham - Born: Gloucester - Age: 20 - Released on 19 May 1815.
Thomas, Stephen - Seaman - Nbr: 1263 - Prize: Adeline - Ship Type: MV - How taken: HMS Magicienne - When taken: 16 Mar 1814 - Where taken: off Cape Finisterre - Date Received: 14 Jun 1814 - From what ship: Mill Prison (Plymouth, England) - Born: Dover - Age: 26 - Released on 28 Apr 1815.
Thomas, Theodore - Seaman - Nbr: 6346 - Prize: Prince de Neufchatel - Ship Type: P - How taken: Leander (Newcastle Acasta) - When taken: 20 Dec 1814 - Where taken: Lat 38 Long 56 - Date Received: 19 Feb 1815 - From what ship: HMT Ganges, Plymouth - Born: Bordeaux - Age: 17 - Released on 6 Jul 1815.
Thomas, Thomas - Seaman - Nbr: 2623 - How taken: Gave himself up from HMS Bellerophon - When taken: 7 Nov 1812 - Date Received: 21 Aug 1814 - From what ship: HMT Freya, Chatham - Born: Long Island - Age: 32 - Race: Black - Released on 26 Apr 1815.
Thomas, Thomas - Seaman - Nbr: 2937 - How taken: Gave himself up from HMS Prince of Wales - Date Received: 24 Aug 1814 - From what ship: HMT Alpheus, Chatham - Born: New York - Age: 23 - Race: Black - Released on 19 May 1815.
Thomas, Timothy - Mate - Nbr: 3528 - Prize: Sabine - Ship Type: P - How taken: HMS Conquistador - When taken: 3 Aug 1814 - Where taken: off the Banks - Date Received: 28 Sep 1814 - From what ship: HMT Salvador del Mundo - Born: Connecticut - Age: 28 - Released on 4 Jun 1815.
Thomas, William - Seaman - Nbr: 4376 - How taken: Gave himself up from HMS Malta - When taken: 20 Oct 1812 - Date Received: 8 Oct 1814 - From what ship: HMT Leyden, Chatham - Born: North Carolina - Age: 20 - Released on 27 Apr 1815.

American Prisoners of War Held at Dartmoor during the War of 1812

## Alphabetical listing of names

Thomas, William - Seaman - Nbr: 5355 - How taken: Impressed at London - When taken: 17 Jan 1814 - Date Received: 31 Oct 1814 - From what ship: HMT Leyden, Chatham - Born: Boston - Age: 30 - Race: Black - Released on 1 Jul 1815.

Thomas, William - Seaman - Nbr: 175 - Prize: Hope - Ship Type: MV - How taken: Chance (Privateer) - When taken: 15 Feb 1813 - Where taken: off Bordeaux - Date Received: 2 Apr 1813 - From what ship: Plymouth - Born: New Orleans - Age: 35 - Sent to Mill Prison (Plymouth, England) on 21 Jun 1814.

Thomas, William - Seaman - Nbr: 1986 - How taken: Gave himself up from HMS Colossus - When taken: Apr 1813 - Date Received: 3 Aug 1814 - From what ship: HMS Lyffey, Chatham Depot - Born: Philadelphia - Age: 23 - Released on 2 May 1815.

Thomas, William - Seaman - Nbr: 2206 - Prize: Hussar - Ship Type: P - How taken: HMS Saturn - When taken: 25 May 1814 - Where taken: off Sandy Hook - Date Received: 16 Aug 1814 - From what ship: HMS Dublin, Halifax - Born: Maryland - Age: 21 - Released on 2 May 1815.

Thomas, William - Seaman - Nbr: 422 - Prize: Viper - Ship Type: MV - How taken: HMS Superb - When taken: 15 Apr 1813 - Where taken: Bay of Biscay - Date Received: 1 Jul 1813 - From what ship: Plymouth - Born: Delaware - Age: 28 - Sent to Plymouth on 7 Dec 1813.

Thomas, William - Seaman - Nbr: 4869 - Prize: Rolla - Ship Type: P - How taken: HMS Loire - When taken: 10 Dec 1813 - Where taken: off Bull Island (South Carolina) - Date Received: 24 Oct 1814 - From what ship: Royal Hospital, Plymouth - Born: Rhode Island - Age: 27 - Released on 21 Jun 1815.

Thompson, Abraham - Seaman - Nbr: 1504 - Prize: Paul Jones - Ship Type: P - How taken: HMS Leonidas - When taken: 22 May 1813 - Where taken: Channel - Date Received: 23 Jun 1814 - From what ship: Stapleton - Born: New Haven - Age: 32 - Race: Negro - Died on 23 Jul 1814 from pneumonia.

Thompson, Charles - Seaman - Nbr: 1576 - Prize: Price - Ship Type: MV - How taken: HMS Pyramus - When taken: 6 Apr 1813 - Where taken: Bay of Biscay - Date Received: 23 Jun 1814 - From what ship: Stapleton - Born: New York - Age: 26 - Released on 1 May 1815.

Thompson, Courtney - Seaman - Nbr: 1974 - Prize: Orbit - Ship Type: MV - How taken: HMS Achates - When taken: 29 Jan 1813 - Where taken: Lat 44N Long 13W - Date Received: 3 Aug 1814 - From what ship: HMS Lyffey, Chatham Depot - Born: New York - Age: 19 - Released on 2 May 1815.

Thompson, George - Seaman - Nbr: 3184 - How taken: Gave himself up from HMS Hibernia - When taken: 27 Jul 1813 - Date Received: 11 Sep 1814 - From what ship: HMT Freya, Chatham - Born: New York - Age: 27 - Released on 28 May 1815.

Thompson, Henry - Boy - Nbr: 593 - Prize: Paul Jones - Ship Type: P - How taken: HMS Leonidas - When taken: 23 May 1813 - Where taken: Channel - Date Received: 8 Sep 1813 - From what ship: Plymouth - Born: Connecticut - Age: 11 - Sent to Ashburton (England) on 11 Dec 1813.

Thompson, Henry - Seaman - Nbr: 3470 - How taken: Gave himself up from HMS Prince - When taken: 12 Sep 1814 - Date Received: 19 Sep 1814 - From what ship: HMT Salvador del Mundo - Born: New York - Age: 28 - Race: Black - Died on 21 Feb 1815 from variola.

Thompson, Isaac - Seaman - Nbr: 5101 - Prize: David Porter - Ship Type: LM - How taken: HMS Pylades - When taken: 12 Sep 1814 - Where taken: Georges Bank - Date Received: 28 Oct 1814 - From what ship: HMT Alkbar, Halifax - Born: Bristol - Age: 18 - Released on 29 Jun 1815.

Thompson, James - Seaman - Nbr: 3858 - Prize: Leicester - Ship Type: Prize - How taken: HMS Prometheus - When taken: Jun 1814 - Where taken: off Halifax - Date Received: 5 Oct 1814 - From what ship: HMT Orpheus, Halifax - Born: Marblehead - Age: 19 - Released on 9 Jun 1815.

Thompson, James - Seaman - Nbr: 1968 - Prize: Orbit - Ship Type: MV - How taken: HMS Achates - When taken: 29 Jan 1813 - Where taken: Lat 44N Long 13W - Date Received: 3 Aug 1814 - From what ship: HMS Lyffey, Chatham Depot - Born: Boston - Age: 36 - Released on 2 May 1815.

Thompson, James - Seaman - Nbr: 751 - How taken: Impressed at Liverpool - Date Received: 3 Nov 1813 - From what ship: Plymouth - Born: Nonith - Age: 28 - Released on 20 Apr 1815.

Thompson, James - Seaman - Nbr: 1818 - How taken: Gave himself up from HMS Swiftsure - When taken: 26 Dec 1813 - Date Received: 20 Jul 1814 - From what ship: HMS Milford, Plymouth - Born: Hudson - Age: 32 - Released on 26 Apr 1815.

Thompson, James - Seaman - Nbr: 2887 - How taken: Gave himself up from HMS Shearwater - When taken: 28 May 1813 - Date Received: 24 Aug 1814 - From what ship: HMT Alpheus, Chatham - Born: Salem - Age: 34 - Released on 19 May 1815.

Thompson, James - Seaman - Nbr: 4028 - Prize: US Brig Rattlesnake - Ship Type: MW - How taken: HMS Leander - When taken: 13 Jul 1814 - Where taken: off Shelburne - Date Received: 6 Oct 1814 - From what ship:

American Prisoners of War Held at Dartmoor during the War of 1812

## Alphabetical listing of names

HMT Chesapeake, Halifax - Born: Pennsylvania - Age: 24 - Released on 9 Jun 1815.
Thompson, John - Seaman - Nbr: 4488 - Prize: Enterprize - Ship Type: P - How taken: HMS Tenedos - When taken: 21 May 1813 - Where taken: off Cape Cod - Date Received: 8 Oct 1814 - From what ship: HMT Leyden, Chatham - Born: Massachusetts - Age: 22 - Released on 15 Jun 1815.
Thompson, John - Seaman - Nbr: 1292 - Prize: Traveler - Ship Type: Prize - How taken: HM Schooner Canso - When taken: 11 May 1814 - Where taken: off Cape Clear - Date Received: 14 Jun 1814 - From what ship: Mill Prison (Plymouth, England) - Born: Brandywine - Age: 45 - Released on 28 Apr 1815.
Thompson, John - Seaman - Nbr: 1116 - Prize: Bunker Hill - Ship Type: P - How taken: HMS Pomone & HMS Cadmus - When taken: 4 Mar 1814 - Where taken: Bay of Biscay - Date Received: 10 May 1814 - From what ship: Plymouth - Born: Kent County - Age: 30 - Released on 20 Mar 1815.
Thompson, John - Seaman - Nbr: 3306 - Prize: Thomas - Ship Type: P - How taken: HMS Nymphe - When taken: 28 Jun 1813 - Where taken: off Halifax - Date Received: 13 Sep 1814 - From what ship: HMT Niobe, Chatham - Born: Massachusetts - Age: 28 - Released on 28 May 1815.
Thompson, John - Seaman - Nbr: 3525 - Prize: Sabine - Ship Type: P - How taken: HMS Conquistador - When taken: 3 Aug 1814 - Where taken: off the Banks - Date Received: 28 Sep 1814 - From what ship: HMT Salvador del Mundo - Born: Baltimore - Age: 30 - Released on 4 Jun 1815.
Thompson, John - Seaman - Nbr: 3371 - How taken: Impressed at London - When taken: 3 Oct 1813 - Date Received: 13 Sep 1814 - From what ship: HMT Niobe, Chatham - Born: Virginia - Age: 24 - Released on 28 May 1815.
Thompson, John - Seaman - Nbr: 3033 - Prize: Achille - Ship Type: MV - How taken: HMS Belvidera - When taken: 23 Jul 1813 - Where taken: off Sandy Hook - Date Received: 2 Sep 1814 - From what ship: HMT Sultan - Born: Connecticut - Age: 22 - Released on 28 May 1815.
Thompson, Joseph - Seaman - Nbr: 6548 - How taken: Gave himself up from HMS Sceptre - When taken: Jul 1814 - Date Received: 7 Mar 1815 - From what ship: HMT Ganges, Plymouth - Born: Connecticut - Age: 30 - Released on 11 Jul 1815.
Thompson, Joseph - Seaman - Nbr: 1372 - Prize: Grand Napoleon - Ship Type: P - How taken: HMS Goldfinch - When taken: 17 Apr 1813 - Where taken: Bay of Biscay - Date Received: 19 Jun 1814 - From what ship: Stapleton - Born: Massachusetts - Age: 22 - Released on 28 Apr 1815.
Thompson, Joseph - Seaman - Nbr: 2615 - How taken: Gave himself up from HMS Romulus - When taken: 14 Aug 1812 - Date Received: 21 Aug 1814 - From what ship: HMT Freya, Chatham - Born: New Castle - Age: 27 - Released on 26 Apr 1815.
Thompson, Lawrence - Seaman - Nbr: 4242 - Prize: Bunker Hill - Ship Type: P - How taken: HMS Pomone - When taken: 4 Mar 1814 - Where taken: Channel - Date Received: 7 Oct 1814 - From what ship: HMT Niobe, Chatham - Born: Sweden - Age: 26 - Released on 13 Jun 1815.
Thompson, Martin - Seaman - Nbr: 135 - Prize: Criterion - Ship Type: MV - How taken: HMS Belle Poule - When taken: 14 Feb 1813 - Where taken: Bay of Biscay - Date Received: 12 Apr 1813 - From what ship: Plymouth - Born: Denmark - Age: 36 - Died on 16 Dec 1813.
Thompson, Michael - Seaman - Nbr: 5160 - Prize: Volante - Ship Type: P - How taken: HMS Curlew - When taken: 25 Nov 1813 - Where taken: off Halifax - Date Received: 31 Oct 1814 - From what ship: HMT Mermaid, Chatham - Born: Kingston - Age: 23 - Released on 29 Jun 1815.
Thompson, Nathaniel - Seaman - Nbr: 2807 - Prize: Prompt - Ship Type: MV - How taken: Change (Privateer) - When taken: 28 May 1813 - Where taken: Bay of Biscay - Date Received: 24 Aug 1814 - From what ship: HMT Liverpool, Chatham - Born: Virginia - Age: 31 - Released on 19 May 1815.
Thompson, Thomas - Boatswain's Mate - Nbr: 565 - Prize: US Brig Argus - Ship Type: MW - How taken: HMS Pelican - When taken: 14 Apr 1813 - Where taken: Irish Channel - Date Received: 8 Sep 1813 - From what ship: Plymouth - Born: Rhode Island - Age: 24 - Sent to Dartmouth on 2 Nov 1814.
Thompson, Thomas - Seaman - Nbr: 5314 - Prize: Thomas - Ship Type: P - How taken: HMS Nymphe - When taken: 24 Jun 1813 - Where taken: off Halifax - Date Received: 31 Oct 1814 - From what ship: HMT Leyden, Chatham - Born: Brooklyn - Age: 32 - Race: Black - Died on 16 Jun 1816 from enteritis.
Thompson, Whitney - Landsman - Nbr: 102 - Prize: St. Martin's Planter - Ship Type: P - How taken: HMS Dublin - When taken: 9 Feb 1813 - Where taken: Lat 43 N, Long 33 50 W - Date Received: 2 Apr 1813 - From what ship: Plymouth - Born: Hartford - Age: 23 - Released on 20 Apr 1815.
Thompson, William - Seaman - Nbr: 4360 - How taken: Not legible - When taken: 1 Oct 1814 - Date Received: 7 Oct 1814 - What ship: HMT Salvador del Mundo, Halifax - Born: New York - Age: 32 - Released on 14 Jun 1815.

American Prisoners of War Held at Dartmoor during the War of 1812

## Alphabetical listing of names

Thompson, William - Seaman - Nbr: 6278 - How taken: Gave himself up from HMS Minden - Date Received: 17 Feb 1815 - From what ship: HMT Ganges, Plymouth - Born: New York - Age: 27 - Race: Mulatto - Released on 5 Jul 1815.

Thompson, William - Seaman - Nbr: 5171 - Prize: Volante - Ship Type: P - How taken: HMS Curlew - When taken: 25 Nov 1813 - Where taken: off Halifax - Date Received: 31 Oct 1814 - From what ship: HMT Mermaid, Chatham - Born: Hancock - Age: 21 - Released on 29 Jun 1815.

Thompson, William - Seaman - Nbr: 3995 - Prize: US Brig Rattlesnake - Ship Type: MW - How taken: HMS Leander - When taken: 13 Jul 1814 - Where taken: off Shelburne - Date Received: 6 Oct 1814 - From what ship: HMT Chesapeake, Halifax - Born: New Orleans - Age: 22 - Released on 9 Jun 1815.

Thompson, William - Seaman - Nbr: 1478 - Prize: Hebe - Ship Type: MV - How taken: HMS Stag - When taken: 18 Apr 1813 - Where taken: Bay of Biscay - Date Received: 19 Jun 1814 - From what ship: Stapleton - Born: Pennsylvania - Age: 24 - Released on 1 May 1815.

Thompson, William - Cook - Nbr: 953 - Prize: Siro - Ship Type: LM - How taken: HMS Pelican - When taken: 13 Jan 1814 - Where taken: at sea - Date Received: 31 Jan 1814 - From what ship: Plymouth - Born: Port au Prince - Age: 25 - Race: Negro - Died on 18 Apr 1815 from variola.

Thompson, William - Seaman - Nbr: 187 - Prize: Star - Ship Type: MV - How taken: HMS Superb - When taken: 9 Feb 1813 - Where taken: Bay of Biscay - Date Received: 2 Apr 1813 - From what ship: Plymouth - Born: Copenhagen - Age: 23 - Sent to Mill Prison (Plymouth, England) on 10 Jul 1813.

Thompson, William - Seaman - Nbr: 2625 - How taken: Gave himself up from HMS Freya - When taken: 1 Nov 1812 - Date Received: 21 Aug 1814 - From what ship: HMT Freya, Chatham - Born: Pittsfield - Age: 29 - Released on 26 Apr 1815.

Thomson, Andrew - Seaman - Nbr: 898 - Prize: Squirrel - Ship Type: MV - How taken: HMS Belle Poule - When taken: 14 Dec 1813 - Where taken: at sea - Date Received: 31 Jan 1814 - From what ship: Plymouth - Born: Providence - Age: 20 - Released on 27 Apr 1815.

Thomson, Peter - Seaman - Nbr: 917 - Prize: Squirrel - Ship Type: MV - How taken: HMS Belle Poule - When taken: 14 Dec 1813 - Where taken: at sea - Date Received: 31 Jan 1814 - From what ship: Plymouth - Born: Maryland - Age: 26 - Released on 27 Apr 1815.

Thomson, William - Seaman - Nbr: 407 - Prize: Viper - Ship Type: MV - How taken: HMS Superb - When taken: 15 Apr 1813 - Where taken: Bay of Biscay - Date Received: 1 Jul 1813 - From what ship: Plymouth - Born: Rochester - Age: 19 - Released on 20 Apr 1815.

Thomson, William - Seaman - Nbr: 914 - How taken: Gave himself up from HMS Lyra - Date Received: 31 Jan 1814 - From what ship: Plymouth - Born: Santee - Age: 32 - Released on 27 Apr 1815.

Thomson, William - Boatswain - Nbr: 3669 - Prize: Alfred - Ship Type: P - How taken: HMS Epervier - When taken: 23 Feb 1812 - Where taken: off Newfoundland - Date Received: 30 Sep 1814 - From what ship: HMT President - Born: Marblehead - Age: 28 - Released on 4 Jun 1815.

Thorndike, Hall - Pilot - Nbr: 3101 - Prize: Mary - Ship Type: Prize - How taken: Taken by pilot boat - When taken: 27 Aug 1814 - Where taken: Bristol Channel - Date Received: 9 Sep 1814 - From what ship: HMS Abercrombie - Born: Massachusetts - Age: 45 - Released on 28 May 1815.

Thorndike, Robert - Seaman - Nbr: 3813 - Prize: Nimble - Ship Type: Prize - How taken: HMS Arab - When taken: 5 Apr 1814 - Where taken: Lat 37 Long 65 - Date Received: 5 Oct 1814 - From what ship: HMT Orpheus, Halifax - Born: Portland - Age: 29 - Released on 9 Jun 1815.

Thornhill, R. - Seaman - Nbr: 5719 - How taken: Gave himself up from HMS Pelican - When taken: 13 Sep 1813 - Date Received: 26 Dec 1814 - From what ship: HMT Argo - Born: Philadelphia - Age: 31 - Released on 9 Jan 1815.

Thornton, Benjamin - Seaman - Nbr: 2197 - Prize: Olio - Ship Type: Prize - How taken: HMS Cyane - When taken: 22 May 1814 - Where taken: off NY - Date Received: 16 Aug 1814 - From what ship: HMS Dublin, Halifax - Born: Chester - Age: 26 - Released on 2 May 1815.

Thornton, David - Seaman - Nbr: 3232 - Prize: Sampson - Ship Type: MV - How taken: Rebuff - When taken: 12 May 1813 - Where taken: off Cape St. Vincent - Date Received: 11 Sep 1814 - From what ship: HMT Freya, Chatham - Born: Virginia - Age: 24 - Released on 28 May 1815.

Thornton, John - Seaman - Nbr: 4494 - Prize: Enterprize - Ship Type: P - How taken: HMS Tenedos - When taken: 21 May 1813 - Where taken: off Cape Cod - Date Received: 8 Oct 1814 - From what ship: HMT Leyden, Chatham - Born: Massachusetts - Age: 19 - Released on 15 Jun 1815.

Thornton, William - Seaman - Nbr: 926 - Prize: Growler - Ship Type: MV - How taken: HMS Wolf - When taken: 11 Aug 1813 - Where taken: at sea - Date Received: 31 Jan 1814 - From what ship: Plymouth - Born:

American Prisoners of War Held at Dartmoor during the War of 1812

## Alphabetical listing of names

Richmond - Age: 19 - Sent to Dartmouth on 19 Oct 1814.

Thrasher, John - Seaman - Nbr: 3392 - Prize: Wasp - Ship Type: P - How taken: HMS Bream - When taken: 10 Jun 1813 - Where taken: off Halifax - Date Received: 13 Sep 1814 - From what ship: HMT Niobe, Chatham - Born: Maryland - Age: 24 - Released on 28 May 1815.

Thrasher, Stephen - Seaman - Nbr: 4343 - Prize: Union - Ship Type: LM - How taken: HMS Curlew - When taken: 1 Apr 1814 - Where taken: off Halifax - Date Received: 7 Oct 1814 - From what ship: HMT Salvador del Mundo, Halifax - Born: Pennsylvania - Age: 30 - Released on 14 Jun 1815.

Threshow, James - Seaman - Nbr: 3383 - Prize: Growler - Ship Type: P - How taken: HMS Electra - When taken: 7 Jul 1813 - Where taken: at sea - Date Received: 13 Sep 1814 - From what ship: HMT Niobe, Chatham - Born: Marblehead - Age: 26 - Released on 28 May 1815.

Tibbett, Henry - Prize Master - Nbr: 526 - Prize: Friends - Ship Type: Bey of Pool - How taken: HMS Whiting - When taken: 15 Jul 1813 - Where taken: Lat 67 N, Long 8 W - Date Received: 8 Sep 1813 - From what ship: Plymouth - Born: Salem - Age: 28 - Released on 14 Apr 1815.

Tilbrook, Barney - Seaman - Nbr: 2592 - How taken: Gave himself up from Fourter - When taken: 29 Dec 1813 - Date Received: 21 Aug 1814 - From what ship: HMT Freya, Chatham - Born: Portsmouth - Age: 29 - Released on 3 May 1815.

Tildon, Robert - Seaman - Nbr: 4489 - Prize: Enterprize - Ship Type: P - How taken: HMS Tenedos - When taken: 21 May 1813 - Where taken: off Cape Cod - Date Received: 8 Oct 1814 - From what ship: HMT Leyden, Chatham - Born: Massachusetts - Age: 21 - Released on 15 Jun 1815.

Tilley, Bernard - Seaman - Nbr: 1165 - How taken: Gave himself up from HMS Hebrius - Date Received: 10 May 1814 - From what ship: Plymouth - Born: Philadelphia - Age: 38 - Released on 28 Apr 1815.

Tilman, Joseph - Seaman - Nbr: 5562 - How taken: Gave himself up from HMS Avon - When taken: 14 Jul 1814 - Date Received: 17 Dec 1814 - From what ship: HMT Loire, Halifax - Born: Virginia - Age: 36 - Race: Negro - Released on 1 Jul 1815.

Timmerman, Mathew - Seaman - Nbr: 743 - Prize: Tom Thumb - Ship Type: MV - How taken: Lyon (Privateer) - When taken: 17 Feb 1813 - Where taken: Bay of Biscay - Date Received: 3 Nov 1813 - From what ship: Plymouth - Born: New York - Age: 30 - Died on 26 Sep 1814.

Tindall, Thomas - Seaman - Nbr: 1682 - How taken: Gave himself up from HMS Orestes - Date Received: 2 Jul 1814 - From what ship: Plymouth - Born: Staten Island - Age: 36 - Released on 1 May 1815.

Tink, Henry - Seaman - Nbr: 1943 - How taken: Gave himself up from HMS Pembroke - When taken: 9 Feb 1813 - Date Received: 3 Aug 1814 - From what ship: HMS Alceste, Chatham Depot - Born: Salem - Age: 24 - Released on 2 May 1815.

Tinker, James - Seaman - Nbr: 5792 - How taken: Sent into custody from MV Kanow - Date Received: 26 Dec 1814 - From what ship: HMT Argo - Born: New York - Age: 27 - Released on 3 Jul 1815.

Tinkham, John - Seaman - Nbr: 4272 - Prize: Stock - Ship Type: P - How taken: HMS Belviders - When taken: 18 Nov 1813 - Where taken: Grand Banks - Date Received: 7 Oct 1814 - From what ship: HMT Niobe, Chatham - Born: Wiscasset - Age: 22 - Released on 13 Jun 1815.

Titus, John - Seaman - Nbr: 3641 - Prize: Bordeaux Packet - Ship Type: LM - How taken: HMS Niemen - When taken: 28 Jan 1814 - Where taken: off Delaware - Date Received: 30 Sep 1814 - From what ship: HMT Sybella - Born: Baltimore - Age: 31 - Race: Negro - Released on 4 Jun 1815.

Toby, John - Seaman - Nbr: 5790 - How taken: Impressed at London - When taken: Nov 1814 - Date Received: 26 Dec 1814 - What ship: HMT Argo - Born: New Bedford - Age: 20 - Race: Mulatto - Released on 3 Jul 1815.

Toby, Peter - Seaman - Nbr: 401 - Prize: Young Holkar - Ship Type: MV - How taken: HMS Superb - When taken: 10 Apr 1813 - Where taken: off Belle Isle, France - Date Received: 1 Jul 1813 - From what ship: Plymouth - Born: New Orleans - Age: 30 - Sent to Plymouth on 8 Jul 1814.

Todd, Robert - Seaman - Nbr: 3028 - Prize: US Sloop Frolic - Ship Type: MW - How taken: HMS Orpheus - When taken: 20 Apr 1814 - Where taken: off Cuba - Date Received: 2 Sep 1814 - From what ship: Naval Hospital, Plymouth - Born: Chester - Age: 52 - Released on 3 May 1815.

Todd, Samuel - Seaman - Nbr: 3394 - Prize: Thomas - Ship Type: P - How taken: HMS Nymphe - When taken: 24 Jun 1813 - Where taken: off Halifax - Date Received: 13 Sep 1814 - From what ship: HMT Niobe, Chatham - Born: Massachusetts - Age: 32 - Released on 28 May 1815.

Toley, Elisha - Seaman - Nbr: 794 - Prize: Collin - Ship Type: MV - How taken: HMS Helicon and HMS Whiting - When taken: 26 Oct 1813 - Where taken: Channel - Date Received: 3 Nov 1813 - From what ship: Plymouth - Born: New York - Age: 24 - Died on 9 Mar 1814 from pneumonia.

Tolman, William - Seaman - Nbr: 4054 - Prize: US Brig Rattlesnake - Ship Type: MW - How taken: HMS Leander -

American Prisoners of War Held at Dartmoor during the War of 1812

## Alphabetical listing of names

When taken: 13 Jul 1814 - Where taken: off Shelburne - Date Received: 6 Oct 1814 - From what ship: HMT Chesapeake, Halifax - Born: Bath - Age: 24 - Released on 13 Jun 1815.

Tolpie, Jonathan - Seaman - Nbr: 5179 - Prize: Vivid - Ship Type: P - How taken: HMS Nymphe - When taken: 20 Apr 1813 - Where taken: off Cape Cod - Date Received: 31 Oct 1814 - From what ship: HMT Mermaid, Chatham - Born: New York - Age: 25 - Released on 29 Jun 1815.

Tolton, Moses - Seaman - Nbr: 5086 - Prize: Ida - Ship Type: LM - How taken: HMS Newcastle - When taken: 9 Aug 1814 - Where taken: Long 34 - Date Received: 28 Oct 1814 - From what ship: HMT Alkbar, Halifax - Born: Hampton - Age: 20 - Released on 29 Jun 1815.

Tolver, Joseph - Marine - Nbr: 2501 - Prize: US Sloop Frolic - Ship Type: MW - How taken: HMS Orpheus - When taken: 20 Apr 1814 - Where taken: off Cuba - Date Received: 16 Aug 1814 - From what ship: HMT Queen, Halifax - Born: Philadelphia - Age: 38 - Released on 3 May 1815.

Tolvett, John - Seaman - Nbr: 617 - Prize: US Brig Argus - Ship Type: MW - How taken: HMS Pelican - When taken: 14 Aug 1813 - Where taken: Irish Channel - Date Received: 8 Sep 1813 - From what ship: Plymouth - Born: Coaseto - Age: 33 - Sent to Dartmouth on 2 Nov 1814.

Tombly, John - Marine - Nbr: 5937 - Prize: Harlequin - Ship Type: P - How taken: HMS Bulwark - When taken: 23 Nov 1814 - Where taken: off Halifax - Date Received: 27 Dec 1814 - From what ship: HMT Penelope - Born: Dover - Age: 23 - Released on 3 Jul 1815.

Tomkins, Abraham - Seaman - Nbr: 3896 - Prize: Governor Shelby - Ship Type: MV - How taken: HMS Saturn - When taken: 9 Jul 1814 - Where taken: off Long Island - Date Received: 5 Oct 1814 - From what ship: HMT Orpheus, Halifax - Born: New York - Age: 36 - Died on 3 Nov 1814.

Tomkins, Ephraim - Seaman - Nbr: 2068 - How taken: Impressed at Hull - When taken: 1 Mar 1813 - Date Received: 3 Aug 1814 - From what ship: HMS Bittern, Chatham Depot - Born: Rhode Island - Age: 27 - Released on 2 May 1815.

Tomlinson, G. W. - Marine - Nbr: 2689 - How taken: Gave himself up from HMS Plantagenet - When taken: 20 Jun 1813 - Date Received: 21 Aug 1814 - From what ship: HMT Freya, Chatham - Born: New York - Age: 25 - Released on 19 May 1815.

Tomus, Andrew - Seaman - Nbr: 4625 - Prize: Argus - Ship Type: MV - How taken: HMS San Domingo - When taken: 1 Mar 1814 - Where taken: off Savannah - Date Received: 9 Oct 1814 - From what ship: HMT Leyden, Chatham - Born: Newport - Age: 22 - Race: Black - Released on 15 Jun 1815.

Tonkin, R. - Seaman - Nbr: 5615 - How taken: Gave himself up from HMS Lion - When taken: 18 Jun 1813 - Date Received: 24 Dec 1814 - From what ship: HMT Tay - Born: New York - Age: 40 - Released on 27 Apr 1815.

Tooley, William - Seaman - Nbr: 3550 - Prize: Hawk - Ship Type: P - How taken: HMS Pique - When taken: 26 Apr 1814 - Where taken: off Bermuda - Date Received: 30 Sep 1814 - From what ship: HMT Sybella - Born: North Carolina - Age: 24 - Released on 4 Jun 1815.

Tophouse, Samuel - Soldier - Nbr: 5604 - How taken: Apprehended at Washington, DC - When taken: 14 Sep 1814 - Date Received: 17 Dec 1814 - From what ship: HMT Loire, Halifax - Born: Washington - Age: 32 - Died on 11 Feb 1815 from an abscess.

Tordey, Edward - Seaman - Nbr: 6181 - Prize: Lion - Ship Type: P - How taken: HMS Granicus - When taken: 2 Dec 1814 - Where taken: off Lisbon - Date Received: 21 Jan 1815 - From what ship: HMT Impregnable - Born: New York - Age: 17 - Released on 5 Jul 1815.

Torgnet, Abel - Seaman - Nbr: 2871 - How taken: Gave himself up from HMS Scorpion - When taken: 27 May 1813 - Date Received: 24 Aug 1814 - From what ship: HMT Alpheus, Chatham - Born: Massachusetts - Age: 25 - Race: Mulatto - Released on 19 May 1815.

Torry, Henry - Seaman - Nbr: 4648 - Prize: Enterprize - Ship Type: P - How taken: HMS Tenedos - When taken: 21 May 1813 - Where taken: off Cape Cod - Date Received: 9 Oct 1814 - From what ship: HMT Leyden, Chatham - Born: Massachusetts - Age: 21 - Released on 10 Apr 1815.

Toskins, Frederick - Captain's clerk - Nbr: 5667 - Prize: Harlequin - Ship Type: P - How taken: HMS Bulwark - When taken: 23 Nov 1814 - Where taken: off Halifax - Date Received: 24 Dec 1814 - From what ship: HMT Penelope - Born: New Hampshire - Age: 18 - Released on 3 Jul 1815.

Toumblin, John - Seaman - Nbr: 3055 - Prize: Viper - Ship Type: MV - How taken: HMS Brilliant - When taken: 25 Dec 1813 - Date Received: 2 Sep 1814 - From what ship: HMT Hydra, Chatham - Born: Salem - Age: 45 - Released on 15 Apr 1815.

Toutz, Vievel - Seaman - Nbr: 5764 - Prize: Young William - Ship Type: Prize - How taken: HMS Plover - When taken: 10 Sep 1814 - Where taken: off Newfoundland - Date Received: 26 Dec 1814 - From what ship: HMT Argo - Born: Saint Malo - Age: 36 - Released on 3 Jul 1815.

American Prisoners of War Held at Dartmoor during the War of 1812

## Alphabetical listing of names

Towan, Robert - Seaman - Nbr: 5687 - Prize: McDonough - Ship Type: P - How taken: HMS Bacchante - When taken: 7 Nov 1814 - Where taken: Lat 42 Long 67 - Date Received: 24 Dec 1814 - From what ship: HMT Penelope - Born: Arundel - Age: 20 - Released on 3 Jul 1815.

Tower, Michael - Seaman - Nbr: 640 - Prize: US Brig Argus - Ship Type: MW - How taken: HMS Pelican - When taken: 14 Aug 1813 - Where taken: Irish Channel - Date Received: 8 Sep 1813 - From what ship: Plymouth - Born: Hingham - Age: 37 - Sent to Dartmouth on 30 Jul 1813.

Tower, Michael - Seaman - Nbr: 1 - How taken: Impressed at Greenock - When taken: 8 Jan 1813 - Date Received: 2 Apr 1813 - What ship: Plymouth - Born: Hingham - Age: 37 - Sent to Dartmouth on 30 Jul 1813.

Towns, Asa - Seaman - Nbr: 4221 - Prize: Liberty - Ship Type: MV - How taken: Surrendered - When taken: 30 Dec 1813 - Where taken: Stromess - Date Received: 7 Oct 1814 - From what ship: HMT Niobe, Chatham - Born: New Hampshire - Age: 25 - Released on 13 Jun 1815.

Townsend, Jeremiah - Soldier - Nbr: 5845 - Prize: Colonel Hopkins' Militia - Ship Type: Troops - How taken: British Army - When taken: 17 Sep 1814 - Where taken: Fort Erie - Date Received: 26 Dec 1814 - From what ship: HMT Argo - Born: New York - Age: 20 - Released on 29 Jun 1815.

Townsend, Thomas - Seaman - Nbr: 1168 - How taken: Gave himself up from HMS Hebrius - Date Received: 10 May 1814 - From what ship: Plymouth - Born: Biddeford - Age: 36 - Released on 28 Apr 1815.

Townson, William - Seaman - Nbr: 5686 - Prize: McDonough - Ship Type: P - How taken: HMS Bacchante - When taken: 7 Nov 1814 - Where taken: Lat 42 Long 67 - Date Received: 24 Dec 1814 - From what ship: HMT Penelope - Born: Biddeford - Age: 25 - Released on 3 Jul 1815.

Towson, William - Seaman - Nbr: 350 - Prize: Courier - Ship Type: LM - How taken: HMS Andromache - When taken: 14 Mar 1813 - Where taken: Bay of Biscay - Date Received: 28 Jun 1813 - From what ship: Plymouth - Born: Maryland - Age: 25 - Released on 20 Apr 1815.

Towson, William - Carpenter - Nbr: 2565 - Prize: Rattlesnake - Ship Type: P - How taken: HMS Hyperion - When taken: 25 Jun 1814 - Where taken: off Cape Finisterre - Date Received: 21 Aug 1814 - From what ship: HMS Hyperion - Born: Maryland - Age: 26 - Released on 3 May 1815.

Tradwell, James - Seaman - Nbr: 5884 - Prize: Harlequin - Ship Type: P - How taken: HMS Bulwark - When taken: 23 Nov 1814 - Where taken: off Halifax - Date Received: 27 Dec 1814 - From what ship: HMT Penelope - Born: Boston - Age: 28 - Released on 3 Jul 1815.

Traphagan, Peter - Seaman - Nbr: 4637 - How taken: Gave himself up from HMS Scorpion - When taken: 27 May 1813 - Date Received: 9 Oct 1814 - From what ship: HMT Leyden, Chatham - Born: New Jersey - Age: 26 - Released on 15 Jun 1815.

Trash, James - Seaman - Nbr: 195 - Prize: Star - Ship Type: MV - How taken: HMS Superb - When taken: 9 Feb 1813 - Where taken: Bay of Biscay - Date Received: 2 Apr 1813 - From what ship: Plymouth - Born: New Castle - Age: 24 - Released on 10 Apr 1815.

Trask, Charles - Seaman - Nbr: 3749 - Prize: Leader - Ship Type: Packet - How taken: HMS Borer - When taken: 17 Sep 1813 - Where taken: off Long Island - Date Received: 30 Sep 1814 - From what ship: HMT President, Halifax - Born: Boston - Age: 19 - Released on 4 Jun 1815.

Trask, Osborne - Seaman - Nbr: 3365 - How taken: Impressed at Hull - When taken: 18 Oct 1813 - Date Received: 13 Sep 1814 - From what ship: HMT Niobe, Chatham - Born: Beverly - Age: 29 - Released on 28 May 1815.

Trask, William - Seaman - Nbr: 2761 - How taken: Gave himself up from HMS Bruin - When taken: 19 Jul 1813 - Date Received: 24 Aug 1814 - From what ship: HMT Liverpool, Chatham - Born: Boston - Age: 25 - Released on 19 May 1815.

Tratt, Lemuel - Seaman - Nbr: 6079 - Prize: Daedalus - Ship Type: MV - How taken: HMS Niemen - When taken: 20 Sep 1814 - Where taken: off Delaware - Date Received: 28 Dec 1814 - From what ship: HMT Penelope - Born: Dorchester - Age: 20 - Released on 3 Jul 1815.

Travers, Thomas - Seaman - Nbr: 4711 - How taken: Impressed at Gravesend - When taken: 20 Dec 1813 - Date Received: 9 Oct 1814 - From what ship: HMT Freya, Chatham - Born: Maryland - Age: 28 - Race: Black - Released on 11 Jul 1815.

Treadwell, Nathaniel - Seaman - Nbr: 4487 - Prize: Enterprize - Ship Type: P - How taken: HMS Tenedos - When taken: 21 May 1813 - Where taken: off Cape Cod - Date Received: 8 Oct 1814 - From what ship: HMT Leyden, Chatham - Born: Massachusetts - Age: 28 - Released on 15 Jun 1815.

Treadwell, Nathaniel - Seaman - Nbr: 3774 - Prize: Fame - Ship Type: P - How taken: HMS Thistle - When taken: 10 Apr 1814 - Where taken: after being cast ashore on a seal island - Date Received: 30 Sep 1814 - From what ship: HMT President, Halifax - Born: Boston - Age: 24 - Released on 9 Jun 1815.

Treadwell, Samuel - Seaman - Nbr: 5893 - Prize: Harlequin - Ship Type: P - How taken: HMS Bulwark - When

American Prisoners of War Held at Dartmoor during the War of 1812

## Alphabetical listing of names

taken: 23 Nov 1814 - Where taken: off Halifax - Date Received: 27 Dec 1814 - From what ship: HMT Penelope - Born: Ipswich - Age: 31 - Released on 3 Jul 1815.

Tempering, Joseph - Seaman - Nbr: 6149 - Prize: Robert - Ship Type: Prize - How taken: HMS Rhin - When taken: Mar 1814 - Where taken: off Charleston - Date Received: 17 Jan 1815 - From what ship: HMT Impregnable - Born: Philadelphia - Age: 19 - Died on 4 Jun 1815 from enteritis.

Trifle, Jasper - Seaman - Nbr: 1628 - Prize: Tom - Ship Type: LM - How taken: HMS Surveillante - When taken: 24 Apr 1813 - Where taken: Bay of Biscay - Date Received: 23 Jun 1814 - From what ship: Stapleton - Born: Massachusetts - Age: 22 - Released on 1 May 1815.

Tripe, W. H. - Captain - Nbr: 4844 - Prize: Hawke - Ship Type: P - How taken: HMS Pique - When taken: 26 Apr 1814 - Where taken: off Porto Rico - Date Received: 9 Oct 1814 - From what ship: HMT Freya, Halifax - Born: Portsmouth - Age: 27 - Released on 21 Jun 1815.

Tripp, Adam - Seaman - Nbr: 2574 - Prize: Rattlesnake - Ship Type: P - How taken: HMS Hyperion - When taken: 25 Jun 1814 - Where taken: off Cape Finisterre - Date Received: 21 Aug 1814 - From what ship: HMS Hyperion - Born: Philadelphia - Age: 17 - Released on 3 May 1815.

Tripper, Robert - Seaman - Nbr: 3407 - Prize: Thomas - Ship Type: P - How taken: HMS Nymphe - When taken: 24 Jun 1813 - Where taken: off Halifax - Date Received: 13 Sep 1814 - From what ship: HMT Niobe, Chatham - Born: Portsmouth - Age: 24 - Released on 8 May 1815.

Trips, O. - Seaman - Nbr: 4092 - Prize: Tyger - Ship Type: Prize - How taken: HMS Bulwark - When taken: 20 Jul 1814 - Where taken: Georges Bank - Date Received: 6 Oct 1814 - From what ship: HMT Chesapeake, Halifax - Born: North Carolina - Age: 22 - Released on 13 Jun 1815.

Troth, George - Master's Mate - Nbr: 3520 - Prize: Snap Dragon - Ship Type: P - How taken: HMS Martin - When taken: 30 Jun 1814 - Where taken: off Halifax - Date Received: 28 Sep 1814 - From what ship: HMT Salvador del Mundo - Born: Maryland - Age: 23 - Released on 3 May 1815.

Trout, Nathaniel - Seaman - Nbr: 4137 - Prize: Bordeaux Packet - Ship Type: LM - How taken: HMS Niemen - When taken: 28 Jun 1814 - Where taken: off Delaware - Date Received: 6 Oct 1814 - From what ship: HMT Chesapeake, Halifax - Born: Salem - Age: 54 - Released on 13 Jun 1815.

Trout, William - Seaman - Nbr: 690 - How taken: Impressed at Liverpool - When taken: 17 Jul 1813 - Date Received: 27 Sep 1813 - From what ship: Plymouth - Born: Boston - Age: 25 - Released on 26 Apr 1815.

Trowbridge, George - Seaman - Nbr: 5587 - Prize: Young William - Ship Type: Prize - How taken: HMS Plover - When taken: 8 Sep 1814 - Where taken: off St. Johns - Date Received: 17 Dec 1814 - From what ship: HMT Loire, Halifax - Born: New Haven - Age: 23 - Released on 1 Jul 1815.

Trowbridge, J. T. - Captain - Nbr: 5125 - How taken: Apprehended at Calcutta - When taken: Mar 1814 - Date Received: 28 Oct 1814 - From what ship: HMT Salvador del Mundo - Born: New Hampshire - Age: 33 - Released on 10 Apr 1815.

Truelove, John - Seaman - Nbr: 6521 - How taken: Gave himself up from HMS Ashea - Date Received: 3 Mar 1815 - From what ship: HMT Ganges, Plymouth - Born: Salem - Age: 32 - Released on 11 Jul 1815.

Truffle, John - Seaman - Nbr: 2445 - Prize: US Sloop Frolic - Ship Type: MW - How taken: HMS Orpheus - When taken: 20 Apr 1814 - Where taken: off Cuba - Date Received: 16 Aug 1814 - From what ship: HMT Queen, Halifax - Born: New Bedford - Age: 22 - Released on 3 May 1815.

Truffle, Joel - Seaman - Nbr: 5801 - Prize: US Frigate Superior (Gig) - Ship Type: MW - How taken: British gunboats - When taken: 26 Aug 1814 - Where taken: Lake Ontario - Date Received: 26 Dec 1814 - From what ship: HMT Argo - Born: Marblehead - Age: 26 - Released on 16 Jun 1815.

Truman, John - Seaman - Nbr: 6296 - Prize: John - Ship Type: Prize - How taken: Leander (Newcastle Acasta) - When taken: 30 Jan 1815 - Date Received: 19 Feb 1815 - From what ship: HMT Ganges, Plymouth - Born: New Haven - Age: 22 - Race: Black - Released on 5 Jul 1815.

Truman, Samuel - Seaman - Nbr: 6197 - Prize: Prince de Neufchatel - Ship Type: P - How taken: Leander (Newcastle Acasta) - When taken: 20 Dec 1814 - Where taken: Lat 38 Long 56 - Date Received: 30 Jan 1815 - From what ship: HMT Pheasant - Born: Boston - Age: 19 - Released on 5 Jul 1815.

Trails, John - Seaman - Nbr: 2306 - Prize: Hussar - Ship Type: P - How taken: HMS Saturn - When taken: 25 May 1814 - Where taken: off Sandy Hook - Date Received: 16 Aug 1814 - From what ship: HMS Dublin, Halifax - Born: Philadelphia - Age: 22 - Released on 3 May 1815.

Trusty, Henry - 2nd Mates - Nbr: 1787 - Prize: Ferox - Ship Type: MV - How taken: HMS Medusa & HMS Lyra - When taken: 28 Mar 1813 - Where taken: off Cape Ortegal (Spain) - Date Received: 20 Jul 1814 - From what ship: HMS Milford, Plymouth - Born: Philadelphia - Age: 29 - Released on 1 May 1815.

Tub, Samuel - Seaman - Nbr: 5555 - Prize: Levant - Ship Type: MV - How taken: HMS Forester - When taken: 4

American Prisoners of War Held at Dartmoor during the War of 1812

## Alphabetical listing of names

Jan 1814 - Where taken: Bahamas Banks - Date Received: 17 Dec 1814 - From what ship: HMT Loire, Halifax - Born: Beverly - Age: 21 - Released on 1 Jul 1815.

Tubbs, Martin - Seaman - Nbr: 967 - Prize: Siro - Ship Type: LM - How taken: HMS Pelican - When taken: 13 Jan 1814 - Where taken: at sea - Date Received: 31 Jan 1814 - From what ship: Plymouth - Born: New York - Age: 24 - Released on 27 Apr 1815.

Tucker, Andrew - Prize master - Nbr: 4541 - Prize: Growler - Ship Type: P - How taken: HMS Electra - When taken: 7 Jul 1813 - Where taken: at sea - Date Received: 8 Oct 1814 - From what ship: HMT Leyden, Chatham - Born: Marblehead - Age: 36 - Released on 15 Jun 1815.

Tucker, Edward - Seaman - Nbr: 3703 - Prize: Alfred - Ship Type: P - How taken: HMS Epervier - When taken: 23 Feb 1814 - Where taken: off Newfoundland - Date Received: 30 Sep 1814 - From what ship: HMT President, Halifax - Born: Salem - Age: 19 - Released on 4 Jun 1815.

Tucker, Henry - Mate - Nbr: 4243 - Prize: Requin - Ship Type: LM - How taken: HMS Venus - When taken: 6 Mar 1814 - Where taken: off Bordeaux - Date Received: 7 Oct 1814 - From what ship: HMT Niobe, Chatham - Born: Charleston - Age: 21 - Released on 13 Jun 1815.

Tucker, James - Seaman - Nbr: 4219 - Prize: Liberty - Ship Type: MV - How taken: Surrendered - When taken: 30 Dec 1813 - Where taken: Stromess - Date Received: 7 Oct 1814 - From what ship: HMT Niobe, Chatham - Born: Long Island - Age: 24 - Died on 28 Apr 1815 from abscess.

Tucker, James - Seaman - Nbr: 5501 - Prize: US Gunboat Nbr 2 - Ship Type: MW - How taken: British forces - When taken: 22 Aug 1814 - Where taken: Chesapeake Bay - Date Received: 17 Dec 1814 - From what ship: HMT Loire, Halifax - Born: New Jersey - Age: 42 - Released on 29 Jun 1815.

Tucker, Levi - Seaman - Nbr: 5939 - Prize: Harlequin - Ship Type: P - How taken: HMS Bulwark - When taken: 23 Nov 1814 - Where taken: off Halifax - Date Received: 27 Dec 1814 - From what ship: HMT Penelope - Born: Lee - Age: 31 - Released on 3 Jul 1815.

Tucker, N. - Seaman - Nbr: 3233 - Prize: Sampson - Ship Type: MV - How taken: Rebuff - When taken: 12 May 1813 - Where taken: off Cape St. Vincent - Date Received: 11 Sep 1814 - From what ship: HMT Freya, Chatham - Born: New Hampshire - Age: 20 - Released on 28 May 1815.

Tucker, Nathaniel - Seaman - Nbr: 4545 - Prize: Growler - Ship Type: P - How taken: HMS Electra - When taken: 7 Jul 1813 - Where taken: at sea - Date Received: 8 Oct 1814 - From what ship: HMT Leyden, Chatham - Born: Marblehead - Age: 22 - Released on 15 Jun 1815.

Tucker, Samuel - Seaman - Nbr: 5170 - Prize: Volante - Ship Type: P - How taken: HMS Curlew - When taken: 25 Nov 1813 - Where taken: off Halifax - Date Received: 31 Oct 1814 - From what ship: HMT Mermaid, Chatham - Born: Salem - Age: 16 - Released on 29 Jun 1815.

Tucker, Samuel - Master - Nbr: 3422 - Prize: Industry - Ship Type: P - How taken: HMS Heron - When taken: 3 Nov 1813 - Where taken: off Halifax - Date Received: 13 Sep 1814 - From what ship: HMT Niobe, Chatham - Born: Marblehead - Age: 21 - Released on 28 May 1815.

Tucker, William - Seaman - Nbr: 4503 - Prize: Enterprize - Ship Type: P - How taken: HMS Tenedos - When taken: 21 May 1813 - Where taken: off Cape Cod - Date Received: 8 Oct 1814 - From what ship: HMT Leyden, Chatham - Born: Massachusetts - Age: 17 - Released on 15 Jun 1815.

Tucker, Zebulon. - Seaman - Nbr: 2212 - Prize: Hussar - Ship Type: P - How taken: HMS Saturn - When taken: 25 May 1814 - Where taken: off Sandy Hook - Date Received: 16 Aug 1814 - From what ship: HMS Dublin, Halifax - Born: Massachusetts - Age: 31 - Released on 2 May 1815.

Tuckerman, William - Seaman - Nbr: 5281 - Prize: Governor Plumer - Ship Type: P - How taken: HMS Shannon - When taken: 26 May 1813 - Where taken: at sea - Date Received: 31 Oct 1814 - From what ship: HMT Leyden, Chatham - Born: New Hampshire - Age: 37 - Released on 11 Jul 1815.

Tufts, Zaria - Seaman - Nbr: 47 - Prize: Terrible - Ship Type: MV - How taken: HMS Foxhound - When taken: 8 Feb 1813 - Where taken: Channel - Date Received: 2 Apr 1813 - From what ship: Plymouth - Born: New Hampton - Age: 28 - Sent to Dartmouth on 10 Jul 1813.

Tuilford, Joseph - Seaman - Nbr: 2354 - Prize: Snap Dragon - Ship Type: P - How taken: HMS Martin - When taken: 10 Jun 1814 - Where taken: off Halifax - Date Received: 16 Aug 1814 - From what ship: HMT Queen, Halifax - Born: North Carolina - Age: 22 - Died on 27 Jan 1815 from variola.

Tulloch, Thomas - Seaman - Nbr: 6072 - Prize: Daedalus - Ship Type: MV - How taken: HMS Niemen - When taken: 20 Sep 1814 - Where taken: off Delaware - Date Received: 28 Dec 1814 - From what ship: HMT Penelope - Born: New York - Age: 38 - Released on 3 Jul 1815.

Tully, Hugh - Seaman - Nbr: 1196 - Prize: Imperatrice Reine - Ship Type: P - How taken: HMS Hotspur - When taken: 13 Jan 1813 - Date Received: 4 Jun 1814 - From what ship: Dartmouth - Born: North Carolina - Age:

American Prisoners of War Held at Dartmoor during the War of 1812

## Alphabetical listing of names

25 - Released on 28 Apr 1815.

Turnbull, James - Seaman - Nbr: 2612 - How taken: Gave himself up from HMS Romulus - When taken: 14 Aug 1812 - Date Received: 21 Aug 1814 - From what ship: HMT Freya, Chatham - Born: Charleston - Age: 33 - Released on 26 Apr 1815.

Turnbull, John - Boy - Nbr: 1881 - Prize: Elbridge Gerry - Ship Type: P - How taken: HMS Crescent - When taken: 16 Sep 1813 - Where taken: off St. George's - Date Received: 29 Jul 1814 - From what ship: HMS Ville de Paris, Chatham Depot - Born: Portland - Age: 16 - Released on 28 May 1815.

Turner, Daniel - Seaman - Nbr: 1057 - How taken: Impressed at Cork - Date Received: 10 May 1814 - From what ship: Plymouth - Born: Maryland - Age: 44 - Released on 27 Apr 1815.

Turner, David - Seaman - Nbr: 5376 - Prize: Derby - Ship Type: MV - How taken: HMS Nereus - When taken: 4 Feb 1813 - Where taken: off Cape of Good Hope - Date Received: 31 Oct 1814 - From what ship: HMT Leyden, Chatham - Born: Boston - Age: 23 - Died on 17 Mar 1815 from phthisis.

Turner, Gardner - Seaman - Nbr: 4765 - Prize: Valentine - Ship Type: MV - How taken: HMS Minden - When taken: 16 Nov 1812 - Where taken: Malager Bay - Date Received: 9 Oct 1814 - From what ship: HMT Freya, Chatham - Born: Tiverton - Age: 20 - Released on 27 Apr 1815.

Turner, George - Seaman - Nbr: 2353 - Prize: Snap Dragon - Ship Type: P - How taken: HMS Martin - When taken: 10 Jun 1814 - Where taken: off Halifax - Date Received: 16 Aug 1814 - From what ship: HMT Queen, Halifax - Born: North Carolina - Age: 24 - Released on 3 May 1815.

Turner, James - Seaman - Nbr: 5343 - Prize: Leda - Ship Type: Packet - How taken: HMS Epervier - When taken: 28 Oct 1813 - Where taken: off Block Island - Date Received: 31 Oct 1814 - From what ship: HMT Leyden, Chatham - Born: Charleston - Age: 22 - Released on 1 Jul 1815.

Turner, James - Seaman - Nbr: 5636 - Prize: Lion - Ship Type: P - How taken: HMS Granicus - When taken: 2 Dec 1814 - Where taken: off Lisbon - Date Received: 24 Dec 1814 - From what ship: HMT Hanover, Gibraltar - Born: Charleston - Age: 44 - Released on 3 Jul 1815.

Turner, John - Seaman - Nbr: 3985 - Prize: US Brig Rattlesnake - Ship Type: MW - How taken: HMS Leander - When taken: 13 Jul 1814 - Where taken: off Shelburne - Date Received: 6 Oct 1814 - From what ship: HMT Chesapeake, Halifax - Born: Massachusetts - Age: 23 - Died on 5 Apr 1815 from variola.

Turner, Joseph - Seaman - Nbr: 6455 - Prize: Jemmell - Ship Type: MV - How taken: HMS Rhin - When taken: 28 May 1814 - Where taken: off Bermuda - Date Received: 3 Mar 1815 - From what ship: HMT Ganges, Plymouth - Born: Massachusetts - Age: 33 - Released on 11 Jul 1815.

Turner, Samuel - Captain - Nbr: 4680 - Prize: Elbridge Gerry - Ship Type: P - How taken: HMS Crescent - When taken: 16 Sep 1813 - Where taken: at sea - Date Received: 9 Oct 1814 - From what ship: HMT Leyden, Chatham - Born: New York - Age: 26 - Released on 2 Apr 1815.

Turner, Silas - Seaman - Nbr: 5293 - Prize: Porcupine - Ship Type: LM - How taken: HMS Wasp - When taken: 17 Jun 1814 - Where taken: off Cape Sable Island (Canada) - Date Received: 31 Oct 1814 - From what ship: HMT Leyden, Chatham - Born: Montville - Age: 24 - Released on 29 Jun 1815.

Turner, Thomas - Seaman - Nbr: 2316 - Prize: Hussar - Ship Type: P - How taken: HMS Saturn - When taken: 25 May 1814 - Where taken: off Sandy Hook - Date Received: 16 Aug 1814 - From what ship: HMS Dublin, Halifax - Born: Charlestown - Age: 32 - Race: Mulatto - Released on 3 May 1815.

Turner, William - Seaman - Nbr: 2111 - How taken: Gave himself up from HMS Akbar - When taken: 20 Jun 1814 - Date Received: 5 Aug 1814 - From what ship: Bristol - Born: Maryland - Age: 28 - Race: Black - Released on 2 May 1815.

Turner, William - Seaman - Nbr: 281 - Prize: Cannoneer - Ship Type: MV - How taken: HMS Warspite - When taken: 14 Mar 1813 - Where taken: Bay of Biscay - Date Received: 28 Jun 1813 - From what ship: Plymouth - Born: Philadelphia - Age: 22 - Released on 20 Apr 1815.

Turney, B. - Seaman - Nbr: 3926 - Prize: Rolla - Ship Type: P - How taken: HMS Loire - When taken: 10 Dec 1813 - Where taken: off Bull Island (South Carolina) - Date Received: 5 Oct 1814 - From what ship: HMT President, Halifax - Born: Newport - Age: 26 - Released on 9 Jun 1815.

Turrell, Ebenezer - Seaman - Nbr: 2262 - Prize: Grand Turk - Ship Type: P - How taken: HMS Martin - When taken: 26 May 1814 - Where taken: off Cape Sable Island (Canada) - Date Received: 16 Aug 1814 - From what ship: HMS Dublin, Halifax - Born: Salem - Age: 24 - Released on 3 May 1815.

Tuttle, French - Seaman - Nbr: 1489 - Prize: Leo - Ship Type: LM - How taken: HMS Magicienne - When taken: 4 Jun 1813 - Where taken: off France - Date Received: 19 Jun 1814 - From what ship: Stapleton - Born: Falmouth - Age: 26 - Died on 24 Nov 1814.

Tuttle, Joseph - Seaman - Nbr: 4212 - Prize: Devon - Ship Type: Prize - How taken: HMS Fly - When taken: 2 Jan

American Prisoners of War Held at Dartmoor during the War of 1812

## Alphabetical listing of names

1814 - Where taken: at sea - Date Received: 7 Oct 1814 - From what ship: HMT Niobe, Chatham - Born: Freeport - Age: 27 - Released on 20 Mar 1815.

Twycross, Samuel - Mate - Nbr: 2132 - How taken: Impressed at London - When taken: 13 Jan 1813 - Date Received: 8 Aug 1814 - From what ship: HMT Raven, Chatham - Born: Dresden - Age: 35 - Released on 27 Apr 1815.

Tyler, Daniel - Seaman - Nbr: 3934 - Prize: Rolla - Ship Type: P - How taken: HMS Loire - When taken: 10 Dec 1813 - Where taken: off Bull Island (South Carolina) - Date Received: 5 Oct 1814 - From what ship: HMT President, Halifax - Born: Norfolk - Age: 21 - Released on 9 Jun 1815.

Tyler, Joseph - Seaman - Nbr: 4609 - Prize: Argus - Ship Type: MV - How taken: HMS San Domingo - When taken: 1 Mar 1814 - Where taken: off Savannah - Date Received: 9 Oct 1814 - From what ship: HMT Leyden, Chatham - Born: Newburyport - Age: 19 - Released on 15 Jun 1815.

Tyler, Louis - Seaman - Nbr: 2774 - How taken: Gave himself up from HMS Puipant - Date Received: 24 Aug 1814 - From what ship: HMT Liverpool, Chatham - Born: Bedford - Age: 23 - Released on 19 May 1815.

Typhon, Louis - Seaman - Nbr: 4230 - Prize: Liberty - Ship Type: MV - How taken: Surrendered - When taken: 30 Dec 1813 - Where taken: Stromess - Date Received: 7 Oct 1814 - From what ship: HMT Niobe, Chatham - Born: Glastonbury - Age: 16 - Race: Mulatto - Released on 13 Jun 1815.

Tyre, William - Seaman - Nbr: 412 - Prize: Viper - Ship Type: MV - How taken: HMS Superb - When taken: 15 Apr 1813 - Where taken: Bay of Biscay - Date Received: 1 Jul 1813 - From what ship: Plymouth - Born: Wendell - Age: 21 - Died on 25 Feb 1814 from dysentery.

Uen, Joseph - Seaman - Nbr: 5954 - Prize: Harlequin - Ship Type: P - How taken: HMS Bulwark - When taken: 23 Nov 1814 - Where taken: off Halifax - Date Received: 27 Dec 1814 - From what ship: HMT Penelope - Born: Africa - Age: 54 - Race: Black - Released on 3 Jul 1815.

Underwood, B. - Boy - Nbr: 5918 - Prize: Harlequin - Ship Type: P - How taken: HMS Bulwark - When taken: 23 Nov 1814 - Where taken: off Halifax - Date Received: 27 Dec 1814 - From what ship: HMT Penelope - Born: Portsmouth - Age: 17 - Released on 2 May 1815.

Underwood, John - Seaman - Nbr: 519 - Prize: Friends - Ship Type: Bey of Pool - How taken: HMS Whiting - When taken: 15 Jul 1813 - Where taken: Lat 67 N, Long 8 W - Date Received: 8 Sep 1813 - From what ship: Plymouth - Born: Massachusetts - Age: 46 - Released on 9 Apr 1815.

Urey, Peter - Seaman - Nbr: 1911 - How taken: Gave himself up from HMS Mars - When taken: 9 Dec 1812 - Date Received: 3 Aug 1814 - From what ship: HMS Alceste, Chatham Depot - Born: New York - Age: 22 - Released on 26 Apr 1815.

Ustick, William - Prize Master - Nbr: 2995 - Prize: St. Lawrence - How taken: HMS Aquilon - When taken: 9 Aug 1814 - Where taken: off Western Islands (England) - Date Received: 29 Aug 1814 - From what ship: HMT Bittern - Born: Philadelphia - Age: 29 - Released on 19 May 1815.

Vail, Jeremiah - Seaman - Nbr: 2921 - How taken: Gave himself up from HMS Dwarf - When taken: 28 Jul 1813 - Date Received: 24 Aug 1814 - From what ship: HMT Alpheus, Chatham - Born: Long Island - Age: 32 - Released on 19 May 1815.

Valentine, John - Seaman - Nbr: 200 - Prize: Star - Ship Type: MV - How taken: HMS Superb - When taken: 9 Feb 1813 - Where taken: Bay of Biscay - Date Received: 2 Apr 1813 - From what ship: Plymouth - Born: New York - Age: 17 - Released on 20 Apr 1815.

Valentine, John - Seaman - Nbr: 3203 - How taken: Gave himself up from HMS America - When taken: 16 Jul 1813 - Date Received: 11 Sep 1814 - From what ship: HMT Freya, Chatham - Born: Boston - Age: 37 - Race: Black - Released on 28 May 1815.

Valiant, Richard - Seaman - Nbr: 827 - Prize: Chesapeake - Ship Type: LM - How taken: HMS Hotspur & HMS Pyramus - When taken: 26 Oct 1813 - Where taken: off Nantes - Date Received: 29 Nov 1813 - From what ship: Plymouth - Born: Maryland - Age: 26 - Escaped on 5 Sep 1814.

Van Horn, James - Seaman - Nbr: 3005 - How taken: Gave himself up from HMS Galatea - Date Received: 29 Aug 1814 - From what ship: HMT Bittern - Born: New Orleans - Age: 26 - Released on 28 May 1815.

Vandeburg, Adam - Seaman - Nbr: 1698 - Prize: Fanny - Ship Type: Prize - How taken: HMS Sceptre - When taken: 12 May 1814 - Date Received: 2 Jul 1814 - From what ship: Plymouth - Born: Malmo, Sweden - Age: 29 - Released on 8 Nov 1814.

Vanderburgh, Cornelius - Seaman - Nbr: 3002 - Prize: St. Lawrence - How taken: HMS Aquilon - When taken: 9 Aug 1814 - Where taken: off Western Islands (England) - Date Received: 29 Aug 1814 - From what ship: HMT Bittern - Born: New York - Age: 32 - Released on 28 May 1815.

Vanderford, Benjamin - Lieutenant - Nbr: 3729 - Prize: Alfred - Ship Type: P - How taken: HMS Epervier - When

American Prisoners of War Held at Dartmoor during the War of 1812

## Alphabetical listing of names

taken: 23 Feb 1814 - Where taken: off Newfoundland - Date Received: 30 Sep 1814 - From what ship: HMT President, Halifax - Born: Salem - Age: 25 - Released on 4 Jun 1815.

Vanderhaven, Mathias - Seaman - Nbr: 1902 - Prize: Quebec - Ship Type: Prize - How taken: HMS Derwent - When taken: 29 Jan 1813 - Where taken: off Lisbon - Date Received: 3 Aug 1814 - From what ship: HMS Alceste, Chatham Depot - Born: New Jersey - Age: 21 - Released on 2 May 1815.

Vanderventer, John - Seaman - Nbr: 2927 - How taken: Gave himself up from HMS Sapion - Date Received: 24 Aug 1814 - What ship: HMT Alpheus, Chatham - Born: Londonderry - Age: 22 - Released on 19 May 1815.

Vangorbet, Cato - Seaman - Nbr: 2808 - Prize: Weazel - Ship Type: MV - How taken: HMS Foxhound - When taken: 25 Mar 1813 - Where taken: Bay of Biscay - Date Received: 24 Aug 1814 - From what ship: HMT Liverpool, Chatham - Born: New Jersey - Age: 22 - Released on 19 May 1815.

Vankirk, Joseph - Seaman - Nbr: 37 - How taken: Apprehended at Gibraltar - When taken: 8 Aug 1813 - Date rec'd: 2 Apr 1813 - What ship: Plymouth - Born: Wheeland - Age: 23 - Sent to Dartmouth on 30 Jul 1813.

Vanmetre, Henry - Seaman - Nbr: 5859 - Prize: Lion - Ship Type: P - How taken: HMS Granicus - When taken: 2 Dec 1814 - Where taken: off Lisbon - Date Received: 26 Dec 1814 - From what ship: HMT Impregnable - Born: Virginia - Age: 39 - Race: Black - Released on 3 Jul 1815.

Vanrant, John - Seaman - Nbr: 5248 - Prize: York Town - Ship Type: P - How taken: HMS Maidstone - When taken: 18 Jul 1813 - Where taken: Grand Banks - Date Received: 31 Oct 1814 - From what ship: HMT Mermaid, Chatham - Born: Savannah - Age: 25 - Released on 29 Jun 1815.

Vansykle, Ralph - Seaman - Nbr: 2319 - Prize: Hussar - Ship Type: P - How taken: HMS Saturn - When taken: 25 May 1814 - Where taken: off Sandy Hook - Date Received: 16 Aug 1814 - From what ship: HMS Dublin, Halifax - Born: New Jersey - Age: 24 - Released on 3 May 1815.

Vanvort, Richard - Seaman - Nbr: 3657 - Prize: Young Teager - Ship Type: P - How taken: HMS La Hogue - When taken: 26 Jul 1813 - Where taken: off Nova Scotia - Date Received: 30 Sep 1814 - From what ship: HMT President - Born: New York - Age: 39 - Released on 4 Jun 1815.

Varmrod, Samuel - Master's mate - Nbr: 5671 - Prize: Harlequin - Ship Type: P - How taken: HMS Bulwark - When taken: 23 Nov 1814 - Where taken: off Halifax - Date Received: 24 Dec 1814 - From what ship: HMT Penelope - Born: Portsmouth - Age: 35 - Released on 3 Jul 1815.

Varney, David - Seaman - Nbr: 5928 - Prize: Harlequin - Ship Type: P - How taken: HMS Bulwark - When taken: 23 Nov 1814 - Where taken: off Halifax - Date Received: 27 Dec 1814 - From what ship: HMT Penelope - Born: Massachusetts - Age: 39 - Released on 3 Jul 1815.

Varney, John - Seaman - Nbr: 3300 - Prize: Governor Plumer - Ship Type: P - How taken: Sent into custody from a privateer - When taken: 1 Jun 1813 - Where taken: off Cape Ann - Date Received: 13 Sep 1814 - From what ship: HMT Niobe, Chatham - Born: Massachusetts - Age: 21 - Released on 28 May 1815.

Vassina, John B. - Seaman - Nbr: 3850 - Prize: Dominique - Ship Type: LM - How taken: HMS Dotterel - When taken: 21 May 1814 - Where taken: off Charleston - Date Received: 5 Oct 1814 - From what ship: HMT Orpheus, Halifax - Born: Bordeaux - Age: 23 - Released on 9 Jun 1815.

Vaughan, Nicholas - Seaman - Nbr: 306 - Prize: Ducornau - Ship Type: MV - How taken: HMS Pheasant - When taken: 15 Mar 1813 - Where taken: Bay of Biscay - Date Received: 28 Jun 1813 - From what ship: Plymouth - Born: Newport - Age: 27 - Died on 31 Aug 1814 from fever.

Vaughan, Robert - Seaman - Nbr: 1987 - How taken: Gave himself up from HMS Peruvian - When taken: Apr 1813 - Date Received: 3 Aug 1814 - From what ship: HMS Lyffey, Chatham Depot - Born: Boston - Age: 24 - Escaped on 19 Sep 1814.

Vaughan, Thomas - Seaman - Nbr: 193 - Prize: Star - Ship Type: MV - How taken: HMS Superb - When taken: 9 Feb 1813 - Where taken: Bay of Biscay - Date Received: 2 Apr 1813 - From what ship: Plymouth - Born: New York - Age: 21 - Released on 20 Apr 1815.

Veal, Charles - Boatswain's Mate - Nbr: 3972 - Prize: US Brig Rattlesnake - Ship Type: MW - How taken: HMS Leander - When taken: 13 Jul 1814 - Where taken: off Shelburne - Date Received: 6 Oct 1814 - From what ship: HMT Chesapeake, Halifax - Born: Marblehead - Age: 55 - Released on 9 Jun 1815.

Veal, Pierre - Seaman - Nbr: 1485 - Prize: Napoleon - Ship Type: LM - How taken: HMS Belle Poule - When taken: 3 Apr 1813 - Where taken: off Cape Ortegal (Spain) - Date Received: 19 Jun 1814 - From what ship: Stapleton - Born: Connecticut - Age: 24 - Released on 1 May 1815.

Veitch, William - Seaman - Nbr: 1333 - Prize: Paul Jones - Ship Type: P - How taken: HMS Leonidas - When taken: 23 May 1813 - Where taken: Channel - Date Received: 19 Jun 1814 - From what ship: Stapleton - Born: New York - Age: 19 - Released on 28 Apr 1815.

Velpley, Israel - Seaman - Nbr: 5546 - Prize: Phaeton - Ship Type: MV - How taken: Cast Away - When taken: 7

American Prisoners of War Held at Dartmoor during the War of 1812

## Alphabetical listing of names

Feb 1814 - Where taken: Old Heights - Date Received: 17 Dec 1814 - From what ship: HMT Loire, Halifax - Born: New York - Age: 22 - Released on 1 Jul 1815.

Veney, George - Seaman - Nbr: 2059 - Prize: Tom Thumb - Ship Type: MV - How taken: Lyon (Privateer) - When taken: 15 Feb 1812 - Where taken: Bay of Biscay - Date Received: 3 Aug 1814 - From what ship: HMS Bittern, Chatham Depot - Born: Philadelphia - Age: 22 - Released on 26 Apr 1815.

Vennard, George - Seaman - Nbr: 4466 - Prize: Governor Plumer - Ship Type: P - How taken: HMS Shamrock - When taken: 4 Mar 1813 - Where taken: at sea - Date Received: 8 Oct 1814 - From what ship: HMT Leyden, Chatham - Born: New Hampshire - Age: 21 - Released on 15 Jun 1815.

Verdon, Louis - Seaman - Nbr: 2179 - Prize: Rattlesnake - Ship Type: P - How taken: HMS Hyperion - When taken: 26 Jun 1814 - Where taken: off Cape Finisterre - Date Received: 16 Aug 1814 - From what ship: HMS Dublin, Halifax - Born: Rochelle - Age: 23 - Released on 2 May 1815.

Very, Samuel - Prize Master - Nbr: 1898 - Prize: Quebec - Ship Type: Prize - How taken: HMS Derwent - When taken: 29 Jan 1813 - Where taken: off Lisbon - Date Received: 3 Aug 1814 - From what ship: HMS Alceste, Chatham Depot - Born: Salem - Age: 28 - Released on 2 May 1815.

Vesan, Weslant - Seaman - Nbr: 5040 - Prize: Betsey - Ship Type: Prize - How taken: HMS Pylades - When taken: 7 Sep 1814 - Where taken: Canten - Date Received: 28 Oct 1814 - From what ship: HMT Alkbar, Halifax - Born: New York - Age: 28 - Released on 29 Jun 1815.

Vesplash, Nicholas - Seaman - Nbr: 5867 - Prize: Lion - Ship Type: P - How taken: HMS Granicus - When taken: 2 Dec 1814 - Where taken: off Lisbon - Date Received: 26 Dec 1814 - From what ship: HMT Impregnable - Born: Massachusetts - Age: 29 - Released on 3 Jul 1815.

Vicary, Richard - Seaman - Nbr: 1835 - Prize: Polly - Ship Type: MV - How taken: HMS Surveillante - When taken: 23 Mar 1813 - Where taken: Bay of Biscay - Date Received: 21 Jul 1814 - From what ship: HMT Redbeard & Pincher, Chatham Depot - Born: Beverly - Age: 17 - Released on 1 May 1815.

Vickery, Thomas - Seaman - Nbr: 3691 - Prize: Alfred - Ship Type: P - How taken: HMS Epervier - When taken: 23 Feb 1814 - Where taken: off Newfoundland - Date Received: 30 Sep 1814 - From what ship: HMT President, Halifax - Born: Marblehead - Age: 18 - Released on 4 Jun 1815.

Vickery, William - Seaman - Nbr: 3984 - Prize: US Brig Rattlesnake - Ship Type: MW - How taken: HMS Leander - When taken: 13 Jul 1814 - Where taken: off Shelburne - Date Received: 6 Oct 1814 - From what ship: HMT Chesapeake, Halifax - Born: Manchester - Age: 21 - Released on 9 Jun 1815.

Vincent, John - Seaman - Nbr: 1981 - How taken: Gave himself up from HMS Barham - When taken: Apr 1813 - Date Received: 3 Aug 1814 - From what ship: HMS Lyffey, Chatham Depot - Born: Philadelphia - Age: 29 - Released on 2 May 1815.

Vindine, Gar. - Seaman - Nbr: 1314 - How taken: Sent into custody from HMS Minden - Date Received: 14 Jun 1814 - From what ship: Mill Prison (Plymouth, England) - Born: Philadelphia - Age: 32 - Released on 28 Apr 1815.

Vinson, Lewis - Boy - Nbr: 367 - Prize: Two Brothers - Ship Type: MV - How taken: Beetle (LM) - When taken: 18 Mar 1813 - Where taken: off Western Islands (England) - Date Received: 1 Jul 1813 - From what ship: Plymouth - Born: Boston - Age: 18 - Sent to Liverpool on 25 Aug 1813.

Virgin, Robert - Seaman - Nbr: 2573 - Prize: Rattlesnake - Ship Type: P - How taken: HMS Hyperion - When taken: 25 Jun 1814 - Where taken: off Cape Finisterre - Date Received: 21 Aug 1814 - From what ship: HMS Hyperion - Born: Philadelphia - Age: 25 - Race: Negro - Released on 3 May 1815.

Voight, Henry - Seaman - Nbr: 1817 - How taken: Gave himself up from HMS Swiftsure - When taken: 26 Dec 1813 - Date Received: 20 Jul 1814 - From what ship: HMS Milford, Plymouth - Born: Pennsylvania - Age: 31 - Released on 26 Apr 1815.

Vorhis, James - Seaman - Nbr: 2792 - Prize: Weazel - Ship Type: MV - How taken: HMS Foxhound - When taken: 25 Mar 1813 - Where taken: Bay of Biscay - Date Received: 24 Aug 1814 - From what ship: HMT Liverpool, Chatham - Born: New Jersey - Age: 20 - Released on 19 May 1815.

Vorhis, Peter - Seaman - Nbr: 5528 - Prize: Sparks - Ship Type: LM - How taken: HMS Maidstone - When taken: 28 Sep 1814 - Where taken: off Nantucket - Date Received: 17 Dec 1814 - From what ship: HMT Loire, Halifax - Born: New York - Age: 27 - Released on 11 Jul 1815.

Voustyten, Christopher - Seaman - Nbr: 2157 - Prize: Rattlesnake - Ship Type: P - How taken: HMS Hyperion - When taken: 26 Jun 1814 - Where taken: off Cape Finisterre - Date Received: 16 Aug 1814 - From what ship: HMS Dublin, Halifax - Born: New York - Age: 23 - Released on 2 May 1815.

Vowdy, William - Mate - Nbr: 3603 - Prize: Tyren - Ship Type: MV - How taken: HMS Barbados - When taken: 20 Jan 1814 - Where taken: off St. Bartholomew - Date Received: 30 Sep 1814 - From what ship: HMT Sybella

American Prisoners of War Held at Dartmoor during the War of 1812

## Alphabetical listing of names

- Born: New York - Age: 32 - Released on 4 Jun 1815.
Wadden, Isaac - Seaman - Nbr: 3716 - Prize: Alfred - Ship Type: P - How taken: HMS Epervier - When taken: 23 Feb 1814 - Where taken: off Newfoundland - Date Received: 30 Sep 1814 - From what ship: HMT President, Halifax - Born: Marblehead - Age: 20 - Released on 4 Jun 1815.
Wadden, Jacob - Seaman - Nbr: 68 - Prize: Spitfire - Ship Type: MV - How taken: HMS Achates - When taken: 14 Feb 1813 - Where taken: off Ushant (France) - Date Received: 2 Apr 1813 - From what ship: Plymouth - Born: Marblehead - Age: 39 - Released on 20 Apr 1815.
Wade, John - Seaman - Nbr: 2308 - Prize: Hussar - Ship Type: P - How taken: HMS Saturn - When taken: 25 May 1814 - Where taken: off Sandy Hook - Date Received: 16 Aug 1814 - From what ship: HMS Dublin, Halifax - Born: Pennsylvania - Age: 35 - Released on 3 May 1815.
Wade, John - Seaman - Nbr: 828 - Prize: Chesapeake - Ship Type: LM - How taken: HMS Hotspur & HMS Pyramus - When taken: 26 Oct 1813 - Where taken: off Nantes - Date Received: 29 Nov 1813 - From what ship: Plymouth - Born: New York - Age: 21 - Released on 26 Apr 1815.
Wade, Nathan - Seaman - Nbr: 5826 - Prize: US Schooner Tigress - Ship Type: MW - How taken: British gunboats - When taken: 2 Sep 1814 - Where taken: Lake Erie - Date Received: 26 Dec 1814 - From what ship: HMT Argo - Born: New London - Age: 18 - Released on 3 Jul 1815.
Wade, Richard H. - Seaman - Nbr: 760 - How taken: Impressed at Liverpool - Date Received: 3 Nov 1813 - From what ship: Plymouth - Born: Wollish - Age: 22 - Released on 26 Apr 1815.
Wadge, John - Seaman - Nbr: 3819 - Prize: Nimble - Ship Type: Prize - How taken: HMS Arab - When taken: 5 Apr 1814 - Where taken: Lat 37 Long 65 - Date Received: 5 Oct 1814 - From what ship: HMT Orpheus, Halifax - Born: Massachusetts - Age: 18 - Released on 9 Jun 1815.
Wadsworth, Thomas - Seaman - Nbr: 1712 - How taken: Gave himself up at Fort Depot - Date Received: 8 Jul 1814 - From what ship: Labrador - Born: Salem - Age: 65 - Released on 1 May 1815.
Waide, Nathaniel - Boy - Nbr: 4043 - Prize: US Brig Rattlesnake - Ship Type: MW - How taken: HMS Leander - When taken: 13 Jul 1814 - Where taken: off Shelburne - Date Received: 6 Oct 1814 - From what ship: HMT Chesapeake, Halifax - Born: Boston - Age: 17 - Released on 9 Jun 1815.
Wainwood, William - Seaman - Nbr: 3954 - Prize: Rolla - Ship Type: P - How taken: HMS Loire - When taken: 10 Dec 1813 - Where taken: off Bull Island (South Carolina) - Date Received: 5 Oct 1814 - From what ship: HMT President, Halifax - Born: Rhode Island - Age: 21 - Race: Negro - Released on 9 Jun 1815.
Waite, Abel - Quartermaster - Nbr: 562 - Prize: US Brig Argus - Ship Type: MW - How taken: HMS Pelican - When taken: 14 Apr 1813 - Where taken: Irish Channel - Date Received: 8 Sep 1813 - From what ship: Plymouth - Born: Rhode Island - Age: 26 - Sent to Dartmouth on 2 Nov 1814.
Waite, Jacob - Seaman - Nbr: 3705 - Prize: Alfred - Ship Type: P - How taken: HMS Epervier - When taken: 23 Feb 1814 - Where taken: off Newfoundland - Date Received: 30 Sep 1814 - From what ship: HMT President, Halifax - Born: Marblehead - Age: 63 - Released on 4 Jun 1815.
Waite, Joshua - Prize Master - Nbr: 2269 - Prize: Success - Ship Type: Prize - How taken: HMS Charybdis - When taken: 29 May 1814 - Where taken: off Cape Logan - Date Received: 16 Aug 1814 - From what ship: HMS Dublin, Halifax - Born: Massachusetts - Age: 50 - Released on 3 May 1815.
Wakefield, John - Seaman - Nbr: 3720 - Prize: Alfred - Ship Type: P - How taken: HMS Epervier - When taken: 23 Feb 1814 - Where taken: off Newfoundland - Date Received: 30 Sep 1814 - From what ship: HMT President, Halifax - Born: Salem - Age: 54 - Released on 4 Jun 1815.
Wakefield, Nathaniel - Seaman - Nbr: 955 - Prize: Siro - Ship Type: LM - How taken: HMS Pelican - When taken: 13 Jan 1814 - Where taken: at sea - Date Received: 31 Jan 1814 - From what ship: Plymouth - Born: Beverly - Age: 52 - Released on 27 Apr 1815.
Walden, James - Seaman - Nbr: 2632 - How taken: Gave himself up from HMS Leviathan - When taken: 22 Oct 1813 - Date Received: 21 Aug 1814 - From what ship: HMT Freya, Chatham - Born: New London - Age: 32 - Released on 3 May 1815.
Waldron, Hiram - Seaman - Nbr: 1655 - How taken: Apprehended at Bristol - When taken: 9 Nov 1813 - Date Received: 23 Jun 1814 - From what ship: Stapleton - Born: New Hampshire - Age: 27 - Released on 1 May 1815.
Walker, Armstrong - Seaman - Nbr: 1381 - Prize: Courier - Ship Type: P - How taken: HMS Rover - When taken: 14 May 1813 - Where taken: Bay of Biscay - Date Received: 19 Jun 1814 - From what ship: Stapleton - Born: Baltimore - Age: 16 - Released on 28 Apr 1815.
Walker, Benjamin - Seaman - Nbr: 3221 - How taken: Gave himself up from HMS Woolwich - When taken: 20 Jun 1813 - Date Received: 11 Sep 1814 - From what ship: HMT Freya, Chatham - Born: Maryland - Age: 27 -

American Prisoners of War Held at Dartmoor during the War of 1812

## Alphabetical listing of names

Race: Mulatto - Released on 28 May 1815.
Walker, Daniel - Seaman - Nbr: 6445 - Prize: Jemmell - Ship Type: MV - How taken: HMS Rhin - When taken: 28 May 1814 - Where taken: off Bermuda - Date Received: 3 Mar 1815 - From what ship: HMT Ganges, Plymouth - Born: Baltimore - Age: 25 - Race: Black - Released on 11 Jul 1815.
Walker, James - Seaman - Nbr: 5757 - Prize: Hope - Ship Type: MV - How taken: HMS Nereus - When taken: 14 May 1814 - Where taken: Rio de la Plata - Date Received: 26 Dec 1814 - From what ship: HMT Argo - Born: Boston - Age: 41 - Released on 3 Jul 1815.
Walker, James - Prize Master - Nbr: 775 - Prize: Avon - Ship Type: MV - How taken: HMS Eurotas - When taken: 27 Oct 1813 - Where taken: off Ushant (France) - Date Received: 3 Nov 1813 - From what ship: Plymouth - Born: Portland - Age: 26 - Released on 10 Apr 1815.
Walker, James Scott - Clerk - Nbr: 4828 - Prize: President - Ship Type: P - How taken: HMS Pique - When taken: 7 May 1814 - Where taken: off Porto Rico - Date Received: 9 Oct 1814 - From what ship: HMT Freya, Halifax - Born: Montrose - Age: 23 - Sent to HMS Impregnable on 27 Dec 1814.
Walker, John - Seaman - Nbr: 2834 - How taken: Gave himself up from HMS Royal George - When taken: 29 Oct 1812 - Date Received: 24 Aug 1814 - From what ship: HMT Liverpool, Chatham - Born: Virginia - Age: 19 - Released on 26 Apr 1815.
Walker, John - Seaman - Nbr: 4070 - Prize: Hussar - Ship Type: P - How taken: HMS Saturn - When taken: 24 May 1814 - Where taken: off Sandy Hook - Date Received: 6 Oct 1814 - From what ship: HMT Chesapeake, Halifax - Born: Philadelphia - Age: 23 - Released on 13 Jun 1815.
Walker, Nathan - Seaman - Nbr: 5689 - Prize: McDonough - Ship Type: P - How taken: HMS Bacchante - When taken: 7 Nov 1814 - Where taken: Lat 42 Long 67 - Date Received: 24 Dec 1814 - From what ship: HMT Penelope - Born: Arundel - Age: 22 - Released on 3 Jul 1815.
Walker, Peter - Seaman - Nbr: 2449 - Prize: US Sloop Frolic - Ship Type: MW - How taken: HMS Orpheus - When taken: 20 Apr 1814 - Where taken: off Cuba - Date Received: 16 Aug 1814 - From what ship: HMT Queen, Halifax - Born: Taunton - Age: 26 - Released on 3 May 1815.
Walker, Richard - Seaman - Nbr: 1650 - How taken: Apprehended at Bristol - When taken: 16 Aug 1813 - Date Received: 23 Jun 1814 - From what ship: Stapleton - Born: Philadelphia - Age: 27 - Race: Negro - Released on 1 May 1815.
Walker, Samuel - Seaman - Nbr: 4491 - Prize: Enterprize - Ship Type: P - How taken: HMS Tenedos - When taken: 21 May 1813 - Where taken: off Cape Cod - Date Received: 8 Oct 1814 - From what ship: HMT Leyden, Chatham - Born: Massachusetts - Age: 20 - Released on 15 Jun 1815.
Walker, Seth - Prize master - Nbr: 4299 - Prize: Zephyr - Ship Type: MV - How taken: HMS Surveillante - When taken: 6 Jan 1814 - Where taken: Bay of Biscay - Date Received: 7 Oct 1814 - From what ship: HMT Niobe, Chatham - Born: Portugal - Age: 28 - Released on 28 May 1815.
Wall, William - Seaman - Nbr: 4167 - How taken: Gave himself up from HMS Leda - When taken: 8 Jul 1814 - Date Received: 6 Oct 1814 - From what ship: HMT Niobe, Chatham - Born: Boston - Age: 23 - Released on 29 Jun 1815.
Wallace, James - Seaman - Nbr: 1831 - Prize: Lightning - Ship Type: MV - How taken: HMS Medusa - When taken: 2 Apr 1813 - Where taken: Bay of Biscay - Date Received: 21 Jul 1814 - From what ship: HMT Redbeard & Pincher, Chatham Depot - Born: St. Michael - Age: 31 - Released on 1 May 1815.
Wallace, Thomas - Master - Nbr: 4250 - Prize: Argus - Ship Type: MV - How taken: HMS San Domingo - When taken: 1 Mar 1814 - Where taken: off Savannah - Date Received: 7 Oct 1814 - From what ship: HMT Niobe, Chatham - Born: Boston - Age: 52 - Escaped on 1 Jun 1815.
Wallace, William - Seaman - Nbr: 5949 - Prize: Harlequin - Ship Type: P - How taken: HMS Bulwark - When taken: 23 Nov 1814 - Where taken: off Halifax - Date Received: 27 Dec 1814 - From what ship: HMT Penelope - Born: Not legible - Age: 15 - Released on 3 Jul 1815.
Wallace, William - Caulker - Nbr: 4169 - How taken: Gave himself up from HMS Alexander - When taken: 11 Feb 1813 - Date Received: 7 Oct 1814 - From what ship: HMT Niobe, Chatham - Born: Baltimore - Age: 38 - Released on 13 Jun 1815.
Wallace, William - Seaman - Nbr: 478 - Prize: Zebra - Ship Type: LM - How taken: HMS Pyramus - When taken: 20 Apr 1813 - Where taken: Bay of Biscay - Date Received: 8 Sep 1813 - From what ship: Plymouth - Born: Washington - Age: 20 - Released on 20 Apr 1815.
Walleman, William - Seaman - Nbr: 527 - How taken: Impressed at Belfast - When taken: 29 Jun 1813 - Date Received: 8 Sep 1813 - From what ship: Plymouth - Born: Boston - Age: 31 - Released on 26 Apr 1815.
Walling, James - Seaman - Nbr: 4660 - How taken: Gave himself up from HMS Bold - When taken: 12 Jul 1813 -

American Prisoners of War Held at Dartmoor during the War of 1812

## Alphabetical listing of names

Date Received: 9 Oct 1814 - From what ship: HMT Leyden, Chatham - Born: New Jersey - Age: 32 - Released on 15 Jun 1815.

Wallis, John - Seaman - Nbr: 2442 - Prize: US Sloop Frolic - Ship Type: MW - How taken: HMS Orpheus - When taken: 20 Apr 1814 - Where taken: off Cuba - Date Received: 16 Aug 1814 - From what ship: HMT Queen, Halifax - Born: New Bedford - Age: 29 - Released on 3 May 1815.

Walsh, John - Marine - Nbr: 4033 - Prize: US Brig Rattlesnake - Ship Type: MW - How taken: HMS Leander - When taken: 13 Jul 1814 - Where taken: off Shelburne - Date Received: 6 Oct 1814 - From what ship: HMT Chesapeake, Halifax - Born: Philadelphia - Age: 20 - Released on 9 Jun 1815.

Walters, Philip - Seaman - Nbr: 1266 - Prize: Adeline - Ship Type: MV - How taken: HMS Magicienne - When taken: 16 Mar 1814 - Where taken: off Cape Finisterre - Date Received: 14 Jun 1814 - From what ship: Mill Prison (Plymouth, England) - Born: Baltimore - Age: 25 - Race: Black - Released on 28 Apr 1815.

Walton, Christopher - Seaman - Nbr: 2855 - Prize: Governor Middleton - Ship Type: MV - How taken: HMS Thetis - When taken: 2 May 1813 - Where taken: Bay of Biscay - Date Received: 24 Aug 1814 - From what ship: HMT Alpheus, Chatham - Born: Danzig - Age: 38 - Released on 28 May 1815.

Walton, John - Seaman - Nbr: 4982 - Prize: Invincible - Ship Type: LM - How taken: HMS Armide - When taken: 15 Aug 1814 - Where taken: off Nantucket - Date Received: 28 Oct 1814 - From what ship: HMT Alkbar, Halifax - Born: Pennsylvania - Age: 23 - Released on 21 Jun 1815.

Wanson, James - Seaman - Nbr: 3471 - How taken: Gave himself up from HMS Prince - When taken: 12 Sep 1814 - Date Received: 19 Sep 1814 - From what ship: HMT Salvador del Mundo - Born: New Haven - Age: 23 - Released on 4 Jun 1815.

Wanton, Samuel - Seaman - Nbr: 3747 - Prize: Mariner - Ship Type: Prize - How taken: HMS Poictiers - When taken: 30 Aug 1813 - Where taken: off Newfoundland - Date Received: 30 Sep 1814 - From what ship: HMT President, Halifax - Born: Newport - Age: 18 - Race: Negro - Released on 4 Jun 1815.

Wanton, William - Seaman - Nbr: 3281 - Prize: Montgomery - Ship Type: P - How taken: HMS Nymphe - When taken: 5 May 1813 - Where taken: off Cape Ann - Date Received: 13 Sep 1814 - From what ship: HMT Niobe, Chatham - Born: Marblehead - Age: 32 - Released on 28 May 1815.

Ward, Alfred - Seaman - Nbr: 4726 - How taken: Gave himself up from HMS Owen Glendower - When taken: 28 Jun 1813 - Date Received: 9 Oct 1814 - From what ship: HMT Freya, Chatham - Born: Middlesex - Age: 28 - Released on 15 Jun 1815.

Ward, Benjamin - Seaman - Nbr: 3277 - Prize: Cossack - Ship Type: P - How taken: HMS Amelia - When taken: 14 Apr 1813 - Where taken: off St. Johns - Date Received: 13 Sep 1814 - From what ship: HMT Niobe, Chatham - Born: Hancock - Age: 21 - Released on 28 May 1815.

Ward, Ebenezer - Mate - Nbr: 2107 - Prize: Thetis - Ship Type: MV - How taken: HMS Flora - When taken: 11 Jun 1814 - Where taken: Lat 30N Long 40W - Date Received: 3 Aug 1814 - From what ship: Plymouth - Born: Salem - Age: 21 - Released on 2 May 1815.

Ward, Israel - Seaman - Nbr: 5997 - Prize: McDonough - Ship Type: P - How taken: HMS Bacchante - When taken: 1 Nov 1814 - Where taken: Lat 42 Long 67 - Date Received: 28 Dec 1814 - From what ship: HMT Penelope - Born: Massachusetts - Age: 18 - Released on 3 Jul 1815.

Ward, John - Seaman - Nbr: 3464 - How taken: Gave himself up from HMS Cyane - When taken: 14 Sep 1814 - Date Received: 19 Sep 1814 - From what ship: HMT Salvador del Mundo - Born: Connecticut - Age: 32 - Released on 4 Jun 1815.

Ward, Nathaniel - Carpenter - Nbr: 5679 - Prize: McDonough - Ship Type: P - How taken: HMS Bacchante - When taken: 7 Nov 1814 - Where taken: Lat 42 Long 67 - Date Received: 24 Dec 1814 - From what ship: HMT Penelope - Born: Arundel - Age: 25 - Released on 3 Jul 1815.

Ward, Peter - Seaman - Nbr: 3363 - How taken: Impressed at London - When taken: 21 Oct 1813 - Date Received: 13 Sep 1814 - From what ship: HMT Niobe, Chatham - Born: Hudson - Age: 20 - Released on 28 May 1815.

Ward, Samuel - Seaman - Nbr: 3012 - How taken: Gave himself up from HMS Zealous - When taken: 27 Aug 1814 - Date Received: 2 Sep 1814 - From what ship: HMT Centaur - Born: Portsmouth - Age: 39 - Released on 28 May 1815.

Ward, Thomas - Prize master - Nbr: 4451 - Prize: Prize to Prince - Ship Type: Prize - How taken: HMS Ethalion - When taken: 14 Dec 1813 - Where taken: at sea - Date Received: 8 Oct 1814 - From what ship: HMT Leyden, Chatham - Born: Baltimore - Age: 32 - Released on 1 May 1815.

Warden, Nathaniel - Seaman - Nbr: 5061 - Prize: Ida - Ship Type: LM - How taken: HMS Newcastle - When taken: 9 Aug 1814 - Where taken: Long 34 - Date Received: 28 Oct 1814 - From what ship: HMT Alkbar, Halifax - Born: Boston - Age: 26 - Released on 29 Jun 1815.

American Prisoners of War Held at Dartmoor during the War of 1812

## Alphabetical listing of names

Ware, John - Seaman - Nbr: 3928 - Prize: Rolla - Ship Type: P - How taken: HMS Loire - When taken: 10 Dec 1813 - Where taken: off Bull Island (South Carolina) - Date Received: 5 Oct 1814 - From what ship: HMT President, Halifax - Born: Baltimore - Age: 38 - Released on 9 Jun 1815.

Ware, William - Seaman - Nbr: 220 - Prize: Pert - Ship Type: MV - How taken: HMS Warspite - When taken: 1 Mar 1813 - Where taken: off Basque Roads (France) - Date Received: 2 Apr 1813 - From what ship: Plymouth - Born: Philadelphia - Age: 38 - Released on 20 Apr 1815.

Warner, Benjamin - Seaman - Nbr: 3545 - Prize: Hawk - Ship Type: P - How taken: HMS Pique - When taken: 26 Apr 1814 - Where taken: off Bermuda - Date Received: 30 Sep 1814 - From what ship: HMT Sybella - Born: North Carolina - Age: 22 - Released on 4 Jun 1815.

Warner, Henry - Seaman - Nbr: 29 - How taken: Apprehended at Gibraltar - When taken: 8 Aug 1813 - Date Received: 2 Apr 1813 - From what ship: Plymouth - Born: New Haven - Age: 28 - Sent to Dartmouth on 30 Jul 1813.

Warner, John - Seaman - Nbr: 4267 - How taken: Gave himself up from HMS Prince of Wales - When taken: 28 May 1813 - Date Received: 7 Oct 1814 - From what ship: HMT Niobe, Chatham - Born: Roxbury - Age: 23 - Released on 26 Apr 1815.

Warner, Stephen - Seaman - Nbr: 2169 - Prize: Rattlesnake - Ship Type: P - How taken: HMS Hyperion - When taken: 26 Jun 1814 - Where taken: off Cape Finisterre - Date Received: 16 Aug 1814 - From what ship: HMS Dublin, Halifax - Born: Delaware - Age: 21 - Released on 2 May 1815.

Warner, Thomas - Seaman - Nbr: 1847 - How taken: Gave himself up from HMS Cordelia - When taken: 28 May 1813 - Date Received: 21 Jul 1814 - From what ship: HMT Redbeard & Pincher, Chatham Depot - Born: New York - Age: 51 - Released on 2 May 1815.

Warney, Benjamin - Seaman - Nbr: 5065 - Prize: Ida - Ship Type: LM - How taken: HMS Newcastle - When taken: 9 Aug 1814 - Where taken: Long 34 - Date Received: 28 Oct 1814 - From what ship: HMT Alkbar, Halifax - Born: Boston - Age: 20 - Released on 29 Jun 1815.

Warnick, Robert - Seaman - Nbr: 2835 - How taken: Gave himself up from HMS Royal George - When taken: 29 Oct 1812 - Date Received: 24 Aug 1814 - From what ship: HMT Liverpool, Chatham - Born: Williston - Age: 34 - Released on 26 Apr 1815.

Warrands, John - Seaman - Nbr: 1829 - Prize: Lightning - Ship Type: MV - How taken: HMS Medusa - When taken: 2 Apr 1813 - Where taken: Bay of Biscay - Date Received: 21 Jul 1814 - From what ship: HMT Redbeard & Pincher, Chatham Depot - Born: Philadelphia - Age: 19 - Released on 1 May 1815.

Warren, David - Seaman - Nbr: 4515 - Prize: Grand Turk - Ship Type: P - How taken: HMS Tenedos - When taken: 26 May 1813 - Where taken: off Cape Cod - Date Received: 8 Oct 1814 - From what ship: HMT Leyden, Chatham - Born: Massachusetts - Age: 18 - Released on 15 Jun 1815.

Warren, David - Seaman - Nbr: 1357 - Prize: Paul Jones - Ship Type: P - How taken: HMS Leonidas - When taken: 23 May 1813 - Where taken: Channel - Date Received: 19 Jun 1814 - From what ship: Stapleton - Born: Massachusetts - Age: 36 - Released on 28 Apr 1815.

Warren, Nathaniel W. - Seaman - Nbr: 2100 - How taken: Gave himself up from HMS Leonidas - When taken: 13 Jun 1813 - Date Received: 3 Aug 1814 - From what ship: HMS Bittern, Chatham Depot - Born: Durham - Age: 33 - Released on 2 May 1815.

Warren, Samuel - Seaman - Nbr: 5969 - Prize: Halifax Packet - Ship Type: Prize - How taken: HMS Bulwark - When taken: 22 Sep 1814 - Where taken: Georges Bank - Date Received: 27 Dec 1814 - From what ship: HMT Penelope - Born: Berwick - Age: 23 - Released on 4 Jun 1815.

Warrimore, Henry - Seaman - Nbr: 2227 - Prize: Hussar - Ship Type: P - How taken: HMS Saturn - When taken: 25 May 1814 - Where taken: off Sandy Hook - Date Received: 16 Aug 1814 - From what ship: HMS Dublin, Halifax - Born: Philadelphia - Age: 23 - Released on 2 May 1815.

Warrimore, Philip - Seaman - Nbr: 2226 - Prize: Hussar - Ship Type: P - How taken: HMS Saturn - When taken: 25 May 1814 - Where taken: off Sandy Hook - Date Received: 16 Aug 1814 - From what ship: HMS Dublin, Halifax - Born: Philadelphia - Age: 17 - Released on 2 May 1815.

Washburn, Edward - Seaman - Nbr: 198 - Prize: Star - Ship Type: MV - How taken: HMS Superb - When taken: 9 Feb 1813 - Where taken: Bay of Biscay - Date Received: 2 Apr 1813 - From what ship: Plymouth - Born: New York - Age: 20 - Sent to Mill Prison (Plymouth, England) on 10 Jul 1813.

Washburn, Jeremiah - Seaman - Nbr: 5996 - Prize: McDonough - Ship Type: P - How taken: HMS Bacchante - When taken: 1 Nov 1814 - Where taken: Lat 42 Long 67 - Date Received: 28 Dec 1814 - From what ship: HMT Penelope - Born: Arundel - Age: 18 - Released on 3 Jul 1815.

Washburn, Thomas - Seaman - Nbr: 5929 - Prize: Harlequin - Ship Type: P - How taken: HMS Bulwark - When

American Prisoners of War Held at Dartmoor during the War of 1812

## Alphabetical listing of names

taken: 23 Nov 1814 - Where taken: off Halifax - Date Received: 27 Dec 1814 - From what ship: HMT Penelope - Born: Arundel - Age: 20 - Released on 3 Jul 1815.

Washbury, Elbert - Seaman - Nbr: 1267 - Prize: Adeline - Ship Type: MV - How taken: HMS Magicienne - When taken: 16 Mar 1814 - Where taken: off Cape Finisterre - Date Received: 14 Jun 1814 - From what ship: Mill Prison (Plymouth, England) - Born: Portland - Age: 17 - Released on 28 Apr 1815.

Washington, John - Seaman - Nbr: 3936 - Prize: Rolla - Ship Type: P - How taken: HMS Loire - When taken: 10 Dec 1813 - Where taken: off Bull Island (South Carolina) - Date Received: 5 Oct 1814 - From what ship: HMT President, Halifax - Born: Savannah - Age: 25 - Died on 6 Apr 1815 from gunshot wound in prison.

Watch, Samuel - Seaman - Nbr: 1096 - Prize: Lyon - Ship Type: P - How taken: Brilliant (Privateer) - When taken: 12 Aug 1814 - Where taken: off Charlestown - Date Received: 10 May 1814 - From what ship: Plymouth - Born: Wales - Age: 27 - Released on 27 Apr 1815.

Waterford, J. H. - Carpenter's mate - Nbr: 3568 - Prize: Hawk - Ship Type: P - How taken: HMS Pique - When taken: 26 Apr 1814 - Where taken: off Bermuda - Date Received: 30 Sep 1814 - From what ship: HMT Sybella - Born: Virginia - Age: 33 - Released on 4 Jun 1815.

Waterhouse, Joseph - Seaman - Nbr: 4754 - How taken: Gave himself up from HMS Africa - When taken: 4 Oct 1813 - Date Received: 9 Oct 1814 - From what ship: HMT Freya, Chatham - Born: Not legible - Age: 24 - Race: Mulatto - Released on 15 Jun 1815.

Waterhouse, Moses - Seaman - Nbr: 3302 - Prize: Theresa - Ship Type: P - How taken: Moor - When taken: 14 Jun 1813 - Where taken: off Cape Ann - Date Received: 13 Sep 1814 - From what ship: HMT Niobe, Chatham - Born: Massachusetts - Age: 23 - Released on 28 May 1815.

Waterman, Cato - Seaman - Nbr: 4802 - Prize: President - Ship Type: P - How taken: HMS Pique - When taken: 7 May 1814 - Where taken: off Porto Rico - Date Received: 9 Oct 1814 - From what ship: HMT Freya, Halifax - Born: Rhode Island - Age: 32 - Race: Black - Released on 21 Jun 1815.

Waterman, John - 2nd Mate - Nbr: 4181 - Prize: Tigre - Ship Type: MV - How taken: HMS Scylla - When taken: 22 Mar 1813 - Where taken: Bay of Biscay - Date Received: 7 Oct 1814 - From what ship: HMT Niobe, Chatham - Born: Nantucket - Age: 21 - Released on 13 Jun 1815.

Waterman, Thomas - Master - Nbr: 4178 - Prize: Robert & Ann - How taken: Delacna - When taken: 14 Apr 1813 - Where taken: London - Date Received: 7 Oct 1814 - From what ship: HMT Niobe, Chatham - Born: New York - Age: 23 - Sent to Ashburton (England) on 8 Dec 1814.

Waterman, William - Seaman - Nbr: 3872 - Prize: Tickler - Ship Type: MV - How taken: HMS Saturn - When taken: 13 Jul 1814 - Where taken: off America - Date Received: 5 Oct 1814 - From what ship: HMT Orpheus, Halifax - Born: Nantucket - Age: 19 - Released on 9 Jun 1815.

Waters, Abraham - Seaman - Nbr: 1532 - Prize: Essex - Ship Type: MV - How taken: HMS Pyramus - When taken: 20 Apr 1813 - Where taken: Bay of Biscay - Date Received: 23 Jun 1814 - From what ship: Stapleton - Born: Boston - Age: 17 - Released on 1 May 1815.

Waters, George - Seaman - Nbr: 6180 - Prize: Lion - Ship Type: P - How taken: HMS Granicus - When taken: 2 Dec 1814 - Where taken: off Lisbon - Date Received: 21 Jan 1815 - From what ship: HMT Impregnable - Born: Portsmouth - Age: 50 - Released on 5 Jul 1815.

Waters, John - Seaman - Nbr: 5491 - Prize: General Putnam - Ship Type: P - How taken: HMS Leander - When taken: 8 Nov 1814 - Where taken: Long 65 Lat 42 - Date Received: 17 Dec 1814 - From what ship: HMT Loire, Halifax - Born: Salem - Age: 23 - Released on 1 Jul 1815.

Waters, Louis - Steward - Nbr: 6441 - Prize: Chance - Ship Type: P - How taken: HMS Statira - When taken: 1 Apr 1814 - Where taken: Lat 38 Long 24 - Date Received: 3 Mar 1815 - From what ship: HMT Ganges, Plymouth - Born: Maryland - Age: 22 - Race: Black - Released on 11 Jul 1815.

Waters, Robert - Seaman - Nbr: 3808 - Prize: Stark - Ship Type: P - How taken: HMS Sophie - When taken: 20 Apr 1814 - Where taken: off Bermuda - Date Received: 5 Oct 1814 - From what ship: HMT Orpheus, Halifax - Born: Salem - Age: 22 - Released on 9 Jun 1815.

Waters, Robert - Seaman - Nbr: 639 - Prize: US Brig Argus - Ship Type: MW - How taken: HMS Pelican - When taken: 14 Aug 1813 - Where taken: Irish Channel - Date Received: 8 Sep 1813 - From what ship: Plymouth - Born: Massachusetts - Age: 25 - Race: Black - Sent to Dartmouth on 2 Nov 1814.

Watkins, George - Seaman - Nbr: 2754 - How taken: Gave himself up from HMS Blake - When taken: 10 Dec 1812 - Date Received: 24 Aug 1814 - From what ship: HMT Liverpool, Chatham - Born: Newport - Age: 32 - Race: Black - Released on 26 Apr 1815.

Watkins, Thomas - Seaman - Nbr: 504 - Prize: Revenge - Ship Type: LM - How taken: HMS Belle Poule - When taken: 10 May 1813 - Where taken: off Cornwall - Date Received: 8 Sep 1813 - From what ship: Plymouth -

American Prisoners of War Held at Dartmoor during the War of 1812

## Alphabetical listing of names

Born: Daton - Age: 38 - Released on 26 Apr 1815.

Watkins, William - Seaman - Nbr: 3623 - Prize: Monarch - Ship Type: MV - How taken: HMS Dotterel - When taken: 14 Dec 1813 - Where taken: off Charleston - Date Received: 30 Sep 1814 - From what ship: HMT Sybella - Born: Richmond - Age: 18 - Released on 4 Jun 1815.

Watson, Daniel - Seaman - Nbr: 3192 - How taken: Gave himself up from HMS Swiftsure - When taken: 26 Dec 1812 - Date Received: 11 Sep 1814 - From what ship: HMT Freya, Chatham - Born: Rhode Island - Age: 24 - Race: Mulatto - Released on 27 Apr 1815.

Watson, George - Seaman - Nbr: 3779 - Prize: Fame - Ship Type: P - How taken: HMS Thistle - When taken: 10 Apr 1814 - Where taken: after being cast ashore on a seal island - Date Received: 30 Sep 1814 - From what ship: HMT President, Halifax - Born: Washington - Age: 28 - Released on 9 Jun 1815.

Watson, Henry - Seaman - Nbr: 5861 - Prize: Lion - Ship Type: P - How taken: HMS Granicus - When taken: 2 Dec 1814 - Where taken: off Lisbon - Date Received: 26 Dec 1814 - From what ship: HMT Impregnable - Born: New York - Age: 24 - Race: Black - Released on 3 Jul 1815.

Watson, James - Seaman - Nbr: 1482 - Prize: Hebe - Ship Type: MV - How taken: HMS Stag - When taken: 18 Apr 1813 - Where taken: Bay of Biscay - Date Received: 19 Jun 1814 - From what ship: Stapleton - Born: New York - Age: 24 - Released on 1 May 1815.

Watson, James - Seaman - Nbr: 3148 - Prize: Kitty - Ship Type: Prize - How taken: Dart of Guernsey (Privateer) - When taken: 20 Jun 1813 - Where taken: off the Western Isles (England) - Date Received: 11 Sep 1814 - From what ship: HMT Freya, Chatham - Born: Boston - Age: 18 - Sent to Dartmouth on 23 Sep 1814.

Watson, Robert - Seaman - Nbr: 3454 - Prize: Mary - Ship Type: Prize - How taken: Not listed - When taken: Jul 1814 - Where taken: off St. Michaels - Date Received: 19 Sep 1814 - From what ship: HMT Salvador del Mundo - Born: Virginia - Age: 23 - Released on 11 Jun 1815.

Watson, William - Seaman - Nbr: 4206 - Prize: Minerva - Ship Type: MV - How taken: HMS Conquistador - When taken: 19 Jan 1814 - Where taken: Bay of Biscay - Date Received: 7 Oct 1814 - From what ship: HMT Niobe, Chatham - Born: Scarborough - Age: 20 - Released on 13 Jun 1815.

Watt, Samuel - Seaman - Nbr: 6401 - Prize: Nellenville - Ship Type: MV - How taken: HMS Onyx - When taken: 25 Dec 1814 - Where taken: off San Domingo - Date Received: 3 Mar 1815 - From what ship: HMT Ganges, Plymouth - Born: Delaware - Age: 22 - Released on 11 Jul 1815.

Watts, Hiram - Seaman - Nbr: 288 - Prize: Cannoneer - Ship Type: MV - How taken: HMS Warspite - When taken: 14 Mar 1813 - Where taken: Bay of Biscay - Date Received: 28 Jun 1813 - From what ship: Plymouth - Born: Newburgh - Age: 18 - Released on 20 Apr 1815.

Watts, James - 2nd Mate - Nbr: 331 - Prize: Pert - Ship Type: MV - How taken: HMS Warspite - When taken: 1 Mar 1813 - Where taken: off Basque Roads (France) - Date Received: 28 Jun 1813 - From what ship: Plymouth - Born: Philadelphia - Age: 43 - Released on 20 Apr 1815.

Watts, John - Seaman - Nbr: 2708 - Prize: Dick - Ship Type: MV - How taken: HMS Dispatch - When taken: 15 Mar 1813 - Where taken: near Corcova Lights - Date Received: 21 Aug 1814 - From what ship: HMT Freya, Chatham - Born: New York - Age: 24 - Released on 19 May 1815.

Watts, William - Seaman - Nbr: 2342 - Prize: Snap Dragon - Ship Type: P - How taken: HMS Martin - When taken: 10 Jun 1814 - Where taken: off Halifax - Date Received: 16 Aug 1814 - From what ship: HMS Dublin, Halifax - Born: North Carolina - Age: 20 - Released on 3 May 1815.

Wavernce, James - Seaman - Nbr: 1229 - How taken: Sent into custody from HMS Furieuse - When taken: 23 Sep 1812 - Date Received: 14 Jun 1814 - From what ship: Mill Prison (Plymouth, England) - Born: Baltimore - Age: 20 - Race: Mulatto - Released on 26 Apr 1815.

Wayne, James - Seaman - Nbr: 6277 - How taken: Gave himself up from HMS Prince - When taken: 3 Feb 1814 - Date Received: 17 Feb 1815 - From what ship: HMT Ganges, Plymouth - Born: Not legible - Race: Black - Released on 5 Jul 1815.

Weaphor, Andrew - Seaman - Nbr: 294 - Prize: Ducornau - Ship Type: MV - How taken: HMS Pheasant - When taken: 15 Mar 1813 - Where taken: Bay of Biscay - Date Received: 28 Jun 1813 - From what ship: Plymouth - Born: Philadelphia - Age: 53 - Released on 20 Apr 1815.

Weare, Ebenezer - Seaman - Nbr: 3393 - Prize: Wasp - Ship Type: P - How taken: HMS Bream - When taken: 10 Jun 1813 - Where taken: off Halifax - Date Received: 13 Sep 1814 - From what ship: HMT Niobe, Chatham - Born: New York - Age: 18 - Released on 28 May 1815.

Wearney, Richard - Seaman - Nbr: 3570 - Prize: Hawk - Ship Type: P - How taken: HMS Pique - When taken: 26 Apr 1814 - Where taken: off Bermuda - Date Received: 30 Sep 1814 - From what ship: HMT Sybella - Born: Washington - Age: 50 - Released on 4 Jun 1815.

American Prisoners of War Held at Dartmoor during the War of 1812

## Alphabetical listing of names

Weaver, Michael - Seaman - Nbr: 4129 - Prize: Bordeaux Packet - Ship Type: LM - How taken: HMS Niemen - When taken: 28 Jun 1814 - Where taken: off Delaware - Date Received: 6 Oct 1814 - From what ship: HMT Chesapeake, Halifax - Born: Philadelphia - Age: 20 - Released on 13 Jun 1815.

Webb, Alexander - Seaman - Nbr: 759 - How taken: Impressed at Liverpool - Date Received: 3 Nov 1813 - From what ship: Plymouth - Born: Philadelphia - Age: 20 - Released on 26 Apr 1815.

Webb, John - Seaman - Nbr: 4499 - Prize: Enterprize - Ship Type: P - How taken: HMS Tenedos - When taken: 21 May 1813 - Where taken: off Cape Cod - Date Received: 8 Oct 1814 - From what ship: HMT Leyden, Chatham - Born: Massachusetts - Age: 27 - Released on 15 Jun 1815.

Webb, John - Seaman - Nbr: 1980 - Prize: Orbit - Ship Type: MV - How taken: HMS Achates - When taken: 29 Jan 1813 - Where taken: Lat 44N Long 13W - Date Received: 3 Aug 1814 - From what ship: HMS Lyffey, Chatham Depot - Born: New York - Age: 16 - Released on 2 May 1815.

Webb, John - Seaman - Nbr: 2727 - Prize: Tickler - Ship Type: LM - How taken: HMS Magicienne - When taken: 5 Jun 1813 - Where taken: Bay of Biscay - Date Received: 23 Aug 1814 - From what ship: Exeter hospital - Born: Edgecombe - Age: 23 - Released on 19 May 1815.

Webb, Michael - Seaman - Nbr: 5755 - Prize: Hope - Ship Type: MV - How taken: HMS Nereus - When taken: 14 May 1814 - Where taken: Rio de la Plata - Date Received: 26 Dec 1814 - From what ship: HMT Argo - Born: Portland - Age: 21 - Released on 3 Jul 1815.

Webb, Nathaniel - Gunner - Nbr: 4425 - Prize: Elbridge Gerry - Ship Type: LM - How taken: HMS Crescent - When taken: 16 Sep 1813 - Where taken: at sea - Date Received: 8 Oct 1814 - From what ship: HMT Leyden, Chatham - Born: Massachusetts - Age: 26 - Released on 14 Jun 1815.

Webb, Thomas - Seaman - Nbr: 5768 - Prize: Rambler - Ship Type: MV - How taken: Morley (Transport) - When taken: 10 Feb 1813 - Where taken: off Isle of France - Date Received: 26 Dec 1814 - From what ship: HMT Argo - Born: New York - Age: 32 - Released on 3 Jul 1815.

Webber, Frederick - Seaman - Nbr: 1019 - Prize: Rachel & Ann - Ship Type: P - How taken: HMS Cydnus - When taken: 6 Jan 1814 - Where taken: at sea - Date Received: 31 Jan 1814 - From what ship: Plymouth - Born: New Orleans - Age: 20 - Race: Black - Released on 27 Apr 1815.

Webber, John - Seaman - Nbr: 767 - How taken: Sent into custody from HMS Pallas - Date Received: 3 Nov 1813 - From what ship: Plymouth - Born: Brunswick - Age: 28 - Released on 11 Jul 1815.

Webster, Asa - Seaman - Nbr: 4485 - Prize: Enterprize - Ship Type: P - How taken: HMS Tenedos - When taken: 21 May 1813 - Where taken: off Cape Cod - Date Received: 8 Oct 1814 - From what ship: HMT Leyden, Chatham - Born: Massachusetts - Age: 23 - Released on 15 Jun 1815.

Webster, David - Seaman - Nbr: 5151 - Prize: Thorn - Ship Type: P - How taken: HMS Shannon - When taken: 7 Nov 1813 - Where taken: off Newfoundland - Date Received: 31 Oct 1814 - From what ship: HMT Mermaid, Chatham - Born: Salisbury - Age: 19 - Released on 29 Jun 1815.

Webster, George - Boy - Nbr: 2214 - Prize: Fiere Facia - Ship Type: P - How taken: HMS Ramillies - When taken: 27 Feb 1814 - Where taken: off NY - Date Received: 16 Aug 1814 - From what ship: HMS Dublin, Halifax - Born: New York - Age: 14 - Released on 2 May 1815.

Webster, James - Seaman - Nbr: 419 - Prize: Viper - Ship Type: MV - How taken: HMS Superb - When taken: 15 Apr 1813 - Where taken: Bay of Biscay - Date Received: 1 Jul 1813 - From what ship: Plymouth - Born: Edenton - Age: 32 - Released on 20 Apr 1815.

Webster, Michael - Seaman - Nbr: 6426 - Prize: Farmer's Daughter - Ship Type: MV - How taken: HMS Leviathan - When taken: 29 May 1814 - Date Received: 3 Mar 1815 - From what ship: HMT Ganges, Plymouth - Born: Annapolis - Age: 32 - Released on 11 Jul 1815.

Webster, William - Seaman - Nbr: 1820 - How taken: Gave himself up from HMS Leviathan - When taken: 28 Oct 1811 - Date Received: 20 Jul 1814 - From what ship: HMS Milford, Plymouth - Born: New York - Age: 28 - Escaped on 21 Apr 1815.

Wedgwood, William - Seaman - Nbr: 2750 - Prize: Dart - Ship Type: MV - How taken: HMS Peterel - When taken: 15 Mar 1813 - Where taken: at sea - Date Received: 24 Aug 1814 - From what ship: HMT Liverpool, Chatham - Born: Porterfield - Age: 29 - Released on 19 May 1815.

Weeble, Robert - Seaman - Nbr: 3078 - How taken: Gave himself up from HMS Warrior - When taken: 21 Apr 1813 - Date Received: 3 Sep 1814 - From what ship: HMT Hydra, Chatham - Born: Newbury - Age: 33 - Released on 28 May 1815.

Weeden, Richard - Cook - Nbr: 3161 - How taken: Impressed at Gravesend - When taken: 20 Aug 1813 - Date Received: 11 Sep 1814 - From what ship: HMT Freya, Chatham - Born: Rhode Island - Age: 30 - Race: Black - Released on 28 May 1815.

American Prisoners of War Held at Dartmoor during the War of 1812

## Alphabetical listing of names

Weeding, Isaac - Seaman - Nbr: 3499 - How taken: Taken in a boat of the Fox - Date Received: 24 Sep 1814 - From what ship: HMT Salvador del Mundo - Born: Rhode Island - Age: 19 - Released on 4 Jun 1815.

Weeks, Benjamin - Seaman - Nbr: 514 - How taken: Impressed at Liverpool - When taken: 25 Apr 1813 - Date Received: 8 Sep 1813 - From what ship: Plymouth - Born: Philadelphia - Age: 30 - Released on 26 Apr 1815.

Weeks, David - Seaman - Nbr: 1398 - Prize: Zebra - Ship Type: LM - How taken: HMS Pyramus - When taken: 20 Apr 1813 - Where taken: Bay of Biscay - Date Received: 19 Jun 1814 - From what ship: Stapleton - Born: New Jersey - Age: 21 - Released on 28 Apr 1815.

Weeks, James - Seaman - Nbr: 1321 - How taken: Impressed at Bristol - When taken: 14 May 1813 - Date Received: 19 Jun 1814 - From what ship: Stapleton - Born: Maryland - Age: 32 - Race: Negro - Released on 28 Apr 1815.

Weeks, James - Seaman - Nbr: 952 - Prize: Siro - Ship Type: LM - How taken: HMS Pelican - When taken: 13 Jan 1814 - Where taken: at sea - Date Received: 31 Jan 1814 - From what ship: Plymouth - Born: Palermo - Age: 21 - Released on 27 Apr 1815.

Weeks, Joseph - Captain - Nbr: 5572 - Prize: McDonough - Ship Type: P - How taken: HMS Bacchante - When taken: 1 Nov 1814 - Where taken: Lat 42 Long 67 - Date Received: 17 Dec 1814 - From what ship: HMT Loire, Halifax - Born: Portland - Age: 29 - Released on 10 Apr 1815.

Weeks, Louis - Seaman - Nbr: 3362 - How taken: Impressed at London - When taken: 21 Oct 1813 - Date Received: 13 Sep 1814 - From what ship: HMT Niobe, Chatham - Born: Hudson - Age: 32 - Released on 28 May 1815.

Weeks, Thomas - Seaman - Nbr: 3887 - Prize: US Brig Rattlesnake - Ship Type: MW - How taken: HMS Leander - When taken: 11 Jul 1814 - Where taken: off Shelburne - Date Received: 5 Oct 1814 - From what ship: HMT Orpheus, Halifax - Born: Portsmouth - Age: 32 - Released on 9 Jun 1815.

Weeks, Thomas N. - Boy - Nbr: 5520 - Prize: James - Ship Type: Prize - How taken: HMS Galatea - When taken: 7 Sep 1814 - Where taken: Channel - Date Received: 17 Dec 1814 - From what ship: HMT Loire, Halifax - Born: Virginia - Age: 19 - Released on 1 Jul 1815.

Weiden, Charles - Boatswain's mate - Nbr: 5822 - Prize: US Schooner Ohio - Ship Type: MW - How taken: British gunboats - When taken: 12 Jun 1814 - Where taken: Fort Erie - Date Received: 26 Dec 1814 - From what ship: HMT Argo - Born: New London - Age: 28 - Released on 3 Jul 1815.

Weimon, John - Seaman - Nbr: 3961 - Prize: Rolla - Ship Type: P - How taken: HMS Loire - When taken: 10 Dec 1813 - Where taken: off Bull Island (South Carolina) - Date Received: 5 Oct 1814 - From what ship: HMT President, Halifax - Born: Philadelphia - Age: 35 - Released on 9 Jun 1815.

Weisse, Philip - Seaman - Nbr: 4248 - Prize: Requin - Ship Type: LM - How taken: HMS Venus - When taken: 6 Mar 1814 - Where taken: off Bordeaux - Date Received: 7 Oct 1814 - From what ship: HMT Niobe, Chatham - Born: Boston - Age: 27 - Released on 13 Jun 1815.

Welbor, John - Seaman - Nbr: 5625 - Prize: Rose - Ship Type: MV - How taken: HMS Harpy - When taken: 9 Feb 1813 - Date rec'd: 24 Dec 1814 - From what ship: HMT Tay - Born: Vermont - Age: 26 - Released on 3 Jul 1815.

Welch, Benjamin - Seaman - Nbr: 1519 - Prize: Essex - Ship Type: MV - How taken: HMS Pyramus - When taken: 20 Apr 1813 - Where taken: Bay of Biscay - Date Received: 23 Jun 1814 - From what ship: Stapleton - Born: Massachusetts - Age: 22 - Released on 1 May 1815.

Welch, John - Seaman - Nbr: 6355 - Prize: Prince de Neufchatel - Ship Type: P - How taken: Leander (Newcastle Acasta) - When taken: 20 Dec 1814 - Where taken: Lat 38 Long 56 - Date Received: 19 Feb 1815 - From what ship: HMT Ganges, Plymouth - Born: Boston - Age: 36 - Released on 6 Jul 1815.

Welch, John - Quartermaster - Nbr: 4889 - How taken: Gave himself up from HMS Centaur - When taken: 16 Oct 1814 - Date Received: 24 Oct 1814 - From what ship: HMT Salvador del Mundo - Born: Baltimore - Age: 25 - Released on 19 May 1815.

Welch, John - Seaman - Nbr: 2859 - How taken: Gave himself up from HMS Leonidas - Date Received: 24 Aug 1814 - From what ship: HMT Alpheus, Chatham - Born: Bridgetown - Age: 28 - Race: Black - Released on 19 May 1815.

Welch, Samuel - Seaman - Nbr: 2539 - Prize: Sea Nymphe - Ship Type: MV - How taken: HMS Thraster - When taken: 4 Mar 1814 - Where taken: North Sea - Date Received: 16 Aug 1814 - From what ship: HMT Salvador del Mundo - Born: Baltimore - Age: 23 - Released on 3 May 1815.

Welch, Thomas - Seaman - Nbr: 3630 - Prize: Monarch - Ship Type: MV - How taken: HMS Dotterel - When taken: 14 Dec 1813 - Where taken: off Charleston - Date Received: 30 Sep 1814 - From what ship: HMT Sybella - Born: Pennsylvania - Age: 22 - Released on 4 Jun 1815.

Welch, William - Seaman - Nbr: 2035 - How taken: Gave himself up from HMS Impeteux - When taken: 2 Dec

American Prisoners of War Held at Dartmoor during the War of 1812

## Alphabetical listing of names

1812 - Date Received: 3 Aug 1814 - From what ship: HMS Lyffey, Chatham Depot - Born: New York - Age: 22 - Released on 26 Apr 1815.

Weldon, Asa - Seaman - Nbr: 5134 - Prize: Calabria - Ship Type: MV - How taken: HMS Castilian - When taken: 29 Sep 1814 - Where taken: off Ireland - Date Received: 31 Oct 1814 - From what ship: HMT Castillian - Born: Worcester - Age: 21 - Released on 29 Jun 1815.

Wellander, Adam - Seaman - Nbr: 292 - Prize: Cannoneer - Ship Type: MV - How taken: HMS Warspite - When taken: 14 Mar 1813 - Where taken: Bay of Biscay - Date Received: 28 Jun 1813 - From what ship: Plymouth - Born: Sweden - Age: 28 - Released on 27 Dec 1813 (Swedish citizen).

Wells, James - Seaman - Nbr: 3652 - Prize: Thorn - Ship Type: P - How taken: HMS Tenedos - When taken: 30 Oct 1812 - Where taken: off Newfoundland - Date Received: 30 Sep 1814 - From what ship: HMT President - Born: Massachusetts - Age: 36 - Released on 11 Jul 1815.

Welsh, Ezekiel - Seaman - Nbr: 1676 - How taken: Apprehended at Liverpool - Date Received: 2 Jul 1814 - From what ship: Plymouth - Born: Bath - Age: 21 - Released on 1 May 1815.

Welsh, Richard - Seaman - Nbr: 6466 - Prize: Decatur - Ship Type: P - How taken: HMS Rhin - When taken: 5 Jun 1814 - Where taken: off San Domingo - Date Received: 3 Mar 1815 - From what ship: HMT Ganges, Plymouth - Born: New York - Age: 24 - Released on 11 Jul 1815.

Welsh, William - Seaman - Nbr: 954 - Prize: Siro - Ship Type: LM - How taken: HMS Pelican - When taken: 13 Jan 1814 - Where taken: at sea - Date Received: 31 Jan 1814 - From what ship: Plymouth - Born: Port Royal - Age: 17 - Released on 27 Apr 1815.

Welter, William - Seaman - Nbr: 1205 - Prize: Charles - Ship Type: P - How taken: HMS Amelia - When taken: 8 Oct 1810 - Where taken: Channel - Date Received: 4 Jun 1814 - From what ship: Dartmouth - Born: Philadelphia - Age: 26 - Sent to Plymouth on 8 Jul 1814.

Wentworth, John - Seaman - Nbr: 6317 - Prize: Prince de Neufchatel - Ship Type: P - How taken: Leander (Newcastle Acasta) - When taken: 20 Dec 1814 - Where taken: Lat 38 Long 56 - Date Received: 19 Feb 1815 - From what ship: HMT Ganges, Plymouth - Born: Canton, China - Age: 20 - Released on 11 Jul 1815.

Werell, Francis - Seaman - Nbr: 5158 - Prize: Volante - Ship Type: P - How taken: HMS Curlew - When taken: 25 Nov 1813 - Where taken: off Halifax - Date Received: 31 Oct 1814 - From what ship: HMT Mermaid, Chatham - Born: Marblehead - Age: 22 - Released on 29 Jun 1815.

Werman, John - Seaman - Nbr: 5356 - Prize: Commodore Perry - Ship Type: Schooner - How taken: Sent into custody from a cutter - When taken: 25 Feb 1814 - Where taken: off Bordeaux - Date Received: 31 Oct 1814 - From what ship: HMT Leyden, Chatham - Born: Philadelphia - Age: 18 - Released on 1 Jul 1815.

Wescott, Edward - Seaman - Nbr: 5439 - How taken: Apprehended at Cork - When taken: 27 Oct 1814 - Date Received: 11 Nov 1814 - From what ship: HMT Impregnable - Born: Virginia - Age: 23 - Died on 5 Dec 1814.

Weslemyer, John - Seaman - Nbr: 4264 - Prize: Argus - Ship Type: MV - How taken: HMS San Domingo - When taken: 1 Mar 1814 - Where taken: off Savannah - Date Received: 7 Oct 1814 - From what ship: HMT Niobe, Chatham - Born: Charleston - Age: 27 - Released on 13 Jun 1815.

Wessel, Samuel - Seaman - Nbr: 340 - Prize: Good Friends - Ship Type: MV - How taken: HMS Andromache - When taken: 2 Apr 1813 - Where taken: Bay of Biscay - Date Received: 28 Jun 1813 - From what ship: Plymouth - Born: New Jersey - Age: 20 - Released on 25 Mar 1815.

Wessels, Robert - Seaman - Nbr: 710 - Prize: Ned - Ship Type: LM - How taken: HMS Royalist - When taken: 6 Sep 1813 - Where taken: Bay of Biscay - Date Received: 27 Sep 1813 - From what ship: Plymouth - Born: New York - Age: 17 - Sent to Plymouth on 7 Dec 1813.

Wessels, William - Seaman - Nbr: 2116 - How taken: Gave himself up from HMS Prince William - When taken: 13 Oct 1812 - Date Received: 8 Aug 1814 - From what ship: HMT Raven, Chatham - Born: New York - Age: 23 - Released on 26 Apr 1815.

West, Abiel - Soldier - Nbr: 5844 - Prize: Crosby's Regiment - Ship Type: Troops - How taken: British Army - When taken: 17 Sep 1814 - Where taken: Fort Erie - Date Received: 26 Dec 1814 - From what ship: HMT Argo - Born: Pennsylvania - Age: 28 - Released on 29 Jun 1815.

West, Benjamin - Seaman - Nbr: 400 - Prize: Young Holkar - Ship Type: MV - How taken: HMS Superb - When taken: 10 Apr 1813 - Where taken: off Belle Isle, France - Date Received: 1 Jul 1813 - From what ship: Plymouth - Born: Charlestown - Age: 19 - Released on 20 Apr 1815.

West, Dennis - Seaman - Nbr: 1562 - Prize: Caroline - Ship Type: MV - How taken: HMS Medusa - When taken: 12 Apr 1813 - Where taken: Bay of Biscay - Date Received: 23 Jun 1814 - From what ship: Stapleton - Born: Massachusetts - Age: 26 - Released on 1 May 1815.

American Prisoners of War Held at Dartmoor during the War of 1812

## Alphabetical listing of names

West, Edward - Seaman - Nbr: 3868 - Prize: Lewis Warrington - Ship Type: MV - How taken: HMS Loire - When taken: 23 May 1814 - Where taken: off VA - Date Received: 5 Oct 1814 - From what ship: HMT Orpheus, Halifax - Born: Charleston - Age: 50 - Released on 10 May 1815.

West, George - Seaman - Nbr: 2094 - How taken: Gave himself up from HMS Malta - When taken: 1 Jan 1813 - Date Received: 3 Aug 1814 - From what ship: HMS Bittern, Chatham Depot - Born: Baltimore - Age: 24 - Died on 27 Jan 1815.

West, George - Seaman - Nbr: 5140 - How taken: Gave himself up from HMS Crescent - When taken: 20 Oct 1814 - Date Received: 31 Oct 1814 - From what ship: HMT Castillian - Born: Delaware - Age: 48 - Died on 27 Jan 1815 from gastritis.

West, Jacob - Seaman - Nbr: 6377 - Prize: US Brig Syren - Ship Type: MW - How taken: HMS Medway - When taken: 12 Jul 1814 - Where taken: off Cape of Good Hope - Date Received: 24 Feb 1815 - From what ship: HMT Ganges, Plymouth - Born: Albany - Age: 30 - Released on 11 Jul 1815.

West, John - Seaman - Nbr: 301 - Prize: Ducornau - Ship Type: MV - How taken: HMS Pheasant - When taken: 15 Mar 1813 - Where taken: Bay of Biscay - Date Received: 28 Jun 1813 - From what ship: Plymouth - Born: New York - Age: 21 - Sent to Plymouth on 7 Dec 1813.

West, John - Seaman - Nbr: 5393 - Prize: Gotland - Ship Type: MV - How taken: HMS Barbados - When taken: 31 Jan 1814 - Where taken: off St. Bartholomew - Date Received: 31 Oct 1814 - From what ship: HMT Leyden, Chatham - Born: Virginia - Age: 19 - Released on 1 Jul 1815.

West, John - Seaman - Nbr: 3844 - Prize: Dominique - Ship Type: LM - How taken: HMS Dotterel - When taken: 21 May 1814 - Where taken: off Charleston - Date Received: 5 Oct 1814 - From what ship: HMT Orpheus, Halifax - Born: Charleston - Age: 21 - Released on 9 Jun 1815.

West, Reuben - 2nd Mate - Nbr: 21 - How taken: Apprehended at Gibraltar - When taken: 8 Aug 1813 - Date rec'd: 2 Apr 1813 - What ship: Plymouth - Born: Falmouth - Age: 21 - Sent to Dartmouth on 30 Jul 1813.

West, Richard - Seaman - Nbr: 4029 - Prize: US Brig Rattlesnake - Ship Type: MW - How taken: HMS Leander - When taken: 13 Jul 1814 - Where taken: off Shelburne - Date Received: 6 Oct 1814 - From what ship: HMT Chesapeake, Halifax - Born: New York - Age: 20 - Race: Negro - Released on 9 Jun 1815.

West, Samuel - Seaman - Nbr: 2176 - Prize: Rattlesnake - Ship Type: P - How taken: HMS Hyperion - When taken: 26 Jun 1814 - Where taken: off Cape Finisterre - Date Received: 16 Aug 1814 - From what ship: HMS Dublin, Halifax - Born: New York - Age: 41 - Race: Negro - Released on 2 May 1815.

West, Simon - Seaman - Nbr: 1412 - Prize: Governor Gerry - Ship Type: MV - How taken: HMS Lyra - When taken: 29 May 1813 - Where taken: Bay of Biscay - Date Received: 19 Jun 1814 - From what ship: Stapleton - Born: Rhode Island - Age: 23 - Released on 28 Apr 1815.

West, William - Prize Master - Nbr: 2199 - Prize: Hussar - Ship Type: P - How taken: HMS Saturn - When taken: 25 May 1814 - Where taken: off Sandy Hook - Date Received: 16 Aug 1814 - From what ship: HMS Dublin, Halifax - Born: Philadelphia - Age: 28 - Released on 2 May 1815.

Westcott, Charles - Seaman - Nbr: 2289 - Prize: Diomede - Ship Type: P - How taken: HMS Rifleman - When taken: 28 May 1814 - Where taken: off Sable Island - Date Received: 16 Aug 1814 - From what ship: HMS Dublin, Halifax - Born: Newburyport - Age: 27 - Released on 3 May 1815.

Westcott, William - 1st Lieutenant - Nbr: 6457 - Prize: Decatur - Ship Type: P - How taken: HMS Rhin - When taken: 5 Jun 1814 - Where taken: off San Domingo - Date Received: 3 Mar 1815 - From what ship: HMT Ganges, Plymouth - Born: Baltimore - Age: 45 - Released on 11 Jul 1815.

Westerbest, William - Seaman - Nbr: 1785 - Prize: Tigre - Ship Type: MV - How taken: HMS Scylla - When taken: 22 Mar 1813 - Where taken: Bay of Biscay - Date Received: 20 Jul 1814 - From what ship: HMS Milford, Plymouth - Born: New York - Age: 22 - Released on 1 May 1815.

Western, Asa - Prize master - Nbr: 6082 - Prize: David Porter - Ship Type: LM - How taken: HMS Pylades - When taken: 13 Sep 1814 - Where taken: Georges Bank - Date Received: 28 Dec 1814 - From what ship: HMT Penelope - Born: Massachusetts - Age: 29 - Released on 3 Jul 1815.

Westler, John - Seaman - Nbr: 2907 - Prize: Matilda - Ship Type: MV - How taken: HMS Revolutionnaire - When taken: 25 Jul 1813 - Where taken: off L'Orient (France) - Date Received: 24 Aug 1814 - From what ship: HMT Alpheus, Chatham - Born: New Jersey - Age: 18 - Released on 2 Nov 1814.

Weston, David - Seaman - Nbr: 1516 - Prize: Governor Gerry - Ship Type: LM - How taken: HMS Royalist - When taken: 31 May 1813 - Where taken: Bay of Biscay - Date Received: 23 Jun 1814 - From what ship: Stapleton - Born: Baltimore - Age: 27 - Released on 1 May 1815.

Weston, Nathaniel - Seaman - Nbr: 3275 - Prize: Cossack - Ship Type: P - How taken: HMS Amelia - When taken: 14 Apr 1813 - Where taken: off St. Johns - Date Received: 13 Sep 1814 - From what ship: HMT Niobe,

American Prisoners of War Held at Dartmoor during the War of 1812

## Alphabetical listing of names

Chatham - Born: Woburn - Age: 20 - Released on 28 May 1815.

Weston, William - Seaman - Nbr: 2716 - How taken: Gave himself up from HMS Ajax - When taken: 13 Aug 1812 - Date Received: 21 Aug 1814 - From what ship: HMT Freya, Chatham - Born: New York - Age: 21 - Released on 26 Apr 1815.

Wetherby, Calvin - Seaman - Nbr: 5432 - How taken: Apprehended at Bristol - When taken: 13 Oct 1814 - Date Received: 11 Nov 1814 - From what ship: Bristol, 66th Regiment - Born: Boston - Age: 43 - Released on 1 Jul 1815.

Weymer, John - Seaman - Nbr: 1856 - How taken: Gave himself up from HMS Cressy - When taken: Jul 1812 - Date Received: 29 Jul 1814 - From what ship: HMS Ville de Paris, Chatham Depot - Born: Boston - Age: 24 - Released on 26 Apr 1815.

Wheadon, Anthony - Seaman - Nbr: 1405 - Prize: Eliza - Ship Type: MV - How taken: HMS Surveillante - When taken: 27 Mar 1813 - Where taken: Bay of Biscay - Date Received: 19 Jun 1814 - From what ship: Stapleton - Born: Rhode Island - Age: 31 - Released on 28 Apr 1815.

Wheeler, Benjamin - Seaman - Nbr: 728 - Prize: Ned - Ship Type: LM - How taken: HMS Royalist - When taken: 6 Sep 1813 - Where taken: Bay of Biscay - Date Received: 27 Sep 1813 - From what ship: Plymouth - Born: Le Havre - Age: 27 - Released on 26 Apr 1815.

Wheeler, Charles - Seaman - Nbr: 206 - Prize: Criterion - Ship Type: MV - How taken: HMS Belle Poule - When taken: 14 Feb 1813 - Where taken: Bay of Biscay - Date Received: 2 Apr 1813 - From what ship: Plymouth - Born: Smithtown - Age: 34 - Released on 20 Apr 1815.

Wheeler, Henry - Seaman - Nbr: 777 - Prize: Betsy - Ship Type: MV - How taken: HMS Eurotas - When taken: 26 Oct 1813 - Where taken: off Ushant (France) - Date Received: 3 Nov 1813 - From what ship: Plymouth - Born: Salem - Age: 22 - Released on 26 Apr 1815.

Wheeler, John - Seaman - Nbr: 895 - Prize: General Kempt - Ship Type: P - How taken: HMS Foxhound - When taken: 18 Dec 1813 - Where taken: Lat 48'60" Long 5'7" - Date Received: 31 Jan 1814 - From what ship: Plymouth - Born: Salem - Age: 21 - Released on 27 Apr 1815.

Wheeler, Michael - Seaman - Nbr: 3062 - Prize: Polly - Ship Type: MV - How taken: HMS Maidstone - When taken: 20 Jul 1813 - Where taken: Grand Banks - Date Received: 2 Sep 1814 - From what ship: HMT Hydra, Chatham - Born: Boston - Age: 27 - Released on 28 May 1815.

Wheeler, Richard - Cook - Nbr: 56 - Prize: Spitfire - Ship Type: MV - How taken: HMS Achates - When taken: 14 Feb 1813 - Where taken: off Ushant (France) - Date Received: 2 Apr 1813 - From what ship: Plymouth - Born: Marblehead - Age: 23 - Released on 20 Apr 1815.

Wheeler, William - Seaman - Nbr: 1972 - Prize: Orbit - Ship Type: MV - How taken: HMS Achates - When taken: 29 Jan 1813 - Where taken: Lat 44N Long 13W - Date Received: 3 Aug 1814 - From what ship: HMS Lyffey, Chatham Depot - Born: Newburyport - Age: 20 - Released on 2 May 1815.

Wherry, Daniel - Passenger - Nbr: 5220 - Prize: Rolla - Ship Type: P - How taken: HMS Victorious - When taken: 2 Jun 1813 - Where taken: off Halifax - Date Received: 31 Oct 1814 - From what ship: HMT Mermaid, Chatham - Born: Philadelphia - Age: 18 - Released on 29 Jun 1815.

Whettan, John - Seaman - Nbr: 5895 - Prize: Harlequin - Ship Type: P - How taken: HMS Bulwark - When taken: 23 Nov 1814 - Where taken: off Halifax - Date Received: 27 Dec 1814 - From what ship: HMT Penelope - Born: Portsmouth - Age: 20 - Died on 18 Jan 1815 from phthisis pulmonalis.

Whiffle, Joseph - Seaman - Nbr: 5398 - Prize: Perfect - Ship Type: P - How taken: HMS Grinder - When taken: 6 Jul 1814 - Where taken: off St. Bartholomew - Date Received: 31 Oct 1814 - From what ship: HMT Leyden, Chatham - Born: Providence - Age: 27 - Released on 1 Jul 1815.

Whilday, Joseph - Seaman - Nbr: 6134 - Prize: US Gunboat 121 - Ship Type: MW - How taken: HMS Martin - When taken: Aug 1813 - Where taken: off Delaware - Date Received: 17 Jan 1815 - From what ship: HMT Impregnable - Born: Virginia - Age: 18 - Released on 5 Jul 1815.

Whippen, W. C. - Mate - Nbr: 5707 - Prize: Romp - Ship Type: MV - How taken: HMS Prometheus - When taken: 2 Jan 1814 - Where taken: West Indies - Date Received: 24 Dec 1814 - From what ship: HMT Penelope - Born: Massachusetts - Age: 22 - Released on 1 May 1815.

Whipple, John - Seaman - Nbr: 6434 - Prize: Chance - Ship Type: P - How taken: HMS Statira - When taken: 1 Apr 1814 - Where taken: Lat 38 Long 24 - Date Received: 3 Mar 1815 - From what ship: HMT Ganges, Plymouth - Born: New London - Age: 19 - Released on 11 Jul 1815.

Whistler, Miles - Seaman - Nbr: 5391 - Prize: Union - Ship Type: MV - How taken: HMS Malabar - When taken: 17 Jan 1814 - Where taken: off Calcutta - Date Received: 31 Oct 1814 - From what ship: HMT Leyden, Chatham - Born: Richmond - Age: 27 - Released on 1 Jul 1815.

American Prisoners of War Held at Dartmoor during the War of 1812

## Alphabetical listing of names

Whitamore, William - Seaman - Nbr: 6117 - Prize: Johannes - Ship Type: Prize - How taken: Lacine - When taken: 26 Oct 1814 - Where taken: Grand Banks - Date Received: 6 Jan 1814 - From what ship: HMT Impregnable - Born: Lynn - Age: 22 - Released on 5 Jul 1815.

White, Charles - Seaman - Nbr: 3462 - How taken: Gave himself up from HMS Nimble - Date Received: 19 Sep 1814 - From what ship: HMT Salvador del Mundo - Born: Melton - Age: 33 - Released on 11 Jul 1815.

White, Charles - Seaman - Nbr: 2634 - How taken: Gave himself up from HMS Leviathan - When taken: 22 Oct 1813 - Date Received: 21 Aug 1814 - From what ship: HMT Freya, Chatham - Born: New York - Age: 38 - Race: Mulatto - Released on 3 May 1815.

White, Clement - Seaman - Nbr: 2305 - Prize: Hussar - Ship Type: P - How taken: HMS Saturn - When taken: 25 May 1814 - Where taken: off Sandy Hook - Date Received: 16 Aug 1814 - From what ship: HMS Dublin, Halifax - Born: Portsmouth - Age: 44 - Released on 3 May 1815.

White, Davis - Boy - Nbr: 1385 - Prize: Courier - Ship Type: P - How taken: HMS Rover - When taken: 14 May 1813 - Where taken: Bay of Biscay - Date Received: 19 Jun 1814 - From what ship: Stapleton - Born: Baltimore - Age: 15 - Released on 28 Apr 1815.

White, Edward - Seaman - Nbr: 4446 - Prize: York Town - Ship Type: P - How taken: HMS Maidstone - When taken: 18 Jul 1813 - Where taken: Grand Banks - Date Received: 8 Oct 1814 - From what ship: HMT Leyden, Chatham - Born: New York - Age: 33 - Released on 14 Jun 1815.

White, George - Seaman - Nbr: 4617 - Prize: Argus - Ship Type: MV - How taken: HMS San Domingo - When taken: 1 Mar 1814 - Where taken: off Savannah - Date Received: 9 Oct 1814 - From what ship: HMT Leyden, Chatham - Born: Boston - Age: 19 - Released on 15 Jun 1815.

White, George - Seaman - Nbr: 2551 - Prize: Caromaned - Ship Type: Prize - How taken: HMS Eridanus - When taken: 13 Aug 1814 - Where taken: Lat 40, Long 16 - Date Received: 16 Aug 1814 - From what ship: HMT Salvador del Mundo - Born: New York - Age: 20 - Race: Mulatto - Released on 3 May 1815.

White, Gilman - Seaman - Nbr: 5974 - Prize: Halifax Packet - Ship Type: Prize - How taken: HMS Bulwark - When taken: 22 Sep 1814 - Where taken: Georges Bank - Date Received: 27 Dec 1814 - From what ship: HMT Penelope - Born: Not legible - Released on 3 Jul 1815.

White, Henry - Seaman - Nbr: 209 - Prize: Criterion - Ship Type: MV - How taken: HMS Belle Poule - When taken: 14 Feb 1813 - Where taken: Bay of Biscay - Date Received: 2 Apr 1813 - From what ship: Plymouth - Born: Maryland - Age: 21 - Released on 20 Apr 1815.

White, Henry - Shipwright - Nbr: 2124 - How taken: Gave himself up from HMS Mulgrave - When taken: 23 Nov 1812 - Date Received: 8 Aug 1814 - From what ship: HMT Raven, Chatham - Born: Hancock - Age: 43 - Released on 26 Apr 1815.

White, Henry - Seaman - Nbr: 1528 - Prize: Essex - Ship Type: MV - How taken: HMS Pyramus - When taken: 20 Apr 1813 - Where taken: Bay of Biscay - Date Received: 23 Jun 1814 - From what ship: Stapleton - Born: Marblehead - Age: 24 - Released on 1 May 1815.

White, Isaac - Seaman - Nbr: 1370 - Prize: Grand Napoleon - Ship Type: P - How taken: HMS Goldfinch - When taken: 17 Apr 1813 - Where taken: Bay of Biscay - Date Received: 19 Jun 1814 - From what ship: Stapleton - Born: Connecticut - Age: 19 - Released on 28 Apr 1815.

White, James - Seaman - Nbr: 6487 - Prize: John - Ship Type: MV - How taken: Variable - When taken: 11 Aug 1814 - Where taken: off Cuba - Date Received: 3 Mar 1815 - From what ship: HMT Ganges, Plymouth - Born: Charleston - Age: 32 - Released on 11 Jul 1815.

White, James - Seaman - Nbr: 793 - Prize: Betsy - Ship Type: MV - How taken: HMS Eurotas - When taken: 26 Oct 1813 - Where taken: off Ushant (France) - Date Received: 3 Nov 1813 - From what ship: Plymouth - Born: New York - Age: 27 - Released on 26 Apr 1815.

White, James - Seaman - Nbr: 5781 - Prize: William Penn - Ship Type: MV - How taken: HMS Acorn - When taken: 27 Oct 1812 - Where taken: Lat 14 - Date Received: 26 Dec 1814 - From what ship: HMT Argo - Born: Massachusetts - Age: 19 - Race: Negro - Released on 27 Apr 1815.

White, James - Seaman - Nbr: 5405 - How taken: Impressed at London - When taken: 6 Oct 1814 - Date Received: 31 Oct 1814 - From what ship: HMT Leyden, Chatham - Born: Hartford - Age: 26 - Released on 1 Jul 1815.

White, John - Seaman - Nbr: 2370 - Prize: Snap Dragon - Ship Type: P - How taken: HMS Martin - When taken: 10 Jun 1814 - Where taken: off Halifax - Date Received: 16 Aug 1814 - From what ship: HMT Queen, Halifax - Born: New York - Age: 17 - Released on 3 Apr 1815.

White, John - Seaman - Nbr: 1457 - Prize: Revenge - Ship Type: LM - How taken: HMS Belle Poule - When taken: 10 Mar 1813 - Where taken: off Cornwall - Date Received: 19 Jun 1814 - From what ship: Stapleton - Born: Boston - Age: 24 - Released on 28 Apr 1815.

American Prisoners of War Held at Dartmoor during the War of 1812

## Alphabetical listing of names

White, John - Marine - Nbr: 4058 - Prize: US Brig Rattlesnake - Ship Type: MW - How taken: HMS Leander - When taken: 13 Jul 1814 - Where taken: off Shelburne - Date Received: 6 Oct 1814 - From what ship: HMT Chesapeake, Halifax - Born: North Carolina - Age: 19 - Released on 13 Jun 1815.

White, Kitt - Seaman - Nbr: 3702 - Prize: Alfred - Ship Type: P - How taken: HMS Epervier - When taken: 23 Feb 1814 - Where taken: off Newfoundland - Date Received: 30 Sep 1814 - From what ship: HMT President, Halifax - Born: Charleston - Age: 63 - Race: Creole - Released on 4 Jun 1815.

White, Peter - Seaman - Nbr: 5697 - Prize: McDonough - Ship Type: P - How taken: HMS Bacchante - When taken: 7 Nov 1814 - Where taken: Lat 42 Long 67 - Date Received: 24 Dec 1814 - From what ship: HMT Penelope - Born: Not listed - Released on 11 Jul 1815.

White, Philip - Prize Master - Nbr: 1282 - Prize: Indian Lass - Ship Type: Prize - How taken: Not listed - When taken: 29 Apr 1814 - Date Received: 14 Jun 1814 - From what ship: Mill Prison (Plymouth, England) - Born: Marblehead - Age: 30 - Released on 28 Apr 1815.

White, Philip - Seaman - Nbr: 122 - Prize: Governor McKean - Ship Type: LM - How taken: HMS Rover - When taken: 26 Jan 1813 - Where taken: off Bordeaux - Date Received: 2 Apr 1813 - From what ship: Plymouth - Born: Pennsylvania - Age: 17 - Sent to Dartmouth on 30 Jul 1813.

White, Richard - Cook - Nbr: 2903 - How taken: Gave himself up from HMS Confounder - When taken: 10 Jun 1813 - Date Received: 24 Aug 1814 - From what ship: HMT Alpheus, Chatham - Born: Salem - Age: 27 - Released on 19 May 1815.

White, W. K. - Prize master - Nbr: 3810 - Prize: Nimble - Ship Type: Prize - How taken: HMS Arab - When taken: 5 Apr 1814 - Where taken: Lat 37 Long 65 - Date Received: 5 Oct 1814 - From what ship: HMT Orpheus, Halifax - Born: New York - Age: 31 - Released on 9 Jun 1815.

White, William - Seaman - Nbr: 3807 - Prize: Stark - Ship Type: P - How taken: HMS Sophie - When taken: 20 Apr 1814 - Where taken: off Bermuda - Date Received: 5 Oct 1814 - From what ship: HMT Orpheus, Halifax - Born: Norfolk - Age: 25 - Race: Black - Released on 9 Jun 1815.

White, William - Seaman - Nbr: 5312 - Prize: Thomas - Ship Type: P - How taken: HMS Nymphe - When taken: 24 Jun 1813 - Where taken: off Halifax - Date Received: 31 Oct 1814 - From what ship: HMT Leyden, Chatham - Born: Portsmouth - Age: 19 - Released on 11 Jul 1815.

White, William 1st - Seaman - Nbr: 1802 - How taken: Gave himself up from HMS Clarence - Date Received: 20 Jul 1814 - From what ship: HMS Milford, Plymouth - Born: New York - Age: 25 - Released on 1 May 1815.

White, William 2nd - Seaman - Nbr: 1803 - How taken: Gave himself up from HMS Clarence - Date Received: 20 Jul 1814 - From what ship: HMS Milford, Plymouth - Born: Boston - Age: 23 - Released on 1 May 1815.

Whiteby, Samuel - Seaman - Nbr: 4974 - Prize: Invincible - Ship Type: LM - How taken: HMS Armide - When taken: 15 Aug 1814 - Where taken: off Nantucket - Date Received: 28 Oct 1814 - From what ship: HMT Alkbar, Halifax - Born: Portsmouth - Age: 19 - Released on 21 Jun 1815.

Whitecombe, John - Seaman - Nbr: 3982 - Prize: US Brig Rattlesnake - Ship Type: MW - How taken: HMS Leander - When taken: 13 Jul 1814 - Where taken: off Shelburne - Date Received: 6 Oct 1814 - From what ship: HMT Chesapeake, Halifax - Born: Boston - Age: 22 - Released on 9 Jun 1815.

Whitehead, James - Seaman - Nbr: 1497 - Prize: Margaret - Ship Type: Prize - How taken: HMS Foxhound - When taken: 27 May 1814 - Where taken: off Isles of Scilly - Date Received: 20 Jun 1814 - From what ship: Mill Prison (Plymouth, England) - Born: Norfolk - Age: 19 - Released on 1 May 1815.

Whitehead, John - Seaman - Nbr: 6428 - How taken: Sent into custody from Prize Patty - When taken: 12 Apr 1814 - Date Received: 3 Mar 1815 - From what ship: HMT Ganges, Plymouth - Born: Bridgewater - Age: 29 - Released on 11 Jul 1815.

Whitehead, Reuben - Prize Master - Nbr: 3564 - Prize: Hawk - Ship Type: P - How taken: HMS Pique - When taken: 26 Apr 1814 - Where taken: off Bermuda - Date Received: 30 Sep 1814 - From what ship: HMT Sybella - Born: North Carolina - Age: 22 - Released on 4 Jun 1815.

Whitehouse, Asay - Seaman - Nbr: 161 - How taken: Impressed at Dublin - When taken: 7 Jan 1813 - Date Received: 2 Apr 1813 - From what ship: Plymouth - Born: Brookfield - Age: 23 - Sent to Dartmouth on 30 Jul 1813.

Whitehouse, Daniel - Novice - Nbr: 1194 - Prize: Imperatrice Reine - Ship Type: P - How taken: HMS Hotspur - When taken: 13 Jan 1813 - Date Received: 4 Jun 1814 - From what ship: Dartmouth - Born: New York - Age: 18 - Released on 28 Apr 1815.

Whiteman, Peter - Seaman - Nbr: 5075 - Prize: Ida - Ship Type: LM - How taken: HMS Newcastle - When taken: 9 Aug 1814 - Where taken: Long 34 - Date Received: 28 Oct 1814 - From what ship: HMT Alkbar, Halifax - Born: Trinidad - Age: 28 - Released on 29 Jun 1815.

American Prisoners of War Held at Dartmoor during the War of 1812

## Alphabetical listing of names

Whitewood, Charles - Seaman - Nbr: 216 - Prize: Mars - Ship Type: MV - How taken: HMS Warspite - When taken: 26 Feb 1813 - Where taken: off Basque Roads (France) - Date Received: 2 Apr 1813 - From what ship: Plymouth - Born: New York - Age: 28 - Released on 20 Apr 1815.

Whitehead, Daniel - Seaman - Nbr: 3116 - Prize: Harmony - Ship Type: Prize - How taken: HMS Brisk - When taken: 24 Aug 1814 - Where taken: Bristol Channel - Date Received: 11 Sep 1814 - From what ship: HMT Salvador del Mundo - Born: New York - Age: 27 - Released on 28 May 1815.

Whiting, Samuel - Seaman - Nbr: 1338 - Prize: Paul Jones - Ship Type: P - How taken: HMS Leonidas - When taken: 23 May 1813 - Where taken: Channel - Date Received: 19 Jun 1814 - From what ship: Stapleton - Born: Rhode Island - Age: 22 - Released on 28 Apr 1815.

Whitmore, John - Seaman - Nbr: 1308 - How taken: Sent into custody from HMS Cadmus - Date Received: 14 Jun 1814 - From what ship: Mill Prison (Plymouth, England) - Born: Petersburg - Age: 24 - Released on 28 Apr 1815.

Whitney, Thomas - Prize Master - Nbr: 1152 - Prize: Bunker Hill - Ship Type: P - How taken: HMS Pomone - When taken: 8 Mar 1814 - Where taken: at sea - Date Received: 10 May 1814 - From what ship: Plymouth - Born: Fairfield - Age: 28 - Released on 30 Mar 1815.

Whitning, William - Seaman - Nbr: 6450 - Prize: Jemmell - Ship Type: MV - How taken: HMS Rhin - When taken: 28 May 1814 - Where taken: off Bermuda - Date Received: 3 Mar 1815 - From what ship: HMT Ganges, Plymouth - Born: Boston - Age: 25 - Released on 11 Jul 1815.

Whitten, Elijah - Seaman - Nbr: 1673 - Prize: Vivid - Ship Type: Prize - How taken: HMS Ceres - When taken: 5 Jun 1814 - Where taken: Channel - Date Received: 2 Jul 1814 - From what ship: Plymouth - Born: Barrington - Age: 34 - Released on 1 May 1815.

Whitter, J. W. - Seaman - Nbr: 1464 - Prize: Revenge - Ship Type: LM - How taken: HMS Belle Poule - When taken: 10 Mar 1813 - Where taken: off Cornwall - Date Received: 19 Jun 1814 - From what ship: Stapleton - Born: Connecticut - Age: 26 - Released on 28 Apr 1815.

Whittington, George - Seaman - Nbr: 1885 - Prize: Mariner - Ship Type: Prize - How taken: HMS Poictiers - When taken: 3 Sep 1813 - Where taken: off Sable Island - Date Received: 29 Jul 1814 - From what ship: HMS Ville de Paris, Chatham Depot - Born: Providence - Age: 20 - Released on 2 May 1815.

Whittington, Lemuel - Seaman - Nbr: 6335 - Prize: Prince de Neufchatel - Ship Type: P - How taken: Leander (Newcastle Acasta) - When taken: 20 Dec 1814 - Where taken: Lat 38 Long 56 - Date Received: 19 Feb 1815 - From what ship: HMT Ganges, Plymouth - Born: Providence - Age: 20 - Released on 6 Jul 1815.

Whormsley, R. - Seaman - Nbr: 5798 - Prize: US Frigate Superior (Gig) - Ship Type: MW - How taken: British gunboats - When taken: 26 Aug 1814 - Where taken: Lake Ontario - Date Received: 26 Dec 1814 - From what ship: HMT Argo - Born: Germantown - Age: 22 - Released on 16 Jun 1815.

Wickham, John - Seaman - Nbr: 2650 - How taken: Gave himself up from HMS Goliath - When taken: 8 Aug 1813 - Date Received: 21 Aug 1814 - From what ship: HMT Freya, Chatham - Born: Connecticut - Age: 25 - Released on 3 May 1815.

Wickham, Thaddeus - Seaman - Nbr: 2759 - How taken: Gave himself up from HMS Bruin - When taken: 19 Jul 1813 - Date Received: 24 Aug 1814 - From what ship: HMT Liverpool, Chatham - Born: Bridgewater - Age: 28 - Race: Mulatto - Released on 19 May 1815.

Wicks, Samuel - Seaman - Nbr: 1693 - Prize: Nonsuch - Ship Type: MV - How taken: HMS Dotterel - When taken: 14 Dec 1813 - Where taken: off Charlestown - Date Received: 2 Jul 1814 - From what ship: Plymouth - Born: Philadelphia - Age: 18 - Released on 5 Apr 1815.

Wickwall, Joseph - Seaman - Nbr: 2742 - How taken: Gave himself up from HMS Ceres - When taken: 20 Dec 1812 - Date Received: 24 Aug 1814 - From what ship: HMT Liverpool, Chatham - Born: Washington - Age: 25 - Released on 26 Apr 1815.

Widger, John - Seaman - Nbr: 3289 - Prize: Enterprize - Ship Type: P - How taken: HMS Tenedos - When taken: 21 May 1813 - Where taken: off Cape Cod - Date Received: 13 Sep 1814 - From what ship: HMT Niobe, Chatham - Born: Marblehead - Age: 27 - Released on 28 May 1815.

Widger, Joseph - Seaman - Nbr: 3376 - Prize: Growler - Ship Type: P - How taken: HMS Electra - When taken: 7 Jul 1813 - Where taken: at sea - Date Received: 13 Sep 1814 - From what ship: HMT Niobe, Chatham - Born: Marblehead - Age: 21 - Died on 6 Jan 1815 from phthisis.

Wieman, Laurence - Soldier - Nbr: 745 - Prize: 21st US Infantry - Ship Type: Troops - How taken: British Army - When taken: 28 May 1813 - Where taken: Lake Ontario - Date Received: 3 Nov 1813 - From what ship: Plymouth - Born: Berlin - Age: 37 - Sent to Dartmouth on 2 Nov 1814.

Wiggent, John - Seaman - Nbr: 3453 - How taken: Sent as prisoner from St. Michaels - Date Received: 19 Sep 1814

American Prisoners of War Held at Dartmoor during the War of 1812

## Alphabetical listing of names

- From what ship: HMT Salvador del Mundo - Born: Konigsberg - Age: 25 - Released on 11 Jun 1815.

Wiggins, Richard - Seaman - Nbr: 5192 - Prize: Lark - Ship Type: MV - How taken: HMS Brearn - When taken: 12 Apr 1813 - Where taken: off Cape Sable Island (Canada) - Date Received: 31 Oct 1814 - From what ship: HMT Mermaid, Chatham - Born: Salem - Age: 30 - Released on 29 Jun 1815.

Wiggins, Samuel - Seaman - Nbr: 5239 - Prize: Rose - Ship Type: MV - How taken: HMS Racehorse - When taken: 4 Feb 1813 - Where taken: at sea - Date Received: 31 Oct 1814 - From what ship: HMT Mermaid, Chatham - Born: Salem - Age: 30 - Released on 29 Jun 1815.

Wilbourg, David - Seaman - Nbr: 6453 - Prize: Jemmell - Ship Type: MV - How taken: HMS Rhin - When taken: 28 May 1814 - Where taken: off Bermuda - Date Received: 3 Mar 1815 - From what ship: HMT Ganges, Plymouth - Born: Falmouth - Age: 24 - Race: Black - Released on 11 Jul 1815.

Wilcox, Caesar - Seaman - Nbr: 3722 - Prize: Alfred - Ship Type: P - How taken: HMS Epervier - When taken: 23 Feb 1814 - Where taken: off Newfoundland - Date Received: 30 Sep 1814 - From what ship: HMT President, Halifax - Born: Salem - Age: 42 - Race: Negro - Released on 4 Jun 1815.

Wilcox, Ephraim - Seaman - Nbr: 2920 - How taken: Gave himself up from HMS Acteon - Date Received: 24 Aug 1814 - From what ship: HMT Alpheus, Chatham - Born: Stanton - Age: 23 - Released on 19 May 1815.

Wilcox, Lewis - Seaman - Nbr: 1376 - Prize: Grand Napoleon - Ship Type: P - How taken: HMS Goldfinch - When taken: 17 Apr 1813 - Where taken: Bay of Biscay - Date Received: 19 Jun 1814 - From what ship: Stapleton - Born: Rhode Island - Age: 32 - Released on 28 Apr 1815.

Wilcox, William - Seaman - Nbr: 4781 - How taken: Gave himself up from HMS Quebec - When taken: 21 Sep 1814 - Date Received: 9 Oct 1814 - From what ship: HMT Freya, Chatham - Born: Connecticut - Age: 43 - Released on 11 Jul 1815.

Wild, Thomas - Seaman - Nbr: 1337 - Prize: Paul Jones - Ship Type: P - How taken: HMS Leonidas - When taken: 23 May 1813 - Where taken: Channel - Date Received: 19 Jun 1814 - From what ship: Stapleton - Born: Delaware - Age: 27 - Released on 28 Apr 1815.

Wildon, Thomas - Seaman - Nbr: 1703 - Prize: Fanny - Ship Type: Prize - How taken: HMS Sceptre - When taken: 12 May 1814 - Date Received: 2 Jul 1814 - From what ship: Plymouth - Born: New York - Age: 32 - Released on 1 May 1815.

Wiley, David - Seaman - Nbr: 2710 - How taken: Gave himself up from HMS Unicorn - When taken: 2 Dec 1812 - Date Received: 21 Aug 1814 - From what ship: HMT Freya, Chatham - Born: Wakefield - Age: 37 - Released on 26 Apr 1815.

Wiley, John - Seaman - Nbr: 3894 - Prize: Governor Shelby - Ship Type: MV - How taken: HMS Saturn - When taken: 9 Jul 1814 - Where taken: off Long Island - Date Received: 5 Oct 1814 - From what ship: HMT Orpheus, Halifax - Born: Philadelphia - Age: 36 - Released on 9 Jun 1815.

Wilkey, Timothy - Seaman - Nbr: 3236 - Prize: Maydock - Ship Type: MV - How taken: Rebuff - When taken: 16 Jun 1813 - Where taken: off St. Marys - Date Received: 11 Sep 1814 - From what ship: HMT Freya, Chatham - Born: Dartmouth - Age: 18 - Released on 28 May 1815.

Wilkins, William - Seaman - Nbr: 1609 - Prize: Fox - Ship Type: LM - How taken: HMS Pheasant - When taken: 23 Apr 1813 - Where taken: Bay of Biscay - Date Received: 23 Jun 1814 - From what ship: Stapleton - Born: New Jersey - Age: 23 - Sent to Mill Prison (Plymouth, England) on 30 Jun 1814.

Wilkins, William - Seaman - Nbr: 3542 - Prize: Hawk - Ship Type: P - How taken: HMS Pique - When taken: 26 Apr 1814 - Where taken: off Bermuda - Date Received: 30 Sep 1814 - From what ship: HMT Sybella - Born: Norfolk - Age: 20 - Released on 4 Jun 1815.

Willett, John - Seaman - Nbr: 1812 - How taken: Gave himself up from HMS Rosario - When taken: 28 May 1813 - Date Received: 20 Jul 1814 - From what ship: HMS Milford, Plymouth - Born: New Castle - Age: 32 - Race: Black - Released on 1 Jul 1815.

Willett, Robert - Seaman - Nbr: 1965 - How taken: Gave himself up from HMS Andromache - When taken: 24 Dec 1812 - Date Received: 3 Aug 1814 - From what ship: HMS Lyffey, Chatham Depot - Born: Newburgh - Age: 21 - Released on 11 Jul 1815.

Willey, Ebenezer - Seaman - Nbr: 784 - Prize: Betsy - Ship Type: MV - How taken: HMS Eurotas - When taken: 26 Oct 1813 - Where taken: off Ushant (France) - Date Received: 3 Nov 1813 - From what ship: Plymouth - Born: New Hampshire - Age: 34 - Released on 26 Apr 1815.

Willey, James - Soldier - Nbr: 5262 - Prize: 13th US Infantry - Ship Type: Troops - How taken: British Army - When taken: 13 Oct 1812 - Where taken: Canada - Date Received: 31 Oct 1814 - From what ship: HMT Leyden, Chatham - Born: Down - Age: 41 - Released on 29 Jun 1815.

Williams, Joseph - Seaman - Nbr: 2136 - How taken: Gave himself up from HMS Frederickstein - When taken: 14

American Prisoners of War Held at Dartmoor during the War of 1812

## Alphabetical listing of names

Oct 1812 - Date Received: 8 Aug 1814 - From what ship: HMT Raven, Chatham - Born: New York - Age: 29 - Released on 26 Apr 1815.

Williams, Abraham - Seaman - Nbr: 2041 - How taken: Gave himself up from HMS Acteon - When taken: 12 Jun 1812 - Date Received: 3 Aug 1814 - From what ship: HMS Lyffey, Chatham Depot - Born: Philadelphia - Age: 44 - Race: Black - Released on 26 Apr 1815.

Williams, Alexander - Seaman - Nbr: 1021 - Prize: Rachel & Ann - Ship Type: P - How taken: HMS Cydnus - When taken: 6 Jan 1814 - Where taken: at sea - Date Received: 31 Jan 1814 - From what ship: Plymouth - Born: New York - Age: 22 - Race: Black - Released on 27 Apr 1815.

Williams, Ambrose - Seaman - Nbr: 71 - Prize: Rolla - Ship Type: MV - How taken: HMS Surveillante - When taken: 11 Feb 1813 - Where taken: Bay of Biscay - Date Received: 2 Apr 1813 - From what ship: Plymouth - Born: Salem - Age: 19 - Sent to Plymouth on 25 Jul 1814.

Williams, Ambrose - Seaman - Nbr: 681 - Prize: Rolla - Ship Type: MV - How taken: HMS Surveillante - When taken: 11 Feb 1813 - Where taken: Bay of Biscay - Date Received: 27 Sep 1813 - From what ship: Plymouth - Born: Salem - Age: 19 - Released on 26 Apr 1815.

Williams, Benjamin - Seaman - Nbr: 4559 - How taken: Gave himself up from MV Emily - When taken: 28 Aug 1814 - Date Received: 8 Oct 1814 - From what ship: HMT Leyden, Chatham - Born: Boston - Age: 24 - Released on 15 Jun 1815.

Williams, Benjamin - Seaman - Nbr: 1000 - How taken: Impressed at Liverpool - When taken: 13 Nov 1813 - Date Received: 31 Jan 1814 - From what ship: Plymouth - Born: New Jersey - Age: 20 - Released on 27 Apr 1815.

Williams, Charles - Seaman - Nbr: 4236 - Prize: Pilot - Ship Type: LM - How taken: Victoria (Privateer) - When taken: 28 Jan 1814 - Where taken: off Bordeaux - Date Received: 7 Oct 1814 - From what ship: HMT Niobe, Chatham - Born: New London - Age: 22 - Race: Black - Died on 9 Mar 1815 from febris.

Williams, Charles - Seaman - Nbr: 4801 - Prize: President - Ship Type: P - How taken: HMS Pique - When taken: 7 May 1814 - Where taken: off Porto Rico - Date Received: 9 Oct 1814 - From what ship: HMT Freya, Halifax - Born: Providence - Age: 25 - Race: Mulatto - Released on 21 Jun 1815.

Williams, Charles - Seaman - Nbr: 1858 - Prize: Volante - Ship Type: LM - How taken: HMS Valiant - When taken: 25 Mar 1813 - Where taken: Georges Bank - Date Received: 29 Jul 1814 - From what ship: HMS Ville de Paris, Chatham Depot - Born: Marblehead - Age: 21 - Released on 2 May 1815.

Williams, Darius - Seaman - Nbr: 2152 - How taken: Gave himself up from HMS Ruby - When taken: 15 Aug 1812 - Date Received: 8 Aug 1814 - From what ship: HMT Raven, Chatham - Born: Seabrook - Age: 27 - Released on 26 Apr 1815.

Williams, David - Seaman - Nbr: 3011 - How taken: Gave himself up from HMS Milford - When taken: 1 Sep 1814 - Date Received: 2 Sep 1814 - From what ship: HMT Salvador del Mundo - Born: Salem - Age: 29 - Released on 11 Jul 1815.

Williams, Edward - Seaman - Nbr: 1054 - How taken: Impressed at Cork - When taken: 23 Jan 1814 - Date Received: 10 May 1814 - From what ship: Plymouth - Born: Virginia - Age: 22 - Died on 21 Mar 1815 from variola.

Williams, Elisha - Seaman - Nbr: 1598 - Prize: Shadow - Ship Type: LM - How taken: HMS Reindeer - When taken: 6 Apr 1813 - Where taken: Bay of Biscay - Date Received: 23 Jun 1814 - From what ship: Stapleton - Born: Delaware - Age: 35 - Race: Negro - Released on 1 May 1815.

Williams, Frederick - Seaman - Nbr: 3294 - Prize: Enterprize - Ship Type: P - How taken: HMS Tenedos - When taken: 21 May 1813 - Where taken: off Cape Cod - Date Received: 13 Sep 1814 - From what ship: HMT Niobe, Chatham - Born: Marblehead - Age: 21 - Released on 28 May 1815.

Williams, George - Seaman - Nbr: 6279 - How taken: Gave himself up from HMS Gloucester - Date Received: 17 Feb 1815 - From what ship: HMT Ganges, Plymouth - Born: Philadelphia - Age: 29 - Race: Black - Released on 13 Jun 1815.

Williams, George - Seaman - Nbr: 4245 - Prize: Requin - Ship Type: LM - How taken: HMS Venus - When taken: 6 Mar 1814 - Where taken: off Bordeaux - Date Received: 7 Oct 1814 - From what ship: HMT Niobe, Chatham - Born: Charleston - Age: 32 - Race: Black - Released on 13 Jun 1815.

Williams, George - Seaman - Nbr: 239 - Prize: Charlotte - Ship Type: MV - How taken: HMS Warspite - When taken: 3 Mar 1813 - Where taken: Bay of Biscay - Date Received: 2 Apr 1813 - From what ship: Plymouth - Born: Maryland - Age: 24 - Sent to Mill Prison (Plymouth, England) on 10 Jul 1813.

Williams, George - Seaman - Nbr: 3553 - Prize: Hawk - Ship Type: P - How taken: HMS Pique - When taken: 26 Apr 1814 - Where taken: off Bermuda - Date Received: 30 Sep 1814 - From what ship: HMT Sybella - Born: New York - Age: 24 - Released on 4 Jun 1815.

American Prisoners of War Held at Dartmoor during the War of 1812

## Alphabetical listing of names

Williams, George - Seaman - Nbr: 5770 - How taken: Gave himself up from HMS Tremendous - Date Received: 26 Dec 1814 - From what ship: HMT Argo - Born: Moscow - Age: 39 - Released on 3 Jul 1815.

Williams, George - Boatswain - Nbr: 4922 - Prize: Ulysses - Ship Type: LM - How taken: HMS Majestic - When taken: 29 Jun 1813 - Where taken: off Western Islands (England) - Date Received: 28 Oct 1814 - From what ship: HMT Alkbar, Halifax - Born: Norfolk - Age: 35 - Released on 21 Jun 1815.

Williams, George - Seaman - Nbr: 2642 - How taken: Gave himself up from HMS Caledonia - When taken: 5 Dec 1812 - Date Received: 21 Aug 1814 - From what ship: HMT Freya, Chatham - Born: Massachusetts - Age: 39 - Released on 26 Apr 1815.

Williams, George F. - Seaman - Nbr: 3094 - Prize: Argus - Ship Type: MV - How taken: HMS San Domingo - When taken: 1 Mar 1814 - Where taken: off Savannah - Date Received: 3 Sep 1814 - From what ship: HMT Hydra, Chatham - Born: Nottingham - Age: 29 - Released on 2 Apr 1815.

Williams, Henry - Seaman - Nbr: 204 - Prize: Star - Ship Type: MV - How taken: HMS Superb - When taken: 9 Feb 1813 - Where taken: Bay of Biscay - Date Received: 2 Apr 1813 - From what ship: Plymouth - Born: New Orleans - Age: 35 - Race: Black - Sent to Plymouth on 8 Jul 1814.

Williams, Henry - Boy - Nbr: 848 - Prize: Chesapeake - Ship Type: LM - How taken: HMS Hotspur & HMS Pyramus - When taken: 26 Oct 1813 - Where taken: off Nantes - Date Received: 29 Nov 1813 - From what ship: Plymouth - Born: Maryland - Age: 16 - Released on 26 Apr 1815.

Williams, Hezekiel - Seaman - Nbr: 3736 - Prize: Lizard - Ship Type: P - How taken: HMS Prometheus - When taken: 5 May 1814 - Where taken: off Halifax - Date Received: 30 Sep 1814 - From what ship: HMT President, Halifax - Born: Salem - Age: 18 - Released on 4 Jun 1815.

Williams, Jacques - Seaman - Nbr: 715 - Prize: Ned - Ship Type: LM - How taken: HMS Royalist - When taken: 6 Sep 1813 - Where taken: Bay of Biscay - Date Received: 27 Sep 1813 - From what ship: Plymouth - Born: Baltimore - Age: 18 - Race: Black - Released on 26 Apr 1815.

Williams, James - Seaman - Nbr: 4622 - Prize: Argus - Ship Type: MV - How taken: HMS San Domingo - When taken: 1 Mar 1814 - Where taken: off Savannah - Date Received: 9 Oct 1814 - From what ship: HMT Leyden, Chatham - Born: Baltimore - Age: 21 - Race: Black - Released on 15 Jun 1815.

Williams, James - Boy - Nbr: 4095 - Prize: Tyger - Ship Type: Prize - How taken: HMS Bulwark - When taken: 20 Jul 1814 - Where taken: Georges Bank - Date Received: 6 Oct 1814 - From what ship: HMT Chesapeake, Halifax - Born: New York - Age: 16 - Released on 13 Jun 1815.

Williams, James - Seaman - Nbr: 6253 - How taken: Sent into custody from MV Sophia - When taken: 10 Dec 1814 - Date Received: 4 Feb 1815 - From what ship: HMT Ganges - Born: Staten Island - Age: 21 - Race: Black - Released on 5 Jul 1815.

Williams, James - Seaman - Nbr: 1759 - How taken: Gave himself up from HMS Badger - When taken: 15 Nov 1812 - Date Received: 20 Jul 1814 - From what ship: HMS Milford, Plymouth - Born: Baltimore - Age: 46 - Released on 26 Apr 1815.

Williams, James - Seaman - Nbr: 1066 - Prize: Fair American - Ship Type: MV - How taken: HMS Andromache - When taken: 19 Jan 1814 - Where taken: Bay of Biscay - Date Received: 10 May 1814 - From what ship: Plymouth - Born: Gacra, Italy - Age: 30 - Sent to Plymouth on 8 Jul 1814.

Williams, James - Seaman - Nbr: 3115 - Prize: Harmony - Ship Type: Prize - How taken: HMS Brisk - When taken: 24 Aug 1814 - Where taken: Bristol Channel - Date Received: 11 Sep 1814 - From what ship: HMT Salvador del Mundo - Born: New York - Age: 31 - Released on 28 May 1815.

Williams, James - Seaman - Nbr: 2940 - How taken: Gave himself up from HMS Leander - When taken: 6 Jul 1813 - Date Received: 24 Aug 1814 - From what ship: HMT Alpheus, Chatham - Born: Boston - Age: 27 - Released on 19 May 1815.

Williams, James - Seaman - Nbr: 2683 - Prize: Hannah & Eliza - Ship Type: MV - How taken: HMS Lyra - When taken: 29 May 1813 - Where taken: off Bayonne - Date Received: 21 Aug 1814 - From what ship: HMT Freya, Chatham - Born: New London - Age: 20 - Released on 19 May 1815.

Williams, James - Seaman - Nbr: 2717 - How taken: Gave himself up from HMS Loire - When taken: 13 Aug 1812 - Date Received: 21 Aug 1814 - From what ship: HMT Freya, Chatham - Born: Taunton - Age: 26 - Released on 26 Apr 1815.

Williams, Jesse - Seaman - Nbr: 5820 - Prize: US Schooner Scorpion - Ship Type: MW - How taken: British gunboats - When taken: 6 Sep 1814 - Where taken: Lake Erie - Date Received: 26 Dec 1814 - From what ship: HMT Argo - Born: Pennsylvania - Age: 42 - Race: Black - Released on 3 Jul 1815.

Williams, John - Seaman - Nbr: 6472 - Prize: Decatur - Ship Type: P - How taken: HMS Rhin - When taken: 5 Jun 1814 - Where taken: off San Domingo - Date Received: 3 Mar 1815 - From what ship: HMT Ganges,

American Prisoners of War Held at Dartmoor during the War of 1812

# Alphabetical listing of names

Plymouth - Born: Roxborough - Age: 42 - Released on 11 Jul 1815.
Williams, John - Seaman - Nbr: 5953 - Prize: Harlequin - Ship Type: P - How taken: HMS Bulwark - When taken: 23 Nov 1814 - Where taken: off Halifax - Date Received: 27 Dec 1814 - From what ship: HMT Penelope - Born: Staten Island - Age: 23 - Released on 4 Jun 1815.
Williams, John - Seaman - Nbr: 5960 - Prize: Harlequin - Ship Type: P - How taken: HMS Bulwark - When taken: 23 Nov 1814 - Where taken: off Halifax - Date Received: 27 Dec 1814 - From what ship: HMT Penelope - Born: Santiago - Age: 26 - Race: Mulatto - Released on 3 Jul 1815.
Williams, John - Seaman - Nbr: 5972 - Prize: Halifax Packet - Ship Type: Prize - How taken: HMS Bulwark - When taken: 22 Sep 1814 - Where taken: Georges Bank - Date Received: 27 Dec 1814 - From what ship: HMT Penelope - Born: Baltimore - Age: 38 - Released on 3 Jul 1815.
Williams, John - Seaman - Nbr: 6103 - How taken: Gave himself up from HMS Glasgow - When taken: 31 Dec 1814 - Date Received: 6 Jan 1814 - From what ship: HMT Impregnable - Born: Rhode Island - Age: 33 - Race: Mulatto - Released on 5 Jul 1815.
Williams, John - Seaman - Nbr: 6483 - Prize: Wolf - Ship Type: P - How taken: Jaiouse - When taken: 1 Aug 1814 - Date Received: 3 Mar 1815 - From what ship: HMT Ganges, Plymouth - Born: Norfolk - Age: 17 - Released on 11 Jul 1815.
Williams, John - Seaman - Nbr: 6122 - Prize: Betsey - Ship Type: Prize - How taken: HMS Bellerophon - When taken: 2 Nov 1814 - Where taken: Long 61 - Date Received: 6 Jan 1814 - From what ship: HMT Impregnable - Born: Baltimore - Age: 25 - Race: Black - Released on 5 Jul 1815.
Williams, John - Seaman - Nbr: 4338 - Prize: Venus - Ship Type: P - How taken: HMS Loire - When taken: 18 Feb 1814 - Where taken: off St. Thomas - Date Received: 7 Oct 1814 - From what ship: HMT Salvador del Mundo, Halifax - Born: Baltimore - Age: 26 - Race: Negro - Released on 14 Jun 1815.
Williams, John - Seaman - Nbr: 413 - Prize: Viper - Ship Type: MV - How taken: HMS Superb - When taken: 15 Apr 1813 - Where taken: Bay of Biscay - Date Received: 1 Jul 1813 - From what ship: Plymouth - Born: New Orleans - Age: 22 - Race: Black - Released on 20 Apr 1815.
Williams, John - Seaman - Nbr: 626 - Prize: US Brig Argus - Ship Type: MW - How taken: HMS Pelican - When taken: 14 Aug 1813 - Where taken: Irish Channel - Date Received: 8 Sep 1813 - From what ship: Plymouth - Born: New York - Age: 22 - Race: Black - Sent to Dartmouth on 2 Nov 1814.
Williams, John - Seaman - Nbr: 2519 - Prize: Snap Dragon - Ship Type: P - How taken: HMS Martin - When taken: 10 Jun 1814 - Where taken: off Halifax - Date Received: 16 Aug 1814 - From what ship: HMT Queen, Halifax - Born: Philadelphia - Age: 28 - Race: Negro - Released on 3 May 1815.
Williams, John - Seaman - Nbr: 854 - Prize: Amiable - Ship Type: LM - How taken: HMS Magnificent - When taken: 30 Oct 1813 - Where taken: off Bordeaux - Date Received: 4 Dec 1813 - From what ship: Plymouth - Born: New Jersey - Age: 45 - Released on 26 Apr 1815.
Williams, John - Seaman - Nbr: 256 - Prize: William Bayard - Ship Type: MV - How taken: HMS Warspite - When taken: 3 Mar 1813 - Where taken: Bay of Biscay - Date Received: 28 Jun 1813 - From what ship: Plymouth - Born: Baltimore - Age: 25 - Entered British service on 13 Sep 1813.
Williams, John - Seaman - Nbr: 611 - Prize: US Brig Argus - Ship Type: MW - How taken: HMS Pelican - When taken: 14 Aug 1813 - Where taken: Irish Channel - Date Received: 8 Sep 1813 - From what ship: Plymouth - Born: Menzel, Prussia - Age: 20 - Sent to Mill Prison (Plymouth, England) on 28 Feb 1814.
Williams, John - Seaman - Nbr: 1306 - How taken: Sent into custody from HMS Minden - When taken: 17 May 1813 - Date Received: 14 Jun 1814 - From what ship: Mill Prison (Plymouth, England) - Born: Virginia - Age: 33 - Race: Black - Released on 28 Apr 1815.
Williams, John - Seaman - Nbr: 411 - Prize: Viper - Ship Type: MV - How taken: HMS Superb - When taken: 15 Apr 1813 - Where taken: Bay of Biscay - Date Received: 1 Jul 1813 - From what ship: Plymouth - Born: New Orleans - Age: 21 - Race: Black - Released on 20 Apr 1815.
Williams, John - Seaman - Nbr: 1590 - Prize: Eliza - Ship Type: MV - How taken: HMS Surveillante - When taken: 27 Mar 1813 - Where taken: Bay of Biscay - Date Received: 23 Jun 1814 - From what ship: Stapleton - Born: Philadelphia - Age: 30 - Released on 1 May 1815.
Williams, John - Seaman - Nbr: 471 - Prize: Hebe - Ship Type: MV - How taken: HMS Stag - When taken: 13 Apr 1813 - Where taken: Bay of Biscay - Date Received: 8 Sep 1813 - From what ship: Plymouth - Born: Amsterdam - Age: 31 - Sent to Mill Prison (Plymouth, England) on 21 Jun 1814.
Williams, John - Seaman - Nbr: 2221 - Prize: Hussar - Ship Type: P - How taken: HMS Saturn - When taken: 25 May 1814 - Where taken: off Sandy Hook - Date Received: 16 Aug 1814 - From what ship: HMS Dublin, Halifax - Born: Massachusetts - Age: 23 - Released on 2 May 1815.

American Prisoners of War Held at Dartmoor during the War of 1812

## Alphabetical listing of names

Williams, John - Seaman - Nbr: 717 - Prize: Ned - Ship Type: LM - How taken: HMS Royalist - When taken: 6 Sep 1813 - Where taken: Bay of Biscay - Date Received: 27 Sep 1813 - From what ship: Plymouth - Born: Baltimore - Age: 26 - Released on 7 May 1814.

Williams, John - Seaman - Nbr: 2181 - Prize: Rattlesnake - Ship Type: P - How taken: HMS Hyperion - When taken: 26 Jun 1814 - Where taken: off Cape Finisterre - Date Received: 16 Aug 1814 - From what ship: HMS Dublin, Halifax - Born: Lancaster County - Age: 39 - Released on 2 May 1815.

Williams, John - Seaman - Nbr: 1559 - Prize: Caroline - Ship Type: MV - How taken: HMS Medusa - When taken: 12 Apr 1813 - Where taken: Bay of Biscay - Date Received: 23 Jun 1814 - From what ship: Stapleton - Born: Connecticut - Age: 25 - Died on 14 Jan 1814 from variola.

Williams, John - Seaman - Nbr: 1085 - How taken: Impressed at Falmouth - When taken: 20 Mar 1814 - Date Received: 10 May 1814 - From what ship: Plymouth - Born: Philadelphia - Age: 30 - Race: Mulatto - Released on 27 Apr 1815.

Williams, John - Seaman - Nbr: 2559 - How taken: Gave himself up from HMS Bittern - When taken: 6 Aug 1814 - Date Received: 16 Aug 1814 - From what ship: HMT Salvador del Mundo - Born: Philadelphia - Age: 27 - Released on 3 May 1815.

Williams, John - Seaman - Nbr: 3778 - Prize: Fame - Ship Type: P - How taken: HMS Thistle - When taken: 10 Apr 1814 - Where taken: after being cast ashore on a seal island - Date Received: 30 Sep 1814 - From what ship: HMT President, Halifax - Born: Nantucket - Age: 28 - Race: Mulatto - Released on 9 Jun 1815.

Williams, John - Seaman - Nbr: 2831 - How taken: Gave himself up from HMS Leloir - When taken: 25 May 1813 - Date Received: 24 Aug 1814 - From what ship: HMT Liverpool, Chatham - Born: Worcester - Age: 32 - Released on 2 Nov 1814.

Williams, John - Seaman - Nbr: 3096 - How taken: Gave himself up from HMS Norge - Date Received: 3 Sep 1814 - From what ship: HMT Bristol - Born: Boston - Age: 40 - Race: Black - Released on 28 May 1815.

Williams, John - Seaman - Nbr: 2828 - How taken: Gave himself up from HMS Monmouth - When taken: 15 Jul 1813 - Date Received: 24 Aug 1814 - From what ship: HMT Liverpool, Chatham - Born: Kennebec - Age: 33 - Released on 19 May 1815.

Williams, John - Seaman - Nbr: 2478 - Prize: US Sloop Frolic - Ship Type: MW - How taken: HMS Orpheus - When taken: 20 Apr 1814 - Where taken: off Cuba - Date Received: 16 Aug 1814 - From what ship: HMT Queen, Halifax - Born: Pennsylvania - Age: 20 - Released on 3 May 1815.

Williams, John - Seaman - Nbr: 2691 - How taken: Gave himself up from HMS Prince of Wales - Date Received: 21 Aug 1814 - From what ship: HMT Freya, Chatham - Born: New Jersey - Age: 26 - Released on 19 May 1815.

Williams, John - Seaman - Nbr: 4910 - Prize: Hussar - Ship Type: P - How taken: HMS Saturn - When taken: 24 May 1814 - Where taken: off Sandy Hook - Date Received: 28 Oct 1814 - From what ship: HMT Alkbar, Halifax - Born: Baltimore - Age: 19 - Released on 21 Jun 1815.

Williams, John - Seaman - Nbr: 4323 - Prize: Fiere Facia - Ship Type: P - How taken: HMS Ramillies - When taken: 26 Feb 1814 - Where taken: off NY - Date Received: 7 Oct 1814 - From what ship: HMT Salvador del Mundo, Halifax - Born: New Orleans - Age: 40 - Race: Negro - Released on 14 Jun 1815.

Williams, John - Seaman - Nbr: 4324 - Prize: Fiere Facia - Ship Type: P - How taken: HMS Ramillies - When taken: 26 Feb 1814 - Where taken: off NY - Date Received: 7 Oct 1814 - From what ship: HMT Salvador del Mundo, Halifax - Born: New York - Age: 24 - Race: Negro - Released on 14 Jun 1815.

Williams, Jonathan - Seaman - Nbr: 2139 - How taken: Gave himself up from HMS Raleigh - When taken: 7 Jan 1813 - Date Received: 8 Aug 1814 - From what ship: HMT Raven, Chatham - Born: Boston - Age: 33 - Released on 10 Apr 1815.

Williams, Joseph - Seaman - Nbr: 2736 - How taken: Gave himself up from HMS Comet - When taken: 25 Nov 1812 - Date Received: 24 Aug 1814 - From what ship: HMT Liverpool, Chatham - Born: Boston - Age: 36 - Race: Black - Released on 26 Apr 1815.

Williams, Joseph - Seaman - Nbr: 3409 - Prize: Thomas - Ship Type: P - How taken: HMS Nymphe - When taken: 24 Jun 1813 - Where taken: off Halifax - Date Received: 13 Sep 1814 - From what ship: HMT Niobe, Chatham - Born: Massachusetts - Age: 23 - Released on 28 May 1815.

Williams, Joseph - Seaman - Nbr: 3442 - How taken: Gave himself up from HMS Clorinda - When taken: 18 Dec 1813 - Date Received: 13 Sep 1814 - From what ship: HMT Niobe, Chatham - Born: Martha's Vineyard - Age: 26 - Race: Black - Died on 1 Feb 1815 from phthisis pulmonalis.

Williams, Joseph - Seaman - Nbr: 4883 - Prize: Saratoga - Ship Type: P - How taken: HMS Barracouta - When taken: 9 Oct 1814 - Where taken: off Western Islands (England) - Date Received: 24 Oct 1814 - From what

American Prisoners of War Held at Dartmoor during the War of 1812

## Alphabetical listing of names

ship: HMT Salvador del Mundo - Born: New Orleans - Age: 34 - Race: Mulatto - Released on 21 Jun 1815.
Williams, Joseph - Seaman - Nbr: 5611 - How taken: Gave himself up from HMS Astora - When taken: 13 May 1814 - Date Received: 24 Dec 1814 - From what ship: HMT Tay - Born: Boston - Age: 34 - Escaped on 1 Jun 1815.
Williams, Moses - Seaman - Nbr: 4238 - Prize: Prize to the Diomede - Ship Type: Prize - How taken: HMS Sapphire - When taken: 27 Feb 1814 - Where taken: at sea - Date Received: 7 Oct 1814 - From what ship: HMT Niobe, Chatham - Born: Middleton - Age: 25 - Race: Mulatto - Released on 13 Jun 1815.
Williams, Noble - Seaman - Nbr: 5431 - Prize: York Town - Ship Type: P - How taken: HMS Maidstone - When taken: 18 Jul 1813 - Where taken: Grand Banks - Date Received: 31 Oct 1814 - From what ship: HMT Leyden, Chatham - Born: Jersey - Age: 47 - Race: Negro - Released on 1 Jul 1815.
Williams, Peter - Seaman - Nbr: 4357 - Prize: Fiere Facia - Ship Type: P - How taken: HMS Ramillies - When taken: 26 Feb 1814 - Where taken: off NY - Date Received: 7 Oct 1814 - From what ship: HMT Salvador del Mundo, Halifax - Born: Trinidad - Age: 25 - Race: Negro - Released on 14 Jun 1815.
Williams, Peter - Seaman - Nbr: 450 - Prize: Magdalen - Ship Type: MV - How taken: HMS Superb - When taken: 15 Apr 1813 - Where taken: off Belle Isle, France - Date Received: 1 Jul 1813 - From what ship: Plymouth - Born: New Orleans - Age: 33 - Sent to Plymouth on 8 Jul 1814.
Williams, Peter - Seaman - Nbr: 1211 - Prize: Imperatrice Reine - Ship Type: P - How taken: HMS Hotspur - When taken: 15 Feb 1813 - Where taken: off St. Antonio - Date Received: 4 Jun 1814 - From what ship: Dartmouth - Born: New York - Age: 25 - Released on 28 Apr 1815.
Williams, Peter - Seaman - Nbr: 5747 - Prize: US Schooner Ohio - Ship Type: MW - How taken: British gunboats - When taken: 12 Aug 1814 - Where taken: Fort Erie - Date Received: 26 Dec 1814 - From what ship: HMT Argo - Born: New York - Age: 20 - Released on 3 Jul 1815.
Williams, Peter - Seaman - Nbr: 3496 - How taken: Impressed at Liverpool - When taken: Jul 1814 - Date Received: 24 Sep 1814 - From what ship: HMT Salvador del Mundo - Born: New York - Age: 30 - Released on 4 Jun 1815.
Williams, Richard - Seaman - Nbr: 2073 - How taken: Impressed at London - When taken: 24 Jun 1813 - Date Received: 3 Aug 1814 - From what ship: HMS Bittern, Chatham Depot - Born: New Jersey - Age: 21 - Race: Mulatto - Released on 2 May 1815.
Williams, Robert - Seaman - Nbr: 2236 - Prize: General Hart - Ship Type: P - How taken: HMS Sophie - When taken: 24 Apr 1814 - Where taken: off Bermuda - Date Received: 16 Aug 1814 - From what ship: HMS Dublin, Halifax - Born: New York - Age: 28 - Released on 2 May 1815.
Williams, Robert - Seaman - Nbr: 3354 - How taken: Gave himself up from HMS Scorpion - When taken: 2 Dec 1812 - Date Received: 13 Sep 1814 - From what ship: HMT Niobe, Chatham - Born: New York - Age: 24 - Race: Black - Released on 27 Apr 1815.
Williams, Robert - Seaman - Nbr: 3359 - How taken: Gave himself up from HMS Union - When taken: 9 Dec 1812 - Date Received: 13 Sep 1814 - From what ship: HMT Niobe, Chatham - Born: New York - Age: 23 - Released on 27 Apr 1815.
Williams, Samuel - Seaman - Nbr: 5811 - Prize: US Schooner Scorpion - Ship Type: MW - How taken: British gunboats - When taken: 6 Sep 1814 - Where taken: Lake Erie - Date Received: 26 Dec 1814 - From what ship: HMT Argo - Born: Massachusetts - Age: 31 - Died on 15 Mar 1815 from variola.
Williams, Stephen - Seaman - Nbr: 3160 - Prize: Fame (Whaler) - Ship Type: MV - How taken: HMS Cressy - When taken: 20 Jul 1813 - Where taken: at sea - Date Received: 11 Sep 1814 - From what ship: HMT Freya, Chatham - Born: Albany - Age: 27 - Race: Black - Released on 28 May 1815.
Williams, Thomas - Seaman - Nbr: 5862 - Prize: Lion - Ship Type: P - How taken: HMS Granicus - When taken: 2 Dec 1814 - Where taken: off Lisbon - Date Received: 26 Dec 1814 - From what ship: HMT Impregnable - Born: Haiti - Age: 24 - Race: Black - Released on 3 Jul 1815.
Williams, Thomas - Soldier - Nbr: 5268 - Prize: 6th US Infantry - Ship Type: Troops - How taken: British Army - When taken: 13 Oct 1812 - Where taken: Canada - Date Received: 31 Oct 1814 - From what ship: HMT Leyden, Chatham - Born: Monmouth - Age: 40 - Released on 29 Jun 1815.
Williams, Thomas - Seaman - Nbr: 1816 - How taken: Gave himself up from HMS Unicorn - When taken: 17 Jun 1812 - Date Received: 20 Jul 1814 - From what ship: HMS Milford, Plymouth - Born: Connecticut - Age: 23 - Released on 26 Apr 1815.
Williams, Thomas - Seaman - Nbr: 1702 - Prize: Fanny - Ship Type: Prize - How taken: HMS Sceptre - When taken: 12 May 1814 - Date Received: 2 Jul 1814 - From what ship: Plymouth - Born: New York - Age: 37 - Released on 1 May 1815.

American Prisoners of War Held at Dartmoor during the War of 1812

## Alphabetical listing of names

Williams, Thomas - Seaman - Nbr: 414 - Prize: Viper - Ship Type: MV - How taken: HMS Superb - When taken: 15 Apr 1813 - Where taken: Bay of Biscay - Date Received: 1 Jul 1813 - From what ship: Plymouth - Born: Connecticut - Age: 23 - Died on 20 Mar 1814 from diarrhea.

Williams, Thomas - Seaman - Nbr: 675 - Prize: Henry Clements - Ship Type: MV - How taken: HMS Orestes - When taken: 13 Apr 1813 - Where taken: Bay of Biscay - Date Received: 8 Sep 1813 - From what ship: Plymouth - Born: Charlestown - Age: 24 - Released on 26 Apr 1815.

Williams, Thomas - Seaman - Nbr: 1526 - Prize: Essex - Ship Type: MV - How taken: HMS Pyramus - When taken: 20 Apr 1813 - Where taken: Bay of Biscay - Date Received: 23 Jun 1814 - From what ship: Stapleton - Born: New York - Age: 22 - Race: Colored - Sent to Mill Prison (Plymouth, England) on 30 Jun 1814.

Williams, Thomas - Seaman - Nbr: 2486 - Prize: US Sloop Frolic - Ship Type: MW - How taken: HMS Orpheus - When taken: 20 Apr 1814 - Where taken: off Cuba - Date Received: 16 Aug 1814 - From what ship: HMT Queen, Halifax - Born: Philadelphia - Age: 19 - Race: Mulatto - Released on 3 May 1815.

Williams, Thomas - Seaman - Nbr: 3240 - Prize: Hepsey - Ship Type: MV - How taken: HMS Tenedos - When taken: 22 Jun 1813 - Where taken: off Lisbon - Date Received: 11 Sep 1814 - From what ship: HMT Freya, Chatham - Born: Baltimore - Age: 40 - Released on 28 May 1815.

Williams, Thomas - Seaman - Nbr: 2636 - How taken: Gave himself up from HMS Leviathan - When taken: 22 Oct 1813 - Date Received: 21 Aug 1814 - From what ship: HMT Freya, Chatham - Born: Maryland - Age: 27 - Race: Mulatto - Released on 3 May 1815.

Williams, William - Boy - Nbr: 4417 - Prize: Fire Fly - Ship Type: LM - How taken: HMS Revolutionnaire - When taken: 19 Oct 1813 - Where taken: at sea - Date Received: 8 Oct 1814 - From what ship: HMT Leyden, Chatham - Born: Massachusetts - Age: 14 - Released on 14 Jun 1815.

Williams, William - Seaman - Nbr: 3910 - Prize: Thorn - Ship Type: MV - How taken: HMS Bulwark - When taken: 9 Jul 1814 - Where taken: off Nantucket - Date Received: 5 Oct 1814 - From what ship: HMT President, Halifax - Born: Connecticut - Age: 22 - Race: Negro - Released on 9 Jun 1815.

Williams, William - Seaman - Nbr: 4939 - Prize: Herald - Ship Type: P - How taken: HMS Endymion - When taken: 15 Aug 1814 - Where taken: off Nantucket - Date Received: 28 Oct 1814 - From what ship: HMT Alkbar, Halifax - Born: Charleston - Age: 38 - Released on 21 Jun 1815.

Williams, William - Seaman - Nbr: 402 - Prize: Young Holkar - Ship Type: MV - How taken: HMS Superb - When taken: 10 Apr 1813 - Where taken: off Belle Isle, France - Date Received: 1 Jul 1813 - From what ship: Plymouth - Born: Virginia - Age: 25 - Race: Black - Released on 20 Apr 1815.

Williams, William - Seaman - Nbr: 1945 - How taken: Gave himself up from HMS Sultan - When taken: 3 Feb 1813 - Date Received: 3 Aug 1814 - From what ship: HMS Alceste, Chatham Depot - Born: Wilmington - Age: 31 - Released on 2 May 1815.

Williams, William - Seaman - Nbr: 689 - How taken: Impressed at Liverpool - When taken: 10 Sep 1813 - Date Received: 27 Sep 1813 - From what ship: Plymouth - Born: New York - Age: 29 - Released on 26 Apr 1815.

Williams, William - Seaman - Nbr: 1747 - How taken: Gave himself up from HMS Malta - When taken: 28 Dec 1812 - Date Received: 20 Jul 1814 - From what ship: HMS Milford, Plymouth - Born: Delaware - Age: 24 - Released on 26 Apr 1815.

Williams, William - Seaman - Nbr: 3141 - How taken: Impressed at London - When taken: 22 Sep 1813 - Date Received: 11 Sep 1814 - From what ship: HMT Freya, Chatham - Born: Yorktown - Age: 22 - Died on 27 Oct 1814 from pneumonia.

Williams, William - Seaman - Nbr: 4047 - Prize: US Brig Rattlesnake - Ship Type: MW - How taken: HMS Leander - When taken: 13 Jul 1814 - Where taken: off Shelburne - Date Received: 6 Oct 1814 - From what ship: HMT Chesapeake, Halifax - Born: Marblehead - Age: 20 - Released on 13 Jun 1815.

Williamson, William - Seaman - Nbr: 5459 - How taken: Impressed at Dublin - When taken: Oct 1814 - Date Received: 10 Dec 1814 - From what ship: HMT Impregnable - Born: Portsmouth - Age: 30 - Released on 1 Jul 1815.

Williams, William - Seaman - Nbr: 5372 - Prize: James - Ship Type: MV - How taken: HMS Harpy - When taken: 16 Dec 1812 - Where taken: off Isle of France - Date Received: 31 Oct 1814 - From what ship: HMT Leyden, Chatham - Born: Eastport - Age: 28 - Released on 27 Apr 1815.

Williamson, Charles - Seaman - Nbr: 2377 - Prize: US Sloop Frolic - Ship Type: MW - How taken: HMS Orpheus - When taken: 20 Apr 1814 - Where taken: off Cuba - Date Received: 16 Aug 1814 - From what ship: HMT Queen, Halifax - Born: Maryland - Age: 29 - Sent to Dartmouth on 19 Oct 1814.

Williamson, David - Seaman - Nbr: 4197 - Prize: Zephyr - Ship Type: MV - How taken: HMS Surveillante - When taken: 4 Jan 1814 - Where taken: Bay of Biscay - Date Received: 7 Oct 1814 - From what ship: HMT Niobe,

American Prisoners of War Held at Dartmoor during the War of 1812

## Alphabetical listing of names

Chatham - Born: Philadelphia - Age: 25 - Released on 13 Jun 1815.

Williamson, John - Seaman - Nbr: 4447 - Prize: Dart - Ship Type: LM - How taken: HMS Niger - When taken: 13 Nov 1815 - Where taken: at sea - Date Received: 8 Oct 1814 - From what ship: HMT Leyden, Chatham - Born: Boston - Age: 37 - Released on 14 Jun 1815.

Williamson, John - Seaman - Nbr: 2591 - How taken: Gave himself up from HMS Electra - When taken: 20 Sep 1812 - Date Received: 21 Aug 1814 - From what ship: HMT Freya, Chatham - Born: New Haven - Age: 28 - Released on 26 Apr 1815.

Williamson, John - Seaman - Nbr: 3048 - How taken: Gave himself up from HMS Pembroke - When taken: 16 Aug 1814 - Date Received: 2 Sep 1814 - From what ship: HMT Sultan - Born: Brooklyn - Age: 27 - Race: Mulatto - Released on 28 Apr 1815.

Williamson, Joseph - Seaman - Nbr: 1933 - How taken: Gave himself up from HMS Desiree - When taken: 8 Jan 1814 - Date Received: 3 Aug 1814 - From what ship: HMS Alceste, Chatham Depot - Born: Philadelphia - Age: 30 - Released on 26 Apr 1815.

Williamson, Richard - Seaman - Nbr: 3877 - Prize: Fame - Ship Type: MV - How taken: HMS Niemen - When taken: 21 May 1814 - Where taken: off Cape Hatteras- Date Received: 5 Oct 1814 - From what ship: HMT Orpheus, Halifax - Born: North Carolina - Age: 27 - Released on 9 Jun 1815.

Williamson, William - Boy - Nbr: 51 - Prize: Terrible - Ship Type: MV - How taken: HMS Foxhound - When taken: 8 Feb 1813 - Where taken: Channel - Date Received: 2 Apr 1813 - From what ship: Plymouth - Born: Bridgeport - Age: 15 - Sent to Dartmouth on 30 Jul 1813.

Willis, John - Seaman - Nbr: 5756 - Prize: Hope - Ship Type: MV - How taken: HMS Nereus - When taken: 14 May 1814 - Where taken: Rio de la Plata - Date Received: 26 Dec 1814 - From what ship: HMT Argo - Born: Salem - Age: 24 - Released on 11 Jul 1815.

Willison, John - 2nd Lieutenant - Nbr: 5462 - Prize: General Putnam - Ship Type: P - How taken: HMS Leander - When taken: 8 Nov 1814 - Where taken: Long 65 Lat 42 - Date Received: 17 Dec 1814 - From what ship: HMT Loire, Halifax - Born: Marblehead - Age: 34 - Released on 10 Apr 1815.

Wills, John - Seaman - Nbr: 1567 - Prize: Messenger - Ship Type: MV - How taken: HMS Iris - When taken: 10 Mar 1813 - Where taken: Bay of Biscay - Date Received: 23 Jun 1814 - From what ship: Stapleton - Born: Philadelphia - Age: 48 - Released on 1 May 1815.

Wills, Nathaniel - Boy - Nbr: 2977 - Prize: Frolic - Ship Type: P - How taken: HMS Heron - When taken: 25 Jan 1814 - Where taken: off St. Thomas - Date Received: 29 Aug 1814 - From what ship: HMT Bittern - Born: Salem - Age: 15 - Released on 19 May 1815.

Wills, Peter - Landsman - Nbr: 107 - Prize: St. Martin's Planter - Ship Type: P - How taken: HMS Dublin - When taken: 9 Feb 1813 - Where taken: Lat 43 N, Long 33 50 W - Date Received: 2 Apr 1813 - From what ship: Plymouth - Born: Norwich - Age: 29 - Race: Black - Released on 20 Apr 1815.

Willson, Charles - Seaman - Nbr: 1167 - How taken: Gave himself up from HMS Hebrius - Date Received: 10 May 1814 - From what ship: Plymouth - Born: Rhode Island - Age: 41 - Race: Black - Released on 28 Apr 1815.

Willy, John - Seaman - Nbr: 1860 - Prize: Volante - Ship Type: LM - How taken: HMS Valiant - When taken: 25 Mar 1813 - Where taken: Georges Bank - Date Received: 29 Jul 1814 - From what ship: HMS Ville de Paris, Chatham Depot - Born: Baltimore - Age: 24 - Released on 2 May 1815.

Wilson, Alexander - Cook - Nbr: 6178 - Prize: Lion - Ship Type: P - How taken: HMS Granicus - When taken: 2 Dec 1814 - Where taken: off Lisbon - Date Received: 21 Jan 1815 - From what ship: HMT Impregnable - Born: Providence - Age: 43 - Race: Colored - Released on 5 Jul 1815.

Wilson, Andrew - Seaman - Nbr: 6046 - Prize: Levi - Ship Type: MV - How taken: HMS Albion - When taken: 30 Jan 1814 - Where taken: off NY - Date Received: 28 Dec 1814 - From what ship: HMT Penelope - Born: New Jersey - Age: 43 - Released on 11 Jul 1815.

Wilson, Charles - Seaman - Nbr: 1249 - How taken: Sent into custody from HMS Edinburgh - When taken: 28 Oct 1812 - Date Received: 14 Jun 1814 - From what ship: Mill Prison (Plymouth, England) - Born: Lewistown - Age: 35 - Race: Black - Released on 26 Apr 1815.

Wilson, Charles - Seaman - Nbr: 974 - Prize: Siro - Ship Type: LM - How taken: HMS Pelican - When taken: 13 Jan 1814 - Where taken: at sea - Date Received: 31 Jan 1814 - From what ship: Plymouth - Born: Baltimore - Age: 19 - Race: Negro - Released on 27 Apr 1815.

Wilson, Daniel - Seaman - Nbr: 4173 - How taken: Gave himself up from HMS Royal William - When taken: 4 Mar 1813 - Date Received: 7 Oct 1814 - From what ship: HMT Niobe, Chatham - Born: Boston - Age: 39 - Released on 13 Jun 1815.

Wilson, Francis - Seaman - Nbr: 352 - Prize: Amphitrite - Ship Type: MV - How taken: HMS Gleaner - When

## Alphabetical listing of names

 taken: 27 Feb 1813 - Where taken: Bay of Biscay - Date Received: 1 Jul 1813 - From what ship: Plymouth - Born: New York - Age: 24 - Released on 20 Apr 1815.

Wilson, Francis - Seaman - Nbr: 5408 - How taken: Gave himself up from HMS Dover - Date Received: 31 Oct 1814 - What ship: HMT Leyden, Chatham - Born: New York - Age: 31 - Race: Black - Released on 13 Jun 1815.

Wilson, George - Seaman - Nbr: 5712 - Prize: Regent - Ship Type: LM - How taken: HMS Forth - When taken: 19 Sep 1814 - Where taken: off Egg Harbor (New Jersey) - Date Received: 24 Dec 1814 - From what ship: HMT Penelope - Born: Not listed - Released on 3 Jul 1815.

Wilson, George - Seaman - Nbr: 580 - Prize: US Brig Argus - Ship Type: MW - How taken: HMS Pelican - When taken: 14 Apr 1813 - Where taken: Irish Channel - Date Received: 8 Sep 1813 - From what ship: Plymouth - Born: New York - Age: 14 - Sent to Dartmouth on 2 Nov 1814.

Wilson, George - Seaman - Nbr: 4560 - How taken: Gave himself up from HMS Blenheim - When taken: 28 Aug 1814 - Date Received: 8 Oct 1814 - From what ship: HMT Leyden, Chatham - Born: Hollis - Age: 25 - Released on 15 Jun 1815.

Wilson, Henry - Seaman - Nbr: 2662 - How taken: Gave himself up from HMS Fortune - When taken: 16 Jan 1813 - Date Received: 21 Aug 1814 - From what ship: HMT Freya, Chatham - Born: Beverly - Age: 43 - Race: Black - Released on 19 May 1815.

Wilson, James - Seaman - Nbr: 4655 - How taken: Gave himself up from HMS Colossus - When taken: 17 Oct 1812 - Date Received: 9 Oct 1814 - From what ship: HMT Leyden, Chatham - Born: Hampshire - Age: 37 - Released on 27 Apr 1815.

Wilson, James - Seaman - Nbr: 1327 - Prize: Paul Jones - Ship Type: P - How taken: HMS Leonidas - When taken: 23 May 1813 - Where taken: Channel - Date Received: 19 Jun 1814 - From what ship: Stapleton - Born: Philadelphia - Age: 30 - Released on 28 Apr 1815.

Wilson, James - Seaman - Nbr: 3391 - Prize: Wasp - Ship Type: P - How taken: HMS Bream - When taken: 10 Jun 1813 - Where taken: off Halifax - Date Received: 13 Sep 1814 - From what ship: HMT Niobe, Chatham - Born: Massachusetts - Age: 50 - Released on 28 May 1815.

Wilson, James - Seaman - Nbr: 3326 - Prize: York Town - Ship Type: P - How taken: HMS Nimrod - When taken: 17 Jul 1813 - Where taken: off St. Johns - Date Received: 13 Sep 1814 - From what ship: HMT Niobe, Chatham - Born: Portsmouth - Age: 31 - Released on 28 May 1815.

Wilson, James - Seaman - Nbr: 2494 - Prize: US Sloop Frolic - Ship Type: MW - How taken: HMS Orpheus - When taken: 20 Apr 1814 - Where taken: off Cuba - Date Received: 16 Aug 1814 - From what ship: HMT Queen, Halifax - Born: New Jersey - Age: 23 - Released on 3 May 1815.

Wilson, James - Seaman - Nbr: 2934 - How taken: Gave himself up from HMS Berwick - Date Received: 24 Aug 1814 - What ship: HMT Alpheus, Chatham - Born: Maryland - Age: 38 - Race: Black - Released on 19 May 1815.

Wilson, John - Seaman - Nbr: 3726 - Prize: Alfred - Ship Type: P - How taken: HMS Epervier - When taken: 23 Feb 1814 - Where taken: off Newfoundland - Date Received: 30 Sep 1814 - From what ship: HMT President, Halifax - Born: Salem - Age: 31 - Race: Negro - Released on 4 Jun 1815.

Wilson, John - Seaman - Nbr: 5649 - Prize: Lion - Ship Type: P - How taken: HMS Granicus - When taken: 2 Dec 1814 - Where taken: off Lisbon - Date Received: 24 Dec 1814 - From what ship: HMT Tay - Born: Charleston - Age: 21 - Released on 3 Jul 1815.

Wilson, John - Seaman - Nbr: 1495 - Prize: Tom - Ship Type: MV - How taken: HMS Surveillante - When taken: 27 Apr 1813 - Where taken: Bay of Biscay - Date Received: 19 Jun 1814 - From what ship: Mill Prison (Plymouth, England) - Born: New York - Age: 19 - Released on 1 May 1815.

Wilson, John - Seaman - Nbr: 2025 - Prize: Governor Middleton - Ship Type: MV - How taken: Thetis (Privateer) - When taken: 2 May 1813 - Where taken: Bay of Biscay - Date Received: 3 Aug 1814 - From what ship: HMS Lyffey, Chatham Depot - Born: New York - Age: 36 - Released on 2 May 1815.

Wilson, John - Seaman - Nbr: 1290 - Prize: Traveler - Ship Type: Prize - How taken: HM Schooner Canso - When taken: 11 May 1814 - Where taken: off Cape Clear - Date Received: 14 Jun 1814 - From what ship: Mill Prison (Plymouth, England) - Born: Baltimore - Age: 23 - Released on 28 Apr 1815.

Wilson, John - Seaman - Nbr: 814 - How taken: Impressed at Cork - When taken: 25 Oct 1813 - Date Received: 3 Nov 1813 - From what ship: Plymouth - Born: Bath - Age: 22 - Released on 26 Apr 1815.

Wilson, John - Seaman - Nbr: 370 - How taken: Impressed at sea - When taken: 29 Mar 1813 - Date Received: 1 Jul 1813 - From what ship: Plymouth - Born: New York - Age: 23 - Released on 20 Apr 1815.

Wilson, John - Seaman - Nbr: 912 - Prize: Zephyr - Ship Type: MV - How taken: HMS Pyramus - When taken: 30

American Prisoners of War Held at Dartmoor during the War of 1812

## Alphabetical listing of names

Nov 1813 - Where taken: off L'Orient (France) - Date Received: 31 Jan 1814 - From what ship: Plymouth - Born: New York - Age: 22 - Released on 27 Apr 1815.

Wilson, John - Seaman - Nbr: 1994 - How taken: Impressed at Yarmouth - When taken: 6 Mar 1813 - Date Received: 3 Aug 1814 - From what ship: HMS Lyffey, Chatham Depot - Born: Pennsylvania - Age: 35 - Released on 2 May 1815.

Wilson, John - Seaman - Nbr: 2253 - Prize: Otario - Ship Type: Prize - How taken: HMS Curlew - When taken: 2 May 1814 - Where taken: off Halifax - Date Received: 16 Aug 1814 - From what ship: HMS Dublin, Halifax - Born: Pennsylvania - Age: 20 - Released on 3 May 1815.

Wilson, John - Seaman - Nbr: 5236 - Prize: Teazer - Ship Type: P - How taken: HMS Maidstone - When taken: 18 Jul 1813 - Where taken: at sea - Date Received: 31 Oct 1814 - From what ship: HMT Mermaid, Chatham - Born: Newport - Age: 25 - Released on 29 Jun 1815.

Wilson, John - Seaman - Nbr: 2671 - Prize: Orders in Council - Ship Type: MV - How taken: HMS Surveillante - When taken: 1 Jun 1813 - Where taken: off Cape Ortegal (Spain) - Date Received: 21 Aug 1814 - From what ship: HMT Freya, Chatham - Born: Charlestown - Age: 23 - Released on 19 May 1815.

Wilson, John - Seaman - Nbr: 5120 - Prize: William - Ship Type: MV - How taken: HMS Ringdove - When taken: 12 Aug 1814 - Where taken: off Portland - Date Received: 28 Oct 1814 - From what ship: HMT Alkbar, Halifax - Born: Philadelphia - Age: 23 - Released on 29 Jun 1815.

Wilson, Nicholas - 1st Mate - Nbr: 5527 - Prize: Sparks - Ship Type: LM - How taken: HMS Maidstone - When taken: 28 Sep 1814 - Where taken: off Nantucket - Date Received: 17 Dec 1814 - From what ship: HMT Loire, Halifax - Born: Newport - Age: 24 - Sent to Ashburton (England) on 4 Jan 1815.

Wilson, Peter - Boy - Nbr: 380 - Prize: VA Planter - Ship Type: MV - How taken: HMS Pyramus - When taken: 18 Mar 1813 - Where taken: off Nantes - Date Received: 1 Jul 1813 - From what ship: Plymouth - Born: New York - Age: 25 - Race: Black - Released on 14 Jun 1815.

Wilson, Robert - Passenger - Nbr: 5350 - Prize: Atlantic - Ship Type: MV - How taken: HMS Swiftsure - Where taken: off Corsica - Date Received: 31 Oct 1814 - From what ship: HMT Leyden, Chatham - Born: New York - Age: 25 - Released on 1 Jul 1815.

Wilson, Robert - Seaman - Nbr: 5514 - Prize: Mary - Ship Type: Prize - How taken: HMS Wasp - When taken: 6 Oct 1814 - Date Received: 17 Dec 1814 - From what ship: HMT Loire, Halifax - Born: New Hampshire - Age: 28.

Wilson, Robert - Seaman - Nbr: 3201 - How taken: Gave himself up from HMS America - When taken: 16 Jul 1813 - Date Received: 11 Sep 1814 - From what ship: HMT Freya, Chatham - Born: Connecticut - Age: 45 - Released on 28 May 1815.

Wilson, Samuel - Seaman - Nbr: 3465 - How taken: Gave himself up from HMS Minorca - When taken: 14 Sep 1814 - Date Received: 19 Sep 1814 - From what ship: HMT Salvador del Mundo - Born: Philadelphia - Age: 34 - Race: Black - Released on 5 Jul 1815.

Wilson, Thomas - Seaman - Nbr: 2091 - How taken: Gave himself up from HMS Hyperion - When taken: 28 Mar 1813 - Date Received: 3 Aug 1814 - From what ship: HMS Bittern, Chatham Depot - Born: Alexandria - Age: 34 - Released on 2 May 1815.

Wilson, Thomas - Seaman - Nbr: 3533 - Prize: Rover - Ship Type: Prize - How taken: HMS Conquistador - When taken: 22 Aug 1814 - Where taken: Long 19 Lat 107 - Date Received: 28 Sep 1814 - From what ship: HMT Salvador del Mundo - Born: Baltimore - Age: 26 - Released on 4 Jun 1815.

Wilson, William - Seaman - Nbr: 4177 - How taken: Gave himself up from HMS Blake - When taken: 28 Dec 1812 - Date Received: 7 Oct 1814 - From what ship: HMT Niobe, Chatham - Born: Boston - Age: 28 - Released on 27 Apr 1815.

Wilson, William - Seaman - Nbr: 711 - Prize: Ned - Ship Type: LM - How taken: HMS Royalist - When taken: 6 Sep 1813 - Where taken: Bay of Biscay - Date Received: 27 Sep 1813 - From what ship: Plymouth - Born: Newport - Age: 23 - Sent to Plymouth on 7 Dec 1813.

Wilson, William - Seaman - Nbr: 3107 - Prize: Gotland - Ship Type: MV - How taken: HMS Barbados - When taken: 22 Jun 1814 - Where taken: off St. Bartholomew - Date Received: 11 Sep 1814 - From what ship: HMT Salvador del Mundo - Born: Virginia - Age: 20 - Released on 28 May 1815.

Wilson, William - Seaman - Nbr: 4160 - How taken: Gave himself up from HMS Minos - When taken: 28 Dec 1813 - Date Received: 6 Oct 1814 - From what ship: HMT Niobe, Chatham - Born: Boston - Age: 32 - Released on 13 Jun 1815.

Wilson, William - Seaman - Nbr: 2653 - Prize: Jane - Ship Type: MV - How taken: HMS Crescent - When taken: 28 Jun 1813 - Where taken: off Newfoundland - Date Received: 21 Aug 1814 - From what ship: HMT Freya,

American Prisoners of War Held at Dartmoor during the War of 1812

## Alphabetical listing of names

Chatham - Born: Philadelphia - Age: 31 - Released on 19 May 1815.

Winand, John - Seaman - Nbr: 5717 - Prize: Nancy - Ship Type: MV - How taken: HMS Forth - When taken: 7 Sep 1814 - Where taken: off Sandy Hook - Date Received: 24 Dec 1814 - From what ship: HMT Penelope - Born: Staten Island - Age: 21 - Released on 3 Jul 1815.

Winchester, Ebenezer - Seaman - Nbr: 4155 - Prize: Argus - Ship Type: MV - How taken: HMS San Domingo - When taken: 1 Mar 1814 - Where taken: off Savannah - Date Received: 6 Oct 1814 - From what ship: HMT Niobe, Chatham - Born: Boston - Age: 21 - Released on 13 Jun 1815.

Windell, Isaac - Seaman - Nbr: 314 - Prize: Pallas - Ship Type: MV - How taken: Rebuff - When taken: 23 Dec 1812 - Where taken: off Cadiz - Date Received: 28 Jun 1813 - From what ship: Plymouth - Born: Boston - Age: 35 - Sent to Dartmouth on 30 Jul 1813.

Windham, John - Seaman - Nbr: 730 - Prize: Ned - Ship Type: LM - How taken: HMS Royalist - When taken: 6 Sep 1813 - Where taken: Bay of Biscay - Date Received: 27 Sep 1813 - From what ship: Plymouth - Born: New York - Age: 25 - Escaped on 30 Aug 1814.

Wing, Joshua - Seaman - Nbr: 356 - Prize: Two Brothers - Ship Type: MV - How taken: Beetle (LM) - When taken: 18 Mar 1813 - Where taken: off Western Islands (England) - Date Received: 1 Jul 1813 - From what ship: Plymouth - Born: Boston - Age: 22 - Released on 20 Apr 1815.

Wing, Judah - Seaman - Nbr: 2452 - Prize: US Sloop Frolic - Ship Type: MW - How taken: HMS Orpheus - When taken: 20 Apr 1814 - Where taken: off Cuba - Date Received: 16 Aug 1814 - From what ship: HMT Queen, Halifax - Born: Rochester - Age: 34 - Released on 3 May 1815.

Wing, Nathaniel - Seaman - Nbr: 2682 - Prize: Hannah & Eliza - Ship Type: MV - How taken: HMS Lyra - When taken: 29 May 1813 - Where taken: off Bayonne - Date Received: 21 Aug 1814 - From what ship: HMT Freya, Chatham - Born: Massachusetts - Age: 26 - Released on 19 May 1815.

Wingate, David - Seaman - Nbr: 4652 - How taken: Gave himself up from HMS Gorgon - When taken: 1 Nov 1812 - Date Received: 9 Oct 1814 - From what ship: HMT Leyden, Chatham - Born: Hampshire - Age: 23 - Released on 27 Apr 1815.

Wingate, Jacob - Seaman - Nbr: 5550 - Prize: Hornet - Ship Type: MV - How taken: HMS Surprize - When taken: 19 Aug 1814 - Where taken: Lat 35 Long 24 - Date Received: 17 Dec 1814 - From what ship: HMT Loire, Halifax - Born: Northampton - Age: 48 - Released on 1 Jul 1815.

Winkinpor, Andrew - Seaman - Nbr: 4107 - Prize: Resolution - Ship Type: MV - How taken: HMS Junon - When taken: 6 Jun 1814 - Where taken: off Cape Ann - Date Received: 6 Oct 1814 - From what ship: HMT Chesapeake, Halifax - Born: Massachusetts - Age: 27 - Released on 13 Jun 1815.

Winkley, W. B. - Prize master - Nbr: 5662 - Prize: Harlequin - Ship Type: P - How taken: HMS Bulwark - When taken: 23 Nov 1814 - Where taken: off Halifax - Date Received: 24 Dec 1814 - From what ship: HMT Penelope - Born: Massachusetts - Age: 19 - Released on 20 Mar 1815.

Winn, Joseph R. - Mater's mate - Nbr: 3731 - Prize: Alfred - Ship Type: P - How taken: HMS Epervier - When taken: 23 Feb 1814 - Where taken: off Newfoundland - Date Received: 30 Sep 1814 - From what ship: HMT President, Halifax - Born: Salem - Age: 33 - Released on 4 Jun 1815.

Winn, Theodore - Seaman - Nbr: 5580 - Prize: William - Ship Type: MV - How taken: HMS Barbados - When taken: 19 Jan 1814 - Where taken: off St. Bartholomew - Date Received: 17 Dec 1814 - From what ship: HMT Loire, Halifax - Born: Massachusetts - Age: 26 - Released on 1 Jul 1815.

Winslow, Elijah - Seaman - Nbr: 406 - Prize: Viper - Ship Type: MV - How taken: HMS Superb - When taken: 15 Apr 1813 - Where taken: Bay of Biscay - Date Received: 1 Jul 1813 - From what ship: Plymouth - Born: Salem - Age: 18 - Released on 20 Apr 1815.

Winter, Andrew - Seaman - Nbr: 1651 - How taken: Apprehended at Bristol - When taken: 7 Sep 1813 - Date Received: 23 Jun 1814 - From what ship: Stapleton - Born: New York - Age: 32 - Released on 1 May 1815.

Winter, Peter - Seaman - Nbr: 123 - Prize: Governor McKean - Ship Type: LM - How taken: HMS Rover - When taken: 26 Jan 1813 - Where taken: off Bordeaux - Date Received: 2 Apr 1813 - From what ship: Plymouth - Born: New Castle - Age: 25 - Race: Negro - Sent to Dartmouth on 30 Jul 1813.

Wintory, Antony - Seaman - Nbr: 417 - Prize: Viper - Ship Type: MV - How taken: HMS Superb - When taken: 15 Apr 1813 - Where taken: Bay of Biscay - Date Received: 1 Jul 1813 - From what ship: Plymouth - Born: New Orleans - Age: 28 - Released on 20 Apr 1815.

Wise, John - Seaman - Nbr: 3173 - How taken: Gave himself up from HMS Impeccable - When taken: 10 Sep 1813 - Date Received: 11 Sep 1814 - From what ship: HMT Freya, Chatham - Born: New York - Age: 26 - Race: Mulatto - Released on 28 May 1815.

Witham, Burrell - Boatswain - Nbr: 3255 - Prize: Growler - Ship Type: P - How taken: HMS Electra - When taken:

American Prisoners of War Held at Dartmoor during the War of 1812

## Alphabetical listing of names

7 Jul 1813 - Where taken: off St. Johns - Date Received: 11 Sep 1814 - From what ship: HMT Freya, Chatham - Born: Marblehead - Age: 23 - Released on 28 May 1815.
Witherett, Charles - Seaman - Nbr: 5357 - Prize: Amity - Ship Type: MV - How taken: HMS Achates - When taken: 22 Dec 1813 - Where taken: Bay of Biscay - Date Received: 31 Oct 1814 - From what ship: HMT Leyden, Chatham - Born: Portsmouth - Age: 35 - Released on 1 Jul 1815.
Witherham, Nicholas - Landsman - Nbr: 64 - Prize: Spitfire - Ship Type: MV - How taken: HMS Achates - When taken: 14 Feb 1813 - Where taken: off Ushant (France) - Date Received: 2 Apr 1813 - From what ship: Plymouth - Born: Marblehead - Age: 27 - Released on 20 Apr 1815.
Witherington, James - Seaman - Nbr: 6219 - Prize: Prince de Neufchatel - Ship Type: P - How taken: Leander (Newcastle Acasta) - When taken: 20 Dec 1814 - Where taken: Lat 38 Long 56 - Date Received: 30 Jan 1815 - From what ship: HMT Pheasant - Born: Boston - Age: 17 - Released on 5 Jul 1815.
Wittlebanks, Edward - Seaman - Nbr: 2148 - Prize: Nancy - Ship Type: MV - How taken: HMS Parthian - When taken: 1 Aug 1812 - Where taken: Needles - Date Received: 8 Aug 1814 - From what ship: HMT Raven, Chatham - Born: Portsmouth - Age: 21 - Released on 11 Jul 1815.
Woglon, Peter - Seaman - Nbr: 5716 - Prize: Nancy - Ship Type: MV - How taken: HMS Forth - When taken: 7 Sep 1814 - Where taken: off Sandy Hook - Date Received: 24 Dec 1814 - From what ship: HMT Penelope - Born: New York - Age: 20 - Released on 3 Jul 1815.
Wolfe, William - Seaman - Nbr: 718 - Prize: Ned - Ship Type: LM - How taken: HMS Royalist - When taken: 6 Sep 1813 - Where taken: Bay of Biscay - Date Received: 27 Sep 1813 - From what ship: Plymouth - Born: Baltimore - Age: 20 - Released on 26 Apr 1815.
Wolsey, M. - Seaman - Nbr: 5720 - How taken: Gave himself up from HMS Pelican - When taken: 13 Sep 1813 - Date Received: 26 Dec 1814 - From what ship: HMT Argo - Born: New York - Age: 25 - Released on 17 Jan 1815.
Wood, Benjamin - Boy - Nbr: 5303 - Prize: Yankee - Ship Type: P - How taken: HMS Shannon - When taken: 20 Aug 1813 - Where taken: at sea - Date Received: 31 Oct 1814 - From what ship: HMT Leyden, Chatham - Born: Warren - Age: 11 - Released on 29 Jun 1815.
Wood, George - Seaman - Nbr: 6028 - Prize: Regent - Ship Type: LM - How taken: HMS Forth - When taken: 19 Sep 1814 - Where taken: off Egg Harbor (New Jersey) - Date Received: 28 Dec 1814 - From what ship: HMT Penelope - Born: New York - Age: 28 - Released on 3 Jul 1815.
Wood, John - Seaman - Nbr: 1389 - Prize: Courier - Ship Type: P - How taken: HMS Rover - When taken: 14 May 1813 - Where taken: Bay of Biscay - Date Received: 19 Jun 1814 - From what ship: Stapleton - Born: Virginia - Age: 22 - Released on 28 Apr 1815.
Wood, John - Marine - Nbr: 4059 - Prize: US Brig Rattlesnake - Ship Type: MW - How taken: HMS Leander - When taken: 13 Jul 1814 - Where taken: off Shelburne - Date Received: 6 Oct 1814 - From what ship: HMT Chesapeake, Halifax - Born: Newburyport - Age: 23 - Released on 13 Jun 1815.
Wood, John - Seaman - Nbr: 5449 - Prize: Prize to the Lawrence - Ship Type: P - How taken: HMS Glasgow - When taken: 2 Nov 1814 - Where taken: Channel - Date Received: 10 Dec 1814 - From what ship: HMT Impregnable - Born: Baltimore - Age: 24 - Released on 11 Jul 1815.
Wood, Joseph - Seaman - Nbr: 3254 - Prize: Wiley Reynard - Ship Type: P - How taken: HMS Shannon - When taken: 15 Aug 1812 - Where taken: off Halifax - Date Received: 11 Sep 1814 - From what ship: HMT Freya, Chatham - Born: Baltimore - Age: 27 - Released on 27 Apr 1815.
Wood, Robert - Seaman - Nbr: 6280 - How taken: Gave himself up from HMS Iris - Date Received: 17 Feb 1815 - From what ship: HMT Ganges, Plymouth - Born: New York - Age: 28 - Race: Black - Released on 5 Jul 1815.
Wood, Samuel - Seaman - Nbr: 1736 - Prize: Hugh Jones - Ship Type: MV - How taken: HMS Bittern - When taken: 14 Jun 1814 - Where taken: at sea - Date Received: 18 Jul 1814 - From what ship: HMT Salvador del Mundo, Plymouth - Born: New London - Age: 22 - Released on 1 May 1815.
Wood, Samuel - Seaman - Nbr: 5451 - Prize: Prize to the Lawrence - Ship Type: P - How taken: HMS Glasgow - When taken: 2 Nov 1814 - Where taken: Channel - Date Received: 10 Dec 1814 - From what ship: HMT Impregnable - Born: Savannah - Age: 21 - Race: Black - Released on 1 Jul 1815.
Wood, Samuel - Prize master - Nbr: 3798 - Prize: Hannah - Ship Type: Prize - How taken: HMS Martin - When taken: 29 Apr 1814 - Where taken: off Cape Lopez - Date Received: 5 Oct 1814 - From what ship: HMT President, Halifax - Born: New Bedford - Age: 22 - Released on 9 Jun 1815.
Wood, Sylvester - Seaman - Nbr: 1437 - Prize: Leo - Ship Type: LM - How taken: HMS Magicienne - When taken: 4 Jun 1813 - Where taken: off France - Date Received: 19 Jun 1814 - From what ship: Stapleton - Born: New

American Prisoners of War Held at Dartmoor during the War of 1812

## Alphabetical listing of names

York - Age: 23 - Released on 28 Apr 1815.
Wood, Thomas - Seaman - Nbr: 5020 - Prize: Landrail - Ship Type: Prize - How taken: HMS Wasp - When taken: 27 Jul 1814 - Where taken: Georges Bank - Date Received: 28 Oct 1814 - From what ship: HMT Alkbar, Halifax - Born: Baltimore - Age: 22 - Released on 21 Jun 1815.
Wood, Thomas - Seaman - Nbr: 8 - Prize: Cashier - Ship Type: LM - How taken: HMS Reindeer - When taken: 3 Feb 1813 - Where taken: Bay of Biscay - Date Received: 2 Apr 1813 - From what ship: Plymouth - Born: Prince George's County - Age: 21 - Sent to Dartmouth on 30 Jul 1813.
Wood, Thomas - Seaman - Nbr: 647 - Prize: US Brig Argus - Ship Type: MW - How taken: HMS Pelican - When taken: 14 Aug 1813 - Where taken: Irish Channel - Date Received: 8 Sep 1813 - From what ship: Plymouth - Born: Prince George's County - Age: 21 - Sent to Dartmouth on 30 Jul 1813.
Wood, Thomas - Seaman - Nbr: 3477 - How taken: Gave himself up from HMS Prince - When taken: 12 Sep 1814 - Date Received: 19 Sep 1814 - From what ship: HMT Salvador del Mundo - Born: New York - Age: 27 - Race: Mulatto - Released on 4 Jun 1815.
Wood, William - Seaman - Nbr: 1272 - How taken: Sent into custody from HMS Seylla - Date Received: 14 Jun 1814 - From what ship: Mill Prison (Plymouth, England) - Born: Philadelphia - Age: 50 - Race: Black - Released on 28 Apr 1815.
Woodbury, Dixey - Seaman - Nbr: 4296 - Prize: Commodore Perry - Ship Type: Schooner - How taken: Sent into custody from a cutter - When taken: 25 Feb 1814 - Where taken: off Bordeaux - Date Received: 7 Oct 1814 - From what ship: HMT Niobe, Chatham - Born: Beverly - Age: 27 - Released on 14 Jun 1815.
Woodbury, John - Seaman - Nbr: 2676 - Prize: Joseph - Ship Type: MV - How taken: HMS Iris - When taken: 8 Jun 1813 - Where taken: off Spain - Date Received: 21 Aug 1814 - From what ship: HMT Freya, Chatham - Born: Gloucester - Age: 24 - Released on 10 Apr 1815.
Woodcraft, John - Seaman - Nbr: 518 - Prize: Fox - Ship Type: Packet prize to the Fox - How taken: Superior - When taken: 25 Jun 1813 - Where taken: Lat 50 N, Long 21 W - Date Received: 8 Sep 1813 - From what ship: Plymouth - Born: Great Island - Age: 18 - Released on 26 Apr 1815.
Woodford, James - Seaman - Nbr: 4678 - Prize: Blockade - Ship Type: P - How taken: HMS Recruit - When taken: 17 Aug 1813 - Where taken: off America - Date Received: 9 Oct 1814 - From what ship: HMT Leyden, Chatham - Born: Baltimore - Age: 19 - Released on 15 Jun 1815.
Woodford, William - Seaman - Nbr: 4456 - Prize: Juliana Smith - Ship Type: P - How taken: HMS Nymphe - When taken: 11 May 1813 - Where taken: off Cape Sable Island (Canada) - Date Received: 8 Oct 1814 - From what ship: HMT Leyden, Chatham - Born: New York - Age: 23 - Race: Black - Released on 15 Jun 1815.
Woodman, Thomas - Seaman - Nbr: 3536 - Prize: Hawk - Ship Type: P - How taken: HMS Pique - When taken: 26 Apr 1814 - Where taken: off Bermuda - Date Received: 30 Sep 1814 - From what ship: HMT Sybella - Born: Norfolk - Age: 19 - Released on 4 Jun 1815.
Woods, Charles - Seaman - Nbr: 6228 - Prize: Prince de Neufchatel - Ship Type: P - How taken: Leander (Newcastle Acasta) - When taken: 20 Dec 1814 - Where taken: Lat 38 Long 56 - Date Received: 30 Jan 1815 - From what ship: HMT Pheasant - Born: Philadelphia - Age: 23 - Released on 5 Jul 1815.
Woods, Edward - Seaman - Nbr: 996 - Prize: Apparencen - Ship Type: MV - How taken: HMS Castilian - When taken: 27 Jan 1814 - Where taken: off Ushant (France) - Date Received: 31 Jan 1814 - From what ship: Plymouth - Born: Gloucester - Age: 21 - Released on 27 Apr 1815.
Woods, Merrill - Seaman - Nbr: 2019 - How taken: Gave himself up from HMS Chatham - When taken: 1 May 1813 - Date Received: 3 Aug 1814 - From what ship: HMS Lyffey, Chatham Depot - Born: Heaton - Age: 30 - Released on 2 May 1815.
Woodson, Isaac - Seaman - Nbr: 5458 - How taken: Gave himself up from HMS Arackne - When taken: 30 Nov 1814 - Date Received: 10 Dec 1814 - From what ship: HMT Impregnable - Born: Cape Cod - Age: 25 - Released on 1 Jul 1815.
Woodward, Anthony - Seaman - Nbr: 95 - Prize: St. Martin's Planter - Ship Type: P - How taken: HMS Dublin - When taken: 9 Feb 1813 - Where taken: Lat 43 N, Long 33 50 W - Date Received: 2 Apr 1813 - From what ship: Plymouth - Born: Philadelphia - Age: 17 - Released on 20 Apr 1815.
Woodward, Francis - Seaman - Nbr: 5791 - Prize: Jane - Ship Type: Prize - How taken: HMS Rhin - When taken: Jun 1814 - Where taken: West Indies - Date Received: 26 Dec 1814 - From what ship: HMT Argo - Born: Hagerstown - Age: 35 - Released on 3 Jul 1815.
Woodward, Thomas - Seaman - Nbr: 5516 - Prize: Robert - When taken: 20 Feb 1814 - Date Received: 17 Dec 1814 - From what ship: HMT Loire, Halifax - Born: New Hampshire - Age: 27.
Wooldridge, Thomas - Seaman - Nbr: 61 - Prize: Spitfire - Ship Type: MV - How taken: HMS Achates - When

American Prisoners of War Held at Dartmoor during the War of 1812

## Alphabetical listing of names

taken: 14 Feb 1813 - Where taken: off Ushant (France) - Date Received: 2 Apr 1813 - From what ship: Plymouth - Born: Marblehead - Age: 19 - Released on 20 Apr 1815.

Woolf, Andrew - Seaman - Nbr: 1913 - How taken: Gave himself up from HMS Mars - When taken: 9 Dec 1812 - Date Received: 3 Aug 1814 - From what ship: HMS Alceste, Chatham Depot - Born: Baltimore - Age: 29 - Released on 26 Apr 1815.

Woolridge, Robert - Seaman - Nbr: 3696 - Prize: Alfred - Ship Type: P - How taken: HMS Epervier - When taken: 23 Feb 1814 - Where taken: off Newfoundland - Date Received: 30 Sep 1814 - From what ship: HMT President, Halifax - Born: Marblehead - Age: 19 - Released on 4 Jun 1815.

Woolridge, W. - Boy - Nbr: 2979 - Prize: Frolic - Ship Type: P - How taken: HMS Heron - When taken: 25 Jan 1814 - Where taken: off St. Thomas - Date Received: 29 Aug 1814 - From what ship: HMT Bittern - Born: Marblehead - Age: 16 - Released on 19 May 1815.

Woolridge, William - Seaman - Nbr: 2706 - How taken: Gave himself up from HMS Barfleur - When taken: 12 Dec 1813 - Date Received: 21 Aug 1814 - From what ship: HMT Freya, Chatham - Born: Boston - Age: 38 - Released on 19 May 1815.

Works, Alford - Seaman - Nbr: 1283 - Prize: Indian Lass - Ship Type: Prize - How taken: Not listed - When taken: 29 Apr 1814 - Date Received: 14 Jun 1814 - From what ship: Mill Prison (Plymouth, England) - Born: Stafford - Age: 21 - Released on 28 Apr 1815.

Worthey, Samuel - Seaman - Nbr: 1876 - Prize: Elbridge Gerry - Ship Type: P - How taken: HMS Crescent - When taken: 16 Sep 1813 - Where taken: off St. George's - Date Received: 29 Jul 1814 - From what ship: HMS Ville de Paris, Chatham Depot - Born: North Yarmouth - Age: 30 - Released on 2 May 1815.

Wright, Daniel - Seaman - Nbr: 2526 - Prize: Diomede - Ship Type: P - How taken: HMS Rifleman - When taken: 28 Jul 1814 - Where taken: off Halifax - Date Received: 16 Aug 1814 - From what ship: HMT Queen, Halifax - Born: New York - Age: 22 - Released on 3 May 1815.

Wright, Edward - Seaman - Nbr: 3207 - How taken: Gave himself up from HMS Achille - When taken: 8 Sep 1813 - Date Received: 11 Sep 1814 - From what ship: HMT Freya, Chatham - Born: New Jersey - Age: 25 - Race: Black - Released on 28 May 1815.

Wright, Amos - Seaman - Nbr: 1048 - How taken: Sent into custody from Mary (Transport) - Date Received: 10 May 1814 - From what ship: Plymouth - Born: Philadelphia - Age: 20 - Released on 27 Apr 1815.

Wright, George - Seaman - Nbr: 1554 - Prize: Caroline - Ship Type: MV - How taken: HMS Medusa - When taken: 12 Apr 1813 - Where taken: Bay of Biscay - Date Received: 23 Jun 1814 - From what ship: Stapleton - Born: Delaware - Age: 30 - Race: Negro - Released on 1 May 1815.

Wright, Israel - Seaman - Nbr: 4635 - How taken: Gave himself up from HMS Ajax - Date Received: 9 Oct 1814 - From what ship: HMT Leyden, Chatham - Born: Philadelphia - Age: 33 - Released on 15 Jun 1815.

Wright, James - Seaman - Nbr: 1732 - How taken: Gave himself up from HMS Union - When taken: 12 Jul 1814 - Date Received: 18 Jul 1814 - From what ship: HMT Salvador del Mundo, Plymouth - Born: Charleston - Age: 38 - Released on 1 May 1815.

Wright, John - Seaman - Nbr: 6334 - Prize: Prince de Neufchatel - Ship Type: P - How taken: Leander (Newcastle Acasta) - When taken: 20 Dec 1814 - Where taken: Lat 38 Long 56 - Date Received: 19 Feb 1815 - From what ship: HMT Ganges, Plymouth - Born: New York - Age: 19 - Released on 6 Jul 1815.

Wright, John - Seaman - Nbr: 4166 - How taken: Gave himself up from HMS Leda - When taken: 8 Jul 1814 - Date Received: 6 Oct 1814 - From what ship: HMT Niobe, Chatham - Born: New Jersey - Age: 23 - Released on 13 Jun 1815.

Wright, John - Seaman - Nbr: 1575 - Prize: Price - Ship Type: MV - How taken: HMS Pyramus - When taken: 6 Apr 1813 - Where taken: Bay of Biscay - Date Received: 23 Jun 1814 - From what ship: Stapleton - Born: New York - Age: 22 - Released on 1 May 1815.

Wright, Jonathan - Seaman - Nbr: 6135 - How taken: Gave himself up from HMS Woolark - When taken: 3 Jan 1814 - Date Received: 17 Jan 1815 - From what ship: HMT Impregnable - Born: Philadelphia - Age: 40 - Released on 1 May 1815.

Wright, Mathew - 2nd Master - Nbr: 3675 - Prize: Alfred - Ship Type: P - How taken: HMS Epervier - When taken: 23 Feb 1812 - Where taken: off Newfoundland - Date Received: 30 Sep 1814 - From what ship: HMT President - Born: Ipswich - Age: 26 - Released on 4 Jun 1815.

Wright, William - Seaman - Nbr: 947 - Prize: Siro - Ship Type: LM - How taken: HMS Pelican - When taken: 13 Jan 1814 - Where taken: at sea - Date Received: 31 Jan 1814 - From what ship: Plymouth - Born: Baltimore - Age: 37 - Released on 27 Apr 1815.

Wright, William - Seaman - Nbr: 1134 - Prize: Lyon - Ship Type: P - How taken: Brilliant (Privateer) - When taken:

American Prisoners of War Held at Dartmoor during the War of 1812

## Alphabetical listing of names

12 Aug 1814 - Where taken: off Charlestown - Date Received: 10 May 1814 - From what ship: Plymouth - Born: Portland - Age: 23 - Released on 28 Apr 1815.

Wright, Thomas - Seaman - Nbr: 2005 - Prize: Governor Middleton - Ship Type: MV - How taken: Thetis (Privateer) - When taken: 2 May 1813 - Where taken: Bay of Biscay - Date Received: 3 Aug 1814 - From what ship: HMS Lyffey, Chatham Depot - Born: Wilmington - Age: 29 - Released on 2 May 1815.

Wyatt, Frederick - Seaman - Nbr: 575 - Prize: US Brig Argus - Ship Type: MW - How taken: HMS Pelican - When taken: 14 Apr 1813 - Where taken: Irish Channel - Date Received: 8 Sep 1813 - From what ship: Plymouth - Born: Baltimore - Age: 25 - Sent to Dartmouth on 2 Nov 1814.

Wyatt, Jesse - Seaman - Nbr: 4676 - Prize: Portsmouth Packet - Ship Type: Prize - How taken: HMS Fantome - When taken: 5 Oct 1813 - Where taken: off Portland - Date Received: 9 Oct 1814 - From what ship: HMT Leyden, Chatham - Born: New York - Age: 20 - Released on 15 Jun 1815.

Wyatt, William - Seaman - Nbr: 4952 - Prize: Herald - Ship Type: P - How taken: HMS Endymion - When taken: 15 Aug 1814 - Where taken: off Nantucket - Date Received: 28 Oct 1814 - From what ship: HMT Alkbar, Halifax - Born: Newport - Age: 24 - Escaped on 1 Jun 1815.

Wyer, Joseph - Seaman - Nbr: 891 - Prize: General Kempt - Ship Type: P - How taken: HMS Foxhound - When taken: 18 Dec 1813 - Where taken: Lat 48'60" Long 5'7" - Date Received: 31 Jan 1814 - From what ship: Plymouth - Born: Beverly - Age: 17 - Released on 27 Apr 1815.

Wyman, William - Seaman - Nbr: 4305 - Prize: Argus - Ship Type: MV - How taken: HMS San Domingo - When taken: 1 Mar 1814 - Where taken: off Savannah - Date Received: 7 Oct 1814 - From what ship: HMT Niobe, Chatham - Born: Boston - Age: 16 - Released on 4 Jun 1815.

Wynn, John - Seaman - Nbr: 4774 - How taken: Gave himself up from HMS Amaranith - When taken: 17 Jan 1814 - Date Received: 9 Oct 1814 - From what ship: HMT Freya, Chatham - Born: Boston - Age: 28 - Released on 15 Jun 1815.

Yard, William - Seaman - Nbr: 1408 - Prize: Miranda - Ship Type: Prize - How taken: HMS Unicorn - When taken: 21 May 1813 - Where taken: off Ushant (France) - Date Received: 19 Jun 1814 - From what ship: Stapleton - Born: New Jersey - Age: 26 - Released on 28 Apr 1815.

Yarnell, William - Seaman - Nbr: 3622 - Prize: Monarch - Ship Type: MV - How taken: HMS Dotterel - When taken: 14 Dec 1813 - Where taken: off Charleston - Date Received: 30 Sep 1814 - From what ship: HMT Sybella - Born: Pennsylvania - Age: 55 - Released on 4 Jun 1815.

Yeaton, Charles - Seaman - Nbr: 3410 - Prize: Thomas - Ship Type: P - How taken: HMS Nymphe - When taken: 24 Jun 1813 - Where taken: off Halifax - Date Received: 13 Sep 1814 - From what ship: HMT Niobe, Chatham - Born: Massachusetts - Age: 22 - Released on 28 May 1815.

Yeaton, James - Seaman - Nbr: 3213 - How taken: Gave himself up from HMS Crebenis - When taken: 8 Sep 1813 - Date Received: 11 Sep 1814 - From what ship: HMT Freya, Chatham - Born: New Hampshire - Age: 30 - Released on 28 May 1815.

Yeoman, John - Seaman - Nbr: 4131 - Prize: Bordeaux Packet - Ship Type: LM - How taken: HMS Niemen - When taken: 28 Jun 1814 - Where taken: off Delaware - Date Received: 6 Oct 1814 - From what ship: HMT Chesapeake, Halifax - Born: New Jersey - Age: 22 - Released on 13 Jun 1815.

Yoing, William - Seaman - Nbr: 554 - How taken: Gave himself up from HMS Pompeii - When taken: 12 Jul 1813 - Date Received: 8 Sep 1813 - From what ship: Plymouth - Born: Portsmouth - Age: 32 - Race: Black - Released on 26 Apr 1815.

Yorkman, Archibald - Seaman - Nbr: 3590 - Prize: Fairy - Ship Type: MV - How taken: HMS Hardy - When taken: 26 Mar 1814 - Where taken: West Indies - Date Received: 30 Sep 1814 - From what ship: HMT Sybella - Born: Baltimore - Age: 21 - Race: Negro - Released on 4 Jun 1815.

Young, Cook - Seaman - Nbr: 3905 - Prize: Eliza - Ship Type: Packet - How taken: HMS Loire - When taken: 10 Jun 1814 - Where taken: Chesapeake Bay - Date Received: 5 Oct 1814 - From what ship: HMT Orpheus, Halifax - Born: Baltimore - Age: 21 - Race: Negro - Released on 9 Jun 1815.

Young, Ebenezer - 2nd Mate - Nbr: 2665 - Prize: Confidence - Ship Type: MV - How taken: HMS Erebus - When taken: 25 Jun 1813 - Where taken: off Gottenburgh - Date Received: 21 Aug 1814 - From what ship: HMT Freya, Chatham - Born: Chatham - Age: 21 - Released on 19 May 1815.

Young, James - Seaman - Nbr: 5596 - Prize: Albion - Ship Type: Prize - How taken: HMS Jaseur - When taken: 21 Sep 1814 - Where taken: off Halifax - Date Received: 17 Dec 1814 - From what ship: HMT Loire, Halifax - Born: New Providence - Age: 19 - Released on 1 Jul 1815.

Young, John - Seaman - Nbr: 6267 - Prize: Plutarch - Ship Type: MV - How taken: HMS Helicon - When taken: 5 Feb 1815 - Where taken: off Bordeaux - Date Received: 10 Feb 1815 - From what ship: HMT Ganges,

American Prisoners of War Held at Dartmoor during the War of 1812

## Alphabetical listing of names

Plymouth - Born: Wiscasset - Age: 22 - Released on 5 Jul 1815.
Young, John - Captain's mate - Nbr: 1073 - How taken: Sent from Mill Prison (Plymouth, England) - Date Received: 10 May 1814 - From what ship: Plymouth - Born: Virginia - Age: 28 - Released on 27 Apr 1815.
Young, John - Quartermaster - Nbr: 1316 - Prize: US Brig Argus - Ship Type: MW - How taken: HMS Pelican - When taken: 14 Aug 1813 - Where taken: Irish Channel - Date Received: 14 Jun 1814 - From what ship: Mill Prison (Plymouth, England) - Born: Philadelphia - Age: 25 - Sent to Dartmouth on 19 Oct 1814.
Young, John - Seaman - Nbr: 408 - Prize: Viper - Ship Type: MV - How taken: HMS Superb - When taken: 15 Apr 1813 - Where taken: Bay of Biscay - Date Received: 1 Jul 1813 - From what ship: Plymouth - Born: Kent County - Age: 36 - Released on 20 Apr 1815.
Young, John - Seaman - Nbr: 552 - How taken: Gave himself up from HMS Tremendous - When taken: 1 Jul 1813 - Date Received: 8 Sep 1813 - What ship: Plymouth - Born: Albany - Age: 34 - Released on 26 Apr 1815.
Young, John - Seaman - Nbr: 5409 - How taken: HMS Wasp - When taken: 17 Jun 1813 - Where taken: off Cape Sable Island (Canada) - Date Received: 31 Oct 1814 - From what ship: HMT Leyden, Chatham - Born: Milford - Age: 21 - Released on 1 Jul 1815.
Young, John - Seaman - Nbr: 5216 - Prize: Thomas - Ship Type: P - How taken: HMS Nymphe - When taken: 24 Jun 1813 - Where taken: off Halifax - Date Received: 31 Oct 1814 - From what ship: HMT Mermaid, Chatham - Born: Barnstable - Age: 44 - Released on 29 Jun 1815.
Young, John - Seaman - Nbr: 4341 - Prize: Lizard - Ship Type: P - How taken: HMS Barbados - When taken: 1 Mar 1814 - Where taken: off Halifax - Date Received: 7 Oct 1814 - From what ship: HMT Salvador del Mundo, Halifax - Born: New Hampshire - Age: 35 - Released on 14 Jun 1815.
Young, Joseph - Seaman - Nbr: 522 - Prize: Friends - Ship Type: Bey of Pool - How taken: HMS Whiting - When taken: 15 Jul 1813 - Where taken: Lat 67 N, Long 8 W - Date Received: 8 Sep 1813 - From what ship: Plymouth - Born: Plymouth - Age: 28 - Released on 26 Apr 1815.
Young, Moses - Seaman - Nbr: 2141 - How taken: Gave himself up from HMS Ruby - When taken: 1 Aug 1812 - Date Received: 8 Aug 1814 - From what ship: HMT Raven, Chatham - Born: Chatham - Age: 35 - Released on 26 Apr 1815.
Young, Nathaniel - Seaman - Nbr: 3128 - Prize: Garter Wester - Ship Type: MV - How taken: HMS Elizabeth - When taken: 15 Aug 1813 - Where taken: off St. Bartholomew - Date Received: 11 Sep 1814 - From what ship: HMT Freya, Chatham - Born: Baltimore - Age: 30 - Race: Black - Released on 28 May 1815.
Young, Philip - Seaman - Nbr: 3791 - Prize: James - Ship Type: Prize - How taken: Rebecca (LM) - When taken: 13 Apr 1814 - Where taken: off Delaware - Date Received: 5 Oct 1814 - From what ship: HMT President, Halifax - Born: Philadelphia - Age: 20 - Released on 9 Jun 1815.
Young, Richard - Prize master - Nbr: 3763 - Prize: Yankee Lass - Ship Type: P - How taken: HMS Surprize - When taken: 1 May 1814 - Where taken: off Western Islands (England) - Date Received: 30 Sep 1814 - From what ship: HMT President, Halifax - Born: Connecticut - Age: 29 - Released on 9 Jun 1815.
Young, Rufus - Seaman - Nbr: 701 - Prize: Ned - Ship Type: LM - How taken: HMS Royalist - When taken: 6 Sep 1813 - Where taken: Bay of Biscay - Date Received: 27 Sep 1813 - From what ship: Plymouth - Born: Providence - Age: 18 - Released on 29 Jun 1815.
Young, Thomas - Seaman - Nbr: 618 - Prize: US Brig Argus - Ship Type: MW - How taken: HMS Pelican - When taken: 14 Aug 1813 - Where taken: Irish Channel - Date Received: 8 Sep 1813 - From what ship: Plymouth - Born: Providence - Age: 35 - Sent to Dartmouth on 2 Nov 1814.
Young, Thomas - Seaman - Nbr: 881 - Prize: Charlotte - Ship Type: MV - How taken: HMS Dwarf - When taken: 4 Nov 1813 - Where taken: off Bordeaux - Date Received: 31 Jan 1814 - From what ship: Plymouth - Born: Wilmington - Age: 24 - Released on 27 Apr 1815.
Young, Thomas - Seaman - Nbr: 2877 - How taken: Gave himself up from HMS Barfleur - When taken: 27 May 1813 - Date Received: 24 Aug 1814 - From what ship: HMT Alpheus, Chatham - Born: Salem - Age: 45 - Released on 19 May 1815.
Young, William - Gunner - Nbr: 3882 - Prize: US Brig Rattlesnake - Ship Type: MW - How taken: HMS Leander - When taken: 11 Jul 1814 - Where taken: off Shelburne - Date Received: 5 Oct 1814 - From what ship: HMT Orpheus, Halifax - Born: Norfolk - Age: 50 - Released on 9 Jun 1815.
Young, William - Seaman - Nbr: 5559 - Prize: Levant - Ship Type: MV - How taken: HMS Forester - When taken: 4 Jan 1814 - Where taken: Bahamas Banks - Date Received: 17 Dec 1814 - From what ship: HMT Loire, Halifax - Born: Beverly - Age: 29 - Died on 21 Jan 1815 from perissneuoniria.
Young, William - Seaman - Nbr: 684 - Prize: US Brig Argus - Ship Type: MW - How taken: HMS Pelican - When taken: 14 Aug 1813 - Where taken: Irish Channel - Date Received: 27 Sep 1813 - From what ship: Plymouth

American Prisoners of War Held at Dartmoor during the War of 1812

## Alphabetical listing of names

- Born: Portsmouth - Age: 38 - Sent to Dartmouth on 2 Nov 1814.
Young, William - Seaman - Nbr: 1580 - Prize: Price - Ship Type: MV - How taken: HMS Pyramus - When taken: 6 Apr 1813 - Where taken: Bay of Biscay - Date Received: 23 Jun 1814 - From what ship: Stapleton - Born: New York - Age: 22 - Released on 1 May 1815.
Young, William - Seaman - Nbr: 357 - Prize: Two Brothers - Ship Type: MV - How taken: Beetle (LM) - When taken: 18 Mar 1813 - Where taken: off Western Islands (England) - Date Received: 1 Jul 1813 - From what ship: Plymouth - Born: New York - Age: 22 - Released on 20 Apr 1815.
Younger, Louis - Seaman - Nbr: 3177 - How taken: Gave himself up from HMS Pompeii - When taken: 27 May 1813 - Date Received: 11 Sep 1814 - From what ship: HMT Freya, Chatham - Born: Massachusetts - Age: 27 - Released on 3 May 1815.
Younger, William - Seaman - Nbr: 1273 - How taken: Sent into custody from HMS Seylla - Date Received: 14 Jun 1814 - From what ship: Mill Prison (Plymouth, England) - Born: New York - Age: 36 - Released on 28 Apr 1815.
Zellie, Thomas - Seaman - Nbr: 1396 - Prize: Zebra - Ship Type: LM - How taken: HMS Pyramus - When taken: 20 Apr 1813 - Where taken: Bay of Biscay - Date Received: 19 Jun 1814 - From what ship: Stapleton - Born: Newport - Age: 33 - Race: Negro - Released on 28 Apr 1815.

American Prisoners of War Held at Dartmoor during the War of 1812

## Numeric listing by prison number

| | | | | | |
|---|---|---|---|---|---|
| 1 | Tower, Michael | 47 | Tufts, Zarara | 93 | Jones, William P. |
| 2 | Smith, Richard | 48 | Leach, William | 94 | Gray, John |
| 3 | Spears, Samuel | 49 | Smith, William | 95 | Woodward, Anthony |
| 4 | Mullan, John | 50 | Cook, James | 96 | Ellis, William |
| 5 | Deal, William | 51 | Williamson, William | 97 | Covell, Isaac |
| 6 | Halfpenny, Robert | 52 | Freeman, Prince | 98 | Allen, John |
| 7 | Schaeman, Frederick | 53 | Lamson, Charles | 99 | Mains, Daius |
| 8 | Wood, Thomas | 54 | Carton, Thomas | 100 | Russell, Isaac |
| 9 | Gallaway, Joseph | 55 | Bishop, William | 101 | Hulet, Michael |
| 10 | Birch, Andrew | 56 | Wheeler, Richard | 102 | Thompson, Whitney |
| 11 | Jentile, Joseph | 57 | Bartlett, Thomas | 103 | Merritt, Almon |
| 12 | Riceo, Joachim | 58 | Jones, Francis | 104 | Taylor, George |
| 13 | Blackston, Edward | 59 | Jones, F. V. | 105 | Gilbert, John |
| 14 | Gilbert, Thomas | 60 | Jervis, John | 106 | Morrison, Davis |
| 15 | Drake, Henry | 61 | Wooldridge, Thomas | 107 | Wills, Peter |
| 16 | Hodge, Rufus | 62 | Dodd, Samuel | 108 | Athroun, Samuel |
| 17 | Evans, Henry | 63 | Bridge, Francis | 109 | Osten, John |
| 18 | Morel, Thomas | 64 | Witherham, Nicholas | 110 | Anderson, Robert |
| 19 | Kimmins, John | 65 | Jiett, Benjamin | 111 | Evans, James |
| 20 | Madden, Frederick | 66 | Dolliver, Francis | 112 | Parsons, Ignatius |
| 21 | West, Reubin | 67 | Lovett, William | 113 | Grey, Thomas |
| 22 | Newell, John | 68 | Wadden, Jacob | 114 | Boggs, James |
| 23 | Atkins, Uriah | 69 | Fletcher, William B. | 115 | Baptist, John |
| 24 | Hubbard, John | 70 | Miller, Henry | 116 | Lehens, Andrew |
| 25 | Leeds, Leon | 71 | Williams, Ambrose | 117 | Capewill, Bartholomew |
| 26 | Blanchard, Carvan | 72 | Johnson, John | 118 | Regens, Jonathan |
| 27 | Hough, Ebenezer | 73 | Morris, John | 119 | Hook, John |
| 28 | Odeen, John | 74 | Ryan, William | 120 | Farrell, Francis |
| 29 | Warner, Henry | 75 | Ireland, James | 121 | Hayes, Simon |
| 30 | Berry, Brook | 76 | Armstrong William | 122 | White, Philip |
| 31 | Conway, William | 77 | Bright, Samuel | 123 | Winter, Peter |
| 32 | Higgins, George | 78 | Rotner, James W. | 124 | Humphries, Jacob |
| 33 | Mires, John | 79 | Reeves, James | 125 | Augustus, Amos |
| 34 | Glover, John | 80 | Augustin, Anthony | 126 | Davis, Benjamin S. |
| 35 | Hovey, Joseph | 81 | Johnston, Edward | 127 | Sargant, Samuel H. |
| 36 | Foster, Joseph | 82 | Lessall, Francis | 128 | Mellett, James |
| 37 | Vankirk, Joseph | 83 | Colcocha, Anthony | 129 | Andrews, Thomas |
| 38 | Paterson, Hanse | 84 | McCauley, George | 130 | Dobins, John |
| 39 | Barnes, William | 85 | Bloom, Joseph | 131 | Brewster, Jacob |
| 40 | Burnes, Robert | 86 | Longworthy, John | 132 | Johnson, John |
| 41 | Lovel, John | 87 | Harris, John | 133 | Thomas, Archibald |
| 42 | Smith, John | 88 | Inbritson, Nicholas | 134 | Miller, John |
| 43 | Martin, Peter | 89 | Bannister, George | 135 | Thompson, Martin |
| 44 | Kromkout, Baney | 90 | Cooper, Daniel | 136 | Royal, John |
| 45 | Root, Eleazer F. | 91 | Pack, Abraham | 137 | Marble, Jabez |
| 46 | Smith, Eldridge | 92 | Shamtan, Samuel | 138 | Griffiths, Thomas |

American Prisoners of War Held at Dartmoor during the War of 1812

## Numeric listing by prison number

| | | | | | |
|---|---|---|---|---|---|
| 139 | Dickson, Charles | 185 | Romain, Samuel | 231 | Forbes, William |
| 140 | Chambers, Henry | 186 | Johnstone, Gersham | 232 | Christy, Alexander |
| 141 | Neel, David | 187 | Thompson, William | 233 | Davis, Charles |
| 142 | Schultz, John | 188 | Gordon, Thomas | 234 | Goff, James |
| 143 | Mannett, Richard | 189 | Richman, Joshua | 235 | Pike, John |
| 144 | Baron, William | 190 | Isaac, Moses | 236 | Taylor, Jacob |
| 145 | Bidson, Thomas | 191 | Clark, William | 237 | Mitchell, Ezekiel |
| 146 | Sage, Moses | 192 | Borgin, Gabriel | 238 | Akens, William |
| 147 | Dewett, John | 193 | Vaughan, Thomas | 239 | Williams, George |
| 148 | Reeves, William | 194 | Sullivan, Hampton | 240 | Cox, Miles |
| 149 | Brown, Henry | 195 | Trash, James | 241 | Blanchard, George |
| 150 | Ludlow, Charles | 196 | Clements, John C. | 242 | Roe, John |
| 151 | Harry, John | 197 | Joseph, Fois | 243 | Averell, Loring |
| 152 | Fawcett, William | 198 | Washburn, Edward | 244 | Mather, John |
| 153 | Fossendor, William | 199 | Louring, Henry | 245 | Barlow, Robert |
| 154 | McLelland, William | 200 | Valentine, John | 246 | Lewes, John |
| 155 | Staitmand, James | 201 | Mix, William A. | 247 | Linnott, John |
| 156 | Mathews, Joseph | 202 | Bisley, Horace | 248 | Jones, Benjamin |
| 157 | Foster, Cato | 203 | Cox, John | 249 | Finch, William |
| 158 | Pierce, Amos | 204 | Williams, Henry | 250 | Gorling, George C. |
| 159 | Johnson, William | 205 | Lewes, John | 251 | Groward, Peter |
| 160 | Starbuck, Thaddeus B. | 206 | Wheeler, Charles | 252 | Foster, Joseph |
| 161 | Whitehouse, Asay | 207 | Monks, John | 253 | Ryley, William |
| 162 | Brown, George | 208 | Porter, William | 254 | Jocelyn, Robert |
| 163 | Fox, Washington | 209 | White, Henry | 255 | Bellinger, William |
| 164 | Rock, Oliver | 210 | Lawson, James | 256 | Williams, John |
| 165 | Allen, John | 211 | Jones, Benjamin | 257 | Bunker, James |
| 166 | Check, Stephen | 212 | Parker, George | 258 | Jackson, John |
| 167 | Lingall, George | 213 | Barrett, Anthony | 259 | Sherwood, John |
| 168 | Eddey, John | 214 | Baker, John | 260 | Kellem, John C. |
| 169 | Gilbert, Thomas | 215 | Caldwell, James | 261 | Hoskins, James |
| 170 | Flood, David | 216 | Whitewood, Charles | 262 | Francis, Benjamin |
| 171 | Allen, William | 217 | Allen, Henry | 263 | Steel, John |
| 172 | Rosett, Samuel | 218 | Saunders, William | 264 | Leach, William |
| 173 | Peterson, Alexander | 219 | Smith, William | 265 | Reley, Michael |
| 174 | Rhodrick, Joseph | 220 | Ware, William | 266 | Davis, James |
| 175 | Thomas, William | 221 | Freddle, John | 267 | Dorrell, John |
| 176 | Crete, John Isaac | 222 | Lawson, Thomas | 268 | Rollands, Joseph |
| 177 | Smith, Charles | 223 | Gellens, William | 269 | Jones, John |
| 178 | Holts, Daniel | 224 | Court, Robert | 270 | Richardson, Francis |
| 179 | Mallery, William | 225 | Hull, William | 271 | Elm, John |
| 180 | Hughes, John | 226 | Stoddard, Reuben | 272 | Jennings, Nathaniel |
| 181 | Right, Joshua | 227 | Norman, Peter | 273 | Coffin, Samuel |
| 182 | Erwin, William | 228 | Brown, William | 274 | Landerman, Joseph |
| 183 | Dennison, Judah | 229 | Holmes, Andrew | 275 | Blackler, Bernard |
| 184 | Duston, Peter | 230 | Sammers, Henry | 276 | Grant, Peter |

American Prisoners of War Held at Dartmoor during the War of 1812

## Numeric listing by prison number

| | | | | | |
|---|---|---|---|---|---|
| 277 | Greenwood, Thales | 323 | Smith, George | 369 | Bagley, William |
| 278 | Eaton, George | 324 | Hockman, William | 370 | Wilson, John |
| 279 | Hock, J. N. | 325 | Edwards, Price | 371 | Erwin, Elijah |
| 280 | Jones, William P. | 326 | Moore, Laurence | 372 | Boivie, James |
| 281 | Turner, William | 327 | Lewis, Edward | 373 | Griffin, William |
| 282 | Holmes, Caleb R. | 328 | Meigs, John | 374 | Howell, Thomas |
| 283 | Erlstroom, John | 329 | Lilliford, Jacob | 375 | Robinson, Josiah |
| 284 | Palmer, John | 330 | Everley, John | 376 | Hodgkins, Daniel |
| 285 | Colvelli, John | 331 | Watts, James | 377 | Harvey, Joseph |
| 286 | Stoddart, Robert | 332 | Blake, Philip | 378 | Marsh, Jesse |
| 287 | Divers, Charles | 333 | Talgart, Shephred | 379 | Rice, Francis |
| 288 | Watts, Hiram | 334 | Sherman, Reuben | 380 | Wilson, Peter |
| 289 | Heath, Henry | 335 | Brown, Jesse | 381 | Andrews, Charles |
| 290 | Leroy, Alexander | 336 | Conklin, William | 382 | Coombes, William |
| 291 | Richardson, William | 337 | Sproson, James | 383 | Harding, John |
| 292 | Wellander, Adam | 338 | Hogabets, John | 384 | Hartar, Henry |
| 293 | Philen, Richard | 339 | Swain, Darius | 385 | Carnes, Joseph |
| 294 | Weaphor, Andrew | 340 | Wessel, Samuel | 386 | Joseph, Thomas |
| 295 | Bennett, James | 341 | Morris, John | 387 | Miller, Stephen |
| 296 | Norton, George | 342 | Moss, John | 388 | Griffin, John |
| 297 | Lemercier, Peter | 343 | Johnson, Samuel | 389 | Rogers, James |
| 298 | Robinson, Samuel | 344 | Harris, George | 390 | Mumford, Thomas |
| 299 | Brownell, Richard | 345 | Morse, Henry | 391 | Cammon, Robert |
| 300 | Hurd, William | 346 | Dowell, Isaac | 392 | Higgins, James |
| 301 | West, John | 347 | Hall, David | 393 | Howard, Samuel |
| 302 | Heydon, Elie | 348 | Hall, Richard | 394 | Monturn, Francis |
| 303 | Brown, Benjamin | 349 | Powers, James | 395 | Sturges, Bradley |
| 304 | Garret, James | 350 | Towson, William | 396 | Lebaith, John P. |
| 305 | Brown, Ludwig | 351 | Smith, John | 397 | Burr, Isaac |
| 306 | Vaughan, Nicholas | 352 | Wilson, Francis | 398 | James, Thomas |
| 307 | Brown, Samuel | 353 | Davidson, Henry | 399 | Blake, Alexander |
| 308 | Evans, Jacob | 354 | Somerville, Charles | 400 | West, Benjamin |
| 309 | Cooper, James | 355 | Stanton, Henry | 401 | Toby, Peter |
| 310 | Farrett, George | 356 | Wing, Joshua | 402 | Williams, William |
| 311 | Burdge, Samuel | 357 | Young, William | 403 | Thomas, John |
| 312 | McCormick, Simon | 358 | Leas, Anthony | 404 | Cowen, Robert |
| 313 | Ceaser, Joseph | 359 | Sheppard, William | 405 | Morgan, Joseph |
| 314 | Windell, Isaac | 360 | Paulm, Nas. | 406 | Winslow, Elijah |
| 315 | Ratoon, Thomas | 361 | Atkinson, William | 407 | Thomson, William |
| 316 | Calfax, William | 362 | Thomas, John | 408 | Young, John |
| 317 | Edgerley, George | 363 | Bridges, John | 409 | Allen, Thomas |
| 318 | Robinson, William | 364 | Dear, Andrew | 410 | Foster, David |
| 319 | Burton, John | 365 | Picking, George | 411 | Williams, John |
| 320 | Reaudon, Andrew | 366 | Jackson, Daniel | 412 | Tyre, William |
| 321 | Brown, James | 367 | Vinson, Lewis | 413 | Williams, John |
| 322 | Douarte, Angelo | 368 | Gifford, Robert | 414 | Williams, Thomas |

American Prisoners of War Held at Dartmoor during the War of 1812

## Numeric listing by prison number

| | | | | | |
|---|---|---|---|---|---|
| 415 | Smalley, Thomas | 461 | Forester, Francis | 508 | Taylor, Thomas |
| 416 | Hinkley, Aaron | 462 | Perney, William | 509 | Carpenter, Henry |
| 417 | Wintory, Antony | 463 | Schole, John | 510 | Fry, John |
| 418 | Farrell, John | 464 | Kennedy, John | 511 | Roman, John |
| 419 | Webster, James | 465 | Hill, Leonard | 512 | Stewart, William |
| 420 | McNelly, Thomas | 466 | Snate, George | 513 | Page, John |
| 421 | Baker, John | 467 | Burn, Reuben | 514 | Weeks, Benjamin |
| 422 | Thomas, William | 468 | Perne, George | 515 | Jenkins, Joseph |
| 423 | Brady, Hugh | 469 | Eastland, James | 516 | Jackson, William |
| 424 | Hunter, George | 470 | Bancroft, Samuel | 517 | Dymoss, Peter |
| 425 | Arnold, Alfred | 471 | Williams, John | 518 | Woodcroft, John |
| 426 | Robinson, Charles | 472 | Conway, Charles | 519 | Underwood, John |
| 427 | Small, Richard | 473 | Denham, John | 520 | Green, Moses |
| 428 | Pain, Joseph | 474 | Chiney, Amos | 521 | Currien, Stephen |
| 429 | Hoye, Cornelius | 475 | Garcia, Francis | 522 | Young, Joseph |
| 430 | Jones, George | 476 | Jones, Charles | 523 | Curtis, Francis |
| 431 | Crouch, Richard | 477 | Mercer, Benjamin | 524 | Curtis, Henry |
| 432 | Campbell, John | 478 | Wallace, William | 525 | Giles, Edward |
| 433 | Belfast, Richard | 479 | Dillin, Pierce | 526 | Tibbett, Henry |
| 434 | Barkman, Henry | 480 | Cook, Samuel | 527 | Walleman, William |
| 435 | Money, Henry | 481 | Huggins, Daniel | 528 | D'Olivera, Manuel |
| 436 | Leversage, William | 482 | Resmabin, Benjamin | 529 | Grace, Allen |
| 437 | Foster, Thomas | 483 | Irwin, Mathew | 530 | Denham, Cornelius |
| 438 | Howe, John | 484 | Gilmore, William H. | 531 | Robinson, John |
| 439 | Kennedy, John | 485 | Gallibrandt, Bernard | 532 | Roberts, Hugh |
| 440 | Holmes, James | 486 | Roberts, John | 533 | Gray, Isaac |
| 441 | Lewis, Francis B. | 487 | Grover, Edmund | 534 | Bell, James |
| 442 | Jackson, Thomas | 488 | Lippart, Thomas D. | 535 | Drew, William |
| 443 | Harris, Simeon | 489 | Kewen, Edward | 536 | Curtis, John |
| 444 | Adams, John | 490 | Sandford, John William | 537 | Little, John |
| 445 | Reynolds, Owen | 491 | Hydra, Dempey | 538 | Gibbs, William |
| 446 | Harts, William | 492 | Parsons, Samuel D. | 539 | Brown, William |
| 447 | Prince, Benjamin | 494 | Long, R. S. | 540 | Harter, Henry |
| 448 | Johnstone, John | 495 | Jamison, Peter | 541 | Hall, James |
| 449 | Johnstone, Peter | 496 | Gandell, Epne | 542 | Russell, James |
| 450 | Williams, Peter | 497 | Rozier, William | 543 | Abbott, Timothy |
| 451 | Jackson, James | 498 | Brown, Aaron | 544 | Amos, Cheney |
| 452 | Brown, John | 499 | Helair, Gasper | 545 | Jordan, Richard |
| 453 | Casey, Henry | 500 | Fox, Washington | 546 | Patten, James |
| 454 | Alley, Samuel | 501 | Laurenceau, John | 547 | Manuel, Diego |
| 455 | Keen, Nathaniel | 502 | Kenny, Harris | 548 | Sutter, John G. |
| 456 | Dore, Charles | 503 | Charter, James | 549 | Hartfield, James |
| 457 | McKenzie, Alexander | 504 | Watkins, Thomas | 550 | Johnson, Samuel |
| 458 | Shairs, Samuel | 505 | Hume, George | 551 | Hubbard, John |
| 459 | Smith, William | 506 | Dinsmore, John | 552 | Young, John |
| 460 | Latimore, Mathew | 507 | Pickrage, George | 553 | Marshall, Thomas |

American Prisoners of War Held at Dartmoor during the War of 1812

## Numeric listing by prison number

| | | | | | |
|---|---|---|---|---|---|
| 554 | Yoing, William | 599 | Dillon, William | 645 | Ash, Sampson |
| 555 | Freely, Henry | 600 | Brown, Alexander | 646 | Daniels, John |
| 556 | Brown, Francis | 601 | Donham, Ebenezer | 647 | Wood, Thomas |
| 557 | Peckham, Henry | 602 | Groger, Henry | 648 | Gallaway, Joseph |
| 558 | Pratt, Paris | 603 | Barry, James | 649 | Bennett, Charles |
| 559 | Clerke, George | 604 | Preston, Jonas | 650 | Rodwin, Walter |
| 560 | Fleming, John J. | 605 | Figasson, Henry | 651 | Silva, Francis |
| 561 | Day, Lyles C. | 606 | Russell, Frederick | 652 | Allen, Philip |
| 562 | Waite, Abel | 607 | Couret, Francis | 653 | Fedie, John |
| 563 | Salley, James | 608 | Nash, Manuel | 654 | Moore, Henry |
| 564 | Place, John | 609 | Allen, Joseph | 655 | Start, Joshua |
| 565 | Thompson, Thomas | 610 | Benson, James | 656 | Roberson, Henry |
| 566 | Sears, Abraham | 611 | Williams, John | 657 | Sanis, Lewis |
| 567 | Curtis, Joseph | 612 | Blisset, James | 658 | Charles, Samuel |
| 568 | Phillips, Benjamin | 613 | Rosah, Samuel | 659 | Menard, Augustus |
| 569 | Hopkins, William | 614 | Clark, John | 660 | Coader, Anthony |
| 570 | Allister, Isaac | 615 | Smith, Adam | 661 | Peters, Aaron |
| 571 | Lee, Ely | 616 | Noyes, Charles | 662 | Allen, James |
| 572 | Henry, James | 617 | Tolvett, John | 663 | Stevens, Hugh |
| 573 | Givell, Oliver | 618 | Young, Thomas | 664 | Hart, Bartholew |
| 574 | Barlow, John | 619 | Leach, Josem | 665 | Palmer, Pero |
| 575 | Wyatt, Frederick | 620 | Bevin, William | 666 | Griffin, William |
| 576 | Nagle, George | 621 | Collins, Andrew | 667 | Morell, Benjamin |
| 577 | Scott, Robert | 622 | McDonald, John | 668 | Berdick, Simon |
| 578 | Graham, David | 623 | Paris, William | 669 | Chipman, Christian |
| 579 | Bogart, K. W. | 624 | Johns, Thomas | 670 | Johnson, William |
| 580 | Wilson, George | 625 | Davis, William | 671 | Black, George |
| 581 | Carbenett, John | 626 | Williams, John | 672 | Durvolf, Stephen |
| 582 | Smith, James | 627 | Brown, Asher | 673 | Brown, Ebenezer |
| 583 | Bladen, John | 628 | Jenkins, William | 674 | Hall, James |
| 584 | Jones, Charles | 629 | Conely, Samuel | 675 | Williams, Thomas |
| 585 | Garner, James | 630 | Hanscom, Moses | 676 | Murray, James |
| 586 | Bowes, David | 631 | Tafe, William | 677 | Eskinson, George |
| 587 | Barren, Thomas | 632 | Griffiths, Benjamin | 678 | Andrews, Charles |
| 588 | Lounsbury, Benjamin F. | 633 | Henry, John | 679 | Jackson, James |
| 589 | Lovett, Henry | 634 | Ayres, Robert | 680 | Jones, Richard |
| 590 | Appene | 635 | Cobbs, James | 681 | Williams, Ambrose |
| 591 | Hunt, Charles | 636 | Fadden, John | 682 | Reeves, James |
| 592 | Bradt, Francis | 637 | Godfrey, Edward | 683 | Martin, Thomas |
| 593 | Thompson, Henry | 638 | Noney, Peter | 684 | Young, William |
| 593 | Stevens, John | 639 | Waters, Robert | 685 | Bell, James |
| 594 | Himes, Walter | 640 | Tower, Michael | 686 | Harrington, William |
| 595 | Benedict, William | 641 | Smith, Richard | 687 | Freeman, John |
| 596 | Blacklidge, John | 642 | Croker, Nathaniel | 688 | Jeffery, Robert |
| 597 | Richards, George | 643 | Mullan, John | 689 | Williams, William |
| 598 | Chugler, John | 644 | Butnell, William | 690 | Trout, William |

American Prisoners of War Held at Dartmoor during the War of 1812

## Numeric listing by prison number

| | | | | | |
|---|---|---|---|---|---|
| 691 | McFall, William | 737 | Baday, Edward | 783 | Bienfaux, Allen |
| 692 | Brown, William | 738 | Fardell, Thomas | 784 | Willey, Ebenezer |
| 693 | Newby, Lambert | 739 | Addigo, Henry | 785 | Meeden, William |
| 694 | Brown, William | 740 | Smith, William | 786 | Morris, John |
| 695 | May, Henry | 741 | Scott, John | 787 | Carland, Lewis |
| 696 | Branger, John | 742 | Gunby, James | 788 | Dyer, Jonathan |
| 697 | Jackson, David | 743 | Timmerman, Mathew | 789 | Smith, John |
| 698 | Roath, Stephen | 744 | Johansen, Johan | 790 | McDowell, John |
| 699 | Bradford, James | 745 | Wieman, Laurence | 791 | Olseen, Elias |
| 700 | Ranson, Joseph | 746 | Oakley, Jacob | 792 | Johnson, Elias |
| 701 | Young, Rufus | 747 | Anderson, James | 793 | White, James |
| 702 | Smith, Elisha | 748 | Killam, James | 794 | Toley, Elisha |
| 703 | Robinson, Nicholas | 749 | Tamer, John | 795 | Lindburgh, Charles |
| 704 | Moore, William | 750 | Odgen, James | 796 | Livermore, Arthur |
| 705 | Niel, Henry | 751 | Thompson, James | 797 | Jones, Walter |
| 706 | Buchan, John | 752 | March, Beverley | 798 | Luce, Prosper |
| 707 | Duncan, George | 753 | Davis, George | 799 | Dennison, Thomas |
| 708 | Lewis, Thomas | 754 | Montgomery, John | 800 | Garrett, William |
| 709 | Mathias, Henry | 755 | Shipley, Daniel | 801 | Humphries, Warren |
| 710 | Wessels, Robert | 756 | Blandel, Jonathan | 802 | Holden, George |
| 711 | Wilson, William | 757 | Carter, James Wilson | 803 | Kennay, Jacob |
| 712 | Sickless, J. M. | 758 | Peters, Peter | 804 | Steward, Isaac |
| 713 | Snowdon, Jacob | 759 | Webb, Alexander | 805 | Moore, George |
| 714 | Norris, Robert | 760 | Wade, Richard H. | 806 | Allen, Eveshia |
| 715 | Williams, Jacques | 761 | Jansen, Knute | 807 | Prior, Asa |
| 716 | Ashmore, Edward | 762 | Nicholson, James | 808 | Small, Enoch |
| 717 | Williams, John | 763 | Bannister, Joshua | 809 | Ross, Philip |
| 718 | Wolfe, William | 764 | Broughton, Glover | 810 | Dubois, Alexander |
| 719 | Mines, Artemas | 765 | Sandburn, John | 811 | Mitchell, Faisly |
| 720 | Morweek, Andrew | 766 | Muggin, John | 812 | Andress, George |
| 721 | Minifee, Charles | 767 | Webber, John | 813 | Horn, Abner |
| 722 | Coffin, John | 768 | Mason, J. Bude | 814 | Wilson, John |
| 723 | Lettington, James | 769 | Holmes, Charles | 815 | Beymer, George |
| 724 | Hebron, John | 770 | Pyneo, John Rogers | 816 | Edgar, William |
| 725 | Spires, Thomas | 771 | Sares, Charles | 817 | Marshall, John |
| 726 | Dalloway, John | 772 | Nickerson, James | 818 | Stone, Edward |
| 727 | Churchill, Men'd. | 773 | Elliott, Benjamin | 819 | Symons, William |
| 728 | Wheeler, Benjamin | 774 | Landford, John | 820 | Campbell, Alexander |
| 729 | Anderson, William | 775 | Walker, James | 821 | Brown, Richard |
| 730 | Windham, John | 776 | Bunker, Isaiah | 822 | Haskell, Thomas |
| 731 | McNab, John | 777 | Wheeler, Henry | 823 | Franks, Francis |
| 732 | Barker, Israel | 778 | Nugent, John | 824 | Pearce, Alexander |
| 733 | Landerkin, Daniel | 779 | Selston, Issac | 825 | Bebe, Joseph |
| 734 | Spangle, Frederick | 780 | Skelder, Samuel | 826 | Gustave, Peter |
| 735 | Esdale, Thomas | 781 | Green, Rebuen | 827 | Valient, Richard |
| 736 | Moulds, Benjamin | 782 | Harrington, John | 828 | Wade, John |

American Prisoners of War Held at Dartmoor during the War of 1812

## Numeric listing by prison number

| | | | | | |
|---|---|---|---|---|---|
| 829 | Roberts, Francis | 875 | Ingles, Daniel | 921 | Anthony, John |
| 830 | Bassonet, Charles | 876 | Kennisen, William | 922 | Lingrin, Peter |
| 831 | Pearer, John | 877 | Cartwright, George | 923 | Clark, Jacob |
| 832 | Goldsbury, William | 878 | Smith, James | 924 | Taylor, James |
| 833 | Johnson, John | 879 | Taylor, James | 925 | Roberts, Joel |
| 834 | Dusheels, Arthur | 880 | Redman, William | 926 | Thornton, William |
| 835 | McDonald, John | 881 | Young, Thomas | 927 | Dunn, William |
| 836 | Palmer, John | 882 | Booth, Charles | 928 | Lambert, William |
| 837 | Moore, John | 883 | Gallin, John | 929 | McFadon, James |
| 838 | Cornish, Charles | 884 | Conray, William M. | 930 | Hoselquist, John |
| 839 | Oakley, Joseph | 885 | Symonds, M. John | 931 | Hayes, Benjamin |
| 840 | Cook, Benjamin | 886 | Archer, Joseph | 932 | Cheney, Daniel |
| 841 | Gillstone, George | 887 | Fisher, John | 933 | Abbey, Obadiah |
| 842 | Homell, Christopher | 888 | Morie, Denis | 934 | Sereder, Martin |
| 843 | Barton, Isaac | 889 | Parley, William | 935 | Dematra, George |
| 844 | Marshall, Emil | 890 | Dolliver, Richard | 936 | Routeman, Benedict |
| 845 | Crawford, William | 891 | Wyer, Joseph | 937 | Drinkwater, Peter |
| 846 | Martin, Stephen | 892 | Pierce, Samuel | 938 | Brown, Stephen |
| 847 | Fletcher, John | 893 | Rogers, Robert | 939 | Neal, John |
| 848 | Williams, Henry | 894 | Perkins, Elijah | 940 | Dennison, George |
| 849 | McLeod, M. | 895 | Wheeler, John | 941 | Prestley, James |
| 850 | Jameson, Robert | 896 | Andrews, Benjamin | 942 | Hutchins, William |
| 851 | Schew, Richard | 897 | Hughes, John | 943 | Gordnow, S. B. |
| 852 | Green, Peter | 898 | Thomson, Andrew | 944 | Gall, Michael |
| 853 | Crepo, William | 899 | Gilchrist, John | 945 | Niles, George |
| 854 | Williams, John | 900 | Chain, John | 946 | Perkins, John |
| 855 | Smith, John | 901 | Road, John | 947 | Wright, William |
| 856 | Crosby, George | 902 | Martin, John | 948 | Patterson, Andrew |
| 857 | Perkins, Thomas | 903 | McFarlane, John | 949 | Sheardon, Elison |
| 858 | Nelson, Thomas | 904 | Long, Charles | 950 | Harris, Bradley |
| 859 | Astropp, Hans Christopher | 905 | Pagechin, Henry | 951 | Stout, John |
| 860 | Peterson, Laurence | 906 | Gable, John | 952 | Weeks, James |
| 861 | Campbell, John | 907 | Postlock, William | 953 | Thompson, William |
| 862 | Peane, Henry | 908 | Hencock, William | 954 | Welsh, William |
| 863 | Rodyman, John | 909 | Graham, James | 955 | Wakefield, Nathaniel |
| 864 | Kitchen, Richard | 910 | Coffin, Alexander | 956 | Pinckham, Amos |
| 865 | Prymvoy, Richard | 911 | Mathews, Joseph | 957 | Pain, George |
| 866 | Blake, Thomas | 912 | Wilson, John | 958 | Bodfish, William |
| 867 | Newell, Nathaniel | 913 | Jones, Thomas | 959 | Nye, Cornelius |
| 868 | Jackson, William | 914 | Thomson, William | 960 | Nanson, Henry |
| 869 | Nicholas, John Buess | 915 | Dixon, Benjamin | 961 | Nanson, Peter |
| 870 | Pearce, Joseph | 916 | Mathews, Henry | 962 | Soderback, Andrew |
| 871 | Lehart, Jacob | 917 | Thomson, Peter | 963 | Patterson, John |
| 872 | Jackson, Curtis | 918 | Laborde, Peter | 964 | Samuel, Nathaniel |
| 873 | Adams, John | 919 | Cross, Peter | 965 | Brown, Sandy |
| 874 | Hoppins, William | 920 | Barker, Charles G. | 966 | Mason, William |

American Prisoners of War Held at Dartmoor during the War of 1812

## Numeric listing by prison number

| | | | | | |
|---|---|---|---|---|---|
| 967 | Tubbs, Martin | 1013 | Brutus, Mario | 1059 | Gardner, Edward |
| 968 | Coffee, William | 1014 | Deparvier, John | 1060 | Burts, Joseph |
| 969 | Botellio, Anthony | 1015 | Erighton, Manuel | 1061 | Richardson, Cheney |
| 970 | Mann, John | 1016 | Rogers, Henry | 1062 | Clapp, George |
| 971 | Ringold, Peregrin | 1017 | Smith, John | 1063 | Myers, Daniel |
| 972 | Miller, William | 1018 | Morgan, John | 1064 | Abbott, Ephraim |
| 973 | Moore, Alexander | 1019 | Webber, Frederick | 1065 | Baldwin, John |
| 974 | Wilson, Charles | 1020 | Ferris, Joseph | 1066 | Williams, James |
| 975 | Cooper, John | 1021 | Williams, Alexander | 1067 | Mathews, Williams |
| 976 | Guilmot, Richard | 1022 | Brady, Jason | 1068 | Homer, Michael |
| 977 | Smiles, John | 1023 | Hammond, William | 1069 | Still, David |
| 978 | Jackson, Samuel | 1024 | Gardner, Edward | 1070 | Gerard, William |
| 979 | Furlong, William | 1025 | Stacey, William | 1071 | Sparks, Robert |
| 980 | Hobart, William | 1026 | Besson, Phillipe | 1072 | Morgan, John |
| 981 | Popal, Richard | 1027 | Chambers, Elias | 1073 | Young, John |
| 982 | David, William | 1028 | Crafford, Robert | 1074 | Nicholls, John |
| 983 | Duff, James | 1029 | Conklin, Robert | 1075 | Freeman, David |
| 984 | Moore, John | 1030 | Thomas, David | 1076 | Elliott, Robert |
| 985 | Leger, Andrew | 1031 | Smith, John | 1077 | Philling, Nathaniel |
| 986 | Ramsden, John | 1032 | Cooper, James | 1078 | Bird, Comford |
| 987 | Charles, John | 1033 | Price, William | 1079 | Carrot, Charles |
| 988 | Poland, Thomas | 1034 | Robinson, James | 1080 | Cassam, Michael |
| 989 | LeCore, Peter M. | 1035 | Leary, Daniel | 1081 | Davey, Charles |
| 990 | Belavoine, L. | 1036 | Gilbert, Gurdonel | 1082 | Coleman, William |
| 991 | Pelhem, Robert | 1037 | Bron, James | 1083 | Cooper, Tannick |
| 992 | Anderson, Edward | 1038 | Shaw, William | 1084 | Simons, William |
| 993 | Neal, Daniel | 1039 | Smith, William | 1085 | Williams, John |
| 994 | Barbadoes, Robert | 1040 | Hill, Thomas | 1086 | Lewis, John |
| 995 | Bunkerson, John | 1041 | Kennedy, Richard | 1087 | Brooks, Thomas |
| 996 | Woods, Edward | 1042 | Anderson, George | 1088 | Handfield, Robert |
| 997 | Raquinis, Francis | 1043 | Starbuck, George | 1089 | McConnell, John |
| 998 | Guiho, Pierre | 1044 | Gennodo, Samuel H. | 1090 | Briggs, William |
| 999 | Castana, John Baptist | 1045 | Ham, John | 1091 | Moore, James |
| 1000 | Williams, Benjamin | 1046 | Berto, John | 1092 | Phillips, James |
| 1001 | Garrison, Richard | 1047 | Poswell, John | 1093 | Fletcher, John W. |
| 1002 | McNeil, John | 1048 | Wright, Emos | 1094 | Brooks, Russell |
| 1003 | Donnelson, Joseph | 1049 | Hudson, James | 1095 | Conner, Edward |
| 1004 | Carr, James | 1050 | Frederick, Charles | 1096 | Watch, Samuel |
| 1005 | Golandre, John | 1051 | Gasey, Raymond | 1097 | Hanson, J. A. |
| 1006 | Lepberg, John P. | 1052 | Lunberg, Berg | 1098 | Jackson, James |
| 1007 | Edwards, John | 1053 | Graham, John | 1099 | Major, James |
| 1008 | Holden, Andrew | 1054 | Williams, Edward | 1100 | Downe, William |
| 1009 | Morrison, John Baptist | 1055 | Bensted, John | 1101 | Meech, David |
| 1010 | Ranger, Desire | 1056 | Norcross, Phillipe | 1102 | Potter, John |
| 1011 | Grenaux, Yves | 1057 | Turner, Daniel | 1103 | Cowder, James |
| 1012 | Robins, John | 1058 | Stafford, John | 1104 | Armstead, Edward |

American Prisoners of War Held at Dartmoor during the War of 1812

## Numeric listing by prison number

| | | | | | |
|---|---|---|---|---|---|
| 1105 | Oytiel, John | 1151 | Daggett, Robert | 1197 | Harburn, Thomas |
| 1106 | Orne, W. B. | 1152 | Whitney, Thomas | 1198 | Smith, Shapley |
| 1107 | Leggett, William | 1153 | Stephenson, John | 1199 | Baker, John |
| 1108 | Hedenburg, Jacob | 1154 | Catwood, Zenos | 1200 | Lloyd, David |
| 1109 | Gardner, Joseph | 1155 | Fuller, Enoch | 1201 | Ravett, William |
| 1110 | Swasey, John | 1156 | Nesbit, Richard | 1202 | Lang, William |
| 1111 | Graves, Ebenezer | 1157 | Soreby, Robert | 1203 | Meadows, Timothy |
| 1112 | Hilbert, Henry | 1158 | Paulin, Nicholas | 1204 | Egbert, Peter |
| 1113 | Gatewood, James | 1159 | Monturne, Jais | 1205 | Welter, William |
| 1114 | Channing, John | 1160 | Minois, Pierre | 1206 | Alexander, J. B. |
| 1115 | Stephens, Amos | 1161 | Crosby, John | 1207 | Cowing, Charles |
| 1116 | Thompson, John | 1162 | Jackson, Thomas | 1208 | Colton, Charles |
| 1117 | Brightman, Joseph | 1163 | Adams, Peter W. | 1209 | Monsieur, Peter |
| 1118 | Nooney, William | 1164 | Jonson, Edward | 1210 | Dugaretz, Augusta |
| 1119 | Smith, John D. | 1165 | Tilley, Bernard | 1211 | Williams, Peter |
| 1120 | Gillett, Francis | 1166 | Shepherd, William | 1212 | Basson, John |
| 1121 | Elves, Manuel | 1167 | Willson, Charles | 1213 | Orwick, Thomas C. |
| 1122 | Elves, Jospeh | 1168 | Townsend, Thomas | 1214 | Boyd, William |
| 1123 | Mitchell, John | 1169 | Fisher, James | 1215 | Rice, John |
| 1124 | Drake, John | 1170 | Miles, William | 1216 | Murray, Nathaniel |
| 1125 | Blasse, John M. | 1171 | Murphy, William | 1217 | Shorts, Clement |
| 1126 | Lewis, James | 1172 | Brown, David | 1218 | Ley, John |
| 1127 | Garreai, Anthony | 1173 | Gillia, Joseph | 1219 | Adams, John |
| 1128 | Lopez, Joseph | 1174 | Barnes, Joseph | 1220 | Greenland, Stephen |
| 1129 | Cotterell, Henry C. | 1175 | Scary, Joseph | 1221 | Potter, Richard H. |
| 1130 | Chessebrough, Ben. F. | 1176 | Dunbar, Luther | 1222 | Holstein, Peter |
| 1131 | Badson, Jacob | 1177 | Gray, Thomas | 1223 | Akerman, William |
| 1132 | Smith, John | 1178 | Springstern, Abraham | 1224 | Holstade, Joseph |
| 1133 | Parker, Peter | 1179 | Preston, John | 1225 | Brown, Thomas |
| 1134 | Wright, William | 1180 | Fornver, Virgil | 1226 | Jarvis, James |
| 1135 | Grant, James | 1181 | Lamar, Edward | 1227 | Brown, Jesse |
| 1136 | Lerwich, John | 1182 | Maxen, Joseph | 1228 | Masick, Joseph |
| 1137 | Dean, Moses | 1183 | Frederique, John | 1229 | Wavernce, James |
| 1138 | Neel, William | 1184 | Lavasseur, Rene | 1230 | Rose, William |
| 1139 | Ray, David | 1185 | Neyren, John | 1231 | Taylor, William |
| 1140 | Diamond, William | 1186 | Dillingham, F. A. | 1232 | Smith, William |
| 1141 | Spark, Samuel | 1187 | Scott, John | 1233 | McCarthy, John |
| 1142 | Rideout, James | 1188 | Phillips, Peter | 1234 | Green, Solomon |
| 1143 | Davis, John | 1189 | Myers, George | 1235 | Hartwell, Berry |
| 1144 | Bowmer, Isaac | 1190 | Bassington, John | 1236 | Ford, Philip |
| 1145 | Christian, John | 1191 | Spince, William | 1237 | Johnson, Andrew |
| 1146 | Fisher, John | 1192 | Roberts, George | 1238 | Eyers, John |
| 1147 | Day, Benjamin | 1193 | Commous, John | 1239 | Craig, William |
| 1148 | Borsdell, Justice | 1194 | Whitehouse, Daniel | 1240 | Stephens, Obadiah |
| 1149 | Drew, Charles | 1195 | Dale, Hercules | 1241 | Day, John |
| 1150 | Joseph, Francis | 1196 | Tully, Hugh | 1242 | Merchant, William |

American Prisoners of War Held at Dartmoor during the War of 1812

## Numeric listing by prison number

| | | | | | |
|---|---|---|---|---|---|
| 1243 | Daniel, Robert | 1289 | Church, William | 1335 | Gibbs, Henry |
| 1244 | Hooper, William | 1290 | Wilson, John | 1336 | Tardy, Andy |
| 1245 | Cox, John | 1291 | Flinn, Pearce | 1337 | Wild, Thomas |
| 1246 | Holbery, Emil | 1292 | Thompson, John | 1338 | Whiting, Samuel |
| 1247 | Driver, John | 1293 | Long, Joseph | 1339 | Newell, Isaac |
| 1248 | Chase, Joseph | 1294 | Harker, John | 1340 | Tassier, Louis |
| 1249 | Wilson, Charles | 1295 | McKertre, Abraham | 1341 | Anderson, Joseph |
| 1250 | Ford, Charles | 1296 | Bitters, John | 1342 | Cato, John |
| 1251 | Morris, Isaac | 1297 | Phillips, John | 1343 | Cramstead, James |
| 1252 | Johnston, Edward | 1298 | Simpson, William | 1344 | Prince, George |
| 1253 | Cole, John | 1299 | Sone, John | 1345 | Sentille, Francis |
| 1254 | Newman, Daniel | 1300 | Farmer, George | 1346 | Martin, Daniel |
| 1255 | Pottigal, John | 1301 | Johnson, Henry | 1347 | Johnson, Joseph Toker |
| 1256 | Conklin, Edward | 1302 | Skinner, Tileman | 1348 | Cook, William |
| 1257 | Lorenze, Thomas | 1303 | Kimball, Nathaniel | 1349 | Ruffield, Samuel |
| 1258 | Bertol, Samuel | 1304 | Day, James | 1350 | Brown, Charles |
| 1259 | Flatt, Rob | 1305 | Bannell, James | 1351 | Green, William |
| 1260 | Sole, Edward | 1306 | Williams, John | 1352 | Cook, Charles Howe |
| 1261 | Holbrook, Elias | 1307 | Cadwell, John | 1353 | Allen, John |
| 1262 | Sole, Elias | 1308 | Whitmore, John | 1354 | Dibble, Rebuen |
| 1263 | Thomas, Stephen | 1309 | McQuillan, James | 1355 | Roley, John |
| 1264 | Teggett, John | 1310 | Norman, Michael | 1356 | Kister, John |
| 1265 | Sole, Thomas | 1311 | Leckler, Anthony | 1357 | Warren, David |
| 1266 | Walters, Philip | 1312 | Parker, Edward | 1358 | Bernard, John |
| 1267 | Washbury, Elbert | 1313 | Gregory, Elijah | 1359 | Mooney, Peter |
| 1268 | Roberts, Moses | 1314 | Vindine, Gar. | 1360 | Guillard, Louis |
| 1269 | Benny, David | 1315 | Snyder, William | 1361 | Martin, Isaac |
| 1270 | Jackson, William | 1316 | Young, John | 1362 | Hamilton, Alexander M. |
| 1271 | Atwood, John | 1317 | Stewart, Adam | 1363 | Irvin, Arthur |
| 1272 | Wood, William | 1318 | Harris, John | 1364 | Gross, William |
| 1273 | Younger, William | 1319 | Martin, John | 1365 | Colman, David |
| 1274 | Gangler, George | 1320 | Judson, Obediah | 1366 | Stevens, William |
| 1275 | Gilpin, William | 1321 | Weeks, James | 1367 | Little, George |
| 1276 | Alton, Peter | 1322 | Bryant, Stephen | 1368 | Colton, Walter |
| 1277 | Armstrong, Joseph | 1323 | Cotton, Samuel | 1369 | Barnes, Nathaniel |
| 1278 | Knox, John | 1324 | Bourton, George | 1370 | White, Isaac |
| 1279 | Burford, William | 1325 | Baker, Stephen | 1371 | Lane, James |
| 1280 | Hoyt, James | 1326 | Coburn, Abel | 1372 | Thompson, Joseph |
| 1281 | Chane, Daniel | 1327 | Wilson, James | 1373 | Pote, Jeremiah |
| 1282 | White, Philip | 1328 | Byard, Joseph | 1374 | Poland, John |
| 1283 | Works, Elford | 1329 | Muley, Etienne | 1375 | Redding, John |
| 1284 | Bassett, Gorham | 1330 | Godfrey, William | 1376 | Wilcox, Lewis |
| 1285 | Holland, James | 1331 | Louis, Nicholas | 1377 | Cornwall, Arthur |
| 1286 | Stone, Samuel | 1332 | Peck, Thomas | 1378 | Ostand, Michael |
| 1287 | Merrill, John | 1333 | Veitch, William | 1379 | Atkins, Joseph |
| 1288 | Simons, John | 1334 | Smith, John | 1380 | Hutchins, Edward |

American Prisoners of War Held at Dartmoor during the War of 1812

## Numeric listing by prison number

| | | | | | |
|---|---|---|---|---|---|
| 1381 | Walker, Armstrong | 1427 | McKinney, John | 1473 | Armstrong, William |
| 1382 | Shaw, Richard | 1428 | Evelish, William | 1474 | Jones, Thomas |
| 1383 | Biss, Daniel W. | 1429 | Flinn, Abraham | 1475 | Fisher, Lewis |
| 1384 | Mullins, James | 1430 | Price, Samuel | 1476 | Maine, William |
| 1385 | White, Davis | 1431 | Lethorpe, James | 1477 | Mode, David |
| 1386 | Roles, John | 1432 | Holland, Richard | 1478 | Thompson, William |
| 1387 | Logan, William | 1433 | Bartlett, Caleb | 1479 | Page, Thomas |
| 1388 | Martin, James | 1434 | Codman, Richard | 1480 | Fish, Joseph |
| 1389 | Wood, John | 1435 | Mason, Nathaniel | 1481 | Chiseldine, John |
| 1390 | Mills, William | 1436 | Davies, John | 1482 | Watson, James |
| 1391 | Hopkins, Daniel | 1437 | Wood, Sylvester | 1483 | Bradford, Charles |
| 1392 | Boyer, Joseph | 1438 | Doughty, Jesse | 1484 | Grey, Moorhouse |
| 1393 | Durand, John | 1439 | Foss, Edward | 1485 | Veal, Pierre |
| 1394 | Carter, Edward | 1440 | Anderson, Daniel | 1486 | Prett, Benjamin |
| 1395 | Carter, Daniel | 1441 | Carey, John | 1487 | Hobson, Abraham |
| 1396 | Zellie, Thomas | 1442 | Foss, Joseph | 1488 | Clepp, Abraham |
| 1397 | Barber, William | 1443 | Forster, Thomas | 1489 | Tuttle, French |
| 1398 | Weeks, David | 1444 | Christie, James | 1490 | Perry, Charles |
| 1399 | Pitts, William | 1445 | McKinnon, Nathaniel | 1491 | Burnham, Enoch |
| 1400 | Reynolds, James | 1446 | Dougall, Thomas | 1492 | Hickman, Joseph |
| 1401 | Martin, Henry | 1447 | Best, Robert | 1493 | Scovel, John |
| 1402 | Boriesa, Pierre | 1448 | Isles, Robert | 1494 | Jones, Thomas |
| 1403 | Lerna, John | 1449 | Meurinose, John | 1495 | Wilson, John |
| 1404 | Hudson, Daniel | 1450 | Batman, C. P. | 1496 | Burnet, Charles |
| 1405 | Wheadon, Anthony | 1451 | English, David | 1497 | Whitehead, James |
| 1406 | Berryman, John | 1452 | Armstrong, John | 1498 | Armstrong, David |
| 1407 | Neale, Dennis | 1453 | Norris, August G. | 1499 | Grubb, Andrew |
| 1408 | Yard, William | 1454 | Scott, John | 1500 | Banks, Perry |
| 1409 | Robertson, Robert | 1455 | Gayer, Joseph | 1501 | Jones, Kenny |
| 1410 | Barraso, John | 1456 | Putman, Charles | 1502 | Howe, Phineas |
| 1411 | Littlefield, Rufus | 1457 | White, John | 1503 | Simmons, Robert |
| 1412 | West, Simon | 1458 | Killenger, John | 1504 | Thompson, Abraham |
| 1413 | Goodwin, William | 1459 | Mezich, Elisha | 1505 | Risings, John |
| 1414 | Butler, George | 1460 | Hely, John | 1506 | Guillard, Peter |
| 1415 | Doughty, Levi | 1461 | Brown, Benjamin | 1507 | Parish, Samuel |
| 1416 | Haskett, Robert | 1462 | Merritt, Jonah | 1508 | Hart, James |
| 1417 | Robinson, Elias | 1463 | Bunnel, Benjamin | 1509 | Manson, William |
| 1418 | Nicherson, Joseph | 1464 | Whitter, J. W. | 1510 | Manson, William |
| 1419 | Merritt, Robert | 1465 | Gardiner, John | 1511 | Dickenson, Chester |
| 1420 | Cudsworth, Henry | 1466 | Gabriel, John | 1512 | Carter, Edward |
| 1421 | Stewart, William | 1467 | Killer, John | 1513 | Edsom, John |
| 1422 | Smith, John | 1468 | Harris, James | 1514 | Smith, Chester |
| 1423 | Cross, Oliver | 1469 | Smith, James | 1515 | Anthony, Stephen |
| 1424 | Hudson, Thomas | 1470 | Bosset, David | 1516 | Weston, David |
| 1425 | Lamond, John | 1471 | Linsey, Alexander | 1517 | Hill, John |
| 1426 | English, Edward | 1472 | Lopez, B. Eldridge | 1518 | Amerson, Charles |

American Prisoners of War Held at Dartmoor during the War of 1812

## Numeric listing by prison number

| | | | | | |
|---|---|---|---|---|---|
| 1519 | Welch, Benjamin | 1565 | Johnson, Joseph | 1611 | Canabon, J. |
| 1520 | Blodget, Caleb | 1566 | Metley, Thomas | 1612 | Beard, Francis |
| 1521 | Ripley, Eden | 1567 | Wills, John | 1613 | Andress, Daniel |
| 1522 | Diluo, Benjamin | 1568 | Broadwater, Samuel | 1614 | Innes, J. |
| 1523 | Chandler, Simon | 1569 | Ingle, John | 1615 | Muller, Edward |
| 1524 | Parker, J. A. | 1570 | Stuff, Francis | 1616 | Jeffrey, Francis |
| 1525 | Russell, Patton | 1571 | Morgan, John | 1617 | Brights, George |
| 1526 | Williams, Thomas | 1572 | Simpson, Thomas | 1618 | Hollinger, William |
| 1527 | Davis, John | 1573 | Hale, Shederick | 1619 | Rowley, Henry |
| 1528 | White, Henry | 1574 | Clothey, Thomas | 1620 | Davis, John |
| 1529 | Moss, Thomas | 1575 | Wright, John | 1621 | Brown, John W. |
| 1530 | Stockham, W. B. | 1576 | Thompson, Charles | 1622 | Layfield, Littleton |
| 1531 | Libley, Moses | 1577 | May, Walter | 1623 | McIntire, Samuel |
| 1532 | Waters, Abraham | 1578 | Robertson, James | 1624 | Smith, Andrew |
| 1533 | Lowe, John | 1579 | Price, Jacob | 1625 | Nelly, Richard J. |
| 1534 | Mathews, Richard | 1580 | Young, William | 1626 | Petton, Alexander |
| 1535 | Harrington, Simon | 1581 | Ingerson, Michael | 1627 | Bourdon, John |
| 1536 | Jackson, Joseph | 1582 | Francis, John | 1628 | Trifle, Jasper |
| 1537 | Payer, Walter | 1583 | Hunter, William | 1629 | Mansfield, James |
| 1538 | Mills, William | 1584 | Myers, John | 1630 | Siffers, Stuben |
| 1539 | Hardingbrook, Theop. | 1585 | Dean, Jonas | 1631 | Davis, John |
| 1540 | Avis, James | 1586 | Brown, Jean | 1632 | Rowe, John |
| 1541 | Logan, Charles | 1587 | Blanchet, Simon | 1633 | Doolittle, H. |
| 1542 | Laurence, Jean | 1588 | Sparrow, James | 1634 | Cummings, James |
| 1543 | Faye, Samuel | 1589 | Haye, Moses | 1635 | Trefry, Peter |
| 1544 | Montbelly, William | 1590 | Williams, John | 1636 | Jenkins, Nathaniel |
| 1545 | Brant, Thomas | 1591 | Murray, Jacob | 1637 | Dean, N. B. |
| 1546 | Phipps, Joseph | 1592 | Longford, Samuel | 1638 | Beck, Steward |
| 1547 | Davis, William | 1593 | Hammond, Joseph | 1639 | Smith, Thomas |
| 1548 | Stagg, J. Boulton | 1594 | Strube, John | 1640 | Smith, Thomas |
| 1549 | Merritt, Thomas | 1595 | Phillips, John | 1641 | Edwards, John |
| 1550 | Pomp, William | 1596 | Marshall, Alexander | 1642 | Friday, John |
| 1551 | Hanford, William | 1597 | Hensell, John | 1643 | Fink, Jonas |
| 1552 | Banker, Robert | 1598 | Williams, Elisha | 1644 | Robinson, John |
| 1553 | Plumer, John | 1599 | Doliver, William | 1645 | Abman, John |
| 1554 | Wright, George | 1600 | Lilley, Simon | 1646 | Lacour, John Baptist |
| 1555 | Edwards, David | 1601 | Hane, W. B. | 1647 | Sandford, William |
| 1556 | Martin, Anthony | 1602 | Rowle, Benjamin | 1648 | Keg, Philip |
| 1557 | Anderson, David | 1603 | Mingle, Thomas | 1649 | Ricks, Thomas |
| 1558 | Fitts, Joseph | 1604 | Moore, Richard | 1650 | Walker, Richard |
| 1559 | Williams, John | 1605 | Harris, John | 1651 | Winter, Andrew |
| 1560 | Evans, Moses | 1606 | Slocombe, Abraham | 1652 | Jefferson, Edward |
| 1561 | Barrester, Peter | 1607 | James, Daniel | 1653 | North, Thomas |
| 1562 | West, Dennis | 1608 | Baldwin, John | 1654 | Lake, Charles |
| 1563 | Burns, Charles | 1609 | Wilkins, William | 1655 | Waldron, Hiram |
| 1564 | Peregrine, Taggert | 1610 | McKey, James Abercomby | 1656 | Bickwith, Benjamin |

American Prisoners of War Held at Dartmoor during the War of 1812

## Numeric listing by prison number

| | | | | | |
|---|---|---|---|---|---|
| 1657 | Haney, Edward | 1703 | Wildon, Thomas | 1749 | Silverlock, John |
| 1658 | Gore, William | 1704 | Hedley, John | 1750 | Frizzle, David |
| 1659 | Edwards, John | 1705 | Miller, Thomas | 1751 | Moore, James |
| 1660 | Littlefield, James | 1706 | Mack, Theron | 1752 | Marshall, John |
| 1661 | Parinele, Benjamin | 1707 | Penman, Richard | 1753 | Homes, Ensign E. |
| 1662 | Gage, Isaac | 1708 | Hall, Thomas | 1754 | Dalton, Samuel |
| 1663 | Cooper, Andrew A. | 1709 | Brown, William | 1755 | Dorchester, Preston |
| 1664 | Hutchins, Henry | 1710 | Dunning, William | 1756 | Parker, Richard |
| 1665 | Brown, Samuel | 1711 | Peal, Andrew | 1757 | Stewart, Scipio |
| 1666 | Manson, Jeremiah | 1712 | Wadsworth, Thomas | 1758 | Enoch, Joseph |
| 1667 | Murrell, Mark | 1713 | Davenport, Russell | 1759 | Williams, James |
| 1668 | Taylor, Samuel E. | 1714 | Mingalls, Robert | 1760 | Lane, W. S. |
| 1669 | Payne, Josiah Smith | 1715 | Lake, Daniel | 1761 | Smith, Turpin |
| 1670 | Freeman, John | 1716 | Lewis, John | 1762 | Lambert, Samuel |
| 1671 | Hayes, Simon | 1717 | Green, John M. | 1763 | Maret, Ebenezer |
| 1672 | Ricker, James | 1718 | Symes, John | 1764 | Dowling, John |
| 1673 | Whitten, Elijah | 1719 | Richardson, John | 1765 | Soie, Henry |
| 1674 | Miller, John | 1720 | Crawford, George | 1766 | Fuller, Gideon |
| 1675 | McLeod, Colin | 1721 | Curry, Samuel | 1767 | Farrell, James |
| 1676 | Welsh, Ezekiel | 1722 | Penn, William | 1768 | Larkin, Amos |
| 1677 | Russell, Richard | 1723 | Bond, Peter | 1769 | Geline, John |
| 1678 | Elvell, Robert | 1724 | Jackson, John | 1770 | Knight, George |
| 1679 | Fellows, Isaac | 1725 | Nugent, John | 1771 | Dolabar, John |
| 1680 | Hudson, James | 1726 | Sniffon, John | 1772 | Knight, William |
| 1681 | Chapman, Abraham | 1727 | Brown, Wiliam | 1773 | Sweet, Francis |
| 1682 | Tindall, Thomas | 1728 | Bowers, Jonathan | 1774 | Ireson, Robert B. |
| 1683 | Evans, Robert | 1729 | Adams, James | 1775 | Graves, Samuel |
| 1684 | Allen, Henry | 1730 | Clark, Philip | 1776 | Symonds, Israel |
| 1685 | Carter, Jesse | 1731 | Calhoun, Richard | 1777 | Arnold, James |
| 1686 | Golden, Edward | 1732 | Wright, James | 1778 | Blue, Peter |
| 1687 | Parr, James | 1733 | Perlie, Abraham | 1779 | Johnstone, Thomas |
| 1688 | Stephens, John | 1734 | Dias, Joseph | 1780 | Henderson, Alexander |
| 1689 | Blair, John R. | 1735 | Pipe, Henry | 1781 | Gault, William |
| 1690 | Holsley, John | 1736 | Wood, Samuel | 1782 | Hawkins, John |
| 1691 | Bradbury, John H. | 1737 | Salisbury, John | 1783 | Stow, Jeremiah |
| 1692 | Hardy, Robert | 1738 | Saunders, Cornelius | 1784 | Fernald, John |
| 1693 | Wicks, Samuel | 1739 | Louis, John | 1785 | Westerbest, William |
| 1694 | McIntire, A. | 1740 | Jane, William | 1786 | Layton, William |
| 1695 | Howard, Benjamin | 1741 | Jack, John | 1787 | Trusty, Henry |
| 1696 | Cottle, John | 1742 | Green, John | 1788 | Bailey, Samuel |
| 1697 | Martling, Abraham | 1743 | Mathews, William | 1789 | Lee, George |
| 1698 | Vandeburg, Ad'm | 1744 | Lander, John | 1790 | Meyers, David |
| 1699 | Hollis, Nathaniel | 1745 | Lane, John | 1791 | Durham, Charles |
| 1700 | Boose, Abraham | 1746 | Keeffe, Alexander | 1792 | Deal, John |
| 1701 | Davis, James | 1747 | Williams, William | 1793 | Crandell, John |
| 1702 | Williams, Thomas | 1748 | Rice, William | 1794 | Freeman, Charles |

American Prisoners of War Held at Dartmoor during the War of 1812

## Numeric listing by prison number

| | | | | | |
|---|---|---|---|---|---|
| 1795 | Cadwell, James | 1841 | McBride, James | 1887 | Lane, William |
| 1796 | Reynolds, Frederick | 1842 | Richards, James | 1888 | Dibble, Zachariah |
| 1797 | Southwich, Israel | 1843 | Harvey, John | 1889 | Platt, Daniel |
| 1798 | Rogers, Nathaniel | 1844 | Gardner, Peter | 1890 | Crow, John |
| 1799 | Jackson, Daniel | 1845 | Bryant, James | 1891 | Duncan, George |
| 1800 | Kennedy, William | 1846 | Egoss, Joseph | 1892 | Norcross, Abel |
| 1801 | Jones, Peter | 1847 | Warner, Thomas | 1893 | Domeree, John |
| 1802 | White, William | 1848 | Lawrence, Robert | 1894 | Brower, John |
| 1803 | White, William | 1849 | Rich, Elisha | 1895 | Mortice, William |
| 1804 | Roach, Reuben | 1850 | Carter, Thomas | 1896 | Prentis, James |
| 1805 | Austin, Jonathan | 1851 | Fields, Alexander | 1897 | Hunberly, Elisha |
| 1806 | Freeman, Prince | 1852 | Armstrong, Thomas | 1898 | Very, Samuel |
| 1807 | Evans, James | 1853 | Brown, Joseph | 1899 | Lane, John |
| 1808 | Parsons, Eugene | 1854 | Nicholls, William | 1900 | Lockwood, Caleb |
| 1809 | Claw, Morris | 1855 | Brown, Benjamin | 1901 | Bovey, Benjamin |
| 1810 | Hamel, William | 1856 | Weymer, John | 1902 | Vanderhaven, Mathias |
| 1811 | Dunchellier, Isaac | 1857 | Quince, Peter | 1903 | Beattie, James |
| 1812 | Willett, John | 1858 | Williams, Charles | 1904 | Richardson, James |
| 1813 | Clark, Abraham | 1859 | Bean, John | 1905 | Broden, Norman |
| 1814 | Dunn, James | 1860 | Willy, John | 1906 | Arnold, William |
| 1815 | Rideout, J. J. | 1861 | How, William | 1907 | Nunns, William |
| 1816 | Williams, Thomas | 1862 | Abbott, Daniel | 1908 | Robinson, Edward |
| 1817 | Voight, Henry | 1863 | Butler, George | 1909 | Courtis, Thomas |
| 1818 | Thompson, James | 1864 | Sanburn, James | 1910 | Dildure, Samuel |
| 1819 | Codding, Caleb | 1865 | Cleveland, Ebenezer | 1911 | Urey, Peter |
| 1820 | Webster, William | 1866 | Lamson, William | 1912 | Saunders, Thomas |
| 1821 | Hamilton, Richard | 1867 | Mason, Daniel | 1913 | Woolf, Andrew |
| 1822 | Moffett, John | 1868 | Bowden, John | 1914 | Ferris, Jacob |
| 1823 | Enderson, James | 1869 | Kirk, William | 1915 | Silsby, Nathaniel |
| 1824 | Sims, Oliver | 1870 | Baptiste, John | 1916 | Sidebottom, John |
| 1825 | Jones, Benjamin | 1871 | Drake, Daniel | 1917 | Poole, John |
| 1826 | Ruddick, William | 1872 | Barton, James | 1918 | McDonald, John |
| 1827 | Jordan, Peter | 1873 | Homer, Henry | 1919 | Patterson, Peter |
| 1828 | Garthy, James | 1874 | Clark, Joseph | 1920 | Johnson, William |
| 1829 | Warrands, John | 1875 | Lee, Michael | 1921 | Church, Benjamin |
| 1830 | Crawford, Nelson | 1876 | Worthey, Samuel | 1922 | Babb, Benjamin |
| 1831 | Wallace, James | 1877 | Cree, William | 1923 | Covall, Ephraim |
| 1832 | Bustin, John | 1878 | Luffie, Warren | 1924 | Houseman, John |
| 1833 | Harris, William | 1879 | Foot, Benjamin | 1925 | Atwood, Edward |
| 1834 | Brooks, John | 1880 | Plummer, William | 1926 | Brenton, York |
| 1835 | Vicary, Richard | 1881 | Trumbull, John | 1927 | Booth, Thomas |
| 1836 | Mazal, James | 1882 | Blossom, Seth | 1928 | Malis, John |
| 1837 | Scott, John | 1883 | Osborn, Stephen | 1929 | Johnson, Frederick |
| 1838 | Nelson, Richard | 1884 | Lake, George | 1930 | Smith, Caesar |
| 1839 | Patton, Robert | 1885 | Whittington, George | 1931 | Scribner, William |
| 1840 | Farley, John | 1886 | Coas, Samuel | 1932 | Davis, William |

American Prisoners of War Held at Dartmoor during the War of 1812

## Numeric listing by prison number

| | | | | | |
|---|---|---|---|---|---|
| 1933 | Williamson, Joseph | 1979 | Sweetman, Samuel | 2025 | Wilson, John |
| 1934 | Beck, William | 1980 | Webb, John | 2026 | Lesle, Richard |
| 1935 | Mitchell, Thomas | 1981 | Vincent, John | 2027 | Fogerty, Archibald |
| 1936 | Robinson, Benjamin | 1982 | Scott, Alexander | 2028 | Baley, John |
| 1937 | Davis, George | 1983 | McFree, John | 2029 | Bourns, George |
| 1938 | Best, John | 1984 | Reynolds, Amos | 2030 | Litchfield, Erick |
| 1939 | Francis, Prince | 1985 | Benson, George | 2031 | Starboard, William |
| 1940 | Barrett, James | 1986 | Thomas, William | 2032 | Sturdwent, David |
| 1941 | Burnham, David | 1987 | Vaughan, Robert | 2033 | Rodgers, John |
| 1942 | Lethorpe, James | 1988 | Cannon, William | 2034 | Allen, Isaac |
| 1943 | Tink, Henry | 1989 | Squires, Prince | 2035 | Welch, William |
| 1944 | Albert, John | 1990 | Paine, R. B. | 2036 | Henshaw, George |
| 1945 | Williams, William | 1991 | Irwin, Andrew | 2037 | Cooper, James |
| 1946 | Amos, Isaac | 1992 | Brown, George | 2038 | Roberts, David |
| 1947 | Allen, John | 1993 | Kaller, Joseph | 2039 | Lynch, Elisha |
| 1948 | Green, John | 1994 | Wilson, John | 2040 | Stevens, Joseph |
| 1949 | Davis, Daniel | 1995 | Butler, William | 2041 | Williams, Abraham |
| 1950 | Deverter, William | 1996 | Groves, Pierce | 2042 | Benson, Leven |
| 1951 | Lent, Joseph | 1997 | Simons, John | 2043 | Rumlet, Ebenezer |
| 1952 | Morris, Louis | 1998 | Paul, Jonathan | 2044 | Clark, William |
| 1953 | Merle, John | 1999 | Horsey, J. H. | 2045 | Benson, James |
| 1954 | Martin, Henry | 2000 | Cushman, Orson | 2046 | Rowe, Richard |
| 1955 | Sparrow, John | 2001 | Dodge, Joseph | 2047 | Muckleroy, Samuel |
| 1956 | Hutchinson, Townsend | 2002 | Doer, James | 2048 | Pearce, Thomas |
| 1957 | Thomas, Francis | 2003 | Driver, Thomas | 2049 | Reed, James |
| 1958 | Sutton, Prince | 2004 | Fry, Peter | 2050 | Davis, William |
| 1959 | Haydon, William | 2005 | Wrigth, Thomas | 2051 | Hughes, John |
| 1960 | McGeorge, William | 2006 | Johnson, John | 2052 | Lee, Joseph |
| 1961 | Payne, James | 2007 | Jerralls, Abraham | 2053 | Babcock, Charles |
| 1962 | Nolton, John | 2008 | Newell, George | 2054 | Chappel, Edward |
| 1963 | Clark, Elisha | 2009 | Bourns, John | 2055 | Pendergrast, Morris |
| 1964 | Kinsley, Benjamin | 2010 | Lewis, Gabriel | 2056 | Rae, William Iron |
| 1965 | Willett, Robert | 2011 | Stevenson, George | 2057 | Thomas, Henry |
| 1966 | Huntress, Robert | 2012 | Ingram, John | 2058 | Burrell, Ryal |
| 1967 | Brown, William | 2013 | Isdale, James | 2059 | Veney, George |
| 1968 | Thompson, James | 2014 | Stinchcombe, George | 2060 | Patten, John |
| 1969 | Rape, Nicholas | 2015 | Posey, Valentine | 2061 | Parsons, Rufus |
| 1970 | Kennedy, Peter | 2016 | Slebar, Samuel | 2062 | Phillips, Thomas |
| 1971 | Cleveland, Davis | 2017 | Briant, Moses | 2063 | Gould, John |
| 1972 | Wheeler, William | 2018 | Francis, John | 2064 | Hibben, Jeremiah |
| 1973 | Ripton, Solomon | 2019 | Woods, Merril | 2065 | Adivoe, Henry |
| 1974 | Thompson, Courtney | 2020 | Gilbert, Thomas | 2066 | Meker, James |
| 1975 | Greenleaf, Thomas | 2021 | Reid, Thomas | 2067 | Church, Jeremiah |
| 1976 | Johnson, James | 2022 | Gunn, Charles William | 2068 | Tomkins, Ephraim |
| 1977 | Johnson, Jacob | 2023 | Prince, Wiliam | 2069 | Carpenter, Nathaniel |
| 1978 | Larrabee, William | 2024 | Rossuros, Philip | 2070 | Hofslider, Jessy |

American Prisoners of War Held at Dartmoor during the War of 1812

## Numeric listing by prison number

| | | | | | |
|---|---|---|---|---|---|
| 2071 | Fogg, Noel | 2117 | Johnson, Joseph | 2163 | Clarke, Peter |
| 2072 | Lewis, Solomon | 2118 | Carr, Jonathan | 2164 | Frederick, Charles |
| 2073 | Williams, Richard | 2119 | Darran, Daniel | 2165 | Shorts, Richard |
| 2074 | Forster, Joseph | 2120 | Cain, John | 2166 | McCormick, Daniel |
| 2075 | Elisha, Thomas | 2121 | Canada, William | 2167 | Listrades, Joseph |
| 2076 | Ingersoll, Abraham | 2122 | Smith, James | 2168 | Stone, Samuel |
| 2077 | Briggs, Boileau | 2123 | Hoffman, James | 2169 | Warner, Stephen |
| 2078 | Mathias, Louis | 2124 | White, Henry | 2170 | Strouse, Martin |
| 2079 | Christie, John | 2125 | Pearson, Benjamin | 2171 | Bann, John |
| 2080 | Taylor, Peter | 2126 | Prince, Jeffery | 2172 | Fackney, John |
| 2081 | Moore, John | 2127 | McCormic, William | 2173 | Grough, Jacob |
| 2082 | Peckham, H. | 2128 | Fargo, Elijah | 2174 | Shaw, Thomas |
| 2083 | Allen, William | 2129 | Taggart, John | 2175 | Peck, Thomas |
| 2084 | Richards, John | 2130 | Jacobson, William | 2176 | West, Samuel |
| 2085 | Follingsby, William | 2131 | Stewart, James | 2177 | Allerd, Erick |
| 2086 | Brown, Seth | 2132 | Twycross, Samuel | 2178 | Smith, John |
| 2087 | Huntley, Charles | 2133 | Calentin, Samuel | 2179 | Verdon, Louis |
| 2088 | Cole, William | 2134 | Edmonds, Francis | 2180 | Parker, Colby |
| 2089 | Kellogg, Amos | 2135 | Saul, Francis | 2181 | Williams, John |
| 2090 | Taylor, James | 2136 | Williams, Joseph | 2182 | Gordon, Philip |
| 2091 | Wilson, Thomas | 2137 | Freeborn, Joseph | 2183 | Gorham, Emanuel |
| 2092 | Palmer, William | 2138 | Macure, Angelo | 2184 | Jewitt, Samuel |
| 2093 | Steward, John | 2139 | Williams, Jonathan | 2185 | Ringe, John |
| 2094 | West, George | 2140 | Cummins, Edward | 2186 | Giddens, John |
| 2095 | Harris, William | 2141 | Young, Moses | 2187 | Caban, Samuel |
| 2096 | Bean, William | 2142 | Magrath, Samuel | 2188 | Soderbury, John |
| 2097 | Johnson, Thomas | 2143 | Armstrong, Elisha | 2189 | Greenleaf, William |
| 2098 | Richardson, Edward | 2144 | Knight, Elisha | 2190 | Gramber, Gustoff |
| 2099 | Avery, Charles | 2145 | Stanwood, Timothy | 2191 | Slater, Ludowick |
| 2100 | Warren, Nathaniel W. | 2146 | Nichols, John | 2192 | Cooke, John |
| 2101 | Kemp, James | 2147 | Golliver, William | 2193 | Reid, Thomas |
| 2102 | Gilbert, George | 2148 | Wittlebanks, Edward | 2194 | Ryder, George |
| 2103 | Spear, Joseph | 2149 | Dunn, Henry G. | 2195 | Brown, Thomas |
| 2104 | Lowdie, Samuel | 2150 | Hogan, William | 2196 | Peterson, William |
| 2105 | Smith, John | 2151 | Thimonier, Pr. | 2197 | Thornton, Benjamin |
| 2106 | Nicholas, John | 2152 | Williams, Darius | 2198 | Ashby, William |
| 2107 | Ward, Edenezer | 2153 | Shadwick, John | 2199 | West, William |
| 2108 | Hammond, William | 2154 | Atkinson, Henry | 2200 | Sutton, Richard |
| 2109 | Hill, Thomas | 2155 | Main, Charles | 2201 | Coffin, Frederick |
| 2110 | Conway, James | 2156 | Smith, Christopher | 2202 | Penn, John |
| 2111 | Turner, William | 2157 | Voustyten, Christopher | 2203 | Loveland, Daniel |
| 2112 | Henman, Elisha | 2158 | Mordaunt, John | 2204 | Minks, Jacob |
| 2113 | Robinet, Samuel | 2159 | Dalmark, William | 2205 | Armstrong, George |
| 2114 | Green, James | 2160 | Baker, Benjamin | 2206 | Thomas, William |
| 2115 | Ballard, John | 2161 | Abbert, Solomon | 2207 | Leech, William |
| 2116 | Wessels, William | 2162 | Brannen, Alexander | 2208 | Clark, William |

American Prisoners of War Held at Dartmoor during the War of 1812

## Numeric listing by prison number

| | | | | | |
|---|---|---|---|---|---|
| 2209 | Argent, William | 2255 | Douglass, Dover | 2301 | Mathews, Pedro |
| 2210 | Long, Charles | 2256 | Hooke, George | 2302 | Barker, John |
| 2211 | Lester, Nathaniel R. | 2257 | McLean, Thomas | 2303 | Denyer, Richard |
| 2212 | Tucker, Zebulon | 2258 | Griffin, John | 2304 | Desharn, George |
| 2213 | Holmes, Robert | 2259 | Markins, John | 2305 | White, Clement |
| 2214 | Webster, George | 2260 | Barnes, Henry | 2306 | Trumills, John |
| 2215 | Byers, Sharp | 2261 | Cogwell, James | 2307 | Jones, William |
| 2216 | Bennett, Stephen | 2262 | Turrell, Ebenezer | 2308 | Wade, John |
| 2217 | Karnes, James | 2263 | Pellow, William | 2309 | Dickinson, Francis |
| 2218 | Adams, James | 2264 | Downing, Benjamin | 2310 | Barrett, George |
| 2219 | Raven, James | 2265 | Henderson, James | 2311 | Haines, Daniel |
| 2220 | McKray, John W. | 2266 | Griffin, William | 2312 | Beans, James |
| 2221 | Williams, John | 2267 | Sutton, Benjamin | 2313 | Smith, Charles |
| 2222 | Tar, Zedediah | 2268 | Adams, Samuel | 2314 | Gould, Samuel |
| 2223 | Chapel, John | 2269 | Waite, Joshua | 2315 | Lewis, George |
| 2224 | Rowe, Cornelius | 2270 | Seaman, Francis | 2316 | Turner, Thomas |
| 2225 | Anderson, Jacob | 2271 | Hamilton, Elijah | 2317 | Dunham, John |
| 2226 | Warrimore, Philip | 2272 | Goodall, William | 2318 | Collins, Thomas |
| 2227 | Warrimore, Henry | 2273 | Rauzelin, Jacob | 2319 | Vansykle, Ralph |
| 2228 | McEvoy, John | 2274 | Downey, Richard | 2320 | Johnson, Frederick |
| 2229 | Hanson, Henry | 2275 | Dempsey, John | 2321 | Doyle, James |
| 2230 | Rusto, Peter | 2276 | Porston, Joseph | 2322 | Badger, Peter |
| 2231 | Lackman, Isaac | 2277 | Page, Reuben | 2323 | Grimes, William |
| 2232 | Quincy, Benjamin | 2278 | Shurman, Thomas | 2324 | Dandy, Hamilton |
| 2233 | Springle, James | 2279 | Burns, Alexander | 2325 | Graham, Mourice |
| 2234 | Haley, John | 2280 | Flint, John | 2326 | Morgan, Jacob |
| 2235 | Moses, John | 2281 | Millet, Joseph | 2327 | Saunders, Jospeh |
| 2236 | Williams, Robert | 2282 | Cliff, Peter | 2328 | Miller, Isaac |
| 2237 | Pendleton, Charles | 2283 | Harrison, James | 2329 | Oliver, Samuel C. |
| 2238 | Hupton, Henry | 2284 | Downey, Charles | 2330 | Smith, James |
| 2239 | Atwood, P. | 2285 | Cowley, Samuel | 2331 | Fetch, Theodore |
| 2240 | Larkin, Lewis | 2286 | Ronning, Mathew | 2332 | Atwood, John |
| 2241 | Martin, Henry | 2287 | Catona, Manuel | 2333 | Piner, Thomas |
| 2242 | Armstrong, Andrew | 2288 | Pierce, Moses | 2334 | Colhoun, William |
| 2243 | Ainsley, William | 2289 | Westcott, Charles | 2335 | Dizere, John |
| 2244 | Parker, Thomas | 2290 | Dunbar, William | 2336 | Mazely, William |
| 2245 | Snade, William | 2291 | Chambers, Jospeh | 2337 | Hosson, John |
| 2246 | Bartrom, Lewis | 2292 | Myers, Edward | 2338 | Johnson, Thomas |
| 2247 | Hamilton, John | 2293 | Cooper, Ezekiel | 2339 | Miller, Thomas |
| 2248 | Norbury, Joseph | 2294 | Keitch, Jonah | 2340 | Stone, William |
| 2249 | Neptune, Daniel | 2295 | Oliver, Stephen | 2341 | Thomas, Allan |
| 2250 | Dorr, Edward | 2296 | Dunn, Charles | 2342 | Watts, William |
| 2251 | Clark, John | 2297 | Seaberry, John | 2343 | Alexander, Joseph |
| 2252 | McDonald, John | 2298 | Boys, Hugh | 2344 | Connor, Fenton |
| 2253 | Wilson, John | 2299 | Pidgeon, Jacques | 2345 | Dill, William |
| 2254 | Columbia, John | 2300 | Francis, John | 2346 | Gribble, George |

American Prisoners of War Held at Dartmoor during the War of 1812

## Numeric listing by prison number

| | | | | | |
|---|---|---|---|---|---|
| 2347 | Frobus, Henry | 2393 | Rogers, Francis | 2439 | Delong, Zaba |
| 2348 | Howland, Solomon | 2394 | Clark, John | 2440 | Johnson, John |
| 2349 | Golf, John | 2395 | Murray, Alexander | 2441 | Bond, William |
| 2350 | Boziman, Ralph | 2396 | Andrews, Asa | 2442 | Wallis, John |
| 2351 | Miller, Richard | 2397 | Dalton, Fred W. | 2443 | Kales, John |
| 2352 | Budwell, Alpheus | 2398 | Dominigue, Jospeh | 2444 | Christie, Robert |
| 2353 | Turner, George | 2399 | Felt, John | 2445 | Truffe, John |
| 2354 | Tulford, Joseph | 2400 | Pedrick, George | 2446 | Stevens, Leman |
| 2355 | McFarlane, Robert | 2401 | Hammond, Benjamin | 2447 | Stetson, Mathew |
| 2356 | Taylor, Alexander | 2402 | Horton, G. A. | 2448 | Morris, Manuel |
| 2357 | Johnson, John | 2403 | Rogers, Nathaniel | 2449 | Walker, Peter |
| 2358 | Fletcher, Henry | 2404 | Curtis, Stacey | 2450 | Ramsey, Samuel |
| 2359 | Fisher, James | 2405 | Logan, James | 2451 | Briennis, John |
| 2360 | Gibley, Thomas | 2406 | Holden, John | 2452 | Wing, Judah |
| 2361 | Studley, Simon | 2407 | Collett, John | 2453 | Lyman, Saul |
| 2362 | Cox, Caleb | 2408 | Benson, Joseph | 2454 | Colhoun, Joseph |
| 2363 | Secundie, James | 2409 | Keeting, John | 2455 | Sargeant, Charles |
| 2364 | Dyer, Israel | 2410 | Carroll, Michael | 2456 | Peterson, James |
| 2365 | Connor, Joseph | 2411 | Fields, George | 2457 | Smith, Isaac |
| 2366 | Petion, Samuel | 2412 | Pitman, John | 2458 | Elvin, John |
| 2367 | Clarke, Simon | 2413 | Ricker, Edward | 2459 | Marthoup, John |
| 2368 | Graham, W. R. | 2414 | Edwards, John | 2460 | Denny, James |
| 2369 | Place, Aaron | 2415 | Parsons, Andrew | 2461 | Sherlock, Charles |
| 2370 | White, John | 2416 | Lear, Alexander | 2462 | Johnson, John |
| 2371 | Arthur, John | 2417 | Boss, Charles | 2463 | Haskins, Joseph |
| 2372 | Jordan, Christopher | 2418 | Bella, Darius | 2464 | Small, William |
| 2373 | Dorman, John | 2419 | Osborne, Henry | 2465 | Anderson, Edward |
| 2374 | Foster, James | 2420 | Hoff, Charles | 2466 | Chapple, John |
| 2375 | Briggs, John | 2421 | Harris, David | 2467 | Holmes, Christopher |
| 2376 | Blesdale, Jacob | 2422 | Coffin, George | 2468 | Gregory, Joseph |
| 2377 | Williamson, Charles | 2423 | Brown, Edward | 2469 | Fraste, Samuel |
| 2378 | Hill, James A. | 2424 | Field, Samuel | 2470 | Parker, John |
| 2379 | Johnson, John | 2425 | Appleton, Daniel | 2471 | Handell, Henry |
| 2380 | Spencer, Robert | 2426 | Justice, John | 2472 | Snowdon, John |
| 2381 | Smith, John | 2427 | Dempston, Daniel | 2473 | Hart, George |
| 2382 | Mariner, Joseph | 2428 | Campbell, James | 2474 | Ronse, Joseph |
| 2383 | Gerard, Henry | 2429 | Meek, Thomas | 2475 | Miller, Henry |
| 2384 | Jamieson, Daniel | 2430 | Burgess, John | 2476 | Dobson, Joseph |
| 2385 | Johnson, James | 2431 | Scribble, Henry | 2477 | Hall, Charles |
| 2386 | Bell, George | 2432 | Smith, Samuel | 2478 | Williams, John |
| 2387 | Reynolds, W. B. | 2433 | Mason, Richard | 2479 | Bostell, James |
| 2388 | Nye, Stephen | 2434 | Dunn, Henry | 2480 | Brown, William |
| 2389 | Dickenson, Henry | 2435 | Taylor, George | 2481 | Davis, Hamilton |
| 2390 | Harvey, Edward | 2436 | Chambers, William | 2482 | Nivers, James |
| 2391 | Burrell, Michael | 2437 | Atkins, William | 2483 | Rogers, Asa |
| 2392 | Price, John | 2438 | Brown, William | 2484 | Butler, Benjamin |

American Prisoners of War Held at Dartmoor during the War of 1812

## Numeric listing by prison number

| | | | | | |
|---|---|---|---|---|---|
| 2485 | Blair, John | 2531 | Morrison, Samuel | 2577 | Ellis, John |
| 2486 | Williams, Thomas | 2532 | Baker, Basil | 2578 | Ray, Charles |
| 2487 | Doddard, John | 2533 | Robinson, Wililam | 2579 | Burke, David |
| 2488 | Abbott, Benjamin | 2534 | Perry, John | 2580 | Chine, Samuel |
| 2489 | Brazel, James | 2535 | Douglass, John | 2581 | Carney, William |
| 2490 | McManus, Michael | 2536 | Campbell, Thomas | 2582 | Headley, John |
| 2491 | Fowler, Isaac | 2537 | Lewis, Thomas | 2583 | Spratt, Thomas |
| 2492 | Randolph, George | 2538 | Irvine, Andrew | 2584 | Osgood, David |
| 2493 | Rawl, Elias B. | 2539 | Welch, Samuel | 2585 | Church, Richard |
| 2494 | Wilson, James | 2540 | Allen, Edward | 2586 | Lister, Louis |
| 2495 | Norton, Edward | 2541 | Gayler, James | 2587 | Marshall, Francis |
| 2496 | Richardson, John | 2542 | Jackson, Thomas | 2588 | Shipley, Charles |
| 2497 | Depero, Henry | 2543 | Day, James | 2589 | Syeway, Peter |
| 2498 | Hadley, James | 2544 | Jenkins, Samuel | 2590 | Johnson, John |
| 2499 | Robins, Philip | 2545 | Griffee, Thomas | 2591 | Williamson, John |
| 2500 | Mangin, John B. | 2546 | Forbes, James | 2592 | Tilbrook, Barney |
| 2501 | Tolver, Joseph | 2547 | Middleton, John W. | 2593 | Mason, John |
| 2502 | Spellard, Robert | 2548 | Frenair, Domingo | 2594 | Burnham, John |
| 2503 | Frasher, Daniel | 2549 | Dawes, W. B. | 2595 | Delabar, Joseph |
| 2504 | Harris, John | 2550 | Jarey, Barthomlew | 2596 | Howman, John |
| 2505 | Porter, Samuel | 2551 | White, George | 2597 | Mitchell, James M. |
| 2506 | Davis, Elisha | 2552 | Lambert, William | 2598 | Blagcon, Stephen |
| 2507 | Ariskins, Samuel | 2553 | Duffy, James | 2599 | Palmer, Peter |
| 2508 | Stanley, Redmond | 2554 | James, Benjamin | 2600 | Cooper, Alfred |
| 2509 | Hall, Lewis | 2555 | Lock, James | 2601 | Cain, Enos |
| 2510 | Hendrickson, T. | 2556 | Kelly, Henry | 2602 | Kanada, Joseph |
| 2511 | Hannon, Alexander | 2557 | Mills, Joseph | 2603 | Pearce, William |
| 2512 | Glover, Allen | 2558 | Jones, Reuben | 2604 | Lapham, Cushman |
| 2513 | Jones, James | 2559 | Williams, John | 2605 | Dennison, Laurence |
| 2514 | Sullivam, P. | 2560 | Brown, Charles | 2606 | Benjamin, Joseph |
| 2515 | Harrison, John | 2561 | Moffatt, David | 2607 | Anderson, Joseph |
| 2516 | Stanbeck, Jacob | 2562 | Solden, Chaarles | 2608 | Carleau, Daniel |
| 2517 | Pendleton, Samuel | 2563 | Glashon, Richard | 2609 | Baddington, Asa |
| 2518 | Porter, Joseph | 2564 | Sharp, John | 2610 | Roberts, Robert |
| 2519 | Williams, John | 2565 | Towson, William | 2611 | Brantre, James |
| 2520 | Crosbeck, Nathaniel | 2566 | Philman, Adam | 2612 | Turnbull, James |
| 2521 | Bailey, Aaron | 2567 | Ewing, Thomas | 2613 | Johnson, Oliver |
| 2522 | Dyer, Samuel | 2568 | Ferris, Jonathan | 2614 | Staggs, Henry |
| 2523 | Mason, William | 2569 | Coleby, Ebenezer | 2615 | Thompson, Joseph |
| 2524 | Folwett, Jospeh | 2570 | Pane, William | 2616 | Marks, Peter |
| 2525 | Nye, William | 2571 | Cooper, John | 2617 | Clark, Amos |
| 2526 | Wright, Daniel | 2572 | Curtis, Henry | 2618 | Howman, Jonas |
| 2527 | Briggs, Samuel | 2573 | Virgin, Robert | 2619 | Meinier, Benjamin |
| 2528 | Anderson, William | 2574 | Tripp, Adam | 2620 | Smith, Peter |
| 2529 | Miller, Jonathan | 2575 | Boisseau, Joseph | 2621 | Colquhoun, William |
| 2530 | Butler, Benjamin | 2576 | Hough, Samuel | 2622 | Cromwell, Glacio |

American Prisoners of War Held at Dartmoor during the War of 1812

## Numeric listing by prison number

| | | | | | |
|---|---|---|---|---|---|
| 2623 | Thomas, Thomas | 2669 | Martin, William | 2715 | Harris, James |
| 2624 | Connelly, James | 2670 | Garner, William | 2716 | Weston, William |
| 2625 | Thompson, William | 2671 | Wilson, John | 2717 | Williams, James |
| 2626 | Clay, John | 2672 | Allen, Joseph | 2718 | Lee, Edward |
| 2627 | Merckett, John | 2673 | Davis, Moses | 2719 | Haywood, John |
| 2628 | Miller, William | 2674 | Bartlett, John | 2720 | Edwards, Isaac |
| 2629 | Levin, Alexander | 2675 | Parsons, Daniel | 2721 | Howell, John |
| 2630 | Smith, John | 2676 | Woodbury, John | 2722 | Smith, W. L. |
| 2631 | Richardson, John | 2677 | Doliver, Joseph | 2723 | Lemon, N. C. |
| 2632 | Walden, James | 2678 | Thomas, Spencer | 2724 | Cooper, Thomas |
| 2633 | Brown, Mark | 2679 | Rodgers, George | 2725 | Ross, Isaac |
| 2634 | White, Charles | 2680 | Cove, John | 2726 | Marlborough, Francis |
| 2635 | Porter, Samuel | 2681 | Mitchell, Carr | 2727 | Webb, John |
| 2636 | Williams, Thomas | 2682 | Wing, Nathaniel | 2728 | Shaw, Edward |
| 2637 | Morrison, William | 2683 | Williams, James | 2729 | Clark, John D. |
| 2638 | Story, William | 2684 | Boggs, James | 2730 | Mellim, George |
| 2639 | Rodgers, Samuel | 2685 | Shepherd, James | 2731 | Davis, Daniel |
| 2640 | LeBaron, Peter | 2686 | Hall, John | 2732 | Pendleton, Evan |
| 2641 | Mayo, Nathaniel | 2687 | Jackson, William | 2733 | Beverly, Henry |
| 2642 | Williams, George | 2688 | McUmber, Jacob | 2734 | Prichard, William |
| 2643 | Stewart, John | 2689 | Tomlinson, G. W. | 2735 | Caroline, Tobias |
| 2644 | Brown, James | 2690 | Johnson, Jacob | 2736 | Williams, Joseph |
| 2645 | Hansey, Peter | 2691 | Williams, John | 2737 | Gunnell, William |
| 2646 | McIver, John | 2692 | Forster, Thomas | 2738 | Mold, John |
| 2647 | Campbell, James | 2693 | Brown, Thomas | 2739 | Raymond, George |
| 2648 | Taylor, Joseph | 2694 | Brush, Thomas | 2740 | Chult, David |
| 2649 | Simson, John | 2695 | Lee, John | 2741 | Dawson, John |
| 2650 | Wickham, John | 2696 | Read, William | 2742 | Wickwall, Joseph |
| 2651 | Rodgers, Ephraim | 2697 | Melvin, John | 2743 | Perkins, Henry |
| 2652 | Sandor, James | 2698 | Heaton, William | 2744 | Fellebrown, William |
| 2653 | Wilson, William | 2699 | Shilling, Morris | 2745 | Butler, George |
| 2654 | Eastlake, James | 2700 | Day, Thomas | 2746 | Brightman, Joseph |
| 2655 | McIntire, William | 2701 | Sutton, John | 2747 | Pierce, Edward |
| 2656 | Lindsay, Nathaniel | 2702 | Angel, Silvester | 2748 | Sanderson, William |
| 2657 | Lee, Nathaniel | 2703 | Read, William | 2749 | Paul, Jacob |
| 2658 | Hitchcock, Moses | 2704 | McInley, James | 2750 | Wedgwood, William |
| 2659 | Grush, Joseph | 2705 | Brown, James | 2751 | Tarlton, J. P. |
| 2660 | Davis, Andrew | 2706 | Woolridge, William | 2752 | Coombs, John |
| 2661 | Reynolds, Stephen | 2707 | Fingall, William | 2753 | Kent, James |
| 2662 | Wilson, Henry | 2708 | Watts, John | 2754 | Watkins, George |
| 2663 | Jennings, Luther | 2709 | Jeffreys, Henry | 2755 | Stacey, Perry |
| 2664 | Mason, John | 2710 | Wiley, David | 2756 | Albro, George |
| 2665 | Young, Ebenezer | 2711 | Grey, Thomas | 2757 | Burton, William |
| 2666 | Northcote, George | 2712 | Banta, John | 2758 | Moulden, William |
| 2667 | Norton, Richard | 2713 | Flood, John | 2759 | Wickham, Thaddeus |
| 2668 | Slocum, William | 2714 | Jackson, William | 2760 | Johnson, Edward |

American Prisoners of War Held at Dartmoor during the War of 1812

## Numeric listing by prison number

| | | | | | |
|---|---|---|---|---|---|
| 2761 | Trask, William | 2807 | Thompson, Nathaniel | 2853 | Chip, Charles |
| 2762 | Malone, James | 2808 | Vangorbet, Cato | 2854 | Potter, John |
| 2763 | Clifford, L. L. | 2809 | Ingerson, James B. | 2855 | Walton, Christopher |
| 2764 | Morell, John | 2810 | Moses, James | 2856 | Shaw, John |
| 2765 | Sims, Clement | 2811 | Carlton, William N. | 2857 | Gordan, William |
| 2766 | Thomas, Charles | 2812 | Codshell, Joseph | 2858 | Frazier, James |
| 2767 | Brown, Elisha | 2813 | Davis, Joseph | 2859 | Welch, John |
| 2768 | Hall, William | 2814 | Moore, John | 2860 | Donaldson, Joseph |
| 2769 | Pinkham, Allen | 2815 | Smith, John | 2861 | Bishop, Edward |
| 2770 | Gardner, Jonathan | 2816 | McMiller, Andrew | 2862 | Lopans, William |
| 2771 | Forrest, James | 2817 | Rightman, John | 2863 | Bartis, John |
| 2772 | Gooley, James | 2818 | Stibbins, Thomas | 2864 | Chase, Nathaniel |
| 2773 | Anderson, James | 2819 | Brown, Samuel | 2865 | Atwood, Thomas |
| 2774 | Tyler, Louis | 2820 | Johnson, Mathew | 2866 | Hutson, Peter |
| 2775 | Plowman, Joseph | 2821 | Butler, John | 2867 | Farmer, Joseph |
| 2776 | Spencer, Leonard | 2822 | Kelley, John | 2868 | Alexander, George |
| 2777 | Calder, J. H. | 2823 | Bird, James | 2869 | Penny, Richard |
| 2778 | Churchill, Henry | 2824 | Jeuvmeson, John | 2870 | Gordon, James |
| 2779 | Sulton, John | 2825 | Chase, Jacob | 2871 | Torgnet, Abel |
| 2780 | Mars, George | 2826 | Brown, Jacob | 2872 | Thomas, John |
| 2781 | Hill, Joseph | 2827 | Blumbhouser, Samuel | 2873 | Thomas, John |
| 2782 | Cadwell, James | 2828 | Williams, John | 2874 | Murray, James |
| 2783 | Nicholls, John | 2829 | Brown, Reuben | 2875 | Cook, John |
| 2784 | Sampson, Jacob | 2830 | Flood, John | 2876 | Kellum, Smith |
| 2785 | Hill, Ephraim | 2831 | Williams, John | 2877 | Young, Thomas |
| 2786 | Hubbard, Alfred | 2832 | Hill, Timothy | 2878 | Bordley, George |
| 2787 | Cole, William | 2833 | Grey, John | 2879 | McKenzie, John |
| 2788 | Beecher, William | 2834 | Walker, John | 2880 | Schanck, William |
| 2789 | Allen, John | 2835 | Warnick, Robert | 2881 | Lindsay, Samuel |
| 2790 | Gibbs, Daniel | 2836 | Augustus, Benjamin | 2882 | Holbrook, David |
| 2791 | Hull, Edward | 2837 | Elliott, Robert | 2883 | Brown, Thomas |
| 2792 | Vorhis, James | 2838 | Jackson, Sidney | 2884 | Blake, William |
| 2793 | Ashfield, Henry | 2839 | Emming, Thomas | 2885 | Jetson, Samuel |
| 2794 | Cappel, John | 2840 | Smith, Paul | 2886 | Mutch, James |
| 2795 | King, John | 2841 | Hawker, Edward | 2887 | Thompson, James |
| 2796 | Bailey, John | 2842 | Pool, John | 2888 | Latham, John |
| 2797 | Jeffs, Joseph | 2843 | Howland, William | 2889 | Scott, William |
| 2798 | Brill, John | 2844 | Handley, John | 2890 | Porter, Joseph |
| 2799 | Carter, John | 2845 | Jackson, William | 2891 | Smith, John |
| 2800 | Dunn, David | 2846 | Pardett, Charles | 2892 | Baker, Daniel |
| 2801 | Hussey, Thomas | 2847 | Johnson, Samuel | 2893 | Bailey, William |
| 2802 | Dennis, Thomas | 2848 | Hawley, Frederick | 2894 | Hutson, John |
| 2803 | Holmes, John | 2849 | Smith, John | 2895 | Stevenson, Levi |
| 2804 | Munster, Isaac | 2850 | Forrest, Wiliam | 2896 | Richards, Henry |
| 2805 | Allen, John D. | 2851 | Benjamin, P. | 2897 | Price, John |
| 2806 | Nasson, Joseph | 2852 | Smith, Thomas | 2898 | Murray, Peter |

American Prisoners of War Held at Dartmoor during the War of 1812

## Numeric listing by prison number

| | | | | | |
|---|---|---|---|---|---|
| 2899 | Bond, Samuel | 2945 | Davies, Nathaniel | 2991 | Pettingell, James |
| 2900 | Peterson, William | 2946 | Duncan, Jesse | 2992 | Newberry, John |
| 2901 | Brown, John | 2947 | Fitzgerald, John | 2993 | Johnson, Thomas |
| 2902 | Penrose, Abraham | 2948 | Garnder, Francis | 2994 | Buchannon, Robert |
| 2903 | White, Richard | 2949 | Robinson, Tony | 2995 | Ustick, William |
| 2904 | Filch, John | 2950 | Brush, Samuel | 2996 | Lupton, John |
| 2905 | Degget, Hamel | 2951 | Ruling, Arnold | 2997 | Frost, John |
| 2906 | Sparrow, Joseph | 2952 | Sline, Samuel | 2998 | Hampton, Thomas |
| 2907 | Westler, John | 2953 | Hunter, Alexander | 2999 | Moore, John |
| 2908 | Hatch, William | 2954 | Jackson, George | 3000 | Stirrell, George |
| 2909 | Legos, Philip | 2955 | Thomas, John | 3001 | Lowrie, William |
| 2910 | Hanson, Peter | 2956 | Hughes, Thomas | 3002 | Vanderburgh, Cornelius |
| 2911 | Smith, Esop | 2957 | Gladding, Joseph | 3003 | Brazier, Thomas |
| 2912 | Hamcomb, Thomas | 2958 | Page, Cato | 3004 | Colton, Joseph |
| 2913 | Jasmine, Paul | 2959 | Grindall, Joseph | 3005 | Van Horn, James |
| 2914 | Miller, George | 2960 | Bailey, Thomas | 3006 | Smith, John |
| 2915 | Jones, William | 2961 | Foxey, Thomas | 3007 | Downing, John |
| 2916 | Leonard, John | 2962 | Shute, Thomas | 3008 | Harris, Benjamin |
| 2917 | Chapley, John | 2963 | Miller, John | 3009 | Crowell, Seth |
| 2918 | Richardson, Samuel | 2964 | Richardson, John | 3010 | Mercey, Thomas |
| 2919 | Gannett, Mathew | 2965 | Peterson, John | 3011 | Williams, David |
| 2920 | Wilcox, Ephraim | 2966 | Shaw, John | 3012 | Ward, Samuel |
| 2921 | Vail, Jeremiah | 2967 | Robinson, Edward | 3013 | Grandison, William |
| 2922 | MacKenzie, George | 2968 | Smith, Thomas | 3014 | Cooper, John |
| 2923 | Cummings, James | 2969 | Housewife, Mathew | 3015 | McKeige, Denis |
| 2924 | Gordon, Abraham | 2970 | Badger, John | 3016 | Thomas, Andrew |
| 2925 | Saunders, Peter | 2971 | Swaine, Thomas | 3017 | Cairns, Thomas |
| 2926 | Cox, Abraham | 2972 | Smith, Charles | 3018 | Hatkins, Henry |
| 2927 | Vanderventer, John | 2973 | Clexton, R. | 3019 | Butler, Charles |
| 2928 | Nixon, Charles | 2974 | Andrews, Nathaniel | 3020 | Cole, Benjamin |
| 2929 | Taylor, William | 2975 | Parr, John | 3021 | Fowler, William |
| 2930 | Hall, James | 2976 | Shott, John | 3022 | Austin, Joseph |
| 2931 | Robinson, Henry | 2977 | Wills, Nathaniel | 3023 | McKinnon, Noel |
| 2932 | Greaves, John | 2978 | Murray, M. | 3024 | Elger, Richard |
| 2933 | Sims, William | 2979 | Woolridge, W. | 3025 | Lad, Daniel |
| 2934 | Wilson, James | 2980 | Hartsoln, John | 3026 | Dyke, Stuard |
| 2935 | Adams, Thomas | 2981 | Grey, Francis | 3027 | Dorsey, John |
| 2936 | Backman, Charles | 2982 | Pittman, Samuel | 3028 | Todd, Robert |
| 2937 | Thomas, Thomas | 2983 | McCarthy, Henry | 3029 | Bunker, Robert |
| 2938 | Johnson, Richard | 2984 | Jackson, George | 3030 | Brown, William |
| 2939 | Curtis, Enoch | 2985 | Holden, Charles | 3031 | Smallpiece, John |
| 2940 | Williams, James | 2986 | Richardson, Andrew | 3032 | James, Joseph |
| 2941 | Mann, Samuel | 2987 | Clements, Samuel | 3033 | Thompson, John |
| 2942 | Baker, Charles | 2988 | Leach, George | 3034 | Stanwood, James |
| 2943 | Gilchrist, Samuel | 2989 | Southwick, George | 3035 | Studdy, Richard |
| 2944 | Fletcher, B. | 2990 | Rowland, William | 3036 | Murray, William |

American Prisoners of War Held at Dartmoor during the War of 1812

## Numeric listing by prison number

| | | | | | |
|---|---|---|---|---|---|
| 3037 | Quinn, William | 3083 | Clark, Titus | 3129 | Laws, Peter |
| 3038 | Hankey, Frederick | 3084 | Huston, James | 3130 | Shomodia, John |
| 3039 | Charlotte, William | 3085 | Pratt, Lester | 3131 | Ely, Abraham |
| 3040 | Pain, Mark | 3086 | Sims, Joseph | 3132 | Miller, John |
| 3041 | Campbell, John | 3087 | Burrow, Charles | 3133 | Phillips, William |
| 3042 | Fowler, Isaac | 3088 | Crumpton, William | 3134 | Haywood, John |
| 3043 | Carroll, John | 3089 | Lee, Isaac | 3135 | Howell, John |
| 3044 | Smith, John | 3090 | Black, Philip | 3136 | Douglass, John |
| 3045 | Babcock, Daniel | 3091 | Francis, John | 3137 | Adams, Thomas |
| 3046 | Seymond, Abner | 3092 | Collins, George | 3138 | Eldridge, William |
| 3047 | Cornish, John | 3093 | Ferris, James | 3139 | Brown, David |
| 3048 | Williamson, John | 3094 | Williams, George F. | 3140 | Robinson, James |
| 3049 | Osbourne, Archibald | 3095 | Henry, Robert | 3141 | Williams, William |
| 3050 | Higginson, David | 3096 | Williams, John | 3142 | Deagle, James |
| 3051 | Johnson, John | 3097 | Parker, William | 3143 | Taylor, Thomas |
| 3052 | Gordon, John | 3098 | Hall, Daniel | 3144 | Ayres, Henry |
| 3053 | Stevens, Ira | 3099 | Hall, Hammon | 3145 | Ferguson, John |
| 3054 | Holding, Henry | 3100 | Pinkham, Jacob | 3146 | Jackson, Isaac |
| 3055 | Toumblin, John | 3101 | Thorndike, Hall | 3147 | Davis, George |
| 3056 | Jarvis, George | 3102 | Brown, Anthony | 3148 | Watson, James |
| 3057 | Chivers, Cantab | 3103 | Collins, John | 3149 | Huse, Ebenezer |
| 3058 | Lincoln, Solomon | 3104 | Cole, Perry | 3150 | Rodgers, Samuel |
| 3059 | Barrett, George | 3105 | Brown, James | 3151 | Chase, Nathaniel |
| 3060 | Putnam, A. | 3106 | Parr, Samuel | 3152 | Greenwood, Joseph |
| 3061 | Brown, Thomas | 3107 | Wilson, William | 3153 | Mason, Joseph J. |
| 3062 | Wheeler, Michael | 3108 | Grahl, Charles | 3154 | Swain, James |
| 3063 | Stilman, John | 3109 | Forsdick, Joseph H. | 3155 | Nicholls, Thomas |
| 3064 | Sloane, William | 3110 | Savage, John H. | 3156 | Bunker, Thomas |
| 3065 | Johnson, Samuel | 3111 | Packhouse, William | 3157 | Marsh, Hercules |
| 3066 | Neil, John | 3112 | Hull, Homer | 3158 | Theal, Bristo |
| 3067 | Mead, Ezekiel | 3113 | Langley, Richard | 3159 | Ostman, David |
| 3068 | Black, Charles | 3114 | King, John | 3160 | Williams, Stephen |
| 3069 | Beck, Willaim | 3115 | Williams, James | 3161 | Weeden, Richard |
| 3070 | Stitman, A. W. | 3116 | Whithead, Daniel | 3162 | Suckley, Richard |
| 3071 | Jardine, Samuel | 3117 | Navar, Francis | 3163 | Rodgers, Edward |
| 3072 | Jackson, Elias | 3118 | Hall, David | 3164 | Burdock, Enos |
| 3073 | Potter, Jacob | 3119 | Benson, Mingo | 3165 | Mack, John |
| 3074 | Perkins, William | 3120 | Morton, William | 3166 | Henry, Henry |
| 3075 | Hinton, John | 3121 | Hitchcock, Edward | 3167 | Hadley, George |
| 3076 | Thayer, Ebenezer | 3122 | Lovell, Robert | 3168 | Homer, John |
| 3077 | Hopkins, Samuel | 3123 | Naborne, Nicholas | 3169 | Stage, William |
| 3078 | Weeble, Robert | 3124 | Phillips, George W. | 3170 | Durval, N. D. |
| 3079 | Davis, James | 3125 | Dailey, George W. | 3171 | Lamb, Jack |
| 3080 | Paris, John | 3126 | Clements, William | 3172 | Elfs, James |
| 3081 | Smith, John | 3127 | Boillet, John | 3173 | Wise, John |
| 3082 | Middleton, Reuben | 3128 | Young, Nathaniel | 3174 | Sherrif, Benjamin P. |

American Prisoners of War Held at Dartmoor during the War of 1812

## Numeric listing by prison number

| | | | | | |
|---|---|---|---|---|---|
| 3175 | Branch, Anthony | 3221 | Walker, Benjamin | 3267 | Foster, Joseph |
| 3176 | Claussen, John | 3222 | Barry, Peter | 3268 | Robinson, John |
| 3177 | Younger, Louis | 3223 | Perry, William | 3269 | Holdridge, Hector |
| 3178 | Potter, John | 3224 | Hurd, Abel | 3270 | Stamford, James |
| 3179 | Quarterman, William | 3225 | Hoskins, John | 3271 | Mead, William |
| 3180 | Thaley, William | 3226 | Darrow, Aaron | 3272 | Roddy, W. B. |
| 3181 | Jameson, George | 3227 | McAlpin, Charles | 3273 | George W. W. |
| 3182 | Duncan, Edward | 3228 | Price, Charlton | 3274 | Pinder, George |
| 3183 | Baker, Robert | 3229 | Gibbs, Valentine | 3275 | Weston, Nathaniel |
| 3184 | Thompson, George | 3230 | Leonard, Robert | 3276 | Green, Charles |
| 3185 | Smith, John | 3231 | Coffin, James | 3277 | Ward, Benjamin |
| 3186 | Gardner, James | 3232 | Thornton, David | 3278 | Hill, Benjamin |
| 3187 | Sherridan, Henry | 3233 | Tucker, N. | 3279 | Hanfield, Enos |
| 3188 | Lucas, Martin | 3234 | Pitts, Charles | 3280 | Clark, William |
| 3189 | Thomas, Moses | 3235 | Peckham, Isaac | 3281 | Wanton, William |
| 3190 | Folger, Frederick | 3236 | Wilkey, Timothy | 3282 | Forbes, John |
| 3191 | Glover, Samuel | 3237 | Fleming, Alexander | 3283 | Sparkes, Thomas |
| 3192 | Watson, Daniel | 3238 | Harwood, William | 3284 | Phenny, John |
| 3193 | Richardson, Robert | 3239 | Boyd, Andrew | 3285 | Higgins, Asa |
| 3194 | Smith, Henry | 3240 | Williams, Thomas | 3286 | Lawrence, George |
| 3195 | Bates, Joseph | 3241 | McKensie, William | 3287 | Snow, Thomas |
| 3196 | Davis, Osborne | 3242 | Boussell, Samuel | 3288 | Cloutman, Joseph |
| 3197 | Heaton, Henry | 3243 | Noel, Stephen C. | 3289 | Widger, John |
| 3198 | Fool, Garret | 3244 | Johnson, Henry | 3290 | Melzard, Peter |
| 3199 | Robinson, John | 3245 | Frazier, William | 3291 | Fuller, Nathaniel |
| 3200 | Anders, John | 3246 | Brown, Francis | 3292 | Clothy, John |
| 3201 | Wilson, Robert | 3247 | Jackson, John | 3293 | Russell, Robert |
| 3202 | Green, George | 3248 | Butler, Henry | 3294 | Williams, Frederick |
| 3203 | Valentine, John | 3249 | Carles, John | 3295 | Goss, Jesse |
| 3204 | Anders, John | 3250 | McKinnon, John | 3296 | Clothy, William |
| 3205 | Perry, Samuel | 3251 | Kirkpatrick, William | 3297 | Pettingall, Joseph |
| 3206 | Mayers, James | 3252 | Hall, Perry | 3298 | Tarlton, John |
| 3207 | Wright, Edward | 3253 | Lindsey, William | 3299 | McKinney, Isaac |
| 3208 | Fry, Thomas | 3254 | Wood, Joseph | 3300 | Varney, John |
| 3209 | Garder, George | 3255 | Witham, Burrell | 3301 | Moore, Samuel |
| 3210 | Pinne, John | 3256 | Brown, Sawyer | 3302 | Waterhouse, Moses |
| 3211 | Stinton, James | 3257 | Pith, William | 3303 | Francis, Abraham |
| 3212 | McKennie, Barney | 3258 | Conner, John | 3304 | Lucas, Daniel |
| 3213 | Yeaton, James | 3259 | Holding, Nathaniel | 3305 | Andrews, James |
| 3214 | Phillips, Benjamin | 3260 | Kingman, Charles | 3306 | Thompson, John |
| 3215 | Rankin, William | 3261 | Dissmore, Abraham | 3307 | Card, John |
| 3216 | Ross, John | 3262 | Dexter, Charles | 3308 | Holbrook, Robert |
| 3217 | Cottrell, Henry | 3263 | Dexter, George W. | 3309 | Pitman, Henry |
| 3218 | Redman, David | 3264 | Smith, Kilby | 3310 | Marshall, John |
| 3219 | Ranlot, John | 3265 | Dunham, Daniel | 3311 | McIntire, Petty |
| 3220 | Brown, Thomas | 3266 | Parker, Robert | 3312 | Cross, Ephraim |

American Prisoners of War Held at Dartmoor during the War of 1812

## Numeric listing by prison number

| | | | | | |
|---|---|---|---|---|---|
| 3313 | Brown, Robert | 3359 | Williams, Robert | 3405 | Knight, Daniel |
| 3314 | Ferguson, Thomas | 3360 | Bale, Charles | 3406 | Johnson, Thomas |
| 3315 | Hunter, James | 3361 | Horsfalt, William | 3407 | Tripper, Robert |
| 3316 | Forsyth, Robert | 3362 | Weeks, Louis | 3408 | Jones, James |
| 3317 | Cooper, Edward | 3363 | Ward, Peter | 3409 | Williams, Joseph |
| 3318 | Spouding, Joseph | 3364 | Bone, Charles | 3410 | Yeaton, Charles |
| 3319 | Haddart, Robert | 3365 | Trask, Osborne | 3411 | Harbrook, Richard |
| 3320 | Hamilton, George W. | 3366 | Cullett, William | 3412 | Smith, Henry |
| 3321 | Jessamin, John | 3367 | Stevens, James | 3413 | Marble, Samuel |
| 3322 | David, John | 3368 | Swain, David | 3414 | McGill, Robert |
| 3323 | Blake, Charles | 3369 | Dole, Henry | 3415 | Bailey, Peter |
| 3324 | Goulding, Samuel | 3370 | Shaw, Andrew | 3416 | Mallard, James |
| 3325 | Gilbert, Isaac | 3371 | Thompson, John | 3417 | Brown, Thomas |
| 3326 | Wilson, James | 3372 | Palmer, George H. | 3418 | Sawyer, Peter |
| 3327 | Rodgers, William | 3373 | Maloon, Bryant | 3419 | Harvey, Joseph |
| 3328 | Eddy, Richard | 3374 | Bartlett, Robert | 3420 | Rice, Thomas |
| 3329 | Mackey, John | 3375 | Bowden, William | 3421 | Eaton, Israel |
| 3330 | Brown, Charles | 3376 | Widger, Joseph | 3422 | Tucker, Samuel |
| 3331 | Storm, Daniel | 3377 | Russell, William | 3423 | Burridge, Robert |
| 3332 | Brown, William | 3378 | Smith, Benjamin | 3424 | Inglas, John |
| 3333 | Cook, John | 3379 | Crocker, Silvester | 3425 | Gowalter, John |
| 3334 | Hawkins, Isaac | 3380 | Orne, Israel | 3426 | Bartell, William |
| 3335 | Everard, Joseph | 3381 | Bowden, William | 3427 | Russell, Louis |
| 3336 | Shorts, Samuel | 3382 | Garrett, L. P. | 3428 | Morris, Jacob |
| 3337 | Colborne, John L. | 3383 | Threshow, James | 3429 | Snow, James |
| 3338 | Gardner, John | 3384 | Roundy, Thomas | 3430 | Stacey, John |
| 3339 | Johnson, Richard | 3385 | Devereux, Benjamin | 3431 | Puffer, John |
| 3340 | Dunham, John | 3386 | Griffin, William | 3432 | Aldridge, Richard |
| 3341 | Cadwell, Abraham | 3387 | Lewis, Peter | 3433 | Hoper, Joseph |
| 3342 | Noble, Charles | 3388 | Rowe, Seth | 3434 | Jones, Thomas |
| 3343 | James, John | 3389 | Robinson, John | 3435 | Hutes, Cyrus |
| 3344 | Cook, Isaac | 3390 | Preston, John | 3436 | Newell, Paul |
| 3345 | Owen, Burden | 3391 | Wilson, James | 3437 | Lucas, Robert |
| 3346 | Boyd, John | 3392 | Thrasher, John | 3438 | Hunn, John |
| 3347 | Campbell, Nathaniel | 3393 | Weare, Ebenezer | 3439 | Smith, John |
| 3348 | Stone, Henry | 3394 | Todd, Samuel | 3440 | Clark, Samuel |
| 3349 | Lowe, Thomas | 3395 | Penn, William | 3441 | Austen, William |
| 3350 | Theyer, James | 3396 | Perkins, John | 3442 | Williams, Joseph |
| 3351 | Jordan, Artemas | 3397 | Billings, Richard | 3443 | Dyer, Thomas |
| 3352 | Powell, Joseph | 3398 | Pettigrene, William | 3444 | Newman, Henry |
| 3353 | Butts, Joseph | 3399 | Bailey, Daniel | 3445 | Monroe, James |
| 3354 | Williams, Robert | 3400 | Langford, Samuel | 3446 | Rolla, William |
| 3355 | Brown, Isaac | 3401 | Brown, Samuel | 3447 | Peake, J. W. |
| 3356 | Hayday, Larsay | 3402 | Beverley, Richard | 3448 | Laskey, Benjamin |
| 3357 | Brown, George | 3403 | Norton, Joseph | 3449 | Penny, James |
| 3358 | Osborne, Louis | 3404 | Davis, William | 3450 | Mitchell, William |

American Prisoners of War Held at Dartmoor during the War of 1812

## Numeric listing by prison number

| | | | | | |
|---|---|---|---|---|---|
| 3451 | Florance, Charles | 3497 | James, John | 3543 | Savage, Robert |
| 3452 | Pearson, Charles | 3498 | Smith, George | 3544 | Lovely, Placid |
| 3453 | Wiggent, John | 3499 | Weeding, Isaac | 3545 | Warner, Benjamin |
| 3454 | Watson, Robert | 3500 | Simes, Henry | 3546 | Marshall, Levy |
| 3455 | Coffin, Caleb | 3501 | Simes, Stephen | 3547 | Coleman, William |
| 3456 | Frizle, John | 3502 | Johnson, Samuel | 3548 | Bolding, Garret |
| 3457 | Smith, John | 3503 | Small, Joseph | 3549 | Brien, Lewis |
| 3458 | Brown, George | 3504 | Nixon, Primos | 3550 | Tooley, William |
| 3459 | Bateman, John | 3505 | Roach, Pedro | 3551 | Grist, William |
| 3460 | Fenhouse, John | 3506 | Schneider, John A. | 3552 | Paul, Daniel |
| 3461 | Joseph, John | 3507 | Rice, James | 3553 | Williams, George |
| 3462 | White, Charles | 3508 | Briggs, Benjamin | 3554 | Jervis, Isaac |
| 3463 | Prince, Lubin | 3509 | Mitchell, Henry | 3555 | Ringrove, William |
| 3464 | Ward, John | 3510 | Gray, Ephraim | 3556 | Brook, David |
| 3465 | Wilson, Samuel | 3511 | Cornish, Thomas | 3557 | Cain, Joseph |
| 3466 | Badcock, Bradley | 3512 | Legur, Ephaim | 3558 | Buckley, William |
| 3467 | Hall, Stephen | 3513 | Martin, Peter | 3559 | Jordan, Simon |
| 3468 | Howell, Thomas | 3514 | Johnson, George | 3560 | Eldon, James |
| 3469 | Perkington, John | 3515 | Hesty, John | 3561 | Eburn, Aaron |
| 3470 | Thompson, Henry | 3516 | Smith, William | 3562 | Davis, John |
| 3471 | Wanson, James | 3517 | Morton, Samuel | 3563 | Bill, Zachariah |
| 3472 | Anderson, Thomas | 3518 | Swain, Edward | 3564 | Whitehead, Reuben |
| 3473 | Jones, Saul | 3519 | Freeman, Nathaniel | 3565 | Sole, John |
| 3474 | Smith, Henry | 3520 | Troth, George | 3566 | Counthon, John |
| 3475 | Foster, James | 3521 | Spanow, W. T. | 3567 | Cowen, Thomas |
| 3476 | Lindsey, James | 3522 | Davenport, John | 3568 | Waterford, J. H. |
| 3477 | Wood, Thomas | 3523 | Martin, Philip | 3569 | Pannison, Henry |
| 3478 | Bingham, Little | 3524 | Simmons, Richard | 3570 | Wearney, Richard |
| 3479 | Peterson, John | 3525 | Thompson, John | 3571 | Harrison, Silas |
| 3480 | Martin, Andrew | 3526 | Hanson, Stephen | 3572 | Osborne, John |
| 3481 | Phillips, Burton | 3527 | Proctor, Henry | 3573 | McQueen, Charles |
| 3482 | Holbrook, Benjamin | 3528 | Thomas, Timothy | 3574 | Johnson, William |
| 3483 | Davis, John | 3529 | Meeds, Joseph | 3575 | Higgins, Francis |
| 3484 | Rameson, Thomas | 3530 | McDaniel, John | 3576 | Callow, William |
| 3485 | Nash, Daniel | 3531 | Sandy, W. P. | 3577 | Jones, David |
| 3486 | Sanders, Gabriel | 3532 | Doyle, John | 3578 | Stanley, Samuel |
| 3487 | Johnson, William | 3533 | Wilson, Thomas | 3579 | Smith, William |
| 3488 | Barker, Thomas | 3534 | Jones, Ezekiel | 3580 | Card, James |
| 3489 | Lemon, James | 3535 | Hanson, John | 3581 | Hutchins, John |
| 3490 | Cousins, Samuel | 3536 | Woodman, Thomas | 3582 | Burbage, Henry |
| 3491 | Sewell, Lot | 3537 | Days, Thomas | 3583 | Read, Thomas |
| 3492 | Smith, James | 3538 | Jones, Paul | 3584 | Ray, William |
| 3493 | Jones, John | 3539 | Laurent, Jonathan | 3585 | Berry, Anton |
| 3494 | Brown, Charles | 3540 | Elwell, Samuel | 3586 | Groves, George W. |
| 3495 | Harris, James | 3541 | Gilbert, Henry | 3587 | Michaels, John |
| 3496 | Williams, Peter | 3542 | Wilkins, William | 3588 | Peterson, Jacob |

American Prisoners of War Held at Dartmoor during the War of 1812

## Numeric listing by prison number

| | | | | | |
|---|---|---|---|---|---|
| 3589 | Chapman, James | 3635 | Davis, George | 3681 | Conway, Samuel |
| 3590 | Yorkman, Archibald | 3636 | Holmes, John | 3682 | Techand, John |
| 3591 | Hancock, James | 3637 | Curtis, Zebediah | 3683 | Pierce, John |
| 3592 | Odiorne, John | 3638 | Barnett, Thomas | 3684 | Barker, Joseph |
| 3593 | Swaine, Thomas | 3639 | Jefferson, Jacob | 3685 | Savary, Peter |
| 3594 | Stevens, Thomas | 3640 | Franics, Michael | 3686 | Carswell, John |
| 3595 | Heny, Daniel | 3641 | Titus, John | 3687 | Stacey, Osmond C. |
| 3596 | Simon, Mark | 3642 | Miller, Thomas | 3688 | Bessop, John |
| 3597 | Perkins, Jonathan | 3643 | Gordon, William | 3689 | Smith, Joseph |
| 3598 | Brown, Benjamin | 3644 | Jones, Charles | 3690 | Hammond, Edward |
| 3599 | Stone, Benjamin | 3645 | Jackson, William | 3691 | Vickery, Thomas |
| 3600 | Maddison, Alexander | 3646 | Miller, John | 3692 | Brown, William |
| 3601 | Jackson, David | 3647 | Allen, Charles | 3693 | Bumbelcomb, Sewel |
| 3602 | Savage, James | 3648 | Staples, William | 3694 | Preston, Samuel |
| 3603 | Vowdy, William | 3649 | Philbrook, Cyrus | 3695 | Cox, Daniel |
| 3604 | Ashton, William | 3650 | Mullin, Robert | 3696 | Woolridge, Robert |
| 3605 | Conner, Galen | 3651 | Miles, Thomas | 3697 | Roundy, Samuel |
| 3606 | Bradie, Thomas | 3652 | Wells, James | 3698 | Corbett, Michael |
| 3607 | Barker, A. L. | 3653 | Shields, John | 3699 | Proctor, John |
| 3608 | Johnson, James | 3654 | Grush, John | 3700 | Prichard, Robert |
| 3609 | Hooper, John | 3655 | Lewis, John | 3701 | Lane, Henry |
| 3610 | Clark, Peleg | 3656 | Thomas, John | 3702 | White, Kitt |
| 3611 | Johnson, Peter | 3657 | Vanvort, Richard | 3703 | Tucker, Edward |
| 3612 | Paine, Charles | 3658 | Taylor, William | 3704 | Nowland, Andrew |
| 3613 | Garrett, Robert | 3659 | Northey, Joseph | 3705 | Waite, Jacob |
| 3614 | Montcalm, Henry | 3660 | Lowman, David | 3706 | Prince, Prince |
| 3615 | Bennett, Peleg | 3661 | Ryan, Patrick | 3707 | Francis, Oliver |
| 3616 | James, John | 3662 | Phip, William | 3708 | Leech, Ezekiel |
| 3617 | Green, John | 3663 | Blackler, William | 3709 | Gilbert, David |
| 3618 | Prindall, Samuel | 3664 | Goss, Richard | 3710 | Storey, John |
| 3619 | Lachame, Thomas | 3665 | Graham, Benjamin | 3711 | Hawks, Benjamin |
| 3620 | Marshall, George | 3666 | Green, Joseph W. | 3712 | Fuller, Nathaniel |
| 3621 | Coles, Charles | 3667 | Richardson, William | 3713 | Pritchard, John |
| 3622 | Yarnell, William | 3668 | Broughton, John | 3714 | Cox, Peter J. |
| 3623 | Watkins, William | 3669 | Thomson, William | 3715 | Rogers, Ambrose |
| 3624 | Block, Edward | 3670 | Roff, Samuel | 3716 | Wadden, Isaac |
| 3625 | Peterson, James | 3671 | Bowden, Frederick | 3717 | Chapman, Daniel |
| 3626 | Bush, George | 3672 | Thomas, Jonah | 3718 | Bunyan, John |
| 3627 | Blasker, Charles | 3673 | Ryan, William | 3719 | Davis, John |
| 3628 | Brown, James | 3674 | Patten, John U. | 3720 | Wakefield, John |
| 3629 | Peterson, Lawrence | 3675 | Wright, Mathew | 3721 | Lane, Titus |
| 3630 | Welch, Thomas | 3676 | Gale, Benjamin G. | 3722 | Wilcox, Ceasar |
| 3631 | Bankstone, William | 3677 | Roundy, Francis | 3723 | Storey, George |
| 3632 | Blanchard, Samuel | 3678 | Hunt, Job | 3724 | Carroll, Samuel |
| 3633 | Hughes, William | 3679 | Brumblecom, Merrit | 3725 | Stacey, William |
| 3634 | Butler, Robert | 3680 | Ronder, Stephen C. | 3726 | Wilson, John |

American Prisoners of War Held at Dartmoor during the War of 1812

## Numeric listing by prison number

| | | | | | |
|---|---|---|---|---|---|
| 3727 | Ross, Richard | 3773 | Forrest, John | 3819 | Wadge, John |
| 3728 | Dennis, Jonas | 3774 | Treadwell, Nathaniel | 3820 | Eldridge, Ephraim |
| 3729 | Vanderford, Benjamin | 3775 | Jennings, Henry | 3821 | Amos, Peter |
| 3730 | Cook, Benjamin | 3776 | Bray, Zachariah | 3822 | Smith, John |
| 3731 | Winn, Joseph R. | 3777 | Simpson, William | 3823 | Halfpenny, William |
| 3732 | Hulin, Abraham | 3778 | Williams, John | 3824 | Higginbotham, William |
| 3733 | Curren, Nathaniel | 3779 | Watson, George | 3825 | Palmer, Robert |
| 3734 | Done, Elisha | 3780 | Stoddard, Thomas | 3826 | Glynn, Hanson |
| 3735 | Simons, John C. | 3781 | Steele, Mathew | 3827 | Mingo, Albert |
| 3736 | Williams, Hezekial | 3782 | Linton, Joseph | 3828 | Kirby, Anthony |
| 3737 | Barrentt, John | 3783 | Hensow, John | 3829 | McNeel, Philip |
| 3738 | Seagell, William | 3784 | Shaw, Benjamin | 3830 | Shean, John |
| 3739 | Fabin, James | 3785 | Steele, William | 3831 | Fowler, Robert |
| 3740 | Graham, Robert | 3786 | Harding, John | 3832 | Cantelo, James |
| 3741 | May, Joseph | 3787 | Smith, Samuel | 3833 | Lynch, James |
| 3742 | Richards, George | 3788 | Elmick, Joseph | 3834 | Gilliken, Daniel |
| 3743 | Pedang, John | 3789 | Raton, Peter | 3835 | St. Martin, Joseph |
| 3744 | Daigney, Jasper | 3790 | Pierce, John | 3836 | Pike, James |
| 3745 | Pitts, Thomas | 3791 | Young, Philip | 3837 | Riviere, Toussaint |
| 3746 | Murray, George | 3792 | Cameron, William | 3838 | Jones, Lawrence |
| 3747 | Wanton, Samuel | 3793 | Heath, James | 3839 | Hughes, Abijah |
| 3748 | Hadley, Andrew | 3794 | Starberne, John | 3840 | Jackson, Henry |
| 3749 | Trask, Charles | 3795 | Raysden, John | 3841 | Mitchell, John |
| 3750 | Peck, Benjamin | 3796 | Hoyt, David | 3842 | Parker, Thomas |
| 3751 | Swain, Luke | 3797 | Cooper, James | 3843 | Honner, John |
| 3752 | Green, Tobias | 3798 | Wood, Samuel | 3844 | West, John |
| 3753 | Spencer, Job | 3799 | Parkinson, Stephen | 3845 | Harley, Joshua |
| 3754 | Pittman, Benjamin | 3800 | Strouts, Joseph | 3846 | Studley, John |
| 3755 | Richards, Benjamin | 3801 | Higley, John | 3847 | King, Uriel |
| 3756 | Kelley, John | 3802 | Randall, Charles | 3848 | Dunn, Edward |
| 3757 | Bowden, Benjamin | 3803 | Hampstead, Cambridge | 3849 | Lamataille, John |
| 3758 | Lee, Richard | 3804 | Brown, John | 3850 | Vassina, John B. |
| 3759 | Cole, John | 3805 | Johnston, Michael | 3851 | Newons, Francis |
| 3760 | Hammond, Edward | 3806 | Lourie, Solomon | 3852 | Allison, Thomas |
| 3761 | Downing, George | 3807 | White, William | 3853 | Collison, Elliot |
| 3762 | Russell, Thomas | 3808 | Waters, Robert | 3854 | Joseph, John |
| 3763 | Young, Richard | 3809 | Branblen, Andrew | 3855 | Dunnings, Charles |
| 3764 | Ganell, James | 3810 | White, W. K. | 3856 | Durham, Samuel |
| 3765 | Pierce, Edward | 3811 | Loveland, Leonard | 3857 | Myers, John |
| 3766 | Blyth, Jonathan | 3812 | Haynes, Stephen | 3858 | Thompson, James |
| 3767 | Archer, Samuel | 3813 | Thorndike, Robert | 3859 | Houghton, Timothy |
| 3768 | Bird, Thomas | 3814 | Green, Timothy | 3860 | Gray, James |
| 3769 | Palmer, Peter | 3815 | Ryder, Amos | 3861 | Allison, J. B. |
| 3770 | Kenny, George | 3816 | Michaels, John | 3862 | Barnett, James |
| 3771 | Peabody, John | 3817 | Potter, Titus | 3863 | Mathews, P. |
| 3772 | Kelley, James | 3818 | Call, George | 3864 | Segeant, John |

American Prisoners of War Held at Dartmoor during the War of 1812

## Numeric listing by prison number

| | | | | | |
|---|---|---|---|---|---|
| 3865 | Dobson, Robert | 3911 | Neil, David A. | 3957 | Loring, George |
| 3866 | McGilmore, John | 3912 | Bernard, William | 3958 | McGee, John |
| 3867 | Dorrell, John | 3913 | Cole, H. A. | 3959 | Maigot, Abner |
| 3868 | West, Edward | 3914 | Dixon, Archibald | 3960 | Briggs, B. |
| 3869 | Cecil, Francis | 3915 | Lyons, Henry | 3961 | Weimon, John |
| 3870 | Carr, Thomas | 3916 | Shutcliffe, John | 3962 | Ellison, Benjamin |
| 3871 | Bignell, Peter | 3917 | Simson, John | 3963 | Clark, John |
| 3872 | Waternamn, William | 3918 | Burt, Thomas | 3964 | Cooper, Lodwig |
| 3873 | Green, Walter | 3919 | Bailey, Warren | 3965 | Justin, William |
| 3874 | Stove, Lewis | 3920 | Holden, James | 3966 | Cole, Nathaniel |
| 3875 | Brush, Ceasar | 3921 | Day, William | 3967 | Barber, John H. |
| 3876 | Brown, Thomas | 3922 | Setchell, Jonathan | 3968 | Hoston, Cato |
| 3877 | Williamson, Richard | 3923 | Baird, James | 3969 | Dixon, John |
| 3878 | Perkham, Peter | 3924 | Hobday, William | 3970 | Davis, John |
| 3879 | Monte, Charles | 3925 | Huntingdon, Edward | 3971 | Burnham, James |
| 3880 | Lewis, Henry | 3926 | Turney, B. | 3972 | Veal, Charles |
| 3881 | Hutchinson, Lamore | 3927 | Crofts, John | 3973 | Ross, William |
| 3882 | Young, William | 3928 | Ware, John | 3974 | Rue, William |
| 3883 | Baker, Richard | 3929 | Little, William | 3975 | Stafford, John |
| 3884 | Bastard, Walter | 3930 | Castagne, John | 3976 | Gale, Amos |
| 3885 | Hurt, John A. | 3931 | Johnson, James | 3977 | Hayes, Edward |
| 3886 | Dyer, Benjamin | 3932 | Barnes, James | 3978 | Campbell, James |
| 3887 | Weeks, Thomas | 3933 | Medker, Charles D. | 3979 | Patterson, John |
| 3888 | McLaughlan, M. | 3934 | Tyler, Daniel | 3980 | Garney, John |
| 3889 | Page, Joseph | 3935 | Baird, Martin | 3981 | Gibson, William |
| 3890 | Crellin, William | 3936 | Washington, John | 3982 | Whitecombe, John |
| 3891 | Rowley, James | 3937 | Horwell, John | 3983 | Gill, John |
| 3892 | Greaves, John | 3938 | Hill, Pompey | 3984 | Vickery, William |
| 3893 | Shanks, William | 3939 | Coston, Thomas | 3985 | Turner, John |
| 3894 | Wiley, John | 3940 | Richardson, Perry | 3986 | Ritchie, James |
| 3895 | Oliver, John | 3941 | Brownell, Paul | 3987 | Ballard, John |
| 3896 | Tomkins, Abraham | 3942 | Fellows, Nathaniel | 3988 | Lampriere, David |
| 3897 | Halton, William | 3943 | Phelps, Elijah | 3989 | Guard, Caleb |
| 3898 | Hussey, Edward | 3944 | Palmer, Benjamin | 3990 | Griffin, David |
| 3899 | Giddons, Andrew | 3945 | Jones, Edward | 3991 | Smith, John |
| 3900 | Birch, Peter | 3946 | Shaw, Robert | 3992 | Nichols, Silas |
| 3901 | Cateret, James | 3947 | Cogent, Benjamin | 3993 | Gale, Edward |
| 3902 | Bonny, John | 3948 | Brown, George | 3994 | Davis, Thomas |
| 3903 | Johnson, John | 3949 | Bennett, Abisa | 3995 | Thompson, William |
| 3904 | Johnson, Thomas | 3950 | Hammond, Samuel | 3996 | Buckley, Joseph |
| 3905 | Young, Cook | 3951 | Brown, William | 3997 | Taylor, John |
| 3906 | Dallison, James | 3952 | Anderson, Francis | 3998 | Baxter, David |
| 3907 | Marshall, John | 3953 | Gardner, Timothy | 3999 | Hughes, Peter |
| 3908 | Lanstone, Peter | 3954 | Wainwood, William | 4000 | George, Thomas |
| 3909 | Esterstone, John | 3955 | Stoddard, Isaac | 4001 | Ashton, Joseph |
| 3910 | Williams, William | 3956 | Lewis, Daniel | 4002 | Antoine, John |

American Prisoners of War Held at Dartmoor during the War of 1812

## Numeric listing by prison number

| | | | | | |
|---|---|---|---|---|---|
| 4003 | Gratton, Edward | 4049 | Morgan, William | 4095 | Williams, James |
| 4004 | Foran, William | 4050 | Payne, Joseph | 4096 | Harman, Mathew |
| 4005 | Alley, Jacob | 4051 | Baker, Henry | 4097 | Elwell, Lewis |
| 4006 | Huckber, Venus | 4052 | Chapple, Samuel | 4098 | Croilt, Aaron |
| 4007 | Stafford, Thomas | 4052 | Cole, Nathaniel | 4099 | Pendleton, George |
| 4008 | Newport, Mathew | 4054 | Tolman, William | 4100 | Russell, James |
| 4009 | Allen, Ambrose | 4055 | Gabbett, Henry | 4101 | Cayler, Jacob |
| 4010 | Dennis, William | 4056 | Brayden, Theodore | 4102 | Cayler, Charles |
| 4011 | Gladding, William | 4057 | Masser, Enoch | 4103 | Cayler, Henry |
| 4012 | Jones, Jeremiah | 4058 | White, John | 4104 | Bennett, James |
| 4013 | Goldsmith, Nathaniel | 4059 | Wood, John | 4105 | Benyman, James |
| 4014 | Logere, H. C. | 4060 | O'Hara, Terrence | 4106 | Buckling, Warren |
| 4015 | Scott, William | 4061 | Ham, Robert | 4107 | Winkinpor, Andrew |
| 4016 | Drew, Samuel | 4062 | Frenage, Michael | 4108 | Miller, Peter |
| 4017 | Stevens, Charles | 4063 | East, John | 4109 | Howard, Frederick |
| 4018 | Curtis, William | 4064 | Buttman, Nathaniel | 4110 | Dexter, David |
| 4019 | Goodman, Caleb | 4065 | Sprague, Henry | 4111 | Buttman, Thomas |
| 4020 | Bouton, John | 4066 | Dean, William | 4112 | Hoyt, David |
| 4021 | Capps, Denis | 4067 | Burke, James | 4113 | Henry, Nathanial |
| 4022 | Saywood, John | 4068 | Read, David | 4114 | Sullivan, Florence |
| 4023 | Cooney, Samuel | 4069 | Read, David | 4115 | Currant, Michael |
| 4024 | Bigsby, Samuel | 4070 | Walker, John | 4116 | Thomas, John |
| 4025 | Perkins, Joseph | 4071 | Garrison, Charles | 4117 | Johnson, Jacob |
| 4026 | Ranford, Henry | 4072 | Patterson, John | 4118 | McFadden, John |
| 4027 | Mann, Richard | 4073 | Jenkins, Samuel | 4119 | Haro, John |
| 4028 | Thompson, James | 4074 | Dashells, Richard | 4120 | Headen, John |
| 4029 | West, Richard | 4075 | Banks, James | 4121 | Parnell, Hugh |
| 4030 | Lilley, Samuel | 4076 | Dunn, Robert | 4122 | Hendrickson, John |
| 4031 | Abbott, Enoch | 4077 | Olliver, John | 4123 | Rockhill, John |
| 4032 | Prindell, Daniel | 4078 | Johnson, James | 4124 | Grimes, George |
| 4033 | Walsh, John | 4079 | Boyle, John | 4125 | Mason, Thomas |
| 4034 | Bradford, Elisha | 4080 | Dave, David | 4126 | Burnham, Benjamin |
| 4035 | Rogers, George | 4081 | Leggins, Ebenezer | 4127 | Barton, Samuel |
| 4036 | Read, John | 4082 | Fulgar, William | 4128 | Morris, Joseph |
| 4037 | Kean, Robert | 4083 | Selman, Edward | 4129 | Weaver, Michael |
| 4038 | Holmes, John | 4084 | Morris, John | 4130 | Harley, Edward |
| 4039 | Bruce, Joseph | 4085 | Dutton, Thomas | 4131 | Yeoman, John |
| 4040 | Peterson, Christopher | 4086 | Budd, John | 4132 | Keane, Daniel |
| 4041 | Holiday, Francis | 4087 | Burn, Rufus | 4133 | Rogers, Henry |
| 4042 | Florence, John | 4088 | Andrews, Peter | 4134 | Ford, Stan |
| 4043 | Waide, Nathaniel | 4089 | Chase, Allen | 4135 | Law, Thomas |
| 4044 | Anderson, Ebenezer | 4090 | Parsons, Simon | 4136 | Sutherland, John |
| 4045 | Lonie, James | 4091 | Crapeman, Henry | 4137 | Trout, Nathaniel |
| 4046 | Bowden, Thomas | 4092 | Tripps, O. | 4138 | Cline, Lewis |
| 4047 | Williams, William | 4093 | Harman, Peter | 4139 | Hudson, William |
| 4048 | Goldsmith, Nathaniel | 4094 | Legatt, Charles | 4140 | Kitchen, Robert |

American Prisoners of War Held at Dartmoor during the War of 1812

## Numeric listing by prison number

| | | | | | |
|---|---|---|---|---|---|
| 4141 | Norris, Philip | 4187 | Lubery, John C. | 4233 | Bassett, William |
| 4142 | Duncan, Nathaniel | 4188 | Henzeman, Christopher | 4234 | Manley, David |
| 4143 | Clements, William | 4189 | Hamilton, Robert | 4235 | Harding, William |
| 4144 | Campbell, William | 4190 | Lynch, William | 4236 | Williams, Charles |
| 4145 | Arnold, Benjamin | 4191 | Reed, Abraham | 4237 | Doosenbery, R. |
| 4146 | Shoe, Bernard | 4192 | Hanner, Daniel | 4238 | Williams, Moses |
| 4147 | Leno, Frederick | 4193 | Spern, Elijah | 4239 | Melbourne, William |
| 4148 | Finn, John | 4194 | Smith, Richard | 4240 | Hermanden, Peter |
| 4149 | Griffin, John | 4195 | Plumer, William | 4241 | O'Sheffey, Jacobus |
| 4150 | Green, John | 4196 | Modre, John | 4242 | Thompson, Lawrence |
| 4151 | Jones, William | 4197 | Williamson, David | 4243 | Tucker, Henry |
| 4152 | Crosby, John | 4198 | Keen, Joseph | 4244 | Haley, John |
| 4153 | Lincoln, Ephraim | 4199 | Hanes, Simon | 4245 | Williams, George |
| 4154 | Harding, John | 4200 | Pritchard, Israel | 4246 | Bartlett, Scipro |
| 4155 | Winchester, Ebenezer | 4201 | Selman, John | 4247 | Kenny, William |
| 4156 | Gale, William | 4202 | Harvey, Joseph | 4248 | Weisse, Philip |
| 4157 | Porter, Calvin | 4203 | Carter, Henry | 4249 | Davis, Henry |
| 4158 | Jones, Theodore | 4204 | Merritt, Enoch | 4250 | Wallace, Thomas |
| 4159 | Molloy, Peter | 4205 | Harris, Ebenezer | 4251 | Kitchen, Daniel |
| 4160 | Wilson, William | 4206 | Watson, William | 4252 | Chase, Constant |
| 4161 | James, John | 4207 | Conner, Michael | 4253 | Skudder, Alexander |
| 4162 | Calebs, Lewis | 4208 | Hill, Daniel | 4254 | Shepherd, Samuel |
| 4163 | Penfold, John | 4209 | Murray, Richard | 4255 | Peters, Benjamin |
| 4164 | Gosling, Joseph | 4210 | Robins, William | 4256 | Lewis, Henry |
| 4165 | Hammond, Isaac | 4211 | Newton, John | 4257 | Stetson, Hiram |
| 4166 | Wright, John | 4212 | Tuttle, Joseph | 4258 | Norcross, Thomas |
| 4167 | Wall, William | 4213 | Carter, Enoch | 4259 | Bridge, Jeremiah |
| 4168 | Boyce, Abraham | 4214 | Bloventon, John | 4260 | Simmonds, Charles |
| 4169 | Wallace, William | 4215 | Ewell, Edward | 4261 | Gage, Lot |
| 4170 | Chalk, John | 4216 | Port, John | 4262 | Simmonds, Joel |
| 4171 | Johnson, Alexander | 4217 | Hilby, Thomas | 4263 | Furness, Jesse |
| 4172 | Nicholson, Thomas | 4218 | Defray, Edward | 4264 | Weslemyer, John |
| 4173 | Wilson, Daniel | 4219 | Tucker, James | 4265 | Harris, Elpheus |
| 4174 | Langroth, Francis | 4220 | Douglass, William | 4266 | Griffiths, Joseph |
| 4175 | Frees, James | 4221 | Towns, Asa | 4267 | Warner, John |
| 4176 | Fitch, William | 4222 | Nichols, John | 4268 | Jones, Anthony |
| 4177 | Wilson, William | 4223 | Hamilton, John | 4269 | Johnston, Charles |
| 4178 | Waterman, Thomas | 4224 | Hall, David | 4270 | Fields, Jacob |
| 4179 | Minor, David | 4225 | Creiger, John | 4271 | Crosby, Andrew |
| 4180 | Small, George D. | 4226 | Ned, Deaf | 4272 | Tinkham, John |
| 4181 | Waterman, John | 4227 | Green, Levenet | 4273 | Cannon, Thomas |
| 4182 | Bliss, Frederick | 4228 | Hill, Justice | 4274 | Butler, Thomas |
| 4183 | Ray, James | 4229 | Jackson, Charles | 4275 | Dixon, Peter |
| 4184 | Crasey, James | 4230 | Typhon, Louis | 4276 | Fiske, Cyrus |
| 4185 | Harrison, Henry | 4231 | Brisons, John | 4277 | Dean, James B. |
| 4186 | Thomas, John | 4232 | Scott, Henry | 4278 | Barnes, Edward L. |

American Prisoners of War Held at Dartmoor during the War of 1812

## Numeric listing by prison number

| | | | | | |
|---|---|---|---|---|---|
| 4279 | Howitt, William | 4325 | Joseph, Michael | 4371 | Davis, William |
| 4280 | Downs, William | 4326 | Carson, John | 4372 | Phippen, Israel |
| 4281 | Downs, John | 4327 | Poor, David | 4373 | Cash, John |
| 4282 | Livingston, Henry | 4328 | Palfrey, John | 4374 | Foster, Samuel |
| 4283 | Coffin, Edward | 4329 | Baker, James | 4375 | Day, Frederick |
| 4284 | Thomas, John | 4330 | Fuller, Thomas | 4376 | Thomas, William |
| 4285 | Simmonds, Henry | 4331 | Falkern, W. | 4377 | Reed, John |
| 4286 | Mids, Michael | 4332 | Langley, Richard | 4378 | Brown, John |
| 4287 | Demarlow, Francis | 4333 | Smithers, Richard | 4379 | Chandler, Enoch |
| 4288 | Stafford, John | 4334 | Mainey, John | 4380 | Samerton, George |
| 4289 | Jackson, Frederick | 4335 | Swaby, Jonathan | 4381 | Sawyer, James |
| 4290 | Taylor, George | 4336 | Dykes, Leven | 4382 | Fortune, John |
| 4291 | Gray, William | 4337 | Craig, John | 4383 | Dew, Frederick |
| 4292 | Hugg, Jacob | 4338 | Williams, John | 4384 | Stanfield, George |
| 4293 | Carman, George | 4339 | Peterson, P. | 4385 | Shelton, Samuel |
| 4294 | Reeves, Asa | 4340 | Marral, Ephraim | 4386 | Seith, John |
| 4295 | Killingsworth, John | 4341 | Young, John | 4387 | Le More, John |
| 4296 | Woodbury, Dixey | 4342 | Lovell, Samuel | 4388 | Bacchus, John |
| 4297 | Edwards, William | 4343 | Thrasher, Stephen | 4389 | Canfield, Arthur |
| 4298 | Gregory, George | 4344 | Atwood, Jesse | 4390 | Chase, Nathaniel |
| 4299 | Walker, Seth | 4345 | Kendrick, Benjamin | 4391 | Appleton, John |
| 4300 | Curtis, Ephraim | 4346 | Boyd, William | 4392 | Golf, Peter |
| 4301 | Bennett, John | 4347 | Holbrook, Richard | 4393 | Pike, Jeremiah |
| 4302 | Johnson, George | 4348 | Jereb, Duly | 4394 | Russell, Moses |
| 4303 | Shrousy, William | 4349 | Peak, James | 4395 | King, Solomon |
| 4304 | Caldwell, Charles | 4350 | Knight, Samuel | 4396 | Robinson, Thomas |
| 4305 | Wyman, William | 4351 | Reed, William | 4397 | Hanson, William |
| 4306 | Bisbee, J. D. | 4352 | Brown, Amos | 4398 | Tarlton, George |
| 4307 | Loring, Samuel | 4353 | Homan, Edwad | 4399 | Perkins, John |
| 4308 | Higby, James | 4354 | Fitzsmond, Jeremiah | 4400 | Stone, Robert |
| 4309 | St. Vincent, Stephen | 4355 | Pigger, Stephen | 4401 | Norcross, Archibald |
| 4310 | Murray, David | 4356 | Bradly, Samuel | 4402 | Perkinson, James |
| 4311 | Maine, John | 4357 | Williams, Peter | 4403 | Leach, Daniel |
| 4312 | Congleton, A. | 4358 | Bridson, Thomas | 4404 | Larabee, Thomas |
| 4313 | Porter, Ephraim | 4359 | Camp, Tobias | 4405 | Perkins, Benjamin |
| 4314 | Calaman, John | 4360 | Thompson, William | 4406 | Dennison, Andrew |
| 4315 | Franklin, William | 4361 | Black, Nicholas | 4407 | Thomas, Elisha |
| 4316 | Haster, John | 4362 | Swinbourne, George | 4408 | Perkins, James |
| 4317 | Rye, William | 4363 | Franklin, Edward | 4409 | Grant, Samuel |
| 4318 | Thomas, James | 4364 | Ricker, Thomas | 4410 | Silverthorn, James |
| 4319 | Robinson, Robert | 4365 | Smith, Richard | 4411 | Rowe, William |
| 4320 | Ralph, David | 4366 | Garrison, Charles | 4412 | Tar, Caleb |
| 4321 | Swadey, Samuel | 4367 | Mayeau, Morris | 4413 | Pearson, Samuel |
| 4322 | Smith, David | 4368 | Eldridge, Nathaniel | 4414 | Allen, David |
| 4323 | Williams, John | 4369 | Driscol, Jeremiah | 4415 | Millett, Joseph |
| 4324 | Williams, John | 4370 | Smith, John | 4416 | Jones, Stephen |

American Prisoners of War Held at Dartmoor during the War of 1812

## Numeric listing by prison number

| | | | | | |
|---|---|---|---|---|---|
| 4417 | Williams, William | 4463 | Fall, James | 4509 | Richardson, William |
| 4418 | Day, John | 4464 | Keen, Benjamin | 4510 | Pierce, Joseph |
| 4419 | Shaw, Henry | 4465 | Sonnett, James | 4511 | Hubbard, William |
| 4420 | Jeffreys, Henry | 4466 | Vennard, George | 4512 | Reed, George |
| 4421 | Cook, Silvanus | 4467 | Place, Thomas | 4513 | Kent, William |
| 4422 | Parsons, John | 4468 | Fitherly, Robert | 4514 | Philips, Timothy |
| 4423 | Harman, Isaac | 4469 | Randall, Henry | 4515 | Warren, David |
| 4424 | Brown, John | 4470 | Ducat, William | 4516 | Malloy, William |
| 4425 | Webb, Nathaniel | 4471 | Fernald, William | 4517 | Hall, James |
| 4426 | Davis, John | 4472 | Sides, Samuel | 4518 | Beatty, John |
| 4427 | Primas, James | 4473 | Smith, Thomas | 4519 | Pitts, George |
| 4428 | Jackson, William | 4474 | Price, John | 4520 | Porter, Louis |
| 4429 | Kingbutton, John | 4475 | Richardson, William | 4521 | Dunningberg, Henry |
| 4430 | Robertson, William | 4476 | Hamson, William | 4522 | Allen, Edward D. |
| 4431 | Porter, Charles | 4477 | Battes, John | 4523 | Devol, Alexander |
| 4432 | Staples, James | 4478 | Bassett, John | 4524 | Forstman, John |
| 4433 | Johnson, Robert | 4479 | Simmons, David | 4525 | Hall, Silvester |
| 4434 | Hill, John | 4480 | Giddins, Hiram | 4526 | Hill, James |
| 4435 | Rodgers, Abraham | 4481 | Horne, John | 4527 | Chase, Oliver |
| 4436 | Lehman, Francis | 4482 | Duncan, Thomas | 4528 | Tarlton, James |
| 4437 | Brown, Joseph | 4483 | Hamson, John | 4529 | Cloutman, Robert |
| 4438 | Gatchell, John G. | 4484 | Lamson, Amos | 4530 | Phippen, John |
| 4439 | Odiam, Joseph H. | 4485 | Webster, Asa | 4531 | Stacey, Benjamin |
| 4440 | Easten, Ephraim | 4486 | Lackey, Joseph | 4532 | Stacey, Samuel |
| 4441 | Emerson, David | 4487 | Treadwell, Nathaniel | 4533 | Lyons, Charles |
| 4442 | Kegs, Zenas | 4488 | Thompson, John | 4534 | Grant, William |
| 4443 | Nicholson, James | 4489 | Tildon, Robert | 4535 | Alexander, R. |
| 4444 | Malcomb, Alexander | 4490 | Clarke, Isaac | 4536 | Chivers, Joseph |
| 4445 | Sillick, Thomas | 4491 | Walker, Samuel | 4537 | Pearce, David |
| 4446 | White, Edward | 4492 | Kelley, John | 4538 | Knap, Samuel |
| 4447 | Williamson, John | 4493 | Findley, Thomas | 4539 | Quiner, Stephen |
| 4448 | Miller, James | 4494 | Thornton, John | 4540 | Pearse, Prince |
| 4449 | Phipps, Richard | 4495 | McIntire, John | 4541 | Tucker, Andrew |
| 4450 | Smith, Thomas | 4496 | Kindan, Ephraim | 4542 | Pittman, Benjamin |
| 4451 | Ward, Thomas | 4497 | Lufkin, William | 4543 | Coombe, Michael |
| 4452 | Lane, James | 4498 | Robinson, William | 4544 | Forster, George |
| 4453 | Holmes, Almoran | 4499 | Webb, John | 4545 | Tucker, Nathaniel |
| 4454 | Behon, Simon | 4500 | Millett, John | 4546 | Gursh, Nathaniel |
| 4455 | Atwood, Nathaniel | 4501 | Oakes, George | 4547 | Campbell, John |
| 4456 | Woodford, William | 4502 | Johnson, George | 4548 | Chase, John |
| 4457 | Briggs, Thomas | 4503 | Tucker, William | 4549 | Cooper, Thomas |
| 4458 | Norton, Solomon | 4504 | Blaney, Stephen | 4550 | Henry, Isaac |
| 4459 | Smith, Elisha | 4505 | Crowel, Mathew | 4551 | Grundy, Edward |
| 4460 | Long, Joseph | 4506 | Carswell, William | 4552 | Brown, Michael |
| 4461 | Porter, Edward | 4507 | Salmon, Archibald | 4553 | Anderson, John |
| 4462 | Rowland, Abner | 4508 | Blanchard, John | 4554 | Spicer, Alexander |

American Prisoners of War Held at Dartmoor during the War of 1812

## Numeric listing by prison number

| | | | | | |
|---|---|---|---|---|---|
| 4555 | Cooper, William | 4601 | Boston, John | 4647 | Owen, Zachariah |
| 4556 | Jones, Isaac | 4602 | Smith, W. B. | 4648 | Torry, Henry |
| 4557 | Allen, John | 4603 | Crasus, Richard | 4649 | Richardson, James |
| 4558 | Evans, Hezekial | 4604 | Johnson, James | 4650 | Hult, Thomas |
| 4559 | Williams, Benjamin | 4605 | Stickney, Abraham | 4651 | Clarke, John |
| 4560 | Wilson, George | 4606 | Gavet, James | 4652 | Wingate, David |
| 4561 | Ellis, Cornelius | 4607 | Ringgold, Thomas | 4653 | Spencer, John |
| 4562 | Snow, Colger | 4608 | Dixon, John | 4654 | Reed, John |
| 4563 | Bushfield, James | 4609 | Tyler, Joseph | 4655 | Wilson, James |
| 4564 | Francis, John | 4610 | Lee. John | 4656 | Nicholson, John |
| 4565 | Forbes, Robert | 4611 | Thinney, Elvin | 4657 | Derring, William F. |
| 4566 | Pines, Isaac | 4612 | Howe, Jacob | 4658 | Smith, Jeremiah |
| 4568 | Kanes, William | 4613 | Choete, Thomas | 4659 | Rust, John |
| 4569 | Davis, Stephen | 4614 | Bisbee, Elijah | 4660 | Walling, James |
| 4569 | Liscomb, William | 4615 | Hallet, William | 4661 | Stephens, William |
| 4570 | Jackson, Henry | 4616 | Mains, Henry | 4662 | Piles, John |
| 4571 | Mark, James | 4617 | White, George | 4663 | Bridges, John |
| 4572 | Stone, Isaac | 4618 | Hodges, Hercules | 4664 | Pratt, Philip |
| 4573 | Hall, Richard | 4619 | Dill, William | 4665 | Holmes, Elisha |
| 4574 | Spiers, Nathaniel | 4620 | Randall, Jacob | 4666 | Lawson, Thomas |
| 4575 | Morrison, Thomas | 4621 | Robinson, Stephen | 4667 | Blair, Robert |
| 4576 | Richards, James | 4622 | Williams, James | 4668 | Cheslie, Amos |
| 4577 | Shaw, William | 4623 | Edwards, John | 4669 | Burrell, Jesse |
| 4578 | Hazard, Charles | 4624 | Morriss, Andrew | 4670 | Melzard, George |
| 4579 | Arnold, Obediah | 4625 | Tomus, Andrew | 4671 | Ross, David |
| 4580 | Kelly, Samuel | 4626 | Francis, John | 4672 | Robes, Edward |
| 4581 | McFarlane, Daniel | 4627 | Freeman, Plim | 4673 | Glover, John |
| 4582 | Moore, Edward | 4628 | Brown, William | 4674 | Lapish, Andrew |
| 4583 | Deane, Peter | 4629 | Rennel, States | 4675 | Anderson, David |
| 4584 | Livesley, Thomas | 4630 | Jennings, John | 4676 | Wyatt, Jesse |
| 4585 | Brown, Frederick | 4631 | Clark, Samuel | 4677 | Spenny, Nathaniel |
| 4586 | Shephard, James | 4632 | Albert, Hezekieh | 4678 | Woodford, James |
| 4587 | Osborne, Samuel | 4633 | Smith, Thomas | 4679 | Hazard, Prince |
| 4588 | Butler, William | 4634 | Baptiste, John | 4680 | Turner, Samuel |
| 4589 | Lawton, William | 4635 | Wright, Israel | 4681 | Hitchins, William |
| 4590 | Gordon, Richard | 4636 | Farrell, Andrew | 4682 | Claby, Martin |
| 4591 | Oliver, Joseph | 4637 | Traphagan, Peter | 4683 | Craft, Richard |
| 4592 | Allison, William R. | 4638 | Poverley, Henry | 4684 | Lane, William |
| 4593 | Coffin, Abel | 4639 | McGee, Robert | 4685 | Hall, John |
| 4594 | Baxter, David | 4640 | Guire, Andrew | 4686 | Hatheray, Philip |
| 4595 | Jones, Lewis | 4641 | Peters, John | 4687 | Coffin, Joseph |
| 4596 | Golding, Abijah | 4642 | Reed, Joseph | 4688 | Jones, Thomas |
| 4597 | Gotier, Charles J. | 4643 | Fadden, Charles | 4689 | Meath, Samuel |
| 4598 | Bartlett, John | 4644 | Lambert, Thomas | 4690 | Peadon, William |
| 4599 | Sterns, Joseph | 4645 | Gordon, John | 4691 | Roweth, William |
| 4600 | Baisley, Abraham | 4646 | Peters, Thomas | 4692 | Davis, Nicholas |

American Prisoners of War Held at Dartmoor during the War of 1812

## Numeric listing by prison number

| | | | | | |
|---|---|---|---|---|---|
| 4693 | Harvey, Anthony | 4739 | Gardner, Jerry | 4785 | Graves, Thomas |
| 4694 | Roundy, Jeremiah | 4740 | Congleton, Smith | 4786 | Hubbard, John G. |
| 4695 | Rust, John | 4741 | Harris, John | 4787 | Ross, George |
| 4696 | Bradford, George | 4742 | Baxter, Francis | 4788 | Sawyer, Jacob |
| 4697 | Shaw, Samuel | 4743 | Barnard, John | 4789 | Jones, Samuel B. |
| 4698 | Smith, John | 4744 | Cooper, Thomas | 4790 | Dennis, William |
| 4699 | Kennedy, Dennis | 4745 | Phillips, John | 4791 | Pilot, Samuel G. |
| 4700 | Marlow, Owen | 4746 | Bradie, John | 4792 | Rice, M. |
| 4701 | Beckwith, James | 4747 | Little, Thomas | 4793 | Halbert, Henry |
| 4702 | Morgan, James | 4748 | Scott, Anthony | 4794 | Gair, Thomas |
| 4703 | Maria, Hosea | 4749 | Jonathan, Jonathan | 4795 | Sprague, Stephen |
| 4704 | Jupiter, james | 4750 | Christian, Tyrel | 4796 | Barron, John |
| 4705 | Cussar, Jacob O. | 4751 | Peckham, Ezekiel | 4797 | Ross, William |
| 4706 | Martin, Francis | 4752 | Lovell, William | 4798 | Knabbs, James |
| 4707 | Jones, Stephen | 4753 | Bordage, Raymond | 4799 | Stephens, William |
| 4708 | Hill, Jeremiah | 4754 | Waterhouse, Joseph | 4800 | Barnett, Tobias |
| 4709 | Maria, Josea | 4755 | Adams, William | 4801 | Williams, Charles |
| 4710 | Gibson, Samuel | 4756 | Lewis, John | 4802 | Waterman, Cato |
| 4711 | Travers, Thomas | 4757 | Morris, Samuel | 4803 | Marie, John |
| 4712 | Callam, John | 4758 | Jones, John | 4804 | Lorang, M. |
| 4713 | Donnell, Samuel | 4759 | Cameron, Daniel | 4805 | Sissons, Pedro |
| 4714 | Bowen, Sylvester | 4760 | Senter, Noah | 4806 | Fry, A. L. |
| 4715 | Blaird, David | 4761 | Roberts, Nathaniel | 4807 | Johnson, John |
| 4716 | Rust, John | 4762 | Coffin, Valentine | 4808 | Adrianne, Jose |
| 4717 | Locker, Michael | 4763 | Luther, Cromwell | 4809 | Ebier, Joseph |
| 4718 | Jones, James | 4764 | Crapon, George | 4810 | Joseph, Pedro |
| 4719 | Bressy, Charles | 4765 | Turner, Gardner | 4811 | Mendoza, Caesar |
| 4720 | Johnson, Henry | 4766 | Ingersole, John | 4812 | Jones, Thomas |
| 4721 | Paris, Peter | 4767 | Davis, Solomon | 4813 | Falcon, J. B. |
| 4722 | Northey, Joseph | 4768 | Felt, George | 4814 | Hidalgo, Valentine |
| 4723 | Starkweather, James | 4769 | Phillips, George | 4815 | Complaro, A. A. |
| 4724 | Kilham, John | 4770 | Monk, Joseph | 4816 | Achro, Joseph |
| 4725 | Marvel, David | 4771 | Feilman, John | 4817 | Lopez, Domingo |
| 4726 | Ward, Alfred | 4772 | Merrill, Enoch | 4818 | Parks, A. D. La Enoh |
| 4727 | Jewitt, Jasper | 4773 | Swain, Obediah | 4819 | Azumus, Jerome |
| 4728 | Glover, Benjamin | 4774 | Wynn, John | 4820 | Campreche, St. Jago |
| 4729 | Brown, Abraham | 4775 | Tarr, James | 4821 | Palmo, Vincent |
| 4730 | Fowler, Joshua | 4776 | Stilwell, William | 4822 | Roberts, John |
| 4731 | Boggart, John | 4777 | Laingan, Daniel | 4823 | De La Canto Sepo, I. |
| 4732 | Middleton, L. | 4778 | Robinson, William | 4824 | Leurand, Ambrose |
| 4733 | Burke, John | 4779 | Simpson, Smith | 4825 | Almeno, Jose |
| 4734 | Hazard, Robert | 4780 | Jones, Thomas | 4826 | Aubry, John Martinalias |
| 4735 | Cutler, Thomas | 4781 | Wilcox, William | 4827 | Flinn, John F. |
| 4735 | Smith, Thomas | 4782 | Griffin, John | 4828 | Walker, James Scott |
| 4737 | Porter, Gideon | 4783 | Green, Thomas | 4829 | Craig, John E. |
| 4738 | Nye, William | 4784 | Drayton, John | 4830 | Montgomery, William |

American Prisoners of War Held at Dartmoor during the War of 1812

## Numeric listing by prison number

| | | | | | |
|---|---|---|---|---|---|
| 4831 | Mathy, James | 4877 | Newcomb, Tilton | 4923 | Herrimond, Hezekiel |
| 4832 | Bretade, E. F. | 4878 | Johnson, Peter | 4924 | Mackey, Richard |
| 4833 | Ivy, John | 4879 | Muller, John | 4925 | Bull, Henry |
| 4834 | Hydalgo, Vincent | 4880 | Carter, George | 4926 | Campbell, Moses |
| 4835 | Bauld, J. P. | 4881 | Pembroke, James | 4927 | Bird, John |
| 4836 | Johannes, John | 4882 | Armandez, Justine | 4928 | Rowe, Abraham |
| 4837 | Milborne, William | 4883 | Williams, Joseph | 4929 | Cochrane, William |
| 4838 | Harman, John | 4884 | Leverage, William | 4930 | Gwynn, Josiah |
| 4839 | George, John | 4885 | Anderson, John | 4931 | Densey, Peter |
| 4840 | Davis, George | 4886 | Fisher, Charles | 4932 | Hunter, William |
| 4841 | Taylor, Samuel | 4887 | Sands, Henry | 4933 | Gibbs, Thomas |
| 4842 | Fog, John | 4888 | Adams, Jesse | 4934 | Nisbett, James |
| 4843 | Smith, J. W. | 4889 | Welch, John | 4935 | Greenfield, James |
| 4844 | Tripe, W. H. | 4890 | Davis, John | 4936 | Gould, John |
| 4845 | Franklin, W. H. | 4891 | Houghman, John | 4937 | Flynn, John |
| 4846 | Jennings, John | 4892 | Stibbens, Stephen | 4938 | Chariton, James |
| 4847 | Jones, John | 4893 | Congdon, James | 4939 | Williams, William |
| 4848 | Adams, William | 4894 | Steward, William | 4940 | Colbourn, James |
| 4849 | Jackson, Joseph | 4895 | Smith, John | 4941 | Carty, Joseph |
| 4850 | Simmons, Ebenezer | 4896 | Long, John | 4942 | Dennis, Francis |
| 4851 | Adams, James | 4897 | Knapp, Joseph | 4943 | Ball, William |
| 4852 | Johnson, Luke | 4898 | Cross, Nathaniel | 4944 | Leverage, Zachariah |
| 4853 | Evans, Edward | 4899 | Hecock, William | 4945 | Freeman, Halkins |
| 4854 | Newell, Charles | 4900 | Frack, Daniel | 4946 | Higgins, Isaac |
| 4855 | Lester, James | 4901 | Russell, Benjamin | 4947 | Calpry, Joseph |
| 4856 | Rogers, Luke | 4902 | Kelley, James | 4948 | Doane, Joshua |
| 4857 | Lynch, William | 4903 | Skenny, Francis | 4949 | Norris, Henry |
| 4858 | Nichols, William | 4904 | Jack, John | 4950 | McCalla, David |
| 4859 | Dale, John | 4905 | Lane, Henry | 4951 | Hall, Zachariah |
| 4860 | Godfrey, John | 4906 | Shaw, Joseph | 4952 | Wyatt, William |
| 4861 | King, Joseph | 4907 | Hazard, D. | 4953 | Morrison, James |
| 4862 | Latham, Giles | 4908 | Howland, J. W. | 4954 | Edwards, John |
| 4863 | Evans, William | 4909 | Craig, William | 4955 | Shoot, William |
| 4864 | Herring, William | 4910 | Williams, John | 4956 | Allan, Asa |
| 4865 | Adams, Isaac | 4911 | Crump, Joseph R. | 4957 | Barnes, Nathaniel |
| 4866 | Thomas, James | 4912 | Hopkins, John | 4958 | Smith, Nicholas |
| 4867 | Greswold, Truman | 4913 | Thomas, Isaac | 4959 | Prime, James |
| 4868 | Dill, Samuel | 4914 | Cook, Samuel | 4960 | Steel, Mark |
| 4869 | Thomas, William | 4915 | Fenderson, Nathaniel | 4961 | Hobert, James |
| 4870 | Johnson, John | 4916 | Bowland, William | 4962 | Jacob, Peter |
| 4871 | Morrell, Jacob | 4917 | Menillo, John | 4963 | Merchant, Elijah |
| 4872 | Charles, Philip | 4918 | Swaine, Joseph | 4964 | Emas, Comfort |
| 4873 | Payne, John | 4919 | Clarke, William | 4965 | Bushnell, Joseph |
| 4874 | Lee, Samuel | 4920 | Munsey, Daniel | 4966 | Lockwood, William |
| 4875 | Brain, William | 4921 | Childs, Adam | 4967 | Allen, John Baptiste |
| 4876 | Gibson, Richard | 4922 | Williams, George | 4968 | Rogers, Gaff |

American Prisoners of War Held at Dartmoor during the War of 1812

## Numeric listing by prison number

| No. | Name | No. | Name | No. | Name |
|---|---|---|---|---|---|
| 4969 | Moore, J. B. | 5015 | Davis, Jacob | 5061 | Warden, Nathaniel |
| 4970 | Kempt, Eric | 5016 | Mason, Moses | 5062 | Hadlock, Nathaniel |
| 4971 | Thomas, Richard | 5017 | Fulton, William | 5063 | DeBates, Amos |
| 4972 | Clark, Thomas | 5018 | Chase, William | 5064 | Perigo, Joel |
| 4973 | Nowland, Edward | 5019 | Marshall, Solomon | 5065 | Warney, Benjamin |
| 4974 | Whiteby, Samuel | 5020 | Wood, Thomas | 5066 | Furs, Theodore |
| 4975 | Sweetland, William | 5021 | Ellis, John | 5067 | Anderson, Alexander |
| 4976 | Peterson, Samuel | 5022 | Kembourn, John | 5068 | Bray, Andrew |
| 4977 | Sherry, Peter | 5023 | Baker, Henry | 5069 | Knowlton, Enos |
| 4978 | Haycock, John | 5024 | Hickson, John | 5070 | Babson, John |
| 4979 | Davis, John | 5025 | Gennison, Michael | 5071 | Sands, John |
| 4980 | Swinley, John | 5026 | Ball, Peter | 5072 | Kellem, Frederick |
| 4981 | Forbes, William | 5027 | Mackay, Joseph | 5073 | DeRichardson, Comte |
| 4982 | Walton, John | 5028 | Gardner, John | 5074 | Denham, Silas |
| 4983 | Fields, Michael | 5029 | Carnass, William | 5075 | Whiteman, Peter |
| 4984 | Coleman, Joseph | 5030 | Anderson, John | 5076 | Miller, George |
| 4985 | Jefferson, Richard | 5031 | Green, Elijah | 5077 | Pyne, Joseph |
| 4986 | Seywood, James | 5032 | Evans, Benjamin | 5078 | Lewis, Gitt |
| 4987 | Murray, Colton | 5033 | Myers, Joseph | 5079 | Russell, John |
| 4988 | Morgan, Henry | 5034 | Hanson, John | 5080 | Alexander, James |
| 4989 | Smith, Daniel | 5035 | Allen, Arthur | 5081 | Terry, Daniel |
| 4990 | Kempt, George | 5036 | Gould, Samuel | 5082 | Bacon, Elisha |
| 4991 | Morier, John | 5037 | Payne, Alexander | 5083 | Daverille, William |
| 4992 | Celutan, Alexander | 5038 | Robinson, Thomas | 5084 | Baker, Edward E. |
| 4993 | Michell, Charles | 5039 | Sharpless, Robert | 5085 | Keough, Edward |
| 4994 | Sobier, Bernard | 5040 | Vesan, Weslant | 5086 | Tolton, Moses |
| 4995 | Simondson, Isaac | 5041 | Hardman, John | 5087 | Churchill, Manuel |
| 4996 | Becamp, Jaques | 5042 | Jones, John | 5088 | Bancroft, James |
| 4997 | DeSilvin, Jean | 5043 | Campbell, James | 5089 | Adams, John |
| 4998 | Hale, John | 5044 | Simmonds, Alexander | 5090 | Lowe, Frederick |
| 4999 | Jouranne, Louis | 5045 | Creamer, George | 5091 | Davies, William |
| 5000 | Merlo, Cristopher | 5046 | Rand, Robert | 5092 | Cashman, Ansel |
| 5001 | Foss, Supply | 5047 | Lewis, William | 5093 | Belcour, James |
| 5002 | Huzzey, John | 5048 | Neumann, Gustavus | 5094 | Restrum, Andrew |
| 5003 | Garrison, Cornelius | 5049 | Sturgess, Major | 5095 | Jose, Emanuel |
| 5004 | Green, John | 5050 | Beets, Thomas | 5096 | Saunders, John |
| 5005 | Bolton, John | 5051 | Cotton, William | 5097 | Bent, Joseph |
| 5006 | Babb, Nathaniel | 5052 | Polland, John | 5098 | Sholtz, Charles |
| 5007 | Florence, John | 5053 | Bray, Isacher | 5099 | Green, Samuel |
| 5008 | Lamb, Anthony | 5054 | Palmer, Joseph | 5100 | Ayres, Parker |
| 5009 | King, Peter | 5055 | Hambleton, John | 5101 | Thompson, Isaac |
| 5010 | Frewan, Nathaniel | 5056 | Cody, James | 5102 | Colley, Thomas |
| 5011 | Loring, William | 5057 | Haffer, Robert | 5103 | Johnson, Samuel |
| 5012 | Garrison, Babet | 5058 | Fowling, Jeremiah | 5104 | Henderson, John |
| 5013 | Johnson, Stephen | 5059 | Johnson, William | 5105 | Taylor, David |
| 5014 | Miller, Edward | 5060 | Moss, N. J. | 5106 | Frank, Joshua |

American Prisoners of War Held at Dartmoor during the War of 1812

## Numeric listing by prison number

| | | | | | |
|---|---|---|---|---|---|
| 5107 | Highby, William | 5153 | Green, John | 5199 | Hanely, Thomas |
| 5108 | Andrews, Joshua | 5154 | Melcher, John | 5200 | Peters, John |
| 5109 | Backman, Isaac | 5155 | Bracket, John | 5201 | Phillips, John |
| 5110 | Damrell, William | 5156 | Alston, Richard | 5202 | Cross, Stephen |
| 5111 | Hannah, Thomas | 5157 | Dotts, Cornelius | 5203 | Hagen, Joel |
| 5112 | Davis, David | 5158 | Werell, Francis | 5204 | Prissey, John |
| 5113 | Roach, James | 5159 | Conley, Cornelius | 5205 | Allen, George |
| 5114 | Saunders, James | 5160 | Thompson, Michael | 5206 | Stiles, Jodick |
| 5115 | Smith, Robert | 5161 | McLane, John | 5207 | Smith, Charles |
| 5116 | Harris, William | 5162 | James, George | 5208 | Sumerg, Benjamin |
| 5117 | McCanon, John | 5163 | Jacob, Louis | 5209 | Davis, John |
| 5118 | Coffee, Ram'l | 5164 | Bailey, Isaac | 5210 | Brown, Benjamin |
| 5119 | Baldwin, James | 5165 | Dow, Henry | 5211 | Tapley, Isaac |
| 5120 | Wilson, John | 5166 | Boyle, Joseph | 5212 | Hill, Hannel |
| 5121 | Anderson, John | 5167 | Anthony, John | 5213 | Girdler, James |
| 5122 | Gier, Alexander | 5168 | Ropes, David | 5214 | Terry, Joseph |
| 5123 | Mott, Thomas | 5169 | Lamson, Noah | 5215 | Mitchell, John |
| 5124 | Minor, John | 5170 | Tucker, Samuel | 5216 | Young, John |
| 5125 | Trowbridge, J. T. | 5171 | Thompson, William | 5217 | Bodkin, William |
| 5126 | Delaney, Mathew | 5172 | Freeman, Isaac | 5218 | Spick, John |
| 5127 | Grafton, James | 5173 | Nelson, Thomas | 5219 | Champlin, Thomas |
| 5128 | Sault, Manuel | 5174 | Snow, Daniel | 5220 | Wherry, Daniel |
| 5129 | Alexander, Pedro | 5175 | Bateman, Michael | 5221 | Adams, Henry |
| 5130 | Francisco, Thomas | 5176 | Lambert, Ephraim | 5222 | Dealing, Elisha |
| 5131 | Hamilton, William | 5177 | Avery, John | 5223 | Soley, Nataniel |
| 5132 | Flood, Francis | 5178 | Higgins, John | 5224 | Thomas, John |
| 5133 | Sullivan, John | 5179 | Tolpie, Jonathan | 5225 | Berry, William |
| 5134 | Weldon, Asa | 5180 | Bragden, James | 5226 | Brush, Able |
| 5135 | Jones, William | 5181 | Thomas, John | 5227 | Fray, James |
| 5136 | Pearcey, John | 5182 | Briggs, William | 5228 | Lewis, William |
| 5137 | Rawlinson, Thomas | 5183 | Brooks, John | 5229 | Mason, Aaron |
| 5138 | Gould, Obadiah | 5184 | Riswell, Palmer | 5230 | Moor, Abraham |
| 5139 | Hogan, John | 5185 | Henley, John | 5231 | Hubble, James |
| 5140 | West, George | 5186 | Murray, Richard | 5232 | Roth, James |
| 5141 | Scott, Samuel | 5187 | Goodwin, John | 5233 | Self, Thomas |
| 5142 | Johnston, Thomas | 5188 | Barker, George | 5234 | Merrish, Joseph |
| 5143 | Parker, George | 5189 | Hay, John | 5235 | Stoddard, Conrad |
| 5144 | Holstein, Richard | 5190 | Harding, Joseph | 5236 | Wilson, John |
| 5145 | Patterson, William | 5191 | Bids, Thomas | 5237 | Moffatt, Hugh |
| 5146 | Burger, Francis | 5192 | Wiggins, Richard | 5238 | Brown, Thomas |
| 5147 | Archer, James | 5193 | Drisco, James | 5239 | Wiggins, Samuel |
| 5148 | Hammond, John | 5194 | Mullett, Joseph | 5240 | Davis, John |
| 5149 | Gross, James | 5195 | Saundry, Nathaniel | 5241 | Shotsenberg, Manuel |
| 5150 | Holden, John | 5196 | Molley, William | 5242 | Smith, George M. |
| 5151 | Webster, David | 5197 | Abbott, William | 5243 | Ayres, William |
| 5152 | Saunders, Richard | 5198 | Didderas, William | 5244 | Manning, George |

American Prisoners of War Held at Dartmoor during the War of 1812

## Numeric listing by prison number

| | | | | | |
|---|---|---|---|---|---|
| 5245 | Marshall, Benjamin | 5291 | Didler, Henry | 5337 | Perish, William |
| 5246 | Gaskin, William | 5292 | Jacobs, William | 5338 | Ring, Andrew |
| 5247 | Mista, Sullivan | 5293 | Turner, Silas | 5339 | Humphrey, Asa |
| 5248 | Vanrant, John | 5294 | Donison, William | 5340 | Shilling, James |
| 5249 | Lowe, George | 5295 | Killerman, M. | 5341 | Davis, Lott |
| 5250 | Munro, Harry | 5296 | Black, James | 5342 | Rogers, Pely |
| 5251 | Lewis, John | 5297 | Mason, James | 5343 | Turner, James |
| 5252 | McCannon, Joseph | 5298 | Domerell, Edward | 5344 | Briggs, Frank |
| 5253 | Bin, Peter | 5299 | Elliott, Andrew | 5345 | Atwood, James |
| 5254 | McBuchey, Patrick | 5300 | Robinson, Robert | 5346 | Lambert, Calvin |
| 5255 | Gill, James | 5301 | Mackay, Charles | 5347 | Black, Ruddick |
| 5256 | McGowen, Joseph | 5302 | Fletcher, John | 5348 | Levy, Charles |
| 5257 | Dalton, John | 5303 | Wood, Benjamin | 5349 | King, Peter |
| 5258 | Mooney, Mathew | 5304 | Boswell, John | 5350 | Wilson, Robert |
| 5259 | Doniner, John | 5305 | Sherdon, Joseph | 5351 | Hunt, Samuel |
| 5260 | Blarney, Henry | 5306 | Allen, William | 5352 | Thaya, Lubin |
| 5261 | Condon, Michael | 5307 | Miller, James | 5353 | Smith, Titus |
| 5262 | Willey, James | 5308 | Richards, Sandy | 5354 | Raddick, Ebenezer |
| 5263 | Donelly, Anthony | 5309 | Giles, John | 5355 | Thomas, William |
| 5264 | Fitzgerald, James | 5310 | Linnett, William | 5356 | Werman, John |
| 5265 | Clark, John | 5311 | Brown, John | 5357 | Witherett, Charles |
| 5266 | Kelley, Henry | 5312 | White, William | 5358 | Peters, John |
| 5267 | Ganganes, Edward | 5313 | Cotton, Edward | 5359 | Edwards, Thomas |
| 5268 | Williams, Thomas | 5314 | Thompson, Thomas | 5360 | Collins, John |
| 5269 | Johnson, George | 5315 | Bell, Richard | 5361 | Lunt, Daniel |
| 5270 | Hearns, Patrick | 5316 | Diamond, George | 5362 | Hughes, John |
| 5271 | Sheilds, Michael | 5317 | Edmond, John | 5363 | Kimbull, Samuel |
| 5272 | Dole, Andrew | 5318 | Ellis, John | 5364 | Avlyn, Lawrence |
| 5273 | Anderson, Henry | 5319 | Fulton, James | 5365 | Montgomery, William |
| 5274 | Strong, William | 5320 | Dullivan, James | 5366 | Harris, George |
| 5275 | Frederick, John | 5321 | Jarvis, Thomas | 5367 | Silvey, John |
| 5276 | Boyd, Stephen | 5322 | Lewis, John | 5368 | Atkinson, Charles |
| 5277 | Jessamine, John | 5323 | Artis, William | 5369 | Cowen, William |
| 5278 | Joseph, Nicholas | 5324 | Blair, David | 5370 | Steele, Thomas |
| 5279 | Anthony, James | 5325 | Blair, Benjamin | 5371 | Sankey, Ceasar |
| 5280 | La Roche, Jean | 5326 | Peach, John | 5372 | Williams, William |
| 5281 | Tuckerman, William | 5327 | Jackson, James | 5373 | Alford, William |
| 5282 | Harris, Joseph | 5328 | Hooper, Samuel | 5374 | Shepherd, Henry |
| 5283 | Gornerson, James | 5329 | Childs, William | 5375 | Parker, William |
| 5284 | Guages, Charles | 5330 | Byron, Ebenezer | 5376 | Turner, David |
| 5285 | Christian, John | 5331 | Douglass, Samuel | 5377 | Evans, William |
| 5286 | Payne, William | 5332 | Parrott, Ebenezer | 5378 | Fife, Thomas |
| 5287 | Carnes, John | 5333 | Boyd, Andrew | 5379 | Bennett, Robert |
| 5288 | Hatton, Peter | 5334 | Smith, Henry | 5380 | Read, Charles |
| 5289 | Anderson, George | 5335 | Eagerly, Ely | 5381 | Mathews, Cornelius |
| 5290 | Hammon, William | 5336 | Smith, Monday | 5382 | Robertson, Thomas |

American Prisoners of War Held at Dartmoor during the War of 1812

## Numeric listing by prison number

| | | | | | |
|---|---|---|---|---|---|
| 5383 | Groves, Thomas | 5429 | Kemble, John | 5475 | Pike, Benjamin |
| 5384 | Buskell, William | 5430 | Oliver, Anthony | 5476 | Richards, Willis |
| 5385 | Gaston, John | 5431 | Williams, Noble | 5477 | Gently, Thomas |
| 5386 | Gould, Henry | 5432 | Wetherby, Calvin | 5478 | Shipling, Peter |
| 5387 | Collier, Thomas | 5433 | Mills, John | 5479 | Colvell, John B. |
| 5388 | Straye, George L. | 5434 | Miller, William | 5480 | Fuller, Benjamin |
| 5389 | Bates, Josiah | 5435 | Hayes, John | 5481 | Cummings, Nathaniel |
| 5390 | Piles, James | 5436 | Storey, William | 5482 | Shade, Nathaniel |
| 5391 | Whistler, Miles | 5437 | Price, John | 5483 | Benson, Samuel |
| 5392 | Clover, Louis | 5438 | Bean, Amos | 5484 | Ingerson, Nathaniel |
| 5393 | West, John | 5439 | Wescott, Edward | 5485 | Killett, William |
| 5394 | Catley, William | 5440 | Dillon, Richard | 5486 | Crescoll, Joseph |
| 5395 | Dame, John | 5441 | Gardner, Daniel | 5487 | Osten, John |
| 5396 | Davis, John | 5442 | Morrison, Michael | 5488 | Dourville, Francis |
| 5397 | Percival, John | 5443 | Marchant, Isaac | 5489 | Frank, Louis Ville |
| 5398 | Whiffle, Joseph | 5444 | Brady, John | 5490 | Golliver, Thomas |
| 5399 | Lent, Samuel | 5445 | Bump, William | 5491 | Waters, John |
| 5400 | Dodge, Joseph | 5446 | Morrison, Joseph | 5492 | Andrews, James |
| 5401 | Noble, Daniel | 5447 | Peterson, John | 5493 | Francis, Hugh |
| 5402 | Sprague, John | 5448 | Jones, Richard | 5494 | Grant, Thomas |
| 5403 | Lewis, Thomas | 5449 | Wood, John | 5495 | Burgen, William |
| 5404 | Lacey, Henry | 5450 | Marshall, William | 5496 | Manning, Samuel |
| 5405 | White, James | 5451 | Wood, Samuel | 5497 | Forster, Samuel |
| 5406 | Beckford, John | 5452 | Ogle, Charles | 5498 | Harvey, Dover |
| 5407 | Adams, Abijah | 5453 | Langhane, Bell'm | 5499 | Cunninghan, John |
| 5408 | Wilson, Francis | 5454 | Scott, William | 5500 | Mitchell, Reuben |
| 5409 | Young, John | 5455 | Robinson, James | 5501 | Tucker, James |
| 5410 | McLane, George | 5456 | Daniels, John | 5502 | Baxter, John |
| 5411 | Brickman, John | 5457 | Jacobs, Hans | 5503 | Orcroft, Lewis |
| 5412 | Sergeant, Daniel | 5458 | Woodson, Isaac | 5504 | Edwards, Paul |
| 5413 | Mason, Richard | 5459 | Williams, William | 5505 | Doogood, Abraham |
| 5414 | Hicks, Ogershill | 5460 | Bell, Joshua | 5506 | McGuire, John |
| 5415 | Leach, Charles | 5461 | Cloutman, Thomas | 5507 | Pope, E. |
| 5416 | Kyler, John | 5462 | Willison, John | 5508 | Lattimore, John |
| 5417 | Anon, John | 5463 | Gale, Samuel | 5509 | Morris, Benjamin |
| 5418 | Queen, Daniel | 5464 | Farford, Samuel | 5510 | Petterson, Frederick |
| 5419 | Kennedy, James | 5465 | Balch, Samuel | 5511 | Alder, Clough |
| 5420 | Fosset, Rodwell | 5466 | Reed, Robert | 5512 | Shepherd, Isaac D. |
| 5421 | Strand, Peter | 5467 | Cloutman, Samuel | 5513 | Frost, Dephne |
| 5422 | Hicks, James | 5468 | Richards, Isaac | 5514 | Wilson, Robert |
| 5423 | Card, Israel | 5469 | Green, William | 5515 | Fresher, Henry |
| 5424 | Bowen, Oliver | 5470 | Dunham, George | 5516 | Woodward, Thomas |
| 5425 | Antoni, Francis | 5471 | Boardman, John | 5517 | Cook, Jeremiah |
| 5426 | Arnold, James | 5472 | Stroud, Joshua | 5518 | Clintic, Ralph |
| 5427 | Littleford, L. | 5473 | Bowden, William | 5519 | Ellis, Thomas |
| 5428 | Averill, Samuel | 5474 | Kellem, Daniel | 5520 | Weeks, Thomas N. |

American Prisoners of War Held at Dartmoor during the War of 1812

## Numeric listing by prison number

| | | | | | |
|---|---|---|---|---|---|
| 5521 | Green, John | 5567 | Parrish, William | 5613 | Porter, Frederick |
| 5522 | Fitzgerald, Edward | 5568 | Moor, William | 5614 | Caulfied, E. |
| 5523 | Jones, Edward | 5569 | Griffen, Andrew | 5615 | Tonkin, R. |
| 5524 | Merchant, John | 5570 | Curlis, Aaron | 5616 | Reily, Lewis |
| 5525 | Scobe, William | 5571 | Simmons, Thomas | 5617 | Furguson, L. |
| 5526 | Reynard, John | 5572 | Weeks, Joseph | 5618 | Coates, Russell |
| 5527 | Wilson, Nicholas | 5573 | Burbalk, Moses | 5619 | Conner, Edward |
| 5528 | Vorhis, Peter | 5574 | Robinson, Joshua | 5620 | Richards, Lawrence |
| 5528 | Coven, Clement | 5575 | Mason, Daniel | 5621 | Richards, Thomas |
| 5530 | Smith, Charles P. | 5576 | Landsbury, John | 5622 | Mugford, William |
| 5531 | Lenham, John | 5577 | Laurence, John | 5623 | Goldsmith, Thomas |
| 5532 | Creek, Frederick | 5578 | Ratcliffe, Joseph | 5624 | Brooks, John |
| 5533 | Hier, Andrew | 5579 | Ruby, Isaiah | 5625 | Welbor, John |
| 5534 | Phoenix, John | 5580 | Winn, Theodore | 5626 | Butler, John |
| 5535 | Morris, Manuel | 5581 | Dellanon, L. | 5627 | Long, David |
| 5536 | Glass, John | 5582 | Merrill, Abraham | 5628 | Hart, Emuis |
| 5537 | Lee, Richard Robert | 5583 | Smith, Thomas | 5629 | Queenwell, Peter |
| 5538 | Howard, Bethnel | 5584 | Blakely, Michael | 5630 | Bartlett, Samuel |
| 5539 | Antonia, Peter | 5585 | Harker, William | 5631 | Murray, Oliver |
| 5540 | Ellingwood, Joshua | 5586 | Taylor, James | 5632 | Marshall, B. |
| 5541 | Chandler, William | 5587 | Trowbridge, George | 5633 | Adams, J. |
| 5542 | Chandler, Samuel | 5588 | Ewer, David | 5634 | Fletcher, P. |
| 5543 | Hastings, Ephraim | 5589 | Prince, Silvester | 5635 | Miflin, Richard |
| 5544 | Johnston, William | 5590 | Jones, Henry | 5636 | Turner, James |
| 5545 | Golding, William | 5591 | Deroche, John Baptiste | 5637 | Thomas, Joseph |
| 5546 | Velpley, Israel | 5592 | Miller, John | 5638 | Muss, Joseph |
| 5547 | Robinson, Joseph | 5593 | Cops, Darius | 5639 | Smith, Charles |
| 5548 | Barrett, Thomas | 5594 | Boyd, Ephraim | 5640 | Gustable, Joseph |
| 5549 | Staines, Samuel | 5595 | Bennett, David | 5641 | Portlock, William |
| 5550 | Wingate, Jacob | 5596 | Young, James | 5642 | Lemorrett, John P. |
| 5551 | Briggs, John | 5597 | Joscelyn, Ambrose | 5643 | Shoals, Christopher |
| 5552 | Loring, Rufus | 5598 | Francois | 5644 | Monroe, John |
| 5553 | Perkins, Joseph | 5599 | Hall, Reuben | 5645 | Brownell, Jonathan |
| 5554 | Stickney, Amos | 5600 | Ramsden, William | 5646 | Brown, James |
| 5555 | Tubb, Samuel | 5601 | Rawlins, Nicholas | 5647 | Sutton, Martin |
| 5556 | Ayres, John | 5602 | Ridland, Ephraim | 5648 | Bell, Jacob |
| 5557 | Smith, George | 5603 | Carroll, William | 5649 | Wilson, John |
| 5558 | Stoddard, Samuel | 5604 | Tophouse, Samuel | 5650 | Duncan, Abel |
| 5559 | Young, William | 5605 | Davis, Richard | 5651 | Hutchins, Samuel |
| 5560 | Mansfield, George | 5606 | Cole, Nathaniel | 5652 | Lay, Lee |
| 5561 | Saunderson, Jacob | 5607 | Andrews, Jeremiah | 5653 | Caslow, John |
| 5562 | Tilman, Joseph | 5608 | Smith, Gerrard | 5654 | Hellin, John P. |
| 5563 | Hull, George | 5609 | Morrison, D. | 5655 | Gray, William |
| 5564 | Greenough, James | 5610 | Clim, George | 5656 | Adams, Robert |
| 5565 | Bowles, John | 5611 | Williams, Joseph | 5657 | Conklin, Vertius |
| 5566 | Balch, George W. | 5612 | Brown, George | 5658 | Smith, Joseph |

American Prisoners of War Held at Dartmoor during the War of 1812

## Numeric listing by prison number

| | | | | | |
|---|---|---|---|---|---|
| 5659 | Michaels, Henry | 5705 | Elliott, Stephen | 5751 | Carnell, George |
| 5660 | Rennard, George | 5706 | Allen, Archibald | 5752 | Price, Francis |
| 5661 | Morell, Samuel J. | 5707 | Whippen, W. C. | 5753 | Maynard, Humphrey |
| 5662 | Winkley, W. B. | 5708 | Reed, John | 5754 | Thomas, John |
| 5663 | Kennard, N. | 5709 | Thomas, James | 5755 | Webb, Michael |
| 5664 | Beck, Henry | 5710 | Hamilton, Clayton | 5756 | Willis, John |
| 5665 | Gordon, Joseph | 5711 | Nickerson, Warren | 5757 | Walker, James |
| 5666 | Moor, Samuel | 5712 | Crowell, Elijah | 5758 | Coleman, Andrew |
| 5667 | Toskins, Frederick | 5712 | Wilson, George | 5759 | Taylor, James |
| 5668 | Simpson, John | 5714 | Hart, Samuel | 5760 | Green, Robert |
| 5669 | Brown, John | 5715 | Henderson, Benjamin | 5761 | Lockwood, Benjamin |
| 5670 | Rickard, Elijah | 5716 | Woglon, Peter | 5762 | Gamble, James |
| 5671 | Varmrod, Samuel | 5717 | Winand, John | 5763 | Eldred, Peter |
| 5672 | Horney, Gilbert | 5718 | Leggatt, Charles | 5764 | Toutz, Vievel |
| 5673 | Taft, Henry | 5719 | Thornhill, R. | 5765 | Kevel, Alexander |
| 5674 | Horney, Charles | 5720 | Wolsey, M. | 5766 | Ballent, John |
| 5675 | Glaridge, Stephen | 5721 | Mainwarring, John | 5767 | Dolavice, Henry |
| 5676 | Stone, Abner | 5722 | Jackson, James | 5768 | Webb, Thomas |
| 5677 | Perkins, Joseph | 5723 | Jackson, Benjamin | 5769 | Gier, John |
| 5678 | Fairfield, James | 5724 | Preston, Thomas | 5770 | Williams, George |
| 5679 | Ward, Nathaniel | 5725 | Morris, Peter | 5771 | Harris, Moses |
| 5680 | Lord, Joseph | 5726 | Carney, Samuel | 5772 | Earle, James |
| 5681 | Lord, Dormer | 5727 | Smith, James | 5773 | Ebby, Samuel |
| 5682 | Lord, John | 5728 | Hall, James | 5774 | McKinzie, Daniel |
| 5683 | Lord, Benjamin | 5729 | Taylor, David | 5775 | Shore, Ceasar |
| 5684 | Patten, Robert | 5730 | Forrier, John | 5776 | Allen, John |
| 5685 | Perkins, Samuel | 5731 | Conner, William | 5777 | Rush, Samuel |
| 5686 | Townson, William | 5732 | Coleman, James | 5778 | Robinson, Garner |
| 5687 | Towan, Robert | 5733 | Hawkins, Parker | 5779 | Hussey, Joseph |
| 5688 | Mitchell, James | 5734 | Colston, John | 5780 | Cairns, Thomas |
| 5689 | Walker, Nathan | 5735 | Griffiths, Henry | 5781 | White, James |
| 5690 | Perkins, George | 5736 | McJugen, Robert | 5782 | Robinson, George |
| 5691 | March, Jesse | 5737 | DeCoine, John | 5783 | Morris, George |
| 5692 | Stone, Jonathan | 5738 | Peters, John | 5784 | Dickson, Abraham |
| 5693 | Dorman, Israel | 5739 | Hubbard, Joseph | 5785 | Jackson, James |
| 5694 | Roberts, Nathaniel | 5740 | Newman, John | 5786 | Davis, Peter |
| 5695 | Patten, Nathaniel | 5741 | Bryan, John | 5787 | Ellis, George |
| 5696 | Avril, Ebenezer | 5742 | Otiel, Charles | 5788 | Brown, George |
| 5697 | White, Peter | 5743 | Devinas, John | 5789 | Shiers, Vincent |
| 5698 | Archer, Daniel | 5744 | Denning, Joseph | 5790 | Toby, John |
| 5699 | Currie, James | 5745 | Manuel, John | 5791 | Woodward, Francis |
| 5700 | Francis, Thomas | 5746 | Atwood, Elisha | 5792 | Tinker, James |
| 5701 | Bradley, William | 5747 | Williams, Peter | 5793 | Holbrook, Ebenezer |
| 5702 | Place, James | 5748 | Chargne, Robert | 5794 | Addy, Francis |
| 5703 | Sherbon, Henry | 5749 | Johnson, Mark | 5795 | Lander, Warren |
| 5704 | Hobart, Samuel B. | 5750 | Stacey, Stephen | 5796 | Cramprey, James |

American Prisoners of War Held at Dartmoor during the War of 1812

## Numeric listing by prison number

| | | | | | |
|---|---|---|---|---|---|
| 5797 | Rich, Francis | 5843 | McGowen, Patrick | 5889 | Seapatch, John |
| 5798 | Whormsley, R. | 5844 | West, Abiel | 5890 | Paul, John |
| 5799 | Atkins, Henry | 5845 | Townsend, Jeremiah | 5891 | Batts, William |
| 5800 | Oliver, Griffith | 5846 | Pier, Henry | 5892 | Pritchard, T. G. |
| 5801 | Truffey, Joel | 5847 | Shelton, Smith | 5893 | Treadwell, Samuel |
| 5802 | Curgan, John | 5848 | Crawford, John | 5894 | Billings, Daniel |
| 5803 | Ridden, John | 5849 | Chasson, Thomas | 5895 | Whettan, John |
| 5804 | Chews, Thomas | 5850 | Mackay, Thomas | 5896 | Goodwin, Simon |
| 5805 | Gustavus, John | 5851 | Jones, Samuel | 5897 | Burdeen, Joseph |
| 5806 | Thomas, John | 5852 | Latimer, John | 5898 | Patch, George |
| 5807 | Felton, John | 5853 | Holford, Elisha | 5899 | Stover, David |
| 5808 | Lowton, John | 5854 | Cook, Edward | 5900 | Horne, E. L. |
| 5809 | Bennett, D. C. | 5855 | Bissen, John | 5901 | Ricker, Wentworth |
| 5810 | Fisher, William | 5856 | Beeston, Robert | 5902 | Marshall, Joseph |
| 5811 | Williams, Samuel | 5857 | Pacher, John | 5903 | Cowing, Samuel |
| 5812 | Morgan, William | 5858 | Nelson, William | 5904 | Lear, George |
| 5813 | Rowe, William | 5859 | Vanmetre, Henry | 5905 | Taylor, Paul |
| 5814 | Rutter, William | 5860 | Rick, Francis | 5906 | Hutchins, Joseph |
| 5815 | Rea, Charles | 5861 | Watson, Henry | 5907 | Seyeant, Henry |
| 5816 | Read, Thomas | 5862 | Williams, Thomas | 5908 | Cranley, James |
| 5817 | Bourdinon, Elijah | 5863 | Steel, John | 5909 | Deering, M. H. |
| 5818 | Lord, Samuel | 5864 | Hall, Robert | 5910 | Butler, Michael |
| 5819 | Bailey, Moses | 5865 | Pendleton, Jonathan | 5911 | Prely, Charles |
| 5820 | Williams, Jesse | 5866 | Nolen, William | 5912 | Lush, John G. |
| 5821 | Brown, Henry | 5867 | Vesplash, Nicholas | 5913 | Baker, Edward |
| 5822 | Weiden, Charles | 5868 | Garrison, James | 5914 | Cherry, David |
| 5823 | McStarbuck, Jordan | 5869 | Jacobson, Mathew | 5915 | Hosgood, B. |
| 5824 | Farrell, Richard | 5870 | Grinard, Bray | 5916 | Davis, Joseph |
| 5825 | Stanford, Nicholas | 5871 | Frederick, John | 5917 | Goodings, Thomas |
| 5826 | Wade, Nathan | 5872 | Roille, Oliver | 5918 | Underwood, B. |
| 5827 | Mastin, James | 5873 | Roderigo, Joseph | 5919 | Durgens, John |
| 5828 | Packett, John | 5874 | Gerard, Pierre | 5920 | Haley, Samuel |
| 5829 | Palmer, Thomas | 5875 | Guillaume, Jean Marie | 5921 | Grant, Samuel |
| 5830 | Ely, Daniel | 5876 | Evans, Pierrs | 5922 | Drinkwater, Daniel |
| 5831 | Roderick, John | 5877 | Dame, Esop | 5923 | Frazer, William |
| 5832 | Norton, Andrew | 5878 | Barton, Robert | 5924 | Cate, Joseph |
| 5833 | Griffin, William | 5879 | Mason, Joshua | 5925 | Smart, Richard |
| 5834 | Cedus, Francis | 5880 | Simpson, Joseph | 5926 | Adams, William |
| 5835 | Phillips, Augusta | 5881 | Hill, Timothy | 5927 | Spring, Seth |
| 5836 | Morey, Ezekiel | 5882 | Hutchins, John | 5928 | Varney, David |
| 5837 | Campbell, Jesse | 5883 | Modge, Daniel | 5929 | Washburn, Thomas |
| 5838 | Henderson, George | 5884 | Tradwell, James | 5930 | Hutchins, Theodore |
| 5839 | Blasdon, Philip | 5885 | Thomas, George | 5931 | Pukiam, E. L. |
| 5840 | Mount, James | 5886 | Jackson, Henry | 5932 | Crassey, Benjamin |
| 5841 | Hardy, Andrew H. | 5887 | Cornivalt, James | 5933 | Butler, James |
| 5842 | Snell, Shaderick | 5888 | Stone, John | 5934 | Clamberg, Peter |

American Prisoners of War Held at Dartmoor during the War of 1812

## Numeric listing by prison number

| | | | | | |
|---|---|---|---|---|---|
| 5935 | Moore, James | 5981 | Rider, John | 6027 | Johnson, Albert |
| 5936 | Farquhar, John | 5982 | Pickett, Josiah | 6028 | Wood, George |
| 5937 | Tombly, John | 5983 | Cole, David | 6029 | Christian, Nicholas |
| 5938 | Preniergath, Nicholas | 5984 | Stoves, Daniel | 6030 | Stanfield, Thomas |
| 5939 | Tucker, Levi | 5985 | Allen, Thomas | 6031 | Lee, John |
| 5940 | Averill, Joseph | 5986 | Burns, Benjamin | 6032 | Lynes, Amos |
| 5941 | Hoff, Abraham | 5987 | Merrill, Jacob | 6033 | Moody, John |
| 5942 | Clarke, Aaron | 5988 | Harding, Jonathan | 6034 | Jenkins, James |
| 5943 | Libbey, John | 5989 | Charles, Samuel | 6035 | Solomon, John |
| 5944 | Ellison, Joseph | 5990 | Middlefield, Abraham | 6036 | Merrick, William |
| 5945 | Carpenter, C. R. | 5991 | Gatchell, Benjamin | 6037 | Barrows, John |
| 5946 | Bartlett, Enoch | 5992 | Emlyn, Isaac | 6038 | Collins, George |
| 5947 | Holiday, John H. | 5993 | Hoff, Nicholas | 6039 | Patterson, Peter |
| 5948 | Clay, David | 5994 | Fisk, George | 6040 | Bruce, Peter |
| 5949 | Wallace, William | 5995 | Fletcher, Robert | 6041 | Curden, John |
| 5950 | Jordan, Samuel | 5996 | Washburn, Jeremiah | 6042 | Ellis, John |
| 5951 | Jennings, Samuel | 5997 | Ward, Israel | 6043 | Jones, John |
| 5952 | Bacock, Mathew | 5998 | Hutchins, Nathaniel | 6044 | Smith, Cornelius |
| 5953 | Williams, John | 5999 | Barnes, Thomas | 6045 | Littlefield, Nicholas |
| 5954 | Uen, Joseph | 6000 | Fairfield, Asa | 6046 | Wilson, Andrew |
| 5955 | Cutts, Ceaser | 6001 | Clough, Obed | 6047 | Berry, Jesse |
| 5956 | Logan, William | 6002 | Clough, Enoch | 6048 | Rogers, Daniel |
| 5957 | Pearce, Charles | 6003 | Clough, Shadrick | 6049 | Smith, Oakley |
| 5958 | Pickrase, Daniel | 6004 | Foss, Abijah | 6050 | Baxter, Marion |
| 5959 | Ridley, Nathaniel | 6005 | Richards, Richard | 6051 | Amos, Elijah |
| 5960 | Williams, John | 6006 | Noble, Joseph | 6052 | Sprust, Anthony |
| 5961 | Franics, John | 6007 | Marshall, M. | 6053 | Milner, Benjamin |
| 5962 | Bunharm, Jeremiah | 6008 | Marshall, Thomas | 6054 | Sacalogos, Nicholas |
| 5963 | Billings, Robert | 6009 | Ferrin, Daniel | 6055 | Fletcher, James |
| 5964 | Bailey, Joseph | 6010 | Murphy, Samuel | 6056 | Marchand, Angelo |
| 5965 | Byron, John | 6011 | Lewis, Benjamin | 6057 | Bird, Joseph |
| 5966 | Horman, Edward | 6012 | Cleeves, Jonathan | 6058 | Allen, Lark |
| 5967 | Plummer, John | 6013 | Littlefiled, Oliver | 6059 | Bakeman, Robert |
| 5968 | Dailey, Daniel | 6014 | Rhodes, Daniel | 6060 | Barden, Abel |
| 5969 | Warren, Samuel | 6015 | Linscott, Josiah | 6061 | McDougall, Hugh |
| 5970 | Long, John | 6016 | Arnold, John | 6062 | Southcombe, Kemp |
| 5971 | Allen, William | 6017 | Emery, John | 6063 | Dolphin, Francis |
| 5972 | Williams, John | 6018 | Littlefield, Israel | 6064 | Duffel, Barnett |
| 5973 | Simson, John | 6019 | Stone, Adam | 6065 | Doue, Alexander |
| 5974 | White, Gilman | 6020 | Derrill, Benjamin | 6066 | Nichols, John |
| 5975 | Hughes, Eben | 6021 | Hodsdale, Daniel | 6067 | Marion, John |
| 5976 | Squibb, Silas | 6022 | Eden, Mack | 6068 | Moore, George |
| 5977 | Puss, Nathaniel | 6023 | Stackpole, Andrew | 6069 | Johnson, Thomas |
| 5978 | Stacy, Samuel S. | 6024 | Parsons, Junius | 6070 | Carberry, Thomas |
| 5979 | Pendess, Lewis | 6025 | Clough, David | 6071 | Selley, Miles |
| 5980 | Hammon, Abner | 6026 | Collins, John | 6072 | Tullock, Thomas |

American Prisoners of War Held at Dartmoor during the War of 1812

## Numeric listing by prison number

| | | | | | |
|---|---|---|---|---|---|
| 6073 | Holmes, John | 6119 | Gracio, Anthony | 6165 | Pearse, John |
| 6074 | Moore, T. T. | 6120 | Smith, Ezkiel | 6166 | Baker, Jesse |
| 6075 | Seaman, M. C. | 6121 | Lee, Charles | 6167 | Snalt, George |
| 6076 | Sevier, William | 6122 | Williams, John | 6168 | Silkes, Peter |
| 6077 | Batterson, John | 6123 | Fetters, Robert | 6169 | Flowers, John |
| 6078 | Mendez, Joseph | 6124 | Roderique, Joseph | 6170 | Palmer, Peter |
| 6079 | Tratt, Lemuel | 6125 | Dellson, John | 6171 | Lewis, John |
| 6080 | Patterson, Joseph | 6126 | Lee, Richard | 6172 | Morris, Richard |
| 6081 | Cleverley, William | 6127 | Archer, Benjamin | 6173 | Bunkerson, John |
| 6082 | Western, Asa | 6128 | Ferry, William | 6174 | Hulen, Edward |
| 6083 | Craig, John | 6129 | Ballace, Daniel | 6175 | Goodrich, Edward |
| 6084 | Alston, Andrew | 6130 | Prescott, Abraham | 6176 | Black, Charles |
| 6085 | Morris, John | 6131 | Rutter, Thomas | 6177 | Robinson, Michael |
| 6086 | Bartholemew, B. | 6132 | Dolling, Gannet | 6178 | Wilson, Alexander |
| 6087 | Latham, Amel | 6133 | Mann, Charles | 6179 | Martin, John |
| 6088 | Davis, John | 6134 | Whilday, Jospeh | 6180 | Waters, George |
| 6089 | Farmer, Thomas | 6135 | Wright, Jonathan | 6181 | Tordey, Edward |
| 6090 | Johnson, Joseph | 6136 | Pass, Benjamin | 6182 | Cannon, William |
| 6091 | McCormick, Simon | 6137 | Shackley, John | 6183 | Barber, William |
| 6092 | Desley, Charles | 6138 | Pedro, Francisco | 6184 | Richards, Edward |
| 6093 | Bartlett, Henry | 6139 | Smith, Abraham | 6185 | Dole, William |
| 6094 | Horrow, Ansel | 6140 | Gorton, William | 6186 | Randall, James |
| 6095 | Platt, John | 6141 | Rigg, Robert | 6187 | Roberts, Nathaniel |
| 6096 | Hubbard, Daniel | 6142 | Beckett, William | 6188 | Miln, Nicholas |
| 6097 | Rivers, Thomas | 6143 | Bagley, William | 6189 | Fluken, William |
| 6098 | Jenkins, Elijah | 6144 | Sale, Thomas | 6190 | Sinclair, Thomas |
| 6099 | Huish, John | 6145 | Reninel, William | 6191 | Berry, Joseph |
| 6100 | Huane, Benjamin | 6146 | Clarke, William | 6192 | Colby, Benjamin |
| 6101 | Miller, William | 6147 | Hall, Joseph | 6193 | Bacon, James |
| 6102 | Murray, John | 6148 | Billows, Charles | 6194 | Harrison, George |
| 6103 | Williams, John | 6149 | Tremerin, Joseph | 6195 | Nelson, David |
| 6104 | Parker, Samuel | 6150 | Freeman, John | 6196 | O'Neil, Joseph |
| 6105 | Mitchell, John | 6151 | Tendreth, Robert | 6197 | Truman, Samuel |
| 6106 | Chace, George | 6152 | Depuyster, Pierre | 6198 | Lamb, Joseph |
| 6107 | Peterson, Jacob | 6153 | Allen, Henry | 6199 | Doggett, James |
| 6108 | Carey, Liberty | 6154 | Conner, William | 6200 | Rand, Thomas |
| 6109 | Smith, John | 6155 | Taylor, David A. | 6201 | Rand, Charles |
| 6110 | Fields, Robert | 6156 | Maxine, Joseph | 6202 | Davis, William |
| 6111 | Newman, John | 6157 | Jones, Abraham | 6203 | Adams, Samuel |
| 6112 | Fernald, William | 6158 | Cole, Charles | 6204 | Newell, Joseph |
| 6113 | Pike, Benjamin | 6159 | Brooks, Philip | 6205 | Condon, Samuel |
| 6114 | Rogers, Robert | 6160 | Fedy, Philip | 6206 | Balch, William |
| 6115 | Dixey, William | 6161 | Charles, John | 6207 | Barrett, Thomas |
| 6116 | Sherry, John | 6162 | Chase, Samuel | 6208 | Nowland, Thomas |
| 6117 | Whitamore, William | 6163 | Henry, William | 6209 | Keith, James |
| 6118 | James, Jeremiah | 6164 | Burnham, David | 6210 | Pritchard, George |

American Prisoners of War Held at Dartmoor during the War of 1812

## Numeric listing by prison number

| | | | | | |
|---|---|---|---|---|---|
| 6211 | Barker, Thomas | 6257 | Callaghan, James | 6303 | Coffin, Theodore |
| 6212 | Burchstad, John | 6258 | Clerk, James | 6304 | Martin, John |
| 6213 | Brown, John | 6259 | James, William | 6305 | Greenlaw, Jeremiah |
| 6214 | Dodge, David | 6260 | Ray, Richard | 6306 | Boyleston, Zebediah |
| 6215 | Fields, Charles | 6261 | Pitts, George | 6307 | Bang, George |
| 6216 | Creighton, William | 6262 | Cooper, Thomas | 6308 | Door, Ebenezer |
| 6217 | Gardner, Andrew | 6263 | Lambert, Andres | 6309 | Burnham, Abraham |
| 6218 | Parsons, William | 6264 | Robinson, William | 6310 | Blake, Ebenezer |
| 6219 | Witherington, James | 6265 | Coville, Nathaniel | 6311 | Studley, Warren |
| 6220 | Clapp, George | 6266 | Hautell, Stephen | 6312 | Jarvis, Peter |
| 6221 | Newell, Joseph | 6267 | Young, John | 6313 | Obrey, Mathew |
| 6222 | Richardson, Nathan | 6268 | Marshall, Anthony | 6314 | David, James |
| 6223 | May, John | 6269 | Peterson, John | 6315 | Davis, Frederick |
| 6224 | Loring, Caleb | 6270 | Peterson, Alexander | 6316 | Perry, Ebenezer |
| 6225 | Saverio, John | 6271 | Bourdon, Amond | 6317 | Wentworth, John |
| 6226 | Haskett, Luther | 6272 | Clark, Francis Marie | 6318 | Delaware, John |
| 6227 | King, Seth | 6273 | Nichelle, Pierie | 6319 | Jones, Edward |
| 6228 | Woods, Charles | 6274 | Frost, Thomas Bell | 6320 | Raymond, John |
| 6229 | Avies, Samuel | 6275 | Frazier, Charles | 6321 | Giddings, Johh |
| 6230 | Holmes, Samuel | 6276 | Hepburn, James | 6322 | Downs, Jesse |
| 6231 | Ball, Stephen | 6277 | Wayne, James | 6323 | Hawes, Caleb |
| 6232 | Newcomb, John | 6278 | Thompson, William | 6324 | Moor, Joshua |
| 6233 | Orkitt, Horea | 6279 | Williams, George | 6325 | Levit, Caleb |
| 6234 | Cobb, Josiah | 6280 | Wood, Robert | 6326 | Haley, Samuel |
| 6235 | Clowes, Philip | 6281 | Evans, Thomas | 6327 | Tarbor, Samuel |
| 6236 | Newall, John | 6282 | Conton, Philip | 6328 | Lewis, Jesse |
| 6237 | Hall, Stephen | 6283 | David, John M. | 6329 | Newall, J. B. L. |
| 6238 | Hingston, Richard | 6284 | Coffee, Jacob | 6330 | Snow, Daniel |
| 6239 | Lowe, Isaac | 6285 | Smith, John | 6331 | Frain, Joshua |
| 6240 | Johnston, Perry | 6286 | Burton, William | 6332 | Ganster, John |
| 6241 | Lee, Abraham | 6287 | Dine, William | 6333 | Bramant, Laurence |
| 6242 | Mounts, John | 6288 | Bryant, Benjamin | 6334 | Wright, John |
| 6243 | Stone, Jerry | 6289 | Banning, Peter | 6335 | Whittington, Lemuel |
| 6244 | Swaney, William | 6290 | Adams, Robert | 6336 | Haywood, Samuel |
| 6245 | Douglass, Mathew | 6291 | Judah, David | 6337 | Alexander, James |
| 6246 | Johnson, Jesse | 6292 | Miller, Jeremiah | 6338 | Tarr, William |
| 6247 | Green, Peter | 6293 | Barton, Mathew | 6339 | Hall, John |
| 6248 | Kendel, Franklin | 6294 | Pierce, Samuel | 6340 | Demedorff, John |
| 6249 | Pitts, Francis | 6295 | Finney, James | 6341 | Leonard, Thomas |
| 6250 | Stenson, David | 6296 | Truman, John | 6342 | Constant, William |
| 6251 | Oliver, Mathew | 6297 | Fisher, Robert | 6343 | Carleton, Christopher |
| 6252 | Davis, Joseph | 6298 | Barker, Charles | 6344 | Peterson, David |
| 6253 | Williams, James | 6299 | Samuels, Samuel | 6345 | Moore, Warren |
| 6254 | Stevens, William | 6300 | Monk, Philip | 6346 | Thomas, Theodore |
| 6255 | Limbourg, Charles | 6301 | Hawkins, John | 6347 | Goodhall, Joseph |
| 6256 | Coren, Hugh | 6302 | Symonds, Moses | 6348 | Fernandez, George |

American Prisoners of War Held at Dartmoor during the War of 1812

## Numeric listing by prison number

| | | | | | |
|---|---|---|---|---|---|
| 6349 | Nash, Alexander | 6395 | Kell, Francis | 6441 | Waters, Louis |
| 6350 | Hose, Richard | 6396 | Kitton, Abraham | 6442 | Salisbury, Joseph |
| 6351 | Lyon, John | 6397 | Blackston, William | 6443 | Pockmitt, Jackness |
| 6352 | Thomas, Charles | 6398 | Cooper, Peter | 6444 | Diverall, John |
| 6353 | Churchill, Joseph | 6399 | Madden, John | 6445 | Walker, Daniel |
| 6354 | Carnes, William | 6400 | Jones, James | 6446 | Holmes, Isaac |
| 6355 | Welch, John | 6401 | Watt, Samuel | 6447 | Clark, John |
| 6356 | Miller, John | 6402 | Johannes, Logan | 6448 | Skipper, John |
| 6357 | Glover, William | 6403 | Pray, Samuel | 6449 | Gowner, Joseph |
| 6358 | Spiller, Moses | 6404 | Bracket, James | 6450 | Whitning, William |
| 6359 | Jackson, Ebenezer | 6405 | Norris, Benjamin | 6451 | Henley, Jacob |
| 6360 | Gurney, John | 6406 | Osborne, John L. | 6452 | Perkins, Thomas |
| 6361 | Dunklin, Jesse | 6407 | Dolphin, Joseph | 6453 | Wilbourg, David |
| 6362 | Payton, James | 6408 | Perkins, Clement | 6454 | Dyer, Joseph |
| 6363 | Martyn, John | 6409 | Jones, John | 6455 | Turner, Joseph |
| 6364 | Black, William F. | 6410 | Jones, Calvin | 6456 | McCormick, James |
| 6365 | Baptiste, John | 6411 | Moore, Robert | 6457 | Westcott, William |
| 6366 | Marshall, William | 6412 | Plumber, James | 6458 | Bonfonce, Anthony |
| 6367 | Coates, Samuel M. | 6413 | Randell, Thomas | 6459 | Bidbee, Joseph |
| 6368 | Morris, John | 6414 | Doing, Denis O. | 6460 | Steel, John |
| 6369 | Sparkes, John | 6415 | Bisett, Robert | 6461 | Taylor, William |
| 6370 | Clyde, John | 6416 | Butcher, Jacob | 6462 | McNeil, Dennis |
| 6371 | Haycock, Joseph | 6417 | Jennings, Francis | 6463 | Phillips, William |
| 6372 | Thomas, John | 6418 | Jack, John | 6464 | Lacey, Zachariah |
| 6373 | Chick, Moses | 6419 | Hunt, James | 6465 | Garbonne, John |
| 6374 | Cloutman, Ephraim | 6420 | Benton, Samuel | 6466 | Welsh, Richard |
| 6375 | Andrews, Samuel | 6421 | Dawson, John | 6467 | Johnson, Charles |
| 6376 | Dunham, William | 6422 | Edwards, William | 6468 | Johnson, William |
| 6377 | West, Jacob | 6423 | Kingdom, John | 6469 | Coombs, John |
| 6378 | Hodge, Edward | 6424 | Johnson, Thomas | 6470 | Beck, Francis |
| 6379 | Millow, John | 6425 | Carle, James | 6471 | Anderson, John |
| 6380 | Kitchen, George | 6426 | Webster, Michael | 6472 | Williams, John |
| 6381 | Pope, Oliver | 6427 | Griffin, Heathcote | 6473 | Bowen, John |
| 6382 | Lewis, Winslow | 6428 | Whitehead, John | 6474 | Day, Samuel |
| 6383 | Rust, Zebulin | 6429 | Smith, John | 6475 | Moore, William |
| 6384 | Davis, Elias S. | 6430 | Armstrong, Charles | 6476 | Hicks, Benjmain |
| 6385 | Casper, William | 6431 | Henson, Peter | 6477 | Hitchell, George |
| 6386 | Kirby, Robert | 6432 | Butler, David | 6478 | Lumsby, Frederick |
| 6387 | Taylor, John | 6433 | Little, William | 6479 | Brown, William |
| 6388 | Anderson, John | 6434 | Whipple, John | 6480 | Favish, John |
| 6389 | Baker, Thomas | 6435 | Smith, Samuel | 6481 | Backley, Walter |
| 6390 | Southcomb, Peter | 6436 | Maceman, Thomas | 6482 | Hunter, G. S. |
| 6391 | Jenkins, Thomas | 6437 | Macklin, John | 6483 | Williams, John |
| 6392 | Harvey, James | 6438 | Gale, Russel | 6484 | Henderson, Joseph |
| 6393 | Hindman, John | 6439 | Hill, Francis | 6485 | Beter, John |
| 6394 | Baptiste, John | 6440 | Dominico, John | 6486 | Farrell, John |

American Prisoners of War Held at Dartmoor during the War of 1812

## Numeric listing by prison number

| | | | |
|---|---|---|---|
| 6487 | White, James | 6533 | Nicholas, Jacob |
| 6488 | Hinchman, George | 6534 | Deamon, Thomas |
| 6489 | Ramsdell, Charles | 6535 | Blackford, Henry |
| 6490 | McCarthy, Samuel | 6536 | Bicker, Charles |
| 6491 | Finch, Etienne | 6537 | Prout, Henry |
| 6492 | Jones, Henry | 6538 | Daniels, Henry |
| 6493 | Martyn, John | 6539 | Spear, Joseph |
| 6494 | Bonner, John | 6540 | Henory, James |
| 6495 | Rake, Martin | 6541 | Fuller, Bernard |
| 6496 | Lemon, John | 6542 | Roseberry, Charles |
| 6497 | Bailey, S. | 6543 | Price, Peter |
| 6498 | Fox, Edward | 6544 | Heasey, John |
| 6499 | Lawrence, Francis | 6545 | Sewell, John |
| 6500 | Detandes, Joseph | 6546 | Fisher, James |
| 6501 | Huxtable, William | 6547 | Alexander, George |
| 6502 | Johnson, Abraham | 6548 | Thompson, Joseph |
| 6503 | Phillips, John | 6549 | Gibson, Francis |
| 6504 | Campbell, Henry | 6550 | Hogerberth, John |
| 6505 | Hendry, William | 6551 | Saundes, Cornelius |
| 6506 | Battle, John | 6552 | Jackson, Joseph |
| 6507 | Finch, Abraham | 6553 | Robinett, Samuel |
| 6508 | Parsons, James | | |
| 6509 | Little, Charles | | |
| 6510 | Moss, Richard | | |
| 6511 | Roberts, James | | |
| 6512 | Benson, John | | |
| 6513 | Moulston, Nathaniel | | |
| 6514 | Jack, John | | |
| 6515 | Peterson, John | | |
| 6516 | Ash, Oliver | | |
| 6517 | Norman, William | | |
| 6518 | Pender, John | | |
| 6519 | Davis, James | | |
| 6520 | Jackson, Thomas | | |
| 6521 | Truelove, John | | |
| 6522 | Peterson, Thomas | | |
| 6523 | Richmond, Caleb | | |
| 6524 | Brown, Henry | | |
| 6525 | Baldwin, Theophilius | | |
| 6526 | Howdy, Joseph | | |
| 6527 | Davis, Richman | | |
| 6528 | Butler, Henry | | |
| 6529 | Stevens, John | | |
| 6530 | Groveman, Frederick | | |
| 6531 | Kellock, Stephen | | |
| 6532 | Stone, John | | |

American Prisoners of War held at Dartmoor during the War of 1812

## Prisoner listing by ship or regiment

Unknown

Adams, J.
Adams, Jesse
Adams, Peter W.
Adams, Robert
Adams, Thomas
Adams, William
Adivoe, Henry
Akerman, William
Albert, John
Albro, George
Alexander, George
Allen, Isaac
Allen, John
Allen, John D.
Alton, Peter
Amos, Isaac
Anders, John
Anderson, James
Anderson, John
Anderson, Joseph
Anderson, Robert
Anderson, Thomas
Angel, Silvester
Anon, John
Armstrong, Elisha
Armstrong, John
Armstrong, Joseph
Armstrong, Thomas
Arnold, James
Ash, Oliver
Atkins, Uriah
Atwood, Edward
Atwood, John
Augustus, Benjamin
Austen, William
Austin, Jonathan
Avery, Charles
Avlyn, Lawrence
Ayres, William
Babb, Benjamin
Babcock, Charles

Babcock, Daniel
Backman, Charles
Badcock, Bradley
Baddington, Asa
Badger, John
Bagley, William
Bailey, William
Baker, Daniel
Baker, Robert
Baker, Stephen
Baldwin, Theophilius
Bale, Charles
Baley, John
Ballard, John
Bannell, James
Banning, Peter
Bannister, Joshua
Banta, John
Barker, Charles G.
Barnard, John
Barnes, William
Barrett, George
Barrett, James
Barry, Peter
Barton, Mathew
Bates, Joseph
Battle, John
Baxter, Francis
Bean, Amos
Bean, William
Beck, Willaim
Beck, William
Beckett, William
Beckford, John
Beckwith, James
Bell, Joshua
Benjamin, Joseph
Bennett, John
Benson, George
Benson, James
Benson, John

American Prisoners of War held at Dartmoor during the War of 1812

## Prisoner listing by ship or regiment

Benson, Leven
Benson, Mingo
Bensted, John
Berry, Anton
Berry, Brook
Best, John
Beverly, Henry
Beymer, George
Billows, Charles
Bingham, Little
Bird, Comford
Bird, James
Black, Nicholas
Blake, William
Blanchard, Carvan
Blandel, Jonathan
Bliss, Frederick
Blossom, Seth
Blumbhouser, Samuel
Boggart, John
Boillet, John
Bond, Peter
Bond, Samuel
Bone, Charles
Booth, Thomas
Bordage, Raymond
Bordley, George
Bourns, George
Bourton, George
Boyce, Abraham
Boyd, John
Boyd, William
Bradbury, John H.
Bradie, John
Brady, Jason
Branch, Anthony
Brantre, James
Brenton, York
Bressy, Charles
Briant, Moses
Briggs, William

Brightman, Joseph
Brooks, Thomas
Broughton, Glover
Brown, Abraham
Brown, Charles
Brown, David
Brown, Elisha
Brown, Francis
Brown, George
Brown, Henry
Brown, Isaac
Brown, Jacob
Brown, James
Brown, Jesse
Brown, John
Brown, Joseph
Brown, Mark
Brown, Michael
Brown, Reuben
Brown, Richard
Brown, Samuel
Brown, Sawyer
Brown, Thomas
Brown, William
Bryant, Benjamin
Bryant, Stephen
Bump, William
Burdock, Enos
Burford, William
Burke, David
Burke, John
Burnes, Robert
Burnham, David
Burnham, John
Burrell, Ryal
Burton, William
Butler, Charles
Butler, George
Butler, John
Butler, Thomas
Butts, Joseph

American Prisoners of War held at Dartmoor during the War of 1812

## Prisoner listing by ship or regiment

Cadwell, Abraham
Cadwell, John
Cain, Enos
Cain, John
Cairns, Thomas
Calaman, John
Calebs, Lewis
Calentin, Samuel
Callaghan, James
Campbell, Henry
Campbell, James
Campbell, John
Campbell, Nathaniel
Canada, William
Cannon, Thomas
Cannon, William
Card, Israel
Carleau, Daniel
Carlton, William N.
Carney, Samuel
Carney, William
Caroline, Tobias
Carpenter, Henry
Carpenter, Nathaniel
Carr, Jonathan
Carroll, John
Carrot, Charles
Carter, James Wilson
Carter, Jesse
Cartwright, George
Cassam, Michael
Catwood, Zenos
Caulfied, E.
Chane, Daniel
Chapley, John
Chapman, Abraham
Chappel, Edward
Charles, John
Charles, Philip
Chase, Jacob
Chase, Joseph

Chase, Oliver
Check, Stephen
Chivers, Cantab
Christian, Tyrel
Chult, David
Church, Benjamin
Church, Jeremiah
Church, Richard
Clark, Amos
Clark, Elisha
Clark, Jacob
Clark, John D.
Clark, Philip
Clark, Samuel
Clark, William
Clarke, John
Clarke, William
Claussen, John
Claw, Morris
Clay, John
Clements, William
Clerk, James
Clifford, L. L.
Clim, George
Coburn, Abel
Codding, Caleb
Coffee, Jacob
Coffin, Joseph
Cole, Benjamin
Coleman, James
Collier, Thomas
Collins, George
Colquhoun, William
Congleton, A.
Congleton, Smith
Connelly, James
Conner, Edward
Conner, John
Conner, William
Conton, Philip
Conway, James

American Prisoners of War held at Dartmoor during the War of 1812

## Prisoner listing by ship or regiment

Conway, William
Cook, Isaac
Cook, John
Coombs, John
Cooper, James
Cooper, John
Cooper, Tannick
Cooper, Thomas
Cooper, William
Coren, Hugh
Cornish, John
Cornish, Thomas
Cotton, Samuel
Cottrell, Henry
Courtis, Thomas
Covall, Ephraim
Cowen, William
Cox, Abraham
Cox, John
Craig, William
Crawford, George
Cromwell, Glacio
Crosby, John
Crow, John
Cullett, William
Cummings, James
Cummins, Edward
Curry, Samuel
Curtis, Enoch
Cushman, Orson
Cutler, Thomas
Dalton, Samuel
Daniel, Robert
Daniels, John
Darran, Daniel
Darrow, Aaron
Davenport, Russell
Davey, Charles
David, John M.
Davis, Andrew
Davis, Daniel

Davis, George
Davis, James
Davis, John
Davis, Joseph
Davis, Osborne
Davis, Peter
Davis, Richman
Davis, Stephen
Davis, William
Dawson, John
Day, James
Day, John
Day, Thomas
Deagle, James
Degget, Hamel
Delaney, Mathew
Denham, Cornelius
Dennis, Thomas
Dennison, Laurence
Deverter, William
Dibble, Zachariah
Dickson, Abraham
Dildure, Samuel
Dillon, Richard
Dine, William
Dixon, Benjamin
Dixon, Peter
Dobins, John
Dole, Henry
D'Olivera, Manuel
Dorchester, Preston
Douglass, John
Downing, John
Drayton, John
Driver, John
Duncan, Edward
Duncan, George
Dunham, John
Dunn, Henry G.
Dunn, James
Dunning, William

American Prisoners of War held at Dartmoor during the War of 1812

## Prisoner listing by ship or regiment

Eddey, John
Edmonds, Francis
Edwards, Isaac
Edwards, John
Edwards, Thomas
Eldridge, Nathaniel
Elisha, Thomas
Elliott, Benjamin
Elliott, Robert
Ellis, Cornelius
Ellis, George
Ellis, John
Ely, Abraham
Emming, Thomas
Enderson, James
English, David
Enoch, Joseph
Erwin, Elijah
Evans, Hezekial
Evans, James
Evans, Thomas
Eyers, John
Fargo, Elijah
Farley, John
Farmer, Joseph
Farrell, Andrew
Fawcett, William
Fellebrown, William
Fernald, John
Ferris, Jacob
Ferris, James
Fields, Alexander
Filch, John
Finch, Abraham
Finney, James
Fisher, James
Fiske, Cyrus
Fletcher, P.
Flood, David
Flood, Francis
Flood, John

Fogerty, Archibald
Fogg, Noel
Folger, Frederick
Fool, Garret
Forbes, Robert
Ford, Charles
Ford, Philip
Forrest, James
Forrest, Wiliam
Forrier, John
Forster, Joseph
Forster, Thomas
Fossendor, William
Fosset, Rodwell
Foster, Cato
Foster, James
Foster, Joseph
Fowler, Isaac
Fowler, Joshua
Foxey, Thomas
Francis, John
Francis, Prince
Franklin, Edward
Franklin, William
Frazier, James
Frazier, William
Frederick, Charles
Freeborn, Joseph
Freely, Henry
Freeman, David
Freeman, Prince
Frizzle, David
Fry, Thomas
Fuller, Bernard
Fuller, Enoch
Furguson, L.
Gamble, James
Gandell, Epne
Gangler, George
Gannett, Mathew
Garder, George

American Prisoners of War held at Dartmoor during the War of 1812

## Prisoner listing by ship or regiment

| | |
|---|---|
| Gardner, Daniel | Greaves, John |
| Gardner, James | Green, George |
| Gardner, Jerry | Green, James |
| Gardner, Jonathan | Green, John |
| Gardner, Peter | Green, John M. |
| Garner, James | Green, Solomon |
| Garrison, Charles | Green, Thomas |
| Garrison, Richard | Gregory, Elijah |
| Gasey, Raymond | Grey, John |
| Gaskin, William | Grey, Thomas |
| Gaston, John | Griffin, John |
| Gayler, James | Groves, George W. |
| Geline, John | Groward, Peter |
| Gibbs, Valentine | Grundy, Edward |
| Gibbs, William | Grush, Joseph |
| Gibson, Francis | Guire, Andrew |
| Gibson, Samuel | Gunnell, William |
| Gier, Alexander | Hall, Daniel |
| Gifford, Robert | Hall, James |
| Gilbert, Thomas | Hall, John |
| Gilpin, William | Hall, Joseph |
| Glover, Benjamin | Hall, Richard |
| Glover, John | Hall, Stephen |
| Glover, Samuel | Hall, William |
| Golding, Abijah | Halton, William |
| Golliver, William | Hamel, William |
| Gooley, James | Hamilton, William |
| Gordan, William | Hammond, Isaac |
| Gordon, Abraham | Handfield, Robert |
| Gordon, James | Handley, John |
| Gordon, John | Haney, Edward |
| Gore, William | Hansey, Peter |
| Gorton, William | Harris, James |
| Gould, Henry | Harris, John |
| Grace, Allen | Harris, William |
| Graham, John | Harrison, Henry |
| Grandison, William | Harry, John |
| Grant, James | Hartfield, James |
| Graves, Thomas | Hartwell, Berry |
| Gray, Ephraim | Harvey, John |
| Gray, William | Hatkins, Henry |

# American Prisoners of War held at Dartmoor during the War of 1812

## Prisoner listing by ship or regiment

Hawker, Edward
Hawkins, John
Hawkins, Parker
Hawley, Frederick
Hayday, Larsay
Haydon, William
Haywood, John
Hazard, Robert
Headley, John
Heasey, John
Heaton, Henry
Heaton, William
Hendry, William
Henman, Elisha
Henry, Isaac
Henry, Robert
Henshaw, George
Hibben, Jeremiah
Hicks, James
Higgins, George
Higginson, David
Hill, James
Hill, Timothy
Hinton, John
Hitchcock, Edward
Hitchcock, Moses
Hoffman, James
Hofslider, Jessy
Hogan, William
Holbery, Emil
Holbrook, Benjamin
Holbrook, David
Holding, Henry
Holford, Elisha
Holmes, Charles
Holmes, John
Holstade, Joseph
Holstein, Peter
Homes, Ensign E.
Hooper, William
Hopkins, Samuel

Horn, Abner
Horsey, J. H.
Horsfalt, William
Hoskins, John
Hough, Ebenezer
Houghman, John
Houseman, John
Housewife, Mathew
Hovey, Joseph
Howdy, Joseph
Howe, Phineas
Howell, John
Howell, Thomas
Howland, William
Hoyt, James
Hubbard, John
Hubbard, John G.
Hudson, James
Hughes, John
Hughes, Thomas
Hunn, John
Hurd, Abel
Hussey, Thomas
Hutchins, Samuel
Hutchinson, Townsend
Hutes, Cyrus
Hutson, John
Huxtable, William
Ingerson, James B.
Irvine, Andrew
Irwin, Andrew
Jack, John
Jackson, Benjamin
Jackson, Daniel
Jackson, Elias
Jackson, Henry
Jackson, Isaac
Jackson, James
Jackson, John
Jackson, Sidney
Jackson, Thomas

American Prisoners of War held at Dartmoor during the War of 1812

## Prisoner listing by ship or regiment

Jackson, William
Jacobs, Hans
James, John
Jameson, George
Jamison, Peter
Jansen, Knute
Jardine, Samuel
Jarvis, George
Jarvis, James
Jefferson, Edward
Jenkins, Samuel
Jennings, John
Jetson, Samuel
Jeuvmeson, John
Jewitt, Jasper
Johnson, Abraham
Johnson, Andrew
Johnson, Edward
Johnson, Frederick
Johnson, George
Johnson, Henry
Johnson, John
Johnson, Joseph
Johnson, Mathew
Johnson, Oliver
Johnson, Richard
Johnson, Samuel
Johnson, Thomas
Johnson, William
Jonathan, Jonathan
Jones, Anthony
Jones, Charles
Jones, Isaac
Jones, James
Jones, John
Jones, Peter
Jones, Saul
Jones, Theodore
Jones, Thomas
Jones, William
Jonson, Edward

Jordan, Artemas
Judah, David
Judson, Obediah
Kanada, Joseph
Kanes, William
Keeffe, Alexander
Keg, Philip
Kelley, John
Kellogg, Amos
Kellum, Smith
Kemp, James
Kennedy, James
Kennedy, William
Kennisen, William
Kent, James
Kilham, John
Kimball, Nathaniel
Kimbull, Samuel
King, Peter
Kinsley, Benjamin
Knight, Elisha
Knox, John
Kyler, John
Lacey, Henry
Laingan, Daniel
Lake, Charles
Lake, Daniel
Lambert, Thomas
Lander, John
Landsbury, John
Lane, John
Lane, William
Langroth, Francis
Lapham, Cushman
Larkin, Amos
Latham, John
Lawrence, Robert
Laws, Peter
Lay, Lee
Leckler, Anthony
Lee, Edward

American Prisoners of War held at Dartmoor during the War of 1812

## Prisoner listing by ship or regiment

Lee, John
Lee, Joseph
Leeds, Leon
Lent, Joseph
Leonard, John
Lesle, Richard
Lethorpe, James
Levin, Alexander
Lewis, John
Lewis, Solomon
Lewis, Thomas
Limbourg, Charles
Lincoln, Solomon
Lindsay, Samuel
Lindsey, James
Lingall, George
Liscomb, William
Lister, Louis
Litchfield, Erick
Little, Charles
Little, Thomas
Locker, Michael
Lovell, William
Lowdie, Samuel
Lowe, Thomas
Lucas, Martin
Lucas, Robert
Lunberg, Berg
Lynch, Elisha
Mack, John
Mackay, Thomas
MacKenzie, George
Macure, Angelo
Magrath, Samuel
Maine, John
Mainwarring, John
Malis, John
Malone, James
Maloon, Bryant
Manning, George
March, Beverley

Marchant, Isaac
Maria, Josea
Mark, James
Marks, Peter
Marlborough, Francis
Marshall, Anthony
Marshall, Benjamin
Marshall, Francis
Marshall, John
Marshall, Thomas
Martin, Andrew
Martin, Henry
Martin, John
Martin, Thomas
Marvel, David
Masick, Joseph
Mason, J. Bude
Mason, John
Mathews, Henry
Mathews, Joseph
Mathews, Richard
Mayers, James
Mazal, James
McAlpin, Charles
McBride, James
McCarthy, John
McConnell, John
McCormic, William
McDonald, John
McFall, William
McFree, John
McGee, Robert
McGeorge, William
McInley, James
McIver, John
McKeige, Denis
McKennie, Barney
McKenzie, John
McLelland, William
McNeil, John
McUmber, Jacob

American Prisoners of War held at Dartmoor during the War of 1812

## Prisoner listing by ship or regiment

Meath, Samuel
Meker, James
Mellim, George
Melvin, John
Mercey, Thomas
Merchant, William
Merckett, John
Merle, John
Merrill, Enoch
Michaels, Henry
Michaels, John
Middleton, L.
Miller, Jeremiah
Miller, John
Miller, William
Mills, John
Mingalls, Robert
Minor, David
Mires, John
Mitchell, Henry
Mitchell, John
Mitchell, Thomas
Mold, John
Molloy, Peter
Monroe, James
Montgomery, John
Moor, Abraham
Moore, James
Moore, John
Morell, John
Morris, George
Morris, Louis
Morris, Peter
Morris, Samuel
Morrison, D.
Morrison, Michael
Morrison, William
Mortice, William
Morton, William
Moses, James
Moss, Richard

Moulden, William
Moulston, Nathaniel
Muckleroy, Samuel
Muggin, John
Munster, Isaac
Murray, James
Murray, John
Murray, Peter
Mutch, James
Nash, Daniel
Neil, John
Nelson, Richard
Newby, Lambert
Newell, John
Newell, Paul
Newman, Henry
Nicholas, John
Nicholls, John
Nichols, John
Nicholson, James
Nickerson, James
Nixon, Charles
Nixon, Primos
Noble, Charles
Noble, Daniel
Nolton, John
Norcross, Abel
Norcross, Phillipe
Norman, Michael
Norman, William
Norris, August G.
North, Thomas
Northcote, George
Northey, Joseph
Norton, Richard
Nunns, William
Nye, William
Oakley, Jacob
Odeen, John
Odgen, James
Oliver, Mathew

American Prisoners of War held at Dartmoor during the War of 1812

## Prisoner listing by ship or regiment

Osborne, Louis
Osbourne, Archibald
Osgood, David
Osten, John
Owen, Burden
Packhouse, William
Page, Cato
Paine, R. B.
Palmer, George H.
Palmer, William
Pardett, Charles
Paris, John
Paris, Peter
Parker, Edward
Parker, George
Parker, Richard
Parker, Samuel
Parker, William
Parr, Samuel
Parsons, Eugene
Parsons, Ignatius
Parsons, James
Pass, Benjamin
Paterson, Hanse
Patterson, Peter
Patterson, William
Patton, Robert
Paul, Jonathan
Payne, James
Peadon, William
Peake, J. W.
Peal, Andrew
Pearce, Thomas
Pearce, William
Pearson, Benjamin
Pearson, Charles
Peckham, Ezekiel
Pender, John
Pendergrast, Morris
Pendleton, Evan
Penfold, John

Penn, William
Penny, Richard
Penrose, Abraham
Perkington, John
Perkins, Henry
Perkins, William
Perry, Samuel
Perry, William
Peters, John
Peters, Peter
Peters, Thomas
Peterson, Alexander
Peterson, Jacob
Peterson, John
Peterson, Thomas
Peterson, William
Phillips, Benjamin
Phillips, Burton
Phillips, James
Phillips, John
Phillips, Peter
Phillips, William
Pierce, Amos
Pierce, Edward
Pierce, Samuel
Piles, James
Pines, Isaac
Pinkham, Allen
Pinne, John
Pith, William
Plowman, Joseph
Poland, Thomas
Pool, John
Poole, John
Porter, Ephraim
Porter, Frederick
Porter, Gideon
Porter, Joseph
Porter, Samuel
Potter, Jacob
Potter, John

American Prisoners of War held at Dartmoor during the War of 1812

## Prisoner listing by ship or regiment

Poverley, Henry
Powell, Joseph
Prentis, James
Preston, Thomas
Price, Charlton
Price, John
Price, Peter
Prichard, William
Prince, Jeffery
Prince, Lubin
Prince, Wiliam
Putnam, A.
Pyneo, John Rogers
Quarterman, William
Queen, Daniel
Raddick, Ebenezer
Rae, William Iron
Rameson, Thomas
Rankin, William
Ranlot, John
Ray, Charles
Ray, James
Ray, William
Raymond, George
Read, Thomas
Read, William
Redman, David
Redman, William
Reed, James
Reed, John
Reid, Thomas
Reily, Lewis
Reninel, William
Rennel, States
Reynolds, Amos
Reynolds, Stephen
Rice, William
Rich, Elisha
Rich, Francis
Richards, Henry
Richards, James

Richards, Lawrence
Richards, Thomas
Richardson, Edward
Richardson, James
Richardson, John
Richardson, Robert
Richardson, Samuel
Richmond, Caleb
Ricker, Thomas
Ricks, Thomas
Rideout, J. J.
Rideout, James
Rigg, Robert
Roach, Reuben
Roberts, David
Roberts, James
Roberts, John
Roberts, Robert
Robinet, Samuel
Robinett, Samuel
Robinson, Benjamin
Robinson, Edward
Robinson, George
Robinson, Henry
Robinson, James
Robinson, John
Robinson, Michael
Robinson, Robert
Robinson, William
Rodgers, Edward
Rodgers, John
Rolla, William
Rose, William
Roseberry, Charles
Rosett, Samuel
Ross, George
Ross, John
Rowe, Richard
Roweth, William
Rumlet, Ebenezer
Ryley, William

American Prisoners of War held at Dartmoor during the War of 1812

## Prisoner listing by ship or regiment

| | |
|---|---|
| Sale, Thomas | Simpson, Smith |
| Sandburn, John | Sims, Clement |
| Sanders, Gabriel | Sims, Oliver |
| Sanderson, William | Sims, William |
| Sankey, Ceasar | Simson, John |
| Sares, Charles | Slebar, Samuel |
| Saul, Francis | Sloane, William |
| Saunders, Peter | Small, Joseph |
| Saunders, Thomas | Smith, Abraham |
| Sawyer, Jacob | Smith, Caesar |
| Schanck, William | Smith, Charles |
| Scott, Alexander | Smith, George |
| Scott, Anthony | Smith, Gerrard |
| Scott, John | Smith, Henry |
| Scott, Samuel | Smith, James |
| Scott, William | Smith, John |
| Scribner, William | Smith, Joseph |
| Self, Thomas | Smith, Paul |
| Sewell, John | Smith, Peter |
| Sewell, Lot | Smith, Richard |
| Seymond, Abner | Smith, Thomas |
| Shaw, Andrew | Smith, William |
| Shepherd, James | Snow, Colger |
| Shepherd, William | Southwich, Israel |
| Sherridan, Henry | Sparrow, John |
| Sherrif, Benjamin P. | Sparrow, Joseph |
| Shiers, Vincent | Spear, Joseph |
| Shilling, Morris | Spencer, John |
| Shipley, Charles | Spicer, Alexander |
| Shipley, Daniel | Sprague, John |
| Shomodia, John | Spratt, Thomas |
| Shotsenberg, Manuel | Squires, Prince |
| Shrousy, William | Stacey, Perry |
| Shute, Thomas | Stafford, John |
| Sidebottom, John | Staggs, Henry |
| Silsby, Nathaniel | Staitmand, James |
| Silverlock, John | Stamford, James |
| Simes, Henry | Stanwood, Timothy |
| Simmons, Ebenezer | Starboard, William |
| Simmons, Robert | Starbuck, Thaddeus B. |
| Simons, William | Starkweather, James |

American Prisoners of War held at Dartmoor during the War of 1812

## Prisoner listing by ship or regiment

Stenson, David
Stephens, Obadiah
Stephens, William
Stephenson, John
Stevens, Ira
Stevens, Joseph
Stevens, William
Stevenson, Levi
Steward, John
Steward, William
Stewart, James
Stewart, John
Stewart, Scipio
Stibbins, Thomas
Stilwell, William
Stinton, James
Stone, Henry
Stone, Isaac
Stow, Jeremiah
Strand, Peter
Sturdwent, David
Suckley, Richard
Sutter, John G.
Sutton, John
Sutton, Prince
Swinbourne, George
Syeway, Peter
Symes, John
Taggart, John
Tarr, James
Taylor, James
Taylor, John
Taylor, Joseph
Taylor, Thomas
Taylor, William
Thaley, William
Thayer, Ebenezer
Theyer, James
Thimonier, Pr.
Thomas, Andrew
Thomas, Charles

Thomas, Francis
Thomas, Henry
Thomas, John
Thomas, Moses
Thomas, Thomas
Thomas, William
Thompson, George
Thompson, Henry
Thompson, James
Thompson, John
Thompson, Joseph
Thompson, William
Thomson, William
Thornhill, R.
Tilbrook, Barney
Tilley, Bernard
Tilman, Joseph
Tindall, Thomas
Tink, Henry
Tinker, James
Toby, John
Tomkins, Ephraim
Tomlinson, G. W.
Tonkin, R.
Tophouse, Samuel
Torgnet, Abel
Tower, Michael
Townsend, Thomas
Traphagan, Peter
Trask, Osborne
Trask, William
Travers, Thomas
Trout, William
Trowbridge, J. T.
Truelove, John
Turnbull, James
Turner, Daniel
Turner, William
Twycross, Samuel
Tyler, Louis
Urey, Peter

American Prisoners of War held at Dartmoor during the War of 1812

## Prisoner listing by ship or regiment

Vail, Jeremiah
Valentine, John
Van Horn, James
Vanderventer, John
Vankirk, Joseph
Vaughan, Robert
Vincent, John
Vindine, Gar.
Voight, Henry
Wade, Richard H.
Wadsworth, Thomas
Walden, James
Waldron, Hiram
Walker, Benjamin
Walker, John
Walker, Richard
Wall, William
Wallace, William
Walleman, William
Walling, James
Wanson, James
Ward, Alfred
Ward, John
Ward, Peter
Ward, Samuel
Warner, Henry
Warner, John
Warner, Thomas
Warnick, Robert
Warren, Nathaniel W.
Waterhouse, Joseph
Watkins, George
Watson, Daniel
Wavernce, James
Wayne, James
Webb, Alexander
Webber, John
Webster, William
Weeble, Robert
Weeden, Richard
Weeding, Isaac

Weeks, Benjamin
Weeks, James
Weeks, Louis
Welch, John
Welch, William
Welsh, Ezekiel
Wescott, Edward
Wessels, William
West, George
West, Reubin
Weston, William
Wetherby, Calvin
Weymer, John
White, Charles
White, Henry
White, James
White, Richard
White, William
Whitehead, John
Whitehouse, Asay
Whitmore, John
Wickham, John
Wickham, Thaddeus
Wickwall, Joseph
Wiggent, John
Wilcox, Ephraim
Wilcox, William
Wiley, David
Willett, John
Willett, Robert
Williams, Abraham
Williams, Benjamin
Williams, Darius
Williams, David
Williams, Edward
Williams, George
Williams, James
Williams, John
Williams, Jonathan
Williams, Joseph
Williams, Peter

American Prisoners of War held at Dartmoor during the War of 1812

## Prisoner listing by ship or regiment

| | | |
|---|---|---|
| Williams, Richard | | Younger, Louis |
| Williams, Robert | | Younger, William |
| Williams, Thomas | 13th US Infantry | Bin, Peter |
| Williams, William | | Blarney, Henry |
| Williamson, John | | Clark, John |
| Williamson, Joseph | | Condon, Michael |
| Willson, Charles | | Dalton, John |
| Wilson, Charles | | Dole, Andrew |
| Wilson, Daniel | | Donelly, Anthony |
| Wilson, Francis | | Doniner, John |
| Wilson, George | | Fitzgerald, James |
| Wilson, Henry | | Gill, James |
| Wilson, James | | Kelley, Henry |
| Wilson, John | | McBuchey, Patrick |
| Wilson, Robert | | McCannon, Joseph |
| Wilson, Samuel | | McGowen, Joseph |
| Wilson, Thomas | | Mooney, Mathew |
| Wilson, William | | Sheilds, Michael |
| Wingate, David | | Willey, James |
| Winter, Andrew | 14th US Infantry | Johansen, Johan |
| Wise, John | 1st Militia Regiment | Pier, Henry |
| Wolsey, M. | 1st US Artillery | Hearns, Patrick |
| Wood, Robert | 1st US Rifles | Snell, Shaderick |
| Wood, Thomas | 21st US Infantry | Wieman, Laurence |
| Wood, William | 22nd US Infantry | Mount, James |
| Woods, Merril | 25th US Infantry | Hardy, Andrew H. |
| Woodson, Isaac | 38th US Infantry | Cunninghan, John |
| Woolf, Andrew | 4th US Rifles | Blasdon, Philip |
| Woolridge, William | | Campbell, Jesse |
| Wright, Edward | | Henderson, George |
| Wright, Emos | 6th US Infantry | Ganganes, Edward |
| Wright, Israel | | Johnson, George |
| Wright, James | | Williams, Thomas |
| Wright, John | a recaptured MV | Ingram, John |
| Wright, Jonathan | | Isdale, James |
| Wynn, John | | Stevenson, George |
| Yeaton, James | Abel | Simes, Stephen |
| Yoing, William | Achille | James, Joseph |
| Young, John | | Thompson, John |
| Young, Moses | Adamant | Hunter, G. S. |
| Young, Thomas | Adams | Baldwin, James |

# American Prisoners of War held at Dartmoor during the War of 1812

## Prisoner listing by ship or regiment

| Ship | Prisoner | Ship | Prisoner |
|---|---|---|---|
| Adeline | Ruby, Isaiah | | Boyd, Ephraim |
| | Benny, David | | Cops, Darius |
| | Bertol, Samuel | | Deroche, John Baptiste |
| | Cole, John | | Douglass, Mathew |
| | Conklin, Edward | | Green, Peter |
| | Flatt, Rob | | Hall, Stephen |
| | Holbrook, Elias | | Hingston, Richard |
| | Hughes, John | | Johnson, Jesse |
| | Johnston, Edward | | Johnston, Perry |
| | Lorenze, Thomas | | Kendel, Franklin |
| | Morris, Isaac | | Lee, Abraham |
| | Newman, Daniel | | Lowe, Isaac |
| | Pottigal, John | | Miller, John |
| | Roberts, Moses | | Mounts, John |
| | Roberts, Nathaniel | | Stone, Jerry |
| | Senter, Noah | | Swaney, William |
| | Sole, Edward | | Young, James |
| | Sole, Elias | Alexander | Campbell, Alexander |
| | Sole, Thomas | | Carroll, William |
| | Teggett, John | | Chase, William |
| | Thomas, Stephen | | Dellanon, L. |
| | Walters, Philip | | Fulton, William |
| | Washbury, Elbert | | Marshall, Solomon |
| Aeolus | Hesty, John | | Merrill, Abraham |
| | Johnson, George | | Ramsden, William |
| | Legur, Ephaim | | Rawlins, Nicholas |
| | Martin, Peter | | Ridland, Ephraim |
| | Mason, William | | Smith, Thomas |
| | Morton, Samuel | | Stone, Edward |
| | Smith, William | | Symons, William |
| | Swain, Edward | | Tendreth, Robert |
| Agnes | Adams, John | Alfred | Barker, Joseph |
| | Conray, William M. | | Bessop, John |
| | Hoppins, William | | Blackler, William |
| | Jackson, Curtis | | Bowden, Benjamin |
| | Lehart, Jacob | | Bowden, Frederick |
| | Pearce, Joseph | | Broughton, John |
| Aigle | Dillingham, F. A. | | Brown, William |
| Ajax | Anderson, Henry | | Brumblecom, Merrit |
| Alatanta | Mista, Sullivan | | Bumbelcomb, Sewel |
| Albion | Bennett, David | | Bunyan, John |

American Prisoners of War held at Dartmoor during the War of 1812

## Prisoner listing by ship or regiment

| | | |
|---|---|---|
| Carroll, Samuel | | Richardson, William |
| Carswell, John | | Roff, Samuel |
| Chapman, Daniel | | Rogers, Ambrose |
| Cole, John | | Ronder, Stephen C. |
| Conway, Samuel | | Ross, Richard |
| Cook, Benjamin | | Roundy, Francis |
| Corbett, Michael | | Roundy, Samuel |
| Cox, Daniel | | Russell, Benjamin |
| Cox, Peter J. | | Ryan, Patrick |
| Curren, Nathaniel | | Ryan, William |
| Davis, John | | Savary, Peter |
| Dennis, Jonas | | Simons, John C. |
| Done, Elisha | | Smith, Joseph |
| Francis, Oliver | | Stacey, Osmond C. |
| Fuller, Nathaniel | | Stacey, William |
| Gale, Benjamin G. | | Storey, George |
| Gilbert, David | | Storey, John |
| Goss, Richard | | Techand, John |
| Graham, Benjamin | | Thomas, Jonah |
| Green, Joseph W. | | Thomson, William |
| Hammond, Edward | | Tucker, Edward |
| Hawks, Benjamin | | Vanderford, Benjamin |
| Hulin, Abraham | | Vickery, Thomas |
| Hunt, Job | | Wadden, Isaac |
| Kelley, James | | Waite, Jacob |
| Kelley, John | | Wakefield, John |
| Lane, Henry | | White, Kitt |
| Lane, Titus | | Wilcox, Ceasar |
| Lee, Richard | | Wilson, John |
| Leech, Ezekiel | | Winn, Joseph R. |
| Nowland, Andrew | | Woolridge, Robert |
| Patten, John U. | | Wright, Mathew |
| Phip, William | Alligator | Higby, James |
| Pierce, John | | Ingersole, John |
| Pittman, Benjamin | Amazon | Allen, Archibald |
| Preston, Samuel | | Bailey, Joseph |
| Prichard, Robert | | Billings, Robert |
| Prince, Prince | | Bunharm, Jeremiah |
| Pritchard, John | | Byron, John |
| Proctor, John | | Dailey, Daniel |
| Richards, Benjamin | | Elliott, Stephen |

American Prisoners of War held at Dartmoor during the War of 1812

## Prisoner listing by ship or regiment

|  |  |  |  |
|---|---|---|---|
|  | Horman, Edward | Ann | Antonia, Peter |
|  | Plummer, John |  | Ganell, James |
|  | Sherbon, Henry |  | Glass, John |
| Amelia | Dale, John |  | Howard, Bethnel |
| America | Blanchard, John |  | Jackson, David |
|  | Carswell, William |  | Lee, Richard Robert |
|  | Didderas, William | Ann Dorothy | Alder, Clough |
|  | Hubbard, William |  | Lattimore, John |
|  | Kent, William |  | McGuire, John |
|  | Pierce, Joseph |  | Morris, Benjamin |
|  | Reed, George |  | Petterson, Frederick |
|  | Richardson, William |  | Pope, E. |
|  | Salmon, Archibald | Ann Packet | Mitchell, William |
| Amiable | Astropp, Hans Christopher | Apparencen | Anderson, Edward |
|  | Blake, Thomas |  | Barbadoes, Robert |
|  | Campbell, John |  | Bunkerson, John |
|  | Crepo, William |  | Guiho, Pierre |
|  | Crosby, George |  | Neal, Daniel |
|  | Franks, Francis |  | Pelhem, Robert |
|  | Green, Peter |  | Raquinis, Francis |
|  | Jackson, William |  | Woods, Edward |
|  | Kitchen, Richard | Argner | Stilman, John |
|  | Nelson, Thomas | Argus | Bisbee, Elijah |
|  | Newell, Nathaniel |  | Bisbee, J. D. |
|  | Peane, Henry |  | Bridge, Jeremiah |
|  | Perkins, Thomas |  | Brower, John |
|  | Peterson, Laurence |  | Chase, Constant |
|  | Prymvoy, Richard |  | Choete, Thomas |
|  | Rodyman, John |  | Collins, John |
|  | Schew, Richard |  | Crosby, John |
|  | Smith, John |  | Davis, Solomon |
|  | Williams, John |  | Dill, William |
| Amity | David, William |  | Dixon, John |
|  | Duff, James |  | Edwards, John |
|  | Hobart, William |  | Francis, John |
|  | Leger, Andrew |  | Freeman, Plim |
|  | Moore, John |  | Furness, Jesse |
|  | Popal, Richard |  | Gage, Lot |
|  | Witherett, Charles |  | Gale, William |
| Amphitrite | Wilson, Francis |  | Gavet, James |
|  | Shairs, Samuel |  | Griffiths, Joseph |

# American Prisoners of War held at Dartmoor during the War of 1812

## Prisoner listing by ship or regiment

|  | | |
|---|---|---|
| | Hallet, William | Gilchrist, Samuel |
| | Harding, John | Savage, James |
| | Harris, Elpheus — Atlantic | Cross, Nathaniel |
| | Hodges, Hercules | Frack, Daniel |
| | Howe, Jacob | Hecock, William |
| | Jones, William | Knapp, Joseph |
| | Kitchen, Daniel | Wilson, Robert |
| | Lee. John — Atlas | Hancock, James |
| | Lewis, Henry | Long, John |
| | Lincoln, Ephraim — Avon | Carland, Lewis |
| | Loring, Samuel | Dyer, Jonathan |
| | Lunt, Daniel | Johnson, Elias |
| | Mains, Henry | McDowell, John |
| | Morriss, Andrew | Morris, John |
| | Norcross, Thomas | Olseen, Elias |
| | Peters, Benjamin | Smith, John |
| | Porter, Calvin | Walker, James |
| | Randall, Jacob — Baroness Longueville | Holstein, Richard |
| | Ringgold, Thomas | Jackson, William |
| | Robinson, Stephen | Primas, James |
| | Shepherd, Samuel — Barracuda | Snyder, William |
| | Simmonds, Charles — Betsey | Allen, Arthur |
| | Simmonds, Joel | Archer, Benjamin |
| | Skudder, Alexander | Ballace, Daniel |
| | Smith, George M. | Dellson, John |
| | Stetson, Hiram | Ferry, William |
| | Stickney, Abraham | Fetters, Robert |
| | Stitman, A. W. | Gould, Samuel |
| | Thinney, Elvin | Hanson, John |
| | Tomus, Andrew | Hardman, John |
| | Tyler, Joseph | Lee, Charles |
| | Wallace, Thomas | Lee, Richard |
| | Weslemyer, John | Payne, Alexander |
| | White, George | Robinson, Thomas |
| | Williams, George F. | Roderique, Joseph |
| | Williams, James | Sharpless, Robert |
| | Winchester, Ebenezer | Smith, Ezkiel |
| | Wyman, William | Vesan, Weslant |
| Atalanta | Baker, Charles | Williams, John |
| | Davies, Nathaniel — Betsy | Bevin, William |
| | Fletcher, B. | Bienfaux, Allen |

American Prisoners of War held at Dartmoor during the War of 1812

## Prisoner listing by ship or regiment

|  |  |  |  |
|---|---|---|---|
|  | Bunker, Isaiah |  | Curtis, Zebediah |
|  | Couret, Francis |  | Ford, Stan |
|  | Croker, Nathaniel |  | Franics, Michael |
|  | Daniels, John |  | Gordon, William |
|  | Green, Rebuen |  | Grimes, George |
|  | Harrington, John |  | Halfpenny, William |
|  | Landford, John |  | Harley, Edward |
|  | Meeden, William |  | Haro, John |
|  | Noyes, Charles |  | Headen, John |
|  | Nugent, John |  | Hendrickson, John |
|  | Paris, William |  | Higginbotham, William |
|  | Preston, Jonas |  | Hudson, William |
|  | Selston, Issac |  | Jefferson, Jacob |
|  | Skelder, Samuel |  | Johnson, Jacob |
|  | Wheeler, Henry |  | Keane, Daniel |
|  | White, James |  | Kitchen, Robert |
|  | Willey, Ebenezer |  | Law, Thomas |
| Black Swan | Bartlett, Henry |  | Mason, Thomas |
|  | Desley, Charles |  | McFadden, John |
|  | Farmer, Thomas |  | Miller, Thomas |
|  | Hastings, Ephraim |  | Morris, Joseph |
|  | Horrow, Ansel |  | Munsey, Daniel |
|  | Hubbard, Daniel |  | Norris, Philip |
|  | Johnson, Joseph |  | Palmer, Robert |
|  | McCormick, Simon |  | Parnell, Hugh |
|  | Platt, John |  | Rockhill, John |
|  | Rivers, Thomas |  | Rogers, Henry |
| Blanche | Ashton, William |  | Smith, John |
|  | Bailey, Thomas |  | Sullivan, Florence |
| Blockade | Anthony, James |  | Sutherland, John |
|  | Greenwood, Joseph |  | Thomas, John |
|  | Hazard, Prince |  | Titus, John |
|  | Silverthorn, James |  | Trout, Nathaniel |
|  | Woodford, James |  | Weaver, Michael |
| Bordeaux Packet | Barnett, Thomas |  | Yeoman, John |
|  | Barton, Samuel | Boros | Crasey, James |
|  | Bridson, Thomas | Brazilian | Allen, William |
|  | Burnham, Benjamin | Bunker Hill | Arnold, Obediah |
|  | Childs, Adam |  | Black, Charles |
|  | Cline, Lewis |  | Borsdell, Justice |
|  | Currant, Michael |  | Bowmer, Isaac |

American Prisoners of War held at Dartmoor during the War of 1812

## Prisoner listing by ship or regiment

|  |  |  |
|---|---|---|
| Brightman, Joseph |  | Swasey, John |
| Brisons, John |  | Thompson, John |
| Brooks, Russell |  | Thompson, Lawrence |
| Brown, Frederick |  | Whitney, Thomas |
| Butler, William | Buzi | Barnett, James |
| Channing, John |  | Mathews, P. |
| Christian, John |  | Segeant, John |
| Conner, Edward | Calabria | Alexander, Pedro |
| Daggett, Robert |  | Francisco, Thomas |
| Davis, John |  | Gould, Obadiah |
| Day, Benjamin |  | Hogan, John |
| Deane, Peter |  | Jones, William |
| Drew, Charles |  | Pearcey, John |
| Fisher, John |  | Rawlinson, Thomas |
| Gardner, Joseph |  | Sault, Manuel |
| Gatewood, James |  | Sullivan, John |
| Gordon, Richard |  | Weldon, Asa |
| Graves, Ebenezer | Cannoneer | Blackler, Bernard |
| Hazard, Charles |  | Coffin, Samuel |
| Hedenburg, Jacob |  | Colvelli, John |
| Hilbert, Henry |  | Divers, Charles |
| Joseph, Francis |  | Eaton, George |
| Kelly, Samuel |  | Erlstroom, John |
| Lawton, William |  | Grant, Peter |
| Leggett, William |  | Greenwood, Thales |
| Leno, Frederick |  | Heath, Henry |
| Livesley, Thomas |  | Hock, J. N. |
| McFarlane, Daniel |  | Holmes, Caleb R. |
| Miles, William |  | Jennings, Nathaniel |
| Moore, Edward |  | Jones, William P. |
| Morrison, Thomas |  | Landerman, Joseph |
| Nooney, William |  | Leroy, Alexander |
| Oliver, Joseph |  | Palmer, John |
| Orne, W. B. |  | Richardson, William |
| Osborne, Samuel |  | Stoddart, Robert |
| O'Sheffey, Jacobus |  | Turner, William |
| Richards, James |  | Watts, Hiram |
| Shaw, William |  | Wellander, Adam |
| Shephard, James | Carbineer | Ainsley, William |
| Spiers, Nathaniel |  | Bartrom, Lewis |
| Stephens, Amos |  | Hamilton, John |

# American Prisoners of War held at Dartmoor during the War of 1812

## Prisoner listing by ship or regiment

|  |  |  |  |
|---|---|---|---|
| | Neptune, Daniel | | Gallaway, Joseph |
| | Norbury, Joseph | | Gilbert, Thomas |
| | Parker, Thomas | | Halfpenny, Robert |
| | Snade, William | | Harris, John |
| Caroline | Anderson, David | | Hodge, Rufus |
| | Barrester, Peter | | Inbritson, Nicholas |
| | Baxter, David | | Jentile, Joseph |
| | Burns, Charles | | Kimmins, John |
| | Edwards, David | | Longworthy, John |
| | Evans, Moses | | Madden, Frederick |
| | Fitts, Joseph | | Morel, Thomas |
| | Jones, Lewis | | Mullan, John |
| | Martin, Anthony | | Pack, Abraham |
| | Minois, Pierre | | Riceo, Joachim |
| | Peregrine, Taggert | | Schaeman, Frederick |
| | Plumer, John | | Shamtan, Samuel |
| | West, Dennis | | Smith, Richard |
| | Williams, John | | Spears, Samuel |
| | Wright, George | | Wood, Thomas |
| Caromaned | Dawes, W. B. | Catherine | Barker, George |
| | Duffy, James | | Goodwin, John |
| | Forbes, James | Cato | Smith, Oakley |
| | Frenair, Domingo | Centurion | Mackay, Charles |
| | James, Benjamin | Chance | Armstrong, Charles |
| | Jarey, Barthomlew | | Butler, David |
| | Jones, Reuben | | Dominico, John |
| | Kelly, Henry | | Gale, Russel |
| | Lambert, William | | Henson, Peter |
| | Lock, James | | Hill, Francis |
| | Middleton, John W. | | Little, William |
| | Mills, Joseph | | Maceman, Thomas |
| | White, George | | Macklin, John |
| Cashier | Bannister, George | | Smith, Samuel |
| | Birch, Andrew | | Waters, Louis |
| | Blackston, Edward | | Whipple, John |
| | Bloom, Joseph | Charles | Welter, William |
| | Cooper, Daniel | Charlotte | Akens, William |
| | Deal, William | | Babb, Nathaniel |
| | Drake, Henry | | Blake, Philip |
| | Evans, Henry | | Booth, Charles |
| | Foster, Joseph | | Bowers, Jonathan |

American Prisoners of War held at Dartmoor during the War of 1812

## Prisoner listing by ship or regiment

|  |  |  |  |
|---|---|---|---|
|  | Brown, William |  | Crawford, William |
|  | Christy, Alexander |  | Dusheels, Arthur |
|  | Davis, Charles |  | Fletcher, John |
|  | Davis, Jacob |  | Gillstone, George |
|  | Florence, John |  | Goldsbury, William |
|  | Forbes, William |  | Gustave, Peter |
|  | Frewan, Nathaniel |  | Homell, Christopher |
|  | Gallin, John |  | Johnson, John |
|  | Garrison, Babet |  | Marshall, Emil |
|  | Goff, James |  | Martin, Stephen |
|  | Holmes, Andrew |  | McDonald, John |
|  | Johnson, Stephen |  | Moore, John |
|  | King, Peter |  | Oakley, Joseph |
|  | Lamb, Anthony |  | Palmer, John |
|  | Loring, William |  | Pearer, John |
|  | Mason, Moses |  | Roberts, Francis |
|  | Miller, Edward |  | Shaw, Edward |
|  | Mitchell, Ezekiel |  | Valient, Richard |
|  | Norman, Peter |  | Wade, John |
|  | Pike, John |  | Williams, Henry |
|  | Sammers, Henry | Christiana | Fellows, Isaac |
|  | Stoddard, Reuben | Circe | Blair, John R. |
|  | Taylor, Jacob |  | Holsley, John |
|  | Williams, George |  | Miller, William |
|  | Young, Thomas | Collin | Allen, Eveshia |
| Chasseur | Briggs, Benjamin |  | Dennison, Thomas |
|  | Dolphin, Francis |  | Garrett, William |
|  | Doue, Alexander |  | Holden, George |
|  | Duffel, Barnett |  | Humphries, Warren |
|  | Latham, Giles |  | Jones, Walter |
|  | Marion, John |  | Kennay, Jacob |
|  | McDougall, Hugh |  | Lindburgh, Charles |
|  | Moore, George |  | Livermore, Arthur |
|  | Nichols, John |  | Luce, Prosper |
|  | Rice, James |  | Moore, George |
|  | Southcombe, Kemp |  | Prior, Asa |
| Chesapeake | Barton, Isaac |  | Steward, Isaac |
|  | Bassonet, Charles |  | Toley, Elisha |
|  | Bebe, Joseph | Colonel Hopkins' Mililita | Townsend, Jeremiah |
|  | Cook, Benjamin | Columbia | Hawkins, Isaac |
|  | Cornish, Charles |  | Merrill, John |

American Prisoners of War held at Dartmoor during the War of 1812

## Prisoner listing by ship or regiment

| Ship/Regiment | Name | | Name |
|---|---|---|---|
| Comet | Colton, Charles | | Atkins, Joseph |
| | Monsieur, Peter | | Berryman, John |
| Commerce | Charlotte, William | | Biss, Daniel W. |
| | Hankey, Frederick | | Boyer, Joseph |
| | Murray, William | | Dickenson, Chester |
| | Pain, Mark | | Dore, Charles |
| | Quinn, William | | Dowell, Isaac |
| | Stanwood, James | | Hall, David |
| | Studdy, Richard | | Hall, Richard |
| Commodore Perry | Carman, George | | Hart, James |
| | Edwards, William | | Helair, Gasper |
| | Gosling, Joseph | | Hopkins, Daniel |
| | Hamilton, Robert | | Hutchins, Edward |
| | Hanner, Daniel | | Keen, Nathaniel |
| | Henzeman, Christopher | | Logan, William |
| | Hugg, Jacob | | Lowe, John |
| | Killingsworth, John | | Manson, William |
| | Lubery, John C. | | McKenzie, Alexander |
| | Lynch, William | | Mills, William |
| | Murray, Richard | | Mullins, James |
| | Newton, John | | Neale, Dennis |
| | Reed, Abraham | | Powers, James |
| | Reeves, Asa | | Roles, John |
| | Robins, William | | Shaw, Richard |
| | Werman, John | | Towson, William |
| | Woodbury, Dixey | | Walker, Armstrong |
| Confidence | Young, Ebenezer | | White, Davis |
| Corinthian | Rake, Martin | | Wood, John |
| Cossack | Cole, David | Creole | Scary, Joseph |
| | Green, Charles | Criterion | Baron, William |
| | Hammon, Abner | | Bidson, Thomas |
| | Hill, Benjamin | | Brown, Henry |
| | Pendess, Lewis | | Chambers, Henry |
| | Pickett, Josiah | | Dewett, John |
| | Pinder, George | | Dickson, Charles |
| | Puss, Nathaniel | | Griffiths, Thomas |
| | Rider, John | | Henderson, Alexander |
| | Stacy, Samuel S. | | Johnson, John |
| | Ward, Benjamin | | Johnstone, Thomas |
| | Weston, Nathaniel | | Ludlow, Charles |
| Courier | Anthony, Stephen | | Mannett, Richard |

American Prisoners of War held at Dartmoor during the War of 1812

## Prisoner listing by ship or regiment

|  |  |  |  |
|---|---|---|---|
|  | Marble, Jabez |  | Gould, John |
|  | Miller, John |  | Lewis, John |
|  | Monks, John |  | Miller, James |
|  | Neel, David |  | Nicholas, John Buess |
|  | Porter, William |  | Parsons, Rufus |
|  | Reeves, William |  | Patten, John |
|  | Royal, John |  | Paul, Jacob |
|  | Sage, Moses |  | Pearce, Alexander |
|  | Schultz, John |  | Phillips, Thomas |
|  | Sherman, Reuben |  | Phipps, Richard |
|  | Thomas, Archibald |  | Pierce, Samuel |
|  | Thompson, Martin |  | Rodgers, Abraham |
|  | Wheeler, Charles |  | Smith, Thomas |
|  | White, Henry |  | Tarlton, J. P. |
| Crosby's Regiment | West, Abiel |  | Wedgwood, William |
| Cupidon | Barnes, Joseph |  | Williamson, John |
| Cygnet | Dowling, John | Dash | Ratcliffe, Joseph |
|  | Farrell, James | David Porter | Alston, Andrew |
|  | Fuller, Gideon |  | Andrews, Joshua |
|  | Lambert, Samuel |  | Ayres, Parker |
|  | Lane, W. S. |  | Backman, Isaac |
|  | Maret, Ebenezer |  | Belcour, James |
|  | Smith, Turpin |  | Bent, Joseph |
|  | Soie, Henry |  | Bradley, William |
| Daedalus | Batterson, John |  | Cashman, Ansel |
|  | Carberry, Thomas |  | Cleverley, William |
|  | Griffen, Andrew |  | Colley, Thomas |
|  | Holmes, John |  | Craig, John |
|  | Mendez, Joseph |  | Currie, James |
|  | Moor, William |  | Francis, Thomas |
|  | Moore, T. T. |  | Frank, Joshua |
|  | Parrish, William |  | Goodrich, Edward |
|  | Patterson, Joseph |  | Green, Samuel |
|  | Reed, John |  | Henderson, John |
|  | Seaman, M. C. |  | Highby, William |
|  | Selley, Miles |  | Johnson, Samuel |
|  | Sevier, William |  | Jose, Emanuel |
|  | Tratt, Lemuel |  | Morris, John |
|  | Tullock, Thomas |  | Restrum, Andrew |
| Darby | Jerralls, Abraham |  | Saunders, John |
| Dart | Clark, Samuel |  | Sholtz, Charles |

American Prisoners of War held at Dartmoor during the War of 1812

## Prisoner listing by ship or regiment

|  |  |  |  |
|---|---|---|---|
|  | Taylor, David |  | Fingall, William |
|  | Thompson, Isaac |  | Jeffreys, Henry |
|  | Western, Asa |  | Lilliford, Jacob |
| Decatur | Mayeau, Morris |  | Watts, John |
| Decature | Anderson, John | Dictator | Bourns, John |
|  | Beck, Francis |  | Lewis, Gabriel |
|  | Bidbee, Joseph |  | Newell, George |
|  | Bonfonce, Anthony | Diomede | Anderson, William |
|  | Bowen, John |  | Briggs, Samuel |
|  | Coombs, John |  | Burns, Alexander |
|  | Day, Samuel |  | Caban, Samuel |
|  | Garbonne, John |  | Catona, Manuel |
|  | Johnson, Charles |  | Cliff, Peter |
|  | Johnson, William |  | Cowley, Samuel |
|  | Lacey, Zachariah |  | Dempsey, John |
|  | McNeil, Dennis |  | Downey, Charles |
|  | Phillips, William |  | Downey, Richard |
|  | Steel, John |  | Flint, John |
|  | Taylor, William |  | Giddens, John |
|  | Welsh, Richard |  | Gorham, Emanuel |
|  | Westcott, William |  | Gramber, Gustoff |
|  | Williams, John |  | Greenleaf, William |
| Derby | Dexter, Charles |  | Harrison, James |
|  | Dexter, George W. |  | Jewitt, Samuel |
|  | Dyer, Thomas |  | Lane, Henry |
|  | Holbrook, Ebenezer |  | Millet, Joseph |
|  | Parker, William |  | Neil, David A. |
|  | Smith, Kilby |  | Page, Reuben |
|  | Turner, David |  | Pierce, Moses |
| Devon | Bloventon, John |  | Porston, Joseph |
|  | Carter, Enoch |  | Ringe, John |
|  | Ewell, Edward |  | Robinson, Wililam |
|  | Hubble, James |  | Ronning, Mathew |
|  | Tuttle, Joseph |  | Shurman, Thomas |
| Diamond | Armstead, Edward |  | Slater, Ludowick |
|  | Cowder, James |  | Soderbury, John |
|  | Orwick, Thomas C. |  | Westcott, Charles |
|  | Oytiel, John |  | Wright, Daniel |
| Dick | Bishop, Edward | Dispatch | Atwood, P. |
|  | Chip, Charles | Dolores | Bailey, S. |
|  | Donaldson, Joseph | Dolphin | Ayres, Henry |

American Prisoners of War held at Dartmoor during the War of 1812

## Prisoner listing by ship or regiment

|  |  |  |  |
|---|---|---|---|
| | Swain, Luke | | Robinson, Samuel |
| Dominique | Allison, Thomas | | Vaughan, Nicholas |
| | Cecil, Francis | | Weaphor, Andrew |
| | Collison, Elliot | | West, John |
| | Cook, Samuel | Eagle | Phoenix, John |
| | Dunn, Edward | Echo | Davis, Nicholas |
| | Dunnings, Charles | El Patrick | Peterson, John |
| | Harley, Joshua | Elbridge Gerry | Allen, William |
| | Honner, John | | Baptiste, John |
| | Hughes, Abijah | | Barnes, Edward L. |
| | Jackson, Henry | | Bartlett, John |
| | Jones, Lawrence | | Barton, James |
| | Joseph, John | | Boyd, Andrew |
| | King, Uriel | | Brown, John |
| | Lamataille, John | | Clark, Joseph |
| | Mitchell, John | | Cook, Silvanus |
| | Newons, Francis | | Cree, William |
| | Parker, Thomas | | Davis, John |
| | Pike, James | | Davis, Lott |
| | Riviere, Toussaint | | Dean, James B. |
| | St. Martin, Joseph | | Donnell, Samuel |
| | Studley, John | | Douglass, Samuel |
| | Vassina, John B. | | Drake, Daniel |
| | West, John | | Eagerly, Ely |
| Dorothea | Nesbit, Richard | | Foot, Benjamin |
| | Soreby, Robert | | Harman, Isaac |
| | Detandes, Joseph | | Hitchins, William |
| | Lawrence, Francis | | Homer, Henry |
| Ducornau | Bennett, James | | Humphrey, Asa |
| | Brown, Benjamin | | Lee, Michael |
| | Brown, Ludwig | | Lewis, John |
| | Brown, Samuel | | Luffie, Warren |
| | Brownell, Richard | | Miller, James |
| | Burdge, Samuel | | Parrott, Ebenezer |
| | Garret, James | | Parsons, John |
| | Heydon, Elie | | Perish, William |
| | Hurd, William | | Plummer, William |
| | Lemercier, Peter | | Richards, Sandy |
| | McCormick, Simon | | Ring, Andrew |
| | Norton, George | | Shilling, James |
| | Philen, Richard | | Smith, Henry |

American Prisoners of War held at Dartmoor during the War of 1812

## Prisoner listing by ship or regiment

|  |  |  |
|---|---|---|
|  | Smith, Monday | Butcher, Jacob |
|  | Sterns, Joseph | Christian, John |
|  | Thaya, Lubin | Clarke, Isaac |
|  | Thomas, John | Clothy, John |
|  | Trumbull, John | Clothy, William |
|  | Turner, Samuel | Cloutman, Joseph |
|  | Webb, Nathaniel | Crowel, Mathew |
|  | Worthey, Samuel | Doing, Denis O. |
| Eliza | Alley, Samuel | Duncan, Thomas |
|  | Casey, Henry | Findley, Thomas |
|  | Dunchellier, Isaac | Fuller, Nathaniel |
|  | Hammond, Joseph | Giddins, Hiram |
|  | Hudson, Daniel | Gornerson, James |
|  | Longford, Samuel | Goss, Jesse |
|  | Mason, John | Guages, Charles |
|  | Murray, Jacob | Hamson, John |
|  | Smith, Chester | Hamson, William |
|  | Strube, John | Harris, Joseph |
|  | Wheadon, Anthony | Horne, John |
|  | Williams, John | Jack, John |
|  | Young, Cook | Jackson, James |
| Eliza (Whaler) | Burrow, Charles | Jennings, Francis |
|  | Clark, Titus | Johnson, George |
|  | Crumpton, William | Kelley, John |
|  | Frederick, John | Kindan, Ephraim |
|  | Holdridge, Hector | Lackey, Joseph |
|  | Huston, James | Lamson, Amos |
|  | Lee, Isaac | Lufkin, William |
|  | Pratt, Lester | McIntire, John |
|  | Sims, Joseph | Melzard, Peter |
| Ellen and Elizabeth | Hunter, Alexander | Millett, John |
|  | Sline, Samuel | Molley, William |
| Emmeline | Dallison, James | Mullett, Joseph |
| Emperor Napeleon | Johnson, Thomas | Oakes, George |
| Enterprize | Abbott, William | Peach, John |
|  | Amos, Elijah | Peck, Benjamin |
|  | Bassett, John | Pettingall, Joseph |
|  | Battes, John | Price, John |
|  | Bisett, Robert | Randell, Thomas |
|  | Blair, Benjamin | Richardson, William |
|  | Blaney, Stephen | Robinson, William |

# American Prisoners of War held at Dartmoor during the War of 1812

## Prisoner listing by ship or regiment

| | | |
|---|---|---|
| | Russell, Robert | Trefry, Peter |
| | Sanburn, James | Waters, Abraham |
| | Saundry, Nathaniel | Welch, Benjamin |
| | Simmons, David | White, Henry |
| | Smith, Thomas | Williams, Thomas |
| | Thompson, John | Experiment | Adams, Samuel |
| | Thornton, John | Fair American | Abbott, Ephraim |
| | Tildon, Robert | Baldwin, John |
| | Torry, Henry | Burts, Joseph |
| | Treadwell, Nathaniel | Clapp, George |
| | Tucker, William | Downe, William |
| | Walker, Samuel | Gardner, Edward |
| | Webb, John | Gerard, William |
| | Webster, Asa | Homer, Michael |
| | Widger, John | Major, James |
| | Williams, Frederick | Mathews, Williams |
| Enterprize (Saratoga) | Hall, James | Morgan, John |
| | Taylor, David | Myers, Daniel |
| Espadron | Gray, Thomas | Richardson, Cheney |
| Essex | Amerson, Charles | Sparks, Robert |
| | Bancroft, Samuel | Still, David |
| | Blodget, Caleb | Williams, James |
| | Brown, Samuel | Fairy | Lester, James |
| | Burnham, Enoch | Lynch, William |
| | Chandler, Simon | Nichols, William |
| | Clothey, Thomas | Rogers, Luke |
| | Davis, John | Yorkman, Archibald |
| | Diluo, Benjamin | Falcon | Hadley, George |
| | Doliver, William | Henry, Henry |
| | Eastland, James | Homer, John |
| | Fisher, Lewis | Fame | Archer, Samuel |
| | Jeffrey, Francis | Bird, Thomas |
| | Libley, Moses | Blyth, Jonathan |
| | Maine, William | Bray, Zachariah |
| | Merritt, Thomas | Brown, Thomas |
| | Moss, Thomas | Forrest, John |
| | Parker, J. A. | Jenkins, Samuel |
| | Ripley, Eden | Jennings, Henry |
| | Russell, Patton | Kelley, James |
| | Scovel, John | Kenny, George |
| | Stockham, W. B. | Lewis, Henry |

American Prisoners of War held at Dartmoor during the War of 1812

## Prisoner listing by ship or regiment

| | | | |
|---|---|---|---|
| | Long, Joseph | Farmer's Daughter | Benton, Samuel |
| | Monte, Charles | | Carle, James |
| | Palmer, Peter | | Dawson, John |
| | Peabody, John | | Edwards, William |
| | Perkham, Peter | | Griffin, Heathcote |
| | Pierce, Edward | | Hunt, James |
| | Porter, Edward | | Johnson, Thomas |
| | Simpson, William | | Kingdom, John |
| | Stoddard, Thomas | | Webster, Michael |
| | Treadwell, Nathaniel | Ferox | Bailey, Samuel |
| | Watson, George | | Cadwell, James |
| | Williams, John | | Crandell, John |
| | Williamson, Richard | | Deal, John |
| Fame (Whaler) | Bunker, Thomas | | Durham, Charles |
| | Dunham, Daniel | | Freeman, Charles |
| | Johnson, Richard | | Gilbert, George |
| | Marsh, Hercules | | Lee, George |
| | Nicholls, Thomas | | Meyers, David |
| | Ostman, David | | Trusty, Henry |
| | Swain, James | Fiere Facia | Barker, A. L. |
| | Theal, Bristo | | Bradly, Samuel |
| | Williams, Stephen | | Byers, Sharp |
| Fanny | Abbey, Obadiah | | Camp, Tobias |
| | Boose, Abraham | | Carson, John |
| | Cheney, Daniel | | Cooke, John |
| | Cottle, John | | Fitzsmond, Jeremiah |
| | Davis, James | | Howard, Benjamin |
| | Dematra, George | | Johnson, James |
| | Griffee, Thomas | | Joseph, Michael |
| | Hayes, Benjamin | | Logan, James |
| | Hedley, John | | Pigger, Stephen |
| | Hollis, Nathaniel | | Webster, George |
| | Hoselquist, John | | Williams, John |
| | Lambert, William | | Williams, Peter |
| | Martling, Abraham | Fire Fly | Allen, David |
| | McFadon, James | | Day, John |
| | Routeman, Benedict | | Haskell, Thomas |
| | Sereder, Martin | | Jeffreys, Henry |
| | Vandeburg, Ad'm | | Jones, Stephen |
| | Wildon, Thomas | | Millett, Joseph |
| | Williams, Thomas | | Pearson, Samuel |

American Prisoners of War held at Dartmoor during the War of 1812

## Prisoner listing by ship or regiment

|       |               |              |                      |
|-------|---------------|--------------|----------------------|
|       | Rowe, William |              | Mason, James         |
|       | Shaw, Henry   |              | McKey, James Abercomby |
|       | Tar, Caleb    |              | Mingle, Thomas       |
|       | Williams, William |          | Moore, Richard       |
| Flash | Bradie, Thomas |             | Morrell, Jacob       |
|       | Dexter, David |              | Perry, Charles       |
|       | Falkern, W.   |              | Price, John          |
|       | Howard, Frederick |          | Resmabin, Benjamin   |
|       | Langley, Richard |           | Rowle, Benjamin      |
|       | Smith, David  |              | Sawyer, Peter        |
|       | Smithers, Richard |          | Slocombe, Abraham    |
|       | Starberne, John |            | Wilkins, William     |
| Flath | Garnder, Francis |           | Woodcroft, John      |
| Fly   | Adams, Henry  | Francis      | Ferguson, John       |
|       | Dealing, Elisha | Francis and Ann | Coffin, John     |
|       | Hatton, Peter |              | Hebron, John         |
| Foria | Bartholemew, B. |            | Lettington, James    |
|       | Latham, Amel  | Franklin     | Mason, Richard       |
| Fox   | Andress, Daniel | Frederick  | Clark, John          |
|       | Baldwin, John |              | Dorr, Edward         |
|       | Beard, Francis | Frederick Augusta | Roath, Stephen  |
|       | Black, James  | Friends      | Currien, Stephen     |
|       | Brights, George |            | Curtis, Francis      |
|       | Brown, Thomas |              | Curtis, Henry        |
|       | Calhoun, Richard |           | Giles, Edward        |
|       | Canabon, J.   |              | Green, Moses         |
|       | Domerell, Edward |           | Tibbett, Henry       |
|       | Dymoss, Peter |              | Underwood, John      |
|       | Elliott, Andrew |            | Young, Joseph        |
|       | Fox, Edward   | Frolic       | Andrews, Nathaniel   |
|       | Green, John   |              | Brown, Benjamin      |
|       | Hammon, William |            | Buchannon, Robert    |
|       | Hane, W. B.   |              | Clements, Samuel     |
|       | Harris, John  |              | Clexton, R.          |
|       | Harvey, Joseph |             | Dennis, William      |
|       | Irwin, Mathew |              | Elger, Richard       |
|       | Jackson, William |           | Grey, Francis        |
|       | James, Daniel |              | Hartsoln, John       |
|       | Jenkins, Joseph |            | Heny, Daniel         |
|       | Kirk, William |              | Holden, Charles      |
|       | Mack, Theron  |              | Jackson, David       |

American Prisoners of War held at Dartmoor during the War of 1812

## Prisoner listing by ship or regiment

|  |  |  |  |
|---|---|---|---|
|  | Jackson, George |  | Hupton, Henry |
|  | Johnson, Thomas |  | Lackman, Isaac |
|  | Leach, George |  | Moses, John |
|  | Maddison, Alexander |  | Pendleton, Charles |
|  | McLane, George |  | Quincy, Benjamin |
|  | Murray, M. |  | Rusto, Peter |
|  | Newberry, John |  | Springle, James |
|  | Odiorne, John |  | Williams, Robert |
|  | Parr, John | General Kempt | Archer, Joseph |
|  | Perkins, Jonathan |  | Dolliver, Richard |
|  | Pettingell, James |  | Domeree, John |
|  | Pittman, Samuel |  | Fisher, John |
|  | Richardson, Andrew |  | Morie, Denis |
|  | Rowland, William |  | Parley, William |
|  | Shott, John |  | Perkins, Elijah |
|  | Simon, Mark |  | Rogers, Robert |
|  | Smith, Charles |  | Smith, Richard |
|  | Southwick, George |  | Spern, Elijah |
|  | Stevens, Thomas |  | Symonds, M. John |
|  | Stone, Benjamin |  | Wheeler, John |
|  | Swaine, Thomas |  | Wyer, Joseph |
|  | Wills, Nathaniel | General Patch's Volunteers | McGowen, Patrick |
|  | Woolridge, W. | General Putnam | Andrews, James |
| Garland | Carey, Liberty |  | Balch, Samuel |
|  | Dixey, William |  | Benson, Samuel |
|  | Dole, William |  | Boardman, John |
|  | Fernald, William |  | Bowden, William |
|  | Fields, Robert |  | Burgen, William |
|  | Newman, John |  | Cloutman, Samuel |
|  | Pike, Benjamin |  | Cloutman, Thomas |
|  | Prescott, Abraham |  | Colvell, John B. |
|  | Randall, James |  | Cramprey, James |
|  | Roberts, Nathaniel |  | Crescoll, Joseph |
|  | Rogers, Robert |  | Cummings, Nathaniel |
|  | Smith, John |  | Davis, Richard |
| Garter Wester | Young, Nathaniel |  | Dourville, Francis |
| Gazelle | Cowing, Charles |  | Dunham, George |
|  | Dunbar, Luther |  | Farford, Samuel |
|  | Scott, John |  | Forster, Samuel |
| General Hart | Haley, John |  | Francis, Hugh |
|  | Hanson, Henry |  | Frank, Louis Ville |

American Prisoners of War held at Dartmoor during the War of 1812

## Prisoner listing by ship or regiment

|  |  |  |  |
|---|---|---|---|
|  | Fuller, Benjamin |  | Pratt, Paris |
|  | Gale, Samuel | Gold Coiner | Meigs, John |
|  | Gently, Thomas | Good Friends | Boriesa, Pierre |
|  | Golliver, Thomas |  | Harris, George |
|  | Grant, Thomas |  | Hogabets, John |
|  | Green, William |  | Hogerberth, John |
|  | Harvey, Dover |  | Johnson, Samuel |
|  | Ingerson, Nathaniel |  | Lerna, John |
|  | Kellem, Daniel |  | Martin, Henry |
|  | Killett, William |  | Morris, John |
|  | Lander, Warren |  | Morse, Henry |
|  | Manning, Samuel |  | Moss, John |
|  | Osten, John |  | Swain, Darius |
|  | Pike, Benjamin |  | Wessel, Samuel |
|  | Reed, Robert | Good Intent | Andrews, Thomas |
|  | Richards, Isaac |  | Davis, Benjamin S. |
|  | Richards, Willis |  | Mellett, James |
|  | Shade, Nathaniel |  | Sargant, Samuel H. |
|  | Shipling, Peter | Goree | Congdon, James |
|  | Stroud, Joshua |  | Stibbens, Stephen |
|  | Waters, John | Gothland | Conner, Galen |
|  | Willison, John |  | West, John |
| General Stark | Ariskins, Samuel |  | Wilson, William |
|  | Budwell, Alpheus | Gotley | Duncan, Jesse |
|  | Mason, Daniel | Governor Gerry | Armstrong, William |
| George | Green, Tobias |  | Bosset, David |
| Gleaner | Layton, William |  | Charter, James |
| Globe | Brown, Robert |  | Cross, Oliver |
|  | Dew, Frederick |  | Cudsworth, Henry |
|  | Ferguson, Thomas |  | Gabriel, John |
|  | Forsyth, Robert |  | Gage, Isaac |
|  | Godfrey, John |  | Harris, James |
|  | Hunter, James |  | Hill, John |
|  | Le More, John |  | Jones, Thomas |
|  | Seith, John |  | Killer, John |
|  | Shelton, Samuel |  | Linsey, Alexander |
|  | Stanfield, George |  | Lopez, B. Eldridge |
|  | Taylor, William |  | Merritt, Robert |
| Gloucester | Davis, John |  | Smith, James |
| Godfrey and Mary | Brown, Francis |  | Smith, John |
|  | Peckham, Henry |  | West, Simon |

American Prisoners of War held at Dartmoor during the War of 1812

## Prisoner listing by ship or regiment

|  |  |  |  |
|---|---|---|---|
|  | Weston, David |  | Tomkins, Abraham |
| Governor McKean | Augustus, Amos |  | Wiley, John |
|  | Baptist, John | Grand Napoleon | Brown, Aaron |
|  | Boggs, James |  | Cornwall, Arthur |
|  | Capewill, Bartholomew |  | Hutchins, Henry |
|  | Farrell, Francis |  | Lane, James |
|  | Hayes, Simon |  | Ostand, Michael |
|  | Hook, John |  | Parish, Samuel |
|  | Humphries, Jacob |  | Poland, John |
|  | Lehens, Andrew |  | Pote, Jeremiah |
|  | Regens, Jonathan |  | Redding, John |
|  | White, Philip |  | Rozier, William |
|  | Winter, Peter |  | Thompson, Joseph |
| Governor Middleton | Dodge, Joseph |  | White, Isaac |
|  | Doer, James |  | Wilcox, Lewis |
|  | Driver, Thomas | Grand Turk | Downing, Benjamin |
|  | Fry, Peter |  | Hall, James |
|  | Johnson, John |  | Henderson, James |
|  | Shaw, John |  | Malloy, William |
|  | Walton, Christopher |  | Pellow, William |
|  | Wilson, John |  | Philips, Timothy |
|  | Wrigth, Thomas |  | Soley, Nataniel |
| Governor Plumer | Butler, George |  | Sutton, Benjamin |
|  | Drisco, James |  | Turrell, Ebenezer |
|  | Ducat, William |  | Warren, David |
|  | Fall, James | Grecian | Downing, George |
|  | Fernald, William |  | Evans, William |
|  | Fitherly, Robert |  | Herring, William |
|  | Keen, Benjamin | Greyhound | Adams, James |
|  | McKinney, Isaac |  | Burbage, Henry |
|  | Moore, Samuel |  | Hutchins, John |
|  | Place, Thomas | Growler | Bartlett, Robert |
|  | Randall, Henry |  | Bowden, William |
|  | Sides, Samuel |  | Bradford, George |
|  | Sonnett, James |  | Brown, John |
|  | Tarlton, John |  | Brown, Joseph |
|  | Tuckerman, William |  | Cash, John |
|  | Varney, John |  | Chandler, Enoch |
|  | Vennard, George |  | Coombe, Michael |
| Governor Shelby | Oliver, John |  | Craft, Richard |
|  | Shanks, William |  | Crocker, Silvester |

American Prisoners of War held at Dartmoor during the War of 1812

## Prisoner listing by ship or regiment

| | | |
|---|---|---|
| Davis, William | | Tucker, Andrew |
| Day, Frederick | | Tucker, Nathaniel |
| Derring, William F. | | Widger, Joseph |
| Devereux, Benjamin | | Witham, Burrell |
| Dunn, William | Halifax Packet | Allen, William |
| Easten, Ephraim | | Hobart, Samuel B. |
| Emerson, David | | Hughes, Eben |
| Fletcher, John | | Long, John |
| Florance, Charles | | Place, James |
| Forster, George | | Simson, John |
| Foster, Samuel | | Squibb, Silas |
| Garrett, L. P. | | Warren, Samuel |
| Gatchell, John G. | | White, Gilman |
| Grush, John | | Williams, John |
| Gursh, Nathaniel | Hall | Castana, John Baptist |
| Harvey, Anthony | Hannah | Blaird, David |
| Hoper, Joseph | | Cooper, James |
| Jones, Thomas | | Harvey, Joseph |
| Joseph, Nicholas | | Hoyt, David |
| Kegs, Zenas | | Pritchard, Israel |
| Lee, Nathaniel | | Ramsden, John |
| Lehman, Francis | | Raysden, John |
| Lindsay, Nathaniel | | Selman, John |
| Malcomb, Alexander | | Wood, Samuel |
| Nicholson, James | Hannah & Eliza | Baday, Edward |
| Nicholson, John | | Fry, John |
| Odiam, Joseph H. | | Laurenceau, John |
| Orne, Israel | | Roman, John |
| Phippen, Israel | | Williams, James |
| Pittman, Benjamin | | Wing, Nathaniel |
| Roberts, Joel | Hannibal | Harris, James |
| Roundy, Jeremiah | Harford | Elwell, Lewis |
| Roundy, Thomas | | Pendleton, George |
| Russell, William | Harlequin | Adams, William |
| Rust, John | | Averill, Joseph |
| Shaw, Samuel | | Bacock, Mathew |
| Smith, Benjamin | | Baker, Edward |
| Smith, Jeremiah | | Balch, George W. |
| Smith, John | | Bartlett, Enoch |
| Thornton, William | | Barton, Robert |
| Threshow, James | | Batts, William |

American Prisoners of War held at Dartmoor during the War of 1812

## Prisoner listing by ship or regiment

Beck, Henry
Billings, Daniel
Bowles, John
Brown, John
Burdeen, Joseph
Butler, James
Butler, Michael
Carpenter, C. R.
Cate, Joseph
Cherry, David
Clamberg, Peter
Clarke, Aaron
Clay, David
Cornivalt, James
Cowing, Samuel
Cranley, James
Crassey, Benjamin
Cutts, Ceaser
Dame, Esop
Davis, Joseph
Deering, M. H.
Drinkwater, Daniel
Durgens, John
Ellison, Joseph
Farquhar, John
Franics, John
Frazer, William
Glaridge, Stephen
Goodings, Thomas
Goodwin, Simon
Gordon, Joseph
Grant, Samuel
Greenough, James
Haley, Samuel
Hill, Timothy
Hoff, Abraham
Holiday, John H.
Horne, E. L.
Horney, Charles
Horney, Gilbert
Hosgood, B.

Huane, Benjamin
Huish, John
Hull, George
Hutchins, John
Hutchins, Joseph
Hutchins, Theodore
Jackson, Henry
Jenkins, Elijah
Jennings, Samuel
Jordan, Samuel
Kennard, N.
Lear, George
Libbey, John
Logan, William
Lush, John G.
Marshall, Joseph
Martyn, John
Mason, Joshua
Modge, Daniel
Moor, Samuel
Moore, James
Morell, Samuel J.
Patch, George
Paul, John
Pearce, Charles
Pickrase, Daniel
Prely, Charles
Preniergath, Nicholas
Pritchard, T. G.
Pukiam, E. L.
Rennard, George
Rickard, Elijah
Ricker, Wentworth
Ridley, Nathaniel
Seapatch, John
Seyeant, Henry
Simpson, John
Simpson, Joseph
Smart, Richard
Spring, Seth
Stone, John

American Prisoners of War held at Dartmoor during the War of 1812

## Prisoner listing by ship or regiment

|  |  |  |  |
|---|---|---|---|
|  | Stover, David |  | Brien, Lewis |
|  | Taft, Henry |  | Brook, David |
|  | Taylor, Paul |  | Buckley, William |
|  | Thomas, George |  | Cain, Joseph |
|  | Tombly, John |  | Coleman, William |
|  | Toskins, Frederick |  | Counthon, John |
|  | Tradwell, James |  | Cowen, Thomas |
|  | Treadwell, Samuel |  | Davis, John |
|  | Tucker, Levi |  | Days, Thomas |
|  | Uen, Joseph |  | Eburn, Aaron |
|  | Underwood, B. |  | Eldon, James |
|  | Varmrod, Samuel |  | Elwell, Samuel |
|  | Varney, David |  | Gilbert, Henry |
|  | Wallace, William |  | Grist, William |
|  | Washburn, Thomas |  | Harrison, Silas |
|  | Whettan, John (Whitorn) |  | Jennings, John |
|  | Williams, John |  | Jervis, Isaac |
|  | Winkley, W. B. |  | Jones, Paul |
| Harmony | Hall, David |  | Jordan, Simon |
|  | Hull, Homer |  | Laurent, Jonathan |
|  | King, John |  | Lovely, Placid |
|  | Langley, Richard |  | Marshall, Levy |
|  | Navar, Francis |  | McQueen, Charles |
|  | Whithead, Daniel |  | Osborne, John |
|  | Williams, James |  | Pannison, Henry |
| Harriett | Baptiste, John |  | Paul, Daniel |
|  | Mathews, Cornelius |  | Ringrove, William |
|  | Read, Charles |  | Savage, Robert |
| Hartford | Croilt, Aaron |  | Sole, John |
| Harvest | Belavoine, L. |  | Tooley, William |
|  | Carter, Henry |  | Warner, Benjamin |
|  | Grenaux, Yves |  | Waterford, J. H. |
|  | Holden, Andrew |  | Wearney, Richard |
|  | LeCore, Peter M. |  | Whitehead, Reuben |
|  | Morrison, John Baptist |  | Wilkins, William |
|  | Murphy, William |  | Williams, George |
|  | Port, John |  | Woodman, Thomas |
|  | Ranger, Desire | Hawke | Catley, William |
| Hawk | Adams, William |  | Dame, John |
|  | Bill, Zachariah |  | Fitzgerald, John |
|  | Bolding, Garret |  | Franklin, W. H. |

447

American Prisoners of War held at Dartmoor during the War of 1812

## Prisoner listing by ship or regiment

|  |  |  |  |
|---|---|---|---|
|  | Jones, John |  | Carty, Joseph |
|  | Percival, John |  | Chariton, James |
|  | Tripe, W. H. |  | Cochrane, William |
| Hazard | Collett, John |  | Colbourn, James |
|  | Smallpiece, John |  | Conklin, Vertius |
| Hebe | Chiseldine, John |  | Dennis, Francis |
|  | Conway, Charles |  | Densey, Peter |
|  | Denham, John |  | Doane, Joshua |
|  | Fish, Joseph |  | Edwards, John |
|  | Mode, David |  | Emas, Comfort |
|  | Page, Thomas |  | Flynn, John |
|  | Thompson, William |  | Freeman, Halkins |
|  | Watson, James |  | Gibbs, Thomas |
|  | Williams, John |  | Gould, John |
| Helen and Emeline | Adams, James |  | Gray, William |
|  | Gallaway, Joseph |  | Greenfield, James |
| Henrietta | Gunn, Charles William |  | Gwynn, Josiah |
|  | Rossuros, Philip |  | Hall, Zachariah |
| Henry Clements | Beck, Steward |  | Higgins, Isaac |
|  | Gallibrandt, Bernard |  | Hobert, James |
|  | Smith, Thomas |  | Hunter, William |
|  | Williams, Thomas |  | Jacob, Peter |
| Henry Guilder | Bonny, John |  | Lemon, John |
|  | Johnson, John |  | Leverage, Zachariah |
|  | Johnson, Thomas |  | Lockwood, William |
| Hepsa | Edgar, William |  | Mackey, Richard |
| Hepsey | Boussell, Samuel |  | McCalla, David |
|  | Boyd, Andrew |  | Merchant, Elijah |
|  | McKensie, William |  | Morrison, James |
|  | Noel, Stephen C. |  | Nisbett, James |
|  | Williams, Thomas |  | Norris, Henry |
| Herald | Adams, Robert |  | Prime, James |
|  | Allan, Asa |  | Rogers, Gaff |
|  | Allen, John Baptiste |  | Rowe, Abraham |
|  | Ball, William |  | Shoot, William |
|  | Barnes, Nathaniel |  | Smith, Nicholas |
|  | Bird, John |  | Steel, Mark |
|  | Bull, Henry |  | Storey, William |
|  | Bushnell, Joseph |  | Williams, William |
|  | Calpry, Joseph |  | Wyatt, William |
|  | Campbell, Moses | Hero | Bradford, James |

American Prisoners of War held at Dartmoor during the War of 1812

## Prisoner listing by ship or regiment

| Ship/Regiment | Name | | Name |
|---|---|---|---|
| High Flyer | Barden, Abel | | Green, John |
| Hindortar | Boyd, Stephen | | Jack, John |
| | Durval, N. D. | | Jane, William |
| | Elfs, James | | Louis, John |
| | Stage, William | | Mathews, William |
| Hindortar | Lamb, Jack | | Pipe, Henry |
| Holstein | Buttman, Thomas | | Salisbury, John |
| Hope | Blasse, John M. | | Saunders, Cornelius |
| | Blue, Peter | | Saundes, Cornelius |
| | Coleman, Andrew | | Wood, Samuel |
| | Crete, John Isaac | Hussar | Adams, James |
| | Drake, John | | Anderson, Jacob |
| | Elves, Jospeh | | Argent, William |
| | Elves, Manuel | | Armstrong, George |
| | Gillett, Francis | | Ashby, William |
| | Green, Robert | | Badger, Peter |
| | Maynard, Humphrey | | Baker, Basil |
| | Mitchell, John | | Barker, John |
| | Price, Francis | | Barrett, George |
| | Rhodrick, Joseph | | Beans, James |
| | Smith, John D. | | Bennett, Stephen |
| | Taylor, James | | Boys, Hugh |
| | Thomas, John | | Butler, Benjamin |
| | Thomas, William | | Campbell, Thomas |
| | Walker, James | | Cantelo, James |
| | Webb, Michael | | Chambers, Jospeh |
| | Willis, John | | Chapel, John |
| Hope & Anchor | Codshell, Joseph | | Clark, William |
| Hornby | Clark, Peleg | | Coffin, Frederick |
| | Dixon, Archibald | | Collins, Thomas |
| | Garrett, Robert | | Cooper, Ezekiel |
| | Johnson, Peter | | Dandy, Hamilton |
| | Lyons, Henry | | Denyer, Richard |
| | McIntire, A. | | Desharn, George |
| | Montcalm, Henry | | Dickinson, Francis |
| | Paine, Charles | | Doyle, James |
| Hornet | Barrett, Thomas | | Dunbar, William |
| | Jones, Samuel | | Dunham, John |
| | Staines, Samuel | | Dunn, Charles |
| | Wingate, Jacob | | Folwett, Jospeh |
| Hugh Jones | Dias, Joseph | | Fowler, Robert |

American Prisoners of War held at Dartmoor during the War of 1812

## Prisoner listing by ship or regiment

| | | |
|---|---|---|
| Francis, John | | Sutton, Richard |
| Freeman, Nathaniel | | Tar, Zedediah |
| Garrison, Charles | | Thomas, William |
| Gibson, Richard | | Trumills, John |
| Gould, Samuel | | Tucker, Zebulon |
| Graham, Mourice | | Turner, Thomas |
| Grimes, William | | Vansykle, Ralph |
| Haines, Daniel | | Wade, John |
| Holmes, Robert | | Walker, John |
| Johnson, Frederick | | Warrimore, Henry |
| Jones, William | | Warrimore, Philip |
| Karnes, James | | West, William |
| Keitch, Jonah | | White, Clement |
| Leech, William | | Williams, John |
| Lester, Nathaniel R. | Ida | Adams, John |
| Lewis, George | | Alexander, James |
| Lewis, Thomas | | Anderson, Alexander |
| Long, Charles | | Babson, John |
| Loveland, Daniel | | Bacon, Elisha |
| Mathews, Pedro | | Baker, Edward E. |
| McKinnon, Noel | | Bancroft, James |
| McKray, John W. | | Beets, Thomas |
| McNeel, Philip | | Brady, John |
| Miller, Isaac | | Bray, Andrew |
| Miller, Jonathan | | Bray, Isacher |
| Minks, Jacob | | Campbell, James |
| Morgan, Jacob | | Churchill, Manuel |
| Morrison, Samuel | | Cody, James |
| Myers, Edward | | Cotton, William |
| Nye, William | | Creamer, George |
| Oliver, Samuel C. | | Daverille, William |
| Oliver, Stephen | | Davies, William |
| Patterson, John | | DeBates, Amos |
| Penn, John | | Denham, Silas |
| Pidgeon, Jacques | | DeRichardson, Comte |
| Raven, James | | Fowling, Jeremiah |
| Rowe, Cornelius | | Furs, Theodore |
| Saunders, Jospeh | | Hadlock, Nathaniel |
| Seaberry, John | | Haffer, Robert |
| Shean, John | | Hambleton, John |
| Smith, Charles | | Johnson, William |

American Prisoners of War held at Dartmoor during the War of 1812

## Prisoner listing by ship or regiment

|  |  |  |  |
|---|---|---|---|
|  | Jones, John |  | Holland, James |
|  | Kellem, Frederick |  | Stone, Samuel |
|  | Keough, Edward |  | White, Philip |
|  | Knowlton, Enos |  | Works, Elford |
|  | Lewis, Gitt | Indostan | Lerwich, John |
|  | Lewis, William | Industry | Bartell, William |
|  | Lowe, Frederick |  | Blair, Robert |
|  | Miller, George |  | Brush, Thomas |
|  | Moss, N. J. |  | Burrell, Jesse |
|  | Neumann, Gustavus |  | Burridge, Robert |
|  | Palmer, Joseph |  | Cheslie, Amos |
|  | Perigo, Joel |  | Dullivan, James |
|  | Polland, John |  | Eaton, Israel |
|  | Pyne, Joseph |  | Ellis, John |
|  | Rand, Robert |  | Fulton, James |
|  | Russell, John |  | Glover, John |
|  | Sands, John |  | Gowalter, John |
|  | Simmonds, Alexander |  | Inglas, John |
|  | Sturgess, Major |  | Jarvis, Thomas |
|  | Terry, Daniel |  | Laskey, Benjamin |
|  | Tolton, Moses |  | Melzard, George |
|  | Warden, Nathaniel |  | Morris, Jacob |
|  | Warney, Benjamin |  | Northey, Joseph |
|  | Whiteman, Peter |  | Rice, Thomas |
| Imperatrice Reine | Basson, John |  | Robes, Edward |
|  | Commous, John |  | Ross, David |
|  | Dale, Hercules |  | Russell, Louis |
|  | Harburn, Thomas |  | Snow, James |
|  | Preston, John |  | Stacey, John |
|  | Springstern, Abraham |  | Tucker, Samuel |
|  | Tully, Hugh | Invincible | Becamp, Jaques |
|  | Whitehouse, Daniel |  | Bolton, John |
|  | Williams, Peter |  | Celutan, Alexander |
| Independence | Boivie, James |  | Clark, Thomas |
|  | Griffin, William |  | Coleman, Joseph |
|  | Harvey, Joseph |  | Davis, John |
|  | Hodgkins, Daniel |  | DeSilvin, Jean |
|  | Howell, Thomas |  | Fields, Michael |
|  | Marsh, Jesse |  | Forbes, William |
|  | Robinson, Josiah |  | Foss, Supply |
| Indian Lass | Bassett, Gorham |  | Garrison, Cornelius |

# American Prisoners of War held at Dartmoor during the War of 1812

## Prisoner listing by ship or regiment

| | | | |
|---|---|---|---|
| | Green, John | | Harding, John |
| | Hale, John | | Heath, James |
| | Haycock, John | | Hensow, John |
| | Hayes, John | | Jones, Edward |
| | Huzzey, John | | Ley, John |
| | Jefferson, Richard | | Linton, Joseph |
| | Jouranne, Louis | | Monk, Joseph |
| | Kempt, Eric | | Murray, Nathaniel |
| | Kempt, George | | Pierce, John |
| | Merlo, Cristopher | | Raton, Peter |
| | Michell, Charles | | Rice, John |
| | Moore, J. B. | | Shaw, Benjamin |
| | Morgan, Henry | | Shepherd, Henry |
| | Morier, John | | Shorts, Clement |
| | Murray, Colton | | Smith, Samuel |
| | Nowland, Edward | | Steele, Mathew |
| | Peterson, Samuel | | Steele, William |
| | Seywood, James | | Weeks, Thomas N. |
| | Sherry, Peter | | Williams, William |
| | Simondson, Isaac | | Young, Philip |
| | Smith, Daniel | Jane | Eastlake, James |
| | Sobier, Bernard | | Johnson, John |
| | Sweetland, William | | McIntire, William |
| | Swinley, John | | Pedro, Francisco |
| | Thomas, Richard | | Penny, James |
| | Walton, John | | Reed, Joseph |
| | Whiteby, Samuel | | Rodgers, Ephraim |
| James | Addy, Francis | | Sandor, James |
| | Alford, William | | Wilson, William |
| | Bonner, John | | Woodward, Francis |
| | Cameron, William | Jane Barns | Manuel, Diego |
| | Chasson, Thomas | Jason | Brown, Benjamin |
| | Crawford, John | | Nicholls, William |
| | Dolavice, Henry | Jemmell | Clark, John |
| | Ellis, Thomas | | Diverall, John |
| | Elmick, Joseph | | Dyer, Joseph |
| | Evans, William | | Gowner, Joseph |
| | Feilman, John | | Henley, Jacob |
| | Fife, Thomas | | Holmes, Isaac |
| | Fitzgerald, Edward | | Perkins, Thomas |
| | Green, John | | Pockmitt, Jackness |

# American Prisoners of War held at Dartmoor during the War of 1812

## Prisoner listing by ship or regiment

|  |  |  |  |
|---|---|---|---|
|  | Salisbury, Joseph |  | Carnes, Joseph |
|  | Skipper, John |  | Farmer, George |
|  | Turner, Joseph |  | Griffin, John |
|  | Walker, Daniel |  | Higgins, James |
|  | Whitning, William |  | Howard, Samuel |
|  | Wilbourg, David |  | Johnson, Henry |
| Jenny | Bassington, John |  | Joseph, Thomas |
| Joel Barlow | Allen, James |  | McKertre, Abraham |
|  | Bell, James |  | Miller, Stephen |
|  | Berdick, Simon |  | Monturn, Francis |
|  | Black, George |  | Monturne, Jais |
|  | Brown, Ebenezer |  | Mumford, Thomas |
|  | Chipman, Christian |  | Phillips, John |
|  | Durvolf, Stephen |  | Rogers, James |
|  | Griffin, William |  | Simpson, William |
|  | Hart, Bartholew |  | Skinner, Tileman |
|  | Jeffery, Robert |  | Sone, John |
|  | Johnson, William | John & Mary | Atwood, James |
|  | Morell, Benjamin |  | Briggs, Frank |
|  | Palmer, Pero |  | Hicks, Ogershill |
|  | Peters, Aaron |  | Rogers, Pely |
|  | Stevens, Hugh | John of Salem | Lemon, N. C. |
| Johannes | Gracio, Anthony |  | Mason, Joseph J. |
|  | James, Jeremiah | Joseph | Bartlett, John |
|  | Sherry, John |  | Besson, Phillipe |
|  | Whitamore, William |  | Chambers, Elias |
| John | Barker, Charles |  | Cove, John |
|  | Beter, John |  | Davis, Moses |
|  | Coffin, Theodore |  | Doliver, Joseph |
|  | Farrell, John |  | Gardner, Edward |
|  | Fisher, Robert |  | Hammond, William |
|  | Hawkins, John |  | Mitchell, Carr |
|  | Henderson, Joseph |  | Parsons, Daniel |
|  | Hinchman, George |  | Rodgers, George |
|  | Monk, Philip |  | Stacey, William |
|  | Samuels, Samuel |  | Thomas, Spencer |
|  | Symonds, Moses |  | Woodbury, John |
|  | Truman, John | Joseph Ricketson | Jacobson, William |
|  | White, James | Julia | Smith, John |
| John & Frances | Bitters, John | Juliana Smith | Artis, William |
|  | Cammon, Robert |  | Atwood, Nathaniel |

American Prisoners of War held at Dartmoor during the War of 1812

## Prisoner listing by ship or regiment

|  |  |  |  |
|---|---|---|---|
|  | Behon, Simon |  | Hickson, John |
|  | Blair, David |  | Kembourn, John |
|  | Briggs, Thomas |  | Mackay, Joseph |
|  | Eldridge, William |  | Myers, Joseph |
|  | Harding, Joseph |  | Wood, Thomas |
|  | Hay, John | Lark | Bids, Thomas |
|  | Higgins, Asa |  | Rowland, Abner |
|  | Holmes, Almoran |  | Wiggins, Richard |
|  | Lawrence, George | Lash | Beatty, John |
|  | Lewis, John | Leader | Trask, Charles |
|  | Norton, Solomon | Leander | Johnston, William |
|  | Phenny, John | Leda | Turner, James |
|  | Smith, Elisha | Leicester | Burke, James |
|  | Snow, Thomas |  | Dean, William |
|  | Woodford, William |  | Durham, Samuel |
| Julianne | Greaves, John |  | Myers, John |
|  | Rowley, James |  | Read, David |
| Kergettos | Butler, William |  | Sprague, Henry |
|  | Groves, Pierce |  | Thompson, James |
|  | Kaller, Joseph | Leo | Anderson, Daniel |
|  | Simons, John |  | Bartlett, Caleb |
| King George | Cooper, James |  | Carey, John |
|  | Evans, Jacob |  | Codman, Richard |
|  | Farrett, George |  | Davies, John |
|  | Munro, Harry |  | Doughty, Jesse |
| Kitty | Chase, Nathaniel |  | Foss, Edward |
|  | Huse, Ebenezer |  | Foss, Joseph |
|  | McMiller, Andrew |  | Hume, George |
|  | Rightman, John |  | Mason, Nathaniel |
|  | Rodgers, Samuel |  | Nasson, Joseph |
|  | Watson, James |  | Tuttle, French |
| L'Aigle | Spince, William |  | Wood, Sylvester |
| Landrail | Anderson, John | Levant | Ayres, John |
|  | Baker, Henry |  | Mansfield, George |
|  | Ball, Peter |  | Saunderson, Jacob |
|  | Carnass, William |  | Smith, George |
|  | Ellis, John |  | Stickney, Amos |
|  | Evans, Benjamin |  | Stoddard, Samuel |
|  | Gardner, John |  | Tubb, Samuel |
|  | Gennison, Michael |  | Young, William |
|  | Green, Elijah | Levi | Wilson, Andrew |

American Prisoners of War held at Dartmoor during the War of 1812

## Prisoner listing by ship or regiment

| | | |
|---|---|---|
| Lewis Warrington | Dobson, Robert | Brownell, Jonathan |
| | Dorrell, John | Bunkerson, John |
| | McGilmore, John | Burnham, David |
| | West, Edward | Cannon, William |
| Liberty | Creiger, John | Chandler, Samuel |
| | Curtis, Ephraim | Chandler, William |
| | Defray, Edward | Charles, John |
| | Douglass, William | Chase, Samuel |
| | Green, Levenet | Clark, Francis Marie |
| | Hall, David | Cole, Charles |
| | Hamilton, John | Conner, William |
| | Hilby, Thomas | Cook, Edward |
| | Hill, Justice | Depuyster, Pierre |
| | Jackson, Charles | Ellingwood, Joshua |
| | Mead, Ezekiel | Evans, Pierrs |
| | Minor, John | Fedy, Philip |
| | Ned, Deaf | Flowers, John |
| | Nichols, John | Frederick, John |
| | Peters, John | Garrison, James |
| | Towns, Asa | Gerard, Pierre |
| | Tucker, James | Grinard, Bray |
| | Typhon, Louis | Guillaume, Jean Marie |
| Lightning | Brooks, John | Gustable, Joseph |
| | Bustin, John | Hall, Robert |
| | Crawford, Nelson | Henry, William |
| | Garthy, James | Hulen, Edward |
| | Harris, William | Jacobson, Mathew |
| | Jones, Richard | Jones, Abraham |
| | Jordan, Peter | Lemorrett, John P. |
| | Ruddick, William | Lewis, John |
| | Wallace, James | Martin, John |
| | Warrands, John | Maxine, Joseph |
| Lion | Allen, Henry | Monroe, John |
| | Baker, Jesse | Morris, Richard |
| | Barber, William | Muss, Joseph |
| | Beeston, Robert | Nelson, William |
| | Bell, Jacob | Nichelle, Pierie |
| | Bissen, John | Nolen, William |
| | Bourdon, Amond | Pacher, John |
| | Brooks, Philip | Palmer, Peter |
| | Brown, James | Pearse, John |

# American Prisoners of War held at Dartmoor during the War of 1812

## Prisoner listing by ship or regiment

|  |  |  |  |
|---|---|---|---|
|  | Pendleton, Jonathan |  | McCarthy, Samuel |
|  | Portlock, William |  | Ramsdell, Charles |
|  | Richards, Edward | Louisiana | Golden, Edward |
|  | Rick, Francis | Lucy | Perkins, Joseph |
|  | Roderigo, Joseph | Lyon | Coleman, William |
|  | Roille, Oliver |  | Hanson, J. A. |
|  | Shoals, Christopher |  | Jackson, James |
|  | Silkes, Peter |  | Lang, William |
|  | Smith, Charles |  | Ravett, William |
|  | Snalt, George |  | Watch, Samuel |
|  | Steel, John |  | Wright, William |
|  | Sutton, Martin | Magdalen | Adams, John |
|  | Taylor, David A. |  | Barkman, Henry |
|  | Thomas, Joseph |  | Foster, Thomas |
|  | Tordey, Edward |  | Harris, Simeon |
|  | Turner, James |  | Harts, William |
|  | Vanmetre, Henry |  | Holmes, James |
|  | Vesplash, Nicholas |  | Howe, John |
|  | Waters, George |  | Jackson, Thomas |
|  | Watson, Henry |  | Johnstone, John |
|  | Williams, Thomas |  | Johnstone, Peter |
|  | Wilson, Alexander |  | Kennedy, John |
|  | Wilson, John |  | Leversage, William |
| Little Belt | Craig, William |  | Lewis, Francis B. |
|  | Littlefield, Nicholas |  | Money, Henry |
| Lizard | Barrentt, John |  | Prince, Benjamin |
|  | Daigney, Jasper |  | Reynolds, Owen |
|  | Fabin, James |  | Williams, Peter |
|  | Graham, Robert | Man | Fletcher, John W. |
|  | Marral, Ephraim | Marengo | Thomas, James |
|  | May, Joseph | Margaret | Armstrong, David |
|  | Pedang, John |  | Banks, Perry |
|  | Ralph, David |  | Brown, George |
|  | Richards, George |  | Fox, Washington |
|  | Seagell, William |  | Grubb, Andrew |
|  | Skenny, Francis |  | Jones, Kenny |
|  | Swadey, Samuel |  | McQuillan, James |
|  | Williams, Hezekial |  | Osborn, Stephen |
|  | Young, John |  | Rock, Oliver |
| Louisa | Finch, Etienne |  | Whitehead, James |
|  | Jones, Henry | Margetits | Hampstead, Cambridge |

American Prisoners of War held at Dartmoor during the War of 1812

## Prisoner listing by ship or regiment

|  |  |  |  |
|---|---|---|---|
|  | Higley, John | Martin | Barron, John |
|  | Randall, Charles |  | Ross, William |
|  | Strouts, Joseph | Mary | Aldridge, Richard |
| Maria | Roach, Pedro |  | Andrews, Jeremiah |
|  | Schneider, John A. |  | Backley, Walter |
|  | Smith, W. L. |  | Bartlett, Samuel |
| Marie Christaina | Garreai, Anthony |  | Bennett, James |
|  | Lewis, James |  | Brown, Anthony |
|  | Lopez, Joseph |  | Brown, William |
| Mariner | Murray, George |  | Cateret, James |
|  | Pitts, Thomas |  | Cayler, Charles |
|  | Wanton, Samuel |  | Cayler, Henry |
|  | Whittington, George |  | Cayler, Jacob |
| Marmion | Allen, Philip |  | Coates, Russell |
|  | Bennett, Charles |  | Cole, Perry |
|  | Charles, Samuel |  | Collins, John |
|  | Coader, Anthony |  | Dean, Moses |
|  | Fedie, John |  | Diamond, William |
|  | Menard, Augustus |  | Favish, John |
|  | Moore, Henry |  | Fresher, Henry |
|  | Roberson, Henry |  | Frost, Dephne |
|  | Rodwin, Walter |  | Griffin, William |
|  | Sanis, Lewis |  | Hall, Hammon |
|  | Silva, Francis |  | Hicks, Benjmain |
|  | Start, Joshua |  | Hitchell, George |
| Mars | Allen, Henry |  | Lumsby, Frederick |
|  | Baker, John |  | Marshall, B. |
|  | Barrett, Anthony |  | Meech, David |
|  | Brewster, Jacob |  | Moore, William |
|  | Caldwell, James |  | Murray, Oliver |
|  | Jones, Benjamin |  | Neel, William |
|  | Lawson, James |  | Pinkham, Jacob |
|  | Parker, George |  | Ray, David |
|  | Saunders, William |  | Roth, James |
|  | Whitewood, Charles |  | Shepherd, Isaac D. |
| Martha | Baker, James |  | Spark, Samuel |
|  | Fuller, Thomas |  | Thorndike, Hall |
|  | Hooper, John |  | Watson, Robert |
|  | Palfrey, John |  | Wilson, Robert |
|  | Poor, David | Mary (True Blooded Yankee) | Lewis, Peter |
|  | Sprague, Stephen | Matilda | Brown, William |

# American Prisoners of War held at Dartmoor during the War of 1812

## Prisoner listing by ship or regiment

|  |  |  |
|---|---|---|
|  | Eskinson, George | Hodsdale, Daniel |
|  | Hamcomb, Thomas | Hoff, Nicholas |
|  | Hanson, Peter | Hutchins, Nathaniel |
|  | Harter, Henry | Lewis, Benjamin |
|  | Hatch, William | Linscott, Josiah |
|  | Jasmine, Paul | Littlefield, Israel |
|  | Legos, Philip | Littlefiled, Oliver |
|  | Miller, George | Lord, Benjamin |
|  | Smith, Esop | Lord, Dormer |
|  | Westler, John | Lord, John |
| Maydock | Fleming, Alexander | Lord, Joseph |
|  | Harwood, William | March, Jesse |
|  | Peckham, Isaac | Marshall, M. |
|  | Pitts, Charles | Marshall, Thomas |
|  | Wilkey, Timothy | Mason, Daniel |
| McDonough | Allen, Thomas | Merrill, Jacob |
|  | Archer, Daniel | Middlefield, Abraham |
|  | Arnold, John | Mitchell, James |
|  | Avril, Ebenezer | Murphy, Samuel |
|  | Barnes, Thomas | Noble, Joseph |
|  | Burbalk, Moses | Parsons, Junius |
|  | Burns, Benjamin | Patten, Nathaniel |
|  | Charles, Samuel | Patten, Robert |
|  | Cleeves, Jonathan | Perkins, George |
|  | Clough, David | Perkins, Joseph |
|  | Clough, Enoch | Perkins, Samuel |
|  | Clough, Obed | Rhodes, Daniel |
|  | Clough, Shadrick | Richards, Richard |
|  | Derrill, Benjamin | Roberts, Nathaniel |
|  | Dorman, Israel | Robinson, Joshua |
|  | Eden, Mack | Stackpole, Andrew |
|  | Emery, John | Stone, Abner |
|  | Emlyn, Isaac | Stone, Adam |
|  | Fairfield, Asa | Stone, Jonathan |
|  | Fairfield, James | Stoves, Daniel |
|  | Ferrin, Daniel | Towan, Robert |
|  | Fisk, George | Townson, William |
|  | Fletcher, Robert | Walker, Nathan |
|  | Foss, Abijah | Ward, Israel |
|  | Gatchell, Benjamin | Ward, Nathaniel |
|  | Harding, Jonathan | Washburn, Jeremiah |

American Prisoners of War held at Dartmoor during the War of 1812

## Prisoner listing by ship or regiment

|  |  |  |  |
|---|---|---|---|
|  | Weeks, Joseph |  | Lepberg, John P. |
|  | White, Peter |  | Little, John |
| Melanie | Egbert, Peter |  | Loring, Rufus |
|  | Fornver, Virgil |  | Merritt, Enoch |
|  | Frederique, John |  | Watson, William |
|  | Lamar, Edward | Miquelonnaiss | Roberts, George |
|  | Lavasseur, Rene | Miranda | Barraso, John |
|  | Maxen, Joseph |  | Grover, Edmund |
|  | Neyren, John |  | Littlefield, Rufus |
| Melville | Jackson, Joseph |  | Parinele, Benjamin |
| Mentor | Cole, Nathaniel |  | Robertson, Robert |
|  | Francois |  | Yard, William |
|  | Hall, Reuben | Modelle | Kirby, Anthony |
|  | Joscelyn, Ambrose | Monarch | Bankstone, William |
| Messenger | Broadwater, Samuel |  | Bennett, Peleg |
|  | Hale, Shederick |  | Blanchard, Samuel |
|  | Ingle, John |  | Blasker, Charles |
|  | Johnson, Joseph |  | Block, Edward |
|  | Metley, Thomas |  | Brown, James |
|  | Morgan, John |  | Bush, George |
|  | Murray, James |  | Butler, Robert |
|  | Posey, Valentine |  | Coles, Charles |
|  | Simpson, Thomas |  | Davis, George |
|  | Stuff, Francis |  | Green, John |
|  | Wills, John |  | Holmes, John |
| Meteor | Burn, Reuben |  | Hughes, William |
|  | Forester, Francis |  | James, John |
|  | Hill, Leonard |  | Lachame, Thomas |
|  | Kennedy, John |  | Marshall, George |
|  | Latimore, Mathew |  | Peterson, James |
|  | Perne, George |  | Peterson, Lawrence |
|  | Perney, William |  | Prindall, Samuel |
|  | Schole, John |  | Watkins, William |
|  | Snate, George |  | Welch, Thomas |
| Minerva | Briggs, John |  | Yarnell, William |
|  | Campbell, William | Montecello | Greenland, Stephen |
|  | Conner, Michael | Montgomery | Bragden, James |
|  | Davis, John |  | Branger, John |
|  | Fardell, Thomas |  | Briggs, William |
|  | Harris, Ebenezer |  | Brooks, John |
|  | Hill, Daniel |  | Callam, John |

# American Prisoners of War held at Dartmoor during the War of 1812

## Prisoner listing by ship or regiment

|  |  |  |  |
|---|---|---|---|
|  | Clark, William |  | Gunby, James |
|  | Forbes, John |  | Landerkin, Daniel |
|  | Hanfield, Enos |  | Lewis, Thomas |
|  | Henley, John |  | Mathias, Henry |
|  | May, Henry |  | McNab, John |
|  | Murray, Richard |  | Mines, Artemas |
|  | Riswell, Palmer |  | Minifee, Charles |
|  | Sparkes, Thomas |  | Moore, William |
|  | Thomas, John |  | Morweek, Andrew |
|  | Wanton, William |  | Niel, Henry |
| Monticello | Miflin, Richard |  | Norris, Robert |
| Moranda | McCarthy, Henry |  | Ranson, Joseph |
| Nancy | Bracket, James |  | Robinson, Nicholas |
|  | Dolphin, Joseph |  | Sickless, J. M. |
|  | Frost, Thomas Bell |  | Smith, Elisha |
|  | Jones, Calvin |  | Snowdon, Jacob |
|  | Jones, John |  | Spangle, Frederick |
|  | Moore, Robert |  | Spires, Thomas |
|  | Norris, Benjamin |  | Wessels, Robert |
|  | Osborne, John L. |  | Wheeler, Benjamin |
|  | Perkins, Clement |  | Williams, Jacques |
|  | Plumber, James |  | Williams, John |
|  | Pray, Samuel |  | Wilson, William |
|  | Winand, John |  | Windham, John |
|  | Wittlebanks, Edward |  | Wolfe, William |
|  | Woglon, Peter |  | Young, Rufus |
| Napoleon | Bradford, Charles | Nellenville | Baker, Thomas |
|  | Clepp, Abraham |  | Baptiste, John |
|  | Grey, Moorhouse |  | Blackston, William |
|  | Hobson, Abraham |  | Cooper, Peter |
|  | Moffett, John |  | Harvey, James |
|  | Prett, Benjamin |  | Hindman, John |
|  | Reynolds, Frederick |  | Jenkins, Thomas |
|  | Veal, Pierre |  | Johannes, Logan |
| Ned | Anderson, William |  | Jones, James |
|  | Ashmore, Edward |  | Kell, Francis |
|  | Barker, Israel |  | Kitton, Abraham |
|  | Buchan, John |  | Madden, John |
|  | Churchill, Men'd. |  | Southcomb, Peter |
|  | Dalloway, John |  | Watt, Samuel |
|  | Duncan, George | Nelly | Andrews, Peter |

American Prisoners of War held at Dartmoor during the War of 1812

## Prisoner listing by ship or regiment

|  |  |  |  |
|---|---|---|---|
|  | Burn, Rufus |  | Lloyd, David |
| New Zealander | Banks, James |  | Rye, William |
|  | Boyle, John |  | Smith, Shapley |
|  | Budd, John | Old Friend | James, John |
|  | Dashells, Richard |  | Mann, Samuel |
|  | Dave, David | Olio | Brown, Thomas |
|  | Dunn, Robert |  | Peterson, William |
|  | Dutton, Thomas |  | Reid, Thomas |
|  | Fulgar, William |  | Ryder, George |
|  | Johnson, James |  | Thornton, Benjamin |
|  | Leggins, Ebenezer | Olive Branch | Adams, Isaac |
|  | Morris, John | Orbit | Albert, Hezekieh |
|  | Olliver, John |  | Brown, William |
|  | Selman, Edward |  | Cleveland, Davis |
| Nimble | Amos, Peter |  | Cox, Miles |
|  | Call, George |  | Frees, James |
|  | Eldridge, Ephraim |  | Greenleaf, Thomas |
|  | Green, Timothy |  | Huntress, Robert |
|  | Haynes, Stephen |  | Johnson, Jacob |
|  | Loveland, Leonard |  | Johnson, James |
|  | Michaels, John |  | Kennedy, Peter |
|  | Potter, Titus |  | Larrabee, William |
|  | Ryder, Amos |  | Rape, Nicholas |
|  | Thorndike, Robert |  | Ripton, Solomon |
|  | Wadge, John |  | Sweetman, Samuel |
|  | White, W. K. |  | Thompson, Courtney |
| Nonsuch | Bernard, William |  | Thompson, James |
|  | Blakely, Michael |  | Webb, John |
|  | Hardy, Robert |  | Wheeler, William |
|  | McCormick, James | Orders in Council | Abbott, Timothy |
|  | Peterson, P. |  | Allen, Joseph |
|  | Wicks, Samuel |  | Amos, Cheney |
| Norfolk | Bacchus, John |  | Butler, George |
|  | Canfield, Arthur |  | Doughty, Levi |
| North Star | Card, James |  | Garner, William |
|  | Evans, Edward |  | Goodwin, William |
|  | Johnson, Luke |  | Hall, James |
|  | Newell, Charles |  | Haskett, Robert |
| New York Militia | Shelton, Smith |  | Jordan, Richard |
| Ocean | Baker, John |  | Kenny, Harris |
|  | Coffin, Valentine |  | Martin, William |

American Prisoners of War held at Dartmoor during the War of 1812

## Prisoner listing by ship or regiment

|  |  |  |
|---|---|---|
|  | Moulds, Benjamin | Cooper, Andrew A. |
|  | Nicherson, Joseph | Cramstead, James |
|  | Patten, James | Dibble, Rebuen |
|  | Robinson, Elias | Edwards, John |
|  | Russell, James | Fink, Jonas |
|  | Slocum, William | Freeman, John |
|  | Wilson, John | Friday, John |
| Otario | Columbia, John | Gibbs, Henry |
|  | Douglass, Dover | Godfrey, William |
|  | Hooke, George | Green, William |
|  | McDonald, John | Gross, William |
|  | McLean, Thomas | Guillard, Louis |
|  | Wilson, John | Guillard, Peter |
| Pallas | Benjamin, P. | Hamilton, Alexander M. |
|  | Brown, James | Hydra, Dempey |
|  | Burton, John | Irvin, Arthur |
|  | Calfax, William | Jackson, James |
|  | Douarte, Angelo | Johnson, Joseph Toker |
|  | Edgerley, George | Kewen, Edward |
|  | Fitch, William | Kister, John |
|  | Hadley, Andrew | Lacour, John Baptist |
|  | Hockman, William | Lippart, Thomas D. |
|  | Ratoon, Thomas | Little, George |
|  | Reaudon, Andrew | Littlefield, James |
|  | Robinson, William | Long, R. S. |
|  | Smith, George | Louis, Nicholas |
|  | Windell, Isaac | Martin, Daniel |
| Pandour | Dugaretz, Augusta | Martin, Isaac |
| Patriot | Hutchinson, Lamore | Martin, James |
| Paul Jones | Abman, John | Miller, Thomas |
|  | Allen, John | Mooney, Peter |
|  | Anderson, Joseph | Muley, Etienne |
|  | Barnes, Nathaniel | Newell, Isaac |
|  | Bernard, John | Parsons, Samuel D. |
|  | Brown, Charles | Peck, Thomas |
|  | Byard, Joseph | Prince, George |
|  | Cato, John | Risings, John |
|  | Colman, David | Robinson, John |
|  | Colton, Walter | Roley, John |
|  | Cook, Charles Howe | Ruffield, Samuel |
|  | Cook, William | Sandford, John William |

American Prisoners of War held at Dartmoor during the War of 1812

## Prisoner listing by ship or regiment

|  |  |  |  |
|---|---|---|---|
|  | Sandford, William | Pilot | Anderson, John |
|  | Sentille, Francis |  | Brickman, John |
|  | Smith, John |  | Coffin, Edward |
|  | Smith, Thomas |  | Demarlow, Francis |
|  | Stevens, William |  | Downs, John |
|  | Stewart, William |  | Downs, William |
|  | Tardy, Andy |  | Gotier, Charles J. |
|  | Tassier, Louis |  | Harding, William |
|  | Thompson, Abraham |  | Howitt, William |
|  | Thompson, Henry |  | Jackson, Frederick |
|  | Veitch, William |  | Livingston, Henry |
|  | Warren, David |  | Manley, David |
|  | Whiting, Samuel |  | Mids, Michael |
|  | Wild, Thomas |  | Simmonds, Henry |
|  | Wilson, James |  | Smith, Titus |
| Paulina | Ceaser, Joseph |  | Stafford, John |
|  | Edwards, Price |  | Taylor, George |
|  | Lewis, Edward |  | Thomas, John |
|  | Moore, Laurence |  | Williams, Charles |
| Pearl | Platt, Daniel | Plutarch | Cooper, Thomas |
| Peggy | Alexander, J. B. |  | Coville, Nathaniel |
|  | Lowman, David |  | Frazier, Charles |
| Penn | Harris, George |  | Hautell, Stephen |
|  | Silvey, John |  | Hepburn, James |
| Perfect | Dodge, Joseph |  | James, William |
|  | Whiffle, Joseph |  | Lambert, Andres |
| Perry | Berry, Jesse |  | Pitts, George |
| Pert | Bellinger, William |  | Ray, Richard |
|  | Court, Robert |  | Robinson, William |
|  | Everley, John |  | Young, John |
|  | Freddle, John | Plutus | Brown, Amos |
|  | Gellens, William |  | Homan, Edwad |
|  | Hull, William |  | Hoyt, David |
|  | Lawson, Thomas |  | Knight, Samuel |
|  | Smith, William |  | Peak, James |
|  | Ware, William |  | Reed, William |
|  | Watts, James | Polly | Allen, Henry |
| Perverance | Lent, Samuel |  | Brown, Thomas |
| Phaeton | Golding, William |  | Brown, William |
|  | Robinson, Joseph |  | Bryant, James |
|  | Velpley, Israel |  | Cook, John |

American Prisoners of War held at Dartmoor during the War of 1812

## Prisoner listing by ship or regiment

|  |  |
|---|---|
| | Day, James |
| | Donison, William |
| | Egoss, Joseph |
| | Elvell, Robert |
| Portsmouth Packet | |
| | Evans, Robert |
| | Herrimond, Hezekiel |
| | Mason, Aaron |
| | Robinson, Robert |
| | Rogers, Nathaniel |
| | Russell, Richard |
| | Vicary, Richard |
| | Wheeler, Michael |
| Pomona | La Roche, Jean |
| Porcupine | Allen, George |
| | Andrews, James |
| | Brown, Benjamin |
| | Cross, Peter |
| | Cross, Stephen |
| | Davis, John |
| | Francis, Abraham |
| | Girdler, James |
| | Hagen, Joel |
| | Hanely, Thomas |
| | Hill, Hannel |
| | Johnson, Jacob |
| | Lucas, Daniel |
| | Peters, John |
| | Phillips, John |
| | Prissey, John |
| | Smith, Charles |
| | Stiles, Jodick |
| | Sumerg, Benjamin |
| | Tapley, Isaac |
| | Terry, Joseph |
| | Turner, Silas |
| Portsmouth | Coffee, Ram'l |
| | Damrell, William |
| | Davis, David |
| | Hannah, Thomas |
| | Harris, William |
| | McCanon, John |

Roach, James
Saunders, James
Smith, Robert
Anderson, David
Appleton, John
Bowen, Oliver
Byron, Ebenezer
Chase, Nathaniel
Childs, William
Dennison, Andrew
Golf, Peter
Grant, Samuel
Hanson, William
Hooper, Samuel
King, Solomon
Lapish, Andrew
Larabee, Thomas
Leach, Daniel
Norcross, Archibald
Perkins, Benjamin
Perkins, James
Perkins, John
Perkinson, James
Pike, Jeremiah
Robinson, Thomas
Russell, Moses
Spenny, Nathaniel
Stone, Robert
Tarlton, George
Tarlton, James
Thomas, Elisha
Wyatt, Jesse

Preseverance  Harris, Benjamin
            Savage, John H.
President   Achro, Joseph
            Adrianne, Jose
            Almeno, Jose
            Aubry, John Martinalias
            Azumus, Jerome
            Barnett, Tobias
            Bauld, J. P.

464

American Prisoners of War held at Dartmoor during the War of 1812

## Prisoner listing by ship or regiment

| | | |
|---|---|---|
| | Bretade, E. F. | Burnet, Charles |
| | Campreche, St. Jago | Cook, Samuel |
| | Complaro, A. A. | Dean, Jonas |
| | Craig, John E. | Dillin, Pierce |
| | Davis, George | Francis, John |
| | De La Canto Sepo, I. | Haye, Moses |
| | Ebier, Joseph | Hickman, Joseph |
| | Falcon, J. B. | Hunter, William |
| | Flinn, John F. | Ingerson, Michael |
| | Fog, John | May, Walter |
| | Fry, A. L. | Myers, John |
| | George, John | Price, Jacob |
| | Harman, John | Robertson, James |
| | Hidalgo, Valentine | Sparrow, James |
| | Hydalgo, Vincent | Stinchcombe, George |
| | Ivy, John | Thompson, Charles |
| | Johannes, John | Wright, John |
| | Johnson, John | Young, William |
| | Jones, Thomas | Prince de Neufchatel | Adams, Samuel |
| | Joseph, Pedro | Alexander, James |
| | Knabbs, James | Avies, Samuel |
| | Leurand, Ambrose | Bacon, James |
| | Lopez, Domingo | Balch, William |
| | Lorang, M. | Ball, Stephen |
| | Marie, John | Bang, George |
| | Mathy, James | Barker, Thomas |
| | Mendoza, Caesar | Barrett, Thomas |
| | Milborne, William | Berry, Joseph |
| | Montgomery, William | Blake, Ebenezer |
| | Palmo, Vincent | Boyleston, Zebediah |
| | Parks, A. D. La Enoh | Bramant, Laurence |
| | Roberts, John | Brown, John |
| | Sissons, Pedro | Burchstad, John |
| | Smith, J. W. | Burnham, Abraham |
| | Stephens, William | Carleton, Christopher |
| | Taylor, Samuel | Carnes, William |
| | Walker, James Scott | Churchill, Joseph |
| | Waterman, Cato | Clapp, George |
| | Williams, Charles | Clowes, Philip |
| Price | Blanchet, Simon | Cobb, Josiah |
| | Brown, Jean | Colby, Benjamin |

American Prisoners of War held at Dartmoor during the War of 1812

## Prisoner listing by ship or regiment

Condon, Samuel
Constant, William
Creighton, William
David, James
Davis, Frederick
Davis, William
Delaware, John
Demedorff, John
Dodge, David
Doggett, James
Door, Ebenezer
Downs, Jesse
Dunklin, Jesse
Fernandez, George
Fields, Charles
Fluken, William
Frain, Joshua
Ganster, John
Gardner, Andrew
Giddings, Johh
Glover, William
Goodhall, Joseph
Greenlaw, Jeremiah
Gurney, John
Haley, Samuel
Hall, John
Harrison, George
Haskett, Luther
Hawes, Caleb
Haywood, Samuel
Holmes, Samuel
Hose, Richard
Jackson, Ebenezer
Jarvis, Peter
Jones, Edward
Keith, James
King, Seth
Lamb, Joseph
Leonard, Thomas
Levit, Caleb
Lewis, Jesse

Loring, Caleb
Lyon, John
Martin, John
Martyn, John
May, John
Miller, John
Miln, Nicholas
Moor, Joshua
Moore, Warren
Nash, Alexander
Nelson, David
Newall, J. B. L.
Newall, John
Newcomb, John
Newell, Joseph
Nowland, Thomas
Obrey, Mathew
O'Neil, Joseph
Orkitt, Horea
Parsons, William
Payton, James
Perry, Ebenezer
Peterson, David
Pritchard, George
Rand, Charles
Rand, Thomas
Raymond, John
Richardson, Nathan
Saverio, John
Sinclair, Thomas
Snow, Daniel
Spiller, Moses
Studley, Warren
Tarbor, Samuel
Tarr, William
Thomas, Charles
Thomas, Theodore
Truman, Samuel
Welch, John
Wentworth, John
Whittington, Lemuel

American Prisoners of War held at Dartmoor during the War of 1812

## Prisoner listing by ship or regiment

|  |  |  |  |
|---|---|---|---|
|  | Witherington, James |  | Shoe, Bernard |
|  | Woods, Charles |  | Williams, Moses |
|  | Wright, John | Prize to the Hunter | Owen, Zachariah |
| Prince of Wales | Berto, John | Prize to the Lawrence | Jones, Richard |
|  | Gennodo, Samuel H. |  | Langhane, Bell'm |
|  | Ham, John |  | Marshall, William |
|  | Poswell, John |  | Morrison, Joseph |
| Print | Chine, Samuel |  | Ogle, Charles |
|  | Dolabar, John |  | Peterson, John |
|  | Graves, Samuel |  | Wood, John |
|  | Ireson, Robert B. |  | Wood, Samuel |
|  | Knight, George | Prize to the Scourge | Barnes, Henry |
|  | Knight, William |  | Cogwell, James |
|  | Sweet, Francis |  | Griffin, John |
|  | Symonds, Israel |  | Markins, John |
| Prize of the Blockage | Dailey, George W. | Prize to the Wiley Reynan | Lane, William |
|  | Lovell, Robert | Prize to Yankee | Cole, H. A. |
|  | Naborne, Nicholas | Procupine | Cleveland, Ebenezer |
|  | Phillips, George W. |  | Lamson, William |
| Prize of the Bunker Hill | Dorsey, John | Prompt | Allen, John |
| Prize of the Chasseur | Duncan, Abel |  | Atwood, Thomas |
| Prize of the Enterprize | Latimer, John |  | Bartis, John |
| Prize of the President | Colborne, John L. |  | Beecher, William |
|  | Gardner, John |  | Chase, Nathaniel |
|  | Shorts, Samuel |  | Cole, William |
| Prize of the Yankee | Spencer, Job |  | Hubbard, Alfred |
| Prize of the Young Wasp | Perry, John |  | Lopans, William |
| Prize to Chasseur | Bateman, John |  | Thompson, Nathaniel |
|  | Brown, George | Prosperity | Birch, Peter |
|  | Coffin, Caleb | Quebec | Beattie, James |
|  | Fenhouse, John |  | Bovey, Benjamin |
|  | Frizle, John |  | Chalk, John |
|  | Joseph, John |  | Hunberly, Elisha |
|  | Smith, John |  | Johnson, Alexander |
| Prize to Prince | Lane, James |  | Lane, John |
|  | Ward, Thomas |  | Lockwood, Caleb |
| Prize to the Diomede | Allison, William R. |  | Vanderhaven, Mathias |
|  | Arnold, Benjamin |  | Very, Samuel |
|  | Coffin, Abel | Quiz | Glynn, Hanson |
|  | Doosenbery, R. |  | Mingo, Albert |
|  | Leach, Charles | Raccoon | Caldwell, Charles |

American Prisoners of War held at Dartmoor during the War of 1812

## Prisoner listing by ship or regiment

| | | |
|---|---|---|
| Rachael | Broden, Norman | Cooper, John |
| | Delabar, Joseph | Crowell, Seth |
| | George W. W. | Curtis, Henry |
| | Howman, John | Dalmark, William |
| | Howman, Jonas | Ewing, Thomas |
| Rachel & Ann | Brutus, Mario | Fackney, John |
| | Deparvier, John | Ferris, Jonathan |
| | Erighton, Manuel | Forsdick, Joseph H. |
| | Ferris, Joseph | Fowler, William |
| | Morgan, John | Frederick, Charles |
| | Robins, John | Gair, Thomas |
| | Rogers, Henry | Gladding, Joseph |
| | Smith, John | Glashon, Richard |
| | Webber, Frederick | Gordon, Philip |
| | Williams, Alexander | Grahl, Charles |
| Rambler | Adams, John | Grough, Jacob |
| | Atkinson, Charles | Groves, Thomas |
| | Brooks, John | Halbert, Henry |
| | Dissmore, Abraham | Hough, Samuel |
| | Felt, George | Listrades, Joseph |
| | Gier, John | Littleford, L. |
| | Goldsmith, Thomas | Main, Charles |
| | Kemble, John | McCormick, Daniel |
| | Mugford, William | Moffatt, David |
| | Phillips, George | Mordaunt, John |
| | Stevens, James | Pane, William |
| | Webb, Thomas | Parker, Colby |
| Rattlesnake | Abbert, Solomon | Peck, Thomas |
| | Adams, Abijah | Perlie, Abraham |
| | Allerd, Erick | Philman, Adam |
| | Arnold, James | Rice, M. |
| | Atkinson, Henry | Robertson, Thomas |
| | Averill, Samuel | Robinson, Tony |
| | Baker, Benjamin | Ruling, Arnold |
| | Bann, John | Shadwick, John |
| | Boisseau, Joseph | Sharp, John |
| | Brannen, Alexander | Shaw, Thomas |
| | Brush, Samuel | Shorts, Richard |
| | Buskell, William | Smith, Christopher |
| | Clarke, Peter | Smith, John |
| | Coleby, Ebenezer | Solden, Chaarles |

# American Prisoners of War held at Dartmoor during the War of 1812

## Prisoner listing by ship or regiment

| Ship | Prisoner | Ship | Prisoner |
|---|---|---|---|
| | Steele, Thomas | | Griffin, John |
| | Stone, Samuel | | Haley, John |
| | Strouse, Martin | | Johnson, James |
| | Thomas, John | | Kenny, William |
| | Towson, William | | Tucker, Henry |
| | Tripp, Adam | | Weisse, Philip |
| | Verdon, Louis | | Williams, George |
| | Virgin, Robert | Resolution | Miller, Peter |
| | Voustyten, Christopher | | Winkinpor, Andrew |
| | Warner, Stephen | Revenge | Bracket, John |
| | West, Samuel | | Brown, Benjamin |
| | Williams, John | | Bunnel, Benjamin |
| Recovery | Hussey, Edward | | English, Edward |
| Regent | Barrows, John | | Evelish, William |
| | Bruce, Peter | | Flinn, Abraham |
| | Christian, Nicholas | | Gardiner, John |
| | Clintic, Ralph | | Gayer, Joseph |
| | Collins, George | | Hely, John |
| | Collins, John | | Holland, Richard |
| | Curden, John | | Hudson, Thomas |
| | Ellis, John | | Johnston, Thomas |
| | Hamilton, Clayton | | Killenger, John |
| | Jenkins, James | | Lamond, John |
| | Johnson, Albert | | Lethorpe, James |
| | Lee, John | | McKinney, John |
| | Lynes, Amos | | Melcher, John |
| | Merrick, William | | Merritt, Jonah |
| | Moody, John | | Mezich, Elisha |
| | Nickerson, Warren | | Page, John |
| | Patterson, Peter | | Price, Samuel |
| | Solomon, John | | Putman, Charles |
| | Stanfield, Thomas | | Watkins, Thomas |
| | Thomas, James | | White, John |
| | Wilson, George | | Whitter, J. W. |
| | Wood, George | Rising States | Jones, Charles |
| Renommee | Meadows, Timothy | Robert | Cook, Jeremiah |
| Requin | Bartlett, Scipro | | Giddons, Andrew |
| | Crasus, Richard | | Shackley, John |
| | Davis, Henry | | Tremerin, Joseph |
| | Finn, John | | Woodward, Thomas |
| | Green, John | Robert & Ann | Waterman, Thomas |

American Prisoners of War held at Dartmoor during the War of 1812

## Prisoner listing by ship or regiment

Rolla

| | |
|---|---|
| Anderson, Francis | Howland, J. W. |
| Armstrong William | Huntingdon, Edward |
| Armstrong, Andrew | Ireland, James |
| Athroun, Samuel | Johnson, James |
| Augustin, Anthony | Johnson, John |
| Bailey, Warren | Johnston, Edward |
| Baird, James | Jones, Edward |
| Baird, Martin | Justin, William |
| Barber, John H. | Larkin, Lewis |
| Barnes, James | Lessall, Francis |
| Bennett, Abisa | Lewis, Daniel |
| Briggs, B. | Little, William |
| Bright, Samuel | Loring, George |
| Brown, George | Maigot, Abner |
| Brown, William | Martin, Henry |
| Brownell, Paul | McCauley, George |
| Burnham, James | McGee, John |
| Burt, Thomas | Medker, Charles D. |
| Castagne, John | Morris, John |
| Champlin, Thomas | Palmer, Benjamin |
| Clark, John | Phelps, Elijah |
| Cogent, Benjamin | Reeves, James |
| Colcocha, Anthony | Richardson, Perry |
| Cole, Nathaniel | Rotner, James W. |
| Cooper, Lodwig | Ryan, William |
| Coston, Thomas | Setchell, Jonathan |
| Crofts, John | Shaw, Joseph |
| Davis, John | Shaw, Robert |
| Day, William | Shutcliffe, John |
| Dixon, John | Simson, John |
| Ellison, Benjamin | Spick, John |
| Fellows, Nathaniel | Stoddard, Isaac |
| Gardner, Timothy | Thomas, William |
| Grey, Thomas | Turney, B. |
| Hammond, Samuel | Tyler, Daniel |
| Hazard, D. | Wainwood, William |
| Hill, Pompey | Ware, John |
| Hobday, William | Washington, John |
| Holden, James | Weimon, John |
| Horwell, John | Wherry, Daniel |
| Hoston, Cato | Williams, Ambrose |

American Prisoners of War held at Dartmoor during the War of 1812

## Prisoner listing by ship or regiment

| Ship | Prisoner | Ship | Prisoner |
|---|---|---|---|
| Romp | Whippen, W. C. | | Carter, George |
| Rose | Bennett, Robert | | Fisher, Charles |
| | Welbor, John | | Johnson, Peter |
| | Wiggins, Samuel | | Leverage, William |
| Rover | Doyle, John | | Muller, John |
| | Hanson, John | | Newcomb, Tilton |
| | Jones, Ezekiel | | Pembroke, James |
| | McDaniel, John | | Sands, Henry |
| | Meeds, Joseph | | Simmons, Thomas |
| | Reynard, John | | Williams, Joseph |
| | Sandy, W. P. | Scourge | Anderson, John |
| | Wilson, Thomas | | Griffin, William |
| Sabine | Davenport, John | | Russell, James |
| | Hanson, Stephen | Sea Nymphe | Allen, Edward |
| | Martin, Philip | | Nicholson, Thomas |
| | Proctor, Henry | | Smith, John |
| | Simmons, Richard | | Welch, Samuel |
| | Spanow, W. T. | Shadow | Esdale, Thomas |
| | Thomas, Timothy | | Hensell, John |
| | Thompson, John | | Huggins, Daniel |
| Sally | Black, Ruddick | | Innes, J. |
| | Boston, John | | Lilley, Simon |
| | Crellin, William | | Marshall, Alexander |
| | Hermanden, Peter | | Muller, Edward |
| | Lambert, Calvin | | Murrell, Mark |
| | Melbourne, William | | Williams, Elisha |
| | Page, Joseph | Siro | Bodfish, William |
| | Rogers, Daniel | | Botellio, Anthony |
| | Rust, John | | Brown, Sandy |
| | Scobe, William | | Brown, Stephen |
| | Smith, W. B. | | Coffee, William |
| Sally & Betsey | Jackson, George | | Cooper, John |
| | Chapman, James | | Dennison, George |
| Sampson | Coffin, James | | Drinkwater, Peter |
| | Leonard, Robert | | Furlong, William |
| | Thornton, David | | Gall, Michael |
| | Tucker, N. | | Gordnow, S. B. |
| Sarah | Laurence, John | | Guilmot, Richard |
| Sarah & Ann | Pilot, Samuel G. | | Harris, Bradley |
| Saratoga | Anderson, John | | Hutchins, William |
| | Armandez, Justine | | Jackson, Samuel |

# American Prisoners of War held at Dartmoor during the War of 1812

## Prisoner listing by ship or regiment

|  |  |  |
|---|---|---|
|  | Mann, John | Cox, Caleb |
|  | Mason, William | Crosbeck, Nathaniel |
|  | Miller, William | Crump, Joseph R. |
|  | Moore, Alexander | Davis, Elisha |
|  | Nanson, Henry | Dill, William |
|  | Nanson, Peter | Dizere, John |
|  | Neal, John | Dyer, Israel |
|  | Niles, George | Dyer, Samuel |
|  | Nye, Cornelius | Fetch, Theodore |
|  | Pain, George | Fisher, James |
|  | Patterson, Andrew | Fletcher, Henry |
|  | Patterson, John | Frobus, Henry |
|  | Perkins, John | Gibley, Thomas |
|  | Pinckham, Amos | Gilliken, Daniel |
|  | Prestley, James | Glover, Allen |
|  | Ringold, Peregrin | Golf, John |
|  | Samuel, Nathaniel | Graham, W. R. |
|  | Sheardon, Elison | Gribble, George |
|  | Smiles, John | Hall, Lewis |
|  | Soderback, Andrew | Hannon, Alexander |
|  | Stewart, Adam | Harrison, John |
|  | Stout, John | Hendrickson, T. |
|  | Thompson, William | Hosson, John |
|  | Tubbs, Martin | Howland, Solomon |
|  | Wakefield, Nathaniel | Johnson, John |
|  | Weeks, James | Johnson, Thomas |
|  | Welsh, William | Jones, James |
|  | Wilson, Charles | Lynch, James |
|  | Wright, William | Mazely, William |
| Sister | Bates, Josiah | McFarlane, Robert |
|  | Straye, George L. | Miller, Richard |
| Snap Dragon | Alexander, Joseph | Miller, Thomas |
|  | Arthur, John | Pendleton, Samuel |
|  | Atwood, John | Petion, Samuel |
|  | Bailey, Aaron | Piner, Thomas |
|  | Barker, Thomas | Place, Aaron |
|  | Boziman, Ralph | Porter, Joseph |
|  | Clarke, Simon | Secundie, James |
|  | Colhoun, William | Smith, James |
|  | Connor, Fenton | Stanbeck, Jacob |
|  | Connor, Joseph | Stanley, Redmond |

American Prisoners of War held at Dartmoor during the War of 1812

## Prisoner listing by ship or regiment

|  |  |  |  |
|---|---|---|---|
|  | Stone, William |  | Hughes, John |
|  | Studley, Simon |  | Long, Charles |
|  | Sullivam, P. |  | Martin, John |
|  | Taylor, Alexander |  | McFarlane, John |
|  | Thomas, Allan |  | Pagechin, Henry |
|  | Troth, George |  | Road, John |
|  | Tulford, Joseph |  | Thomson, Andrew |
|  | Turner, George |  | Thomson, Peter |
|  | Watts, William | St. Johanna | Baptiste, John |
|  | White, John |  | Black, William F. |
|  | Williams, John |  | Marshall, William |
| Sparks | Coven, Clement | St. Lawrence | Brazier, Thomas |
|  | Creek, Frederick |  | Colton, Joseph |
|  | Hier, Andrew |  | Frost, John |
|  | Lenham, John |  | Hampton, Thomas |
|  | Smith, Charles P. |  | Lowrie, William |
|  | Vorhis, Peter |  | Lupton, John |
|  | Wilson, Nicholas |  | Moore, John |
| Spitfire | Bartlett, Thomas |  | Stirrell, George |
|  | Bishop, William |  | Ustick, William |
|  | Bridge, Francis |  | Vanderburgh, Cornelius |
|  | Carton, Thomas | St. Martin's Planter | Allen, John |
|  | Dodd, Samuel |  | Brown, John |
|  | Dolliver, Francis |  | Covell, Isaac |
|  | Fletcher, William B. |  | Ellis, William |
|  | Jervis, John |  | Gilbert, John |
|  | Jiett, Benjamin |  | Gray, John |
|  | Jones, F. V. |  | Hulet, Michael |
|  | Jones, Francis |  | Jones, William P. |
|  | Lamson, Charles |  | Mains, Daius |
|  | Lovett, William |  | Merritt, Almon |
|  | Miller, Henry |  | Morrison, Davis |
|  | Wadden, Jacob |  | Russell, Isaac |
|  | Wheeler, Richard |  | Taylor, George |
|  | Witherham, Nicholas |  | Thompson, Whitney |
|  | Wooldridge, Thomas |  | Wills, Peter |
| Squirrel | Andrews, Benjamin |  | Woodward, Anthony |
|  | Chain, John | Star | Bisley, Horace |
|  | Gable, John |  | Borgin, Gabriel |
|  | Gilchrist, John |  | Brown, Jesse |
|  | Gregory, George |  | Clark, William |

American Prisoners of War held at Dartmoor during the War of 1812

## Prisoner listing by ship or regiment

|  |  |  |  |
|---|---|---|---|
|  | Clements, John C. |  | Hamilton, Elijah |
|  | Cox, John |  | Rauzelin, Jacob |
|  | Dennison, Judah |  | Seaman, Francis |
|  | Duston, Peter |  | Waite, Joshua |
|  | Erwin, William | Swift | Baxter, Marion |
|  | Gordon, Thomas |  | Crowell, Elijah |
|  | Holts, Daniel | Sword Fish | Middleton, Reuben |
|  | Hughes, John |  | Strong, William |
|  | Isaac, Moses | Sybelle | Andress, George |
|  | Jocelyn, Robert |  | Dubois, Alexander |
|  | Johnstone, Gersham |  | Mitchell, Faisly |
|  | Joseph, Fois |  | Ross, Philip |
|  | Lewes, John |  | Small, Enoch |
|  | Louring, Henry | Taken off an English whaler | Claby, Martin |
|  | Mallery, William |  | Hill, John |
|  | Mix, William A. |  | Johnson, Robert |
|  | Parkinson, Stephen |  | Kingbutton, John |
|  | Richman, Joshua |  | Porter, Charles |
|  | Right, Joshua |  | Robertson, William |
|  | Romain, Samuel |  | Staples, James |
|  | Sullivan, Hampton | Teazer | Bassett, William |
|  | Thompson, William |  | Fortune, John |
|  | Trash, James |  | Merrish, Joseph |
|  | Valentine, John |  | Mitchell, John |
|  | Vaughan, Thomas |  | Moffatt, Hugh |
|  | Washburn, Edward |  | Piles, John |
|  | Williams, Henry |  | Samerton, George |
| Stark | Benyman, James |  | Sawyer, James |
|  | Branblen, Andrew |  | Stoddard, Conrad |
|  | Brown, John |  | Wilson, John |
|  | Buckling, Warren | Terrible | Cook, James |
|  | Giles, John |  | Kromkout, Baney |
|  | Johnston, Michael |  | Leach, William |
|  | Lourie, Solomon |  | Lovel, John |
|  | Waters, Robert |  | Martin, Peter |
|  | White, William |  | Root, Eleazer F. |
| Starks | Clarke, William |  | Smith, Eldridge |
| Steven Getard | Pitts, Francis |  | Smith, John |
| Stock | Tinkham, John |  | Smith, William |
| Stockholm | Jackson, William |  | Tufts, Zarara |
| Success | Goodall, William |  | Williamson, William |

American Prisoners of War held at Dartmoor during the War of 1812

## Prisoner listing by ship or regiment

| Ship/Regiment | Name | | Name |
|---|---|---|---|
| Theodore | Henderson, Benjamin | | Pitman, Henry |
| | Merchant, John | | Sergeant, Daniel |
| | Morris, Manuel | | Shields, John |
| Theresa | Waterhouse, Moses | | Smith, Henry |
| Thetis | Hammond, William | | Thompson, John |
| | Ward, Edenezer | | Thompson, Thomas |
| Thomas | Bailey, Daniel | | Todd, Samuel |
| | Bailey, Peter | | Tripper, Robert |
| | Bell, Richard | | White, William |
| | Beverley, Richard | | Williams, Joseph |
| | Billings, Richard | | Yeaton, Charles |
| | Bodkin, William | | Young, John |
| | Brown, Samuel | Thorn | Allison, J. B. |
| | Card, John | | Archer, James |
| | Cotton, Edward | | Burger, Francis |
| | Cross, Ephraim | | Esterstone, John |
| | Davis, William | | Gray, James |
| | Diamond, George | | Gross, James |
| | Driscol, Jeremiah | | Hammond, John |
| | Fadden, Charles | | Henry, Nathanial |
| | Harbrook, Richard | | Holden, John |
| | Holbrook, Robert | | Houghton, Timothy |
| | Holmes, Elisha | | Lanstone, Peter |
| | Hult, Thomas | | Marshall, John |
| | Jennings, Luther | | Quince, Peter |
| | Johnson, Thomas | | Saunders, Richard |
| | Jones, James | | Webster, David |
| | Knight, Daniel | | Wells, James |
| | Lake, George | | Williams, William |
| | Langford, Samuel | Thrasher | Davidson, Henry |
| | Lawson, Thomas | | Somerville, Charles |
| | Lowe, George | Three Brothers | Elliott, Robert |
| | Mallard, James | | Philling, Nathaniel |
| | Marble, Samuel | Tickler | Batman, C. P. |
| | Marshall, John | | Best, Robert |
| | McGill, Robert | | Bignell, Peter |
| | McIntire, Petty | | Brush, Ceasar |
| | Norton, Joseph | | Christie, James |
| | Penn, William | | Dinsmore, John |
| | Perkins, John | | Dougall, Thomas |
| | Pettigrene, William | | Forster, Thomas |

# American Prisoners of War held at Dartmoor during the War of 1812

## Prisoner listing by ship or regiment

|  |  |  |  |
|---|---|---|---|
|  | Green, Walter |  | Payne, Josiah Smith |
|  | Hart, Samuel |  | Penman, Richard |
|  | Hutson, Peter |  | Petton, Alexander |
|  | Isles, Robert |  | Rowe, John |
|  | Jones, John |  | Rowley, Henry |
|  | McKinnon, Nathaniel |  | Siffers, Stuben |
|  | Meurinose, John |  | Smith, Andrew |
|  | Pickrage, George |  | Taylor, Samuel E. |
|  | Smith, Cornelius |  | Trifle, Jasper |
|  | Stove, Lewis |  | Wilson, John |
|  | Waternamn, William | Tom Thumb | Arnold, William |
|  | Webb, John |  | Blagcon, Stephen |
| Tiger | Cadwell, James |  | Cooper, Alfred |
|  | Calder, J. H. |  | Jenkins, Nathanile |
|  | Churchill, Henry |  | Mitchell, James M. |
|  | Conklin, William |  | Palmer, Peter |
|  | Hill, Ephraim |  | Timmerman, Mathew |
|  | Hill, Joseph |  | Veney, George |
|  | Mars, George | Traveller | Church, William |
|  | Nicholls, John |  | Flinn, Pearce |
|  | Sampson, Jacob |  | Hall, Thomas |
|  | Spencer, Leonard |  | Harker, John |
|  | Sproson, James |  | Long, Joseph |
|  | Sulton, John |  | Simons, John |
| Tigre | Small, George D. |  | Thompson, John |
|  | Waterman, John |  | Wilson, John |
|  | Westerbest, William | True Blooded Yankee | Allen, William |
| Tom | Bickwith, Benjamin |  | Briggs, Boileau |
|  | Bourdon, John |  | Brown, Seth |
|  | Brown, John W. |  | Carter, Thomas |
|  | Brown, Wiliam |  | Christie, John |
|  | Cummings, James |  | Clark, Abraham |
|  | Davis, John |  | Cole, William |
|  | Dean, N. B. |  | Follingsby, William |
|  | Doolittle, H. |  | Hamilton, Richard |
|  | Gilmore, William H. |  | Huntley, Charles |
|  | Hollinger, William |  | Ingersoll, Abraham |
|  | Layfield, Littleton |  | LeBaron, Peter |
|  | Mansfield, James |  | Mathias, Louis |
|  | McIntire, Samuel |  | Mayo, Nathaniel |
|  | Nelly, Richard J. |  | Miller, John |

American Prisoners of War held at Dartmoor during the War of 1812

## Prisoner listing by ship or regiment

|  |  |  |  |
|---|---|---|---|
|  | Moore, John |  | Cooper, Thomas |
|  | Peckham, H. |  | Curtis, John |
|  | Richards, John |  | Drew, William |
|  | Rodgers, Samuel |  | Finch, William |
|  | Story, William |  | Gray, Isaac |
|  | Taylor, Peter |  | Haster, John |
| Two Brothers | Atkinson, William |  | Holbrook, Richard |
|  | Bridges, John |  | Jereb, Duly |
|  | Dear, Andrew |  | Kendrick, Benjamin |
|  | Jackson, Daniel |  | Lovell, Samuel |
|  | Leas, Anthony |  | Meinier, Benjamin |
|  | Paulin, Nicholas |  | Roberts, Hugh |
|  | Paulm, Nas. |  | Robinson, John |
|  | Picking, George |  | Thrasher, Stephen |
|  | Sheppard, William |  | Whistler, Miles |
|  | Stanton, Henry | Unity | Curlis, Aaron |
|  | Thomas, John |  | Swaine, Joseph |
|  | Vinson, Lewis | US Brig Argus | Addigo, Henry |
|  | Wing, Joshua |  | Allen, Joseph |
|  | Young, William |  | Allister, Isaac |
| Two Sisters | King, Joseph |  | Anderson, George |
| Tyger | Chase, Allen |  | Appene |
|  | Crapeman, Henry |  | Ash, Sampson |
|  | Harman, Mathew |  | Ayres, Robert |
|  | Harman, Peter |  | Barlow, John |
|  | Legatt, Charles |  | Barren, Thomas |
|  | Leggatt, Charles |  | Barry, James |
|  | Parsons, Simon |  | Benedict, William |
|  | Tripps, O. |  | Benson, James |
|  | Williams, James |  | Blacklidge, John |
| Tyren | Vowdy, William |  | Bladen, John |
| Ulysses | Allen, Charles |  | Blisset, James |
|  | Miles, Thomas |  | Bogart, K. W. |
|  | Mullin, Robert |  | Bowes, David |
|  | Philbrook, Cyrus |  | Bradt, Francis |
|  | Staples, William |  | Bron, James |
|  | Williams, George |  | Brown, Alexander |
| Union | Atwood, Jesse |  | Brown, Asher |
|  | Bell, James |  | Butnell, William |
|  | Boyd, William |  | Carbenett, John |
|  | Clover, Louis |  | Chugler, John |

American Prisoners of War held at Dartmoor during the War of 1812

## Prisoner listing by ship or regiment

| | |
|---|---|
| Clark, John | Lovett, Henry |
| Clerke, George | McDonald, John |
| Cobbs, James | McLeod, Colin |
| Collins, Andrew | McLeod, M. |
| Conely, Samuel | Mullan, John |
| Conklin, Robert | Nagle, George |
| Cooper, James | Nash, Manuel |
| Crafford, Robert | Noney, Peter |
| Curtis, Joseph | Nugent, John |
| Davis, William | Phillips, Benjamin |
| Day, Lyles C. | Place, John |
| Dillon, William | Price, William |
| Donham, Ebenezer | Richards, George |
| Fadden, John | Robinson, James |
| Figasson, Henry | Rosah, Samuel |
| Fleming, John J. | Russell, Frederick |
| Freeman, John | Salley, James |
| Gilbert, Gurdonel | Scott, John |
| Givell, Oliver | Scott, Robert |
| Godfrey, Edward | Sears, Abraham |
| Graham, David | Shaw, William |
| Griffiths, Benjamin | Smith, Adam |
| Groger, Henry | Smith, James |
| Hall, James | Smith, John |
| Hanscom, Moses | Smith, Richard |
| Harrington, William | Smith, William |
| Henry, James | Sniffon, John |
| Henry, John | Starbuck, George |
| Hill, Thomas | Stevens, John |
| Himes, Walter | Tafe, William |
| Hopkins, William | Tamer, John |
| Hunt, Charles | Thomas, David |
| Jameson, Robert | Thompson, Thomas |
| Jenkins, William | Tolvett, John |
| Johns, Thomas | Tower, Michael |
| Kennedy, Richard | Waite, Abel |
| Killam, James | Waters, Robert |
| Leach, Josem | Williams, John |
| Leary, Daniel | Wilson, George |
| Lee, Ely | Wood, Thomas |
| Lounsbury, Benjamin F. | Wyatt, Frederick |

American Prisoners of War held at Dartmoor during the War of 1812

## Prisoner listing by ship or regiment

|  |  |  |
|---|---|---|
|  | Young, John | Gabbett, Henry |
|  | Young, Thomas | Gale, Amos |
|  | Young, William | Gale, Edward |
| US Brig Rattlesnake | Abbott, Enoch | Garney, John |
|  | Allen, Ambrose | George, Thomas |
|  | Alley, Jacob | Gibson, William |
|  | Anderson, Ebenezer | Gill, John |
|  | Antoine, John | Gladding, William |
|  | Ashton, Joseph | Goldsmith, Nathaniel |
|  | Baker, Henry | Goodman, Caleb |
|  | Baker, Richard | Gratton, Edward |
|  | Ballard, John | Griffin, David |
|  | Bastard, Walter | Guard, Caleb |
|  | Baxter, David | Ham, Robert |
|  | Bigsby, Samuel | Hayes, Edward |
|  | Bouton, John | Holiday, Francis |
|  | Bowden, Thomas | Holmes, John |
|  | Bowland, William | Huckber, Venus |
|  | Bradford, Elisha | Hughes, Peter |
|  | Brain, William | Hurt, John A. |
|  | Brayden, Theodore | Jones, Jeremiah |
|  | Bruce, Joseph | Kean, Robert |
|  | Buckley, Joseph | Lampriere, David |
|  | Buttman, Nathaniel | Lee, Samuel |
|  | Campbell, James | Lilley, Samuel |
|  | Capps, Denis | Logere, H. C. |
|  | Caslow, John | Lonie, James |
|  | Chapple, Samuel | Mann, Richard |
|  | Cole, Nathaniel | Masser, Enoch |
|  | Cooney, Samuel | McLaughlan, M. |
|  | Curtis, William | Menillo, John |
|  | Davis, Thomas | Morgan, William |
|  | Dennis, William | Newport, Mathew |
|  | Dill, Samuel | Nichols, Silas |
|  | Drew, Samuel | O'Hara, Terrence |
|  | Dyer, Benjamin | Patterson, John |
|  | East, John | Payne, John |
|  | Fenderson, Nathaniel | Payne, Joseph |
|  | Florence, John | Perkins, Joseph |
|  | Foran, William | Peterson, Christopher |
|  | Frenage, Michael | Prindell, Daniel |

American Prisoners of War held at Dartmoor during the War of 1812

## Prisoner listing by ship or regiment

|  |  |  |  |
|---|---|---|---|
|  | Ranford, Henry |  | Haycock, Joseph |
|  | Read, John |  | Henory, James |
|  | Ritchie, James |  | Hodge, Edward |
|  | Rogers, George |  | Kellock, Stephen |
|  | Ross, William |  | Kirby, Robert |
|  | Rue, William |  | Kitchen, George |
|  | Saywood, John |  | Lewis, Winslow |
|  | Scott, William |  | Millow, John |
|  | Smith, John |  | Morris, John |
|  | Stafford, John |  | Nicholas, Jacob |
|  | Stafford, Thomas |  | Pope, Oliver |
|  | Stevens, Charles |  | Prout, Henry |
|  | Taylor, John |  | Rust, Zebulin |
|  | Thompson, James |  | Sparkes, John |
|  | Thompson, William |  | Spear, Joseph |
|  | Tolman, William |  | Stevens, John |
|  | Turner, John |  | Stone, John |
|  | Veal, Charles |  | Thomas, John |
|  | Vickery, William |  | West, Jacob |
|  | Waide, Nathaniel | US Frigate Chesapeake | Anthony, John |
|  | Walsh, John |  | Douglass, John |
|  | Weeks, Thomas |  | Lingrin, Peter |
|  | West, Richard | US Frigate Superior | Atkins, Henry |
|  | White, John |  | Chews, Thomas |
|  | Whitecombe, John |  | Curgan, John |
|  | Williams, William |  | Gustavus, John |
|  | Wood, John |  | Oliver, Griffith |
|  | Young, William |  | Mann, Charles |
| US Brig Syren | Andrews, Samuel |  | Ridden, John |
|  | Bicker, Charles |  | Thomas, John |
|  | Blackford, Henry |  | Truffey, Joel |
|  | Casper, William |  | Whormsley, R. |
|  | Chick, Moses | US Gunboat Number 121 | Whilday, Jospeh |
|  | Cloutman, Ephraim | US Gunboat Number 2 | Baxter, John |
|  | Clyde, John |  | Doogood, Abraham |
|  | Coates, Samuel M. |  | Edwards, Paul |
|  | Daniels, Henry |  | Mitchell, Reuben |
|  | Davis, Elias S. |  | Tucker, James |
|  | Deamon, Thomas | US Gunboat Number 51 | Carr, Thomas |
|  | Dunham, William | US Schooner Growler | McEvoy, John |
|  | Groveman, Frederick | US Schooner Ohio | Atwood, Elisha |

# American Prisoners of War held at Dartmoor during the War of 1812

## Prisoner listing by ship or regiment

| | | | |
|---|---|---|---|
| | Carnell, George | | Griffin, William |
| | Chargne, Robert | | Mastin, James |
| | Denning, Joseph | | Norton, Andrew |
| | Devinas, John | | Packett, John |
| | Johnson, Mark | | Palmer, Thomas |
| | Manuel, John | | Phillips, Augusta |
| | McStarbuck, Jordan | | Roderick, John |
| | Stacey, Stephen | | Stanford, Nicholas |
| | Weiden, Charles | | Wade, Nathan |
| | Williams, Peter | US Scorpion (Commodore Barney) | Orcroft, Lewis |
| US Schooner Scorpion | Bailey, Moses | US Sloop Frolic | Abbott, Benjamin |
| | Bennett, D. C. | | Anderson, Edward |
| | Bourdinon, Elijah | | Andrews, Asa |
| | Brown, Henry | | Appleton, Daniel |
| | Felton, John | | Atkins, William |
| | Fisher, William | | Austin, Joseph |
| | Lord, Samuel | | Bell, George |
| | Lowton, John | | Bella, Darius |
| | Morgan, William | | Benson, Joseph |
| | Rea, Charles | | Blair, John |
| | Read, Thomas | | Blesdale, Jacob |
| | Rowe, William | | Bond, William |
| | Rutter, Thomas | | Boss, Charles |
| | Rutter, William | | Bostell, James |
| | Williams, Jesse | | Brazel, James |
| | Williams, Samuel | | Briennis, John |
| | Black, Charles | | Briggs, John |
| | Bryan, John | | Brown, Edward |
| | Colston, John | | Brown, William |
| | DeCoine, John | | Bunker, Robert |
| | Dolling, Gannet | | Burgess, John |
| | Griffiths, Henry | | Burrell, Michael |
| | Hubbard, Joseph | | Butler, Benjamin |
| | McJugen, Robert | | Campbell, James |
| | Newman, John | | Carroll, Michael |
| | Otiel, Charles | | Chambers, William |
| | Peters, John | | Chapple, John |
| | Peterson, Jacob | | Christie, Robert |
| US Schooner Tigress | Cedus, Francis | | Clark, John |
| | Ely, Daniel | | Coffin, George |
| | Farrell, Richard | | Colhoun, Joseph |

American Prisoners of War held at Dartmoor during the War of 1812

## Prisoner listing by ship or regiment

| | |
|---|---|
| Cousins, Samuel | Jamieson, Daniel |
| Curtis, Stacey | Johnson, James |
| Dalton, Fred W. | Johnson, John |
| Davis, Hamilton | Jordan, Christopher |
| Delong, Zaba | Justice, John |
| Dempston, Daniel | Kales, John |
| Denny, James | Keeting, John |
| Depero, Henry | Lad, Daniel |
| Dickenson, Henry | Lear, Alexander |
| Dobson, Joseph | Lemon, James |
| Doddard, John | Lyman, Saul |
| Dominigue, Jospeh | Mangin, John B. |
| Dorman, John | Mariner, Joseph |
| Dunn, Henry | Marthoup, John |
| Dyke, Stuard | Mason, Richard |
| Edwards, John | McManus, Michael |
| Elvin, John | Mead, William |
| Felt, John | Meek, Thomas |
| Field, Samuel | Miller, Henry |
| Fields, George | Morris, Manuel |
| Foster, James | Murray, Alexander |
| Fowler, Isaac | Nivers, James |
| Frasher, Daniel | Norton, Edward |
| Fraste, Samuel | Nye, Stephen |
| Gerard, Henry | Osborne, Henry |
| Gregory, Joseph | Parker, John |
| Grindall, Joseph | Parsons, Andrew |
| Hadley, James | Pedrick, George |
| Hall, Charles | Peterson, James |
| Hammond, Benjamin | Pitman, John |
| Handell, Henry | Porter, Samuel |
| Harris, David | Price, John |
| Harris, John | Ramsey, Samuel |
| Hart, George | Randolph, George |
| Harvey, Edward | Rawl, Elias B. |
| Haskins, Joseph | Reynolds, W. B. |
| Hill, James A. | Richardson, John |
| Hoff, Charles | Ricker, Edward |
| Holden, John | Robins, Philip |
| Holmes, Christopher | Roddy, W. B. |
| Horton, G. A. | Rogers, Asa |

American Prisoners of War held at Dartmoor during the War of 1812

## Prisoner listing by ship or regiment

|  |  |  |  |
|---|---|---|---|
|  | Rogers, Francis |  | Mainey, John |
|  | Rogers, Nathaniel |  | Swaby, Jonathan |
|  | Ronse, Joseph |  | Williams, John |
|  | Sargeant, Charles | Vice Admiral Martin | Myers, George |
|  | Scribble, Henry | Ville de Milan | Gillia, Joseph |
|  | Sherlock, Charles | Viper | Allen, Thomas |
|  | Small, William |  | Arnold, Alfred |
|  | Smith, Isaac |  | Baker, John |
|  | Smith, John |  | Belfast, Richard |
|  | Smith, Samuel |  | Brady, Hugh |
|  | Snowdon, John |  | Campbell, John |
|  | Spellard, Robert |  | Crouch, Richard |
|  | Spencer, Robert |  | Farrell, John |
|  | Stetson, Mathew |  | Foster, David |
|  | Stevens, Leman |  | Hinkley, Aaron |
|  | Taylor, George |  | Hoye, Cornelius |
|  | Todd, Robert |  | Hunter, George |
|  | Tolver, Joseph |  | Jones, George |
|  | Truffe, John |  | McNelly, Thomas |
|  | Walker, Peter |  | Morgan, Joseph |
|  | Wallis, John |  | Pain, Joseph |
|  | Williams, John |  | Robinson, Charles |
|  | Williams, Thomas |  | Small, Richard |
|  | Williamson, Charles |  | Smalley, Thomas |
|  | Wilson, James |  | Thomas, William |
|  | Wing, Judah |  | Thomson, William |
| US Sloop Growler | Duncan, Nathaniel |  | Toumblin, John |
| US Sloop-of-War Frolic | Grafton, James |  | Tyre, William |
|  | Morey, Ezekiel |  | Webster, James |
| USRC Surveyor | Bowden, John |  | Williams, John |
| Valentine | Crapon, George |  | Williams, Thomas |
|  | Hellin, John P. |  | Winslow, Elijah |
|  | Jones, Samuel B. |  | Wintory, Antony |
|  | Lockwood, Benjamin |  | Young, John |
|  | Luther, Cromwell | Virginia Planter | Andrews, Charles |
|  | Potter, Richard H. |  | Coombes, William |
|  | Turner, Gardner |  | Harding, John |
| Vengeance | Montgomery, William |  | Hartar, Henry |
|  | Parker, Peter |  | Rice, Francis |
| Venus | Craig, John |  | Wilson, Peter |
|  | Dykes, Leven | Vivid | Abbott, Daniel |

American Prisoners of War held at Dartmoor during the War of 1812

## Prisoner listing by ship or regiment

|  | | | |
|---|---|---|---|
| | Avery, John | Volunteer | Anderson, George |
| | Bateman, Michael | | Antoni, Francis |
| | Fields, Jacob | | Bushfield, James |
| | Freeman, Isaac | | Campbell, John |
| | Hayes, Simon | | Chase, John |
| | Higgins, John | | Cussar, Jacob O. |
| | How, William | | Hunt, Samuel |
| | Jacobs, William | | Jones, Stephen |
| | Lambert, Ephraim | | Jupiter, james |
| | Miller, John | | Kennedy, Dennis |
| | Murray, David | | Maria, Hosea |
| | Nelson, Thomas | | Marlow, Owen |
| | Ricker, James | | Martin, Francis |
| | Snow, Daniel | | Morgan, James |
| | St. Vincent, Stephen | Vulture | Butler, Henry |
| | Tolpie, Jonathan | Walker | Queenwell, Peter |
| | Whitten, Elijah | | Swain, David |
| Vixen | Stephens, John | Wasp | Bridges, John |
| Volante | Alston, Richard | | Cooper, Edward |
| | Anthony, John | | Edmond, John |
| | Bailey, Isaac | | Haddart, Robert |
| | Bean, John | | Hamilton, George W. |
| | Boyle, Joseph | | Pratt, Philip |
| | Conley, Cornelius | | Preston, John |
| | Didler, Henry | | Richardson, James |
| | Dotts, Cornelius | | Robinson, John |
| | Dow, Henry | | Rowe, Seth |
| | Jacob, Louis | | Spouding, Joseph |
| | James, George | | Thrasher, John |
| | Laborde, Peter | | Weare, Ebenezer |
| | Lamson, Noah | | Wilson, James |
| | McLane, John | Watson | Coas, Samuel |
| | Ropes, David | | Hill, Jeremiah |
| | Scott, Henry | Weazel | Ashfield, Henry |
| | Taylor, Thomas | | Bailey, John |
| | Thompson, Michael | | Brill, John |
| | Thompson, William | | Cappel, John |
| | Tucker, Samuel | | Carter, John |
| | Werell, Francis | | Dunn, David |
| | Williams, Charles | | Gibbs, Daniel |
| | Willy, John | | Hull, Edward |

American Prisoners of War held at Dartmoor during the War of 1812

## Prisoner listing by ship or regiment

| | | | |
|---|---|---|---|
| | Jeffs, Joseph | | Gault, William |
| | King, John | | Gorling, George C. |
| | Vangorbet, Cato | | Hoskins, James |
| | Vorhis, James | | Jackson, John |
| Wig | Thomas, John | | Jones, Benjamin |
| Wiley Reynard | Butler, Henry | | Jones, John |
| | Carles, John | | Kellem, John C. |
| | Hall, Perry | | Leach, William |
| | Holding, Nathaniel | | Lewes, John |
| | Jackson, John | | Linnott, John |
| | Kingman, Charles | | Mather, John |
| | Kirkpatrick, William | | Reley, Michael |
| | Lindsey, William | | Richardson, Francis |
| | McKinnon, John | | Roe, John |
| | Wood, Joseph | | Rollands, Joseph |
| William | Allen, Lark | | Sherwood, John |
| | Bakeman, Robert | | Steel, John |
| | Bird, Joseph | | Talgart, Shephred |
| | Callow, William | | Williams, John |
| | Fletcher, James | William Penn | Allen, John |
| | Higgins, Francis | | Cairns, Thomas |
| | Hopkins, John | | Chace, George |
| | Johnson, William | | Earle, James |
| | Jones, David | | Ebby, Samuel |
| | Marchand, Angelo | | Freeman, John |
| | Milner, Benjamin | | Harris, Moses |
| | Sacalogos, Nicholas | | Hart, Emuis |
| | Smith, William | | Hussey, Joseph |
| | Sprust, Anthony | | Long, David |
| | Stanley, Samuel | | McKinzie, Daniel |
| | Thomas, Isaac | | Robinson, Garner |
| | Wilson, John | | Rush, Samuel |
| | Winn, Theodore | | Shore, Ceasar |
| William Bayard | Averell, Loring | | Swain, Obediah |
| | Barlow, Robert | | White, James |
| | Blanchard, George | William Wilson | Parr, James |
| | Bunker, James | Wily Reynard | Foster, Joseph |
| | Davis, James | | Parker, Robert |
| | Dorrell, John | | Swaine, Thomas |
| | Elm, John | Wolf | Williams, John |
| | Francis, Benjamin | Wolf Cove | Alexander, R. |

# American Prisoners of War held at Dartmoor during the War of 1812

## Prisoner listing by ship or regiment

|  |  |  |  |
|---|---|---|---|
| | Carnes, John | | Davis, John |
| | Chivers, Joseph | | Eddy, Richard |
| | Cloutman, Robert | | Everard, Joseph |
| | Grant, William | | Fray, James |
| | Knap, Samuel | | Gilbert, Isaac |
| | Lyons, Charles | | Goulding, Samuel |
| | Pearce, David | | Jessamin, John |
| | Pearse, Prince | | Jessamine, John |
| | Phippen, John | | Johnston, Charles |
| | Quiner, Stephen | | Killerman, M. |
| | Stacey, Benjamin | | Lewis, William |
| | Stacey, Samuel | | Mackey, John |
| Yankee | Allen, Edward D. | | Mott, Thomas |
| | Boswell, John | | Oliver, Anthony |
| | Bowen, Sylvester | | Puffer, John |
| | Devol, Alexander | | Rodgers, William |
| | Dunningberg, Henry | | Sillick, Thomas |
| | Forstman, John | | Storm, Daniel |
| | Hall, John | | Vanrant, John |
| | Hall, Silvester | | White, Edward |
| | Hatheray, Philip | | Williams, Noble |
| | Levy, Charles | | Wilson, James |
| | Linnett, William | Young Dixon | Badson, Jacob |
| | Payne, William | | Chessebrough, Benjamin F. |
| | Pitts, George | | Cotterell, Henry C. |
| | Porter, Louis | Young Holkar | Blake, Alexander |
| | Sherdon, Joseph | | Burr, Isaac |
| | Wood, Benjamin | | Cowen, Robert |
| Yankee Lass | Greswold, Truman | | James, Thomas |
| | Russell, Thomas | | Lebaith, John P. |
| | Thomas, James | | Sturges, Bradley |
| | Young, Richard | | Thomas, John |
| York Town | Baisley, Abraham | | Toby, Peter |
| | Berry, William | | West, Benjamin |
| | Black, Philip | | Williams, William |
| | Blake, Charles | Young Teager | Vanvort, Richard |
| | Brown, Charles | Young William | Ballent, John |
| | Brush, Able | | Eldred, Peter |
| | Cameron, Daniel | | Ewer, David |
| | Crosby, Andrew | | Harker, William |
| | David, John | | Jones, Henry |

American Prisoners of War held at Dartmoor during the War of 1812

## Prisoner listing by ship or regiment

|  |  |  |
|---|---|---|
|  | Kevel, Alexander | Donnelson, Joseph |
|  | Prince, Silvester | Golandre, John |
|  | Taylor, James | Graham, James |
|  | Toutz, Vievel | Hanes, Simon |
|  | Trowbridge, George | Hencock, William |
| Zebra | Avis, James | Ingles, Daniel |
|  | Banker, Robert | Jones, Thomas |
|  | Barber, William | Keen, Joseph |
|  | Brant, Thomas | Mathews, Joseph |
|  | Carter, Daniel | Modre, John |
|  | Carter, Edward | Plumer, William |
|  | Chiney, Amos | Postlock, William |
|  | Davis, William | Walker, Seth |
|  | Durand, John | Williamson, David |
|  | Edsom, John | Wilson, John |
|  | Faye, Samuel |  |
|  | Garcia, Francis |  |
|  | Hanford, William |  |
|  | Hardingbrook, Theop. |  |
|  | Harrington, Simon |  |
|  | Jackson, Joseph |  |
|  | Jones, Charles |  |
|  | Laurence, Jean |  |
|  | Logan, Charles |  |
|  | Manson, Jeremiah |  |
|  | Mercer, Benjamin |  |
|  | Mills, William |  |
|  | Montbelly, William |  |
|  | Payer, Walter |  |
|  | Phipps, Joseph |  |
|  | Pitts, William |  |
|  | Pomp, William |  |
|  | Reynolds, James |  |
|  | Ross, Isaac |  |
|  | Stagg, J. Boulton |  |
|  | Wallace, William |  |
|  | Weeks, David |  |
|  | Zellie, Thomas |  |
| Zephyr | Brown, David |  |
|  | Carr, James |  |
|  | Coffin, Alexander |  |

# American Prisoners of War held at Dartmoor during the War of 1812

## Definitions

## Prisoner of War Terms

**Barracks**
A permanent structure for housing or quartering prisoners of war within a prison depot.

**Cartel**
A cartel is a military agreement between hostile powers, which regulates the conduct of warfare.

**Cartel Ship**
A vessel chartered to carry prisoners or impressed to carry prisoners taken by a vessel of war to the nearest port; also, formerly a vessel which carried proposals between belligerent powers. A chartered cartel must carry no arms, and is liable to seizure by the enemy if she attempts to trade.

**Depot**
Depot is a British term for a prisoner of war camp. Depots were actually forts which contained up to seven prisons within their walls.

**Goal**
Goal is the British word for 'jail.'

**Letter of Marque** (or Letter of Marque and Reprisal)
A Letter of Marque is a license granted by a nation to a private citizen which permits that person to arm a ship and seize merchant vessels from an enemy nation.

**Non-Combatants**
Non-combatants were military surgeons, surgeons' mates, pursers, secretaries, chaplains and schoolmasters. All passengers on warships and merchant vessels were classified as non-combatants, which included women and girls, and all boys under the age of twelve.

**Parole**
Prisoners of war who were on parole status were given permission to return to their homes if they signed an agreement not to take up arms until they were exchanged. Enemy officers were granted another type parole in which they agreed not to escape and in turn they were permitted to walk freely around a town or a given area.

**Parole Station**
A village or city in which prisoners of war were housed.

**Prison Ship**
Prison ships were obsolete warships and transports, ranging in size from ships-of-the-line to frigates and cargo ships used to house prisoners of war and convicts.

**Prisoner of War**
Soldiers or sailors who had been captured by an enemy and confined until either exchanged or paroled.

**Transport Board**
His Majesty's Transport Board of the Royal Navy was responsible for transporting Royal Army regiments and their supplies throughout the British Empire. The board was also responsible for handling the prisoner of war facilities and the transportation of these prisoners.

## Definitions

## Naval Terms

**Armourer**
An armourer repairs the ship's small arms.

**Boatswain**
A warrant officer who has charge of the work of the seamen, the general oversight of the cleanliness of the ship, and of the work pertaining to the boats, spars, rigging, etc., anchoring and the mooring and unmooring of the ship.

**Boatswain's Mate**
Assists the boatswain in his duties.

**Boy**
A male minor who is in naval training to become an officer. Also called a cabin boy, a powder monkey and ship's boy.

**Brig**
A brig is a two-mast vessel with square sails.

**Captain**
The highest commissioned officer rank in the U.S. Navy during the War of 1812. The captain's rank was above a master commandant.

**Captain's Clerk**
The commanding officer of a vessel of war was formerly allowed a civilian clerk who was frequently a male member of his own family.

**Carpenter**
A carpenter is responsible for maintenance of the ship's hull and masts.

**Carpenter's Mate**
Assists the carpenter in his duties.

**Chaplain**
A chaplain provides pastoral, spiritual and emotional support for the ship's personnel

**Clerk** – see Captain's Clerk

**Commandant**
The officer in command of a navy yard or station.

**Commodore**
A title in the U.S. Navy given by the Navy Secretary for certain commissioned officers who were in command of a naval base, a squadron or a flotilla. The rank of commodore would not be created in the navy until 1862.

**Cook**
A cook handles the preparation of food for the ship's personnel.

**Coxswain**
A coxswain is in charge of the crew of a small boat from a vessel of war.

## Definitions

**Frigate**
A three mast sailing ship with square sails which had a gun deck. The number of guns varied from 28 to 44 and sometimes more guns. Guns were also placed on the spar deck.

**Gunboat**
A gunboat was the smallest warship in the U.S. Navy. They could be propelled by oars or sails, and were usually designed for coastal waters or the Great Lakes. Many gunboats carried only one or two guns on swivel mounts.

**Gunner**
A gunner was responsible for the care and maintenance of the ship's guns and gunpowder.

**Gunner's Mate**
Assists the gunner in his duties.

**Landsman**
A recruit with no sea experience.

**Lieutenant**
A commissioned officer who was below the rank of master commandant and above a warrant officer.

**Marines**
Naval soldiers who were used as guards aboard ships and provided musket support during naval battles. They also assisted in shore actions.

**Master (Naval)**
The master was the senior warrant officer on board a ship who was a qualified navigator and experienced seaman who set the sails, maintained the ship's log and advised the captain on the seaworthiness of the ship and crew.

**Master (Merchant Vessel)**
The commander of a merchant vessel was called a master. Other terms used were sea captain, captain and shipmaster.

**Masters-at-Arms**
Masters-at-Arms were in charge of keeping the swords, pistols, carbines and muskets in good working order.

**Master Commandant**
A commissioned officer's rank which was below a captain and above a lieutenant. The name of this rank was changed to 'commander' in 1838.

**Mate**
The assistant or subordinate of a warrant officer.

**Merchant Vessel or Merchantman**
A ship that transports cargo and/or passengers.

**Midshipman**

## Definitions

The lowest commissioned officer's rank in the U.S. Navy was a midshipman. The rank was below a lieutenant. This rank was abolished in 1845 and term was then used to describe cadets at the U.S. Naval Academy.

**Privateer**
An armed vessel, owned by private parties, licensed to prey on an enemy's commerce in time of war. In the War of 1812 a number of the merchant vessels of the United States, which were debarred from their usual trade, were fitted out as armed cruisers and created much havoc among British shipping.

**Prize**
A captured vessel or other property taken by a naval vessel in war. The circumstances of the capture and of the ownership of the property are taken under consideration by a court which awards the proportionate share of the money accruing from the sale of a prize. Provision is made by statute for distribution of prize money from the Treasury in cases of destruction of the vessels of an enemy.

**Provincial Marine**
A naval force operated by the Canadian provinces to man the warships on the Great Lakes, Lake Champlain and the St. Lawrence River. The naval force was established in 1778 and lasted until the Royal Navy absorbed the Provincial Marine in May 1813. It was also the main transport service for the government in western Upper Canada.

**Purser**
The financial officer responsible for supplies, provisions, and pay for the crew.

**Sailmaker**
A sailmaker makes and repairs the ship's sails.

**Sailmaker's Mate**
Assists the sailmaker in his duties.

**Sailing Master** – see Master
Another term for master.

**Schoolmaster**
Schoolmasters were involved in the education of boys, midshipmen and others aboard ship.

**Schooner**
A two-mast vessel with triangular sails before and after the mast is called a schooner.

**Seamen**
The lowest skilled enlisted rank in the navy.

**Ship**
A sailing vessel with three masts and rigged with square sails. A captain normally commanded a ship. The ship did not have a gun deck and it was rated between a frigate and a brig.

**Sloop**
A small vessel with one mast equipped with triangular sails before and after the mast.

**Steward**
A steward organizes the mess (meals) aboard ship working with the cook and the purser.

## Definitions

**Store Ship**
A vessel attached to a navy and used to transport supplies to distant naval depots.

**Supercargo**
A supercargo is a person employed on board a vessel by the owner of cargo carried on the ship. The duties of a supercargo include managing the cargo owner's trade, selling the merchandise in ports to which the vessel is sailing, and buying and receiving goods to be carried on the return voyage.

**Surgeon**
A surgeon was a warrant officer in charge of the medical department. On larger ships, the surgeon had one or more assistant surgeons, also called a surgeon's mate.

**Warrant Officer**
Warrant officers were ship officers who had a warrant and not a commission. Warrants were issued for a specific trade, that is, a purser, a carpenter, a sailmaker, etc. These men were not line officers and could not command a ship. However, masters (sailing masters) could and did command small vessels for the navy.

## Bibliography

*American Vessels Captured by the British during the Revolution and War of 1812*, (The Essex Institute: Salem, MA 1911).

Coggeshall, George, *History of the American Privateers, and Letters-of-Marque, during our War with England in the Years 1812, 1813 and 1814*, (New York, 1856).

*General Entry Book of American Prisoners of War*, British Admiralty, Public Record Office, London, Great Britain (Series ADM 103 / Ledgers 87 through 91 and 511), General Entry Book of American prisoners of war at Dartmoor Prison and American alphabetical book.

*General Entry Book of American Prisoners of War*, British Admiralty, Public Record Office, London, Great Britain (Series ADM 103 / Ledgers 465 and 640), Miscellaneous Lists and Records (Certificates of Death).

Harris, Captain Vernon, *Dartmoor Prison: Past and Present*, (William Brendon and Son: Plymouth, England, 1888).

Lewis, Lt. Col. George C., and Capt. John Mewha, *History of Prisoner of War Utilization by the United States Army 1776-1945*, Department of the Army Pamphlet number 20-213, (Washington, DC: Department of the Army, 1955), Part One – the Early Wars, Chapter 2 – Prisoners of War as Instruments of Retaliation and Parole, The War of 1812, pp. 22-25.

Maclay, Edgar Stanton, *A History of American Privateers*, (D. Appleton and Company: New York 1900).

*Records Relating to American Prisoners of War, 1812-1815*, British Records Relating to America Microform (BRRAM) Series, (Microform Academic Publishers: Wakefield, West Yorkshire, United Kingdom, 1980).

Scott, Colonel H. L., *Military Dictionary comprising Technical Definitions; information of raising and keeping troops, including makeshifts and improved materiel; and law, government, regulation, and administration relating to land forces*, (D. Van Nostrand, New York, New York: 1861).

Thomas, Basil, *The Story of Dartmoor Prison*, (William Heinemann: London, England 1907).

Waterhouse, Benjamin, *A Journal of a Young Man of Massachusetts*, (Rowe & Hopper: Boston, 1816).

www.ingramcontent.com/pod-product-compliance
Lightning Source LLC
Chambersburg PA
CBHW080933300426
44115CB00017B/2798